Pain Management in Palliative Care for Patients with Cancer

Pain Management in Palliative Care for Patients with Cancer

Edited by

Marin Golčić
Tumor Clinic, Clinical Hospital Center Rijeka, Rijeka, Croatia; Faculty of Medicine, University of Rijeka, Rijeka, Croatia

Tomi Kovačević
Faculty of Medicine, University of Novi Sad, Novi Sad, Serbia; Institute for Pulmonary Diseases of Vojvodina, Sremska Kamenica, Serbia

Amir Radfar
Former Faculty Member, College of Medicine, University of Central Florida, Orlando, FL, United States

ELSEVIER

ACADEMIC PRESS
An imprint of Elsevier

Academic Press is an imprint of Elsevier
125 London Wall, London EC2Y 5AS, United Kingdom
50 Hampshire Street, 5th Floor, Cambridge, MA 02139, United States

ISBN: 978-0-443-30050-9

For Information on all Academic Press publications
visit our website at https://www.elsevier.com/books-and-journals

Publisher: Megan Ball
Acquisitions Editor: Sangeeta Yadav
Editorial Project Manager: Pat Gonzalez
Production Project Manager: Gomathi Sugumar
Cover Designer: Greg Harris

Typeset by MPS Limited, Chennai, India

Contents

List of contributors

Ashique Ahamed Department of Supportive and Palliative Care, The Christie NHS Foundation Trust, Manchester, United Kingdom

Imatullah Akyar Faculty of Nursing, Hacettepe University, Ankara, Türkiye

Ahmad Samir Alfaar Al-Moriat for eHealth Solutions, Berlin, Germany; Vistarlin Center for Integrated Medicine, Berlin, Germany; Roya Eye Center, Stendal, Sachsen-Anhalt, Germany

Andrej Belančić Department of Basic and Clinical Pharmacology and Toxicology, Faculty of Medicine, University of Rijeka, Rijeka, Croatia

Borislav Belev Department of Medical Oncology, Clinical Hospital Center Zagreb, Zagreb, Croatia; Faculty of Medicine, University of Zagreb, Zagreb, Croatia

Krešimir Blažičević University Hospital Centre Zagreb, Zagreb, Croatia

Gorana Matovina Brko Oncology Institute of Vojvodina, Sremska Kamenica, Serbia

Dino Budrovac Faculty of Medicine Osijek, Josip Juraj Strossmayer University of Osijek, Osijek, Croatia; Department of Anaesthesiology, Reanimatology and Intensive Care, University Hospital Centre Osijek, Osijek, Croatia

Ajaz Bulbul Hematology/Oncology, Nancy N. and J.C. Lewis Cancer & Research Pavilion, Hinesville, GA, United States

Daliborka Bursać Faculty of Medicine, University of Novi Sad, Novi Sad, Serbia; Institute for Pulmonary Diseases of Vojvodina, Sremska Kamenica, Serbia

Anastasia Constantinidou Medical School, University of Cyprus, Nicosia, Cyprus; Bank of Cyprus Oncology Centre, Nicosia, Cyprus

Luis Felipe Calle Cruz Primary Care Centre of Castilleja de la Cuesta, Seville, Spain

Irina Filip Diplomate of the American Board of Psychiatry and Neurology, Avicenna Medical and Clinical Research Institute, Mission Viejo, CA, United States

Srdjan Gavrilović Faculty of Medicine, University of Novi Sad, Novi Sad, Serbia; Institute for Pulmonary Diseases of Vojvodina, Sremska Kamenica, Serbia

Jagoš Golubović Faculty of Medicine, University of Novi Sad, Novi Sad, Serbia; Department of Neurosurgery, Clinical Centre of Vojvodina, Novi Sad, Serbia

Marin Golčić Tumor Clinic, Clinical Hospital Center Rijeka, Rijeka, Croatia; Faculty of Medicine, University of Rijeka, Rijeka, Croatia

Davorin Herceg Department of Medical Oncology, Clinical Hospital Center Zagreb, Zagreb, Croatia; Faculty of Dental Medicine, University of Zagreb, Zagreb, Croatia

Ryouhei Ishii Department of Occupational Therapy, Graduate School of Rehabilitation Science, Osaka Metropolitan University, Osaka, Osaka, Japan

Poorni Priya Jaganathan Chelsea and Westminster Hospital NHS Foundation Trust, London, United Kingdom

Urška Janžič Medical Oncology Unit, University Clinic Golnik, Golnik, Slovenia; Faculty of Medicine, University of Ljubljana, Ljubljana, Slovenia

Bojan Jelača Faculty of Medicine, University of Novi Sad, Novi Sad, Serbia; Department of Neurosurgery, Clinical Centre of Vojvodina, Novi Sad, Serbia

Nikola Jojić Faculty of Pharmacy, University Business Academy, Novi Sad, Serbia

Robin Jones Royal Marsden NHS Foundation Trust, London, United Kingdom

Gordana Kenđel Jovanović Faculty of Medicine, University of Rijeka, Rijeka, Croatia; Department of Health Ecology, Teaching Institute of Public Health of Primorsko-goranska County, Rijeka, Croatia

Laila Kamal Royal Marsden NHS Foundation Trust, London, United Kingdom

Vlasta Orlić Karbić Department of Anaesthesiology, Intensive and Pain Medicine, University Hospital Rijeka, Rijeka, Croatia

Farzana Khan Fasiuddin Khan Research Foundation, Dhaka, Bangladesh; Knowledge Mobilisation Hub, NIHR Global Health Research Unit on Respiratory Health (RESPIRE), Usher Institute, University of Edinburgh, United Kingdom

Aleksandar Knežević Faculty of Medicine, University of Novi Sad, Novi Sad, Serbia; Medical Rehabilitation Clinic, University Clinical Center of Vojvodina, Novi Sad, Serbia

Alojzije Košić Tumor Clinic, Clinical Hospital Center Rijeka, Rijeka, Croatia; Faculty of Medicine, University of Rijeka, Rijeka, Croatia

Tomi Kovačević Faculty of Medicine, University of Novi Sad, Novi Sad, Serbia; Institute for Pulmonary Diseases of Vojvodina, Sremska Kamenica, Serbia

Ana Mišir Krpan Special Hospital Radiochirurgia Zagreb, Zagreb, Croatia; Faculty of Dental Medicine and Health Osijek, University of Josip Juraj Strossmayer, Osijek, Croatia

Ivana Kukec Department of Medical Oncology, Clinical Hospital Center Zagreb, Zagreb, Croatia

Knut Liepe Department of Nuclear Medicine, Klinikum Frankfurt (Oder), Frankfurt (Oder), Germany

Auxiliadora Fernández López Home Support Team for Palliative Care, Virgen del Rocío University Hospital, Seville, Spain; Faculty of Medicine, University of Seville, Seville, Spain

Nevenka Lukovac Janjić Clinic for Medical Oncology, Institute for Oncology, Clinical Centre of Montenegro, Podgorica, Montenegro

Dragana Marić Faculty of Medicine, University of Belgrade, Belgrade, Serbia; Clinic of Pulmonology, University Clinical Center of Serbia, Belgrade, Serbia

Antonio Markotić Department of Physiology, School of Medicine, University of Mostar, Mostar, Bosnia and Herzegovina; Department of Clinical Pharmacology, University Clinical Hospital Mostar, Mostar, Bosnia and Herzegovina

Hedieh Matinrad UCSF Division of Palliative Medicine, San Francisco, CA, United States

Ivana Mikolašević Tumor Clinic, Clinical Hospital Center Rijeka, Rijeka, Croatia; Faculty of Medicine, University of Rijeka, Rijeka, Croatia

Nikola Milašević Clinic for Medical Oncology, Institute for Oncology, Clinical Centre of Montenegro, Podgorica, Montenegro

Boris Milijašević Department of Pharmacology, Toxicology and Clinical Pharmacology, University of Novi Sad, Novi Sad, Serbia

Mihael Mišir Department of Neurology and Neurosurgery, University Hospital Centre Osijek, Osijek, Croatia; Faculty of Medicine Osijek, Josip Juraj Strossmayer University of Osijek, Osijek, Croatia

Ivana Mudnić Department of Basic and Clinical Pharmacology, School of Medicine, University of Split, Split, Croatia

Sergej Nadalin Department of Psychiatry, General Hospital "Dr. Josip Benčević", Slavonski Brod, Croatia; Faculty of Medicine, Catholic University of Croatia, Zagreb, Croatia

Mario Nalbani General Hospital Dubrovnik, Dubrovnik, Croatia

Kazuyuki Niki Department of Clinical Pharmacy Research and Education, Graduate School of Pharmaceutical Sciences, The University of Osaka, Suita, Osaka, Japan

Dragan Nikolić Faculty of Medicine, University of Novi Sad, Novi Sad, Serbia; Department of Vascular and Endovascular Surgery, Clinical Centre of Vojvodina, Novi Sad, Serbia

Marko Pavlović Department of Psychiatry, School of Medicine, University of Mostar, Mostar, Bosnia and Herzegovina; Department of Psychiatry, University Clinical Hospital Mostar, Mostar, Bosnia and Herzegovina

Hayda Perdue Hospice Savannah Inc., Savannah, GA, United States

Branka Petrić Miše Department of Oncology and Radiotherapy, University Hospital of Split, Split, Croatia; Faculty of Medicine, University of Split, Split, Croatia

Pampina Pilavaki Barking, Havering and Redbridge University Hospitals NHS Trust, London, United Kingdom

Kristina Pilipović Department of Basic and Clinical Pharmacology and Toxicology, Faculty of Medicine, University of Rijeka, Rijeka, Croatia

Jerolima Pinjatela Department of Internal Intensive Medicine and Clinical Pharmacology and Toxicology, University Hospital of Split, Split, Croatia

Sanja Pleština Department of Pulmonology, University Hospital Centre Zagreb, Zagreb, Croatia; Faculty of Medicine, University of Rijeka, Rijeka, Croatia

Lazar Popović Faculty of Medicine, University of Novi Sad, Novi Sad, Serbia; Oncology Institute of Vojvodina, Sremska Kamenica, Serbia

Maja Popović Faculty of Medicine, University of Novi Sad, Novi Sad, Serbia; Oncology Institute of Vojvodina, Sremska Kamenica, Serbia

Juraj Prejac Department of Oncology, University Hospital Centre Zagreb, Kišpatićeva, Zagreb, Croatia; School of Dental Medicine, University of Zagreb, Gundulićeva, Zagreb, Croatia

Dirk Rades Radiation Oncology Department, University of Lübeck/UKSH-CL, Lübeck, Germany

Amir Radfar Former Faculty Member, College of Medicine, University of Central Florida, Orlando, FL, United States

Ivan Radoš Faculty of Medicine Osijek, Josip Juraj Strossmayer University of Osijek, Osijek, Croatia; Department of Anaesthesiology, Reanimatology and Intensive Care, University Hospital Centre Osijek, Osijek, Croatia

Angelo Restivo Faculty of Medicine, University of Cagliari, Cagliari, Italy

Zeeshan M. Rizwan Department of Pharmacy, Mayo Clinic, Rochester, MN, United States

Jelena Roganović Department of Pediatric Oncology and Hematology, Children's Hospital Zagreb, Zagreb, Croatia; Faculty of Biotechnology and Drug Development, University of Rijeka, Rijeka, Croatia

Viktor Šabarić Department of Medical Oncology, Clinical Hospital Center Zagreb, Zagreb, Croatia

Ehsan Shamsi Gooshki Monash Bioethics Centre, Monash University, Melbourne, VIC, Australia; Medical Ethics and History of Medicine Research Centre, Tehran University of Medical Sciences, Tehran, Iran

Velizar Stefanov Shivarov Department of Experimental Research, Medical University Pleven, Pleven, Bulgaria

Luka Simetić Department of Medical Oncology, Clinical Hospital Center Zagreb, Zagreb, Croatia

Nataša Skočibušić Department of Clinical Pharmacology, University Hospital Center Rijeka, Rijeka, Croatia; Faculty of Medicine, University of Rijeka, Rijeka, Croatia

Iva Skočilić Tumor Clinic, Clinical Hospital Center Rijeka, Rijeka, Croatia; Faculty of Medicine, University of Rijeka, Rijeka, Croatia

Tamer Soror Radiation Oncology Department, University of Lübeck/UKSH-CL, Lübeck, Germany

Majana Soče Department of Medical Oncology, Clinical Hospital Center Zagreb, Zagreb, Croatia; University of Applied Health Sciences, University of Zagreb, Croatia

Jelena Spasić Clinic for Medical Oncology, Institute for Oncology and Radiology of Serbia, Belgrade, Serbia

Dragana P. Srebro Faculty of Medicine, Institute of Pharmacology, Clinical Pharmacology and Toxicology, University of Belgrade, Belgrade, Serbia

Nuša Matijašić Stjepović Department of Pediatric Oncology and Hematology, Children's Hospital Zagreb, Zagreb, Croatia

Vladimir Stojšić Faculty of Medicine, University of Novi Sad, Novi Sad, Serbia; Institute for Pulmonary Diseases of Vojvodina, Sremska Kamenica, Serbia

Mamak Tahmasebi Palliative Medicine Unit, Radiation Oncology Department, Tehran University of Medical Sciences, Tehran, Iran

Andrew Tweddle Royal Marsden NHS Foundation Trust, London, United Kingdom

Marijana Vučković Department of Internal Medicine, Division of Nephrology and Dialysis, University Hospital of Split, Split, Croatia

Muhammad Atif Waqar Section of Palliative Medicine, Department of Oncology, The Aga Khan University Hospital, Karachi, Pakistan

Christopher M. Wilson School of Health Sciences, Oakland University, Rochester, MI, United States; Corewell Health, Southfield, MI, United States

Bojan Zarić Faculty of Medicine, University of Novi Sad, Novi Sad, Serbia; Institute for Pulmonary Diseases of Vojvodina, Sremska Kamenica, Serbia

Andrej Žist Division of Medical Oncology, Institute of Oncology Ljubljana, Ljubljana, Slovenia

Feđa Đorđević Department of Medical Oncology, Institute for Oncology and Radiology of Serbia, Belgrade, Serbia

About the editors

Amir Radfar, MD, MPH, MSc, DHSc, ESCO Fellow, is a former faculty member at the University of Central Florida College of Medicine in Orlando, Florida. He received his medical degree from the Shiraz University of Medical Sciences, Iran, and completed postdoctoral training in internal medicine at the University of Southern California and Los Angeles County hospitals.

He completed advanced oncology training in Germany, earning a Master of Science in Oncology from the University of Ulm, with clinical training at Ludwig Maximilian University Hospitals, Munich, and palliative care education through the European Palliative Care Academy. He further completed postgraduate medical education at Harvard Medical School, focused on cancer biology, precision diagnostics, and next-generation therapies, as well as specialized training in cancer immunotherapy through the Society for Immunotherapy of Cancer. He is a Fellow of the College of the European School of Oncology (ESCO).

Dr. Radfar also holds a Master of Public Health from Florida International University and a Doctor of Health Science from A.T. Still University, Arizona. He is a member of the American Society of Clinical Oncology and the European Society for Medical Oncology and has collaborated with academic institutions and biopharmaceutical companies on therapeutic development and supportive care initiatives. He serves on the medical student admissions screening and interview committee at the UCF College of Medicine.

Dr. Radfar has authored more than 100 peer-reviewed publications and has served as an academic editor for *PLOS One* and the *Journal of Oncology*. His research interests include innovative cancer therapies, oncogenesis, digital health, and palliative care.

Tomi Kovačević, MD, PhD, palliative medicine specialist, geriatrician and Engineering Management Specialist, serves as Assistant Professor at the Faculty of Medicine, University of Novi Sad, in the Department of General Medicine and Geriatrics and the Department of Palliative Medicine. After graduation at the Faculty of Medicine, University of Novi Sad, his medical education in the field of palliative care includes palliative medicine specialization at the Faculty of Medicine, University of Belgrade, ESMO, IASLC, and other PC fellowships and mentorships (Canada, Germany, Belgium, Switzerland, and others), as well as specialized European Palliative Care Academy (EUPCA) education in Germany. His current clinical engagement includes employment at the Institute for Pulmonary Diseases of Vojvodina as chief of the Palliative Care Unit and as a PC consultant. He is also the president of the Palliative Medicine Section at the Medical Chamber of Serbia and served as the member of the Ministry of Health's committee for palliative care from 2019 to 2023. His dedication toward palliative care is visible through active participation as a presenter, invited lecturer and course instructor in numerous professional meetings in the country and abroad, with the aim to promote early integration of palliative care into regular clinical practice. His scientific contribution is reflected through the authorship and coauthorship in scientific papers and international collaborations, including projects, actions, and clinical trials in the field of palliative care and oncology.

Marin Golčić, MD, PhD, is a faculty member at the Faculty of Medicine, University of Rijeka, Croatia, and serves as a consultant medical oncologist at the Clinical Hospital Center in Rijeka, with a particular focus on gastrointestinal and skin cancers. He is also the head of the Committee for Medicinal Products at Marija K. Kozulić, the first Croatian hospice, where he works to optimize pharmacotherapy and advance research in palliative care. He completed clinical ESMO fellowships at the Royal Marsden in London and the Churchill Hospital in Oxford and has participated in virtual mentorship programs through ESMO and ASCO, as well as several ESMO preceptorships. He has published over 50 scientific papers and conference proceedings, acts as a scientific reviewer for more than a dozen medical journals, and has mentored several graduate theses. From 2019 to 2024, he served as the elected representative of his county at the Croatian Medical Association.

Preface and how to read this book

This book provides a thorough resource for managing pain in patients with cancer receiving palliative care. Contrary to the common misconception that palliative care is solely limited to end-of-life care, it plays a vital role in early cancer treatment, beginning at diagnosis.

Pain management in cancer care must be multifaceted, timely, and individualized. While pharmacological treatment, including opioids and supportive medications, remains foundational, it is only one part of a broader strategy. The WHO three-step analgesic ladder provides a useful starting point, but clinical decision-making requires flexibility and awareness of adjuvant options, especially in neuropathic or refractory pain syndromes. The concept of "total pain" reminds us that psychological, social, and spiritual suffering often compound physical distress.

Nonpharmacological strategies—such as physical therapy, psychological support, and innovative tools like virtual reality—are increasingly vital in managing complex or chronic cancer-related pain. These methods, combined with spiritual care, are complementary interventions for comprehensive patient care.

Cancer-targeted therapies can also help manage pain, alongside other supportive treatments. For many patients—especially those with bone metastases or malignant obstructions—radiotherapy or surgery can offer meaningful, long-term relief. Systemic cancer therapies, including chemotherapy, targeted agents, and immunotherapy, are increasingly recognized for their potential to reduce tumor burden and alleviate pain. A significant reduction in tumor size, for instance, often results in meaningful pain relief by decompressing nerves, organs, or bone structures. Reflecting this link, newer trials increasingly incorporate patient-reported outcomes—such as pain scores and daily functioning—to better capture the full impact of treatment. Clinicians must leverage the dual roles of these therapies in disease control and symptom relief to optimize patient outcomes.

Effective pain management requires collaboration across oncology, palliative care, and allied health disciplines. This book provides a framework to support that collaboration through clinical guidance, interdisciplinary tools, and evidence-informed strategies.

This book is divided into five main parts. An overview of palliative care, cancer pain mechanisms, and evolving cancer therapies is provided in the Introduction. The following two sections cover pharmacological and interventional strategies, offering tools to integrate them into daily practice. The core of this book focuses on cancer-type-specific chapters, each examining how disease biology, treatments, and pain syndromes intersect—and how cancer therapies influence both survival and symptom control.

The final section addresses special populations and care settings—including pediatric and geriatric patients, the ICU, and emergency departments. Additionally, it addresses ethical dilemmas, such as balancing appropriate pain relief with the risk of oversedation, addressing family concerns about opioid use and addiction, or navigating patient refusal of pain medication due to cultural beliefs. Clinical expertise and cultural sensitivity are both needed to meet these real-world challenges.

Who this book is for

This book is intended for all healthcare professionals involved in the care of patients with cancer: (1) Oncologists and palliative care specialists will find strategies for integrating therapeutic modalities with symptom relief. (2) Pain management physicians, surgeons, radiation therapists, and interventionalists will benefit from cancer-specific procedural insights. (3) Primary care providers, psychologists, nurses, and allied health professionals will gain tools for coordinating interdisciplinary care and managing comorbidities, such as addressing chronic conditions alongside cancer pain. (4) Students and trainees will find a clear foundation for understanding the evolving field of cancer pain management.

Note for Clinicians: This book serves as both an educational and a practical resource for clinicians. This is not clinical advice and should not be used as a substitute for clinical judgment. Treatment strategies described here must be applied in accordance with local, institutional, and national guidelines and regulations for clinicians and other healthcare practitioners.

Why it matters

For patients, pain is not just a symptom, it is often the most disruptive and feared part of the cancer experience. Compassionate, evidence-based pain management restores dignity, improves function, and enhances quality of life. Even modest pain relief empowers patients to engage in their care, connect with loved ones, maintain control, and rediscover meaning in their daily lives. Our goal with this book is to equip clinicians with the tools and knowledge to deliver care throughout the course of disease and across disciplines.

Acknowledgments

We, the editors of *Pain Management in Palliative Care for Patients with Cancer*, express our sincere gratitude to the individuals and organizations whose support and expertise made this volume possible.

We are especially grateful to our families for their unwavering encouragement and understanding during the preparation of this work. Their patience and support provided the foundation for our efforts throughout the writing and editorial process.

To the patients and families affected by cancer, we extend our most profound appreciation. Your courage and lived experiences have been a profound source of motivation. This book is dedicated to you, with the hope that it contributes meaningfully to improving the quality of life through better palliative care practices.

We thank our coauthors—distinguished colleagues from 18 countries—for their invaluable contributions. Their commitment to evidence-based, patient-centered care and their willingness to share clinical and research expertise have enriched the content and scope of this book.

We acknowledge the support of Avicenna Medical and Clinical Research Institute, whose resources and staff assistance significantly contributed to the development of this work. Special thanks to Mr. Ali Derakhshani for his contribution to the glossary and research assistance, which helped make complex concepts more accessible to a broad audience.

We are grateful to our peer reviewers, including Dr. Irina Filip, MD, DHSc, and Dr. Niseno Terhuja, MBBS, MD, for their insightful feedback and critical review, which enhanced the clarity and accuracy of the final manuscript.

We also extend our appreciation to **Elsevier** for commissioning and supporting the development of this title. We thank the editorial and production teams for their guidance and professionalism throughout the publishing process.

This volume represents a collaborative effort toward improving pain management and supportive care in oncology and palliative medicine. We hope it serves as a valuable resource for clinicians, researchers, and healthcare professionals worldwide.

Marin Golčić, MD, PhD
Tomi Kovačević, MD, PhD
Amir Radfar, MD, MPH, MSc, DHSc, ESCO Fellow

Part I

Introduction

Chapter 1

A new era in cancer treatment

Marin Golčić[1,2]
[1]*Tumor Clinic, Clinical Hospital Center Rijeka, Rijeka, Croatia,* [2]*Faculty of Medicine, University of Rijeka, Rijeka, Croatia*

Key learning objectives

- Understand the historical evolution of cancer treatment from ancient times to modern targeted therapy and immunotherapy.
- Recognize how advances in cancer treatment have blurred the distinction between curative and palliative approaches.
- Identify the importance of individualized, multidimensional cancer care that integrates disease modification with comprehensive symptom management.

The history of cancer and cancer treatment is intertwined with the history of humankind. Even mummies from ancient Egypt suffered from metastatic cancer, and due to the preservation of ancient papyrus scrolls, we know that surgery was used to try to treat cancer since around 1600 BCE. Different plant remedies were also used to treat cancer in the same time period, albeit both treatments had very limited success [1,2]. The ancient Greeks and Romans further improved on the protocol for antisepsis and used remedies such as mandragoras (with scopolamine as an active ingredient) and various sedatives for analgesia during the operations. Some also consider the treatment of Queen Atossa, in the 5th century BCE, to be the first recorded successful treatment of breast cancer, although other authors consider it to be an inflammatory mastitis [3–5].

However, there wasn't a significant breakthrough in cancer management during the ancient times mainly due to a deeply flawed understanding of cancer biology. From today's perspective, it almost seems unbelievable that it is over 3400 years since the first surgical attempts and herbal prescriptions to Virchow's advancement in the understanding of cancer cells in the 19th century [6]. A period of rapid scientific advances ensued, from the discovery of modern anesthesia for surgical procedures to X-rays in 1895, which were soon used as an attempt to cure cancer [7]. At the same time, Dr. Coley was famous for using streptococcal organisms to treat patients with inoperable cancers since 1891 [8]. However, this field was hampered for more than a century due to a lack of understanding of microbial organisms and the immune system, and the inability to produce reproducible, consistent results. Modern chemotherapy was not developed until the 1940s, and it was only in the 1950s that a metastatic cancer was successfully cured by chemotherapy [7,9]. It then took almost 20 years for the application of tamoxifen as the first targeted cancer treatment, while the use of monoclonal antibodies started in the 1980s [10,11].

At the same time, it took much longer to develop modern palliative care, and some consider the year 1967, just before tamoxifen was put into practice, to mark the start of the modern palliative care era. However, data suggest that various ointments such as cannabis were used for analgesia long before any attempt to treat cancer was even recorded in human history [12]. Despite the fact that symptom management was much more successful than cancer management for a long period of human history, palliative care has long been stigmatized as care only for the dying. A change in the perception began with the groundbreaking work of several researchers such as Temel et al. in 2010 [13]. The premise was simple: patients with newly diagnosed metastatic non–small-cell lung cancer would have a better quality of life if palliative care was started early, parallel with the diagnosis of the cancer. The study showed not only that early palliative care improves quality of life but also that it was associated with longer survival than those in the standard care group [13]. While some analyses confirmed the results by Temel et al. that a good symptom control can lead to longer survival, others did not. Furthermore, some trials have shown that home-based palliative care was associated with longer survival than hospital-based care [14–17]. Such results suggest that not all palliative care is the same and that an umbrella of different measures and approaches is grouped under the same definition.

Pain Management in Palliative Care for Patients With Cancer. DOI: https://doi.org/10.1016/B978-0-443-30050-9.00036-9

But why was that patient cohort recruited by Temel and colleagues [13]? One of the reasons was grim—metastatic non–small-cell lung cancer was lethal in patients with advanced disease. Data from the beginning of the 21st century report a median survival of up to 10 months, 1-year survival of less than 40%, with long-term survival deemed "improbable." Not only that, but less than a quarter of patients over the age of 65 actually received any kind of chemotherapy due to galloping disease and rapidly declining performance status. A particularly poor prognosis was reported for patients with brain metastases, less than 20% of whom survived 1 year following the diagnosis [18]. Hence, providing palliative care in such lethal, rapidly progressing disease was entirely appropriate and was often the only treatment available.

However, less than 15 years later, the distinction between the incurable and curable is less clear, and recent data show remarkable advances in cancer treatment. For example, metastatic non–small-cell lung cancers with ALK mutations treated with lorlatinib in the first-line setting achieved a staggering response—60% of the patients had no disease progression after 5 years of follow-up, with 71% of the patients achieving an intracranial complete response—basically an eradication of metastases. Remarkably, only 8% of patients had intracranial progression during the 5-year follow-up, compared with 79% of patients treated with the previous gold standard, crizotinib [19]. Similar remarkable results were shown for patients treated with targeted therapy whose cancers harbor mutations in *EGFR*, *KRAS G12C*, *ROS1*, *BRAF*, *MET*, *RET*, *NTRK*, and *HER2* genes [20]. Even multitarget kinase inhibitors (TKIs), which are not specific to a single mutation, changed the approach in the treatment of various metastatic diseases. Patients with metastatic gastrointestinal stromal tumors survived less than 2 years at the beginning of the 21st century, compared to the median survival of up to 80 months less than 20 years later, thanks to imatinib and other TKIs [21]. Targeted therapy is not the only mainstay of treatment, as immunotherapy has revolutionized oncology as well. Metastatic melanoma was considered a lethal cancer at the beginning of the 21st century, as fewer than 10% of the patients survived 3 years after the diagnosis. The discovery of checkpoint immunotherapy (anti-PD-1 and anti-CTLA-4) has led to recently published median survival of 72 months, with 43% of patients alive 10 years after the diagnosis of a metastatic disease [22]. Similarly, remarkable results were shown for patients with metastatic colorectal cancer with microsatellite instability. A previously lethal disease can now be treated with a combination immunotherapy, which allowed 74% of patients to be without signs of progression after 2 years, with an objective response rate of up to 69% [23].

Hence, for many patients with advanced cancer, a dramatic improvement has been reached in only one decade, and talking about a long-term remission or even cure is no longer a taboo topic nowadays. Furthermore, many of the modern cancer treatments not only improve survival but also improve quality of life and reduce pain levels, which will be demonstrated in the following chapters of this book. Such data suggest that even modern cancer treatments can be part of palliative care in the sense that they can lessen the cancer burden and pain levels. The same is true for palliative care as well, and since the distinction between the curable and incurable is constantly being erased, palliative care should not only be reserved for those patients who are deemed unlikely to survive for a substantial period of time. There is an ongoing debate on whether palliative care should be rebranded, as the name itself is viewed as one of the barriers to applying proper palliative care early [24].

Regardless of the names, caring for patients with cancer should undoubtedly incorporate treatments aimed at prolonging or even curing the disease and treatments aimed at alleviating the symptoms, particularly approaching the total pain with a multidisciplinary and multidimensional approach. Although we still do not have appropriate prognostic factors to precisely evaluate which patients will have a significant benefit from modern cancer treatments, patients with a significant tumor burden can have dramatic responses, particularly when a targetable mutation is present [18–20]. In addition, while many people primarily equate classical analgesics with pain management, the advances of different branches of medicine and the technological sector have led to an increased arsenal of options that can be used in pain management in patients with metastatic or inoperable cancer. Hence, a simplified three-step ladder from the WHO might not be the gold standard in pain management in cancer patients in the 21st century [25]. A complete and individualized cancer care, incorporating oncological, pharmacological, surgical, psychological, spiritual, emotional, nutritional, physical therapy, and other adjuvant measures, should be the approach used in everyday clinical practice.

References

[1] David AR, Zimmerman MR. Cancer: an old disease, a new disease or something in between? Nat Rev Cancer 2010;10(10):728–33. Available from: https://doi.org/10.1038/nrc2914.

[2] Nunn JF. Ancient Egyptian medicine. Norman (OK): University of Oklahoma Press; 2002.

[3] Chidiac EJ, Kaddoum RN, Fuleihan SF. Special article: mandragora: anesthetic of the ancients. Anesth Analg 2012;115(6):1437–41. Available from: https://doi.org/10.1213/ANE.0b013e318259ee4d.

[4] Sandison AT. The first recorded case of inflammatory mastitis–Queen Atossa of Persia and the physician Democedes. Med Hist 1959;3(4):317–22. Available from: https://doi.org/10.1017/s0025727300024820.

[5] Key TJ, Verkasalo PK, Banks E. Epidemiology of breast cancer. Lancet Oncol 2001;2(3):133–40. Available from: https://doi.org/10.1016/S1470-2045(00)00254-0.

[6] Virchow R. Cellular pathology. As based upon physiological and pathological histology. Lecture XVI–Atheromatous affection of arteries. 1858. Nutr Rev 1989;47(1):23–5. Available from: https://doi.org/10.1111/j.1753-4887.1989.tb02747.x.

[7] McAleer S. A history of cancer and its treatment: Presidential Address to the Ulster Medical Society. 7th October 2021. Ulster Med J 2022;91(3):124–9.

[8] Coley WB. The treatment of malignant tumors by repeated inoculations of erysipelas. With a report of ten original cases. 1893. Clin Orthop Relat Res 1991;262:3–11.

[9] American Cancer Society. A brief history of cancer [Internet]. Atlanta (GA): American Cancer Society; 2019 [cited 2024 Oct 7]. Available from: https://www.cancer.org/cancer/understanding-cancer/history-of-cancer.html.

[10] Jordan VC. Tamoxifen (ICI46,474) as a targeted therapy to treat and prevent breast cancer. Br J Pharmacol 2006;147(Suppl. 1):S269–76. Available from: https://doi.org/10.1038/sj.bjp.0706399.

[11] Oldham RK. Monoclonal antibodies in cancer therapy. J Clin Oncol 1983;1(9):582–90. Available from: https://doi.org/10.1200/JCO.1983.1.9.582.

[12] Pickett M, Cooley ME, Gordon DB. Palliative care: past, present, and future perspectives. Semin Oncol Nurs 1998;14(2):86–94. Available from: https://doi.org/10.1016/s0749-2081(98)80014-4.

[13] Temel JS, Greer JA, Muzikansky A, et al. Early palliative care for patients with metastatic non-small-cell lung cancer. N Engl J Med 2010;363(8):733–42. Available from: https://doi.org/10.1056/NEJMoa1000678.

[14] Sullivan DR, Chan B, Lapidus JA, et al. Association of early palliative care use with survival and place of death among patients with advanced lung cancer receiving care in the Veterans Health Administration. JAMA Oncol 2019;5(12):1702–9. Available from: https://doi.org/10.1001/jamaoncol.2019.3105.

[15] Huo B, Song Y, Chang L, et al. Effects of early palliative care on patients with incurable cancer: a meta-analysis and systematic review. Eur J Cancer Care 2022;31(6):e13620. Available from: https://doi.org/10.1111/ecc.13620.

[16] Haun MW, Estel S, Rücker G, et al. Early palliative care for adults with advanced cancer. Cochrane Database Syst Rev 2017;6(6):CD011129. Available from: https://doi.org/10.1002/14651858.CD011129.pub2.

[17] Hamano J, Takeuchi A, Mori M, et al. Comparison of survival times of advanced cancer patients with palliative care at home and in hospital. PLoS One 2023;18(4):e0284147. Available from: https://doi.org/10.1371/journal.pone.0284147.

[18] Schiller JH. Current standards of care in small-cell and non-small-cell lung cancer. Oncology 2001;61(Suppl. 1):3–13. Available from: https://doi.org/10.1159/000055386.

[19] Shaw AT, Bauer TM, de Marinis F, et al. First-line lorlatinib or crizotinib in advanced ALK-positive lung cancer. N Engl J Med 2020;383(21):2018–29. Available from: https://doi.org/10.1056/NEJMoa2027187.

[20] National Comprehensive Cancer Network. NCCN Clinical Practice Guidelines in Oncology: Non-Small Cell Lung Cancer. Version 11.2024. [Internet]. Available from: https://www.nccn.org/professionals/physician_gls/pdf/nscl.pdf [accessed 7 Oct 2024].

[21] Li B, Chen H, Yang S, et al. Advances in immunology and immunotherapy for mesenchymal gastrointestinal cancers. Mol Cancer 2023;22(1):71. Available from: https://doi.org/10.1186/s12943-023-01770-6.

[22] Wolchok JD, Chiarion-Sileni V, Gonzalez R, et al. Long-term outcomes with nivolumab plus ipilimumab or nivolumab alone versus ipilimumab in patients with advanced melanoma. J Clin Oncol 2022;40(2):127–37. Available from: https://doi.org/10.1200/JCO.21.02229.

[23] Lenz HJ, Van Cutsem E, Limon ML, et al. First-line nivolumab plus low-dose ipilimumab for microsatellite instability-high/mismatch repair-deficient metastatic colorectal cancer: the phase II CheckMate 142 study. J Clin Oncol 2022;40(2):161–70. Available from: https://doi.org/10.1200/JCO.21.01015.

[24] Dalal S, Palla S, Hui D, et al. Association between a name change from palliative to supportive care and the timing of patient referrals at a comprehensive cancer center. Oncologist 2011;16(1):105–11. Available from: https://doi.org/10.1634/theoncologist.2010-0161.

[25] Ventafridda V, Saita L, Ripamonti C, et al. WHO guidelines for the use of analgesics in cancer pain. Int J Tissue React 1985;7(1):93–6.

Chapter 2

Evolution of palliative care

Amir Radfar
Former Faculty Member, College of Medicine, University of Central Florida, Orlando, FL, United States

Key learning objectives

By the end of this chapter, readers will be able to:

- Define palliative care and outline different levels of palliative care services.
- Identify the turning points and developments that led to contemporary palliative care services.
- Explain how palliative care has become an integral part of cancer treatment across all stages of illness.
- Describe the structure and function of multidisciplinary teams in delivering comprehensive palliative care.
- Anticipate new developments, including research directions, personalized care planning, and digital health.

Introduction

The evolution of palliative care reflects a growing recognition that managing cancer and other serious illnesses requires more than disease-focused treatment. Palliative care expands the goals of medicine to include relief of suffering and support across physical, emotional, psychological, and spiritual domains [1]. From its roots in the mid–20th-century hospice movement to its integration within oncology and other specialties, palliative care has developed into a team-based discipline that benefits patients and families throughout the course of illness. Evidence from randomized controlled trials demonstrates that early palliative care interventions improve quality of life and clinical outcomes for patients with advanced cancer [2,3]. Guideline updates have reinforced this approach, highlighting the importance of early and routine integration of palliative care into oncology practice [4–6].

Defining palliative care and levels of service

The WHO defines palliative care as an approach that improves the quality of life of patients and families facing problems associated with life-threatening illness. It prevents and relieves suffering through early identification, correct assessment, and treatment of pain and other problems, whether physical, psychosocial, or spiritual [1].

Primary palliative care typically involves core palliative care services by primary care physicians. These include core pain management, emotional and psychosocial support, care coordination, and advance care planning. In contrast, palliative care specialists are trained physicians who offer specialist (secondary) level palliative care services for seriously ill and dying patients. These specialists work in outpatient clinics or as hospital-based consultants [7].

Tertiary palliative care services address rare, complex, or treatment-refractory needs. They are usually provided by highly specialized interdisciplinary teams, often including physicians with fellowship training in palliative medicine. In addition to patient care, these teams may contribute to education, quality improvement, clinical research, and policy development [8].

Historical background

Over the centuries, medical care has been influenced and shaped by philosophies, schools of thought, social norms, and scientific discoveries, and palliative care is no exception. Understanding these early ideas puts the current palliative care approach to the cancer treatment paradigm in perspective.

Pain Management in Palliative Care for Patients With Cancer. DOI: https://doi.org/10.1016/B978-0-443-30050-9.00007-2

Early historical roots

Ancient and medieval practices

The concern for relieving pain, suffering, and distress has deep historical roots in ancient and medieval medical traditions, even though palliative care was not yet established as a distinct medical discipline. Earlier medical traditions associated with figures such as Hippocrates, Avicenna, and Rhazes emphasized clinical care, symptom relief, and ethical responsibility toward patients. Avicenna's Canon of Medicine examined pain in detail, including its causes, types, and treatment methods, and described approaches to pain relief using analgesic agents, anesthetics, and antispasmodics [9]. This systematic concern with pain relief may be viewed as an early antecedent to later palliative principles.

During the medieval period, hospices and hospitals were often faith-based or charitable institutions that cared for the poor, sick, and vulnerable. Their care emphasized humanitarian and spiritual support, addressing not only bodily illness but also existential, emotional, and personal concerns at the end of life [10].

Renaissance to early modern era

During the Renaissance and early modern period, palliative care was still not established as a formal medical specialty. However, broader medical traditions continued to develop practices relevant to later palliative care, including symptom observation, medicinal treatment, wound care, lifestyle guidance, and ethical responsibility toward patients. These traditions should be viewed as indirect historical antecedents rather than as palliative care in the modern clinical sense.

From the mid-19th century, religious and philanthropic organizations began establishing dedicated homes for the dying. These early hospice-like institutions offered rudimentary but holistic support, including shelter, nursing attention, spiritual care, emotional comfort, and relief of distressing symptoms. They provided an important foundation for the modern hospice and palliative care movement [11].

Modern origins of palliative care

Dame Cicely Saunders and the hospice movement

Dame Cicely Saunders is widely considered the founder of the hospice movement. Beginning her professional career as a nurse (St. Thomas' Hospital), then a medical social worker, and finally as a physician, she brought a unique perspective to dealing with terminally ill patients. In the 1960s, Saunders was the first to introduce the concept of "total pain," which includes all aspects of pain: physical, emotional, social, and spiritual. Her holistic approach was unheard of, as it was considered essential to look after all aspects of a person's well-being [12].

Development of St Christopher's Hospice

In 1967, Cicely Saunders founded St Christopher's Hospice in London. This establishment pioneered the modern palliative care concept. The facility provided comprehensive end-of-life care, including pain management, recognizing that pain extends beyond the physical. Its model integrated psychological and spiritual support, with medical staff trained to offer counseling and guidance to both patients and their families [11].

Recognition by the World Health Organization

In 1990, the WHO formally defined palliative care as a key component of care for patients with life-threatening illnesses, emphasizing pain relief and symptom control alongside curative and preventive strategies [13]. This marked a major shift in global health policy, positioning palliative care as an essential part of comprehensive cancer management. The WHO definition helped the integration of palliative services into national health systems and encouraged countries to adopt supportive policies and training programs [13].

Integration of palliative care in cancer treatment

Early integration and timing

Palliative care is now recognized as a core component of cancer treatment—providing symptom relief, psychosocial support, and goal-aligned communication throughout the illness trajectory [14,15].

A landmark study by Temel et al. demonstrated that early integration of palliative care into the treatment of patients with non–small-cell lung cancer (NSCLC) significantly improves clinical outcomes. Patients receiving early palliative care, compared to those receiving standard care, had better quality-of-life scores (98.0 vs 91.5; $P = .03$), fewer depressive symptoms (16% vs 38%; $P = .01$), lower rates of aggressive end-of-life care (33% vs 54%; $P = .05$), and longer median survival (11.6 vs 8.9 months; $P = .02$) [3].

These findings challenged the traditional view of palliative care as solely end-of-life support and helped reframe it as a proactive intervention that enhances overall cancer care. Early integration also supports advance care planning and shared decision-making—essential elements of patient-centered oncology. These clinical benefits have reshaped how professional societies view the timing and role of palliative care in oncology. The American Society of Clinical Oncology (ASCO) recommendations support the integration of palliative care in the earlier stages of treatment, particularly for patients with advanced solid tumors or hematologic malignancies [14]. The European Society for Medical Oncology (ESMO) also advocates for patient-centered care, including supportive and palliative interventions, from diagnosis through end of life or survivorship [15].

Stages of integration and shared decision-making

Key transition points

Effective implementation of palliative care requires recognizing "key transition points" across the cancer trajectory—moments when patient needs shift significantly. These include the initial diagnosis of a life-limiting illness, evidence of disease progression, changes in prognosis, or referral to hospice. At each of these inflection points, clinicians should reassess goals of care, update symptom management strategies, and adjust the palliative care plan to reflect evolving clinical realities and patient preferences [6,14,15].

For patients with metastatic NSCLC, palliative care should be integrated early alongside oncology treatment [3,14]. Early interventions should include systematic assessment and guideline-based management of pain, dyspnea, cough, fatigue, anxiety, depression, and psychosocial distress [6]. Early palliative care in metastatic NSCLC has been shown to improve quality of life, reduce depressive symptoms, reduce aggressive end-of-life care, and improve survival [3]. Open discussions about prognosis, treatment goals, and the risks and benefits of chemotherapy versus supportive care are essential [6,14]. Advance care planning, including documentation of advance directives and preferences for end-of-life care, should also be initiated early in the treatment course [6,14].

Shared decision-making

Shared decision-making is a collaborative process in which patients and their families engage with clinicians to make treatment choices that align with the patient's goals, values, and quality-of-life preferences. Rather than relying on unilateral medical decisions, clinicians guide patients through the risks, benefits, and trade-offs of various available options. For example, when an oncologist recommends chemotherapy for a patient with advanced cancer, the care team should also explore the patient's priorities—such as minimizing hospital visits or preserving functional independence. If the burden of treatment outweighs the potential benefit from the patient's perspective, they may choose supportive care focused on comfort and symptom relief [16].

Shared decision-making is also important in chronic or progressive illnesses, where patients may eventually lose capacity to express their wishes. Early discussions about future care preferences allow patients to define priorities while they are still able to communicate, helping ensure that later decisions remain consistent with their values [16].

Development of a multidisciplinary team in palliative care

The multidisciplinary team is instrumental in palliative care practices. A multidisciplinary team uses specialized skills and patient-centered care to deliver holistic palliative care. A typical team is comprised of different providers and specialists. Some of the job definitions of the interdisciplinary team are listed here:

Palliative care physicians: Manage complex symptoms such as pain, dyspnea, fatigue, and anxiety; guide advance care planning; and coordinate interdisciplinary care.

Oncologists: Direct cancer-specific treatment plans and collaborate with the palliative team to balance disease-modifying therapy with symptom control.

Nurses: Monitor symptoms, administer medications, provide patient education, and serve as a communication bridge between patients, families, and clinicians.

Social workers: Address emotional, psychosocial, and practical concerns; facilitate advance directives and resource coordination; and support caregivers.

Therapists (psychologists, counselors): Provide mental health support, teach coping strategies, and assist with grief and existential distress.

Pharmacists: Optimize medication regimens for symptom control, monitor drug interactions, and support deprescribing where appropriate.

Dietitians: Assess nutritional status, recommend dietary modifications for symptom relief (e.g., nausea, anorexia, constipation), and counsel patients and families.

Chaplains or spiritual care providers: Support patients' spiritual and existential needs, offer rituals and counseling, and assist with meaning-making at the end of life.

Ethicists: Provide guidance in complex decision-making situations, clarify patient values, and mediate conflicts around goals of care.

Volunteers: Offer companionship, emotional support, and practical assistance to patients and families, particularly for those with limited social support.

Family members/caregivers: Serve as core partners in care, often managing daily needs, advocating for patient preferences, and participating in decision-making.

This team-based approach is consistent with international guidelines, including those from the ESMO [15].

Current trends and future directions

Palliative care is evolving rapidly in response to advances in personalized care planning, digital health, and global health policy. These developments aim not only to improve symptom control but also to expand access, equity, and quality of care across diverse patient populations.

Personalized care planning allows clinicians to tailor care plans based on disease trajectory, treatment response, symptom burden, tolerability, available resources, and patient goals [16]. In oncology, treatment decisions should be aligned with quality of life, prognosis, toxicity, access, and patient preferences, including when supportive and palliative interventions are integrated from diagnosis through survivorship or end-of-life care [15]. This individualized approach informs palliative decisions, particularly in advanced cancer care [15,16].

Digital health and telemedicine are emerging tools for palliative care delivery. Remote monitoring tools, virtual consultations, and mobile applications may support access to specialist input, particularly for patients in rural or underserved areas [17]. Artificial intelligence and wearable technologies are emerging areas for symptom monitoring and risk prediction, but they require further validation, privacy safeguards, and equity assessment before routine implementation [17,18].

Beyond technology, addressing global disparities remains a critical priority. Access to palliative care remains severely limited in many low- and middle-income countries; worldwide, only about 14% of people who need palliative care receive it [1]. The Lancet Commission frames palliative care and pain relief as an imperative of universal health coverage [19]. WHO guidance also emphasizes integrating palliative care and symptom relief into humanitarian emergency and crisis responses [20], and the COVID-19 pandemic highlighted the key role of palliative care in responding to large-scale suffering [21]. Expanding universal health coverage to include palliative care and investing in continuing education for physicians, nurses, and allied health professionals are essential steps toward equity [19,22]. Structured CAPC (Center to Advance Palliative Care) education supports workforce development and integration of palliative care into clinical practice [22].

Educational resources and training opportunities to support this integration are listed at the end of this chapter.

Research in palliative care

Despite advances in clinical care, palliative care research remains underfunded and underrepresented compared to other medical disciplines. Nonetheless, its importance is growing, particularly in addressing patient-centered outcomes, strengthening research infrastructure, and improving care quality [23].

Recent efforts have moved beyond proving palliative care's efficacy to determining optimal delivery models, particularly in low-resource settings, across disease types, and in low- and middle-income countries [23,24]. One research priority is the development and validation of patient-reported outcome measures to guide care. For example, the Edmonton Symptom Assessment System (ESAS) is widely used in oncology and palliative care to track symptoms such as pain, fatigue, anxiety, and nausea. It has been validated across multiple populations and is integral to real-time symptom monitoring and care adjustments [25].

Other research areas include the integration of palliative care in noncancer illnesses, caregiver burden, and implementation science focused on translating guidelines into practice. However, critical gaps persist—especially in pediatric care, community-based interventions, and equity-driven models for diverse cultural and socioeconomic contexts [23]. Strengthening the global research infrastructure is essential to ensure all patients have access to high-quality, evidence-based palliative care.

Conclusion

Palliative care has evolved well beyond its historical roots in hospice care. Once primarily associated with end-of-life support, it is now recognized as a vital component of comprehensive cancer management. Today, palliative care addresses the needs of patients at any stage of a serious illness, offering holistic, interdisciplinary support that enhances quality of life for both patients and their families. Looking ahead, innovations in personalized care planning, digital health, and global education are poised to expand the reach and impact of palliative care—ensuring that comfort, dignity, and patient-centered care remain integral to the cancer journey.

Additional resources

There is a collection of books and articles that will be useful to anyone considering deepening the palliative care practice; it contains important aspects of palliative care. Possible readings include:

- *The Oxford Textbook of Palliative Medicine* by Geoffrey Hanks, Nathan I. Cherny, Nicholas A. Christakis, Marie Fallon, Stein Kaasa, and Russell K. Portenoy covers a wide range of topics relevant to palliative care.
- *Palliative Care: Core Skills and Clinical Competencies* by Linda L. Emanuel and S. Lawrence Librach offers a comprehensive overview of essential skills for effective palliative care.
- Key articles such as *Early Palliative Care for Patients With Metastatic Non-Small-Cell Lung Cancer* by Jennifer S. Temel et al. underscore the benefits of integrating palliation early in the treatment process of terminal illnesses.

Online courses and workshops

Various online courses and workshops are available for healthcare professionals seeking to expand their palliative care knowledge and skills. These educational opportunities offer flexibility and access to expert instruction. Notable programs include:

- The ELNEC courses provide specialized training for nurses in palliative care.
- The palliative care education and practice program offered by Harvard Medical School is designed for healthcare professionals to enhance their palliative care skills.
- Online modules from the CAPC cover pain management, communication strategies, and symptom management.

Professional organizations and journals

The commitment to seeking information and acquiring knowledge within professional organizations and their journals is important in addressing the problems related to research and practice within the area of palliative care. Pivotal organizations are:

- The International Association for Hospice and Palliative Care (IAHPC) promotes global access to palliative care.
- The American Academy of Hospice and Palliative Medicine (AAHPM) offers resources, conferences, and networking opportunities for professionals in the field.
- Leading journals, such as *the Journal of Pain and Symptom Management* and *Palliative Medicine*, publish cutting-edge research and reviews on palliative care topics.

Acknowledgments

This chapter was prepared with the assistance of AI-based tools for literature synthesis, formatting, reference management, and language refinement. All content was reviewed, verified, and curated by the author in accordance with established publication ethics and authorship guidelines.

References

[1] World Health Organization. Palliative care. 5 August 2020. Available from: https://www.who.int/news-room/fact-sheets/detail/palliative-care.

[2] Bakitas M, Lyons KD, Hegel MT, et al. Effects of a palliative care intervention on clinical outcomes in patients with advanced cancer: the project ENABLE II randomized controlled trial. JAMA 2009;302(7):741–9. Available from: https://doi.org/10.1001/jama.2009.1198.

[3] Temel JS, Greer JA, Muzikansky A, et al. Early palliative care for patients with metastatic non–small-cell lung cancer. N Engl J Med 2010;363(8):733–42. Available from: https://doi.org/10.1056/NEJMoa1000678.

[4] Ferrell BR, Temel JS, Temin S, et al. Integration of palliative care into standard oncology care: American Society of Clinical Oncology clinical practice guideline update. J Clin Oncol 2017;35(1):96–112. Available from: https://doi.org/10.1200/JCO.2016.70.1474.

[5] American Society of Clinical Oncology. Integration of palliative care into standard oncology care: ASCO guideline; 2023. Available from: https://www.asco.org/guidelines/clinical-practice-guidelines/palliative-care.

[6] National Comprehensive Cancer Network. NCCN clinical practice guidelines in oncology: palliative care, Version 2.2024; 2024. Available from: https://www.nccn.org/professionals/physician_gls/pdf/palliative.pdf.

[7] Radfar A. Integrating palliative care into primary care: an educational project to meet an unmet need. In: Schmidt-Straßburger U, editor. Improving oncology worldwide: education, clinical research and global cancer care. Cham: Springer; 2022. p. 117–21. Available from: https://doi.org/10.1007/978-3-030-96053-7_15.

[8] Carmont SAS. An integrated approach to palliative care in the community: exploring a model of multidisciplinary collaboration for frail older people; 2020 Doctoral dissertation, University of Queensland. Available from: https://doi.org/10.14264/uql.2020.125.

[9] Çetkin M, Ergin G, Mehmetoğlu G, et al. The pain in the Canon of Medicine: types, causes, and treatment. Eur J Ther 2019;25(3):164–9. Available from: http://doi.org/10.5152/EurJTher.2018.551.

[10] Rawcliffe C. Medicine and society in later medieval England. Stroud (UK): Sutton Publishing; 1997.

[11] Clark D. To comfort always: a history of palliative medicine since the nineteenth century. Oxford University Press; 2016.

[12] Monroe DB. Cicely Saunders: A life and legacy. J Palliat Med. 2019;22(2):234–35. Available from: https://doi.org/10.1089/jpm.2018.0634.

[13] World Health Organization. Cancer pain relief and palliative care: report of a WHO expert committee (technical report series No. 804); 1990. Available from: https://iris.who.int/handle/10665/39524.

[14] Sanders JJ, Temin S, Ghoshal A, et al. Palliative care for patients with cancer: ASCO guideline update. J Clin Oncol 2024;42(19):2336–57. Available from: https://doi.org/10.1200/JCO.24.00542.

[15] Jordan K, Aapro M, Kaasa S, et al. European Society for Medical Oncology (ESMO) position paper on supportive and palliative care. Ann Oncol 2018;29(1):36–43. Available from: https://doi.org/10.1093/annonc/mdx757.

[16] Coulter A, Entwistle VA, Eccles A, et al. Personalised care planning for adults with chronic or long-term health conditions. Cochrane Database Syst Rev 2016;3:CD010523. Available from: https://doi.org/10.1002/14651858.CD010523.pub2.

[17] Nwosu AC, Stanley S, Norris J, et al. Digital health and palliative care. In: Technologies in palliative care for older people. Cham: Springer Nature Switzerland; 2026. p. 17–36. Available from: https://doi.org/10.1007/978-3-032-04277-4_3.

[18] Topol E. Deep medicine: how artificial intelligence can make healthcare human again. New York, NY: Basic Books; 2019.

[19] Knaul FM, Farmer PE, Krakauer EL, et al. Alleviating the access abyss in palliative care and pain relief—an imperative of universal health coverage: the Lancet Commission report. Lancet 2018;391(10128):1391–454. Available from: https://doi.org/10.1016/S0140-6736(17)32513-8.

[20] World Health Organization. Integrating palliative care and symptom relief into the response to humanitarian emergencies and crises: a WHO guide. 2018. Available from: https://www.who.int/publications/i/item/9789241514460.

[21] Radbruch L, Knaul FM, de Lima L, et al. The key role of palliative care in response to the COVID-19 tsunami of suffering. Lancet 2020;395(10235):1467–9. Available from: https://doi.org/10.1016/S0140-6736(20)30964-8.

[22] Center to Advance Palliative Care. National palliative care continuing education programs: building workforce capacity; 2021. Available from: https://www.capc.org.

[23] Morrison RS, Aldridge M, Block J, et al. The National Palliative Care Research Center: ten years of promoting and developing research in palliative care. J Palliat Med 2018;21(11):1548–57. Available from: https://doi.org/10.1089/jpm.2018.0204.

[24] Pramesh CS, Badwe RA, Bhoo-Pathy N, et al. Priorities for cancer research in low- and middle-income countries: a global perspective. Nat Med 2022;28(4):649–57. Available from: https://doi.org/10.1038/s41591-022-01738-x.

[25] Bruera E, Kuehn N, Miller MJ, et al. The Edmonton Symptom Assessment System (ESAS): a simple method for the assessment of palliative care patients. J Palliat Care 1991;7(2):6–9. Available from: https://doi.org/10.1177/082585979100700202.

Chapter 3

Early palliative care

Laila Kamal and Andrew Tweddle
Royal Marsden NHS Foundation Trust, London, United Kingdom

Key learning objectives

- To understand the role of early palliative care in supporting patients through the illness journey.
- To understand some of the evidence that exists to support the practice of early palliative care.
- To appreciate some of the different models of delivering early palliative care.
- To understand some of the challenges faced in the delivery of early palliative care alongside standard treatments.

Introduction

The origins of the field of palliative care are rooted in the hospice movement associated with Dame Cicely Saunders and Elisabeth Kubler-Ross, championing the care of patients with terminal illnesses with a multidisciplinary approach to alleviating, or at least acknowledging, "total pain" and suffering [1]. Over the decades, the field has evolved into expertise in symptom control with a holistic approach in looking after the physical, psychosocial, and spiritual needs of patients and their families [2]. Modern-day palliative care aspires to work side by side with other specialties, including oncology, rather than take over from them when there are no further "curative" treatment options available [3]. Despite the evolution of this specialty, the stigma of the association with death and dying remains at the forefront of most people's minds, including healthcare professionals as well as patients and their families.

Well-designed studies have consistently evidenced that early palliative care delivered concurrently with oncology treatments enables patients to have a better quality of life (QoL), reduced symptom burden, and enhanced communication [4–10]. Furthermore, early palliative care has been shown to reduce the use of acute care services, including inpatient hospital admissions and emergency department attendances [11,12], as well as improve QoL and end-of-life care [13].

The 2010 study by Temel et al. was a landmark randomized controlled trial that provided evidence for the benefit of early palliative care versus standard oncological care amongst patients with a new diagnosis of metastatic non–small-cell lung cancer [4]. The study included 151 patients with a performance status of 0–2, within 8 weeks of diagnosis of cancer, who were randomized to oncological care with or without specialist palliative care. The early palliative care intervention provided by specialist palliative care physicians and advanced nurse practitioners reported clinically meaningful improvements in QoL (Trial Outcome Index, 59 vs 53; $P = .009$) and fewer depressive symptoms (16% vs 38%, $P = .01$) as well as better end-of-life care (33% vs 54%, $P = .05$) and longer overall survival (11.6 vs 8.9 months, $P = .02$). Patients in the intervention arm also had a better understanding of their disease as well as a lower rate of aggressive chemotherapy at the end of life, suggesting a link between these two phenomena.

In 2014, Zimmermann et al. published their findings on a cluster randomized controlled trial of early palliative care versus routine oncology care for 461 patients from 24 clinics treating lung, gastrointestinal (GI), genitourinary, breast, and gynecological cancers [6]. The palliative care intervention provided by a specialist palliative care physician and a nurse in a free-standing clinic demonstrated improved QoL (Functional Assessment of Chronic Illness Therapy–Spiritual Well-Being [FACIT-Sp] Scale and Quality of Life at the End of Life [QUAL-E] Scale), symptom severity for patients (Edmonton Symptom Assessment System [ESAS]), and satisfaction of care (FAMCARE-P16). Four hundred and sixty-one patients completed baseline measures (228 intervention, 233 control), and 393 completed at least one follow-up assessment. Using five different assessment scales, significant changes in scores were observed at 4 months for all outcomes apart from

Pain Management in Palliative Care for Patients With Cancer. DOI: https://doi.org/10.1016/B978-0-443-30050-9.00030-8

problems with medical interventions (Cancer Rehabilitation Evaluation System Medical Interaction Subscale [CARES-MIS]).

Maltoni et al. (2016) reported on a multicenter randomized controlled trial of systematic early versus on-demand palliative care for patients with metastatic or locally advanced inoperable pancreatic cancer [14]. Two hundred and seven patients were recruited and randomized to receive "standard cancer care plus on-demand early palliative care" ($n = 100$) or "standard cancer care plus systematic early palliative care" ($n = 107$). The primary outcome was change in QoL evaluated through the Functional Assessment of Cancer Therapy—hepatobiliary questionnaire between baseline (T0) and after 12 weeks (T1), particularly the integration of physical, functional, and Hepatic Cancer Subscale (HCS) combined in the Trial Outcome Index (TOI). The palliative care intervention delivered by a specialist palliative care physician demonstrated an improved QoL at 3 months with QoL scores at T1 of the TOI scale and HCS of 84.4 versus 78.1 ($P = .022$) and 52.0 versus 48.2 ($P = .008$), respectively, for the interventional and standard arms. There was no difference in overall survival between treatment arms.

In 2017, Temel et al. carried out a further randomized controlled trial of early palliative care for advanced lung and noncolorectal GI cancers delivered by palliative care physicians and advanced nurse practitioners [15]. Patients newly diagnosed with incurable lung or noncolorectal GI cancer were randomized to receive either early integrated palliative care and oncology care ($n = 175$) or usual care ($n = 175$) between May 2011 and July 2015. The primary end point was change in QoL from baseline to week 12, per scoring by the Functional Assessment of Cancer Therapy–General Scale. Secondary end points included change in QoL from baseline to week 24, change in depression per the Patient Health Questionnaire-9, and differences in end-of-life communication. Intervention patients (vs usual care) reported greater improvement in QoL from baseline to week 24 (1.59 vs −3.40; $P = .010$) but not week 12 (0.39 vs −1.13; $P = .339$). Intervention patients also reported lower depression at week 24, controlling for baseline scores (adjusted mean difference, −1.17 [95% CI, −2.33 to −0.01]; $P = .048$). Intervention effects varied by cancer type, such that intervention patients with lung cancer reported improvements in QoL and depression at 12 and 24 weeks, whereas usual care patients with lung cancer reported deterioration. Patients with GI cancers in both study groups reported improvements in QoL and mood by week 12. Intervention patients versus usual care patients were more likely to discuss their wishes with their oncologist if they were dying (30.2% vs 14.5%; $P = .004$). The conclusions drawn were that for patients with newly diagnosed incurable cancers, early integrated palliative care improved QoL and other pertinent outcomes, with variable effects by cancer type. Early integrated palliative care was felt to be the most effective when it was targeted to the specific needs of each patient population.

In 2018, Vanbutsele et al. reported on a randomized controlled trial of early systematic palliative care for patients with advanced solid tumor diagnoses [16]. One hundred and eighty-six patients were enrolled and randomly assigned to the early and systematic palliative care group (92 patients) or the standard oncological care group (94 patients). The primary end point was change in the Global Health Status/QoL scale assessed by the 30-item European Organisation for Research and Treatment of Cancer Quality of Life Questionnaire Core (EORTC QLQ-C30) at 12 weeks. The overall QoL score at 12 weeks, by the EORTC QLQ-C30, was 54.39 (95% CI, 49.23–59.56) in the standard oncological care group versus 61.98 (95% CI, 57.02–66.95) in the early and systematic palliative care group (difference, 7.60 [95% CI, 0.59–14.60]; $P = .03$). The intervention was delivered by palliative care nurses and specialized palliative care physicians available on demand and demonstrated improved QoL compared to standard care.

The Lancet Commission endorsed palliative care as "an essential component of comprehensive care" [17]. The WHO, the European Society for Medical Oncology (ESMO), and the American Society of Clinical Oncology (ASCO) all advocate palliative care integration into standard oncology care early in the disease course, alongside active cancer treatment [18–21].

Challenges

Despite the body of evidence supporting early palliative care, referrals to this specialty continue to come late in the patients' journeys, thus depriving them of the opportunity to benefit from a thorough assessment and management of their palliative care needs. Referrals to palliative care based on estimated prognosis are problematic, as such estimates are notoriously inaccurate, especially in the era of novel therapeutic agents. Referrals based solely on the stage of disease risk miss patients with early-stage disease and unmet palliative care needs. Oncologist-driven referrals can be subjective based on resources and variable thresholds for reporting of symptoms and supportive care needs.

Timing of referrals: early or *timely* palliative care

A significant challenge faced by healthcare professionals remains the timing of palliative care interventions [22]. There is no single model of palliative care delivery that is suitable for all patients and settings. Needs-based, personalized patient-centered palliative care delivered at the optimal time for each individual is more favorable as compared to a blanket approach with an early but fixed timeline for palliative care intervention. Additionally, the palliative care framework required would be led by resources including but not limited to access to education, training in the specialty, cultural acceptability of palliative care, and national health policies.

Palliative care≠no treatment≠no hope≠death

In addition to the timing of referrals, other challenges that palliative care professionals face are (1) the impact and understanding of the term "palliative care" and the association with "death and dying" and (2) the association with absence of treatment and hope [22,23].

The semantics of the term "palliative care" remains perhaps the biggest barrier to the acceptance of this field of modern medicine. Healthcare professionals as well as patients and caregivers find it challenging to grasp the ethos of the specialty, especially when they have limited or no experience of palliative care [24,25]. There is a leaning toward alternative names like "supportive care" and "comfort care," which makes it more acceptable and less alarming. There does also remain a perception that palliative care is appropriate in the last days and months of life because of a perceived reduction in care that does not work concurrently with active treatment being delivered aimed at curing the disease. These misconceptions about the scope of the specialty are shared by patients and caregivers.

Qualitative studies exploring healthcare professionals and patients and caregiver perceptions after interactions with palliative care services have a different perspective. They describe an improved understanding of the specialty as well as an appreciation for the improvement in the QoL, the individualized symptom control input, the elements of holistic care, and the support in planning for the future and in decisions related to ongoing care and treatment [26–29]. It was interesting to note that the term "palliative care" remained a source of concern, and the rebranding of the specialty remained a significant theme, given the emotional impact and negative connotations of palliative care and its association with end of life.

Resources

While early palliative care integration is encouraged and even welcomed in some countries, basic palliative care services are a scant resource for over half the world's population. Even when palliative care services are readily available, in practical terms, it is unlikely to be feasible, efficient, or even necessary for every patient with cancer to be reviewed by a specialist palliative care team [30,31]. Sensible and pragmatic allocation of specialist resources is required. A major challenge faced by palliative care as a specialty is how to address the increased workload brought about by proactive early identification of patients for palliative care referral. This is coupled with forecasts of rapidly increasing volumes of patients needing palliative care for end-of-life care in the next 20 years, which adds to the challenge of resource allocation [32,33]. Multifaceted solutions have been proposed, including prioritizing the availability of existing specialist palliative care resources for those with the greatest needs [34–36], the use of telephone or virtual consultations [37–39], and upskilling oncologists and students in basic and generalist palliative care approaches, competencies, and communication [40].

Models for early palliative care in the outpatient setting

Data supporting the benefits of early palliative care and taking the challenges into account supports consideration and development of multiple models of early palliative care delivery in the outpatient setting [41]. Just as there is no single right time point for the introduction of palliative care for all patients, there is no single model that is appropriate for all settings.

Embedded palliative care clinics with variable team makeup

In the embedded model, palliative care is embedded into oncology clinics and delivers specialist palliative care on the same day and in the same space either via palliative care physicians or via advanced nurse practitioners only, or an interdisciplinary team including physicians and allied health professionals. The advantages of this include improved opportunities for integration, support with communication and symptom control, and normalization of palliative care as part of standard cancer care.

TABLE 3.1 Palliative care referral criteria.

Severe physical symptoms
Severe emotional symptoms
Request for hastened death
Spiritual or existential crisis
Assistance with decision-making or care-planning
Patient request for referral
Delirium
Spinal cord compression
Brain or leptomeningeal metastases
Within 3 months of advanced cancer diagnosis for patients with median survival of 1 year or less
Progressive disease despite second-line therapy

Adapted from: Hui D, Mori M, Watanabe SM, et al. Referral criteria for outpatient specialty palliative cancer care: an international consensus. Lancet Oncol 2016;17(12):e552–9.

TABLE 3.2 Early palliative care referral criteria—Royal Marsden Hospital.

Metastatic cancer progressing after first-line therapy
Performance status Eastern Cooperative Oncology Group Scale (ECOG) 2 and deteriorating
Acute oncology or unplanned admission
Severe or overwhelming symptoms
Anorexia, hypercalcaemia, or any effusion
Moderate or severe psychological or existential distress
Complex social issues

Adapted from: Kamal L, Kano Y, Stevens AM, et al. Assessing the sensitivity and acceptability of the Royal Marsden palliative care referral "triggers" tool for outpatients with cancer. Support Care Cancer 2024;32(11):730.

One type of embedded model of delivery of early palliative care uses bespoke referral criteria to identify those patients who may benefit from early palliative care review. Multiple sets of referral criteria or screening tools are in various stages of development, including but not limited to the 11 major criteria published by Hui et al. (Table 3.1) [42] and the locally designed "triggers" tool from the Royal Marsden Hospital in the United Kingdom (Table 3.2) [43]. These tools usually encompass several physical and psychosocial domains, prompting an early palliative care review in the oncology clinic. The benefits of using shared clinical space alongside oncologists to present the involvement as part of standard care have been clear to see and have strengthened relationships with oncologists to be able to deliver the best care possible to patients.

Free-standing palliative care clinics

Free-standing palliative care clinics can provide a more comprehensive offering and operate from dedicated clinic space within the hospital, which can offer elective and emergency appointments. In this model, the palliative care team, comprising clinicians as well as nursing and allied health professionals, provides holistic care encompassing physical, psychological, spiritual, and social domains.

Referral methods can include oncologist-initiated judgement or criteria or using referral criteria or screening tools. There are potential disadvantages of this kind of approach. The stand-alone setup does not lend itself to a tightly integrated way of working and, as such, can be seen as separate from the "normal" cancer journey with a stepped approach *from* cancer treatment *to* palliative care.

Telehealth palliative care clinics

Palliative care clinics run by telephone or by video link have become more common after their use during the COVID-19 pandemic. A Cochrane review in 2020 concluded that telephone interventions provided a convenient way of supporting self-management of cancer-related symptoms for patients with additional benefits, including resource containment and the potential for support from the voluntary sector [44]. The review authors were unable to determine whether telephone input alone or in combination with other interventions provided optimal reduction in symptoms.

The advantages for patients are clear: being able to attend a consultation in the comfort of the home without the need to travel. For the clinician, the lack of physical clinic space can be a challenge, and the need for only a private desk space to conduct the consultation can be convenient. However, the lack of face-to-face contact can make clinical assessment difficult and introduces the risk of missing important signs or features alongside the potential negative impact upon the doctor–patient relationship. Despite its disadvantages, the telehealth model continues to be an attractive option for patients as well as healthcare providers [45], and there may be additional evidence supporting this with the ongoing digital evolution.

Conclusions

Palliative care integrated early on with cancer care and treatments provides a wide range of benefits, including improved patient outcomes, improved symptom control, improved QoL, and a better understanding of disease, as evidenced by multiple randomized controlled trials and metaanalyses.

Close communications between oncologists and specialist palliative care professionals, as well as between healthcare professionals and patients, can help overcome challenges, including the appropriateness and timeliness of palliative care interventions, a better understanding of the scope of the palliative care specialty, and improved resource utilization to help the maximum number of patients with the most severe palliative care needs.

There are several different models of early or timely palliative care provision, which can be targeted to the needs of different groups of patients or healthcare organizations with no evidence of superiority of one model over the others at this stage of research.

New avenues of healthcare provision that need further research are virtual or telehealth models. These could increase the reach of palliative care provision across wider geographical domains but may diminish the essence of true holistic care, which remains the primary essence of good and effective palliative care.

References

[1] Hui D, Kim YJ, Park JC, et al. Integration of oncology and palliative care: a systematic review. Oncologist 2015;20(1):77–83. Available from: https://doi.org/10.1634/theoncologist.2014-0312.

[2] World Health Organization. Definition of palliative care; 2012. Available from: https://www.who.int/cancer/palliative/definition/en/#.XJ68GpU9hPU.mendeley.

[3] Higginson I. Palliative care: a review of past changes and future trends. J Public Health Med 1993;15(1):3–8. Available from: https://doi.org/10.1093/oxfordjournals.pubmed.a042817.

[4] Temel JS, Greer JA, Muzikansky A, et al. Early palliative care for patients with metastatic non-small-cell lung cancer. N Engl J Med 2010;363(8):733–42. Available from: https://doi.org/10.1056/NEJMoa1000678.

[5] Ferrell B, Sun V, Hurria A, et al. Interdisciplinary palliative care for patients with lung cancer. J Pain Symptom Manage 2015;50(6):758–67. Available from: https://doi.org/10.1016/j.jpainsymman.2015.07.005.

[6] Zimmermann C, Swami N, Krzyzanowska M, et al. Early palliative care for patients with advanced cancer: a cluster-randomised controlled trial. Lancet 2014;383(9930):1721–30. Available from: https://doi.org/10.1016/S0140-6736(13)62416-2.

[7] Bakitas MA, Tosteson TD, Li Z, et al. Early versus delayed initiation of concurrent palliative oncology care: patient outcomes in the ENABLE III randomized controlled trial. J Clin Oncol 2015;33(13):1438–45. Available from: https://doi.org/10.1200/JCO.2014.58.6362.

[8] King JD, Eickhoff J, Traynor A, et al. Integrated onco-palliative care associated with prolonged survival compared to standard care for patients with advanced lung cancer: a retrospective review. J Pain Symptom Manage 2016;51(6):1027–32. Available from: http://doi.org/10.1016/j.jpainsymman.2016.01.003.

[9] Hannon B, Swami N, Rodin G, et al. Experiences of patients and caregivers with early palliative care: a qualitative study. Palliat Med 2017;31(1):72–81. Available from: https://doi.org/10.1177/0269216316649126.

[10] Fulton JJ, LeBlanc TW, Cutson TM, et al. Integrated outpatient palliative care for patients with advanced cancer: a systematic review and meta-analysis. Palliat Med 2019;33(2):123–34. Available from: https://doi.org/10.1177/0269216318812633.

[11] Brumley R, Enguidanos S, Jamison P, et al. Increased satisfaction with care and lower costs: results of a randomized trial of in-home palliative care. J Am Geriatr Soc 2007;55(7):993–1000. Available from: https://doi.org/10.1111/j.1532-5415.2007.01234.x.
[12] McNamara BA, Rosenwax LK, Murray K, et al. Early admission to community-based palliative care reduces use of emergency departments in the ninety days before death. J Palliat Med 2013;16(7):774–9. Available from: https://doi.org/10.1089/jpm.2012.0403.
[13] Ziegler LE, Craigs CL, West RM, et al. Is palliative care support associated with better quality end-of-life care indicators for patients with advanced cancer? A retrospective cohort study. BMJ Open 2018;8(1):e018284. Available from: https://doi.org/10.1136/bmjopen-2017-018284.
[14] Maltoni M, Scarpi E, Dall'Agata M, et al. Systematic versus on-demand early palliative care: results from a multicentre, randomised clinical trial. Eur J Cancer 2016;65:61–8. Available from: https://doi.org/10.1016/j.ejca.2016.06.007.
[15] Temel JS, Greer JA, El-Jawahri A, et al. Effects of early integrated palliative care in patients with lung and GI cancer: a randomized clinical trial. J Clin Oncol 2017;35(8):834–41. Available from: https://doi.org/10.1200/JCO.2016.70.5046.
[16] Vanbutsele G, Pardon K, Van Belle S, et al. Effect of early and systematic integration of palliative care in patients with advanced cancer: a randomised controlled trial. Lancet Oncol 2018;19(3):394–404. Available from: https://doi.org/10.1016/S1470-2045(18)30060-3.
[17] Knaul FM, Garcia PJ, Gospodarowicz M, et al. The Lancet Commission on cancer and health systems: harnessing synergies to achieve solutions. Lancet 2021;398(10306):1114–6. Available from: https://doi.org/10.1016/S0140-6736(21)01895-X.
[18] Schrijvers D, Cherny NI. ESMO clinical practice guidelines on palliative care: advanced care planning. Ann Oncol 2014;25(Suppl. 3):iii138–42. Available from: https://doi.org/10.1093/annonc/mdu241.
[19] Nickolich MS, El-Jawahri A, Temel JS, et al. Discussing the evidence for upstream palliative care in improving outcomes in advanced cancer. Am Soc Clin Oncol Educ Book 2016(35):e534–8. Available from: https://doi.org/10.1200/EDBK_159224.
[20] Ferrell BR, Temel JS, Temin S, et al. Integration of palliative care into standard oncology care: American Society of Clinical Oncology clinical practice guideline update. J Clin Oncol 2017;35(1):96–112. Available from: https://doi.org/10.1200/JCO.2016.70.1474.
[21] World Health Organization. Palliative care factsheet; 2020. Available from: https://www.who.int/news-room/fact-sheets/detail/palliative-care.
[22] Rugno FC, Paiva BSR, Nunes JS, et al. "There won't' be anything else…it's over": perceptions of women referred to palliative care only. Eur J Oncol Nurs 2014;18(3):261–6. Available from: https://doi.org/10.1016/j.ejon.2014.01.003.
[23] Collins A, McLachlan SA, Philip J. Initial perceptions of palliative care: an exploratory qualitative study of patients with advanced cancer and their family caregivers. Palliat Med 2017;31(9):825–32. Available from: https://doi.org/10.1177/0269216317696420.
[24] Fadul N, Elsayem A, Palmer JL, et al. Supportive versus palliative care: what's in a name?: a survey of medical oncologists and midlevel providers at a comprehensive cancer center. Cancer 2009;115(9):2013–21. Available from: http://doi.org/10.1002/cncr.24206.
[25] Tricou C, Munier S, Phan-Hoang N, et al. Haematologists and palliative care: a multicentric qualitative study. BMJ Support Palliat Care 2019;12(e6):e798–802. Available from: https://doi.org/10.1136/bmjspcare-2018-001714.
[26] Hausner D, Tricou C, Mathews J, et al. Timing of palliative care referral before and after evidence from trials supporting early palliative care. Oncologist 2021;26(4):332–40. Available from: https://doi.org/10.1002/onco.13625.
[27] Babu D, Dionne-Odom JN, Zubkoff L, et al. Oncology providers' perceptions of early/concurrent palliative care (S727). J Pain Symptom Manage 2016;51(2):421–2. Available from: http://doi.org/10.1016/j.jpainsymman.2015.12.030.
[28] Bakitas M, Lyons KD, Hegel MT, et al. Oncologists' perspectives on concurrent palliative care in a National Cancer Institute-designated comprehensive cancer center. Palliat Support Care 2013;11(5):415–23. Available from: https://doi.org/10.1017/S1478951512000673.
[29] Bakitas M. ES22.06 early integration of palliative care is the answer to improving quality of life. J Thorac Oncol 2021;16(3 Supplement):S88–9. Available from: https://doi.org/10.1016/j.jtho.2021.01.049.
[30] Zimmermann C, Pope A, Hannon B, et al. Phase II trial of symptom screening with targeted early palliative care for patients with advanced cancer. J Natl Compr Canc Netw 2022;20(4):361–370.e3. Available from: https://doi.org/10.6004/jnccn.2020.7803.
[31] Glare PA, Semple D, Stabler SM, et al. Palliative care in the outpatient oncology setting: evaluation of a practical set of referral criteria. J Oncol Pract 2011;7(6):366–70. Available from: https://doi.org/10.1200/JOP.2011.000367.
[32] Etkind SN, Bone AE, Gomes B, et al. How many people will need palliative care in 2040? Past trends, future projections and implications for services. BMC Med 2017;15(1):102. Available from: https://doi.org/10.1186/s12916-017-0860-2.
[33] Bone AE, Gomes B, Etkind SN, et al. What is the impact of population ageing on the future provision of end-of-life care? Population-based projections of place of death. Palliat Med 2018;32(2):329–36. Available from: https://doi.org/10.1177/0269216317734435.
[34] Hui D, Hannon BL, Zimmermann C, et al. Improving patient and caregiver outcomes in oncology: team-based, timely, and targeted palliative care. CA Cancer J Clin 2018;68(5):356–76. Available from: https://doi.org/10.3322/caac.21490.
[35] Hui D, Heung Y, Bruera E. Timely palliative care: personalizing the process of referral. Cancers (Basel) 2022;14(4):1047. Available from: https://doi.org/10.3390/cancers14041047.
[36] Kaasa S, Loge JH, Aapro M, et al. Integration of oncology and palliative care: a Lancet Oncology Commission. Lancet Oncol 2018;19(11):e588–653. Available from: https://doi.org/10.1016/S1470-2045(18)30415-7.
[37] Hannon B, Mak E, Al Awamer A, et al. Palliative care provision at a tertiary cancer center during a global pandemic. Support Care Cancer 2021;29(5):2501–7. Available from: https://doi.org/10.1007/s00520-020-05767-5.
[38] Dalal S, Palla S, Hui D, et al. Association between a name change from palliative to supportive care and the timing of patient referrals at a comprehensive cancer center. Oncologist 2011;16(1):105–11. Available from: https://doi.org/10.1634/theoncologist.2010-0161.
[39] Lee J, Mc Carthy O, Ryan S, et al. Video consultations in community palliative care – patient satisfaction: mixed methods study. BMJ Support Palliat Care 2023;13(e3):e1212–6. Available from: https://doi.org/10.1136/spcare-2022-003757.

[40] Block SD, Billings JA. A need for scalable outpatient palliative care interventions. Lancet 2014;383(9930):1699–700. Available from: https://doi.org/10.1016/S0140-6736(13)62676-8.

[41] Hui D. Palliative cancer care in the outpatient setting: which model works best? Curr Treat Options Oncol 2019;20(2):17. Available from: https://doi.org/10.1007/s11864-019-0615-8.

[42] Hui D, Mori M, Watanabe SM, et al. Referral criteria for outpatient specialty palliative cancer care: an international consensus. Lancet Oncol 2016;17(12):e552–9. Available from: https://doi.org/10.1016/S1470-2045(16)30577-0.

[43] Kamal L, Kano Y, Stevens AM, et al. Assessing the sensitivity and acceptability of the Royal Marsden palliative care referral "triggers" tool for outpatients with cancer. Support Care Cancer 2024;32(11):730. Available from: https://doi.org/10.1007/s00520-024-08921-5.

[44] Ream E, Hughes AE, Cox A, et al. Telephone interventions for symptom management in adults with cancer. Cochrane Database Syst Rev 2020;6(6):CD007568. Available from: https://doi.org/10.1002/14651858.CD007568.pub2.

[45] van Gurp J, van Selm M, Vissers K, et al. How outpatient palliative care teleconsultation facilitates empathic patient-professional relationships: a qualitative study. PLoS One 2015;10(4):e0124387. Available from: https://doi.org/10.1371/journal.pone.0124387.

Chapter 4

What is pain?

Ashique Ahamed
Department of Supportive and Palliative Care, The Christie NHS Foundation Trust, Manchester, United Kingdom

Key learning objectives

- Define pain in palliative care and explore the physiological mechanisms that lead to different types of pain.
- Distinguish between nociceptive and neuropathic pain, and background and breakthrough pain.
- Understand the concept of "total pain" and its implications for palliative care, addressing the physical, emotional, social, and spiritual dimensions of suffering.
- Analyze the epidemiology and burden of pain in cancer patients, focusing on the physical, social, and economic impact on patients, caregivers, and healthcare systems.

Introduction

Pain in palliative care is complex and multifactorial. It encompasses not only physical but also psychological, emotional, and spiritual characteristics that can worsen the patient's experience of the symptom. Patients with palliative illness can experience pain at various stages of their disease. For cancer patients, the terminal stage is associated with a higher pain burden compared to early or intermediate stages of cancer [1].

In 1979, the International Association for the Study of Pain (IASP) initially defined pain as "an unpleasant sensory and emotional experience associated with actual or potential tissue damage." However, critics argued that pain is subjective and can also occur in the absence of tissue damage and due to psychological reasons. It was recognized that the definition did not account for the subjective and emotional nature of pain. To address this, in 2018, a 14-member committee with expertise in clinical and basic science related to pain undertook a review of the definition. They recommended a new definition of pain as "an unpleasant sensory and emotional experience associated with, or resembling that associated with, actual or potential tissue damage." This was unanimously accepted by the IASP council in 2020 as the new definition for pain [2].

Epidemiology and burden of pain in cancer patients

Pain is one of the most common and distressing symptoms experienced by cancer patients, particularly in the advanced stages of the disease. Approximately 50% of cancer patients report experiencing pain, with the prevalence rising to 70%–90% in those with advanced disease [3]. Cancer pain may result from tumor invasion of bones, nerves, or other tissues, as well as from the effects of cancer treatments such as surgery, chemotherapy, or radiation. The burden of pain extends beyond physical suffering. Cancer-related pain often leads to reduced mobility, loss of independence, and a diminished quality of life. Socially, chronic pain impacts patients' ability to maintain relationships and perform daily activities, leading to isolation and increased reliance on caregivers. Economically, pain contributes to prolonged hospital stays, increased healthcare costs, and a higher demand for palliative care services. Uncontrolled pain can lead to frequent emergency room visits, adding to the strain on healthcare systems [4].

Pathophysiology of pain

There are four major processes that are involved in the processing of pain-related information in the brain. They are transduction, transmission, modulation, and perception. Transduction refers to the processes by which tissue-damaging

Pain Management in Palliative Care for Patients With Cancer. DOI: https://doi.org/10.1016/B978-0-443-30050-9.00043-6

stimuli activate nerve endings. Transmission refers to the process by which the message is carried from the site of tissue injury to the brain regions underlying perception. Perception is subjective awareness produced by the sensory signals [5].

Transduction, as a process, starts with nociceptors converting harmful stimuli into electrical signals. For example, in cancer pain, stimuli such as tumor growth, inflammation, or ischemia activate nociceptors in the affected tissues. Inflammatory mediators such as prostaglandins and bradykinin are released at the site of injury or tumor invasion, sensitizing nociceptors and lowering their threshold for activation [6]. This heightened sensitivity, known as peripheral sensitization, is a hallmark of cancer pain. It leads to increased pain perception (hyperalgesia) and abnormal responses to normally nonpainful stimuli (allodynia) [7].

After nociceptors are activated, pain signals are transmitted via afferent nerve fibers to the spinal cord. A-delta and C-fibers carry these signals from the periphery to the dorsal horn of the spinal cord, where they synapse with second-order neurons. Neurotransmitters such as glutamate and substance P are released in the spinal cord, facilitating the transmission of pain signals to higher centers in the brain. The pain signal ascends through the spinothalamic tract to the thalamus, which serves as a relay station before the signal reaches the somatosensory cortex, where pain is perceived.

Pain modulation refers to the process by which the body alters a pain signal as it is transmitted along the pain pathways. It occurs in both the spinal cord and brain, where the intensity of pain signals can be amplified or dampened. Descending inhibitory pathways originating in the brainstem—specifically the periaqueductal gray and rostral ventromedial medulla—play a critical role in pain modulation. These pathways release neurotransmitters such as serotonin, norepinephrine, and endogenous opioids to inhibit the transmission of pain signals in the spinal cord. The modulatory structures in the brain can often mediate both "facilitatory" and "inhibitory" modulation of pain, which can lead to both analgesia and hyperalgesia [8]. Hyperalgesia is commonly seen in cancer pain, where ongoing nerve damage and inflammation can lead to central sensitization, whereby the nervous system becomes hyperresponsive to stimuli.

Perception refers to the brain's interpretation of pain signals. Once pain signals reach the thalamus, they are relayed to the somatosensory cortex for localization and intensity assessment and to the limbic system for emotional processing. Pain is therefore experienced as both a sensory and affective phenomenon, influencing the patient's emotional state and response to pain. Prolonged and repetitive exposure to noxious stimuli (e.g., tissue injury from cancer) can lead to enhanced sensitivity of primary afferent nociceptors. It also results in increased release of inflammatory mediators such as inflammatory cytokines and prostaglandins. Repetitive stimulation of peripheral nociceptors in turn leads to the amplification of these pain signals in the dorsal horn of the spinal cord, leading to central sensitization. As a result, the pain becomes more widespread and persistent. Central sensitization can lead to increased sensitivity to even mild pain, known as hyperalgesia. In extreme cases, even nonnoxious stimuli such as light touch can cause pain, which is called allodynia. These sensitization mechanisms can develop in chronic pain conditions, including chronic cancer pain. The resulting amplification and persistence of pain significantly complicate its management and treatment.

Receptors and neurotransmitters involved in cancer pain

Pain perception in cancer involves several types of receptors, each playing a specific role in the detection and modulation of pain signals. Nociceptors and opioid receptors play a key role in the transmission and modulation of pain signals. Nociceptors are specialized receptors that detect harmful stimuli and initiate the pain response. The nociceptors are present in the viscera, skin, joints, bone, and muscle. Brain and spinal parenchyma lack nociceptors; pain in the head largely arises from meninges and vessels. Nociceptors are classified into mechanoreceptors that respond to physical damage such as pressure or distortion, thermoreceptors that respond to extreme temperatures, whether heat or cold, and polymodal nociceptors that respond to a combination of mechanical, thermal, and chemical stimuli [9]. In cancer, nociceptors are often sensitized by inflammatory mediators released by tumors or surrounding tissues, resulting in an exaggerated pain response [6]. Opioid receptors are located throughout the nervous system and are critical for the modulation of pain. These receptors, including mu (μ), delta (δ), and kappa (κ) receptors, mediate the effects of endogenous opioids (such as endorphins) and opioid medications. Activation of these receptors inhibits the release of excitatory neurotransmitters and reduces pain perception.

Several neurotransmitters play key roles in the transmission and modulation of pain signals in cancer patients. The key ones among them are glutamate, substance P, and endogenous opioids. Glutamate is the primary excitatory neurotransmitter involved in pain transmission. Glutamate is the most abundant neurotransmitter in the vertebrate nervous system. Glutamate plays a crucial role in modulating nociception and neurotransmitter release from the terminal afferents [7]. It binds to N-methyl-D-aspartate and α-amino-3-hydroxy-5-methyl-4-isoxazolepropionic acid receptors in the spinal cord, facilitating the propagation of pain signals to higher brain centers. In chronic cancer pain, excessive glutamate release can lead to central sensitization, where neurons in the spinal cord become hyperresponsive to even minor stimuli. Substance P is a neuropeptide released from nociceptors that enhances pain transmission by binding to neurokinin-1 receptors in the spinal cord. It is involved in the transmission of pain signals, particularly those related to chronic and persistent pain. Endogenous opioids, such as enkephalins and endorphins, are naturally produced by the body to regulate pain. These opioid receptors, μ, κ, and δ, are present throughout the peripheral and central nervous system and play an active role in the modulation of pain perception by inhibiting pain transmission and reducing the overall perception of pain [10].

Types of pain

Pain in cancer patients can be broadly divided into nociceptive and neuropathic pain. Nociceptive pain arises from the activation of nociceptors due to tissue injury or inflammation, while neuropathic pain results from nerve damage. Patients often experience mixed pain, which includes both nociceptive and neuropathic components, further complicating treatment.

Nociceptive pain is the most frequent type of pain. It is a physiological protective mechanism against noxious stimuli. Nociceptive pain arises in response to actual or potential harmful factors, triggering an evasive action or protective behavior to prevent further harm [11]. Nociceptive pain can be further classified into two main categories: somatic pain and visceral pain, which differ based on the location and nature of the tissue involved. Somatic pain originates from the skin, muscles, bones, or connective tissues. It is typically well localized and can be described as sharp, throbbing, or aching. Somatic nociceptors detect harmful stimuli that cause mechanical, thermal, or chemical damage to these tissues. The pain is usually well defined and can be exacerbated by movement or pressure on the affected area [9].

Somatic pain is transmitted by two types of nerve fibers, namely A-delta fibers and C-fibers. A-delta fibers are myelinated, enabling them to conduct pain signals rapidly, resulting in sharp, well-localized pain. On the other hand, C-fibers are unmyelinated and transmit pain signals more slowly, leading to a dull, aching sensation that is longer-lasting [6]. Visceral pain originates from the internal organs and surrounding tissues. It is typically diffuse, poorly localized, and can be described as deep, cramping, or pressure-like. Visceral nociceptors have a broad receptive field, which makes it difficult to pinpoint the exact source of pain [6]. Visceral pain is generally mediated by C-fibers, and as a result, visceral pain often presents as a persistent, aching sensation that is harder to relieve.

Neuropathic pain results from damage or dysfunction of the peripheral or central nervous system and is often described as burning, shooting, or electric-like. This type of pain is particularly challenging to treat in palliative care because it is often resistant to traditional analgesics such as opioids. Certain cancer treatments can also cause neuropathic pain. Many chemotherapy agents, such as paclitaxel, vincristine, and platinum compounds, can damage peripheral nerves, resulting in chemotherapy-induced neuropathic pain. Patients report painful sensations in the hands and feet ("glove-and-stocking" distribution) and describe tingling, numbness, or burning pain, often exacerbated by cold or movement. This type of pain may persist long after chemotherapy is completed and can severely impact a patient's mobility and quality of life [12]. Radiotherapy that covers the areas that are rich with nerve innervation, such as the brachial plexus and lumbosacral plexus, can also lead to treatment-induced neuropathic pain.

Pain in cancer is often of mixed etiology. For instance, a patient with breast cancer and bone metastasis can experience nociceptive and neuropathic pain from the tumor. They can suffer from nociceptive pain from the tissue damage around the cancer. The damage from the surrounding areas, such as the brachial plexus, can also result in neuropathic pain from the damage to the surrounding nerves. Additionally, the patient may also be suffering from inflammatory pain from the primary tumor and especially from the spinal metastases. The patient can also experience posttreatment pain, such as postmastectomy neuropathic pain or chemo-induced peripheral neuropathy, that can further complicate their pain management. As illustrated, pain in palliative conditions such as cancer is multifactorial and of mixed etiology. Hence, a multimodal approach is required to manage these symptoms effectively.

Pain in palliative care can also be categorized based on its temporal characteristics, particularly into background and breakthrough pain (BTP). Background pain is the persistent, ongoing pain that a patient experiences consistently throughout the day. It often results from chronic conditions such as cancer, where the disease or its treatment causes continuous tissue damage or nerve dysfunction. An example is a patient with bone metastases who requires long-acting drugs to manage a constant, dull pain in the lower back or pelvis. Proper management of background pain is essential to maintain a stable level of comfort for the patient. BTP is defined as a transient exacerbation of pain, with a rapid onset, which occurs while baseline pain is controlled and the patient receives a stable background analgesic regimen [13]. Its incidence is estimated to be between 40% and 80% and is particularly elevated in advanced disease [14]. BTP can arise due to excess stimuli acting on injured tissue or pathological tissue affecting adjacent structures (peripheral sensitization) or by lowering of the nociceptor threshold or the development of ectopic sites associated with allodynia and hyperalgesia.

Transient increases in sensitivity of spinal neurons (central sensitization) can also give rise to BTP [15]. It is typically sudden in onset and spontaneous (stimulus independent) and can be volitional (e.g., walking, coughing) or nonvolitional (e.g., ischemia). For example, a patient receiving long-acting opioids for background pain due to bone metastases may still experience sudden, sharp pain when turning in bed or standing, requiring immediate-release opioids such as morphine or fentanyl for rapid relief. Understanding the need to treat both background pain and BTP is critical in palliative care, as uncontrolled BTP can significantly impact a patient's quality of life.

The concept of "total pain"

Cicely Saunders, the founder of modern hospice and palliative care, introduced the concept of total pain in the early 1960s to describe patients' lived experience of suffering due to a combination of aspects of physical, psychological, social, and spiritual pain [16]. The concept of total pain reflects the cumulative burden of suffering experienced by individuals facing terminal illness, especially advanced cancer. This understanding goes beyond the simple categorization of pain as nociceptive or neuropathic, recognizing that pain is influenced by factors such as psychological distress, social isolation, financial hardship, and spiritual anguish [17]. In palliative care, total pain refers to not just the physical symptoms but also the emotional, psychological, and existential distress.

Emotional pain involves feelings of fear, anxiety, and depression, which often accompany a cancer or terminal diagnosis. Patients often worry about the emotional impact of a cancer diagnosis on their families and loved ones. These anxieties can further exacerbate their distress. Psychological interventions such as cognitive-behavioral therapy can help patients manage anxiety and depression, which often exacerbate physical pain. Pain in palliative conditions can lead to social isolation. In some countries, conditions such as cancer still carry a stigma and can result in social isolation. The patients can also feel as if they are a burden to their families. This can be further impacted by the financial burden of cancer treatments, especially if the patient is the main breadwinner for the family [18]. Social interventions might include family counseling, financial planning assistance, or social support groups to alleviate social burdens.

One of the most challenging aspects of total pain is spiritual pain. Patients with terminal cancer often grapple with the question of "Why me?" as they search for the meaning of their illness. Sometimes this is interpreted as a punishment for their past actions. In other instances, it can lead to patients questioning their faith. It can also lead to anger and feelings of hopelessness, which can further exacerbate their suffering. Spiritual interventions could involve chaplaincy services or discussions about the patient's values and beliefs to help them find meaning in their experience.

The coexisting psychological, social, and spiritual distress can often worsen or even manifest as physical pain. Lack of recognition of these can often underestimate the holistic nature of pain and complicate its management. Often total pain can manifest as a worsening of physical pain. This can lead to inappropriate management with drugs such as high-dose opioids and lead to a worse outcome. There are validated tools such as the Integrated Palliative Outcome Scale (IPOS), Edmonton Symptom Assessment Scale (ESAS), and unidimensional Visual Analog Scale (VAS) that can reveal important aspects about the patient's total pain [18]. The concept of total pain highlights the need for a multidisciplinary team and a multimodal approach to managing pain.

Conclusion

Pain is a common feature in patients suffering from palliative conditions, often intensifying as patients approach terminal stages of the disease, such as cancer. The pathophysiology and the various pain types are well researched and help in managing the pain more effectively. Pain is all-encompassing and is multifaceted (total pain). Therefore, a multidisciplinary and holistic approach is crucial to its effective management and to help reduce patient suffering.

References

[1] St Marie B. Pain management in patients receiving palliative care. Oncol Nurse Advis 2013;2013:e1–6.
[2] Raja SN, Carr DB, Cohen M, et al. The revised IASP definition of pain: concepts, challenges, and compromises. Pain 2020;161(9):1976–82. Available from: https://doi.org/10.1097/j.pain.0000000000001939.
[3] van den Beuken-van Everdingen MHJ, de Rijke JM, Kessels AG, et al. Prevalence of pain in patients with cancer: a systematic review of the past 40 years. Ann Oncol 2007;18(9):1437–49. Available from: https://doi.org/10.1093/annonc/mdm056.
[4] Breivik H, Collett B, Ventafridda V, et al. Survey of chronic pain in Europe: prevalence, impact on daily life, and treatment. Eur J Pain 2006;10(4):287–333. Available from: https://doi.org/10.1016/j.ejpain.2005.06.009.

[5] Osterweis M, Kleinman A, Mechanic D, editors. Pain and disability: clinical, behavioural, and public policy perspectives. Washington (DC): National Academies Press (US); 1987.
[6] Julius D, Basbaum AI. Molecular mechanisms of nociception. Nature 2001;413(6852):203–10. Available from: https://doi.org/10.1038/35093019.
[7] Yam MF, Loh YC, Tan CS, et al. General pathways of pain sensation and the major neurotransmitters involved in pain regulation. Int J Mol Sci 2018;19(8):2164. Available from: https://doi.org/10.3390/ijms19082164.
[8] Kirkpatrick DR, McEntire DM, Hambsch ZJ, et al. Therapeutic basis of clinical pain modulation. Clin Transl Sci 2015;8(6):848–56. Available from: https://doi.org/10.1111/cts.12282.
[9] Baron R. Mechanisms of disease: neuropathic pain–a clinical perspective. Nat Clin Pract Neurol 2006;2(2):95–106. Available from: https://doi.org/10.1038/ncpneuro0113.
[10] Lubejko ST, Graham RD, Livrizzi G, et al. The role of endogenous opioid neuropeptides in neurostimulation-driven analgesia. Front Syst Neurosci 2022;16:1044686. Available from: https://doi.org/10.3389/fnsys.2022.1044686.
[11] Cao B, Xu Q, Shi Y, et al. Pathology of pain and its implications for therapeutic interventions. Signal Transduct Target Ther 2024;9(1):155. Available from: https://doi.org/10.1038/s41392-024-01845-w.
[12] Hou S, Huh B, Kim HK, et al. Treatment of chemotherapy-induced peripheral neuropathy: systematic review and recommendations. Pain Physician 2018;21(6):571–92.
[13] Mercadante S, Portenoy RK. Breakthrough cancer pain: twenty-five years of study. Pain 2016;157(12):2657–63. Available from: https://doi.org/10.1097/j.pain.0000000000000721.
[14] Mestdagh F, Steyaert A, Lavand'homme P. Cancer pain management: a narrative review of current concepts, strategies, and techniques. Curr Oncol 2023;30(7):6838–58. Available from: https://doi.org/10.3390/curroncol30070500.
[15] Zucco F, Bonezzi C, Fornasari D. Breakthrough cancer pain (BTcP): a synthesis of taxonomy, pathogenesis, therapy, and good clinical practice in adult patients in Italy. Adv Ther 2014;31(7):657–82. Available from: https://doi.org/10.1007/s12325-014-0130-z.
[16] Saunders C. The evolution of hospices. In: Mann RD, editor. The history of the management of pain. Boca Raton: CRC Press; 1988. p. 167–78.
[17] Jespersen E, Minet LR, Nissen N. Symptoms of total pain experienced by older people with advanced gastrointestinal cancer receiving palliative chemotherapy. Eur J Cancer Care 2022;31(6):e13674. Available from: https://doi.org/10.1111/ecc.13674.
[18] Gomes-Ferraz CA, Rezende G, Fagundes AA, et al. Assessment of total pain in people in oncologic palliative care: integrative literature review. Palliat Care Soc Pract 2022;16:1–12. Available from: https://doi.org/10.1177/26323524221125244.

Chapter 5

Diagnosis, assessment, and initial treatment of pain

Tomi Kovačević[1,2]

[1]*Faculty of Medicine, University of Novi Sad, Novi Sad, Serbia,* [2]*Institute for Pulmonary Diseases of Vojvodina, Sremska Kamenica, Serbia*

Key learning objectives

- Importance and principles of pain assessment: to understand why the detailed pain assessment is crucial for successful pain management.
- Different models of pain assessment: to learn about key components of pain assessment using the different models.
- Pain assessment tools: to give insight into various pain scales.
- How to assess pain in people unable to communicate: to learn how to assess the pain when communication is disabled.
- Initial treatment of pain in patients with cancer: to learn the basic principles of the pharmacological approach in pain management, especially for moderate–severe pain.

Introduction

Pain is one of the most common symptoms in patients with cancer and often has a negative impact on patients' functional status and quality of life (QoL). Studies from the beginning of the 21st century showed that roughly 80% of patients with advanced-stage cancer had moderate–severe pain and that more than half of the patients had pain [1,2]. New data, according to the International Association for the Study of Pain, are somewhat encouraging and provide information that approximately one-third of adults who are actively receiving treatment for cancer and two-thirds of those with advanced malignant disease experience pain [3]. Recent metaanalysis also pointed out the decrease in pain prevalence in the group of patients with cancer to 44.5% and provided data that moderate–severe pain was experienced by 30.6% of this population [4]. Even though the overall prevalence of pain in patients with cancer decreased, attention to the assessment and management of pain is still an unmet need in most of the world. Effective pain management can generally be accomplished by paying attention to the details. Proper diagnosis of the pain, giving detailed information that identifies the underlying pathophysiology, can and will significantly influence its management. The crucial significance for further decision on pain management is to conduct a detailed pain assessment and provide answers to many questions: Is the pain acute or chronic? Is the pain due to cancer or secondary to cancer? Is it due to cancer treatment or some other causes? Is it somatic, visceral, neuropathic, or mixed? What is the impact of pain on a patient's life? Is there any psychosocial issue? and so forth [5]. One of the most important tools and fundamental skills in pain assessment is good communication and a pain diagnosis structure, which needs to be followed throughout the disease trajectory. In this chapter, readers will be introduced to up-to-date modalities and models of pain diagnosis and assessment tools, helping them to implement this important task in regular clinical practice.

Principles of diagnosis and assessment of pain

Each patient is unique, and so is the pain, and this should be kept in mind when we diagnose and assess the pain. "Pain is what the patient says it is," and the complexity of pain experience is modulated by individual differences. The multifaceted nature of pain is the main reason for many challenges when communicating with patients and assessing the pain [5–7]. In order to assess pain and to avoid many barriers and obstacles, it is important to keep in mind that pain is,

Pain Management in Palliative Care for Patients With Cancer. DOI: https://doi.org/10.1016/B978-0-443-30050-9.00025-4

first of all, a subjective experience, and, due to that, pain assessment should be relied on patient self-reported assessment, which is still a gold standard for this important task in pain management. Many patients with cancer experience different types of pain at the same time. Due to various patients' cultural differences and educational level, the complexity of pain assessment is more than a simple task, and insufficient pain assessment can result in significant errors in pain management [8,9].

Therefore, the pain assessment should be comprehensive, following the main principles: to assess physical domain, to assess pain manifestation, to assess effect of pain on a functional status, to assess psychosocial domain influencing on and being influenced by pain, to assess spiritual domain influencing on and being influenced by pain, and to assess modulators of pain expression [9].

Critical point of optimal pain management is pain diagnosis and assessment, and the quality of this assessment directly influences further pain management and strategy [10,11]. If the quality of a pain assessment is insufficient, it will lead to poor pain control and further suffering in all domains of patients' lives. Managing the pain in patients with cancer in a palliative care setting is complex, and, in many cases, the primary task of healthcare professionals (HCPs) during the disease trajectory. Hereof, the efforts should be made from all team members to complete the pain assessment successfully using their knowledge, empathy, and experience in order to respond to patients' needs and develop an adequate pain management plan [9,10].

Pain assessment tools

One of the systematic approach models for assessing pain is the Symptom Assessment Acronym **OPQRSTUV**—a practical and easy-to-use mnemonic tool which can be used as a guide for a consistent and comprehensive pain assessment [12] (Fig. 5.1).

Another, also a mnemonic tool, is the **SOCRATES** model—a structured framework used by many HCPs to assess pain [13]. This structured model is used by many HCPs and is also an easy-to-use tool in regular clinical practice. This model is described in Fig. 5.2.

There are many other alternatives such as: OLDCARTS (Onset, Location/Radiation, Duration, Character, Aggravating factors, Relieving factors, Timing, Severity), SCHOLAR (Symptoms, Characteristics, History, Onset, Location, Aggravating factors, Remitting factors), COLDER AS (Characteristics, Onset, Location, Duration, Exacerbation, Remitting factors, Associated symptoms, Severity), CLEARAST (Characteristics, Location, Exacerbation, Alleviation, Radiation, Associated symptoms, Severity, Time course), and LIQOR AAA (Location, Intensity, Quality, Radiation, Alleviating and Aggravating factors, Associated signs) [13].

Pain assessment model of choice depends on HCPs' personal decision. It is important to highlight that it is of great importance, regardless of model choice, to gather as much detailed pain history as possible, which will help in the decision of further pain management. The more specific the pain assessment, the better. In order to achieve the best possible pain diagnosis, regardless of the method, it is important to observe and gather data about key components/questions:

- Onset and duration of pain.
- Description of localization, irradiation, characteristic, pattern, and pain quality.
- What is the current and what were the lowest and highest pain intensities?
- What are the factors that potentiate and/or relieve pain?
- What are the other symptoms and signs associated with pain ("total pain concept": physical, psychological, social, and spiritual domain)?
- How does the pain influence activities of daily living (ADL) and QoL?
- The past and current pain management plan and its effect.
- What are the patient's goals of treatment? [9]

Pain assessment scales

The useful and commonly used tool for pain assessment is pain scale. Today, there are many pain scales in use, and there is no consensus about the gold standard, but the scale must be easy to use, sensitive to treatment effect, validated, and reliable. It is needed to highlight that scales are not a substitution for a complex pain assessment and cannot be used alone when assessing the pain, but they are a necessary and inevitable tool for proper pain assessment. Which scale should be used during the assessment depends on patients and their abilities to understand and provide the answer(s). It is important to bear in mind the existence of different forms of pain scales and to tailor the usage depending on patients' needs and abilities [5,9].

	Description	Questions to ask	Goal
O	Onset	When did the pain start? How long does it last? How often does it occur?	To identify the onset of pain (when active, or resting) or whether pain is acute/chronic.
P	Provoking/palliating	Does your pain increase with movement or activity? Are the symptoms relieved with rest? Were any previous treatments effective? What brings it on? What makes it better? What makes it worse?	To identify quality of pain and differentiate between nociceptive and neuropathic pain mechanisms. To identify alleviating and aggravating factors. To evaluate effectiveness of current treatment.
Q	Quality	What does your pain feel like? What words describe your pain? What does it feel like? Can you describe it?	To identify mechanism of pain (terms such as "throbbing," "aching," "shooting," and "dull" may provide clues).
R	Region/radiation	Where is your pain? Does the pain radiate, or move to other areas? Where is it? Does it spread anywhere?	To identify one or more areas of the body that are affected by pain.
S	Severity	How would you rate your pain on an intensity scale? What is the intensity of this symptom (on a scale of 0–10 with 0 being none and 10 being the worst possible)? Right now? At best? At worst? On average? How bothered are you by this symptom? Are there any other symptom(s) that accompany this symptom?	To identify intensity (refer to various intensity scales). To identify degree of impairment and effect on quality of life or ability to perform activities of daily living (ADLs).
T	Treatment	What treatments have worked for you in the past? Is it a constant, dull, or intermittent pain? What medications and treatments are you currently using? How effective are these? Do you have any side effects from the medications and treatments? What medications and treatments have you used in the past?	To identify treatments which have been successful in the past. To identify the timing of the pain so that treatment can be focused on spikes in pain.
U	Understanding/impact	What do you believe is causi ng the pain? What do you believe is causing this symptom? How is this symptom affecting you and/or your family?	To understand patient history of pain. To be able to set achievable pain and function goals when reviewing the plan of care.
V	Values	What is your acceptable level for this pain? Is there anything else that you would like to say about your pain? What is your goal for this symptom? What is your comfort goal or acceptable level for this symptom (on a scale of 0 to 10 with 0 being none a nd 10 being worst possible)? Are there any other views or feelings about this symptom that are important to you or your family? Are there any other symptoms related to the pain?	To understand and discuss other stressors, spiritual pain.

FIGURE 5.1 OPQRSTUV symptom assessment tool.
Adapted from Fraser Health Authority [12].

	Description	Questions to ask	Goal
S	Site	Where is the pain? Can you point to where exactly the pain is?	To define the part of body where the pain is.
O	Onset	When did you start feeling pain? Did the pain start suddenly or gradually? What were you doing when you started feeling the pain? How long have you been feeling the pain?	To establish how and when the pain started. To define duration of pain To find out the change of intensity during the time.
C	Characteristic	How would you describe the pain?	To understand the type of pain: somatic, visceral, or mixed.
R	Radiation	Does the pain radiate? If yes where? How often does it happen?	To identify one or more areas of the body that are affected by pain.
A	Associated symptoms	Are you experiencing other symptoms that might be connected to the pain? Is there anything you think is associated with the pain?	To identify possible other symptoms influencing on or influenced by pain.
T	Time course	Ever since you started experiencing the pain, has it gotten better, worse, or is it the same? Does the pain follow a pattern?	To identify degree of impairment and effect on quality of life or ability to perform activities of daily living (ADLs).
E	Exacerbating/relieving factors	What makes the pain worse or better? If you don't do anything, will you still experience pain?	To identify treatments which have been successful in the past. To identify the movements or activities which influence on pain.
S	Severity	On a scale of 1 to 10, with 10 being the highest, how bad is the pain?	To identify intensity (refer to various intensity scales).

FIGURE 5.2 SOCRATES mnemonic structured framework for pain assessment.
Adapted from SOCRATES: Pain Assessment Acronym and Mnemonic, skills training group [13].

The main division of pain scales is:

1. Uni(single)dimensional pain scales: provide information about the patient's self-reported severity of pain (Visual Analog Scale, Verbal [Categorical] Rating Scales, Numeric Rating Scales [NRS], Wong–Baker Faces Pain Rating Scale [14]) (Fig. 5.3).
2. Multidimensional pain scales: provide information not only about intensity but also about other aspects of pain: location, duration, quality, and other pain components (e.g., McGill Pain Questionnaire, Brief Pain Inventory, etc.).

The pain scales should be used at the initial assessment and in each subsequent pain assessment. They are helpful in many aspects: provide insight into the trend of response to therapy, encourage communication during visits, facilitate the understanding of pain complexity, and so forth [9].

The detailed pain assessment, as mentioned many times, should be carried out at each visit of a patient with cancer. It should not be assumed that a patient does not have pain if it was not present previously; does not have "new pain"; does not have a change in characteristics of known pain; the provided therapy is successful; and so forth. It is recommended to encourage patients to keep records about the pain on a daily basis, to write a "pain diary"—a simple and practical tool that will help patients and HCPs to have a direct insight about success and other events during day-by-day pain management. It is usual that patients, when asked, rate the pain intensity based on their memory, taking into account the daily fluctuation, which is usual in patients with cancer. It was shown that recall accuracy depends on the stability of pain, and most of the time, patients rely on the worst experience they had. Also, the pain diary helps more than half of patients to cope with pain [15]. This tool can be used both in the hospital and at home. The data which should be followed in the diary should include at least:

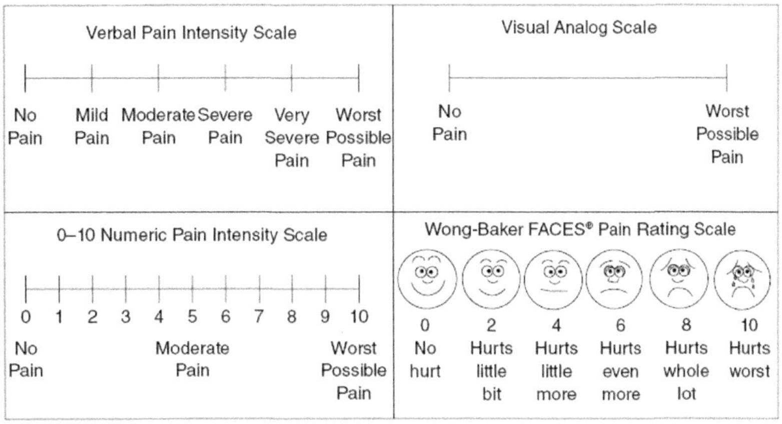

FIGURE 5.3 Uni(single)dimensional pain scales.

DATE:					
Time	**Pain before (0–10)**	**Medication**	**Dose**	**Pain after (0–10)**	**Observations**
0					
....					
24					

FIGURE 5.4 Example of a simple pain diary.

- Date
- Time of day
- Severity of pain before using any pain medication
- Type and dosage of medication in use
- Severity of pain after using any pain medication
- Time of breakthrough pain
- Other important observation (side effects, new pain characteristics, new pain site, etc.) (Fig. 5.4)

Another scale which can be used in order to assess not only pain but also other symptoms of patients in need of palliative care is the Integrated Palliative Care Outcome Scale (IPOS). It measures the symptoms and outcomes through 10 questions scored from 0 to 4, assessing symptoms and physical, social, psychological, and spiritual needs, and provides a better understanding of patients' issues as well as effects of prescribed therapy over time. This scale is based on self-report and can be used at home, completed by the patient with or without the support of family, caregivers, or HCPs. The first symptom on this scale is pain, followed by other symptoms which can be of great influence on pain experience, and it should also be repeated daily [16]. The IPOS was developed in 2013 and has evolved since then, translated and validated in many languages, and nowadays is used worldwide [17,18].

How to assess pain in people unable to communicate?

The pain is often undertreated in patients with cognitive impairment and communication disabilities, mainly due to difficulties in pain assessment. Accuracy of pain diagnosis and treatment success is a major challenge in these patients due to the inability for self-assessment. Indicators that are commonly used in pain assessment may not effectively measure distress in patients who are unable to communicate their discomfort, resulting in pain that is often unnoticed and/or undertreated [19].

In order to provide better pain assessment in this group of patients, the Nonverbal Pain Scale (NVPS) was developed. It is based on several categories which need to be observed when communication with the patient is aggravated or impossible: face expression, body movements (activities), guarding, vital signs, and respiratory signs, and each of the categories needs to be assessed partially on the scale 0–2 by its severity (Fig. 5.5) [19]. It was shown in research that NVPS improves diagnosing not only the presence but also the severity of pain in patients with communication difficulties

Category	0	1	2
Face expression	No particular expression No smile	Occasional grimace Occasional tearing Occasional frowning Occasional forehead wrinkling	Frequent grimace Frequent tearing Frequent frowning Frequent forehead wrinkling
Body movement	Lying quietly Normal position	Seeking attention through movement Slow, cautious movements	Restless Excessive activity Withdrawal reflexes
Guarding	No positioning of hands over body areas	Splinting areas of the body Tense	Rigid Stiff
Vital signs	Stable	Change in SBP for >20mmHg Change in HR for >20/min	Change in SBP for >30mmHg Change in HR for >25/min
Respiratory signs	Baseline RR and SpO^2 Compliant with ventilator	RR >10 above baseline or 5% ↓SpO_2 Mild asynchrony with ventilator	RR >20 above baseline or 10% ↓SpO_2 Severe asynchrony with ventilator

FIGURE 5.5 Nonverbal Pain Scale. *HR*, Heart rate; *RR*, respiratory rate; *SBP*, systolic blood pressure.
Adapted from Pain Assessment and Management Initiative [19].

in settings such as the ICU and in palliative care [20,21]. As in all other diagnostic uses of scales, this scale cannot be used alone and needs to be coupled with physical examination. Use of NVPS for pain diagnosis and assessment of the presence and severity of pain can provide consistent follow-up in patients unable to communicate and also enable revision of the pain management plan [21,22].

How to assess pain in older people with cancer?

The elderly are characterized by the coexistence of frailty and functional decline, which complicates both assessment and treatment of pain in older patients with cancer. For learning more about the importance and use of tools tailored to cognitive status and functional ability in the elderly, as well as specificities of diagnosis, assessment, and pain management in older patients with cancer, please refer to Chapter 48.

Initial treatment of pain

Basic WHO principles for pain management in patients with cancer

The most widely used approach in pain management is the one introduced by pain experts in 1986, extensively validated across the globe—the WHO analgesic ladder [23]. This approach consists of three simple steps, structured but flexible, in which selection of analgesics is based on appropriate assessment of pain intensity. Depending on pain intensity (mild, 1–2/10 NRS; moderate, 3–6/10 NRS; severe, 7–10/10 NRS; common cutoff points, 1–3/4–6/7–10), the following principles guide the choice of pain medication:

STEP 1—Nonopioids ± coanalgesics (adjuvants)

STEP 2—Opioids for mild–moderate pain+nonopioid ± coanalgesics (adjuvants)

STEP 3—Opioids for moderate–severe pain+nonopioid ± coanalgesics (adjuvants [23])

Despite this approach, which alleviates pain in 70%–80% of patients with cancer [23], the revision was made and included one more step with a suggestion to integrate this step alongside other steps and not as an alternative after failure of introduced measures [24, 25].

STEP 4—Interventional treatment (e.g., invasive and minimally invasive techniques, radiotherapy, etc.)

This step provided an added option for pain management in patients with cancer not responding adequately to the prescribed medication and also introduced a bidirectional approach in pain management, which implies de-escalation on the ladder depending on pain intensity and going back to step 2 or even step 1 (Fig. 5.6).

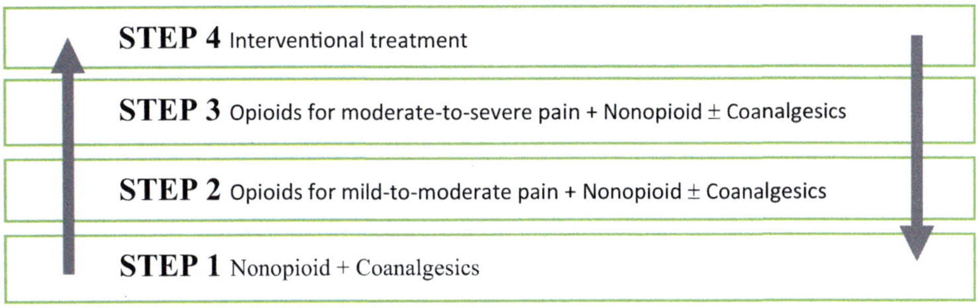

FIGURE 5.6 Modified, bidirectional WHO analgesic ladder including interventional treatment.

Addressing the limitation of the stepwise WHO analgesic ladder, especially in the case of acute pain, led to a recommendation proposed by the European Association for Palliative Care, which was later revised [26–28]. This approach suggests the introduction of strong opioids for moderate–severe pain as the first-line treatment and is also known as "*Analgesic elevator*" [29].

How to start the pharmacological treatment for cancer pain?

Since the cancer pain is (most of the time) chronic, the most important, regardless of the choice of which analgesic will be used, is to follow the basic principles of pharmacological pain treatment:

1. ***By the mouth***—preferred route for all analgesics
2. ***By the clock***—at regular intervals
3. ***By the ladder*** —following the proposed three-step approach
4. ***For the individual***—tailored based on regular assessment and the patient's preferences
5. ***Attention to detail***—assessing all details about pain (e.g., character, onset, dependence of movements, adverse effects, etc. [23,26])

Choosing the right nonopioid analgesic

The choice of nonopioid analgesics should be based on the type of pain and risk/benefit basis (e.g., comorbidities, other used drugs, age, etc.). For more information about nonopioid analgesics, please refer to Chapter 7.

Choosing the right opioid analgesic

The choice of opioid analgesic should be based first of all on pain intensity (opioids for mild–moderate pain or opioids for severe pain) as well as on a risk/benefit basis (e.g., comorbidities, other used drugs, age, etc.). Choice of opioid is of major importance, especially due to heterogeneity and complexities of patients with cancer pain and the goal to achieve a balance between desired analgesic effects and unwanted adverse effects [30]. For more information about opioids, please refer to Chapter 6.

Introducing and titrating opioids for severe cancer pain

There are several reasons for the titration of opioids for moderate–severe pain: management of moderate–severe pain in opioid-naive patients, in patients unsuccessfully and/or undertreated with opioids for mild–moderate pain, and in patients who need rapid intervention due to pain severity [31]. The incremental and rapid dose adjustment based on the patients' needs and response for adequate and satisfactory pain relief with the smallest dose needed and fewest side effects is the main pillar of introduction and titration of opioids for moderate–severe pain [30,32]. Individualized titration should start at the minimum recommended drug dose and should be increased with the goal of reaching the optimum analgesic effect and without serious and unacceptable side effects [30]. According to the European Society for Medical Oncology (ESMO) and the American Society of Clinical Oncology (ASCO) guidelines, the first-choice opioid for moderate–severe cancer pain is oral morphine (level of evidence I, A; strength of recommendation: strong), and oral administration of immediate- and modified-release formulations can be used. Transdermal route is usually contraindicated for the opioid titration phase [29,32]. After the titration phase, when optimum pain control is reached, the long-acting opioid for moderate–severe pain should be introduced together with medication for breakthrough pain in adjusted dose [29,33].

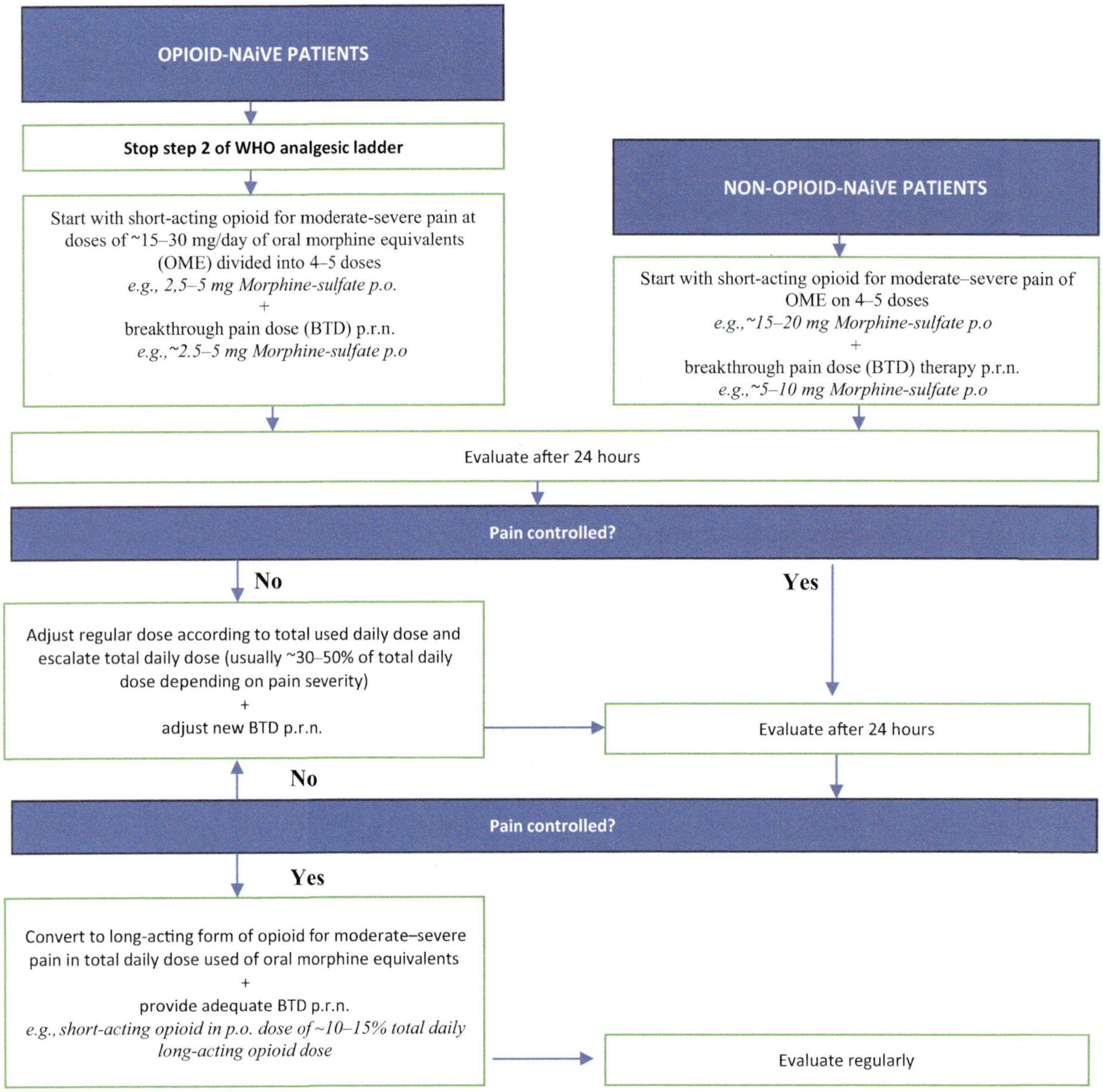

FIGURE 5.7 Titration of opioids for moderate–severe pain.
Adapted and adjusted from Watson M, Ward S, Vallath N, et al. Oxford handbook of palliative care. 3rd ed. Oxford (UK): Oxford University Press; 2019, and Paice J, Bohlke K, Barton D, et al. Use of opioids for adults with pain from cancer or cancer treatment: ASCO guideline. J Clin Oncol 2023;41(4):914–30. Available from: https://doi.org/10.1200/JCO.22.02198.

The titration of opioids for moderate–severe pain depends on the fact whether patients were or were not on opioid therapy (non–opioid-naive vs opioid-naive patients) (Fig. 5.7).

Dose escalation

Opioid dose escalation should be considered when the use of breakthrough pain dose (BTD) is used more than four times daily, and if the patient tolerates opioids well (with no or minimal sedation), the treatment should be adjusted [32,34]. When opioid escalation is indicated, the following principles must be followed: Calculate the total used dose of opioids for 24 hours (daily dose of long-acting+all BTD used) in OME → Escalate the dose of long-acting opioid based on equianalgesic table → Adjust BTD of short-acting opioid (~10%–15% of total daily dose of prescribed long-acting opioid; i.e., 4-hourly dose) → Monitor and reevaluate regularly [34]. For more information about opioids, please refer to Chapter 6.

Dose de-escalation

For the dose de-escalation principles, please refer to Chapter 6.

Opioid rotation

If the prescribed opioid fails to provide adequate antidolorous effect or unmanageable adverse effects occur, the used opioid should be discontinued and another should be prescribed (evidence level III, C) [9,32]. The process in which one opioid is substituted with another is called opioid rotation (opioid switching) [9,29]. Data shows that in up to 44% of patients with cancer, opioid rotation (opioid switching) may be required, and in >50% of patients, improvement of analgesia is seen when opioid rotation is performed [26]. Decision for opioid rotation should be made only for the patient's benefit and in the following cases:

- Due to adverse effects or intolerance of the opioid used.
- In case of a change in clinical status (e.g., liver impairment, renal impairment, bowel obstruction, malabsorption, etc.).
- Due to the inability of oral intake (e.g., reduced level of consciousness, dysphagia, nausea, vomiting [9,32]).

When opioid rotation is indicated, the following principles must be followed: Convert to oral morphine equivalents (OME) using the opioid equivalence tables → Reduce the dose of OME for ~25%–50% due to incomplete cross-tolerance → Prescribe the BTD of short-acting opioid (~10%–15% of the total daily long-acting opioid [i.e., 4-hourly dose]) → Review and reassess regularly [9,32].

Due to insufficient evidence, do not combine two strong (long-acting) opioids simultaneously [35].

Coanalgesics

Although opioids are the drugs of first choice for moderate–severe cancer pain treatment, the nature of underlying pain guides the use of additional medication when managing pain in patients with cancer. Aimed to better control some types of pain, coanalgesics play an important role in pain management, especially in cases of insufficient analgesia with opioids or their intolerable side effects [9,29]. The main role when choosing the right adjuvant analgesic is detailed pain assessment. Using the right coanalgesic provides better tailored pain control and leads to less often escalation of opioids. The group of most used adjuvant analgesics is shown in Fig. 5.8.

Further implications

The aim of palliative care, by WHO definition, among others, is *"to prevent and relieve suffering through the early identification, correct assessment and treatment of pain,"* and therefore all HCPs need to have education about pain assessment [36]. Palliative care needs to be provided by an interdisciplinary and multiprofessional team, and that means that patients with a life-threatening illness will face a large number of people caring about their condition with the same goal—to relieve suffering. This necessarily means, in order to fulfill this goal, that all professionals need to communicate. Incorporation of pain assessment scales, but also their results, into electronic health records would improve and promote better communication. It is shown that incorporation of these tools in hospital electronic systems enables timely reassessment of pain as well as interventions [21]. Today, there are many "pain tracking" applications which can be useful for HCPs when assessing treatment success, and it needs to be encouraged for wider use.

Conclusions (clinical implications)

- Pain diagnosis and assessment are crucial in pain management plan.
- Pain is a complex and first-of-all subjective symptom, and therefore detailed, personalized pain assessment should be made for each patient, considering not only the physical aspect of pain but also psychological, social, and spiritual:

  When did pain occur for the first time, and what is the duration of pain?
  What are localization, irradiation, characteristic, pattern, and pain quality?
  What is the current, the lowest, and the highest pain intensity?
  What are the factors that potentiate and/or relieve pain?
  What are the other symptoms and signs associated with pain?
  How does the pain influence ADL and QoL?
  The past and current pain treatment and its effect.
  What are the patient's goals of treatment?

Coanalgesic Group	Indication	Refer to
Antidepressants	Neuropathic pain	***Chapter 8***
Anticonvulsants	Neuropathic pain	***Chapter 10***
Bisphosphonate	Bone metastases pain	***Chapter 11***
Corticosteroids	Pain due to edema, nerve compression, and bone metastases	***Chapter 12***
Muscle relaxants	Muscle spasm or muscle cramp pain	***Chapter 13***
Antispasmodics	Pain due to colic	***Chapter 13***

FIGURE 5.8 Coanalgesics: the most used adjuvant therapy in pain management of patients with cancer.

- The choice of pain scale should be based on the patient's abilities to understand, complementary to pain assessment and combined with physical exam.
- The choice of nonopioid analgesics should be based on the type of pain and risk/benefit basis.
- The choice of opioid analgesic should be based on pain intensity (opioids for mild–moderate pain or opioids for severe pain) as well as on risk/benefit basis (e.g., comorbidities, other used drugs, age, etc.) and is of major importance, especially due to heterogeneity and complexities of patients with cancer pain.
- Introducing and titrating opioids for moderate–severe cancer pain should be based on the "start low and go slow" principle.
- The goal of pain management in patients with cancer is to achieve a balance between desired analgesic effects and unwanted adverse effects.
- Coanalgesics play an important role in pain management, especially in cases of insufficient analgesia by opioids or their intolerable adverse effects.
- Assessment, monitoring, and evaluation of pain and (side)effects of provided therapy should be introduced into regular clinical practice using the appropriate tools (in outpatient clinics, hospitals, and at home).

AI disclosure

During the preparation of this work, the author used ChatGPT to assist with proofreading, formatting, and grammar checking. After using this tool/service, the author reviewed and edited the content as needed and takes full responsibility for the content of the publication.

References

[1] van den Beuken-van Everdingen MHJ, de Rijke JM, Kessels AG, et al. Prevalence of pain in patients with cancer: a systematic review of the past 40 years. Ann Oncol 2007;18(9):1437–49. Available from: https://doi.org/10.1093/annonc/mdm056.
[2] Bruera E, Kim HN. Cancer pain. JAMA 2003;290(18):2476–9. Available from: https://doi.org/10.1001/jama.290.18.2476.
[3] International Association for the Study of Pain (IASP). Advocacy/Global Year/Cancer Pain. Retrieved October 3, 2024. https://www.iasp-pain.org/advocacy/global-year/cancer-pain/
[4] Snijders RAH, Brom L, Theunissen M, et al. Update on prevalence of pain in patients with cancer 2022: a systematic literature review and meta-analysis. Cancers (Basel) 2023;15(3):591. Available from: https://doi.org/10.3390/cancers15030591.
[5] Bruera E, Higginson IJ, von Gunten CF, et al. Multidimensional patient assessment. Textbook of palliative medicine and supportive care. 3rd ed. Boca Raton: CRC Press; 2021. p. 149–65.
[6] Longo R. "Pain is what the patient says it is": a secondary analysis of nurses' reflections on the term pain catastrophizing. 2022. [Master Thesis]. Western University Open Repository. Available from: https://doi.org/10.71858/123456.
[7] Miller LE, Eldredge SA, Dalton ED. "Pain Is What the Patient Says It Is": nurse-patient communication, information seeking, and pain management. Am J Hosp Palliat Care 2016;34(10):966–76. Available from: https://doi.org/10.1177/1049909116661815.
[8] Oldenmenger WH, Geerling JI, Mostovaya I, et al. A systematic review of the effectiveness of patient-based educational interventions to improve cancer-related pain. Cancer Treat Rev 2018;63:96–103. Available from: https://doi.org/10.1016/j.ctrv.2017.12.005.
[9] Watson M, Campbell R, Vallath N, et al. Pain management. Oxford handbook of palliative care. 3rd ed. Oxford (UK): Oxford University Press; 2019. p. 237–316.

[10] Fink R. Pain assessment: the cornerstone to optimal pain management. Proc (Bayl Univ Med Cent) 2000;13(3):236–9. Available from: https://doi.org/10.1080/08998280.2000.11927681.

[11] Kishner S, Paulk KL. Pain assessment from Medscape clinical procedures. Chief Editor: chraga ED. Updated Oct 2022. Retrieved October 3, 2024. https://emedicine.medscape.com/article/1948069-overview?form=fpf.

[12] Pain from BC. Inter-professional palliative symptom management guidelines. New Westminster: BC Centre for Palliative Care; 2017. Available from: https://bc-cpc.ca/wp-content/uploads/2019/10/2-BCPC-Clinical-Best-Practices-colour-Pain.pdf.

[13] Clayton HA, Reschak GL, Gaynor SE, et al. A novel program to assess and manage pain. Medsurg Nurs 2000;9(6):318–21, 317.

[14] Jacques E. Pain scales: types of scales and using them to explain pain from Verywellhealth. Updated Aug 30, 2023. Retrieved October, 2024. https://www.verywellhealth.com/pain-scales-assessment-tools-4020329

[15] de Wit R, van Dam F, Hanneman M, et al. Evaluation of the use of a pain diary in chronic cancer pain patients at home. Pain 1999;79(1):89–99. Available from: https://doi.org/10.1016/S0304-3959(98)00158-4.

[16] Murtagh FE, Ramsenthaler C, Firth A, et al. A brief, patient- and proxy-reported outcome measure in advanced illness: validity, reliability and responsiveness of the Integrated Palliative Care Outcome Scale (IPOS). Palliat Med 2019;33(8):1045–57. Available from: https://doi.org/10.1177/0269216319854264.

[17] Rugno FC, Carlo MM. The Palliative Outcome Scale (POS) applied to clinical practice and research: an integrative review. Rev Lat Am Enfermagem 2016;24:e2764. Available from: https://doi.org/10.1590/1518-8345.0993.2764.

[18] Nakajima N. Palliative Care Outcome Scale assessment for cancer patients eligible for palliative care: perspectives on the relationship between patient-reported outcome and objective assessments. Curr Oncol. 2022;29(10):7140–7. Available from: https://doi.org/10.3390/curroncol29100561.

[19] Odhner M, Wegman D, Freeland N, et al. Assessing pain control in nonverbal critically ill adults. Dimens Crit Care Nurs 2003;22(6):260–7. Available from: https://doi.org/10.1097/00003465-200311000-00010.

[20] Ghayem H, Heidari MR, Aghaei B, et al. The effect of training the Nonverbal Pain Scale (NVPS) on the ability of nurses to monitor the pain of patients in the intensive care unit. Indian J Crit Care Med 2023;27(3):195–200. Available from: https://doi.org/10.5005/jp-journals-10071-24425.

[21] McGuire DB, Kaiser KS, Haisfield-Wolfe ME, et al. Pain assessment in noncommunicative adult palliative care patients. Nurs Clin North Am 2016;51(3):397–431. Available from: https://doi.org/10.1016/j.cnur.2016.05.009.

[22] Rababa M, Al-Sabbah S, Abu Zahra T. Importance of pain assessment, documentation, and education among nurses caring for critically Ill patients: a cross-sectional study. Electron J Gen Med 2023;20(5):em510. Available from: https://doi.org/10.29333/ejgm/13292.

[23] WHO Guidelines for the Pharmacological and Radiotherapeutic Management of Cancer Pain in Adults and Adolescents. Geneva: World Health Organization; 2018. ANNEX 1, Evaluationof Pain. Available from: https://www.ncbi.nlm.nih.gov/books/NBK537489/.

[24] Vargas-Schaffer G. Is the WHO analgesic ladder still valid? Twenty-four years of experience. Can Fam Physician 2010;56(6):514–7.

[25] Hochberg U, Minerbi A, Boucher LM, et al. Interventional pain management for cancer pain: an analysis of outcomes and predictors of clinical response. Pain Physician 2020;23(5):E451–60.

[26] Hanks GW, Conno F, Cherny N, et al. Expert Working Group of the Research Network of the European Association for Palliative Care. Morphine and alternative opioids in cancer pain: the EAPC recommendations. Br J Cancer 2001;84(5):587–93. Available from: https://doi.org/10.1054/bjoc.2001.1680.

[27] Klepstad P, Kaasa S, Borchgrevink PC. Starting step III opioids for moderate to severe pain in cancer patients: dose titration: a systematic review. Palliat Med 2011;25(5):424–30. Available from: https://doi.org/10.1177/0269216310386280.

[28] Caraceni A, Pigni A, Brunelli C. Is oral morphine still the first choice opioid for moderate to severe cancer pain? A systematic review within the European Palliative Care Research Collaborative guidelines project. Palliat Med 2011;25(5):402–9. Available from: https://doi.org/10.1177/0269216310392102.

[29] Fallon M, Giusti R, Aielli F, et al. ESMO Guidelines Committee. Management of cancer pain in adult patients: ESMO Clinical Practice Guidelines. Ann Oncol 2018;29(4):iv166–91. Available from: https://doi.org/10.1093/annonc/mdy152.

[30] Mercadante S. Opioid titration in cancer pain: a critical review. Eur J Pain 2007;11(8):823–30. Available from: https://doi.org/10.1016/j.ejpain.2007.01.003.

[31] Mercadante S. Opioid dose titration for cancer pain. Eur J Pain 2024;28(3):359–68. Available from: https://doi.org/10.1002/ejp.2194.

[32] Paice JA, Bohlke K, Barton D, et al. Use of opioids for adults with pain from cancer or cancer treatment: ASCO guideline. J Clin Oncol 2023;41(4):914–30. Available from: https://doi.org/10.1200/JCO.22.02198.

[33] Fallon M. Principles of opioid titration. In: Forbes K, editor. Opioids in cancer pain. 1st ed. Oxford, UK: Oxford University Press; 2008.

[34] Weissman DE, Kamell A. Opioid dose escalation. Retrieved June 20, 2025, from: https://www.mypcnow.org/fast-fact/opioid-dose-escalation/; 2020.

[35] Fallon M, Laird BJ. A systematic review of combination step III opioid therapy in cancer pain: an EPCRC opioid guideline project. Palliat Med 2011;25(5):597–603. Available from: https://doi.org/10.1177/0269216310392101.

[36] WHO. Retrieved October 3, 2024, from: Palliative care. https://www.who.int/news-room/fact-sheets/detail/palliative-care; 2020.

Part II

Pharmacological strategies in pain management

Chapter 6

Opioids

Kristina Pilipović

Department of Basic and Clinical Pharmacology and Toxicology, Faculty of Medicine, University of Rijeka, Rijeka, Croatia

Key learning objectives

- To provide insight into the current evidence on the use of opioids in the treatment of pain in patients with cancer.
- To describe and explain the findings of clinical trials investigating the use of opioid analgesics in patients with cancer.
- To provide information about specific drugs that are often used or recommended by guidelines for the treatment of pain or related conditions.
- To give a brief overview of the current guidelines and opinions regarding the usage of opioid medicines and drugs in cancer pain management.

Introduction

In the treatment of pain, the use of opioids goes back thousands of years. The opium poppy (*Papaver somniferum*) was cultivated as early as 3400 BCE in Mesopotamia by the Sumerians, who called it the joy plant [1]. The name opium itself comes from the Greek word for juice, "*opos*," used for the liquid that appears on the unripe seed capsule when it is scratched. This raw opium is a complex mixture of chemical substances, including up to 25 different alkaloids, which are categorized into two main chemical classes, phenanthrenes and benzylisoquinolines [2]. Morphine, the main active ingredient in opium, was isolated in 1806 by F.W.A. Sertürner, while codeine was isolated a few years later, in 1832, by P. Robiquet [3].

In terms of terminology (Table 6.1), morphine, codeine, and structurally related compounds found in opium are referred to as opiates. However, a broader term—opioids—is used for any agent that binds to the ligand-binding site of any receptor of the opioid family. Opioids thus include the naturally occurring neurotransmitters such as enkephalins, endorphins, and dynorphins; semisynthetic opiate derivatives such as oxycodone, heroin, and buprenorphine; as well as synthetic opioids such as fentanyl, methadone, and pethidine. Opioid drugs are also commonly referred to as narcotics or narcotic analgesics, which comes from the Greek word "*narkōtikós*," with the meaning being "making something numb" or "causing stupor" due to their sedative properties and ability to induce sleep.

Opioid receptors

Opioid drugs exert their effect through binding to three types of classical opioid receptors: mu (μ), delta (δ), and kappa (κ) [4]. There is evidence for the existence of the subtypes of the μ, δ, and κ receptors, and it is believed that the differences in the effects of certain opioids are related to their selectivity to a particular receptor subtype. Another receptor is the most recently discovered member of the opioid receptor family—nociceptin/orphanin FQ—and, together with its endogenous ligand, N/OFQ, it makes up the fourth member of the opioid receptor and opioid peptide family [5] (Table 6.2).

Pharmacokinetics of opioids

The majority of opioids are quickly and fully absorbed through different places and surfaces, although some formulations like transdermal patches exhibit a deliberately slow release. They can be applied by various application routes and are most commonly administered orally, transdermally, subcutaneously, intramuscularly, and intravenously. Most opioids are

Pain Management in Palliative Care for Patients With Cancer. DOI: https://doi.org/10.1016/B978-0-443-30050-9.00008-4

TABLE 6.1 Commonly used terms related to opioids.

Opium	Borrowed from Latin word *opium* and ancient Greek word *ὄπιον (ópion)*. Mixture of alkaloids from the poppy seed.
Opiates	Naturally occurring alkaloids such as morphine or codeine.
Opioid	All compounds that work at the opioid receptors.
Narcotic	Term originally used to describe medications for sleep, then used to describe opioids, but now is a legal term for drugs of abuse.

TABLE 6.2 Opioid receptor classification, distribution, function, and endogenous and exogenous ligands.

Currently accepted name	μ	κ	δ	ORL1
Tissue expression	Brain Spinal cord Peripheral nerves Gastrointestinal (GI) tract Immune cells	Brain Spinal cord Peripheral nerves Heart Kidney	Brain Spinal cord Peripheral nerves GI Heart Immune cells	Brain Spinal cord Peripheral nerves Immune cells GI tract Cardiovascular system
Physiological function	Supraspinal analgesia, euphoria and reward, sedation, respiratory function, GI motility, hormone secretion, immune functions, miosis	Antinociception, diuresis, food intake, neuroendocrine and immune system functions, dysphoria and aversion, response to stress	Antinociception, GI motility, mood and behavior, cardiovascular regulation, immune function	Pain modulation Mood regulation Stress response

readily absorbed after oral application; however, their oral bioavailability tends to be poor (20%–40% for morphine) as the first-pass metabolism in the liver is responsible for the extensive metabolism.

In the treatment of cancer pain, oral administration remains the most common, the least invasive, and the easiest route for the application of opioid drugs and should be the chosen route if possible. Some opioids, like codeine and oxycodone, are more effective when taken orally because they undergo less first-pass metabolism. Bypassing this process, methods like nasal insufflation can quickly achieve therapeutic blood levels. Other administration routes include oral mucosa (e.g., lozenges) and transdermal patches, the latter providing continuous delivery of potent analgesics over several days. For chronic treatment, transdermal application is often a good choice. Among other special routes of administration, intrathecal and epidural applications offer an alternative if the maximum analgesic effect is desired at the principal site of pain transmission, achieving high drug concentrations while minimizing the risk of adverse drug effects.

The distribution of opioids throughout the body depends on both physiological factors, such as blood flow, and chemical properties of the opioid agents, like lipid solubility and protein binding. Although opioids vary in their affinity for plasma proteins—from 7% (codeine) to around 93% (sufentanil)—they quickly leave the bloodstream and concentrate most heavily in highly perfused organs, including the brain, lungs, liver, kidneys, and spleen [6]. Even though opioid concentrations are low in the skeletal muscles, it serves as a key reservoir due to its substantial mass in proportion to other tissues. Fatty tissues, although less perfused, can store significant amounts of opioids, especially after repeated high doses or prolonged infusions of highly lipophilic opioids (e.g., fentanyl and sufentanil), which metabolize slowly and can accumulate over time.

Opioids are biotransformed by phase 1 metabolism through the cytochrome P450 (CYP) pathway, phase 2 metabolism by conjugation, or both. Phase 1 metabolism primarily utilizes the CYP3A4 and CYP2D6 enzymes, and as CYP3A4 is responsible for metabolizing over 50% of all medications, this presents a significant risk for drug–drug interactions. In contrast, CYP2D6 handles a smaller number of drugs and carries a moderate risk of interactions. Conversely, opioids that undergo phase 2 conjugation, with little to no involvement of the CYP system, generally exhibit minimal interaction potential [7].

Opioids are largely metabolized into polar compounds, such as glucuronides, which are easily excreted by the kidneys [6,7]. Morphine, for example, is primarily converted to morphine-3-glucuronide (M3G), a neuroexcitatory metabolite, and to a lesser extent to morphine-6-glucuronide (M6G), which is a potent analgesic. Despite their potency, these polar metabolites have limited blood–brain barrier (BBB) penetration and typically have no effect on the central nervous system (CNS) after a single dose. However, in cases of renal failure or prolonged high-dose use, M3G can cause CNS excitation (e.g., seizures), and M6G can lead to prolonged opioid effects, and coadministration of certain drugs (e.g., probenecid or drugs that inhibit the P-glycoprotein drug transporter) may increase their CNS uptake.

Effects of opioid agonists on organs and organ systems

Opioid agonists and agents with mixed receptor effects possess various effects on organs and organ systems. Their primary site of action is the CNS, where their main effects are analgesia, euphoria, sedation, and respiratory depression.

Central effects of opioids

Analgesia

Opioid receptor-binding sites have been identified in various regions of the CNS, including the dorsal horn of the spinal cord, where all three major receptor types are present on pain transmission neurons and primary afferents. Opioid agonists not only inhibit dorsal horn pain transmission neurons but also reduce the release of excitatory transmitters from primary afferents. As opioids have a potent analgesic effect directly on the spinal cord, in clinical practice, opioid agonists can be administered directly to this site, providing localized pain relief while minimizing unwanted side effects that can occur with systemically administered opioids. When given systemically, opioids can affect multiple sites responsible for pain modulation, including the ascending pain transmission pathways and descending modulatory pathways. This complex interaction ultimately enhances the overall analgesic effect of opioid agonists [8–10].

Both animal and human studies have shown that endogenous and exogenous opioids can produce analgesia at sites outside the CNS, particularly in cases of inflammation, where pain seems especially sensitive to these peripheral actions. Functional μ receptors on the peripheral terminals of sensory neurons support the hypothesis that activation of these receptors decreases sensory neuron activity and transmitter release, with β-endorphin released by immune cells in injured or inflamed tissue contributing to this activation. Accordingly, in the treatment of pain with a strong inflammatory component, opioids that target peripheral sites may serve as valuable complementary options [11].

Sedation

Morphine and other μ-opioid receptor agonists can cause drowsiness and cognitive impairment, most commonly when starting opioid therapy or increasing the dosage. Fortunately, tolerance to sedation typically develops within a few days, but certain patient factors (such as dementia, encephalopathies, brain tumors, and the use of other depressants like sleep aids, antihistamines, alcohol, antidepressants, and anxiolytics) amplify these effects [12,13].

Respiratory system effects

All the opioid analgesics can cause significant respiratory depression, particularly in patients with conditions like asthma, chronic obstructive pulmonary disease, or preexisting respiratory issues. This depression affects all phases of respiratory activity, leading to irregular breathing patterns and a decrease in respiratory rate and volume. The mechanism involves a direct depressant effect on respiratory rhythm generation and reduced responsiveness to CO_2 and hypoxia. While maximal respiratory depression typically occurs within minutes to hours of administration, this effect can persist for several hours, especially with long-acting agents like methadone. Fortunately, respiratory depression can be reversed with opioid antagonists, though the risk of "renarcotization" must be carefully managed due to the shorter action of antagonists compared to the agonists. In the treatment of severe pain, respiratory depression presents a significant clinical challenge, and ongoing research is focused on efforts to overcome this therapeutic hurdle [12,14,15]. Preventing respiratory depression requires a combination of strategies, including patient and caregiver education, avoidance of concomitant use of other CNS depressants, naloxone access, avoidance of risky drug combinations, and regular follow-up visits [16,17].

Cough suppression is another well-known effect that the opioids have on the respiratory system. Morphine, codeine, and related opioids achieve this effect by directly affecting the cough center in the medulla. However, by suppressing the cough reflex, opioids may precipitate accumulation of secretions, leading to airway obstruction [18,19].

Miosis

Virtually all opioid agonists are known to cause pupillary constriction (miosis) by increasing parasympathetic outflow. The marked constriction is a diagnostic sign of opioid overdose and can be reversed with opioid antagonists or blocked by atropine [20,21].

Nausea and vomiting

Opioid analgesics can stimulate the brainstem chemoreceptor trigger zone, leading to nausea and vomiting. This effect is more pronounced in ambulatory patients, likely due to a vestibular component, with about 40% experiencing nausea and 15% vomiting after analgesic doses. Morphine also increases vestibular sensitivity and contributes to gastric stasis, further exacerbating postoperative nausea [22,23].

Peripheral effects of opioids

Cardiovascular system

Most opioids have minimal direct effects on the heart, aside from bradycardia. Meperidine can cause tachycardia due to its antimuscarinic action. Blood pressure is generally stable but can drop under stress due to peripheral vasodilation, often linked to central vasomotor depression and histamine release [24].

In angina pectoris and acute myocardial infarction treatment, morphine can be effective by reducing preload, inotropy, and chronotropy, which positively influences myocardial oxygen consumption. Additionally, it has cardioprotective effects and can mimic ischemic preconditioning, where a brief period of ischemia paradoxically helps protect the heart from subsequent ischemic events [25].

Skeletal muscle tone

At therapeutic analgesic doses, opioid agonists generally have minimal impact on motor tone and function. However, high doses, such as those used for anesthetic induction, can lead to muscular rigidity. With the involvement of truncal muscles, muscle rigidity can reduce thoracic compliance and interfere with ventilation. This is most apparent when high doses of the highly lipid-soluble opioids, such as fentanyl, are rapidly administered intravenously and can be overcome with concomitant application of muscle relaxants [26–28].

Gastrointestinal and biliary tract

Opioids are well-known for causing constipation, an effect that does not diminish with long-term use due to the high density of opioid receptors in the gastrointestinal system. In patients using opioid agonists, constipation and changes in bowel function develop in 40%–95% of individuals and can occur even with a single dose of morphine [23]. In contrast to the majority of other opioid side effects, constipation is not likely to improve with time. As a result, it must be anticipated, monitored, and managed throughout the duration of opioid treatment. Opioids can also contract biliary smooth muscle, causing biliary colic and potential constriction of the sphincter of Oddi, which may lead to reflux of secretions and elevated pancreatic enzyme levels [29].

Renal system

Opioids can impair renal function primarily by decreasing renal plasma flow. They also exhibit an antidiuretic effect and enhance sodium reabsorption in renal tubules. Opioids increase ureteral and bladder tone, which may cause urinary retention, especially postoperatively. Additionally, ureteral colic from kidney stones may worsen due to increased ureteral tone [30–32].

Neuroendocrine effects

Opioids stimulate the release of pituitary hormones such as antidiuretic hormone, prolactin, and somatotropin, while inhibiting luteinizing hormone release [33]. Chronic opioid use can lead to lower testosterone levels, resulting in decreased libido, energy, and mood, while women may experience menstrual irregularities [34].

Pruritus

Opioids like morphine and codeine can cause skin flushing, sweating, urticaria, and itching, primarily due to peripheral histamine release, but also through central actions on itch pathways [35]. This side effect is especially pronounced with spinal and epidural administration, where the incidence may reach 70%–100%.

Immune system

Opioids influence immune function by affecting lymphocyte activity and the release of opioid peptides that modulate inflammatory pain. While opioids may inhibit certain immune responses, such as natural killer cell activity, they can also contribute to pain modulation [23].

Precautions and the contraindications for the use of opioid agents

Opioid analgesics cause a similar spectrum of adverse effects, as they are mostly directly related to their mechanism of action. This is particularly true for opioids that act on the μ-opioid receptor, even though there are some individual differences between specific agents. However, except for constipation, the majority of opioid adverse effects generally improve over time.

Regarding the contraindications and precautions, opioids should be avoided in acute respiratory depression, acute alcoholism, and patients at high risk of paralytic ileus [36]. In high intracranial pressure or head injury patients, opioids lead to fatal alterations in brain functions as these substances, through respiratory depression, cause carbon dioxide retention with subsequent cerebral vasodilation [37,38]. Precautions are necessary in patients with hepatic and renal impairment, as it may cause drug or active accumulation and might require dosage modification. Opioid doses should also be reduced or even avoided in older and debilitated patients. Endocrine system conditions such as hypothyroidism and adrenal insufficiency may cause prolonged and exaggerated responses to opioid agents. Other conditions that also require caution are convulsive disorders, decreased respiratory reserve and acute asthma, hypotension, and prostatic hypertrophy.

With opioids being mostly used orally in patients with cancer pain, medical practitioners also have to consider food interactions. Specifically, alcoholic beverages or ethanol-containing products should be avoided, as they are known to disrupt extended-release capsule formulations, resulting in rapid release of the entire morphine dose. Administration of certain opioid formulations with food might increase the bioavailability of the drug. For example, taking morphine in the form of an oral solution after a high-fat meal has been reported to result in a 34% increase in its AUC [39]. Interactions of opioids with other medications are evaluated in depth in another chapter of this book.

Adverse effects, contraindications, important interactions, and specific comments for individual opioid drugs are given in Table 6.3.

Opioid tolerance and dependence

Opioid tolerance occurs when repeated doses of opioids lead to a reduced analgesic effect, often requiring higher doses to achieve the same pain relief. This tolerance can develop after just a few weeks of regular use, though ultrapotent opioids like remifentanil can induce tolerance in hours [40]. While tolerance develops to many of the effects of opioids, such as analgesia and respiratory depression, some effects like constipation and miosis are less affected. Cross-tolerance is common among μ-receptor agonists, meaning tolerance to one opioid may reduce the effectiveness of others, which has led to the practice of opioid rotation in pain management. Adjunctive treatments, such as N-methyl-D-aspartate (NMDA) receptor antagonists like ketamine, are being explored to mitigate opioid-induced tolerance and enhance pain relief in tolerant patients.

Opioid dependence is characterized by a specific withdrawal syndrome, with notable variations in psychological dependence and withdrawal severity among different opioids. The development of physical dependence is a common result of repeated administration of μ-opioid receptor agonists, leading to a withdrawal syndrome upon cessation. Withdrawal symptoms include rhinorrhea, lacrimation, yawning, chills, muscle aches, vomiting, diarrhea, anxiety, and hostility, with the severity dependent on the level of dependence. The onset, intensity, and duration of withdrawal vary by substance. In the case of fentanyl, heroin, or morphine, symptoms can begin within 6–10 hours, peaking at 36–48 hours and generally subsiding within 5 days, whereas meperidine's symptoms often resolve within 24 hours. Methadone withdrawal is characterized by a delayed peak and can last up to 2 weeks, making it useful for heroin detoxification despite persistent cravings. Additionally, a transient, explosive abstinence syndrome—antagonist-precipitated withdrawal—can occur rapidly with naloxone or buprenorphine, producing intense symptoms that peak within 10–20 minutes and subside in about an hour.

TABLE 6.3 Comparison of opioid agents for palliative care, containing key characteristics of various opioid agents, with a focus on their initial dosing for opioid-naive patients, maximum daily doses, clinical applications, and important precautions.

Opioid agent	Initial dose in opioid-naive patients[a,b]	Comments	Precautions
Morphine (oral): short-acting long-acting	–5 to 10 mg every 4 h –15 mg every 12 h or 20 or 30 mg PO once daily	The opioid of first choice for moderate-to-severe cancer pain. Available as tablet or liquid preparation.	Avoid use in renal dysfunction; use with caution in liver dysfunction.
Morphine (parenteral)	–2 to 4 mg every 4 h (IV, short-acting)	Typically used for patients requiring rapid pain relief.	
Fentanyl (TD)[c,d]	–12 mcg/h during 48–72 h	For patient with stable opioid requirements. Useful for patients with GI tract issues. Safe in patients with chronic kidney disease stage 4 or 5.	Do *not* cut patch, apply heat, and use cautiously in patients who develop fever. After patch removal, continued absorption of fentanyl occurs from the skin.
Hydrocodone (PO)	–5 to 10 mg every 6 h (IR) –10 mg every 12 h (ER) –20 mg every 24 h (ER)	Often used in combination with nonopioids. Generally well-tolerated, less sedation, and respiratory depression.	Doses greater than 160 mg/day of hydrocodone ER have been associated with increased risk of QTc prolongation. Use with caution in renal and/or hepatic dysfunction.
Oxycodone (PO)	–5 to 10 mg every 4 h (short-acting) –10 mg every 12 h (long-acting)	Alternative to oral morphine. Available as tablet, liquid preparation, or in combination with acetaminophen.	Use with caution in renal and/or liver dysfunction.
Codeine (PO)	–15 to 60 mg every 4 h	For mild-to-moderate pain.	Contraindicated in children younger than 12 or in rapid metabolizers of CYP2D6. Avoid or reduce dose in renal impairment.
Hydromorphone (PO)	–2 to 4 mg every 4 h (short-acting) –8 mg every 24 h	Alternative to oral morphine. Available as tablet or liquid preparation.	Use with caution in renal and/or liver dysfunction.
Hydromorphone (IV, SC)	–0.5 to 1 mg every 4 h		
Methadone (PO)	–2.5 to 10 mg every 6–8 h	Has the potential to control pain that does not respond to morphine or other opioids.	Long half-life, higher risk of adverse effects. Dose should be titrated slowly, usually every 5–7 days to achieve steady state.
Tramadol (PO)	–50 to 100 mg every 4–6 h (IR) or every 12 h (ER)	Useful for moderate pain. Provides dual-act on pain relief. Generally well-tolerated with a lower risk of serious side effects.	Risk of seizures with overdose. Caution in patients with renal or liver impairment. Avoid with MAOI.
Tapentadol (PO)	–50 to 100 mg every 4–6 h (IR) or 50 mg every 12 h (ER)	Useful for moderate pain. Provides dual-action pain relief. Favorable side effect profile.	Caution in patients with a history of seizures, renal impairment, or liver dysfunction. Risk of serotonin syndrome.
Buprenorphine (TD)[d]	20–35 mcg/h, up to every 72–96 h	For patient with stable opioid requirements Safe in chronic kidney disease stage 4 or 5.	May precipitate withdrawal in patients on full agonists.

CYP, Cytochrome P450; *ER*, extended-release; *GI*, gastrointestinal; *IR*, immediate-release; *IV*, intravenous; *MAOIs*, monoaminooxidase inhibitors; *PO*, peroral; *SC*, subcutaneous; *TD*, transdermal.

[a] *For the majority of opioid agents, there is no established "maximum daily dose" for opioid-tolerant individuals, as doses are titrated upwards based on the patient's response to pain relief and side effects. However, there are suggestions limiting tramadol to 400 mg/daily, tapentadol 600 mg/daily, and codeine 240 mg/daily*

[b] *Different markets could have different dose formulations.*

[c] *For the dose conversion and the calculation of equivalent analgesic strength of TD fentanyl, see Tables 6.4 and 6.5.*

[d] *Generally recommended only in opioid-tolerant patients with stable opioid requirements after conversion.*

TABLE 6.4 Dose conversion table—multiplication factors for converting the previous daily dose of opioid into an equianalgesic 24-h dose of oral morphine (mg/day of previous opioid x factor = equianalgesic 24-h dose of oral morphine).

Current opioid	Route of administration	Multiplication factor
Morphine	Peroral	1[a]
	Parenteral	3
Buprenorphine	Sublingual	75
	Parenteral	100
Codeine	Peroral	0.15
	Parenteral	0.23[b]
Fentanyl	Peroral	–
	Parenteral	300
Hydromorphone	Peroral	4
	Parenteral	20[b]
Methadone	Peroral	1.5
	Parenteral	3[b]
Oxycodone	Peroral	1.5
	Parenteral	3
Tapentadol	Peroral	0.4
	Parenteral	–
Tramadol	Peroral	0.25
	Parenteral	0.3

[a] *The strength ratio of oral and intramuscular morphine is based on clinical experience in patients with chronic pain.*
[b] *Based on single-dose studies in which an intramuscular dose of each of the listed active substances was compared with morphine to determine relative potency. Oral doses are those recommended when switching from parenteral to oral route of administration.*
Source: Taken and adapted from Foley KM. The treatment of cancer pain. N Engl J Med 1985;313:84–95. Available from: https://doi.org/10.1056/NEJM198507113130205 [48]; McPherson ML. Introduction to opioid conversion calculations. In: Demystifying opioid conversion calculations. Bethesda, MD: American Society of Health-System Pharmacists; 2018. p. c2010 [49].

Clinical pharmacology of individual opioids

Morphine

Considered the "golden standard" for opioid analgesics, morphine is a full agonist of opioid receptors that achieves its effects by binding to the μ-opioid receptor, with weaker affinity toward the κ-opioid receptor and almost no activity via the δ-opioid receptor. Although morphine is well absorbed after oral application, it is extensively inactivated in the liver in the process of first-pass metabolism. Consequently, it is more readily applied parenterally, that is, by subcutaneous (SC), intramuscular (IM), or intravenous (IV) injection. Its peak analgesic activity is achieved at 15 minutes following IV application and 45 minutes after SC injection, while the duration of action is around 1–2 hours following IV injection and up to 6 hours in the case of SC or IM administration. However, the pharmacological activity of morphine is poorly correlated with its blood concentration. This is considered the consequence of delayed entry into the cerebrospinal fluid (CSF) through the BBB. Similarly, morphine effects in the CNS persevere even with decreasing blood concentrations.

According to the latest update from the Cochrane Database of Systematic Reviews from 2016, and based on an analysis of 62 studies with 4241 subjects, the consensus is that oral morphine is effective in the treatment of pain in cancer patients [41]. Over 90% of subjects receiving oral morphine experienced pain relief to mild or no pain. Furthermore, more than 60% of respondents expressed satisfaction with morphine treatment, describing it as very good or excellent, and only about 5% of subjects discontinued treatment due to intolerable side effects. Since the previous update in 2007, it is important to note that

TABLE 6.5 Recommended starting dose of transdermal fentanyl based on daily doses of oral morphine for the patients requiring opioid switching or who are less clinically stable and for the patients on a stable and well-tolerated opioid regimen.

Patients requiring opioid switching or who are less clinically stable (conversion ratio 150:1)		Patients on a stable and well-tolerated opioid regimen (conversion ratio 100:1)	
24-h dose of oral morphine (mg/day)	Fentanyl dose (μg/h)	24-h dose of oral morphine (mg/day)	Fentanyl dose (μg/h)
<90	12	<44	12
90–134	25	45–89	25
135–224	50	90–149	50
225–314	75	150–209	75
315–404	100	210–269	100
405–494	125	270–329	125
495–584	150	330–389	150
585–674	175	390–449	175
675–764	200	450–509	200
765–854	225	510–569	225
855–944	250	570–629	250
945–1034	275	630–689	275
1035–1124	300	690–749	300

most studies that were added are focused on comparing different formulations of oral morphine or comparing oral morphine with other opioids, resulting in a lack of comprehensive evidence for certain types of comparisons [42–47].

Since the latest Cochrane update, additional randomized controlled trials (RCTs) were conducted with similar premises as previously mentioned. Such was the RCT reported by Sun et al. in 2022 [52]. In a multicenter, 7-day randomized controlled study involving 108 patients with moderate-to-severe cancer pain, participants were assigned to either a rapid titration group (receiving both sustained-release and SC morphine) or a group receiving only SC morphine for dose titration. The rapid titration group had significantly fewer rescue analgesic uses (0.4 vs 2.3 times, $P = .000$) with no observed differences in opioid-related symptoms between the groups. Rapid titration was found to be safe and more effective in reducing the need for rescue analgesics and improving pain control.

In another RCT published in 2021, it was evaluated if combining a low dose of methadone with morphine could reduce opioid consumption in cancer-related pain management [53]. Involving 41 patients in palliative care, one group received methadone (2.5 mg/12 h) with morphine, while the other group received morphine alone. The outcomes measured included total morphine use, pain intensity, and adverse effects, and the results showed no difference in morphine dose or adverse effects between groups. However, the methadone/morphine group experienced significantly lower pain intensity after 2 weeks, though not at other time points. The combination provided faster pain control without notable differences in adverse effects.

Cancer pain management guidelines tend to consider morphine and morphine-like opioids comparable and interchangeable in the treatment of chronic cancer pain, but individual responses can vary. As the main outcome of pain management for the physicians treating pain in patients with cancer is to relieve the pain over time, they tend to adapt the therapy whenever pain tends to increase. This might require a dose change, adding another opioid or an adjuvant drug, switching to another molecule, discontinuing opioids altogether, and considering the use of other analgesic techniques. In a multicenter randomized phase IV "real-life" trial, morphine and three other morphine-like opioids were compared for their analgesic efficacy, changes of therapy, and safety profile over time [54]. It was found that 22% of patients receiving morphine switched to another opioid (compared with 12% of those on oxycodone). Compared to other opioids, patients on morphine required lower dose escalation, adjuvants were scantily used, but the frequency of drug switches and discontinuations was the highest, as was the prevalence of neurotoxic effects. See Table 6.6 for morphine titration for severe cancer pain.

TABLE 6.6 Opioid titration (dose finding) for severe cancer pain.

Opioid	Initial dose	Following dosage	Rescue opioid
IV morphine	1.5 mg bolus every 10 min until pain relief (or adverse effects)	IR PO morphine every 4 h, on the basis of the previous IV requirements (IV to PO conversion 1:3)	IR PO morphine (10%–15% of the total daily dose)
PO morphine (or equivalent dose of oxycodone or hydrocodone)	IR morphine 5 mg every 4 h in opioid-naive patients, 10 mg in patients already on weak opioids	Follow the same scheme	IR PO morphine (up to hourly)

IR, Immediate-release; *IV*, intravenous; *PO*, peroral.
Source: Adapted from Fallon M, Giusti R, Aielli F, et al. Management of cancer pain in adult patients: ESMO clinical practice guidelines†. Ann Oncol 2018;29:iv166–91. Available from: https://doi.org/10.1093/annonc/mdy152; Caraceni A, Hanks G, Kaasa S, et al. Use of opioid analgesics in the treatment of cancer pain: evidence-based recommendations from the EAPC. Lancet Oncol 2012.13:e58–68. Available from: https://doi.org/10.1016/S1470-2045(12)70040-2 [51].

In a recent RCT, morphine was compared with hydromorphone using the patient-controlled subcutaneous analgesia (PCSA) protocol, and both drugs effectively and safely relieved short-term moderate-to-severe cancer pain [55]. In this randomized, double-blind trial, 60 patients were assigned to receive either hydromorphone or morphine through PCSA. Pain intensity, breakthrough pain, quality of life, opioid consumption, and adverse events were measured, and the results showed that both hydromorphone and morphine effectively reduced pain over 72 hours. Hydromorphone provided faster pain relief, with significantly lower pain scores at 30 minutes compared to morphine. Breakthrough pain occurrence and quality of life improved similarly in both groups, and no significant differences were observed in opioid consumption or adverse events such as nausea, vomiting, dizziness, or constipation. Overall, both drugs were effective and safe, with hydromorphone acting more quickly. In their study published in 2016, Bandieri et al. challenged the WHO's three-step analgesic ladder, specifically focusing on the effectiveness of step II opioids [56]. To achieve this, they designed an RCT to assess the efficacy and tolerability of low doses of morphine compared to standard doses of weak opioids for managing moderate cancer pain in opioid-naive patients. In a 28-day, multicenter, open-label RCT, 240 adults with moderate cancer pain were assigned to receive either weak opioids or low-dose morphine. Results showed that 88.2% of patients in the low-dose morphine group achieved a 20% reduction in pain intensity, compared to 57.7% in the weak opioid group. Pain reduction was greater in the morphine group as early as 1 week, with more patients achieving clinically significant pain relief. Treatment changes due to inadequate pain control were more common in the weak opioid group. Both groups had similar tolerability and adverse effects, but patients on low-dose morphine experienced better overall symptom improvement. Overall, this study demonstrated that low-dose morphine, compared to step II opioids, achieved an earlier and more effective level of analgesia for moderate cancer pain while maintaining a favorable tolerability profile and positively impacting overall well-being [56].

Morphine remains the gold standard for managing moderate-to-severe pain. However, assistance may be required to address common side effects, such as constipation and nausea, which are not morphine-exclusive but are issues related to most opioid analgesics.

Fentanyl

Fentanyl is a synthetic phenylpiperidine derivative with potent μ-opioid receptor activity and some δ-opioid receptor and κ-opioid receptor activity. It has high lipid solubility, allowing for administration through various routes, including parenteral, spinal, transdermal, transmucosal, buccal, and intranasal. Compared to morphine, fentanyl is around 80 times more potent but has significantly lower oral bioavailability with a shorter duration of analgesia of 1–1.5 hours.

When used via transdermal and transmucosal routes of delivery, its bioavailability ranges between 50% (oral transmucosal fentanyl citrate [OTFC]) and 90% (intranasal fentanyl). Options like fentanyl pectin nasal spray, buccal fentanyl, and sublingual fentanyl offer 60%–65% availability and can be chosen based on clinical conditions and patient preferences. It also reaches the CNS within minutes after the application [57]. For chronic pain, fentanyl is typically used as a transdermal formulation. There is a lag time of 6–12 hours before the onset of action following fentanyl application, and it generally achieves a steady state within 3–6 days. The duration of the analgesic effect of a single patch is usually 72 hours. However, in patients who experience significant end-of-dose failure, they can be replaced every 48 hours. After removing the patch, a SC reservoir persists, and the drug may take up to 24 hours to be fully cleared from the system [58–60]. Transdermal fentanyl is an especially useful option for patients with trouble swallowing as well as for patients with gastrointestinal tract (GIT) issues since transdermal fentanyl was reported to have a lower incidence of constipation compared to morphine.

Additionally, fentanyl can be recommended to patients with poor tolerance to morphine and in cases of inadequate patient compliance. Adverse effects commonly associated with transdermal preparations, such as rash and pruritus, were infrequent and improved over time. According to the National Comprehensive Cancer Network (NCCN) guidelines, fentanyl patches are recommended to be used only in patients tolerant to opioid therapy and are not to be used for treating unstable pain that requires frequent dose changes or titration [61]. This is due to various factors, including that the delayed analgesia onset makes it unsuitable for rapid pain control in unstable or acute pain scenarios and the fact that dose adjustments require 72 hours (patch replacement interval) to take full effect. Furthermore, prolonged elimination half-life and existence of the SC reservoir mean that the residual drug persists even after patch removal, increasing overdose risk if pain or opioid requirements decrease suddenly. In case of the need for breakthrough pain management, patients using transdermal fentanyl require immediate-release opioids (e.g., oral morphine) for breakthrough pain, which is impractical if baseline pain is not first stabilized with shorter-acting agents. All of these—delayed kinetics and inflexible dosing—make transdermal fentanyl appropriate only for stable, chronic cancer pain after initial stabilization with faster-acting opioids [62–66].

Fentanyl's high lipophilicity enables absorption through various routes, leading to the development of new, noninvasive formulations. Fentanyl has different well-established formulations which have proven effective in managing breakthrough cancer pain, emergencies, and pediatric care. Transmucosal fentanyl has been shown to be effective in treating breakthrough pain (transitory exacerbation of pain in cancer patients with otherwise persistent pain). In a double-blind, randomized crossover trial that compared OTFC with morphine sulfate immediate-release (MSIR) for managing breakthrough cancer pain in patients on a fixed opioid regimen, 134 cancer patients were enrolled, and OTFC was titrated to provide effective pain relief with manageable side effects. OTFC showed significantly better pain relief (pain intensity, pain relief, and global performance) compared to MSIR at all time points, and more patients preferred OTFC after the study [67]. Portenoy et al. [68] reported the results of an RCT conducted on 77 opioid-treated patients randomized to receive fentanyl buccal tablets or placebo for managing cancer-related breakthrough pain.

Fentanyl buccal tablets significantly improved pain relief and pain intensity differences at all time points compared to placebo. Adverse events were typical of opioids, with poor oral tolerability reported in two patients. Another long-term, open-label safety study evaluated the safety and tolerability of fentanyl buccal tablets in opioid-tolerant cancer patients with breakthrough pain [69]. Of the 232 patients enrolled, 79 found a successful fentanyl buccal tablet dose after titration, and 197 patients entered the maintenance phase, lasting 12 months or more. Adverse events, mostly typical of opioid use, led to therapy discontinuation in 33% of patients, and no unexpected adverse effects occurred, and the safety profile was consistent with short-term studies. For transmucosal fentanyl preparations, proper titration is crucial for optimizing pain management, as individual factors, such as patient characteristics, comorbidities, and other treatments, affect drug absorption and response, requiring dose adjustments [70]. Clinical studies show effective fentanyl buccal tablet doses range from 100 to 800 μg per episode, underscoring the importance of titration. Once an effective dose is determined, fentanyl buccal tablets can provide significant pain relief within 10 minutes, resulting in high patient satisfaction. A nasal fentanyl formulation for breakthrough cancer pain was studied in a randomized, crossover study that compared fentanyl pectin nasal spray and oral morphine in patients on regular opioids [71].

Nasal fentanyl demonstrated superior pain relief over oral morphine, with significantly greater reductions in pain intensity at 15 minutes, and by 15 and 30 minutes, more patients treated with nasal preparation experienced ≥33% and ≥50% pain reductions compared to morphine. In another observational, prospective cohort study done by the same group of investigators, the long-term safety and efficacy of intranasal fentanyl for managing breakthrough cancer pain in opioid-tolerant patients were assessed [72]. Among 75 patients, intranasal fentanyl was well-tolerated with no serious adverse effects, and efficacy was rated as good to excellent, with significant improvements in pain control and quality of sleep, particularly during the first 3–4 months.

Hydrocodone

Hydrocodone is a μ-opioid receptor and δ-opioid receptor agonist with approximately equipotent, if slightly lower, potency as oral morphine. It is not typically used in cancer patients with severe pain as a first-choice opioid but is suggested to be used as a mild, initial-use opioid, even though effective doses may vary. Additionally, in cases where patients may not tolerate morphine or oxycodone, or when those medications are ineffective, hydrocodone may be considered as part of an opioid rotation strategy. Hydrocodone is available in immediate-release formulations in combinations with acetaminophen or ibuprofen, but extended-release preparations (without added nonopioid analgesics) are also available. The bioavailability of hydrocodone after oral administration is high, and the elimination half-life is 4–6 hours for immediate-release and prolonged to 7–9 hours in extended-release formulations [73–75].

Oxycodone

Oxycodone is a semisynthetic opioid with an elimination half-life of 2–4 hours. Compared to morphine, it has 1.5–2 times higher potency. Oxycodone is available as immediate-release and modified-release tablets, capsules, and oral liquid and is used for moderate-to-severe persistent pain. Oral bioavailability of oxycodone is high at around 60%–80% as it is readily absorbed in the GIT. Although oxycodone is metabolized in the liver, the first-pass effect is not so pronounced as to significantly reduce the amount of drug that reaches the systemic circulation. The European Association for Palliative Care guidelines recommend oxycodone as a potential first-line option for managing moderate-to-severe cancer pain, serving as an alternative to morphine [51]. In the WHO's three-step analgesic ladder, oxycodone is considered a step III opioid but is available for use in step II when used at low doses (e.g., ≤20 mg/day), alone or in combination with paracetamol [76].

Despite morphine traditionally being the first-choice opioid, evidence suggests that oxycodone provides similar pain relief without significant differences in tolerability. Although a slight benefit was observed in some studies favoring morphine, the overall findings indicate that oxycodone is a viable option for patients, especially when considering individual patient factors such as side effect profiles, preferences, and availability [77,78]. A Cochrane review from 2022 included 42 studies involving 4485 participants and found no significant difference between oxycodone and other strong opioids, such as morphine, in terms of pain relief, cancer pain management, and side effects in adults with cancer [79]. While controlled-release morphine showed a slight benefit over controlled-release oxycodone in pain relief, this difference disappeared after sensitivity analysis and was deemed clinically insignificant. The Cochrane review noted a higher frequency of hallucinations with controlled-release morphine (7.8%) compared to controlled-release oxycodone (4%), though oxycodone had a slightly higher incidence of myoclonus. The evidence for these side effects, however, is of low certainty [79]. Oxycodone, like other opioid analgesics, often induces euphoria, relaxation, and reduced anxiety in occasional users, contributing to its widespread abuse, particularly in the United States [80]. Opioid misuse has contributed significantly to the opioid epidemic, with opioids causing 49,000 of the 72,000 drug overdose deaths in the United States in 2017. In response to the crisis, the FDA updated labeling for long-acting opioids in 2013, limiting their use to severe pain [81].

Codeine

A weak μ-opioid receptor and δ-opioid receptor agonist, codeine has very little direct analgesic effect. Namely, it is a prodrug that is metabolized in the liver via the CYP2D6 enzyme to several metabolites: codeine-6-glucuronide, norcodeine, morphine, M3G, M6G, and normorphine. For codeine to exert its analgesic activity, it therefore has to be metabolized to morphine through the process of O-methylation since the 3-methoxy group drastically reduces affinity for the μ-opioid receptor. This also means that the individuals who are poor CYP2D6 metabolizers will have no or weak analgesic effect when taking codeine, and the rapid metabolizers might experience toxic effects upon codeine application [82].

Hydromorphone

Hydromorphone is a potent, semisynthetic opioid derived from morphine. It is primarily a μ-opioid receptor agonist and a weak δ-opioid receptor agonist with a potency 5–8 times greater than that of morphine [83]. Hydromorphone is available in different formulations—oral tablets, liquids, suppositories, and parenteral preparations. Immediate-release oral hydromorphone has an onset of action of about 30 minutes and lasts around 4 hours. Its oral bioavailability is relatively low, around 38%–48%, due to the significant effect of the extensive liver metabolism, with 62% of an oral dose eliminated during the first pass [83,84]. Sustained and controlled-release tablets of hydromorphone are also available, with around 12 hours duration of action. The controlled-release hydromorphone achieves 90% bioavailability of the immediate-release tablets; thus the dose adjustment when switching from one type of tablet to the other is not necessary, although careful monitoring is still suggested [85].

When it comes to adverse effects, hydromorphone shares similar side effects with morphine but tends to cause less pruritus, nausea, vomiting, sedation, and cognitive impairment [83]. In the 2021 Cochrane review on the use of hydromorphone in cancer pain [86], eight studies involving 1283 adults with moderate-to-severe cancer pain, comparing hydromorphone with oxycodone, morphine, or fentanyl, were analyzed. No studies included a placebo or children. Results showed no significant differences in pain intensity across treatment groups, with all opioids effectively reducing pain. Hydromorphone performed similarly to morphine, oxycodone, and fentanyl. Side effects like nausea, vomiting, dizziness, and constipation were reported, but there was no clear difference in the frequency of these effects between hydromorphone and the other opioids [87,88]. The evidence is currently insufficient to confirm or deny the effectiveness of hydromorphone for cancer pain compared to other analgesics based on reported outcomes.

Methadone

Methadone is a long-acting μ-opioid receptor agonist similar to morphine but with extended duration and unique pharmacokinetics. Methadone also has an effect on the neurotransmission, as it blocks the NMDA receptors and monoaminergic reuptake transporters, which might explain its effectiveness in difficult-to-treat pain, especially if treatment with other opioids has failed [89–91]. Methadone is extremely well absorbed orally, and it can also be administered by IV, SC, and rectal routes, as well as by spinal injection. However, there are very high individual differences in the pharmacokinetics and half-life, which is long and highly variable (up to 120 hours) [92], but the variations can be even higher. Thus, its use, especially in cancer patients, can be complex and requires experience in pain management and palliative therapy [93]. Methadone is primarily used for opioid detoxification and maintenance therapy but is also effective for chronic pain. Its respiratory depressant effects can outlast its analgesic action, requiring careful dosage management, and the main side effects are similar to morphine, but it is uniquely associated with prolonged QT syndrome. RCT data indicate that methadone, although more challenging to manage than morphine, offers similar efficacy and tolerability for cancer pain treatment [94]. Studies show that outpatient initiation and rotation to methadone in cancer patients can be done safely without significant adverse effects [95]. Additionally, retrospective studies suggest that low-dose methadone may enhance pain control when used alongside another opioid analgesic in cancer patients [96,97].

Tramadol

Tramadol is a centrally acting analgesic agent with a complex mechanism of action. It produces analgesia through weak μ-opioid receptor activity and the inhibition of norepinephrine and serotonin reuptake [4,98]. Tramadol is effective for mild-to-moderate pain, comparable to morphine, but is less effective for severe pain, as its potency is only one-tenth that of morphine. It also has a favorable safety profile, causing fewer respiratory and cardiovascular issues than morphine. Common side effects include nausea, dizziness, and dry mouth, while the risk of seizures exists, particularly in susceptible individuals. After a single oral dose, tramadol has 68% bioavailability; it is extensively metabolized in the liver, with a half-life of approximately 6 hours for tramadol and 7.5 hours for its active metabolite. Tramadol-induced analgesia starts within an hour upon oral application with peak effect within 2–3 hours, and the duration of analgesia is about 6 hours. Tramadol is available as immediate-release and extended-release oral formulations. In a nonrandomized, observational study in cancer patients, high-dose tramadol was found to have comparable analgesic efficacy with low-dose morphine, but in patients receiving low-dose morphine, higher rates of constipation, neuropsychological symptoms, and pruritus were observed [99]. Conversely, in a double-blind study involving cancer patients, tramadol was associated with a higher incidence of adverse effects, such as vomiting, dizziness, and weakness, compared to hydrocodone and codeine [100]. A Cochrane review on tramadol, whether used alone or in combination with acetaminophen, determined that there is limited evidence supporting tramadol's effectiveness for cancer pain treatment, suggesting that it is probably not as effective as morphine in this context [101].

Tapentadol

Similar to tramadol, tapentadol is an atypical opioid with a mixed mechanism of action—it binds to the μ-opioid receptor and additionally inhibits norepinephrine reuptake [102–104]. Because of its effect on the neurotransmitter uptake, it should be used with caution or avoided in patients taking other serotonergic medications, as well as monoaminooxidase inhibitors, tricyclic antidepressants, selective serotonin reuptake inhibitors, and serotonin–norepinephrine reuptake inhibitors. The plasma half-life of tapentadol is around 4 hours after oral administration, with peak effect attained after 1 hour, while the duration of action is 4–6 hours. The drug undergoes extensive first-pass hepatic metabolism [103,105]. This indicates that hepatic function appears to strongly affect tapentadol clearance, and the dose should be lowered in patients with moderate hepatic impairment. Tapentadol received the US Food and Drug Administration (FDA) approval in 2008 for treating moderate-to-severe pain in adults in the United States [106]. It is available in an immediate-release form for acute pain and as an extended-release formulation for chronic pain management when continuous analgesia is needed.

It has been extensively researched in pain management, mostly for the treatment of noncancer pain, with a limited number of studies evaluating the use of tapentadol for cancer pain [103]. However, it has been demonstrated that tapentadol's effectiveness is comparable to morphine and oxycodone [107–109]. It was shown to be beneficial for both opioid-naive patients and those already receiving opioid therapy [110]. Due to its lower binding affinity for the μ-opioid receptor, tapentadol tends to result in fewer opioid-related side effects, such as constipation and nausea [107,108]. A recent randomized trial comparing tapentadol alone to a combination of tapentadol and duloxetine in patients with chemotherapy-induced peripheral neuropathy found similar improvements in pain relief and quality-of-life measures in both groups, with comparable adverse effects. In a recent prospective, open-label, multicenter trial [111], a total of 752 patients with moderate-to-severe chronic cancer pain were observed during a 6-month period. Patients were opioid-naive or opioid-resistant, and out of 650 who completed the study, 349 were cancer patients. Extended-release tapentadol significantly reduced the mean pain intensity, including neuropathic pain, during the observation period with improvements in the quality-of-life measurements and very few drug-related adverse reactions.

Buprenorphine

A partial μ-opioid receptor agonist, buprenorphine has been approved for the treatment of mild-to-moderate pain in opioid-naive and opioid-tolerant patients. Buprenorphine also acts as a δ- and κ-opioid antagonist and has weak partial agonist activity at NOP receptors. It is highly lipophilic in comparison to morphine; its potency is much higher, but its analgesic effect has a ceiling due to partial agonism. It is used primarily for managing opioid use disorder and as an analgesic, often administered sublingually or via injection. Buprenorphine is also available as a sublingual tablet, oral lyophilizate, and transdermal patch. For moderate-to-severe chronic pain, buccal film and transdermal patch formulations are usually used due to a slower onset of action, longer elimination half-life, and the duration of action. Buprenorphine offers several benefits compared to other opioids, including a lower risk of misuse, reduced development of analgesic tolerance, lower likelihood of constipation, and less impact on the tone of the sphincter of Oddi. Additionally, it has a ceiling effect on respiratory depression, making it safer at higher doses; it tends to cause less cognitive impairment, lower potential for drug dependence, and milder withdrawal symptoms.

In the 2015 Cochrane review, 19 studies with 1421 patients were analyzed, comparing buprenorphine (any formulation and any route of administration) with a placebo or an active drug (including buprenorphine) for cancer background pain in adults and children [112]. Among the studies comparing buprenorphine to other drugs, five found it superior, three showed no difference, and three found it inferior based on side effects or patient preference. Pain intensity was similar between IM buprenorphine and suppository forms, but the IM group experienced more dizziness and nausea. Sublingual buprenorphine provided faster pain relief than subdermal, with similar adverse event rates. Transdermal buprenorphine was found superior to placebo in two studies but showed no dose–response relationship. The overall evidence quality was rated as very low due to bias, small sample sizes, and incomplete reporting. Regardless of the lack of strong evidence from clinical studies on the superiority or at least noninferiority in the chronic treatment of cancer pain compared to other opioid medicines, buprenorphine has high potential for expanded use in the treatment of adult cancer pain and especially in particular populations of patients, like older adults, patients with renal failure, those with preexisting or high risk for substance dependence, or those not tolerating other opioids [113].

Other opioid agents with mixed actions

Butorphanol, pentazocine, and nalbuphine are κ- and μ-opioid receptor partial agonists. Pentazocine and nalbuphine are also opioid analgesics with mixed agonist–antagonist properties [114]. Nalbuphine acts as a strong agonist at κ- and a partial antagonist at μ-opioid receptors, and it is administered parenterally. Unlike morphine, nalbuphine exhibits a ceiling effect on respiratory depression at higher doses. However, if respiratory depression does occur, it may be difficult to reverse with naloxone due to nalbuphine's higher receptor affinity. Pentazocine is the oldest mixed agent available, and it is a κ-opioid receptor agonist with weak μ-antagonist or partial agonist properties. It may be used orally or parenterally, but SC application is not recommended due to its irritant properties. Mixed agonists–antagonists, such as butorphanol and pentazocine, should not be used in conjunction with opioid agonists for cancer pain management. Switching from a pure opioid agonist to an agonist–antagonist can precipitate withdrawal symptoms in patients who are physically dependent on opioid agonists.

Opioid antagonists

Naloxone, naltrexone, and nalmefene are pure opioid antagonists derived from morphine with relatively high affinity for μ-opioid binding sites and lower affinity for κ- and δ-opioid receptors [114]. Their main use is for treating opioid overdose, but they are also utilized in managing opioid-induced constipation and alcohol abuse. Typically, opioid antagonists have minimal effects when there is no external agonist present.

Naloxone is usually applied by injection and autoinjector, but nasal sprays are also available for emergency administration. Oral application is ineffective, as naloxone undergoes significant first-pass metabolism and has very low oral bioavailability. Following injection, the duration of action of naloxone is short at around 1–2 hours, and renarcotization may occur with long-lasting agonists as naloxone is metabolized [4,114]. Naloxone rapidly reverses central and peripheral opioid effects within 1–3 minutes, and in overdosed patients effectively normalizes respiration, reverses disturbances of consciousness and miosis, and normalizes bowel activity as well as nociception. In opioid-dependent subjects, administration of naloxone and other μ-antagonists almost instantaneously induces withdrawal symptoms. Low-dose naloxone (0.04 mg) is increasingly used to treat adverse effects like itching, nausea, and vomiting from IV or epidural opioids while preserving analgesia through careful titration. Oral naloxone and modified analogs, such as methylnaltrexone bromide and naldemedine, have been approved for opioid-induced constipation in chronic pain patients [115]. Naloxegol and naldemedine have also been used for the treatment of opioid-induced constipation, while alvimopan is approved for managing postoperative ileus [116]. These agents primarily work as peripheral μ-receptor antagonists with limited CNS effects due to poor BBB penetration.

Naltrexone is efficiently absorbed when taken orally, although it can be rapidly metabolized during first-pass metabolism. It has a half-life of 10 hours, and a single oral dose of 100 mg can block the effects of injected heroin for as long as 48 hours.

Nalmefene, the most recent addition to this class of drugs, is a derivative of naltrexone. Similar to naloxone, nalmefene is utilized for treating opioid overdose and has a longer half-life of 8–10 hours [117].

Opioid de-escalation

Opioid de-escalation (tapering) is a procedure recommended to reduce the risks of overdose, dependence, and adverse effects, especially when the risks of continued high-dose opioid therapy outweigh the benefits. It is to be done gradually, with patient-centered planning and close monitoring. Key recommendations for opioid de-escalation are presented in Table 6.7. In essence, opioid de-escalation is a gradual, patient-tailored process balancing safety, efficacy, and patient comfort, supported by education, monitoring, and multidisciplinary care when needed [118,119].

TABLE 6.7 Key recommendations for opioid de-escalation.

Recommendation for opioid de-escalation	
Assess benefits and risks	During the ongoing opioid therapy, benefits and risks should be assessed regularly and the use of opioid tapered if necessary.
Engage in shared decision-making	Continuously communicate with the patient, discuss goals, concerns, and expectations.
Taper slowly and individually	Typical dose reductions are about 5%–25% per week or month, tailored to patient tolerance and duration of prior use. Faster tapers may be needed in urgent situations (e.g., overdose risk, misuse).
Avoid abrupt discontinuation	Abrupt discontinuation of opioids may precipitate withdrawal symptoms and adverse outcomes.
Monitor closely	Patients need to be observed for the symptoms and signs of withdrawal symptoms, efficacy of pain control, and mental health status. Taper speed should be adjusted accordingly.
Provide education	Patients need to be familiarized with overdose risks, especially after tapering due to loss of tolerance, and prescribing naloxone should be considered.
Use validated tools and protocols	Opioid tapering should be done by following the established protocols and flowcharts which are helpful to guide the process.
Refer to specialists	If the opioid use disorder is suspected or if tapering is complex, refer the patients to pain management or addiction medicine specialists.
How to perform opioid tapering	
1. Convert to a single opioid if multiple are used. **2.** Reduce dose by 10%–25% per week or month depending on patient factors and response. **3.** Allow flexibility: pauses or slower tapering if withdrawal or pain worsens. **4.** Support patients with nonopioid pain management and self-management strategies. **5.** Regularly reassess benefits, risks, and patient readiness throughout the taper.	

Conclusion and future perspectives on the use of opioids in cancer pain management

The consensus on the use of opioids for managing cancer pain highlights their effectiveness and necessity, particularly for patients experiencing moderate-to-severe pain. Pain management should be individualized based on the patient's specific cancer type, pain characteristics, treatment goals, and potential side effects, with regular assessment and adjustment of opioid dosages as necessary. A multimodal pain management strategy that includes nonopioid medications, adjuvant therapies, and psychological support is encouraged to optimize pain control. Additionally, integrating palliative care early in the treatment process focuses on symptom management and improvements in the quality of life.

Looking to the future, ongoing research into the pharmacogenomics of opioids, the development of less addictive analgesics and improved guidelines for their use in cancer pain management are essential. There is a pressing need for innovative pain management strategies that minimize the risk of opioid dependency while maintaining effective pain control. Enhanced education and training for healthcare providers about the complexities of opioid use, as well as patient-centered approaches to pain management, will be vital in addressing the evolving challenges in cancer care, including palliative care settings.

References

[1] Veiga P. Opium: was it used as a recreational drug in ancient Egypt? Cultural and linguistic transition explored. Proceedings of the ATrA closing workshop Trieste, May 25–26, 2016, Trieste, Italy; 2017. p. 199–215.

[2] Kalant H. Opium revisited: a brief review of its nature, composition, non-medical use and relative risks. Addiction 1997;92:267–77. Available from: https://doi.org/10.1111/j.1360-0443.1997.tb03197.x.

[3] Zarin MKZ, Dehaen W, Salehi P, et al. Synthesis and modification of morphine and codeine, leading to diverse libraries with improved pain relief properties. Pharmaceutics 2023;15:1779. Available from: https://doi.org/10.3390/pharmaceutics15061779.

[4] Jutkiewicz EM, Traynor JR. Opioid analgesics. In: Brunton LL, Knollmann BC, editors. Goodman & Gilman's: the pharmacological basis of therapeutics. New York, NY: McGraw-Hill Education; 2023.

[5] Toll L, Bruchas MR, Calo' G, et al. Nociceptin/orphanin fq receptor structure, signaling, ligands, functions, and interactions with opioid systems. Pharmacol Rev 2016;68:419–57. Available from: https://doi.org/10.1124/pr.114.009209.

[6] Davies G, Kingswood C, Street M. Pharmacokinetics of opioids in renal dysfunction. Clin Pharmacokinet 1996;31:410–22. Available from: https://doi.org/10.2165/00003088-199631060-00002.

[7] Smith HS. Opioid metabolism. Mayo Clin Proc 2009;84(7):613–24. Available from: https://doi.org/10.1016/S0025-6196(11)60750-7.

[8] Kohno T, Kumamoto E, Higashi H, et al. Actions of opioids on excitatory and inhibitory transmission in substantia gelatinosa of adult rat spinal cord. J Physiol 1999;518:803–13. Available from: https://doi.org/10.1111/j.1469-7793.1999.0803p.x.

[9] Kondo I, Marvizon JCG, Song B, et al. Inhibition by spinal μ- and δ-opioid agonists of afferent-evoked substance P release. J Neurosci 2005;25:3651–60. Available from: https://doi.org/10.1523/JNEUROSCI.0252-05.2005.

[10] Mizoguchi H, Watanabe C, Sakurada T, et al. New vistas in opioid control of pain. Curr Opin Pharmacol 2012;12:87–91. Available from: https://doi.org/10.1016/j.coph.2011.10.020.

[11] Smith HS. Peripherally-acting opioids. Pain Physician 2008;11(2 Suppl.):S121–32.

[12] Elliott A. Reducing the incidence of opioid-induced unintended advanced sedation and respiratory depression in hospitalized patients. Doctor of Nursing Practice Projects. 88. https://digitalcommons.jsu.edu/etds_nursing/88; 2023.

[13] Garrett J, Vanston A, Ogola G, et al. Predicting opioid-induced oversedation in hospitalised patients: a multicentre observational study. BMJ Open 2021;11(11):e051663. Available from: https://doi.org/10.1136/bmjopen-2021-051663.

[14] Boom M, Niesters M, Sarton E, et al. Non-analgesic effects of opioids: opioid-induced respiratory depression. Curr Pharm Des 2012;18:5994–6004. Available from: https://doi.org/10.2174/138161212803582469.

[15] Gupta K, Prasad A, Nagappa M, et al. Risk factors for opioid-induced respiratory depression and failure to rescue: a review. Curr Opin Anaesthesiol 2018;31:110–9. Available from: https://doi.org/10.1097/ACO.0000000000000541.

[16] European Monitoring Centre for Drugs and Drug Addiction (EU body or agency), Strang J, McDonald R, Simon R, Hedrich D. Preventing opioid overdose deaths with take-home naloxone. Publications Office of the European Union; 2016.

[17] van der Schier R, Roozekrans M, van Velzen M, et al. Opioid-induced respiratory depression: reversal by non-opioid drugs. F1000Prime Rep 2014;6:79. Available from: https://doi.org/10.12703/P6-79.

[18] Joskova M, Durdik P, Sutovska M, et al. Negative impact of anesthesia with midazolam, sufentanil, and propofol used in pediatric flexible bronchoscopy on the tracheal ciliary beat frequency in guinea pigs. J Pharmacol Sci 2020;142:165–71. Available from: https://doi.org/10.1016/j.jphs.2020.01.005.

[19] Reynolds SM, Mackenzie AJ, Spina D, et al. The pharmacology of cough. Trends Pharmacol Sci 2004;25:569–76. Available from: https://doi.org/10.1016/j.tips.2004.09.009.

[20] Collett BJ. Opioid tolerance: the clinical perspective. Br J Anaesth 1998;81:58–68. Available from: https://doi.org/10.1093/bja/81.1.58.

[21] Kongsgaard UE, Høiseth G. Dynamic assessment of the pupillary reflex in patients on high-dose opioids. Scand J Pain 2019;19:465–71. Available from: https://doi.org/10.1515/sjpain-2019-0032.

[22] Porreca F, Ossipov MH. Nausea and vomiting side effects with opioid analgesics during treatment of chronic pain: mechanisms, implications, and management options. Pain Med 2009;10:654–62. Available from: https://doi.org/10.1111/j.1526-4637.2009.00583.x.

[23] Benyamin R, Trescot AM, Datta S, et al. Opioid complications and side effects. Pain Physician 2008;11:S105–20.

[24] Chen A, Ashburn MA. Cardiac effects of opioid therapy. Pain Med 2015;16:S27–31. Available from: https://doi.org/10.1111/pme.12915.

[25] Schultz JEJ, Hsu AK, Gross GJ. Morphine mimics the cardioprotective effect of ischemic preconditioning via a glibenclamide-sensitive mechanism in the rat heart. Circ Res 1996;78:1100–4. Available from: https://doi.org/10.1161/01.RES.78.6.1100.

[26] Çoruh B, Tonelli MR, Park DR. Fentanyl-induced chest wall rigidity. Chest 2013;143:1145–6. Available from: https://doi.org/10.1378/chest.12-2131.

[27] McNicol E, Horowicz-Mehler N, Fisk RA, et al. Management of opioid side effects in cancer-related and chronic noncancer pain: a systematic review. J Pain 2003;4:231–56. Available from: https://doi.org/10.1016/S1526-5900(03)00556-X.

[28] Struppler A, Burgmayer B, Ochs GB, et al. The effect of epidural application of opioids on spasticity of spinal origin. Life Sci 1983;33:607–10. Available from: https://doi.org/10.1016/0024-3205(83)90576-3.

[29] Thompson DR. Narcotic analgesic effects on the sphincter of Oddi: a review of the data and therapeutic implications in treating pancreatitis. Am J Gastroenterol 2001;96:1266–72. Available from: https://doi.org/10.1111/j.1572-0241.2001.03536.x.

[30] Davenport K, Timoney AG, Keeley FX. Conventional and alternative methods for providing analgesia in renal colic. BJU Int 2005;95:297–300. Available from: https://doi.org/10.1111/j.1464-410X.2005.05286.x.

[31] Baldini G, Bagry H, Aprikian A, et al. Postoperative urinary retention: anesthetic and perioperative considerations. Anesthesiology 2009;110:1139–57. Available from: https://doi.org/10.1097/ALN.0b013e31819f7aea.

[32] Mercadante S, Arcuri E. Opioids and renal function. J Pain 2004;5:2–19. Available from: https://doi.org/10.1016/j.jpain.2003.09.007.

[33] Armario A. Activation of the hypothalamic-pituitary-adrenal axis by addictive drugs: different pathways, common outcome. Trends Pharmacol Sci 2010;31:318–25. Available from: https://doi.org/10.1016/j.tips.2010.04.005.

[34] Bawor M, Bami H, Dennis BB, et al. Testosterone suppression in opioid users: a systematic review and meta-analysis. Drug Alcohol Depend 2015;149:1–9. Available from: https://doi.org/10.1016/j.drugalcdep.2015.01.038.

[35] Ganesh A, Maxwell LG. Pathophysiology and management of opioid-induced pruritus. Drugs 2007;67:2323–33. Available from: https://doi.org/10.2165/00003495-200767160-00003.

[36] World Health Organization. WHO guidelines for the pharmacological and radiotherapeutic management of cancer pain in adults and adolescents, <https://www.who.int/publications/i/item/9789241550390>; [accessed 10.03.24].

[37] Hocker SE, Fogelson J, Rabinstein AA. Refractory intracranial hypertension due to fentanyl administration following closed head injury. Front Neurol 2013;4:3. Available from: https://doi.org/10.3389/fneur.2013.00003.

[38] Wiener J, McIntyre A, Janzen S. Opioids and cerebral physiology in the acute management of traumatic brain injury: a systematic review. Brain Inj 2019;33:559–66. Available from: https://doi.org/10.1080/02699052.2019.1574328.

[39] Gourlay GK, Plummer JL, Cherry DA, et al. Influence of a high-fat meal on the absorption of morphine from oral solutions. Clin Pharmacol Ther 1989;46:463–8. Available from: https://doi.org/10.1038/clpt.1989.166.

[40] Morgan MM, Christie MJ. Analysis of opioid efficacy, tolerance, addiction and dependence from cell culture to human. Br J Pharmacol 2011;164:1322–34. Available from: https://doi.org/10.1111/j.1476-5381.2011.01335.x.

[41] Wiffen PJ, Wee B, Moore RA. Oral morphine for cancer pain. Cochrane Database Syst Rev 2016;2016:CD003868. Available from: https://doi.org/10.1002/14651858.CD003868.pub4.

[42] Hanna M, Thipphawong J. 118 Study Group. A randomized, double-blind comparison of OROS(R) hydromorphone and controlled-release morphine for the control of chronic cancer pain. BMC Palliat Care 2008;7:17. Available from: https://doi.org/10.1186/1472-684X-7-17.

[43] Homsi J, Walsh D, Lasheen W, et al. A comparative study of 2 sustained-release morphine preparations for pain in advanced cancer. Am J Hosp Palliat Care 2010;27:99–105. Available from: https://doi.org/10.1177/1049909109345146.

[44] Mercadante S, Porzio G, Ferrera P, et al. Sustained-release oral morphine versus transdermal fentanyl and oral methadone in cancer pain management. Eur J Pain 2008;12:1040–6. Available from: https://doi.org/10.1016/j.ejpain.2008.01.013.

[45] Mercadante S, Tirelli W, David F, et al. Morphine versus oxycodone in pancreatic cancer pain: a randomized controlled study. Clin J Pain 2010;26:794–7. Available from: https://doi.org/10.1097/ajp.0b013e3181ecd895.

[46] Oztürk T, Karadibak K, Catal D, et al. [Comparison of TD-fentanyl with sustained-release morphine in the pain treatment of patients with lung cancer]. Agri 2008;20(3):20–5.

[47] Ridgway D, Sopata M, Burneckis A, et al. Clinical efficacy and safety of once-daily dosing of a novel, prolonged-release oral morphine tablet compared with twice-daily dosing of a standard controlled-release morphine tablet in patients with cancer pain: a randomized, double-blind, exploratory crossover study. J Pain Symptom Manage 2010;39:712–20. Available from: https://doi.org/10.1016/j.jpainsymman.2009.08.013.

[48] Foley KM. The treatment of cancer pain. N Engl J Med 1985;313:84–95. Available from: https://doi.org/10.1056/NEJM198507113130205.

[49] McPherson ML. Introduction to opioid conversion calculations. In: Demystifying opioid conversion calculations. Bethesda, MD: American Society of Health-System Pharmacists; 2018:c2010.

[50] Fallon M, Giusti R, Aielli F, et al. Management of cancer pain in adult patients: ESMO clinical practice guidelines†. Ann Oncol 2018;29:iv166–91. Available from: https://doi.org/10.1093/annonc/mdy152.

[51] Caraceni A, Hanks G, Kaasa S, et al. Use of opioid analgesics in the treatment of cancer pain: evidence-based recommendations from the EAPC. Lancet Oncol. 2012;13:e58–68. Available from: https://doi.org/10.1016/S1470-2045(12)70040-2.
[52] Sun Y, Wei H, Yu M, et al. Rapid titration with oral sustained-release morphine plus subcutaneous morphine in a multi-center, randomized control study of cancer patients with moderate to severe cancer pain. Jpn J Clin Oncol 2022;52:1303–10. Available from: https://doi.org/10.1093/jjco/hyac128.
[53] Duarte FCN, Ferraro LHDC, Ferreira A, et al. A randomized controlled trial evaluating the analgesic effect of the combination of methadone with morphine for cancer related pain. Clin J Pain 2021;37:664–8. Available from: https://doi.org/10.1097/AJP.0000000000000959.
[54] Corli O, Floriani I, Roberto A, et al. Are strong opioids equally effective and safe in the treatment of chronic cancer pain? A multicenter randomized phase IV 'real life' trial on the variability of response to opioids. Ann Oncol 2016;27:1107–15. Available from: https://doi.org/10.1093/annonc/mdw097.
[55] Zeng X, Zhu J, Li J, et al. Patient controlled subcutaneous analgesia of hydromorphone versus morphine to treat moderate and severe cancer pain: a randomized double-blind controlled trial. J Pain Symptom Manage 2024;67:50–8. Available from: https://doi.org/10.1016/j.jpainsymman.2023.09.018.
[56] Bandieri E, Romero M, Ripamonti CI, et al. Early Strong Opioid Treatment Study (ESOT) investigators. Randomized trial of low-dose morphine versus weak opioids in moderate cancer pain. J Clin Oncol 2016;34:436–42. Available from: https://doi.org/10.1200/JCO.2015.61.0733.
[57] Lötsch J, Walter C, Parnham MJ, et al. Pharmacokinetics of non-intravenous formulations of fentanyl. Clin Pharmacokinet 2013;52:23–36. Available from: https://doi.org/10.1007/s40262-012-0016-7.
[58] Grond S, Radbruch L, Lehmann KA. Clinical pharmacokinetics of transdermal opioids: focus on transdermal fentanyl. Clin Pharmacokinet 2000;38:59–89. Available from: https://doi.org/10.2165/00003088-200038010-00004.
[59] Kuip EJM, Zandvliet ML, Koolen SLW, et al. A review of factors explaining variability in fentanyl pharmacokinetics; focus on implications for cancer patients. Br J Clin Pharmacol 2017;83:294–313. Available from: https://doi.org/10.1111/bcp.13129.
[60] Varvel JR, Shafer SL, Hwang SS, et al. Absorption characteristics of transdermally administered fentanyl. Anesthesiology 1989;70:928–34. Available from: https://doi.org/10.1097/00000542-198906000-00008.
[61] NCCN Clinical Practice Guidelines in Oncology. Adult cancer pain, <https://www.nccn.org/guidelines/guidelines-detail?category=3&id=1413>; 2024 [accessed 10.03.24].
[62] Allan L, Richarz U, Simpson K, et al. Transdermal fentanyl versus sustained release oral morphine in strong-opioid naïve patients with chronic low back pain. Spine 2005;30:2484–90. Available from: https://doi.org/10.1097/01.brs.0000186860.23078.a8.
[63] Dowell D. CDC clinical practice guideline for prescribing opioids for pain—United States, 2022. MMWR Recomm Rep 2022;71(3):1–95. Available from: https://doi.org/10.15585/mmwr.rr7103a1.
[64] Fu Q, Han N, Li N, et al. Guidelines for rational clinical use of fentanyl transdermal patch. Drug Des Devel Ther 2024;18:233–55. Available from: https://doi.org/10.2147/DDDT.S414318.
[65] Gourlay GK. Treatment of cancer pain with transdermal fentanyl. Lancet Oncol 2001;2:165–72. Available from: https://doi.org/10.1016/S1470-2045(00)00258-8.
[66] Kornick CA, Santiago-Palma J, Schulman G, et al. A safe and effective method for converting patients from transdermal to intravenous fentanyl for the treatment of acute cancer-related pain. Cancer 2003;97:3121–4. Available from: https://doi.org/10.1002/cncr.11457.
[67] Coluzzi PH, Schwartzberg L, Conroy JD, et al. Breakthrough cancer pain: a randomized trial comparing oral transmucosal fentanyl citrate (OTFC) and morphine sulfate immediate release (MSIR). Pain 2001;91:123–30. Available from: https://doi.org/10.1016/s0304-3959(00)00427-9.
[68] Portenoy RK, Taylor D, Messina J, et al. A randomized, placebo-controlled study of fentanyl buccal tablet for breakthrough pain in opioid-treated patients with cancer. Clin J Pain 2006;22:805–11. Available from: https://doi.org/10.1097/01.ajp.0000210932.27945.4a.
[69] Weinstein SM, Messina J, Xie F. Fentanyl buccal tablet for the treatment of breakthrough pain in opioid-tolerant patients with chronic cancer pain: a long-term, open-label safety study. Cancer 2009;115:2571–9. Available from: https://doi.org/10.1002/cncr.24279.
[70] Kleeberg UR, Filbet M, Zeppetella G. Fentanyl buccal tablet for breakthrough cancer pain: why titrate? Pain Pract 2011;11:185–90. Available from: https://doi.org/10.1111/j.1533-2500.2010.00414.x.
[71] Mercadante S, Aielli F, Adile C, et al. Fentanyl pectin nasal spray versus oral morphine in doses proportional to the basal opioid regimen for the management of breakthrough cancer pain: a comparative study. J Pain Symptom Manage 2016;52:27–34. Available from: https://doi.org/10.1016/j.jpainsymman.2016.01.010.
[72] Mercadante S, Vellucci R, Cuomo A, et al. Long-term efficacy and tolerability of intranasal fentanyl in the treatment of breakthrough cancer pain. Support Care Cancer 2015;23:1349–54. Available from: https://doi.org/10.1007/s00520-014-2491-x.
[73] Cardia L, Calapai G, Quattrone D, et al. Preclinical and clinical pharmacology of hydrocodone for chronic pain: a mini review. Front Pharmacol 2018;9:1122. Available from: https://doi.org/10.3389/fphar.2018.01122.
[74] Charles Argoff MD, Paul Arnstein RN, Steven Stanos DO, et al. Relationship between change in pain intensity and functional outcomes in patients with chronic pain receiving twice daily extended-release hydrocodone bitartrate. J Opioid Manag 2015;11:417–24. Available from: https://doi.org/10.5055/jom.2015.0291.
[75] Cofano S, Patel P, Yellon R. Hydrocodone. In: StatPearls [Internet]. Treasure Island (FL): StatPearls Publishing; 2025.
[76] World Health Organization. Cancer pain relief: with a guide to opioid availability. World Health Organization; 1996.
[77] Parris WC-V, Johnson BW, Croghan MK, et al. The use of controlled-release oxycodone for the treatment of chronic cancer pain: a randomized, double-blind study. J Pain Symptom Manage 1998;16:205–11. Available from: https://doi.org/10.1016/S0885-3924(98)00064-5.

[78] Reid CM, Martin RM, Sterne JAC, et al. Oxycodone for cancer-related pain: meta-analysis of randomized controlled trials. Arch Intern Med 2006;166:837–43. Available from: https://doi.org/10.1001/archinte.166.8.837.

[79] Schmidt-Hansen M, Bennett MI, Arnold S, et al. Oxycodone for cancer-related pain. Cochrane Database Syst Rev 2022;2022:CD003870. Available from: https://doi.org/10.1002/14651858.CD003870.pub7.

[80] U.S. Department of Health and Human Services. Opioid abuse in the United States and Department of Health and Human Services actions to address opioid-drug-related overdoses and deaths. J Pain Palliat Care Pharmacother 2015;29:133–9. Available from: https://doi.org/10.3109/15360288.2015.1037530.

[81] Barlas S. FDA requires new labeling for some opioids: modest changes walk narrow political line. P T 2013;38(11):644.

[82] Kirchheiner J, Schmidt H, Tzvetkov M, et al. Pharmacokinetics of codeine and its metabolite morphine in ultra-rapid metabolizers due to CYP2D6 duplication. Pharmacogenomics J 2007;7:257–65. Available from: https://doi.org/10.1038/sj.tpj.6500406.

[83] Sarhill N, Walsh D, Nelson KA. Hydromorphone: pharmacology and clinical applications in cancer patients. Support Care Cancer 2001;9:84–96. Available from: https://doi.org/10.1007/s005200000183.

[84] Parab PV, Ritschel WA, Coyle DE, et al. Pharmacokinetics of hydromorphone after intravenous, peroral and rectal administration to human subjects. Biopharm Drug Dispos 1988;9:187–99. Available from: https://doi.org/10.1002/bod.2510090207.

[85] Walsh D, Doona M, Molnar M, et al. Symptom control in advanced cancer: important drugs and routes of administration. Semin Oncol 2000;27:69–83.

[86] Li Y, Ma J, Lu G, et al. Hydromorphone for cancer pain. Cochrane Database Syst Rev 2021;8:CD011108. Available from: https://doi.org/10.1002/14651858.CD011108.pub3.

[87] Abi-Aad KR, Derian A. Hydromorphone. In: StatPearls [Internet]. Treasure Island (FL): StatPearls Publishing; 2025.

[88] Rapp SE, Egan KJ, Ross BK. A multidimensional comparison of morphine and hydromorphone patient-controlled analgesia. Anesth Analg 1996;82:1043–8. Available from: https://doi.org/10.1097/00000539-199605000-00029.

[89] Benítez-Rosario MA, Salinas-Martín A, Aguirre-Jaime A, et al. Morphine-methadone opioid rotation in cancer patients: analysis of dose ratio predicting factors. J Pain Symptom Manage 2009;37:1061–8. Available from: https://doi.org/10.1016/j.jpainsymman.2008.05.016.

[90] McLean S, Twomey F. Methods of rotation from another strong opioid to methadone for the management of cancer pain: a systematic review of the available evidence. J Pain Symptom Manage 2015;50:248–59. Available from: https://doi.org/10.1016/j.jpainsymman.2015.02.029. e1.

[91] Tan C, Wong JF, Yee CM, et al. Methadone rotation for cancer pain: an observational study. BMJ Support Palliat Care 2022;12:e736–9. Available from: https://doi.org/10.1136/bmjspcare-2019-002175.

[92] Davis MP, Walsh D. Methadone for relief of cancer pain: a review of pharmacokinetics, pharmacodynamics, drug interactions and protocols of administration. Support Care Cancer 2001;9:73–83. Available from: https://doi.org/10.1007/s005200000180.

[93] Stringer J, Welsh C, Tommasello A. Methadone-associated Q-T interval prolongation and torsades de pointes. Am J Health Syst Pharm 2009;66:825–33. Available from: https://doi.org/10.2146/ajhp070392.

[94] Nicholson AB, Watson GR, Derry S, et al. Methadone for cancer pain. Cochrane Database Syst Rev 2017;2:CD003971. Available from: https://doi.org/10.1002/14651858.CD003971.pub4.

[95] Parsons HA, de la Cruz M, El Osta B, et al. Methadone initiation and rotation in the outpatient setting for patients with cancer pain. Cancer 2010;116:520–8. Available from: https://doi.org/10.1002/cncr.24754.

[96] Courtemanche F, Dao D, Gagné F, et al. Methadone as a coanalgesic for palliative care cancer patients. J Palliat Med 2016;19:972–8. Available from: https://doi.org/10.1089/jpm.2015.0525.

[97] Fürst P, Lundström S, Klepstad P, et al. Improved pain control in terminally ill cancer patients by introducing low-dose oral methadone in addition to ongoing opioid treatment. J Palliat Med 2018;21:177–81. Available from: https://doi.org/10.1089/jpm.2017.0157.

[98] Schumacher MA, Basbaum AI, Naidu RK. Opioid agonists & antagonists. In: Vanderah TW, editor. Katzung's basic & clinical pharmacology. New York, NY: McGraw-Hill; 2024.

[99] Grond S, Radbruch L, Meuser T, et al. High-dose tramadol in comparison to low-dose morphine for cancer pain relief. J Pain Symptom Manage 1999;18:174–9. Available from: https://doi.org/10.1016/s0885-3924(99)00060-3.

[100] Rodriguez RF, Bravo LE, Castro F, et al. Incidence of weak opioids adverse events in the management of cancer pain: a double-blind comparative trial. J Palliat Med 2007;10:56–60. Available from: https://doi.org/10.1089/jpm.2006.0117.

[101] Wiffen PJ, Derry S, Moore RA. Tramadol with or without paracetamol (acetaminophen) for cancer pain. Cochrane Database Syst Rev 2017;5:CD012508. Available from: https://doi.org/10.1002/14651858.CD012508.pub2.

[102] Hartrick CT, Rodríguez Hernandez JR. Tapentadol for pain: a treatment evaluation. Expert Opin Pharmacother 2012;13:283–6. Available from: https://doi.org/10.1517/14656566.2012.648616.

[103] Jain D, Basniwal PK. Tapentadol, a novel analgesic: review of recent trends in synthesis, related substances, analytical methods, pharmacodynamics and pharmacokinetics. Bull Fac Pharm Cairo Univ 2013;51:283–9. Available from: https://doi.org/10.1016/j.bfopcu.2013.04.003.

[104] Wade WE, Spruill WJ. Tapentadol hydrochloride: a centrally acting oral analgesic. Clin Ther 2009;31:2804–18. Available from: https://doi.org/10.1016/j.clinthera.2009.12.003.

[105] Homma M, Kokubun H, Okuwaki K, et al. Pharmacokinetic analysis, analgesic effects, and adverse effects of tapentadol in cancer patients with pain. Biol Pharm Bull 2020;43:1000–6. Available from: https://doi.org/10.1248/bpb.b20-00084.

[106] Hartrick CT. Tapentadol immediate release for the relief of moderate-to-severe acute pain. Expert Opin Pharmacother 2009;10:2687–96. Available from: https://doi.org/10.1517/14656560903313734.

[107] Imanaka K, Tominaga Y, Etropolski M, et al. Ready conversion of patients with well-controlled, moderate to severe, chronic malignant tumor-related pain on other opioids to tapentadol extended release. Clin Drug Investig 2014;34:501–11. Available from: https://doi.org/10.1007/s40261-014-0204-3.
[108] Imanaka K, Tominaga Y, Etropolski M, et al. Efficacy and safety of oral tapentadol extended release in Japanese and Korean patients with moderate to severe, chronic malignant tumor-related pain. Curr Med Res Opin 2013;29:1399–409. Available from: https://doi.org/10.1185/03007995.2013.831816.
[109] Kress HG, Koch ED, Kosturski H, et al. Tapentadol prolonged release for managing moderate to severe, chronic malignant tumor-related pain. Pain Physician 2014;17(4):329–43.
[110] Mercadante S, Porzio G, Ferrera P, et al. Tapentadol in cancer pain management: a prospective open-label study. Curr Med Res Opin 2012;28:1775–9. Available from: https://doi.org/10.1185/03007995.2012.739151.
[111] Jung JY, Chon HJ, Choi YJ, et al. A prospective, multicenter, open-label study of the clinical efficacy of tapentadol extended-release in the treatment of cancer-related pain and improvement in the quality of life of opioid-naïve or opioid-resistant patients. Support Care Cancer 2022;30:6103–12. Available from: https://doi.org/10.1007/s00520-022-06992-w.
[112] Schmidt-Hansen M, Bromham N, Taubert M, et al. Buprenorphine for treating cancer pain. Cochrane Database Syst Rev 2015;2015:CD009596. Available from: https://doi.org/10.1002/14651858.CD009596.pub4.
[113] Jackson LK, Poon IO, Garcia MA, et al. Buprenorphine use for analgesia in palliative care. Pharmacy 2024;12:78. Available from: https://doi.org/10.3390/pharmacy12030078.
[114] Trescot AM, Datta S, Lee M, et al. Opioid pharmacology. Pain Physician 2008;11(2 Suppl.):S133–53.
[115] Kistemaker KRJ, Sijani F, Brinkman DJ, et al. Pharmacological prevention and treatment of opioid-induced constipation in cancer patients: a systematic review and meta-analysis. Cancer Treat Rev 2024;125:102704. Available from: https://doi.org/10.1016/j.ctrv.2024.102704.
[116] Rekatsina M, Paladini A, Drewes AM, et al. Efficacy and safety of peripherally acting μ-opioid receptor antagonist (PAMORAs) for the management of patients with opioid-induced constipation: a systematic review. Cureus 2021;13:e16201. Available from: https://doi.org/10.7759/cureus.16201.
[117] Wang DS, Sternbach G, Varon J. Nalmefene: a long-acting opioid antagonist. clinical applications in emergency medicine. J Emerg Med 1998;16:471–5. Available from: https://doi.org/10.1016/S0736-4679(98)00019-5.
[118] Dowell D, Compton WM, Giroir BP. Patient-centered reduction or discontinuation of long-term opioid analgesics: the HHS guide for clinicians. JAMA 2019;322:1855–6. Available from: https://doi.org/10.1001/jama.2019.16409.
[119] Kral LA, Jackson K, Uritsky TJ. A practical guide to tapering opioids. Ment Health Clin 2015;5:102–8. Available from: https://doi.org/10.9740/mhc.2015.05.102.

Chapter 7

NSAID and acetaminophen (paracetamol)

Ajaz Bulbul
Hematology/Oncology, Nancy N. and J.C. Lewis Cancer & Research Pavilion, Hinesville, GA, United States

Key learning objectives

- Understand the mechanisms of action of nonopioid analgesics, specifically nonsteroidal antiinflammatory drugs (NSAIDs) and acetaminophen, in cancer pain management.
- Recognize the constraints in existing evidence on the effectiveness of acetaminophen and NSAIDs in cancer pain management, with many studies lacking robust methodologies.
- Differentiate the analgesic benefit of NSAIDs based on pain type, noting their potential efficacy for bone and inflammation-related pain, but limited effectiveness for neuropathic pain.
- Integrate NSAIDs and acetaminophen into a comprehensive pain management plan for cancer patients, considering the complexity of pain sources, potential side effects, and drug interactions.

Introduction

Pain is a common experience for over half of the patients undergoing anticancer treatment for metastatic disease [1]. For mild-to-moderate cancer pain, acetaminophen (paracetamol) and nonsteroidal antiinflammatory drugs (NSAIDs) provide a nonopioid pain relief option. These medications can complement opioids, allowing for a lower opioid dose and potentially fewer side effects [2].

While traditionally positioned as first-line agents in pain management approaches, current understanding recognizes that in cases of moderate-to-severe cancer pain, the initial step using nonopioid analgesics alone may be bypassed in favor of more appropriate pain control strategies including steroids, antidepressants, antiseizure medications, or gamma-aminobutyric acid agonists [3]. The comprehensive pain management approach is described in detail in other chapters of this book. Most studies on the role of nonopioid analgesics for the management of cancer pain were performed between the 1970s and the early 2000s after the mechanism of action related to the inhibition of prostaglandin synthesis was discovered [4].

Identifying two cyclooxygenase (COX) isoforms, COX-1 and COX-2, was a crucial breakthrough in understanding the side effects of NSAIDs. COX-1 exhibits constitutive expression across most human tissues, while COX-2 expression is induced during inflammatory processes and tumorigenesis. This distinction has demonstrably linked the inhibition of COX-1 to the manifestation of many of their adverse effects. The development of COX-2 selective inhibitors emerged as a promising strategy for pain management, achieving widespread use. However, this approach faced a setback in 2004 with the discovery of an association between these drugs and an increased relative risk of cardiovascular and cerebrovascular events [5].

Optimizing pain management in cancer patients prioritizes achieving a tolerable pain level that supports an acceptable quality of life. This approach necessitates balancing the benefits of pain relief against the potential risks of adverse effects and overdose, particularly respiratory depression. Data from low- and middle-income countries highlights the significant burden of untreated pain, with estimates suggesting that roughly 80% of individuals with terminal cancer experience moderate-to-severe pain for an average of 90 days [6]. These findings underscore the critical public health concern posed by inadequate cancer pain management.

Pain Management in Palliative Care for Patients With Cancer. DOI: https://doi.org/10.1016/B978-0-443-30050-9.00052-7

Acetaminophen (paracetamol)

Mechanism of action

Despite its widespread use as an analgesic, the precise mechanisms underlying acetaminophen's efficacy remain elusive. While some evidence suggests that it might inhibit prostaglandin synthesis within the central nervous system, other contributing factors are likely at play. Its analgesic effect may inhibit the same COX enzyme and prostaglandin H_2 synthase inhibited by NSAIDs. However, it inhibits the enzyme at its peroxidase catalytic rather than at the COX catalytic site. Acetaminophen is sensitive to tissue peroxide levels and higher peroxidase concentrations, as seen in the inflammatory milieu, and activated leukocytes and platelets block the effect of acetaminophen on inflammation in the periphery. Still, it can inhibit prostaglandins in the central nervous system, thus providing relief of pain and fever [7]. Studies investigating the coadministration of serotonin receptor antagonists, tropisetron or granisetron, with acetaminophen have revealed a blockade of the drug's analgesic effect. This finding supports the hypothesis that the serotonergic system may be integral to the mechanism of action underlying acetaminophen's analgesia [8], although others posit that an active metabolite of acetaminophen might exert its influence at least partially through modulation of the cannabinoid receptor system [9].

Clinical data

Despite its widespread use, there is a scarcity of trials evaluating the effectiveness of acetaminophen in patients with cancer (Table 7.1). In a randomized, blinded clinical trial of 112 hospitalized cancer patients on high-dose opioids, acetaminophen did not improve pain control or decrease total opioid use [10]. A separate study included patients diagnosed with cancer who reported a pain score of ≥4 on a 0–10 Numeric Rating Scale. All participants were required to use at least step 1 (acetaminophen) of the WHO analgesic ladder for pain management. Following randomization, patients were assigned to either the control arm receiving a weak opioid (step 2 of the WHO ladder, e.g., codeine) or the experimental arm receiving a potent opioid (step 3). The primary outcome measure was time to achieve stable pain control, defined as three consecutive days with a pain score of ≤3. The findings demonstrate that there was no significant difference in time to stable pain control between the arms, demonstrating the two-step approach as a viable alternative for managing cancer pain [11]. However, there is a lack of proof using acetaminophen as the first-step approach. A 2017 metaanalysis evaluated the use of oral acetaminophen on cancer pain, including randomized, double-blind studies of 5 days' duration or longer, with a minimum of 25 participants. The results were very modest, with only 122 participants included, and with no studies reporting reduction of pain 30% or 50% from baseline. The results suggest no difference between acetaminophen and placebo when added to another treatment. There was no convincing evidence of acetaminophen achieving a more significant effect than placebo with regard to quality of life, use of rescue medication, or participant satisfaction or preference. The authors concluded that there is limited high-quality evidence to support or refute the use of acetaminophen alone or in combination with opioids for the first two steps of the three-step WHO cancer pain ladder [11,12].

TABLE 7.1 Summary of acetaminophen use in patients with cancer.

Indication: Mild cancer pain (NRS 1–3).
Prescription: 325 mg–1 g every 4–6 h as needed. IV acetaminophen may be used temporarily if not able to use oral medications.
Max daily dose: 3–4 g (for long-term use exceeding 7 days).
Warning: For patients with risk factors, including increasing age or history of liver disease, monitoring liver function and reducing the dose to 2–3 g daily are crucial. For patients with severe renal impairment, extending the dosing interval to every 6 h is suggested. In patients undergoing immunotherapy, use acetaminophen, cautiously; limited emerging data suggest possible impact, although clinical relevance remains uncertain [16]. Monitor closely for fever, as acetaminophen may mask it in immunocompromised patients.
Rationale: This approach aligns with step 1 of the WHO analgesic ladder, providing a nonopioid option for initial pain management.

IV, Intravenous; *NRS*, Numeric Rating Scale.

Despite this, there is some evidence for its use in the mild acute postoperative setting. A 2022 US Department of Health and Human Services–led task force consensus statement advocates for the routine use of acetaminophen (unless contraindicated) as part of multimodal analgesia for acute perioperative pain [13]. The magnitude of its impact on pain scores and opioid consumption following major surgery remains unclear, with conflicting evidence emerging from metaanalyses. Notably, one metaanalysis observed a modest reduction in morphine use within the first 24 hours postoperatively after major abdominal surgery, while another did not report a clinically significant difference in pain scores after hip arthroplasty with acetaminophen [14].

Many patients combine acetaminophen to weak opioids, but even this combination has very limited data to support it [15]. Acetaminophen may not offer additional pain control or reduce opioid consumption in patients with advanced cancer pain who are already receiving a strong opioid regimen. These findings contribute to the growing body of evidence that questions the routine use of acetaminophen as an adjuvant analgesic in this specific patient population experiencing moderate-to-severe cancer pain [10].

Application and toxicity

Although there is even less evidence for using intravenous (IV) than peroral acetaminophen [17], IV acetaminophen offers a more rapid and predictable onset of action compared to rectal or oral administration. Following IV administration, the onset of effect occurs within 5–10 minutes, with peak concentration achieved within 15 minutes. Conversely, rectal and oral routes exhibit a delayed onset, typically ranging from 10 to 60 minutes or longer.

Acetaminophen exhibits a relatively narrow therapeutic window, meaning a small range between an effective dose and a potentially harmful dose. Due to limited evidence demonstrating additional benefit beyond 325 mg per dosage unit while conversely increasing the risk of liver injury, the US Food and Drug Administration implemented a policy change in 2011 mandating drug manufacturers to limit the amount of acetaminophen in prescription combination products to a maximum of 325 mg per dosage unit. Furthermore, regulatory bodies have advocated restricting the maximum single dose to 650 mg and the daily maximum dose to 4 g for adults (applicable for 5 days or less) [18]. Building upon consensus guidelines from the National Comprehensive Cancer Network (NCCN) for managing adult cancer pain [19], some experts advocate for a reduced maximum daily acetaminophen dosage of 3 g for patients with normal liver function. This recommendation is particularly relevant for long-term use exceeding 7 days. It is crucial to consider all sources of acetaminophen, including prescription medications over the counter. NCCN also recommends caution in immunotherapy for the possibility of a negative impact on its efficacy and over the effects of masking fever in immunocompromised patients [16,19].

Caution is warranted in patients weighing less than 50 kg or those with preexisting conditions such as chronic alcoholism, dehydration, prolonged fasting or anorexia, and concurrent use of medications that impact acetaminophen metabolism, such as St. John's wort and phenobarbital. Beyond exceeding recommended dosages, several additional factors elevate the risk of acetaminophen-induced hepatotoxicity such as advanced age and compromised nutritional status. The maximum daily dose should be limited to 2–3 g in these populations [18,20]. Complete avoidance is recommended for patients with alcoholic hepatitis, severe hepatic impairment, or severe active liver disease [21]. When used for a short duration, the acetaminophen and placebo groups exhibited similar rates of clinically significant adverse events [17,22]. Patients with ≤30 mL/min/1.73 m^2 creatinine clearance could receive similar doses at less frequent intervals, once every 6 hours rather than every 4 hours.

Synergistic analgesia of acetaminophen and NSAIDs

NSAIDs combined with acetaminophen show efficacy primarily in postoperative settings, particularly ENT and orthopedic surgeries. They appear more effective for bone pain and inflammatory lesions but less so for neuropathic pain [23]. The evidence for synergistic effects between acetaminophen and NSAIDs in cancer pain management remains limited and inconclusive based on current research. Findings from systematic reviews and metaanalyses indicate that while NSAIDs are frequently combined with opioids in cancer pain management, studies show mixed results, with five of seven trials demonstrating that adding NSAIDs to step 3 opioids improved analgesia (26%–51% pain reduction) or reduced opioid requirements in some cases [24–26]. However, no studies directly examined NSAID–acetaminophen synergism without opioid involvement. A 2017 Cochrane review found no high-quality evidence supporting acetaminophen's efficacy either alone or in combination with opioids for cancer pain, with only one of five trials showing marginal benefit when combining acetaminophen with opioids [12]. Methodological limitations further complicate interpretation, as most trials (42 studies, 3084 patients) focused on NSAID–opioid combinations rather than NSAID–acetaminophen pairings, with heterogeneous methodologies, short durations, and lack of standardized outcomes preventing definitive conclusions [27].

Nonsteroidal antiinflammatory drugs

Mechanism of action

Due to their analgesic and antiinflammatory properties, NSAIDs are a cornerstone of multimodal pain management strategies. Their mechanism of action centers on inhibiting COX, a key enzyme responsible for prostaglandin synthesis in both peripheral and central tissues. Two main COX isoforms exist: COX-1, constitutively expressed in various tissues and essential for physiological functions mediated by prostaglandins, and COX-2, inducibly produced at sites of inflammation. While most NSAIDs inhibit both COX-1 and COX-2 enzymes, they exhibit varying degrees of selectivity. NSAIDs with greater COX-2 selectivity offer a potential therapeutic advantage. They can provide analgesia and reduce inflammation, which is particularly beneficial for bone pain [28], while minimizing the risk of gastrointestinal (GI) complications associated with COX-1 inhibition.

Clinical data

Following the pioneering study by Moertel et al., which demonstrated the analgesic superiority of aspirin (650 mg) compared to placebo for pancreatic cancer pain, NSAIDs emerged as a treatment option [29]. Subsequent research has extensively explored the efficacy of NSAIDs, particularly in comparison to placebo and other analgesic agents. However, only a handful of studies since 2015 have looked at NSAIDs/acetaminophen for cancer pain. One study suggested that COX-2 inhibition with celecoxib exhibited additional antidepressive activity in a cohort of 52 breast cancer patients [30]. Another smaller study showed that adding ibuprofen (600 mg) to oxycodone/acetaminophen enhances analgesia and reduces narcotic use in chronic cancer pain management [31]. Another study of 100 inpatients with cancer showed similar effectiveness of ketorolac 10 mg intramuscularly (IM) compared to ketorolac 30 mg and diclofenac 75 mg after a single IM injection [32].

A metaanalysis in 2017 evaluated the effect of NSAID for cancer pain including randomized, double-blind, single-blind, or open-label studies of 5 days' duration or longer, with a minimum study size of 25 participants. Despite the high prevalence of both cancer and NSAID, only 11 studies met the inclusion criteria, of which only 8 were double blinded. Hence, the evidence was defined as very low quality, while showing that with NSAID, initially moderate or severe pain was reduced to no worse than mild pain after 1 or 2 weeks in four studies, with a range of estimates between 26% and 51% in individual studies. There is no high-quality evidence to support or refute the use of NSAIDs, alone or in combination with opioids, at any step of the three-step WHO cancer pain ladder [26].

Another review study in 2019 investigated 30 studies, and despite NSAID being a key component of the WHO analgesic ladder, very little solid evidence was found. This analysis showed that six out of seven studies specifically commented on the fact that no opioid analgesia was permitted during the study period, which could be ethically doubtful [24].

Research has particularly focused on the effect of NSAID on bone metastases, which involve complex pathophysiological processes that remain incompletely understood. These mechanisms include tumor-induced osteolytic bone destruction, the production of pain-promoting growth factors and cytokines, stimulation of neuronal ion channels, and direct nerve infiltration by tumor cells. Mechanistically, the pain associated with bone metastases can be categorized as primary (direct skeletal destruction) or secondary (release of chemical pain mediators or compression of nerve roots) [19,28]. In a study of patients with bone metastases, the combination of dual NSAIDs with IV morphine demonstrated superior pain control compared to single NSAID use [33]. However, NSAID dosing may have been suboptimal in that study. Such an approach would not be employed today as corticosteroids are readily available, and dual NSAID approach would not be the standard of care due to higher toxicity and no proven benefit. There is some but limited evidence that IV ketorolac might be particularly useful for patients with bone metastasis who have already been treated [32], but larger studies are lacking. One of the only recent trials comparing NSAID to sublingual fentanyl was for patients with pain due to bone metastasis, demonstrating that both medications are successful, but fentanyl had more rapid onset of analgesia [34]. Current management strategies for metastatic bone pain typically incorporate a multimodal approach, including opioids, NSAIDs, and adjuvant therapies such as corticosteroids and bisphosphonates tailored to the patient's risk profile and needs.

TABLE 7.2 Summary of the use of NSAID in patients with cancer.

NSAID	Usual dose	Maximum dose/day	Preferred scenario	Avoid
Ibuprofen	400 mg every 4–6 h	2400 mg	Shorter-acting alternative Preferred in patients without cardiovascular risks Higher analgesic: antiinflammatory ratio	Take with food, consider PPI Avoid in patients with bleeding risk and thrombocytopenia
Naproxen	Naproxen sodium: 275–550 mg every 12 h or 275 mg every 6–8 h	1100–1650 mg	Preferred due to lower cardiovascular toxicity than comparable NSAIDs Naproxen sodium has a faster onset	Similar GI, renal toxicity as other NSAIDs
Meloxicam	7.5–15 mg once daily	15 mg (conventional tablet, oral suspension)	Long half-life Relatively COX-2 selective	Avoid if quick reversal required (pre surgery)
Celecoxib	200 mg daily or 100 mg every 12 h	400 mg	Less risk of GI toxicity Minimal effect on platelet function Preferred in AERD, asthma, or allergic reactions	Cardiovascular and kidney risks are dose related
Salsalate	1000 mg every 8–12 h or 1500 mg every 12 h	3000 mg	Lower bleeding risk	Consider lower daily doses in adults with AERD or allergic reactions
Aspirin	325–1000 mg every 4–6 h	4000 mg	Possible cancer prevention benefit	Limited antiinflammatory property Irreversible platelet inhibition
Diclofenac	50 mg every 8–12 h	150 mg	Dosing differs for free acid versus sodium/potassium salt Topical alternative for joint pain	Higher cardiovascular risk
Etodolac	200–400 mg every 6–8 h	1000 mg	Relatively COX-2 selective Minimal platelet effect	Cardiovascular risk
Indomethacin	25–50 mg every 8–12 h	150–200 mg	Preferred for gout exacerbation	Significant renal risk CNS side effects (AMS)
Sulindac	150–200 mg every 12 h	400 mg	Rarely used for desmoid tumors	Hepatic inflammation Renal calculi

AERD, Aspirin-exacerbated respiratory disease; *AMS*, altered mental status; *CNS*, central nervous system; *COX*, cyclooxygenases; GI, gastrointestinal; *NSAID*, nonsteroidal antiinflammatory drug; *PPI*, proton pump inhibitor.

Application and toxicity

Various different NSAIDs are used as adjunctive therapy alongside other analgesics. A systematic review of the literature identified about 30 studies investigating the use of NSAIDs for cancer pain management (Table 7.2). However, these studies exhibited limitations that hinder definitive conclusions regarding the analgesic efficacy of NSAIDs in this context. Notably, a lack of high-quality evidence was observed, with methodological shortcomings, including brief study durations and substantial heterogeneity, in the outcome measures. These limitations restrict the generalizability of the findings and impede the ability to draw robust conclusions about the effectiveness of NSAIDs for cancer pain control [24]. On the other hand, there is no evidence that supports the use of NSAID in neuropathic pain [23,35].

A recent Italian Delphi study on the use of NSAID in cancer pain reached a consensus that assessment of liver function for acetaminophen and kidney function for NSAID is crucial and that acetaminophen is the analgesic of first choice for patients with mild chronic pain while NSAID can play a role in pain management as well. However, no consensus or agreement was found for the recommendations on routine prescriptions [23,35,36]. On the other hand, a survey of UK palliative care physicians demonstrated that most physicians consider NSAIDs improve cancer pain either "sometimes" (57.7%) or "often" (40%), with ibuprofen being the first-line oral NSAID of choice (42.6%) [37]. However, following the WHO recommendations, for patients experiencing mild-to-moderate cancer-related pain, NSAIDs may be used as a first-line treatment option, particularly for bone pain from metastases, short-term pain relief following cancer-related procedures, and as an adjuvant to opioid treatment.

The therapeutic benefits of NSAIDs come with a range of severe side effects, which we must be aware of, particularly cardiovascular, GI, bleeding, nephrotoxicity, and hepatotoxicity complications that can have significant implications for the cancer patient. Immunoglobulin E–mediated immunologic reactions, including urticaria, angioedema, generalized pruritus, tachycardia, and hypotension, can be seen rarely [38].

Cardiovascular risk

NSAIDs, both traditional and COX-2 inhibitors, are associated with an elevated risk of major adverse cardiovascular events like stroke, major coronary events (nonfatal myocardial infarction or coronary death), heart failure, and death. This concerning trend holds regardless of the drug's COX-2 selectivity. Most nonselective NSAIDs, including ibuprofen, naproxen, and diclofenac, inhibit both COX-1 and COX-2 enzymes to varying degrees rather than selectively targeting one isoform. Differences in their relative COX-1 and COX-2 inhibition may contribute to variations in gastrointestinal and cardiovascular risk profiles, with greater COX-2 inhibition thought to play a role in the increased cardiovascular risk observed with selective COX-2 inhibitors. However, that trend is unclear in smaller celecoxib doses of 200 mg daily [39]. The metaanalysis, conducted by the COXib and traditional NSAID Trialists' Collaboration, found that all NSAIDs doubled the risk of heart failure hospitalizations. The risk is amplified more so in patients with preexisting cardiovascular disease (eight events per 1000 person-years), with about a 30%–40% higher relative risk compared to placebo in all patients [40]. Recommendations from the American Heart Association discourage the use of NSAIDs in patients with established heart disease as their use was associated with an increased risk of bleeding and excess thrombotic events with no safe therapeutic window for concomitant NSAID use [41]. Even a 3-day treatment was associated with increased risk with diclofenac, higher doses of celecoxib, and ibuprofen [42]. Mechanistically, this could be explained by inhibition of COX-2–mediated prostacyclin vasodilation [43]. Naproxen has not been associated with an elevated vascular event risk [42].

Gastrointestinal toxicity (peptic ulcer disease)

Prostaglandins play a crucial role in maintaining the integrity of the gastroduodenal mucosa by stimulating factors such as mucus synthesis and secretion, mucosal bicarbonate secretion, mucosal blood flow, and cellular repair making the mucosa more susceptible to luminal agents like acid, pepsin, ethanol, and steroids [44,45]. There is a shift in the arachidonic acid metabolism from protective prostaglandins, which promotes the production of injurious leukotrienes. NSAIDs can interact directly with the mucosa to increase permeability and impair mucosal repair, driven primarily by COX-1 inhibition. Therefore the COX-2 inhibitors have limited GI toxicity. Age over 60, previous history of peptic ulcer or bleeding, *Helicobacter pylori*, higher doses, and corticosteroids all increase the risk of GI toxicity about fivefold. The concurrent use of blood thinners can increase the risk up to 15×. GI risk depends on the NSAID's antiinflammatory/analgesic activity ratio. NSAIDs for which the doses required for antiinflammatory effects are comparable to those needed for analgesia (e.g., piroxicam and ketorolac) confer a heightened risk of GI toxicity and should be judiciously prescribed or avoided, particularly in patients with preexisting risk factors for NSAID-related GI complications [46].

Bleeding

The nonselective NSAIDs inhibit platelet aggregation by blocking thromboxane A2 [47]. Aspirin has an irreversible effect through COX inhibition for the lifespan of the platelet. Other NSAIDs have a temporary effect based on dosing and half-life. This effect is less significant with nonacetylated salicylates (up to 3 g/day) and selective COX-2 inhibitors. Patients with preexisting platelet dysfunction (qualitative or quantitative defects, e.g., uremia, severe thrombocytosis, or hereditary von Willebrand disease) or thrombocytopenia (platelet count <50,000/μL) should exercise caution when using NSAIDs due to increased bleeding risk. It should also be withheld for up to five half-lives before surgery (generally 3–5 days). Platelet function tends to normalize within 24 hours after cessation of ibuprofen use in healthy individuals (600 mg orally every 8 hours) [48].

Nephrotoxicity

NSAID use has been linked to renal complications through renal vasoconstriction via COX enzyme inhibition in glomerular, vascular, collecting, and interstitial cells. Acute interstitial nephritis and predisposition to acute tubular necrosis have been seen, particularly in patients with low renal perfusion. Both COX unselective and COX-2 selective drugs can decrease glomerular filtration rate, and prolonged use may lead to irreversible renal toxicity [49,50]. Patients should avoid NSAIDs if their estimated glomerular filtration rate (eGFR) is <30 mL/min/1.73 m^2 (stage 4–5 chronic kidney disease [CKD]) and use them with caution if eGFR is 30–60 mL/min/1.73 m^2 (stage 3 CKD). Regular monitoring of serum creatinine and blood urea nitrogen is recommended before initiating therapy and within 1–2 weeks after starting NSAIDs in at-risk patients. A rise in serum creatinine >0.3 mg/dL or >20% from baseline warrants NSAID discontinuation. Patients with volume depletion, nephrotic syndrome, CKD, heart failure, cirrhosis, hypercalcemia, or those taking diuretics or angiotensin-converting enzyme inhibitors should also avoid using NSAIDs concurrently [51].

Hepatotoxicity

Mild elevation of liver enzymes is commonly seen with NSAIDs. These tend to be seen in the first 6 weeks of treatment, with resolution within 8 weeks of stopping. Frank liver failure is uncommon (1 case per 100,000). Sulindac has a greater risk than other NSAIDs (27/100,000), which is usually reversible after holding the dose [52]. Diclofenac has been reported to cause antinuclear antibody–positive immune hepatitis syndrome. Use in patients with concomitant rheumatological conditions increases the risk of liver toxicity. Toxicity should be suspected in cases of elevated liver enzymes, typically aspartate aminotransferase ≥3× normal with low albumin and prolonged international normalized ratio.

Pulmonary effects and allergy

Nonselective NSAIDs may precipitate acute exacerbations of bronchospasm, especially in those with aspirin-exacerbated respiratory disease. This may be akin to allergy in the form of inflammation of the bronchial tree, flushing, and nasal congestion; it is less likely to be seen with COX-2 inhibitors [53].

Potential future implications of NSAID

Despite being the cornerstone of the WHO pain ladder, there has been a scarcity of trials evaluating NSAID use in cancer patients in the last decade. Predominant research in cancer setting has evaluated the repurposing of NSAIDs for cancer prevention and treatment. For example, in patients with familial adenomatous polyposis, low-dose aspirin (100 mg daily) was found to suppress colorectal adenoma development and reduce polyp recurrence [54], while for Lynch syndrome patients, long-term use of high-dose aspirin (600 mg daily) significantly reduced cancer risk [55]. There is also research showing that in advanced gastric cancer, celecoxib (400 mg daily) as an adjuvant to chemotherapy improved overall survival, disease-free survival, and quality of life, particularly in COX-2–positive patients [56]. However, not all trials showed positive results. In stage 3 colon cancer patients, adding celecoxib (400 mg daily) to standard adjuvant chemotherapy did not significantly improve outcomes compared to placebo [57]. The anticancer effects of NSAIDs are thought to be mediated through inhibiting COX enzymes, reducing prostaglandin synthesis, and suppressing inflammation. Preclinical studies suggest that inhibiting prostaglandin E2 can overcome tumor immune evasion [58]. However, serious adverse effects, particularly cardiovascular and gastrointestinal, limit the development of further research in cancer patients, where any pharmacological pain treatment is often long term and usually escalating in dose. Hence, most of the newer studies focus either on the use of NSAID as a cancer prevention measure or on the prevention of secondary relapse adjuvant to systemic treatment [59,60]. However, there is also a clear need for a newer trial evaluating and offering evidence to support the WHO three-step ladder of pain management in patients with cancer, particularly in the light of the development of new opioid drugs and advances of the systemic therapy. Developing NSAIDs with improved safety profiles that minimize cardiovascular risks while retaining anticancer efficacy is also crucial for their broader application in clinical practice. Nitric oxide–releasing NSAIDs and dual COX/lipoxygenase NSAIDs are being studied in clinical trials to evaluate their role in mitigating GI side effects and enhancing cancer-protective effects and could provide a potential option in the future [61,62].

Conclusion

Cancer pain management is a complex issue that requires a comprehensive approach, incorporating nonopioid analgesics like acetaminophen and NSAIDs, as well as opioids, in a stepwise manner. While NSAIDs and acetaminophen may provide modest analgesic benefits, their use is limited by cardiovascular risks and various other systemic toxicities. Modern trials primarily evaluate their use as a cancer-preventive treatment, or as enhancers to systemic therapy, although new drugs from the same classes are being developed and could provide a viable treatment in the future years.

References

[1] World Health Organization. WHO guidelines for the pharmacological and radiotherapeutic management of cancer pain in adults and adolescents. Geneva: World Health Organization; 2018.

[2] Huang R, Jiang L, Cao Y, et al. Comparative efficacy of therapeutics for chronic cancer pain: a Bayesian network meta-analysis. J Clin Oncol 2019;37(20):1742–52. Available from: https://doi.org/10.1200/JCO.18.01567.

[3] Portenoy RK. A practical approach to using adjuvant analgesics in older adults. J Am Geriatr Soc 2020;68(4):691–8. Available from: https://doi.org/10.1111/jgs.16340.

[4] Vane JR. Inhibition of prostaglandin synthesis as a mechanism of action for aspirin-like drugs. Nat N Biol 1971;231(25):232–5. Available from: https://doi.org/10.1038/newbio231232a0.

[5] Sun SX, Lee KY, Bertram CT, et al. Withdrawal of COX-2 selective inhibitors rofecoxib and valdecoxib: impact on NSAID and gastroprotective drug prescribing and utilization. Curr Med Res Opin 2007;23(8):1859–66. Available from: https://doi.org/10.1185/030079907X210561.

[6] Knaul FM, Farmer PE, Krakauer EL, et al. Alleviating the access abyss in palliative care and pain relief – an imperative of universal health coverage: the Lancet Commission report. Lancet 2018;391(10128):1391–454. Available from: https://doi.org/10.1016/S0140-6736(17)32513-8.

[7] Aronoff DM, Oates JA, Boutaud O. New insights into the mechanism of action of acetaminophen: its clinical pharmacologic characteristics reflect its inhibition of the two prostaglandin H2 synthases. Clin Pharmacol Ther 2006;79(1):9–19. Available from: https://doi.org/10.1016/j.clpt.2005.09.009.

[8] Pickering G, Loriot MA, Libert F, et al. Analgesic effect of acetaminophen in humans: first evidence of a central serotonergic mechanism. Clin Pharmacol Ther 2006;79(4):371–8. Available from: https://doi.org/10.1016/j.clpt.2005.12.307.

[9] Anderson BJ. Paracetamol (acetaminophen): mechanisms of action. Paediatr Anaesth 2008;18(10):915–21. Available from: https://doi.org/10.1111/j.1460-9592.2008.02764.x.

[10] Leppert W, Zajaczkowska R, Wordliczek J, et al. Is acetaminophen beneficial in patients with cancer pain who are on strong opioids? A randomized controlled trial. J Pain Symptom Manage 2023;66(3):e216–26. Available from: https://doi.org/10.1016/j.jpainsymman.2023.05.002.

[11] Fallon M, Dierberger K, Leng M, et al. An international, open label, randomised trial comparing a two-step approach versus the standard three-step approach of the WHO analgesic ladder in patients with cancer. Ann Oncol 2022;33(12):1296–303. Available from: https://doi.org/10.1016/j.annonc.2022.08.083.

[12] Wiffen PJ, Derry S, Moore RA, et al. Oral paracetamol (acetaminophen) for cancer pain. Cochrane Database Syst Rev 2017;7:CD012637. Available from: https://doi.org/10.1002/14651858.CD012637.pub2.

[13] Mariano ER, Dickerson DM, Szokol JW, et al. Multisociety Organizational Consensus Group. A multisociety organizational consensus process to define guiding principles for acute perioperative pain management. Reg Anesth Pain Med 2022;47(2):118–27. Available from: https://doi.org/10.1136/rapm-2021-103083.

[14] De Oliveira GS Jr, Castro-Alves LJ, McCarthy RJ. Single-dose systemic acetaminophen to prevent postoperative pain: a meta-analysis of randomized controlled trials. Clin J Pain 2015;31(1):86–93. Available from: https://doi.org/10.1097/AJP.0000000000000081.

[15] Straube S, Derry S, Jackson KC, et al. Codeine, alone and with paracetamol (acetaminophen), for cancer pain. Cochrane Database Syst Rev 2014;2014(9):CD006601. Available from: https://doi.org/10.1002/14651858.CD006601.pub4.

[16] Bessede A, Marabelle A, Guégan JP, et al. Impact of acetaminophen on the efficacy of immunotherapy in cancer patients. Ann Oncol 2022;33(9):969–80. Available from: https://doi.org/10.1016/j.annonc.2022.05.010.

[17] Subramaniam K, Esper SA, Mallikarjun K, et al. The effect of scheduled intravenous acetaminophen in an enhanced recovery protocol pathway in patients undergoing major abdominal procedures: a prospective, randomized, and placebo-controlled clinical trial. Pain Med 2022;23(1):23–32. Available from: https://doi.org/10.1093/pm/pnab272.

[18] U.S. Food and Drug Administration. Acetaminophen. FDA Drug Database; 2024.

[19] National Comprehensive Cancer Network (NCCN). NCCN clinical practice guidelines in oncology (NCCN Guidelines®) for adult cancer pain, Version 2.2024. Available from: https://doi.org/10.6004/jnccn.2025.0032.

[20] Twycross R, Pace V, Mihalyo M, et al. Acetaminophen (paracetamol). J Pain Symptom Manage 2013;46(5):747–55. Available from: https://doi.org/10.1016/j.jpainsymman.2013.08.001.

[21] Larson AM, Neuberger JM, Liver E, et al. Acute liver injury with therapeutic doses of acetaminophen: a prospective study. Hepatology 2021;73(5):1890–900. Available from: https://doi.org/10.1002/hep.31678.

[22] Liang L, Cai Y, Li A, et al. The efficiency of intravenous acetaminophen for pain control following total knee and hip arthroplasty: a systematic review and meta-analysis. Medicine 2017;96(46):e8586. Available from: https://doi.org/10.1097/MD.0000000000008586.

[23] Moore RA, Chi C-C, Wiffen PJ, et al. Oral nonsteroidal anti-inflammatory drugs for neuropathic pain. Cochrane Database Syst Rev 2015;2015(10):CD010902. Available from: https://doi.org/10.1002/14651858.CD010902.pub2.

[24] Magee DJ, Jhanji S, Poulogiannis G, et al. Nonsteroidal anti-inflammatory drugs and pain in cancer patients: a systematic review and reappraisal of the evidence. Br J Anaesth 2019;123(2):e412–23. Available from: https://doi.org/10.1016/j.bja.2019.02.028.

[25] Nabal M, Librada S, Redondo MJ, et al. The role of paracetamol and nonsteroidal anti-inflammatory drugs in addition to WHO Step III opioids in the control of pain in advanced cancer. A systematic review of the literature. Palliat Med 2012;26(4):305–12. Available from: https://doi.org/10.1177/0269216311428528.

[26] Derry S, Wiffen PJ, Moore RA, et al. Oral nonsteroidal anti-inflammatory drugs (NSAIDs) for cancer pain in adults. Cochrane Database Syst Rev 2017;2017(7):CD012638. Available from: https://doi.org/10.1002/14651858.CD012638.pub2.

[27] McNicol ED, Strassels SA, Goudas LC, et al. NSAIDS or paracetamol, alone or combined with opioids, for cancer pain. Cochrane Database Syst Rev 2005;1:CD005180. Available from: https://doi.org/10.1002/14651858.CD005180.

[28] Rodriguez D, Serrano SE, Alfaro JL. Treatment of painful bone metastases. Nat Rev Clin Oncol 2010;7(4):206–14. Available from: https://doi.org/10.1038/nrclinonc.2010.17.

[29] Moertel CG, Ahmann DL, Taylor WF, et al. Aspirin and pancreatic cancer pain. Gastroenterology 1971;60(4):552–3.

[30] Maghsoudi A, Mokhtari A, Ghaffari JP, et al. Celecoxib versus diclofenac in mild to moderate depression management among breast cancer patients: a double-blind, placebo-controlled, randomized trial. Ann Pharmacother 2015;49(9):953–9. Available from: https://doi.org/10.1177/1060028015592215.

[31] Stambaugh Jr JE, Drew J. The combination of ibuprofen and oxycodone/acetaminophen in the management of chronic cancer pain. Clin Pharmacol Ther 1988;44(6):665–9. Available from: https://doi.org/10.1038/clpt.1988.209.

[32] Minotti V, Betti M, Ciccarese G, et al. A double-blind study comparing two single-dose regimens of ketorolac with diclofenac in pain due to cancer. Pharmacotherapy 1998;18(3):504–8. Available from: https://doi.org/10.1002/j.1875-9114.1998.tb03113.x.

[33] Liu Z, Xu Y, Liu ZL, et al. Combined application of diclofenac and celecoxib with an opioid yields superior efficacy in metastatic bone cancer pain: a randomized controlled trial. Int J Clin Oncol 2017;22(5):980–5. Available from: https://doi.org/10.1007/s10147-017-1133-y.

[34] Yousef AA, Alzeftawy AE. The efficacy of oral piroxicam fast-dissolving tablets versus sublingual fentanyl in incident breakthrough pain due to bone metastases: a double-blinded randomized study. Support Care Cancer 2019;27(6):2171–7. Available from: https://doi.org/10.1007/s00520-018-4469-6.

[35] Zhen G, Fu Y, Zhang C, et al. Mechanisms of bone pain: progress in research from bench to bedside. Bone Res 2022;10(1):44. Available from: https://doi.org/10.1038/s41413-022-00217-w.

[36] Varrassi G, Coluzzi F, Fornasari D, et al. New perspectives on the adverse effects of NSAIDs in cancer pain: an Italian Delphi study from the Rational Use of Analgesics (RUA) Group. J Clin Med 2022;11(24):7451. Available from: https://doi.org/10.3390/jcm11247451.

[37] Page AJ, Mulvey MR, Bennett MI. Designing a clinical trial of non-steroidal anti-inflammatory drugs for cancer pain: a survey of UK palliative care physicians. BMJ Support Palliat Care 2023;13(e1):e55–8. Available from: https://doi.org/10.1136/bmjspcare-2020-002792.

[38] Stevenson DD, Sanchez-Borges M, Szczeklik A. Classification of allergic and pseudoallergic reactions to drugs that inhibit cyclooxygenase enzymes. Ann Allergy Asthma Immunol 2001;87(3):177–80. Available from: https://doi.org/10.1016/S1081-1206(10)62221-1.

[39] García Rodríguez LA, Tacconelli S, Patrignani P. Role of dose potency in the prediction of risk of myocardial infarction associated with nonsteroidal anti-inflammatory drugs in the general population. J Am Coll Cardiol 2008;52(20):1628–36. Available from: https://doi.org/10.1016/j.jacc.2008.08.041.

[40] Bhala N, Emberson J, Merhi A, et al. Vascular and upper gastrointestinal effects of non-steroidal anti-inflammatory drugs: meta-analyses of individual participant data from randomised trials. Lancet 2013;382(9894):769–79. Available from: https://doi.org/10.1016/S0140-6736(13)60900-9.

[41] Antman EM, Bennett JS, Daugherty A, et al. American Heart Association. Use of nonsteroidal antiinflammatory drugs: an update for clinicians: a scientific statement from the American Heart Association. Circulation 2007;115(12):1634–42. Available from: https://doi.org/10.1161/CIRCULATIONAHA.106.181424.

[42] Schjerning Olsen A, Gislason GH, McGettigan P, et al. Association of NSAID use with risk of bleeding and cardiovascular events in patients receiving antithrombotic therapy after myocardial infarction. JAMA 2015;313(8):805–14. Available from: https://doi.org/10.1001/jama.2015.0809.

[43] Grosser T, Yu Y, Fitzgerald GA. Emotion recollected in tranquility: lessons learned from the COX-2 saga. Annu Rev Med 2010;61:17–33. Available from: https://doi.org/10.1146/annurev-med-011209-153129.

[44] Wilson DE. Role of prostaglandins in gastroduodenal mucosal protection. J Clin Gastroenterol 1991;13(Suppl 1):S65–71. Available from: https://doi.org/10.1097/00004836-199112001-00011.

[45] Chan FKL, Ching JYL, Tse YK, et al. Gastrointestinal safety of celecoxib versus naproxen in patients with cardiothrombotic diseases and arthritis after upper gastrointestinal bleeding (CONCERN): an industry-independent, double-blind, double-dummy, randomised trial. Lancet 2017;389(10087):2375–82. Available from: https://doi.org/10.1016/S0140-6736(17)30981-9.
[46] Hunter TS, Robison C, Gerbino PP. Emerging evidence in NSAID pharmacology: important considerations for product selection. Am J Manag Care 2015;21(7 Suppl):S139–47.
[47] Schafer AI. Effects of nonsteroidal antiinflammatory drugs on platelet function and systemic hemostasis. J Clin Pharmacol 1995;35(3):209–19. Available from: https://doi.org/10.1002/j.1552-4604.1995.tb04050.x.
[48] Goldenberg NA, Jacobson L, Manco-Johnson MJ. Brief communication: duration of platelet dysfunction after a 7-day course of ibuprofen. Ann Intern Med 2005;142(7):506–9. Available from: https://doi.org/10.7326/0003-4819-142-7-200504050-00009.
[49] Schneider V, Lévesque LE, Zhang B, et al. Association of selective and conventional nonsteroidal antiinflammatory drugs with acute renal failure: a population-based, nested case-control analysis. Am J Epidemiol 2006;164(9):881–9. Available from: https://doi.org/10.1093/aje/kwj331.
[50] Perazella MA. COX-2 selective inhibitors: analysis of the renal effects. Expert Opin Drug Saf 2002;1(1):53–64. Available from: https://doi.org/10.1517/14740338.1.1.53.
[51] Baker M, Perazella MA. NSAIDs in CKD: are they safe? Am J Kidney Dis 2020;76(4):546–57. Available from: https://doi.org/10.1053/j.ajkd.2020.03.023.
[52] García Rodríguez LA, Williams R, Derby LE, et al. Acute liver injury associated with nonsteroidal anti-inflammatory drugs and the role of risk factors. Arch Intern Med 1994;154(3):311–6. Available from: https://doi.org/10.1001/archinte.1994.00420030117012.
[53] Lo PC, Tsai YT, Lin SK, et al. Risk of asthma exacerbation associated with nonsteroidal anti-inflammatory drugs in childhood asthma: a nationwide population-based cohort study in Taiwan. Medicine (Baltimore) 2016;95(41):e5109. Available from: https://doi.org/10.1097/MD.0000000000005109.
[54] Ishikawa H, Mutoh M, Sato Y, et al. Chemoprevention with low-dose aspirin, mesalazine, or both in patients with familial adenomatous polyposis without previous colectomy (J-FAPP Study IV): a multicentre, double-blind, randomised, two-by-two factorial design trial. Lancet Gastroenterol Hepatol 2021;6(6):474–81. Available from: https://doi.org/10.1016/S2468-1253(21)00018-2.
[55] Burn J, Sheth H, Elliott F, et al. CAPP2 Investigators. Cancer prevention with aspirin in hereditary colorectal cancer (Lynch syndrome), 10-year follow-up and registry-based 20-year data in the CAPP2 study: a double-blind, randomised, placebo-controlled trial. Lancet 2020;395(10240):1855–63. Available from: https://doi.org/10.1016/S0140-6736(20)30366-4.
[56] Guo Q, Li Q, Wang J, et al. A comprehensive evaluation of clinical efficacy and safety of celecoxib in combination with chemotherapy in metastatic or postoperative recurrent gastric cancer patients: a preliminary, three-center, clinical trial study. Medicine (Baltimore) 2019;98(27):e16234. Available from: https://doi.org/10.1097/MD.0000000000016234.
[57] Meyerhardt JA, Shi Q, Fuchs CS, et al. Effect of celecoxib vs placebo added to standard adjuvant therapy on disease-free survival among patients with stage III colon cancer: the CALGB/SWOG 80702 (Alliance) Randomized Clinical Trial. JAMA 2021;325(13):1277–86. Available from: https://doi.org/10.1001/jama.2021.2454.
[58] Zelenay S, van der Veen AG, Bottcher JP, et al. Cyclooxygenase-dependent tumor growth through evasion of immunity. Cell 2015;162(6):1257–70. Available from: https://doi.org/10.1016/j.cell.2015.08.015.
[59] Liao X, Lochhead P, Nishihara R, et al. Aspirin use, tumor PIK3CA mutation, and colorectal-cancer survival. N Engl J Med 2012;367(17):1596–606. Available from: https://doi.org/10.1056/NEJMoa1207756.
[60] Hamada T, Ogino S, Nishihara R, et al. Aspirin use and colorectal cancer survival according to tumor CD274 (programmed cell death 1 ligand 1) expression status. J Clin Oncol 2017;35(16):1836–44. Available from: https://doi.org/10.1200/JCO.2016.70.7547.
[61] Fiorucci S, Meli R, Bucci M, et al. Dual inhibitors of cyclooxygenase and 5-lipoxygenase. A new avenue in anti-inflammatory therapy? Biochem Pharmacol 2001;62(11):1433–8. Available from: https://doi.org/10.1016/s0006-2952(01)00747-x.
[62] Fiorucci S, Santucci L, Distrutti E. NSAIDs, coxibs, CINOD and H2S-releasing NSAIDs: what lies beyond the horizon. Dig Liver Dis 2007;39(12):1043–51. Available from: https://doi.org/10.1016/j.dld.2007.09.001.

Chapter 8

Anxiolytics, sedatives, and antidepressants

Antonio Markotić[1,2] and Marko Pavlović[3,4]
[1]*Department of Physiology, School of Medicine, University of Mostar, Mostar, Bosnia and Herzegovina,* [2]*Department of Clinical Pharmacology, University Clinical Hospital Mostar, Mostar, Bosnia and Herzegovina,* [3]*Department of Psychiatry, School of Medicine, University of Mostar, Mostar, Bosnia and Herzegovina,* [4]*Department of Psychiatry, University Clinical Hospital Mostar, Mostar, Bosnia and Herzegovina*

Key learning objectives

- To give insight into current evidence regarding the usage of anxiolytics, sedatives, and antidepressants for the treatment of pain in cancer patients.
- To describe and explain data from clinical trials conducted specifically in cancer patients.
- To provide information related to specific individual drugs that are commonly used or recommended by guidelines for pain management or related conditions.
- To give a brief overview of current guidelines and opinions regarding the usage of these drugs in pain management.

Introduction

Anxiolytics, sedatives, and antidepressants are central nervous system (CNS)-acting medications used for various disorders. In terms of pain treatment in cancer patients, these drugs are usually designated as adjuvant analgesics [1]. Adjuvant analgesics represent a diverse group of drugs that are used primarily for indications other than pain but exhibit analgesic properties under certain circumstances. As the usage of these drugs increased over time, and some of them got approved for pain management, the term *adjuvant* became a misnomer. The term *coanalgesic* is sometimes used in the sense of their usual administration with other drugs (such as with opioids in cancer pain management) [1,2]. These drugs are mainly used to treat neuropathic pain that affects approximately 40% of patients with cancer [3]. The International Association for the Study of Pain defined neuropathic pain as "pain that arises as a direct consequence of a lesion or disease affecting the somatosensory system" [4]. Neuropathic pain in oncological patients may be caused directly by cancer (through damage or injury of the nervous system), cancer treatment (surgery, chemotherapy, and radiotherapy), or comorbid diseases (such as diabetes) [5,6]. Chemotherapy-induced peripheral neuropathy (CIPN) is a condition that may even be considered as a common side effect of oral or intravenous chemotherapeutics, such as taxanes, platinum compounds, and vinca alkaloids [5,6]. Therefore, neuropathic pain in oncological patients exhibits a complex pathophysiology. The heterogeneous nature of pain in cancer, including the neuropathic component itself, and the complex characteristics of cancer patients hamper the effort of healthcare professionals and the scientific community in terms of both providing good quality data from clinical trials and achieving effective pain management in patients [7,8]. In this chapter, we will give insight into current evidence regarding the usage of anxiolytics, sedatives, and antidepressants for cancer pain treatment and related conditions. This chapter will also include the presumed mechanism of action, recommendations for dosage, common side effects, contraindications, and interactions of several drug class representatives that are most commonly used, and provide a brief overview of current guidelines and recommendations.

Anxiolytics and sedatives

Anxiolytics are a group of medications used for the treatment of various anxiety disorders, for short-term treatment of anxiety symptoms in other psychiatric conditions, as well as for many other purposes [9]. Sedatives are a group of drugs also known as hypnotics, which are primarily used for the treatment of insomnia but may also be used to reduce anxiety or

Pain Management in Palliative Care for Patients With Cancer. DOI: https://doi.org/10.1016/B978-0-443-30050-9.00012-6

excitement [10,11]. Although anxiolytics and sedatives are used for different purposes in clinical practice, there is a lot of overlap between these medications since many anxiolytics also cause sedation and drowsiness. Anxiolytics and sedatives/hypnotics collectively include benzodiazepines, barbiturates, 5-hydroxytryptamine 1A receptor partial agonists (e.g., buspirone), Z-drugs (drugs with novel chemical structures for use in sleep disorders; e.g., zolpidem), newer hypnotic drugs (e.g., melatonin receptor agonists, orexin antagonists), and other drugs (e.g., meprobamate, chloral hydrate) [12,13]. Benzodiazepines, barbiturates, Z-drugs, and other drugs bind to the $GABA_A$ receptor, which functions as a chloride ion channel and is activated by GABA (γ-aminobutyric acid), the principal inhibitory neurotransmitter in the CNS [13]. Although several classes of drugs fit into this broad category, in this chapter, we will primarily focus on the role of benzodiazepines, one of the most commonly prescribed psychotropic drugs for cancer patients worldwide [14].

Benzodiazepines

Benzodiazepines that are frequently prescribed in clinical practice include diazepam, flurazepam, alprazolam, clonazepam, lorazepam, oxazepam, and others [15]. Although benzodiazepines can be used in the treatment of different anxiety disorders, they are most commonly used for the treatment of panic disorder and generalized anxiety disorder. Other indications include the treatment of insomnia, catatonia, seizure disorders, and alcohol and benzodiazepine withdrawal syndromes [16]. They can also be used for preoperative relief of anxiety, for conscious sedation, and as skeletal muscle relaxants [9]. Benzodiazepines enhance GABA's inhibitory effects by positive allosteric modulation of postsynaptic $GABA_A$ receptors, which increases the frequency of channel opening events [13,17]. Most benzodiazepines are metabolized through the liver by the cytochrome P450 (CYP) enzymes (especially CYP3A4), and many metabolites are pharmacologically active. The metabolites are conjugated with glucuronide and excreted almost entirely in the urine [13,15]. Based on the elimination half-life, benzodiazepines can be classified as short-acting, intermediate-acting, or long-acting benzodiazepines, as shown in Table 8.1 [15].

Some of the most important adverse effects of benzodiazepines include drowsiness, fatigue, and impaired memory and psychomotor function [15,18]. Contraindications, adverse effects, important interactions, and specific comments for individual drugs are given in Table 8.2. If used in higher dosages, benzodiazepines can lead to impaired motor coordination, dizziness, vertigo, slurred speech, blurry vision, mood swings, and euphoria. Long-term use of benzodiazepines increases the risk of addiction [15].

Benzodiazepines and Z-drugs as adjuvant analgesics

The effectiveness of benzodiazepines in pain management as adjuvant analgesics has long been a controversial issue. Previous studies do not provide sufficient evidence to support the claim that benzodiazepines have intrinsic analgesic

TABLE 8.1 Administration, dosing, and pharmacokinetic characteristics of benzodiazepines.

Benzodiazepines	Available forms	Usual daily dose (mg)	Half-life (h)	Tapering[a]
Long-acting				
Diazepam	Oral, intramuscular, intravenous, intranasal, rectal	20–40	20–80	Taper by 2 mg every 3 days
Short- to intermediate-acting				
Alprazolam	Oral	0.25–4	12–15	Taper by 0.5 mg every 3–5 days
Clonazepam	Oral	0.5–4	18–50	Taper by 0.25 mg every 3–5 days
Lorazepam	Oral, intramuscular, intravenous	0.5–5	10–20	Taper by 0.5 mg every 3–5 days
Oxazepam	Oral	30–120	10–20	Taper by 15 mg every 3–5 days
Midazolam	Oral, sublingual, intramuscular, intravenous, intranasal, rectal	0.25–20	1–3	If administration was prolonged, do not stop abruptly

[a]*Potential clinical recommendations, other schedules may be appropriate based on different patient characteristics.*
Source: Data are compiled from Refs. [13,15,18,19].

TABLE 8.2 Adverse effects and contraindications of benzodiazepines [19,20].

Diazepam	
Contraindications:	Hypersensitivity to benzodiazepines, angle-closure glaucoma
Adverse effects:	Sedation, fatigue, ataxia, dysarthria, dizziness, forgetfulness
Drug interactions:	Central nervous system (CNS) depressants—additive effect
Comments:	Use with caution in patients with pulmonary disease
Alprazolam	
Contraindications:	Hypersensitivity, angle-closure glaucoma, concomitant use with ketoconazole/itraconazole
Adverse effects:	Sedation, fatigue, depression, ataxia, dysarthria, dizziness, forgetfulness
Drug interactions:	CNS depressants—additive effect; inhibitors/inducers of CYP3A4 may impact clearance
Comments:	Use with caution if a patient has obstructive sleep apnea (OSA) or pulmonary disease
Clonazepam	
Contraindications:	Hypersensitivity to benzodiazepines, angle-closure glaucoma, severe liver disease
Adverse effects:	Sedation, dizziness, fatigue, ataxia, forgetfulness
Drug interactions:	CNS depressants—additive effect
Comments:	Use with caution if a patient has OSA or pulmonary disease
Lorazepam	
Contraindications:	Hypersensitivity to benzodiazepines, angle-closure glaucoma
Adverse effects:	Sedation, dizziness, fatigue, ataxia, forgetfulness
Drug interactions:	CNS depressants—additive effect
Comments:	May be used in liver impairment; use with caution in patients with OSA or pulmonary disease
Midazolam	
Contraindications:	Hypersensitivity to benzodiazepines, angle-closure glaucoma
Adverse effects:	Oversedation, impaired recall, agitation, involuntary movements, headache, nausea, vomiting
Drug interactions:	CNS depressants—additive effect; inhibitors of CYP3A4 may decrease midazolam clearance
Comments:	Primarily used preoperatively as an anxiolytic and sedative; not intended for long-term use

Source: Data are compiled from Stahl SM. Prescriber's guide: Stahl's essential psychopharmacology. 7th ed. Cambridge: Cambridge University Press; 2021; Bounds CG, Patel P. Benzodiazepines. Treasure Island (FL): StatPearls; 2024.

properties in the treatment of pain in most clinical circumstances, as well as in cancer patients [21]. However, on the basis of reports in the literature or on clinical experience, benzodiazepines are widely accepted as adjuvants in the management of several specific pain states [21,22]. Also, cancer patients are frequently prescribed benzodiazepines to reduce anxiety, insomnia, phobia, and anticipatory nausea. The analgesic effect of benzodiazepines in these patients is thought to be particularly related to the state of high anxiety, muscle spasms, and deafferentation [23]. Moreover, neuropathic cancer pain often does not respond to opioids, and adjuvant analgesics are often used to complement opioid therapy in the treatment of neuropathic pain [24,25]. It has been postulated that benzodiazepines may alleviate neuropathic pain by inhibiting ectopic firing in peripheral nerves [21].

To date, only a few studies have investigated the use of individual benzodiazepines as adjuvant analgesics in patients with cancer-related pain. To investigate the adjuvant analgesic properties of clonazepam, Hugel et al. undertook a prospective case series involving five patients with cancer-related neuropathic pain. The patients were monitored for pain control during 5 days and titrated onto clonazepam up to a dose of 2 mg per day, while the dose of opioids remained unchanged. The pain scores, measured by a four-point rating scale that describes increasing levels of pain intensity, were improved in all patients [26].

Although the results of this study suggest that clonazepam may have beneficial effects on pain relief in cancer patients, in order to establish a cause–effect relationship between the use of clonazepam and its analgesic activity in cancer patients, a randomized controlled trial (RCT) should be performed. Fernandez et al. investigated the effect of alprazolam for the treatment of psychiatric disturbances in 39 patients with malignancies. They found decreased reports of pain in cancer patients who were, based on their clinical presentation, classified as experiencing neuropathic or mixed pain. In contrast, no improvement was observed in patients classified as having somatic pain. However, this was an open-label uncontrolled study, and relations between analgesia during treatment with alprazolam and psychiatric disorders were not investigated [27]. Yosselson-Superstine et al. investigated the clinical efficacy of several antianxiety medications as adjuvant analgesic treatment in a placebo-controlled, double-blind, crossover study. The study included nine patients with chronic malignant and arthritic pain, and no effect was found in reducing pain levels for chlordiazepoxide or other antianxiety medications [28].

Nonbenzodiazepine hypnotics or Z-drugs (zolpidem, zaleplon, zopiclone, and eszopiclone) bind preferentially to $\alpha 1$ subunits of $GABA_A$ receptors [17]. Z-drugs have become the most prescribed medications for insomnia management worldwide [29]. To the best of our knowledge, only one previous study has been conducted to address the effect of Z-drugs on pain in cancer patients. Dimsdale et al. examined the effect of eszopiclone on pain in a randomized, double-blind, placebo-controlled study. The study included 45 inpatients with hematological malignancies and acute pain due to mucositis. Eszopiclone has been shown to significantly reduce reports of pain throughout the day [30]. This study is limited by its small sample size and a short observation period of 2 days, although it is indicative of a favorable effect of eszopiclone on pain. However, considering that impaired sleep lowers the pain threshold [31], the effect of eszopiclone may have occurred secondarily due to improved sleep.

Given the limited data from the small number of studies conducted so far, meaningful conclusions about the effectiveness of benzodiazepines and Z-drugs as adjuvant analgesics in pain treatment for cancer patients are difficult to draw. Nevertheless, based on current evidence, these medications cannot be recommended for treating cancer pain. If benzodiazepines and Z-drugs are used based on clinical experience, it is important to note that their use should take into account the potential for the development of cognitive impairment, dependence, overdose, and other adverse effects [21,32,33]. Special attention should be paid when using benzodiazepines or Z-drugs in elderly patients due to the increased risk of falls and fractures [18,34]. Likewise, the US Food and Drug Administration (FDA) expanded its warnings against using opioids and benzodiazepines concomitantly due to the risk of serious adverse effects, including slowed or difficult breathing and death [35]. Taken together, careful consideration of the risk–benefit ratio prior to treatment with benzodiazepines or Z-drugs in cancer patients should be evaluated.

Benzodiazepines are also frequently used in terminally ill cancer patients who may experience a variety of symptoms refractory to standard palliative therapy, including pain. In these patients, benzodiazepines are used for palliative sedation therapy to reduce consciousness and alleviate unnecessary suffering before death [36]. Although there is insufficient evidence from prospective clinical studies to confirm the effectiveness of palliative sedation in symptom control, it has become a widely accepted medical practice, despite being primarily based on expert opinion. Palliative sedation requires a tailored approach for each cancer patient, considering their different clinical manifestations and symptom burden [37]. According to most guidelines, benzodiazepines (primarily midazolam) should be the first-choice drugs in palliative sedation therapy. Midazolam is preferred over other benzodiazepines due to its rapid onset of action and short elimination half-life, which allows rapid titration [38]. However, a variety of ethical challenges exist when using sedatives and other medications in palliative sedation. The basic ethical principles of palliative sedation should include autonomy, beneficence, nonmaleficence, and justice [39,40]. In the context of pain management, the proportionality principle is essential and should be applied with care. The primary goal is to control refractory pain, starting with a low level of sedation and gradually increasing it if needed [41].

Tolerance, dependence, and abuse are important adverse effects of benzodiazepines in palliative care [22]. Tolerance and dependence are associated with long-term use of benzodiazepines or with short-term use if high doses of benzodiazepines are administered [9]. Tolerance often develops over a few weeks and requires an increase in dose to maintain the therapeutic effect. Given the poor prognosis of patients with advanced cancer, benzodiazepine dependence and abuse should not be a reason for discontinuation of benzodiazepines [22].

To the best of our knowledge, no study to date has investigated the effect of other sedatives (e.g., barbiturates, buspirone, melatonin receptor agonists, and orexin antagonists) on pain management in palliative care for cancer patients.

Potential applications

There is insufficient evidence to confirm the effectiveness of anxiolytics and sedatives in pain management in palliative care for cancer patients. Based on clinical experience, benzodiazepines and Z-drugs can be used in palliative care if, in addition to pain, patients suffer from severe anxiety, insomnia, or psychological distress, with an appropriate evaluation of the risk–benefit ratio for each patient individually. If palliative sedation is administered, benzodiazepines can be used as the first-line treatment.

Antidepressants

Antidepressants comprise several drug subclasses named by their chemical structure or presumed mechanism of action: tricyclic antidepressants (TCAs), serotonin–norepinephrine reuptake inhibitors (SNRIs), selective serotonin reuptake inhibitors (SSRIs), monoamine oxidase inhibitors, and others [42,43]. In terms of pain and especially neuropathic pain treatment, TCAs and SNRIs are most commonly used [44]. The proposed mechanism for pain alleviation in neuropathic pain is an increase in the availability of monoamines in neural pathways in the spinal cord and locus coeruleus [45]. It seems that the inhibition of norepinephrine uptake is the most important mechanism, but serotonin and dopamine may also contribute to the analgesic effect. Furthermore, the analgesic effect is observed even in patients without mood disorders, and it is observed before the antidepressant effect, suggesting that there could be differences in mechanisms that mediate these effects [45–47]. However, there are still controversies about their usage, as a novel Cochrane review and network metaanalysis found only low-certainty evidence for the majority of examined drugs, and specific data about their effectiveness in treating cancer pain are scarce [48,49]. Ferreira et al. gave an overview of systematic reviews addressing the efficacy of antidepressants in pain management in a recent paper published in *BMJ* [50]. They found that moderate-certainty evidence exists only for SNRIs in the treatment of neuropathic pain and a few other conditions. However, they also raised several concerns about industry ties of trials (45%) and the clinical relevance of treatment effects, concluding that a more nuanced approach is both essential and warranted when antidepressants are prescribed for pain [50].

Tricyclic antidepressants

This subclass includes drugs such as amitriptyline and imipramine, which are tertiary amines, and nortriptyline and desipramine, which are secondary amines. They are indicated for the treatment of major depressive disorder, but clinicians also use them off-label to treat other disorders such as anxiety, pain, migraine prophylaxis, insomnia, and enuresis [42,43,51]. The mechanism of action is inhibition of norepinephrine and serotonin reuptake through the blockade of their transporters (NET and SERT) at presynaptic terminals. Plasma half-lives range from 8 to 80 hours. Therefore, most of them are used once daily, usually at bedtime, due to their sedating effect. They are metabolized in the liver (substrates of CYP2D6), and some of them have active metabolites—for example, amitriptyline is converted to nortriptyline. Due to their antagonism of histamine, muscarinic, and alpha-1 adrenergic receptors, typical side effects include sedation, dry mouth, constipation, urinary retention, orthostatic hypotension, and serious cardiovascular complications. To avoid adverse effects, TCAs are usually initiated in a low-dose manner, with slow dose escalation, and tapered when discontinuing to prevent withdrawal [42,43,51]. Dosage considerations, contraindications, side effects, drug interactions, and comments regarding specific drugs from this subclass are provided in Tables 8.3 and 8.4.

Amitriptyline has been used over the years for the treatment of neuropathic pain. It is widely recommended as a first-line treatment and even approved in the United Kingdom [50,52]. A systematic review and metaanalysis provided by Finnerup et al. analyzed the effectiveness of TCAs (mainly amitriptyline) in neuropathic pain. Based on the Grading of Recommendations Assessment, Development and Evaluation (GRADE) evaluation system, they recommended TCAs as one of the possible first-line treatments, with a number needed to treat (NNT) of 3.6 for TCAs [44]. However, a Cochrane review raised concerns about overestimating the treatment effects of amitriptyline [53]. As with the majority of other adjuvant analgesics, recommendations for amitriptyline use in cancer-related neuropathic pain are based mainly on the data from noncancer patients and conditions, although there are several studies that examined its analgesic properties in oncological patients.

Mishra et al. conducted an RCT on the comparative efficacy of amitriptyline and anticonvulsants [54]. The study enrolled 120 patients with neuropathic cancer pain. The amitriptyline was started at 50 mg/day at bedtime with dose escalation, and patients rated their pain on the 10-cm Visual Analog Scale (VAS). The authors reported a significant decrease in VAS value in all groups (including placebo) compared to baseline values, and the pregabalin group showed better results than amitriptyline and gabapentin groups. As rescue morphine was required by 56.7% of patients on amitriptyline and 16.7% in the pregabalin group, the authors concluded that all examined drugs were effective in relieving pain, with pregabalin showing the best opioid-sparing effect [54]. Other trials demonstrate only a marginal difference in the change of VAS value between the amitriptyline and placebo groups, with low quality of evidence [55]. Similar results were reported in a randomized, open-label study that compared the efficacy and safety of amitriptyline and gabapentin in

TABLE 8.3 Dosage considerations for major antidepressant drug representatives in cancer pain management.

Tricyclic antidepressants	
Amitriptyline oral use	**Dosage considerations**[a]: *Start with a low dose:* 10–25 mg at bedtime to prevent/minimize adverse effects. *Escalate dose:* every few days or weekly by 10–25 mg (based on efficacy and side effects). *Usual effective dose*: 50–100 mg at bedtime (once daily) or in two divided doses. *Maximal dose*: 150 mg/day if tolerated (note that for depression, even higher doses may be used). *ESMO guidelines* suggest doses <75 mg/day in cancer-related neuropathic pain for tricyclic antidepressants [68]. *Analgesic effect* is usually observed in 2–4 weeks. **Additional considerations:** Avoid or use in low doses in the elderly (due to anticholinergic side effects). Taper is recommended—gradually reduce the dose in 2–4 weeks (i.e., by 50% every 2 weeks or by 25% every week); no dosage reduction in kidney failure, use cautiously; reduce doses in patients with impaired liver function.
Nortriptyline oral use	**Dosage considerations**[a]: See Amitriptyline. Plasma levels may be measured—the optimum range is 50–150 ng/mL (there is no specific data regarding optimal levels in neuropathic pain); better tolerated than tertiary amines.
Serotonin–norepinephrine reuptake inhibitors (SNRIs)	
Duloxetine oral use	**Dosage considerations**[a]: *Start with a low dose:* 20–30 mg to prevent/minimize adverse effects. *Escalate dose:* every 1 week by 20–30 mg (based on efficacy and side effects). *Usual effective dose:* 60 mg/day (can be taken once daily). *Maximal dose:* 60 mg/day (higher doses up to 120 mg/day may be used, although there is no evidence that higher doses yield analgesic benefits in cancer patients). *Analgesic effect* may be observed even after 1 week [69], whereas trial duration of 8 weeks is recommended for painful diabetic neuropathy. **Additional considerations:** May be used in older patients in standard doses (use with caution). Taper is recommended due to possible discontinuation syndrome—gradually reduce the dose in 2–4 weeks (i.e., by 50% every 2 weeks or by 25% every week). Manufacturer recommends avoidance if creatinine clearance is <30 mL/min. Due to rare hepatotoxicity, avoidance is also recommended in hepatic disorders or alcohol abuse.
Venlafaxine ER oral use	**Dosage considerations**[a]: *Extended-release (ER)* hydrochloride is preferred over immediate-release (IR) form—ER is used once per day (better compliance) with better tolerance. *Start with a low dose:* 37.5 mg of ER formulation. *Escalate dose:* every 4–7 days by 37.5 mg (taken once daily) if tolerated. *Usual effective dose:* 150–225 mg/day, a dose of 150 mg is presumed to achieve SNRI-mediated analgesia [70]. *Maximal dose:* 225 mg/day (once daily). **Additional considerations:** May be used in older patients in standard doses (use with caution). Taper is recommended due to possible discontinuation syndrome—gradually reduce the dose over 2–4 weeks by 37.5–75 mg/week (preferably 4 weeks due to relatively short half-life) [71]. In patients with creatinine clearance <30 mL/min or those on hemodialysis treatment, start with a low dose (37.5 mg) and do not use more than 50% of the maximum recommended dose. Use cautiously. Reduce the dose by 50% in mild/moderate liver impairment or higher in severe disorders.

[a]*Potential clinical recommendations, other schedules may be appropriate based on different patient characteristics.*
Source: Data were obtained through UpToDate Lexidrug drug information packages, in table labeled and listed sources [1,25,52,68–76].

88 patients receiving opioid treatment for neuropathic cancer pain [56]. While both treatments provided significant pain relief and a similar number of side effects, the major limitation of this study was the open-label trial, which was thus unable to control for the placebo effect [56].

A small, crossover RCT examined the effect of amitriptyline on neuropathic pain following surgical treatment of breast cancer in Finnish females [57]. Even though authors concluded that amitriptyline may reduce neuropathic pain, the adverse effects were a major issue, as the majority of patients did not continue therapy due to the adverse effects [57]. A

TABLE 8.4 Contraindications, warnings, major adverse effects, and drug interactions with specific comments for major antidepressant drug representatives in cancer pain management.

Tricyclic antidepressants	
Amitriptyline	
Contraindications:	Hypersensitivity, recent myocardial infarction, conduction abnormalities, concurrent or recent therapy with monoamine oxidase inhibitors (MAOI). Generally avoided if QTc > 500 ms.
Adverse effects:	Anticholinergic effects (dry mouth, urinary retention, constipation), sedation, orthostatic hypotension, QT prolongation, arrhythmias.
Drug interactions:	MAOIs, selective serotonin reuptake inhibitors (SSRIs), serotonin–norepinephrine reuptake inhibitors (SNRIs)—risk of serotonin syndrome (caution with tramadol); anticholinergic drugs; central nervous system (CNS) depressants—additive sedation; drugs that prolong QT interval.
Comments:	Before initiation, ask about heart disease or symptoms, check potassium levels and QTc in patients at risk, check for drug interactions; box warning—suicidality.
Nortriptyline	
Comments:	Contraindications, adverse effects, and interactions generally the same as with amitriptyline.
SNRI	
Duloxetine	
Contraindications:	Hypersensitivity, concurrent or recent therapy with MAOI. Avoided in severe renal or liver impairment or uncontrolled hypertension.
Adverse effects:	Nausea, dry mouth, constipation, headache, drowsiness, sexual dysfunction. Rare but serious: hepatotoxicity, increased risk of bleeding, hyponatremia.
Drug interactions:	MAOIs, tricyclic antidepressants (TCAs), SSRIs—risk of serotonin syndrome (caution with tramadol); anticoagulants or antiplatelet agents—increased risk of bleeding; CNS depressants—additive sedation; CYP1A2 and 2D6 inhibitors/inducers may affect duloxetine effectiveness.
Comments:	It is a moderate CYP2D6 inhibitor that may hamper the effectiveness of drugs that are converted to active metabolites by this enzyme (tamoxifen, codeine [76,81]); box warning—suicidality.
Venlafaxine ER	
Contraindications	Hypersensitivity, concurrent or recent therapy with MAOI; caution is advised in patients with cardiac disease, renal or liver impairment, seizure disorders.
Adverse effects:	Nausea, vomiting, diarrhea, headache, hypertension (with higher doses), sexual dysfunction. Rare but serious: increased risk of bleeding, hyponatremia, serotonin syndrome, seizures.
Drug interactions:	MAOIs, TCAs, SSRIs—risk of serotonin syndrome (caution with tramadol); anticoagulants or antiplatelet agents—increased risk of bleeding; CNS depressants—additive sedative effects.
Comments:	Check blood pressure; optimal for patients on tamoxifen [76]; box warning—suicidality.

Source: Data were obtained through UpToDate Lexidrug drug information packages, in table labeled and listed sources [1,52,69–76,81].

significantly higher number of adverse effects was also reported by Mercadante et al., but without benefit in terms of analgesia when compared to placebo. The study included 16 advanced cancer patients with neuropathic pain [58].

A group of authors from Finland published two studies that examined the effectiveness of amitriptyline in the prevention or treatment of CIPN. Unfortunately, amitriptyline failed to provide prevention from CIPN [59] or to improve neuropathic symptoms in cancer patients in these randomized, placebo-controlled trials [60]. However, a recent article suggested a different approach in terms of high-dose topical amitriptyline for patients with CIPN [61]. The 25 patients with CIPN were treated with a cream for 1 month, which resulted in significant pain relief without adverse effects. Even though this approach sounds interesting, well-designed RCTs are warranted to make further conclusions.

Secondary amines such as **nortriptyline** and desipramine are more selective in the inhibition of norepinephrine uptake, which appears to be a major mechanism of how antidepressants alleviate pain [42,45]. Due to that and a better adverse effect profile, some experts suggest that these drugs should be preferred over amitriptyline, especially in older patients [62–64]. A recent metaanalysis quantitatively analyzed the analgesic effect and safety of nortriptyline over various pain types and suggested its effectiveness [65]. However, a previous Cochrane review found little evidence to suggest its usage in neuropathic pain [66]. Data are particularly scarce when it comes to cancer pain treatment. We found only one trial that evaluated the efficacy of nortriptyline in cisplatin-induced CIPN sufferers [63]. It was a crossover RCT that enrolled 57 patients. Authors found no significant differences in pain, but the effect of pain on activities was slightly improved along with hours of sleep in the nortriptyline arm. Dry mouth, dizziness, and constipation were more common with the nortriptyline. A recent systematic review concluded that there is insufficient evidence regarding nortriptyline use in CIPN [67].

Serotonin–norepinephrine reuptake inhibitors

This subclass includes several chemically unrelated drugs, such as *venlafaxine*, and its metabolites, *desvenlafaxine*, *duloxetine*, and *milnacipran*. These medications are used mainly for the treatment of depression, anxiety disorders, and pain [43]. However, there are some differences in their indications. For instance, milnacipran is indicated by the FDA only for fibromyalgia, while duloxetine was the first drug indicated for the treatment of diabetic neuropathy, and it is approved for fibromyalgia and chronic musculoskeletal pain. Venlafaxine shares the indications regarding depression and generalized anxiety with duloxetine, and it is approved for panic and social anxiety disorders. However, in terms of pain syndromes, it is used off-label [43,76,77]. The mechanism of action of SNRIs is described in their name—they inhibit the reuptake of serotonin and norepinephrine, thereby stimulating serotonergic and noradrenergic neurotransmission [42,78]. There are some differences in their affinity for transporters, with milnacipran having higher selectivity for norepinephrine uptake [78,79] and venlafaxine showing differential, dose-dependent inhibition, with low doses selectively inhibiting serotonin transporters and high doses inhibiting both transporters [80]. The plasma half-life range is shorter compared to TCAs (6–12 hours). Venlafaxine and duloxetine are extensively metabolized in the liver by CYP2D6, and duloxetine is a moderate CYP2D6 inhibitor, due to which it may exhibit relevant pharmacokinetic interactions with several clinically important drugs, as shown in Table 8.4 [42,43,76,77]. The most common side effects include nausea, dizziness, constipation, headaches, and sexual dysfunction, and venlafaxine may increase blood pressure at higher doses [43,71,76]. In order to avoid adverse effects, it is recommended to start with a low dose and then escalate by one or two increments. The starting and maximum dose may vary depending on the indication, albeit it seems that the analgesic effect occurs with lower doses and is faster compared to the antidepressant effect [8,62,81]. Tapering is recommended in order to avoid discontinuation syndrome [76].

Duloxetine is approved for the treatment of pain over several conditions, as noted in the beginning. Two Cochrane metaanalyses found moderate certainty evidence that duloxetine is effective in the treatment of neuropathic and chronic pain, with a strong GRADE recommendation for its usage as a first-line treatment option in neuropathic pain conditions [44,49,82]. Nonetheless, a recent Bayesian adaptive, open-label, randomized trial reported that duloxetine might outperform pregabalin when pain and adverse effects are combined as an end point in cryptogenic neuropathy [83]. However, there are only a few studies that examined the effect of duloxetine on pain in cancer patients, and recommendations for duloxetine use are, as noted before, mainly based on data from trials with noncancer patients, with the exception regarding CIPN. We will give a short narrative review of these studies.

A randomized, double-blind, placebo-controlled trial examined if adding duloxetine is effective in patients with cancer-related neuropathic pain who are nonresponsive or intolerant to opioid-pregabalin therapy [84]. The study enrolled 70 patients with neuropathic pain whose average pain intensity was measured by the Brief Pain Inventory (BPI)—item 5 on day 10. The duloxetine group (40 mg/day) exhibited a lower average pain score compared to placebo. Furthermore, the responder analysis revealed that 44% and 32% of patients in the duloxetine arm reported moderately (decrease by 30%) and substantially important (decrease by 50%) improvement, respectively, both significantly higher in comparison with the placebo group [84]. The main limitations were a short observation period and moderate pain intensity of enrolled patients, so it is questionable if the results may apply to patients with severe pain. It is also not clear if duloxetine with an opioid but without pregabalin would confer significant and satisfying pain relief. The latter was in part addressed by Curry et al., who concluded that combined methadone-duloxetine therapy reduces cancer-related pain better than either drug in monotherapy [85]. However, due to a retrospective design (chart review) and a relatively small sample size, these results should be taken cautiously.

A prospective, randomized, open-label study showed a comparable efficacy of duloxetine and pregabalin in the treatment of 44 lung cancer patients with neuropathic pain [86]. Leeds Assessment of Neuropathic Symptoms and Signs (LANSS) and VAS scores, which were used for pain assessment, were significantly lower after 3 months in both groups, suggesting that both drugs are effective [86]. The lack of placebo control with an open-label design and the small sample size are some of the major drawbacks of the study. Additionally, one large RCT reported that duloxetine may improve joint pain in aromatase inhibitor-associated arthralgias along with significant improvements in quality of life (QoL) of 255 breast cancer patients [87]. Interestingly, the analgesic effect of duloxetine was more pronounced in obese patients [88].

Several systematic reviews supported the use of duloxetine for CIPN, suggesting moderate benefit [67,89]. Furthermore, American Society of Clinical Oncology (ASCO) guidelines recommended it for painful CIPN as the only agent with appropriate evidence to support usage [90]. We identified four RCTs that addressed duloxetine's effectiveness in CIPN.

A highly cited double-blind, placebo-controlled, crossover RCT by Smith et al. was the first large trial that elucidated a possibly effective treatment for painful CIPN [91]. The study included 231 patients who had grade 1 or higher sensory neuropathy after taxane or oxaliplatin treatment. The average pain level was assessed by BPI. Patients treated with duloxetine (60 mg/day) described a significantly better pain alleviation compared to placebo. Approximately 33% of duloxetine-treated patients exhibited a moderate improvement, and 20.7% reported a substantially important improvement, both significantly higher compared to placebo, with an NNT of 3.7. The dropout rate due to adverse effects was around 11% in the duloxetine group. Furthermore, duloxetine also reduced the interference of pain with daily function significantly better than placebo. The reduction in pain level and its interference with daily activities, combined with a low occurrence of adverse effects, appears to be particularly important for these patients, as they also reported a significant improvement in QoL with duloxetine treatment [91]. Given that cancer pain is strongly correlated with deterioration in QoL [92], appropriate pain management may contribute to overall QoL improvement.

The study by Smith et al. shed light on duloxetine in CIPN and was followed by several smaller studies. An open-label, randomized study comparing duloxetine and vitamin B12 further suggested that duloxetine might be effective in treating pain and numbness of CIPN caused by various chemotherapeutics [93]. The other two RCTs were comparative trials that compared duloxetine to venlafaxine and to pregabalin in CIPN [94,95]. In a double-blind RCT by Salefihar and associates, 82 patients with breast cancer who received taxanes were treated with either pregabalin or duloxetine [95]. Both treatments were effective in reducing pain and improving sensory neuropathy compared to baseline measurements, but pregabalin seemed superior to duloxetine. There was no placebo-treated group. The second RCT compared the effects of venlafaxine and duloxetine on CIPN and reported that both treatments are effective when compared to placebo, but duloxetine is superior in terms of decreasing the symptoms of CIPN [94].

Additionally, two recent double-blind RCTs suggested that duloxetine might even be effective in preventing oxaliplatin- and paclitaxel-induced neuropathies [96,97]. Further RCTs with larger sample sizes are particularly warranted, as there are currently no effective treatments or interventions to recommend for the prevention of CIPN [90].

Even though largely "neglected" and excluded from trials, pregnant patients may also experience cancer-related pain, including neuropathic pain. One comparative advantage of duloxetine in these settings is that evidence suggests that duloxetine is unlikely to be a major teratogen and might be used in pregnant and breastfeeding patients if clearly needed. However, findings of increased risk for postpartum hemorrhage and a slight increase in the risk of congenital cardiac malformation warrant precaution [98,99,100].

Venlafaxine is also suggested for the symptomatic treatment of neuropathic pain, although one Cochrane review found little evidence to support its use [101,102]. Kus et al. reported that venlafaxine is effective in the treatment of taxane and oxaliplatin acute neurotoxicity [103]. However, it was a retrospective, case–control study that lacked randomization and placebo control, and therefore, the level of confidence in their findings is very low. There is also one RCT that examined the efficacy of venlafaxine in oxaliplatin-induced acute neurotoxicity [104]. The analysis revealed that a significantly higher number of patients experienced full relief in the venlafaxine group, but adverse effects were also more frequent in the venlafaxine arm. The study that compared venlafaxine to duloxetine was described before [94].

A recent systematic review suggested promising evidence regarding the effectiveness of **milnacipran** in chronic pain, but we found no relevant papers regarding its use in cancer patients [49].

Selective serotonin reuptake inhibitors and other antidepressants

SSRIs are generally considered as a first-line treatment for depression and several other disorders and symptoms in cancer patients [76]. However, they are not recommended as a first-line treatment for chronic pain conditions, including neuropathic pain [25,52,68,76,105]. Due to a lack of evidence and relevant papers that assessed their use in cancer-related (neuropathic) pain, we will not give a comprehensive overview of SSRIs. There is one open-label, randomized trial that investigated if fluvoxamine combined with oxycodone adds any benefit in terms of pain management in moderate-to-severe cancer pain [106]. A conclusion was made that adding fluvoxamine could provide better pain control and reduce the dosage of opioids. However, the open-label design with a lack of placebo control hampers these findings. Nevertheless, it should be noted that emotional disorders may exacerbate chronic pain, including both intensity and duration [105]. Therefore, a successful treatment of depression, for example, may contribute to pain alleviation. *Mirtazapine* is a tetracyclic agent with a good safety and tolerability profile that may be helpful in pain syndromes [42,107]. However, data from a small number of nonrandomized and/or noncontrolled trials in cancer patients that mainly addressed pain as a secondary outcome do not provide evidence that mirtazapine is efficient in treating cancer-related pain [108].

Potential application—to "antidepressant" or not?

Strong opioids are the drugs of choice for patients with advanced cancer and moderate or severe cancer pain [48,109]. However, opioid analgesia is sometimes insufficient to control pain, especially in the setting of neuropathic pain, and additional agents such as adjuvant analgesics may then step in [110]. A review article describing the possibilities of nonopioid drugs in the treatment of cancer pain concluded that the evidence for the use of these drugs, including antidepressants, remains limited, as most of the studies lack adequate sample size and long-term follow-up [110]. Hence, there is no clear consensus regarding antidepressant use. While some experts argue that the advantages of analgesic action outweigh the side effects and favor the use of SNRIs and TCAs for neuropathic pain in a patient-tailored fashion, others highlight that due to limited evidence, the benefit-to-risk balance remains unclear [48,111]. And what do the guidelines say?

WHO suggests that adjuvant analgesics are beneficial when used with opioids for cancer pain but are currently underutilized. However, they make no recommendations for or against the use of antidepressants and anticonvulsants in the treatment of cancer-associated neuropathic pain but consider that individual trials with antidepressants may be taken for patients with neuropathic pain where opioids fail to provide adequate pain relief [55]. A similar recommendation for an opioid-adjuvant combination in cancer-related neuropathic pain is given by European Society for Medical Oncology (ESMO) guidelines. However, these guidelines also strongly recommend gabapentin, pregabalin, duloxetine, and TCAs as single agents for first-line treatment of neuropathic pain [68]. National Comprehensive Cancer Network (NCCN) guidelines also find SNRIs, TCAs, and gabapentinoids as first-line adjuvant analgesics in the treatment of neuropathic pain that can also be combined with opioids if the pain is not adequately controlled [81]. For patients with painful CIPN, ASCO guidelines provide an evidence-based, moderately strong recommendation for duloxetine prescription. Conversely, no recommendation was given for TCAs and anticonvulsants [90]. This is slightly different from the ESMO–European Oncology Nursing Society (EONS)–European Association of Neuro-Oncology (EANO) joint guidelines that consider duloxetine as the only agent with level I evidence but also suggest that venlafaxine, gabapentinoids, and amitriptyline may be considered as treatment options [112]. Guidelines generally do not suggest anxiolytics or sedatives for pain alleviation, while in ESMO guidelines, sedation is proposed as maybe the only therapeutic option to provide pain relief in rare circumstances of end-of-life refractory pain [68]. WHO guidelines point out that sedatives and antidepressants may be necessary to cope with issues such as insomnia, anxiety, and depression [55].

We agree that data and evidence are scarce and limited, but in view of that, almost one-third of patients are still undertreated [113], a trial of adjuvant analgesics in patients with cancer pain and insufficient response or intolerance to opioids sounds like a reasonable approach. That is especially relevant to mixed or purely neuropathic pain that might benefit the most. Which drug to use and when should be decided in a patient-tailored fashion on a case-by-case basis. The clinician in charge should take into consideration several factors, including comorbidities, drug interactions, kidney and liver function, and side effects, and discuss the approach with the patient. The fundamental importance of tailoring treatment in comparison with a one-size-fits-all approach is that it considers individual patient characteristics rather than solely the diagnosis, thereby ensuring better pain management while minimizing adverse effects. For example, patients with pain and concurrent anxiety and/or depression may benefit more from antidepressants than anticonvulsants, and in patients suffering from insomnia, a sedating drug is a valid option. If an antidepressant is to be used, SNRIs, or more specifically, duloxetine, are probably the first drugs that should be considered. SNRIs are better tolerated than TCAs and have a better safety profile, and duloxetine is highlighted by a recent Cochrane network metaanalysis as "the only antidepressant the authors were certain about for the treatment of chronic pain" with moderate efficacy across outcomes [49]. This may be particularly applied to the treatment of CIPN, as several RCTs presented in this chapter reported moderate but significant effects on pain relief.

Future implications

In two words: future research. Current evidence is weak, and emphasis should be placed on adequately powered RCTs that will prove or refute the use of these drugs in practice. Additionally, comparative trials that will assess not only the efficacy of a particular drug but also compare the effect size to other antidepressants or anticonvulsants in cancer-related neuropathic pain are of major interest. The use of anxiolytics and sedatives in pain management within palliative care remains a controversial issue due to the lack of evidence of their benefit for this indication and the significant burden of adverse effects. High-quality RCTs to address the efficacy of anxiolytics and sedatives in alleviating pain in cancer patients, compared to nonsedated patients, are required to provide evidence on this issue.

Conclusions

- Antidepressants are a valid option in treating cancer-related neuropathic pain. Even though data is still insufficient and the quality of evidence is low, we find that benefits outweigh potential harms in selected patients.
- There is no substantial evidence for the use of anxiolytics and sedatives in the treatment of cancer pain.
- We endorse a patient-tailored approach, meaning several factors must be addressed prior to prescribing, such as comorbidities, drug interactions, adverse effect profiles, and patient concerns. For example, patients with depression disorder and/or anxiety and cancer-related neuropathic pain will probably benefit more from an antidepressant than anticonvulsants.
- If anxiolytics and sedatives are used in palliative care based on clinical experience, a careful assessment of the risk–benefit ratio is required.
- SNRIs are preferred over TCAs based on their better safety and tolerance profiles. Due to a lack of data in cancer patients and limited efficacy in noncancer patients, SSRIs are not recommended.
- Duloxetine (SNRI) is a first choice in CIPN treatment, even though it confers modest benefits.
- Midazolam is a first-line drug in palliative sedation therapy.

References

[1] Lussier D, Huskey AG, Portenoy RK. Adjuvant analgesics in cancer pain management. Oncologist 2004;9(5):571–91. Available from: https://doi.org/10.1634/theoncologist.9-5-571.

[2] Glare P, Aubrey K, Gulati A, et al. Pharmacologic management of persistent pain in cancer survivors. Drugs 2022;82(3):275–91. Available from: https://doi.org/10.1007/s40265-022-01675-6.

[3] Bennett MI, Rayment C, Hjermstad M, et al. Prevalence and aetiology of neuropathic pain in cancer patients: a systematic review. Pain 2012;153(2):359–65. Available from: https://doi.org/10.1016/j.pain.2011.10.028.

[4] International Association for the Study of Pain (IASP). Neuropathic pain. Retrieved August 18, 2024, from: https://www.iasp-pain.org/advocacy/global-year/neuropathic-pain; 2021.

[5] Bennett MI, Kaasa S, Barke A, et al. The IASP classification of chronic pain for ICD-11: chronic cancer-related pain. Pain 2019;160(1):38–44. Available from: https://doi.org/10.1097/j.pain.0000000000001363.

[6] Edwards HL, Mulvey MR, Bennett MI. Cancer-related neuropathic pain. Cancers (Basel) 2019;11(3):373. Available from: https://doi.org/10.3390/cancers11030373.

[7] Shkodra M, Caraceni A. Treatment of neuropathic pain directly due to cancer: an update. Cancers (Basel) 2022;14(8):1992. Available from: https://doi.org/10.3390/cancers14081992.

[8] Mulvey MR, Paley CA, Schuberth A, et al. Neuropathic pain in cancer: what are the current guidelines? Curr Treat Options Oncol 2024;25(9):1193–202. Available from: https://doi.org/10.1007/s11864-024-01248-7.

[9] Fuentes D, Ray SD, Holstege CP. Anxiolytics. In: Wexler P, editor. Encyclopedia of toxicology. 3rd ed. Cambridge: Academic Press; 2014. p. 280–6.

[10] Sedatives and hypnotics. LiverTox: clinical and research information on drug-induced liver injury. Bethesda (MD): National Institute of Diabetes and Digestive and Kidney Diseases; 2012.

[11] Moini J, Koenitzer J, LoGalbo A. Substance-related disorders. In: Moini J, Koenitzer J, LoGalbo A, editors. Global emergency of mental disorders. Cambridge: Academic Press; 2021. p. 273–301.

[12] Argyropoulos SV, Sandford JJ, Nutt DJ. The psychobiology of anxiolytic drug. Part 2: pharmacological treatments of anxiety. Pharmacol Ther 2000;88(3):213–27. Available from: https://doi.org/10.1016/s0163-7258(00)00083-8.

[13] Trevor AJ. Sedative-hypnotic drugs. In: Katzung B, editor. Katzung's basic & clinical pharmacology. 14th ed. New York: McGraw-Hill Education; 2018. p. 381–95.

[14] Mohamed F, Uvais NA, Moideen S, et al. Psychotropic medication prescriptions for home-based palliative care oncology patients. Prim Care Companion CNS Disord 2024;26(2):23m03668. Available from: https://doi.org/10.4088/PCC.23m03668.

[15] Griffin CE 3rd, Kaye AM, Bueno FR, et al. Benzodiazepine pharmacology and central nervous system-mediated effects. Ochsner J 2013;13(2):214–23.

[16] Edinoff AN, Nix CA, Hollier J, et al. Benzodiazepines: uses, dangers, and clinical considerations. Neurol Int 2021;13(4):594–607. Available from: https://doi.org/10.3390/neurolint13040059.

[17] Stahl SM. Anxiety, trauma, and treatment. In: Stahl SM, editor. Stahl's essential psychopharmacology: neuroscientific basis and practical applications. Cambridge: Cambridge University Press; 2021. p. 359–78.

[18] Dubovsky SL, Marshall D. Benzodiazepines remain important therapeutic options in psychiatric practice. Psychother Psychosom 2022;91(5):307–34. Available from: https://doi.org/10.1159/000524400.

[19] Stahl SM. Prescriber's guide: Stahl's essential psychopharmacology. 7th ed. Cambridge: Cambridge University Press; 2021.

[20] Bounds CG, Patel P. Benzodiazepines. Treasure Island (FL): StatPearls; 2024.

[21] Reddy S, Patt RB. The benzodiazepines as adjuvant analgesics. J Pain Symptom Manage 1994;9(8):510–4. Available from: https://doi.org/10.1016/0885-3924(94)90112-0.

[22] Stiefel F, Berney A, Mazzocato C. Psychopharmacology in supportive care in cancer: a review for the clinician. I. Benzodiazepines. Support Care Cancer 1999;7(6):379–85. Available from: https://doi.org/10.1007/s005200050296.

[23] Greenberg DB. Strategic use of benzodiazepines in cancer patients. Oncology (Williston Park) 1991;5(4):83–8; discussion 88, 90, 95.

[24] Farrar JT, Portenoy RK. Neuropathic cancer pain: the role of adjuvant analgesics. Oncology (Williston Park) 2001;15(11):1435–42, 1445; discussion 1445, 1450–3.

[25] Yoon SY, Oh J. Neuropathic cancer pain: prevalence, pathophysiology, and management. Korean J Intern Med 2018;33(6):1058–69. Available from: https://doi.org/10.3904/kjim.2018.162.

[26] Hugel H, Ellershaw JE, Dickman A. Clonazepam as an adjuvant analgesic in patients with cancer-related neuropathic pain. J Pain Symptom Manage 2003;26(6):1073–4. Available from: https://doi.org/10.1016/j.jpainsymman.2003.09.005.

[27] Fernandez F, Adams F, Holmes VF. Analgesic effect of alprazolam in patients with chronic, organic pain of malignant origin. J Clin Psychopharmacol 1987;7(3):167–9.

[28] Yosselson-Superstine S, Lipman AG, Sanders SH. Adjunctive antianxiety agents in the management of chronic pain. Isr J Med Sci 1985;21(2):113–7.

[29] Scharner V, Hasieber L, Sönnichsen A, et al. Efficacy and safety of Z-substances in the management of insomnia in older adults: a systematic review for the development of recommendations to reduce potentially inappropriate prescribing. BMC Geriatr 2022;22(1):87. Available from: https://doi.org/10.1186/s12877-022-02757-6.

[30] Dimsdale JE, Ball ED, Carrier E, et al. Effect of eszopiclone on sleep, fatigue, and pain in patients with mucositis associated with hematologic malignancies. Support Care Cancer 2011;19(12):2015–20. Available from: https://doi.org/10.1007/s00520-010-1052-1.

[31] Frohnhofen H. Pain and sleep: a bidirectional relationship. Z Gerontol Geriatr 2018;51(8):871–4. Available from: https://doi.org/10.1007/s00391-018-01461-8.

[32] Atkin T, Comai S, Gobbi G. Drugs for insomnia beyond benzodiazepines: pharmacology, clinical applications, and discovery. Pharmacol Rev 2018;70(2):197–245. Available from: https://doi.org/10.1124/pr.117.014381.

[33] Schifano F, Chiappini S, Corkery JM, et al. An insight into Z-drug abuse and dependence: an examination of reports to the European medicines agency database of suspected adverse drug reactions. Int J Neuropsychopharmacol 2019;22(4):270–7. Available from: https://doi.org/10.1093/ijnp/pyz007.

[34] Treves N, Perlman A, Kolenberg Geron L, et al. Z-drugs and risk for falls and fractures in older adults – a systematic review and meta-analysis. Age Ageing 2018;47(2):201–8. Available from: https://doi.org/10.1093/ageing/afx167.

[35] U.S. Food and Drug Administration. FDA warns about serious risks and death when combining opioid pain or cough medicines with benzodiazepines; requires its strongest warning. Retrieved September 4, 2024, from: https://www.fda.gov/media/127688/download?attachment; 2017.

[36] Beller EM, van Driel ML, McGregor L, et al. Palliative pharmacological sedation for terminally ill adults. Cochrane Database Syst Rev 2015;1(1):CD010206. Available from: https://doi.org/10.1002/14651858.CD010206.pub2.

[37] Arantzamendi M, Belar A, Payne S, et al. Clinical aspects of palliative sedation in prospective studies. A systematic review. J Pain Symptom Manage 2021;61(4) 831–44e10. Available from: https://doi.org/10.1016/j.jpainsymman.2020.09.022.

[38] Schildmann EK, Schildmann J, Kiesewetter I. Medication and monitoring in palliative sedation therapy: a systematic review and quality assessment of published guidelines. J Pain Symptom Manage 2015;49(4):734–46. Available from: https://doi.org/10.1016/j.jpainsymman.2014.08.013.

[39] Tomczyk M, Jaques C, Jox RJ. Ethical challenges in palliative sedation of adults: protocol for a systematic review of current clinical practice guidelines. BMJ Open 2022;12(7):e059189. Available from: https://doi.org/10.1136/bmjopen-2021-059189.

[40] Akdeniz M, Yardımcı B, Kavukcu E. Ethical considerations at the end-of-life care. SAGE Open Med 2021;9:20503121211000918. Available from: https://doi.org/10.1177/20503121211000918.

[41] Raho JA. Palliative sedation: a review of the ethical debate. Health Care Ethics USA [Online]; 2014. https://www.chausa.org/publications/health-care-ethics-usa.

[42] DeBattista C. Antidepressant agents. In: Katzung B, editor. Katzung's basic & clinical pharmacology. 14th ed. New York: McGraw-Hill Education; 2018. p. 532–52.

[43] O'Donnell JM, Bies RR, Williams AJ. Drug therapy of depression and anxiety disorders. In: Brunton LL, Knollmann BC, editors. Goodman & Gilman's the pharmacological basis of therapeutics. 14th ed. New York: McGraw-Hill Education; 2023. p. 343–56.

[44] Finnerup NB, Attal N, Haroutounian S, et al. Pharmacotherapy for neuropathic pain in adults: a systematic review and meta-analysis. Lancet Neurol 2015;14(2):162–73. Available from: https://doi.org/10.1016/S1474-4422(14)70251-0.

[45] Obata H. Analgesic mechanisms of antidepressants for neuropathic pain. Int J Mol Sci 2017;18(11):2483. Available from: https://doi.org/10.3390/ijms18112483.

[46] Fishbain DA, Cutler RB, Rosomoff HL, et al. Do antidepressants have an analgesic effect in psychogenic pain and somatoform pain disorder? A meta-analysis. Psychosom Med 1998;60(4):503–9. Available from: https://doi.org/10.1097/00006842-199807000-00019.

[47] Onghena P, Van Houdenhove B. Antidepressant-induced analgesia in chronic non-malignant pain: a meta-analysis of 39 placebo-controlled studies. Pain 1992;49(2):205–19. Available from: https://doi.org/10.1016/0304-3959(92)90144-Z.

[48] Chapman EJ, Edwards Z, Boland JW, et al. Practice review: evidence-based and effective management of pain in patients with advanced cancer. Palliat Med 2020;34(4):444–53. Available from: https://doi.org/10.1177/0269216319896955.

[49] Birkinshaw H, Friedrich CM, Cole P, et al. Antidepressants for pain management in adults with chronic pain: a network meta-analysis. Cochrane Database Syst Rev 2023;5(5):CD014682. Available from: https://doi.org/10.1002/14651858.CD014682.pub2.

[50] Ferreira GE, Abdel-Shaheed C, Underwood M, et al. Efficacy, safety, and tolerability of antidepressants for pain in adults: overview of systematic reviews. BMJ 2023;380:e072415. Available from: https://doi.org/10.1136/bmj-2022-072415.

[51] Moraczewski J, Awosika AO, Aedma KK. Tricyclic antidepressants. Treasure Island, FL: StatPearls; 2024.

[52] Attal N, Bouhassira D. Pharmacotherapy of neuropathic pain: which drugs, which treatment algorithms? Pain 2015;156(Suppl. 1):S104–14. Available from: https://doi.org/10.1097/01.j.pain.0000460358.01998.15.

[53] Moore RA, Derry S, Aldington D, et al. Amitriptyline for neuropathic pain in adults. Cochrane Database Syst Rev 2015;2015(7):CD008242. Available from: https://doi.org/10.1002/14651858.CD008242.pub3.

[54] Mishra S, Bhatnagar S, Goyal GN, et al. A comparative efficacy of amitriptyline, gabapentin, and pregabalin in neuropathic cancer pain: a prospective randomized double-blind placebo-controlled study. Am J Hosp Palliat Care 2012;29(3):177–82. Available from: https://doi.org/10.1177/1049909111412539.

[55] World Health Organization. WHO guidelines for the pharmacological and radiotherapeutic management of cancer pain in adults and adolescents. WHO guidelines approved by the guidelines review committee. Geneva: WHO; 2018.

[56] Banerjee M, Pal S, Bhattacharya B, et al. A comparative study of efficacy and safety of gabapentin versus amitriptyline as coanalgesics in patients receiving opioid analgesics for neuropathic pain in malignancy. Indian J Pharmacol 2013;45(4):334–8. Available from: https://doi.org/10.4103/0253-7613.115000.

[57] Eija K, Tiina T, Pertti JN, et al. Amitriptyline effectively relieves neuropathic pain following treatment of breast cancer. Pain 1996;64(2):293–302. Available from: http://doi.org/10.1016/0304-3959(95)00138-7.

[58] Mercadante S, Arcuri E, Tirelli W, et al. Amitriptyline in neuropathic cancer pain in patients on morphine therapy: a randomized placebo-controlled, double-blind crossover study. Tumori 2002;88(3):239–42. Available from: https://doi.org/10.1177/030089160208800310.

[59] Kautio AL, Haanpää M, Leminen A, et al. Amitriptyline in the prevention of chemotherapy-induced neuropathic symptoms. Anticancer Res 2009;29(7):2601–6.

[60] Kautio AL, Haanpää M, Saarto T, et al. Amitriptyline in the treatment of chemotherapy-induced neuropathic symptoms. J Pain Symptom Manage 2008;35(1):31–9. Available from: https://doi.org/10.1016/j.jpainsymman.2007.02.043.

[61] Genevois AL, Ruel J, Penalba V, et al. Analgesic effects of topical amitriptyline in patients with chemotherapy-induced peripheral neuropathy: mechanistic insights from studies in mice. J Pain 2021;22(4):440–53. Available from: https://doi.org/10.1016/j.jpain.2020.11.002.

[62] Portenoy RK, Ahmed E, Keilson YY. UpToDate. In: Connor RF, Abrahm J, editors. Cancer pain management: role of adjuvant analgesics (coanalgesics). Waltham, MA: Wolters Kluwer; 2024.

[63] Hammack JE, Michalak JC, Loprinzi CL, et al. Phase III evaluation of nortriptyline for alleviation of symptoms of cis-platinum-induced peripheral neuropathy. Pain 2002;98(1–2):195–203. Available from: https://doi.org/10.1016/s0304-3959(02)00047-7.

[64] Nersesyan H, Slavin KV. Current aproach to cancer pain management: availability and implications of different treatment options. Ther Clin Risk Manag 2007;3(3):381–400.

[65] Hashemzadeh S, Mortazavi M, Abdi Dezfouli R. Quantitative analysis of nortriptyline's analgesic properties: a comparative systematic review and meta-analysis. BMJ Open 2024;14(8):e085438. Available from: https://doi.org/10.1136/bmjopen-2024-085438.

[66] Derry S, Wiffen PJ, Aldington D, et al. Nortriptyline for neuropathic pain in adults. Cochrane Database Syst Rev 2015;1(1):CD011209. Available from: https://doi.org/10.1002/14651858.CD011209.pub2.

[67] Hou S, Huh B, Kim HK, et al. Treatment of chemotherapy-induced peripheral neuropathy: systematic review and recommendations. Pain Physician 2018;21(6):571–92.

[68] Fallon M, Giusti R, Aielli F, et al. Management of cancer pain in adult patients: ESMO clinical practice guidelines. Ann Oncol 2018;29(Suppl. 4):iv166–91. Available from: https://doi.org/10.1093/annonc/mdy152.

[69] Bellingham GA, Peng PW. Duloxetine: a review of its pharmacology and use in chronic pain management. Reg Anesth Pain Med 2010;35(3):294–303. Available from: https://doi.org/10.1097/AAP.0b013e3181df2645.

[70] Sansone RA, Sansone LA. Pain, pain go away: antidepressants and pain management. Psychiatry (Edgmont) 2008;5(12):16–9.

[71] Nelson C. UpToDate. In: Connor RF, Roy-Byrne PP, editors. Serotonin-norepinephrine reuptake inhibitors: pharmacology, administration, and side effects. Waltham, MA: Wolters Kluwer; 2024.

[72] UpToDate. Amitriptyline. In: Lexidrug database. Wolters Kluwer; 2024a. [Retrieved September 12, 2024].

[73] UpToDate. Nortriptyline. In: Lexidrug database. Wolters Kluwer; 2024b. [Retrieved September 12, 2024].

[74] UpToDate. Duloxetine. In: Lexidrug database. Wolters Kluwer; 2024c. [Retrieved September 13, 2024].

[75] UpToDate. Venlafaxine. In: Lexidrug database. Wolters Kluwer; 2024d. [Retrieved September 13, 2024].
[76] Grassi L, Nanni MG, Rodin G, et al. The use of antidepressants in oncology: a review and practical tips for oncologists. Ann Oncol 2018;29(1):101–11. Available from: https://doi.org/10.1093/annonc/mdx526.
[77] Sansone RA, Sansone LA. Serotonin norepinephrine reuptake inhibitors: a pharmacological comparison. Innov Clin Neurosci 2014;11(3–4):37–42.
[78] Sheffler ZM, Patel P, Abdijadid S. Antidepressants. Treasure Island (FL): StatPearls; 2024.
[79] Asnis GM, Henderson MA. Levomilnacipran for the treatment of major depressive disorder: a review. Neuropsychiatr Dis Treat 2015;11:125–35. Available from: https://doi.org/10.2147/NDT.S54710.
[80] Debonnel G, Saint-Andre E, Hebert C, et al. Differential physiological effects of a low dose and high doses of venlafaxine in major depression. Int J Neuropsychopharmacol 2007;10(1):51–61. Available from: https://doi.org/10.1017/S1461145705006413.
[81] National Comprehensive Cancer Network. NCCN clinical practice guidelines in oncology: adult cancer pain (version 2.2024). Retrieved September 6, 2024, from: https://www.nccn.org/guidelines/guidelines-detail?category=3&id=14132024; 2024.
[82] Lunn MP, Hughes RA, Wiffen PJ. Duloxetine for treating painful neuropathy, chronic pain or fibromyalgia. Cochrane Database Syst Rev 2014;2014(1):CD007115. Available from: https://doi.org/10.1002/14651858.CD007115.pub3.
[83] Barohn RJ, Gajewski B, Pasnoor M, et al. Patient assisted intervention for neuropathy: comparison of treatment in real life situations (PAIN-CONTRoLS): Bayesian adaptive comparative effectiveness randomized trial. JAMA Neurol 2021;78(1):68–76. Available from: https://doi.org/10.1001/jamaneurol.2020.2590.
[84] Matsuoka H, Iwase S, Miyaji T, et al. Additive duloxetine for cancer-related neuropathic pain nonresponsive or intolerant to opioid-pregabalin therapy: a randomized controlled trial (JORTC-PAL08). J Pain Symptom Manage 2019;58(4):645–53. Available from: https://doi.org/10.1016/j.jpainsymman.2019.06.020.
[85] Curry ZA, Dang MC, Sima AP, et al. Combination therapy with methadone and duloxetine for cancer-related pain: a retrospective study. Ann Palliat Med 2021;10(3):2505–11. Available from: https://doi.org/10.21037/apm-20-1455.
[86] Gül SK, Tepetam H, Gül HL. Duloxetine and pregabalin in neuropathic pain of lung cancer patients. Brain Behav 2020;10(3):e01527. Available from: https://doi.org/10.1002/brb3.1527.
[87] Henry NL, Unger JM, Schott AF, et al. Randomized, multicenter, placebo-controlled clinical trial of duloxetine versus placebo for aromatase inhibitor-associated arthralgias in early-stage breast cancer: SWOG S1202. J Clin Oncol 2018;36(4):326–32. Available from: https://doi.org/10.1200/JCO.2017.74.6651.
[88] Henry NL, Unger JM, Till C, et al. Association between body mass index and response to duloxetine for aromatase inhibitor-associated musculoskeletal symptoms in SWOG S1202. Cancer 2019;125(12):2123–9. Available from: https://doi.org/10.1002/cncr.32024.
[89] D'Souza RS, Alvarez GAM, Dombovy-Johnson M, et al. Evidence-based treatment of pain in chemotherapy-induced peripheral neuropathy. Curr Pain Headache Rep 2023;27(5):99–116. Available from: https://doi.org/10.1007/s11916-023-01107-4.
[90] Loprinzi CL, Lacchetti C, Bleeker J, et al. Prevention and management of chemotherapy-induced peripheral neuropathy in survivors of adult cancers: ASCO guideline update. J Clin Oncol 2020;38(28):3325–48. Available from: https://doi.org/10.1200/JCO.20.01399.
[91] Smith EML, Pang H, Cirrincione C, et al. Effect of duloxetine on pain, function, and quality of life among patients with chemotherapy-induced painful peripheral neuropathy: a randomized clinical trial. JAMA 2013;309(13):1359–67. Available from: https://doi.org/10.1001/jama.2013.2813.
[92] Costa MFFD, Bilobran MA, de Oliveira LC. Correlation between cancer pain and quality of life in patients with advanced cancer admitted to a palliative care unit. Am J Hosp Palliat Care 2024;41:882–8. Available from: https://doi.org/10.1177/10499091231195318.
[93] Hirayama Y, Ishitani K, Sato Y, et al. Effect of duloxetine in Japanese patients with chemotherapy-induced peripheral neuropathy: a pilot randomized trial. Int J Clin Oncol 2015;20(5):866–71. Available from: https://doi.org/10.1007/s10147-015-0810-y.
[94] Farshchian N, Alavi A, Heydarheydari S, et al. Comparative study of the effects of venlafaxine and duloxetine on chemotherapy-induced peripheral neuropathy. Cancer Chemother Pharmacol 2018;82(5):787–93. Available from: https://doi.org/10.1007/s00280-018-3664-y.
[95] Salehifar E, Janbabaei G, Hendouei N, et al. Comparison of the efficacy and safety of pregabalin and duloxetine in taxane-induced sensory neuropathy: a randomized controlled trial. Clin Drug Investig 2020;40(3):249–57. Available from: https://doi.org/10.1007/s40261-019-00882-6.
[96] Rokhsareh S, Haghighi S, Tavakoli-Ardakani M. Evaluating the effects of duloxetine on prophylaxis of oxaliplatin-induced peripheral neuropathy in patients with gastrointestinal cancer: a randomized double-blind placebo controlled clinical trial. J Oncol Pharm Pract 2023;29(1):60–5. Available from: https://doi.org/10.1177/10781552211052646.
[97] Aghili M, Taherioun M, Jafari F, et al. Duloxetine to prevent neuropathy in breast cancer patients under paclitaxel chemotherapy (a double-blind randomized trial). Support Care Cancer 2024;32(8):493. Available from: https://doi.org/10.1007/s00520-024-08669-y.
[98] Huybrechts KF, Bateman BT, Pawar A, et al. Maternal and fetal outcomes following exposure to duloxetine in pregnancy: cohort study. BMJ 2020;368:m237. Available from: https://doi.org/10.1136/bmj.m237.
[99] Zerfas I, McGinn R, Smith MA. Pharmacologic management of cancer-related pain in pregnant patients. Drugs 2023;83(12):1067–76. Available from: https://doi.org/10.1007/s40265-023-01906-4.
[100] Larsen ER, Damkier P, Pedersen LH, et al. Use of psychotropic drugs during pregnancy and breast-feeding. Acta Psychiatr Scand Suppl 2015;445:1–28. Available from: https://doi.org/10.1111/acps.12479.
[101] Aiyer R, Barkin RL, Bhatia A. Treatment of neuropathic pain with venlafaxine: a systematic review. Pain Med 2017;18(10):1999–2012. Available from: https://doi.org/10.1093/pm/pnw261.

[102] Gallagher HC, Gallagher RM, Butler M, et al. Venlafaxine for neuropathic pain in adults. Cochrane Database Syst Rev 2015;2015(8):CD011091. Available from: https://doi.org/10.1002/14651858.CD011091.pub2.

[103] Kus T, Aktas G, Alpak G, et al. Efficacy of venlafaxine for the relief of taxane and oxaliplatin-induced acute neurotoxicity: a single-center retrospective case-control study. Support Care Cancer 2016;24(5):2085–91. Available from: https://doi.org/10.1007/s00520-015-3009-x.

[104] Durand JP, Deplanque G, Montheil V, et al. Efficacy of venlafaxine for the prevention and relief of oxaliplatin-induced acute neurotoxicity: results of EFFOX, a randomized, double-blind, placebo-controlled phase III trial. Ann Oncol 2012;23(1):200–5. Available from: https://doi.org/10.1093/annonc/mdr045.

[105] Bonilla-Jaime H, Sanchez-Salcedo JA, Estevez-Cabrera MM, et al. Depression and pain: use of antidepressants. Curr Neuropharmacol 2022;20(2):384–402. Available from: https://doi.org/10.2174/1570159X19666210609161447.

[106] Xiao Y, Liu J, Huang XE, et al. Clinical study on fluvoxamine combined with oxycodone prolonged-release tablets in treating patients with moderate to severe cancer pain. Asian Pac J Cancer Prev 2014;15(23):10445–9. Available from: https://doi.org/10.7314/apjcp.2014.15.23.10445.

[107] Hassanein EHM, Althagafy HS, Baraka MA, et al. Pharmacological update of mirtazapine: a narrative literature review. Naunyn Schmiedebergs Arch Pharmacol 2024;397(5):2603–19. Available from: https://doi.org/10.1007/s00210-023-02818-6.

[108] Economos G, Lovell N, Johnston A, et al. What is the evidence for mirtazapine in treating cancer-related symptomatology? A systematic review. Support Care Cancer 2020;28(4):1597–606. Available from: https://doi.org/10.1007/s00520-019-05229-7.

[109] Paice JA, Bohlke K, Barton D, et al. Use of opioids for adults with pain from cancer or cancer treatment: ASCO guideline. J Clin Oncol 2023;41(4):914–30. Available from: https://doi.org/10.1200/JCO.22.02198.

[110] Vardy J, Agar M. Nonopioid drugs in the treatment of cancer pain. J Clin Oncol 2014;32(16):1677–90. Available from: https://doi.org/10.1200/JCO.2013.52.8356.

[111] van den Beuken-van Everdingen MHJ, de Graeff A, Jongen JLM, et al. Pharmacological treatment of pain in cancer patients: the role of adjuvant analgesics, a systematic review. Pain Pract 2017;17(3):409–19. Available from: https://doi.org/10.1111/papr.12459.

[112] Jordan B, Margulies A, Cardoso F, et al. Systemic anticancer therapy-induced peripheral and central neurotoxicity: ESMO-EONS-EANO clinical practice guidelines for diagnosis, prevention, treatment and follow-up. Ann Oncol 2020;31(10):1306–19. Available from: https://doi.org/10.1016/j.annonc.2020.07.003.

[113] Greco MT, Roberto A, Corli O, et al. Quality of cancer pain management: an update of a systematic review of undertreatment of patients with cancer. J Clin Oncol 2014;32(36):4149–54. Available from: https://doi.org/10.1200/JCO.2014.56.0383.

Chapter 9

Cannabinoids, psychostimulants, and N-methyl-D-aspartate antagonists

Nikola Jojić
Faculty of Pharmacy, University Business Academy, Novi Sad, Serbia

Key learning objectives

- Understand the pharmacological mechanisms of cannabis compounds, including the interaction between phytocannabinoids (Δ-9-tetrahydrocannabinol and cannabidiol) and the endocannabinoid system, as well as their effects on various physiological processes in palliative care patients.
- Identify appropriate patient selection criteria for cannabinoid therapy, including contraindications.
- Recognize clinically significant drug–drug interactions between cannabinoids and commonly prescribed medications.
- Evaluate the current evidence base for cannabinoids in treating cancer-related symptoms.
- Evaluate the effect of psychostimulants and N-methyl-D-aspartate antagonists in palliative care patients.

Introduction

The rise in the use of the cannabis plant for medical purposes is increasing worldwide, and patients often seek advice from healthcare professionals regarding the efficacy and safety of a cannabis-based treatment. In spite of an increase in research on medicinal cannabis for a wide range of indications, there were few scientific data supporting its efficacy. Patients' access to and usage of cannabis and/or other cannabinoids has surpassed the body of research proving their therapeutic value [1]. Cancer, chronic cancer pain, chemotherapy-induced nausea and vomiting (CINV), and cachexia-anorexia cancer syndrome (CACS) have been the most frequently investigated indications for medical cannabis use in palliative care settings.

Botany and phytochemistry

Cannabis sativa L. (the cannabis plant) is a flowering herb of the Cannabaceae family, and since time immemorial, the cannabis plant has been used for medical and mind-altering purposes, but also for its fiber and oil (vegetable oil and oilseed products) [2]. *C. sativa* is a source of more than 500 compounds, of which at least 150 are known to be cannabinoids called phytocannabinoids. The most abundant phytocannabinoids present in the female plants of *C. sativa* are Δ-9-tetrahydrocannabinol (THC), with psychoactive activity, and cannabidiol (CBD), which is nonpsychoactive. Apart from THC and CBD, several other cannabinoids were isolated from *C. sativa*. In addition to phytocannabinoids, cannabis plant also contains flavonoids and terpenes [3].

With the exception of THC and CBD, two major pharmacologically relevant compounds of cannabis, which have been extensively studied for their potential therapeutic applications, most of these other compounds have been understudied. The chemical diversity of cannabinoids and the notion that these substances may cooperate through inhibitory and synergistic interactions coined the term "entourage effect."

Cannabis sativa: difference between hemp and marijuana

Classification of cannabis plants is based on their primary purpose (textile, food, or medicinal/recreational purposes) and chemical phenotypes (levels of plant cannabinoids, mainly THC). Hemp (industrial *C. sativa*—fiber type) contains only

Pain Management in Palliative Care for Patients With Cancer. DOI: https://doi.org/10.1016/B978-0-443-30050-9.00051-5

low levels of THC and high levels of CBD, unlike marijuana (medical *C. sativa*—marijuana/drug type) which contains both THC and CBD [4].

The threshold for differentiating between hemp and marijuana *C. sativa* is based on the concentration of THC. In North America, plants that accumulate more than 0.3% THC are classified as marijuana/drug-type *C. sativa*, while those that accumulate less than this amount are classified as hemp. The threshold in the European Union (EU) was harmonized in recent years, so the maximum level for THC has been increased from 0.2% to 0.3%. Compared to drug-type *C. sativa*, industrial hemp grown today is primarily CBD dominant and has a lower overall cannabinoid concentration, although hemp can also be used for pharmaceutical purposes.

Cannabinoids and cannabinoid receptors

Cannabinoids are compounds that interact with the endocannabinoid system. Though our understanding of the endocannabinoid system is still incomplete, the main components of the endocannabinoid system include endocannabinoids, enzymes/transporters responsible for the synthesis and degradation of endocannabinoids, and endocannabinoid receptors.

There are three different types of cannabinoids: endogenous cannabinoids (also known as endocannabinoids), such as anandamide and 2-arachidonoylglycerol, which are lipid-based molecules produced in many organs and cell types; phytocannabinoids (plant-based), such as THC and CBD; and many other synthetic cannabinoids, such as dronabinol and nabilone [2].

The primary targets of the cannabinoids are cannabinoid receptors 1 (CB1) and 2 (CB2), which are G-protein–coupled receptors. As a component of the endocannabinoid system, cannabinoid receptors are essential in many physiological and pathological processes. CB1 receptors are widely distributed throughout the central nervous system (CNS) and are primarily expressed by the axons and presynaptic termini of neurons in the amygdala, hippocampus, cortex, basal ganglia outflow tracts, and cerebellum. Both gamma-aminobutyric acid and glutamate release are inhibited by activation of CB1 [5]. The immunomodulatory effects of cannabinoids were partially explained by the discovery of CB2 receptors, which are predominantly expressed on immune system cells, astrocytes, and microglia in the CNS [6–8].

The pharmacological mechanism of the effects of the phytocannabinoids is still not fully explained. It is very likely that, contrary to popular belief, the effects of phytocannabinoids are mediated by a variety of targets, not just the cannabinoid receptors, specifically CB1 and CB2. The antiemetic and increased appetite effects are at least partially modulated by antagonistic activity on 5-HT3A receptors and by potential additional effects on the serotonergic system [9].

Moreover, the receptor profile has been studied mainly for THC and CBD, and we know that THC acts as a partial agonist at CB1. It is worth mentioning that a ligand being partial agonist also has a mixed agonist–antagonist activity, depending on the presence of full agonists. On the other hand, CBD has low affinity for CB1/CB2; at CB1 it functions primarily as a negative allosteric modulator and at CB2 it behaves as an antagonist/inverse agonist depending on context [10].

Cannabis safety considerations

Compared to many other medicines, safety data from clinical trials for medical cannabis are less clear. Before starting cannabis-based treatment, the overall risk versus benefit of cannabis use should be assessed for each patient individually. Clinicians who recommend cannabis need to take into account potential contraindications, precautions, and drug–drug interactions (DDIs) prior to initiating cannabis use.

Contraindications

Unstable cardiovascular disease: Due to endocannabinoid receptors being ubiquitous throughout the cardiovascular system, hemodynamic effects of THC include blood pressure and heart rate changes due to THC's sympathomimetic properties. In the presence of health conditions such as acute congestive heart failure, critical aortic stenosis, poorly controlled atrial fibrillation, and coronary artery disease, concomitant use of cannabis should be avoided. The most common way to use cannabis is to smoke, which further increases the risk of cardiovascular disease [11,12].

Respiratory disease: Smoking marijuana exposes a user to a variety of other substances. Chemicals including acetaldehyde, ammonia, benzene, carbon monoxide, hydrogen cyanide, and polycyclic hydrocarbons are inhaled during the combustion of cannabis [13–16]. These substances may also further exacerbate respiratory symptoms (e.g., cough, sputum production, wheezing, and chest tightness) [11,12].

Psychosis and bipolar disorder: Longer exposure to cannabis is linked to a dose-dependent increase in the risk of psychotic disorder, so people with bipolar disorder and/or current psychosis may experience worsening symptoms. Although the precise neurobiological mechanisms underlying these mental health effects are not fully understood and likely involve multiple contributing factors, the use of THC, the primary psychoactive component of cannabis, is regarded as a major risk factor [12,17].

Pregnancy and breastfeeding: Phytocannabinoids cross the placenta and pass into breast milk, resulting in fetal and neonatal exposure. Though evidence is limited, there is possibly an association between cannabis use and adverse perinatal outcomes, especially with heavy use; therefore, women should be advised to refrain from using cannabis during pregnancy and lactation [12,18].

Patients treated with calcineurin and checkpoint inhibitors: The main concern regarding the use of cannabinoids in immunocompromised patients is the potential for DDIs, which can result in an increased risk of toxicity or decreased effectiveness. Concomitant use of calcineurin inhibitors (e.g., tacrolimus) and CBD may increase calcineurin inhibitor toxicity due to increased serum concentrations of calcineurin inhibitors [19]. Decreased effectiveness of programmed cell death protein 1 inhibitors when used concomitantly with CBD is possible due to the immunosuppressive mechanism of CBD. The potential interaction between cannabis use and immune checkpoint inhibitors has yet to be studied, so caution is warranted [20]. The lack of controlled studies restricts the conclusiveness of this finding.

Elderly patients: As aging is associated with several biological changes in human organs, the possible use of cannabinoids may result in altered medication pharmacokinetics and pharmacodynamics. Additionally, in this population, there is an increased risk of DDIs, so more frequent monitoring is usually recommended.

Drug–drug interactions with cannabidiol

Information related to the clinical relevance of interaction between CBD and other drugs is limited. CBD has potential for pharmacokinetic and pharmacodynamic DDIs [21,22].

The potential for DDIs with commonly used medications is high because CBD has effects on biological targets that are implicated in drug metabolism (e.g., CYP3A4/2C19) and excretion (e.g., P-glycoprotein). CBD competitively inhibits the enzymes CYP3A4, CYP2B6, CYP2C9, CYP2C19, CYP2D6, and CYP2E1, so CBD–drug interactions mainly occur in cases where CBD alters the metabolic activity of some isoenzymes of the cytochrome P450 (CYP) family [23].

In literature, there is evidence for numerous CBD–drug interactions, although clinical relevance for most of those interactions remains unclear. Still, several clinically relevant interactions have been reported. For instance, CBD–warfarin, CBD–buprenorphine, and CBD–clopidogrel interactions have all been classified as clinically relevant. Concomitant use of CBD and warfarin could lead to increased international normalized ratio (INR) due to inhibition of the metabolic enzymes of warfarin. The effectiveness and safety of warfarin treatment rely on keeping the INR within the target range. For most indications, dosing should be adjusted to achieve and maintain a target INR of 2.5, with an acceptable range of 2.0–3.0. In case of concomitant use of warfarin and CBD, clinicians should consider the possibility of additional INR elevations or adverse bleeding and adjust the dose accordingly to the INR values [24,25]. Therefore, when used concomitantly, weekly INR checks for the first 4 weeks are a reasonable approach.

Concomitant use of CBD and buprenorphine may increase the plasma levels of buprenorphine due to the ability of the CBD to inhibit CYP3A4, an enzyme responsible for buprenorphine metabolism. If the simultaneous administration cannot be avoided, buprenorphine dosage adjustments and plasma level monitoring of buprenorphine should be considered [26].

Clopidogrel, an antiplatelet prodrug, requires CYP enzymes, including CYP2C19, CYP3A, CYP2B6, CYP1A2, and CYP2C9, for conversion to a pharmacologically active metabolite. CBD, being a mild inhibitor of several P450 enzymes, may compromise clopidogrel's antiplatelet activity, increasing the risk of cardiac adverse effects [26]. Based on limited pharmacodynamic and pharmacokinetic studies, it is essential to closely monitor laboratory assessments of platelet response and aggregation in patients using clopidogrel and possibly avoid using CBD concomitantly with the P2Y12 inhibitor clopidogrel. Prasugrel, being the alternative, may be considered.

Drug–drug interactions with Δ-9-tetrahydrocannabinol

Generally, THC is less linked to DDIs than CBD, but at clinically meaningful concentrations, it is still likely to increase this risk.

Broad inhibitory effects of THC on CYP enzymes, such as CYP3A, CYP2D6, CYP2C9, CYP2C19, CYP2A6, CYP2B6, CYP1A1/2, and CYP2J2, have been reported. Consequently, concomitant use of THC with medications metabolized by enzymes that THC inhibits could result in increased medication bioavailability and increased risk of adverse drug reactions. Furthermore, recent data indicates that THC and other cannabinoids have potent inhibitory effects on carboxylesterase 1, a key enzyme in the metabolism of numerous drugs [27].

Cannabis pharmacogenomics

The research in the field of pharmacogenomics has made meaningful progress and can assist in improving the outcomes of cannabinoid therapy, but the widespread integration remains a work in progress. Screening for polymorphisms in CYP2C9 and CYP3A4 should be considered for patients using cannabinoids alongside other medications, particularly when managing conditions with narrow therapeutic windows or when polypharmacy is involved. CYP2C9 plays a significant role in metabolizing cannabinoids like THC and CBD. Variants such as CYP2C9*2 and CYP2C9*3 result in reduced enzyme activity and increased risk of sedation and adverse effects [28]. Considering pharmacogenomic testing for CYP2C9 and CYP3A4, polymorphisms can aid in personalizing cannabinoid therapy, optimizing drug efficacy, and minimizing adverse effects, especially in patients on multiple medications. For instance, in patients categorized as poor CYP2C9 metabolizers, the starting dose of THC could be 1.25 mg/day, with the "start low and go slow" approach being critical.

Cannabis in palliative care

There is a lot of information available online, much of it inconsistent with scientific research, regarding the benefits of cannabinoid use. Such false information has the potential to create irrational expectations for cannabis as a natural treatment devoid of side effects. Pain, cachexia, loss of appetite, cancer, and weight loss, as well as CINV, have been the most frequently investigated indications for medical cannabis use in palliative care settings.

Medical cannabis as an anticancer agent

Numerous studies have investigated the effectiveness of cannabis use in oncology. Although preclinical data suggest cannabinoids may be beneficial as anticancer agents, real clinical data on cannabinoids' antitumorous properties are still lacking [29,30]. Moreover, when used as a cancer-directed treatment, outside of evidence-based indications, cannabinoids may cause several side effects with no clinical benefit [31].

Another worrying possibility is that adults with cancer receiving immunotherapy may have worse clinical outcomes, including overall and progression-free survival, due to the use of cannabinoids. While the precise nature of the relationship between cannabinoids and anticancer immunotherapy remains unclear, clinicians should exercise caution when recommending cannabis-based treatments to adults undergoing immunotherapy, especially among adults with cancer receiving a checkpoint inhibitor [1,20,32,33].

Medicinal cannabis for the treatment of chronic cancer pain

A number of cannabis-based products have been proposed as pain relievers, and cannabinoids could be an option for patients suffering from neuropathic, inflammatory, nociplastic, and mixed pain, with the most compelling evidence supporting their use in the management of neuropathic pain. The Canada Drug and Health Technology Agency (CADTH) reports that there is evidence indicating possible advantages of cannabis-derived medications in managing neuropathic pain [34]. However, the effect sizes are of questionable importance for the abovementioned types of pain.

The WHO advises using morphine-like medications for moderate-to-severe cancer pain. However, in spite of these recommendations, morphine-like medications may inefficiently relieve the pain [35]. Additionally, the opioid treatment increases the risk of addiction and causes side effects, especially at the higher doses required to relieve cancer pain.

Strong preclinical animal data and a significant body of observational evidence were found to be in favor of cannabis' potential utility in the treatment of cancer pain. Clinical trial data, however, only sporadically support clinical efficacy. A limited number of randomized controlled clinical trials have investigated the use of cannabinoids as a therapeutic option, yielding inconsistent results.

Comprehensive systematic reviews with metaanalyses have found little evidence supporting the effectiveness of cannabinoids as an additional treatment or pain reliever for cancer-related pain. Studies suggest that cannabinoids reduced chronic pain when compared with placebo; however, whether this effect is clinically important seems questionable [36]. Overall, there is not enough data to make a recommendation for or against cannabis and/or cannabinoids in the management of chronic cancer pain.

A 2021 systematic review evaluated noninhaled medical cannabis or cannabinoids for chronic cancer or noncancer pain. In all the randomized controlled trials (RCTs) included in this systematic review, cannabinoids were administered primarily to address pain, with the main outcome being the change in pain score or pain intensity. A study found no statistically significant effect on cancer pain in 4 RCTs of nabiximols (cannabinoid-based medication consisting of a 1:1 THC:CBD), but a small, statistically significant improvement in noncancer pain in 23 RCTs. This systematic review concluded that cannabinoids do not have a role in cancer-related pain.

This aligns with the 2023 Multinational Association of Supportive Care in Cancer (MASCC) guideline, which does not recommend cannabinoids for cancer pain outside of a clinical trial [33,37]. Furthermore, the International Association for the Study of Pain (IASP) has recognized significant research deficiencies. In light of insufficient high-quality clinical evidence, IASP does not presently endorse the use of cannabis and cannabinoids for pain management, while acknowledging the pressing need for preclinical and clinical studies to fill the research gap [38].

Due to the limited number and generally inadequate quality of studies available to assist clinicians, it is advisable to encourage patients to participate in clinical trials involving medicinal cannabis for palliative care whenever possible. Outside the clinical trial scenario, when the patient insists on cannabinoid-based treatment, it is highly recommended that physicians discuss with the patients the evidence supporting medicinal cannabis use and the potential for adverse effects. In the absence of evidence supporting the first-line or exclusive use of cannabinoid products in palliative care, it is recommended that if the treating physician decides to commence medicinal cannabis therapy for palliative care patients, it should be considered only after standard palliative treatments have been evaluated and implemented. For instance, cannabinoids may become part of an analgesic treatment in adult patients with advanced cancer and persistent pain who no longer respond to the highest tolerated dose of morphine-like drugs [39].

A systematic review and metaanalysis indicated that selective cannabinoids, such as dronabinol, nabilone, and nabiximols, offer a modest reduction in pain scores compared to placebo for neuropathic pain. However, the clinical significance of this reduction remains uncertain, and the use remains controversial [40].

In terms of medical cannabis use in patients with neuropathic pain, there is moderate-quality evidence supported by multiple systematic reviews and metaanalyses that suggests modest benefit. On the other hand, evidence for medical cannabis use for pure nociceptive pain is low/insufficient, showing no consistent benefit over placebo. Therefore, medical cannabis should be considered only in mixed pain or refractory cases [41]. In a clinical context, without robust evidence indicating therapeutic benefits of medical cannabis, patient selection criteria often rely on PainDETECT scores ≥19 or Numeric Rating Scale (NRS) scores ≥7 to identify candidates for cannabinoid treatment in neuropathic pain.

There is a limited number of RCT examining recommendations on how to dose and administer medical cannabis in patients with chronic pain. In order to fulfill the existing gap in clinical guidance regarding the safe and effective application of medical cannabis for chronic pain, the modified Delphi process has been widely utilized in healthcare environments to generate consensus-driven recommendations, although there is a lack of data specifically for palliative oncology.

The Delphi process resulted in a consensus that a patient may commence treatment with a CBD at a dosage of 5 mg twice daily (bid) and increase the dosage by 10 mg per day (5 mg CBD bid) every 2–3 days, up to a maximum of 40 mg of CBD per day. Should the 40 mg/day CBD dosage fail to achieve the desired treatment outcomes, healthcare providers may consider the introduction of 2.5 mg of THC per day, with a titration of 2.5 mg THC every 2–7 days, up to a limit of 40 mg/day, while continuing the same CBD dosage. It is advisable to seek expert guidance if considering an increase beyond 40 mg/day of THC [42].

Future RCT examining how to dose and administer medical cannabis safely and effectively are needed. It is crucial to recognize that each patient is unique, and the treatment with medical cannabis, similar to many other therapies, must be tailored to the individual patient.

Integration protocol

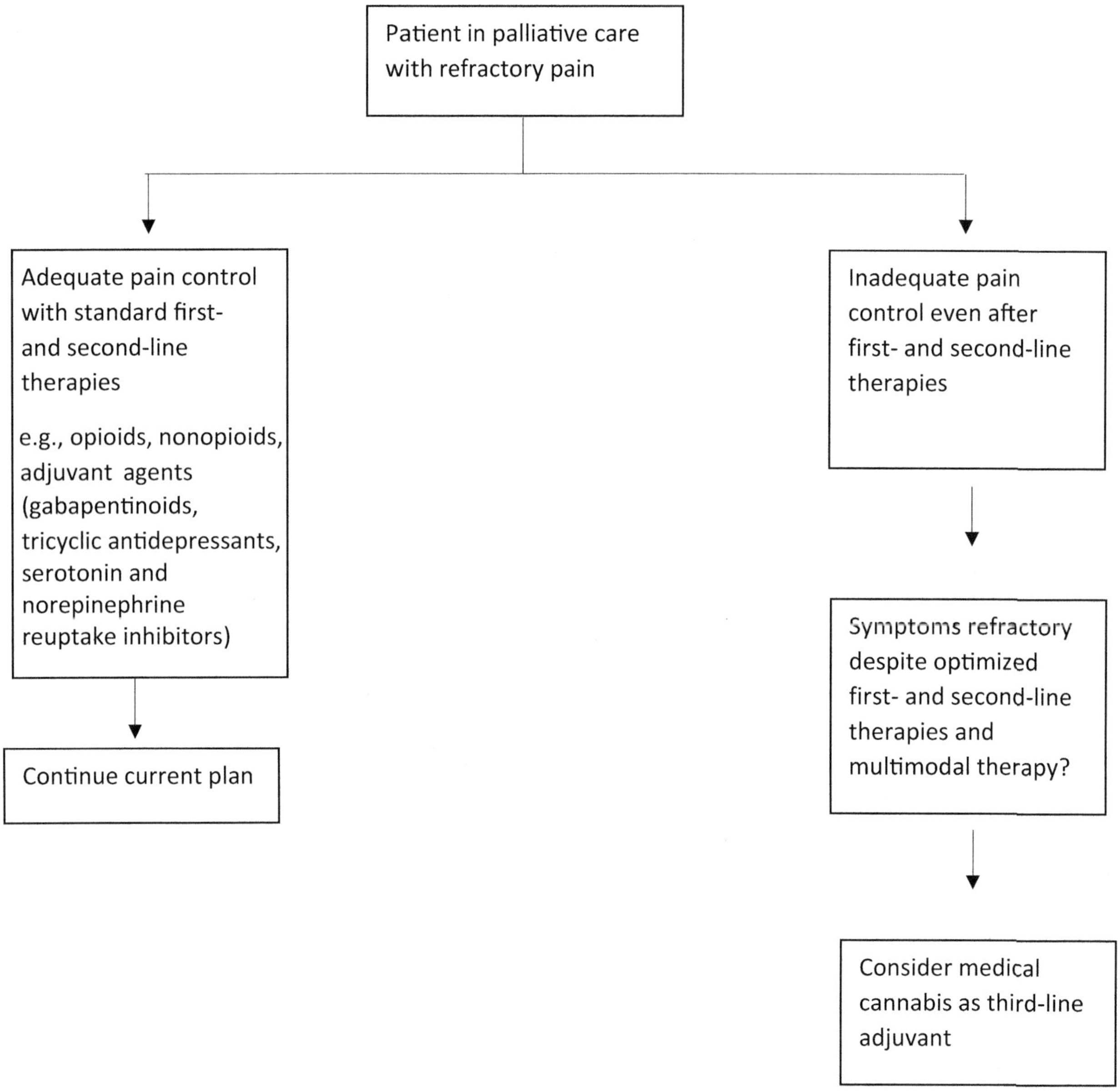

Medicinal cannabis as antiemetic

CINV is a debilitating side effect of cancer treatment. In spite of receiving antiemetic prophylaxis, 40% of patients experience refractory nausea or vomiting [43]. Interplay between the CB1 and CB2 receptors and the cannabinoids THC and CBD has been investigated as a new approach for alleviating these symptoms. According to the American Society of Clinical Oncology (ASCO) guideline, nausea and vomiting that do not improve with conventional antiemetic medications may be treated with the US Food and Drug Administration (FDA)-approved cannabinoids dronabinol and nabilone but not for first-line treatment [1]. According to the MASCC guideline, adults with cancer who are not taking checkpoint inhibitors may consider cannabinoids for refractory CINV [44].

Medical cannabis in the treatment of cachexia-anorexia cancer syndrome

More than half of patients with advanced cancer have CACS [45]. In theory, cannabinoids could be a potential treatment for CACS. A 2022 systematic review and metaanalysis of four RCTs evaluated the effects of cannabis and cannabinoids on cachexia outcomes, finding no statistically significant effect on appetite or weight and a small detriment in quality of life compared to active comparators or placebo [46,47].

Medicinal cannabis—legislative

It is not easy to discern a common approach to the legislation surrounding cannabis use for medicinal purposes. As a result of the growing interest in medicinal cannabis, many countries have introduced specific legislation, regulations, and guidelines regarding the use of medical cannabis.

In the United States, at the federal level, cannabis remains classified as a Schedule I substance under the Controlled Substances Act. In terms of the use of medicinal cannabis, as of February 2024, 47 states, the District of Columbia, and 3 territories (Guam, Puerto Rico, and the US Virgin Islands) allow for the use of cannabis for medical purposes [48].

Similarly to the United States, there is no harmonized EU law on medicinal cannabis use. Some EU countries do not recognize the medicinal potential of cannabis and, as such, actively oppose its use. On the other hand, medicalization as a clinical policy has been embraced in some EU countries. However, there is no harmonized EU law on medicinal cannabis; each member state regulates access nationally.

Use of psychostimulants in palliative care

According to the European Society for Medical Oncology (ESMO), cancer-related fatigue (CRF) is defined as a distressing, persistent, subjective sense of physical, emotional, and/or cognitive tiredness or exhaustion related to cancer or cancer treatment that is not proportional to recent physical activity and that interferes with usual functioning. More than two-thirds of patients experience fatigue during the cancer trajectory from diagnosis to the end of life. Although the exact cause of CRF is unknown, it may involve a number of physiological and biochemical contributing factors: tumor-related factors and complications, comorbid conditions, iatrogenic factors, physical symptoms, medications, and behavioral factors [49,50].

Although the specific therapeutic algorithm is not yet well defined, the primary goal of treatment of CRF should be addressing the underlying causes of fatigue, such as hypothyroidism and electrolyte imbalances if present, before proceeding to general and pharmacological modalities. General nonpharmacological strategies like physical activity, energy conservation, and cognitive behavioral therapy have the strongest evidence for treatment of fatigue, and the National Comprehensive Cancer Network (NCCN) recommends physical activity, counseling, and nutritional support to manage CRF after treatment and in end-of-life care [49,51,52].

In addition to general treatment modalities, CRF can be treated pharmacologically with psychostimulants: methylphenidate, dexmethylphenidate, dexamphetamine, modafinil, and armodafinil. Dexamphetamine, being a sympathomimetic drug, alleviates CRF by acting as an agonist on adrenergic receptors, while others act through mechanisms involving the reuptake transporter and/or the vesicular transporters to cause the release of dopamine and norepinephrine from presynaptic terminals. Theoretically, advantages of psychostimulant use could include increased energy, enhancing the analgesic effect of opioids, preventing opioid-induced sedation, boosting appetite, and enhancing cognitive functions.

Although it may be appropriate to consider these agents in the treatment of CRF, especially in patients with a Fatigue Severity Scale ≥4, there is limited evidence to support the use of pharmacologic interventions. Most studies evaluating the benefits of psychostimulants on CRF showed conflicting or weak evidence [51]. Methylphenidate has the strongest body of evidence, including three RCTs in which a superior effect over placebo has been noticed, although there are RCTs in which methylphenidate was no more effective than placebo at relieving fatigue [51,53]. Despite conflicting evidence, when used in low doses, as is common practice in palliative care settings, methylphenidate is well tolerated and has no or only mild side effects, making it the first choice in CRF. Being proposed as a first-line drug in the treatment of CRF, the recommended dosing regimen for methylphenidate starts at 5 mg orally (PO) bid and increases, based on perceived efficacy and side effects, to a maximum of 20 mg/day.

One of the most important benefits of these medicines is their action generating immediate behavioral effects. With a rapid onset of action, improvement in fatigue is frequently noticeable even after the initial dose. Because of this, the effect might be assessed within a few days and continued only in patients who experience a beneficial effect.

Conversely, little information about the short- and long-term side effects of psychostimulants is available. Despite the favorable safety profile of these drugs, agents with amphetamine-like actions, notably dexamphetamine, are linked to an infrequent increased risk of cardiac adverse effects and temperature control dysregulation [54]. For most psychostimulant drugs used in CRF, the monitoring protocol includes blood pressure and heart rate, as well as mood changes.

To conclude, the evidence supporting the use of psychostimulant drugs in CRF is still weak, and there are several reasons for that. For instance, it is challenging to determine the influence of coexisting factors influencing the effect on CRF, such as depression and emotional status. Furthermore, in a few studies, the placebo has been shown to be equally effective, highlighting the fact that the placebo itself may have beneficial effects in patients with CRF, suggesting further research is needed regarding biological mechanisms affecting fatigue [55]. Furthermore, interpretation of results from trials evaluating the effects of psychostimulants is difficult because of the clinical heterogeneity, differences in disease severity, life expectancy, symptom load, and so on.

Although the strength of evidence remains low to very low for most agents, psychostimulants, particularly methylphenidate, might be the drug of choice for pharmacological treatment of fatigue in palliative cancer care [51,52].

N-methyl-D-aspartate antagonists in palliative care

In cancer pain, morphine and its derivatives are considered first-line opioids. However, 20%–40% of cancer patients do not respond well to conventional therapy including morphine and coanalgesics, due to the disease's complex pathophysiology, which has been described as refractory cancer pain [56,57]. N-methyl-D-aspartate (NMDA) receptors play a role in the development of opioid tolerance, opioid-resistant neuropathic pain, hyperalgesia, and allodynia. Furthermore, the NMDA receptor is at least partially responsible for the development of central sensitization, which is linked to reduced opioid sensitivity [58].

The development of central sensitization of dorsal horn neurons as a result of continuous nociceptive input from inflammation or injury further aggravates hyperalgesia, allodynia, and relative nonresponsiveness to opioids in chronic pain [59]. Therefore, the idea behind the use of NMDA antagonists is the assumption that pain can be better relieved if the NMDA receptors are blocked. In palliative care, NMDA receptor antagonists most commonly used as an add-on to ongoing opioid treatment are ketamine and methadone, which are perhaps best known for their use in anesthesia and heroin withdrawal, respectively.

NMDA antagonists should be considered as an adjuvant to opioids for the treatment of refractory cancer pain when opioids, alone or in combination with appropriate adjuvant analgesics, prove to be ineffective. However, there is not enough data available to evaluate the advantages and disadvantages of using NMDA antagonists as an adjuvant to opioids to treat refractory cancer pain.

Ketamine is a noncompetitive NMDA receptor antagonist, and by reducing the excitatory pathways mediated by the NMDA receptor, ketamine produces somatic analgesia. Ketamine also has an immediate and long-term impact on neuropathic pain. Due to blockage of NMDA-mediated pathways, ketamine can alleviate neuropathic pain and its associated symptoms, including allodynia and hyperalgesia [60].

A substantial body of evidence demonstrates the efficacy of ketamine in managing severe pain, although its use in pain management is still off-label. In pain management, subdissociative dosing, commonly referred to as low-dose ketamine, can be utilized either independently or in conjunction with other analgesics. It is both safe and effective when combined with injectable nonsteroidal antiinflammatory drugs and opioids. Most commonly used for acute pain, the guidelines recommend starting at the lower end and titrating as needed: intermittent bolus doses of 0.1–0.35 mg/kg or continuous infusion of 0.1–0.5 mg/kg/h (max 1 mg/kg/h infusion).

There is insufficient evidence supporting the use of ketamine for chronic pain management, and it is advisable to consult a pain specialist prior to starting treatment. The usefulness of ketamine in the treatment of cancer pain remains up for debate despite the fact that numerous studies on the drug have been carried out.

In addition to acting as an analgesic through μ-receptors, methadone is a serotonin and norepinephrine reuptake inhibitor, as well as an NMDA receptor antagonist. There is evidence that the modulation of neuropathic pain is caused by antagonistic interactions at the NMDA receptor [61].

Having multiple complementary mechanisms, methadone might be helpful in the treatment of refractory cancer pain. However, methadone use is associated with several possible disadvantages, such as a lengthy half-life, individual metabolic rate variation due to significant genetic polymorphism of the CYP3A4 and 2D6, the enzymes that metabolize methadone, possible QT interval prolongation, and risk of torsades de pointes and sudden death. Clinicians should be aware of methadone's possible cardiotoxicities while prescribing medication, and a baseline ECG should be obtained before starting methadone, with follow-up ECGs at 30 days and annually thereafter for stable doses, as recommended by guidelines [62].

Furthermore, a number of the DDIs may occur. Drugs that inhibit or induce CYP enzymes have an impact on methadone concentration and therapeutic effect. Methadone is usually considered as a second-line option for the treatment of cancer pain, and the dosing regimen for methadone is 2.5 mg PO bid, titrated up to 30 mg/day.

Integration protocol

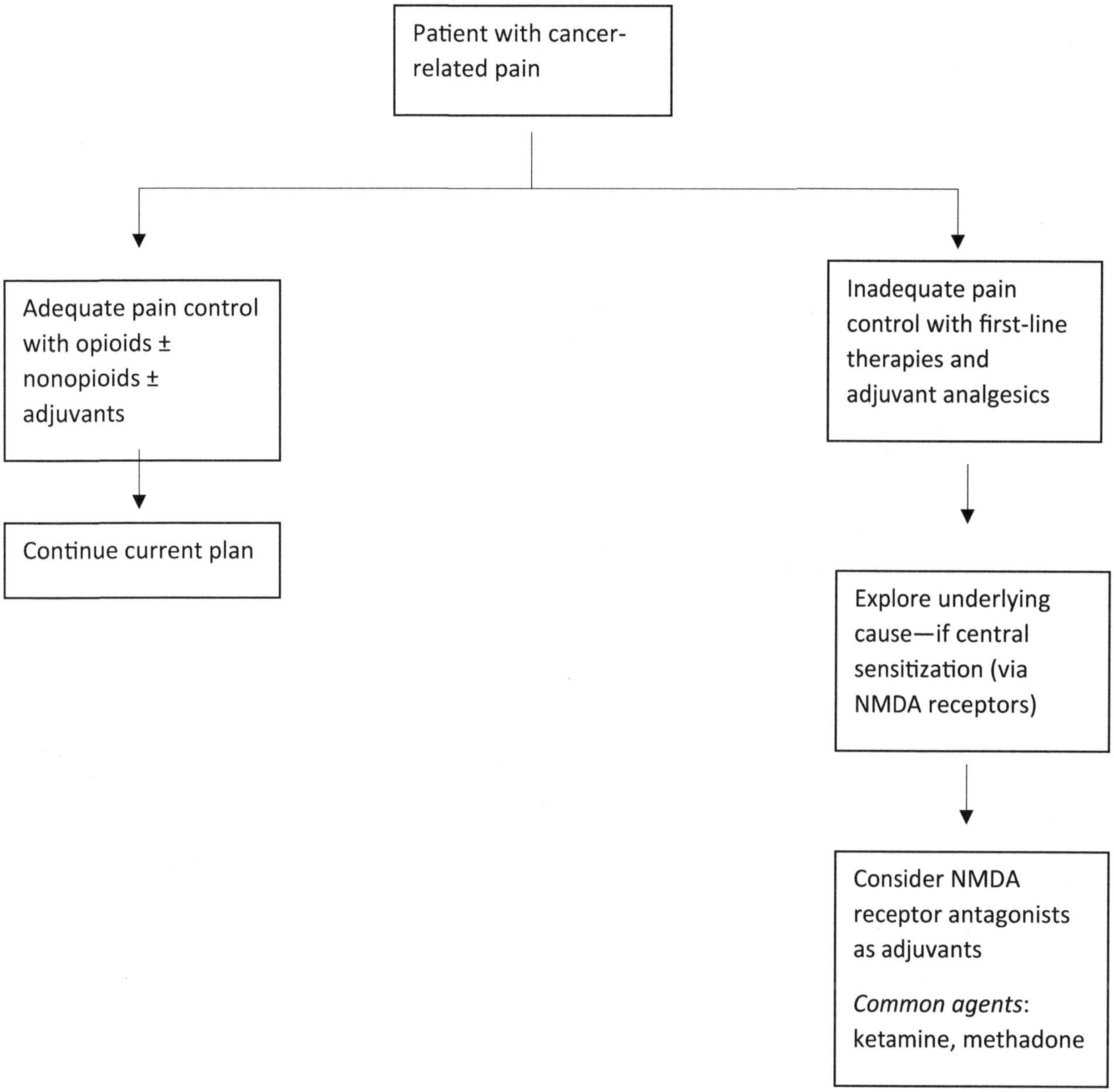

Conclusions

Chronic cancer pain significantly impairs patients' quality of life. Despite pharmacological efforts, with opioids being the primary treatment for chronic cancer pain, many patients do not achieve adequate pain relief. Additionally, opioid-based treatments can cause side effects. Consequently, the medical use of cannabis and cannabinoids has been proposed as an alternative therapeutic option. However, for any therapeutic option to be adopted in clinical practice, its benefits must outweigh its risks. Conflicting results from clinical trials have not provided sufficient evidence to support the use of cannabinoids as therapeutic agents for cancer pain management. Therefore, further high-quality clinical trials are necessary to establish their safety and efficacy.

References

[1] Braun IM, Bohlke K, Abrams DI, et al. Cannabis and cannabinoids in adults with cancer: ASCO guideline. J Clin Oncol 2024;42(13):1575–93. Available from: https://doi.org/10.1200/JCO.23.02596.

[2] Mangal N, Erridge S, Habib N, et al. Cannabinoids in the landscape of cancer. J Cancer Res Clin Oncol 2021;147:2507–34. Available from: https://doi.org/10.1007/s00432-021-03710-7.

[3] Khalsa JH, Bunt G, Blum K, et al. Cannabinoids as medicinals. Curr Addict Rep 2022;9(4):630–46. Available from: https://doi.org/10.1007/s40429-022-00438-3.

[4] Malabadi RB, Kolkar KP, Chalannavar RK. *Cannabis sativa*: ethnobotany and phytochemistry. Int J Innov Sci Res Rev 2023;5(2):3990–8.

[5] Baker D, Pryce G, Giovannoni G, et al. The therapeutic potential of cannabis. Lancet Neurol 2003;2(5):291–8. Available from: https://doi.org/10.1016/s1474-4422(03)00381-8.

[6] Klein TW, Newton C, Larsen K, et al. The cannabinoid system and immune modulation. J Leukoc Biol 2003;74(4):486–96. Available from: https://doi.org/10.1189/jlb.0303101.

[7] Mackie K. Cannabinoid receptors: where they are and what they do. J Neuroendocrinol 2008;20:10–4. Available from: https://doi.org/10.1111/j.1365-2826.2008.01671.x.

[8] An D, Peigneur S, Hendrickx LA, et al. Targeting cannabinoid receptors: current status and prospects of natural products. Int J Mol Sci 2020;21(14):5064. Available from: https://doi.org/10.3390/ijms21145064.

[9] Pagano C, Navarra G, Coppola L, et al. Cannabinoids: therapeutic use in clinical practice. Int J Mol Sci 2022;23(6):3344. Available from: https://doi.org/10.3390/ijms23063344.

[10] Laprairie RB, Bagher AM, Kelly MEM, et al. Cannabidiol is a negative allosteric modulator of the cannabinoid CB1 receptor. Br J Pharmacol 2015;172(20):4790–805. Available from: https://doi.org/10.1111/bph.13250.

[11] National Academies of Sciences, Engineering, and Medicine. The Health Effects of Cannabis and Cannabinoids: The Current State of Evidence and Recommendations for Research. Washington, DC: National Academies Press; 2017. Available from: https://www.nationalacademies.org/read/24625/chapter/1.

[12] MacCallum CA, Lo LA, Boivin M. "Is medical cannabis safe for my patients?" A practical review of cannabis safety considerations. Eur J Intern Med 2021;89:10–8. Available from: https://doi.org/10.1016/j.ejim.2021.05.002.

[13] Maertens RM, White PA, Rickert W, et al. The genotoxicity of mainstream and sidestream marijuana and tobacco smoke condensates. Chem Res Toxicol 2009;22(8):1406–14. Available from: https://doi.org/10.1021/tx9000286.

[14] Moir D, Rickert WS, Levasseur G, et al. A comparison of mainstream and sidestream marijuana and tobacco cigarette smoke produced under two machine smoking conditions. Chem Res Toxicol 2008;21(2):494–502. Available from: https://doi.org/10.1021/tx700275p.

[15] Novotný M, Merli F, Wiesler D, et al. Fractionation and capillary gas chromatographic—mass spectrometric characterization of the neutral components in marijuana and tobacco smoke condensates. J Chromatogr A 1982;238(1):141–50. Available from: https://doi.org/10.1016/S0021-9673(00)82720-X.

[16] Tashkin DP. Effects of marijuana smoking on the lung. Ann Am Thorac Soc 2013;10(3):239–47. Available from: https://doi.org/10.1513/AnnalsATS.201212-127FR.

[17] Walsh Z, Gonzalez R, Crosby K, et al. Medical cannabis and mental health: a guided systematic review. Clin Psychol Rev 2017;51:15–29. Available from: https://doi.org/10.1016/j.cpr.2016.10.002.

[18] Chabarria KC, Racusin DA, Antony KM, et al. Marijuana use and its effects in pregnancy. Am J Obstet Gynecol 2016;215(4):506.e1. Available from: https://doi.org/10.1016/j.ajog.2016.05.044.

[19] Leino AD, Emoto C, Fukuda T, et al. Evidence of a clinically significant drug-drug interaction between cannabidiol and tacrolimus. Am J Transplant 2019;19(10):2944–8. Available from: https://doi.org/10.1111/ajt.15398.

[20] Taha T, Meiri D, Talhamy S, et al. Cannabis impacts tumor response rate to nivolumab in patients with advanced malignancies. Oncologist 2019;24(4):549–54. Available from: https://doi.org/10.1634/theoncologist.2018-0383.

[21] Johannessen SI, Landmark CJ. Antiepileptic drug interactions-principles and clinical implications. Curr Neuropharmacol 2010;8(3):254–67. Available from: https://doi.org/10.2174/157015910792246254.

[22] Landmark CJ, Patsalos PN. Drug interactions involving the new second and third-generation antiepileptic drugs. Expert Rev Neurother 2010;10(1):119–40. Available from: https://doi.org/10.1586/ern.09.136.

[23] Stöllberger C, Finsterer J. Cannabidiol's impact on drug-metabolization. Eur J Intern Med 2023. Available from: https://doi.org/10.1016/j.ejim.2023.07.029.

[24] Grayson L, Vines B, Nichol K, et al. An interaction between warfarin and cannabidiol, a case report. Epilepsy Behav Case Rep 2018;9 10–11. Available from: https://doi.org/10.1016/j.ebcr.2017.10.001.

[25] Cortopassi J. Warfarin dose adjustment required after cannabidiol initiation and titration. Am J Health Syst Pharm 2020;77(22):1846–51. Available from: https://doi.org/10.1093/ajhp/zxaa268.

[26] Brown JD, Winterstein AG. Potential adverse drug events and drug–drug interactions with medical and consumer cannabidiol (CBD) use. J Clin Med 2019;8(7):989. Available from: https://doi.org/10.3390/jcm8070989.

[27] Brown JD. Potential adverse drug events with tetrahydrocannabinol (THC) due to drug–drug interactions. J Clin Med 2020;9(4):919. Available from: https://doi.org/10.3390/jcm9040919.

[28] Babayeva M, Loewy ZG. Cannabis pharmacogenomics: a path to personalized medicine. Curr Issues Mol Biol 2023;45(4):3479–514. Available from: https://doi.org/10.3390/cimb45040228.

[29] Hinz B, Ramer R. Cannabinoids as anticancer drugs: current status of preclinical research. Br J Cancer 2022;127(1):1–13. Available from: https://doi.org/10.1038/s41416-022-01727-4.

[30] Creanga-Murariu I, Filipiuc LE, Cuciureanu M, et al. Should oncologists trust cannabinoids? Front Pharmacol 2023;14:1211506. Available from: https://doi.org/10.3389/fphar.2023.1211506.

[31] Whiting PF, Wolff RF, Deshpande S, et al. Cannabinoids for medical use: a systematic review and meta-analysis. JAMA 2015;313(24):2456–73. Available from: https://doi.org/10.1001/jama.2015.6358.

[32] Bar-Sela G, Cohen I, Campisi-Pinto S, et al. Cannabis consumption used by cancer patients during immunotherapy correlates with poor clinical outcome. Cancers 2020;12(9):2447. Available from: https://doi.org/10.3390/cancers12092447.

[33] To J, Davis M, Sbrana A, et al. MASCC guideline: cannabis for cancer-related pain and risk of harms and adverse events. Supportive Care Cancer 2023;31(4):202. Available from: https://doi.org/10.1007/s00520-023-07662-1.

[34] Banerjee S, McCormack S. Medical cannabis for the treatment of chronic pain: a review of clinical effectiveness and guidelines. CADTH rapid response report: summary with critical appraisal. Ottawa (ON): Canadian Agency for Drugs and Technologies in Health; 2019.

[35] Häuser W, Welsch P, Radbruch L, et al. Cannabis-based medicines and medical cannabis for adults with cancer pain. Cochrane Database Syst Rev 2023;6(6):CD014915. Available from: https://doi.org/10.1002/14651858.CD014915.pub2.

[36] Barakji J, Korang SK, Feinberg J, et al. Cannabinoids versus placebo for pain: a systematic review with meta-analysis and trial sequential analysis. PLoS One 2023;18(1):e0267420. Available from: https://doi.org/10.1371/journal.pone.0267420.

[37] Boland EG, Bennett MI, Allgar V, et al. Cannabinoids for cancer-related pain in adults: systematic review and meta-analysis. BMJ Support Palliat Care 2020;10(1):14–24. Available from: https://doi.org/10.1136/bmjspcare-2019-002032.

[38] IASP Presidential Task Force on Cannabis and Cannabinoid Analgesia. International association for the study of pain presidential task force on cannabis and cannabinoid analgesia position statement. Pain 2021;162(Suppl. 1):S1–2. Available from: https://doi.org/10.1097/j.pain.0000000000002265.

[39] Therapeutic Goods Administration. Guidance for the use of medicinal cannabis in the treatment of palliative care patients in Australia (v1.1). https://www.tga.gov.au/products/unapproved-therapeutic-goods/medicinal-cannabis-hub/medicinal-cannabis-guidance-documents/guidance-use-medicinal-cannabis-treatment-palliative-care-patients-australia; 2024.

[40] Meng H, Johnston B, Englesakis M, et al. Selective cannabinoids for chronic neuropathic pain: a systematic review and meta-analysis. Anesth Analg 2017;125(5):1638–52. Available from: https://doi.org/10.1213/ANE.0000000000002110.

[41] Mücke M, Phillips T, Radbruch L, et al. Cannabis-based medicines for chronic neuropathic pain in adults. Cochrane Database Syst Rev 2018;3(3):CD012182. Available from: https://doi.org/10.1002/14651858.CD012182.pub2.

[42] Bhaskar A, Bell A, Boivin M, et al. Consensus recommendations on dosing and administration of medical cannabis to treat chronic pain: results of a modified Delphi process. J Cannabis Res 2021;3:1–12. Available from: https://doi.org/10.1186/s42238-021-00073-1.

[43] Dranitsaris G, Molassiotis A, Clemons M, et al. The development of a prediction tool to identify cancer patients at high risk for chemotherapy-induced nausea and vomiting. Ann Oncol 2017;28(6):1260–7. Available from: https://doi.org/10.1093/annonc/mdx100.

[44] Ruhlmann CH, Jordan K, Jahn F, et al. 2023 Updated MASCC/ESMO consensus recommendations: prevention of radiotherapy-and chemoradiotherapy-induced nausea and vomiting. Support Care Cancer 2024;32(1):26. Available from: https://doi.org/10.1007/s00520-023-08226-z.

[45] Lasheen W, Walsh D. The cancer anorexia-cachexia syndrome: myth or reality? Support Care Cancer 2010;18:265–72. Available from: https://doi.org/10.1007/s00520-009-0772-6.

[46] Simon L, Baldwin C, Kalea AZ, et al. Cannabinoid interventions for improving cachexia outcomes in cancer: a systematic review and meta-analysis. J Cachexia Sarcopenia Muscle 2022;13(1):23–41. Available from: https://doi.org/10.1002/jcsm.12861.

[47] Hammond S, Erridge S, Mangal N, et al. The effect of cannabis-based medicine in the treatment of cachexia: a systematic review and meta-analysis. Cannabis Cannabinoid Res 2021;6(6):474–87. Available from: https://doi.org/10.1089/can.2021.0048.

[48] State Medical Cannabis Laws. https://www.cdc.gov/cannabis/about/state-medical-cannabis-laws.html.

[49] Suliman I, Bacon M, Drury L, et al. Palliative care symptom management guidelines. Boston, MA, USA: Dana-Farber Cancer Institute/Brigham & Women's Hospital; 2023.

[50] Thong MS, Van Noorden CJF, Steindorf K, et al. Cancer-related fatigue: causes and current treatment options. Curr Treat Options Oncol 2020;21(2):17. Available from: https://doi.org/10.1007/s11864-020-0707-5.

[51] Klasson C, Helde Frankling M, Lundh Hagelin C, et al. Fatigue in cancer patients in palliative care—a review on pharmacological interventions. Cancers 2021;13(5):985. Available from: https://doi.org/10.3390/cancers13050985.

[52] Berger AM, Mooney K, Alvarez-Perez A, et al. Cancer-related fatigue, version 2.2015. J Natl Compr Canc Netw 2015;13(8):1012–39. Available from: https://doi.org/10.6004/jnccn.2015.0122.

[53] Stone PC, Minton O, Richardson A, et al. Methylphenidate versus placebo for treating fatigue in patients with advanced cancer: randomized, double-blind, multicenter, placebo-controlled trial. J Clin Oncol 2024;42(20):2382–92. Available from: https://doi.org/10.1200/JCO.23.02639.

[54] Wei W, Chen L, Zhou H, et al. Safety profiles of methylphenidate, amphetamine, and atomoxetine: analysis of spontaneous reports submitted to the food and drug administration adverse event reporting system. Front Pharmacol 2023;14:1208456. Available from: https://doi.org/10.3389/fphar.2023.1208456.

[55] Roji R, Stone P, Ricciardi F, et al. Placebo response in trials of drug treatments for cancer-related fatigue: a systematic review, meta-analysis and meta-regression. BMJ Support Palliat care 2020;10(4):385–94. Available from: https://doi.org/10.1136/bmjspcare-2019-002163.

[56] Paice JA, Portenoy R, Lacchetti C, et al. Management of chronic pain in survivors of adult cancers: American Society of Clinical Oncology clinical practice guideline. J Clin Oncol 2016;34(27):3325–45. Available from: https://doi.org/10.1200/JCO.2016.68.5206.

[57] Weiss SC, Emanuel LL, Fairclough DL, et al. Understanding the experience of pain in terminally ill patients. Lancet 2001;357(9265):1311–5. Available from: https://doi.org/10.1016/S0140-6736(00)04515-3.

[58] Falk S, Bannister K, Dickenson AH. Cancer pain physiology. Br J Pain 2014;8(4):154–62. Available from: https://doi.org/10.1177/2049463714545136.

[59] Latremoliere A, Woolf CJ. Central sensitization: a generator of pain hypersensitivity by central neural plasticity. J Pain 2009;10(9):895–926. Available from: https://doi.org/10.1016/j.jpain.2009.06.012.

[60] Guimarães Pereira JE, Ferreira Gomes Pereira L, Mercante Linhares R, et al. Efficacy and safety of ketamine in the treatment of neuropathic pain: a systematic review and meta-analysis of randomized controlled trials. J Pain Res. 2022;15:1011–37. Available from: https://doi.org/10.2147/JPR.S358070.

[61] Zhou HY, Chen SR, Pan HL. Targeting N-methyl-D-aspartate receptors for treatment of neuropathic pain. Expert Rev Clin Pharmacol 2011;4(3):379–88. Available from: https://doi.org/10.1586/ecp.11.17.

[62] Swarm RA, Paice JA, Anghelescu DL, et al. Adult cancer pain, version 3.2019, NCCN clinical practice guidelines in oncology. J Natl Compr Canc Netw 2019;17(8):977–1007. Available from: https://doi.org/10.6004/jnccn.2019.0038.

Chapter 10

Anticonvulsants

Boris Milijašević[1] and Feđa Đorđević[2]

[1]Department of Pharmacology, Toxicology and Clinical Pharmacology, University of Novi Sad, Novi Sad, Serbia, [2]Department of Medical Oncology, Institute for Oncology and Radiology of Serbia, Belgrade, Serbia

Key learning objectives

- To give insight into current evidence regarding anticonvulsant usage for the treatment of pain in cancer patients.
- To describe and explain data from clinical trials conducted specifically in cancer patients.
- To provide information related to specific individual drugs that are commonly used or recommended by guidelines for pain management or related conditions.
- To give a brief overview of current guidelines and opinions regarding the usage of anticonvulsants in pain management.

Anticonvulsants

Anticonvulsants are drugs that are primarily used in the treatment of convulsions. This is a heterogeneous group of drugs in terms of the mechanism of action, which, in common, eliminate convulsions. However, some of these drugs can also be used in pain management.

Pharmacology

Carbamazepine and **oxcarbazepine** block voltage-gated sodium channels and potentiate the effects of GABA and are considered the drugs of choice for trigeminal neuralgia [1,2]. **Carbamazepine** tablet is absorbed almost completely but relatively slowly. The bioavailability of carbamazepine in various oral formulations has been shown to be between 85% and 100%. Ingestion of food has no significant influence on the rate and extent of absorption. Carbamazepine is bound to serum proteins to the extent of 70%–80%. Carbamazepine is metabolized in the liver primarily via cytochrome P450 3A4 (CYP3A4). The elimination half-life of unchanged carbamazepine averages approximately 36 hours following a single oral dose, whereas, after repeated administration, it averages only 16–24 hours. After administration of a single oral dose of carbamazepine, approximately 72% of the dose is recovered in urine and 28% in feces, primarily as metabolites, because carbamazepine undergoes extensive hepatic metabolism prior to excretion [3]. **Oxcarbazepine**, following oral administration, is completely absorbed and extensively metabolized to its active metabolite, the monohydroxy derivative (MHD). Oxcarbazepine is cleared from the body mostly (95%) in the form of metabolites which are predominantly excreted by the kidneys. Steady-state plasma concentrations of MHD are reached within 2–3 days in patients when oxcarbazepine is given twice a day [4].

Gabapentinoids, including **gabapentin** and **pregabalin**, are commonly used for treating neuropathic pain. Pharmacologically, they decrease the release of excitatory neurotransmitters glutamate, norepinephrine, and substance P [1]. **Gabapentin** binds with high affinity to the $\alpha 2\delta$ (alpha-2-delta) subunit of voltage-gated calcium channels leading to several different activities that may be responsible for the analgesic activity. The analgesic activities of gabapentin may occur in the spinal cord, as well as at higher brain centers, through interactions with the descending pain inhibitory pathways. Gabapentin is not bound to plasma proteins and does not induce/inhibit microsomal enzymes responsible for the metabolism of other drugs. It is eliminated solely unchanged by renal excretion. The elimination half-life of gabapentin is independent of dose and averages 5–7 hours. In elderly patients and in patients with impaired renal function, gabapentin plasma clearance is reduced [5]. **Pregabalin**, similar to gabapentin, binds to a subunit $\alpha 2\delta$ of voltage-gated calcium

Pain Management in Palliative Care for Patients With Cancer. DOI: https://doi.org/10.1016/B978-0-443-30050-9.00011-4

channels in the central nervous system. The oral bioavailability of pregabalin is estimated to be ≥90% and steady state is achieved within 24–48 hours. Pregabalin is not bound to plasma proteins. It is negligibly metabolized and is excreted in the urine. Dose adjustment in patients with reduced renal function is necessary [6].

Lamotrigine is an agent used in a specialist setting due to the risk of life-threatening rash requiring increased monitoring and careful dose titration. It blocks voltage-gated sodium channels on the membrane of neurons. Lamotrigine is rapidly and completely absorbed from the gut; approximately 55% of the drug is bound to plasma proteins. It is metabolized by glucuronidation and induces its own metabolism in a modest dose-dependent manner. Lamotrigine, as well as its metabolites, are excreted in the urine [1,7].

Topiramate is an agent used most commonly for the treatment of migraine. It blocks action potentials generated by depolarization of neurons, which indicates blockade of sodium channels, among its other mechanisms of action. It also inhibits some isoenzymes of carbonic anhydrase. Topiramate is rapidly and well absorbed, without a significant effect of food on its bioavailability. About 15% of the drug is bound to plasma proteins. It is metabolized to a lesser extent; however, in patients who use enzyme inducers for a long time, topiramate is metabolized up to 50%. The major route of elimination of unchanged topiramate and its metabolites is via the kidneys [1,8].

A summary of contraindications, adverse effects, and major drug interactions of anticonvulsants is given in Table 10.1.

Clinical evidence and role in treating cancer pain

Gabapentinoids

The most robust evidence of usage of anticonvulsants for management of pain is reported for gabapentinoids, gabapentin, and pregabalin, which were primarily used for epileptic seizure prevention and treatment, and restless leg syndrome. However, gabapentinoids are considered first-line agents in the treatment of neuropathic pain by multiple international societies [9–11], as well as neuropathic cancer pain (NCP) [12–14]. Although majority of recommendations were derived from the results extrapolated from the studies of noncancer patients, there is data to support their use in patients with cancer.

A parallel-group study randomized 121 participants with NCP to a maximum dose of gabapentin 1800 mg daily or placebo, during 10 days of treatment. All groups received stable opioid dosing (~100 mg/day oral morphine equivalent). The study showed that average pain intensity was lower with gabapentin than with placebo, but the number of participants described as having pain under control was very similar with both treatments after 6 days, with 50%–60% with pain under control over 6–10 days [15]. Mishra et al. compared gabapentin 1800 mg daily with pregabalin 600 mg daily, amitriptyline 100 mg daily, and placebo in groups of 30 patients with severe NCP for a total of 4 weeks [16]. Results favored pregabalin over all other groups with a reduction in mean Visual Analog Scale (VAS) score from 7.77 to 2.5, and significantly lower scores in comparison with amitriptyline by week 3, and gabapentin and placebo by week 4 ($P = .024$). The pregabalin group also showed the greatest improvements in dysesthesia, burning, and lancinating pain, but not allodynia. They reported significantly better morphine-sparing and functional capacity in the pregabalin group, compared to other groups. In an open-label study, Raptis et al. randomized 120 patients with moderate-to-severe NCP, uncontrolled on mild opioids, to the pregabalin or transdermal fentanyl group. In the pregabalin group, 73.3% achieved at least a 30% reduction in the VAS pain scores, in comparison to 36.7% in the fentanyl group ($P < .0001$). [17] A crossover study compared gabapentin titrated to 2700 mg daily with placebo in chemotherapy-induced neuropathic pain over 3 weeks [18]. No significant difference was found between gabapentin and placebo, although the study was underpowered. Gabapentinoids were also tested against duloxetine; a trial published by Salehifar et al. suggests that pregabalin in doses of 150 mg daily is superior to a 60-mg dose of duloxetine in efficacy in relieving taxane-induced peripheral neuropathic pain in breast cancer patients. The trial demonstrated that both arms were effective, but at week 6, pain scores improved in 38.1% of the patients treated with duloxetine and 92.5% of patients treated with pregabalin ($P < .001$). The majority of the patients in both arms started their treatment while they were receiving chemotherapy, and some of this improvement may have been related to chemotherapy discontinuation [19].

While larger randomized controlled trials are lacking, several studies were conducted to evaluate the potential of gabapentinoids for preventing chemotherapy-associated acute pain in chronic neuropathy [20]. Vondracek et al. addressed the decrease in pain score of chemotherapy-induced peripheral neuropathy (CIPN) with pregabalin despite the suboptimal study design [21]. Another smaller study did not show clinical benefits using pregabalin (75 mg two times per day) to prevent the paclitaxel-associated acute pain syndrome or paclitaxel-induced peripheral neuropathy [22]. Similarly, the results from a double-blind, placebo-controlled trial did not support that preemptive use of pregabalin (150–600 mg daily) during oxaliplatin infusions could decrease the incidence of chronic pain related to oxaliplatin. This study was

TABLE 10.1 Contraindications, adverse effects, and interactions of anticonvulsants.

Carbamazepine [3]	
Contraindications	Hypersensitivity, atrioventricular block, history of bone-marrow depression or a history of hepatic porphyrias. Avoid concomitant use with monoamine oxidase inhibitors.
Adverse effects	Bone-marrow suppression, fluid retention, antidiuretic hormone (ADH)–like effect, urticaria, neurological disorders.
Drug interactions	Potent inducer of CYP3A4, may lower the plasma level of certain drugs (e.g., buprenorphine, methadone, tramadol). Hepatotoxicity when administered long term with paracetamol.
Oxcarbazepine [4]	
Contraindications	Hypersensitivity.
Adverse effects	Weight increase, neurological disorders, hyponatremia, gastrointestinal disorders.
Drug interactions	Weak inducers of cytochrome P450 enzymes (CYP3A4 and CYP3A5) and UDP-glucuronyl transferase. Potential neurotoxicity when used with lithium.
Gabapentin [5]	
Contraindications	Hypersensitivity.
Adverse effects	Increased likelihood of infections, gastrointestinal and neurological disorders, edema, weight gain, skin disorders, fatigue, and sedation.
Drug interactions	Coadministration of gabapentin with CNS depressants, including opioids, can increase the risk of respiratory depression, sedation, and death. Coadministration of gabapentin with antacids containing aluminum and magnesium reduces gabapentin bioavailability up to 24%. Use gabapentin at the earliest 2 h, following antacid administration.
Pregabalin [6]	
Contraindications	Hypersensitivity.
Adverse effects	Neurological disorders and sedation, increase in appetite, weight gain, gastrointestinal adverse effects.
Drug interactions	Pregabalin may potentiate the effects of ethanol and lorazepam. Coadministration of pregabalin with CNS depressants, including opioids, can increase the risk of respiratory depression, sedation, and death.
Lamotrigine [7]	
Contraindications	Hypersensitivity.
Adverse effects	Neurological disorders, gastrointestinal disorders, arthralgia.
Drug interactions	Drugs that induce or inhibit glucuronidation may affect the clearance of lamotrigine. Inducers of the cytochrome P450 3A4 could decrease the concentration of lamotrine. No significant interactions with opiate analgesics or other CNS depressants have been noted.
Topiramate [8]	
Contraindications	Hypersensitivity.
Adverse effects	Neurological and cognitive disorders, fatigue, sedation, respiratory symptoms, gastrointestinal symptoms.
Drug interactions	Phenytoin and carbamazepine decrease the plasma concentration of topiramate. Decreased prothrombin time/international normalized ratio has been reported in patients treated with topiramate in combination with warfarin. No significant interactions with opiate analgesics or other CNS depressants have been noted.

CNS, Central nervous system.

corroborated by nerve conduction studies that also failed to show any protective effect of pregabalin [23]. Aghili et al. randomized women with breast cancer into two groups: (1) paclitaxel chemotherapy with gabapentin 300 mg/three times daily and (2) a placebo. The study shows that gabapentin given with paclitaxel is effective in the prevention of intermediate and high-grade neuropathies both objectively and subjectively [24]. A small pilot trial showed that prophylactic use of gabapentin during chemoradiation for head and neck cancer patients resulted in a decrease in pain, neurosensory symptoms, and general systemic symptoms [25].

A recent metaanalysis found a nonsignificant trend that combining opioids with gabapentinoids improves cancer pain as a whole [26]. The addition of gabapentin or pregabalin to an opioid as combination therapy can increase effectiveness [27] and may enable the dose of opioids to be reduced [28]. Some authors suggest that coadministration of gabapentinoids and a tricyclic antidepressant could have a greater clinical benefit and effect on pain [29]; however, this should be weighed against the potential for increased side effects.

According to the 2020 joint European Society for Medical Oncology (ESMO)–European Oncology Nursing Society (EONS)–European Association of Neuro-Oncology (EANO) guidelines, membrane stabilizing agents like gabapentin and pregabalin might be the reasonable options for symptom control in patients with CIPN if first-line treatment like duloxetine has failed or contraindications are present. From the practical side, they suggest to apply gabapentinoids for neuropathic pain for at least 2 weeks at the appropriate dose in order to assess their efficacy before changing to another option (V,B) [20]. The American Society of Clinical Oncology guidelines do not recommend gabapentinoids as standard treatment for CIPN, although limited evidence suggests they may provide benefit in selected patients. Duloxetine remains the primary guideline-supported pharmacologic option for painful CIPN [30]. On the other hand, ESMO guidelines also recommend duloxetine with level I evidence based on randomized trials but also give pregabalin level II evidence, based on nonrandomized trials [20].

Taken together, the evidence seems in favor of a positive but rather modest effect of gabapentinoids on the intensity of NCP. Moreover, clinicians should always balance the potential benefits of these medications with their potential side

TABLE 10.2 Dosing of anticonvulsants and other considerations.

Name of drugs	Dosage/additional considerations
Carbamazepine	**Dosage considerations:** Neuropathic cancer pain (cancer-induced trigeminal neuralgia), start with **50–100 mg PO b.d.** **Maximum daily dose 2 g (participants in most RCTs received ≤600 mg/24 h).** **Additional considerations:** Caution in patients with renal and hepatic impairment and cardiac disease. Agranulocytosis and aplastic anemia affect about 5 and 2 patients/million/year, respectively.
Oxcarbazepine	**Dosage considerations:** Structurally related to **carbamazepine**, improve trigeminal neuralgia unresponsive to carbamazepine. Starts with 150 mg b.d. (75 mg b.d. in elderly, frail, or in severe renal impairment CrCl ≤30 mL/min). Increase the dose, 75–150 mg increments weekly, maximum 1200 mg b.d. **Additional considerations:** Fewer drug interactions than carbamazepine. Caution in patients with renal and hepatic impairment and cardiac disease and hyponatremia.
Gabapentin	**Dosage considerations:** Start with 300 mg PO at night, if necessary increase by 300 mg/24 h every 2–3 days. In neuropathic cancer pain most effective dose 600 mg t.d.s. Max recommended dose 1.200 mg t.d.s. **Additional considerations:** The starting and maximum dose should be reduced in adults with renal impairment. CrCl 50–79 mL/min, start with 300 mg at bedtime, max 600 mg t.d.s. CrCl 30–49 mL/min, start with 300 mg at bedtime, max 300 mg t.d.s. CrCl 15–29 mL/min, start with 100 mg at bedtime, max 300 mg b.d. <15 mL/min, start with 100 mg on alternate nights, max 300 mg at bedtime.
Pregabalin	**Dosage considerations:** Start with 75 mg PO b.d., or lower in frail patients, if necessary, at intervals of 3–7 days, increase to 150 mg b.d. to 225 mg b.d. to 300 mg b.d. which is max recommended dose. In debilitated patients, start with 25–50 mg b.d. **Additional considerations:** The starting and maximum dose should be reduced in adults with renal impairment. CrCl >60 mL/min, start with 75 mg b.d., max 300 mg b.d. CrCl 31–60 mL/min, start with 25 mg t.d.s., max 150 mg b.d. CrCl 15–30 mL/min, start with 25–50 mg once daily, max 150 mg once daily. <15 mL/min, start with 25 mg once daily, max 75 mg once daily.
Lamotrigine	**Dosage considerations:** Start with 25 mg daily up to maximum of 400 mg. **Additional considerations:** This drug is associated with a small risk of severe cutaneous hypersensitivity, which is relatively higher in younger patients. It should not be used in younger children, and slow titration of the dose from a low initial dose is necessary to reduce the risk of hypersensitivity.
Topiramate	**Dosage considerations:** Second- and third-line therapies in patients with cancer who did not experience adequate pain control on previous regimens start with 25 mg PO at bedtime to 25 mg increments every 1–2 weeks, effective dose <200 mg/24 h daily. **Additional considerations:** Retrospective study showed topiramate to be beneficial; further prospective studies are needed.

b.d., Twice daily; *CrCl*, creatinine clearance; *mg*, milligrams; *mL/min*, milliliters per minute; *PO*, per os (oral administration); *RCTs*, randomized controlled trials; *t.d.s.*, three times daily.

effects. Patients on gabapentinoids frequently complain of dizziness or somnolence, diarrhea, nausea/vomiting, and rash, but both pregabalin and gabapentin are also associated with more serious adverse events, such as misuse and respiratory depression, especially when combined with opioids [31–33].

Other anticonvulsants

Lamotrigine has been effective in studies of central pain, trigeminal neuralgia, painful human immunodeficiency virus polyneuropathy, and postherpetic neuralgia, but it has not shown analgesic effect in other trials, including one conducted in patients with painful CIPN [34]. Given the limited evidence of efficacy, the need for slow titration, and the risk of severe cutaneous hypersensitivity, this drug is not a preferred second-line agent for neuropathic pain and is usually considered only after neuropathic pain has proved to be refractory to other preferred drugs.

Carbamazepine is a first-line drug for focal seizures and trigeminal neuralgia [35]. Because it requires slow titration and particular care with regard to drug interactions, its use in other neuropathic pain is limited. Oxcarbazepine is a metabolite of carbamazepine that has similar anticonvulsant properties and a safer pharmacologic profile. Although the quality of the evidence is limited [36], it has been shown to be analgesic in trigeminal neuralgia and several other types of peripheral neuropathic pain [37]. A trial of this drug may be considered after trials of the gabapentinoids and perhaps one or two of the analgesic antidepressants have proven unhelpful.

Topiramate, an effective drug for the prevention of episodic migraine, has been variably effective in studies of neuropathic pain [38]. Although anecdotal experience suggests that some patients with cancer and neuropathic pain derive benefit, it is generally considered only after other second-line drugs have been proved ineffective.

A summary of dosing recommendations is given in Table 10.2.

Conclusion

Anticonvulsants represent valuable additions to cancer pain management. Both pregabalin and gabapentin have demonstrated efficacy in neuropathic pain, although some studies suggest pregabalin may provide greater analgesic benefit in certain patients. While other anticonvulsants serve specialized roles for specific pain syndromes, their cancer pain applications remain limited. Future research should focus on optimizing patient selection and combination therapies through larger randomized trials. Clinicians must carefully monitor for drug interactions and adverse effects, particularly respiratory depression, when combining with opioids.

References

[1] Schwinghammer TL, DiPiro JT, Ellingrod VL, DiPiro CV, editors. Pharmacotherapy handbook. 11th ed. New York, NY: McGraw-Hill; 2021.

[2] DiPiro JT, Yee GC, Haines ST, Nolin TD, Ellingrod VL, Posey L, editors. DiPiro's pharmacotherapy: a pathophysiologic approach. 12th ed. New York, NY: McGraw-Hill; 2023.

[3] Novartis Pharmaceuticals UK Ltd. Tegretol 200mg tablets - summary of product characteristics [internet]. London: Electronic Medicines Compendium; 2024 [cited 20 June 2025]. Available from: https://www.medicines.org.uk/emc/product/7845/smpc.

[4] Novartis Ireland Limited. Trileptal 300 mg film-coated tablets - summary of product characteristics [internet]. Dublin: Electronic Medicines Compendium; 2024 [cited 20 June 2025]. Available from: https://www.medicines.org.uk/emc/product/7851/smpc.

[5] SmPC gabapentin - summary of product characteristics [internet]. London: Electronic Medicines Compendium; date unknown [cited 20 June 2025]. Available from: https://www.medicines.org.uk/emc/product/2362/smpc.

[6] Dr. Reddy's Laboratories (UK) Ltd. Alzain 100 mg capsules, hard - summary of product characteristics [internet]. Cambridge: Electronic Medicines Compendium; 2024 [cited 20 June 2025]. Available from: https://www.medicines.org.uk/emc/product/1761/smpc.

[7] Milpharm Limited. Lamotrigine Milpharm 100 mg tablets - summary of product characteristics [internet]. Dungannon: Electronic Medicines Compendium; 2024 [cited 20 June 2025]. Available from: https://www.medicines.org.uk/emc/product/4739/smpc.

[8] Janssen-Cilag Limited. Topamax 100 mg film-coated tablets - summary of product characteristics [internet]. High Wycombe: Electronic Medicines Compendium; 2024 [cited 20 June 2025]. Available from: https://www.medicines.org.uk/emc/product/1977/smpc.

[9] Finnerup NB, Attal N, Haroutounian S, et al. Pharmacotherapy for neuropathic pain in adults: a systematic review and meta-analysis. Lancet Neurol 2015;14(2):162–73. Available from: https://doi.org/10.1016/S1474-4422(14)70251-0.

[10] Attal N, Cruccu G, Baron R, et al. EFNS guidelines on the pharmacological treatment of neuropathic pain. Eur J Neurol 2010;17(9):e67–88. Available from: https://doi.org/10.1111/j.1468-1331.2010.02999.x.

[11] Moulin DE, Clark AJ, Gilron I, et al. Pharmacological management of chronic neuropathic pain – consensus statement and guidelines from the Canadian Pain Society. Pain Res Manag 2007;12(1):13–21. Available from: https://doi.org/10.1155/2007/730785.

[12] Piano V, Schalkwijk A, Burgers J, et al. Guidelines for neuropathic pain management in patients with cancer: a European survey and comparison. Pain Pract 2013;13(5):349–57. Available from: https://doi.org/10.1111/j.1533-2500.2012.00602.x.

[13] Ramanjulu R, Thota RS, Ahmed A, et al. Indian society for study of pain, cancer pain special interest group guidelines on pharmacological management of cancer pain (part III). Indian J Palliat Care 2020;26(2):191–7. Available from: https://doi.org/10.4103/0973-1075.285694.

[14] Wordliczek J, Kotlinska-Lemieszek A, Leppert W, et al. Pharmacotherapy of pain in cancer patients - recommendations of the Polish Association for the Study of Pain, Polish Society of Palliative Medicine, Polish Society of Oncology, Polish Society of Family Medicine, Polish Society of Anaesthesiology and Intensive Therapy and Association of Polish Surgeons. Pol Przegl Chir 2018;90(4):55–84. Available from: https://doi.org/10.5604/01.3001.0012.2904.

[15] Caraceni A, Zecca E, Bonezzi C, et al. Gabapentin for neuropathic cancer pain: a randomized controlled trial from the Gabapentin Cancer Pain Study Group. J Clin Oncol 2004;22(14):2909–17. Available from: https://doi.org/10.1200/JCO.2004.08.141.

[16] Mishra S, Bhatnagar S, Goyal GN, et al. A comparative efficacy of amitriptyline, gabapentin, and pregabalin in neuropathic cancer pain: a prospective randomized double-blind placebo-controlled study. Am J Hosp Palliat Care 2012;29(3):177–82. Available from: https://doi.org/10.1177/1049909111412539.

[17] Raptis E, Vadalouca A, Stavropoulou E, et al. Pregabalin vs. opioids for the treatment of neuropathic cancer pain: a prospective, head-to-head, randomized, open-label study. Pain Pract 2014;14(1):32–42. Available from: https://doi.org/10.1111/papr.12045.

[18] Rao RD, Michalak JC, Sloan JA, et al. Efficacy of gabapentin in the management of chemotherapy-induced peripheral neuropathy: a phase 3 randomized, double-blind, placebo-controlled, crossover trial (N00C3). Cancer 2007;110(9):2110–8. Available from: https://doi.org/10.1002/cncr.23008.

[19] Salehifar E, Janbabaei G, Hendouei N, et al. Comparison of the efficacy and safety of pregabalin and duloxetine in taxane-induced sensory neuropathy: a randomized controlled trial. Clin Drug Investig 2020;40(3):249–57. Available from: https://doi.org/10.1007/s40261-019-00882-6.

[20] Jordan B, Margulies A, Cardoso F, et al. Systemic anticancer therapy-induced peripheral and central neurotoxicity: ESMO-EONS-EANO Clinical Practice Guidelines for diagnosis, prevention, treatment and follow-up. Ann Oncol 2020;31(10):1306–19. Available from: https://doi.org/10.1016/j.annonc.2020.07.003.

[21] Vondracek P, Oslejskova H, Kepak T, et al. Efficacy of pregabalin in neuropathic pain in paediatric oncological patients. Eur J Paediatr Neurol 2009;13(4):332–6. Available from: https://doi.org/10.1016/j.ejpn.2008.06.011.

[22] Shinde SS, Seisler D, Soori G, et al. Can pregabalin prevent paclitaxel-associated neuropathy?—An ACCRU pilot trial. Support Care Cancer 2016;24(2):547–53. Available from: https://doi.org/10.1007/s00520-015-2807-5.

[23] De Andrade DC, Jacobsen Teixeira M, Galhardoni R, et al. Pregabalin for the prevention of oxaliplatin-induced painful neuropathy: a randomized, double-blind trial. Oncologist 2017;22(10):1154-e105. Available from: https://doi.org/10.1634/theoncologist.2017-0235.

[24] Aghili M, Zare M, Mousavi N, et al. Efficacy of gabapentin for the prevention of paclitaxel induced peripheral neuropathy: a randomized placebo controlled clinical trial. Breast J 2019;25(2):226–31. Available from: https://doi.org/10.1111/tbj.13196.

[25] Smith DK, Cmelak A, Niermann K, et al. Preventive use of gabapentin to decrease pain and systemic symptoms in patients with head and neck cancer undergoing chemoradiation. Head Neck 2020;42(12):3497–505. Available from: https://doi.org/10.1002/hed.26407.

[26] Kane CM, Mulvey MR, Wright S, et al. Opioids combined with antidepressants or antiepileptic drugs for cancer pain: systematic review and meta-analysis. Palliat Med 2018;32(1):276–86. Available from: https://doi.org/10.1177/0269216317711826.

[27] Guan J, Tanaka S, Kawakami K. Anticonvulsants or antidepressants in combination pharmacotherapy for treatment of neuropathic pain in cancer patients: a systematic review and meta-analysis. Clin J Pain 2016;32(8):719–25. Available from: https://doi.org/10.1097/AJP.0000000000000310.

[28] Szczudlik A, Dobrogowski J, Wordliczek J, et al. Diagnosis and management of neuropathic pain: review of literature and recommendations of the Polish Association for the study of pain and the Polish Neurological Society - part two. Neurol Neurochir Pol 2014;48(6):423–35. Available from: https://doi.org/10.1016/j.pjnns.2014.11.002.

[29] Taverner T. Neuropathic pain in people with cancer (part 2): pharmacological and non-pharmacological management. Int J Palliat Nurs 2015;21(8):380–4. Available from: https://doi.org/10.12968/ijpn.2015.21.8.380.

[30] Loprinzi CL, Lacchetti C, Bleeker J, et al. Prevention and management of chemotherapy-induced peripheral neuropathy in survivors of adult cancers: ASCO guideline update. J Clin Oncol 2020;38(28):3325–48. Available from: https://doi.org/10.1200/JCO.20.01399.

[31] Chang TW, Yang FY, Liu YC, et al. Gabapentinoids for chemotherapy-induced peripheral neuropathy: systematic review and meta-analysis. BMJ Support Palliat Care 2024;14(3):269–78. Available from: https://doi.org/10.1136/spcare-2023-004362.

[32] Driot D, Jouanjus E, Oustric S, et al. Patterns of gabapentin and pregabalin use and misuse: results of a population-based cohort study in France. Br J Clin Pharmacol 2019;85(6):1260–9. Available from: https://doi.org/10.1111/bcp.13892.

[33] Hägg S, Jönsson AK, Ahlner J. Current evidence on abuse and misuse of gabapentinoids. Drug Saf 2020;43(12):1235–54. Available from: https://doi.org/10.1007/s40264-020-00985-6.

[34] Wiffen PJ, Derry S, Moore RA, et al. Lamotrigine for chronic neuropathic pain and fibromyalgia in adults. Cochrane Database Syst Rev 2013;2013(12):CD006044. Available from: https://doi.org/10.1002/14651858.CD006044.pub4.

[35] Wiffen PJ, Derry S, Moore RA, et al. Carbamazepine for chronic neuropathic pain and fibromyalgia in adults. Cochrane Database Syst Rev 2014;2014(4):CD005451. Available from: https://doi.org/10.1002/14651858.CD005451.pub3.

[36] Demant DT, Lund K, Vollert J, et al. The effect of oxcarbazepine in peripheral neuropathic pain depends on pain phenotype: a randomised, double-blind, placebo-controlled phenotype-stratified study. Pain 2014;155(11):2263–73. Available from: https://doi.org/10.1016/j.pain.2014.08.014.

[37] Gomez-Arguelles JM, Dorado R, Sepulveda JM, et al. Oxcarbazepine monotherapy in carbamazepine-unresponsive trigeminal neuralgia. J Clin Neurosci 2008;15(5):516–9. Available from: https://doi.org/10.1016/j.jocn.2007.04.010.

[38] Bendaly EA, Jordan CA, Staehler SS, et al. Topiramate in the treatment of neuropathic pain in patients with cancer. Support Cancer Ther 2007;4(4):241–6. Available from: https://doi.org/10.3816/SCT.2007.n.021.

Chapter 11

Bisphosphonates and denosumab

Feđa Đorđević[1] and Boris Milijašević[2]

[1]*Department of Medical Oncology, Institute for Oncology and Radiology of Serbia, Belgrade, Serbia,* [2]*Department of Pharmacology, Toxicology and Clinical Pharmacology, University of Novi Sad, Novi Sad, Serbia*

Key learning objectives

- Understand the molecular mechanism of bisphosphonate action.
- Identify appropriate clinical indications for bisphosphonate therapy and recognize the dual role of bisphosphonates in preventing skeletal-related events and providing adjuvant analgesia in cancer patients with bone metastases.
- Evaluate risk factors and prevention strategies for bisphosphonate-related complications.
- Apply evidence-based dosing and monitoring principles.

Introduction

Bisphosphonates are medications that affect bone structure and mineralization, most commonly used in osteoporosis therapy, but also have a role in managing pain from bone metastases. Bisphosphonates are analogs of pyrophosphate in which the P–O–P bond has been replaced with a nonhydrolyzable P–C–P bond. The mechanism of their action is that they inhibit the activity of osteoclasts. Preclinical and clinical studies indicate that the absorption of bisphosphonates after oral administration is less than 10%. Since they are not subject to biotransformation, they rarely interact with the metabolism of other drugs. Since they are excreted unchanged by the kidneys, caution should be exercised in patients with impaired renal function. Adverse effects of bisphosphonates include gastrointestinal complaints (esophagitis, gastritis, and ulcer disease), muscle pain, joint pain, and headache, as well as hypocalcemia, impaired kidney function, and flu-like symptoms that may occur after intravenous (IV) administration. Osteonecrosis of the jaw, as an adverse effect of bisphosphonates, was recorded most commonly in patients with cancer, who also received chemotherapy and corticosteroids at the same time. Dentist evaluation and treatment are recommended before use due to the risk of jaw osteonecrosis [1–6].

Currently available bisphosphonates include etidronate, pamidronate, alendronate, risedronate, tiludronate, and the most commonly used ones, ibandronate and zoledronate. Ibandronic acid rapidly binds to bone or is excreted into urine. The amount of dose reaching the bone is estimated to be 40%–50% of the circulating dose. Protein binding in plasma is approximately 87%. Ibandronic acid is removed from the circulation by uptake into bone and renal excretion [5]. Zoledronic acid is a drug administered intravenously that, along with a potent inhibition of bone resorption, exhibits antitumor properties, at least partially through alteration of the bone marrow microenvironment, making it less conducive to tumor cell growth [7]. Zoledronic acid is not metabolized and is excreted unchanged via the kidneys. No relevant role of liver function in the pharmacokinetics of zoledronic acid is reported [6]. Another osteoclast inhibitor includes denosumab, a fully human monoclonal antibody that targets receptor activator of nuclear factor-κB ligand (RANKL), a key mediator of osteoclast formation, function, and survival. By binding to RANKL, denosumab prevents its interaction with the RANK receptor on osteoclast precursors and mature osteoclasts, thereby inhibiting osteoclast-mediated bone resorption. Denosumab is administered by subcutaneous injection and, after administration, exhibits nonlinear pharmacokinetics typical of monoclonal antibodies, with slow absorption and a prolonged half-life. Because denosumab is not eliminated via the kidneys and does not undergo hepatic metabolism, dose adjustment based on renal or hepatic function is generally not required. However, patients with severe renal impairment are at increased risk of hypocalcemia and require appropriate calcium and vitamin D supplementation and careful monitoring during therapy [2–4,8,9].

Pain Management in Palliative Care for Patients With Cancer. DOI: https://doi.org/10.1016/B978-0-443-30050-9.00016-3

Clinical use

Bisphosphonates and denosumab are the treatment of choice for hypercalcemia of malignancy [10]. However, in patients with myeloma, breast cancer, or prostate cancer and bone metastases, osteoclast inhibitors are given to decrease the incidence of skeletal-related events (SREs) [11], which include pathological fracture, spinal cord compression; hence, and due to reduction of pain, they can also be considered as adjuvant analgesics. This is because the bone is one of the most common pain-sensitive structures affected by cancer, and the majority of patients with cancer-related bone pain require analgesia. The usual approach is a combined one, utilizing nonsteroidal antiinflammatory drugs, unless they have a specific contraindication to the use of these agents, plus an opioid, with or without an adjuvant analgesic, used specifically for bone pain. Adjuvant analgesics to be additionally considered in this setting include osteoclast inhibitors (bisphosphonates, denosumab) and bone-seeking radiopharmaceuticals (the topic will be evaluated in another chapter of this book). Osteoclast inhibitors can be used in two ways: prophylactically, over months to years, to reduce the risk/rate of SREs or as adjuvant analgesics for moderate-to-severe bone pain (often following SRE). Bisphosphonates and denosumab, given prophylactically over months to years to reduce the risk and rate of pain as SRE, are most commonly used in breast cancer, myeloma, and metastatic hormone–relapsed prostate cancer [8,11]. However, bone pain was often not the focus of studies evaluating bisphosphonates and denosumab, either as a primary or secondary end point. When pain was assessed, the use of different measures made comparison difficult, primarily as the main analgesic effect is indirect and derived from the prevention of SRE.

As adjuvant agents, there is evidence to support the analgesic potential of all osteoclast inhibitors, although the benefit seems modest. In a Cochrane review, a significant improvement in bone pain was reported after receiving a bisphosphonate in patients with bone metastases from breast cancer in 6 of 11 studies [12]. Bisphosphonates used as adjuvant analgesics for moderate-to-severe bone pain include zoledronic acid and ibandronic acid [9,13]. Doses are given in Table 11.1 and are the same as for the prophylactic treatment of SRE and hypercalcemia in tumor-induced hypercalcemia. Due to their excretion through the kidneys, patients are sensitive to changes in renal function, which should be monitored closely during bisphosphonate application and adjusted as necessary (Table 11.2).

The choice of one agent over another for patients with bone metastases is based on many factors, including cost. For many patients, the dosing interval for zoledronic acid can be extended to every 3 months, rather than monthly, which increases convenience and decreases cost [14].

As previously mentioned, there are two general classes of agents used for osteoclast inhibition in patients with bone metastases from solid tumors: bisphosphonates and denosumab.

The selection of one osteoclast inhibitor over another is influenced by tumor type, patient preference, and other factors, such as route of administration, frequency of administration, tolerance, accessibility, and cost of therapy.

For patients with bone metastases from breast cancer and castration-resistant prostate cancer, denosumab and zoledronic acid reduce SREs. In the case of metastatic breast cancer, international practice guideline groups have not expressed a preference for a particular agent [15–17]. While monthly denosumab may appear preferable to monthly zoledronic acid in terms of a modest improvement in SRE reduction and analgesic effects, and the fact that denosumab can be administered as a subcutaneous injection rather than an IV infusion, its use needs to be balanced by the significantly higher direct drug costs and the data demonstrating that a 3-month dosing interval of zoledronic acid is noninferior to monthly dosing. Additionally, unlike

TABLE 11.1 Pharmacological effects and dosage of ibandronic and zoledronic acids.

	Zoledronic acid	Ibandronic acid
Intravenous dose	4 mg	2–6 mg
Onset of effect	<4 days	<4 days
Maximum effect	4–7 days	7 days
Duration of effect	4 weeks	Up to 4 weeks
Efficacy in tumor-induced hypercalcemia	90%	75%

Ibandronic acid can be given both intravenously and orally. Data is presented for hypercalcemia.

TABLE 11.2 Dose reduction for zoledronic acid in cancer patients and mild-to-moderate renal impairment.

Baseline creatinine clearance (mL/min)	Recommended dose (mg)
>60	4 (no reduction)
50–60	3.5
40–49	3.3
30–39	3

Manufacturer's recommendations for patients with multiple myeloma or bone metastasis. No data exist for severe renal impairment (creatinine clearance <30 mL/min) because these patients were excluded from the studies. Reduced doses are diluted in 100 mL 0.9% saline or 5% glucose and given intravenously over 15 min.

bisphosphonates, denosumab does not require dose adjustment in renal impairment and is safer for use even when creatinine clearance (CrCl) is <30 mL/min, although the risk of severe hypocalcemia is higher. As such, denosumab is preferred over bisphosphonates in patients with significant renal dysfunction [10,14,18,19].

Considerations prior to initiating an osteoclast inhibitor

Dental health: The risk for osteonecrosis of the jaw with osteoclast inhibitor therapy appears to be up to 2.8% in 3 years [20] but rises with shorter intervals between treatments and longer duration of osteoclast inhibitor use [21]. We agree with the joint guidelines recommending that all patients, if possible, undergo a dental examination before initiation of treatment with an osteoclast inhibitor at regular intervals during therapy [22].

Calcium and vitamin D supplementation: For patients in whom an osteoclast inhibitor is indicated, preexisting hypocalcemia and/or vitamin D deficiency should be corrected prior to treatment. Ongoing assessment is needed during therapy due to the risk of secondary hyperparathyroidism. Patients with solid tumors may be vitamin D-deficient, which places them at risk for hypocalcemia if they are treated with an osteoclast inhibitor without additional calcium (and vitamin D) supplementation [23,24]. As an example, in a combined analysis of three randomized trials involving over 5600 women with breast cancer who were randomly assigned to treatment with denosumab or zoledronic acid [23,24], hypocalcemia was more common among patients who did not report using calcium and vitamin D compared with those who did (16% vs 9% for denosumab and 7.2% vs 4.6% for zoledronic acid). Serious (grade 3/4) hypocalcemia within 6 months of starting treatment is more common with denosumab (3% vs 1%). If there are no contraindications (e.g., history of recurrent renal stones), all patients receiving osteoclast inhibitors who do not have adequate dietary intake should receive calcium and vitamin D supplementation to decrease the risk of hypocalcemia.

Conclusion

This chapter evaluates the mechanism of action, clinical applications, and safety considerations of bisphosphonates and denosumab in cancer pain management. Key evidence supports their role as effective agents for preventing SREs and providing modest analgesic benefits in cancer patients with bone metastases. While complications such as osteonecrosis of the jaw require careful monitoring, appropriate patient selection, prophylactic measures, and evidence-based dosing protocols ensure optimal therapeutic outcomes. The integration of bisphosphonates and denosumab into multimodal cancer pain management strategies represents an important advancement in supportive oncology care.

References

[1] Schwinghammer TL, editor. Pharmacotherapy handbook. New York (NY): McGraw-Hill; 2021.

[2] DiPiro JT, editor. Pharmacotherapy: a pathophysiologic approach. New York (NY): McGraw-Hill; 2020.

[3] Katzung BG, editor. Basic & clinical pharmacology. New York (NY): McGraw-Hill; 2020.

[4] Stevens CW. Brenner and Stevens' pharmacology. 6th ed. Philadelphia (PA): Elsevier/Health Sciences; 2022.

[5] Atnahs Pharma UK Limited. Bondronat 2 mg concentrate for solution for infusion SmPC. Basildon: Atnahs Pharma UK Limited; 2024. [cited 31 May 2025]. Available from: https://www.ema.europa.eu/en/documents/product-information/bondronat-epar-product-information_en.pdf.

[6] Phoenix Labs. Zometa 4 mg/100 mL solution for infusion SmPC. Clonee: Phoenix Labs; 2021. [cited 31 May 2025]. Available from: https://www.medicines.org.uk/emc/product/12754/smpc.

[7] Green J, Lipton A. Anticancer properties of zoledronic acid. Cancer Invest. 2010;28(9):944–57. Available from: https://doi.org/10.3109/07357907.2010.512598.

[8] Porta-Sales J, Garzón-Rodríguez C, Llorens-Torromé S, et al. Evidence on the analgesic role of bisphosphonates and denosumab in the treatment of pain due to bone metastases: a systematic review within the European Association for Palliative Care guidelines project. Palliat Med 2017;31(1):5–25. Available from: https://doi.org/10.1177/0269216316639793.

[9] Cleeland CS, Body JJ, Stopeck A, et al. Pain outcomes in patients with advanced breast cancer and bone metastases: results from a randomized, double-blind study of denosumab and zoledronic acid. Cancer 2013;119:832–8. Available from: https://doi.org/10.1002/cncr.27789.

[10] Saunders Y, Ross JR, Broadley KE, et al. Systematic review of bisphosphonates for hypercalcaemia of malignancy. Palliat Med 2004;18(5):418–31. Available from: https://doi.org/10.1191/0269216304pm914ra.

[11] Palmieri C, Fullarton JR, Brown J. Comparative efficacy of bisphosphonates in metastatic breast and prostate cancer and multiple myeloma: a mixed-treatment meta-analysis. Clin Cancer Res 2013;19(24):6863–72. Available from: https://doi.org/10.1158/1078-0432.CCR-13-2275.

[12] O'Carrigan B, Wong MH, Willson ML, et al. Bisphosphonates and other bone agents for breast cancer. Cochrane Database Syst Rev 2017;10:CD003474. Available from: https://doi.org/10.1002/14651858.CD003474.pub4.

[13] Hoskin PJ, Sundar S, Reczko K, et al. A multicenter randomized trial of ibandronate compared with single-dose radiotherapy for localized metastatic bone pain in prostate cancer. J Natl Cancer Inst 2015;107(10):djv197. Available from: https://doi.org/10.1093/jnci/djv197.

[14] Shapiro CL, Moriarty JP, Dusetzina S, et al. Cost-effectiveness analysis of monthly zoledronic acid, zoledronic acid every 3 months, and monthly denosumab in women with breast cancer and skeletal metastases: CALGB 70604 (alliance). J Clin Oncol 2017;35(35):3949–55. Available from: https://doi.org/10.1200/JCO.2017.73.7437.

[15] Cardoso F, Costa A, Norton L, et al. ESO-ESMO 2nd international consensus guidelines for advanced breast cancer (ABC2). Breast 2014;23(5):489–502. Available from: https://doi.org/10.1016/j.breast.2014.08.009.

[16] Van Poznak CH, Somerfield MR, Barlow WE, et al. Role of bone-modifying agents in metastatic breast cancer: an American Society of Clinical Oncology-Cancer Care Ontario focused guideline update. J Clin Oncol 2017;35(35):3978–86. Available from: https://doi.org/10.1200/JCO.2017.75.4614.

[17] Coleman R, Hadji P, Body JJ, et al. Bone health in cancer: ESMO clinical practice guidelines. Ann Oncol 2020;31:1650–63. Available from: https://doi.org/10.1016/j.annonc.2020.07.019.

[18] Amadori D, Aglietta M, Alessi B, et al. Efficacy and safety of 12-weekly versus 4-weekly zoledronic acid for prolonged treatment of patients with bone metastases from breast cancer (ZOOM): a phase 3, open-label, randomised, non-inferiority trial. Lancet Oncol 2013;14(7):663–70. Available from: https://doi.org/10.1016/S1470-2045(13)70174-8.

[19] Hortobagyi GN, Van Poznak C, Harker WG, et al. Continued treatment effect of zoledronic acid dosing every 12 vs 4 weeks in women with breast cancer metastatic to bone: the OPTIMIZE-2 randomized clinical trial. JAMA Oncol 2017;3(7):906–12. Available from: https://doi.org/10.1001/jamaoncol.2016.6316.

[20] Van Poznak CH, Unger JM, Darke AK, et al. Association of osteonecrosis of the jaw with zoledronic acid treatment for bone metastases in patients with cancer. JAMA Oncol 2021;7:246–54. Available from: https://doi.org/10.1001/jamaoncol.2020.6353.

[21] Ng TL, Tu MM, Ibrahim MFK, et al. Long-term impact of bone-modifying agents for the treatment of bone metastases: a systematic review. Support Care Cancer 2021;29(2):925–43. Available from: https://doi.org/10.1007/s00520-020-05556-0.

[22] Yarom N, Shapiro CL, Peterson DE, et al. Medication-related osteonecrosis of the jaw: MASCC/ISOO/ASCO clinical practice guideline. J Clin Oncol 2019;37:2270–90. Available from: https://doi.org/10.1200/JCO.19.01186.

[23] Wang-Gillam A, Miles DA, Hutchins LF. Evaluation of vitamin D deficiency in breast cancer patients on bisphosphonates. Oncologist 2008;13:821–7. Available from: https://doi.org/10.1634/theoncologist.2008-0013.

[24] Body JJ, Bone HG, de Boer RH, et al. Hypocalcemia during denosumab treatment of metastatic bone disease. Eur J Cancer 2015;51:1812–21. Available from: https://doi.org/10.1016/j.ejca.2015.05.016.

Chapter 12

Corticosteroids

Feđa Đorđević[1] and Boris Milijašević[2]
[1]*Department of Medical Oncology, Institute for Oncology and Radiology of Serbia, Belgrade, Serbia,* [2]*Department of Pharmacology, Toxicology and Clinical Pharmacology, University of Novi Sad, Novi Sad, Serbia*

Key learning objectives

- Understand the pharmacological mechanisms by which glucocorticoids provide analgesic effects in cancer patients.
- Identify appropriate clinical applications for glucocorticoid therapy in cancer pain management, including spinal cord compression, brain metastases, malignant bowel obstruction, and neuropathic pain syndromes.
- Recognize the preferred agents and dosing strategies for different cancer pain scenarios.
- Understand the adverse effects and dangers associated with glucocorticoid therapy, focusing on screening for infections, glucose intolerance, osteoporosis, and neuropsychiatric complications, while implementing appropriate preventive measures.

Pharmacology

Corticosteroids are a broad class of steroid hormones produced by the adrenal cortex, which includes both naturally occurring hormones and their synthetic analogs, often classified into two major categories: glucocorticoids (involved in a variety of physiological functions, including glucose metabolism, inflammation, and immune system modulation) and mineralocorticoids (primarily involved in electrolyte balance and blood pressure regulation). This chapter will focus on glucocorticoids.

In the 1950s, glucocorticoids were approved by the US Food and Drug Administration for use in human medicine. Since their introduction into therapy, these drugs have recorded an increasing number of indications. Additionally, due to their impact on numerous bodily functions, glucocorticoids exhibit various unwanted effects [1–5]. Numerous metabolic effects of glucocorticoids in the body have been described as they affect the metabolism of carbohydrates, fats, and proteins. Glucocorticoids have catabolic and antianabolic effects on lymphoid and connective tissue, as well as muscles, peripheral fat, and skin, leading to a decrease in muscle mass, weakness, and skin thinning. Catabolic and antianabolic effects on bones result in a reduction of bone density, leading to the development of osteoporosis. In children, glucocorticoids can reduce growth, which is why growth hormone replacement may be necessary in some instances. In short, these drugs lead to hyperglycemia, cause a specific redistribution of fat tissue, and result in a negative nitrogen balance. Due to their mineralocorticoid action, they also affect the retention of water and sodium in the body [1–5].

The antiinflammatory and immunosuppressive effects of corticosteroids are also well known. The antiinflammatory action of glucocorticoids is attributed to their inhibition of the enzyme phospholipase A2, which ultimately leads to reduced levels of leukotrienes, prostaglandins, and platelet-activating factors, all of which are involved in inflammation. Glucocorticoids also reduce the number of circulating lymphocytes, monocytes, eosinophils, and basophils by withdrawing these cells into the lymphatic tissue. On the other hand, the number of neutrophils in the circulation increases because they are mobilized from the bone marrow. However, the concentration of both neutrophils and lymphocytes at the site of inflammation is reduced. The immunosuppressive action of these drugs, in addition to the above, is also reflected in the fact that they reduce the activity of tissue macrophages and other cells that present antigens [1–5]. Resorption of corticosteroids after oral administration is rapid and complete. About 90% of circulating cortisol is bound to globulins. On the other hand, synthetic glucocorticoids bind to a large percentage of plasma albumins. All

Pain Management in Palliative Care for Patients With Cancer. DOI: https://doi.org/10.1016/B978-0-443-30050-9.00027-8

glucocorticoids are metabolized, and some are even inactive until they become active after metabolism. They are excreted by the kidneys [1–5].

Since they act on numerous functions of the body, this group of drugs has countless unwanted effects. In order to avoid systemic side effects, depending on the indication, these drugs can also be applied locally (intraarticular, inhaled, on the skin, and in the form of drops for the nose, eyes, or ears) [1–5].

Corticosteroids can cause behavioral disturbances in humans, initially insomnia and euphoria, and later depression. Due to the inhibition of the synthesis of protective prostaglandins at the level of the stomach lining, they can cause peptic ulcers. Side effects include Cushing syndrome, suppression of the response to infection, osteoporosis, atrophy of the adrenal gland, as well as numerous metabolic disorders [1–5]. Discontinuation of long-term therapy with these drugs must be carried out gradually and under the supervision of a doctor. The reason for this is that the body already contains endogenous cortisol, the secretion of which is strictly regulated by the hypothalamus and pituitary gland. Exogenously introduced corticosteroids inhibit the activity of the hypothalamus-pituitary-adrenal cortex axis, so that a sudden stoppage of corticosteroid therapy would leave the axis unprepared for further work and, consequently, lower cortisol levels in the body [1–5].

Indications

The indication and role of glucocorticoids in the treatment of cancer-related pain are evidence-based due to existing literature. A 2015 systematic review of six randomized trials found that glucocorticoid therapy resulted in modestly less pain compared with control treatment (placebo or usual treatment/supportive care) at 1 week (mean difference on a 1–10 Numeric Pain Scale of 0.84 (95% CI, 1.38–0.30), although the evidence quality is low to moderate. Factors limiting analysis included the lack of standardized measurements of pain and the use of different agents, dosages, and routes of delivery. Subgroup analysis regarding the cancer type was not possible [6]. Bruera et al. conducted a randomized trial in 40 patients with advanced cancer. Patients received methylprednisolone 16 mg twice daily or placebo for 5 days, followed by a period of washout for 3 days. All the patients were given methylprednisolone for a total of 20 days, and the study showed that Visual Analog Scale scores for pain were lower in the case of the corticosteroid treatment [7]. In a Swedish survey, out of a total of 147 patients with pain, 50 (34%) had a very good effect of the treatment with corticosteroids, 71 (48%) had some effect, 4 (3%) had no effect, and the effect could not be assessed in 22 patients (15%). The positive response came within a week and lasted for more than 4 weeks. Of these patients, 81% used corticosteroids for more than 4 weeks, and 90% were on betamethasone, with a daily dose of less than 3 mg in most patients [8]. Della Cuna et al. [9] reported a randomized double-blind study conducted among 403 patients with advanced cancer who were given either methylprednisolone, 125 mg intravenously daily, or placebo for 8 weeks. Visual Analog scale scores for pain improved in the methylprednisolone-treated group. Tannock et al. [10] reported an uncontrolled study of oral prednisone, 7.5–10 mg daily, in 37 patients with metastatic prostate cancer. Bone pain was relieved in 14 patients. The response appeared to correlate with a decrease in androgen levels.

Clinical trials demonstrated a wide experience suggesting that glucocorticoids can be helpful in the treatment of different types of cancer pain, including neuropathic and bone pain, pain associated with stretching of an organ capsule caused by a space-occupying lesion (e.g., within the liver or kidney) or duct obstruction, pain from malignant bowel obstruction, pain caused by lymphedema, and headache caused by increased intracranial pressure [11]. Dexamethasone is the preferred drug for the management of cancer-related pain because of its long half-life (up to 72 hours) and low mineralocorticoid effects, although prednisone has the advantage of less hyperglycemia or neuropsychiatric adverse effects, especially at higher doses or prolonged use. It has a shorter half-life, which can be beneficial in situations where rapid tapering is required, although it may be associated with increased fluid retention. Methylprednisolone is, on the other hand, particularly useful in acute settings as it is available both as oral and intravenous therapy and has less mineralocorticoid activity compared to prednisone. However, there is no strong empirical evidence that this drug is more effective in the cancer population than any other glucocorticoid. The dose and frequency should be selected based on the cause of pain and its severity; equivalent dosages are given in Table 12.1.

Special situations with cancer pain and use of corticosteroids

Pain crisis

Managing patients with "pain crisis" who have not experienced rapid relief following opioid administration, whose pain is severe and escalating despite opioid treatment, and whose distress is very high may include treatment with a test dose regimen of 8–10 mg dexamethasone to determine if the pain is steroid responsive [6,12]. If there is a dramatic response,

TABLE 12.1 Equivalent doses, antiinflammatory activity, and duration of action of corticosteroids.

	Equivalent doses (mg)[a]	Antiinflammatory activity relative to hydrocortisone	Duration of action (h)
Short-acting			
Hydrocortisone (cortisol)	20	1	8–12
Cortisone acetate	25	0.8	8–12
Intermediate-acting			
Prednisone	5	4	12–36
Prednisolone	5	4	12–36
Methylprednisolone	4	5	12–36
Triamcinolone	4	5	12–36
Long-acting			
Dexamethasone	0.75	30	36–72
Betamethasone	0.6	30	36–72

[a] *Pain-specific dosage is often higher than the abovementioned doses. For example, prednisone 5–10 mg or methylprednisolone 16–32 mg doses can be required for pain.*

then 4 mg bid or 8 mg daily can be tried, and the dose can be downtitrated to the lowest dose that maintains pain relief while other analgesic treatments are added, if indicated.

Spinal cord compression

Malignant spinal cord compression (mSCC) is an oncological emergency. Pain accompanies mSCC in 95% of patients and usually precedes the diagnosis by days to months. Pain can be local (such as back or neck pain), radicular, or both [12,13]. Patients with established neurological deficits have a poor prognosis for recovery; thus, early diagnosis confirmed by magnetic resonance imaging and prompt therapy are critical. Steroids should be given immediately when the clinical and radiological diagnosis of mSCC is suspected. Dexamethasone is the most frequently used drug. There is a lack of high-quality data to guide the choice between high-dose and moderate-dose regimens; dexamethasone (16 mg/day) remains the most often used prescription, although doses ranging from moderate (8 mg/day) to ultrahigh levels (36–96 mg/day preceded by a bolus of 10–100 mg intravenously [IV]) have been advocated. The steroids are usually tapered over 2 weeks [12,13]. Greenberg et al. [14] reported an uncontrolled study of 83 patients with spinal cord compression and pain, who received 100 mg of dexamethasone intravenously, followed by 24 mg, orally, four times daily, tapered over 2 weeks. The patients were treated concurrently with radiotherapy. Pain relief was observed in up to 82% of patients, and in some cases, before the commencement of radiotherapy.

Obstructive syndromes with pain in cancer

Corticosteroids help by reducing inflammation at the site of the obstruction. Dexamethasone, 8 mg/24 h, may help resolve malignant bowel obstruction and palliate associated symptoms; available evidence does not suggest a survival benefit [15]. The corticosteroids do not appear to affect the length of survival. Still, they may palliate symptoms extremely well, enabling the multidisciplinary team caring for the patient to address further issues regarding placement or treatment [15].

Pain in patients with brain metastases

The reduction of peritumoral edema, in response to steroid therapy, can also lead to improvement in analgesia in patients with brain metastases [16]. Dexamethasone is recommended for the treatment of adults with symptomatic brain metastases

[16]. A dose of 4–8 mg/24 h PO (per os) provides temporary symptomatic relief for patients with pain due to raised intracranial pressure from cerebral edema [16]. If there is a risk of herniation, doses of 16 mg/24 h are recommended [16].

Monitoring and treatment of adverse effects

Patients should be routinely asked about adverse effects related to glucocorticoids. The intensity of monitoring will depend on individual risk factors, such as glucocorticoid dose and duration, other medications being used, and comorbidities. However, generally, during treatment with glucocorticoids, particular attention should be given to several significant potential adverse effects.

Patients on high-dose therapy should have screening for latent infections such as tuberculosis. Respiratory symptoms in patients on moderate- to high-dose glucocorticoids should also prompt consideration of workup for *Pneumocystis jirovecii* pneumonia (PJP), especially if patients are on prolonged glucocorticoid courses with other immunosuppressive agents and/or are not taking prophylactic PJP therapy. Patients who may require prolonged courses or high doses of glucocorticoids may benefit from screening for osteoporosis and consideration of treatment. Particular care should be taken to prevent the development of peptic ulcers.

Additionally, screening for glucose intolerance and type 2 diabetes may need to be intensified for patients on moderate- to high-dose glucocorticoids or on prolonged courses of low-dose therapy. When glucocorticoids induce symptomatic diabetes mellitus or asymptomatic but clinically significant hyperglycemia, these conditions are generally treated pharmacologically in the same way as for patients who are not on glucocorticoid therapy. Screening for neuropsychiatric adverse events, such as mood changes (including irritability, anxiety, depression, insomnia, and even mania and psychosis, along with cognitive impairment), is also warranted during corticosteroid use. Other potential mineralocorticoid adverse events, such as fluid retention and edema, hypertension, hypokalemia, and arrhythmia, are also readily available for regular clinical evaluation [17,18].

Several tools are available to measure glucocorticoid toxicity, including the Glucocorticoid Toxicity Index for adults and the Pediatric Glucocorticoid Toxicity Index for children [19–21]. These have primarily been used in clinical research, although they may be helpful in clinical practice. A gradual taper is recommended after prolonged corticosteroid therapy to reduce the risk of adrenal insufficiency, whereas treatment courses of up to 2 weeks can generally be discontinued without tapering. The rate of dose reduction should be individualized based on treatment duration, dose, and patient response, although the most appropriate method of tapering has not been determined [22].

Conclusion

Corticosteroids represent important adjuvant analgesics in comprehensive cancer pain management strategies, offering modest but clinically meaningful pain relief across diverse cancer-related pain syndromes. Dexamethasone emerges as the preferred agent due to its favorable pharmacokinetic profile and reduced mineralocorticoid activity. Clinical applications span from oncological emergencies like spinal cord compression to chronic conditions, including neuropathic pain and malignant bowel obstruction. However, successful implementation requires careful patient selection, appropriate dosing strategies, and systematic monitoring for adverse effects, including infection risk, metabolic disturbances, and bone health complications. Future research should focus on optimizing dosing protocols and identifying patient populations most likely to benefit from glucocorticoid therapy, while minimizing treatment-related morbidity.

References

[1] Kapugi M, Cunningham K. Corticosteroids. Orthop Nurs 2019;38(5):336–9. Available from: https://doi.org/10.1097/NOR.0000000000000595.

[2] Schwinghammer TL, editor. Pharmacotherapy handbook. New York, NY: McGraw-Hill; 2021.

[3] DiPiro JT, editor. Pharmacotherapy: a pathophysiologic approach. New York, NY: McGraw-Hill; 2020.

[4] Katzung BG, editor. Basic & clinical pharmacology. McGraw-Hill: New York, NY; 2020.

[5] Stevens CW. Brenner and Stevens' pharmacology. 6th ed. Philadelphia (PA): Elsevier; 2022.

[6] Haywood A, Good P, Khan S, et al. Corticosteroids for the management of cancer-related pain in adults. Cochrane Database Syst Rev. 2015;2015(4):CD010756. Available from: https://doi.org/10.1002/14651858.CD010756.pub2.

[7] Bruera E, Roca E, Cedaro L, et al. Action of oral methylprednisolone in terminal cancer patients—a prospective, randomized double-blind study. Cancer Treat Rep 1985;69(7–8):751–4.

[8] Lundström SH, Furst CJ. The use of corticosteroids in Swedish palliative care. Acta Oncol 2006;45(4):430–7. Available from: https://doi.org/10.1080/02841860500401167.

[9] Della Cuna GR, Pellegrini A, Piazzi M. Effect of methylprednisolone sodium succinate on quality of life in preterminal cancer patients—a placebo controlled, multicenter study. Eur J Cancer Clin Oncol 1989;25(12):1817–21. Available from: https://doi.org/10.1016/0277-5379(89)90353-2.

[10] Tannock I, Gospodarowicz M, Meakin W, et al. Treatment of metastatic prostatic cancer with low-dose prednisone—evaluation of pain and quality of life as pragmatic indexes of response. J Clin Oncol 1989;7(5):590–7. Available from: https://doi.org/10.1200/JCO.1989.7.5.590.

[11] Hardy JR, Rees E, Ling J, et al. A prospective survey of the use of dexamethasone on a palliative care unit. Palliat Med 2001;15(1):3–8. Available from: https://doi.org/10.1191/026921601673324846.

[12] Lawton AJ, Lee KA, Cheville AL, et al. Assessment and management of patients with metastatic spinal cord compression: a multidisciplinary review. J Clin Oncol 2019;37(1):61–71. Available from: https://doi.org/10.1200/JCO.2018.78.1211.

[13] Cowap J, Hardy JR, A'Hern R. Outcome of malignant spinal cord compression at a cancer center: implications for palliative care services. J Pain Symptom Manage 2000;19(4):257–64. Available from: https://doi.org/10.1016/s0885-3924(00)00110-x.

[14] Greenberg HS, Kim JH, Posner JB. Epidural spinal-cord compression from metastatic tumor—results with a new treatment protocol. Ann Neurol 1980;8(4):361–6. Available from: https://doi.org/10.1002/ana.410080404.

[15] Feuer DJ, Broadley KE. Systematic review and meta-analysis of corticosteroids for the resolution of malignant bowel obstruction in advanced gynaecological and gastrointestinal cancers. Ann Oncol 1999;10(9):1035–41. Available from: https://doi.org/10.1023/a:1008361102808.

[16] Vecht CJ, Hovestadt A, Verbiest HB, et al. Dose-effect relationship of dexamethasone on Karnofsky performance in metastatic brain tumors: a randomized study of doses of 4, 8, and 16 mg per day. Neurology 1994;44(4):675–80. Available from: https://doi.org/10.1212/WNL.44.4.675.

[17] Liu D, Ahmet A, Ward L, et al. A practical guide to the monitoring and management of the complications of systemic corticosteroid therapy. Allergy Asthma Clin Immunol 2013;9(1):30. Available from: https://doi.org/10.1186/1710-1492-9-30.

[18] Kobza AO, Herman D, Papaioannou A, et al. Understanding and managing corticosteroid-induced osteoporosis. Open Access Rheumatol 2021;13:177–90. Available from: https://doi.org/10.2147/OARRR.S282606.

[19] Brogan P, Naden R, Ardoin SP, et al. The pediatric glucocorticoid toxicity index. Semin Arthritis Rheum. 2022;56:152068. Available from: https://doi.org/10.1016/j.semarthrit.2022.152068.

[20] Miloslavsky EM, Naden RP, Bijlsma JWJ, et al. Development of a glucocorticoid toxicity index (GTI) using multicriteria decision analysis. Ann Rheum Dis. 2017;76(3):543–6. Available from: https://doi.org/10.1136/annrheumdis-2016-210002.

[21] Stone JH, McDowell PJ, Jayne DRW, et al. The glucocorticoid toxicity index: measuring change in glucocorticoid toxicity over time. Semin Arthritis Rheum. 2022;55:152010. Available from: https://doi.org/10.1016/j.semarthrit.2022.152010.

[22] Vyvey M. Steroids as pain relief adjuvants. Can Fam Physician 2010;56(12):1295–7.e415.

Chapter 13

Antispasmodics and muscle relaxants

Iva Skočilić[1,2], Alojzije Košić[1,2], and Ivana Mikolašević[1,2]
[1]*Tumor Clinic, Clinical Hospital Center Rijeka, Rijeka, Croatia,* [2]*Faculty of Medicine, University of Rijeka, Rijeka, Croatia*

Key learning objectives

- Recognize the pharmacological mechanisms and clinical indications for antispasmodics and muscle relaxants in palliative care.
- Identify the appropriate use and dosing of hyoscine butylbromide, baclofen, diazepam, and tizanidine for specific pain and spasticity syndromes.
- Apply principles of individualized therapy, considering comorbidities, organ function, and potential drug interactions.
- Understand the importance of nonpharmacological interventions as first-line treatment for spasticity, with pharmacological options reserved for refractory cases.
- Monitor for and manage common adverse effects and withdrawal symptoms associated with these medications in the palliative care population.

Introduction

The treatment of pain in palliative patients not only includes the use of analgesics but also a plethora of medications that synergistically, in combination with analgesics, help decrease pain perception and, consequently, improve the quality of life of palliative patients. This type of medication is often referred to as coanalgesics or adjuvant analgesics [1]. Some coanalgesics act on the smooth muscle tissue, while others act on skeletal muscles. In this chapter, two groups of medications will be presented systematically:

1. Medications that decrease spasticity and spasms of the skeletal muscles (skeletal muscle relaxants).
2. Medications that act on the smooth muscle tissue (antispasmodics).

It is important to note that the coanalgesics described in this chapter—both antispasmodics and muscle relaxants—do not act on pain perception in the same way as traditional analgesics. Instead, they reduce pain through different mechanisms, for instance, by relaxing smooth muscles or alleviating spasms. Classical analgesics, particularly opioids, remain a cornerstone of pain management in patients with cancer who are receiving palliative care, but coanalgesics have a particular place in helping with particular painful situations which will be covered in this chapter.

Antispasmodics (anticholinergics)

Hyoscine and dicyclomine fall under this category. These medications inhibit muscarinic cholinergic receptors of the enteric plexus and smooth muscle [2]. Considering the more frequent use of hyoscine in palliative care, it will be discussed in this chapter.

Hyoscine butylbromide

Hyoscine butylbromide (also known as Buscopan) is a medication that can be used in palliative patients, primarily for the treatment of acute spasms of the gastrointestinal tract [3]. In addition to the antimuscarinic effect, it also has an antisecretory effect. Some scientific papers suggest its potential use in the treatment of death rattle, genitourinary colic, sialorrhea, and inoperable intestinal obstruction [3,4].

Pain Management in Palliative Care for Patients With Cancer. DOI: https://doi.org/10.1016/B978-0-443-30050-9.00026-6

The primary indication for managing inoperable malignant bowel obstruction typically arises in patients with colorectal, ovarian, or gastric malignancies, as these cancers frequently lead to peritoneal dissemination or direct tumor infiltration of the intestinal lumen, culminating in obstruction [5,6]. The obstruction itself may be partial or complete and is most often caused by the primary tumor or metastases. Other causes of intestinal obstruction include intestinal adhesions, intestinal stenosis, hypokalemia, abdominal hernia, and stool incarceration in elderly patients following surgery, radiotherapy, or platinum-based chemotherapy [5]. The development of mechanical bowel obstruction is multifactorial and is based on an imbalance between fluid secretion and absorption in the intestines, which consequently leads to electrolyte imbalance, acid–base disturbances, and a reduction in circulating volume.

Pain associated with intestinal obstruction is most commonly the result of continuous pain from bowel distension, secretion, and motor activity of the obstructed bowel. It presents as dull and continuous pain [5]. It may also result in intestinal spasms characterized by paroxysmal pain [5]. The primary cause of pain is best treated with opioid analgesics, though it is essential to titrate the dosage to prevent side effects, including constipation, which could exacerbate the preexisting obstruction. In cases of paroxysmal intestinal spasms, anticholinergics should be included in the pain management regimen. However, due to the synergistic effect of anticholinergics and opioids, which can sometimes worsen the symptoms of intestinal obstruction, the simultaneous prescription of stool softeners is recommended to mitigate constipation [5]. Details about constipation and its treatment are covered by another chapter of this book in more depth.

The treatment of intestinal obstruction can be divided into surgical and pharmacological approaches. Given the vulnerability of palliative patients, surgical treatment is often not an option, making pharmacological treatment the only viable approach. This treatment includes the use of antisecretory drugs such as hyoscine butylbromide, glucocorticoids, antiemetics, and opioids [5]. The role of antisecretory drugs is to reduce intestinal peristalsis and fluid secretion into the intestinal lumen, and they may also contribute to the control of vomiting and intestinal colics.

The use of hyoscine butylbromide is contraindicated in patients who suffer from myasthenia gravis, toxic megacolon, glaucoma, narrow-angle glaucoma, tachycardia, hyperplastic prostate with urinary retention, and/or paralytic ileus [3,7].

In palliative care, it is predominantly administered parenterally, either intramuscularly, slowly intravenously, or subcutaneously [3,8]. It is also available in oral form as a tablet.

Hyoscine butylbromide operates by binding to muscarinic receptors located on smooth muscle cells and intramural ganglia of abdominal organs. Consequently, muscle cells relax, and motor activity is suppressed, resulting in decreased emptying of the abdominal organs [8].

Due to low absorption in oral administration, parenteral administration is recommended. The onset of action is expected in 5–10 minutes after subcutaneous administration and 1–2 hours after oral administration [7]. The dosage and method of administration, depending on the indication, are shown in Table 13.1.

Information on dose adjustment based on renal and hepatic function is not included in the drug's prescribing information and has not been studied in depth. Therefore, in cases of impaired liver or kidney function, it is recommended to start with lower doses or extend the interval between administrations while carefully monitoring the side effects. There are no clear instructions on the discontinuation of hyoscine butylbromide or its use among older patients.

If used simultaneously with hyoscine butylbromide, a reduced clinical effect of metoclopramide and domperidone is expected [7], which should also guide the treatment plan.

There are currently no studies dealing with the use of hyoscine butylbromide primarily for the treatment of pain in palliative patients. The most significant data for hyoscine butylbromide in palliative care is for the treatment of death rattles. Death rattle occurs in 23%–92% of dying patients, making it one of the most common phenomena in palliative medicine [4]. In the study mentioned previously, the patients were divided into two groups: the first group received hyoscine butylbromide after the onset of death rattle, while the second group received medication in a prophylactic

TABLE 13.1 Dosage and administration of hyoscine butylbromide for bowel obstruction with painful colic [3].

Mode	Dosage
Continuous subcutaneous infusion	Initial dose: 20 mg; continue with 60 mg/24 h via continuous subcutaneous infusion
As needed	20 mg every 2 h; maximum daily dose: 120 mg

manner in the absence of death rattle. All patients were administered 20 mg of hyoscine butylbromide as a subcutaneous bolus or intravenously, followed by 60 mg/24 h (20 mg every 8 hours or continuous intravenous/subcutaneous infusion). The results showed that prophylactic administration of hyoscine butylbromide is an effective method for alleviating death rattles. Late applications lead to limited responses [4].

The most frequent side effects include xerostomia and, less frequently, urine retention, tachycardia, visual disturbances, confusion, and constipation [3]. Currently, there is no international consensus on the treatment of xerostomia in palliative care patients, and treatment is primarily based on practical experience. Based on the frequent fluid imbalance in palliative patients, the aim is to maintain a balance between fluid intake and loss to prevent dehydration and worsening xerostomia. There are also saliva substitutes available on the market in various forms; however, they do not possess the physical, enzymatic, antimicrobial, antifungal, and antiviral properties that saliva does. Additionally, they can also worsen nausea in patients due to their taste and consistency. In patients with preserved salivary secretion, products such as lozenges and chewing gums can be used [9].

Currently, a study is underway examining the topical use of pilocarpine, a muscarinic receptor agonist that stimulates secretion [10]. The results of this study have not yet been published [10]. The effectiveness of pilocarpine is questionable depending on the remaining therapy and possible side effects in combination with other drugs, which should be further investigated.

Skeletal muscle relaxants

Therapeutics that act on skeletal muscles are divided into two groups: neuromuscular blockers, used in surgical procedures and intensive care units, and spasmolytics and antispasmodics, which are used to decrease spasticity and spasms. This chapter will explain the role of the latter group. The goal of these medications is to decrease pain; however, there is no proof that they can be used to increase muscle function [11].

Spasticity is one of the symptoms associated with central nervous system tumors due to damage to the upper motor neurons of the brain parenchyma or spinal cord injury, regardless of whether the lesions are primary or secondary [12]. Subtle muscle weakness is observed in the early phase of central tumors, and it worsens in the later phase. In addition, due to the increasing disruption of the inhibitory modulation of alpha motor neurons, spasticity occurs. Headaches are also common, usually accompanied by nausea and vomiting. Seizures are most often focal, though they can be generalized. In spinal cord tumors, spasticity typically occurs in muscles below the level of the injury, often in the later stages of the disease [12].

The medications presented in this chapter are used in the treatment of clinical spasticity caused by upper motor neuron disorders; however, none are explicitly approved for spasticity caused by cancer. It is recommended to begin treating spasticity with nonpharmacological methods, such as stretching, before introducing pharmacological treatments [12].

Benzodiazepines

The most used medication from the benzodiazepine group in the treatment of spasticity is diazepam [13]. Other medications from the same group are used to treat spasticity, such as midazolam; however, there is not enough clinical evidence about its effects. As the role of benzodiazepines will be discussed in more detail in another chapter, only a brief overview of diazepam's role in treating spasticity will be presented here.

Diazepam, like other benzodiazepines, acts by binding to $GABA_A$ receptors on the membrane of central nervous system neurons. By binding to the subunits of the $GABA_A$ receptor, it leads to an increased effect of gamma-aminobutyric acid, a ligand of this receptor, resulting in increased inhibition of signal transmission in the nervous system [13].

In clinical practice, diazepam is typically used as an adjunct to baclofen in the treatment of spasticity, whereas its use as a monotherapy is less common [13]. Initial treatment may start with 2 mg administered twice daily, with gradual dose escalation based on the clinical response. Alternatively, a single 5-mg dose at night may be helpful to control nocturnal symptoms [13].

Baclofen

Table 13.2 shows the indications and patient groups in which caution is advised while administering baclofen, as well as those in which its administration is contraindicated. Besides sensitivity to the medication and its ingredients, there are no contraindications. This chapter will address the oral form of the medication, although intrathecal administration is possible as well.

TABLE 13.2 Indications and patient groups in which caution is advised.

Indication [7,14]	Precautions/contraindications [7]
Hiccups (off-label)	Epilepsy, psychosis, schizophrenia, depression, mania, gastrointestinal ulceration, alcoholism, diabetes, hypertension, cerebrovascular disease, decreased renal function, urinary retention
Muscle spasms and musculoskeletal pain (off-label)	
Spasticity	

TABLE 13.3 Baclofen dosage depending on the indication and dose adjustments in patients with chronic kidney disease.

Indication		Dosage and administration [17,18]	
Muscle spasms and musculoskeletal pain (off label)		Initial dose: 5–10 mg up to three times per day as needed	
Spasticity		Initial dose: 5 mg once to three times per day; *increase by 5 mg every 3 days*; maximum dose: 80 mg/day	
Creatinine clearance (mL/min)	Starting daily dose (mg)	Dosage regimen	eGFR (mL/min/1.73 m^2)
>80	15	5 mg/8 h	≥90
50–80	10	5 mg/12 h	60–89
30–50	7.5	2.5 mg/8 h	30–59
<30	5, or avoid completely	2.5 mg/12 h	15–29

In contrast to benzodiazepines, baclofen has an affinity for $GABA_B$ receptors [15]. In addition to decreasing excitatory neurotransmitters in the brain and spinal cord, baclofen also reduces pain, likely by inhibiting substance P in the spinal cord [15]. Although there is a need for more data, baclofen does not appear inferior to benzodiazepine for the treatment of spasticity [13].

It is characterized by good and rapid absorption when administered orally, and the peak concentration in the blood is reached between 1 and 2.79 hours after administration [16]. It binds to plasma proteins at approximately 30% [16]. It is predominantly excreted unchanged via urine [16].

The dosage of baclofen, depending on indications and kidney function, is shown in Table 13.3.

Considering the compromised renal function in palliative patients, dosage adjustment may be necessary [7,18]. Also, since 15% of the drug is metabolized in the liver, it is recommended to use it with additional precautions in patients with primary or secondary liver disease or in cases of impaired liver function [12].

A gradual discontinuation of the medication over a span of a week is recommended unless due to severe side effects. A rapid discontinuation can lead to altered mental status, worsening of spasticity, febrility, nausea, and weakness. Considering the possibility of symptom progression after the discontinuation, the first-line treatment would be to continue therapy in the same form as before. It is also important to avoid dehydration to prevent and cure the consequences of potential rhabdomyolysis and provide cardiopulmonary support [17].

The most common side effects include nausea and vomiting, gastrointestinal disturbance, respiratory depression, sedation, somnolence, insomnia, euphoria, headache, hallucinations, hypotension, tremor, and dry mouth [7].

The use of baclofen in palliative patients has been predominantly studied for the treatment of hiccups. Unfortunately, most of the information on this topic has been obtained from the case series. These studies support the use of baclofen for treating hiccups in adult patients, although only a low level of recommendation could be made as the strength of evidence is weak [14]. In addition, the role of baclofen as an analgesic in the treatment of cancer pain has also been investigated. In the aforementioned retrospective study, it was demonstrated that baclofen could help manage pain as an adjuvant analgesic. A significant reduction in pain perception was observed in most patients treated with baclofen [15].

TABLE 13.4 Indications and contraindications for using tizanidine.

Indications [20]	Contraindications [21]
Muscle spasm and/or musculoskeletal pain (off-label)	Concomitant therapy with CYP1A2 inhibitors (e.g., ciprofloxacin)
Spasticity	

TABLE 13.5 The dosage and mode of administration of tizanidine [13,22].

Indications	Dosage and administration
Spasticity	Initial dose: 2-mg tablet once daily at bedtime. Increase the dose, depending on tolerability and effect, by 2 mg weekly to a maximum of 36 mg/day, divided into three or four doses.

Tizanidine

Tizanidine falls into the category of α-2 adrenergic agonists, which decrease spasticity by increasing presynaptic inhibition [13,19]. It acts predominantly in the area of the spinal cord. According to clinical studies, tizanidine does not exhibit an inferior effect compared to diazepam and baclofen; however, the side effects are milder [13]. The indications and contraindications of tizanidine are shown in Table 13.4.

Approximately 30% of the drug is bound to plasma proteins [19]. The liver metabolizes the majority of the medication, while the elimination half-life is 2.5 hours, and it is primarily excreted via urine [19]. The dosage and mode of administration are shown in Table 13.5.

Caution is advised when administering tizanidine in elderly patients due to decreased creatinine clearance [19]; hence, the medication should be used cautiously with dose reduction and close monitoring in patients with severe kidney damage [19]. In patients with moderate damage, it is essential to initiate treatment at a lower dosage and adjust it according to tolerance. A dosage adjustment is also required depending on creatinine clearance [19]. Elevation of liver enzymes can be expected in 5% of patients, and it is recommended to monitor them at the initiation of therapy and intermittently thereafter [12].

Discontinuation is conducted gradually to avoid withdrawal symptoms (e.g., tachycardia, hypertension, hypertonia), especially after prolonged exposure to higher doses [23].

Side effects differ from those of the previously discussed medication and include hepatotoxicity, asthenia, significantly less muscle weakness, drowsiness, hypotension, dry mouth, and dizziness [13,19]. Drowsiness could potentially be controlled by taking the medication before bedtime [24]. In patients receiving antihypertensive therapy or those with insufficient fluid intake, hypotensive episodes or orthostatic hypotension may develop; so the drug should be used with caution [12]. To prevent this side effect, maintaining fluid balance and adjusting chronic therapy is essential.

No recent studies are addressing the use of tizanidine in palliative care, which is a major limitation to any recommendation for the use of tizanidine. A study in the year 2000 concluded that tizanidine can contribute to the treatment of spasticity, particularly in patients with multiple sclerosis. The study also found that tizanidine does not lead to a decrease in functional status, which is a potential concern with other drugs. Compared to baclofen, tizanidine causes less muscle weakness, although dry mouth and somnolence are more common. The authors suggested that the choice of drug should be based on the side effect profile, as the effectiveness of both drugs is comparable. When compared to diazepam, tizanidine demonstrates similar effectiveness while causing less muscle weakness [25].

Conclusion

The treatment of pain in palliative cancer patients, in addition to analgesics, often includes the use of coanalgesic medications, which, with their synergistic effects, reduce pain and contribute to patients' quality of life. Coanalgesics also

TABLE 13.6 The drug of choice based on the indication.

Medication	Hyoscine butylbromide	Baclofen	Tizanidine
Indications	Spasms of the gastrointestinal tract	Muscle spasms and musculoskeletal pain[a] (off-label)	Muscle spasm and/or musculoskeletal pain[a] (off-label)
	Genitourinary spasm	Spasticity[a]	Spasticity[a]
	Inoperable intestinal obstruction (off-label)	Hiccups (off-label)	
	Death rattle (off-label)		
	Sialorrhea (off-label)		

[a] *Choice of the drug should be based on the side effect profile, as the effectiveness of both drugs is comparable.*

include skeletal muscle relaxants, which help reduce pain caused by spasms or spasticity, while anticholinergics act on smooth muscles, alleviating pain in the gastrointestinal, biliary, and urogenital tracts. The use of analgesics, such as opioids, is recommended despite potential side effects and remains a gold standard for the majority of patients with cancer receiving palliative care [1,5,7]. If opioids alone do not achieve the desired clinical response, the use of anticholinergics is advised. For pain caused by spasms, the choice of medication depends on the side effect profile. For instance, baclofen is less likely to cause dry mouth and somnolence, while tizanidine is associated with less muscle weakness. A summary of medications, based on indications discussed in the previous text, is also presented in Table 13.6. An appropriate choice of medication, dosage, mode of administration, and adjustment of treatment, depending on other comorbidities, is necessary to achieve the desired effect.

References

[1] Watson M, Campbell R, Vallath N, Ward S, Wells J. Chapter 8: Pain management. In: Watson M, Campbell R, Vallath N, Ward S, Wells J, editors. Oxford handbook of palliative care. Oxford: Oxford University Press; 2019.

[2] McQuaid KR. Drugs used in the treatment of gastrointestinal diseases. In: Katzung BG, Vanderah TW, editors. Basic & clinical pharmacology. 15th ed. New York (NY): McGraw-Hill; 2021. Available from: https://accessmedicine.mhmedical.com/content.aspx?bookid=2988&sectionid=250604938.

[3] Health New Zealand I Te Whatu Ora. Hyoscine N-BUTYLbromide (Buscopan) subcutaneous—palliative care. Medicine & administration, medicine & infusion protocols (adults)/1d palliative care; 2019. Available from: https://www.waitematadhb.govt.nz/assets/Documents/health-professionals/palliative-care/P-Hyoscine-N-BUTYLbromide-Buscopan-Subcutaneous-Palliative-Care-Jul19.pdf.

[4] Mercadante S, Marinangeli F, Masedu F, et al. Hyoscine butylbromide for the management of death rattle: sooner rather than later. J Pain Symptom Manage 2018;56(6):902–7. Available from: https://doi.org/10.1016/j.jpainsymman.2018.08.018.

[5] Huang X, Xue J, Gao M, et al. Medical management of inoperable malignant bowel obstruction. Ann Pharmacother 2021;55(9):1134–45. Available from: https://doi.org/10.1177/1060028020979773.

[6] Wey W, Mian M, Calabrese R, et al. Palliative medical management of inoperable malignant bowel obstruction with "triple therapy": dexamethasone, octreotide, and metoclopramide. Am J Hosp Palliat Care 2021;38(4):340–5. Available from: https://doi.org/10.1177/1049909120968249.

[7] MacLeod R, Macfarlane S. The palliative care handbook: guidelines for clinical management and symptom control. 9th ed. Sydney (NSW): HammondCare Media; 2018.

[8] Tytgat GN. Hyoscine butylbromide - a review on its parenteral use in acute abdominal spasm and as an aid in abdominal diagnostic and therapeutic procedures. Curr Med Res Opin 2008;24(11):3159–73. Available from: https://doi.org/10.1185/03007990802472700.

[9] Kvalheim SF, Strand GV. A narrative of oral care in palliative patients. Int J Environ Res Public Health 2022;19(10):6306. Available from: https://doi.org/10.3390/ijerph19106306.

[10] van der Meulen AI, Neis EPJG, de Nijs EJM, et al. Dry mouth in patients with a life-limiting condition or frailty: a study protocol for two intervention studies and a nested qualitative sub-study (the Dry mOuth Project DROP). BMC Palliat Care 2023;22(1):120. Available from: https://doi.org/10.1186/s12904-023-01242-0.

[11] Kruidering-Hall M, Campbell L. Skeletal muscle relaxants. In: Katzung BG, Vanderah TW, editors. Basic & clinical pharmacology. 15th ed. McGraw-Hill; 2021. Available from: https://accessmedicine.mhmedical.com/content.aspx?bookid=2988&sectionid=250598374.

[12] Hill AR, Vishwa RS, Heather W. Chapter 27: Management of the cancer patient with spasticity. In: Brashear A, Elovic E, editors. Spasticity. 2nd ed. New York (NY): Demos Medical Publishing; 2015. Available from: https://doi.org/10.1891/9781617052422.0027.

[13] Kita M, Goodkin DE. Drugs used to treat spasticity. Drugs 2000;59(3):487–95. Available from: https://doi.org/10.2165/00003495-200059030-00006.
[14] Adam E. A systematic review of the effectiveness of oral baclofen in the management of hiccups in adult palliative care patients. J Pain Palliat Care Pharmacother 2020;34(1):43–54. Available from: https://doi.org/10.1080/15360288.2019.1705457.
[15] Yomiya K, Matsuo N, Tomiyasu S, et al. Baclofen as an adjuvant analgesic for cancer pain. Am J Hosp Palliat Care 2009;26(2):112–8. Available from: https://doi.org/10.1177/1049909108327968.
[16] Simon N, Franchitto N, Rolland B. Pharmacokinetic studies of baclofen are not sufficient to establish an optimized dosage for management of alcohol disorder. Front Psychiatry 2018;9:485. Available from: https://doi.org/10.3389/fpsyt.2018.00485.
[17] Romito JW, Turner ER, Rosener JA, et al. Baclofen therapeutics, toxicity, and withdrawal: a narrative review. SAGE Open Med 2021;9:20503121211022197. Available from: https://doi.org/10.1177/20503121211022197.
[18] Vlavonou R, Perreault MM, Barrière O, et al. Pharmacokinetic characterization of baclofen in patients with chronic kidney disease: dose adjustment recommendations. J Clin Pharmacol 2014;54(5):584–92. Available from: https://doi.org/10.1002/jcph.247.
[19] Ghanavatian S, Derian A. Tizanidine StatPearls [Internet]. Treasure Island (FL): StatPearls Publishing; 2025 [Updated 28 August 2023] Available from: https://www.ncbi.nlm.nih.gov/books/NBK519505/.
[20] Pareek A, Chandurkar N, Chandanwale AS, et al. Aceclofenac–tizanidine in the treatment of acute low back pain: a double-blind, double-dummy, randomized, multicentric, comparative study against aceclofenac alone. Eur Spine J 2009;18(12):1836–42. Available from: https://doi.org/10.1007/s00586-009-1019-4.
[21] Villa-Zapata L, Gómez-Lumbreras A, Horn J, et al. A disproportionality analysis of drug-drug interactions of tizanidine and CYP1A2 inhibitors from the FDA adverse event reporting system (FAERS). Drug Saf 2022;45(8):863–71. Available from: https://doi.org/10.1007/s40264-022-01200-4.
[22] Nair KP, Marsden J. The management of spasticity in adults. BMJ 2014;349:g4737. Available from: https://doi.org/10.1136/bmj.g4737.
[23] Suárez-Lledó A, Padullés A, Lozano T, et al. Management of tizanidine withdrawal syndrome: a case report. Clin Med Insights Case Rep 2018;11:1179547618758022. Available from: https://doi.org/10.1177/1179547618758022.
[24] Tanaka H, Fukuda I, Miyamoto A, et al. Effects of tizanidine for refractory sleep disturbance in disabled children with spastic quadriplegia. No Hattatsu [Brain Dev] 2004;36(6):455–60. [Japanese].
[25] Smith HS, Barton AE. Tizanidine in the management of spasticity and musculoskeletal complaints in the palliative care population. Am J Hosp Palliat Care 2000;17(1):50–8. Available from: https://doi.org/10.1177/104990910001700111.

Chapter 14

Vitamin D

Ivana Mudnić[1], Jerolima Pinjatela[2], and Branka Petrić Miše[3,4]
[1]Department of Basic and Clinical Pharmacology, School of Medicine, University of Split, Split, Croatia, [2]Department of Internal Intensive Medicine and Clinical Pharmacology and Toxicology, University Hospital of Split, Split, Croatia, [3]Department of Oncology and Radiotherapy, University Hospital of Split, Split, Croatia, [4]Faculty of Medicine, University of Split, Split, Croatia

Key learning objectives

- Identify patients in palliative care who are likely to benefit from vitamin D screening and supplementation.
- Interpret 25-hydroxyvitamin D serum levels in the context of cancer-related pain and supportive care.
- Implement safe, evidence-based dosing strategies with specific vitamin D formulations.
- Integrate vitamin D correction into multimodal pain and symptom management plans in palliative care.

Introduction

Vitamin D exerts immunomodulatory, neuroprotective, and antiinflammatory effects that are relevant in cancer pain management in palliative care settings. While evidence from observational studies is compelling, results from randomized trials remain mixed, often limited by sample size, baseline deficiency severity, and variability in dosing strategies [1]. Nevertheless, the potential for vitamin D supplementation to reduce pain, improve immune competence, and lower opioid requirements in palliative care cancer patient populations justifies continued investigation [2].

Dietary sources, endogenous synthesis, and bioactivation

Vitamin D is, chemically, a fat-soluble secosteroid; biochemically, a pro-hormone that can be endogenously produced by human skin and further metabolized to attain biological activity; and physiologically, an agonist of ubiquitous intracellular steroid receptors responsible for pleiotropic effects.

Calciferol is the common collective name of cholecalciferol, vitamin D_3, and ergocalciferol, vitamin D_2. Both forms can be found in dietary sources: vitamin D_3 mainly in foods of animal origin, whereas vitamin D_2 mainly in foods of plant origin. Foods high in vitamin D_3 include cod liver, tuna, and fish oil. In contrast, plants and mushrooms naturally contain only small amounts of vitamin D_2. However, this content can be significantly increased through specialized cultivation methods, such as exposure to ultraviolet light. Furthermore, extracts derived from dried plants and mushrooms can be used to fortify foods and enhance the vitamin D content of milk and other dairy products [3].

However, only a minor proportion of total vitamin D in the human body is derived from exogenous dietary sources, and the majority of vitamin D is synthesized in the skin through a photochemical reaction triggered by exposure to sunlight (UVB radiation). In this process, 7-dehydrocholesterol is converted into precholecalciferol, which subsequently undergoes a thermal isomerization to form cholecalciferol. Endogenous synthesis of vitamin D_3 via cutaneous production can yield approximately 10,000–20,000 international units (IU) within 30 minutes of full-body exposure to midday sunlight under optimal conditions. The subsequent metabolism of the two forms of vitamin D is fundamentally similar [4]. In the following text, the term *vitamin D*, without a subscript, is used, as it refers to both vitamin D_2 and D_3 and to their metabolites.

Vitamin D is then transported to the liver, where it is hydroxylated to 25-hydroxyvitamin D (25(OH)D, calcifediol) mostly by CYP2R1. Although biologically inactive, calcifediol is the major circulating form of vitamin D and is characterized by its chemical stability and relatively long half-life (~21 days). For these reasons and strong detection

Pain Management in Palliative Care for Patients With Cancer. DOI: https://doi.org/10.1016/B978-0-443-30050-9.00013-8

repeatability, serum 25(OH)D is used as the standard biomarker for assessing vitamin D status [5]. The final step in vitamin D bioactivation occurs in the kidneys, where calcifediol is further hydroxylated by the enzyme 25-hydroxyvitamin D-1α-hydroxylase (CYP27B1) to 1,25-dihydroxyvitamin D (1,25$(OH)_2$D, calcitriol), the hormonally active form with the half-life of approximately 5–8 hours.

In addition to hepatic conversion, calcifediol can also be synthesized locally in various tissues, supporting autocrine signaling pathways. Accordingly, subsequent activation to the biologically active form occurs not only in the kidneys but also in multiple extrarenal tissues [6].

Pharmacokinetics

In healthy individuals, approximately 85% of circulating vitamin D metabolites are tightly bound to vitamin D-binding protein (vDBP), while around 15% are loosely associated with albumin and only a small fraction circulates in the unbound form. Importantly, most tissues can only utilize the free, unbound fraction of calcitriol for cellular uptake and biological activity [7]. The binding affinity of vitamin D metabolites to vDBP also determines their plasma half-life and systemic retention. While adipose tissue serves as a storage reservoir for vitamin D, the dynamics of 25(OH)D mobilization from the lipid pools remain incompletely understood and may have implications for vitamin D-related health outcomes [8].

Vitamin D and its metabolites are taken up by cells via two primary mechanisms: passive diffusion of the unbound fractions across cell membranes and receptor-mediated endocytosis of the complexes with the vDBP via the endocytic coreceptors megalin and cubilin [9].

Pharmacodynamics

Vitamin D mediates its biological effects primarily through binding to the nuclear vitamin D receptor (VDR), which regulates gene transcription. Maestro et al. reported that the VDR is present in most human tissues and regulates over 1000 genes [10]. The active form of vitamin D, 1,25$(OH)_2$D, acts through VDR via genomic and nongenomic pathways. Genomic effects occur when nuclear VDR (nVDR) functions as a ligand-activated transcription factor, modulating gene expression over hours to days. The rapid, nongenomic effects of 1,25$(OH)_2$D are mediated by its interaction with membrane-bound vitamin D receptors [11]. These receptors mediate rapid vitamin D signaling through interaction with membrane scaffolds and signaling molecules, complementing the slower genomic actions of nVDR [12].

Genetic polymorphisms in genes related to vitamin D metabolism, transport, storage, and receptor signaling may significantly affect individual responses to vitamin D exposure [13].

Calcitriol exerts its biological effects by enhancing calcium absorption in the intestines, promoting calcium incorporation into the bone, and exerting additional regulatory roles, including antiinflammatory, immunomodulatory, antineoplastic, and analgesic actions.

Vitamin D and pain in palliative care cancer patients: proposed mechanisms of action on the subcellular level

Emerging evidence suggests that vitamin D supplementation may serve as a beneficial adjunct in managing pain in patients with cancer receiving palliative care. VDR and 1α-hydroxylase are expressed in key central nervous system regions involved in pain processing both in glial and neuronal cells, where local conversion of 25(OH)D to 1,25$(OH)_2$D allows autocrine and paracrine regulation of neuroimmune responses involved in the pathophysiology of chronic pain, as described below.

Vitamin D functions as a neuroactive steroid, modulating neuronal excitability via effects on spontaneous firing, action potential duration, and neurotransmitter sensitivity. It regulates both excitatory and inhibitory transmission by influencing receptor expression and modulates ion channels which contribute to the attenuation of nociceptive signaling [14].

Moreover, it upregulates neurotrophins such as nerve growth factor (NGF), neurotrophin 3, and glial cell line-derived neurotrophic factor, while downregulating neurotrophin 4. These neurotrophic factors are critical for neuronal development, maintenance, and repair. NGF is a key mediator in inflammatory hyperalgesia by increasing nociceptor sensitivity and innervation [15].

Vitamin D suppresses prostaglandin-mediated pain pathways as it regulates inhibition of cyclooxygenase-2 expression and induction of 15-prostaglandin dehydrogenase which degrades prostaglandins. This reduces prostaglandin E2 (PGE_2) and prostaglandin F2 ($PGF_2α$) receptor activation on sensory neurons. Furthermore, vitamin D modulates neuroinflammation by upregulating antiinflammatory cytokines such as transforming growth factor-beta 1 (TGF-β1) and IL-4 and suppressing proinflammatory mediators including tumor necrosis factor-alpha (TNF-α), IL-1β, and IL-2 [16]. It inhibits macrophage colony-stimulating factor (M-CSF) activity in glial cells and macrophages, limiting their release of pronociceptive substances, thus attenuating peripheral and central sensitization in chronic pain states. Vitamin D inhibits inducible nitric oxide synthase expression in macrophages, decreasing nitric oxide production, which is one of the key mediators in central sensitization and neurotoxicity [17]. It also supports astrocytic detoxification mechanisms by upregulating γ-glutamyl transpeptidase involved in glutathione metabolism, contributing to neuroprotection and reduction of pain-related oxidative stress [18].

Vitamin D regulates immune responses relevant to chronic pain by inhibiting neutrophil infiltration and T-helper cell activation [19].

Vitamin D and pain in palliative care cancer patients: overview of clinical research

Multiple observational studies have demonstrated that cancer patients typically exhibit lower vitamin D levels compared to healthy individuals [20,21]. Although a metaanalysis of 52 randomized controlled trials involving 75,454 participants demonstrated that vitamin D supplementation was not associated with a significant reduction in all-cause mortality [22], which contrasts with the previous analysis [23], a decrease in cancer-related mortality was repeatedly reported [22,24].

Focused on palliative care cancer patients, observational data from Sweden involving 100 palliative cancer patients indicated that low serum 25(OH)D levels were correlated with increased pain intensity, higher opioid consumption, greater infectious burden, and reduced quality of life [20].

Furthermore, in a cross-sectional study of 30 advanced cancer patients receiving palliative care, higher 25(OH)D levels were positively associated with reduced fatigue and better physical and functional well-being [25]. In addition, a matched case–control study involving 39 palliative cancer patients found that daily supplementation with 4000 IU of vitamin D significantly decreased fentanyl requirements compared to controls, alongside improvements in quality of life and reduced antibiotic use. Notably, vitamin D was well tolerated in all patients [26].

A planned trial in breast cancer patients receiving doxorubicin and cyclophosphamide, followed by treatment with paclitaxel (AC-T) chemotherapy, investigates vitamin D's role in preventing chemotherapy-induced peripheral neuropathy. Deficient patients are administered 50,000 IU weekly for 8 weeks, followed by 100,000 IU monthly, with the Modified Total Neuropathy Score as the primary end point [27].

Another ongoing trial evaluates high-dose enteral vitamin D (576,000 IU in week 1, followed by 72,000 IU/week for 3 weeks) in advanced cancer patients with pain, aiming to raise serum 25(OH)D to ≥30 ng/mL. Outcomes include changes in oral morphine dose, pain scores, and correlations with inflammatory, immune, and nutritional markers [28].

High-dose or "stoss" vitamin D therapy (applied as single or divided large doses) has been tested in ICU patients with sepsis and severe illness including palliative care cancer patients. Han et al. administered either placebo, 50,000 IU, or 100,000 IU vitamin D daily for 5 days to mechanically ventilated ICU patients. High-dose vitamin D effectively increased plasma 25(OH)D into the sufficient range and was associated with a reduction in hospital stay [29].

Vitamin D and pain in palliative care cancer patients: practical implications

Cancer patients generally exhibit lower serum vitamin D levels compared to healthy individuals. Notably, in palliative care settings, an inverse correlation has been observed between vitamin D concentration and pain intensity [30,31], suggesting that even mild vitamin D deficiency should be addressed with supplementation, although causality and optimal thresholds for analgesic benefit remain uncertain.

Vitamin D deficiency is diagnosed based on laboratory measurements of circulating 25(OH)D. It is essential to consider the measurement units used in laboratory reports. According to the Endocrine Society's consensus papers, a serum 25(OH)D level above 50 nmol/L (equivalent to 20 ng/mL) indicates sufficiency, insufficiency between 30 and 50 nmol/L, and deficiency below 30 nmol/L (12 ng/mL) [32].

However, it has been hypothesized that serum levels above 75 nmol/L may be necessary to achieve the neuroprotective and immunomodulatory benefits of vitamin D [31].

Therapeutic strategies for vitamin D supplementation vary in dosage, administration route, and duration. Oral administration is considered the most effective and practical method, although parenteral formulations are also available. While a typical maintenance dose is around 800–1000 IU/day, in palliative cancer care settings, doses of up to 4000 IU/day are frequently used over a treatment period of approximately 3 months [33]. This dose corresponds to 100 μg or roughly eight drops of standard liquid formulations per day (10 μg corresponds to 400 IU and approximately 0.8–1 drop of vitamin D) [30].

Cholecalciferol and ergocalciferol and their metabolites calcifediol and calcitriol are available for oral administration in different vehicles, such as oil-containing gel capsules, oily drops, and hard-powder tablets, capsules, and liquids in a range of dose strengths of 400–50,000 IU and 2400–50,000 IU per dose unit, respectively. Parenteral administration of intermittent vitamin D boluses (not higher than 100,000 IU) may be indicated in patients with hypovitaminosis D who are not suitable for oral intake. Intramuscular cholecalciferol may be the preferred form of vitamin D to be used in these clinical settings [32,34]. Two analogs of calcitriol, doxercalciferol, and paricalcitol are also used therapeutically. One additional analog of calcitriol (calcipotriol) can be used topically [35].

However, according to the latest guidelines, oral cholecalciferol remains the preferred form of vitamin D for supplementation in the majority of patients, while other vitamin D analogs and parenteral administration should be used in specific conditions [32].

The choice of a specific vitamin D formulation should be tailored to the individual patient's clinical status. For instance, in patients with renal insufficiency, where renal hydroxylation is impaired, calcitriol, the active form of vitamin D, is preferred, whereas calcifediol should be reserved for patients with liver failure or severe intestinal malabsorption syndromes [32,36].

Several medications may interfere with vitamin D metabolism or absorption. For example, enzyme-inducing anticonvulsants (e.g., phenytoin) or antimycobacterial drug rifampin increase its elimination, whereas isoniazid as an inhibitor of the activation enzymes can reduce vitamin D bioactivation and effectiveness. Lipid absorption inhibitors such as orlistat or cholestyramine may impair gastrointestinal absorption. Imidazole antifungals inhibit vitamin D bioactivation, while glucocorticoids accelerate its metabolism. Thiazide diuretics may increase the risk of hypercalcemia by decreasing renal calcium excretion, and magnesium-containing products (e.g., some antacids) should be avoided with high-dose

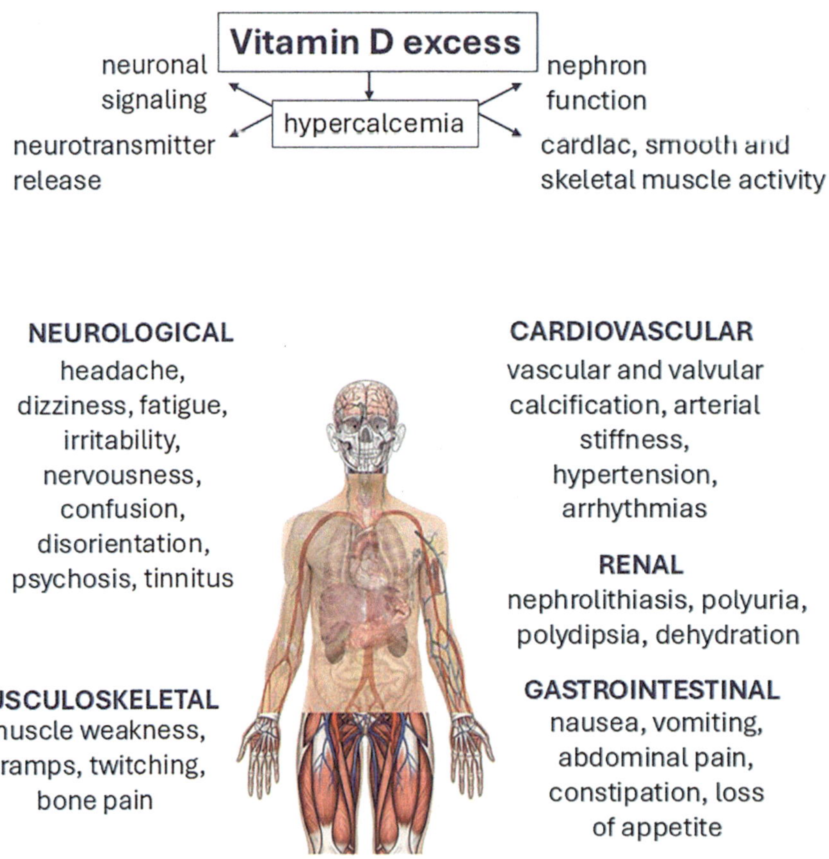

FIGURE 14.1 Signs and symptoms associated with vitamin D excess.
Source: Reprinted via a creative commons attribution-NonCommercial-NoDerivatives license (http://creativecommons.org/licenses/bync-nd/4.0/). From Janoušek J, Pilařová V, Macáková K, et al. Vitamin D: sources, physiological role, biokinetics, deficiency, therapeutic use, toxicity, and overview of analytical methods for detection of vitamin D and its metabolites. Crit Rev Clin Lab Sci 2022;59(8):517–54. doi:10.1080/10408363.2022.2070595. Epub 2022 May 16. PMID: 35575431.

vitamin D due to the risk of hypermagnesemia. In addition, oral vitamin D supplementation can enhance both the efficacy and toxicity of cardiac glycosides like digitalis, increasing the risk of arrhythmias due to elevated calcium levels [37].

Vitamin D toxicity

Vitamin D has a broad therapeutic window, with toxicity occurring only at substantially elevated doses. Hypercalcemia is generally observed when serum 25(OH)D concentrations exceed 150 ng/mL (375 nmol/L) [38]. The Endocrine Society considers serum levels up to 100 ng/mL (250 nmol/L) to be safe and not associated with risk of hypercalcemia [39,40]. However, individuals with certain conditions are more susceptible to vitamin D toxicity due to heightened sensitivity to both endogenous synthesis and exogenous intake. Clinical signs of vitamin D toxicity can vary from asymptomatic to coma and are presented in Fig. 14.1.

In general, vitamin D-fortified foods and over-the-counter supplements have become widely accessible in pharmacies, supermarkets, and online [41]. This increasing availability has led to a rise in vitamin D toxicity cases, often attributed to self-medication, accidental overdosing (e.g., prescribing errors), or use of unregulated or poorly standardized products [42], which is rarely seen in palliative care cancer patients.

Conclusion

In conclusion, while vitamin D is not used as primary analgesic, it offers a potential adjuvant treatment which is both safe and could improve quality of life and reduce opioid burden in advanced cancer patients. Hence, routine screening and timely correction of vitamin D status is recommended.

References

[1] Klasson C, Helde Frankling M, Lundh Hagelin C, et al. Patient experiences of randomised placebo-controlled trial participation during end-of-life palliative cancer care. BMJ Support Palliat Care 2024;15(1):141–8. Available from: https://doi.org/10.1136/spcare-2023-004628.

[2] Helde Frankling M, Klasson C, Björkhem-Bergman L. Successful strategies and areas of improvement-lessons learned from design and conduction of a randomized placebo-controlled trial in palliative care, 'Palliative-D' Life 2021;11(11):1233. Available from: https://doi.org/10.3390/life11111233.

[3] Benedik E. Sources of vitamin D for humans. Int J Vitam Nutr Res 2022;92(2):118–25. Available from: https://doi.org/10.1024/0300-9831/a000733.

[4] Janoušek J, Pilařová V, Macáková K, et al. Vitamin D: sources, physiological role, biokinetics, deficiency, therapeutic use, toxicity, and overview of analytical methods for detection of vitamin D and its metabolites. Crit Rev Clin Lab Sci 2022;59(8):517–54. Available from: https://doi.org/10.1080/10408363.2022.2070595.

[5] Binkley N, Dawson-Hughes B, Durazo-Arvizu R, et al. Vitamin D measurement standardization: the way out of the chaos. J Steroid Biochem Mol Biol 2017;173:117–21. Available from: https://doi.org/10.1016/j.jsbmb.2016.12.002.

[6] Bikle DD. Vitamin D: newer concepts of its metabolism and function at the basic and clinical level. J Endocr Soc 2020;4(2):bvz038. Available from: https://doi.org/10.1210/jendso/bvz038.

[7] Bikle DD, Schwartz J. Vitamin D binding protein, total and free vitamin D levels in different physiological and pathophysiological conditions. Front Endocrinol 2019;10:317. Available from: https://doi.org/10.3389/fendo.2019.00317.

[8] Cominacini M, Fumaneri A, Ballerini L, et al. Unraveling the connection: visceral adipose tissue and vitamin D levels in obesity. Nutrients 2023;15(19):4259. Available from: https://doi.org/10.3390/nu15194259.

[9] Sirajudeen S, Shah I, Al Menhali A. A narrative role of vitamin D and its receptor: with current evidence on the gastric tissues. Int J Mol Sci 2019;20(15):3832. Available from: https://doi.org/10.3390/ijms20153832.

[10] Maestro MA, Molnár F, Mouriño A, et al. Vitamin D receptor 2016: novel ligands and structural insights. Expert Opin Ther Pat 2016;26(11):1291–306. Available from: https://doi.org/10.1080/13543776.2016.1216547.

[11] Hii CS, Ferrante A. The non-genomic actions of vitamin D. Nutrients 2016;8(3):135. Available from: https://doi.org/10.3390/nu8030135.

[12] Bikle DD. Vitamin D metabolism, mechanism of action, and clinical applications. Chem Biol 2014;21(3):319–29. Available from: https://doi.org/10.1016/j.chembiol.2013.12.016.

[13] Uitterlinden AG, Fang Y, Van Meurs JB, et al. Genetics and biology of vitamin D receptor polymorphisms. Gene 2004;338(2):143–56. Available from: https://doi.org/10.1016/j.gene.2004.05.014.

[14] Shipton EA, Shipton EE. Vitamin D and pain: vitamin D and its role in the aetiology and maintenance of chronic pain states and associated comorbidities. Pain Res Treat 2015;2015:904967. Available from: https://doi.org/10.1155/2015/904967.

[15] Habib AM, Nagi K, Thillaiappan NB, et al. Vitamin D and its potential interplay with pain signaling pathways. Front Immunol 2020;11:820. Available from: https://doi.org/10.3389/fimmu.2020.00820.

[16] Leung L, Cahill CM. TNF-alpha and neuropathic pain–a review. J Neuroinflammation 2010;7:27. Available from: https://doi.org/10.1186/1742-2094-7-27.

[17] Almeida Moreira Leal LK, Lima LA, Alexandre de Aquino PE, et al. Vitamin D (VD3) antioxidative and antiinflammatory activities: peripheral and central effects. Eur J Pharmacol 2020;879:173099. Available from: https://doi.org/10.1016/j.ejphar.2020.173099.

[18] Garcion E, Sindji L, Leblondel G, et al. 1,25-Dihydroxyvitamin D3 regulates the synthesis of gamma-glutamyl transpeptidase and glutathione levels in rat primary astrocytes. J Neurochem 1999;73(2):859–66. Available from: https://doi.org/10.1046/j.1471-4159.1999.0730859.x.

[19] Cantorna MT, Snyder L, Lin YD, et al. Vitamin D and 1,25(OH)$_2$D regulation of T cells. Nutrients 2015;7(4):3011–21. Available from: https://doi.org/10.3390/nu7043011.

[20] Bergman P, Sperneder S, Höijer J, et al. Low vitamin D levels are associated with higher opioid dose in palliative cancer patients–results from an observational study in Sweden. PLoS One 2015;10(5):e0128223. Available from: https://doi.org/10.1371/journal.pone.0128223.

[21] Heaney RP. Vitamin D in health and disease. Clin J Am Soc Nephrol 2008;3(5):1535–41. Available from: https://doi.org/10.2215/CJN.01160308.

[22] Zhang Y, Fang F, Tang J, et al. Association between vitamin D supplementation and mortality: systematic review and meta-analysis. BMJ 2019;366:l4673. Available from: https://doi.org/10.1136/bmj.l4673. Erratum in: BMJ. 2020 Sep 22;370:m2329. https://10.1136/bmj.m2329.

[23] Autier P, Gandini S. Vitamin D supplementation and total mortality: a meta-analysis of randomized controlled trials. Arch Intern Med 2007;167(16):1730–7. Available from: https://doi.org/10.1001/archinte.167.16.1730.

[24] Kuznia S, Zhu A, Akutsu T, et al. Efficacy of vitamin D3 supplementation on cancer mortality: systematic review and individual patient data meta-analysis of randomised controlled trials. Ageing Res Rev 2023;87:101923. Available from: https://doi.org/10.1016/j.arr.2023.101923.

[25] Martínez-Alonso M, Dusso A, Ariza G, et al. Vitamin D deficiency and its association with fatigue and quality of life in advanced cancer patients under palliative care: a cross-sectional study. Palliat Med 2016;30(1):89–96. Available from: https://doi.org/10.1177/0269216315601954.

[26] Helde-Frankling M, Höijer J, Bergqvist J, et al. Vitamin D supplementation to palliative cancer patients shows positive effects on pain and infections-results from a matched case-control study. PLoS One 2017;12(8):e0184208. Available from: https://doi.org/10.1371/journal.pone.0184208.

[27] The Effect of Vitamin D Supplementation in Breast Cancer Patients Receiving Taxane-based Chemotherapy [Internet]. ClinicalTrials.gov Identifier: NCT06596122; 2024 [updated 2024-09-19; cited 2025-06-06]. Available from: https://clinicaltrials.gov/study/NCT06596122.

[28] The Role of Vitamin D3 Supplementation in Advanced Cancer Patients With Pain [Internet]. ClinicalTrials.gov Identifier: NCT05450419; 2024 [cited 2025-06-06]. Available from: https://clinicaltrials.gov/study/NCT05450419.

[29] Han JE, Jones JL, Tangpricha V, et al. High dose vitamin D administration in ventilated intensive care unit patients: a pilot double blind randomized controlled trial. J Clin Transl Endocrinol 2016;4:59–65. Available from: https://doi.org/10.1016/j.jcte.2016.04.004

[30] Helde-Frankling M, Björkhem-Bergman L. Vitamin D in pain management. Int J Mol Sci 2017;18(10):2170. Available from: https://doi.org/10.3390/ijms18102170.

[31] Helde Frankling M, Klasson C, Sandberg C, et al. 'Palliative-D'—vitamin D supplementation to palliative cancer patients: a double blind, randomized placebo-controlled multicenter trial. Cancers 2021;13(15):3707. Available from: https://doi.org/10.3390/cancers13153707.

[32] Giustina A, Bilezikian JP, Adler RA, et al. Consensus statement on vitamin D status assessment and supplementation: whys, whens, and hows. Endocr Rev 2024;45(5):625–54. Available from: https://doi.org/10.1210/endrev/bnae009.

[33] Helde Frankling M, Klasson C, Björkhem-Bergman L. 25-Hydroxyvitamin D in cancer patients admitted to palliative care: a post-hoc analysis of the Swedish trial 'Palliative-D' nutrients 2022;14(3):602. Available from: https://doi.org/10.3390/nu14030602.

[34] Agenzia Italiana del Farmaco (AIFA). Nota 96 per la prescrizione di farmaci a base di vitamina D. Rome: AIFA; 2011. Available from: https://www.aifa.gov.it/documents/20142/1728113/nota-96.pdf [accessed 06.06.25].

[35] Gisondi P, Gracia-Cazaña T, Kurzen H, et al. Calcipotriol/betamethasone dipropionate for the treatment of psoriasis: mechanism of action and evidence of efficacy and safety versus topical corticosteroids. J Clin Med 2024;13(15):4484. Available from: https://doi.org/10.3390/jcm13154484.

[36] Sosa Henríquez M, Gómez de Tejada Romero MJ. Cholecalciferol or calcifediol in the management of vitamin D deficiency. Nutrients 2020;12(6):1617. Available from: https://doi.org/10.3390/nu12061617.

[37] Hrvatska agencija za lijekove i medicinske proizvode (HALMED). SmPC D-VITAL 25,000 IU oralna otopina. HALMED, [accessed 06.06.25]. Available from: https://www.halmed.hr/lijekovi/baza-lijekova/D-VITAL-25-000-IU-oralna-otopina/13376/.

[38] Galior K, Grebe S, Singh R. Development of vitamin D toxicity from overcorrection of vitamin D deficiency: a review of case reports. Nutrients 2018;10(8):953. Available from: https://doi.org/10.3390/nu10080953.

[39] Holick MF, Binkley NC, Bischoff-Ferrari HA, et al. Evaluation, treatment, and prevention of vitamin D deficiency: an Endocrine Society clinical practice guideline. J Clin Endocrinol Metab 2011;96(7):1911–30. Available from: https://doi.org/10.1210/jc.2011-0385. Epub 2011 Jun 6. Erratum in: J Clin Endocrinol Metab. 2011;96(12):3908; https://doi.org/10.1210/jcem.96.12.zeg3908. Erratum in: J Clin Endocrinol Metab. 2024;109(10):e1991; https://doi.org/10.1210/clinem/dgae373.

[40] Holick MF. Revisiting vitamin D guidelines: a critical appraisal of the literature. Endocr Pract 2024;30(12):1227–41. Available from: https://doi.org/10.1016/j.eprac.2024.10.011.

[41] Wan M, Patel A, Patel JP, et al. Quality and use of unlicensed vitamin D preparations in primary care in England: retrospective review of national prescription data and laboratory analysis. Br J Clin Pharmacol 2021;87(3):1338–46. Available from: https://doi.org/10.1111/bcp.14521.

[42] Taylor PN, Davies JS. A review of the growing risk of vitamin D toxicity from inappropriate practice. Br J Clin Pharmacol 2018;84(6):1121–7. Available from: https://doi.org/10.1111/bcp.13573.

Chapter 15

Oral rinses and topical agents

Hayda Perdue
Hospice Savannah Inc., Savannah, GA, United States

Key learning objectives

- To provide an in-depth analysis of oral mucositis (OM), focusing on its pathophysiology, incidence, and available treatments.
- To review and explore the effectiveness of various oral rinses and topical agents in managing the symptoms of OM.
- To help healthcare professionals optimize their treatment strategies for cancer patients experiencing this debilitating condition, such as OM and localized cancer pain.
- To evaluate the clinical effectiveness of various topical agents used in the treatment of cancer pain.

Oral mucositis

Oral mucositis (OM) is a painful condition that can be found in patients with different cancers, but affects up to 90% of cancer patients undergoing chemotherapy (CT) and radiotherapy (RT) for head and neck cancers and up to 60%–85% of patients undergoing hematopoietic stem cell transplantation (HSCT) [1]. The reason is that one of the most commonly used drugs in the treatment of head and neck cancers, cisplatin, directly affects mucosal cells, while RT exacerbates damage through localized tissue irradiation. Additionally, human papillomavirus (HPV)-negative head and neck cancer patients and those with elevated salivary cytokine levels (IL-6, IL-1β) are more prone to developing high-grade mucositis [2]. Treatments for cancers other than head and neck cancers can also cause OM. Patients with colorectal cancer treated with 5-fluorouracil (5-FU), capecitabine, or irinotecan in particular are at high risk of OM, as well as gastrointestinal mucositis, due to the impact on gut mucosa [3]. Furthermore, in metastatic breast cancer, the combination of everolimus (a mechanistic target of rapamycin [mTOR] inhibitor) and exemestane has been linked to OM, specifically mTOR-inhibitor-associated stomatitis. This type of stomatitis differs from conventional mucositis, which is characterized by small, painful lesions with a necrotic center and erythematous halo [4,5]. Consequently, patients affected by OM require dose adjustments or treatment discontinuation of chemoradiation with delayed cancer treatment and a worse prognosis.

OM manifests as painful ulcerations in the mouth, impairing the ability to eat, swallow, or even speak, leading to secondary complications such as malnutrition and infections. However, the mucosa of the whole gastrointestinal tract can be affected by mucositis as well. The pathophysiology of OM involves direct damage to the rapidly dividing cells of the oral mucosa, which involves five stages: initiation, upregulation, signal amplification, ulceration, and healing. CT or radiation leads to direct DNA injury in basal epithelial cells, triggering the release of inflammatory mediators like nuclear factor kappa-light-chain-enhancer of activated B cells (NF-κB), which amplify the inflammatory response. The result is tissue damage, reactive oxygen species formation, and apoptosis. The ulcerative phase is marked by severe mucosal breakdown and bacterial colonization. Healing occurs through epithelial proliferation and matrix regeneration once the damaging factors resolve. Buccal mucosa, soft palate, ventral tongue surface, and floor of the mouth are mostly affected. There are several classifications of OM based on the severity of symptoms and the underlying cancer treatment involved. One of the most common classification includes the World Health Organization Scale [6] (Table 15.1).

Preventive measures such as good oral hygiene or sodium bicarbonate rinses and avoiding irritants that exacerbate mucosal damage (e.g., tobacco) are crucial. For instance, regular brushing, flossing, and mouth rinsing are important to reduce bacterial load and prevent secondary infections. Cryotherapy, where ice chips are used during CT, has also shown some promise in reducing OM incidence [7].

Pain Management in Palliative Care for Patients With Cancer. DOI: https://doi.org/10.1016/B978-0-443-30050-9.00020-5

TABLE 15.1 WHO grading of oral mucositis.

Grades/clinical presentation
Grade 1: Erythema and soreness
Grade 2: Ulcers, able to eat solids
Grade 3: Ulcers, requires a liquid diet (due to mucositis)
Grade 4: Ulcer, bleeding, alimentation not possible (due to mucositis)

WHO, World Health Organization.

Despite these preventive measures, managing OM remains a challenge for clinicians. The lack of a gold-standard treatment leaves physicians and patients to rely on various oral rinses and topical agents for symptom relief (Table 15.2). While orally ingested painkillers, primarily opioid-based, can be used to treat pain from OM as well, the risk of common opioid side effects, such as constipation, nausea, and sedation, remains a concern. Additionally, as oral drugs require systemic absorption, targeted and rapid pain relief is often lacking, which can subsequently lead to weight loss, fatigue, and worsening of the clinical condition. This chapter evaluates the clinical effectiveness of these treatments in reducing the severity and duration of OM in patients with cancer.

Types of oral rinses

Oral indomethacin spray

Oral indomethacin spray is a nonsteroidal antiinflammatory drug (NSAID) used to alleviate pain associated with OM in cancer patients. The use of oral indomethacin spray offers a targeted approach to pain management, providing localized relief with minimal systemic side effects. Indomethacin works by inhibiting the cyclooxygenase (COX) enzymes, COX-1 and COX-2, which are involved in the synthesis of prostaglandins (PG). PGs are lipid compounds that play a key role in inflammation and pain. By reducing the production of PG, indomethacin decreases inflammation and alleviates pain [8].

Several randomized controlled trials (RCTs) have demonstrated the efficacy of oral indomethacin spray in managing OM pain in patients with head and neck carcinomas and hematological tumors. In an RCT including 35 patients, the positive effects of indomethacin oral spray on OM pain were significantly greater than those of a placebo. Pain relief was achieved in 93% of patients within 25 minutes of applying the spray, and the analgesic effects were maintained for up to 120 minutes. As systemic exposure was only 1.8% of the dose, the effect of indomethacin is predominantly local and appears safe [8]. Another randomized placebo-controlled study enrolled 60 patients with cancer undergoing RT or CT, demonstrating a significant improvement in pain levels with oral indomethacin use, which persisted for 180 minutes and allowed patients to drink liquids more easily. Indomethacin spray was used four times (a total of 0.8 mg) during the first 240 minutes and later 30 minutes before a meal. No adverse effects were noted, and only patients with a history of adverse reactions to NSAIDs were excluded from the trial [9].

Synbiotic mouthwash

Synbiotic mouthwash, which works by leveraging the benefits of probiotics and prebiotics to maintain a healthy oral microbiome, has emerged as a promising treatment for OM in cancer patients, particularly those undergoing RT. In a double-blind RCT involving 64 oral cancer patients, the efficacy of synbiotic mouthwash was compared to that of normal saline mouthwash. The study found that the synbiotic mouthwash group experienced a significant reduction in the severity of OM compared to the control group using normal saline. The synbiotic mouthwash not only reduced the severity of mucositis but also prevented its progression, highlighting its potential as an effective adjunctive treatment [10].

Topical opioids

While opioids remain a backbone of cancer pain treatment, there is also clinical data regarding the use of topical morphine. One trial compared topical morphine with a routine mouthwash for patients with grade 3 or 4 OM. Thirty patients received morphine sulfate 2% solution or a solution with magnesium aluminum hydroxide, viscous lidocaine, and diphenhydramine, with the same dietary and oral hygiene instructions. Solutions were administered 10 mL every 3 hours, six times a day, for 6 days. Both arms demonstrated a decrease in mucositis severity, with a more significant finding for morphine on day 6 and with greater satisfaction in patients with opioid treatment [11]. A systematic literature review demonstrated clear clinical benefits described in small RCTs for the use of topical opioids in OM, which included diamorphine, morphine (sulfate and hydrochloride), methadone, oxycodone, and meperidine (pethidine). However, the study concluded that there is a deficiency of higher-quality evidence on the role of topical opioids, and more robust primary studies are required to inform practice recommendations [12]. However, the use of topical opioids could potentially allow for a reduction in the use of oral morphine, which in turn could reduce the systemic side effects of opioid treatment and allow for a safe and effective option for pain in OM.

Doxepin suspension

Doxepin is a tricyclic antidepressant used for the management of OM ordered as a compound form. A doxepin suspension (5 mg/mL) can be prepared in an oral rinse containing 0.1% alcohol and sorbitol. A single-dose application of doxepin rinse was found to have an extended duration of pain relief of greater than 50% for 4 hours or longer. In a longitudinal study, a doxepin rinse was repeated over the course of 1 week in patients with cancer-related OM. Patients were advised to continue using the rinse as needed, three to six times per day, for the weeks until their next follow-up assessment. The study demonstrated that repeated dosing of doxepin solution continued to bring significant pain reductions, although the sample size was small [13].

A more recent randomized, double-blind, placebo-controlled trial compared doxepin oral rinse or a placebo for the treatment of radiotherapy-related OM pain, demonstrating a greater reduction in pain for doxepin (–9.1) than for placebo (–4.7; P <.001). Doxepin was, however, associated with more stinging or burning, unpleasant taste, and greater drowsiness than the placebo rinse [14].

Sodium bicarbonate and zinc chloride mouthwash

Sodium bicarbonate and zinc chloride mouthwash were studied in a trial on 144 patients undergoing CT. The severity of mucositis and quality of life (QoL) were examined blindly at the baseline and at a 3-week follow-up. The results showed the severity of OM in the group treated with the sodium bicarbonate and zinc chloride groups decreased from the end of the first week until the third week (P <.001). In addition, there was a significant difference in the severity of OM among the groups at the end of the second (P = .014) and the third weeks (P <.001). Zinc chloride (0.2% mouthwash) and sodium bicarbonate (5% mouthwash), which were used for 2 minutes, three times a day, improved oral health and promoted QoL in this study [15].

Natural remedies

Curcumin, a herbal agent present in the spice turmeric, has antiinflammatory, immunomodulatory, and wound-healing properties. A study was conducted to measure the effectiveness and safety of 0.1% curcumin (using nanoparticles) and 0.15% benzydamine mouthwash on radiation-induced OM (RIOM) among 74 head and neck cancer patients scheduled to receive RT. Neither of the mouthwashes was able to completely prevent the onset of RIOM or reduce the severity of RIOM, although the use of 0.1% curcumin mouthwash (three times a day for 7 days, with no drinking or eating 30 minutes after the rinse) was able to significantly delay the onset of RIOM [16]. However, a network metaanalysis in the year 2024 included eight articles investigating the effect of turmeric for OM, showing a significant difference in the degree of OM at weeks 4 and 6, compared to placebo. A significant improvement in pain levels was only observed at the 6-week follow-up visit, compared to placebo, but surprisingly at 2, 4, 5, and 6-week follow-ups compared to the non placebo group [17].

Another network metaanalysis evaluated the effect of natural remedies on OM from 36 RCTs involving 2083 patients undergoing cancer therapies. Out of all mentioned, honey was shown to be particularly effective, reducing the incidence of OM in both placebo and standard care (RR = 0.48 [95% CI, 0.30–0.75] and RR = 0.56 [95% CI, 0.34–0.93], respectively). The use of propolis, chamomile, and *Plantago major* was all shown to mitigate the incidence, severity, and associated pain of OM in patients undergoing CT and RT [18].

Yarrow (*Achillea millefolium*) distillate is created through a distillation process that captures the plant's essential oils and active compounds. An RCT evaluated cancer patients diagnosed with chemotherapy-induced stomatitis [19]. All 56 evaluated patients received a combination of lidocaine, dexamethasone, sucralfate, and diphenhydramine, while 28 patients in the experimental group received a mixture of the routine mouthwash and yarrow distillate. The experimental group demonstrated a significant decrease in stomatitis score both on days 7 and 14 ($P < .001$) [19].

Aloe vera mouthwash, known for its soothing and antiinflammatory properties, has been explored as a natural treatment for OM in the prevention and treatment of radiotherapy-induced mucositis. Results from a randomized double-blind trial on 75 patients showed no significant difference in OM severity during the first 3 weeks of treatment between the *A. vera* (three times a day for 6 weeks) and placebo groups. However, by weeks 4–6, the *A. vera* group experienced significantly lower OM severity. This suggests that long-term use of *A. vera* mouthwash may help reduce the severity of RIOM during extended RT [20].

Hypochlorite-based mouth rinse

The efficacy of a hypochlorite-based mouth rinse (Granudacyn), assumed to be highly biocompatible due to the compounds being part of the natural pathogen defense as a possible antiseptic agent in cases of OM, was compared to that of an octenidine-based antiseptic mouth rinse (Octenidol md). A monocentric, controlled, randomized, blinded crossover study was conducted on 20 volunteers. Both antiseptic mouth rinses induced a significant reduction of colony-forming units on buccal mucosa and in saliva 1 minute after mouth rinsing. The effect persisted up to 60 minutes. The octenidine-based rinse was significantly superior to the hypochlorite-based rinse up to the last sample 60 minutes after rinsing. While hypochlorite-based rinse (Granudacyn) can support oral hygiene in patients with cancer, high-quality data in cancer settings are lacking [21].

Dexamethasone mouthwash

Studies have shown that dexamethasone mouthwash can reduce the incidence and severity of OM. By dampening the inflammatory response, dexamethasone can prevent the development of painful ulcers and reduce the severity of existing lesions. The growing body of evidence supporting this treatment has made it a valuable tool in OM, as shown in a multicenter, open-label, randomized phase 2 trial that examined the efficacy and safety of dexamethasone mouthwash in preventing chemotherapy-induced OM in women undergoing breast cancer treatment. Patients treated with CT were randomized to prevention with dexamethasone (10 mL of a 0.1 mg/mL dexamethasone solution; swish for 2 minutes and spit, four times daily for 9 weeks) or only oral care without dexamethasone. A total of 117 patients were randomized, and the results demonstrated a numeric reduction in stomatitis incidence in the intervention group with a significantly lower stomatitis severity in the dexamethasone arm [22]. A single-arm, phase 2 trial by Rugo et al. showed that the prophylactic use of dexamethasone oral solution (0.5 mg/5 mL oral solution for 8 weeks) significantly reduced the incidence of grade 2 or worse stomatitis in patients on everolimus and exemestane, from 33% in the control group to 2%. This demonstrates a substantial benefit in preventing a specific adverse event within a targeted patient population, offering a promising approach to managing a known side effect of mTOR inhibitor-based therapy [5].

Orasol Plus

Orasol Plus is an over-the-counter oral pain relief agent that contains benzocaine, a topical anesthetic known for its numbing effect; its benefits are primarily symptomatic and do not influence the underlying inflammatory processes. In a randomized trial, 62 patients with advanced renal cell carcinoma treated with everolimus were assigned to receive either Orasol Plus with oral hygiene or oral hygiene alone (rinsing three times a day). Results showed that the Orasol Plus group experienced significantly fewer stomatitis episodes (8 vs 20, $P = .021$) and had shorter average stomatitis duration (8 vs 11.2 days, $P = .0416$) compared to the control group. The study concluded that using Orasol Plus alongside oral hygiene effectively reduces the severity and duration of everolimus-induced stomatitis in these patients; however, no active control arm was used [23].

Phenytoin

Originally used for treating epilepsy, phenytoin, in its 1% mouthwash form, has been studied for its potential to promote wound healing when applied topically to OM lesions. Some research suggests that phenytoin may enhance mucosal repair by stimulating fibroblast proliferation and collagen synthesis. In a pilot double-blind randomized clinical trial, eight patients in the study group were given 1% phenytoin mouthwash, while eight patients in the control group used normal saline. This study showed that 1% phenytoin mouthwash caused pain relief and improvement in QoL significantly in patients with mucositis due to cancer therapy, but it did not reduce the severity of mucositis on a statistically significant scale [24].

Benzydamine

Benzydamine is an antiinflammatory and analgesic agent that works by inhibiting proinflammatory molecules and interactions involved in the mucositis process. Particularly useful for patients receiving RT, benzydamine mouthwash helps manage OM symptoms by reducing inflammation and pain, thereby improving patient comfort during treatment [25]. However, in an RCT, the effectiveness of benzydamine hydrochloride was tested against honey–lemon spray in preventing radiation-induced OM in 46 patients with head and neck cancer receiving RT. There were no significant differences between the groups in mucositis scores, pain, or QoL. Moreover, the benzydamine hydrochloride group had a higher rate of OM occurrence compared to the honey–lemon group (hazard ratio = 2.1), warranting caution for further trials with benzydamine [26].

Caphosol

A supersaturated calcium phosphate mouthwash, Caphosol, works by lubricating and moisturizing the oral mucosa, helping to maintain the integrity of mucosal tissues while providing symptomatic relief. Caphosol was assessed in a nonblinded matched clinical study with a total of 100 patients undergoing irradiation for head and neck cancers. The results demonstrated a significant difference between the study groups for mucositis in the clinical target volume area, mucositis in the increased dose (boost) area, dysphagia, and xerostomia, with lower values in the Caphasol group [27]. However, when tested in a controlled, double-blinded, and randomized clinical trial for children with cancer treated with CT, no difference in prevention of OM was observed between the use of Caphosol or saline rinses [28]. Additionally, another RCT comparing Caphosol plus standard oral care (intervention) or standard oral care alone (control) during chemoradiotherapy did not show the reduction of incidence of severe OM in the Caphosol arm despite enrolling 215 patients [29].

Gelclair

Polyvinylpyrrolidone-sodium hyaluronate gel (Gelclair) is an oral gel that forms a protective barrier over ulcerated mucosa, in theory reducing pain and discomfort associated with OM, particularly in patients with postallogeneic stem cell transplantation. In an observational cohort study, 22 patients with active OM who received Gelclair and a control group of 15 patients who used other rinsing solutions, including chlorhexidine, benzydamine, and salvia-based solutions, showed that Gelclair had no difference in pain relief, although it had a longer-lasting effect in reducing oral pain. Gelclair provided a significantly longer duration of symptom relief, lasting a median of 3 (range, 0–5) hours compared to 1 (range, 0–3) hour in the control group ($P = .001$), suggesting potential use of Gelclair [30]. Another trial that evaluated 20 patients compared Gelclair to sucralfate and mucaine on radiotherapy-induced OM but found that Gelclair was no more effective than current standard practice [31].

Sucralfate

Sucralfate is a nonabsorbable aluminum salt of sulfated disaccharide, which is sometimes used in the treatment of OM due to its effect of forming a barrier over damaged mucosa. The majority of data on sucralfate use is in prevention along with the treatment of chemotherapy-induced OM, with one study reporting 58% of patients on sucralfate reporting no oral pain compared with 25% receiving placebo [32]. The double-blinded randomized study on radiation mucositis on 111 patients demonstrated a trend of less severe radiation OM that was not statistically significant for patients receiving sucralfate in addition to the combination of viscous lidocaine, diphenhydramine, and antacid; both combinations were used three times a day before meals and at bedtime [33]. However, a recent metaanalysis evaluated 30 randomized clinical trials with 2565 patients, evaluating different formulations for prevention of OM. Out of 19 compounds, sucralfate resulted in the most significant effect for OM prevention compared to placebo (OR = 0.04 [95% CI, 0.01–0.25]; $P = .001$), while curcumin had the most significant effect out of natural substances [34].

MuGard

MuGard is a proprietary viscous liquid mucoadhesive hydrogel that works by creating a palliative barrier over injured mucosa, which was tested in a randomized, double-blind, placebo-controlled trial in 78 patients with cancer undergoing chemoradiotherapy. The comparator arm was sham-controlled. No significant adverse events were reported, and the adverse event incidence was similar between cohorts, with MuGard demonstrating a mitigation of OM symptoms [35]. However, real-world retrospective data on 147 patients did not show any benefit of MuGard or Caphosol [36].

Avasopasem manganese

Avasopasem is a selective small-molecule dismutase mimetic designed to convert superoxide to hydrogen peroxide, which could potentially protect normal cells from and sensitize cancer cells to radiation. It was recently tested in a randomized, double-blind phase trial demonstrating a statistically significant and clinically meaningful reduction of OM compared to placebo [37]. However, the FDA recently rejected the new drug application, as the data from these trials were not substantial enough to prove avasopasem's efficacy and safety [38].

Multimodal oral rinses

Lidocaine, Benadryl, and Maalox ("magic mouthwash")

A commonly used formulation for pain relief in OM is the combination of lidocaine, diphenhydramine (Benadryl), and Maalox, colloquially known as "magic mouthwash." Lidocaine, a local anesthetic, provides immediate numbing of the affected mucosal tissues, while Benadryl, an antihistamine, helps to reduce inflammation. Maalox, an antacid, coats the mucosa, providing a soothing barrier. Although this combination is widely used, its efficacy remains mostly anecdotal, with studies suggesting that it primarily offers symptomatic relief rather than addressing the underlying inflammatory processes of OM [33]. A randomized clinical trial tested the effectiveness of three mouthwashes: salt and soda, chlorhexidine, and a "magic mouthwash" (lidocaine, Benadryl, and Maalox) to treat chemotherapy-induced mucositis. The trial involved 200 patients and found that in 142 of them, mucositis symptoms resolved within 12 days. There were no significant differences in effectiveness between the three mouthwashes. The study concluded that the least expensive option, the salt and soda mouthwash, was just as effective as the others [39].

Diphenhydramine–lidocaine–antacid mouthwash

A phase 3 randomized clinical trial comparing the effectiveness of doxepin mouthwash, diphenhydramine–lidocaine–antacid mouthwash, and a placebo in treating radiotherapy-related OM pain included patients undergoing RT for head and neck cancer. A total of 275 patients were randomized, with the primary goal to evaluate OM pain reduction. OM pain decreased similarly in both active arms, with a similar number of adverse effects (4% of grade 3 in both active arms), although patients on doxepin reported more drowsiness, unpleasant taste, stinging, and burning. Although both active treatment arms demonstrated effectiveness, the effect size was less than the minimal clinically important difference [40].

Dentoxol

Dentoxol is a multicomponent mouthwash (purified water, xylitol, sodium bicarbonate, eugenol, camphor, parachlorophenol, and peppermint essence) whose main mode of action is mechanical sloughing of the superficial epithelial cell layer of the oral mucosa, stimulating local regeneration of the epithelium. An RCT was conducted on 108 patients undergoing RT for cancer, demonstrating a lower incidence of severe OM compared to the standard-care group, although there was no difference in duration of severe OM (11.95 days in the Dentoxol group vs 14.59 days in the control group [$P = .502$]). There was no difference between groups in mouth pain and its impact on function [41].

Nonpharmacological topical approach to OM management

Oral cryotherapy

Oral cryotherapy with an established oral care protocol in autologous cell transplant was compared with an oral care protocol alone (frequent mouth rinsing using sodium bicarbonate solution and gentle brushing of teeth with Toothette sponges [Sage Products Inc., Elgin, IL]). This prospective, pilot, RCT appears to be beneficial in reducing the severity of OM in a patient population treated with high-dose melphalan compared to an oral care protocol alone [42].

Low-level laser therapy

Though not a topical agent, low-level laser therapy (LLLT) has become an established treatment for both the prevention and management of OM. LLLT utilizes photobiomodulation to stimulate cellular repair processes, reduce inflammation, and promote healing of the mucosal tissues. A metaanalysis of LLLT for the treatment of OM after stem cell

Treatment	Type	Effect	Type of evidence	Comment
Indomethacin spray	Oral nonsteroidal antiinflammatory drug (NSAID) spray	Antiinflammatory, pain relief superior to placebo	Double-blind RCT	Both in chemotherapy (CT) and radiotherapy (RT)
Synbiotic mouthwash	Synbiotic component	Change in the microbiome, prevention and pain relief superior to standard treatment	Double-blind RCT	Specific to RT patients
Morphine	Topical opioid	Pain relief, superior to magnesium aluminum hydroxide, viscous lidocaine, and diphenhydramine	Double-blind RCT	After CT, RT, CRT
Doxepin suspension	Tricyclic antidepressant rinse	Pain relief superior to placebo	Double-blind RCT	After RT, smaller trials after CT, CRT
Sodium bicarbonate and zinc chloride	Mouthwash with electrolytes	Pain relief superior to placebo	Double-blind RCT	CT patients
Curcumin	Herbal mouthwash	Antiinflammatory, longer delay to the onset of OM compared to benzydamine	Triple-blind RCT	Specific to RT patients
Hypochlorite-based rinse	Antiseptic mouthwash	Antiseptic effect, oral hygiene, significant reduction of bacterial colony-forming units, similar to octenidine	Monocentric, randomized controlled study	Compared to octenidine-based rinse
Dexamethasone	Steroid	Prevention of OM incidence and severity during CT or everolimus, more significant than control group	RCT, open-label	Most data for breast cancer patients
Yarrow distillate	Plant distillate	Antiinflammatory, reducing severity, compared to routine mouthwash (lidocaine, sucralfate, diphenhydramine)	Triple-blind RCT	CT patients
Orasol Plus	Benzocaine as the main ingredient	Prevention of incidence, severity, and reduction of duration of mucositis compared to oral hygiene alone	Multicenter RCT	Renal cell carcinoma patients on everolimus
Phenytoin	Originally anticonvulsant	Better pain relief and quality of life than normal saline	Pilot double-blind RCT	RT or CT
Benzydamine	Prostaglandin synthetase inhibitor	Antiinflammatory, pain relief, but not significant compared to honey–lemon spray	RCT	RT patients
Caphosol	Calcium and phosphate mouthwash	Reduction of dryness, mixed results for prevention of OM	Nonblinded matched clinical study	RT patients
Gelclair	Polyvinylpyrrolidone-sodium hyaluronate	Not significant pain relief vs oral rinse, but longer duration	Prospective observational, RCT	Most data in HSCT patients or RT patients
Sucralfate	Acid buffer	Nonsignificant trend in pain reduction vs lidocaine, diphenhydramine, and antacid, although positive results from metaanalysis	Double-blind RCT	RT, CT, and HSCT data
Aloe vera	Natural remedy	More significant reduction of OM compared to placebo	Double-blind RCT	RT
MuGard	Mucoadhesive hydrogel	Less OM compared to sham-control	Double-blind RCT	CRT
Lidocaine, Benadryl, and Maalox	Combination therapy	Comparable effectiveness to chlorhexidine and salt and soda	Double-blind RCT	CT
Diphenhydramine–lidocaine–antacid	Combination therapy	Comparable effectiveness to doxepin, more effective than placebo	Double-blind RCT	RT
Dentoxol	Combination therapy	Lower incidence of severe OM, similar pain score	Double-blind RCT	RT

CRT, Chemoradiotherapy; *HSCT*, hematopoietic stem cell transplantation; *NSAID*, nonsteroidal antiinflammatory drug; *OM*, oral mucositis; *RCT*, randomized controlled trial.

transplantation demonstrated a reduction in the degree of mucositis in the patients treated with laser therapy [43]. On the other hand, a systematic review and metaanalysis in pediatric cancer patients demonstrated that LLLT was not found to be effective for the prevention or treatment of OM grade ≥II (RR = 0.99, P = .99) [44]. As a noninvasive option, LLLT can be recommended as a potential option, particularly for patients undergoing head and neck RT or high-dose CT.

Conclusion for oral rinses and nonpharmacological topical approach

Oral indomethacin, topical opioids, doxepin, and diphenhydramine–lidocaine–antacid demonstrated the effect of rapid pain reduction in various, usually small-sized, RCTs. No significant adverse effects were noted, although patients on doxepin often reported local stinging and burning. On the other hand, synbiotic mouthwashes, sodium bicarbonate and zinc chloride mouthwashes, topical dexamethasone, and sucralfate also demonstrated a reduction of severity of OM in the RCT setting but during the prolonged period and should be primarily used for prevention of OM. It needs to be emphasized that most data are available for RT and not CT settings. Metaanalyses also demonstrate the positive effect of honey, curcumin, propolis, chamomile, and *P. major* primarily in the prevention of OM; these compounds could be an attractive option as they are relatively cheap and abundant, allowing easier access for our patients. However, there is a lack of data suggesting the use of one over the other compound, emphasizing the need for randomized clinical trials with active comparators.

There are also positive data with lower levels of evidence for various other compounds. On the other hand, a recent RCT evaluating benzydamine demonstrated worse results compared to honey, while negative trials exist for Caphosol and "magic mouthwash" despite their widespread use. Hence, caution is warranted when using such formulations except in clinical trials. Except for topical agents, oral cryotherapy and LLLT also remain as nonpharmacological options, although more data is required.

There are several guidelines for the treatment of OM, such as the Multinational Association of Supportive Care in Cancer and the International Society of Oral Oncology (MASCC), which are also endorsed by European guidelines [45,46]. The MASCC panel suggested that implementation of multiagent combination oral care protocols is beneficial for prevention of OM (level of evidence [LoE III], along with benzydamine mouthwash, LLLT [LoE I], cryotherapy [LoE II], and honey [LoE II], while advising against combined topical and systemic sucralfate [LoE II]). The only agent recommended for pain is the 0.2% morphine mouthwash [45].

However, since the latest edition of MASCC guidelines, robust data on dexamethasone, metaanalysis supporting sucralfate, and disappointing results of benzydamine compared to honey, among other studies, have been published, so an update of both MASCC and European guidelines is warranted and expected.

Topical (skin) agents in treatment of cancer pain

Introduction

The use of topical agents to alleviate painful dermato-oncologic complications is an essential strategy to offer palliative care in cancer treatment. Systemic therapies, such as antivascular endothelial growth factor receptor and epidermal growth factor receptor (EGFR) inhibitors, as well as radiation therapy, are associated with significant cutaneous side effects that impact patient comfort and treatment adherence. Hand-foot syndrome (HFS), a painful reaction involving the palms and soles, is commonly triggered by agents such as sorafenib and sunitinib [47]. Similarly, radiation dermatitis, which can range from mild erythema to severe desquamation, requires careful management [48]. Anti-EGFR treatments often induce skin toxicities due to the role of EGFR in normal skin cell regulation. Common side effects include papulopustular eruptions, xerosis, and pruritus, which can be painful and adversely affect patients' QoL and their ability to continue therapy [49].

Topical nonsteroidal antiinflammatory drugs

Topical NSAIDs are widely used for their antiinflammatory and analgesic properties. They work by inhibiting COX enzymes, which are involved in the synthesis of PGs that mediate inflammation and pain. Although topical NSAIDs are often used, their effectiveness in cancer patients is anecdotal. In a case report, a cancer patient with bone metastasis pain was treated with a combination of opioids, oral NSAIDs, and topical NSAIDs. The addition of topical NSAIDs, guided by pharmacists' intervention, significantly reduced pain scores and improved the patient's daily activities. The study highlighted the role of pharmacists and the use of topical pain management strategies for complex home cancer pain [50]. While data in patients with cancer is scarce, topical NSAIDs could potentially provide pain relief without the adverse effects commonly associated with systemic NSAIDs, such as gastrointestinal or cardiovascular issues, making them a safer option for localized pain management in metastatic cancer, although further studies are required.

On the other hand, NSAIDs have recently been shown as an important factor in preventing the painful HFS. In a phase 3 randomized, double-blind trial, patients with breast or gastrointestinal cancer received capecitabine-based treatment. The participants were divided into two groups, one receiving topical diclofenac gel and the other a placebo, aiming to assess the efficacy of topical diclofenac in preventing capecitabine-induced HFS. Furthermore, capecitabine dose reduction was less frequent in the diclofenac groups than in the placebo group, with adverse effects like those of the placebo. The study demonstrated the efficacy of topical diclofenac in achieving significant reduction in the incidence of moderate-to-severe HFS and overall incidence as well as the overall incidence of HFS [51].

Topical corticosteroids

Cancer Care Ontario's Supportive Care Guidelines Group (SCGG) evaluated the prevention and management of acute skin reactions related to radiation therapy and suggested that preventative methods, including skin washing, such as gentle washing with water alone with or without mild soap, should be permitted in patients receiving radiation therapy to prevent acute skin reactions. While there is limited evidence, the expert opinion of SCGG recommended that a low-dose (i.e., 1%) corticosteroid cream may be beneficial in the reduction of itching and irritation [52].

Low-dose topical steroid creams were also observed to be effective in dermatologic toxicities during anti-epidermal growth factor receptor (anti-EGFR) monoclonal inhibitor therapy in patients with metastatic colon cancer without rat sarcoma virus oncogene homologue (RAS) gene mutations. In a systemic review of skin-related toxicity occurring during anti-EGFR monoclonal inhibitor therapy in patients with metastatic colon cancer, particularly panitumumab and cetuximab, for metastatic colorectal cancer [49], it was shown that skin toxicities, including papulopustular rash, dry skin, nail inflammation, and itching, are not only common but can also be severe and painful, impacting patient QoL and potentially leading to treatment interruptions. Low-dose topical steroid creams—such as 0.5% hydrocortisone cream—can be used as part of the preemptive treatment regimen for managing dermatologic toxicities. This mild corticosteroid cream was recommended for application to affected areas, such as the face, neck, chest, back, hands, and feet, to help reduce inflammation and alleviate discomfort associated with skin reactions [53].

Moreover, a systematic literature review by Salvo et al. reviewed various interventions for preventing and managing radiation dermatitis. The systemic review included 39 studies, with findings suggesting that topical corticosteroids reduce the severity of radiation-induced skin reactions. Although no single corticosteroid was recommended, other treatments like amifostine (a cytoprotective drug) show some potential in phase 2 trials; further RCTs were recommended to establish clear, evidence-based guidelines [54].

Topical phytosterols and salicylates

In a randomized, prospective, open-label, phase 3 trial, Mebo ointment was evaluated for the management of acute dermatitis during RT for breast cancer. The efficacy of Mebo (β-sitosterol ointment) was tested against Biafine (trolamine cream). Two groups of 161 breast cancer patients requiring radiation therapy were enrolled and randomized. Skin reactions were evaluated weekly, and additional symptoms like pruritus and pain were also monitored. The results suggest that while Mebo and Biafine are comparably effective in preventing and managing the overall severity of radiation dermatitis, Mebo may offer a superior benefit by reducing severe symptoms of pruritus and pain [55].

Topical lidocaine

Lidocaine is a common anesthetic available in various formulations, including patches, gels, and creams, which works by inhibiting sodium channels, essential for the initiation and conduction of nerve impulses. This action effectively blocks pain signals from reaching the brain. Lidocaine patches are widely used for managing neuropathic pain and are particularly effective in conditions such as postherpetic neuralgia and diabetic neuropathy. In cancer patients, lidocaine patches can be applied to areas affected by surgery or radiation to provide localized pain relief.

A randomized, double-blind, crossover design study evaluated the short-term efficacy and safety of a 10% lidocaine ointment for managing allodynia in cancer-related pain. The study involved 25 cancer patients with static allodynia and a pain rating of ≥4, comparing 10% lidocaine ointment with a placebo for effects on allodynia and spontaneous pain. Results indicated that Numeric Rating Scale (NRS) scores for static allodynia were significantly lower in the lidocaine group than in the placebo group 2–8 hours postapplication, with 56% of patients achieving a ≥50% improvement at 8 hours compared to 20% in the placebo group. Although no significant time–group interaction was noted for spontaneous pain, lidocaine demonstrated an overall benefit in reducing pain (P = .027) without adverse events. This suggests that 10% lidocaine ointment can effectively relieve allodynia for 2–8 hours after application. While there were no adverse events associated with lidocaine use, caution is advised, particularly for patients with renal impairment [56].

Another RCT compared a topical lidocaine–prilocaine cream to morphine in patients with cancer wounds, demonstrating that patients treated with the topical lidocaine–prilocaine cream had significantly lower pain scores and heart rates (P <.01) compared to those in the morphine group. Additionally, the comfort level, as measured by the Kolcaba scale, was significantly higher in the cream group (P <.01). These findings suggest that the compound lidocaine–prilocaine cream is a safe and effective alternative to systemic morphine in cancer wound care [57].

Topical lidocaine–prilocaine cream was also shown to be an appropriate alternative to infiltrative lidocaine in reducing pain during thoracentesis and abdominocentesis, as a trial on 69 patients suggested a nonsignificant difference in overall pain perception between the two groups. However, the number of attempts required to perform successful abdominocentesis was significantly higher in the topical group [58].

Lidocaine 5% patches were also reported in a prospective study that examined the short-term efficacy. Patient-reported satisfaction with 5% lidocaine patches for neuropathic cancer pain from painful scars (postthoracotomy and postmastectomy) and chest wall tumors was shown to be statistically significant. Results indicated a significant clinical improvement in pain severity, with no notable opioid dose escalation throughout treatment [59].

Topical cannabidiol

Topical cannabidiol (CBD) is one of many compounds found in the cannabis plant, and unlike tetrahydrocannabinol, CBD does not produce psychoactive effects, making it a more appealing option for patients seeking pain relief without the dangers associated with cannabis.

CBD interacts with the body's endocannabinoid system, particularly with cannabinoid receptors (CB1 and CB2). By modulating these receptors, CBD can reduce inflammation and pain [60]. A randomized, double-blind, crossover study including 29 patients evaluated topical CBD, 250 mg/dose up to four times a day, in neuropathic pain, and reported a larger decrease in ratings of "intense, sharp, cold, and itchy" on the Neuropathic Pain Scale in the CBD groups compared to placebo. This suggests that there may be a role for CBD in neuropathic pain, but again, the study was small and included a variety of etiologies, making it hard to know how transferable this is to advanced cancer-related neuropathic pain [61]. However, new clinical trials are crucial, as current data does not support the use of topical CBD outside of clinical trials.

Tibetree pain-relieving plaster

Tibetree pain-relieving plaster (PRP) is a compound with an active ingredient, including camphor, known for its analgesic and antiinflammatory properties. Other ingredients include lamiophlomis rotata extract, turmeric, oxytropis chiliophylla plant, myricaria paniculata stem, safflower, and Szechuan peppercorn. Camphor, the primary active ingredient, acts as a counterirritant, creating a cooling sensation that helps to distract from deeper pain. The other herbal components contribute to reducing inflammation, improving blood circulation, and promoting healing. A program evaluation of a pilot herbal dispensary at a National Cancer Institute–designated comprehensive cancer center demonstrated that PRP improved pain and health-related outcomes and addressed gaps left by oral analgesics. Some patients, however, noted adverse effects, such as skin irritation. While the study was conducted through interviews, these findings highlight the potential role of herbs and natural products in cancer pain management and support the need for further research [62].

Conclusion for topical agents

While less data is available for the position of topical agents in the treatment of cancer-related pain, there is still a significant need. Topical NSAIDs have been shown to be a significant factor in preventing painful HFS and remain a standard of care, while topical corticosteroids remain a mainstay of treatment for often painful side effects from RT and systemic therapy. While less data is available for lidocaine, it remains a standard of care before thoracentesis and abdominocentesis, whether in infiltrative or topical form, and lidocaine patches are a reasonable option, particularly for local pain from chest wall tumors.

Less data is available for topical phytosterols, salicylates, CBD, and PRP, although initial data remains positive, and further studies are required to establish a proper position of those medications in cancer care.

Future development for oral and topical agents

Along with further large-scale, multicenter clinical trials directly comparing the most efficacious agents, several other developments might help in alleviating pain with topical agents. Improving bioavailability and potency of topical agents through nanoparticle-based formulations can extend treatment duration and efficacy. For instance, curcumin nanoparticles could enhance absorption and prolong antiinflammatory effects for OM management. Similarly, liposomal or other encapsulation methods may improve topical cannabinoid absorption for chemotherapy-induced peripheral neuropathy (CIPN), offering targeted pain relief without systemic effects.

Combining multiple agents in a single topical formulation may enhance pain relief and reduce the need for systemic medication. Multiagent therapies also allow for symptom-specific treatments, such as corticosteroid and antibiotic combinations for EGFR inhibitor–related rashes. Genetic profiling could enable more tailored pain management strategies, optimizing treatment selection based on individual patient responses. This personalized approach would allow clinicians to adjust topical formulations, concentrations, or combination therapies to meet specific needs, thereby improving treatment outcomes and minimizing adverse effects.

Innovative drug delivery systems, including biologic dressings, barrier films, and advanced emulsions, could improve the therapeutic reach and longevity of topical treatments. These systems could protect compromised skin, such as in radiation dermatitis, enhance moisture retention, and ensure sustained release of active ingredients.

References

[1] Villa A, Sonis ST. Pharmacotherapy for the management of cancer regimen-related oral mucositis. Expert Opin Pharmacother 2016;17(13):1801–7. Available from: https://doi.org/10.1080/14656566.2016.1217993.

[2] Yamamoto H, Ishihara K, Takeda Y, et al. Changes in the mucus barrier during cisplatin-induced intestinal mucositis in rats. Biomed Res Int 2013;2013. Available from: https://doi.org/10.1155/2013/276186.

[3] Abdel-Rahman O, El Halawani H, Essam-Eldin S. S-1-based regimens and the risk of oral and gastrointestinal mucosal injury: a metaanalysis with comparison to other fluoropyrimidines. Expert Opin Drug Saf 2016;15(1):5–20. Available from: https://doi.org/10.1517/14740338.2016.1105959.

[4] Vigarios E, Epstein JB, Sibaud V. Oral mucosal changes induced by anticancer targeted therapies and immune checkpoint inhibitors. Support Care Cancer 2017;25(5):1713–39. Available from: https://doi.org/10.1007/s00520-017-3629-4.

[5] Rugo HS, Seneviratne L, Beck JT, et al. Prevention of everolimus-related stomatitis in women with hormone receptor-positive, HER2-negative metastatic breast cancer using dexamethasone mouthwash (SWISH): a single-arm, phase 2 trial. Lancet Oncol 2017;18(5):654–62. Available from: https://doi.org/10.1016/S1470-2045(17)30109-2.

[6] Pandiyan RS. I. Unveiling the multifaceted management of oral mucositis in cancer patients: a narrative review. Cureus 2024;16(2):e55213. Available from: https://doi.org/10.7759/cureus.55213.

[7] Shankar A, Roy S, Bhandari M, et al. Current trends in management of oral mucositis in cancer treatment. Asian Pac J Cancer Prev 2017;18(8):2019–26. Available from: https://doi.org/10.22034/APJCP.2017.18.8.2019.

[8] Momo K, Nagaoka H, Kizawa Y, et al. Assessment of indomethacin oral spray for the treatment of oropharyngeal mucositis-induced pain during anticancer therapy. Support Care Cancer 2017;25(9):2997–3000. Available from: https://doi.org/10.1007/s00520-017-3817-2.

[9] Nagaoka H, Momo K, Hamano J, et al. Effects of an indomethacin oral spray on pain due to oral mucositis in cancer patients treated with radiotherapy and chemotherapy: a double-blind, randomized, placebo-controlled trial (JORTC-PAL04). J Pain Symptom Manage 2021;62(3):537–44. Available from: https://doi.org/10.1016/j.jpainsymman.2021.01.123.

[10] Manifar S, Koopaie M, Jahromi ZM, et al. Effect of synbiotic mouthwash on oral mucositis induced by radiotherapy in oral cancer patients: a double-blind randomized clinical trial. Support Care Cancer 2022;31(1):31. Available from: https://doi.org/10.1007/s00520-022-07521-5.

[11] Sarvizadeh M, Hemati S, Meidani M, et al. Morphine mouthwash for the management of oral mucositis in patients with head and neck cancer. Adv Biomed Res 2015;4:44. Available from: https://doi.org/10.4103/2277-9175.151254.

[12] LeBon B, Zeppetella G, Higginson IJ. Effectiveness of topical administration of opioids in palliative care: a systematic review. J Pain Symptom Manage 2009;37(5):913–7. Available from: https://doi.org/10.1016/j.jpainsymman.2008.06.007.

[13] Epstein JB, Epstein JD, Epstein MS, et al. Management of pain in cancer patients with oral mucositis: follow-up of multiple doses of doxepin oral rinse. J Pain Symptom Manage 2007;33(2):111–4. Available from: https://doi.org/10.1016/j.jpainsymman.2006.11.002.

[14] Leenstra JL, Miller RC, Qin R, et al. Doxepin rinse versus placebo in the treatment of acute oral mucositis pain in patients receiving head and neck radiotherapy with or without chemotherapy: a phase III, randomized, double-blind trial (NCCTG-N09C6 [Alliance]). J Clin Oncol 2014;32(15):1571–7. Available from: https://doi.org/10.1200/JCO.2013.53.2630.

[15] Mohammadi F, Oshvandi K, Kamallan SR, et al. Effectiveness of sodium bicarbonate and zinc chloride mouthwashes in the treatment of oral mucositis and quality of life in patients with cancer under chemotherapy. Nurs Open 2022;9(3):1602–11. Available from: https://doi.org/10.1002/nop2.1168.

[16] Swikant S, Hemamalini R, Gaurav S, et al. Effectiveness of curcumin mouthwash on radiation-induced oral mucositis among head and neck cancer patients: a triple-blind, pilot randomised controlled trial. Indian J Dent Res 2020;31(5):718–27. Available from: https://doi.org/10.4103/ijdr.IJDR_822_18.

[17] Wu CF, Wu HJ, Shih CL, et al. Efficacy of turmeric in the treatment of oral mucositis in patients with head and neck cancer after radiotherapy or chemoradiotherapy: a systematic review and meta-analysis. Front Pharmacol 2024;15:1363202. Available from: https://doi.org/10.3389/fphar.2024.1363202.

[18] Lee CC, Chen YW, Kang YN, et al. Efficacy of natural products in preventing oral mucositis resulting from cancer therapies: a network meta-analysis of randomized controlled trials. Crit Rev Oncol Hematol 2024;199:104373. Available from: https://doi.org/10.1016/j.critrevonc.2024.104373.

[19] Miranzadeh S, Adib-Hajbaghery M, Soleymanpoor L, et al. A new mouthwash for chemotherapy-induced stomatitis. Nurs Midwifery Stud 2014;3(3):e20249. Available from: https://doi.org/10.17795/nmsjournal20249.

[20] Najafizade N, Mobini Dehkordi R, Hemati S. Investigating the effect of *Aloe vera* on the prevention and treatment of radiotherapy-induced oral mucositis in patients with head-and-neck cancer. J Res Med Sci 2024;29:16. Available from: https://doi.org/10.4103/jrms.jrms_115_23.

[21] Zwicker P, Freitag M, Heidel FH, et al. Antiseptic efficacy of two mouth rinses in the oral cavity to identify a suitable rinsing solution in radiation- or chemotherapy-induced mucositis. BMC Oral Health 2023;23(1):176. Available from: https://doi.org/10.1186/s12903-023-02884-5.

[22] Kuba S, Maeda S, Shibata K, et al. Efficacy and safety of a dexamethasone-based mouthwash to prevent chemotherapy-induced stomatitis in women with breast cancer: a multicentre, open-label, randomised phase 2 study. J Evid Based Dent Pract 2023;23(3):101896. Available from: https://doi.org/10.1016/j.jebdp.2023.101896.

[23] Porta C, Verzoni E, Zai S, et al. Use of a natural multicomponent mouthwash plus oral hygiene vs oral hygiene alone to prevent everolimus-induced stomatitis: the STOP multicenter, randomized trial. Tumori 2020;106(3):257–66. Available from: https://doi.org/10.1177/0300891620915786.

[24] Baharvand M, Hamian M, Moosavizadeh MA, et al. Phenytoin mouthwash to treat cancer therapy-induced oral mucositis: a pilot study. Indian J Cancer 2015;52(1):81–5. Available from: https://doi.org/10.4103/0019-509X.175597.

[25] Nicolatou-Galitis O, Bossi P, Orlandi E, et al. The role of benzydamine in prevention and treatment of chemoradiotherapy-induced mucositis. Support Care Cancer 2021;29(10):5701–9. Available from: https://doi.org/10.1007/s00520-021-06048-5.

[26] Ameri A, Poshtmahi S, Heydarirad G, et al. Effect of honey-lemon spray versus benzydamine hydrochloride spray on radiation-induced acute oral mucositis in head and neck cancer patients: a pilot, randomized, double-blind, active-controlled clinical trial. J Altern Complement Med 2021;27(3):255–62. Available from: https://doi.org/10.1089/acm.2020.0468.

[27] Kiprian D, Jarzabski A, Kawecki A. Evaluation of efficacy of Caphosol in prevention and alleviation of acute side effects in patients treated with radiotherapy for head and neck cancers. Contemp Oncol (Pozn) 2016;20(5):389–93. Available from: https://doi.org/10.5114/wo.2016.64600.

[28] Immonen E, Aine L, Nikkilä A, et al. Randomized controlled and double-blinded study of Caphosol versus saline oral rinses in pediatric patients with cancer. Pediatr Blood Cancer 2020;67(10):e28520. Available from: https://doi.org/10.1002/pbc.28520.

[29] Wong KH, Kuciejewska A, Sharabiani MTA, et al. A randomised controlled trial of Caphosol mouthwash in management of radiation-induced mucositis in head and neck cancer. Radiother Oncol 2017;122(2):207–11. Available from: https://doi.org/10.1016/j.radonc.2016.06.015.

[30] Vokurka S, Skardova J, Hruskova R, et al. The effect of polyvinylpyrrolidone-sodium hyaluronate gel (Gelclair) on oral microbial colonization and pain control compared with other rinsing solutions in patients with oral mucositis after allogeneic stem cells transplantation. Med Sci Monit 2011;17(10):CR572–6. Available from: https://doi.org/10.12659/msm.881983.

[31] Barber C, Powell R, Ellis A, et al. Comparing pain control and ability to eat and drink with standard therapy vs Gelclair: a preliminary, double centre, randomised controlled trial on patients with radiotherapy-induced oral mucositis. Support Care Cancer 2007;15(4):427–40. Available from: https://doi.org/10.1007/s00520-006-0171-1.

[32] Shenep JL, Kalwinsky DK, Hutson PR, et al. Efficacy of oral sucralfate suspension in prevention and treatment of chemotherapy-induced mucositis. J Pediatr 1988;113(4):758–63. Available from: https://doi.org/10.1016/s0022-3476(88)80397-4.

[33] Meredith R, Salter M, Kim R, et al. Sucralfate for radiation mucositis: results of a double-blind randomized trial. Int J Radiat Oncol Biol Phys 1997;37(2):275–9. Available from: https://doi.org/10.1016/s0360-3016(96)00531-7.

[34] Coppini M, Caponio VCA, Mauceri R, et al. Efficacy of topical agents in oral mucositis prevention: systematic review and network meta-analysis. Oral Dis 2024;30(7):4126–44. Available from: https://doi.org/10.1111/odi.15046.

[35] Allison RR, Ambrad AA, Arshoun Y, et al. Multi-institutional, randomized, double-blind, placebo-controlled trial to assess the efficacy of a mucoadhesive hydrogel (MuGard) in mitigating oral mucositis symptoms in patients being treated with chemoradiation therapy for cancers of the head and neck. Cancer 2014;120(9):1433–40. Available from: https://doi.org/10.1002/cncr.28553.

[36] Murdock JL, Reeves DJ. Chemotherapy-induced oral mucositis management: a retrospective analysis of MuGard, caphosol, and standard supportive care measures. J Oncol Pharm Pract 2019;26(3):521–8. Available from: https://doi.org/10.1177/1078155219850298.

[37] Anderson CM, Lee CM, Kelley JR, et al. ROMAN: phase 3 trial of avasopasem manganese (GC4419) for severe oral mucositis (SOM) in patients receiving chemoradiotherapy (CRT) for locally advanced, nonmetastatic head and neck cancer (LAHNC). J Clin Oncol 2022;40(16_suppl.):6005. Available from: https://doi.org/10.1200/JCO.2022.40.16_suppl.6005.

[38] FDA rejects NDA for avasopasem in RT-induced severe oral mucositis. Targeted Oncology. 2023, <https://www.targetedonc.com/view/fda-rejects-nda-for-avasopasem-in-rt-induced-severe-oral-mucositis>; 2023 [accessed 21.01.25].

[39] Dodd MJ, Dibble SL, Miaskowski C, et al. Randomized clinical trial of the effectiveness of 3 commonly used mouthwashes to treat chemotherapy-induced mucositis. Oral Surg Oral Med Oral Pathol Oral Radiol Endod 2000;90(1):39–47. Available from: https://doi.org/10.1067/moe.2000.105713.

[40] Sio TT, Le-Rademacher JG, Leenstra JL, et al. Effect of doxepin mouthwash or diphenhydramine-lidocaine-antacid mouthwash vs placebo on radiotherapy-related oral mucositis pain. JAMA 2019;321(15):1481. Available from: https://doi.org/10.1001/jama.2019.3504.

[41] Lalla RV, Solé S, Becerra S, et al. Efficacy and safety of Dentoxol® in the prevention of radiation-induced oral mucositis in head and neck cancer patients (ESDOM): a randomized, multicenter, double-blind, placebo-controlled, phase II trial. Support Care Cancer 2020;28(12):5871–9. Available from: https://doi.org/10.1007/s00520-020-05358-4.
[42] Salvador P, Azusano C, Wang L, et al. A pilot randomized controlled trial of an oral care intervention to reduce mucositis severity in stem cell transplant patients. J Pain Symptom Manage 2012;44(1):64–73. Available from: https://doi.org/10.1016/j.jpainsymman.2011.08.012.
[43] Franco R, Lupi E, Iacomino E, et al. Low-level laser therapy for the treatment of oral mucositis induced by hematopoietic stem cell transplantation: a systematic review with meta-analysis. Medicina (Kaunas) 2023;59(8):1413. Available from: https://doi.org/10.3390/medicina59081413.
[44] Andriakopoulou CS, Yapijakis C, Koutelekos I, et al. Prevention and treatment of oral mucositis in pediatric patients: systematic review and meta-analysis of randomized controlled trials. In Vivo 2024;38(3):1016–29. Available from: https://doi.org/10.21873/invivo.13535.
[45] Elad S, Cheng KKF, Lalla RV, et al. MASCC/ISOO clinical practice guidelines for the management of mucositis secondary to cancer therapy. Cancer 2020;126(19):4423–31. Available from: https://doi.org/10.1002/cncr.33100.
[46] Peterson DE, Boers-Doets CB, Bensadoun RJ, et al. Management of oral and gastrointestinal mucosal injury: ESMO clinical practice guidelines for diagnosis, treatment, and follow-up. Ann Oncol 2015;26(5):v139–51. Available from: https://doi.org/10.1093/annonc/mdv202.
[47] Miller KK, Gorcey L, McLellan BN. Chemotherapy-induced hand-foot syndrome and nail changes: a review of clinical presentation, etiology, pathogenesis, and management. J Am Acad Dermatol 2014;71(4):787–94. Available from: https://doi.org/10.1016/j.jaad.2014.03.019.
[48] Bernier J, Bonner J, Vermorken JB, et al. Consensus guidelines for the management of radiation dermatitis and coexisting acne-like rash in patients receiving radiotherapy plus EGFR inhibitors for the treatment of squamous cell carcinoma of the head and neck. Ann Oncol 2008;19(1):142–9. Available from: https://doi.org/10.1093/annonc/mdm400.
[49] Cury-Martins J, Eris APM, Abdalla CMZ, et al. Management of dermatologic adverse events from cancer therapies: recommendations of an expert panel. An Bras Dermatol 2020;95(2):221–37. Available from: https://doi.org/10.1016/j.abd.2020.01.001.
[50] Terayama K, Hamana M, Tsuchiya M, et al. Effectiveness of topical NSAIDs in a gastric cancer patient with bone metastasis pain during therapy with opioids and oral NSAIDs. Yakugaku Zasshi 2020;140(8):1081–5. Available from: https://doi.org/10.1248/yakushi.19-00260.
[51] Santhosh A, Sharma A, Bakhshi S, et al. Topical diclofenac for prevention of capecitabine-associated hand-foot syndrome: a double-blind randomized controlled trial. J Clin Oncol 2024;42(15):1821–9. Available from: https://doi.org/10.1200/JCO.23.01730.
[52] Bolderston A, Lloyd NS, Wong RK, et al. Supportive Care Guidelines Group of Cancer Care Ontario Program in Evidence-Based Care. The prevention and management of acute skin reactions related to radiation therapy: a systematic review and practice guideline. Support Care Cancer 2006;14(8):802–17. Available from: https://doi.org/10.1007/s00520-006-0063-4.
[53] Koukourakis GV, Kelekis N, Kouvaris J, et al. Therapeutics interventions with anti-inflammatory creams in postradiation acute skin reactions: a systematic review of most important clinical trials. Recent Pat Inflamm Allergy Drug Discov 2010;4(2):149–58. Available from: https://doi.org/10.2174/187221310791163099.
[54] Salvo N, Barnes E, van Draanen J, et al. Prophylaxis and management of acute radiation-induced skin reactions: a systematic review of the literature. Curr Oncol 2010;17(4):94–112. Available from: https://doi.org/10.3747/co.v17i4.493.
[55] Geara FB, Eid T, Zouain N, et al. Randomized, prospective, open-label phase III trial comparing mebo ointment with biafine cream for the management of acute dermatitis during radiotherapy for breast cancer. Am J Clin Oncol 2018;41(12):1257–62. Available from: https://doi.org/10.1097/COC.0000000000000460.
[56] Hasuo H, Sakuma H, Uchitani K, et al. Short-term effects of 10% lidocaine ointment on allodynia in cancer pain: a randomized, double-blind, placebo-controlled crossover study. J Palliat Med 2019;22(11):1364–9. Available from: https://doi.org/10.1089/jpm.2019.0081.
[57] Peng L, Zheng HY, Dai Y. Local dermal application of a compound lidocaine cream in pain management of cancer wounds. Braz J Med Biol Res 2019;52(11):e8567. Available from: https://doi.org/10.1590/1414-431X20198567.
[58] Halili H, Azizkhani R, Tavakoli Garmaseh S, et al. Comparing the effect of lidocaine-prilocaine cream and infiltrative lidocaine on overall pain perception during thoracentesis and abdominocentesis: a randomized clinical trial. Anesth Pain Med 2020;11(1):e106275. Available from: https://doi.org/10.5812/aapm.106275.
[59] Garzón-Rodríguez C, Casals Merchan M, Calsina-Berna A, et al. Lidocaine 5% patches as an effective short-term co-analgesic in cancer pain. Preliminary results. Support Care Cancer 2013;21(11):3153–8. Available from: https://doi.org/10.1007/s00520-013-1948-7.
[60] Lord S, Hardy J, Good P. Does cannabidiol have a benefit as a supportive care drug in cancer? Curr Treat Options Oncol 2022;23(4):514–25. Available from: https://doi.org/10.1007/s11864-021-00934-0.
[61] Xu DH, Cullen BD, Tang M, et al. The effectiveness of topical cannabidiol oil in symptomatic relief of peripheral neuropathy of the lower extremities. Curr Pharm Biotechnol 2020;21(5):390–402. Available from: https://doi.org/10.2174/1389201020666191202111534.
[62] Liou KT, Chen C, Emard N, et al. Herbal topical analgesic for pain management: perspectives from cancer patients. Pain Med 2021;22(6):1435–40. Available from: https://doi.org/10.1093/pm/pnab072.

Chapter 16

Interactions between pain management drugs

Dragana P. Srebro
Faculty of Medicine, Institute of Pharmacology, Clinical Pharmacology and Toxicology, University of Belgrade, Belgrade, Serbia

Key learning objectives

- Understand the main types of clinically significant interactions involving analgesic drugs.
- Distinguish between pharmacokinetic and pharmacodynamic interactions of opioids and explain their relevance in cancer pain management.
- Identify important interactions involving nonopioid analgesics, including NSAIDs and acetaminophen/paracetamol.
- Apply patient-specific factors, such as age, comorbidities, polypharmacy, and organ dysfunction, when evaluating interaction risks

Introduction

The use of multiple drugs is common in clinical practice [1]. The use of multiple drugs is sometimes a consequence of the disease itself (e.g., diabetes, arterial hypertension, malignant diseases), and it can also be a consequence of self-medication, because some drugs are available on the market without a prescription (over-the-counter drugs). Medicines used in the treatment of pain are divided into opioid analgesics, nonopioid analgesics, and adjuvant analgesics [2,3].

As pain accompanies many conditions and diseases, analgesics are often used in addition to other medicines, especially in the elderly population, which puts them at greater risk for clinically significant drug interactions. Opioid analgesics are most often used in the treatment of cancer and postoperative pain [4,5]. In addition, nonopioid analgesics such as paracetamol and nonsteroidal antiinflammatory drugs (NSAIDs) can be bought from a pharmacy without a prescription in many countries. Compared to opioid analgesics and NSAIDs, the potential for interactions with paracetamol is lower.

The interaction of two drugs is a change in the action of one drug caused by the presence of another drug. For practical work, clinically significant drug interactions are important, where the effectiveness or safety of the therapy is significantly reduced. The interaction between drugs most often occurs when two or more drugs are administered simultaneously, but it can also occur when one drug is discontinued and another drug is continued in an unadjusted dose [6]. In addition to drug–drug interactions, there are several other types of interactions: drug–disease, drug–food, drug–alcohol, and drug–herbal products [7]. Factors such as age-related changes in pharmacokinetics and pharmacodynamics and interindividual variability should be taken into account when evaluating drug interactions.

Interactions of opioids with other drugs

Interactions of opioids with other drugs can be pharmacokinetic and pharmacodynamic [8]. Pharmacokinetic drug–drug interactions include effects on absorption, metabolic pathways, drug transport through membranes, and protein binding. The most well-known are interactions at the level of P-glycoprotein transporter and CYP enzyme metabolism. Many commonly used opioids are metabolized by CYP2D6 and CYP3A4 or CYP3A5. Pharmacodynamic interactions of opioids with other drugs can be synergistic, additive, or antagonistic. The most common of these interactions include synergistic or additive effects with other drugs that have analgesic or sedative effects.

Pain Management in Palliative Care for Patients With Cancer. DOI: https://doi.org/10.1016/B978-0-443-30050-9.00003-5

Opioids—pharmacokinetic interactions

Resorption

The absorption of opioids may be enhanced when the intragastric pH is increased. Thus, using drugs such as ranitidine (H2 antagonist) could result in increased morphine plasma concentrations after acute and chronic morphine administration [9].

Opioids slow gastrointestinal tract motility and decrease intestinal secretions. Opioids induce constipation as an adverse effect, but it may delay drug absorption and decrease the C_{max} of other orally administered drugs. Compared to other opioids, tramadol and tapentadol produce less constipation and are less likely to delay the absorption of other drugs.

Transport

P-glycoprotein, the transmembrane efflux transporter, in the intestine, brain capillary endothelium, and renal tubule may reduce the absorption of drugs [10,11]. Some opioids (morphine, codeine, methadone, and fentanyl) are substrates of P-glycoprotein. Morphine glucuronides are not substrates of P-glycoprotein. Itraconazole and quinidine (P-glycoprotein inhibitors), in combination with oral morphine, increase the peak plasma morphine concentrations, but the effects of morphine are unchanged. Itraconazole, verapamil, HIV protease inhibitors, and grapefruit juice are potent P-glycoprotein inhibitors and may increase opioid exposure and effects [12,13]. There is overlap between the substrate specifics of P-glycoprotein and the CYP3A4 enzymes. St. John's wort, a herbal medicine, decreases concentrations of methadone in plasma by induction of P-glycoprotein and/or CYP3A4.

The association between P-glycoprotein and oxycodone is controversial and not completely elucidated, while buprenorphine is not a P-glycoprotein substrate but is a potent competitive P-glycoprotein inhibitor [14]. Tramadol and its metabolite are not P-glycoprotein substrates, but proton-based efflux pumps may be involved in its transepithelial transport [15].

Plasma protein binding

The free, unbound concentration of a drug in the plasma is pharmacologically active and produces the effects of drugs. The concentration of buprenorphine, fentanyl, and methadone, highly protein-bound opioids (more than 88%), may be changed in combination with other drugs that have high protein-binding affinity. Drugs with high protein binding that are often given in combination with opioids are propofol, haloperidol, midazolam, itraconazole, methylprednisolone, and verapamil. For parenteral administration of fentanyl, alfentanil, remifentanil, and sufentanil, dosing regimens may be adjusted according to changes in plasma protein binding. For opioids orally administered and eliminated primarily by the liver, there is no need for dosing adjustments. Buprenorphine does not bind to albumin (it binds to alpha- and beta globulins) and is not in competition for binding in plasma [3].

Metabolism and excretion

Opioids are metabolized in the liver through the reactions of glucuronidation and conjugation and/or demethylation. The clinical significance of the interactions of morphine with amitriptyline and ranitidine, which occurred as a result of the decrease in glucuronidation of morphine, was recorded [16]. Metabolic reactions of opioids are catalyzed by CYP450 isoenzymes, most often CYP2D6 and CYP3A4/5 [17–19]. Pharmacogenetic differences in opioid metabolism contribute to interindividual variation in response to opioids. For morphine, hydromorphone, and buprenorphine, glucuronidation, catalyzed by UDP-glucuronosyltransferases (UGTs), is a major metabolic pathway. The UGTs (UGT1A1 and UGT2B7 in morphine and buprenorphine metabolism) are polymorphic and inducible.

Tapentadol is extensively metabolized by glucuronic acid to inactive metabolites, while a smaller portion is metabolized by cytochrome P450 enzyme by CYP2C9, CYP2C19, and CYP2D6 [20]. CYP isoenzymes may be inducible or polymorphic (CYP2D6 and CYP2C19). Inductors of CYP enzymes (rifampicin and carbamazepine) may reduce morphine analgesia. Drug inducers of CYP2B6 include phenobarbital, phenytoin, and rifampicin. The CYP2D6 isoenzyme is important for the metabolism of codeine and tramadol, which is why drugs that inhibit CYP2D6 (bupropion, paroxetine, fluoxetine, etc.) reduce their biotransformation and effectiveness. Drug inducers of CYP3A4/5/7 include rifampicin, phenytoin, carbamazepine, glucocorticoids, and herbal medicines such as St. John's wort. CYP3A4 inhibitors (clarithromycin, itraconazole, grapefruit juice, etc.) can slow down the breakdown of opioid analgesics and potentiate their effectiveness or side effects (sedation and respiratory depression). Ciprofloxacin, as an inhibitor of CYP3A4/5/7, may interact with methadone and can lead to serious toxicity [21]. CYP3A4/5 inhibitors such as itraconazole, grapefruit, and voriconazole will increase the effects of oxycodone [12,13,22], while inductors of CYP3A4/5 such as rifampicin will reduce the efficacy of oxycodone.

Opioids—pharmacodynamic drug interactions

Opioids have effects on important neurotransmitter systems, such as cholinergic, GABAergic, dopaminergic, and serotonergic, and consequently may have potential interactions with drugs that affect these neurotransmitter systems. These drugs include GABAergic drugs (e.g., benzodiazepines), anticholinergic drugs (e.g., tricyclic antidepressants), anticholinesterase inhibitors (physostigmine, distigmine, and rivastigmine), dopaminergic drugs (e.g., levodopa), and antidopaminergic drugs (e.g., haloperidol) [23]. Drug interactions of opioids and these drugs depend on acute or chronic administration of opioids, characteristics of opioids, individual patient characteristics, and underlying conditions [24].

Opioids can enter into clinically significant interactions with alcohol and drugs such as antidepressants, antihistamines, anxiolytics, and others where, due to the additive effect, there is depression of the central nervous system (CNS) and sedation. Benzodiazepines potentiate the depressive effect of opioids on the CNS—sedation, decreased consciousness, and respiratory depression [25]. Antiepileptic drugs (pregabalin or gabapentin) and antidepressant drugs may act synergistically with opioids, causing more frequent sedation and confusion [26]. Ketamine and morphine or fentanyl act synergistically, and the combination increases analgesia with mild adverse effects [27–29]. The dry mouth, difficulty urinating, or confusion will increase when using the opioids in combination with anticholinergic drugs, tricyclic antidepressants, or traditional antipsychotics. Tramadol causes acute deterioration of tremor in far-advanced Parkinson disease, and fentanyl causes bradykinesia and rigidity [30,31].

Different types of drugs can prolong the QT interval and induce Torsade de pointes (a fatal form of ventricular arrhythmia), such as antiarrhythmics, antibiotics, antipsychotics, antidepressants, antiemetics, antihistamines (e.g., astemizole, terfenadine), gastrointestinal prokinetics (cisapride), and also opioids. Opioids increase this risk at high doses, when used concomitantly with other drugs, and in patients who are genetically predisposed [32,33]. Methadone exhibits the highest risk for QT prolongation and dose-dependent effect, while tramadol, oxycodone, and fentanyl have intermediate risk and dose-dependent effect, and morphine, buprenorphine, and remifentanil are considered safer, although not completely without risk [34]. Weak opioids such as pethidine (meperidine), dextromethorphan, and tramadol can potentially induce serotonin syndrome [35]. Tramadol can interact additively with drugs that lead to the release or inhibition of serotonin and noradrenaline uptake (levodopa, valproic acid, selective serotonin reuptake inhibitors [SSRIs], serotonin and noradrenaline reuptake inhibitors [SNRIs], tricyclic antidepressants, St. John's wort, monoamine oxidase inhibitors [MAOIs], triptans, lithium, etc.). Tramadol and dextromethorphan are metabolized by CYP2D6, which is strongly inhibited by SSRIs (fluoxetine and paroxetine). Serotonin syndrome is an alteration in mental status, autonomic dysfunction, and neuromuscular abnormalities (confusion, disorientation, irritability, anxiety, sweating, tachycardia, tremors, diarrhea, muscle stiffness, and hyperthermia) after hyperstimulation of postsynaptic 5-HT2A and 5-HT1A receptors and noradrenergic hyperactivity in the CNS. Serotonin syndrome is usually caused by interactions involving SSRIs, MAOIs, serotonin precursors (e.g., tryptophan), and other serotonergic drugs (e.g., some 5-HT3 antagonist antiemetics and antimigraine drugs like sumatriptan). Tramadol may also lead to an increase of the anticoagulant effect of warfarin [36]. Coadministration of tramadol with diuretics (furosemide, hydrochlorothiazide, and indapamide) has a higher risk of diuretic side effects/possible reduced effect of diuretics.

Drug–drug interactions with opioids are also possible. Drug–drug interactions are mostly focused on drug combinations that result in increased CNS depression due to the simultaneous use of three CNS depressants. This may be the use of opioids with benzodiazepines and skeletal muscle relaxants (especially carisoprodol) [37–40]. The simultaneous use of sedative-hypnotics, even at low doses of opioids, represents a significantly higher risk of opioid overdose, especially when central muscle relaxants are also used. In one-third of cases, such combinations are the result of misuse or improper use of drugs, and in two-thirds of cases there is a suspicion of a suicide attempt. Recent research shows that this risk increases with the duration of treatment, minimum ≤14 days, maximum >60 days, when using baclofen or carisoprodol and if benzodiazepines and opioids with a daily dose ≥50 mg are used simultaneously.

The synergistic interaction of paracetamol and tramadol increases the effectiveness and prolongs the duration of analgesia. Preclinical and clinical trials have demonstrated a synergistic antinociceptive interaction between opioids and ketamine at lower doses than those used when administered separately. Serum ketamine concentration that did not alter sedation indices enhanced the antinociceptive effect of fentanyl. Ketamine at low stable doses (30–120 ng/mL) enhances the analgesic effect of fentanyl without increasing sedation. Naloxone is an antagonist of opioid analgesics, which is important in opioid overdose (Table 16.1).

Interactions of nonopioid analgesics with other drugs

Paracetamol (acetaminophen)—pharmacokinetic interactions

Resorption

After ingestion, some drugs can alter the absorption of paracetamol (anion exchange resins, such as activated charcoal and cholestyramine—time-dependently reduce paracetamol absorption). This is important in case of paracetamol poisoning.

TABLE 16.1 Some major clinically significant interactions of opioid analgesics. Recommendation: generally avoid or closely monitor therapy and patient.

Opioid(s)	Other drug(s)	Effects
Morphine, fentanyl, tramadol, tapentadol	Other central nervous system (CNS) depressants (e.g., benzodiazepines [lorazepam, diazepam, alprazolam] and nonbenzodiazepine sedatives/hypnotics anxiolytics [zolpidem], muscle relaxants, general anesthetics, antipsychotics [haloperidol, quetiapine], other opioids [hydromorphone, tapentadol, tramadol], anticonvulsant [pregabalin, gabapentin], alcohol, cough medications containing opioids [e.g., codeine, hydrocodone]), (es)ketamine	Profound sedation, respiratory depression, coma, and death
Morphine	Quetiapine	Constipation, intestinal obstruction, fecal impaction
Fentanyl	Ondansetron, tramadol	Serotonin syndrome
Fentanyl	Inhibitors of CYP450 3A4 (amiodarone, aprepitant, chloramphenicol, ketoconazole, verapamil)	Increase the plasma concentrations of fentanyl, may cause potentially fatal respiratory depression
Fentanyl, butorphanol, hydrocodone, oxycodone	Inductors of CYP3A4 (carbamazepine, dexamethasone, rifampicin)	Reduced efficacy or withdrawal symptoms of opioids
Tramadol	Serotonergic agents (serotonin reuptake inhibitors, monoamine oxidase inhibitors [MAOIs], tricyclic antidepressants, 5-HT1 receptor agonists, ergot alkaloids, cyclobenzaprine, lithium, St. John's wort, opioids [fentanyl], dextromethorphan, and tryptophan)	Potentiate the risk of serotonin syndrome
Tramadol	Substance that can reduce the seizure threshold, such as selective serotonin reuptake inhibitors (SSRIs) (SSRI antidepressants or anorectics), MAOIs, neuroleptic agents, CNS stimulants, opioids, tricyclic antidepressants, other tricyclic compounds (e.g., cyclobenzaprine, phenothiazines), carbapenems, cholinergic agents, fluoroquinolones, interferons, chloroquine, mefloquine, lindane, and theophylline	Seizure, convulsion
Tramadol	Class IA (e.g., disopyramide, quinidine, procainamide) and class III (e.g., amiodarone, dofetilide, sotalol) antiarrhythmic agents	Prolongation of the QT interval, ventricular arrhythmias, torsade de pointes
Tramadol	Warfarin	Increase risk of bleeding

Prokinetic agents (metoclopramide) increase, while opioid analgesics can slow the rate of absorption of paracetamol, changing the rate of gastric emptying [41].

Distribution

Paracetamol at therapeutic doses has low protein binding, and there are no clinically significant interactions based on plasma protein binding [3].

Metabolism and excretion

Drugs that are inducers of microsomal liver enzymes (phenytoin, barbiturates, carbamazepine, and rifampicin) can accelerate the metabolism of paracetamol and the formation of hepatotoxic metabolites. Chronic consumption of alcohol leads to the induction of the CYP2E1 isoenzyme, which participates in the production of the hepatotoxic metabolite of paracetamol. Drugs that affect glutathione levels may reduce glucuronidation of paracetamol and excretion and increase the risk of hepatotoxicity of paracetamol [41].

Paracetamol (acetaminophen)—pharmacodynamic interactions

The major drug interactions (Table 16.2) and the risk of hepatotoxicity increase when paracetamol is administered together with other drugs that have a hepatotoxic effect (ketoconazole, levoketoconazole, leflunomide, teriflunomide, lomitapide, and mipomersen). Drug-induced liver injury occurred in 13% of patients using levoketoconazole. Generally, the risk of hepatotoxicity may exist when acetaminophen is used with other hepatotoxic agents. These include alcohol; androgens and anabolic steroids; antituberculous agents; azole antifungal agents; angiotensin-converting enzyme (ACE) inhibitors; cyclosporine (high dosages); cannabidiol; disulfiram; endothelin receptor antagonists; interferons; ketolide and macrolide antibiotics; kinase inhibitors; minocycline; nonsteroidal antiinflammatory agents; nucleoside reverse transcriptase inhibitors; proteasome inhibitors; retinoids; thiazolidinediones; tolvaptan; vincristine; zileuton; anticonvulsants such as carbamazepine, hydantoins, felbamate, and valproic acid; and lipid-lowering medications such as fenofibrate, lomitapide, mipomersen, niacin, and statins [42].

Prolonged ingestion of relatively high paracetamol dosages (greater than 1.3–4 g/day continuously for greater than 1 week) may potentiate the anticoagulant effect of warfarin and other vitamin K antagonists and increase international normalized ratio (risk of bleeding) [43]. The toxic metabolite of paracetamol (N-acetyl-p-benzoquinone-imine [NAPQI]) is thought to interfere with vitamin K-dependent coagulation factors. Paracetamol is metabolized by CYP2E1 to NAPQI [41,42,44].

N-acetylcysteine provides cysteine for glutathione synthesis and attenuates the hepatotoxicity, which is an important antidote in acetaminophen overdose.

NSAIDs—pharmacokinetic interactions

The most significant pharmacodynamic and pharmacokinetic interactions of NSAIDs are with other drugs that affect blood clotting, kidney function, and the gastrointestinal system, and they are especially common in the elderly [3,45–47].

Resorption

A few NSAIDs, such as aspirin, tolfenamic acid, and diclofenac, undergo presystemic or first-pass elimination and reduction in bioavailability [48].

NSAIDs may be absorbed systemically following topical administration, especially if absorption is increased with frequent applications; use of large quantities or over large areas of skin; use on compromised or diseased skin or open wounds; and use of occlusive dressings or heating pads over application areas. Concomitant use of topical with systemic NSAIDs may result in increased adverse effects such as gastrointestinal toxicity (inflammation, bleeding, ulceration, and perforation), cardiovascular thrombotic events, hepatotoxicity (liver enzyme elevations), renal toxicity, fluid retention, edema, hypertension, and inhibition of platelet aggregation [49]. This interaction is moderate, and these patients should avoid or limit the use of systemic NSAIDs.

Distribution

NSAIDs are weak acids that bind to plasma albumin in a high percentage (95%–99%). Thus, they can displace other drugs (warfarin, salicylates, sulfonylurea derivatives, phenytoin, and methotrexate) and increase their plasma concentration and thus toxicity. Aspirin has a high affinity for binding to plasma proteins, and if used in combination with valproic acid, it displaces it from the drug–protein bond, which increases the free valproic acid fraction and the risk of increased side effects [50]. Aspirin at antiinflammatory dosages or higher has been shown to decrease the plasma concentrations of many NSAIDs, including ibuprofen.

The hypoglycemic effect of insulin secretagogues (e.g., sulfonylureas, meglitinides) may be potentiated by NSAIDs (and other drugs such as ACE inhibitors, 4-aminoquinolines, amylin analogs, anabolic steroids, fibrates, salicylates, SSRIs, and sulfonamides) [51–54]. These drugs may increase the risk of hypoglycemia by stimulating insulin secretion

TABLE 16.2 Some major clinically significant interactions of paracetamol (acetaminophen). Recommendation: generally avoid or closely monitor therapy and patient.

Paracetamol	Other drug(s)	Effects
Paracetamol	Ketoconazole/levoketoconazole, mipomersen, leflunomide/teriflunomide	Increase the risk of hepatotoxicity

(salicylates, NSAIDs), or they may increase plasma concentration of insulin secretagogues by displacing them from plasma protein binding sites and/or inhibiting their metabolism (fibrates, NSAIDs, salicylates, and sulfonamides).

Metabolism

NSAIDs interact, in the process of metabolism, with drugs that inhibit isoenzyme CYP2C9, such as fluconazole and amiodarone [55]. By inhibiting the metabolism of NSAIDs, their concentrations in the blood are increased, as is the risk of bleeding. These interactions are of particular importance when the patient also uses warfarin in therapy, which is also a substrate for CYP2C9. In this case, the interaction of NSAIDs and warfarin becomes pharmacodynamic and pharmacokinetic, and the probability of bleeding increases.

Some NSAIDs are prodrugs and need hepatic metabolism to become their active metabolites (sulindac–sulindac sulfide and parecoxib–valdecoxib).

Excretion

Urine alkalization increases the excretion of weak acids (acetylsalicylic acid). Aspirin and other NSAIDs inhibit the secretion of diuretics into the lumen of the tubules, which reduces their diuretic effect, or reduce the excretion of methotrexate, which increases the toxicity of methotrexate [56]. In the interaction between NSAIDs and methotrexate, the mechanism of interaction is based on the reduced production of prostaglandins, resulting in reduced kidney blood flow and impaired methotrexate excretion. This interaction is clinically significant when methotrexate is used in higher doses in hematological and oncological indications but not in rheumatological indications (a dose of up to 25 mg of methotrexate per week). NSAIDs coadministered with lithium can reduce lithium excretion and increase serum lithium levels and induce toxicity in some patients. NSAIDs, through the inhibition of renal prostaglandin synthesis, decreased renal blood flow and lithium excretion.

NSAIDs (indomethacin and ibuprofen) may increase plasma digoxin concentrations and half-life and reduce renal clearance of digoxin.

Elimination

Metabolism by CYP2C9 contributes extensively to the inactivation of NSAIDs mentioned, such as ibuprofen, flurbiprofen, celecoxib, piroxicam, and tenoxicam [57]. NSAIDs (ibuprofen) interact, in the process of metabolism, with drugs that inhibit isoenzyme CYP2C9, such as fluconazole, voriconazole, and amiodarone. By inhibiting the metabolism of NSAIDs, their concentrations in the blood are increased, as is the risk of bleeding. These interactions are of particular importance when the patient also uses warfarin in therapy, which is also a substrate for CYP2C9. In this case, the interaction of NSAIDs and warfarin becomes pharmacodynamic and pharmacokinetic, and the probability of bleeding increases. Aprepitant is an inductor of CYP2C9, and moderate interaction may occur between ibuprofen and aprepitant or its prodrug, fosaprepitant.

Coxibs also undergo hepatic metabolism: rofecoxib primarily by reduction by cytosolic enzymes and celecoxib by the CYP P450 enzyme system. Historically, rofecoxib had fewer confirmed or potential drug interactions than celecoxib [58], although it has since been withdrawn from the market for cardiovascular safety reasons. However, potent inducers of CYP, such as rifampin, may decrease rofecoxib concentrations because of induction of general hepatic metabolic activity. Celecoxib is metabolized by CYP2C9 and may be increased or decreased by CYP2C9 modifiers. It also inhibits CYP2D6 and may affect concentrations of CYP2D6 substrates.

NSAIDs—pharmacodynamic interactions

NSAIDs increase the risk of toxicity when coadministered with drugs that are nephrotoxic, myelosuppressive, or have side effects on the gastrointestinal tract [3,45–47]. The combination of aspirin (via inhibition of biosynthesis of thromboxane A2 in platelets, which can damage stomach lining) and warfarin increases the risk of bleeding, although this combination can be applied safely with careful monitoring. Coadministration of aspirin in cardioprotective doses with nonselective NSAIDs increases the risk of side effects on the gastrointestinal tract. This combination should be avoided, or simultaneous use of proton pump inhibitors is recommended. The simultaneous use of two NSAIDs is not recommended because of additive interactions with each other and serious gastrointestinal toxicity, including inflammation, bleeding, ulceration, and perforation. However, the exception according to the mechanism of action is the interaction of aspirin and ibuprofen. Aspirin irreversibly binds to COX-1, and ibuprofen is reversible but has a higher affinity for the enzyme. Ibuprofen interferes with the binding of aspirin to COX-1, preventing its antiaggregation effect in lower doses. Therefore, if it is necessary to apply both drugs, it is necessary to separate the applications in time. Aspirin is applied first, and after at least 2 hours, ibuprofen can be applied. Diclofenac (75 mg twice daily of delayed-release) may be an alternative for patients requiring routine NSAID therapy with concomitant low-dose aspirin. Other noninterfering alternatives for pain include paracetamol, celecoxib, or opioid analgesics. The risk of gastrointestinal bleeding is increased with simultaneous use of NSAIDs and corticosteroids. Both drugs are ulcerogenic and have additive effects on the gastrointestinal mucosa during coadministration. Caution is advised, and prophylactic antiulcer therapy is recommended (e.g., antacids, H2 antagonists). Also, SSRIs and SNRIs in combination with NSAIDs may increase the patient's risk of bleeding [59]. Platelets physiologically take serotonin from the blood, which is released again during platelet aggregation and activates new platelets [60]. Due to the mechanism of action of SSRIs and SNRIs, serotonin concentrations are increased in the blood, but platelets do not take them up, which further reduces the possibility of platelet aggregation. Simultaneous application of nonselective NSAIDs with oral anticoagulants such as warfarin, acenocoumarol, direct oral anticoagulants (apixaban, rivaroxaban, dabigatran, etc.), heparin, and P2Y12 inhibitors (clopidogrel, ticagrelor, etc.) should be avoided given that an adverse outcome of the interaction may be bleeding as a consequence of reduced capacity for formation of coagulum and/or inhibition of platelet aggregation. The coadministration of NSAIDs and heparin or low-molecular-weight heparin may potentiate the risk of bleeding; these effects are generally slight and of relatively short duration for most NSAIDs (except aspirin) at recommended dosages.

NSAIDs antagonize the effects of many diuretics, vasodilators, and antihypertensives [47]. NSAIDs reduce the effectiveness of ACE inhibitors. This combination of drugs increases the nephrotoxic effect of NSAIDs and the risk of hyperkalemia [61]. NSAIDs interact with antihypertensives. It is known that the use of NSAIDs may increase the patient's systolic blood pressure and thus antagonize the effect of antihypertensive drugs. Also, simultaneous administration of NSAIDs and ACE inhibitors, in addition to reduced antihypertensive efficiency, can lead to reversible damage to kidney function. The mechanism of interaction is associated with the reduced production of prostaglandin E2, which acts as a vasodilator in the kidneys and enables a good blood supply to the kidneys and maintains glomerular filtration rate. Additionally, ACE inhibitor administration prevents the formation of angiotensin II, which acts as a vasoconstrictor, and enables the release of bradykinin, which acts as a vasodilator. Due to the additive vasodilator effect on the kidneys and reduced blood supply to the kidneys, the speed of glomerular filtration decreases, which, in patients with an increased risk, can lead to renal insufficiency. Unlike interactions that may result in bleeding, where the risk of interaction is attributed to

TABLE 16.3 Some major clinically significant interactions of nonsteroidal antiinflammatory drugs (NSAIDs). Recommendation: generally avoid or closely monitor therapy and patient.

NSAIDs	Other drug(s)	Effects
Ibuprofen, indomethacin, naproxen, tiaprofenic acid	Aspirin (low-dose)	The antiplatelet and cardioprotective effect of aspirin may be antagonized
NSAID(s) (ibuprofen, ketorolac)	Aspirin (high-dose)	Increase the potential for serious gastrointestinal toxicity, including inflammation, bleeding, ulceration, and perforation and renal toxicity. Pharmacokinetically, aspirin decreases the plasma concentrations of many NSAIDs (ibuprofen)
NSAID(s) (ibuprofen, diclofenac)	Etodolac, fenoprofen, ketorolac, ketoprofen, naproxen	Increase the potential for gastrointestinal, renal, or cardiovascular toxicity
NSAID(s) (ibuprofen, diclofenac, naproxen)	COX-2 inhibitors (meloxicam, celecoxib, rofecoxib)	Increase the potential for serious gastrointestinal toxicity or cardiovascular toxicity
NSAID(s) (aspirin, ibuprofen, diclofenac, naproxen)	Warfarin, apixaban, rivaroxaban, dabigatran, prasugrel, fondaparinux	Increase the risk of bleeding
NSAID(s) (aspirin, ibuprofen, diclofenac, naproxen)	Methotrexate	Increase the plasma concentrations and toxicities of methotrexate
NSAID(s) (ibuprofen, diclofenac, naproxen)	Sirolimus, tacrolimus	Additive renal toxicity

nonselective NSAIDs, for renal failure the risk exists regardless of the selectivity of drugs toward COX. Combined use of NSAIDs and cyclosporine or tacrolimus can also lead to renal failure. NSAIDs may potentiate the nephrotoxic effect of aminoglycosides (moderate interaction). NSAIDs with fluoroquinolone (ciprofloxacin and moxifloxacin) may potentiate the risk of CNS toxicity, and their concomitant use increases the risk of convulsions. The piperazine ring of fluoroquinolones may inhibit the binding of gamma-aminobutyric acid to brain receptors, and NSAIDs may synergistically add to this effect.

Coadministration of oral bisphosphonates (alendronate) and NSAIDs may result in additive or synergistic gastrointestinal toxicity (mucosa irritation) and nephrotoxic effects (renal failure) [62]. Nephrotoxicity of bisphosphonates, however, was primarily described for intravenous formulations of bisphosphonates (pamidronic acid and zoledronic acid), especially when they are administered too rapidly [63]. The risk of hypocalcemia may also be increased, as drug-induced renal tubular damage can lead to renal loss of calcium and magnesium. Nephrotoxic effects increase with the chronic use of NSAIDs.

Some of the major clinically significant interactions of the most commonly used NSAIDs are summarized in Table 16.3.

Conclusion

Clinically significant drug interactions are important for clinical practice. Before prescription of analgesics, it is important to evaluate in detail the use of comedication. To avoid the risk of adverse effects due to drug interaction, individual dose titration and careful follow-up are necessary.

References

[1] Guthrie B, Makubate B, Hernandez-Santiago V, et al. The rising tide of polypharmacy and drug-drug interactions: population database analysis 1995-2010. BMC Med 2015;13:74. Available from: https://doi.org/10.1186/s12916-015-0322-7.

[2] Alorfi NM. Pharmacological methods of pain management: narrative review of medication used. Int J Gen Med 2023;16:3247–56. Available from: https://doi.org/10.2147/IJGM.S419239.

[3] Vanderah TW, editor. Katzung's basic & clinical pharmacology. 16th ed. United States of America: McGraw-Hill, 2024.

[4] Garimella V, Cellini C. Postoperative pain control. Clin Colon Rectal Surg 2013;26(3):191–6. Available from: https://doi.org/10.1055/s-0033-1351138.

[5] Klepstad P, Kaasa S, Cherny N, et al. Research Steering Committee of the EAPC. Pain and pain treatments in European palliative care units. A cross sectional survey from the European Association for Palliative Care Research Network. Palliat Med 2005;19(6):477–84. Available from: https://doi.org/10.1191/0269216305pm1054oa.

[6] Tatro D. Drug interaction facts 2014: the authority on drug interactions. Philadelphia, United States: Lippincott Williams & Wilkins; 2014.

[7] Izzo AA, Ernst E. Interactions between herbal medicines and prescribed drugs: an updated systematic review. Drugs 2009;69(13):1777–98. Available from: https://doi.org/10.2165/11317010-000000000-00000.

[8] Kotlinska-Lemieszek A, Klepstad P, Haugen DF. Clinically significant drug-drug interactions involving opioid analgesics used for pain treatment in patients with cancer: a systematic review. Drug Des Devel Ther 2015;9:5255–67. Available from: https://doi.org/10.2147/DDDT.S86983.

[9] Aasmundstad TA, Størset P. Influence of ranitidine on the morphine-3-glucuronide to morphine-6-glucuronide ratio after oral administration of morphine in humans. Hum Exp Toxicol 1998;17(6):347–52. Available from: https://doi.org/10.1177/096032719801700611.

[10] Kharasch ED, Hoffer C, Whittington D, et al. Role of P-glycoprotein in the intestinal absorption and clinical effects of morphine. Clin Pharmacol Ther 2003;74(6):543–54. Available from: https://doi.org/10.1016/j.clpt.2003.08.011.

[11] Linnet K, Ejsing TB. A review on the impact of P-glycoprotein on the penetration of drugs into the brain. Focus on psychotropic drugs. Eur Neuropsychopharmacol 2008;18(3):157–69. Available from: https://doi.org/10.1016/j.euroneuro.2007.06.003.

[12] Saari TI, Grönlund J, Hagelberg NM, et al. Effects of itraconazole on the pharmacokinetics and pharmacodynamics of intravenously and orally administered oxycodone. Eur J Clin Pharmacol 2010;66(4):387–97. Available from: https://doi.org/10.1007/s00228-009-0775-8.

[13] Nieminen TH, Hagelberg NM, Saari TI, et al. Grapefruit juice enhances the exposure to oral oxycodone. Basic Clin Pharmacol Toxicol 2010;107(4):782–8. Available from: https://doi.org/10.1111/j.1742-7843.2010.00582.x.

[14] Kress HG. Clinical update on the pharmacology, efficacy and safety of transdermal buprenorphine. Eur J Pain 2009;13(3):219–30. Available from: https://doi.org/10.1016/j.ejpain.2008.04.011.

[15] Kanaan M, Daali Y, Dayer P, et al. Uptake/efflux transport of tramadol enantiomers and O-desmethyl-tramadol: focus on P-glycoprotein. Basic Clin Pharmacol Toxicol 2009;105(3):199–206. Available from: https://doi.org/10.1111/j.1742-7843.2009.00428.x.

[16] McQuay HJ, Carroll D, Faura CC, et al. Oral morphine in cancer pain: influences on morphine and metabolite concentration. Clin Pharmacol Ther 1990;48(3):236–44. Available from: https://doi.org/10.1038/clpt.1990.145.

[17] Collart L, Luthy C, Favario-Constantin C, et al. Dualité de l'effet analgésique du tramadol chez l'homme [Duality of the analgesic effect of tramadol in humans]. Schweiz Med Wochenschr 1993;123(47):2241–3.

[18] Desmeules J, Gascon MP, Dayer P, et al. Impact of environmental and genetic factors on codeine analgesia. Eur J Clin Pharmacol 1991;41(1):23–6. Available from: https://doi.org/10.1007/BF00280101.

[19] Heiskanen T, Olkkola KT, Kalso E. Effects of blocking CYP2D6 on the pharmacokinetics and pharmacodynamics of oxycodone. Clin Pharmacol Ther 1998;64(6):603–11. Available from: https://doi.org/10.1016/S0009-9236(98)90051-0.

[20] Singh DR, Nag K, Shetti AN, et al. Tapentadol hydrochloride: a novel analgesic. Saudi J Anaesth 2013;7(3):322–6. Available from: https://doi.org/10.4103/1658-354X.115319.

[21] Herrlin K, Segerdahl M, Gustafsson LL, et al. Methadone, ciprofloxacin, and adverse drug reactions. Lancet 2000;356(9247):2069–70. Available from: https://doi.org/10.1016/S0140-6736(00)03409-7.

[22] Hagelberg NM, Nieminen TH, Saari TI, et al. Voriconazole drastically increases exposure to oral oxycodone. Eur J Clin Pharmacol 2009;65(3):263–71. Available from: https://doi.org/10.1007/s00228-008-0568-5.

[23] Wehrfritz AP, Ihmsen H, Schmidt S, et al. Interaction of physostigmine and alfentanil in a human pain model. Br J Anaesth 2010;104(3):359–68. Available from: https://doi.org/10.1093/bja/aep372.

[24] Xu XS, Etropolski M, Upmalis D, et al. Pharmacokinetic and pharmacodynamic modeling of opioid-induced gastrointestinal side effects in patients receiving tapentadol IR and oxycodone IR. Pharm Res 2012;29(9):2555–64. Available from: https://doi.org/10.1007/s11095-012-0786-5.

[25] Lintzeris N, Mitchell TB, Bond AJ, et al. Pharmacodynamics of diazepam co-administered with methadone or buprenorphine under high dose conditions in opioid dependent patients. Drug Alcohol Depend 2007;91(2-3):187–94. Available from: https://doi.org/10.1016/j.drugalcdep.2007.05.019.

[26] Gilron I, Bailey JM, Tu D, et al. Morphine, gabapentin, or their combination for neuropathic pain. N Engl J Med 2005;352(13):1324–34. Available from: https://doi.org/10.1056/NEJMoa042580.

[27] Alvarez P, Saavedra G, Hernández A, et al. Synergistic antinociceptive effects of ketamine and morphine in the orofacial capsaicin test in the rat. Anesthesiology 2003;99(4):969–75. Available from: https://doi.org/10.1097/00000542-200310000-00033.

[28] Bell RF, Dahl JB, Moore RA, et al. Perioperative ketamine for acute postoperative pain. Cochrane Database Syst Rev 2006(1):CD004603. Available from: https://doi.org/10.1002/14651858.CD004603.pub2.

[29] Tucker AP, Kim YI, Nadeson R, et al. Investigation of the potentiation of the analgesic effects of fentanyl by ketamine in humans: a double-blinded, randomised, placebo controlled, crossover study of experimental pain [ISRCTN83088383]. BMC Anesthesiol 2005;5(1):2. Available from: https://doi.org/10.1186/1471-2253-5-2.

[30] Mellor K, Ahmed A, Thomson A. Tramadol hydrochloride use and acute deterioration in Parkinson's disease tremor. Mov Disord 2009;24(4):622–3. Available from: https://doi.org/10.1002/mds.22449.

[31] Zesiewicz TA, Hauser RA, Freeman A, et al. Fentanyl-induced bradykinesia and rigidity after deep brain stimulation in a patient with Parkinson disease. Clin Neuropharmacol 2009;32(1):48–50. Available from: https://doi.org/10.1097/WNF.0b013e31817e23e3.

[32] Isbister GK, Brown AL, Gill A, et al. QT interval prolongation in opioid agonist treatment: analysis of continuous 12-lead electrocardiogram recordings. Br J Clin Pharmacol 2017;83(10):2274–82. Available from: https://doi.org/10.1111/bcp.13326.

[33] Hu J, Song Y, Huang X, et al. Opioids-induced long QT syndrome: a challenge to cardiac health. Cardiovasc Toxicol 2024;24(5):472–80. Available from: https://doi.org/10.1007/s12012-024-09853-6.

[34] Behzadi M, Joukar S, Beik A. Opioids and cardiac arrhythmia: a literature review. Med Princ Pract 2018;27(5):401–14. Available from: https://doi.org/10.1159/000492616.

[35] Baldo BA. Opioid analgesic drugs and serotonin toxicity (syndrome): mechanisms, animal models, and links to clinical effects. Arch Toxicol 2018;92(8):2457–73. Available from: https://doi.org/10.1007/s00204-018-2244-6.

[36] Dumo PA, Kielbasa LA. Successful anticoagulation and continuation of tramadol therapy in the setting of a tramadol-warfarin interaction. Pharmacotherapy 2006;26(11):1654–7. Available from: https://doi.org/10.1592/phco.26.11.1654.

[37] Forrester MB. Ingestions of hydrocodone, carisoprodol, and alprazolam in combination reported to Texas poison centers. J Addict Dis 2011;30(2):110–5. Available from: https://doi.org/10.1080/10550887.2011.554778.

[38] Garg RK, Fulton-Kehoe D, Franklin GM. Patterns of opioid use and risk of opioid overdose death among medicaid patients. Med Care 2017;55(7):661–8. Available from: https://doi.org/10.1097/MLR.0000000000000738.

[39] Wang Y, Delcher C, Li Y, et al. Overlapping prescriptions of opioids, benzodiazepines, and carisoprodol: "Holy Trinity" prescribing in the state of Florida. Drug Alcohol Depend 2019;205:107693. Available from: https://doi.org/10.1016/j.drugalcdep.2019.107693.

[40] Watanabe JH, Yang J. Association of combination opioid, benzodiazepine, and muscle relaxant usage with emergency department visits in a nationwide cohort in the United States. Int J Clin Pharm 2021;43(2):358–64. Available from: https://doi.org/10.1007/s11096-020-01012-5.

[41] Toes MJ, Jones AL, Prescott L. Drug interactions with paracetamol. Am J Ther 2005;12(1):56–66. Available from: https://doi.org/10.1097/00045391-200501000-00009.

[42] Oldenburg LIK, Dalhoff KP, Sandoval LØ, et al. The risk of drug-drug interactions with paracetamol in a population of hospitalized geriatric patients. J Pharm 2020;2020:1354209. Available from: https://doi.org/10.1155/2020/1354209.

[43] Caldeira D, Costa J, Barra M, et al. How safe is acetaminophen use in patients treated with vitamin K antagonists? A systematic review and meta-analysis. Thromb Res 2015;135(1):58–61. Available from: https://doi.org/10.1016/j.thromres.2014.10.011.

[44] Kalsi SS, Wood DM, Waring WS, et al. Does cytochrome P450 liver isoenzyme induction increase the risk of liver toxicity after paracetamol overdose? Open Access Emerg Med 2011;3:69–76. Available from: https://doi.org/10.2147/OAEM.S24962.

[45] Brouwers JR, de Smet PA. Pharmacokinetic-pharmacodynamic drug interactions with nonsteroidal anti-inflammatory drugs. Clin Pharmacokinet 1994;27(6):462–85. Available from: https://doi.org/10.2165/00003088-199427060-00005.

[46] Mehuys E, De Backer T, De Keyser F, et al. Prevalence and management of drug interactions between nonsteroidal anti-inflammatory drugs and antithrombotics in ambulatory care. Br J Clin Pharmacol 2022;88(8):3896–902. Available from: https://doi.org/10.1111/bcp.15288.

[47] Wongrakpanich S, Wongrakpanich A, Melhado K, et al. A comprehensive review of non-steroidal anti-inflammatory drug use in the elderly. Aging Dis 2018;9(1):143–50. Available from: https://doi.org/10.14336/AD.2017.0306.

[48] Hinz B, Chevts J, Renner B, et al. Bioavailability of diclofenac potassium at low doses. Br J Clin Pharmacol 2005;59(1):80–4. Available from: https://doi.org/10.1111/j.1365-2125.2005.02226.x.

[49] Makris UE, Kohler MJ, Fraenkel L. Adverse effects of topical nonsteroidal antiinflammatory drugs in older adults with osteoarthritis: a systematic literature review. J Rheumatol 2010;37(6):1236–43. Available from: https://doi.org/10.3899/jrheum.090935.

[50] Orr JM, Abbott FS, Farrell K, et al. Interaction between valproic acid and aspirin in epileptic children: serum protein binding and metabolic effects. Clin Pharmacol Ther 1982;31(5):642–9. Available from: https://doi.org/10.1038/clpt.1982.89.

[51] Chu PY, Edmondson EK, Flory JH, et al. Risk of hypoglycemia associated with concomitant use of insulin secretagogues and ACE inhibitors in adults with type 2 diabetes: a systematic review. Clin Pharmacol Ther 2025;117(4):1005–11. Available from: https://doi.org/10.1002/cpt.3530.

[52] Leonard CE, Bilker WB, Brensinger CM, et al. Severe hypoglycemia in users of sulfonylurea antidiabetic agents and antihyperlipidemics. Clin Pharmacol Ther 2016;99(5):538–47. Available from: https://doi.org/10.1002/cpt.297.

[53] Isaac R, Boura-Halfon S, Gurevitch D, et al. Selective serotonin reuptake inhibitors (SSRIs) inhibit insulin secretion and action in pancreatic β cells. J Biol Chem. 2018;293(12):4577–8. Available from: https://doi.org/10.1074/jbc.M112.408641.

[54] Sone H, Takahashi A, Yamada N. Ibuprofen-related hypoglycemia in a patient receiving sulfonylurea. Ann Intern Med 2001;134(4):344. Available from: https://doi.org/10.7326/0003-4819-134-4-200102200-00027.

[55] Theken KN, Lee CR, Gong L, et al. Clinical pharmacogenetics implementation consortium guideline (CPIC) for CYP2C9 and nonsteroidal anti-inflammatory drugs. Clin Pharmacol Ther 2020;108(2):191–200. Available from: https://doi.org/10.1002/cpt.1830.

[56] Colebatch AN, Marks JL, Edwards CJ. Safety of non-steroidal anti-inflammatory drugs, including aspirin and paracetamol (acetaminophen), in people receiving methotrexate for inflammatory arthritis (rheumatoid arthritis, ankylosing spondylitis, psoriatic arthritis, other spondyloarthritis). Cochrane Database Syst Rev. 2011;(11):CD008872. Available from: https://doi.org/10.1002/14651858.CD008872.pub2.

[57] Kathuria A, Roosan M, Sharma A. CYP2C9 polymorphism and use of oral nonsteroidal anti-inflammatory drugs. U.S. Pharmacist 2021;54(3):23–30.

[58] Schaefer MG, Plowman BK, Morreale AP, et al. Interaction of rofecoxib and celecoxib with warfarin. Am J Health Syst Pharm 2003;60(13):1319–23. Available from: https://doi.org/10.1093/ajhp/60.13.1319.

[59] Haghbin H, Zakirkhodjaev N, Husain FF, et al. Risk of gastrointestinal bleeding with concurrent use of NSAID and SSRI: a systematic review and network meta-analysis. Dig Dis Sci 2023;68(5):1975–82. Available from: https://doi.org/10.1007/s10620-022-07788-y.

[60] Edinoff AN, Raveendran K, Colon MA, et al. Selective serotonin reuptake inhibitors and associated bleeding risks: a narrative and clinical review. Health Psychol Res 2022;10(4):39580. Available from: https://doi.org/10.52965/001c.39580.

[61] Mohamed Pakkir Maideen N, Balasubramanian R, Muthusamy S, et al. An overview of clinically imperative and pharmacodynamically significant drug interactions of renin-angiotensin-aldosterone system (RAAS) blockers. Curr Cardiol Rev 2022;18(6):e110522204611. Available from: https://doi.org/10.2174/1573403X18666220511152330.

[62] Etminan M, Lévesque L, Fitzgerald JM, et al. Risk of upper gastrointestinal bleeding with oral bisphosphonates and non steroidal anti-inflammatory drugs: a case-control study. Aliment Pharmacol Ther 2009;29(11):1188–92. Available from: https://doi.org/10.1111/j.1365-2036.2009.03989.x.

[63] Perazella MA, Markowitz GS. Bisphosphonate nephrotoxicity. Kidney Int 2008;74(11):1385–93. Available from: https://doi.org/10.1038/ki.2008.356.

Part III

Interventional treatment of pain

Chapter 17

Chemotherapy and targeted therapy

Urška Janžič[1,2]
[1]Medical Oncology Unit, University Clinic Golnik, Golnik, Slovenia, [2]Faculty of Medicine, University of Ljubljana, Ljubljana, Slovenia

Key learning objectives

- Understand the role of systemic therapy in pain management for advanced cancer; recognize that pain is a prevalent and debilitating symptom in advanced cancers. Learn that effective systemic anticancer therapy can reduce tumor burden, which directly alleviates pain and other disease-related symptoms.
- Compare the impact of chemotherapy, antibody-drug conjugates (ADCs), and targeted therapies on quality of life (QoL); understand that targeted therapies and ADCs often provide faster and more predictable symptom relief compared to traditional chemotherapy.
- Evaluate the importance of QoL end points in cancer therapy trials; recognize that traditional tumor-related end points (progression-free survival, overall survival, response rate) do not always correlate with improvements in patients' QoL.
- Analyze clinical evidence supporting pain reduction and QoL improvements with modern systemic therapies; review clinical trial data demonstrating that modern systemic therapies significantly delay pain progression and improve QoL compared to older regimens or placebo.

Introduction

Pain is frequently reported during cancer, and it still remains poorly controlled in 40% of patients. Recent advancements in chemotherapeutic agents and combinations and new targeted treatments may offer patients symptom relief and significantly increase survival [1].

The goal of chemotherapy is to inhibit cell proliferation and tumor multiplication, thus avoiding further invasion and metastasis. Combination chemotherapy is a common choice to produce adequate responses, since it prevents the development of resistant clones by promoting cytotoxicity in resting and dividing cells. While some patients may notice improvements within a few weeks, the full effects of chemotherapy are typically evaluated over several months, with periodic assessments to monitor progress [2]. Targeted cancer therapies can vary in their speed and effectiveness, similar to chemotherapy. However, because they are designed to target certain molecules or pathways involved in cancer growth specifically, they often have a more predictable and faster onset of action—many patients report the relief of symptoms in a matter of a few days or weeks [3].

Tumor-related end points, such as progression-free survival (PFS), overall survival (OS), or overall response rate (ORR), do not inevitably translate into significant improvements in the quality of life (QoL). QoL can be considered as a coprimary end point besides OS if no effect of treatment is observed. Furthermore, in Europe, health assessments to determine reimbursement decisions require both clinical and patient-reported end points (such as QoL assessments); treatments are generally considered for reimbursement based on the added value that these treatments provide [4].

Thus in recent years, QoL has become an important end point for studies of patients with cancer, especially at an advanced stage. Caring for cancer patients in the last years or months of their life requires a good understanding of how QoL can be improved when efforts to prolong life become increasingly futile [5] (Table 17.1).

Pain Management in Palliative Care for Patients With Cancer. DOI: https://doi.org/10.1016/B978-0-443-30050-9.00057-6

TABLE 17.1 Examples of clinical and quality-of-life (QoL) improvements with chemotherapy, antibody-drug conjugates, and targeted therapies.

Therapy or combination	Disease setting	Primary end point/benefit	QoL benefit
Chemotherapy and antibody-drug conjugates			
Burris et al. [6] Gemcitabine vs 5-FU	Metastatic pancreatic cancer	mOS 6.5 months vs 4.41 months (*p* = .0025)	Significant improvement in alleviation of pain and disease-related symptoms (24% vs 5% clinical benefit rate)
Gourgou-Bourgade et al. [7] FOLFIRINOX vs gemcitabine	Metastatic pancreatic cancer	ORR 32% vs 9%; mOS 11.1 vs 6.8 months; HR 0.57 mPFS 6.4 vs 3.3 months; HR 0.47	Significantly increased time until 10- or 20-point deterioration in pain scores (both *P* <. 001)
Loibl et al. [8] Sacituzumab govitecan (SG) vs any chemotherapy	Metastatic triple-negative breast cancer	ORR 35% vs 5%; mOS 12.1 vs 6.7 months; HR 0.48 mPFS 5.6 vs 1.7 months; HR 0.41	Median time to first pain worsening longer for SG vs chemotherapy (21.6 vs 9.9 weeks, *P* < .001)
Rosenberg et al. [9] Enfortumab vedotin (EV) vs chemotherapy	Metastatic urothelial cancer	ORR 44%; mOS 12.9 vs 8.9 months; HR 0.70 mPFS 5.6 vs 3.7 months; HR 0.62	Patients in the EV arm had 2.76x higher odds of clinically meaningful reduction in pain vs the chemotherapy arm
Targeted therapies			
Roussel et al. [10] Lenalidomide, bortezomib, and dexamethasone (RVd)+ASCT	Multiple myeloma	Complete response 59%; mPFS 50 months; OS at 4 years 81%	Clinically significant decreases in pain scores of approximately 15 points
Sonneveld et al. [11] Daratumumab, bortezomib, lenalidomide, and dexamethasone vs bortezomib, lenalidomide, and dexamethasone	Multiple myeloma	Complete response 88% vs 70%; PFS at 48 months 84% vs 68%; HR 0.42	Reduction at post–stem cell transplantation consolidation treatment and throughout the maintenance phase (mean change −29.7 points)
Chen et al. [12] Erlotinib vs chemotherapy	EGFR-positive metastatic NSCLC	ORR 82% vs 36%; mPFS 13.1 vs 4.6 months; HR 0.16 mOS 22.8 vs 27.2 months; HR 1.19	Median time to worsening of pain longer (2.9 vs 1.9 months, *P* = .03)
Yang et al. [13] Afatinib vs chemotherapy	EGFR-positive metastatic NSCLC	ORR 56% vs 23%; mPFS 11.1 vs 6.9 months; HR 0.58 mOS 28.2 vs 28.2 months; HR 0.88	Improvements in pain higher for afatinib (*P* = .051)
Lee et al. [14] Osimertinib vs chemotherapy	EGFR-positive metastatic NSCLC	ORR 71%; mPFS of 10.1 vs 4.4 months; HR 0.30	Median time to pain deterioration 12.4 vs 2.1 months (*P* < .001)
Blackhall et al. [15] Crizotinib vs chemotherapy	ALK-positive metastatic NSCLC	ORR 65% vs 20%; mPFS 7.7 vs 3.0 months; HR 0.49 mOS 21.7 vs 21.9 months; HR 0.85	Median TTD was of pain 5.6 vs 1.9 months (*P* = .02)
Harbeck et al. [16] Palbociclib + HT vs HT alone	Hormone receptor-positive, HER2-negative metastatic breast cancer	ORR 41% vs 29%; mPFS 9.2 vs 3.8 months; HR 0.42 mOS 67.6 vs 51.8 months; HR 0.67	Significantly greater improvement in pain (−3.3 vs 2.0 points; *P* = .0011). Delayed time to deterioration in pain symptoms (HR, 0.6; *P* < .001)
		ORR 53% vs 37%;	

(Continued)

TABLE 17.1 (Continued)

Therapy or combination	Disease setting	Primary end point/benefit	QoL benefit
Harbeck et al. [17] Ribociclib + HT vs HT alone	Hormone receptor-positive, HER2-negative metastatic breast cancer	PFS at 18 months 63% vs 42%; OS at 42 months 70% vs 46%	Delay in time to deterioration of pain ⩾10%; HR = 0.65
Kaufman et al. [18] Abemaciclib + HT vs HT alone	Hormone receptor-positive, HER2-negative metastatic breast cancer	ORR 61% vs 45%; mPFS 28.1 vs 14.7 months; HR 0.54; mOS 66.8 vs 53.7 months; HR 0.80	Delay in pain deterioration by 4.9 months + significantly greater time to sustained deterioration and analgesic use (HR = 0.76) and QLQ-C30 pain item (HR = 0.62)
Fizazi et al. [19] Enzalutamide vs placebo	Metastatic castrate-resistant prostate cancer	ORR 29% vs 4%; mPFS 8.3 vs 2.9 months; mOS 18.4 vs 13.6 months; HR 0.63	Improvement in pain severity and pain interference scores from baseline to week 13, $P < .0001$. Significantly fewer patients had pain progression 28% vs 39% ($P = .0018$)
Rush et al. [20] Abiraterone acetate vs docetaxel	Metastatic castrate-sensitive prostate cancer	Failure-free survival 56% vs 87%; HR 0.34 mPFS 33.0 vs 14.8 months; HR 0.47 mOS 46.0 vs 33.0 months; HR 0.60	Difference in pain score by −6.3 points ($P < .001$)
Badrising et al. [21] Radium-223 vs placebo	Metastatic castrate resistant prostate cancer	Time to first skeletal event 15.6 versus 9.8 months; HR 0.66 mOS 14.9 versus 11.3 months; HR 0.70	Complete pain response in 31% and 58% had an integrated overall clinical response. Time to first opioid use for bone pain was significantly delayed with Ra-223 versus placebo (HR = 0.62, $P = .002$)
Strosberg et al. [22] 177 Dotatate vs octreotide LAR	Metastatic midgut neuroendocrine tumors	ORR 18% vs 3% mPFS 40.0 vs 8.4 months; HR 0.21 mOS 48.0 vs 36.3 months; HR 0.84	Longer TTD in the ^{177}Lu-Dotatate arm compared with the octreotide arm and the HR for pain was 0.57 (95% CI, 0.34–0.94; $P = .025$)

ASCT, Autologous stem cell transplantation; EGFR, epidermal growth factor receptor; *HT*, hormone therapy; *mOS*, median overall survival; *mPFS*, median progression-free survival; *NSCLC*, non–small-cell lung cancer; *ORR*, overall response rate; OS, overall survival; *TTD*, time to deterioration.

Chemotherapy and antibody-drug conjugates

Since QoL measurements only became an important end point in recent years, not many historically important trials assessing chemotherapeutic strategies include this data in their reports. Chemotherapeutic drugs are often toxic themselves, thus causing further tissue damage and QoL deterioration, including symptoms like neuropathic pain. Data on pain relief with chemotherapy is scarce and limited to the most recent publications with the most meaningful benefit.

For example, in pancreatic cancer, pain can be debilitating and is caused by pancreatic enzyme insufficiency (indirectly through bloating, malabsorption, and abdominal discomfort), obstruction, or direct mass effect on nerves in the celiac plexus. Providing proper supportive care to decrease pain is an essential aspect of the overall management of these patients. The first trial to show a significant benefit compared single-agent gemcitabine to 5-fluorouracil (5-FU). The primary measure of efficacy in the trial was the clinical benefit rate, defined as improvement in pain (analgesic consumption and pain intensity), Karnofsky Performance Status, and weight, with no worsening in other parameters. Patients in the gemcitabine arm experienced a significant improvement in alleviation of pain and disease-related symptoms (24% clinical benefit rate vs 5% for 5-FU) [6].

Next, a trial evaluating FOLFIRINOX (5-FU, oxaliplatin, irinotecan) or single-agent gemcitabine proved that after 6 months of therapy, more than twice as many patients experienced a decrease in global health status in the gemcitabine arm compared with the FOLFIRINOX arm ($P < .001$). Patients treated with FOLFIRINOX had significantly increased time until definitive deterioration in their pain scores, demonstrating that for pancreatic cancer, chemotherapy represents one of the multifaceted avenues to pain control [7].

In recent years, it has become increasingly common to encapsulate a cytotoxic drug in an antibody carrier—these drugs are known as antibody-drug conjugates (ADCs). Unlike chemotherapy, ADCs are intended to target and kill tumor cells while sparing healthy cells. ADCs are complex molecules composed of an antibody linked to a biologically active cytotoxic (anticancer) payload or drug. They combine the targeting properties of monoclonal antibodies with the cancer-killing capabilities of cytotoxic drugs, designed to discriminate between healthy and diseased tissue.

An example of ADC is sacituzumab govitecan (SG), which directs SN-38 (the active metabolite of irinotecan) to cells expressing Trop-2, which is highly expressed in triple-negative breast cancer. In the open-label phase III trial, SG significantly prolonged PFS and OS compared to single-agent chemotherapy treatment, and it was noninferior for any of the QoL domains. SG was superior for four of the primary-focused health-related quality-of-life (HRQoL) domains (GHS/QoL, physical functioning, fatigue, and pain), with differences both statistically significant and clinically meaningful. The median time to first clinically meaningful worsening was longer for SG than for chemotherapy alone (22.1 vs 12.1 weeks, $P < .001$), as well as for pain (21.6 vs 9.9 weeks, $P < .001$) [8].

Enfortumab vedotin (EV) is another ADC comprising an antibody directed against nectin-4 and monomethyl auristatin E, a microtubule-disrupting agent, attached to the antibody. The phase III trial demonstrated that EV significantly improved OS versus chemotherapy in patients with advanced or metastatic urothelial cancer (UC) with disease progression after platinum-containing chemotherapy and PD-1/PD-L1 inhibitor treatment. The most significant difference in confirmed improvement was reported for pain: patients in the EV arm had 2.76 times higher odds of achieving a clinically meaningful reduction in pain in comparison to the chemotherapy arm. These results demonstrate that ADC, as well as chemotherapy, should be regarded as agents that can significantly impact pain [9]. More details about chemotherapy and ADCs will be found in the corresponding chapters of this book.

Targeted therapies

Targeted therapy has revolutionized the treatment of cancer by targeting proteins that control the growth, division, and spread of cancer cells. It is the foundation of precision medicine, due to its unique effectiveness, but also because it spares patients from all the unnecessary adverse events that follow chemotherapy treatment, although targeted therapies can sometimes exhibit severe toxicities themselves.

An example of a targeted therapy approach is found for multiple myeloma (MM), one of the most devastating hemato-oncological diseases, where patients experience substantial pain and fatigue along with their illness, with significant reductions in physical functioning, role functioning, and overall QoL. Patients receiving treatment with lenalidomide, bortezomib, and dexamethasone displayed clinically significant decreases in pain scores of approximately 15 points, which were sustained throughout the treatment. Longitudinal analyses showed significant improvements in pain, side effects, and disease symptoms starting at the end of induction therapy, as well as during consolidation treatment and follow-up—with all differences statistically significant ($P < .001$) [10]. Similarly, patients treated with the combination of daratumumab, bortezomib, lenalidomide, and dexamethasone (D-RVd) who are transplant-eligible achieved significant reduction in pain symptoms at most time points, with a large reduction (≥20 point change) in pain in favor of D-RVd at post–stem cell transplantation consolidation treatment and throughout the maintenance phase (mean change, –29.7 [–37.3, –22.2] $P =$ .0497) [11].

Another example would be metastatic non–small-cell lung cancer (NSCLC), which remains incurable in the majority of patients, and thus treatment is aimed at palliation. Furthermore, lung cancer is characterized by a high symptom burden, key symptoms being cough, dyspnea, chest pain, fatigue, and appetite loss, which are experienced by at least 90% of patients and have a significant negative effect on disease-specific HRQoL. Thus symptom measures are now extensively incorporated into clinical studies of lung cancer [23].

With the advent of molecular targets on some of the histological subtypes of NSCLC, the era of precision medicine began for these patients, and improvements in QoL measures were significant. Drugs targeting mutated epidermal growth factor receptors (EGFRs), such as erlotinib, gefitinib, afatinib, and osimertinib, significantly prolong median time to PFS and can also lead to improvement of symptoms compared to those receiving BSC or chemotherapy. For erlotinib, the median time to worsening of pain was significantly longer (2.9 vs 1.9 months, $P = .03$), together with a significant improvement in physical functioning (31% vs 19%, $P = .01$) and global QoL (35% vs 26%, $P < .0001$). A correlation between efficacy and QoL response showed that patients who had a complete or partial response were more likely to have an improvement in QoL than those with stable disease or progressive disease ($P < .01$) [12]. Similarly, the proportion of patients with improvements in pain was higher for afatinib compared to chemotherapy, approaching significance ($P = .051$). The delayed deterioration time for pain did not reach statistical significance (HR, 0.83 [95% CI, 0.62–1.10]; $P = .19$), although afatinib did significantly delay worsening of the individual item of pain in the chest [13]. The last EGFR

TKI, osimertinib, significantly prolonged time to deterioration compared with chemotherapy, particularly for chest pain—median time to pain deterioration for osimertinib was 12.4 months (95% CI, 8.2 to NC) versus for chemotherapy was 2.1 months (95% CI, 0.8–4.8; $P < .001$) [14]. Similar significant results were reported for NSCLC with ALK fusion, which is found in about 4% of patients and can be treated with ALK TKIs. Treatments such as crizotinib prolonged time to deterioration (TTD) for pain compared to chemotherapy. Although newer generations of ALK inhibitors offer significant improvements in both PFS and OS, there is no difference in most QoL measurements compared to first-generation therapies, which provide pain alleviation and symptom control. More details about these will be found in the Lung Cancer chapter of this book [24,25].

Breast cancer is the most common cancer in women across the globe and can present with a heavy burden of disease in metastatic settings [26]. Women with breast cancer, even in the advanced setting, now live much longer thanks to the novel therapeutics in this area. One of the most prominent agents in the last decades was CDK 4/6 inhibitors, which are very successful in the treatment of hormone-positive breast cancer patients and also alleviate symptoms in case of metastatic spread to the bones, which causes pain. CDK 4/6 inhibitors such as palbociclib not only improve PFS but also delay a deterioration of QoL and pain symptoms. Significantly greater improvement from baseline in pain was also observed in this group (−3.3 points [95% CI, −5.1 to −1.5] vs 2.0 points [95% CI, −0.6 to 4.6]; $P = .0011$). Treatment with palbociclib plus fulvestrant significantly delayed time to deterioration in pain symptoms versus fulvestrant (HR, 0.6; $P < .001$) [16]. Similar results were found for the other CDK 4/6 inhibitors, ribociclib and abemaciclib [17,18], with more details in the breast cancer chapter.

For patients with prostate cancer, bone is the most common site of spread, accounting for about 90% of metastases and resulting in some of the most painful and functionally compromising complications of the disease. Newer agents, such as enzalutamide and abiraterone, had prespecified QoL analyses or end points that evaluated common symptoms during treatment, including pain, and demonstrated an improvement of both pain severity and pain interference scores [19,20]. For symptomatic metastatic castration-resistant prostate alpha-emitter radium-223 dichloride (Ra-223), a 3.6-month OS benefit and favorable HRQoL were shown. Moreover, 50% of the evaluable patients had a clinically meaningful improvement of the Brief Pain Inventory-Short Form (BPI-SF). Use of analgesics in the evaluable patient sample decreased during Ra-223 treatment and remained low during follow-up. Complete pain response was achieved in 31% of the patients, while 58% had an integrated overall clinical response [21]. The details can be found in the appropriate chapters of this book.

Another trial with the radio-emitting agent ^{177}Lu-Dotatate for patients with advanced neuroendocrine tumors showed significantly longer TTD in the ^{177}Lu-Dotatate arm compared with the octreotide arm, and the HR for pain was 0.57 (95% CI, 0.34–0.94; $P = .025$). In fact, the symptoms that patients experienced the most benefit from included pain and fatigue [22].

Conclusion

In conclusion, effective systemic anticancer therapy remains one of the pillars of effective pain control and symptom alleviation. Good response rates and tumor shrinkage following systemic treatment could also lead to significant improvement of symptoms, particularly pain, although not all trials reported on pain-related outcomes. New randomized controlled trials (RCTs) are now including QoL end points in their strategies, since QoL is becoming increasingly important for cancer patients with advanced disease. Moreover, these data are crucial for the registration and reimbursement purposes of novel anticancer agents and are as important as percentages gained in the overall response rates or months in the OS.

- Pain is one of the most common and debilitating symptoms of advanced cancer.
- Effective systemic anticancer therapy reduces the disease burden, alleviates cancer symptoms, especially pain, and diminishes the need for analgesic therapy.
- QoL end points are crucial for RCTs and informed decision-making.

References

[1] Mestdagh F, Steyaert A, Lavand'homme P. Cancer pain management: a narrative review of current concepts, strategies, and techniques. Curr Oncol 2023;30(7):6838–58. Available from: https://doi.org/10.3390/curroncol30070500.

[2] Anand U, Dey A, Chandel AKS, et al. Cancer chemotherapy and beyond: current status, drug candidates, associated risks and progress in targeted therapeutics. Genes Dis 2023;10(4):1367–401. Available from: https://doi.org/10.1016/j.gendis.2022.02.007.

[3] Min HY, Lee HY. Molecular targeted therapy for anticancer treatment. Exp Mol Med 2022;54(10):1670–94. Available from: https://doi.org/10.1038/s12276-022-00864-3.

[4] Del Paggio JC, Sullivan R, Schrag D, et al. Delivery of meaningful cancer care: a retrospective cohort study assessing cost and benefit with the ASCO and ESMO frameworks. Lancet Oncol 2017;18(7):887–94. Available from: https://doi.org/10.1016/S1470-2045(17)30415-1.
[5] Schandelmaier S, Conen K, von Elm E, et al. Planning and reporting of quality-of-life outcomes in cancer trials. Ann Oncol 2015;26(9):1966–73. Available from: https://doi.org/10.1093/annonc/mdv283.
[6] Burris HA 3rd, Moore MJ, Andersen J, et al. Improvements in survival and clinical benefit with gemcitabine as first-line therapy for patients with advanced pancreas cancer: a randomized trial. J Clin Oncol. 1997;15(6):2403–13. Available from: https://doi.org/10.1200/JCO.1997.15.6.2403.
[7] Gourgou-Bourgade S, Bascoul-Mollevi C, Desseigne F, et al. Impact of FOLFIRINOX compared with gemcitabine on quality of life in patients with metastatic pancreatic cancer: results from the PRODIGE 4/ACCORD 11 randomized trial. J Clin Oncol 2013;31(1):23–9. Available from: https://doi.org/10.1200/jco.2012.44.4869.
[8] Loibl S, Loirat D, Tolaney SM, et al. Health-related quality of life in the phase III ASCENT trial of sacituzumab govitecan versus standard chemotherapy in metastatic triple-negative breast cancer. Eur J Cancer 2023;178:23–33. Available from: https://doi.org/10.1016/j.ejca.2022.10.003.
[9] Rosenberg JE, Mamtani R, Sonpavde GP, et al. Health-related quality of life in patients with previously treated advanced urothelial carcinoma from EV-301: a phase 3 trial of enfortumab vedotin versus chemotherapy. Eur Urol 2024;85(6):574–85. Available from: https://doi.org/10.1016/j.eururo.2024.01.007.
[10] Roussel M, Hebraud B, Hulin C, et al. Health-related quality of life results from the IFM 2009 trial: treatment with lenalidomide, bortezomib, and dexamethasone in transplant-eligible patients with newly diagnosed multiple myeloma. Leuk Lymphoma 2020;61(6):1323–33. Available from: https://doi.org/10.1080/10428194.2020.1719091.
[11] Sonneveld P, Dimopoulos MA, Boccadoro M, et al. Daratumumab, bortezomib, lenalidomide, and dexamethasone for multiple myeloma. N Engl J Med 2024;390(4):301–13. Available from: https://doi.org/10.1056/nejmoa2312054.
[12] Chen G, Feng J, Zhou C, et al. Quality of life (QoL) analyses from OPTIMAL (CTONG-0802), a phase III, randomised, open-label study of first-line erlotinib versus chemotherapy in patients with advanced EGFR mutation-positive non-small-cell lung cancer (NSCLC). Ann Oncol 2013;24(6):1615–22. Available from: https://doi.org/10.1093/annonc/mdt012.
[13] Yang JCH, Hirsh V, Schuler M, et al. Symptom control and quality of life in lux-lung 3: a phase III study of afatinib or cisplatin/pemetrexed in patients with advanced lung adenocarcinoma with EGFR mutations. J Clin Oncol 2013;31(27):3342–50. Available from: https://doi.org/10.1200/JCO.2012.46.1764.
[14] Lee CK, Novello S, Rydén A, et al. Patient-reported symptoms and impact of treatment with osimertinib versus chemotherapy in advanced non–small-cell lung cancer: the AURA3 trial. J Clin Oncol 2018;36(18):1853–60. Available from: https://doi.org/10.1200/jco.2017.77.2293.
[15] Blackhall F, Kim DW, Besse B, et al. Patient-reported outcomes and quality of life in PROFILE 1007: a randomized trial of crizotinib compared with chemotherapy in previously treated patients with ALK-positive advanced non-small-cell lung cancer. J Thorac Oncol 2014;9(11):1625–33. Available from: https://doi.org/10.1097/JTO.0000000000000318.
[16] Harbeck N, Iyer S, Turner N, et al. Quality of life with palbociclib plus fulvestrant in previously treated hormone receptor-positive, HER2-negative metastatic breast cancer: patient-reported outcomes from the PALOMA-3 trial. Ann Oncol 2016;27(6):1047–54. Available from: https://doi.org/10.1093/annonc/mdw139.
[17] Harbeck N, Franke F, Villanueva-Vazquez R, et al. Health-related quality of life in premenopausal women with hormone-receptor-positive, HER2-negative advanced breast cancer treated with ribociclib plus endocrine therapy: results from a phase III randomized clinical trial (MONALEESA-7). Ther Adv Med Oncol 2020;12:1758835920943065. Available from: https://doi.org/10.1177/1758835920943065.
[18] Kaufman PA, Toi M, Neven P, et al. Health-related quality of life in MONARCH 2: abemaciclib plus fulvestrant in hormone receptor-positive, HER2-negative advanced breast cancer after endocrine therapy. Oncologist 2020;25(2):e243. Available from: https://doi.org/10.1634/theoncologist.2019-0551.
[19] Fizazi K, Scher HI, Miller K, et al. Effect of enzalutamide on time to first skeletal-related event, pain, and quality of life in men with castration-resistant prostate cancer: results from the randomised, phase 3 AFFIRM trial. Lancet Oncol 2014;15(10):1147–56. Available from: https://doi.org/10.1016/S1470-2045(14)70303-1.
[20] Rush HL, Murphy L, Morgans AK, et al. Quality of life in men with prostate cancer randomly allocated to receive docetaxel or abiraterone in the STAMPEDE trial. J Clin Oncol 2022;40(8):825–36. Available from: https://doi.org/10.1200/JCO.21.00728.
[21] Badrising SK, Louhanepessy RD, van der Noort V, et al. Integrated analysis of pain, health-related quality of life, and analgesic use in patients with metastatic castration-resistant prostate cancer treated with radium-223. Prostate Cancer Prostatic Dis 2022;25(2):248–55. Available from: https://doi.org/10.1038/s41391-021-00412-6.
[22] Strosberg J, Wolin E, Chasen B, et al. Health-related quality of life in patients with progressive midgut neuroendocrine tumors treated with 177 Lu-Dotatate in the phase III NETTER-1 trial. J Clin Oncol 2018;36(25):2578–84. Available from: https://doi.org/10.1200/jco.2018.78.5865.
[23] Iyer S, Taylor-Stokes G, Roughley A. Symptom burden and quality of life in advanced non-small cell lung cancer patients in France and Germany. Lung Cancer 2013;81(2):288–93. Available from: https://doi.org/10.1016/j.lungcan.2013.03.008.
[24] Pérol M, Pavlakis N, Levchenko E, et al. Patient-reported outcomes from the randomized phase III ALEX study of alectinib versus crizotinib in patients with ALK-positive non-small-cell lung cancer. Lung Cancer 2019;138:79–87. Available from: https://doi.org/10.1016/j.lungcan.2019.10.002.
[25] Garcia Campelo MR, Lin HM, Zhu Y, et al. Health-related quality of life in the randomized phase III trial of brigatinib vs crizotinib in advanced ALK inhibitor–naive ALK + non–small cell lung cancer (ALTA-1L). Lung Cancer 2021;155:68–77. Available from: https://doi.org/10.1016/j.lungcan.2021.03.005.
[26] Sung H, Ferlay J, Siegel RL, et al. Global cancer statistics 2020: GLOBOCAN estimates of incidence and mortality worldwide for 36 cancers in 185 countries. CA Cancer J Clin 2021;71(3):209–49. Available from: https://doi.org/10.3322/caac.21660.

Chapter 18

Immunotherapy

Amir Radfar
Former Faculty Member, College of Medicine, University of Central Florida, Orlando, FL, United States

Key learning objectives

- Describe immunotherapy's indirect and responder-dependent mechanisms of pain relief, including tumor reduction, immune modulation, and its adjunctive role in supportive care.
- Evaluate evidence from landmark clinical trials on pain relief and quality-of-life outcomes.
- Provide practical guidance for incorporating immunotherapy into multidisciplinary pain management, including strategies for managing immune-related adverse events.
- Discuss ethical, economic, and research considerations, emphasizing pain-specific trial end points.

Introduction

As immunotherapy integrates into cancer treatment, its implications for symptom control, including pain, are increasingly relevant to clinical practice. This chapter reviews four classes of FDA-approved immunotherapies with potential implications for symptom management: checkpoint inhibitors, chimeric antigen receptor (CAR) T-cell therapies, therapeutic cancer vaccines such as sipuleucel-T, and the oncolytic virus talimogene laherparepvec (T-VEC). Greater emphasis is placed on checkpoint inhibitors and CAR T-cell therapy, reflecting their broader clinical uptake and stronger evidence base. Vaccines and oncolytic viruses are discussed selectively to illustrate unique mechanisms and symptomatic considerations. Emerging bispecific antibodies and rarely used cytokine-based immunotherapies were excluded due to a lack of pain-related or quality-of-life (QoL) data and higher toxicity. Drawing on clinical trials, observational studies, and guideline recommendations, the aim is to provide a practical, evidence-based perspective on where immunotherapy contributes to pain relief and QoL and where its limitations underscore the need for palliative approaches. Immunotherapy should not be framed as a primary analgesic intervention; its potential pain benefit is usually indirect, delayed, and limited to selected responders.

Mechanisms of action

Immune checkpoint inhibitors

In cancer immunotherapy, immune checkpoints such as programmed death-1 (PD-1) and cytotoxic T-lymphocyte-associated protein 4 (CTLA-4) function as key regulators of T-cell activity. These inhibitory receptors, expressed on activated T cells, interact with ligands on antigen-presenting cells (APCs), tumor cells, and stromal elements within the tumor microenvironment. Under physiologic conditions, these pathways preserve self-tolerance and prevent chronic inflammation by restraining overactive immune responses. However, tumors frequently exploit these mechanisms—particularly through upregulation of checkpoint ligands—to suppress cytotoxic T-cell function and evade immune detection. Therapeutic blockade of PD-1 or CTLA-4 can reverse this immunosuppression and restore antitumor activity in select patients.

Pain Management in Palliative Care for Patients With Cancer. DOI: https://doi.org/10.1016/B978-0-443-30050-9.00004-7

The PD-1/PD-L1 axis suppresses immune responses primarily in peripheral tissues, including the tumor microenvironment. PD-1, expressed on activated T cells, binds to PD-L1—a ligand often upregulated on tumor cells and APCs in response to interferon-gamma or oncogenic signaling [1]. This interaction inhibits T-cell proliferation, cytokine production, and cytolytic function, leading to T-cell exhaustion and impaired immune surveillance [2].

In contrast, CTLA-4 regulates T-cell activation earlier in the immune response, primarily during T-cell activation (priming) in lymphoid tissues. It competes with the costimulatory receptor CD28 for binding B7 ligands (CD80 and CD86) on APCs. CTLA-4 has a much higher affinity for these ligands and delivers inhibitory signals that suppress T-cell activation. T-cell priming (the initial activation of naive T cells through antigen recognition, costimulation, and cytokine signaling) is tightly regulated by CTLA-4 to ensure appropriate immune activation and prevent autoimmunity.

Checkpoint inhibitors, such as pembrolizumab (which blocks PD-1) and ipilimumab (which blocks CTLA-4), help the immune system recognize and attack cancer cells again. They achieve this by blocking the "off switches" that tumors use to disable T cells from functioning, allowing these immune cells to reactivate and kill cancer cells [2].

Clinically, pain benefit from checkpoint blockade is most plausible when tumor regression reduces mass effect, nerve compression, or tumor-associated inflammatory signaling.

Mechanistic pathway summary:

- PD-1/PD-L1 blockade→T-cell reactivation→tumor regression→reduced mass effect→potential pain relief.
- CTLA-4 blockade→enhanced T-cell activation (priming)→tumor regression→reduced mass effect→potential pain relief.

Chimeric antigen receptor T-cell therapy

CAR T-cell therapy is a personalized form of immunotherapy in which a patient's own T cells are collected through leukapheresis and genetically modified outside the body to express a synthetic receptor—called a CAR—that targets tumor-specific antigens such as cluster of differentiation 19 (CD19) or B-cell maturation antigen (BCMA). This receptor links an antigen-binding fragment with intracellular signaling and costimulatory domains, enabling the T cells to recognize and destroy cancer cells without relying on major histocompatibility complex (MHC) presentation [3]. After lymphodepleting chemotherapy, the engineered T cells are reinfused to trigger targeted cytotoxicity, which may help relieve cancer-related pain through tumor reduction. This potential benefit is most relevant in patients with bulky or symptomatic hematologic disease, but early cytokine release syndrome or immune effector cell-associated neurotoxicity syndrome may temporarily worsen symptom burden.

However, resistance remains a challenge. In B-cell acute lymphoblastic leukemia (B-ALL), relapse can occur in 30%–50% of patients due to loss or downregulation of the CD19 target. To overcome this, newer CAR constructs are being designed to recognize multiple antigens—such as CD19 and cluster of differentiation 22 (CD22)—simultaneously, improving durability and reducing escape, as shown in early-phase studies [4]. Currently approved CAR T-cell therapies like axicabtagene ciloleucel and tisagenlecleucel are used primarily for relapsed or refractory hematologic malignancies (see the section "Clinical applications").

Mechanistic pathway summary:

T-cell harvesting (leukapheresis)→genetic modification (CAR expression)→lymphodepleting chemotherapy→reinfusion→tumor cell lysis.

Sipuleucel-T (therapeutic cancer vaccine)

Sipuleucel-T (Provenge) is the first and only FDA-approved autologous cellular immunotherapy for prostate cancer. It involves collecting the patient's APCs and exposing them ex vivo to a recombinant fusion protein composed of prostatic acid phosphatase (PAP) and granulocyte-macrophage colony-stimulating factor (GM-CSF). GM-CSF functions as an immune adjuvant—a substance that boosts, specifies, or directs the body's immune response to an antigen. Once reinfused, these activated APCs act as vectors, stimulating a systemic T-cell-mediated immune response against prostate cancer cells and thereby slowing disease progression [5].

Because sipuleucel-T generally does not produce rapid tumor shrinkage, any pain-related benefit should be described as delayed, indirect, and exploratory.

Mechanistic pathway summary

Leukapheresis (APC collection)→ex vivo activation with PAP–GM-CSF→reinfusion→T-cell priming→targeting of PAP-expressing prostate cancer cells→immune-mediated tumor destruction.

Oncolytic virus therapy (talimogene laherparepvec)

T-VEC, a live genetically modified herpes simplex virus type 1, is used as an intralesional therapy for unresectable stage IIIB–IVM1a melanoma with cutaneous, subcutaneous, or nodal lesions [6,7]. T-VEC effects are localized to the site of injection, distinguishing it from other systemic agents such as checkpoint inhibitors, CAR T-cell therapies, and therapeutic cancer vaccines.

T-VEC operates through dual mechanisms. It selectively infects and lyses cancer cells, directly reducing tumor mass. Simultaneously, it produces GM-CSF, which recruits dendritic cells and amplifies immune recognition of the tumor antigens [8,9].

Any pain improvement with T-VEC is expected to be local and lesion-related rather than systemic.

Mechanistic pathway summary

T-VEC injection→virus multiplies in cancer cells→tumor cell lysis+GM-CSF production→antigen release→dendritic cell recruitment and immune activation→tumor regression

Pain relief mechanisms across immunotherapies

Immunotherapies may alleviate cancer-related pain indirectly in selected responders through tumor regression or modulation of the tumor microenvironment, but consistent pain relief has not been demonstrated across trials. Below, the pain relief mechanisms of four FDA-approved immunotherapies are outlined, supported by clinical and preclinical evidence.

Checkpoint inhibitors: By blocking PD-1/programmed death-ligand 1 (PD-L1) or CTLA-4 pathways, agents such as pembrolizumab and nivolumab reactivate exhausted T cells, leading to tumor destruction. In melanoma, renal cell carcinoma (RCC), and non–small-cell lung cancer (NSCLC), some responders report modest pain relief 4–12 weeks after radiographic response, likely due to reduced mass effect or nerve compression [10]. However, these observations largely derive from QoL or symptom domains rather than trials designed with pain as a primary end point. Conversely, irAEs like arthralgia or myalgia may exacerbate pain [11].

CAR T-cell therapy: Autologous T cells are genetically engineered to express a CD19-directed CAR, enabling rapid and potent tumor cytoreduction in B-cell malignancies. In diffuse large B-cell lymphoma (DLBCL), commercial CAR-T products (axi-cel, tisa-cel) have demonstrated impressive overall survival and nonrelapse mortality outcomes [12]. In addition, prospective health-related QoL assessments from the ZUMA-7 trial show that pain improves markedly after axi-cel infusion; the Functional Assessment of Cancer Therapy–Lymphoma (FACT-Lym) pain item decreased from a median score of 3 (moderate pain) at baseline to 1 (mild pain) by day 90, and overall QoL scores rose significantly [13]. This is among the stronger pain-relevant signals in the chapter, although it remains embedded within a broader lymphoma QoL instrument rather than a dedicated cancer-pain trial.

Cytokine release syndrome (CRS), a frequent adverse effect of CAR-T therapy, often elevates interleukin-6 (IL-6), tumor necrosis factor-alpha (TNF-α), and interleukin-1 beta (IL-1β), causing transient headache or myalgias in ~40% of patients. These symptoms usually resolve as inflammation subsides or after IL-6 blockade (e.g., tocilizumab) [14].

Sipuleucel-T: Clinical trial data indicate a modest palliative effect, potentially mediated through immune modulation of the tumor microenvironment (see the section "Clinical evidence") [15]. However, the mechanisms underlying pain relief remain poorly defined, as pain was not a primary end point in pivotal trials [5], and no studies have confirmed reductions in nociceptive biomarkers such as TNF-α or IL-1β [16]. Therefore, sipuleucel-T should not be described as an analgesic therapy.

Talimogene laherparepvec (T-VEC): As detailed in the section "Clinical evidence," intralesional T-VEC shrinks injectable melanoma lesions (overall response rate ≈26 % and durable-response rate ≈16 %) [17]. Some patients may experience reduced local pain after lesion regression, but this remains anecdotal or underreported, and systemic pain relief data remain sparse.

Clinical evidence

Landmark trials

Clinical trials in biomarker-selected patients reveal a nuanced relationship between tumor response and symptom outcomes, particularly for pain and function, though this association is not universal [10]. Importantly, most pivotal immunotherapy trials assessed global QoL or symptom scales rather than pain as a prespecified primary end point. The KEYNOTE-024 trial demonstrated that pembrolizumab yielded a 44.8% response rate in NSCLC patients with PD-L1 expression ≥50%, outperforming chemotherapy's 27.8% response rate, and resulted in a QoL enhancement with a 6.9-point improvement and a 7.8-point between-arm difference on the EORTC QLQ-C30 Global Health Status (GHS)/QoL scale [10]. This supports global QoL benefit, but not direct analgesic efficacy.

In the CheckMate-141 trial, nivolumab, used for recurrent or metastatic head and neck squamous cell carcinoma (HNSCC), achieved a 13.3% response rate, with patients experiencing maintenance or slower deterioration of symptom burden, including pain-related domains, compared with investigator's choice, alongside better-preserved overall QoL, despite modest survival gains [18].

In men with asymptomatic or minimally symptomatic metastatic castration-resistant prostate cancer (mCRPC), the IMPACT trial demonstrated that sipuleucel-T significantly extended median overall survival by 4.1 months, without inducing substantial prostate-specific antigen (PSA) declines or radiographic tumor regression [5]. While pain relief was not a primary end point, a post hoc analysis of IMPACT found no significant delay in disease-related pain progression between treatment and placebo groups [19]. However, a post hoc pooled analysis of three randomized phase III trials (D9901, D9902A, and IMPACT) revealed a modest but potentially meaningful palliative effect. Patients receiving sipuleucel-T experienced a delay in median time to first opioid use (12.6 vs 9.7 months; HR 0.76), and a greater proportion remained pain-free at 12 months compared to placebo (39% vs 19%) [16].

These findings suggest an exploratory secondary symptom benefit through immune modulation in select patients, though no validated biomarkers of pain (e.g., TNF-α, IL-1β) have been linked to its use, and analgesia remains an exploratory end point.

The OPTiM trial of T-VEC in melanoma (stage IIIB–IVM1a) reported a 26% response rate in injected lesions but provided limited pain or QoL data, precluding its recommendation for palliative symptom control [17]. Thus, T-VEC should not be presented as an established palliative pain intervention.

Table 18.1 summarizes pivotal immunotherapy trials that link tumor response with reported pain and QoL.

TABLE 18.1 Tumor response, pain, and QoL outcomes.

Trial	Therapy	Tumor type	ORR	Pain/QoL outcome
KEYNOTE-024	Pembrolizumab	Non-small-cell lung cancer (NSCLC) (PD-L1 ≥50%)	44.8%	QoL improvement (+6.9-point improvement; between-arm difference +7.8 on EORTC QLQ-C30 GHS/QoL)
CheckMate-141	Nivolumab	Recurrent/metastatic HNSCC	13.3%	Maintenance or slower deterioration of symptom burden, including pain-related domains, compared with investigator's choice; overall QoL better preserved
IMPACT	Sipuleucel-T	mCRPC (minimally symptomatic)	N/A	Post hoc pooled analysis: delayed opioid initiation (12.6 vs 9.7 months with placebo); 39% vs 19% of patients pain-free at 12 months (20% absolute difference)
OPTiM	Talimogene laherparepvec (T-VEC)	Melanoma (stage IIIB–IVM1a)	26%	Limited pain/QoL data; not established as a palliative pain intervention.

Pain and QoL outcomes were heterogeneous across trials; pain was not uniformly assessed as a prespecified primary end point.
EORTC QLQ-C30, European Organisation for Research and Treatment of Cancer Quality of Life Questionnaire Core 30; *GHS*, Global Health Status; *HNSCC*, head and neck squamous cell carcinoma; *mCRPC*, metastatic castration-resistant prostate cancer; *NSCLC*, non-small-cell lung cancer; *ORR*, objective response rate; *QoL*, quality of life.

Clinical applications

The clinical application of immunotherapy is a nuanced decision guided by the patient's performance status, tumor biology, and overall treatment goals [20]. Understanding Eastern Cooperative Oncology Group (ECOG) performance status is critical for determining patient suitability for immunotherapy. Developed by the ECOG, this 0–5 scale assesses daily functioning: 0 (fully active), 1 (ambulatory but restricted in strenuous activity), 2 (ambulatory more than 50% of waking hours but unable to work), 3 (chair- or bed-bound more than 50% of waking hours), 4 (completely disabled), and 5 (dead) [20]. Performance status should be interpreted alongside prognosis, symptom burden, goals of care, and whether functional decline is tumor-related and reversible. While regulatory labels rarely specify ECOG performance-status cut-offs, clinical experience and data from subgroup analyses suggest that patients with an ECOG status of 0–1 derive the most significant benefit [20]. Patients with ECOG ≥3 rarely benefit and should generally receive supportive care unless the decline is clearly tumor-related, rapidly reversible, and the expected response timeline is clinically meaningful [20].

Immune checkpoint inhibitors (ICIs) have the broadest applications. Approximately 40%–45% of US patients with cancer are eligible for checkpoint inhibitors, and up to ~12.5% of all patients with cancer are expected to respond, highlighting the extensive but still selective impact of these agents [22]. Key approvals include pembrolizumab for non–small-cell lung cancer (NSCLC) with high PD-L1 expression tumor proportion score (TPS ≥50%) [10] and for recurrent/metastatic HNSCC with a combined positive score (CPS) ≥1 [23]. A major advance is the tumor-agnostic approval of pembrolizumab for any solid tumor with high microsatellite instability (MSI-H) or deficient mismatch repair (dMMR) [24]. It is important to note that nivolumab's approval for MSI-H tumors is more specific—limited to colorectal cancer, often in combination with ipilimumab [25]. Other major approvals for combination ICI therapy include nivolumab plus ipilimumab for first-line metastatic RCC [26]. The use of ICIs has also expanded to biomarker-defined settings in hepatocellular carcinoma [27,28], cervical cancer [29], and selected gastric, triple-negative breast, urothelial, esophageal, and biliary tract cancers [30]. For patients with low PD-L1 (1%–49%) NSCLC, ICIs are typically combined with chemotherapy [20]. Beyond ECOG status, recent European Society for Medical Oncology (ESMO) guidance underscores that multidisciplinary collaboration to recognize and manage immune-related toxicities is essential during patient selection [21,31].

As noted before, CAR T-cell therapy is a transformative option for relapsed/refractory hematologic malignancies. For instance, axicabtagene ciloleucel is approved for DLBCL [32], tisagenlecleucel is used for acute lymphoblastic leukemia (ALL) and DLBCL in adults [33], and idecabtagene vicleucel or ciltacabtagene autoleucel are approved for multiple myeloma [34,35]. Pivotal trials enrolled patients with ECOG status ≤2, but careful selection is critical because overall nonrelapse mortality is ≈5%–8% across products [36]. Administration is restricted to specialized centers with 24/7 intensive care and neurology support [32–36].

Other immunotherapies have more targeted applications. Sipuleucel-T is approved for asymptomatic or mCRPC in patients with a life expectancy >6 months and an ECOG status of 0–1 [37,38]. The regimen requires multiple leukapheresis procedures, demanding adequate venous access [37,38].

T-VEC is an intralesional therapy for unresectable melanoma metastases, the most effective in patients with low disease burden [17]. Its use is limited to cutaneous, subcutaneous, or nodal lesions and is not recommended for visceral or CNS metastases [17,39].

A critical takeaway is that for patients with very poor performance status (ECOG ≥3), tumor-directed therapies alone are often insufficient for pain control. This makes adjunctive analgesics—such as gabapentinoids, which have a National Comprehensive Cancer Network (NCCN) Category 2A recommendation as first-line adjuvants for neuropathic cancer pain—essential components of care [40,41].

Table 18.2 provides an ECOG-stratified decision matrix to guide immunotherapy selection, detailing clinical indications, biomarkers, toxicity risks, and palliative suitability. In this context, palliative suitability assesses whether a patient's functional status (ECOG 0–2) supports immunotherapy for symptom relief or modest life extension, as patients with ECOG status >2 often experience limited benefits and significant irAEs. This aligns with NCCN and ASCO guidelines emphasizing performance status [20,21].

TABLE 18.2 Eastern Cooperative Oncology Group (ECOG)–based immunotherapy decision matrix.

ECOG	Therapy	Guideline indications	Mechanism	Clinical benefit	Response timeline	Biomarker (key)	Palliative suitability
0–1	Checkpoint inhibitors	NSCLC (pembro 1L TPS ≥50%, 1%–49% chemo-IO), melanoma, renal cell carcinoma (RCC), head and neck squamous cell carcinoma, microsatellite instability (MSI-H)/deficient mismatch repair (dMMR), TMB-high, HCC, cervical, gastric/GEJ, TNBC (CPS ≥10), urothelial, esophageal SCC, biliary tract	Programmed cell death protein 1 (PD-1)/PD-L1 or cytotoxic T-lymphocyte-associated protein 4 blockade	ΔOS 6–30 months; 5-y OS 30%–55% (melanoma, RCC)	8–12 weeks	PD-L1 TPS or CPS (tumor-specific), MSI-H/dMMR, TMB ≥10 mut/Mb	Treat (≥3 months QALY gain)
0–1	Chimeric antigen receptor (CAR) T-cell	R/R B-cell acute lymphoblastic leukemia, diffuse large B-cell lymphoma, multiple myeloma	Autologous T cells with CD19/BCMA CAR	Median OS 26–60 months (no RCT comparator)**	2–4 weeks	CD19, BCMA	Treat (potentially curative; inpatient)
0–1	Sipuleucel-T	Metastatic castration-resistant prostate cancer (mCRPC), LE>6 months	Antigen-presenting cell vaccine	ΔOS ~4 months	Not applicable	Prostatic acid phosphatase expression (no companion diagnostic)	Consider (≥6 months LE required)
0–1	Talimogene laherparepvec (T-VEC)	Melanoma (IIIB–IVM1a)	Oncolytic herpesvirus (HSV-1)	No OS gain; ~16% durable response (CR/PR ≥6 months; median time to response ~4 months)	12–16 weeks	Injectable lesions, low LDH, minimal tumor bulk	Consider (local; minimal systemic burden)
2	Checkpoint inhibitors	NSCLC (PD-L1 ≥50%), melanoma, RCC (off-label multidisciplinary team [MDT] extrapolation; evidence level: real-world NSCLC, expert opinion elsewhere)	As above	ΔOS 3–5 months (NSCLC; data sparse elsewhere)	8–12 weeks	PD-L1 TPS or CPS (tumor-specific)	Consider; reassess at 8 weeks
2	CAR T-cell	B-cell malignancies	As above	Median OS ~9–12 months (no randomized comparator); early nonrelapse mortality appears higher in frail patients, but ECOG-stratified data are sparse; decisions should be individualized in experienced centers	2–4 weeks	As above	Consider (center-specific)

2	Sipuleucel-T[a]	mCRPC (LE ≥6 months)	As above	ΔOS ~4 months (if LE ≥6 months)	—	As above	Consider (≥6 months LE required)
3	Checkpoint inhibitors[b]	Select NSCLC (PD-L1 ≥50%), melanoma, RCC (off-label MDT extrapolation; evidence level: real-world NSCLC, expert opinion elsewhere)	As above	No robust survival benefit (data sparse)	8–12 weeks	PD-L1 TPS or CPS (tumor-specific)	Avoid; reassess at 8 weeks
3	CAR T-cell[b]	B-cell malignancies (compassionate use)	As above	Not established; early non relapse mortality risk appears high and data are very limited; use only in trials or exceptionally selected cases at specialized centers	2–4 weeks	As above	Avoid outside trials
3	T-VEC[b]	Melanoma (IIIB–IVM1a)	As above	No OS gain; 16% durable RR	12–16 weeks	Injectable lesions, low LDH	Rarely feasible; case-by-case
≥4	Best supportive care/clinical trial[c]	None	Varies	Not expected	—	None	Avoid outside trials

Clinical considerations Clinical benefit for Immune checkpoint inhibitors, sipuleucel-T, and T-VEC is ΔOS versus control; CAR T-cell is median OS on therapy (no RCT comparator; phase II trial of axicabtagene ciloleucel in large B-cell lymphoma [ZUMA-1] ΔOS ~15 months vs salvage chemo, CARTITUDE-1 ΔOS ~38 months vs control): Treat ≥6 months, consider 3–6 months, avoid <3 months or no data (KEYNOTE-024/189/826, CheckMate-067/649, ZUMA-1/7, Phase II global trial of tisagenlecleucel in pediatric and young-adult B-cell acute lymphoblastic leukemia [ELIANA], CARTITUDE-1, IMPACT, OPTiM). NSCLC ECOG 2 data from real-world cohorts [42].

CPS, Combined positive score; *LDH*, lactate dehydrogenase; *NSCLC*, non–small-cell lung cancer; *OS*, overall survival; *PD-L1*, programmed death-ligand 1; *QALY*, quality-adjusted life year; *TPS*, tumor proportion score.

[a] *Off-label/extrapolated use beyond trial and label population (IMPACT trial and FDA label restricted to ECOG 0–1 and life expectancy >6 months [37,38]).*

[b] *Exceptions may apply when poor performance status is primarily tumor-driven and rapid clinical response is plausible.*

[c] *For ECOG ≥4, best supportive care should generally be prioritized according to palliative-care guidance [41], unless a trial or highly responsive tumor scenario is appropriate.*

Immune-related adverse events

irAEs are significant toxicities associated with cancer immunotherapies, particularly ICIs. From a pain management perspective, irAEs may cause new or worsening pain syndromes, including inflammatory arthritis, myalgia, neuropathy, colitis-related abdominal pain, headache, and myositis. They impact multiple organ systems and often necessitate treatment modifications [43,44]. Their subacute onset and potential to mimic disease progression require prompt recognition to manage symptoms effectively while preserving therapeutic benefits [44].

General principles of immune-related adverse event management

irAEs are graded using the Common Terminology Criteria for Adverse Events version 5.0 for ICIs. For CAR-T-cell therapy, the 2019 ASTCT consensus criteria are used for CRS and immune effector cell-associated neurotoxicity syndrome (ICANS) [45].

Persistent grade 2 toxicities typically require oral prednisone 0.5–1 mg/kg/day. Grade ≥3 toxicities, however, require 1–2 mg/kg/day (or intravenous) methylprednisolone for unstable patients and are tapered over ≥4 weeks, guided by symptom resolution and biomarker trends [44]. For higher-risk toxicities, taper over ≥6 weeks to reduce recurrence [31].

Supportive treatments, such as nonsteroidal antiinflammatory drugs (NSAIDs) for low-grade inflammatory pain or neuropathic agents (gabapentin 300–600 mg three times daily [TID], pregabalin 75–150 mg twice daily [BID], or duloxetine 30–60 mg daily), do not address the underlying irAEs but help alleviate symptoms [43,44].

Routine monitoring using the Numeric Rating Scale or Brief Pain Inventory (BPI) every 1–2 weeks supports early detection [31]. Persistent or organ-specific irAEs warrant subspecialty referral. Patient education is essential to encourage early reporting of symptoms such as rash, diarrhea, or fatigue [43,44], so provide an irAE wallet card and 24/7 contact information while reviewing red-flag symptoms (new dyspnea, chest pain, severe diarrhea, jaundice, severe headache/visual change) and instructing patients to report them promptly [31]. For patients receiving prednisone ≥20 mg/day for more than 4 weeks or receiving multiple immunosuppressants, consider Pneumocystis jirovecii prophylaxis, bone-protection measures, and gastric protection when clinically indicated, according to local protocol [31]. Monitoring should also continue for at least 6 months after immunotherapy cessation because late-onset irAEs are increasingly recognized [31]. Suggested monitoring includes pain score/BPI, bowel symptoms, rash, dyspnea, fatigue, endocrine symptoms, AST/ALT, bilirubin, TSH/free T4, glucose, and cortisol when clinically indicated.

Timing and clinical presentation

Building on these management principles and understanding the timing and clinical manifestations of irAEs is crucial for timely intervention. Most irAEs manifest within 3–6 months of ICI initiation, though delayed onset, months after discontinuation, has been reported [11,43,44]. Median onset varies by organ system: dermatologic toxicities (2–6 weeks), gastrointestinal (5–8 weeks), hepatitis or pneumonitis (8–14 weeks), and endocrinopathies (12–24 weeks), with adrenal insufficiency or immune-related diabetes occasionally presenting later [11,43,44]. A practical timing summary is as follows: dermatologic toxicities usually occur at 2–6 weeks, gastrointestinal toxicities at 5–8 weeks, hepatitis or pneumonitis at 8–14 weeks, and endocrinopathies at 12–24 weeks.

Common irAEs include arthralgia/myalgia (10%–20%), inflammatory arthritis (1%–5%), immune-mediated colitis (1%–5% with PD-1/PD-L1 monotherapy, 10%–15% with ipilimumab, 25%–35% with ipilimumab plus PD-1 combinations; grade ≥3 in ~10%–15% for combinations), dermatitis (15%–30%), hepatitis (1%–2% with PD-1/PD-L1, 5%–10% with ipilimumab), and pneumonitis (2%–5%) [44]. Rare but severe events, such as primary adrenalitis (~0.7%), immune-related diabetes (0.2%–1%), myocarditis (~0.5%), and Guillain–Barré syndrome (<1%), demand high clinical suspicion and multidisciplinary management [44]. Rheumatologic and neurologic irAEs may persist for months to years (chronic toxicity) and often require long-term management plans [31]. In addition, endocrinopathies, among the most common long-lasting irAEs, often persist beyond treatment cessation and necessitate long-term follow-up. They include hypothyroidism (up to 20%) and hypophysitis (5%–10% with CTLA-4 blockade, <1% with PD-1/PD-L1) [45]. Rare severe forms, such as primary adrenalitis (≈0.7%) and immune-related diabetes (0.2%–1%), also require ongoing care [44,45].

Differentiating immune-related adverse events from disease progression, pseudoprogression, and hyperprogression

Progressive disease typically produces focal, organ-specific symptoms that correspond to tumor burden, often accompanied by rising tumor markers (when applicable) and solid, enlarging lesions on imaging [44,46]. In practice, irAEs usually present with inflammatory organ-specific findings and compatible laboratory abnormalities, true progression with enlarging tumor burden and concordant symptoms, pseudoprogression with transient radiographic enlargement followed by response, and hyperprogression with rapid clinical and radiographic deterioration soon after ICI initiation. In contrast, irAEs manifest as diffuse, systemic inflammation—for example, poly-articular synovitis or diarrhea without radiographic bowel metastases—and are characterized by organ-specific laboratory abnormalities (elevated thyroid-stimulating hormone [TSH], alanine aminotransferase [ALT]/aspartate aminotransferase [AST], or troponin), while tumor markers usually remain stable [31,43,44]. Radiologically, irAEs exhibit inflammatory patterns—segmental bowel-wall thickening in colitis, bilateral ground-glass opacities in pneumonitis, and pituitary enlargement in hypophysitis—distinct from the mass-like or necrotic lesions typical of progression [31,46]. Pseudoprogression denotes a transient increase in lesion size due to immunecell infiltration, resolving on subsequent imaging and observed in ≈5%–10% of ICI-treated patients [47]. Hyperprogression reflects a ≥2-fold acceleration in tumor growth rate within 1–2 months of ICI initiation, reported in 4%–29% of cases and associated with poor outcomes [48]. Both are tumor-response phenomena rather than irAEs [47]. Accurate distinction integrates symptom distribution, laboratory trends, and imaging features. When infection is excluded, a short corticosteroid trial—undertaken under specialist supervision—may aid differentiation: improvement within 48–72 hours favors an irAE, whereas nonresponse suggests true or hyperprogression [43]. Because delayed or overlapping toxicities are common, evaluation should occur within a multidisciplinary team that includes organ-specific experts [31].

Laboratory and diagnostic workup

Accurate differentiation relies on targeted diagnostic approaches to confirm irAEs and guide management. Routine monitoring includes TSH, free T4, morning cortisol (before steroids if adrenal crisis is suspected), adrenocorticotropic hormone, glucose, HbA1c, ALT, AST, bilirubin, and CRP to screen for endocrinopathies, hepatic irAEs, and systemic inflammation [31,43,44]. Ferritin is reserved for suspected hemophagocytic lymphohistiocytosis—a hyperinflammatory syndrome characterized by very high ferritin, cytopenias, and cytokine storm—or for severe CRS in CAR T-cell therapy, but is not recommended for routine ICI monitoring [45,52]. In patients with compatible symptoms, obtain troponin, CK, and an ECG, and involve cardiology/neurology early to exclude the myocarditis–myositis–myasthenia overlap syndrome [31,43,44]. Colonoscopy with biopsy remains the diagnostic gold standard for immune-mediated colitis and is essential to exclude infectious etiologies such as *Clostridium difficile*. It is typically indicated for grade ≥3 diarrhea or persistent grade 2 symptoms, especially when unresponsive to 3–5 days of corticosteroids [31,43,44]. For pneumonitis grade ≥2, bronchoalveolar lavage is recommended to exclude infection [44] (Fig. 18.1).

Therapy-specific toxicity profiles

Checkpoint inhibitors: irAEs demand a structured, multidisciplinary framework encompassing prevention, anticipation, early detection, prompt treatment, and ongoing monitoring. Figure 18.1 outlines the five pillars of ICI toxicity management, emphasizing proactive risk assessment, timely recognition of toxicity, appropriate intervention, and continuous reassessment to optimize therapeutic efficacy while ensuring patient safety during immunotherapy [50].

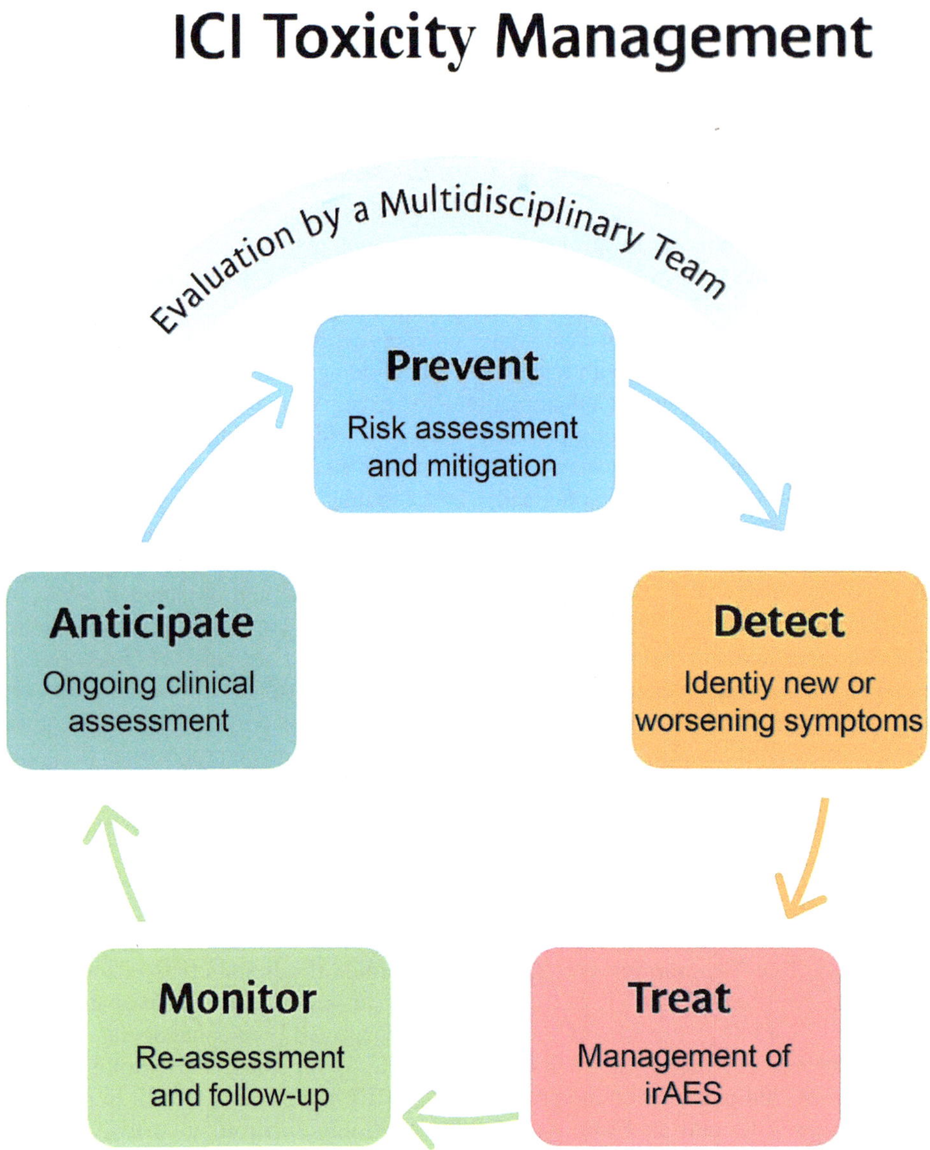

FIGURE 18.1 Framework for immune checkpoint inhibitor toxicity management [50].
Adapted conceptually from Ann Oncol 2016;27(4):559–74. Available from: https://doi.org/10.1093/annonc/mdv623.

Corticosteroids remain the mainstay of treatment, typically oral prednisone ≥1 mg/kg/day or intravenous methylprednisolone 1–2 mg/kg/day [31,44]. If symptoms persist after 48–72 hours, escalation to second-line immunosuppressants such as infliximab, vedolizumab, or mycophenolate is warranted under multidisciplinary team (MDT) supervision [31,44]. Endocrine irAEs—including thyroiditis, hypophysitis, and type 1 diabetes—require lifelong hormone replacement (e.g., levothyroxine, hydrocortisone, or insulin), with high-dose corticosteroids reserved only for adrenal crisis or severe hypophysitis with mass effect [51]. Rechallenge with ICI therapy may be cautiously considered once toxicity resolves to grade ≤1, except in cases of life-threatening or grade 4 events [31].

CAR T-cell therapy: Key toxicities include CRS (fever, hypotension) and ICANS (tremor, confusion, seizures). Grade ≥2 CRS is treated with tocilizumab (8 mg/kg IV, max 800 mg, every 8 hours up to 4 doses within 24 hours) ± corticosteroids if refractory [45,52]. Grade 2 ICANS is managed with dexamethasone 10 mg IV every 6 hours, escalating to 20 mg every 6 hours, or methylprednisolone 1–2 mg/kg/day for grade ≥3, alongside neurologic monitoring [45,52]. Prolonged cytopenias and hypogammaglobulinemia may occur [45].

Sipuleucel-T: Serious adverse events occurred in 24% of patients versus 25% with placebo, and grade ≥3 infusion-related reactions observed in 6.7% of patients [5,38,53]. Common infusion reactions include chills (53%), fever (31%), and fatigue (41%), which are typically self-limited and endocrine irAEs have not been reported [5,38]. Sipuleucel-T is generally well tolerated.

Talimogene laherparepvec (T-VEC): It is generally well-tolerated, with common side effects (fatigue 50%, chills 49%, injection-site pain 28%) resolving within 72 hours [17,39]. Grade 3–4 events are rare (~11%), and autoimmune endocrinopathies are limited to case reports (<1%) [17].

To guide clinical management of major irAEs across immunotherapy types, Table 18.3 provides a comprehensive overview of presentation, initial management strategies, and specialist referral recommendations.

TABLE 18.3 Management of major immune-related adverse events (irAEs).

irAE	Typical presentation	First-line/algorithmic management	Key specialist(s)
Immune-mediated colitis	Diarrhea, cramping ± blood	Hold immune checkpoint inhibitor (ICI); rule out infection (stool cultures, *C. difficile*). Prednisone 1 mg/kg/day PO[c] (or intravenous [IV] methylprednisolone 1–2 mg/kg/day for grade ≥3). If no improvement in 48–72 h, add infliximab 5 mg/kg IV (or vedolizumab if contraindicated). Endoscopy ± biopsy for persistent grade ≥2. Taper steroids ≥4 weeks (≥6 weeks for grade ≥3). MDT[a] oversight.	Gastroenterology
Inflammatory arthritis/ arthralgia	Symmetric/oligo-articular pain, morning stiffness	Grade 1: continue ICI; use NSAIDs or acetaminophen if not contraindicated. Grade 2: hold ICI; prednisone 10–20 mg/day ± intraarticular corticosteroid for limited-joint disease. Grade ≥3: hold ICI; prednisone 0.5–1 mg/kg/day and urgent rheumatology referral. Refractory: add csDMARD[f], such as methotrexate or sulfasalazine, or biologic therapy, such as TNF-α or IL-6 inhibitor. Taper corticosteroids over ≥4 weeks once symptoms improve to grade ≤1 [31,43,44].	Rheumatology
Cytokine-release syndrome (CAR T)	Fever, hypotension, hypoxia, capillary leak	Grade ≥2 CRS or hemodynamic compromise: tocilizumab 8 mg/kg IV, maximum 800 mg, every 8 h, up to 4 doses. If no response after 2 h or grade ≥3 CRS: add dexamethasone 10 mg IV every 6 h or methylprednisolone 1–2 mg/kg/day. Grade ≥3 CRS: ICU-level care and vasopressors as indicated. Monitor for ICANS[d]. Give PJP prophylaxis[b] if corticosteroids continue for >4 weeks [45,52].	Hematology/ oncology, critical care
Peripheral/cranial neuropathy	Paresthesia, weakness, facial palsy	Hold ICI for grade ≥2 neurologic symptoms. Grade 2: prednisone 0.5–1 mg/kg/day. Grade ≥3: IV methylprednisolone 1–2 mg/kg/day and urgent neurology referral. For Guillain–Barré-like presentations, give IVIG 2 g/kg over 2–5 days or plasmapheresis. Bulbar weakness, respiratory compromise, or rapidly progressive weakness requires urgent admission. Taper corticosteroids over ≥4 weeks [31,43,44].	Neurology
Injection-site infection (T-VEC)	Local erythema, vesicles, pain	For grade 1, continue T-VEC with local hygiene ± NSAIDs (or topical acyclovir). If vesicular, obtain HSV[h] PCR; if positive, treat with valacyclovir 1 g PO TID × 7–10 days. For cellulitis or grade ≥2, hold T-VEC and start systemic antibiotics; resume after resolution.	Dermatology, infectious disease

(Continued)

TABLE 18.3 (Continued)

irAE	Typical presentation	First-line/algorithmic management	Key specialist(s)
Pneumonitis	Dyspnea, cough, hypoxia	Hold ICI for grade ≥2 pneumonitis. Start prednisone 1 mg/kg/day or IV methylprednisolone 1–2 mg/kg/day. Perform BAL[e] for grade ≥2 to exclude infection. If no improvement after 48–72 h, add mycophenolate or another guideline-supported immunosuppressant. Taper corticosteroids over ≥4 weeks; give PJP prophylaxis[b] if steroids continue >4 weeks. Severe or recurrent grade 3–4 pneumonitis generally warrants permanent ICI discontinuation [31,43,44].	Pulmonology, critical care
Endocrinopathies (thyroiditis, hypophysitis, adrenalitis, diabetes)	Fatigue, weight change, temperature intolerance, headache, polyuria/polydipsia	Screen with TSH, free T4, morning cortisol, ACTH, and glucose as clinically indicated. Treat hormone deficiencies with replacement therapy, including levothyroxine, hydrocortisone, or insulin. Reserve high-dose corticosteroids for adrenal crisis or hypophysitis with mass effect. ICI may continue once clinically stable on replacement therapy [51].	Endocrinology
Hepatitis	Often asymptomatic; ↑ aspartate aminotransferase (AST)/alanine aminotransferase (ALT)	Hold ICI; prednisone 1 mg/kg/day (or IV methylprednisolone 1–2 mg/kg/day) for grade ≥2. If no response in 48–72 h, add mycophenolate 1 g BID. Avoid infliximab (hepatotoxic). Perform viral serologies and biopsy for grade ≥3. Taper ≥4 weeks.	Hepatology
Myocarditis/myositis/ myasthenia overlap	Chest pain, dyspnea, weakness, ptosis	Admit urgently. Permanently discontinue ICI for grade 2–3 myocarditis, myositis, or myasthenia overlap. Start IV methylprednisolone 1 g/day for 3 days, then taper over ≥4 weeks. If no response within 24–48 h, add abatacept, IVIG, mycophenolate, or another guideline-supported immunosuppressant. Urgent cardiology and neurology involvement is required [31,43,44].	Cardiology, neurology
Dermatologic (rash, pruritus)	Erythema, pruritus, bullae	Hold ICI; topical steroids for grade 1–2; prednisone 0.5–1 mg/kg/day for grade ≥3. Skin biopsy for severe cases. Taper ≥4 weeks.	Dermatology
Renal (nephritis)	Elevated creatinine, proteinuria	Hold ICI for grade ≥2 nephritis. Workup should include creatinine trend, urinalysis, and proteinuria assessment. Start prednisone 1 mg/kg/day for suspected immune-mediated nephritis; add mycophenolate if refractory. Taper corticosteroids over ≥4 weeks once renal function improves. Refer to nephrology for grade ≥2, grade ≥3, or unclear etiology; consider biopsy for grade ≥3 or atypical features [31,43,44].	Nephrology

[a] ***MDT*** *= Multidisciplinary team oversight (e.g., gastroenterology + medical oncology, rheumatology+oncology, etc.).*
[b] ***PJP*** = Pneumocystis jirovecii *pneumonia.*
[c] ***Dosages*** *are expressed in SI units (e.g., mg/kg/day + mg per kg per day).*
[d] ***ICANS*** *= Immune effector cell-associated neurotoxicity syndrome.*
[e] ***BAL*** *= Bronchoalveolar lavage.*
[f] ***csDMARD*** *= Conventional synthetic disease-modifying antirheumatic drug.*
[g] ***HSV*** *= Herpes simplex virus.*

Ethical and economic considerations

The use of immunotherapy in advanced cancer care raises significant ethical and economic challenges, particularly in patients with limited prognosis or access to supportive infrastructure. For individuals with poor performance status (ECOG ≥3) or a <6-month life expectancy, the delayed therapeutic onset and risk of immune-related toxicities may not justify treatment—clinically, ethically, or economically [41,42,54].

From an ethical perspective, initiating immunotherapy in such contexts may conflict with the principles of proportionality and nonmaleficence [56], especially when comfort, dignity, and symptom control are better served through palliative approaches. Current guidelines therefore recommend prioritizing supportive-care measures in patients with advanced frailty or rapid functional decline to align care with patient-centered goals and minimize harm [41,54].

However, exceptions exist. ASCO's Choosing Wisely list notes that highly curable malignancies—such as germ-cell tumors or certain aggressive lymphomas—may warrant a trial of systemic therapy even in ECOG 3–4 patients, provided rapid response (<4 weeks) is plausible and adequate supportive infrastructure is available [55].

Economically, immunotherapies impose substantial direct and indirect costs. PD-1/PD-L1 inhibitors are priced at roughly USD 150,000–180,000 per year [57]; CAR-T products range from USD 410,000 to 475,000 [58], in addition to infusion services, biomarker testing, and the management of irAEs [58]. Persistent or chronic irAEs often require prolonged immunosuppressive therapy and ongoing monitoring, contributing to sustained healthcare resource utilization in palliative care settings [31,59]. These expenditures can strain patients and caregivers, and in low-resource settings, the lack of monitoring capability can convert a potentially life-prolonging treatment into an ethically problematic one. Access to immunotherapy may be limited by geography, insurance status, institutional capability, and availability of biomarker testing [59,60]. Clinicians should practice resource-conscious, shared decision-making, balancing therapeutic intent with realistic prognosis, institutional capacity, and patient's own goals of care. When immunotherapy is unlikely to confer meaningful benefit or cannot be delivered safely, deferring it in favor of high-quality palliative or supportive care is not only reasonable but also ethically and economically responsible [41,54].

Future directions

Future research should define pain, function, and QoL as core outcomes in immunotherapy trials, using standardized tools such as the Brief Pain Inventory and EORTC QLQ-C30. Predictive biomarkers are needed to identify patients most likely to experience symptom benefit, particularly those with limited performance status such as ECOG 2–3 [22]. Studies should also evaluate safe integration of immunotherapy with guideline-based analgesics and investigational symptom-directed approaches, including antinerve growth factor agents and cannabinoids [61]. Broader access to biomarker testing, molecular diagnostics, and safe immunotherapy delivery remains essential in resource-limited settings. Digital health and artificial intelligence may further support symptom monitoring, pain management personalization, and early detection of immune-related toxicities [62].

Conclusion

Immunotherapy has emerged as a cornerstone of systemic cancer treatment, offering survival benefits and, in select biomarker-defined populations, improvements in QoL [1]. As outlined in Table 18.1, these gains are primarily driven by tumor burden reduction and immunologic modulation of the tumor microenvironment [2]. However, consistent evidence of symptom relief—particularly pain reduction—remains limited, as most pivotal immunotherapy trials were not designed with pain as a primary end point.

Checkpoint inhibitors, such as pembrolizumab and nivolumab, have shown durable responses and early QoL signals in PD-L1–positive NSCLC, melanoma, renal RCC, and MSI-H/dMMR tumors [10]. In KEYNOTE-024, PD-L1–high NSCLC patients reported QoL improvement by week 15 [10]. In melanoma, CheckMate-066 demonstrated QoL maintenance in responders, though without early statistically significant improvement [63].

CAR T-cell therapies yield high response rates in hematologic malignancies—approximately 83% in relapsed or refractory large B-cell lymphoma (ZUMA-1 [14]) and comparable benefits reported in real-world DLBCL cohorts [12]; response rates of 73%–98% have been observed in B-ALL [63] and multiple myeloma [34,35]. These approaches may also improve QoL through disease control [12–14]. Yet, pain and QoL outcomes were not consistently prioritized as primary end points in pivotal CAR T-cell trials [14,64]. Moreover, toxicities such as CRS and ICANS can exacerbate symptom burden, particularly in older or functionally impaired patients [45,51].

Sipuleucel-T, approved for asymptomatic or mCRPC in men with ECOG 0–1, confers a 4.1-month survival benefit without measurable tumor shrinkage [5]. Post hoc pooled analyses indicate delayed opioid initiation and a higher proportion of patients remaining pain-free at 12 months, suggesting an exploratory modest palliative benefit [16].

T-VEC, an oncolytic viral therapy for unresectable melanoma, achieves a 26.4% objective response rate and local tumor control [17]. However, evidence remains limited because pivotal trials did not systematically report pain-specific outcomes [17].

Consequently, none of these agents—checkpoint inhibitors, CAR T-cell therapy, Sipuleucel-T, or T-VEC—are currently recommended by major oncology societies, including NCCN [6,65], ESMO [7], or ASCO [66], as primary modalities for palliative pain management [6,7,41,54,65,66]. Symptom relief remains a secondary, inconsistently reported outcome, hindered by the infrequent use of standardized tools such as the BPI or European Organisation for Research and Treatment of Cancer Quality of Life Questionnaire Core 30 (EORTC QLQ-C30) in trial design [61].

Going forward, immunotherapy should be integrated into comprehensive, guideline-concordant cancer care that also addresses symptom control through pharmacologic analgesia, supportive measures, and early palliative care referral [41,54,65]. Decisions to continue, pause, or resume immunotherapy after toxicity should be multidisciplinary, balancing oncologic benefit with symptom burden and the risk of chronic irAEs [31]. Equitable access to biomarker testing and immunotherapies remains a global challenge, particularly in low- and middle-income settings [59]. Future research should prioritize prospective, patient-reported outcomes to better quantify immunotherapy's effects on pain, function, and QoL. Immunotherapy may complement, but should not replace, guideline-based analgesia and early palliative-care involvement.

Acknowledgments

This chapter was developed with material support from the Avicenna Medical and Clinical Research Institute and the use of AI-based tools for literature organization, table and figure formatting, reference management, and language refinement. All content was authored, reviewed, and approved by the author to ensure accuracy and full compliance with ICMJE and COPE authorship and publication ethics guidelines.

References

[1] Topalian SL, Drake CG, Pardoll DM. Immune checkpoint blockade: a common denominator approach to cancer therapy. Cancer Cell 2015;27(4):450–61. Available from: https://doi.org/10.1016/j.ccell.2015.03.001.

[2] Ribas A, Wolchok JD. Cancer immunotherapy using checkpoint blockade. Science 2018;359(6382):1350–5. Available from: https://doi.org/10.1126/science.aar4060.

[3] June CH, O'Connor RS, Kawalekar OU, et al. CAR T cell immunotherapy for human cancer. Science 2018;359(6382):1361–5. Available from: https://doi.org/10.1126/science.aar6711.

[4] Gardner RA, Finney O, Annesley C, et al. Intent-to-treat leukemia remission by CD19 CAR T cells of defined formulation and dose in children and young adults. Blood 2017;129(25):3322–31. Available from: https://doi.org/10.1182/blood-2017-02-769208.

[5] Kantoff PW, Higano CS, Shore ND, et al. Sipuleucel-T immunotherapy for castration-resistant prostate cancer. N Engl J Med 2010;363(5):411–22. Available from: https://doi.org/10.1056/NEJMoa1001294.

[6] National Comprehensive Cancer Network. NCCN Clinical Practice Guidelines in Oncology: Melanoma: Cutaneous. Version 2.2025 [Internet]. Plymouth Meeting (PA): National Comprehensive Cancer Network; 2025 [cited 2025 Dec 22]. Available from: https://www.nccn.org/guidelines/guidelines-detail?id=1484.

[7] ESMO Guidelines Committee. Cutaneous melanoma: ESMO clinical practice guideline for diagnosis, treatment and follow-up. Ann Oncol 2025;36(2):117–33. Available from: https://doi.org/10.1016/j.annonc.2024.11.006.

[8] Harrington KJ, Andtbacka RHI, Collichio F, et al. Efficacy and safety of talimogene laherparepvec versus granulocyte-macrophage colony-stimulating factor in patients with stage IIIB/C and IVM1a melanoma: subanalysis of the phase III OPTiM trial. Onco Targets Ther 2016;9:7081–93. Available from: https://doi.org/10.2147/OTT.S115245.

[9] Kohlhapp FJ, Kaufman HL. Molecular pathways: mechanism of action for talimogene laherparepvec, a new oncolytic virus immunotherapy. Clin Cancer Res 2016;22(5):1048–54. Available from: https://doi.org/10.1158/1078-0432.CCR-15-2667.

[10] Reck M, Rodríguez-Abreu D, Robinson AG, et al. Pembrolizumab versus chemotherapy for PD-L1-positive non-small-cell lung cancer. N Engl J Med 2016;375(19):1823–33. Available from: https://doi.org/10.1056/NEJMoa1606774.

[11] Kennedy LB, Salama AKS. A review of cancer immunotherapy toxicity. CA Cancer J Clin 2020;70(2):86–104. Available from: https://doi.org/10.3322/caac.21596.

[12] Sermer D, Batlevi C, Palomba ML, et al. Outcomes in patients with DLBCL treated with commercial CAR T cells compared with alternate therapies. Blood Adv 2020;4(19):4669–78. Available from: https://doi.org/10.1182/bloodadvances.2020002118.

[13] Jacobson C, Locke FL, Ghobadi A, et al. Long-term survival and gradual recovery of B cells in patients with refractory large B-cell lymphoma treated with axicabtagene ciloleucel (axi-cel). Blood 2020;136(Suppl 1):40–2. Available from: https://doi.org/10.1182/blood-2020-134362.

[14] Neelapu SS, Locke FL, Bartlett NL, et al. Axicabtagene ciloleucel CAR T-cell therapy in refractory large B-cell lymphoma. N Engl J Med 2017;377(26):2531–44. Available from: https://doi.org/10.1056/NEJMoa1707447.

[15] Madan RA, Antonarakis ES, Drake CG, et al. Putting the pieces together: completing the mechanism of action jigsaw for sipuleucel-T. J Natl Cancer Inst 2020;112(6):562–73. Available from: https://doi.org/10.1093/jnci/djaa021.

[16] Small EJ, Higano CS, Kantoff PW, et al. Time to disease-related pain and first opioid use in patients with metastatic castration-resistant prostate cancer treated with sipuleucel-T. Prostate Cancer Prostatic Dis 2014;17(3):259–64. Available from: https://doi.org/10.1038/pcan.2014.21.

[17] Andtbacka RHI, Kaufman HL, Collichio F, et al. Talimogene laherparepvec improves durable response rate in patients with advanced melanoma. J Clin Oncol 2015;33(25):2780–8. Available from: https://doi.org/10.1200/JCO.2014.58.3377.

[18] Ferris RL, Blumenschein G Jr, Fayette J, et al. Nivolumab for recurrent squamous-cell carcinoma of the head and neck. N Engl J Med 2016;375(19):1856–67. Available from: https://doi.org/10.1056/NEJMoa1602252.

[19] Small EJ, Higano CS, Kantoff PW, et al. Time to disease-related pain after sipuleucel-T in asymptomatic patients with metastatic castrate-resistant prostate cancer: results from three randomized phase III trials. J Clin Oncol 2011;29(15_suppl):4661. Available from: https://doi.org/10.1200/JCO.2011.29.15_suppl.4661.

[20] National Comprehensive Cancer Network. NCCN Clinical Practice Guidelines in Oncology: Non-Small Cell Lung Cancer. Version 5.2026 [Internet]. Plymouth Meeting (PA): National Comprehensive Cancer Network; 2026 [cited 2026 Mar 23]. Available from: https://www.nccn.org/guidelines/guidelines-detail?id=1450.

[21] Schneider BJ, Naidoo J, Santomasso BD, et al. Management of immune-related adverse events in patients treated with immune checkpoint inhibitor therapy: ASCO guideline. J Clin Oncol 2021;39(36):4073–126. Available from: https://doi.org/10.1200/JCO.21.01440.

[22] Haslam A, Gill J, Prasad V. Estimation of the percentage of US patients with cancer who are eligible for immune checkpoint inhibitor drugs. JAMA Netw Open 2020;3(3):e200423. Available from: https://doi.org/10.1001/jamanetworkopen.2020.0423.

[23] Burtness B, Harrington KJ, Greil R, et al. Pembrolizumab alone or with chemotherapy versus cetuximab with chemotherapy for recurrent or metastatic squamous-cell carcinoma of the head and neck (KEYNOTE-048). Lancet 2019;394(10212):1915–28. Available from: https://doi.org/10.1016/S0140-6736(19)32591-7.

[24] Le DT, Durham JN, Smith KN, et al. Mismatch repair deficiency predicts response of solid tumors to PD-1 blockade. Science 2017;357(6349):409–13. Available from: https://doi.org/10.1126/science.aan6733.

[25] Overman MJ, Lonardi S, Wong KYM, et al. Durable clinical benefit with nivolumab plus ipilimumab in DNA mismatch repair-deficient/microsatellite instability-high metastatic colorectal cancer. J Clin Oncol 2018;36(8):773–9. Available from: https://doi.org/10.1200/JCO.2017.76.9901.

[26] Motzer RJ, Tannir NM, McDermott DF, et al. Nivolumab plus ipilimumab versus sunitinib in advanced renal-cell carcinoma. N Engl J Med 2018;378(14):1277–90. Available from: https://doi.org/10.1056/NEJMoa1712126.

[27] Abou-Alfa GK, Lau G, Kudo M, et al. Tremelimumab plus durvalumab in unresectable hepatocellular carcinoma. NEJM Evid 2022;1(8). Available from: https://doi.org/10.1056/EVIDoa2100070.

[28] Finn RS, Qin S, Ikeda M, et al. Atezolizumab plus bevacizumab in unresectable hepatocellular carcinoma. N Engl J Med 2020;382(20):1894–905. Available from: https://doi.org/10.1056/NEJMoa1915745.

[29] Tewari KS, Monk BJ, Vergote I, et al. Survival with cemiplimab in recurrent cervical cancer. N Engl J Med 2022;386(6):544–55. Available from: https://doi.org/10.1056/NEJMoa2112187.

[30] National Comprehensive Cancer Network. NCCN clinical practice guidelines in oncology (NCCN Guidelines®). Plymouth Meeting (PA): National Comprehensive Cancer Network; 2025 [cited 2025 Dec 22]. Available from: https://www.nccn.org/guidelines

[31] Haanen J, Obeid M, Spain L, et al. Management of toxicities from immunotherapy: ESMO clinical practice guideline. Ann Oncol 2022;33(12):1217–38. Available from: https://doi.org/10.1016/j.annonc.2022.10.001.

[32] U.S. Food and Drug Administration. Yescarta (axicabtagene ciloleucel) [prescribing information]. Kite Pharma, Inc.; 2024. Available from: https://www.fda.gov/media/108377/download

[33] U.S. Food and Drug Administration. Kymriah (tisagenlecleucel) [prescribing information]. Novartis Pharmaceuticals Corporation; 2024. Available from: https://www.fda.gov/media/107296/download

[34] U.S. Food and Drug Administration. Abecma (idecabtagene vicleucel) [prescribing information]. Celgene Corporation; 2024. Available from: https://www.fda.gov/media/147055/download

[35] U.S. Food and Drug Administration. Carvykti (ciltacabtagene autoleucel) [prescribing information]. Janssen Biotech, Inc.; 2024. Available from: https://www.fda.gov/media/156560/download

[36] Lemoine J, Bachy E, Cartron G, et al. Nonrelapse mortality after CAR T-cell therapy for large B-cell lymphoma: a LYSA study from the DESCAR-T registry. Blood Adv 2023;7(21):6589–98. Available from: https://doi.org/10.1182/bloodadvances.2023010624.

[37] National Comprehensive Cancer Network. NCCN Clinical Practice Guidelines in Oncology: Prostate Cancer. Version 4.2024 [Internet]. Plymouth Meeting (PA): National Comprehensive Cancer Network; 2024. Available from: https://www.nccn.org/guidelines/guidelines-detail?id=1459

[38] U.S. Food and Drug Administration. Provenge (sipuleucel-T) [prescribing information]. Dendreon Pharmaceuticals LLC; 2024. Available from: https://www.fda.gov/media/78511/download

[39] U.S. Food and Drug Administration. Imlygic (talimogene laherparepvec) [prescribing information]. Amgen Inc.; 2024. Available from: https://www.fda.gov/media/94129/download

[40] National Comprehensive Cancer Network. NCCN clinical practice guidelines in oncology: adult cancer pain. J Natl Compr Canc Netw 2019;17(8):977–1007. Available from: https://doi.org/10.6004/jnccn.2019.0038.

[41] National Comprehensive Cancer Network. NCCN Clinical Practice Guidelines in Oncology: Palliative Care. Version 1.2024 [Internet]. Plymouth Meeting (PA): National Comprehensive Cancer Network; 2024 [cited 2025 Dec 22]. Available from: https://www.nccn.org/guidelines/guidelines-detail?id=1454.

[42] Veluswamy R, Hu L, Smith CB, et al. Immunotherapy outcomes in individuals with non-small-cell lung cancer and poor performance status. JNCI Cancer Spectr 2022;6(2):pkac013. Available from: https://doi.org/10.1093/jncics/pkac013.

[43] Brahmer JR, Abu-Sbeih H, Ascierto PA, et al. Society for Immunotherapy of Cancer (SITC) clinical practice guideline on immune checkpoint inhibitor-related adverse events. J Immunother Cancer 2021;9(6):e002435. Available from: https://doi.org/10.1136/jitc-2021-002435.

[44] Brahmer JR, Lacchetti C, Schneider BJ, et al. Management of immune-related adverse events in patients treated with immune checkpoint inhibitor therapy: ASCO clinical practice guideline. J Clin Oncol 2018;36(17):1714–68. Available from: https://doi.org/10.1200/JCO.2017.77.6385.
[45] Lee DW, Santomasso BD, Locke FL, et al. ASTCT consensus grading for cytokine release syndrome and neurologic toxicity associated with immune effector cells. Biol Blood Marrow Transpl 2019;25(4):625–38. Available from: https://doi.org/10.1016/j.bbmt.2018.12.758.
[46] Percik R, Larkin J, Morganstein DL. Endocrinopathies induced by immune checkpoint inhibitors: the need for clear endocrine diagnosis. Lancet Oncol 2021;22(7):905–7. Available from: https://doi.org/10.1016/S1470-2045(21)00305-3.
[47] Puzanov I, Diab A, Abdallah K, et al. Managing toxicities associated with immune checkpoint inhibitors: SITC consensus recommendations. J Immunother Cancer 2017;5(1):95. Available from: https://doi.org/10.1186/s40425-017-0300-z.
[48] Champiat S, Dercle L, Ammari S, et al. Hyperprogressive disease is a new pattern of progression in cancer patients treated by anti-PD-1/PD-L1. Clin Cancer Res 2017;23(8):1920–8. Available from: https://doi.org/10.1158/1078-0432.CCR-16-1741.
[49] Maus MV, Alexander S, Bishop MR, et al. SITC clinical practice guideline on immune effector cell-related adverse events. J Immunother Cancer 2020;8(2):e001511. Available from: https://doi.org/10.1136/jitc-2020-001511.
[50] Champiat S, Lambotte O, Barreau E, et al. Management of immune checkpoint blockade dysimmune toxicities: a collaborative position paper. Ann Oncol 2016;27(4):559–74. Available from: https://doi.org/10.1093/annonc/mdv623.
[51] Percik R, Criseno S, Adam S, et al. Diagnostic criteria and proposed management of immune-related endocrinopathies following immune checkpoint inhibitor therapy. Endocr Connect 2023;12(5):e220513. Available from: https://doi.org/10.1530/EC-22-0513.
[52] Santomasso BD, Nastoupil LJ, Adkins S, et al. Management of immune-related adverse events in patients treated with chimeric antigen receptor T-cell therapy: ASCO guideline. J Clin Oncol 2021;39(35):3978–92. Available from: https://doi.org/10.1200/JCO.21.01992.
[53] Higano CS, Schellhammer PF, Small EJ, et al. Integrated data from two randomized phase 3 trials of sipuleucel-T in advanced prostate cancer. Cancer 2009;115(16):3670–9. Available from: https://doi.org/10.1002/cncr.24429.
[54] Ferrell BR, Sanders JJ, Alesi ER, et al. Palliative care for patients with cancer: ASCO guideline update. J Clin Oncol 2024;42(19):2336–57. Available from: https://doi.org/10.1200/JCO.24.00542.
[55] American Society of Clinical Oncology. Choosing Wisely [Internet]. Alexandria (VA): ASCO; 2021 [cited 2025 Dec 22]. Available from: https://www.choosingwisely.org/societies/american-society-of-clinical-oncology/
[56] Beauchamp TL, Childress JF. Principles of biomedical ethics. 8th ed. New York: Oxford University Press; 2019.
[57] Paul J, Mitchell AP, Kesselheim AS, et al. Trends in prices of checkpoint inhibitors in the US, 2016–2023. J Clin Oncol 2024;42(16_suppl):11075. Available from: https://doi.org/10.1200/JCO.2024.42.16_suppl.11075.
[58] Borgert R. Improving outcomes and mitigating costs associated with CAR T-cell therapy. Am J Manag Care 2021;27(13 Suppl):S253–61. Available from: https://doi.org/10.37765/ajmc.2021.88737.
[59] Zheng Y, Kim R, Yu T, et al. Real-world clinical and economic outcomes in selected immune-related adverse events among patients receiving immune checkpoint inhibitors. Oncologist 2021;26(11):e2002–12. Available from: https://doi.org/10.1002/onco.13918.
[60] Curigliano G, Banerjee S, Cervantes A, et al. Managing cancer patients during the COVID-19 pandemic: ESMO multidisciplinary expert consensus. Ann Oncol 2020;31(10):1320–35. Available from: https://doi.org/10.1016/j.annonc.2020.07.010.
[61] Fallon M, Giusti R, Aielli F, et al. Management of cancer pain in adult patients: ESMO clinical practice guidelines. Ann Oncol 2018;29(Suppl 4):iv166–91. Available from: https://doi.org/10.1093/annonc/mdy152.
[62] Topol E. Deep medicine: how artificial intelligence can make healthcare human again. New York: Basic Books; 2019.
[63] Robert C, Long GV, Brady B, et al. Nivolumab in previously untreated melanoma without BRAF mutation. N Engl J Med 2015;372(4):320–30. Available from: https://doi.org/10.1056/NEJMoa1412082.
[64] Maude SL, Laetsch TW, Buechner J, et al. Tisagenlecleucel in children and young adults with B-cell lymphoblastic leukemia. N Engl J Med 2018;378(5):439–48. Available from: https://doi.org/10.1056/NEJMoa1709866.
[65] National Comprehensive Cancer Network. NCCN clinical practice guidelines in oncology: adult cancer pain. Version 2.2025 [Internet]. Plymouth Meeting (PA): National Comprehensive Cancer Network; 2025 [cited 2025 Dec 22]. Available from: https://www.nccn.org/guidelines/guidelines-detail?category3&id1413
[66] Paice JA, Portenoy R, Lacchetti C, et al. Management of chronic pain in survivors of adult cancers: ASCO clinical practice guideline. J Clin Oncol 2016;34(27):3325–45. Available from: https://doi.org/10.1200/JCO.2016.68.5206.

Chapter 19

Radiotherapy and brachytherapy

Tamer Soror and Dirk Rades
Radiation Oncology Department, University of Lübeck/UKSH-CL, Lübeck, Germany

Key learning objectives

- Understand the role of radiation in managing pain for cancer patients.
- Describe the pathophysiology of pain development and how radiation treatment alleviates pain.
- Discuss the various radiation techniques employed for pain management, including the dosages utilized.
- Identify future applications and research opportunities aimed at enhancing personalized pain treatment through radiation therapy.

Introduction

A significant number of cancer patients experience pain throughout their illness. This could be a result of the progression of the primary tumor locally or distantly. However, it is most commonly observed in patients with distant metastases, especially when the lesions affect the bones [1,2]. Approximately half of all individuals diagnosed with cancer will develop bone metastases, with breast, prostate, and lung cancers accounting for more than 85% of these cases. Other malignancies that often lead to bone metastases include cancers of the urinary bladder, kidney, endometrium, and thyroid, and also malignant melanoma [2].

Patients with bone metastases frequently encounter skeletal-related events, such as pathological fractures, bone pain, hypercalcemia, and spinal cord compression. These complications can lead to a range of physical, functional, and psychosocial challenges, significantly impacting the patients' quality of life (QoL) [3,4]. Bone metastases are frequently linked to poor prognosis. Nonetheless, advancements in systemic and supportive therapies have gradually improved the life expectancy of these patients. Effective pain management can greatly enhance QoL, with studies indicating that radiation therapy (RT) can provide pain relief for up to 80% of patients [1].

The interplay between osteoclasts and osteoblasts is critical in the context of bone metastases. Osteoclasts, responsible for bone resorption, are often activated in the presence of tumor cells, leading to increased bone destruction. In contrast, osteoblasts, which are responsible for bone formation, may be suppressed, resulting in a net loss of bone mass. This imbalance not only leads to structural weakness and increased fracture risk but also contributes to the overall pain experience [5,6].

Pathophysiology of pain in bone metastases

The pathophysiology of pain in cancer patients, especially those with bone metastases, is multifaceted. Bone metastases can disrupt the normal balance between osteoclasts and osteoblasts, leading to bone degradation and resultant pain. Metastatic lesions can induce a local inflammatory response, causing the release of inflammatory mediators, which sensitize nociceptive nerve endings. Furthermore, the presence of tumor cells in the bone microenvironment can provoke neurogenic inflammation, further exacerbating pain perception. This dual mechanism of mechanical pressure from the tumor and biochemical sensitivity complicates the pain experienced by patients. Moreover, the pain is usually aggravated through spasm of the surrounding muscles [5–7].

Pain Management in Palliative Care for Patients With Cancer. DOI: https://doi.org/10.1016/B978-0-443-30050-9.00029-1

Pain evaluation following radiation therapy

A crucial aspect of evaluating the effectiveness of RT for pain relief involves accurately measuring changes in pain levels posttreatment. Relying solely on binary subjective categorical assessments can lead to misleading conclusions about the differences between various RT protocols [8]. In contrast, incorporating additional methods, such as tracking reductions in analgesic use, has been shown to provide a more accurate evaluation of RT's effectiveness [9,10]. The International Bone Metastases Consensus Working Party recommends utilizing a patient-assessed ordinal pain scale ranging from 0 to 10, along with monitoring changes in analgesic consumption [10]. This approach has been increasingly adopted in recent studies evaluating RT end points [11].

Pain relief effect of radiation therapy

The precise mechanisms by which radiation alleviates pain remain largely unclear. However, the analgesic effects are believed to result from several processes, including the promotion of bone formation, reduction in osteoclast activity within the bone microenvironment, and the destruction of cancer cells, all of which contribute to decreased tumor load [5,12]. Some patients experience swift pain relief, sometimes within 24 hours, suggesting a reduction in the activity of inflammatory cells and a decrease in the concentration of chemical mediators associated with pain in the area exposed to radiation [13,14].

Patients typically experience pain relief from RT within 4–6 weeks after treatment, with the relief often stabilizing at that point [15,16]. However, around one-third of patients may experience a temporary increase in pain, known as pain flare, usually occurring within the first 5 days of radiation and lasting for 2–3 days. It is important to note that the occurrence of a pain flare does not indicate the effectiveness of pain relief from the therapy. The likelihood of experiencing a pain flare can be reduced by prophylactic administration of dexamethasone [17,18].

Additionally, the fractionation schedule of RT for patients with bone metastases significantly impacts mineralization, bone density, and the rate of recalcification at the irradiated site [6,12]. Patients receiving single-fraction RT exhibit a significantly lower recalcification rate when compared to those undergoing fractionated RT [14]. Furthermore, a correlation has been identified between the pain relief provided by RT and lower urinary levels of osteoclast activity markers, such as pyridinoline and deoxypyridinoline, measured before and after treatment. Additionally, reirradiation for painful bone metastases also appears to affect osteoclast activity as observed through these urinary biomarkers [19,20].

Inherited radiosensitivity may also play an important role in achieving pain relief following RT. Irradiating painful non–small-cell lung cancer metastases resulted in 46% complete pain relief compared to 83% for prostate cancer metastases or 65% for breast cancer metastases [21].

Fractionation schedules for uncomplicated bone metastases

The question of the ideal fractionation schedule for treating uncomplicated painful bone metastases remains the most extensively investigated single question in radiation oncology. Uncomplicated bone metastases refer to tumor masses that are painful, without impending or existing pathological fracture, and without spinal cord or cauda equina compression, irrespective of size [22]. Over the past decades, numerous randomized and observational studies have investigated the best dosing and fractionation strategies for effective pain relief. Consecutive systematic reviews investigated single-fraction versus multifraction RT trials and demonstrated similar efficacy between both schedules [1,23–26]. The most recent two updates of these systematic reviews published in 2018 by Rich et al. and in 2022 by Behroozian et al. along with the findings of the metaanalysis in 2019 by Chow et al. have consistently concluded that a single dose of 8 Gy provides comparable analgesic benefits to various multifraction regimens such as 30 Gy given in 10 fractions, 24 Gy in 6 fractions, or 20 Gy in 5 fractions [23,24,27].

Rich et al. reported an overall response rate of 61% after single-fraction RT versus 62% after RT with multiple fractions. However, when considering only the assessable patients, the rates were 72% versus 75%, respectively [24]. Behroozian et al. identified no new or ongoing randomized controlled trials investigating the same question [23].

The metaanalysis by Chow et al. that included 25 randomized clinical trials indicated that approximately 80% of patients with uncomplicated bone metastases respond positively to external beam radiotherapy (EBRT), with about 25%–30% achieving complete pain relief within 3–4 weeks after treatment [27]. However, only about 35% of patients experience persistent pain control lasting 3–6 months posttherapy. Interestingly, the incidence of in-field pain recurrence requiring retreatment was higher among patients who received single-fraction treatments compared to those who underwent multiple-fraction schemes (20% vs 8%) [27].

Furthermore, single-fraction RT had significantly fewer acute treatment-related toxicities, such as nausea, vomiting, diarrhea, fatigue, and skin irritation. However, using single instead of multiple-fraction palliative RT was associated with a slightly increased risk of spinal cord compression and pathological fractures during the period of follow-up, though the differences were not statistically significant [27].

Moreover, a number of studies found that the total financial costs, including the need for reirradiation as well as all related nonmedical costs of single-fraction treatments, were significantly lower when compared to multiple-fraction schemes [28–30]. These findings are of special interest for lower-income countries. Single-fraction treatments should also be considered for patients with limited mobility, short life expectancy, and patients living at long distances from the radiation facility, as well as for high-workload cancer centers.

Despite clear evidence supporting the effectiveness of single-fraction treatments compared to multiple-fraction treatments and the treatment recommendations of the most recently published guidelines on uncomplicated bone metastases [22,31], single-fraction treatments are still noticed to be underutilized in clinical practice for managing painful bone metastases. Reasons for this could be the lesser durability of pain relief, especially in the presence of a neuropathic component of the pain [28], the higher rate of retreatment [27], the inherited radiosensitivity of the different histologies [21], and the lower rate of recalcification [14].

Stereotactic body radiation therapy

A considerable number of skeletal metastases are multiple, whereas single lesions make up only about 10% of cases [32]. In clinical settings, an oligometastatic stage is usually defined as 1–5 distant metastases. Stereotactic body radiation therapy (SBRT) consists of delivering a high dose per fraction, either as a single treatment or using a limited number of fractions. Typically, SBRT is employed to treat one to three vertebral metastases that are less than 5 cm in size and do not present with spinal cord compression. SBRT is not considered optimal for symptomatic spinal cord compression and should be used only for appropriately selected patients. It may be associated with an increased risk of neurotoxicity (e.g., radiculopathy) and vertebral fractures, particularly in the case of single-fraction treatment, when compared to fractionated EBRT [33–35]. Moreover, the risk of adjacent level progression needs to be considered after SBRT. Chakravarthy et al. reported a rate of 41% of adjacent level progression, which occurred after a mean interval of 6.1 ± 5.4 months following SBRT and upfront separation surgery [36]. SBRT was administered with a median total dose of 16.0 Gy (range, 15–30 Gy), mainly (92.3%) as single-fraction treatment.

Unlike EBRT, much of the evidence supporting SBRT comes from a large number of retrospective studies [33–41]. However, three randomized trials have compared SBRT to EBRT [39,42,43]. A recent metaanalysis of these trials involving 642 patients found no significant differences in overall pain response between SBRT and conventional EBRT at 3 and 6 months. Two of the studies provided complete data on pain response, revealing that SBRT led to a statistically significant enhancement in complete pain relief compared to EBRT. There were no notable differences in local progression or overall survival rates. Adverse events were comparable, although any grade of radiation dermatitis was significantly less frequent in the SBRT group [44].

Another recent systematic review and metaanalysis that investigated the efficacy and safety of SBRT for spine metastases considered 69 studies evaluating the outcomes of 7236 metastases in 5736 SBRT sessions for patients with spinal metastases [45]. It demonstrated significant efficacy of SBRT, achieving a pooled overall pain response rate of 83%, a pooled complete pain response rate of 36%, and a 1-year local control rate of 94%. However, these results exhibited considerable heterogeneity across the studies. Regarding treatment-related toxicities, SBRT was found to be generally safe, with a pooled vertebral fracture rate of 9%, a pooled rate of radiation-induced myelopathy of 0%, and a pooled pain flare rate of 6%, although levels of heterogeneity varied among the studies [45].

Treatment regimens vary significantly, with doses ranging from single sessions of 5–24 Gy to 18–36 Gy administered over 1–5 fractions [33–45]. The recent European Society for Radiotherapy and Oncology (ESTRO) clinical practice guidelines recommended using doses higher than the biological efficacy of a single fraction of 18 Gy. The suggested dose regimens included 1 × 20 Gy, 1 × 24 Gy, 2 × 12 Gy, 3 × 10 Gy, and 5 × 7 Gy [46]. Prior to SBRT, it is highly recommended to evaluate the need for surgical stabilization using the spinal instability neoplastic score and to assess epidural involvement through the validated Bilsky scale [46].

Compared to vertebral metastases, the use of SBRT to treat nonspine bone metastases is less frequent. A recent meta analysis reviewed seven retrospective studies involving a total of 1048 lesions treated in 807 patients [47]. The most frequently treated sites included the pelvis (39.2%), ribs (25.8%), femur (16.7%), and the humerus/shoulder region (8.7%).

At 1 year, the local failure rate was 7%, while the cumulative incidence of local failure at 2 years was 12.1%. The overall survival rate at 1 year was 82%, while the progression-free survival rate was 33.5%, with a median overall survival duration of 20.2 months. The pooled fracture rate was 5.3%, with a significant correlation observed between planning target volume and fracture rate. The most common sites of fractures were the ribs and femur. Furthermore, the incidence rates of pain flare, fatigue, and dermatitis were reported at 7%, 5.4%, and 0.65%, respectively [47].

Half-body irradiation

Half-body irradiation (HBI) is a form of EBRT designed to treat patients suffering from multiple painful bone metastases. It can be delivered in a single fraction or, less commonly, in a fractionated schedule. Depending on the treatment area, HBI can be categorized into upper HBI, lower HBI, and occasionally mid-body irradiation [48–54]. In the modern era, HBI is not acceptable to many clinicians [55]. Furthermore, concerns about toxicity and the necessity to preserve bone marrow function, which is essential for the use of novel systemic treatments, along with the availability of modern radiation techniques that enable the treatment of complex and large targets, have limited the application of HBI in advanced cancer cases [51,55]. One study found that HBI is cost-effective, reduces overall treatment time, and offers a convenient treatment option for patients and their families [54]. A single fraction of 6–8 Gy HBI resulted in a 70% pain relief rate within 2 days [54]. A recent systematic review reported a high overall pain response rate across all studies, with a median response of 84% and a median complete pain response of 36%. Reported toxicity is generally mild, with most cases classified as grade 1 or 2, and grade 3 cases being extremely rare. Additionally, over 50% of patients were able to reduce their analgesic intake following HBI treatment [55].

Brachytherapy

Brachytherapy (BT) is a specialized form of RT that involves placing radioactive material inside the body, either temporarily or permanently. BT is highly conformal, delivering a very high dose to a small volume while effectively sparing nearby at-risk organs or structures from higher radiation doses due to its unique physical properties, which allow for a rapid decline in radiation intensity. The procedure is typically performed invasively under local or general anesthesia [56].

Temporary applications utilize high-dose-rate or pulsed-dose-rate techniques, where treatment is delivered through an after-loading machine in a shielded room, either in a single session or across a limited number of fractions. In contrast, permanent applications involve the implantation of multiple radioactive sources directly into the tumor volume, such as low-dose-rate iodine seed implants for prostate cancer [56].

The complexity and invasive nature of this technique have restricted its use in patients with metastatic disease, and concerns about higher rates of bone necrosis following BT represent a significant limitation [57–61]. However, a systematic review examining low-dose-rate approaches for treating vertebral tumors included 370 patients across 15 studies, with nearly 80% of participants having metastases. The primary indication for BT was the inability to apply EBRT. Treatment failure rates varied between 13% and 49%, and in seven studies, significant improvements in pain control were documented [57].

For nonspinal bone metastases, several small studies have reported on the use of 125I seed implantation [58–61]. Despite the heterogeneity of these studies, pain control rates were notably high, and reported treatment-related toxicities were generally low [58–61].

Metastatic spinal cord compression

Metastases to the vertebral column or within the epidural space can lead to spinal cord compression, often resulting in neuropathic pain and neurological deficits. Metastatic spinal cord compression (MSCC) is considered a medical emergency that can significantly impact QoL. Up to 60% of MSCC cases are associated with prostate, breast, and lung cancers [62].

Earlier studies suggested that RT was comparable to surgical posterior decompression-laminectomy in terms of pain relief and preservation of motor function [63,64]. However, in a randomized trial by Patchell et al. in the year 2005, patients with MSCC were treated either with surgery followed by RT or with RT alone. The results indicated that surgery was superior to RT in both maintaining and regaining the ability to walk, as well as in reducing the need for corticosteroids and opioid analgesics [65].

A shift in treatment paradigm occurred following the publication of a metaanalysis by Klimo et al. comparing RT to newer surgical techniques such as circumferential decompressive surgery and spinal stabilization [66]. These analyses revealed that there was over a 20% significant difference in overall ambulatory success rates between surgery and RT (85% vs 64%) [66].

Recent ESTRO clinical guidelines recommend that patients with a single area of MSCC, motor deficits lasting less than 48 hours, and a minimum life expectancy of 3 months should be urgently referred for surgical decompression and stabilization followed by postoperative RT. For patients who are not fit or eligible for surgery, RT should be administered as soon as possible, ideally within 24 hours from the patient's first presentation to a radiation oncologist [67]. Patients with poor survival prognosis are generally recommended to receive short-course RT (e.g., with 5 × 4 Gy or 5 × 5 Gy), or even single-fraction RT with 8–10 Gy [68,69]. However, a prospective study involving 224 patients found that longer course fractionation regimens (such as 40 Gy in 20 fractions, 37.5 Gy in 15 fractions, or 30 Gy in 10 fractions) resulted in better 1-year local control when compared to shorter regimens (8 Gy in a single fraction or 20 Gy in 5 fractions) (81% vs 61%). Functional and survival outcomes were comparable between the two groups [70]. Since the risk of an in-field recurrence of spinal cord compression increases with lifetime, patients with more favorable survival prognoses likely benefit from longer-course RT. An individual patient's survival prognosis can be estimated with the help of corresponding prognostic tools [71,72]. Additionally, dexamethasone at 10–16 mg should be initiated immediately for all patients [67].

The role of SBRT in this context remains unclear, with limited literature available [65]. However, precision RT utilizing volumetric modulated arc therapy has been shown to yield higher local progression-free survival rates compared to conventional RT among patients unsuitable for decompression surgery who exhibit motor deficits, without affecting motor function outcomes [73,74].

Survival outcomes and the potential for recovery of ambulatory function after RT for MSCC can be estimated using published scoring systems, which may assist in developing personalized treatment approaches for these patients [71,72].

Fracture and impending fracture

Bone metastases can weaken the affected bone and lead to an increasing risk of pathological fractures. Major indicators of pathological fractures include heightened pain and deformities in the affected bone, such as limb shortening, distortion, or spinal kyphosis. Among long bones, the proximal femur is the most commonly involved site [75].

The management of impending pathological fractures requires a multidisciplinary assessment to determine the appropriateness of prophylactic surgery and/or RT. In cases where patients present with a disease-related poor performance status, RT alone should be considered. For impending fractures that do not require surgical intervention, longer-course RT with 30 Gy in 10 fractions or even higher total doses of 35–40 Gy with doses per fraction of 2.0 Gy or 2.5 Gy are recommended to prevent pathological fractures and promote recalcification [67]. Short-course RT with 6 × 4 Gy may also be a reasonable option [15]. After surgical stabilization for pathological fractures or impending pathological fractures, postoperative RT is typically indicated and should be administered after 1–2 weeks to minimize the risk of wound complications [76]. Selected patients experiencing compression fractures of the vertebrae without neurological deficits can benefit from balloon kyphoplasty or percutaneous vertebroplasty [77–80].

Reirradiation of painful bone metastases or metastatic spinal cord compression

Reirradiation of painful bone metastases is a common clinical scenario. In a study by Chow et al., single-fraction and multiple-fraction treatments were equally effective with respect to pain control; however, a single fraction of 8 Gy was considered to be safer [81].

SBRT is considered an appealing treatment method for reirradiation, especially for spinal metastases [41,82]. Reirradiation of spinal metastases using SBRT has proven effective, with 1-year local control rates ranging from 66% to 90% [82]. Improvement in pain scores varied between 65% and 81%. The median crude rate of vertebral body fractures was 12%, while the incidence of radiation-induced myelopathy was reported to be 1.2% [82].

Reirradiation of MSCC carries the risk of exceeding spinal cord tolerance and the potential development of radiation-induced myelopathy. Pooled results from two randomized studies showed that patients who underwent reirradiation were able to maintain ambulation, with most of them depending on their pretreatment ambulatory status. Notably, no cases of radiation-induced myelopathy were reported [83]. Reirradiation with a cumulative biologically effective dose (BED) of 100 Gy_2 resulted in improvements in motor function in 40% of patients without radiation-induced myelopathy [84].

Furthermore, reirradiation of MSCC is considered safe after 6 months from the initial irradiation, provided that the cumulative BED is less than or equal to 100–135.5 Gy_2 [67].

Painful soft tissue tumors

Primary, recurrent, or metastatic soft tissue tumors may result in significant pain due to local infiltration [41]. Palliative RT achieves high levels of pain relief ranging between 60% and 80% [85–91]. Lung cancer patients often experience severe pain due to tumor invasion of the brachial plexus. A systematic review of 14 randomized clinical trials evaluating palliative RT for lung cancer demonstrated effective symptom relief, particularly pain reduction [85]. The trials employed various fractionation schedules, including a single fraction of 10 Gy, 17 Gy in 2 fractions, 20 Gy in 5 fractions, and longer courses of 30–45 Gy in 10–15 fractions [85].

Palliative localized RT for gastric tumors has been shown to alleviate symptoms, including pain, in more than 70% of patients [86]. Advanced rectal cancer patients did similarly benefit from palliative RT to the pelvis, with both conventional fractionation (45 Gy in 25 fractions) and hypofractionated regimens (30 Gy in 6 fractions) achieving satisfactory pain relief in approximately 70% of cases [87,88].

In pancreatic cancer, up to 40% of patients are diagnosed with locoregionally advanced disease, often resulting in nerve infiltration and severe pain in 43%–72% of cases. The use of SBRT, delivering a median dose of 28 Gy (ranging from 25 to 33 Gy) in five fractions, led to abdominal pain relief in 78% of advanced pancreatic cancer patients [89–91].

Future implications

Prospective phase III randomized studies are underway to determine the optimal use of SBRT for treating newly diagnosed or recurrent painful spinal metastases. The STEREO-OS trial evaluates the efficacy of combining standard systemic treatment with SBRT in patients with oligometastatic prostate cancer, breast cancer, or lung cancer who have up to three bone-only metastases. This study aims to assess progression-free survival at 1 year, with secondary end points including bone progression-free survival, local control, cancer-specific survival, overall survival, toxicity, QoL, pain score analysis, and cost-effectiveness analyses [92]. Another ongoing study, the VERTICAL trial, compares the analgesic effects of SBRT to standard low-dose EBRT in patients with painful bone metastases [93]. Additionally, the IRON-2 trial is investigating the use of intensity-modulated RT with integrated boost for patients with spinal bone metastases using biological planning [94].

The FAST-01 nonrandomized trial investigates the potential role of proton FLASH RT (ultra-high-dose-rate RT delivered at more than 40 Gy/second) for the treatment of symptomatic bone metastases. This is the first in-human use of proton FLASH. The results confirmed that the treatment was clinically feasible. The efficacy and the adverse events were comparable with standard RT [95].

The Palliative Radiotherapy and Inflammation Study (PRAIS) is a multicenter, multinational observational study that aims to identify predictive factors associated with inflammation in patients receiving palliative RT for cancer-induced bone pain. Enrolling 600 adult participants, the study assesses demographic data, clinical variables, osteoclast activity markers, and inflammatory biomarkers before treatment and at various intervals post-RT. The primary goal is to find predictors of pain relief from RT, while secondary objectives include examining predictors for cachexia and depression, as well as the relationship between inflammatory biomarkers and changes in pain intensity. By identifying these predictive factors, PRAIS aims to enhance individualized treatment strategies, reduce unnecessary hospital visits, and improve overall patient outcomes in managing cancer-induced bone pain [96].

These studies aim to provide high-quality data to guide personalized treatment strategies, enhancing QoL and pain relief for patients with painful bone metastases.

Conclusion

A significant proportion of cancer patients experience pain, particularly from bone metastases, which affect approximately half of all diagnosed individuals. Patients with bone metastases frequently suffer from complications such as fractures and spinal cord compression, resulting in physical, functional, and psychosocial challenges. Effective pain management, particularly through RT, has been shown to significantly enhance the QoL of these patients.

RT can relieve pain in 60%–80% of patients, acting through mechanisms that include promoting bone formation and reducing tumor load. Research indicates that single-fraction RT is as effective as multiple-fraction regimens for pain relief in uncomplicated bone metastases. However, single-fraction RT is underutilized in clinical practice, possibly due to concerns about the durability of pain relief and rates of retreatment. SBRT offers high doses of radiation to specific areas, effectively managing pain in patients with limited metastases. Recent studies show comparable pain relief with lower rates of adverse events when compared to conventional RT, making it a promising option for patients with painful bone metastases.

Effective management of pain in cancer patients requires a multidisciplinary approach, including coordination with other specialists, such as palliative care physicians. Integrating various treatment modalities such as RT and supportive and holistic care strategies to address both physical and psychosocial aspects of patient well-being is crucial for optimal pain management.

References

[1] Chow E, Zeng L, Salvo N, et al. Update on the systematic review of palliative radiotherapy trials for bone metastases. Clin Oncol (R Coll Radiol) 2012;24(2):112–24. Available from: https://doi.org/10.1016/j.clon.2011.11.004.

[2] Hayat MJ, Howlader N, Reichman ME, et al. Cancer statistics, trends, and multiple primary cancer analyses from the surveillance, epidemiology, and end results (SEER) program. Oncologist 2007;12(1):20–37. Available from: https://doi.org/10.1634/theoncologist.12-1-20.

[3] Jiang W, Rixiati Y, Zhao B, et al. Incidence, prevalence, and outcomes of systemic malignancy with bone metastases. J Orthop Surg (Hong Kong) 2020;28(2):105989. Available from: https://doi.org/10.1177/2309499020915989.

[4] Tharmalingam S, Chow E, Harris K, et al. Quality of life measurement in bone metastases: a literature review. J Pain Res 2008;1:49–58. Available from: https://doi.org/10.2147/jpr.s4572.

[5] Goblirsch MJ, Zwolak PP, Clohisy DR. Biology of bone cancer pain. Clin Cancer Res 2006;12(20):6231s–5s. Available from: https://doi.org/10.1158/1078-0432.CCR-06-0682.

[6] Johnstone C, Lutz ST. External beam radiotherapy and bone metastases. Ann Palliat Med 2014;3(2):114–22. Available from: https://doi.org/10.3978/j.issn.2224-5820.2014.04.06.

[7] Needham PR, Mithal NP, Hoskin PJ. Radiotherapy for bone pain. J R Soc Med 1994;87(9):503–5.

[8] Tong D, Gillick L, Hendrickson FR. The palliation of symptomatic osseous metastases: final results of the Study by the Radiation Therapy Oncology Group. Cancer 1982;50(5):893–9. Available from: https://doi.org/10.1002/1097-0142(19820901)50:5<893::aid-cncr2820500515>3.0.co;2-y.

[9] Blitzer PH. Reanalysis of the RTOG study of the palliation of symptomatic osseous metastasis. Cancer. 1985;55(7):1468–72. Available from: https://doi.org/10.1002/1097-0142(19850401)55:7<1468::aid-cncr2820550708>3.0.co;2-m.

[10] Chow E, Wu JS, Hoskin P, et al. International consensus on palliative radiotherapy endpoints for future clinical trials in bone metastases. Radiother Oncol 2002;64(3):275–80. Available from: https://doi.org/10.1016/s0167-8140(02)00170-6.

[11] Saito T, Shikama N, Yorozu A, et al. Inconsistencies in assessment of pain endpoints in radiotherapy for painful tumors: analysis of original articles in the green and red journals. Clin Transl Radiat Oncol 2020;24:111–5. Available from: https://doi.org/10.1016/j.ctro.2020.07.003.

[12] Goblirsch M, Mathews W, Lynch C, et al. Radiation treatment decreases bone cancer pain, osteolysis and tumor size. Radiat Res 2004;161(2):228–34. Available from: https://doi.org/10.1667/rr3108.

[13] Wu JSY, Wong RKS, Lloyd NS, et al. Radiotherapy fractionation for the palliation of uncomplicated painful bone metastases—an evidence-based practice guideline. BMC Cancer 2004;4:71. Available from: https://doi.org/10.1186/1471-2407-4-71.

[14] Koswig S, Budach V. Remineralization and pain relief in bone metastases after different radiotherapy fractions (10 times 3Gy vs 1 time 8 Gy): a prospective study. Strahlenther Onkol 1999;175(10):500–8. Available from: https://doi.org/10.1007/s000660050061.

[15] Steenland E, Leer JW, van Houwelingen H, et al. The effect of a single fraction compared to multiple fractions on painful bone metastases: a global analysis of the Dutch bone metastasis study. Radiother Oncol 1999;52(2):101–9. Available from: https://doi.org/10.1016/s0167-8140(99)00110-3.

[16] McDonald R, Ding K, Brundage M, et al. Effect of radiotherapy on painful bone metastases: a secondary analysis of the NCIC clinical trials group symptom control trial SC.23. JAMA Oncol 2017;3(7):953–9. Available from: https://doi.org/10.1001/jamaoncol.2016.6770.

[17] Chow E, Meyer RM, Ding K, et al. Dexamethasone in the prophylaxis of radiation-induced pain flare after palliative radiotherapy for bone metastases: a double-blind, randomised placebo-controlled, phase 3 trial. Lancet Oncol 2015;16(15):1463–72. Available from: https://doi.org/10.1016/S1470-2045(15)00199-0.

[18] Gomez-Iturriaga A, Cacicedo J, Navarro A, et al. Incidence of pain flare following palliative radiotherapy for symptomatic bone metastases: multicenter prospective observational study. BMC Palliat Care 2015;14:48. Available from: https://doi.org/10.1186/s12904-015-0045-8.

[19] Hoskin PJ, Stratford MRL, Folkes LK, et al. Effect of local radiotherapy for bone pain on urinary markers of osteoclast activity. Lancet 2000;355(9213):1428–9. Available from: https://doi.org/10.1016/s0140-6736(00)02144-9.

[20] Chow E, DeAngelis C, Chen BE, et al. Effect of re-irradiation for painful bone metastases on urinary markers of osteoclast activity (NCIC CTG SC.20U). Radiother Oncol 2015;115(2):141–8. Available from: https://doi.org/10.1016/j.radonc.2015.02.025.

[21] Arcangeli G, Giovinazzo G, Saracino B, et al. Radiation therapy in the management of symptomatic bone metastases: the effect of total dose and histology on pain relief and response duration. Int J Radiat Oncol Biol Phys 1998;42(5):1119–26. Available from: https://doi.org/10.1016/s0360-3016(98)00264-8.

[22] van der Velden JM, Willmann J, Spałek M, et al. ESTRO ACROP guidelines for external beam radiotherapy of patients with uncomplicated bone metastases. Radiother Oncol 2022;173:197–206. Available from: https://doi.org/10.1016/j.radonc.2022.05.024.

[23] Behroozian T, Navarro I, Hoskin P, et al. Update on the systematic review/meta-analysis of uncomplicated bone metastases treated with external beam radiation. Radiother Oncol 2022;174:109–10. Available from: https://doi.org/10.1016/j.radonc.2022.07.010.

[24] Rich SE, Chow R, Raman S, et al. Update of the systematic review of palliative radiation therapy fractionation for bone metastases. Radiother Oncol 2018;126(3):547–57. Available from: https://doi.org/10.1016/j.radonc.2018.01.003.

[25] Chow E, Harris K, Fan G, et al. Palliative radiotherapy trials for bone metastases: a systematic review. J Clin Oncol 2007;25(11):1423–36. Available from: https://doi.org/10.1200/JCO.2006.09.5281.

[26] Sze WM, Shelley M, Held I, et al. Palliation of metastatic bone pain: single fraction versus multifraction radiotherapy – a systematic review of the randomised trials. Cochrane Database Syst Rev 2003;15(6):345–52. Available from: https://doi.org/10.1002/14651858.CD004721.

[27] Chow R, Hoskin P, Schild SE, et al. Single vs multiple fraction palliative radiation therapy for bone metastases: cumulative meta-analysis. Radiother Oncol 2019;141:56–61. Available from: https://doi.org/10.1016/j.radonc.2019.06.037.

[28] Van den Hout WB, van der Linden YM, Steenland E, et al. Single-versus multiple-fraction radiotherapy in patients with painful bone metastases: cost-utility analysis based on a randomized trial. J Natl Cancer Inst 2003;95(3):222–9. Available from: https://doi.org/10.1093/jnci/95.3.222.

[29] Konski A, James J, Hartsell W, et al. Economic analysis of radiation therapy oncology group 97-14: multiple versus single fraction radiation treatment of patients with bone metastases. Am J Clin Oncol 2009;32(4):423–8. Available from: https://doi.org/10.1097/COC.0b013e31818da9f7.

[30] Collinson L, Kvizhinadze G, Nair N, et al. Economic evaluation of single-fraction versus multiple-fraction palliative radiotherapy for painful bone metastases in breast, lung, and prostate cancer. J Med Imaging Radiat Oncol 2016;60(5):650–60. Available from: https://doi.org/10.1111/1754-9485.12467.

[31] Alcorn S, Cortés ÁA, Bradfield L, et al. External beam radiation therapy for palliation of symptomatic bone metastases: an ASTRO clinical practice guideline. Pract Radiat Oncol 2024;14(5):377–97. Available from: https://doi.org/10.1016/j.prro.2024.04.018.

[32] Falkmer U, Järhult J, Wersäll P, et al. A systematic overview of radiation therapy effects in skeletal metastases. Acta Oncol 2003;42(5-6):620–33. Available from: https://doi.org/10.1080/02841860310014895.

[33] Osborn VW, Lee A, Yamada Y. Stereotactic body radiation therapy for spinal malignancies. Technol Cancer Res Treat 2018;17. Available from: https://doi.org/10.1177/1533033818802304.

[34] Sahgal A, Weinberg V, Ma L, et al. Probabilities of radiation myelopathy specific to stereotactic body radiation therapy to guide safe practice. Int J Radiat Oncol Biol Phys 2013;85(2):341–7. Available from: https://doi.org/10.1016/j.ijrobp.2012.05.007.

[35] Gong Y, Xu L, Zhuang H, et al. Efficacy and safety of different fractions in stereotactic body radiotherapy for spinal metastases: a systematic review. Cancer Med 2019;8(14):6176–84. Available from: https://doi.org/10.1002/cam4.2546.

[36] Chakravarthy VB, Khan HA, Srivatsa S, et al. Factors associated with adjacent-level tumor progression in patients receiving surgery followed by radiosurgery for metastatic epidural spinal cord compression. Neurosurg Focus 2021;50(5):E15. Available from: https://doi.org/10.3171/2021.2.FOCUS201097.

[37] Chang UK, Cho WI, Kim MS, et al. Local tumor control after retreatment of spinal metastasis using stereotactic body radiotherapy; comparison with initial treatment group. Acta Oncol 2012;51(5):589–95. Available from: https://doi.org/10.3109/0284186X.2012.666637.

[38] Ryu S, Jin R, Jin JY, et al. Pain control by image-guided radiosurgery for solitary spinal metastasis. J Pain Symptom Manage 2008;35(3):292–8. Available from: https://doi.org/10.1016/j.jpainsymman.2007.04.020.

[39] Sprave T, Verma V, Förster R, et al. Randomized phase II trial evaluating pain response in patients with spinal metastases following stereotactic body radiotherapy versus three-dimensional conformal radiotherapy. Radiother Oncol 2018;128(2):274–82. Available from: https://doi.org/10.1016/j.radonc.2018.04.030.

[40] Gibbs IC, Kamnerdsupaphon P, Ryu MR, et al. Image-guided robotic radiosurgery for spinal metastases. Radiother Oncol 2007;82(2):185–90. Available from: https://doi.org/10.1016/j.radonc.2006.11.023.

[41] Sierko E, Hempel D, Zuzda K, et al. Personalized radiation therapy in cancer pain management. Cancers (Basel) 2019;11(3):390. Available from: https://doi.org/10.3390/cancers11030390.

[42] Sahgal A, Myrehaug SD, Siva S, et al. Stereotactic body radiotherapy versus conventional external beam radiotherapy in patients with painful spinal metastases: an open-label, multicentre, randomised, controlled, phase 2/3 trial. Lancet Oncol 2021;22(8):1023–33. Available from: https://doi.org/10.1016/S1470-2045(21)00196-0.

[43] Ryu S, Deshmukh S, Timmerman RD, et al. Stereotactic radiosurgery vs conventional radiotherapy for localized vertebral metastases of the spine: phase 3 results of NRG oncology/RTOG 0631 randomized clinical trial. JAMA Oncol 2023;9(4):489–96. Available from: https://doi.org/10.1001/jamaoncol.2023.0356.

[44] Wong HCY, Lee SF, Chan AW, et al. Stereotactic body radiation therapy versus conventional external beam radiotherapy for spinal metastases: a systematic review and meta-analysis of randomized controlled trials. Radiother Oncol 2023;189:109914. Available from: https://doi.org/10.1016/j.radonc.2023.109914.

[45] Guninski RS, Cuccia F, Alongi F, et al. Efficacy and safety of SBRT for spine metastases: a systematic review and meta-analysis for preparation of an ESTRO practice guideline. Radiother Oncol 2024;190:109969. Available from: https://doi.org/10.1016/j.radonc.2023.109969.

[46] Guckenberger M, Andratschke N, Belka C, et al. ESTRO clinical practice guideline: stereotactic body radiotherapy for spine metastases. Radiother Oncol 2024;190:109966. Available from: https://doi.org/10.1016/j.radonc.2023.109966.

[47] Moraes FY, Gouveia AG, Marta GN, et al. Meta-analysis of stereotactic body radiation therapy in nonspine BONE metastases (MASTER-BONES). Int J Radiat Oncol Biol Phys 2024;119(5):1403–12. Available from: https://doi.org/10.1016/j.ijrobp.2023.12.045.

[48] Nag S, Shah V. Once-a-week lower hemibody irradiation (HBI) for metastatic cancers. Int J Radiat Oncol Biol Phys 1986;12(6):1003–5. Available from: https://doi.org/10.1016/0360-3016(86)90398-6.

[49] Sas-Korczynska B, Rzeznik D, Wojcik M, et al. Half body irradiation (HBI) in palliative treatment of multiple cancer metastases – contemporary evaluation. Contemp Oncol (Pozn) 2002;6(6):395–9.
[50] Skolyszewski J, Sas-Korczynska B, Korzeniowski S, et al. The efficiency and tolerance of half-body irradiation (HBI) in patients with multiple metastases: the Krakow experience. Strahlenther Onkol 2001;177(9):482–6. Available from: https://doi.org/10.1007/PL00002430.
[51] Zamagni A, Buwenge M, Macchia G, et al. Accelerated middle half body radiotherapy in bone metastases from prostate cancer: a phase I study (SHARON project). Anticancer Res 2019;39(9):5065–9. Available from: https://doi.org/10.21873/anticanres.13699.
[52] Furlan C, Trovo M, Drigo A, et al. Half-body irradiation with tomotherapy for pain palliation in metastatic breast cancer. J Pain Symptom Manage 2014;47(1):174–80. Available from: https://doi.org/10.1016/j.jpainsymman.2013.02.022.
[53] Kluska A, Tomasik B, Osadnik A, et al. Retrospective analysis of the effectiveness of volumetric modulated arc therapy half-body irradiation in palliation of pain in patients with multiple bone metastases. Mol Clin Oncol 2022;17(5):90. Available from: https://doi.org/10.3892/mco.2022.2586.
[54] Pal S, Dutta S, Adhikary S, et al. Hemi body irradiation: an economical way of palliation of pain in bone metastasis in advanced cancer. South Asian J Cancer 2014;3(1):28–32. Available from: https://doi.org/10.4103/2278-330X.126513.
[55] Bilski M, Konat-Bąska K, Mastroleo F, et al. Half body irradiation (HBI) for bone metastases in the modern radiotherapy technique era – a systematic review. Clin Transl Radiat Oncol 2024;49:100845. Available from: https://doi.org/10.1016/j.ctro.2024.100845.
[56] Hannoun-Lévi JM, Chargari C, Blanchard P, et al. Best practice in brachytherapy. Cancer Radiother 2022;26(1-2):29–33. Available from: https://doi.org/10.1016/j.canrad.2021.09.003.
[57] Zuckerman SL, Lim J, Yamada Y, et al. Brachytherapy in spinal tumors: a systematic review. World Neurosurg 2018;118:e491–501. Available from: https://doi.org/10.1016/j.wneu.2018.06.166.
[58] Wang W, Liu Z, Zhu J, et al. Brachytherapy with iodine 125 seeds for bone metastases. J Cancer Res Ther 2017;13(5):742–7. Available from: https://doi.org/10.4103/jcrt.JCRT_399_17.
[59] Yao Y, Li Z, Jiao D, et al. Palliative local treatment of bone metastases by 125I seed brachytherapy under DynaCT guidance: single-center experience. Diagn Interv Radiol 2021;27(4):558–63. Available from: https://doi.org/10.5152/dir.2021.20769.
[60] Li H, Wang M, Zhu Z, et al. Application value of the treatment of breast cancer bone metastases with radioactive seed 125I implantation under CT-guidance. BMC Med Imaging 2022;22(1):3. Available from: https://doi.org/10.1186/s12880-021-00726-w.
[61] Liu Y, Zhang C, Xu K, et al. 125I brachytherapy: a useful treatment to control painful osteoblastic bone metastases. Clin Transl Oncol 2023;25(5):1297–306. Available from: https://doi.org/10.1007/s12094-022-03025-0.
[62] Cole JS, Patchell RA. Metastatic epidural spinal cord compression. Lancet Neurol 2008;7(5):459–66. Available from: https://doi.org/10.1016/S1474-4422(08)70089-9.
[63] Martenson JA, Evans RG, Lie MR, et al. Treatment outcome and complications in patients treated for malignant epidural spinal cord compression (SCC). J Neurooncol 1985;3(1):77–84. Available from: https://doi.org/10.1007/BF00165175.
[64] Young RF, Post EM, King GA. Treatment of spinal epidural metastases. Randomized prospective comparison of laminectomy and radiotherapy. J Neurosurg 1980;53(6):741–8. Available from: https://doi.org/10.3171/jns.1980.53.6.0741.
[65] Patchell RA, Tibbs PA, Regine WF, et al. Direct decompressive surgical resection in the treatment of spinal cord compression caused by metastatic cancer: a randomised trial. Lancet 2005;366(9486):643–8. Available from: https://doi.org/10.1016/S0140-6736(05)66954-1.
[66] Klimo P, Thompson CJ, Kestle JRW, et al. A meta-analysis of surgery versus conventional radiotherapy for the treatment of metastatic spinal epidural disease. Neuro Oncol 2005;7(1):64–76. Available from: https://doi.org/10.1215/S1152851704000262.
[67] Oldenburger E, Brown S, Willmann J, et al. ESTRO ACROP guidelines for external beam radiotherapy of patients with complicated bone metastases. Radiother Oncol 2022;173:240–53. Available from: https://doi.org/10.1016/j.radonc.2022.06.002.
[68] Thirion PG, Dunne MT, Kelly PJ, et al. Non-inferiority randomised phase 3 trial comparing two radiation schedules (single vs five fractions) in malignant spinal cord compression. Br J Cancer 2020;122(9):1315–23. Available from: https://doi.org/10.1038/s41416-020-0768-z.
[69] Hoskin PJ, Hopkins K, Misra V, et al. Effect of single-fraction vs multifraction radiotherapy on ambulatory status among patients with spinal canal compression from metastatic cancer: the SCORAD randomized clinical trial. JAMA 2019;322(21):2084–94. Available from: https://doi.org/10.1001/jama.2019.17913.
[70] Rades D, Lange T, Veninga T, et al. Final results of a prospective study comparing the local control of short-course and long-course radiotherapy for metastatic spinal cord compression. Int J Radiat Oncol Biol Phys 2011;79(2):524–30. Available from: https://doi.org/10.1016/j.ijrobp.2009.10.073.
[71] Rades D, Dunst J, Schild SE. The first score predicting overall survival in patients with metastatic spinal cord compression. Cancer 2008;112(1):157–61. Available from: https://doi.org/10.1002/cncr.23150.
[72] Rades D, Cacicedo J, Lomidze D, et al. A new and easy-to-use survival score for patients irradiated for metastatic epidural spinal cord compression. Pract Radiat Oncol 2022;12(4):354–64. Available from: https://doi.org/10.1016/j.prro.2022.03.012.
[73] Rades D, Cacicedo J, Conde-Moreno AJ, et al. Precision radiation therapy for metastatic spinal cord compression: final results of the PRE-MODE trial. Int J Radiat Oncol Biol Phys 2020;106(4):780–9. Available from: https://doi.org/10.1016/j.ijrobp.2019.11.401.
[74] Rades D, Hansen O, Jensen LH, et al. Radiotherapy for metastatic spinal cord compression with increased radiation doses (RAMSES-01): a prospective multicenter study. BMC Cancer 2019;19(1):1163. Available from: https://doi.org/10.1186/s12885-019-6390-x.
[75] Hu YC, Lun DX, Wang H. Clinical features of neoplastic pathological fracture in long bones. Chin Med J (Engl) 2012;125(17):3127–32.
[76] Itshayek J, Yamada Y, Bilsky M, et al. Timing of surgery and radiotherapy in the management of metastatic spine disease: a systematic review. Int J Oncol 2010;36(3):533–44. Available from: https://doi.org/10.3892/ijo_00000527.

[77] Phillips FM, Ho E, Campbell-Hupp M, et al. Early radiographic and clinical results of balloon kyphoplasty for the treatment of osteoporotic vertebral compression fractures. Spine (Phila Pa 1976) 2003;28(19):2260–5. Available from: https://doi.org/10.1097/01.BRS.0000085092.84097.7B.
[78] Gerszten PC, Germanwala A, Burton SA, et al. Combination kyphoplasty and spinal radiosurgery: a new treatment paradigm for pathological fractures. J Neurosurg Spine 2005;3(4):296–301. Available from: https://doi.org/10.3171/spi.2005.3.4.0296.
[79] Gu YF, Li YD, Wu CG, et al. Safety and efficacy of percutaneous vertebroplasty and interventional tumor removal for metastatic spinal tumors and malignant vertebral compression fractures. AJR Am J Roentgenol 2014;202(3):W298–305. Available from: https://doi.org/10.2214/AJR.12.10497.
[80] Hulme PA, Krebs J, Ferguson SJ, et al. Vertebroplasty and kyphoplasty: a systematic review of 69 clinical studies. Spine (Phila Pa 1976) 2006;31(17):1983–2001. Available from: https://doi.org/10.1097/01.brs.0000229254.89952.6b.
[81] Chow E, van der Linden YM, Roos D, et al. Single versus multiple fractions of repeat radiation for painful bone metastases: a randomised, controlled, non-inferiority trial. Lancet Oncol 2014;15(2):164–71. Available from: https://doi.org/10.1016/S1470-2045(13)70556-4.
[82] Myrehaug S, Sahgal A, Hayashi M, et al. Reirradiation spine stereotactic body radiation therapy for spinal metastases: systematic review. J Neurosurg Spine 2017;27(4):428–35. Available from: https://doi.org/10.3171/2017.2.SPINE16976.
[83] Maranzano E, Trippa F, Casale M, et al. Reirradiation of metastatic spinal cord compression: definitive results of two randomized trials. Radiother Oncol 2011;98(2):234–7. Available from: https://doi.org/10.1016/j.radonc.2010.12.011.
[84] Rades D, Stalpers LJ, Veninga T, et al. Spinal reirradiation after short-course RT for metastatic spinal cord compression. Int J Radiat Oncol Biol Phys 2005;63(3):872–5. Available from: https://doi.org/10.1016/j.ijrobp.2005.03.034.
[85] Fairchild A, Harris K, Barnes E, et al. Palliative thoracic radiotherapy for lung cancer: a systematic review. J Clin Oncol 2008;26(24):4001–11. Available from: https://doi.org/10.1200/JCO.2007.15.3312.
[86] Tey J, Choo BA, Leong CN, et al. Clinical outcome of palliative radiotherapy for locally advanced symptomatic gastric cancer in the modern era. Medicine (Baltimore) 2014;93(24):e118. Available from: https://doi.org/10.1097/MD.0000000000000118.
[87] Rominger CJ, Gelber RD, Gunderson LL, et al. Radiation therapy alone or in combination with chemotherapy in the treatment of residual or inoperable carcinoma of the rectum and rectosigmoid or pelvic recurrence following colorectal surgery. Am J Clin Oncol 1985;8(2):118–27. Available from: https://doi.org/10.1097/00000421-198504000-00003.
[88] Bae SH, Park W, Choi DH, et al. Palliative radiotherapy in patients with a symptomatic pelvic mass of metastatic colorectal cancer. Radiat Oncol 2011;6:52. Available from: https://doi.org/10.1186/1748-717X-6-52.
[89] Rosati LM, Herman JM. Role of stereotactic body radiotherapy in the treatment of elderly and poor performance status patients with pancreatic cancer. J Oncol Pract 2017;13(3):157–66. Available from: https://doi.org/10.1200/JOP.2016.020628.
[90] Ng SP, Koay EJ. Current and emerging radiotherapy strategies for pancreatic adenocarcinoma: stereotactic, intensity modulated, and particle radiotherapy. Ann Pancreat Cancer 2018;1:22. Available from: https://doi.org/10.21037/apc.2018.07.03.
[91] Ryan JF, Rosati LM, Groot VP, et al. Stereotactic body radiation therapy for palliative management of pancreatic adenocarcinoma in elderly and medically inoperable patients. Oncotarget 2018;9(22):16427–36. Available from: https://doi.org/10.18632/oncotarget.24713.
[92] Thureau S, Marchesi V, Vieillard MH, et al. Efficacy of extracranial stereotactic body radiation therapy (SBRT) added to standard treatment in patients with solid tumors (breast, prostate and non-small cell lung cancer) with up to 3 bone-only metastases: study protocol for a randomised phase III trial (STEREO-OS). BMC Cancer 2021;21(1):117. Available from: https://doi.org/10.1186/s12885-021-07828-2.
[93] Randomized trial comparing conventional radiotherapy with stereotactic radiotherapy in patients with bone metastases—VERTICAL study (VERTICAL), <https://clinicaltrials.gov/ct2/show/NCT02364115>; 2024 [accessed 08.11.24].
[94] Sprave T, Welte SE, Bruckner T, et al. Intensity-modulated radiotherapy with integrated-boost in patients with bone metastasis of the spine: study protocol for a randomized controlled trial. Trials 2018;19:59. Available from: https://doi.org/10.1186/s13063-018-2452-7.
[95] Mascia AE, Daugherty EC, Zhang Y, et al. Proton FLASH radiotherapy for the treatment of symptomatic bone metastases: the FAST-01 nonrandomized trial. JAMA Oncol 2023;9(1):62–9. Available from: https://doi.org/10.1001/jamaoncol.2022.5843.
[96] Habberstad R, Frøseth TCS, Aass N, et al. The Palliative Radiotherapy and Inflammation Study (PRAIS) – protocol for a longitudinal observational multicenter study on patients with cancer induced bone pain. BMC Palliat Care 2018;17(1):110. Available from: https://doi.org/10.1186/s12904-018-0362-9.

Chapter 20

Nuclear medicine treatment

Knut Liepe
Department of Nuclear Medicine, Klinikum Frankfurt (Oder), Frankfurt (Oder), Germany

Key learning objectives

- To introduce bone pain palliation using bone-seeking agents in the management of pain in patients with cancer.
- To understand the mechanisms of action of different bone-seeking agents.
- To gain an overview of bone-seeking agents currently in use and their effect on pain relief.
- To learn the indication for treatment with bone-seeking agents and their side effects.

Introduction

While radiopharmaceuticals have multiple applications in oncology, this chapter will specifically address their most common use in pain management: the treatment of bone metastases. Skeletal metastases occur in many types of solid malignancies, particularly in advanced stages of prostate, breast, and lung cancer [1]. The resulting bone pain reduces overall survival and quality of life (QoL) through immobilization, isolation, and depression [2,3]. Treatment strategies include pain management, cancer-specific systemic therapy, local radiotherapy, and bone pain palliation using bone-seeking agents.

Osteoblastic bone metastases with evidence of uptake of uptake on bone scintigraphy are suitable for treatment with bone-seeking agents. Typical entities leading to skeletal metastases in advanced stages are prostate cancer with 65%–85% of bone metastases, breast cancer with 65%–75%, and small-cell lung cancer (SCLC) with 34%–50%, respectively [1,3,4]. The pathophysiology of bone pain is not fully understood. However, several key mechanisms are believed to contribute, including tumor-induced osteolysis with infiltration of the bone trabeculae and matrix, the development of microfractures, and mechanical stretching of the periosteum due to tumor growth [2,5,6]. Biochemical mechanisms of pain include the stimulation of nerve endings in the endosteum by a variety of chemical mediators, including bradykinin, prostaglandin, histamine, interleukin, and tumor necrosis factor produced by the osteolytic process [1,3,4].

Systemic radiotherapy with bone-seeking agents is effective in delivering high doses of radiation to widespread metastatic bone lesions and may limit the dose to healthy tissue [8]. Bone-seeking agents can be divided into volume seekers and surface seekers. The alkaline earth elements (e.g., calcium, strontium, radium, and barium) are traditionally considered to be volume seekers, and are readily absorbed into the bone regions where active mineralization is occurring, while surface seekers usually bind to the bone surface at sites of active turnover [9,10].

History of bone-seeking agents

Palliation of bone pain with bone-seeking agents has been used for more than 75 years. The publications showed pain relief of >70% and a pain-free rate of 10%–30% (Table 20.1). The specific uptake of ^{32}P in the red bone marrow results in a high bone marrow toxicity, and so there are only a few publications on that topic [12–14]. From the 1970s to the 1990s, several reports were published on the efficacy of ^{89}Sr in patients with bone pain, especially in patients with breast and prostate cancer. Robinson et al. included 622 patients (466 with prostate cancer, 96 with breast cancer, and 60 others) in their study. A partial response in pain level was documented in 80%, of whom 15% were pain-free [15,16]. A randomized, double-blind "Trans Canada" trial demonstrated the benefits of ^{89}Sr as an adjunct to conventional radiotherapy. The study involved hormonally resistant prostate cancer patients (HRCP; $n = 126$), who all had received local field radiotherapy and

Pain Management in Palliative Care for Patients With Cancer. DOI: https://doi.org/10.1016/B978-0-443-30050-9.00048-5

TABLE 20.1 Summary of pain relief efficacy of the different bone-seeking agents [11].

Bone seekers	References	No. of pts	Activity	Pain relief
^{89}Sr	Pons et al.	76	150 MBq	89%
	Kraeber-Bodere et al.	94	150 MBq	78%
	Dafermou et al.	527	150 MBq	80%
	Liepe et al.	15	150 MBq	72%
	Ma et al.	116	1.5–2.2 MBq/kg	84%
^{223}Ra	Nilsson	100	5–100 kBq/kg	50%–74%
^{32}P	Nair et al.	16	444 MBq (oral)	88%
	Fettich et al.	93		60%–75%
^{186}Re-HEDP	Maxon et al.	43	1295 MBq	60%
	Sciuto et al.	60	1406 MBq	80%
	Dafermou et al.	58	1295 MBq	86%
	Liepe et al.	15	1295 MBq	67%
^{153}Sm-EDTMP	Serafini et al.	118	18.5 or 37.0 MBq/kg	62%–72%
	Tian et al.	105	37 MBq/kg	84%
	Enrique et al.	417	18.5/37.0/55 MBq/kg	73%
	Sartor et al.	152	37 MBq/kg	65%
^{188}Re-HEDP	Palmedo et al.	22	1.3–4.4 GBq	70%
	Liepe et al.	37	3.3 GBq	77%
^{117m}Sn-DTPA	Srivastava et al.	47	2.6–10.6 MBq/kg	75%

were then randomized to placebo or a single injection of 400 MBq (10.8 mCi) of ^{89}Sr. There were no differences in overall survival. However, in the ^{89}Sr group, statistically significant and clinically important effects were reported, including reduced analgesic and radiotherapy requirements [17].

The importance of an early treatment of cancer patients with the systemic radionuclide therapy is underlined by the work of Schmeler et al. In this study the pretreatment factors (Karnofsky Performance Status [KPS]) were correlated with outcome, represented by survival and pain response. Limited survival (mean survival of 12.5 weeks) and poor response (in only 40% of patients) were observed in the group with a KPS of ≤50%. Patients with a pretreatment KPS of 60% had a mean survival of 20.5 weeks after ^{89}Sr, and 42% of patients experienced pain relief. Patients with a KPS ≥ 70% had a response rate of approximately 75%. The authors concluded that patients with a pretreatment KPS ≤ 50% should not be treated with ^{89}Sr, and patients with a pretreatment KPS of 60% should be evaluated on a case-by-case basis [18].

A large Italian multicenter study of 818 patients treated with ^{89}Sr (n = 527) and ^{186}Re-HEDP (n = 83) showed a response rate of 81%, favoring patients with limited skeletal metastases, higher KPS, and radiological osteoblastic metastatic lesions. Retreatments showed a significantly ($P < .01$) worse response rate of 48% compared to initial treatment. The duration of palliation was 5.0 ± 3.5 months and was longer in cases of excellent response, in patients with limited metastases, and when ^{89}Sr was used. Better responses were found in cases of limited skeletal disease, under good clinical conditions, when life expectancy was greater than 3 months, and in radiologic osteoblastic or mixed bone lesions. The only statistically significant predictive factor was life expectancy ($P < .001$). The two radiopharmaceuticals did not show any statistically significant differences in palliative efficacy and toxicity, either in initial treatment or in subsequent treatments [19]. This is supported by other reports that also found no differences between these two bone seekers [20].

Side effects were limited, and while a flare phenomenon was observed in 14% of cases, bone marrow toxicity was mild and reversible, mainly affecting platelets. Thrombocytopenia grade I was reported in 14.0% of patients and grade II in 7.4%. Toxicity did not correlate with pain response [19,21]. ^{89}Sr has similar toxicity to the other common radiopharmaceuticals, such as ^{186}Re-HEDP, ^{188}Re-HEDP, and ^{153}Sm-EDTMP [22]. Due to the long physical half-life, the nadir of limited thrombocytopenia occurs between the fourth and sixth week with a broad recovery by 12 weeks [23].

Currently approved bone-seeking agents

Surface-seeking bone agents include various radionuclide-labeled bisphosphonates, which include agents such as ^{153}Sm-ethylenediaminetetramethylene phosphonate (^{153}Sm-EDTMP) and ^{186}Re-hydroxyethylidene diphosphonate (^{186}Re-HEDP). Recent research has explored the use of zoledronic acid labeled with ^{188}Re or ^{177}Lu [24–28].

Most commonly used radionuclides for bone pain palliation include $^{223}RaCl_2$ and ^{153}Sm-EDTMP [29,30].

^{223}Ra-dichloride

As an alkaline earth metal, ^{223}Ra-dichloride ($^{223}RaCl_2$) mimics the calcium uptake in bone [31]. The α-emitters have a high linear energy transfer and result in a high incidence of DNA double-strand breaks, leading to a strong tumor cell–killing effect [32]. The short track length of 2–10 cell diameters minimizes the cell damage effect in surrounding healthy tissue but also reduces the tumor-killing effect to a very short distance in the bone. However, the bystander effect leads to an effect of the alpha emitter $^{223}RaCl_2$ up to a distance of 1 mm from the irradiation line. The bystander effect is defined as "radiation-induced, signal-mediated effects in unirradiated cells within an irradiated volume" [33,34]. These signal-mediated bystander effects include cell death, DNA damage, chromatid aberrations, genomic instability, transformation, differentiation, proliferation, gene expression, cell cycle, invasion, and radioadaptive responses [35]. The release of signaling molecules and direct intercellular communication via gap junctions are the two primary mechanisms involved [36,37]. Cell proximity has also been shown to be necessary for proliferative bystander responses [38]. Reactive oxygen and nitrogen species together with calcium and cytokines have all been implicated in bystander signaling [39].

Based on the data from the ALSYMPCA trial, the European Medicines Agency (EMA) approved $^{223}RaCl_2$ as a first-line treatment for castration-resistant prostate cancer (CRPC) in November 2013 [30]. $^{223}RaCl_2$ was approved by the FDA for the same indication in May 2013. While the FDA also approved 223RaCl2 in May 2013 for the first-line treatment, EMA later restricted its use to patients who have received at least two prior systemic mCRPC treatments or those unable to receive other treatments. This double-blind, randomized, phase III trial (2008–2011) enrolled 921 CRPC patients. Inclusion criteria included bone pain and at least two bone metastases, without metastases outside the skeleton [40,41]. The study design consisted of six administrations of 50 kBq/kg $^{223}RaCl_2$ with a 4-week interval. Patients were assigned in a 2:1 ratio to receive radium or best standard of care (BSC). The $^{223}RaCl_2$ group provided a longer overall survival (median 14.9 vs 11.3 months; hazard ratio [HR] 0.70, 95% CI 0.58–0.83, $P < .001$) and longer time to first symptomatic skeletal event (median 15.6 vs 9.8 months, HR 0.66, 95% CI 0.52–0.83, $P < .001$) than the placebo group. The short- and long-term safety profiles of patients treated with $^{223}RaCl_2$ were favorable compared to those who received BSC, with improvements in quality of life and a low rate of myelosuppression [42,43].

Smith et al. published the results of the ERA-223 trial in 2019. This randomized, double-blind, placebo-controlled, phase 3 trial analyzed data from 65 oncology and urology centers in 19 countries. A total of 806 chemotherapy-naive CRPC patients were randomized to receive $^{223}RaCl_2$ ($n = 401$) or placebo ($n = 405$) in addition to abiraterone plus prednisone or prednisolone. In the $^{223}RaCl_2$ group, 49% had at least one symptomatic skeletal event or death ($n = 190$) compared to 47% in the placebo group (HR 1.122, $P = .2636$). Unfortunately, the rate of symptomatic pathological fractures was higher in the treatment group at 18% compared to 9% in the placebo group [44]. These data led to a change in the indication for $^{223}RaCl_2$ in Europe. Currently, $^{223}RaCl_2$ is approved as a third-line treatment for patients who have received two previous treatments for metastatic prostate cancer or who cannot receive other treatments. It must also not be used with the medicines Zytiga (abiraterone acetate) and the corticosteroid prednisone or prednisolone [30].

^{153}Sm-ethylenediaminetetramethylene phosphonate

^{153}Sm is produced in high radionuclidic purity in a reactor by neutron bombardment of enriched ^{152}Sm oxide [45]. A dose-escalation study from 10 MBq/kg to 36 MBq/kg of ^{153}Sm-ethylenediaminetetramethylene phosphonate (^{153}Sm-EDTMP) reported a response rate of 65% for pain relief and a duration of response ranging from 4 to 35 weeks [45]. These data were subsequently confirmed in a phase II study by the same authors, which showed a response rate of 61% with a median duration of response of 8 weeks [46].

Larger studies with more than 100 patients showed a median therapeutic efficacy of 80%. In a randomized, double-blind, placebo-controlled study ($n = 152$), pain relief was reported in 65% of patients treated with ^{153}Sm-EDTMP compared to 45% of patients treated with placebo. A significant reduction in pain between ^{153}Sm-EDTMP and placebo was reported after 1 week, and the analgesic intake was significantly reduced at the third and fourth weeks [47]. A trial involving 136 patients with various tumor types, mainly breast and lung cancer, reported a response rate of 93%, with 36% of patients being pain-free [48]. A randomized, dose-controlled study included 114 patients with an applied activity of 17.5 or 37 MBq/kg of ^{153}Sm-EDTMP. A significant difference in pain relief was documented between the 17.5 MBq/kg group (55% of patients at 4 weeks) and the 37 MBq/kg group (70% of patients at 4 weeks; $P = .0476$) [49]. In 118 patients treated with 17.5 MBq/kg of ^{153}Sm-EDTMP ($n = 40$ patients) or 37 MBq/kg of ^{153}Sm-EDTMP or placebo (each $n = 39$ patients), pain relief in 72% of patients treated with 37 MBq/kg of ^{153}Sm-EDTMP was described, with a significant difference ($P = .01$) in reduction of opioid analgesic intake and decrease of pain score compared to placebo [50].

Two large studies with ^{153}Sm-EDTMP involving more than 550 patients reported response rates of 73% and 86% (Table 20.1) [51,52]. Enrique et al. documented pain relief with a duration of response of 4–8 weeks in 50% of patients and more than 8 weeks in the other 50% of patients [52]. In a retrospective study, hematological and nonhematological side effects of patients treated with ^{153}Sm-EDTMP ($n = 103$) and the effect of additional chemotherapy and external beam radiotherapy were observed. Chemotherapy and/or external beam radiotherapy did not affect the extent or duration of myelosuppression. The most common nonhematological adverse events were acute lower limb edema ($n = 27$) and acute and transient neuropathy ($n = 29$ [53]).

Bone-seeking agents not currently in routine use

^{186}Re-1,1-hydroxyethylidene diphosphonate

^{186}Rhenium is a reactor product produced by irradiating enriched ^{185}rhenium. ^{186}Rhenium decays with beta emission and a maximum energy of 1.07 MeV, with a low abundance 137 keV gamma emission.

An early study by Maxon et al. using 1285 MBq of ^{186}Re-1,1-hydroxyethylidene diphosphonate (^{186}Re-HEDP) documented an overall pain relief in 80% of patients, with a mean duration of 7 weeks in HRCP patients (Table 20.1) [54]. A double-blind, placebo-controlled, randomized study (Placorhen study) included 111 HRCP patients. The reduction of the pain level from baseline (scaled from 0% to 100%) was significantly different ($P = .039$) between the two arms. In the placebo group, the main pain index score increased during the study, whereas the main pain index score decreased in the ^{186}Re-HEDP group. The mean percentage of days of response after 6 and 12 weeks of follow-up was also significantly higher in patients treated with ^{186}Re-HEDP (27%) than in the placebo group (13%) [55]. Two studies investigated the effect of ^{186}Re-HEDP in breast cancer patients: Lam et al. included 30 females with breast cancer and used activities ranging from 1285 MBq up to a maximum of 2960 MBq, and improvement in activities of daily living was seen in 58% of patients. Palmedo et al. documented the results of 38 treatments with 1285 MBq of ^{186}Re-HEDP in 30 breast cancer patients, and a pain response was observed in 60% of patients [56,57].

A double-blind, crossover, placebo-controlled trial of ^{186}Re-HEDP reported a response rate of 80% of patients with a mean duration of response of 7 weeks. The only major toxicity was mild reversible thrombocytopenia [58].

^{188}Re-hydroxyethylidene diphosphonate

Radiolabeling of various agents with ^{188}Re offers a variety of different therapeutic options beyond systemic radionuclide therapy with bone-seeking agents [59–61].

When it is labeled with the diphosphonate 1-hydroxyethylidene-1,1-diphosphonic acid (HEDP), it can be used as a surface bone seeker agent. In 27 patients with prostate cancer who received 3.3 GBq of ^{188}Re-hydroxyethylidene diphosphonate (^{188}Re-HEDP), an increase in the Karnofsky Performance Scale from 74% ± 7% before therapy to 85% ± 9% at 12 weeks after therapy was obtained ($P = .001$). In addition, pain relief was achieved in 76% of patients, and 20% of patients were pain-free. The pain score showed a maximum decrease from 4.4% ± 1.8% to 2.7% ± 2.0% between 3 and 8 weeks after therapy ($P = .009$) [62]. ^{188}Re-HEDP is also effective in bone metastases from lung cancer. In a randomized phase 2 trial, 55 patients with previous bisphosphonate and external beam radiotherapy (interval of 12 weeks) were randomized to receive 30, 40, or 50 MBq/kg ^{188}Re-HEDP. Significant reductions in pain scores were observed in the 40 and 50 MBq/kg groups ($P = .033$ and $P = .046$, respectively). Significant improvements in quality of life were also seen in both dose groups. No patient withdrew from the trial due to adverse events in any of the three groups [63].

^{117m}Tin-stannic diethylenetriaminepentaacetic acid

The chelate ^{117m}Tin-diethylenetriaminepentaacetic acid (^{117m}Sn-DTPA) is an experimental radiopharmaceutical produced from an enriched ^{117m}Sn-DTPA target in the Oak Ridge National Laboratory. Preliminary studies have shown that ^{117m}Sn-DTPA is effective in relieving bone pain in patients with breast and prostate cancer and has minimal toxicity. A phase I/II study of ^{117m}Sn-DTPA conducted by Srivastava et al. in 47 patients with painful bone metastases from various cancers found an overall response rate of 75% (range, 60%–83%) in 40 evaluable patients. Patients were assigned to five different dose levels ranging from 2.63 to 10.58 MBq (71–286 μCi) per kilogram of body weight. Twelve (30%) patients experienced complete relief. The time to onset of pain relief was 19 ± 15 days with doses of 5.29 MBq/kg or less and 5 ± 3 days with doses of 6.61 MBq/kg or greater. Myelotoxic effects were minimal with only one patient [64].

Indications for treatment with bone-seeking radiopharmaceuticals

- Painful metastatic bone lesions with osteoblastic response, as confirmed by areas of intense uptake on radionuclide bone scans [65].
- Primary painful bone tumors with an osteoblastic response, as confirmed by areas of intense uptake on radionuclide bone scans. However, this indication is not yet approved [65].
- ^{99m}Tc-HMDP bone scan showing at least three metastases [65].
- Limiting indication for $^{223}RaCl_2$: patients with CRPC with symptomatic bone metastases and no known visceral metastases [66].
- In respect to the ERA-223 trial, the indication of $^{223}RaCl_2$ is limited as a third-line treatment since 2018. The ERA-223 trial showed an increased risk of fracture using abiraterone and prednisolone, so this combination is not allowed [44]. Administration of $^{223}RaCl_2$ should only be performed at least 1 week apart from abiraterone.
- In respect to the latency in the onset of the palliative effect life expectancy >3 months [65] and Karnofsky Performance Scale >50% [65].
- Blood cell counts should be stable before undertaking bone palliation therapy [65].

Contraindications

- Evidence of compromised bone marrow reserve, especially neutropenia and thrombocytopenia (except for $^{223}RaCl_2$): platelet count ≤100×10^9/L; leukocyte count ≤3.0×10^9/L; hemoglobin ≤100 g/L; absolute number of neutrophil granulocytes ≤1.5×10^9/L [65].
- Poor renal function due to reduction of the plasma clearance of bone-seeking radiopharmaceuticals, resulting in a higher whole-body dose and greater risk of myelotoxicity. Therefore, patients had severely reduced renal function, that is, creatinine >180 μmol/L and/or glomerular filtration rate <30 mL/min [65].
- Concomitant or sequential use of radionuclide therapy and bisphosphonates for the treatment of patients with painful osteoblastic or mixed pattern (osteoblastic/osteoclastic) bone metastases is not contraindicated [65].
- Pretreatment hematologic contraindications for $^{223}RaCl_2$:

Before the first administration/cycle: absolute neutrophil count ≤1.5×10^9/L; platelet count ≤100×10^9/L; hemoglobin ≤100 g/L.

Before subsequent administrations/cycles: neutrophil count ≤1.0×10^9/L; platelet count ≤50×10^9/L [66].

Adverse effects

For beta-emitting bone seekers:

- Flare phenomenon for beta-emitting bone seekers: increase in pain symptoms, usually occurring within 14 days of treatment and occurring in 10%–15% of patients [19,62,65]. It is mild and self-limiting in the majority of patients and usually responds to standard analgesics [65]. In general, a flare phenomenon is associated with a good clinical response [49,67–69].
- Bone marrow toxicity is mostly limited up to grade II and reversible 12 weeks after treatment with beta-emitting bone seekers [22,62]. Toxicity mainly affects the platelet counts, with leukopenia and anemia playing a minor role [22]. Depending on the physical half-life of applied radionuclides, the nadir varies from 3 to 6 weeks, shorter for ^{188}Re with a nadir of 3 weeks [62] up to 6 weeks for ^{89}Sr [22,70]. A higher degree of bone marrow toxicity could be observed in patients with previous myelosuppressive therapy [22,65].

TABLE 20.2 Summary of radiation absorbed dose to bone metastases, bone marrow, and total body for bone-seeking beta-emitters.

	Standard applied activity (MBq)	Radiation to bone metastases (mGy/MBq)	Absorbed dose to bone marrow (mGy/MBq)	Whole-body dose (mGy/MBq)	Dose to bone metastases (Gy)	Dose to bone marrow (Gy)	Whole-body dose (Gy)	Tumor-to-marrow dose ratio
^{89}Sr [15]	150		19	16	0.03–50	2.3	0.24	12.3
^{32}P [75]	444		6			2.6		
^{186}Re-HEDP [54,72]	1285		0.8	0.1	40.1	1.8	0.19	
^{153}Sm-EDTMP [64], [68]	37 MBq/kg BW		0.89 ± 0.03		2.9–14.1	2.3		4.4
^{188}Re-HEDP [109]	3300	3.8 ± 2.0	0.61 ± 0.21	0.07 ± 0.03	12.6 ± 6.6	2.3 ± 0.7	0.23 ± 0.07	6.3
^{188}Re zoledronic acid [27]	45 MBq/kg BW				4.7–43.9			
^{177}Lu-EDTMP [76]	37 MBq/kg BW		0.80 ± 0.15					8.3
^{177}Lu zoledronic acid [25]	5780	4.21 ± 2.40	0.36 ± 0.12		24.3 ± 13.9	2.1 ± 0.7		13.9

Because of the mainly feces excretion, patients treated with $^{223}RaCl_2$ reported more gastrointestinal symptoms, for example, nausea and diarrhea, in 36% and 25% of patients, respectively [28,41,43,71]. Other symptoms include fatigue (26%), loss of weight (12%), and peripheral edema (13%). The "flare" phenomenon was described in 50% of treated patients. In contrast to beta-emitters, marrow suppression is most commonly manifested as anemia, which was observed in 31% of patients. Thrombocytopenia (12%) and neutropenia (5%) were less commonly observed [28,41,43,71].

A summary of dose calculations

Dose calculations are useful for estimating the therapeutic effect, the potential toxic effects to critical organs, and for comparing different bone-seeking agents. The total absorbed dose is calculated for standard activities in radio-pharmaceuticals using a weight-adapted activity. A standard patient of 70 kg body weight was underlain (Table 20.2, [15,25,27,54,72–76]).

Combination of systemic and radionuclide therapy

Chemosensitization is a well-known method to improve efficacy of radionuclid treatment. The cytotoxic effect of chemotherapy makes tumor cells more susceptible to radiation damage and can be used to improve the overall efficacy of bone-seeking agents [77]. The first data on the combination of ^{89}Sr and 35 mg/m^2 cisplatin in 18 prostate cancer patients were published by Mertens et al. Good pain relief was documented in 55% with at least a 50% reduction in analgesic use. Toxicity appears to be mild, with no life-threatening complications [78]. Sciuto et al. reported a higher pain relief of 91% for ^{89}Sr plus 50 mg/m^2 cisplatin compared to ^{89}Sr alone [79]. A group from Houston also published results with ^{89}Sr in combination with chemotherapy, including doxorubicin, ketoconazole, paclitaxel, estramustine, and gemcitabine in CRPC patients. A higher rate of PSA response and pain relief was seen with the combination compared to ^{89}Sr alone [79–83]. The combination of 188Re-HEDP plus docetaxel/prednisone showed a prolonged survival compared to docetaxel/prednisone in CRPC metastatic to bone (P = .012) [84]. Further trials showed the same benefit of the combination to radionuclide therapy alone [85–87].

In the 1990s, bisphosphonate therapy was a contraindication for systemic radionuclide therapy. However, data showed that previous bisphosphonate therapy did not affect the uptake of bone-seeking agents [88–90]. More recent cell studies have reported a synergistic effect of bisphosphonates and bone-seeking agents reducing the enzymatic activity of osteoclasts, probably by reducing their ability to recognize and destroy the bone surface and their motility [91,92]. Other effects include a reduction in the number of osteoclasts, inhibition of the differentiation of osteoclastic precursors into mature osteoclasts, and induction of apoptosis in tumor cells [92–94]. Based on the available data, there is no evidence of biological competition between bisphosphonates and bone-seeking agents, and, therefore, they can be used concomitantly [77]. The first human data describing the combination of bisphosphonates and ^{89}Sr were published in 2002. The authors found a significant reduction in the incidence of spinal cord compression in HRCP patients treated with bisphosphonates and ^{89}Sr [95]. Storto et al. evaluated the pain response in patients treated with ^{89}Sr and zoledronic acid (group A) and in patients treated with ^{89}Sr alone (group B) or zoledronic acid (group C). The reduction in total discomfort in group A, as measured by a 10-point Visual Analog Scale, was significantly greater in group A than in group B or C [96].

Combination of external beam radiotherapy and radionuclide therapy

The combination of systemic radiotherapy and external beam radiotherapy may also be an option. A double-blind randomized study controlled the role of ^{89}Sr as an adjuvant to palliative external beam radiotherapy. Ninety-five patients were randomized to receive ^{89}Sr (arm A) or saline (arm B) on the day of external beam radiotherapy. At 3 and 6 months, no difference in progression rate was observed between the several arms. Based on quality-of-life assessments, 50% of patients experienced pain relief, and 32% experienced an improvement in global quality of life [97].

Future developments

New therapeutic options for the treatment of metastatic bone pain, such as endothelin-1 antagonists, osteoprotegerin, TGFβ mAbs, or EGFR inhibitors, offer competition to the therapies under development, although synergistic effects are conceivable [98–100]. This may be particularly true for EGFR , as there is evidence of radiosensitization following EGFR inhibition, and evaluation would certainly be of interest [101,102].

Several ongoing trials started using a combination of ^{223}Ra plus immunotherapy (avelumab, atezolizumab, nivolumab, and pembrolizumab) to control the safety and efficacy [103]. The combination of atezolizumab plus ^{223}Ra showed a greater toxicity without significant benefit for the patients compared to either drug alone. In addition, the combination of pembrolizumab plus ^{223}Ra did not improve efficacy but was well tolerated [104]. In contrast, the combination of sipuleucel showed a higher rate of patients with ≥50% PSA reduction. The COMRADE trial started as a dose escalation study of olaparib with a fixed activity of ^{223}Ra. The observed progression-free survival was 6 months in 58% of patients [105]. The combination of niraparib and ^{223}Ra was reported to be safe [106]. For most of the ongoing trials, we do not have clinical data on the therapeutic efficacy of the combination of ^{223}Ra plus immunotherapy or poly (ADP-ribose) polymerase (PARP) inhibitors [103,107,108]. Another interesting approach may be a direct labeling of the tumor cell by an antibody and an additional indirect radiation by crossfire induced by bone-seeking agents. The AlphaBet trial treated CRPC patients with ^{225}actinium-prostate-specific membrane antigen (PSMA) (direct labeling of the tumor cell) plus ^{153}Sm-EDTMP [107].

Conclusions

Systemic treatment with a bone-seeking agent is an established treatment for metastatic bone pain. To date, only ^{223}Ra has been reported to prolong survival, but we do not have statistically significant survival data for the other bone-seeking agents. ^{223}Ra has limited use as a third-line therapy in CRPC patients; ^{153}Sm-EDTMP can be used as a first-line therapy in all patients with multiple painful osteoblastic skeletal metastases that show uptake on bone scan, depending on country approvals, marrow reserve, performance status, and prior therapies. The treatment is well tolerated with reversible bone marrow toxicity within 3 months posttreatment. To improve the therapeutic efficacy, combination with other anticancer treatment modalities such as chemotherapy, antibodies, and PARP inhibitors is of interest. However, we must await clinical data from several ongoing trials.

References

[1] Liepe K, Shinto A. From palliative therapy to prolongation of survival: (223)$RaCl_2$ in the treatment of bone metastases. Ther Adv Med Oncol 2016;8(4):294–304. Available from: https://doi.org/10.1177/1758834016640494.

[2] Mantyh PW, Clohisy DR, Koltzenburg M, et al. Molecular mechanisms of cancer pain. Nat Rev Cancer 2002;2(3):201–9. Available from: https://doi.org/10.1038/nrc747.

[3] Saad F, Clarke N, Colombel M. Natural history and treatment of bone complications in prostate cancer. Eur Urol 2006;49(3):429–40. Available from: https://doi.org/10.1016/j.eururo.2005.12.045.

[4] Juzeniene A, Stenberg VY, Bruland ØS, et al. Dual targeting with (224)Ra/(212)Pb-conjugates for targeted alpha therapy of disseminated cancers: a conceptual approach. Front Med 2022;9:1051825. Available from: https://doi.org/10.3389/fmed.2022.1051825.

[5] Mundy GR. Metastasis to bone: causes, consequences and therapeutic opportunities. Nat Rev Cancer 2002;2(8):584–93. Available from: https://doi.org/10.1038/nrc867.

[6] Serafini AN. Current status of systemic intravenous radiopharmaceuticals for the treatment of painful metastatic bone disease. Int J Radiat Oncol Biol Phys 1994;30(5):1187–94. Available from: https://doi.org/10.1016/0360-3016(94)90327-1.

[7] Nielsen OS, Munro AJ, Tannock IF. Bone metastases: pathophysiology and management policy. J Clin Oncol 1991;9(3):509–24. Available from: https://doi.org/10.1200/JCO.1991.9.3.509.

[8] Breen SL, Battista JJ. Cavity theory applied to the dosimetry of systemic radiotherapy of bone metastases. Phys Med Biol 2000;45(4):879–96. Available from: https://doi.org/10.1088/0031-9155/45/4/305.

[9] Akabani G. Absorbed dose calculations in Haversian canals for several beta-emitting radionuclides. J Nucl Med 1993;34(8):1361–6.

[10] Spiers FW, Beddoe AH, Whitwell JR. Mean skeletal dose factors for beta-particle emitters in human bone. Part I: volume-seeking radionuclides. Br J Radiol 1978;51(608):622–7. Available from: https://doi.org/10.1259/0007-1285-51-608-622.

[11] Liepe K, Kotzerke J. Internal radiotherapy of painful bone metastases. Methods 2011;55(3):258–70. Available from: https://doi.org/10.1016/j.ymeth.2011.07.001.

[12] Burnet NG, Williams G, Howard N. Phosphorus-32 for intractable bony pain from carcinoma of the prostate. Clin Oncol (R Coll Radiol) 1990;2(4):220–3. Available from: https://doi.org/10.1016/s0936-6555(05)80172-4.

[13] Fettich J, Padhy AK, Nair N. Comparative clinical efficacy and safety of phosphorus and strontium-89 in the palliative treatment of metastatic bone pain: results of an IAEA coordination research project. World J Nucl Med 2003;2:226–31.

[14] Nair N. Relative efficacy of 32P and 89Sr in palliation in skeletal metastases. J Nucl Med 1999;40(2):256–61.

[15] Robinson RG, Blake GM, Preston DF, et al. Strontium-89: treatment results and kinetics in patients with painful metastatic prostate and breast cancer in bone. Radiographics 1989;9(2):271–81. Available from: https://doi.org/10.1148/radiographics.9.2.2467331.

[16] Robinson RG, Preston DF, Schiefelbein M, et al. Strontium 89 therapy for the palliation of pain due to osseous metastases. JAMA 1995;274(5):420–4. Available from: https://doi.org/10.1001/jama.1995.03530050068035.

[17] Porter AT, Davis LP. Systemic radionuclide therapy of bone metastases with strontium-89. Oncology (Willston Park) 1994;8(2):93–6; discussion 96, 99–101.

[18] Schmeler K, Bastin K. Strontium-89 for symptomatic metastatic prostate cancer to bone: recommendations for hospice patients. Hosp J 1996;11(2):1–10. Available from: https://doi.org/10.1080/0742-969x.1996.11882817.

[19] Dafermou A, Colamussi P, Giganti M, et al. A multicentre observational study of radionuclide therapy in patients with painful bone metastases of prostate cancer. Eur J Nucl Med 2001;28(7):788–98. Available from: https://doi.org/10.1007/s002590100533.

[20] Liepe K, Franke WG, Kropp J, et al. Comparison of rhenium-188, rhenium-186-HEDP and strontium-89 in palliation of painful bone metastases. Nuklearmedizin 2000;39(6):146–51.

[21] Cancer pain relief and palliative care. Report of a WHO Expert Committee. World Health Organ Tech Rep Ser 1990:804:1–75.

[22] Liepe K, Kotzerke J. A comparative study of 188Re-HEDP, 186Re HEDP, 153Sm-EDTMP and 89Sr in the treatment of painful skeletal metastases. Nucl Med Commun 2007;28:623–30. Available from: https://doi.org/10.1097/MNM.0b013e32825a6adc.

[23] Liepe K, Runge R, Kotzerke J. Systemic radionuclide therapy in pain palliation. Am J Hosp Palliat Care 2005;22(6):457–64. Available from: https://doi.org/10.1177/104990910502200613.

[24] Erfani M, Tabatabaei M, Doroudi A, et al. Radiolabeling of zoledronic acid with 188 Re as a new palliative agent radiotracer in treatment of bone tumors. J Radioanal Nucl Chem 2018;316:491–500. Available from: https://doi.org/10.1007/s10967-018-5781-9.

[25] Fernandez R, Eppard E, Lehnert W, et al. Evaluation of safety and dosimetry of (177)Lu DOTA-ZOL for therapy of bone metastases. J Nucl Med 2021;62:1. Available from: https://doi.org/10.2967/jnumed.120.255851.

[26] Khawar A, Eppard E, Roesch F, et al. Biodistribution and post-therapy dosimetric analysis of [(177)Lu]Lu-DOTA(ZOL) in patients with osteoblastic metastases: first results. EJNMMI Res 2019;9(1):102. Available from: https://doi.org/10.1186/s13550-019-0566-x.

[27] Kochetova T, Krylov V, Smolyarchuk M, et al. 188Re zoledronic acid in the palliative treatment of painful bone metastases. Int J Nucl Med Res 2017;7:92–100.

[28] Liepe K, Krylov VV, Kochetova TY. [Radionuclide therapy for bone metastases by drugs based on Re-188]. Vopr Onkol 2016;62(4):401–9.

[29] EMA. European medicines agency, <https://ec.europa.eu/health/documents/community-register/html/h057.htm>; 1998.

[30] EMA. European medicines agency, <http://www.ema.europa.eu/ema/index.jsp?curl=pages/medicines/human/medicines/002653/human_med_001692.jsp&mid=WC0b01ac058001d124>; 2013.

[31] Liepe K. Alpharadin, a 223Ra-based alpha-particle-emitting pharmaceutical for the treatment of bone metastases in patients with cancer. Curr Opin Investig Drugs 2009;10(12):1346–58.
[32] Henriksen G, Breistøl K, Bruland ØS, et al. Significant antitumor effect from bone-seeking, alpha-particle-emitting (223)Ra demonstrated in an experimental skeletal metastases model. Cancer Res 2002;62(11):3120–5.
[33] Liepe K. Pros and cons of alpha versus beta bone seeking agents in the treatment of cancer pain. World J Nucl Med 2023;22(4):255–6. Available from: https://doi.org/10.1055/s-0043-1774731.
[34] Blyth BJ, Sykes PJ. Radiation-induced bystander effects: what are they, and how relevant are they to human radiation exposures? Radiat Res 2011;176(2):139–57. Available from: https://doi.org/10.1667/rr2548.1.
[35] Brady D, O'Sullivan JM, Prise KM. What is the role of the bystander response in radionuclide therapies? Front Oncol 2013;3:215. Available from: https://doi.org/10.3389/fonc.2013.00215.
[36] Azzam EI, de Toledo SM, Gooding T, et al. Intercellular communication is involved in the bystander regulation of gene expression in human cells exposed to very low fluences of alpha particles. Radiat Res 1998;150(5):497–504. Available from: https://doi.org/10.2307/3579865.
[37] Mothersill C, Seymour C. Medium from irradiated human epithelial cells but not human fibroblasts reduces the clonogenic survival of unirradiated cells. Int J Radiat Biol 1997;71(4):421–7. Available from: https://doi.org/10.1080/095530097144030.
[38] Gerashchenko BI, Howell RW. Cell proximity is a prerequisite for the proliferative response of bystander cells co-cultured with cells irradiated with gamma-rays. Cytometry A 2003;56(2):71–80. Available from: https://doi.org/10.1002/cyto.a.10092.
[39] Prise KM, O'Sullivan JM. Radiation-induced bystander signalling in cancer therapy. Nat Rev Cancer 2009;9(5):351–60. Available from: https://doi.org/10.1038/nrc2603.
[40] Parker C, Sartor O. Radium-223 in prostate cancer. N Engl J Med 2013;369(17):1659–60. Available from: https://doi.org/10.1056/NEJMc1310231.
[41] Sartor O, Coleman R, Nilsson S, et al. Effect of radium-223 dichloride on symptomatic skeletal events in patients with castration-resistant prostate cancer and bone metastases: results from a phase 3, double-blind, randomised trial. Lancet Oncol 2014;15(7):738–46. Available from: https://doi.org/10.1016/S1470-2045(14)70183-4.
[42] Nilsson S, Cislo P, Sartor O, et al. Patient-reported quality-of-life analysis of radium-223 dichloride from the phase III ALSYMPCA study. Ann Oncol 2016;27(5):868–74. Available from: https://doi.org/10.1093/annonc/mdw065.
[43] Parker C, Nilsson S, Heinrich D, et al. Alpha emitter radium-223 and survival in metastatic prostate cancer. N Engl J Med 2013;369(3):213–23. Available from: https://doi.org/10.1056/NEJMoa1213755.
[44] Smith M, Parker C, Saad F, et al. Addition of radium-223 to abiraterone acetate and prednisone or prednisolone in patients with castration-resistant prostate cancer and bone metastases (ERA 223): a randomised, double-blind, placebo-controlled, phase 3 trial. Lancet Oncol 2019;20(3):408–19. Available from: https://doi.org/10.1016/S1470-2045(18)30860-X.
[45] Turner JH, Claringbold PG, Hetherington EL, et al. A phase I study of samarium-153 ethylenediaminetetramethylene phosphonate therapy for disseminated skeletal metastases. J Clin Oncol 1989;7(12):1926–31. Available from: https://doi.org/10.1200/JCO.1989.7.12.1926.
[46] Turner JH, Claringbold PG. A phase II study of treatment of painful multifocal skeletal metastases with single and repeated dose samarium-153 ethylenediaminetetramethylene phosphonate. Eur J Cancer 1991;27(9):1084–6. Available from: https://doi.org/10.1016/0277-5379(91)90297-q.
[47] Sartor O, Reid RH, Hoskin PJ, et al. Samarium-153-Lexidronam complex for treatment of painful bone metastases in hormone-refractory prostate cancer. Urology 2004;63(5):940–5. Available from: https://doi.org/10.1016/j.urology.2004.01.034.
[48] Deng HF, Luo SF, Tan TF, et al. [153Sm-EDTMP for moderate and severe bone cancer pain]. Hua Xi Yi Ke Da Xue Xue Bao 1995;26(4):391–4.
[49] Resche I, Chatal JF, Pecking A, et al. A dose-controlled study of 153Sm-ethylenediaminetetramethylenephosphonate (EDTMP) in the treatment of patients with painful bone metastases. Eur J Cancer 1997;33(10):1583–91. Available from: https://doi.org/10.1016/s0959-8049(97)00155-x.
[50] Serafini AN, Houston SJ, Resche I, et al. Palliation of pain associated with metastatic bone cancer using samarium-153 lexidronam: a double-blind placebo-controlled clinical trial. J Clin Oncol 1998;16(4):1574–81. Available from: https://doi.org/10.1200/JCO.1998.16.4.1574.
[51] Tian JH, Zhang JM, Hou QT, et al. Multicentre trial on the efficacy and toxicity of single-dose samarium-153-ethylene diamine tetramethylene phosphonate as a palliative treatment for painful skeletal metastases in China. Eur J Nucl Med 1999;26(1):2–7. Available from: https://doi.org/10.1007/s002590050351.
[52] Enrique O, Zhonyun P, Prma EP, et al. Efficacy and toxicity of Sm-153 EDTMP in the palliative treatment of painful bone metastases. World J Nucl Med 2002;1:21–7. Available from: https://doi.org/10.1159/000184746.
[53] Paravati AJ, Russo AL, Aitken C. Adverse events in the long-term follow-up of patients treated with samarium Sm 153 lexidronam for osseous metastases. Int J Radiat Oncol Biol Phys 2011;81(2):506–10. Available from: https://doi.org/10.1016/j.ijrobp.2010.05.066.
[54] Maxon 3rd HR, Schroder LE, Thomas SR, et al. Re-186(Sn) HEDP for treatment of painful osseous metastases: initial clinical experience in 20 patients with hormone-resistant prostate cancer. Radiology 1990;176(1):155–9. Available from: https://doi.org/10.1148/radiology.176.1.1693784.
[55] Han SH, de Klerk JM, Tan S, et al. The PLACORHEN study: a double-blind, placebo-controlled, randomized radionuclide study with (186)Re-etidronate in hormone-resistant prostate cancer patients with painful bone metastases. Placebo controlled rhenium study. J Nucl Med 2002;43(9):1150–6.
[56] Lam MG, de Klerk JM, van Rijk PP. 186Re-HEDP for metastatic bone pain in breast cancer patients. Eur J Nucl Med Mol Imaging 2004;31(1):162–70. Available from: https://doi.org/10.1007/s00259-004-1539-4.
[57] Palmedo H, Bender H, Dierke-Dzierzon C, et al. Pain palliation with rhenium-186 HEDP in breast cancer patients with disseminated bone metastases. Clin Nucl Med 1999;24(9):643–8. Available from: https://doi.org/10.1097/00003072-199909000-00001.

[58] Maxon 3rd HR, Schroder LE, Hertzberg VS, et al. Rhenium-186(Sn)HEDP for treatment of painful osseous metastases: results of a double-blind crossover comparison with placebo. J Nucl Med 1991;32(10):1877–81.

[59] Lambert B, Bacher K, Defreyne L, et al. 188Re-HDD/lipiodol therapy for hepatocellular carcinoma: a phase I clinical trial. J Nucl Med 2005;46(1):60–6.

[60] Liepe K, Brogsitter C, Leonhard J, et al. Feasibility of high activity rhenium-188-microsphere in hepatic radioembolization. Jpn J Clin Oncol 2007;37(12):942–50. Available from: https://doi.org/10.1093/jjco/hym137.

[61] Liepe K, Zaknun JJ, Soroa VE, et al. Radiosynovectomy using yttrium-90, phosphorus-32 and rhenium-188 colloids in rheumatoid arthritis. Eur J Nucl Med 2007;34:476.

[62] Liepe K, Kropp J, Runge R, et al. Therapeutic efficiency of rhenium-188-HEDP in human prostate cancer skeletal metastases. Br J Cancer 2003;89(4):625–9. Available from: https://doi.org/10.1038/sj.bjc.6601158.

[63] Chen P, Li J, Gui J, et al. Efficacy and safety of (188)Re-HEDP in lung cancer patients with bone metastases: a randomized, multicenter, multiple-dose phase IIa study. Int J Clin Oncol 2021;26(7):1212–20. Available from: https://doi.org/10.1007/s10147-021-01906-y.

[64] Srivastava SC, Atkins HL, Krishnamurthy GT, et al. Treatment of metastatic bone pain with tin-117m stannic diethylenetriaminepentaacetic acid: a phase I/II clinical study. Clin Cancer Res 1998;4(1):61–8.

[65] Handkiewicz-Junak D, Poeppel TD, Bodei L, et al. EANM guidelines for radionuclide therapy of bone metastases with beta-emitting radionuclides. Eur J Nucl Med Mol Imaging 2018;45(5):846–59. Available from: https://doi.org/10.1007/s00259-018-3947-x.

[66] Poeppel TD, Handkiewicz-Junak D, Andreeff M, et al. EANM guideline for radionuclide therapy with radium-223 of metastatic castration-resistant prostate cancer. Eur J Nucl Med Mol Imaging 2018;45(5):824–45. Available from: https://doi.org/10.1007/s00259-017-3900-4.

[67] Farhanghi M, Holmes RA, Volkert WA, et al. Samarium-153-EDTMP: pharmacokinetic, toxicity and pain response using an escalating dose schedule in treatment of metastatic bone cancer. J Nucl Med 1992;33(8):1451–8.

[68] Finlay IG, Mason MD, Shelley M. Radioisotopes for the palliation of metastatic bone cancer: a systematic review. Lancet Oncol 2005;6(6):392–400. Available from: https://doi.org/10.1016/S1470-2045(05)70206-0.

[69] Basch E, Autio K, Ryan CJ, et al. Abiraterone acetate plus prednisone versus prednisone alone in chemotherapy-naive men with metastatic castration-resistant prostate cancer: patient-reported outcome results of a randomised phase 3 trial. Lancet Oncol 2013;14(12):1193–9. Available from: https://doi.org/10.1016/S1470-2045(13)70424-8.

[70] Lee CK, Aeppli DM, Unger J, et al. Strontium-89 chloride (Metastron) for palliative treatment of bony metastases. The University of Minnesota experience. Am J Clin Oncol 1996;19(2):102–7. Available from: https://doi.org/10.1097/00000421-199604000-00003.

[71] Liepe K. Radium-223 chloride in bone pain treatment of prostate cancer. Der Nuklearmediziner 2015;38(2):131–7.

[72] Maxon HR, Deutsch EA, Thomas SR, et al. Re-186(Sn) HEDP for treatment of multiple metastatic foci in bone: human biodistribution and dosimetric studies. Radiology 1988;166(2):501–7. Available from: https://doi.org/10.1148/radiology.166.2.3122267.

[73] van Rensburg AJ, Alberts AS, Louw WK. Quantifying the radiation dosage to individual skeletal lesions treated with samarium-153-EDTMP. J Nucl Med 1998;39(12):2110–5.

[74] Bayouth JE, Macey DJ, Kasi LP, et al. Dosimetry and toxicity of samarium-153-EDTMP administered for bone pain due to skeletal metastases. J Nucl Med 1994;35(1):63–9.

[75] Spiers FW, Beddoe AH, King SD. The absorbed dose to bone marrow in the treatment of polycythaemia by 32P. Br J Radiol 1976;49(578):133–40. Available from: https://doi.org/10.1259/0007-1285-49-578-133.

[76] Sharma S, Singh B, Koul A, et al. Comparative therapeutic efficacy of (153)Sm-EDTMP and (177)Lu-EDTMP for bone pain palliation in patients with skeletal metastases: patients' pain score analysis and personalized dosimetry. Front Med 2017;4:46. Available from: https://doi.org/10.3389/fmed.2017.00046.

[77] Paes FM, Serafini AN. Systemic metabolic radiopharmaceutical therapy in the treatment of metastatic bone pain. Semin Nucl Med 2010;40:89–104. Available from: https://doi.org/10.1053/j.semnuclmed.2009.10.003.

[78] Mertens WC, Porter AT, Reid RH, et al. Strontium-89 and low-dose infusion cisplatin for patients with hormone refractory prostate carcinoma metastatic to bone: a preliminary report. J Nucl Med 1992;33(8):1437–43.

[79] Sciuto R, Festa A, Rea S, et al. Effects of low-dose cisplatin on 89Sr therapy for painful bone metastases from prostate cancer: a randomized clinical trial. J Nucl Med 2002;43(1):79–86.

[80] Tu SM, Delpassand ES, Jones RF, et al. Strontium-89 combined with doxorubicin in the treatment of patients with androgen-independent prostate cancer. Urol Oncol 1996;2:191–7. Available from: https://doi.org/10.1016/s1078-1439(97)00013-6.

[81] Tu SM, Kim J, Pagliaro LC, et al. Therapy tolerance in selected patients with androgen-independent prostate cancer following strontium-89 combined with chemotherapy. J Clin Oncol 2005;23(31):7904–10. Available from: https://doi.org/10.1200/JCO.2005.01.2310.

[82] Tu SM, Millikan RE, Mengistu B, et al. Bone-targeted therapy for advanced androgen-independent carcinoma of the prostate: a randomised phase II trial. Lancet 2001;357(9253):336–41. Available from: https://doi.org/10.1016/S0140-6736(00)03639-4.

[83] Pagliaro LC, Delpassand ES, Williams D, et al. A phase I/II study of strontium-89 combined with gemcitabine in the treatment of patients with androgen independent prostate carcinoma and bone metastases. Cancer 2003;97(12):2988. Available from: https://doi.org/10.1002/cncr.11412.

[84] van Dodewaard-de Jong JM, de Klerk JMH, Bloemendal HJ, et al. A randomised, phase II study of repeated rhenium-188-HEDP combined with docetaxel and prednisone versus docetaxel and prednisone alone in castration-resistant prostate cancer (CRPC) metastatic to bone; the Taxium II trial. Eur J Nucl Med Mol Imaging 2017;44(8):1319–27. Available from: https://doi.org/10.1007/s00259-017-3673-9.

[85] Amato RJ, Hernandez-McClain J, Henary H. Bone-targeted therapy: phase II study of strontium-89 in combination with alternating weekly chemohormonal therapies for patients with advanced androgen-independent prostate cancer. Am J Clin Oncol 2008;31(6):532–8. Available from: https://doi.org/10.1097/COC.0b013e318172aa92.

[86] Nilsson S, Strang P, Ginman C, et al. Palliation of bone pain in prostate cancer using chemotherapy and strontium-89. A randomized phase II study. J Pain Symptom Manage 2005;29(4):352–7. Available from: https://doi.org/10.1016/j.jpainsymman.2004.07.008.

[87] Ricci S, Boni G, Pastina I, et al. Clinical benefit of bone-targeted radiometabolic therapy with 153Sm-EDTMP combined with chemotherapy in patients with metastatic hormone-refractory prostate cancer. Eur J Nucl Med Mol Imaging 2007;34(7):1023–30. Available from: https://doi.org/10.1007/s00259-006-0343-8.

[88] Marcus CS, Saeed S, Mlikotic A, et al. Lack of effect of a bisphosphonate (pamidronate disodium) infusion on subsequent skeletal uptake of Sm-153 EDTMP. Clin Nucl Med 2002;27(6):427–30. Available from: https://doi.org/10.1097/00003072-200206000-00008.

[89] Pecherstorfer M, Schilling T, Janisch S, et al. Effect of clodronate treatment on bone scintigraphy in metastatic breast cancer. J Nucl Med 1993;34(7):1039–44.

[90] Roudier MP, Vesselle H, True LD, et al. Bone histology at autopsy and matched bone scintigraphy findings in patients with hormone refractory prostate cancer: the effect of bisphosphonate therapy on bone scintigraphy results. Clin Exp Metastasis 2003;20(2):171–80. Available from: https://doi.org/10.1097/10.1023/a:1022627421000.

[91] Flanagan AM, Chambers TJ. Dichloromethylenebisphosphonate (Cl2MBP) inhibits bone resorption through injury to osteoclasts that resorb Cl2MBP-coated bone. Bone Miner 1989;6(1):33–43. Available from: https://doi.org/10.1016/0169-6009(89)90021-4.

[92] Boonekamp PM, van der Wee-Pals LJ, van Wijk-van Lennep MM, et al. Two modes of action of bisphosphonates on osteoclastic resorption of mineralized matrix. Bone Miner 1986;1(1):27–39.

[93] Fromigue O, Lagneaux L, Body JJ. Bisphosphonates induce breast cancer cell death in vitro. J Bone Miner Res 2000;15(11):2211–21. Available from: https://doi.org/10.1359/jbmr.2000.15.11.2211.

[94] Jagdev SP, Coleman RE, Shipman CM, et al. The bisphosphonate, zoledronic acid, induces apoptosis of breast cancer cells: evidence for synergy with paclitaxel. Br J Cancer 2001;84(8):1126–34. Available from: https://doi.org/10.1054/bjoc.2001.1727.

[95] Soerdjbalie-Maikoe V, Pelger RCM, Lycklama à Nijeholt GAB, et al. Strontium-89 (Metastron) and the bisphosphonate olpadronate reduce the incidence of spinal cord compression in patients with hormone-refractory prostate cancer metastatic to the skeleton. Eur J Nucl Med Mol Imag 2002;29(4):494–8. Available from: https://doi.org/10.1007/s00259-001-0728-7.

[96] Storto G, Klain M, Paone G, et al. Combined therapy of Sr-89 and zoledronic acid in patients with painful bone metastases. Bone 2006;39(1):35–41. Available from: https://doi.org/10.1016/j.bone.2005.12.004.

[97] Smeland S, Erikstein B, Aas M, et al. Role of strontium-89 as adjuvant to palliative external beam radiotherapy is questionable: results of a double-blind randomized study. Int J Radiat Oncol Biol Phys 2003;56(5):1397–404. Available from: https://doi.org/10.1016/s0360-3016(03)00274-8.

[98] Fizazi K, Yang J, Peleg S, et al. Prostate cancer cells-osteoblast interaction shifts expression of growth/survival-related genes in prostate cancer and reduces expression of osteoprotegerin in osteoblasts. Clin Cancer Res 2003;9(7):2587–97.

[99] Kakonen SM, Selander KS, Chirgwin JM, et al. Transforming growth factor-beta stimulates parathyroid hormone-related protein and osteolytic metastases via Smad and mitogen-activated protein kinase signaling pathways. J Biol Chem 2002;277(27):24571–8. Available from: https://doi.org/10.1074/jbc.M202561200.

[100] Nakamura T, Hanada K, Tamura M, et al. Stimulation of endosteal bone formation by systemic injections of recombinant basic fibroblast growth factor in rats. Endocrinology 1995;136(3):1276–84. Available from: https://doi.org/10.1210/endo.136.3.7867582.

[101] Chinnaiyan P, Huang S, Armstrong E, et al. Radiosensitization following EGFR signaling inhibition by erlotinib (Tarceva). Int J Radiat Oncol Biol Phys 2003;57(2):S294.

[102] Harari PM, Huang SM. Epidermal growth factor receptor modulation of radiation response: preclinical and clinical development. Semin Radiat Oncol 2002;12(3 Suppl. 2):21–6. Available from: https://doi.org/10.1053/srao.2002.34865.

[103] Muralidhar A, Potluri HK, Jaiswal T, et al. Targeted radiation and immune therapies-advances and opportunities for the treatment of prostate cancer. Pharmaceutics 2023;15(1):252. Available from: https://doi.org/10.3390/pharmaceutics15010252.

[104] Choudhury AD, Kwak L, Cheung A, et al. Randomized phase II study evaluating the addition of pembrolizumab to radium-223 in metastatic castration-resistant prostate cancer. Cancer Immunol Res 2024;12(6):704–18. Available from: https://doi.org/10.1158/2326-6066.CIR-22-0306.

[105] Pan E, Xie W, Ajmera A, et al. A phase I study of combination olaparib and radium-223 in men with metastatic castration-resistant prostate cancer (mCRPC) with bone metastases (COMRADE). Mol Cancer Ther 2023;22(4):511–8. Available from: https://doi.org/10.1158/1535-7163.MCT-22-0583.

[106] Quinn Z, Leiby B, Sonpavde G, et al. Phase I study of niraparib in combination with radium-223 for the treatment of metastatic castrate-resistant prostate cancer. Clin Cancer Res 2023;29(1):50–9. Available from: https://doi.org/10.1158/1078-0432.CCR-22-2526.

[107] Kostos L, Buteau JP, Yeung T, et al. AlphaBet: combination of radium-223 and [(17) (7)Lu]Lu-PSMA-I&T in men with metastatic castration-resistant prostate cancer (clinical trial protocol). Front Med 2022;9:1059122. Available from: https://doi.org/10.3389/fmed.2022.1059122.

[108] Shore N, Carles J, McDermott R, et al. Treatment intensification with radium-223 plus enzalutamide in patients with metastatic castration-resistant prostate cancer. Front Med 2024;11:1460212. Available from: https://doi.org/10.3389/fmed.2024.1460212.

[109] Liepe K, Hliscs R, Kropp J, et al. Dosimetry of 188Re-hydroxyethylidene diphosphonate in human prostate cancer skeletal metastases. J Nucl Med. 2003;44(6):953–60.

Chapter 21

Procedural interventions

Ivan Radoš[1,2] and Dino Budrovac[1,2]
[1]*Faculty of Medicine Osijek, Josip Juraj Strossmayer University of Osijek, Osijek, Croatia,* [2]*Department of Anaesthesiology, Reanimatology and Intensive Care, University Hospital Centre Osijek, Osijek, Croatia*

Key learning objectives

- Bring the reader closer to the possibilities of invasive pain treatment.
- Understand the indications and contraindications for procedural interventions in pain management.
- Comprehend the most common minimally invasive procedures in the treatment of pain of cancer patients.

Introduction

Often, pharmacotherapeutic treatment of cancer pain is insufficient or is associated with systemic toxicity and side effects that patients cannot tolerate in some cases. For this reason, invasive pain treatment is desirable because it enables the control of such pain with minimal side effects. Invasive pain treatment was previously performed according to specific anatomical markings [1]. Nowadays, with the advancement of imaging methods and the increased availability of devices, it is performed under the control of ultrasound, fluoroscopy (X-ray), or computerized tomography (CT), which significantly increases the success of these methods and reduces the risk of potential complications. The choice of imaging method depends on the type of procedure and the preference of the procedure performer. The thickness and length of the needle vary depending on the procedure itself as well as on the physical characteristics of the individual patient. The solution for injection includes a local anesthetic with or without corticosteroids, and in procedures where there is a high possibility of intravascular administration, particular corticosteroids are avoided in order to avoid erythrocyte aggregation and, consequently, organ ischemia distal to the application site. In some patients, satisfactory analgesia can be achieved by administering a local anesthetic with or without corticosteroids, and in the case of refractory pain, it is possible to perform neurolysis, which is achieved with alcohol or phenol. Neurolysis is performed only after a positive block with a local anesthetic [2].

General contraindications are the patient's refusal to perform the procedure, infection at the site of the procedure, allergy to the drugs required for the procedure, and coagulopathy. Common complications are infection, bleeding, worsening of diabetes mellitus and endocrinopathy due to the action of corticosteroids, nerve injury, and allergic and vasovagal reactions. In addition to the listed general complications that are possible with all procedures, each procedure has a risk of developing some specific complications. Thus, for example, when performing a procedure around the spine, there is a risk of developing a spinal or epidural hematoma, as well as nerve and spinal cord injuries [3].

Invasive procedures in clinical practice

Minimally invasive procedures and neuromodulation techniques are increasing in the field of pain treatment, but neurolytic blocks have an irreplaceable place in the treatment of cancer pain. The goals of neurolytic blocks are protein denaturation and precipitation of Schwann and nerve cell lipoproteins. Alcohol (50%–100%) and phenol (5%–10%) are used for neurolysis, which interrupts the pain signal transmission by a mechanism called Wallerian degeneration distal to the lesion site. The application of alcohol is painful; it quickly spreads into the surrounding tissue, and larger volumes are needed, and the occurrence of neuritis is more common than with phenol. The application of phenol is less painful, and the

Pain Management in Palliative Care for Patients With Cancer. DOI: https://doi.org/10.1016/B978-0-443-30050-9.00010-2

spread to the surrounding tissues is less. Phenol diffuses into the axon and perineural blood vessels, leading to protein denaturation. Expected pain reduction after neurolysis is 3–6 months [2].

Pain resulting from a malignant disease can appear in any part of the body, and wherever there are sensory nerves or nerve plexuses, it is possible to block the transmission of painful impulses. There are numerous minimally invasive procedures that can be used to reduce pain in patients with cancer [3]. In the following, the most often used procedures in clinical practice are described.

Stellate ganglion block

The stellate ganglion is also called the lower cervicothoracic ganglion and is the junction of the lower cervical and the first thoracic ganglion. It is located on the front side of the longus colli muscle at the level of the transverse process of the seventh cervical and first thoracic vertebrae. It is located anteromedially in relation to the vertebral artery and medially in relation to the common carotid artery and jugular vein and laterally in relation to the trachea and esophagus.

Indications

The block of stellate ganglia is traditionally used in the treatment of complex regional pain syndrome of the upper extremities but also for other indications such as the treatment of painful conditions of malignant origin in trigeminal nerve innervation, cervical and upper thoracic dermatomes, and in the treatment of pain resulting from herpes zoster, reflex sympathetic dystrophy, and Raynaud's syndrome of the upper extremities. It has also been shown to be effective in the treatment of anxiety symptoms in patients with posttraumatic stress disorder. The block of stellate ganglia is effective in reducing pain in orofacial cancers, which is caused by the cancer itself, but also pain resulting from chemoradiation therapy and postmastectomy pain. In addition to pain reduction, the quality of speech, swallowing, chewing, and sleep improves [4].

Contraindications

General contraindications are the patient's refusal to perform the procedure, infection at the site of the procedure, allergy to the drugs required for the procedure, and coagulopathy.

Procedure

The block can be performed with an anterior, posterior, and vertebral body approach. The procedure can be performed under the control of a fluoroscope or ultrasound. Fig. 21.1 shows the position of the patient during the block under the control of the fluoroscope. Neurolysis can be achieved chemically and by radiofrequency (RF) ablation. The position of the needle tip on the C6 vertebra is shown in Fig. 21.2, and the confirmation of the correct position of the needle after contrast application is shown in Fig. 21.3. After reaching the desired location, 7–10 mL of local anesthetic is applied [5].

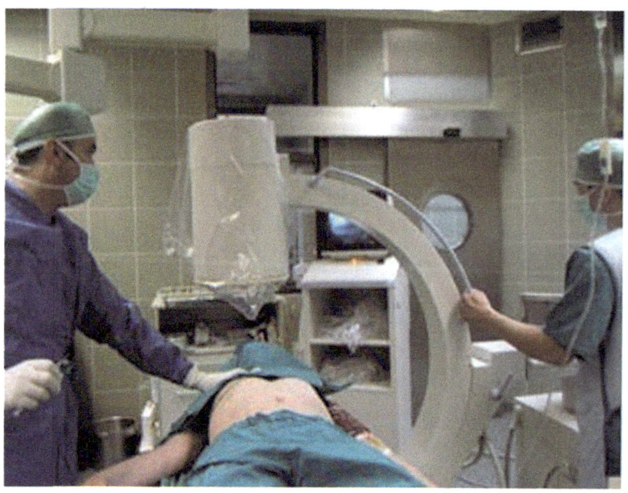

FIGURE 21.1 Patient position for stellate ganglion block in the supine position during the anterior approach.

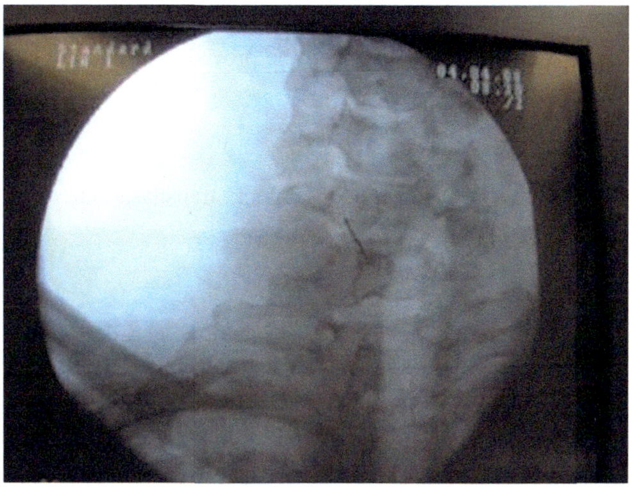

FIGURE 21.2 Stellate ganglion block. The position of the needle tip at C6.

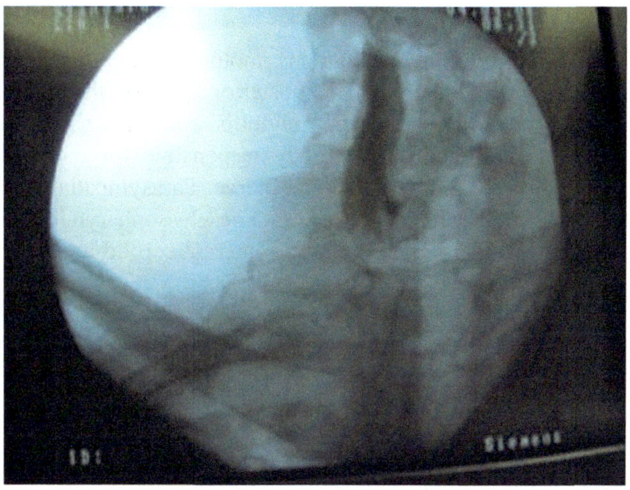

FIGURE 21.3 Stellate ganglion block. The position of the needle tip at C6 and checking the position of the needle after contrast application.

Complications

Considering that this region is well vascularized, it is necessary to watch out for the development of systemic toxicity of the local anesthetic and the development of hematoma. Due to the proximity of the spine, it is possible to apply the drug in the epidural or spinal space, and even a small volume of local anesthetic at this level can lead to the development of total spinal anesthesia. If the needle is placed too low, it is possible to puncture the pleura and develop a pneumothorax. Other side effects of this block include recurrent laryngeal nerve block leading to hoarseness and dysphagia and the development of Horner's syndrome [6,7].

Sphenopalatine ganglion block

The sphenopalatine ganglion (SPG) is also known as the pterygopalatine, nasal, and Meckel's ganglion. It is located in the pterygopalatine fossa and is covered by a thin layer of connective tissue and mucous membrane. It is an extracranial parasympathetic ganglion of the facial nerve, from which fibers go to the Gasser ganglion, trigeminal nerve, carotid plexus, facial nerve, and upper cervical ganglion.

Indications

The SPG block is indicated in the treatment of migraine headaches, cluster headaches, various facial neuralgias, and cancer pain in the face and neck area [8,9].

Contraindications

General contraindications are the patient's refusal to perform the procedure, infection at the site of the procedure, allergy to the drugs required for the procedure, and coagulopathy.

Procedure

There are different anatomical approaches for performing this block: transnasal, lateral, and approach through the greater palatine foramen [9].

Complications

Complications of this block are associated with high vascularization of the nasal mucosa. These include epistaxis, which is a common complication; facial hematoma; hypoesthesia of the nose, palate, and pharynx; ipsilateral lacrimation; and infection. There is also the possibility of significant systemic absorption of the local anesthetic. After the SPG block, it is possible to develop hypotension, which is why pressure monitoring after the procedure is necessary. Local anesthetic is used to perform this block, and if the effect is not satisfactory, neurolytic agents, freezing, and RF lesioning can be applied. Due to the presence of other neural structures in the vicinity, an improperly performed RF lesion can lead to neurological deficits [9,10].

Glossopharyngeal nerve block

The glossopharyngeal nerve is a mixed nerve that contains motor and sensory fibers. Motor fibers innervate the stylopharyngeus muscle, and sensory fibers innervate the posterior third of the tongue, tonsils, soft palate, mucous membrane of the mouth and pharynx, the front side of the epiglottis, and the sense of taste in the posterior third of the tongue. Information from the carotid sinus and carotid body responsible for controlling blood pressure, pulse, and respiration is transmitted by a branch of the glossopharyngeal nerve. Parasympathetic fibers are transmitted to the otic ganglion and the parotid gland via the glossopharyngeal nerve. The glossopharyngeal nerve passes through the jugular foramen near the vagus nerve and the internal jugular vein and internal carotid artery.

Indications

Glossopharyngeal nerve block is indicated in the treatment of cancer pain in the back third of the tongue, hypopharynx, and tonsils and glossopharyngeal neuralgia.

Contraindications

General contraindications are the patient's refusal to perform the procedure, infection at the site of the procedure, allergy to the drugs required for the procedure, and coagulopathy.

Procedure

The key anatomical marking for performing this block is the styloid process of the temporal bone, which is in the middle of the line connecting the mastoid process and the angle of the mandible [10]. The procedure can be performed under the control of ultrasound or fluoroscope [11].

Because of the surrounding structures, caution is required, and no cerebrospinal fluid or blood should appear during aspiration. To achieve the block, a local anesthetic is applied with or without corticosteroids. Longer-lasting analgesia can be achieved by neurolysis with alcohol or phenol and also by RF ablation.

Complications

When performing this block, a possible complication is the puncture of the internal jugular vein or carotid artery and the formation of a hematoma, as well as the intravascular application of a local anesthetic, which can lead to systemic toxicity and seizures. Blockade of the motor part of the glossopharyngeal nerve results in dysphagia, which is a consequence of the weakness of the stylopharyngeus muscle. If the vagus nerve is blocked, dysphonia may develop due to paralysis of the ipsilateral vocal cord. Due to the blockade of the vagus nerve, the development of reflex tachycardia is also possible. In some patients, after chemical neurolysis, dysesthesia may develop in the innervation area, which can be manifested by an unpleasant burning pain, and this is called anesthesia dolorosa. Such pain can be more unpleasant and stronger than the primary pain for which the neurolysis was performed. Infection is also a possible complication. Intraoral access is also described when extraoral access is not possible due to anatomical reasons or illness. In this case, the anatomical marking is the palatine tonsils, and the glossopharyngeal nerve is located submucosally and medial to the tonsils. The internal carotid artery is located posterior and lateral to the nerve, which is why there is a high possibility of puncture [12].

Celiac plexus block and splanchnic nerve block

The celiac plexus is the largest visceral plexus, which is located retroperitoneally and anterolaterally in relation to the abdominal aorta at the point where the celiac artery and superior mesenteric artery separate and anteriorly in relation to the diaphragm. The greater, lesser, and least splanchnic nerves mostly supply the celiac plexus with preganglionic fibers. The greater splanchnic nerve consists of fibers from the Th5–Th10 spinal roots, the lesser splanchnic nerve from Th10 and Th11, and the least from Th11 and Th12, with certain anatomical variations. The celiac plexus supplies sympathetic, parasympathetic, and visceral sensory fibers to the upper abdominal organs, such as the distal esophagus, pancreas, liver, biliary tract, spleen, kidneys, adrenal glands, stomach, small intestine, colon to transverse colon, and mesentery [13]. In 94% of cases, the celiac plexus is located between the Th12 and L1 vertebrae, but there are certain anatomical variations (Fig. 21.4). The size of the ganglia varies from 0.5 to 4.5 cm.

Indications

It was initially designed as a part of an anesthesiology technique, but as such it did not take root in clinical practice but instead found its place in the treatment of abdominal pain caused by pancreatitis and malignant pain.

Contraindications

Contraindications for performing this block are patient rejection, platelet count lower than 50,000, anticoagulant therapy, and coagulopathies. Relative contraindications are altered anatomy due to previous operations or congenital anomalies. Before the procedure, it is necessary to perform laboratory tests, including a complete blood count, and a coagulogram. Local and intraabdominal infections, as well as sepsis, are absolute contraindications for performing this block. CP blockade leads to increased intestinal motility, which is why this block is contraindicated in patients with intestinal obstruction.

Diagnostic celiac plexus block (CPB) with local anesthetic with or without corticosteroids is a procedure that determines whether abdominal pain is mediated by the celiac plexus. If the diagnostic block is positive, neurolysis with alcohol or phenol or RF ablation can be performed [3].

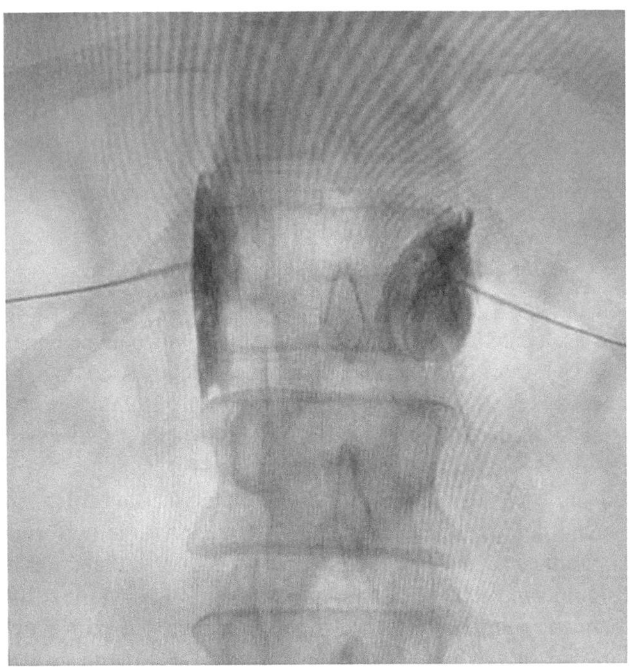

FIGURE 21.4 Celiac plexus block. Bilateral needle position at vertebra Th12 level after contrast application.

Procedure

To increase efficiency and reduce complications, CPB is performed under the control of a fluoroscope, CT, or ultrasound. There are different anatomical approaches to the celiac plexus: retrocrural, transcrural, transaortic, preaortic, and transdiscal. The retrocrural approach is similar to the splanchnic nerve block except that in the case of CPB the needle is positioned lower. The choice of the imaging method used to monitor the progress of the needle and the anatomical approach to the celiac plexus depends on the preference of the doctor performing the procedure, and all techniques are equally effective.

A splanchnic nerve block is preferred in patients in whom it is not possible to perform a CPB due to altered anatomy by a tumor mass or due to lymphadenopathy.

Preparation for the procedure includes adequate intravenous hydration of the patient to prevent hypotension due to nerve block. Patients must be monitored before, during, and after the procedure [14].

Complications

The complication rate of neurolysis of the celiac plexus and splanchnic nerve is 1%–2%. The most common complications are hypotension, pneumothorax, infection, nerve injury, paresthesia, epidural or subarachnoid infection, organ puncture, and drug allergy. Diarrhea and orthostatic hypotension are transient and result from blocking of sympathetic fibers. Intravascular application of drugs into the blood vessel that is responsible for the nutritional supply of the spine can lead to thrombosis of the blood vessel and consequently paraplegia. Retrocrural neurolysis can lead to injury of the lumbar nerve roots and neurological deficits, which occur less frequently with the precrural approach [3].

Both methods, chemical neurolysis and RF ablation, are effective methods in reducing the intensity of pain, reducing the need for opioid analgesics, and improving the quality of life of patients [2,15].

That celiac plexus neurolysis is an effective additional method to pharmacotherapy in the treatment of pain in patients with inoperable pancreatic cancer is also shown by the results of the systematic review and metaanalysis by Okita et al. [16].

Superior hypogastric plexus block

The superior hypogastric plexus transmits visceral nociceptive signals from pelvic organs such as the bladder, prostate, rectum, ovaries, uterus, and vagina.

Indications

Superior hypogastric plexus block (SHPB) is indicated in the treatment of cancerous visceral pain of the pelvic organs, and extended indications include the treatment of pain due to endometriosis and pelvic inflammatory disease [17].

Contraindications

General contraindications are the patient's refusal to perform the procedure, infection at the site of the procedure, allergy to the drugs required for the procedure, and coagulopathy.

Procedure

The superior hypogastric plexus is located retroperitoneally at the level between the L5 and S1 vertebrae near the bifurcation of the iliac vessels. The most common anatomical approaches to this plexus are transdiscal, posterior paravertebral, and anterior. The block is performed under the control of ultrasound, fluoroscopy, or CT. A diagnostic block is performed with local anesthetic with or without corticosteroids, and neurolysis is achieved with alcohol or phenol [18,19].

Abdelghaffar et al. compared the effectiveness of this block under the control of the fluoroscope with an anterior approach compared to an ultrasound-guided block. The results showed that the performance of the procedure under the control of the fluoroscope took significantly less time, which led to greater satisfaction in this group of patients. In both groups of subjects, there was a significant decrease in pain intensity, but there was no difference between the groups [20].

In patients with pelvic and perineal pain associated with pelvic cancer, pulsed RF of the sacral roots in combination with superior hypogastric plexus neurolysis can lead to a greater reduction in pain intensity compared to superior hypogastric plexus (SHP) neurolysis alone [21].

Complications

Complications of SHPB include infection, bleeding, nerve damage, and allergic reactions. Complications such as blood clots in the pelvis, kidney, or ureter injury, and vasovagal reactions may also occur. Perforation and injury to surrounding organs or blood vessels are also possible, although rare with proper X-ray guidance.

Inferior hypogastric plexus block

The inferior hypogastric plexus, also called the pelvic plexus, is a paired group of nerve fibers. In men, it is located posterolaterally in relation to the seminal vesicles, prostate, and at the base of the urinary bladder, and in women, laterally in relation to the cervix, vaginal fornices, and urinary bladder. It is formed from the pelvic and sacral splanchnic nerves and the fibers of the upper hypogastric plexus together with the afferent fibers of the pelvic organs. It contains sympathetic and parasympathetic fibers, and its main function is the innervation of pelvic and perineal structures [22].

Indications

Specific painful conditions that can be treated with this block are pelvic pain, pain due to endometriosis, malignant pelvic pain, vulvodynia, tenesmus caused by radiation, enteritis of the rectum and sigmoid colon, proctalgia fugax, and postherpetic neuralgia of the sacral dermatomes.

Contraindications

General contraindications are the patient's refusal to perform the procedure, infection at the site of the procedure, allergy to the drugs required for the procedure, and coagulopathy.

Procedure

Traditional approaches for performing this block are paravertebral approach, which is accessed anterolaterally in relation to the body of the fifth lumbar vertebra, and the transdiscal approach, which is accessed through the L5/S1 intervertebral disc, which is considered simpler and safer. In addition to these, there is also a transsacral approach that reaches the plexus through the sacral foramen, most often through the second sacral foramen [23].

Complications

When performing this block, possible complications are infection, discitis during the transdiscal approach, bleeding, paresthesia, and rectal puncture during the transsacral approach [24].

Ganglion impar block

The ganglion impar is considered the final part of the sympathetic chain and is located anterior to the sacrococcygeal junction. The ganglion receives sympathetic and parasympathetic fibers from the lumbar and sacral segments and innervates the pelvic and genital organs.

Indications

Ganglion impar block is used in the treatment of sympathetically mediated pain in the perineum, rectum, and genitals. It is most often used to treat pain of malignant origin, and other indications include endometriosis, reflex sympathetic dystrophy, radiation enteritis, and proctalgia fugax.

Contraindications

General contraindications are the patient's refusal to perform the procedure, infection at the site of the procedure, allergy to the drugs required for the procedure, and coagulopathy.

Procedure

The procedure can be performed with the pronation technique through the anococcygeal ligament or with an intercoccygeal and transcoccygeal approach, orienting according to anatomical features, under the control of ultrasound, fluoroscopy, or CT. Fig. 21.5 shows the position of the patient in the pronation position and the placed needle when performing the block under the control of the fluoroscope. Confirmation of correct needle position after contrast application is shown in Fig. 21.6. Diagnostic blocks can be done with local anesthetic with or without corticosteroids, and

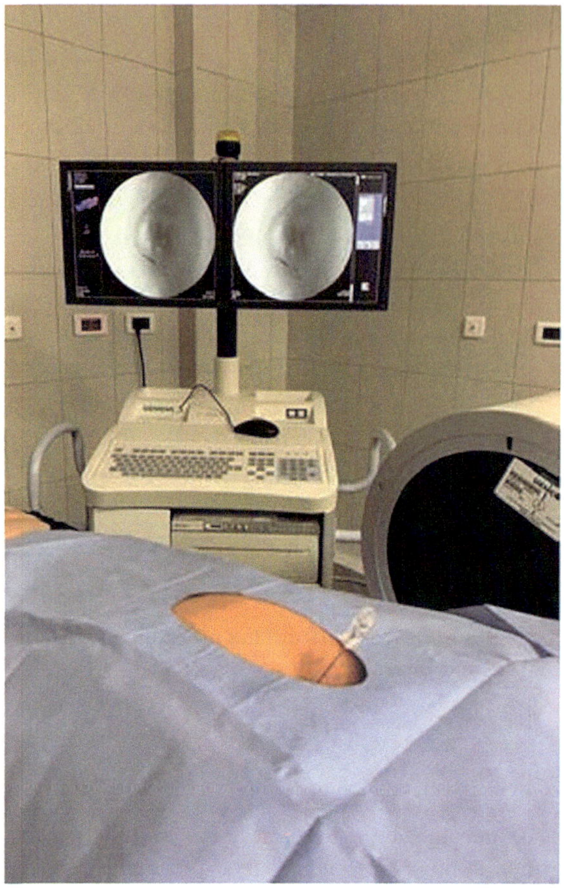

FIGURE 21.5 Ganglion of impar block. Patient position and needle view.

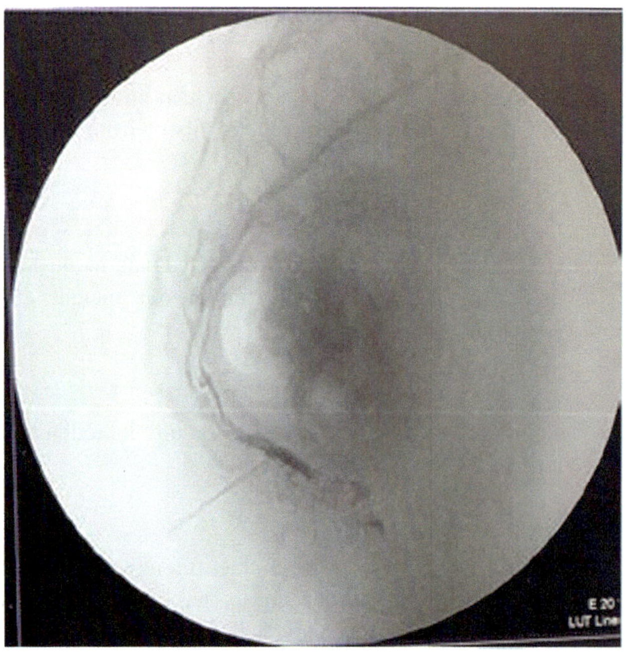

FIGURE 21.6 Ganglion of impar block. Confirmation of needle position after contrast application.

if the diagnostic block is positive, longer-term pain reduction can be achieved with alcohol or phenol neurolysis, RF ablation, or cryoablation [25].

Complications

Due to the proximity of the ganglia impar to the rectum, perforation of the rectum is possible, which is why there is a possibility of contamination and the development of infection. A potential complication, especially in immunocompromised patients and patients who have undergone radiation therapy of the perineum, is the formation of a fistula [26,27].

Treatment of cancer pain is difficult, especially in regions with complex innervation such as the pelvis and perineal region. Impar ganglia blockade is a safe and effective choice in the group of oncology patients [28].

Percutaneous vertebroplasty and balloon kyphoplasty

Indications

Percutaneous vertebroplasty is a procedure that is indicated for patients with a weakened and collapsed vertebral body due to various pathological conditions. The most common conditions are fracture of the vertebral body due to osteoporosis and tumor process of the vertebral body. Indications for percutaneous balloon kyphoplasty are the same as for vertebroplasty. Both methods have their advantages and disadvantages. In balloon kyphoplasty, a kyphoplasty balloon is placed in the body of the vertebra through the trocar, and its inflation restores the height of the collapsed vertebra. The advantage of balloon kyphoplasty is that the balloon creates a cavity inside the vertebral body and enables better control when applying cement, thus reducing the risk of cement leaking into the spinal canal or blood vessels. The best results can be expected if the compression of the vertebral body is partial, if the condition lasts less than 12 months; and if the fracture is older than 12 months, the radionuclide scan of the bone indicates a still active process [29].

Contraindications

General contraindications are the patient's refusal to perform the procedure, infection at the site of the procedure, allergy to the drugs required for the procedure, and coagulopathy.

Procedure

The procedure is performed in the operating room under aseptic conditions under the control of a fluoroscope under local anesthesia and analgosedation.

Under the control of the fluoroscope, the trocar is introduced through the pedicle to the body of the desired vertebra. Alternative approaches are anterolateral, costovertebral, and posterolateral. The target site of the tip of the trocar is the front third of the vertebral body, because that part of the vertebra bears the greatest weight. The correct position of the tip of the trocar is confirmed by the application of a contrast medium, after which cement is applied until the entire body of the vertebra is filled. If the entire vertebral body is not filled with cement, it is possible to introduce a second trocar.

With the correct performance of this procedure, complications are rare, and the risk of intravascular or epidural application of cement is reduced by using a fluoroscope and contrast medium and by preparing the cement according to the manufacturer's instructions [30].

Complications

The trunk of the vertebra is richly supplied with venous plexuses, and knowledge of their anatomy is essential when performing this procedure due to their potential damage by the needle, as well as the possible application of cement into the systemic circulation, which can lead to catastrophic consequences. Unintentional application of cement into the neural foramen and spinal space is possible even if the procedure was performed according to all the rules and results in neurological complications and radicular symptoms that may require urgent surgical decompression. New-onset pain may be the result of a new fracture of the pedicle or another vertebra. The development of infections is also possible, especially in immunocompromised patients on long-term corticosteroid therapy and in cancer patients. Many clinicians performing balloon kyphoplasty place two trocars and two balloons that are inflated simultaneously to prevent herniation of one balloon and to achieve symmetrical height elevation of the vertebral body. After the desired height of the vertebral body is achieved, cement is applied in the formed cavity according to the same procedure as in vertebroplasty. The complications of balloon kyphoplasty are the same as those of vertebroplasty, which were mentioned earlier in the text [30].

Intrathecal and epidural drug delivery

The discovery of opioid receptors in the dorsal horn of the spinal cord made it possible to understand the effectiveness of intrathecal administration of opioid analgesics. Recently, there has been an increase in intrathecal drug delivery (IDD), and the method of administration and the type of drug administered depend on the individual institution, the required time of analgesia, and the life expectancy of the patient.

Indications

This method of analgesia can be used both in patients with noncancer and cancer pain. In patients with cancer, a catheter can be placed intrathecally. Analgesics are most often used intrathecally in cancer patients, which is why the greatest findings are in this group of patients. In patients with cancer, opioids are most often used systemically in high doses, which is why there are frequent and numerous side effects. Intrathecal administration of opioids at a lower dose results in better analgesia with fewer side effects such as sedation, constipation, and respiratory depression. The most commonly used drugs in this way are morphine, hydromorphone, fentanyl, sufentanil, levobupivacaine, ziconotide, and clonidine. There are different modalities of intrathecal administration of drugs. Single-shot injection, multiple bolus injections, and continuous infusion via a pump are possible. Bolus administration of morphine can also be used as a test to assess the effectiveness and safety of this method of administering analgesics. With continuous use of opioids, it is necessary to start with lower doses, which must be carefully titrated every 6–12 hours. To avoid delayed reactions such as respiratory depression, it is preferable that the patient be hospitalized [31].

Contraindications

Contraindications for IDD are systemic infection, coagulopathy, allergy to drugs that are planned to be administered, inadequate previous medication intake, and poor personal hygiene.

Procedure

A psychological assessment of the patient is necessary because this way of administering drugs requires the cooperation of the patient, who must understand the procedure itself and the outcome and have realistic expectations.

Drugs administered epidurally must pass through the dura and arachnoid to reach the site of action, while drugs administered intrathecally diffuse directly to the site of action. Hydrophilic drugs circulate more easily through the cerebrospinal fluid compared to lipophilic drugs; they have a slower onset of action, and the effect lasts longer. With epidural administration, the dose required for the same analgesia is much higher, and systemic absorption is also higher. The doses of morphine administered epidurally compared to intrathecally for the same degree of analgesia are 5–10 times higher. Lipophilic drugs such as fentanyl and sufentanil diffuse more slowly through the cerebrospinal fluid and remain closer to the tip of the catheter. With epidural application of lipophilic drugs, significantly higher serum concentrations are recorded, and the pharmacokinetics are more complex. Morphine is the prototype opioid analgesic and is the only opioid analgesic approved by the FDA for intrathecal use. It is hydrophilic and undergoes minimal changes in the cerebrospinal fluid, which ensures long-term analgesia after intrathecal administration. It achieves its effect through opioid receptors in the spinal cord. Other opioids that are administered intrathecally are hydromorphone, fentanyl, and sufentanil. Other nonopioid drugs that have been described for intrathecal administration include levobupivacaine, ziconotide, clonidine, baclofen, adenosine, and gabapentin [31].

Visceral pain is the most common type of pain in cancer patients, but in 30% of cases, patients also have somatic pain [32]. For the treatment of somatic pain, there are different blocks that are performed depending on the region that is affected. A transversus abdominis plane block with local anesthetic and corticosteroid or ethanol/phenol can be performed to treat postoperative abdominal wall pain, chronic abdominal wall pain, or somatic pain caused by cancer [33]. Pain caused by tumor infiltration or metastasis of the thoracic wall can be reduced by intercostal, paravertebral, or erector spinae plane block, and postmastectomy pain by a serratus plane block [34–36].

Complications

If survival is expected to be longer than a month, intrathecal administration is recommended over epidural administration due to fewer complications. With long-term use of an epidural catheter, loss of function may occur due to fibrosis around the catheter. With an intrathecally placed catheter, it may happen that patients have a mild postpuncture headache, while some may have severe postural headaches with the loss of large amounts of cerebrospinal fluid. The development of an infection is possible, but noninfectious fever can also occur, peaking after 72 hours and presenting as mild stiffness in the neck [31].

Further implications

The wider use of invasive pain management methods requires additional education of healthcare professionals and the inclusion of these procedures in standardized palliative care guidelines. Also, further research is needed to examine the long-term effects, safety, and cost-effectiveness of these methods. A multidisciplinary approach and the availability of technology are key factors for the successful integration of these techniques into everyday oncological practice.

Conclusion

Invasive pain management methods represent an effective approach to cancer pain control in patients in whom pharmacotherapy is insufficient or causes serious side effects. The development of image-guided techniques such as ultrasound, fluoroscopy, and CT has enabled greater precision and safety of the procedure. The use of local anesthetics, corticosteroids, and neurolytic procedures can achieve significant pain reduction and improvement in quality of life in clinical practice.

References

[1] Filippiadis DK, Tselikas L, Bazzocchi A, et al. Percutaneous management of cancer pain. Curr Oncol Rep 2020;22(5):43. Available from: https://doi.org/10.1007/s11912-020-00906-x.

[2] Koyyalagunta D, Engle MP, Yu J, et al. The effectiveness of alcohol versus phenol based splanchnic nerve neurolysis for the treatment of intra-abdominal cancer pain. Pain Physician 2016;19(4):281–92. Available from: https://doi.org/10.36076/ppj/2019.19.281.

[3] Yalamuru B, Weisbein J, Pearson ACS, et al. Minimally-invasive pain management techniques in palliative care. Ann Palliat Med 2022;11(2):947–57. Available from: https://doi.org/10.21037/apm-20-2386.

[4] Abbas DN, Reyad RM. Thermal versus super voltage pulsed radiofrequency of stellate ganglion in post-mastectomy neuropathic pain syndrome: a prospective randomized trial. Pain Physician 2018;21(4):351–62. Available from: https://doi.org/10.36076/ppj.2018.4.351.

[5] Narouze S. Ultrasound-guided stellate ganglion block: safety and efficacy. Curr Pain Headache Rep 2014;18(6):424. Available from: https://doi.org/10.1007/s11916-014-0424-5.

[6] Darabad RR, Kalangara JP, Woodbury A. Case series: cancer-related facial pain treated with stellate ganglion block. Palliat Med Rep 2020;1(1):290–5. Available from: https://doi.org/10.1089/pmr.2020.0042.

[7] Ghai A, Kaushik T, Kumar RWS. Chemical ablation of stellate ganglion for head and neck cancer pain. Acta Anaesthesiol Belg 2016;67(1):6–8.

[8] Candido KD, Massey ST, Sauer R, et al. A novel revision to the classical transnasal topical sphenopalatine ganglion block for the treatment of headache and facial pain. Pain Physician 2013;16(6):769–78.

[9] Ho KWD, Przkora R, Kumar S. Sphenopalatine ganglion: block, radiofrequency ablation and neurostimulation – a systematic review. J Headache Pain 2017;18(1):118. Available from: https://doi.org/10.1186/s10194-017-0826-y.

[10] Ažman J, Stopar Pintarič T, Cvetko E, et al. Ultrasound-guided glossopharyngeal nerve block: a cadaver and a volunteer sonoanatomy study. Reg Anesth Pain Med 2017;42(2):252–8. Available from: https://doi.org/10.1097/AAP.0000000000000561.

[11] Kojima Y, Okamoto S, Hirabayashi K. Ultrasound-guided selective glossopharyngeal nerve block: posterior mandibular ramus approach. Pain Med 2024;25(3):169–72. Available from: https://doi.org/10.1093/pm/pnad143.

[12] Manoharan D, Bharati SJ, Yadav MK. A novel technique of ultrasound-guided glossopharyngeal nerve block to relieve cancer pain. Saudi J Anaesth 2019;13(3):279–80. Available from: https://doi.org/10.4103/sja.SJA_139_19.

[13] Vig S, Bhan SBS. Celiac plexus block – an old technique with new developments. Pain Physician 2021;24(5):379–98.

[14] Yoon WJ, Oh Y, Yoo C, et al. EUS-guided versus percutaneous celiac neurolysis for the management of intractable pain due to unresectable pancreatic cancer: a randomized clinical trial. J Clin Med 2020;9(6):1666. Available from: https://doi.org/10.3390/jcm9061666.

[15] Sachdev AH, Gress FG. Celiac plexus block and neurolysis: a review. Gastrointest Endosc Clin N Am 2018;28(4):579–86. Available from: https://doi.org/10.1016/j.giec.2018.06.004.

[16] Okita M, Otani K, Gibo N, et al. Systematic review and meta- analysis of celiac plexus neurolysis for abdominal pain associated with unresectable pancreatic cancer. Pain Pract 2022;22(7):652–61. Available from: https://doi.org/10.1111/papr.13143.

[17] Mishra S, Bhatnagar S, Rana SP, et al. Efficacy of the anterior ultrasound-guided superior hypogastric plexus neurolysis in pelvic cancer pain in advanced gynecological cancer patients. Pain Med 2013;14(6):837–42. Available from: https://doi.org/10.1111/pme.12106.

[18] Rocha A, Plancarte R, Natarén RGR, et al. Effectiveness of superior hypogastric plexus neurolysis for pelvic cancer pain. Pain Physician 2020;23(2):203–8.

[19] Hou S, Novy D, Felice F, et al. Efficacy of superior hypogastric plexus neurolysis for the treatment of cancer-related pelvic pain. Pain Med 2020;21(6):1255–62. Available from: https://doi.org/10.1093/pm/pnz151.

[20] Abdelghaffar NA, Farahat TE. Fluoroscopic anterior approach versus ultrasound guided superior hypogastric plexus neurolysis in cancer pelvic pain: a randomized controlled study. BMC Anesthesiol 2022;22(1):1–9. Available from: https://doi.org/10.1186/s12871-022-01948-3.

[21] Hetta DF, Mohamed AA, Emam RMA, et al. Pulsed radiofrequency of the sacral roots improves the success rate of superior hypogastric plexus neurolysis in controlling pelvic and perineal cancer pain. Randomized trial. Pain Physician 2020;23(2):149–57. Available from: https://doi.org/10.36076/ppj.2020/23/149.

[22] Yoham AL. Anatomy, abdomen and pelvis: inferior hypogastric plexus. StatPearls. Treasure Island (FL): StatPearls Publishing; 2022.

[23] Urits I, Schwartz R, Bangalore Siddaiah H, et al. Inferior hypogastric block for the treatment of chronic pelvic pain. Anesth Pain Med 2021;11(1):e112225. Available from: https://doi.org/10.5812/aapm.112225.

[24] Schultz DM. Inferior hypogastric plexus blockade: a transsacral approach. Pain Physician 2007;10(6):757–63.

[25] Ferreira F, Pedro A. Ganglion impar neurolysis in the management of pelvic and perineal cancer-related pain. Case Rep Oncol 2020;13(1):29–34. Available from: https://doi.org/10.1159/000505181.

[26] Sindt JE, Brogan SE. Interventional treatments of cancer pain. Anesthesiol Clin 2016;34(2):317–39. Available from: https://doi.org/10.1016/j.anclin.2016.01.004.

[27] Gunduz OH, Kenis-Coskun O. Ganglion blocks as a treatment of pain: current perspectives. J Pain Res 2017;10:2815–26. Available from: https://doi.org/10.2147/JPR.S134775.

[28] Oliveira J, Bem GAA. Ganglion impar block in chronic cancer-related pain – a review of the current literature. Rev Esp Anestesiol Reanim (Engl Ed) 2024;71(8):608–18. Available from: https://doi.org/10.1016/j.redare.2024.04.013.

[29] Kyriakou C, Molloy S, Vrionis F, et al. The role of cement augmentation with percutaneous vertebroplasty and balloon kyphoplasty for the treatment of vertebral compression fractures in multiple myeloma: a consensus statement from the International Myeloma Working Group (IMWG). Blood Cancer J 2019;9(3):27. Available from: https://doi.org/10.1038/s41408-019-0187-7.

[30] Hogan WB, Philips A, Alsoof D, et al. Kyphoplasty and vertebroplasty performed by surgeons versus nonsurgeons: trends in procedure rates, complications, and revisions. World Neurosurg 2022;164:e518–24. Available from: https://doi.org/10.1016/j.wneu.2022.05.004.

[31] Deer TR, Pope JE, Hayek SM, et al. The polyanalgesic consensus conference (PACC): recommendations for intrathecal drug delivery: guidance for improving safety and mitigating risks. Neuromodulation 2017;20(2):155–76. Available from: https://doi.org/10.1111/ner.12579.

[32] Kushchayev SV, Wiener PC, Teytelboym OM, et al. Percutaneous vertebroplasty: a history of procedure, technology, culture, specialty, and economics. Neuroimag Clin N Am 2019;29(4):481–94. Available from: https://doi.org/10.1016/j.nic.2019.07.001.

[33] Abd-Elsayed A, Luo S, Falls C. Transversus abdominis plane block as a treatment modality for chronic abdominal pain. Pain Physician 2020;23(4):405–12.

[34] Yang A, Nadav D, Legler A, et al. An interventional pain algorithm for the treatment of postmastectomy pain syndrome: a single-center retrospective review. Pain Med 2021;22(3):677–86. Available from: https://doi.org/10.1093/pm/pnaa343.

[35] Zhen-Hao L, Wei-Jin H, Xiao-Liang G, et al. Serratus plane block in breast cancer surgery: a systematic review and meta-analysis. Clin Breast Cancer 2023;23:1–13. Available from: https://doi.org/10.1016/j.clbc.2022.10.009.

[36] Shaikh S, Banatwala UESS, Desai P, et al. Analgesic efficacy and safety of erector spinae plane block versus serratus anterior plane block in breast surgery – a meta-analysis and systematic review of randomized controlled trials. J Anesth Analg Crit Care 2024;4(1):82. Available from: https://doi.org/10.1186/s44158-024-00218-7.

Chapter 22

Surgical interventions

Jagoš Golubović[1,2], Angelo Restivo[3], Bojan Jelača[1,2], and Dragan Nikolić[1,4]
[1]*Faculty of Medicine, University of Novi Sad, Novi Sad, Serbia,* [2]*Department of Neurosurgery, Clinical Centre of Vojvodina, Novi Sad, Serbia,* [3]*Faculty of Medicine, University of Cagliari, Cagliari, Italy,* [4]*Department of Vascular and Endovascular Surgery, Clinical Centre of Vojvodina, Novi Sad, Serbia*

Key learning objectives

- Evaluate patient selection and surgical candidacy: This chapter will assess indications for surgical pain interventions in patients with cancer.
- Match pain types to appropriate surgical approaches: This chapter will show how different types of cancer pain should be treated by different interventions.
- Apply core palliative surgical techniques: This chapter will explain key surgical procedures, including their indications and expected outcomes.
- Navigate decision-making in palliative surgery, including ethical considerations.

Introduction

Pain is one of the most feared symptoms among cancer patients, particularly in the palliative care setting, where the focus shifts from curative treatment to comfort care. Although pharmacological approaches, including opioids and adjuvant therapies, form the foundation of pain management, they are not always sufficient, and in some cases, surgical interventions can provide effective pain relief by directly addressing the underlying causes of pain or by interrupting pain pathways [1,2].

The role of surgery in palliative care is distinct from that in curative settings. The primary goals are to reduce pain, manage symptoms, and improve or maintain quality of life rather than to extend life or cure the disease. This chapter delves into the surgical strategies available for managing cancer-related pain, discussing the indications, techniques, and outcomes of various procedures. It also emphasizes the importance of a multidisciplinary approach, ethical considerations, and the integration of emerging technologies in enhancing patient care [3,4].

Indications for surgical interventions

Surgical interventions should be considered when conservative treatments fail to provide adequate pain relief or when there is a clear anatomical basis for the pain that surgery can address. The decision to proceed with surgery in palliative care requires careful consideration of the patient's overall condition, prognosis, and personal preferences [5].

Patient selection

Selecting appropriate candidates for surgical intervention is critical to optimizing outcomes and minimizing harm. Cancer pain specialists must evaluate several factors:

Life expectancy: Accurate estimation of life expectancy is crucial. Surgical interventions are generally more appropriate for patients with a life expectancy of several months or more, where the benefits of pain relief can be sustained over time. For patients with very limited prognoses, the focus may shift toward less invasive methods or nonsurgical pain management strategies [6].

Performance status: The patient's performance status, assessed using tools like the Karnofsky Performance Scale or the Eastern Cooperative Oncology Group performance status, provides insight into their ability to tolerate surgery

Pain Management in Palliative Care for Patients With Cancer. DOI: https://doi.org/10.1016/B978-0-443-30050-9.00049-7

and recover postoperatively. Patients with better performance status are more likely to benefit from surgical interventions [7].

Comorbid conditions: Comorbidities such as cardiovascular disease, diabetes, or renal impairment can increase the risk of surgical complications. A thorough preoperative assessment should include evaluation of these factors, and multidisciplinary discussions can help determine the best course of action [8].

Psychosocial and emotional considerations: The psychological state of the patient, including factors such as anxiety, depression, and overall mental resilience, can influence surgical outcomes. A comprehensive assessment by a psycho-oncologist or a palliative care specialist can help identify patients who might benefit from additional psychological support before and after surgery [9].

Types of pain amenable to surgical intervention

Understanding the pathophysiology of cancer pain is essential for determining the most appropriate surgical intervention. Cancer pain is often categorized as somatic, neuropathic, or visceral, each with distinct characteristics and mechanisms [10]. The details of different types of pain are evaluated in another chapter of this book. Briefly, however, somatic pain is typically well localized, often predictable, and correlates with movement or mechanical stress. For example, a patient with spinal metastases may experience increased pain with axial loading or positional changes [11,12]. Early recognition and precise characterization of this pain are essential for preserving patient autonomy and mobility, which are critical quality-of-life determinants in palliative care [13].

Neuropathic pain arises from injury or dysfunction of the nervous system. In cancer patients, this type of pain is often caused by direct tumor invasion of nerves or plexuses, compression of the spinal cord or nerve roots, or as a side effect of treatments such as surgery, chemotherapy, or radiation. Unlike somatic pain, neuropathic pain often lacks clear mechanical triggers and may be associated with sensory deficits, hyperalgesia (increased pain response), or allodynia (pain from normally nonpainful stimuli). One of the major challenges of neuropathic pain is its variable response to opioids, which are often insufficient in monotherapy. This necessitates a broader pain management approach, frequently involving adjuvants such as anticonvulsants or antidepressants and in certain cases, interventional strategies [14–16].

Visceral pain originates from the internal organs and is mediated by autonomic afferent fibers, which makes it diffuse, poorly localized, and often referred to distant anatomical sites. In cancer patients, visceral pain is frequently encountered in conditions like pancreatic carcinoma, hepatobiliary malignancies, and bowel obstructions. One of the defining features of visceral pain is its complexity and resistance to conventional analgesia, especially when caused by organ distension, ischemia, or inflammatory processes associated with tumor growth or lymphatic obstruction [17]. In many cases, treating the underlying cause of the visceral pain—whether obstruction, pressure, or inflammation—can provide more lasting relief than systemic medications alone.

Distinguishing between somatic, neuropathic, and visceral pain is not merely academic—it carries direct therapeutic implications (Table 22.1). Each type of pain engages different anatomical pathways and neurochemical circuits, which

TABLE 22.1 Indications for surgical interventions in palliative care.

Type of pain	Common causes	Examples of surgical interventions	Expected outcomes
Somatic pain	Bone metastases, pathological fractures, chest wall invasion	Orthopedic stabilization (rods, plates), vertebroplasty/kyphoplasty, rib resection or stabilization	Pain relief, improved mobility, and function
Neuropathic pain	Nerve root or plexus compression, spinal metastases, intercostal nerve invasion	Cordotomy, DREZ lesioning, spinal cord stimulation, intercostal neurectomy, intrathecal drug delivery systems	Significant pain relief, reduced opioid use
Visceral pain	Gastrointestinal or biliary obstruction, pelvic tumor burden	Gastrojejunostomy, colostomy/ileostomy, endoscopic stenting (SEMS/EUS-GE), biliary bypass, celiac plexus block, neurolytic procedures	Substantial pain and symptom relief, improved intake
Ischemic pain	Tumor-induced vascular compression or infiltration	Sympathectomy, angioplasty, and stenting	Pain relief, restored perfusion

DREZ, Dorsal root entry zone; *EUS-GE*, endoscopic ultrasound-guided gastroenterostomy; *SEMS*, self-expanding metal stents.

means that their response to analgesic classes, interventional strategies, and surgical options differs substantially. Misclassification can lead to undertreatment, polypharmacy, or procedural interventions that fail to address the actual pain mechanism [18,19]. Moreover, many patients experience mixed pain syndromes, where multiple mechanisms coexist. For instance, a patient with spinal metastases compressing the cauda equina may simultaneously suffer from both somatic (bone involvement) and neuropathic (nerve compression) pain [20,21].

The decision to proceed with surgery in palliative care is multifactorial and involves a detailed assessment of the potential benefits, risks, and alignment with the patient's overall goals of care. The evaluation should include the expected effectiveness of the surgical intervention in providing pain relief, a comprehensive risk assessment, and detailed discussions with the patient should be undertaken before any surgical procedure. Furthermore, operations in a high-volume center and after multidisciplinary input, which should include oncologists, pain specialists, surgeons, anesthesiologists, and palliative care professionals, are recommended [22–25].

Common surgical procedures

A wide range of surgical procedures is available for managing cancer-related pain, each with its own indications, techniques, and outcomes. Below, we explore these procedures in detail (Table 22.2).

Neurosurgical interventions

Neurosurgical procedures are typically reserved for patients with refractory pain syndromes that have failed or are insufficiently controlled with medical management.

Cordotomy: Cordotomy is a well-established technique for managing unilateral somatic pain, particularly in the lower body. The procedure involves the selective severing of spinothalamic tracts, which transmit pain and temperature sensations. Advances in imaging and percutaneous techniques have made cordotomy safer, with reduced risks of complications such as motor deficits or dysesthesias [26,27]. However, careful patient selection is critical, as the procedure is generally not recommended for patients with bilateral pain or those who are unfit for anesthesia.

Dorsal root entry zone (DREZ) lesioning: DREZ lesioning is a highly specialized procedure particularly useful for patients with severe neuropathic pain that is refractory to other treatments. The procedure involves the targeted destruction of neurons in the DREZ, which are responsible for transmitting pain signals from damaged nerves. While effective, DREZ lesioning carries risks of motor deficits and sensory loss, and its use is generally reserved for patients with severe, intractable pain [28].

Spinal cord stimulation (SCS): SCS involves the implantation of a pulse generator and electrodes in the epidural space, which deliver electrical impulses to modulate pain signals. Recent innovations in SCS technology, such as high-frequency stimulation and burst mode, have shown a potential for improving the outcomes for patients with complex pain syndromes. SCS is particularly useful in patients with pain that is not well controlled by opioids or those who experience significant side effects from systemic medications [29].

Intrathecal drug delivery systems (IDDS): IDDS are used to deliver medications directly into the cerebrospinal fluid, allowing for lower doses and reduced systemic side effects. These systems are particularly useful for patients with severe pain who require high doses of opioids or those with significant side effects from oral or parenteral medications. IDDS can be used to deliver a variety of analgesics, including opioids, local anesthetics, and adjuvant drugs such as clonidine or baclofen. The use of programmable pumps allows for precise titration of medication and adjustment of dosing schedules based on the patient's needs [30,31].

Orthopedic surgical interventions

Orthopedic interventions are primarily aimed at stabilizing skeletal structures compromised by metastases, thereby alleviating pain and preventing fractures. These procedures are particularly beneficial in patients with metastatic bone disease, where the risk of pathological fractures and spinal cord compression is high [32].

Vertebroplasty and kyphoplasty: These minimally invasive procedures involve the injection of bone cement into a collapsed or weakened vertebra, restoring its structural integrity. Kyphoplasty includes balloon inflation before cement injection to restore vertebral height. Both procedures have demonstrated efficacy in reducing pain and improving mobility in patients with spinal metastases [33].

Surgical stabilization of long bones: Patients with metastases to the femur, humerus, or pelvis may benefit from intramedullary nailing, plating, or endoprosthetic replacement. The primary goal is to prevent or treat pathological

TABLE 22.2 Comparative outcomes of common surgical procedures.

Procedure	Indications	Pain relief duration	Complications	Patient selection criteria
Cordotomy	Unilateral somatic or pelvic pain	Up to 6–12 months	Motor deficits, dysesthesias	Limited prognosis, localized unilateral pain
DREZ lesioning	Brachial or lumbosacral plexus pain, root avulsion	Long-term	Neurological deficits, dysesthesia	Intractable neuropathic pain unresponsive to pharmacotherapy
Spinal cord stimulation	Refractory neuropathic pain, lower extremity or trunk	Variable (long-term)	Device migration, infection	Moderate prognosis, opioid-resistant pain
Intrathecal drug delivery	Severe diffuse pain requiring high-dose opioids	Long-term (programmable)	Pump malfunction, infection	Opioid-resistant pain, systemic toxicity from oral opioids
Vertebroplasty/ kyphoplasty	Vertebral metastases with collapse or pain	Immediate to long-term	Cement leakage, infection	Painful compression fractures from metastatic disease
Orthopedic stabilization	Pathologic fractures or risk of fracture (e.g., femur, humerus)	Immediate to long-term	Surgical site infection, implant failure	Ambulatory patients with bone instability and good recovery potential
Prophylactic fixation	Impending fracture due to metastatic bone lesion	Preventive	Hardware complications	Lesions >50% cortex, anticipated survival ≥4–6 weeks
Gastrojejunostomy	Gastric outlet obstruction due to malignancy	Weeks to months	Anastomotic leak, delayed gastric emptying	Good functional status, moderate prognosis
Endoscopic stenting (SEMS/EUS-GE)	Gastric or biliary obstruction in poor surgical candidates	Weeks to several months	Stent migration, reobstruction	Frail patients, short prognosis, need for rapid palliation
Colostomy/ileostomy	Malignant bowel obstruction	Long-term or until death	Stoma complications, infection	Unresectable obstruction, symptomatic relief goal
Biliary bypass/ cholecystojejunostomy	Malignant biliary obstruction	Weeks to months	Anastomotic leak, infection	Failed endoscopy or complex hilar blockages
Pleurodesis	Recurrent malignant pleural effusions	Weeks to months	Fever, pain, empyema	Recurrent effusions, good performance status
Indwelling pleural catheter	Recurrent effusions in poor surgical candidates	Variable (intermittent)	Infection, blockage	Short prognosis, outpatient management preferred
Airway stenting	Central airway obstruction (e.g., trachea, bronchi)	Weeks to months	Stent migration, bleeding, obstruction	Nonsurgical candidates with dyspnea or stridor
Chest wall tumor resection/stabilization	Rib metastases causing localized pain	Long-term	Wound healing issues, respiratory compromise	Good functional status, limited disease burden
Celiac plexus/neurolytic blocks	Severe visceral pain (e.g., pancreatic, upper abdominal tumors)	Weeks to months	Hypotension, transient diarrhea, neuropathy	Severe visceral pain, high opioid burden
Sympathectomy	Refractory ischemic limb pain; selected *cases of CRPS*	Long-term	Hypotension, compensatory *sweating*	Refractory ischemic or sympathetically *mediated pain*
Angioplasty and stenting	Tumor-induced vascular compression (e.g., iliac, SVC)	Immediate to long-term	Vascular injury, restenosis	Clear vascular compromise with clinical symptoms

CRPS, Complex regional pain syndrome; *DREZ*, dorsal root entry zone; *EUS-GE*, endoscopic ultrasound-guided gastroenterostomy; *SVC*, superior vena cava; *SEMS*, self-expanding metal stents.

fractures and maintain weight-bearing function. These interventions improve pain, promote mobility, and can often be performed with palliative intent without extensive recovery periods [34].

Oncological surgical procedures

Oncological surgeries focus on reducing tumor burden and alleviating pain caused by mass effect, nerve compression, or obstruction.

Palliative tumor resections: In cases where tumors are causing significant pain or functional impairment, palliative tumor resection can provide substantial symptom relief. These procedures are typically performed with the goal of debulking the tumor, reducing mass effect, and alleviating pain. Tumor resections may be performed in various locations, including the spine, pelvis, and extremities. Advances in surgical techniques, such as image-guided surgery and intraoperative monitoring, have improved the safety and effectiveness of these procedures [35].

Debulking surgeries: Debulking involves the partial removal of a tumor to alleviate symptoms caused by the mass effect, such as pain, obstruction, or neurological deficits. While not curative, debulking can significantly improve quality of life by reducing tumor burden and relieving pressure on adjacent structures. Debulking surgeries are often performed in conjunction with other therapies, such as radiation or chemotherapy, to maximize symptom relief [36].

Neurolytic procedures: Neurolytic procedures, such as celiac plexus block or splanchnic nerve block, involve the chemical destruction of pain-conducting nerves to provide pain relief. These procedures are particularly effective for visceral pain associated with abdominal or pelvic cancers. The choice of neurolytic agent, such as alcohol or phenol, depends on the desired duration of pain relief and the patient's overall condition. Neurolytic procedures can be performed percutaneously, under imaging guidance, or surgically during an open procedure [16,37].

Vascular surgical procedures

Vascular surgeries may be indicated for pain related to ischemia or vascular compression caused by tumors or other malignancies.

Sympathectomy: Sympathectomy involves the surgical destruction of sympathetic nerves to improve blood flow and relieve ischemic pain. This procedure is particularly useful in patients with peripheral vascular disease or complex regional pain syndrome who experience severe, refractory pain. Sympathectomy can be performed through various approaches, including thoracoscopic, laparoscopic, or open surgery, depending on the location of the sympathetic chain and the patient's overall condition [38].

Angioplasty and stenting: In cases of vascular compression by tumors, angioplasty and stenting can relieve pain by restoring blood flow and reducing ischemia. These procedures are particularly useful in patients with tumors that compress major vessels, such as the superior vena cava or iliac arteries. Angioplasty and stenting can provide immediate symptom relief and improve the patient's quality of life [39].

Abdominal surgical interventions

Abdominal surgeries are frequently used in palliative care to relieve symptoms caused by gastrointestinal or hepatobiliary obstruction, thereby reducing visceral pain and improving nutritional intake and comfort.

Gastrojejunostomy (GJO): GJO is a surgical bypass procedure that connects the stomach directly to the jejunum, effectively relieving gastric outlet obstruction caused by tumor compression or infiltration—commonly seen in pancreatic or gastric cancer. The procedure can be performed via open or laparoscopic techniques, with the latter offering shorter recovery times and fewer complications. GJO is typically considered for patients with moderate life expectancy and good functional status, as it offers durable symptom relief and allows resumption of oral intake [40,41].

Endoscopic stenting: Self-expanding metal stents or endoscopic ultrasound-guided gastroenterostomy are minimally invasive alternatives to GJO for patients with shorter life expectancy or high surgical risk. These interventions provide rapid relief of obstructive symptoms, often within 24–48 hours, and significantly reduce hospital stay. However, stent migration or tumor ingrowth may necessitate repeat procedures in a subset of patients [42].

Colostomy and ileostomy: Diverting stomas are commonly created for patients with malignant bowel obstruction due to colorectal or peritoneal metastases. These procedures can effectively decompress the bowel, relieve abdominal pain, and reduce the risk of perforation. Though not without psychosocial implications, stoma formation can dramatically improve quality of life in appropriately selected patients [43].

Biliary bypass and cholecystojejunostomy: Surgical decompression of the biliary system is indicated in patients with malignant biliary obstruction when endoscopic options have failed or are not feasible. Hepaticojejunostomy is the most durable option, connecting the hepatic duct to the jejunum to bypass the obstructed biliary tree. In select cases where access is limited, cholecystojejunostomy—an anastomosis between the gallbladder and the jejunum—may be used. These procedures can relieve cholestatic pain, pruritus, and jaundice, thus improving patient comfort [44,45].

Thoracic surgical interventions

Thoracic procedures play a critical role in alleviating dyspnea, pleuritic pain, and airway obstruction caused by intrathoracic malignancies, particularly lung cancer, mesothelioma, or metastases to the pleura or mediastinum.

Pleurodesis: Pleurodesis is used to prevent recurrent malignant pleural effusions that cause dyspnea and chest pain. The procedure involves the introduction of a sclerosing agent, most commonly sterile talc, into the pleural space, leading to fibrosis and adhesion of the pleural layers. It can be performed via thoracostomy or thoracoscopy depending on the patient's condition. Pleurodesis can offer durable relief for patients with high-output effusions and can significantly reduce the need for repeated thoracenteses [46].

Thoracostomy with indwelling pleural catheter (IPC): For patients with limited life expectancy or poor functional status, placement of an IPC provides a minimally invasive solution for ongoing pleural drainage. This outpatient-based approach enables symptom control with minimal disruption to patient autonomy and is especially effective in patients with rapid fluid reaccumulation or loculated effusions [47].

Airway stenting: Malignant airway obstruction due to tumor infiltration or external compression can lead to dyspnea, cough, and stridor. Endobronchial stenting with silicone or metallic stents restores airway patency and provides immediate symptom relief. This procedure is often performed under bronchoscopic guidance and is particularly useful in patients who are not candidates for surgical resection or external beam radiation, although these methods can be used in conjunction with stenting [48].

Chest wall tumor resection and stabilization: In cases of painful chest wall involvement due to tumor invasion of the ribs, intercostal nerves, or soft tissue, surgical resection or stabilization can significantly reduce pain and improve respiratory mechanics. Selected patients may benefit from en bloc resection of tumor-affected ribs followed by reconstruction with mesh or titanium plating. These procedures are typically reserved for localized disease with a high symptom burden and stable systemic disease [49].

Preoperative and intraoperative technique considerations

A thorough preoperative assessment is essential to optimize outcomes and minimize risks in palliative surgical interventions. This assessment should include not only the patient's physical condition but also their psychological state and overall goals of care. A multidisciplinary team should always be involved, and psychological assessment and support should be available as anxiety, depression, and fear can all impact the patient's ability to cope with surgery and recover postoperatively. Finally, clear communication is essential to align the patient's expectations with the likely outcomes of the surgery, as in some cases, patients may choose to forego surgery in favor of less invasive options or nonsurgical pain management strategies [50–53].

The intraoperative phase of care is critical to the success of surgical interventions in palliative care. Several factors must be managed to optimize outcomes and minimize complications.

Minimally invasive techniques: Minimally invasive surgical techniques, such as laparoscopy, endoscopy, and percutaneous interventions, are preferred in palliative care settings due to their lower morbidity and faster recovery times. These techniques can be particularly useful in patients with limited life expectancy or those who are at high risk for surgical complications. Advances in imaging and surgical technology have expanded the range of procedures that can be performed minimally invasively, offering new options for patients who may not have been candidates for surgery in the past [54].

Anesthesia considerations: Anesthesia management in palliative care patients presents unique challenges, particularly in those with multiple comorbidities or altered physiology due to advanced disease. The anesthesiologist must carefully balance the need for adequate pain control with the risks of anesthesia-related complications, such as respiratory depression or cardiovascular instability. Regional anesthesia techniques, such as epidural or spinal anesthesia, may be preferred in some cases to reduce the risks associated with general anesthesia. The use of multimodal analgesia, including the use of nonopioid analgesics, local anesthetics, and adjuvant drugs, can help optimize pain control while minimizing opioid-related side effects [55].

Techniques to minimize postoperative complications: Surgical teams should employ strategies to minimize postoperative complications, such as infection prevention, meticulous surgical technique, and close monitoring of the patient's status during and after the procedure. Infection prevention measures may include the use of prophylactic antibiotics, careful skin preparation, and the use of sterile techniques during surgery. Meticulous surgical technique, including careful handling of tissues and precise hemostasis, can help reduce the risk of complications such as bleeding or delayed wound healing. Close monitoring of the patient's status, including vital signs, pain levels, and wound healing, can help identify and address complications early, improving overall outcomes [56].

Postoperative care and pain management

Effective postoperative care is essential to ensure pain relief and address any complications that arise after surgery.

Immediate postoperative pain management strategies: Pain management immediately following surgery should be aggressive and tailored to the individual patient. Options include multimodal analgesia, regional anesthesia techniques, and continuous pain assessments. Multimodal analgesia involves the use of multiple analgesic agents with different mechanisms of action, such as opioids, nonsteroidal antiinflammatory drugs, and local anesthetics, to achieve synergistic pain relief while minimizing side effects. Regional anesthesia techniques, such as epidural or spinal anesthesia, can provide excellent pain control while reducing the need for systemic opioids. Continuous pain assessments, using validated pain scales, can help ensure that pain is effectively managed and that the analgesic regimen is adjusted as needed [57].

Monitoring and managing complications: Close monitoring for surgical complications, such as infection, bleeding, or thromboembolism, is essential in the postoperative period. Surgical teams should be vigilant for signs of complications, such as fever, redness or swelling at the surgical site, or changes in vital signs, and should take prompt action to address any issues that arise [58].

Rehabilitation and supportive care: Postoperative rehabilitation may be necessary to restore function and mobility, particularly in patients who have undergone orthopedic or neurological surgeries. Supportive care, including physical therapy and occupational therapy, plays a critical role in the recovery process. Physical therapy can help improve strength, flexibility, and mobility, while occupational therapy can help patients adapt to any functional limitations and improve their ability to perform activities of daily living. In some cases, additional supportive care, such as nutritional support or psychological counseling, may be necessary to address the patient's overall needs [59].

Role of continuous pain assessment: Continuous pain assessment, using validated pain scales, can help ensure that pain is effectively managed and that the analgesic regimen is adjusted as needed. In some cases, additional interventions, such as nerve blocks or neurostimulation, may be necessary to manage persistent pain [60,61].

Ethical considerations

Ethical considerations are paramount when deciding on surgical interventions in palliative care. While ethical considerations are evaluated in another chapter of this book, briefly, respecting the patient's autonomy is crucial, and no decision should be made without informed consent. The potential benefits of surgery must be carefully weighed against the risks, particularly in patients with limited life expectancy, and any concern of patients, family, or caregivers must be taken into account. In some cases, family members may need additional support, such as counseling or education, to help them understand the risks and benefits of the procedure and to make informed decisions on behalf of the patient. Additionally, in patients nearing the end of life, the focus should be on comfort and quality of life rather than aggressive interventions. Surgical procedures should only be considered if they are likely to provide significant symptom relief with minimal risk [62,63].

Future directions and research

The field of palliative care surgery is continually evolving, with emerging techniques and technologies offering new possibilities for pain management. Advances in minimally invasive surgery, robotics, and image-guided procedures hold promise for reducing the invasiveness of surgical interventions, minimizing recovery time, and improving outcomes in palliative care patients. For example, the use of robotic surgery in palliative tumor resections may allow for more precise removal of tumor tissue while preserving surrounding structures and minimizing morbidity. Similarly, advances in image-guided techniques, such as intraoperative MRI or computed tomography (CT) scan, may improve the accuracy and safety of procedures such as cordotomy or vertebroplasty [64].

Technologies such as augmented reality, artificial intelligence (AI), and precision medicine are increasingly being integrated into surgical practice, offering potential for more personalized and effective pain management strategies. For example, AI algorithms may help identify patients who are most likely to benefit from specific surgical interventions, while precision medicine approaches, such as the use of genetic or molecular profiling, may guide the selection of targeted therapies or neurolytic agents. Augmented reality technology may also be used to enhance surgical planning and execution, improving the accuracy and safety of complex procedures [65,66].

Despite progress, significant gaps remain in the evidence base for surgical interventions in palliative care. Future research should focus on the long-term outcomes of surgical interventions, patient-reported outcomes, and the integration of new technologies into palliative care practice [22,67].

Conclusion

Surgical interventions play a vital role in the multidisciplinary management of pain in patients with cancer in palliative care. While not curative, these procedures can provide significant pain relief, improve function, and enhance the quality of life for patients with advanced cancer. A patient-centered approach, guided by ethical principles, multidisciplinary collaboration, and ongoing research, is essential to optimize outcomes and ensure that surgical interventions align with the patient's overall goals of care. As the field continues to evolve, ongoing research and innovation will be critical for advancing the care of palliative patients and improving their quality of life.

AI disclosure

During the preparation of this work, the author(s) used ChatGPT in order to assist with proofreading, formatting, and grammar checking. After using this tool/service, the author(s) reviewed and edited the content as needed and take(s) full responsibility for the content of the publication.

References

[1] Allam AK, Viswanathan A, Schwalb JM, et al. Editorial: Surgical techniques for the management of pain. Front Pain Res 2023;4:1120174. Available from: https://doi.org/10.3389/fpain.2023.1120174.

[2] Alorfi NM. Pharmacological methods of pain management: narrative review of medication used. Int J Gen Med 2023;16:3247–56. Available from: https://doi.org/10.2147/IJGM.S419239.

[3] Kettyle G. Multidisciplinary approach to cancer pain management. Ulster Med J 2023;92(1):55–8.

[4] Habib MH, Schlögl M, Raza S, et al. Interventional pain management in cancer patients—a scoping review. Ann Palliat Med 2023;12:1198–214. Available from: https://doi.org/10.21037/apm-23-433.

[5] Deo SVS, Kumar N, Rajendra VKJ, et al. Palliative surgery for advanced cancer: clinical profile, spectrum of surgery and outcomes from a tertiary care cancer centre in low-middle-income country. Indian J Palliat Care 2021;27:281–5. Available from: https://doi.org/10.25259/ijpc_399_20.

[6] Lin C-C, Lai Y-L, Lo E-C. Life-extending therapies among patients with advanced cancer: patients' levels of pain and family caregivers' concerns about pain relief. Cancer Nurs 2001;24(6):430–5. Available from: https://doi.org/10.1097/00002820-200112000-00003.

[7] Freeman M, Ennis M, Jerzak KJ. Karnofsky performance status (KPS) ≤60 is strongly associated with shorter brain-specific progression-free survival among patients with metastatic breast cancer with brain metastases. Front Oncol 2022;12:867462. Available from: https://doi.org/10.3389/fonc.2022.867462.

[8] Zhao X, Gao S, Kamarajah SK, et al. Associations of multimorbidity with mortality, hospital stay, and hospitalization costs in Chinese surgical patients: a retrospective cohort study. BMC Anesthesiol 2025;25:407. Available from: https://doi.org/10.1186/s12871-025-03246-0.

[9] Bokhari SFH, Waseem AB, Raza H, et al. Impact of psychiatric disorders on surgical outcomes: a comprehensive review of preoperative screening and interventions. World J Surg Proced 2025;15(1):104178. Available from: https://doi.org/10.5412/wjsp.v15.i1.104178.

[10] Nikitas J, Yanagawa J, Sacks S, et al. Pathophysiology and management of chest wall pain after surgical and non-surgical local therapies for lung cancer. JTO Clin Res Rep 2024;5:100690. Available from: https://doi.org/10.1016/j.jtocrr.2024.100690.

[11] Haroun R, Wood JN, Sikandar S. Mechanisms of cancer pain. Front Pain Res 2023;3:1030899. Available from: https://doi.org/10.3389/fpain.2022.1030899.

[12] Jin Q, Chang Y, Lu C, et al. Referred pain: characteristics, possible mechanisms, and clinical management. Front Neurol 2023;14:1104817. Available from: https://doi.org/10.3389/fneur.2023.1104817.

[13] Pahuta M, Laufer I, Lo S-FL, et al. Defining spine cancer pain syndromes: a systematic review and proposed terminology. Global Spine J 2025;15(1_suppl.):81s–92s. Available from: https://doi.org/10.1177/21925682241259686.

[14] Moisset X. Neuropathic pain: evidence-based recommendations. Presse Med 2024;53:104232. Available from: https://doi.org/10.1016/j.lpm.2024.104232.

[15] Zeng X, Powell R, Woolf CJ. Mechanism-based nonopioid analgesic targets. J Clin Invest 2025;135(11):e191346. Available from: https://doi.org/10.1172/JCI191346.

[16] Dash SS, Khatri NK, Divyateja S, et al. Advances in neuropathic pain management: a review of real-world studies. J Curr Res Sci Med 2024;10(1):15–24. Available from: https://doi.org/10.4103/jcrsm.jcrsm_132_23.

[17] Kuhlmann L, Olesen SS, Drewes AM. Assessment of visceral pain with special reference to chronic pancreatitis. Front Pain Res 2022;3:1067103. Available from: https://doi.org/10.3389/fpain.2022.1067103.

[18] Cao B, Xu Q, Shi Y, et al. Pathology of pain and its implications for therapeutic interventions. Signal Transduct Target Ther 2024;9(1):155. Available from: https://doi.org/10.1038/s41392-024-01845-w.

[19] Leone CM, Truini A. Understanding neuropathic pain: the role of neurophysiological tests in unveiling underlying mechanisms. J Anesth Analg Crit Care 2024;4(1):77. Available from: https://doi.org/10.1186/s44158-024-00212-z.

[20] Hochberg U, Ingelmo P, Solé E, et al. Early interventional treatments for patients with cancer pain: a narrative review. J Pain Res 2023;16:1663–71. Available from: https://doi.org/10.2147/JPR.S405808.

[21] Rugnath R, Orzechowicz C, Newell CR, et al. A literature review: the mechanisms and treatment of neuropathic pain—a brief discussion. Biomedicines 2024;12(1):204. Available from: https://doi.org/10.3390/biomedicines12010204.

[22] Sun V, Krouse RS. Palliative surgery: incidence and outcomes. Semin Oncol Nurs 2014;30(4):234–41. Available from: https://doi.org/10.1016/j.soncn.2014.08.005.

[23] Louie AD, Miner TJ. Palliative surgery and the surgeon's role in the palliative care team: a review. Ann Palliat Med 2021;11(2):907–17. Available from: https://doi.org/10.21037/apm-20-2381.

[24] Aman M, Mahmoud A, Deer TR, et al. The American Society of Pain and Neuroscience (ASPN) best practices and guidelines for the interventional management of cancer-associated pain. J Pain Res 2021;14:2139–64. Available from: https://doi.org/10.2147/JPR.S315585.

[25] Lilley EJ, Farber ON, Cooper Z. Palliative surgery: state of the science and future directions. Br J Surg 2024;111(3):znae068. Available from: https://doi.org/10.1093/bjs/znae068.

[26] Lu K-Y, Lin F-S, Lin C-S, et al. Percutaneous cervical cordotomy for managing refractory pain in a patient with a pancoast tumor: a case report. World J Clin Cases 2024;12:4770–6. Available from: https://doi.org/10.12998/wjcc.v12.i21.4770.

[27] Strauss I, Gabay S, Shofty B. Cordotomy for intractable cancer pain: a historical and technical narrative review with modern perspectives. Stereotact Funct Neurosurg 2025;103(6):511–522. Available from: https://doi.org/10.1159/000548954.

[28] Du T, Ni B, Shu W, et al. Dorsal root entry zone lesioning following unresponsive spinal cord stimulation for post-traumatic neuropathic pain. World Neurosurg 2023;178:e300–6. Available from: https://doi.org/10.1016/j.wneu.2023.07.048.

[29] Zhu C, Esteller R, Block J, et al. Exploratory evaluation of spinal cord stimulation with dynamic pulse patterns: a promising approach to improve stimulation sensation, coverage of pain areas, and expected pain relief. Front Pain Res 2024;4:1339892. Available from: https://doi.org/10.3389/fpain.2023.1339892.

[30] Goel V, Kumar V, Blaes A, Gulati A. Intrathecal drug delivery systems for cancer pain control: insights on current contemporary practices in the US. Neuromodulation. 2023;26(6):1256–62. Available from: https://doi.org/10.1016/j.neurom.2023.05.001.

[31] Chen Q, Yang Z, Liu H, et al. Novel drug delivery systems: an important direction for drug innovation research and development. Pharmaceutics 2024;16:674. Available from: https://doi.org/10.3390/pharmaceutics16050674.

[32] Nowak H, Szwacka DM, Pater M, et al. Holistic approach to the diagnosis and treatment of patients with tumor metastases to the spine. Cancers 2022;14:3480. Available from: https://doi.org/10.3390/cancers14143480.

[33] Giammalva GR, Costanzo R, Paolini F, et al. Management of spinal bone metastases with radiofrequency ablation, vertebral reinforcement and transpedicular fixation: a retrospective single-center case series. Front Oncol 2022;11:818760. Available from: https://doi.org/10.3389/fonc.2021.818760.

[34] Willeumier JJ, van der Linden YM, van de Sande MAJ, et al. Treatment of pathological fractures of the long bones. EFORT Open Rev 2016;1:136–45. Available from: https://doi.org/10.1302/2058-5241.1.000008.

[35] Faier TAS, Queiroz FL, Lacerda-Filho A, et al. Surgical treatment of rectal cancer: prospective cohort study about good oncologic results and low rates of abdominoperineal excision. Rev Col Bras Cir 2023;50:e20233435. Available from: https://doi.org/10.1590/0100-6991e-20233435-en.

[36] Hishida T, Masai K, Kaseda K, et al. Debulking surgery for malignant tumors: the current status, evidence and future perspectives. Jpn J Clin Oncol 2021;51:1349–62. Available from: https://doi.org/10.1093/jjco/hyab107.

[37] Hassan AU, Buch V, Shah P. Advancement in pain relief neurosurgical techniques for pain relief. Univers J Clin Med 2013;1:6–11. Available from: https://doi.org/10.13189/ujcm.2013.010102.

[38] Bandyk DF, Johnson BL, Kirkpatrick AF, et al. Surgical sympathectomy for reflex sympathetic dystrophy syndromes. J Vasc Surg 2002;35:269–77. Available from: https://doi.org/10.1067/mva.2002.121065.

[39] Nguyen NP, Borok TL, Welsh JS, et al. Safety and effectiveness of vascular endoprosthesis for malignant superior vena cava syndrome. Thorax 2009;64(2):174–8. Available from: https://doi.org/10.1136/thx.2007.086017.

[40] Rodríguez J, Kutscher ML, Lemus MB, et al. Palliative gastrojejunostomy in unresectable cancer and gastric outlet obstruction: a retrospective cohort study. Ann R Coll Surg Engl 2021;103(3):197–202. Available from: https://doi.org/10.1308/rcsann.2020.7016.

[41] Papanikolaou IS, Siersema PD. Gastric outlet obstruction: current status and future directions. Gut Liver 2022;16:667–75. Available from: https://doi.org/10.5009/gnl210327.

[42] Maetani I. How to successfully administer palliative treatment with a stent for malignant gastric outlet obstruction? Front Med 2022;9:967740. Available from: https://doi.org/10.3389/fmed.2022.967740.

[43] Pattarajierapan S, Manomayangoon C, Tipsuwannakul P, et al. Comparison of colonic stenting and stoma creation as palliative treatment for incurable malignant colonic obstruction. JGH Open 2022;6:630–6. Available from: https://doi.org/10.1002/jgh3.12800.

[44] Han JK, Choi BI, Chung J-W, et al. Cholecystocholedochostomy with use of a self-expandable metallic stent: treatment method for nonfunctioning cholecystojejunostomy. J Vasc Interv Radiol 1996;7:757–60. Available from: https://doi.org/10.1016/s1051-0443(96)70845-x.
[45] Dora RK, Murmu G, Behera SK, et al. A prospective study on different biliary diversions in common bile duct stones. Ann Int Med Dental Res 2017;3(5):SG36-SG39.
[46] Rodríguez-Panadero F, Antony V. Pleurodesis: state of the art. Eur Respir J 1997;10:1648–54. Available from: https://doi.org/10.1183/09031936.97.10071648.
[47] Lazarus DR. Extending the indications for indwelling pleural catheters: a tube for all seasons. J Thorac Dis 2023;15:3501–4. Available from: https://doi.org/10.21037/jtd-23-765.
[48] Bashour SI, Lazarus DR. Airway stents in interventional pulmonology. J Respir 2024;4:62–78. Available from: https://doi.org/10.3390/jor4010006.
[49] McCormack P, Bains MS, Beattie EJ, et al. New trends in skeletal reconstruction after resection of chest wall tumors. Ann Thorac Surg 1981;31:45–52. Available from: https://doi.org/10.1016/s0003-4975(10)61315-x.
[50] Wada S, Sadahiro R, Matsuoka Y, et al. The association between postoperative recovery and psychosocial factors in cancer patients. J Surg 2020;8:9–15. Available from: https://doi.org/10.11648/j.js.20200801.13.
[51] Doyle CE. Preoperative strategies for managing postoperative pain at home after day surgery. J Perianesth Nurs. 1999;14(6):373-9. Available from: https://doi.org/10.1016/s1089-9472(99)80099-9.
[52] Ziegelmann M, Köhler T, Bailey GC, et al. Surgical patient selection and counseling. Transl Androl Urol 2017;6:609–19. Available from: https://doi.org/10.21037/tau.2017.07.19.
[53] Patil JD, Sefen JAN, Fredericks S. Exploring non-pharmacological methods for pre-operative pain management. Front Surg 2022;9:801742. Available from: https://doi.org/10.3389/fsurg.2022.801742.
[54] Lavoué V, Gotlieb WH. Benefits of minimal access surgery in elderly patients with pelvic cancer. Cancers 2016;8:12. Available from: https://doi.org/10.3390/cancers8010012.
[55] Halaszynski TM. Pain management in the elderly and cognitively impaired patient: the role of regional anesthesia and analgesia. Curr Opin Anaesthesiol 2009;22(5):594–9. Available from: https://doi.org/10.1097/ACO.0b013e32833020dc.
[56] Soliman M, Mowafy K, Elshafae A, et al. The efficacy of BloodSTOP iX, surgicel, and gelfoam in vascular operations: first-in-human head-to-head study. Innov Technol Tech Cardiothorac Vasc Surg 2023;18:276–81. Available from: https://doi.org/10.1177/15569845231176291.
[57] Carter AM, Yost S, Tobin J, et al. Pain management for the neurosurgical patient in spinal procedures: overview of historic and new modalities. Explor Drug Sci 2024;2:268–76. Available from: https://doi.org/10.37349/eds.2024.00046.
[58] Elnicki DM, Shockcor WT. Preventing postoperative medical complications. Postgrad Med 1992;92:189–96, 202. https://doi.org/10.1080/00325481.1992.11701449.
[59] Şahın S, Akel S, Zarif M. Occupational therapy in oncology and palliative care [Internet]. Occupational therapy - occupation-focused holistic practice in rehabilitation. InTech; 2017. Available from: http://dx.doi.org/10.5772/intechopen.68463.
[60] Conn DA. Assessment of acute and chronic pain. Anaesth Intensive Care Med 2005;6:14–15. Available from: https://doi.org/10.1383/anes.6.1.14.57132.
[61] Erskine WA, Raine ER, Lindegger G. Assessment and management of chronic pain. S Afr Med J 1986;69(10):621–5.
[62] Jacobs LG. Managing respiratory symptoms at the end of life. Clin Geriatr Med 2003;19:225–39. Available from: https://doi.org/10.1016/S0749-0690(02)00054-X.
[63] Ross EL, Abrahm JL. Preparation of the patient for palliative procedures. Surg Clin North Am 2005;85:191–207. Available from: https://doi.org/10.1016/j.suc.2004.12.005.
[64] Abidin ZU, Naqvi RA, Haider A, et al. Recent deep learning-based brain tumor segmentation models using multi-modality magnetic resonance imaging: a prospective survey. Front Bioeng Biotechnol 2024;12:1392807. Available from: https://doi.org/10.3389/fbioe.2024.1392807.
[65] Gilard V, Derrey S, Marret S, et al. Precision neurosurgery: a path forward. J Pers Med 2021;11:1019. Available from: https://doi.org/10.3390/jpm11101019.
[66] Rau A, Roelz R, Urbach H, et al. Application of augmented reality in percutaneous procedures—rhizotomy of the gasserian ganglion. Oper Neurosurg 2021;21:160–4. Available from: https://doi.org/10.1093/ons/opab155.
[67] Lilley EJ, Cooper Z, Schwarze ML, et al. Palliative care in surgery. Ann Surg 2018;267(1):66–72. Available from: https://doi.org/10.1097/SLA.0000000000002253.

Chapter 23

Physical therapy and manual therapy interventions

Aleksandar Knežević[1,2] and Christopher M. Wilson[3,4]

[1]*Faculty of Medicine, University of Novi Sad, Novi Sad, Serbia,* [2]*Medical Rehabilitation Clinic, University Clinical Center of Vojvodina, Novi Sad, Serbia,* [3]*School of Health Sciences, Oakland University, Rochester, MI, United States,* [4]*Corewell Health, Southfield, MI, United States*

Key learning objectives

Upon completion of this chapter, the reader will be able to:

- Discuss the impact of physical therapy interventions on pain management in patients with advanced cancer.
- Evaluate the type, intensity, frequency, and duration of physical exercise in palliative care patients.
- Determine the need for further study and potential uses of physical exercise in the palliative population.
- Discuss the relevance of manual therapy, muscle relaxation techniques, and reflexology in palliative care.

Introduction

Physical therapy (also often known as physiotherapy) is a skilled service delivered by physical therapists to help individuals to "develop, maintain, and restore maximum movement and functional ability throughout the lifespan" [1]. Physical therapy consists of a comprehensive, holistic evaluation to develop a patient-centered treatment plan. This plan often includes a wide variety of interventions, such as exercises, soft tissue mobilization (commonly known as massage), joint mobilizations, training in functional mobility, ergonomics, patient/family education, and various biophysical treatments based on the patient's clinical presentation. Nearly all of these physical therapy interventions can directly or indirectly treat biomechanical or nociceptive pain; however, this chapter will focus on exercise and noninvasive techniques to manage pain. A significant portion of physical therapy consists of exercises, with pain relief being one of the main desired outcomes [2]. In addition to its known role in preventing malignant diseases, there is a growing awareness of the importance of physical activity and exercise among individuals already diagnosed with cancer. Evidence even suggests that patients with advanced cancer can benefit from incorporating physical exercise regimens into their overall health management [3].

Pain is one of the most disabling symptoms in the oncological population, impacting roughly two-thirds of patients with cancer. Cancer pain refers to a variety of pain disorders with distinct causes, features, and pathological mechanisms [4]. These patients experience various types of pain, not all of which are directly related to the tumor. To give appropriate treatment for patients with cancer experiencing pain, it is crucial to identify whether the discomfort is caused by the tumor, adverse effects of treatment, or other comorbidities [4], hence the need for a comprehensive examination and evaluation. Research indicates that nonpharmacologic therapy, such as physical therapy, can complement pharmacologic treatments to more effectively manage cancer pain [5–8]. This can not only have positive effects for the patient but also provide relief and support for families and caregivers. Physical therapy represents an important part of a multidisciplinary approach in palliative care [9].

This chapter will examine the critical role of common physical therapy interventions in managing pain within the palliative care context for patients with cancer. The impact of physical exercise, manual therapy, and muscle relaxation techniques will be highlighted. Additionally, concepts of reflexology will be discussed in this patient population.

Pain Management in Palliative Care for Patients With Cancer. DOI: https://doi.org/10.1016/B978-0-443-30050-9.00056-4

Exercise and physical activity

The terms "physical activity" and "exercise" are often used interchangeably and generally refer to physical activities undertaken in leisure time with the primary purpose of improving or maintaining physical fitness, physical performance, or health. However, physical activity is not synonymous with exercise. Exercise or training is characterized as a subcategory of physical activity that is planned, structured, repetitive, and serves the purpose of improving or maintaining one or more components of physical fitness, while physical activity is generally unstructured body movement. In terms of health benefits, the amount of recommended exercise depends on the goal, the type of exercise, and the age and condition of the person exercising [10]. Exercise guidelines are still lacking in the field of palliative medicine [11].

The mechanisms by which exercise reduces pain are complex and multifactorial and involve at least biological, physical, biomechanical, emotional, and psychosocial factors. The extent and certainty of an analgesic effect vary depending on the type of pain (nociceptive, neuropathic, or visceral), the location of the pain (shoulder, hands, feet, breast, etc.), the duration of the pain (acute, subacute, or chronic), and the intensity of the pain [12]. As it relates to physical therapy, exercise can be categorized into general exercises or targeted exercises. General exercises are aimed at improving gross muscle strength and mobility, inducing generalized physiologic benefits (increased heart rate and circulation), and usually targeting large muscles such as biceps and quadriceps in a large range of motion. Conversely, physical therapists can utilize targeted exercises to improve the coordination of motor tasks to control pain by restoring normal biomechanical movement patterns. A key example of this would be targeted muscular training to improve proper biomechanical sequencing for elevation of the arm; this would be achieved by strengthening or inhibiting key scapulothoracic muscles that were causing aberrant movement patterns and pain. This targeted exercise may be accompanied by manual therapy such as joint mobilizations and soft tissue mobilizations to restore normal arthrokinematic motion and muscular length and tone. Most literature on therapeutic exercise in the palliative population focuses on general exercise regimens.

Physical exercise in advanced cancer

Clinically, exercise has been widely used as an adjunct therapy in the early or asymptomatic disease phase in patients diagnosed with cancer. Interestingly, some earlier studies have not found a clear positive effect of physical exercise on pain in palliative care patients [11,13]. A key consideration in reviewing any study for efficacy of exercise is if the article reports key details such as duration, intensity, repetitions, position, dosing of resistance, and, critically, patient compliance with exercise performance. Some early studies on exercise with the palliative population lack these details, which limits their generalizability and clinical applicability as well as weakens the ability to establish the relationship of exercise to pain and symptoms.

Physical exercise effectiveness in cancer pain treatment

More recent studies confirm the effectiveness of physical exercises for treating pain in patients with advanced-stage cancer (Table 23.1). Canamero et al. [3] concluded in their systematic review that the patients with advanced-stage cancer benefit from well-designed, supervised, and guided exercise that includes aerobic and/or strengthening exercises. By engaging in these strength and aerobic training regimens, patients' muscle mass increased, and their fatigue, dyspnea, quality of life, and autonomy were all improved. Additionally, patients experienced improved quantity and quality of their sleep. The training regimen that produced the greatest effects was aerobic/strength training sessions [3]. The impact of physical activity on pain reduction has increasingly been the subject of research and is gaining more significance. A metaanalysis conducted by Tanriverdi et al. [14] investigated the impact of the physical exercise interventions (including walking, resistance training, aerobic and resistance training combined, and balance training) on pain in palliative care patients. Seven studies quantified participants' pain and confirmed that the intensity differed significantly between the exercise group and the control group at the end of the intervention period, favoring the exercise group (standardized mean difference [SMD] = −0.29; [95% CI, −0.54 to −0.03]; I^2 = 0%; 6 studies, 243 participants, moderate certainty evidence). This metaanalysis demonstrated that exercise interventions significantly reduced pain and fatigue, improved exercise capacity, and enhanced quality of life in palliative care recipients [14].

Plinsinga et al. [12] designed the first comprehensive systematic review, which assessed the impact of exercise on pain associated with cancer and considered type of cancer, degree of supervision, length and timing of the intervention (during or postcancer treatment), type of pain, measurement method, and exercise mode. The metaanalysis demonstrated that exercise was associated with reductions in pain. In the majority of subgroup analyses (>82%), the direction of effect was in favor of exercise as opposed to standard care. While trends indicated that exercise can reduce pain in all types of cancer analyzed, only trials including women with breast cancer and those including individuals with more than one cancer type

TABLE 23.1 Recent findings of physical exercise effectiveness in cancer.

Authors (year)	Design	Participants (number, gender, age)	Cancer type	Interventions	Conclusions
Canamero et al. [3]	Systematic review and metaanalysis	1072 patients 60% women, 40% men. No data about mean age for the entire sample	Heterogeneous tumors, breast cancer, lung cancer, colorectal cancer, prostate cancer, or gastrointestinal cancer	**Interventions with aerobic training programs**: the duration of the training: 30–150 min per session three parts: initial warm-up (with an average duration of 5–10 min), main phase/training (45 min), and a cool down 5 min of stretching exercises **Interventions with strength training programs**: in most studies strength training was combined with aerobic exercise The duration of strength training was not precisely described (approximately 12 weeks) The most commonly trained muscle groups were biceps, triceps, abdomen, back, quadriceps, and chest The preferred intensity was between 60% and 90% of 1RM (repetition maximum), followed by 2–3 sets, followed by 10–15 repetitions	Aerobic, strength or combined training that is well-designed, guided, and supervised has a positive impact on advanced-stage cancer patients The training prescription with the reported greatest benefits was aerobic/strength training sessions, lasting 60–90 min, with a medium-vigorous intensity (walk test, resistance circuits, and hypertrophy), structured as a warm-up, training, and cool down
Tanriverdi et al. [14]	Systematic review and metaanalysis	1034 patients 537 men and 453 women (one study). No data about gender on other studies Mean age 52–70 years	Mixed, lung cancer, spinal bone metastases, metastatic colorectal cancer, gastrointestinal cancer, NSCLC[c], metastatic breast cancer	Exercise programs ranged in duration from 2 to 12 weeks **Interventions with aerobic training programs:** the intensity of aerobic exercise was described according to maximum heart rate or heart rate reserve, peak work rate, Borg and modified Borg scale[a] **Interventions with strength training programs:** the intensity of resistance exercise was described according to 1RM, 10RM, and OMNI exertion scale[b] The frequency of interventions: 2–5 times a week The duration of interventions: 15 min–2 h	Exercise training, with aerobic exercise, resistance exercise, or combined aerobic and resistance exercise, helps to maintain or improve exercise capacity, pain, fatigue, and quality of life Exercise training represents important support for cancer patients receiving palliative care
Plinsinga et al. [12]	Systematic review and metaanalysis	5877 patients. No data about gender Mean age 57 years	Breast, mixed, prostate, lung, non-Hodgkin lymphoma, gynecological, gastrointestinal cancer, testicular, thyroid, myeloproliferative neoplasm, multiple myeloma	**Exercise type:** only aerobic exercise interventions, resistance only and mixed mode exercise (i.e., aerobic and resistance exercise), other exercise: yoga interventions **Intervention durations:** 2 weeks–12 months, but most interventions 12 weeks or longer	Benefits for cancer-related pain, with small-to-large effect sizes, across various exercise modalities Benefits were independent of the intervention length (whether less than or more than 12 weeks) and the timing of the intervention relative to cancer treatment (benefits were observed both during and after cancer treatment)

(Continued)

TABLE 23.1 (Continued)

Authors (year)	Design	Participants (number, gender, age)	Cancer type	Interventions	Conclusions
Austin et al. [18]	Metaanalysis of randomized controlled trials	1954 patients 78% women, 22% men Mean age 58 ± 8.5 years	Breast, mixed, head, neck and oral, prostate, leukemia, lung, pancreatic	**Exercise type:** aerobic & resistance, aerobic alone, resistance alone **Typical aerobic interventions**: walking, jogging, aerobic classes, cycling, swimming, aqua-aerobics, or organized sports (e.g., tennis, football) **Typical resistance exercises:** weight training (e.g., dumbbells, barbells), resistance elastic band, participant bodyweight or a combination of both	Aerobic and resistance exercise regimens are effective adjuncts in reducing pain in persons with progressive cancer or who have survived it Most studies that included stretching exercises demonstrated a significant decrease of pain connected to cancer implying that stretching might be a helpful component of exercise programs Aerobic and resistance exercise therapy have a high therapeutic fidelity for lowering cancer pain both during and after anticancer treatments
Peters et al. [15]	Systematic review and metaanalysis	13,806 patients. No data about gender and mean age	Breast, prostate, lung, colorectal, head and neck, hematological, bladder, gastrointestinal, ovarian, combined/ nonspecific sites	**Exercise type:** general physical activity, aerobic exercise, resistance exercise, combined aerobic, and resistance exercise, aquatic exercise, yoga, Pilates, dance, Tai chi, or Qigong	Physical activity led to a reduction in body pain Subgroup analyses showed reduced pain after physical activity interventions for breast cancer and for other cancers or multiple malignancies, but not for prostate or lung cancer Resistance and/or aerobic exercise had greater impacts than overall physical activity While some studies found no difference in benefit for type, others found that aerobic exercise had a greater effect than resistance exercise Individual study findings do not indicate a certain ideal kind of exercise Supervised interventions had greater effects than unsupervised ones Effect estimates were larger for therapies including mixed modalities, supervised interventions, and breast cancer

[a] Borg scale—developed to measure the degree of exertion, dyspnea, and exhaustion experienced during physical exercise. The modified Borg Dyspnea Scale represents a variant of the Borg with rating from 0 to 10.
[b] OMNI scale—a scale used to determine the rating of perceived exertion. Users choose a number from 0 to 10 that best correlates to their perceived intensity of physical exertion.
[c] NSCLC—non–small-cell lung cancer.

showed statistically significant outcomes. Subgroup analyses revealed benefits for cancer-related pain, with small-to-large effect sizes, across various exercise modalities. These benefits were independent of the intervention length, the level of supervision during the intervention, and the timing of the intervention relative to cancer treatment. The results of this metaanalysis showed that exercise can reduce pain when compared to nonexercise or usual care interventions. The overall strength of evidence in this research was very low. The authors concluded that engaging in physical activity does not exacerbate pain associated with cancer and could help with pain management [12].

Are supervised interventions more effective than unsupervised?

To find randomized controlled trials (RCTs), randomized crossover studies, and prospective observational studies that examined the relationship between physical activity and pain outcomes in persons living with or beyond cancer, a comprehensive systematic review and metaanalysis was conducted by Peters et al. [15]. This large metaanalysis revealed a reduction in the severity of pain after therapies involving physical activity. Subgroup analyses showed a reduction in pain after physical activity in breast cancer as well as for studies that investigated other or multiple cancer types (except for lung and prostate cancer). Subgroup analyses revealed higher effect estimates for resistance training–focused therapies or combined modalities. It was also clear that supervised interventions had greater effects than unsupervised ones. Individual study findings did not indicate a certain ideal kind of exercise. According to the review, individuals with cancer may experience less pain sensitivity and severity when engaging in physical activity. High heterogeneity, however, reduced the confidence in these results. Given that several of the included trials did not have pain as their primary outcome, it is likely that the interventions were not optimally designed to reduce pain. This could have diminished the potential benefit of the observed effects on pain reduction. Additionally, the type of exercise and the delivery environment appeared to influence the effects. Subgroup metaanalyses revealed that supervised exercise had greater effects than unsupervised exercise, as well as that resistance and/or aerobic exercise had larger effects than general physical activity. This could point to a greater benefit from exercise—defined as structured and planned movement, such as lifting weights or engaging in continuous cardiorespiratory exercise at a predetermined intensity—than from broader, unstructured physical activity. These findings reinforce the concept that physical activity can help cancer patients experience less pain and that it should be included in cancer-specific pain management guidelines as a nonpharmacological treatment. This research does not provide enough information for recommendation of specific physical exercise for specific forms of cancer-related pain [15].

Are certain types of exercises more effective in pain treatment?

In patients living with and beyond cancer, improvements in physical, psychological, and behavioral symptoms are linked to both aerobic and resistance exercise [16]. These exercises have been demonstrated to have antiinflammatory benefits, including reductions in tumor necrosis factor-alpha, interleukin-6, and C-reactive protein—the factors linked to higher pain levels in people with cancer [17]. While several reviews have shown that exercise can help reduce pain in patients living with and beyond cancer, the results are specific to certain types of cancer, and there is a lack of consensus guidelines or a clear systematic approach. Austin et al. [18] recognized the need for a systematic review of RCTs and evaluation of the impact of aerobic and resistance exercise on cancer pain outcomes. The objective of this research was to examine the effects of aerobic and/or resistance exercise on cancer-related pain in patients undergoing cancer treatment and those with a history of cancer, compared to control groups. They examined baseline pain levels prior to initial exercise sessions and pain scores after the last exercise session in order to account for the impact of delayed onset muscle soreness that would have potentially occurred after the first exercise session. A metaanalysis revealed that intervention groups' pain intensity was considerably lower than that of control groups. This provides further evidence that aerobic and resistance exercise regimens are effective adjuncts in reducing pain in persons with progressive cancer or who have survived the disease. These results regarding the beneficial effects of aerobic and resistance exercise can be generalized across a wide range of cancer settings, although this heterogeneity makes it difficult to interpret the results for specific cancer types and cancer stages. Notably, five out of the six studies that included stretching exercises demonstrated a significant decrease in the degree of pain related to cancer. Intriguingly, these results imply that stretching might be a helpful component of exercise programs; however, there is little evidence that stretching alone or in combination with other multimodal exercise interventions has an analgesic impact in cancer and noncancer contexts. A substantial limitation of this study was that the cancer staging was missing. These results demonstrate the high therapeutic fidelity of aerobic and resistance exercise therapy for lowering cancer pain both during and after anticancer treatments. These interventions are characterized as effective, tolerable, and practical. This review suggests that exercise should be a part of the care plan for individuals experiencing discomfort due to cancer, both during and after the cancer treatment. Future research should also explore the potential benefits of short-term, low-intensity exercise regimens for older patients and those receiving advanced care, particularly when they are closely monitored [18].

Summary of physical exercise used in treatment of pain in palliative care

Through recent studies, it has been shown that exercises in the treatment of patients receiving palliative care bring a greater benefit compared to the standard treatment [3,12,14,15,18]. Although the therapies that included exercises showed statistically significant results in pain reduction only for certain types of cancer (e.g., breast cancer), it was shown that there is a trend that exercise helps lessen cancer pain in all types of disease, even in patients with advanced cancer [3,12,15]. Supervised exercise has superiority over unsupervised exercise when it comes to pain treatment. Although physical activity has been shown to have many beneficial effects in palliative care patients, it is suggested that structured and planned exercises are more effective in pain treatment than unstructured physical activities [3,15]. Aerobic and strength exercises were the most investigated interventions, leading to the conclusion that their application, among other physical exercise interventions, has the best results in cancer pain management. However, it should be kept in mind that other such mind–body exercises also have positive effects. Therefore, aerobic, resistance, and stretching exercises, or a combination of these exercises, should be used, but caution should be taken based on the patient's symptoms and tolerance. Exercise-induced analgesia in older patients or those with advanced cancer may necessitate closer supervision and lower levels of exercise intensity compared to younger patients and those with early-stage disease who are able to complete higher-intensity exercise programs [19]. In other words, providers need to adapt the physical exercise "dose" to every patient individually and maybe even modify the dose on a daily or ongoing basis depending on the patient's clinical presentation.

Many aspects of physical exercise in pain treatment for cancer patients in palliative care remain unclear. There may be additional positive effects of physical exercise in this population (e.g., quality of life, functional status, psychological benefits). Future research should focus on the amount of cancer-related pain relief when modifying exercise session length, dose, frequency, and exercise intensity based on patient condition and cancer stage.

Manual therapy

A key component of treatment for pain management is the use of manual therapy techniques, often provided by physical therapists or other trained providers. When provided by physical therapists, manual therapy is defined as "the synergistic application of movement-oriented strategies integrating exercise and manually applied mobilization and manipulation procedures" [20]. These procedures often involve the physical therapist using their hands or other body parts (e.g., elbow, forearm) to mobilize, stretch, deform, or modify articular or neuromuscular soft tissues in an area affected by a disease or disorder. "Manual therapy procedures are delivered along a continuum of skilled passive and active movement that involves the application of skilled mechanical and movement-oriented stimuli that produce a chain of mechanical and neurophysiological responses, including decreases in pain and symptom modulation, increases in range of motion, modulation of inflammatory mediators, autonomic responses, neuromuscular responses, and endocrine responses that are mediated by individual contextual factors" [20]. There are a wide range of manual therapy techniques, including mobilization or manipulation/thrust techniques to soft tissues as well as spinal and peripheral joints, manual lymphatic drainage, manual traction/distraction, connective tissue massage, therapeutic massage, neural tissue mobilization, and passive range of motion [20]. Soft tissue and/or joint mobilizations/manipulations are often done in conjunction with targeted therapeutic exercises to increase the mobility of stiff or hypomobile joints or tissues while concurrently strengthening or stabilizing other areas that may be weak or hypermobile. As many patients with cancer often experience pain from limited mobility from radiation fibrosis syndrome, joint stiffness, disuse, neuromuscular weakness, or altered biomechanics due to cancer or surgery to treat cancer, manual therapy is appropriate for this patient population. Care must be taken to assure that manual therapy techniques are not inappropriately applied to areas with impaired structural integrity or localized medical issues. Maltser et al. [21] recommended caution with manual therapy in areas with poor sensation or circulation, fragile skin, or open wounds. It was also recommended to avoid manual therapy techniques in areas with acute radiation dermatitis, areas with unmanaged or active cancer, or bony instability due to metastasis or osteoporosis [21].

Manual therapy has been shown to be an effective intervention for cancer pain in a variety of studies, including shoulder and thorax pain after breast cancer [22], chemotherapy-induced peripheral neuropathy [23], neck pain after head and neck cancer [24], trismus (limited mouth opening) [25], and pain after lung cancer [26], to name a few. A systematic review and metaanalysis examining manual therapy on cancer pain included 13 RCTs with 1100 participants [27]. The study demonstrated that manual therapy was effective in reducing pain compared to usual care (n = 592; SMD = 0.63 [95% CI, 0.18–1.08]; P = .006). Furthermore, the authors determined that manual therapy in conjunction with an active therapy program (e.g., physical therapy, kinesiotherapy) was more effective than active therapy treatment alone (n = 123; SMD = 0.79 [95% CI, 0.28–1.30]; P = .002). In addition to the likelihood of skilled manual therapy improving the biomechanics of a body region to reduce pain, there are also positive impacts on a patient's emotional and psychosocial status from the simple act of therapeutic touch that accompanies these hands-on techniques [28].

Reflexology and muscle relaxation

Reflexology is the practice of manually pressing on certain body parts (typically the feet or hands) in order to create a response in distant organs or anatomical body regions [29]. Most reflexology points can be found on a diagram or map of the foot area, with different anatomical locations on the foot corresponding to different organs and body regions; there is wide variation between different reflexology maps, and no scientific consensus has been reached for these anatomical points. Reflexology aims to restore balance, promote relaxation, and alleviate stress symptoms. It is a mild and safe supplementary therapy and a popular choice for people with cancer. Reported benefits include reducing cancer symptoms such as pain, dyspnea, fatigue, anxiety, nausea, and vomiting, as well as potentially improving quality of life [30]. At this time, no clear conclusions can be drawn for the effectiveness of reflexology on pain based on multiple systematic reviews; any effectiveness may be from unknown or unrelated means or via the placebo effect [29,31]. It should be noted that reflexology has a negligible adverse effect profile, and despite an absence of conclusive evidence of effectiveness, this treatment is likely harmless and has the potential to improve pain, relaxation, and other symptoms [31].

Relaxation techniques in palliative care can alleviate anxiety, sleep difficulties, chemotherapy-induced nausea, despair, and pain. Relaxation therapy uses various approaches, including Jacobson's progressive muscle relaxation (PMR), breathing exercises, guided imagery, meditation, and Benson's relaxation response. Relaxation is the most commonly reported complementary and alternative medicine technique among cancer survivors (44.3%) and is generally safe. Regular practice is necessary for optimal results; however, once mastered, relaxation therapy may be included in regular routines [30].

It has been noted that PMR therapy is effective in controlling pain in people with cancer. Edmund Jacobson is considered the founder of this technique, which involves tensing and relaxing different skeletal muscle groups followed by breathing techniques to promote relaxation. While this method has been shown to be beneficial in relieving pain in various populations, there are not many research studies focused on cancer-related pain control using PMR in palliative patients. A study conducted by Anshasi et al. [32] assessed the effectiveness of the PMR approach in reducing cancer-related pain in patients receiving palliative care. They compared mean pain intensity and interference scores between the intervention and control groups before, after 4 weeks of treatment, and at a 1-month follow-up after treatment concluded. The PMR group experienced significantly lower pain intensity levels than the control group at the end of the session (after 4 weeks). Also, the study found that the PMR group had significantly lower pain intensity scores than the control group at 1-month follow-up. The PMR approach appears to have an aftereffect, as evidenced by ongoing observed improvements during the follow-up period. Such results in pain relief are likely due to increased endorphin release and distraction from pain. Therefore the PMR approach might reduce pain intensity and reduce pain interference with daily activities in cancer patients receiving palliative care [32].

There is even research that deals with the comparison of reflexology and relaxation effects on pain, anxiety, depression, and quality of life in patients receiving palliative care [30]. According to the results of this study, after 4 and 6 weeks of treatment, both groups experienced similar levels of pain. However, the reflexology group showed considerable progress over time, while the relaxation group reached a plateau after 4 weeks. This study suggests that for patients with cancer receiving integrative palliative care, both relaxation and reflexology were effective in reducing anxiety and depression, but it seems like reflexology is more effective in pain management. It was also pointed out that these methods are economical, and that is an additional reason why they should be included in regular treatment within palliative care [30,33,34].

Future implications

As physical therapy is delivered by a highly trained specialist who is able to tailor, adjust, and modify a plan of care for each individual patient, best practices for frequency, duration, and dosage of commonly utilized techniques should be established by ongoing research. Additionally, comprehensive research on the effectiveness of treatments such as reflexology and relaxation should be further investigated. Although physical therapy services have historically been designed to be delivered in a rehabilitative setting (i.e., there is an expectation that the patient will get better), the skills and interventions are equally as applicable in patients with advanced disease and in palliative care. As there is often widespread lack of role clarity for physical therapists in palliative care (including by patients, interdisciplinary team members, and even therapists themselves), comprehensive education on the role of the physical therapist and advocacy for their role in palliative care should be emphasized [35]. Finally, insurance providers and healthcare systems should focus on assuring appropriate access and payment for these physical therapist services in palliative care so that patients have unfettered access to the skills of a physical therapist in order to facilitate nonpharmacologic pain management and optimize remaining quality of life and function [35].

Conclusions

1. Cancer patients require multidisciplinary medical support, with physical therapy playing an indispensable role. Supervised, individualized physical exercises are preferred over unsupervised ones.
2. Future research should focus on establishing guidelines defining the type, intensity, and frequency of physical exercise in palliative medicine.
3. Muscle relaxation, reflexology, and manual therapy techniques showed promising results in pain treatment for patients with cancer.

References

[1] World Physiotherapy. Description of physiotherapy: policy statement, <https://world.physio/policy/ps-descriptionPT>; 2023.

[2] InformedHealth.org, In brief: physical therapy, <https://www.ncbi.nlm.nih.gov/books/NBK561514/>; 2022.

[3] Rodríguez-Cañamero S, Cobo-Cuenca AI, Carmona-Torres JM, et al. Impact of physical exercise in advanced-stage cancer patients: systematic review and meta-analysis. Cancer Med 2022;11(19):3714–27. Available from: https://doi.org/10.1002/cam4.4746.

[4] Caraceni A, Shkodra M. Cancer pain assessment and classification. Cancers (Basel) 2019;11(4):510. Available from: https://doi.org/10.3390/cancers11040510.

[5] Deng GE, Rausch SM, Jones LW, et al. Complementary therapies and integrative medicine in lung cancer: diagnosis and management of lung cancer, 3rd ed: AMERICAN COLLEGE OF CHEST Physicians evidence-based clinical practice guidelines. Chest 2013;143(5):e420. Available from: https://doi.org/10.1378/chest.12-2364.

[6] Greenlee H, DuPont-Reyes MJ, Balneaves LG, et al. Clinical practice guidelines on the evidence-based use of integrative therapies during and after breast cancer treatment. CA Cancer J Clin 2017;67(3):194–232. Available from: https://doi.org/10.3322/caac.21397.

[7] Lyman GH, Greenlee H, Bohlke K, et al. Integrative therapies during and after breast cancer treatment: ASCO endorsement of the SIO clinical practice guideline. J Clin Oncol 2018;36(25):2647–55. Available from: https://doi.org/10.1200/JCO.2018.79.2721.

[8] Swarm RA, Paice JA, Anghelescu DL, et al. Adult cancer pain, Version 3.2019, NCCN clinical practice guidelines in oncology. J Natl Compr Canc Netw 2019;17(8):977–1007. Available from: https://doi.org/10.6004/jnccn.2019.0038.

[9] World Health Organization (WHO). Palliative care, <https://www.who.int/news-room/fact-sheets/detail/palliative-care>; 2020.

[10] Caspersen CJ, Powell KE, Christenson GM. Physical activity, exercise, and physical fitness: definitions and distinctions for health-related research. Public Health Rep 1985;100(2):126–31. Available from: https://www.ncbi.nlm.nih.gov/pmc/articles/PMC1424733/.

[11] Heywood R, McCarthy AL, Skinner TL. Efficacy of exercise interventions in patients with advanced cancer: a systematic review. Arch Phys Med Rehabil 2018;99(12):2595–620. Available from: https://doi.org/10.1016/j.apmr.2018.04.008.

[12] Plinsinga ML, Singh B, Rose GL, et al. The effect of exercise on pain in people with cancer: a systematic review with meta-analysis. Sports Med 2023;53(9):1737–52. Available from: https://doi.org/10.1007/s40279-023-01862-9.

[13] Chen YJ, Li XX, Ma HK, et al. Exercise training for improving patient-reported outcomes in patients with advanced-stage cancer: a systematic review and meta-analysis. J Pain Symptom Manage 2020;59(3):734–49.e10. Available from: https://doi.org/10.1016/j.jpainsymman.2019.09.010.

[14] Tanriverdi A, Ozcan Kahraman B, Ergin G, et al. Effect of exercise interventions in adults with cancer receiving palliative care: a systematic review and meta-analysis. Support Care Cancer 2023;31(4):205. Available from: https://doi.org/10.1007/s00520-023-07655-0.

[15] Peters M, Butson G, Mizrahi D, et al. Physical activity and pain in people with cancer: a systematic review and meta-analysis. Support Care Cancer 2024;32(3):145. Available from: https://doi.org/10.1007/s00520-024-08343-3.

[16] Bouillet T, Bigard X, Brami C, et al. Role of physical activity and sport in oncology: scientific commission of the National Federation Sport and Cancer CAMI. Elsevier Ireland Ltd, France. Crit Rev Oncol/Hematol 2015;94(1):74–86. Available from: https://doi.org/10.1016/j.critrevonc.2014.12.012.

[17] Vendrell I, Macedo D, Alho I, et al. Treatment of cancer pain by targeting cytokines. Mediat Inflamm 2015;2015:984570. Available from: https://doi.org/10.1155/2015/984570.

[18] Austin PD, Lee W, Costa DS, et al. Efficacy of aerobic and resistance exercises on cancer pain: a meta-analysis of randomised controlled trials. Heliyon 2024;10(7):e29193. Available from: https://doi.org/10.1016/j.heliyon.2024.e29193.

[19] Avancini A, Pala V, Trestini I, et al. Exercise levels and preferences in cancer patients: a cross-sectional study. Int J Environ Res Public Health 2020;17(15):1–22. Available from: https://doi.org/10.3390/ijerph17155351.

[20] American Physical Therapy Association. Guide to physical therapist practice. 4th edition, Alexandria: American Physical Therapy Association; 2024, <https://guide.apta.org/interventions/categories-interventions/manual-therapy-techniques>.

[21] Maltser S, Cristian A, Silver JK, et al. A focused review of safety considerations in cancer rehabilitation. PM R 2017;9(9):S415. Available from: https://doi.org/10.1016/j.pmrj.2017.08.403.

[22] Pinheiro da Silva F, Moreira GM, Zomkowski K, et al. Manual therapy as treatment for chronic musculoskeletal pain in female breast cancer survivors: a systematic review and meta-analysis. J Manipulative Physiol Ther 2019;42(7):503–13. Available from: https://doi.org/10.1016/j.jmpt.2018.12.007.

[23] Lee E, Kim H. Effects of manual therapy on chemotherapy-induced peripheral neuropathy: a systematic review and meta-analysis of randomized controlled trials. Phys Ther Rehabil Sci 2023;12(1):12–8. Available from: https://doi.org/10.14474/ptrs.2023.12.1.12.

[24] Pattanshetty RB, Patil SN. Role of manual therapy for neck pain and quality of life in head and neck cancer survivors: a systematic review. Indian J Palliat Care 2022;28(1):99–112. Available from: https://doi.org/10.25259/IJPC_10_2021.

[25] McMillan H, Barbon CEA, Cardoso R, et al. Manual therapy for patients with radiation-associated trismus after head and neck cancer. JAMA Otolaryngol Head Neck Surg 2022;148(5):418–25. Available from: https://doi.org/10.1001/jamaoto.2022.0082.

[26] Bellomo RG, Barassi G, Lococo A, et al. Rehabilitation of cancer pain in lung cancer: role of manual therapy. Arch Physiother Glob Res 2015;19:27–38. Available from: https://apgrjournal.com/article/01.3001.0053.9983/en?language=en.

[27] Yao C, Cheng Y, Zhu Q, et al. Clinical evidence for the effects of manual therapy on cancer pain: a systematic review and meta-analysis. Evid Based Complement Alternat Med 2021;2021. Available from: https://doi.org/10.1155/2021/6678184.

[28] Cadet T, Davis C, Wilson P, et al. The experiences of touch therapies in symptom management of rural and regional cancer patients in Australia. Int J Ther Massage Bodywork 2022;15(1):66–71. Available from: https://doi.org/10.3822/ijtmb.v15i1.687.

[29] Klaus, Kutschan, Männle J, et al. Reflexology in oncological treatment – a systematic review. BMC Complement Med Ther 2024;24(1):32. Available from: https://doi.org/10.1186/s12906-023-04220-4.

[30] Mantoudi A, Parpa E, Tsilika E, et al. Complementary therapies for patients with cancer: reflexology and relaxation in integrative palliative care. A randomized controlled comparative study. J Altern Complement Med 2020;26(9):792–8. Available from: https://doi.org/10.1089/acm.2019.0402.

[31] Candy B, Armstrong M, Flemming K, et al. The effectiveness of aromatherapy, massage and reflexology in people with palliative care needs: a systematic review. Palliat Med 2020;34(2):179–94. Available from: https://doi.org/10.1177/0269216319884198.

[32] Anshasi H, Saleh M, Abdalrahim MS, et al. The effectiveness of progressive muscle relaxation technique in reducing cancer-related pain among palliative care patients: a randomized controlled trial. Br J Pain 2023;17(5):501–9. Available from: https://doi.org/10.1177/20494637231190191.

[33] Kwekkeboom KL, Wanta B, Bumpus M. Individual difference variables and the effects of progressive muscle relaxation and analgesic imagery interventions on cancer pain. J Pain Symptom Manage 2008;36(6):604–15. Available from: https://doi.org/10.1016/j.jpainsymman.2007.12.011.

[34] Tarrasch R, Carmel-Neiderman NN, Ben-Ami S, et al. The effect of reflexology on the pain-insomnia-fatigue disturbance cluster of breast cancer patients during adjuvant radiation therapy. J Altern Complement Med 2018;24(1):62–8. Available from: https://doi.org/10.1089/acm.2017.0023.

[35] Wilson CM, Stiller CH, Doherty DJ, et al. Physical therapists in integrated palliative care: a qualitative study. BMJ Support Palliat Care 2022;12(e1):e59. Available from: https://doi.org/10.1136/bmjspcare-2019-002161.

Chapter 24

Noninvasive and supportive interventions

Christopher M. Wilson[1,2] and Aleksandar Knežević[3,4]
[1]*School of Health Sciences, Oakland University, Rochester, MI, United States,* [2]*Corewell Health, Southfield, MI, United States,* [3]*Faculty of Medicine, University of Novi Sad, Novi Sad, Serbia,* [4]*Medical Rehabilitation Clinic, University Clinical Center of Vojvodina, Novi Sad, Serbia*

Key learning objectives

Upon completion of this chapter, the reader will be able to:

- Identify common noninvasive pain management interventions for use in palliative care.
- Determine which interventions have robust evidence for efficacy and which may have less conclusive evidence but a very low adverse effect profile.
- Describe interventions that will modulate the nervous system temporarily or via long-term mechanisms.
- Discuss the applicability of high-intensity focused ultrasound for cancerous bone lesions.
- Identify future applications and research needs for noninvasive interventions in the palliative population.

Introduction

Pharmacologic, invasive, or surgical pain management interventions may not be clinically indicated or may have an unfavorable risk–benefit ratio (e.g., more likely to negatively impact quality of life [QoL]) for patients in palliative care. Therefore, nonpharmacologic, noninvasive modalities for treating pain caused by advanced cancer and its treatments should be considered. This chapter will review common and novel noninvasive pain management techniques, including the current evidence of efficacy, basic principles of application, and physiologic effects. One key consideration with many of these thermal or electrical interventions is that there is often concern for the intervention causing further metastatic spread, usually through means of vasodilation or increasing the metabolic activity of neoplastic tumors [1]. Although there is sparse or inconsistent evidence to support these concerns, hesitancy remains with applying these interventions; however, in the case of already widespread metastatic disease that is life-limiting, the benefit of improving QoL, comfort, and dignity will likely outweigh concerns of spreading an already extensive or terminal disease process [2]. Furthermore, there is often a high amount of anxiety, helplessness, or hopelessness in the presence of a life-limiting disease, and some of the interventions discussed below may not have clear, definitive evidence for their efficacy in managing pain; however, these procedures often come with the perception that a medical intervention is being delivered and therefore may be perceived as efficacious in reducing pain by an individual patient (e.g., placebo effect) [3]. As there is little to no adverse effect profile for most of these interventions, they may be worthwhile to administer despite only limited evidence of definitive clinical efficacy.

Electrotherapy

Transcutaneous electrical nerve stimulation

One of the most common and available electrotherapeutic agents for managing pain is transcutaneous electrical nerve stimulation (TENS). Although there are multiple ways to modify the electrical waveform delivery of TENS, three of the most common techniques are conventional TENS (high frequency, low intensity), acupuncture-like TENS (low frequency, high intensity), and brief intense TENS (high frequency, high intensity) (Table 24.1) [4]. Each of the three common techniques attempts to target different pain pathways, but all include the gate control theory.

Pain Management in Palliative Care for Patients With Cancer. DOI: https://doi.org/10.1016/B978-0-443-30050-9.00022-9

TABLE 24.1 Characteristics of common transcutaneous electrical nerve stimulation (TENS) techniques.

	Clinical purpose	Physiological intention (fiber type)	Desired outcome: patient experience	Optimal electrical characteristics in first instance	Electrode position	Analgesic profile for most patients	Duration of treatment	Main mechanism of analgesic action
Conventional TENS	TENS sensation soothes pain	Selective activation of low threshold nonnoxious afferents, for example, arising from cutaneous mechanoreceptors (A-beta)	Strong comfortable electrical paresthesia with minimal muscle activity	High frequency/low intensity Current amplitude = varies according to patient Pulse pattern = continuous Pulse frequency = 10–200 pps Pulse duration = 50–500 (usually 100–200) μs	Over site of pain Over main nerve bundle Dermatomal	Immediate relief Rapid offset of analgesia often within 30 min after TENS switched off	Whenever in pain—PRN	Gating of peripheral nociceptive input—short-acting neurotransmitters (segmental)
Acupuncture-like-TENS (AL-TENS)	TENS pulsing sensation and muscle twitching may be soothing or may be applied stronger as a counter irritant-accompanied by postTENS relief of pain	(a) Activation of high threshold nonnoxious/ noxious cutaneous afferents (A-beta and A-delta) and/or (b) Activation of low threshold motor efferents to produce muscle twitching, which generates impulses in afferents arising from muscle and deeper tissue (A-delta afferents)	(a) Strong comfortable pulsing without muscle twitching or (b) With muscle twitching	Low frequency/high intensity Current amplitude = varies according to patient (a) Single pulse AL-TENS Pulse pattern = continuous Pulse frequency ≤10 pps Pulse duration = 200–500 μs (b) Burst pattern AL-TENS Pulse pattern = burst (frequency ~2 bursts per second) Pulse frequency = ~100 pps within burst Pulse duration = 100–200 μs	(a) Dermatomal: Close to pain or over main nerve bundle if tolerated Sometimes placed on acupuncture point and described as "Acu-TENS" (b) Myotomal: over motor point/ muscle at site of pain	May be delayed onset of analgesia and maybe up to 30 min after TENS switched on Posttreatment relief may last >1 h after TENS switched off	Treatments of ~30 min a few times per day	Gating of peripheral nociceptive input—short-acting neurotransmitters (segmental) and activation of descending inhibitory pathways—long-acting neuromodulators (extrasegmental)
Intense TENS	TENS sensation is a counter irritant	Generate nerve impulses in afferents arising from high threshold cutaneous afferents (A-delta)	Electrical paresthesias that are slightly uncomfortable with minimal muscle contraction	High frequency/high intensity Current amplitude = discomfort but tolerable Pulse pattern = continuous Pulse frequency = ~50–200 pps Pulse duration >500 μs	Remote body site as a counter irritant	Immediate action Posttreatment relief may last >1 h after TENS switched off	One off treatment for a few minutes during short and painful procedures or breakthrough pain	Blockade of afferent input (peripheral), gating of peripheral nociceptive input—short-acting neurotransmitters (segmental), and activation of descending inhibitory pathways—long-acting neuromodulators (extrasegmental)

pps, Pulses per second.

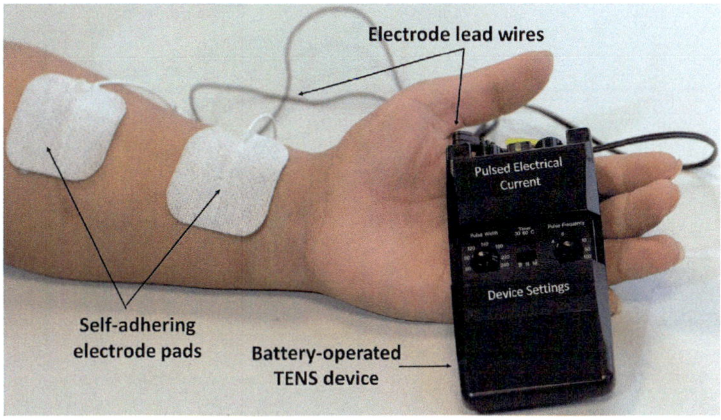

FIGURE 24.1 Transcutaneous electrical nerve stimulation device (TENS) and electrodes.

Advantages of TENS are that it is relatively simple to utilize, has little to no adverse effect profile, and patients and caregivers can be quickly taught to apply and adjust TENS for home or community use; however, the use of TENS in patients with cardiac devices is generally not recommended. Recently, TENS units are commonly commercially available without a prescription and may be acquired for a nominal cost (~40 USD). A key consideration for patients with cancer is that there is an unclear effect of electrotherapy on the growth and spread of cancer cells, with concern of increasing local circulation or cellular metabolism; however, in palliative situations, these considerations may be less of a concern [5] (Fig. 24.1).

Most TENS units have the ability to adjust the frequency (aka pulse rate or pulses per second), pulse duration, and amplitude (intensity). Conventional TENS is usually modulated to achieve a "strong but comfortable" electrical sensation directly over a painful area or over a peripheral nerve proximal to the painful site.

The evidence for TENS remains inconclusive, with very few studies examining pain relief in palliative situations despite its availability for decades. A Cochrane review published in 2012 found that there were only three randomized controlled trials (RCTs) for treating cancer pain, but it was found that none were of sufficient power to achieve conclusive findings of TENS efficacy [6]. More recently, Nakano et al. [7] utilized a single-group crossover design (phases of TENS or no TENS) to examine the use of high-rate TENS delivered at the site of pain for 20 palliative patients. In this study, TENS significantly improved cancer pain by approximately three points ($P < .01$) on an 11-point Numeric Rating Scale (NRS) immediately after administration, and the effect lasted between 60 and 120 minutes. Average pain score during the TENS phase significantly decreased compared to the non-TENS phase (–1.9 vs 0.2, $P = .04$); however, there was no change in maximal pain reported between groups. Additionally, TENS decreased the need for opioid rescue doses (TENS = 11.7 vs non-TENS = 16.3 doses, $P = .04$) during the study. It was concluded that TENS was an effective adjunct for managing pain but should not replace opioids for pain management. Siemens et al. [8] compared intensity-modulated TENS to sham TENS placebo via an RCT on 20 patients in an inpatient palliative care unit. In this study, there were no statistically significant between-group differences in average pain or medication use ($P > .05$). Within the TENS group, worst pain, mood, walking ability, and relationships with others improved by one point on an 11-point NRS, which was felt to be clinically significant. Finally, neither study resulted in any substantial adverse events.

Scrambler therapy

Scrambler therapy (ST), sometimes known as "Calmare" therapy, was introduced in the late 2000s as a neuromodulatory approach that attempts to provide pain relief through the replacement of painful afferent signals with "nonpain" afferent signals from the ST device [9]. Similar to TENS, the ST electrical current is delivered by superficial adhesive electrodes; however, in ST, the electrotherapeutic waveforms are aimed to resemble "normal" or "nonpain" physiologic action potentials [9,10]. The theoretical basis of this treatment for pain is that, over time, afferent pain signals increase the responsiveness of the central nervous system to afferent inputs, resulting in an aberrant response to future painful afferent signals [10]. Employing the concepts of neuroplasticity, repeated scrambler treatments are theorized to retrain the brain, which reduces the sensation of pain over repeated treatments [9].

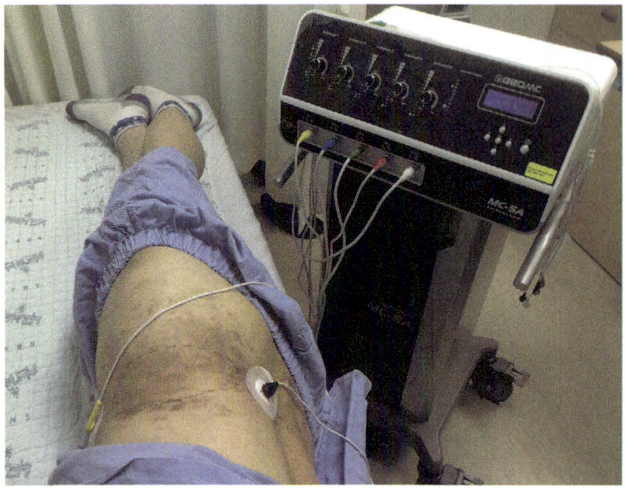

FIGURE 24.2 Scrambler therapy device and delivery.
Reprinted with permission from Lee SC, Park KS, Moon JY, et al. An exploratory study on the effectiveness of "Calmare therapy" in patients with cancer-related neuropathic pain: a pilot study. Eur J Oncol Nurs 2016;21:1–7. https://doi.org/10.1016/j.ejon.2015.12.001.

ST devices have up to five pairs of electrodes from the unit, and electrode pairs are suggested to be placed along a specific dermatome or area of concern (Fig. 24.2) [11]. In order to administer ST, healthcare providers must attend a 3-day in-person course and demonstrate proficiency [12]. Most treatment sessions were 30–60 minutes for 6–10 sessions; however, some studies reported providing "booster" sessions after the initial treatment regimen to enhance or prolong pain relief [13]. Advantages of ST include the negligible adverse effect profile and the potential for longer-term pain reduction, especially in the presence of chronic pain [13]. Some disadvantages include the potentially prohibitively high cost for the unit (~105,000 USD) [14]; that treatment delivery requires the ongoing, direct administration by a licensed, trained healthcare provider; the requirement for multiple in-person treatments; and that the treatment efficacy appears to be related to the experience and expertise of the provider [9].

Although not specifically focused on patients with advanced cancer, Majithia et al. [9] completed a systematic review of 20 reports on ST for chronic pain and found that all but one demonstrated positive results for the effectiveness of ST. Jin et al. [15] performed a metaanalysis of seven RCTs ($n = 287$ patients with chronic pain) and found marginal improvement in pain scores and reduced opioid use. There is a growing body of evidence of the effectiveness of ST in palliative and advanced cancer situations. Kashyap et al. [16] performed an RCT to examine pain, morphine intake, and QoL in 80 patients with advanced head, neck, or thoracic (e.g., breast/lung) cancer. Patients were randomized into groups where they would get conventional pain management via the World Health Organization (WHO) pain ladder versus ST with the WHO pain ladder. Pain via NRS decreased from a baseline mean of ~6.5 for both groups; the ST group decreased to ~0.8, while the usual care group decreased to ~3.1. The ST group demonstrated decreased morphine use ($P < .05$) and improved QoL in all domains ($P < .05$) compared to the usual care group at day 10 and 1 week after ST was completed. Lee et al. [11] performed a single-arm pilot study of 23 patients with cancer with neuropathic, bone, or postsurgical neuropathic pain. ST significantly decreased patients' pain score on an NRS at 1 month from baseline ($P < .001$) in 20 patients. Multiple items on the Brief Pain Inventory also significantly improved ($P < .05$), as well as decreased use of rescue opioids ($P < .05$) [11].

Noninvasive brain stimulation

These procedures attempt to modulate the excitability of brain neurons via magnetic or electrical stimulation via superficial, noninvasive mechanisms through electrical and magnetic stimulation (Fig. 24.3).

Transcranial direct current stimulation

This procedure involves delivery of a steady, low-amplitude direct current via superficial electrodes, which can result in a net negative or positive charge, which can excite or inhibit neurons by depolarizing or hyperpolarizing targeted neurons. Generally, a low-amplitude charge (<2 mA) is delivered for 10–30 minutes over a series of treatments [17]. A recently

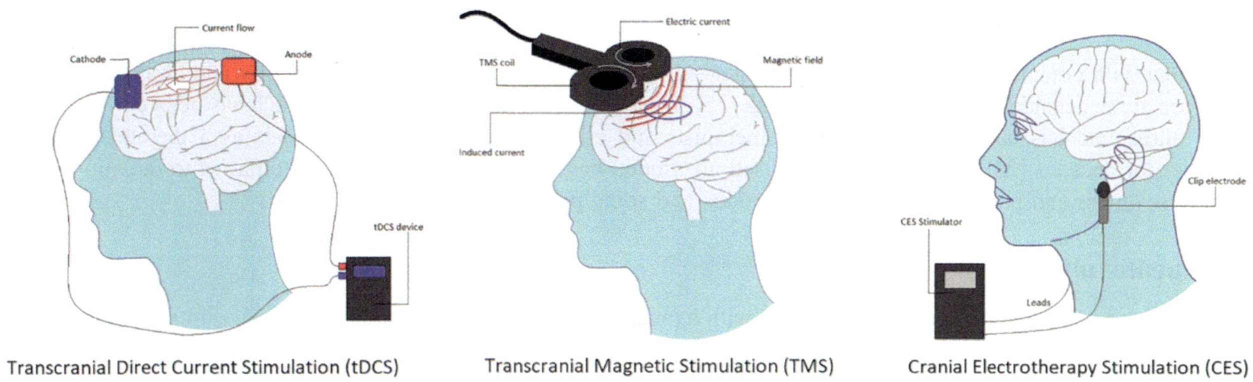

FIGURE 24.3 Noninvasive transcranial brain stimulation methods.
Adapted and reprinted via a Creative Commons 4.0 International License http://creativecommons.org/licenses/by/4.0/. From Szymoniuk M, Chin JH, Domagalski Ł, Biszewski M, Jóźwik K, Kamieniak P. Brain stimulation for chronic pain management: a narrative review of analgesic mechanisms and clinical evidence. Neurosurg Rev 2023;46(1):127. https://doi.org/10.1007/s10143-023-02032-1.

updated Cochrane review of 22 studies found an approximate reduction in chronic pain by 17%, but it was noted that many studies were of low methodological quality [18]. A narrative review by Szymoniuk et al. [17] found studies that demonstrated positive pain relief from transcranial direct current stimulation (tDCS) in knee osteoarthritis, arthralgia from viral infections, fibromyalgia, and multiple sclerosis. Finally, as it relates to cancer pain, a significant difference ($P < .05$) was found between tDCS and sham treatment after 10 sessions that lasted for a month in an RCT for patients with hepatocellular cancer ($n = 40$) [19,20].

Repetitive transcranial magnetic stimulation

This intervention aims to induce neuronal depolarization within the brain. In this intervention, an electrical current is delivered through a circular wire, inducing a magnetic field. This circular wire is placed on the scalp, inducing neuronal depolarization. This same current is repeated to achieve a specific clinical goal. Frequencies can be adjusted for specific clinical goals with higher frequencies of ≥5 Hz or low-frequency repetitive transcranial magnetic stimulation (rTMS) with frequencies ≤1 Hz. In a narrative review of rTMS including 23 articles since 2013, it appears that the main risk is the induction of seizures; however, it is noted that this risk is "rare" [17]. In patients with chronic pain, the pain reduction appears to be approximately 40% [17]. An RCT examined the analgesic effect of rTMS versus sham treatment with malignant visceral pain ($n = 40$). Ten sessions of rTMS over the primary motor cortex resulted in pain relief in malignant visceral pain, which lasted between 15 and 30 days [21].

Cranial electrical stimulation

Cranial electrical stimulation (CES) is a noninvasive electrotherapy technique used to treat pain. An electrical current is delivered through the patient's earlobes, mastoid processes, or forehead (Fig. 24.3) [17]. The intensity of current is usually between 50 μA and 4 mA with a common setting of 2 mA [17]. The specific physiologic mechanism of pain control remains unknown. However, it is theorized that the electrical currents affect the thalamus, limbic system, and reticular activating system [17]. Specifically, modulating the pain-processing network in the thalamus is thought to facilitate pain inhibitory mechanisms to reduce the perception of pain [18]. A secondary theorized means of pain control is through alteration of brain waves (as assessed via electroencephalogram [EEG]); CES appears to increase alpha wave activity, while beta and delta waves are decreased, resulting in an overall decrease in anxiety, negative thoughts, and fatigue, which may further reduce the perception of pain [22].

In a cohort study of patients with cancer ($n = 33$) in palliative care by Yennurajalingam et al. [23], CES demonstrated improved pain on the Brief Pain Inventory ($P = .013$) by one point, as well as reduced medication usage and improvements in sleep, depression, and anxiety ($P < .05$).

There is also preliminary evidence of use of CES via remote monitoring. Ahn et al. [24] performed an RCT of 30 participants with knee osteoarthritis who performed 60 minutes of CES remotely or sham CES and found that the CES group had reduced pain compared to the sham group ($P < .01$). Although CES seems to be safe with a low risk of adverse effects [17,23,24], more investigations are needed, especially with the advanced cancer population.

Ultrasound

Ultrasound is a potentially useful noninvasive intervention due to its wide clinical applicability and ability to be easily delivered at the bedside or other mobile locations [25]. In addition to its diagnostic utility, which has been employed since 1940, ultrasound has been utilized by physiotherapists and other rehabilitation professionals as a therapeutic heating modality for decades [25]. More recently, focused ultrasound has been investigated as an alternative to radiation therapy for the treatment and palliation of metastatic lesions and some primary tumors.

Therapeutic ultrasound

An ultrasound is delivered via a handheld wand with a piezoelectric crystal that converts electricity into ultrasonic waves, which are delivered into a body part [26]. These waves pass through the skin and fat easily, reducing the heating effect on these tissues, and affect tissues with increased collagen content, converting the ultrasound waves into heat by attenuation [25,26]. Therapeutic ultrasound can heat deeper tissues (3–5 cm deep) as compared to superficial heating methods (e.g., hot pack), which generally only changes tissue temperature 1–2 cm deep. Therapeutic ultrasound can be delivered at different duty cycles, with lower duty cycles (e.g., 50% and 20%) providing less thermal and more nonthermal effects. Thermal effects are generally considered to include increased extensibility of collagen (e.g., tendons and joint capsules), decreased joint stiffness and muscle spasm, as well as increased blood flow and mild activation of an inflammatory response [26]. These effects are thought to occur when tissue temperature has been raised to 40°C–45°C for at least 5 minutes [26]. It has been shown that tissues may be damaged when temperatures are raised to 45°C; however, the patient will likely report feeling excessive heat or pain before this threshold is met, resulting in cessation or modification of treatment [26]. Nonthermal ultrasound effects may include cell permeability and fibroblast activity, which may facilitate the soft tissue repair process [26]. Nerve conduction and modulation may also be impacted, which may result in a reduction in local pain. The evidence for therapeutic ultrasound is mixed, with several Cochrane reviews not being able to conclude that therapeutic ultrasound was effective for pain management in a variety of common orthopedic conditions (ankle sprain, low back pain, adhesive capsulitis, carpal tunnel syndrome, or osteoarthritis), and no studies on therapeutic ultrasound could be located for patients with advanced cancer or near the end of life [25].

Focused ultrasound

Heating tissue until damage is actively avoided in therapeutic ultrasound; however, this effect is intentionally utilized in focused ultrasound, sometimes known as high-intensity focused ultrasound (HIFU) or focused ultrasound surgery (FUS) [25]. This procedure seeks to ablate cancerous tissue and has been used to treat breast, prostate, uterine, kidney, liver, and pancreas tumors (Fig. 24.4) [27]. In a metaanalysis of pancreatic tumors treated with HIFU, 567 patients from HIFU and 459 patients from 23 studies experienced partial or complete pain relief, resulting in an estimated 75%–81% of patients expected to experience relief from pancreatic pain [28]. In order to place the HIFU accurately, ultrasound imaging may be

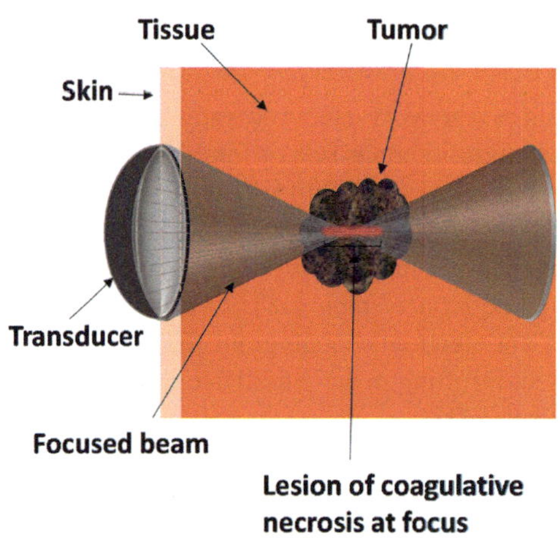

FIGURE 24.4 High-intensity focused ultrasound for tumor ablation.
Adapted and reprinted via a Creative Commons 4.0 International License http://creativecommons.org/licenses/by/4.0/. From Izadifar Z, Izadifar Z, Chapman D, Babyn P. An introduction to high intensity focused ultrasound: systematic review on principles, devices, and clinical applications. J Clin Med 2020;9(2):460. https://doi.org/10.3390/jcm9020460.

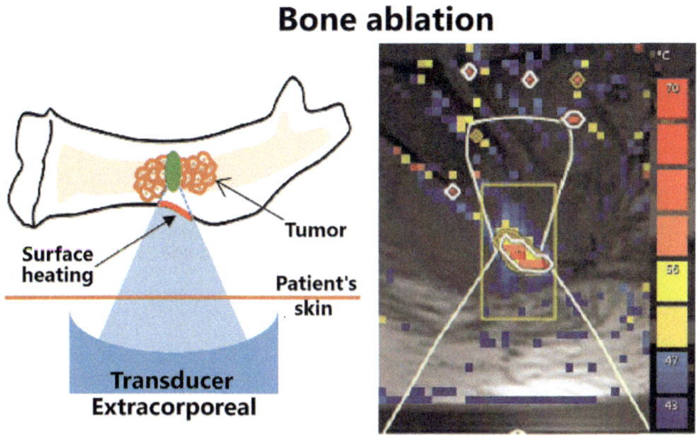

FIGURE 24.5 Schematic illustration and representative magnetic resonance thermometry of bone metastasis ablation.
Adapted and reprinted via a Creative Commons 4.0 International License http://creativecommons.org/licenses/by/4.0/. From Yeo SY, Bratke G, Grüll H. High intensity focused ultrasound for treatment of bone malignancies-20 years of history. Cancers (Basel) 2022;15(1):108. Published 2022 Dec 24. https://doi.org/10.3390/cancers15010108.

used; however, magnetic resonance imaging (MRI) is also utilized, which is called magnetic resonance-guided focused ultrasound surgery (MRgFUS) [29]. The energy and heat delivered to the tissues is much higher than therapeutic ultrasound (65°C–85°C with maximum acoustic energy of 2000 J) with the aim to achieve tumor cell death via coagulative necrosis and mechanical tissue damage [30].

A substantial area of promise for HIFU in the palliative setting is in the treatment of primary or metastatic bone tumors (Fig. 24.5). In an international consensus document published in 2015, more than 70% of patients reported pain reduction after MRgFUS [31]. QoL also improved via reduction in pain interference with functional activities [32]. Although external beam radiation therapy (EBRT) is generally preferred to palliatively treat metastatic bone pain, repeated radiotherapy may not be advisable or feasible in some cases, and treatment effects of radiotherapy may not be appreciated for 4 weeks; HIFU pain reduction can occur within 3 days of treatment [31]. In a randomized sham-controlled phase III multicenter trial ($n = 147$), MRI-guided FUS treatment of bone metastases demonstrated a statistically significant improvement in pain and medication use (FUS = 64%, sham = 20%; $P < .001$) at 3 months; however, there was a high variability to FUS, likely due to the limited ability to directly monitor intratumor temperature [33]. Bitton et al. [34] found that a higher energy density on the bone surface (EDBS), evidence of postprocedural periosteal devascularization around the tumor site, and female sex were correlated ($P < .05$) with improved responsiveness to FUS. It was found that an EDBS larger than 5 J/mm^2 increased the likelihood of a positive response, and further research establishing evidence-based protocols should include a recommended EDBS [33, 34]. In a pilot study, six patients received HIFU 4 days after EBRT, and five of six patients demonstrated a positive response, and no adverse events were noted. This approach may improve the response rate, as 30%–40% of patients have no improvement with EBRT alone [35]. A metaanalysis of 799 patients receiving MRgFUS across 26 studies demonstrated a 0.9% rate of severe treatment-related adverse events (e.g., deep vein thrombosis [DVT], two cases of grade III skin burns, four cases of fracture) and a 5.9% rate of low-grade adverse events (e.g., numbness, edema, posttreatment pain, fever) [36]. Ninety four (11.8%) patients experienced intervention-related pain during MRgFUS treatment [36].

Cognitive retraining approaches

Neurofeedback (electroencephalogram biofeedback)

Patients with acute or chronic pain have been found to have alterations to their resting-state brain oscillations (also known as brain waves or rhythms) as measured via EEG [37]. An EEG can illustrate oscillations of electrical activity that occur at different frequency bands, each of which is associated with certain brain activities and mental states (Fig. 24.6) [38]. An increase in theta, beta, and delta waves and lower alpha waves compared to normal individuals has been observed in patients with chronic pain conditions (e.g., spinal cord injury, migraines, and fibromyalgia) [37]. In addition to the waves detailed in Fig. 24.6, the sensorimotor rhythm (SMR) is important in pain neuromodulation. The SMR consists of 12–15 Hz oscillations around the sensorimotor cortex when the patient is not thinking about or performing a motor task [37]. SMR oscillations decrease when planning or performing a motor task [37].

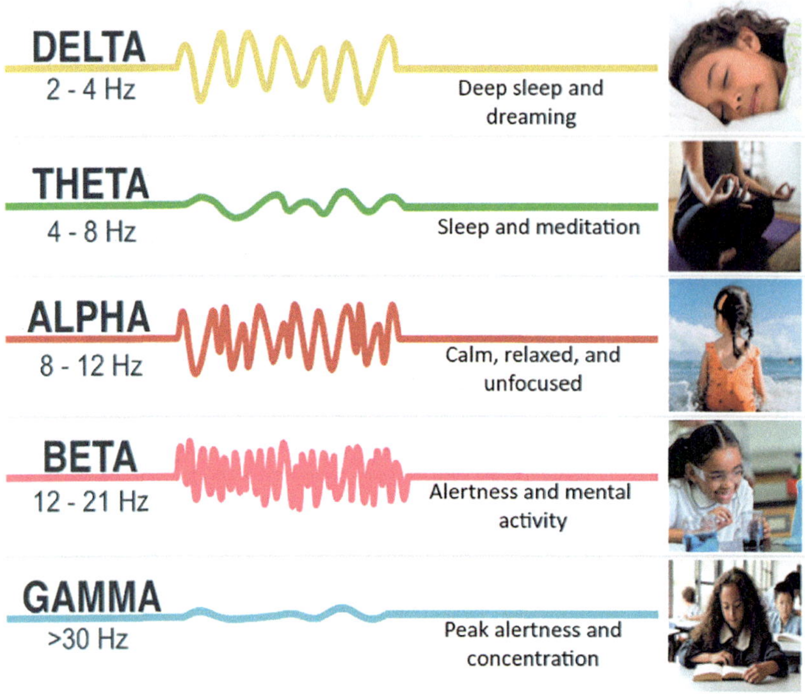

FIGURE 24.6 Electroencephalogram frequency bands and associated cognitive and physiological activities.
Adapted and reprinted via a Creative Commons 4.0 International License http://creativecommons.org/licenses/by/4.0/. From Flanagan K, Saikia MJ. Consumer-grade electroencephalogram and functional near-infrared spectroscopy neurofeedback technologies for mental health and wellbeing. Sensors 2023; 23(20):8482. https://doi.org/10.3390/s23208482.

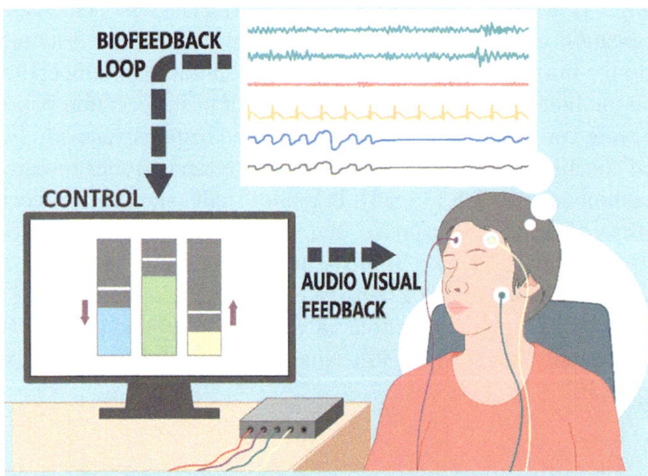

FIGURE 24.7 Depiction of a real-time electroencephalogram (EEG) biofeedback system with an EEG head cap.
Adapted and reprinted via a Creative Commons 4.0 International License http://creativecommons.org/licenses/by/4.0/. From Flanagan K, Saikia MJ. Consumer-grade electroencephalogram and functional near-infrared spectroscopy neurofeedback technologies for mental health and wellbeing. Sensors 2023; 23(20):8482. https://doi.org/10.3390/s23208482.

EEG neurofeedback encourages volitional control of targeted wave oscillations to increase or reduce specific waves to neuromodulate the experience of pain [37,39]. EEG waves will be detected via electrodes placed in specific locations (e.g., sensorimotor cortex), which concentrate and depict brain oscillations in a format where the patient can volitionally act upon them and attempt to modify them. This often includes a video screen or auditory noises and may utilize an interactive videogame-like delivery method (Fig. 24.7) [38]. Patients are provided positive or negative cues when a desired waveform is volitionally altered, allowing the person to provide a level of control of the sensation of pain in the brain, resulting in reduced pain with progressive training [39]. Generally, EEG neurofeedback sessions are 30–45 minutes in duration over 20–40 sessions [37]. Common targets for improving pain control include decreasing the activity of the theta and beta rhythms and/or increasing the activity of alpha waves or the SMR [37].

In a systematic review and metaanalysis by Patel et al. [40] of 18 EEG neurobiofeedback studies, patients reported a clinically significant reduction in pain (>30%) after EEG training in 10 of those studies, with one study reporting an 82% reduction in pain. Pain reports remained decreased for 3–6 months or longer in 7 of the 10 studies, providing evidence of potential long-term efficacy from EEG biofeedback training [40]. Another systematic review and metaanalysis by Hesam–Shariati et al. [39] of nine RCTs for EEG biofeedback found that the intervention may have a clinically meaningful effect on pain intensity in short term; however, the authors found that the confidence level for this efficacy was low. Interestingly, there were only three studies that overlapped between the article by Patel and the RCTs examined by Hesam–Shariati. Additionally, some studies found improvements in depression, anxiety, fatigue, and sleep [39,40]. Studies examining pain relief from EEG biofeedback in advanced cancer or in palliative care were not found. Reported adverse effects were rare but included headache, fatigue, nausea, drowsiness, and irritability, as well as muscle spasms or twitching [39,40]. In addition to the low adverse effect profile and noninvasive nature of EEG biofeedback, there are a number of widely available consumer-grade EEG models that may have future applicability in the palliative care setting due to their low cost, portability, and ease of applicability at the bedside [38].

Hypnosis

The American Psychological Association has defined hypnosis as "a procedure during which a health professional or researcher suggests, while treating someone, that he or she experience changes in sensations, perceptions, thoughts, or behavior" [41]. In hypnosis, the patient receives suggestions for relaxation, calmness, and stillness [42]. There are several variables to achieving a hypnotic state, including a patient's hypnotic susceptibility, expectations, motivation to respond to suggestions, and willingness to have imaginative experiences, as well as fears of achieving a hypnotic state (e.g., the unfounded concern about losing control of oneself) [43]. Most adults (70%) are moderately hypnotizable, and 10%–15% are highly hypnotizable [44,45].

Hypnosis procedures are usually delivered by a trained provider, oftentimes a clinical psychologist or a physician [46]. A hypnosis treatment session generally begins with inducing hypnosis via suggestions for focusing one's attention, attentive breathing, visual imaging, and/or arm levitation [43,46]. Deepening of the hypnotic state includes increased focus of attention, progressive relaxation, and suggestions for increased comfort [46,47]. Finally, posthypnotic suggestions are provided to attempt to maintain the benefits achieved by the hypnosis session [46,47].

No high-quality trials of hypnosis for pain management in palliative care could be located; however, there is some evidence of the efficacy of hypnosis for cancer pain and symptoms [48]. A systematic review by Sine et al. [46] examined 11 controlled clinical trials ($n = 1182$) with a 42-minute median duration of hypnosis intervention. Nine studies examined the effect of pain when receiving hypnosis versus control/sham treatment, and six of the studies demonstrated improvement [46]. Anxiety also demonstrated a statistically significant improvement in five of seven studies examining it [46]. There is also potential for hypnosis delivered via audio recordings. In a randomized trial, Eaton et al. [47] compared 14- to 18-minute hypnosis or relaxation audio recordings over 4 weeks and found moderate to large improvements in pain ($d = 0.88$) as well as pain interference and anxiety ($ds = 0.44$–0.88). Furthermore, the hypnosis group demonstrated an improvement in fatigue ($d = 0.47$) and sleep disturbance ($d = 0.54$) [47].

Future implications

Due to their variable medical stability, patients in palliative care may have a limited tolerance for invasive, elaborate, or uncomfortable procedures. Although there are a wide variety of interventions available, very few have conclusive evidence for use and applicability in the palliative population, and this should be a focus for future studies. As nonhospital-based services are preferred in palliative care, a key focus should be examining the efficacy of consumer products for their use in controlling pain as opposed to opioids or other pharmacologic means. Finally, in order to address geographic disparities in health in palliative situations, development and examination of nonpharmacologic, noninvasive interventions that can be delivered or monitored remotely should be prioritized.

Conclusions

- There are a wide variety of noninvasive interventions, all with a relatively low adverse effect profile. However, some interventions may require sophisticated, expensive equipment (ST, CES, HIFU), while others require little to no equipment (TENS, hypnosis).
- There is a growing body of evidence for the use of scrambler/Calmare therapy in immediate and long-term management of pain in advanced cancer and palliative situations; however, the expense of the unit and advanced training may be a barrier to implementation in palliative care situations.
- In cases where external beam radiotherapy is not an option or demonstrates limited effectiveness, HIFU is a viable option to control pain and ablate bone lesions.
- Interventions for pain in palliative cases should have a low adverse effect profile with limited negative impact on QoL. As patients in palliative care may have limited physical endurance and tolerance to noxious interventions, interventions that are able to be performed at the bedside or in the home or have long-term positive effects are preferred.
- Some interventions do not have conclusive evidence for their efficacy but are readily available for consumer use; due to their low risk of harm and potential for actual or placebo effect, they may be worthwhile to try, including TENS, therapeutic ultrasound, and hypnosis.

References

[1] Zheng JY, Mixon AC, McLarney MD. Safety, precautions, and modalities in cancer rehabilitation: an updated review. Curr Phys Med Rehabil Rep 2021;9(3):142–53. Available from: https://doi.org/10.1007/s40141-021-00312-9.

[2] Rubine W, Jones K. Pain and pain management. In: Cameron MH, editor. Physical agents in rehabilitation. 6th ed. Elsevier; 2022. p. 54.

[3] Lucas V, Booth S. The importance of placebo effects in enhancing palliative care interventions. BMJ Support Palliat Care 2014;4(2). Available from: https://doi.org/10.1136/bmjspcare-2013-000571.

[4] Johnson MI. Resolving long-standing uncertainty about the clinical efficacy of transcutaneous electrical nerve stimulation (TENS) to relieve pain: a comprehensive review of factors influencing outcome. Medicina (Lithuania) 2021;57(4). Available from: https://doi.org/10.3390/medicina57040378.

[5] Wilson CM, Stanczak JF. Palliative pain management using transcutaneous electrical nerve stimulation (TENS). Rehabil Oncol 2020;38(1). Available from: https://doi.org/10.1097/01.REO.0000000000000188.

[6] Hurlow A, Bennett MI, Robb KA, et al. Transcutaneous electric nerve stimulation (TENS) for cancer pain in adults. Cochrane Database Syst Rev 2012;2012(3). Available from: https://doi.org/10.1002/14651858.CD006276.pub3.

[7] Nakano J, Ishii K, Fukushima T, et al. Effects of transcutaneous electrical nerve stimulation on physical symptoms in advanced cancer patients receiving palliative care. Int J Rehabil Res 2020;43(1). Available from: https://doi.org/10.1097/MRR.0000000000000386.

[8] Siemens W, Boehlke C, Bennett MI, et al. Transcutaneous electrical nerve stimulation for advanced cancer pain inpatients in specialist palliative care—a blinded, randomized, sham-controlled pilot cross-over trial. Support Care Cancer 2020;28(11):5323–33. Available from: https://doi.org/10.1007/s00520-020-05370-8.

[9] Majithia N, Smith TJ, Coyne PJ, et al. Scrambler therapy for the management of chronic pain. Support Care Cancer 2016;24(6):1–7. Available from: https://doi.org/10.1007/s00520-016-3177-3.

[10] Smith TJ, Wang EJ, Loprinzi CL. Cutaneous electroanalgesia for relief of chronic and neuropathic pain. N Engl J Med 2023;389(2). Available from: https://doi.org/10.1056/nejmra2110098.

[11] Lee SC, Park KS, Moon JY, et al. An exploratory study on the effectiveness of "Calmare therapy" in patients with cancer-related neuropathic pain: a pilot study. Eur J Oncol Nurs 2016;21. Available from: https://doi.org/10.1016/j.ejon.2015.12.001.

[12] Scrambler Therapy. Training area. Scrambler therapy website, https://www.scramblertherapy.org/training.htm; 2024 [accessed 25.02.24].

[13] Wang EJ, Limerick G, D'Souza RS, et al. Safety of scrambler therapy: a systematic review of complications and adverse effects. Pain Med 2023;24(3). Available from: https://doi.org/10.1093/pm/pnac137.

[14] Smith T, Loprinzi C. Scrambler therapy for treating neuropathic pain. Foundation for peripheral neuropathy website, https://www.foundationforpn.org/scrambler-therapy-for-treating-neuropathic-pain/; 2022 [accessed 24.02.24].

[15] Jin Y, Kim D, Hur J, et al. Efficacy of scrambler therapy for management of chronic pain: a meta-analysis of randomized controlled trials. Pain Physician 2022;25(7):E931–9.

[16] Kashyap K, Singh V, Dwivedi SN, et al. Scrambler therapy enhances quality of life in cancer patients in a palliative care setting: a randomised controlled trial. Indian J Palliat Care 2022;28(3). Available from: https://doi.org/10.25259/IJPC_94_2021.

[17] Szymoniuk M, Chin JH, Domagalski Ł, et al. Brain stimulation for chronic pain management: a narrative review of analgesic mechanisms and clinical evidence. Neurosurg Rev 2023;46(1). Available from: https://doi.org/10.1007/s10143-023-02032-1.

[18] O'Connell NE, Marston L, Spencer S, et al. Non-invasive brain stimulation techniques for chronic pain. Cochrane Database Syst Rev 2018;2018(8). Available from: https://doi.org/10.1002/14651858.cd008208.pub5.

[19] Chien YJ, Chang CY, Wu MY, et al. Noninvasive brain stimulation for cancer pain management in nonbrain malignancy: a meta-analysis. Eur J Cancer Care (Engl) 2023(4):1–10. Available from: https://doi.org/10.1155/2023/5612061.

[20] Ibrahim NM, Abdelhameed KM, Kamal SMM, et al. Effect of transcranial direct current stimulation of the motor cortex on visceral pain in patients with hepatocellular carcinoma. Pain Med 2018;19(3):550–60. Available from: https://doi.org/10.1093/pm/pnx087.
[21] Khedr EM, Kotb HI, Mostafa MG, et al. Repetitive transcranial magnetic stimulation in neuropathic pain secondary to malignancy: a randomized clinical trial. Eur J Pain 2015;19(4):519–27. Available from: https://doi.org/10.1002/ejp.576.
[22] Brunyé TT, Patterson JE, Wooten T, et al. A critical review of cranial electrotherapy stimulation for neuromodulation in clinical and non-clinical samples. Front Hum Neurosci 2021;15:625321. Available from: https://doi.org/10.3389/fnhum.2021.625321.
[23] Yennurajalingam S, Kang DH, Hwu WJ, et al. Cranial electrotherapy stimulation for the management of depression, anxiety, sleep disturbance, and pain in patients with advanced cancer: a preliminary study. J Pain Symptom Manage 2018;55(2):198–206. Available from: https://doi.org/10.1016/j.jpainsymman.2017.08.027.
[24] Ahn H, Galle K, Mathis KB, et al. Feasibility and efficacy of remotely supervised cranial electrical stimulation for pain in older adults with knee osteoarthritis: a randomized controlled pilot study. J Clin Neurosci 2020;77:128–33. Available from: https://doi.org/10.1016/j.jocn.2020.05.003.
[25] Chernack B, Knowlton SE, Kohler MJ. The use of ultrasound in palliative care and hospice. Am J Hosp Palliat Care 2017;34(4):385–91. Available from: https://doi.org/10.1177/1049909115625960.
[26] Kahanov L, Prentice WE, Draper DO. Chapter 10: Therapeutic ultrasound. In: Prentice WE, editor. Therapeutic modalities in rehabilitation. 6th ed. McGraw Hill; 2021.
[27] Al-Bataineh O, Jenne J, Huber P. Clinical and future applications of high intensity focused ultrasound in cancer. Cancer Treat Rev 2012;38(5). Available from: https://doi.org/10.1016/j.ctrv.2011.08.004.
[28] Dababou S, Marrocchio C, Rosenberg J, et al. A meta-analysis of palliative treatment of pancreatic cancer with high intensity focused ultrasound. J Ther Ultrasound 2017;5:9. Available from: https://doi.org/10.1186/s40349-017-0080-4.
[29] Izadifar Z, Izadifar Z, Chapman D, et al. An introduction to high intensity focused ultrasound: systematic review on principles, devices, and clinical applications. J Clin Med 2020;9(2):460. Available from: https://doi.org/10.3390/jcm9020460.
[30] Papalexis N, Parmeggiani A, Peta G, et al. Minimally invasive interventional procedures for metastatic bone disease: a comprehensive review. Curr Oncol 2022;29(6):4155–77. Available from: https://doi.org/10.3390/curroncol29060332.
[31] Huisman M, Ter Haar G, Napoli A, et al. International consensus on use of focused ultrasound for painful bone metastases: current status and future directions. Int J Hyperthermia 2015;31(3):251–9. Available from: https://doi.org/10.3109/02656736.2014.995237.
[32] Chen Z, Wang C, Ma X, et al. Evaluation of quality of life using EORTC QLQ-BM22 in patients with bone metastases after treatment with magnetic resonance guided focused ultrasound. Orthop Surg 2018;10(3):264–71. Available from: https://doi.org/10.1111/os.12383.
[33] Hurwitz MD, Ghanouni P, Kanaev SV, et al. Magnetic resonance-guided focused ultrasound for patients with painful bone metastases: phase III trial results. J Natl Cancer Inst 2014;106(5):dju082. Available from: https://doi.org/10.1093/jnci/dju082.
[34] Bitton RR, Rosenberg J, LeBlang S, et al. MRI-guided focused ultrasound of osseous metastases treatment parameters associated with successful pain reduction. Invest Radiol 2021;56(3):141–6. Available from: https://doi.org/10.1097/RLI.0000000000000721.
[35] Bartels MMTJ, Verpalen IM, Ferrer CJ, et al. Combining radiotherapy and focused ultrasound for pain palliation of cancer induced bone pain; a stage I/IIa study according to the IDEAL framework. Clin Transl Radiat Oncol 2021;27:57–63. Available from: https://doi.org/10.1016/j.ctro.2021.01.005.
[36] Baal JD, Chen WC, Baal U, et al. Efficacy and safety of magnetic resonance-guided focused ultrasound for the treatment of painful bone metastases: a systematic review and meta-analysis. Skeletal Radiol 2021;50(12):2459–69. Available from: https://doi.org/10.1007/s00256-021-03822-8.
[37] Patel K, Sivan M, Henshaw J, et al. Neurofeedback for chronic pain. In: Da-Yin Liao E, editor. Smart biofeedback – perspectives and applications. IntechOpen; 2020. Available from: https://doi.org/10.5772/intechopen.93826.
[38] Flanagan K, Saikia MJ. Consumer-grade electroencephalogram and functional near-infrared spectroscopy neurofeedback technologies for mental health and wellbeing. Sensors 2023;23(20):8482. Available from: https://doi.org/10.3390/s23208482.
[39] Hesam-Shariati N, Chang WJ, Wewege MA, et al. The analgesic effect of electroencephalographic neurofeedback for people with chronic pain: a systematic review and meta-analysis. Eur J Neurol 2022;29(3):921–36. Available from: https://doi.org/10.1111/ene.15189.
[40] Patel K, Sutherland H, Henshaw J, et al. Effects of neurofeedback in the management of chronic pain: a systematic review and meta-analysis of clinical trials. Eur J Pain 2020;24(8):1440–57. Available from: https://doi.org/10.1002/ejp.1612.
[41] American Psychological Association. Hypnosis, https://www.apa.org/topics/psychotherapy/hypnosis; 2008 [accessed 16.03.24].
[42] Goolsby MJ, Dirubbo NE, Phillips BC. Hypnosis in the clinical setting. J Am Assoc Nurse Pract 2020;32(5):351–3. Available from: https://doi.org/10.1097/JXX.0000000000000422.
[43] Cardeña E, Terhune DB. The roles of response expectancies, baseline experiences, and hypnotizability in spontaneous hypnotic experiences. Int J Clin Exp Hypn 2019;67(1):1–27. Available from: https://doi.org/10.1080/00207144.2019.1553759.
[44] Santarcangelo EL, Carli G. Individual traits and pain treatment: the case of hypnotizability. Front Neurosci 2021;15:683045. Available from: https://doi.org/10.3389/fnins.2021.683045.
[45] Jensen MP, Adachi T, Tomé-Pires C, et al. Mechanisms of hypnosis: toward the development of a biopsychosocial model. Int J Clin Exp Hypn 2015;63(1):34–75. Available from: https://doi.org/10.1080/00207144.2014.961875.
[46] Sine H, Achbani A, Filali K. The effect of hypnosis on the intensity of pain and anxiety in cancer patients: a systematic review of controlled experimental trials. Cancer Invest 2022;40(3):235–53. Available from: https://doi.org/10.1080/07357907.2021.1998520.
[47] Eaton LH, Jang MK, Jensen MP, et al. Hypnosis and relaxation interventions for chronic pain management in cancer survivors: a randomized controlled trial. Support Care Cancer 2022;31(1):50. Available from: https://doi.org/10.1007/s00520-022-07498-1.
[48] Booth S. Hypnosis in a specialist palliative care setting – enhancing personalized care for difficult symptoms and situations. Palliat Care Soc Pract 2020;14:2632352420953436. Available from: https://doi.org/10.1177/2632352420953436.

Chapter 25

Acupuncture

Vlasta Orlić Karbić
Department of Anaesthesiology, Intensive and Pain Medicine, University Hospital Rijeka, Rijeka, Croatia

Key learning objectives

Upon completion of this chapter, the reader will be able to:

- Describe the principles and techniques of acupuncture in both traditional and biomedical contexts.
- Explain the proposed mechanisms of acupuncture analgesia.
- Evaluate the evidence for acupuncture in cancer-related pain and symptom management.
- Assess its role as an adjunct within multidisciplinary cancer care.
- Interpret guideline-based recommendations and limitations of current evidence.

Introduction

Acupuncture is a controversial topic as the traditional practice of acupuncture has evolved over many centuries of clinical observation, but there are plenty of aspects of acupuncture that we still cannot explain. Supporters of the traditional approach argue that its concepts of energy flow operate as a knowledge system in their own right, which should not be rejected simply because they do not fit in with current Western understandings of the body's mechanisms. On the other hand, the modern approach declares that acupuncture is a relatively straightforward form of stimulation therapy. Acupuncture has recently gained popularity and is widely used throughout the world. It is described as one of the "complementary and alternative medicine/therapies," showing promising efficacy in the treatment of many conditions and resulting in fewer adverse effects compared with some conventional medical treatments.

Complementary medicine and "natural therapies" have their origins in the civilizations of Babylon, Egypt, and China, dating back approximately 3000 years. The Chinese developed a system of medicine based on an extraordinarily detailed knowledge of herbal remedies combined with acupuncture. Traditional Chinese medicine (TCM) is practiced through a holistic approach, focusing on the unity of the human body with its environment [1]. The TCM view of the human body presents a construction of "energetic functions," as opposed to the traditional Western view of the body, which is based on structure (anatomy) and function (physiology), with the various parts operating together as systems in a mechanical manner. TCM suggests that about 365 acupuncture points are present on the human body, arranged in lines or channels (meridians)—there are 12 main meridians along which energy or "Chi" flows in a coherent and ordered manner (Fig. 25.1). If the flow is interrupted, then illness can occur. It is believed by some that acupuncture is a preventive medicine, enabling individuals to maintain and improve their health, potentially even enhancing their resistance. In treating illness, acupuncture seeks to stimulate the appropriate "point" along the affected channel, allowing the energy to become balanced and flow freely. Treatments are based on the evaluation of the diagnosis to rebalance the Yin and Yang deficiency or excess in the body [1].

Technique

Acupuncture uses ultra-thin needles (0.13–0.15 mm) that are inserted into the epidermis and dermis at specific points based on the principles of TCM (Fig. 25.2). In Western medicine, acupuncture is considered a therapeutic option adapted from TCM using current knowledge of anatomy, physiology, and pathology and following the principles of evidence-based medicine [2]. Acupuncture points and "meridians" can also be activated through the local application of pressure

Pain Management in Palliative Care for Patients With Cancer. DOI: https://doi.org/10.1016/B978-0-443-30050-9.00050-3

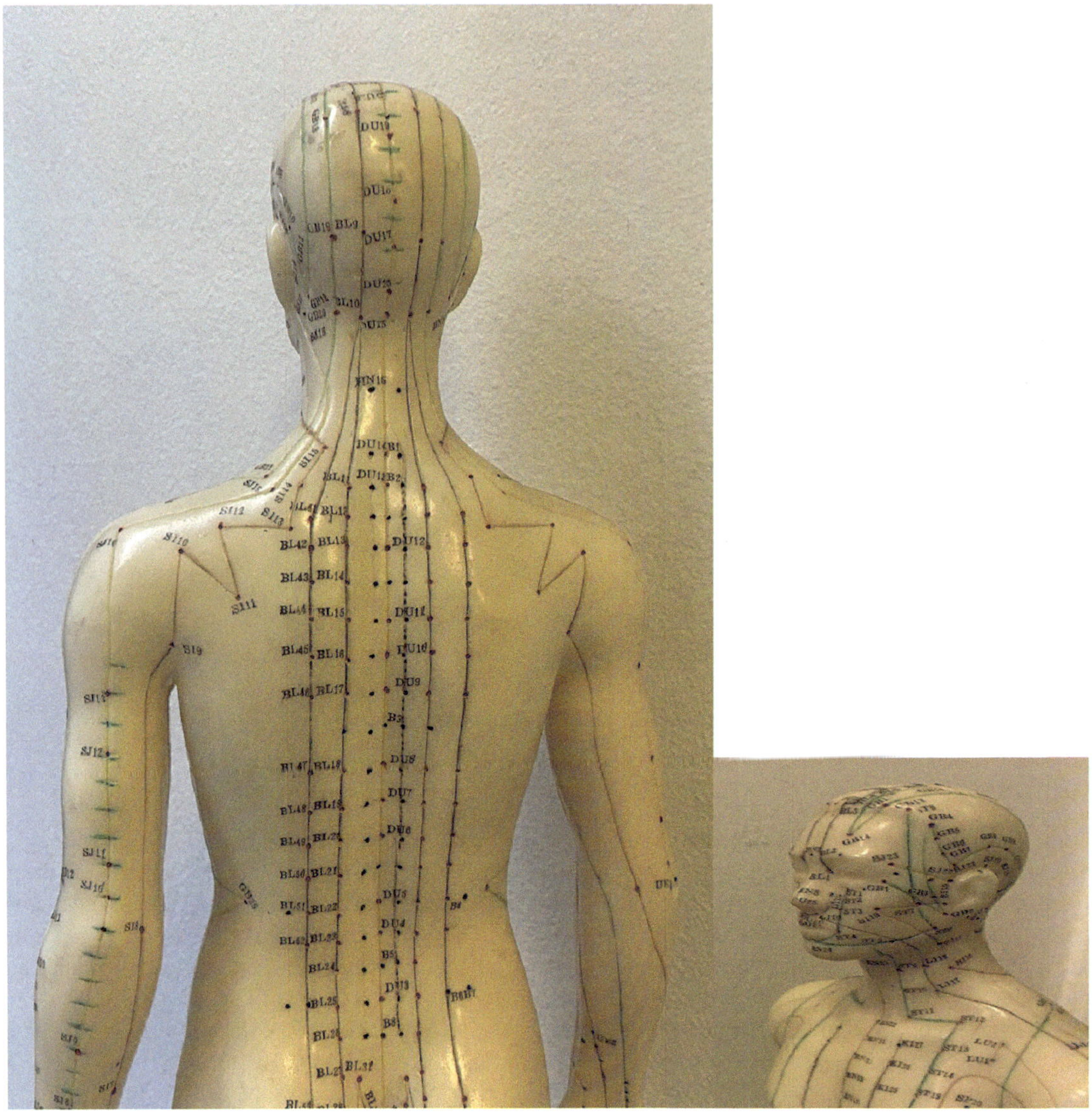

FIGURE 25.1 Meridians and acupuncture points (original points).

(acupressure) or by heat produced by mugwort cigars placed over the point, a technique called "moxibustion." The effectiveness of acupuncture and other methods activating acupuncture points is assessed by evaluating the patient's response to the treatment, as well as the quality of their radial pulse and the appearance of the tongue and its layer of coating, which are believed to reflect the balance of "Chi" in the body's "internal organs" [3,4]. In acupuncture practice, needles can be attached to electrodes (electroacupuncture [EA]) for what is believed to be an enhanced effect. EA involves running a small electrical current through the needles.

Mechanism of action

The proposed mechanisms of action of acupuncture on pain are complex and not fully elucidated. When considering pain, it is essential to distinguish between nociception (the perception of tissue damage) and pain, which can persist even in the absence of tissue damage.

Three types of nerve fibers are involved in nociception (Table 25.1). A-beta fibers are large-diameter myelinated fibers. A-delta fibers are also myelinated but are smaller in diameter. C fibers are unmyelinated small-diameter fibers. Both

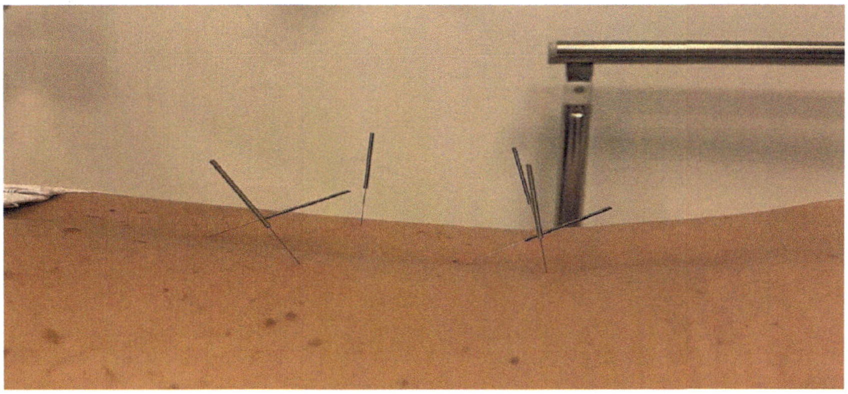

FIGURE 25.2 Acupuncture needle and method of manipulating (original picture).

TABLE 25.1 Classification of sensory nerve fibers in the skin.

Fiber type	Skin	Sensation
Large myelinated	Aβ	Light touch, pressure, vibration
Small myelinated	Aδ	Deep pressure, cold, pinprick
Small unmyelinated	C	Heat, itch, ache

A-delta and C fiber systems are involved in nociception and pain perception, but they play different roles. The A-delta fiber system is responsible for the accurate localization of noxious stimulation without much emotional accompaniment, whereas the C fiber system yields a poorly localized perception of pain accompanied by a stronger emotional effect. The A-delta system mediates rapid-onset pain, while the C fiber system mediates the second pain. A-delta fibers are activated by heavy pressure, heat and cold, and also by pinprick, which is where acupuncture comes in. A-delta fibers are thought to be responsible for acupuncture analgesia; the A-beta fibers are probably involved only in EA analgesia and not in ordinary clinical acupuncture. The A-delta system also has an inhibitory feedback effect on pain transmission because A-delta activity is directed to a part of the hypothalamus (the arcuate nucleus), which in turn activates descending impulses that activate inhibitory systems in the brain stem.

Acupuncture stimulates the immune cells to produce endogenous opioids capable of targeting the nerve fibers. Through this mechanism, two effects are produced: direct analgesia and reduction of hyperalgesia. Locally, the insertion of acupuncture needles results in changes in blood flow (neurogenic inflammation) and cellular remodeling, with reduced mechanical stress in regional mechanoreceptors (needle wrapping) [5]. Then, a chain reaction is triggered, and the induction of A-beta, A-delta, and C afferent fibers, as well as the release of noradrenaline, dopamine, and adenosine, occurs [3]. Neurogenic inflammation is mediated by the release of calcitonin gene-related peptide and substance P, which act directly on vascular endothelial and smooth muscle cells, causing vasodilation and plasma extravasation [6].

Following needle insertion, acupuncture induces mast cell degranulation and the release of histamine, which can result in analgesic effect, particularly in the treatment of neuropathic pain, through the activation of histamine receptors (Fig. 25.3). It also modulates the analgesic action of opioids [7]. Increased mast cell density has been shown to be most significant (by 55%) at the TCM-identified acupuncture point (stomach 36 [ST 36]), and the degranulation of mast cells in acupoints suggests a relationship between degranulation and the analgesic effect of acupuncture [8]. Central effects following the insertion of acupuncture needles include the release of endorphins, serotonin, substance P, and enkephalin [9,10]. Three classes of opioids are involved in analgesia: enkephalins, β-endorphin, and dynorphin. Low-frequency EA triggers the release of β-endorphin and enkephalin in the brain and spinal cord, whereas high-frequency EA increases the release of dynorphin in the spinal cord. Serotonin has been shown in some experiments to increase the effectiveness of

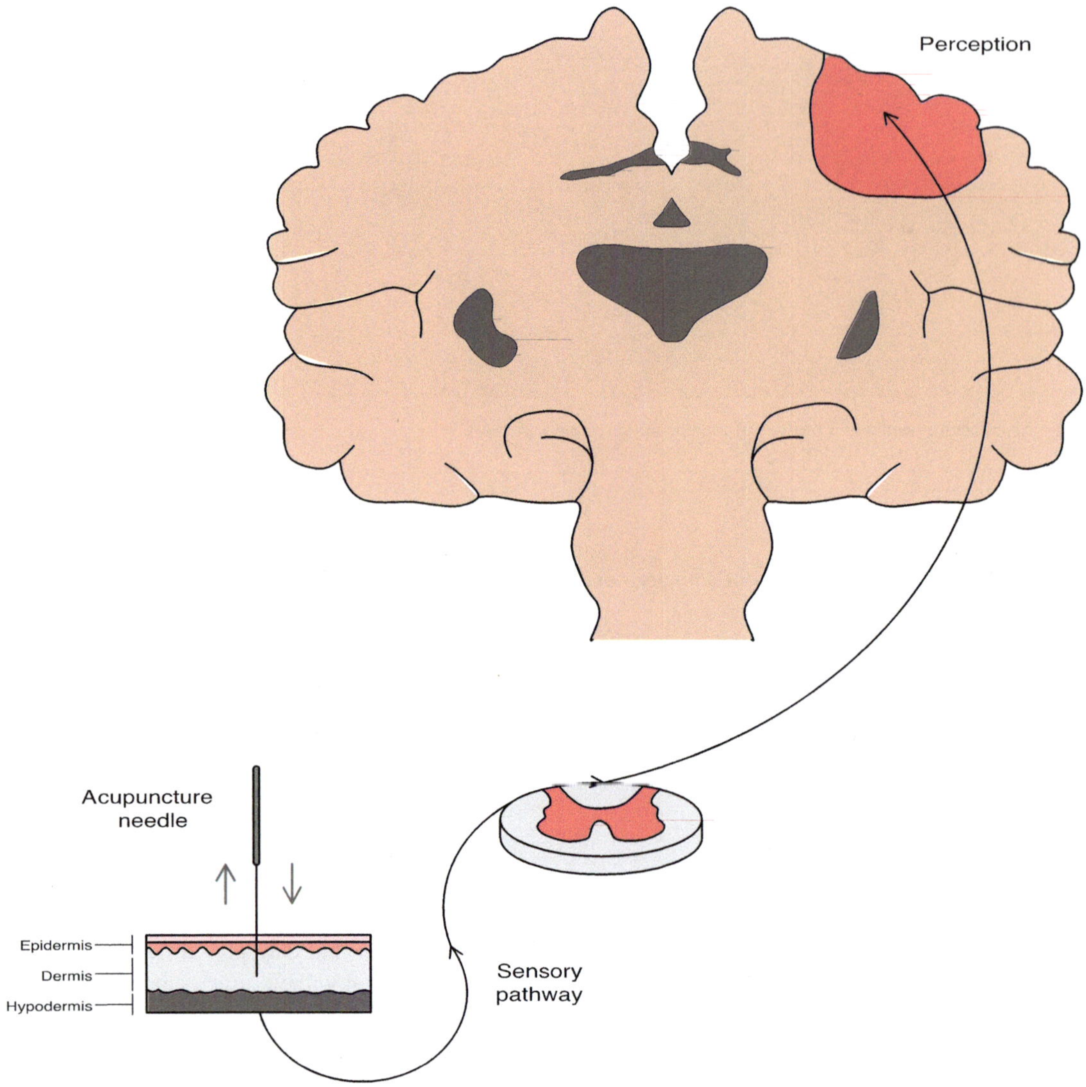

FIGURE 25.3 Acupuncture involvement in pain perception (original picture).

acupuncture analgesia. Oxytocin is another possible candidate for a role in acupuncture. In rats, it causes analgesia and induces feelings of calmness and relaxation, which are commonly observed in patients receiving acupuncture [11].

Acupuncture stimulation activates a variety of sensory receptors located both in the skin and in deeper tissues. This process engages multiple components of afferent nerves, providing several possible sites of interaction among their different constituent fiber types. There are two distinct functional categories of primary afferent axons with cell bodies in the dorsal root ganglion (DRG). Primary afferents include large-diameter myelinated (Aα, Aβ), small-diameter myelinated (Aδ), and unmyelinated (C) axons. All sympathetic postganglionic fibers are unmyelinated. Neurons mediating information of nociception/pain, hard pressure, cutaneous touch, and temperature from the limbs and trunk terminate in the ipsilateral dorsal horn of the spinal cord. From the dorsal horn, the information travels through the anterolateral pathway to the thalamus and hypothalamus. Thalamic neurons that are sensitive to nociceptive or painful stimuli, as well as thermal stimuli, project to the primary somatosensory cortex, the dorsal anterior insular cortex, and the anterior cingulate gyrus [6].

Different studies conducted using diagnostic imaging techniques have highlighted the role of acupuncture in affecting various somatosensory, affective, and cognitive brain regions. In functional magnetic resonance imaging (MRI) studies in human patients, acupuncture has been shown to alter the activity and connectivity associated with pain modulation in areas of the brain comprising the "pain cortex," which includes the insula, limbic areas, and the somatosensory areas S1 and S2 [12]. Moreover, acupuncture suppresses neurodegeneration by interfering with specific signaling pathways that involve the brain-derived neurotrophic factor and the neurotrophin family of proteins [13,14]. It has also been demonstrated that acupuncture

TABLE 25.2 Main opioid peptides.

Peptide	Main site	Receptor
β-endorphin	Midbrain, periaqueductal gray	μ δ
Enkephalin	Dorsal horn of spinal cord	μ δ
Dynorphin	Brain stem, spine	κ
Orphan (nociceptin)	Widespread	NOP (or NOP/ORL1)

NOP, Nociceptin opioid peptide receptor; *ORL1*, opioid-receptor-like 1.

exerts its analgesic effects by inhibiting proinflammatory cytokines, proinflammatory neuropeptides, and neurotrophins through the modulation of transient receptor potential vanilloid (TRPV1) [13,14]. TRPV1 is highly expressed in sensory nerve fibers distributed in the general region of acupoints like ST 36 and bladder 40. However, the contribution of TRPV1 to the activation of sensory nerves by acupuncture leads to the modulation of the pressor response and participates in the activation of their associated cell bodies in the DRG, thereby modulating pressor reflexes. These findings suggest that TRPV1 might act as an acupuncture-responsive channel and participate in acupuncture-related analgesia [15].

Neuromodulators of the opioid peptide group are involved in acupuncture analgesia (Table 25.2). Electrical stimulation at different frequencies stimulates the release of different peptides. Acupuncture can induce widespread relief of pain by stimulating the endogenous system known as descending inhibitory pain control. Opioid peptides have been identified in various regions of the central nervous system. These peptides are referred to as neuromodulators, and they produce a single response and have a sustained effect. The β-endorphin plays an important role in acupuncture analgesia, and its concentration increases in the cerebrospinal fluid of patients with pain. Through the local release of β-endorphin in the periaqueductal gray, at least two descending tracts are activated. They inhibit the transmission of nociceptive pathways in the dorsal horn at every segmental level. The transmitters released locally in the dorsal horn are serotonin and noradrenaline [16].

The benefits of acupuncture

Globally, acupuncture has gained widespread recognition for its role in managing chronic pain, stress, and other health conditions. In recent years, the TCM market, including acupuncture, has experienced significant growth due to its increasing integration into modern healthcare systems. The increasing popularity of acupuncture as a natural and holistic alternative to conventional medicine underscores its economic viability, while its ability to address a wide range of health conditions contributes to better patient outcomes worldwide. For example, in 2022 the TCM market was valued at over USD 400 billion globally, with acupuncture being a primary driver, particularly in pain management and wellness [17]. Acupuncture is increasingly recognized as a cost-effective treatment option, especially for chronic conditions like pain, stress, insomnia, anxiety, and depression. Compared to conventional medical interventions, it offers potential financial advantages without compromising therapeutic outcomes [17].

The integration of technology into TCM and acupuncture is creating new opportunities to modernize and enhance the effectiveness of treatments. In many regions, acupuncture is covered by insurance for specific conditions, further enhancing its accessibility. For example, in the United States, Medicare includes acupuncture for chronic lower back pain. In Germany, Austria, and other European countries, statutory health insurance often reimburses acupuncture for pain-related conditions.

What is the value of investing in complementary therapies such as acupuncture? First, there is a health benefit: improvements in health status lead to cost savings from reduced overall healthcare utilization and expenditures, as well as enhancements in quality of life (QoL), which translate into population welfare gains. Acupuncture is cost-effective in managing chronic pain and mental health conditions. The integration of acupuncture with conventional medicine offers a unique approach that shows promise in enhancing both mental and physical health while also generating cost savings. Second, harm reduction: improvements in pain management could reduce patients' exposure to addictive substances. Third, lifelong productivity: for younger populations, acupuncture can enhance school performance and lead to increasing long-term overall productivity [18].

Acupuncture in patients with cancer

The use of complementary therapies is common among patients with cancer [19]. Complementary and integrative medicine (CIM) modalities are effective and safe for treatment when used in conjunction with conventional medications and under the supervision of a medical professional during cancer therapy.

CIM is a term used to describe a shared work environment in which CIM practices are provided within a conventional medicine setting [20]. Many conventional cancer treatments, such as surgery, chemotherapy, radiotherapy, immunotherapy, or hormonal therapy, can result in both acute and chronic pain conditions. These include aromatase inhibitor (AI)-induced joint pain (somatic, musculoskeletal [MSK] pain) or chemotherapy-induced peripheral neuropathy (CIPN) pain, which is covered in a specific chapter of this book. With improved oncologic treatment, many patients diagnosed with advanced cancer now live longer with symptomatic illness and ongoing oncologic treatment. Therefore effective pain management is crucial throughout the cancer care trajectory.

Integrative medicine is defined as the coordinated use of evidence-based complementary practices and conventional care treatments, including interventions such as acupuncture, massage, meditation, and yoga, which are increasingly available in cancer centers and recommended for symptom and pain management. Approximately 40% of patients with cancer use integrative medicine treatment every year. The principle of integrative medicine is to use these interventions concurrently with conventional pain management approaches, and it is not intended as a replacement for conventional interventions [19–21].

Patients have discovered positive aspects of integrative medicine and perceive that conventional treatments are not always sufficient; they are concerned about the adverse effects of pharmacotherapy and prefer alternative approaches [21]. At the same time, risk–benefit considerations must take into account the ability of acupuncture and other CIM modalities to significantly relieve treatment-related symptoms, with implications for patients' adherence to conventional cancer treatment regimens [22].

Acupuncture can alleviate a wide range of adverse effects associated with cancer treatments and improve overall QoL, although there is a significant variety in the use of acupuncture treatment in patients with cancer (some ranges are between 1.7% and 31%) [23,24]. The National Institutes of Health defined acupuncture as "a family of procedures involving stimulation of anatomical locations on the skin by a variety of techniques" [23,25]. Acupuncture is often used in the oncology setting. It may be beneficial for pain management and alleviation of chemotherapy-induced nausea/vomiting, chronic MSK complaints, treatment- and premature menopause-associated hot flashes, fatigue, stress, anxiety, CIPN, radiation-induced xerostomia, dyspnea, leukopenia, and sleep disorders [26]. Acupuncture research is evolving, and some studies are subject to bias. A particular problem is that different forms of control groups have been used in acupuncture trials, including sham acupuncture that involves a nonpenetrating needle device, sham acupuncture that targets nonacupoints with shallow rather than deep needling, attention controls, and waitlist controls.

Treating pain with acupuncture

Acupuncture-induced analgesic effects can strongly influence the psychological aspect of pain. Tas et al. found that acupuncture is one of the most frequently utilized supportive therapies and has been shown to be effective in patients with cancer [26]. A total of 45 patients were included in the study, and acupuncture was administered to the patients 1 day prior to chemotherapy, on the day of chemotherapy, and 1 day after chemotherapy. Statistically significant decreases ($P < .001$) in pain, nausea, vomiting, insomnia, and anxiety scores were observed after the acupuncture treatment.

Dymackova et al. assessed the function of acupuncture in pain management related to cancer therapy in patients ($n = 60$) with head and neck cancer using a 10-point pain scale and analgesic use [27]. They found that patients who received acupuncture during radiotherapy have reduced average pain (AP) and the worst pain (WP) intensity during radiotherapy (AP median 0.16 vs 1.36, $P < .001$; WP median 0.90 vs 1.96, $P < .001$) and 3 months after radiotherapy (AP median 0.07 vs 0.50, $P < .001$; WP median 0.30 vs 0.83, $P = .002$). The results indicate lower analgesic consumption during radiotherapy ($P = .047$) and 3 months after radiotherapy ($P = .006$) in patients who underwent acupuncture treatment.

Recently, there has been an interest in whether acupuncture can be combined with the three-step approach for cancer pain and whether it has an influence on QoL [28]. The authors concluded that acupuncture combined with three-step analgesic drugs reduced pain intensity and significantly reduced chemotherapy side effects. Nonetheless, the authors noted that acupuncture does not control all cancer pain, and the effect sometimes takes time to be completely evident. They concluded that the optimal pain relief was achieved after 5 days of continuous acupuncture treatment but that despite this relatively slow onset of peak effect, this was then sustained long-term [28].

One trial ($n = 30$) with positive results compared acupuncture and the daily dose of opioid intake according to the Numeric Rating Scale (NRS) measured the intensity of pain with the control group taking only opioids for 1 week with a 2-week follow-up [29]. The acupuncture group showed a significant reduction in pain intensity, with a 2-point decrease in the NRS at the end of treatment, which was achieved in 93% ($n = 14/15$) of the acupuncture group [28]. In addition, there was a reduction in opioid needs and slight side effects.

Ge and colleagues tried to identify studies that support acupuncture and its acceptability in the treatment of patients with moderate-to-severe cancer pain [30]. This cross-sectional survey involved 2503 patients and, among other data, also investigated patients' preferences and attitudes. Most cancer patients thought that acupuncture was worthwhile, important, or effective. Around one-third of cancer pain patients preferred acupuncture over medication for pain management. Of the patients who received acupuncture treatment, most of them (ranging from 70% to 87%) stated that the course of acupuncture met their expectations, and most of them (ranged 90%–100%) were willing to receive acupuncture treatment again and were willing to recommend acupuncture to others [30]. The evidence from the observational study revealed that the adverse events primarily consisted of skin and subcutaneous tissue disorders, as well as mild pain associated with the application of the treatment to the skin [31]. A study involving 454,920 patients who used acupuncture to treat pain showed that 8% of patients experienced mild adverse events, and severe side effects occurred in 13 patients (0.003%). The most frequent mild adverse events included needling pain, hematoma, and bleeding at the point of needle insertion. Severe side reactions were pneumothorax, acute hypertensive or hypotensive crisis, erysipelas, and asthma attacks. A systematic review indicated that acupuncture performed by trained practitioners using clean needle techniques was a generally safe procedure [31].

Cancer-related painful conditions treated by acupuncture

AI-associated arthritis is a common and debilitating adverse effect in cancer patients with hormone-positive breast cancer. AI-related MSK (joint and muscle) pain affects up to 50% of women on this class of drugs and negatively affects the QoL and adherence to hormonal treatment. Crew et al. evaluated the use of acupuncture for the management of AI-associated joint pain in women with breast cancer ($n = 19$). After 6 weeks, acupuncture reduced pain more than the standard of care and provided a clinically meaningful change in pain (a 2-point reduction on a 0–10 scale). The authors concluded that acupuncture reduced joint symptoms and improved functional ability, although larger trials are required [32].

Patients undergoing treatment with taxane drugs (e.g., paclitaxel and docetaxel) and platinum-based drug groups frequently suffer from peripheral nerve damage symptoms, which are, in many cases, irreversible and can lead to long-term disability. Sensory pain–related symptoms include pain, tingling and numbness, "pins-and-needles" sensation, loss of sensation, and a wide range of motor and autonomic symptoms [33]. An integrative medicine program that included acupuncture reduced peripheral neuropathy–related symptoms in female patients with gynecological and breast cancer, as well as providing CIPN-related symptom relief compared with a group of patients receiving standard treatment [34]. In the field of cancer-specific pain management, auricular acupuncture can also be helpful for cancer patients with neuropathic pain [35].

According to the 2022 joint guideline from the Society for Integrative Oncology and the American Society of Clinical Oncology, acupuncture is recommended as an adjunctive treatment for select cancer-related pain syndromes. The recommendation for the use of acupuncture for AI-induced arthralgia and general MSK pain is supported by a moderate

TABLE 25.3 Studies on acupuncture with sufficient evidence to inform recommendations.

Study type	No	Pain system categories
SRs	17	AI-related joint pain CIPN Palliative and survivorship care
RCTs	34	AI-related joint pain General cancer pain CIPN Procedural or surgical pain

AI, Aromatase inhibitor; *CIPN*, chemotherapy-induced peripheral neuropathy; *RCT*, randomized controlled trial; *SR*, systematic review.

TABLE 25.4 Summary of recommendations for the use of acupuncture.

Pain category	Type of recommendation	Quality of evidence	Level of obligation	Benefit/harm	Strength of recommendation
AI-related joint muscle pain	Evidence-based	Intermediate	Should	Benefit	Moderate
Cancer pain/MSK pain	Evidence-based	Intermediate	May	Benefit	Moderate
CIPN	Evidence-based	Low	May	Not assessable	Weak
Surgical pain	Evidence-based	Low	May	Benefit	Weak

AI, Aromatase inhibitor; *CIPN*, chemotherapy-induced peripheral neuropathy; *MSK*, musculoskeletal.

strength of recommendation and intermediate-quality evidence (Table 25.3). For CIPN and postsurgical or procedural pain, acupuncture receives a weak recommendation based on low-quality evidence. Acupuncture can also be considered for reducing anxiety, improving mood disturbance and depressive symptoms, improving posttreatment fatigue, improving hot flashes, and improving QoL in general. In addition, acupressure or EA can be considered as an addition to antiemetic drugs to control nausea and vomiting during chemotherapy. There is also some evidence that acupuncture or acupressure may be offered to patients undergoing cancer surgery or other cancer-related procedures, such as bone marrow biopsy, although with low-quality evidence and weak strength of recommendation.

These distinctions underscore acupuncture's potential utility in supportive cancer care while highlighting the need for continued research and individualized clinical judgment [36,37] (Table 25.4).

Along with the effect on pain, acupuncture administered during a course of radiation may also reduce the incidence of radiation-induced xerostomia, which can also cause pain. Blom and Lundeberg reported a small randomized controlled trial with a placebo acupuncture control, in which acupuncture treatment induced a persistent increase in salivary flow rate among patients with severe xerostomia [38]. A long-term follow-up (up to 32 months) further confirmed his findings. Johnstone and his colleagues used acupuncture for patients with pilocarpine-resistant xerostomia after radiotherapy for head and neck cancer [39]. They found a 70% response rate, a 10% increase from the baseline, and an increase in salivation at both the 3- and 6-month follow-ups. Deng et al. used functional MRI to evaluate changes in saliva production associated with acupuncture. Manual stimulation of LI-2 was found to be associated with neuronal activations that were absent during sham acupuncture [40].

Future direction

While more data and robust clinical trials are required to elucidate the potential use and underlying mechanism, there is data to support the use of acupuncture in patients with different types of cancer and different stages of cancer treatment, particularly for cancer pain. Because of the well-known side effects and complications of pharmacological treatment, cancer patients often prefer acupuncture in pain treatment when possible, although acupuncture can also be used along with modern analgesics. Additional benefits of acupuncture include its favorable safety profile and low rate of complications. Hence, the implementation of acupuncture and CIM principles should become more readily available and affordable in the management of cancer patients.

Conclusion

Acupuncture offers a nonpharmaceutical, evidence-based approach to manage many uncomfortable consequences of cancer treatments. It should be considered as a part of cancer care, as it can be administered safely at any stage of the cancer survivorship pathway.

Pain is one of the most common symptoms in cancer patients and can have complex and different etiologies (e.g., tumor burden, treatment-related, and non-cancer-related) and varying presentations (e.g., neuropathic and MSK) and duration (e.g., acute and chronic). Management of pain requires an interdisciplinary approach and interventions that can alleviate the pain.

The quantity and quality of current economic evaluations of acupuncture interventions are insufficient to meet the needs of health decision-makers. The National Institute for Health and Care Excellence suggests evaluating the applicability of economic evaluations. When evaluating the cost of acupuncture for a healthcare system, several factors should be considered, including space, equipment, practitioners' time, training, and their impact on other healthcare provisions. According to available data, acupuncture is increasingly recognized as a cost-effective treatment option, especially for chronic conditions like pain and stress. Compared to conventional medical interventions, it offers significant financial advantages without compromising therapeutic outcomes [17].

One of the primary challenges for acupuncture and TCM in the global market is ongoing skepticism and scientific debate, particularly in Western medical communities. While acupuncture has gained recognition for its effectiveness in managing chronic pain and stress, there remains mixed evidence regarding its efficacy in treating many other conditions. Critics argue that many studies supporting acupuncture lack rigorous scientific design, such as randomized controlled trials, leading to resistance from some Western practitioners. The placebo effect is often cited as a counterargument, with skeptics questioning whether acupuncture's benefits are due to psychological factors rather than physiological mechanisms [16].

Integration of TCM into evidence-based medical practice, particularly in Western healthcare systems, together with stronger regulatory frameworks and international collaboration, is well positioned to enable acupuncture to achieve sustained long-term growth, thereby enhancing global health and economic development.

References

[1] Cambel, A. Modern acupuncture. Acupuncture in practice; beyond points and meridians. Edinburgh, UK: Butterford-Heinemann; 2001. p. 22–32.

[2] White A. Western medical acupuncture: a definition. Acupunct Med 2009;27(1):33–5. Available from: https://doi.org/10.1136/aim.2008.000372.

[3] Samuels N, Ben-Arye E. Integrative medicine therapies in pain management. In: Cherny NI, Fallon MT, Kaasa S, Portenoy RK, Currow DC, editors. Oxford textbook of palliative medicine. 6th ed. Oxford, UK: Oxford University Press; 2021. p. 471–7.

[4] Samuels N, Ben-Arye E. Integrative medicine for cancer-related pain: a narrative review. Healthcare (Basel) 2024;12(3):403. Available from: https://doi.org/10.3390/healthcare12030403.

[5] Liddle CE, Harris RE. Cellular reorganization plays a vital role in acupuncture analgesia. Med Acupunct 2018;30(1):15–20. Available from: https://doi.org/10.1089/acu.2017.1258.

[6] Lundeberg T, Lund I. Peripheral components of acupuncture stimulation – their contribution to the specific clinical effects of acupuncture. In: Filshie J, White A, Cummings M, Longhurst J, editors. Medical acupuncture; a western scientific approach. Elsevier; 2016. p. 22–58. ISBN: 978-0-7020-4307-9.

[7] Mobarakeh JI, Takahashi K, Yanai K. Enhanced morphine-induced antinociception in histamine H_3 receptor gene knockout mice. Neuropharmacology 2009;57(4):409–14. Available from: https://doi.org/10.1016/j.neuropharm.2009.06.036.

[8] Zhang D, Ding G, Shen X, et al. Role of mast cells in acupuncture effect: a pilot study. Explore 2008;4(3):170–7. Available from: https://doi.org/10.1016/j.explore.2008.02.002.

[9] Lee HJ, Lee JH, Lee EO, et al. Substance P and beta endorphin mediate electroacupuncture induced analgesic activity in mouse cancer pain model. Acupunct Electrother Res 2009;34(1–2):27–40. Available from: https://doi.org/10.3727/036012909803861095.

[10] Zhao Z-Q. Neural mechanism underlying acupuncture analgesia. Prog Neurobiol 2008;85(4):355–75. Available from: https://doi.org/10.1016/j.pneurobio.2008.05.004.

[11] Campbell A. An outline of the traditional system. Acupuncture in practice; beyond points and meridians. Edinburgh, UK: Butterford-Heinemann; 2001. p. 9–21.

[12] Dhond RP, Kettner N, Napadow V. Neuroimaging acupuncture effects in the human brain. J Altern Complement Med 2007;13(6):603–16. Available from: https://doi.org/10.1089/acm.2007.7040.

[13] Evenepoel M, Haenen V, De Baerdemaecker T, et al. Pain prevalence during cancer treatment: a systematic review and meta-analysis. J Pain Symptom Manage 2022;63(3):e317–35. Available from: https://doi.org/10.1016/j.jpainsymman.2021.09.011.

[14] McDonald JL, Cripps AW, Smith PK. Mediators, receptors and signalling pathways in the anti-inflammatory and antihyperalgesic effects of acupuncture. Evid Based Complement Alternat Med. 2015:2015:975632. Available from: https://doi.org/10.1155/2015/975632.

[15] Wu SY, Chen WH, Hsieh CL, et al. Abundant expression and functional participation of TRPV1 at Zusanli acupoint (ST36) in mice: mechanosensitive TRPV1 as an "acupuncture – responding channel". BMC Complement Altern Med 2014;14:96. Available from: https://doi.org/10.1186/1472-6882-14-96.

[16] White A, Cummings M, Filshie J. Neurological mechanisms III: extrasegmental analgesia. An introduction to western medical acupuncture. London: Churchill Livingstone, Elsevier; 2008. p. 41–50.

[17] Litscher G. The global economics of acupuncture and traditional Chinese medicine – a strategic and financial perspective. Sci 2024;7(1):4. Available from: https://doi.org/10.3390/sci7010004.

[18] Lu M, Sharmin S, Tao Y, et al. Economic evaluation of acupuncture in treating patients with pain and mental health concerns: the results of the Alberta Complementary Health Integration Project. Front Public Health 2024;12:e1362751. Available from: https://doi.org/10.3389/fpubh.2024.1362751.

[19] Garcia MK, McQuade J, Haddad R, et al. Systematic review of acupuncture in cancer care: a synthesis of the evidence. J Clin Oncol 2013;31(7):952–60. Available from: https://doi.org/10.1200/JCO.2012.43.5818.

[20] National Centre for Complementary and Integrative Health. Complementary, alternative, or integrative health: what's in a name? <https://www.nccih.nih.gov/health/complementary-alternative-or-integrative-health-whats-in-a-name>; 2018.

[21] Cybularz PA, Brothers K, Singh GM, et al. The safety of acupuncture in patients with cancer therapy–related thrombocytopenia. Med Acupunct 2015;27(4):224–9. Available from: https://doi.org/10.1089/acu.2015.1099.

[22] Lyman GH, Greenlee H, Bohlke K, et al. Integrative therapies during and after breast cancer treatment: ASCO endorsement of the SIO clinical practice guideline. J Clin Oncol 2018;36(25):2647–55. Available from: https://doi.org/10.1200/JCO.2018.79.2721.

[23] NIH Consensus Conference. Acupuncture. JAMA 1998;280(17):1518–24. Available from: https://doi.org/10.1001/jama.280.17.1518.

[24] Lu W, Dean-Clower E, Doherty-Gilman A, Rosenthal DS. The value of acupuncture in cancer care. J Soc Integr Oncol 2008;22(4):631–48. viii. Available from: https://doi.org/10.1016/j.hoc.2008.04.005.

[25] Cascella A, Bimonte S, Schiavo D, et al. Acupuncture application for cancer pain management and its underlying mechanisms. World Cancer Res J 2023;10:e2479. Available from: https://doi.org/10.32113/wcrj_20231_2479.

[26] Tas D, Uncu D, Sendur MA, et al. Acupuncture as a complementary treatment for cancer patients receiving chemotherapy. Asian Pac J Cancer Prev 2014;15(7):3139–44. Available from: https://doi.org/10.7314/apjcp.2014.15.7.3139.

[27] Dymackova R, Selingerova I, Kazda T, et al. Effect of acupuncture in pain management of head and neck cancer radiotherapy: prospective randomized unicentric study. J Clin Med 2021;10(5):1111–27. Available from: https://doi.org/10.3390/jcm10051111.

[28] Li D, Su Y, Fan H, et al. Acupuncture combined with three-step analgesic therapy of cancer pain: a systematic review and meta-analysis of randomised clinical trials. Evid Based Complement Alternat Med 2021;2021(1):5558590. Available from: https://doi.org/10.1155/2021/5558590.

[29] He Y, Zhang H, Li Y, et al. Acupuncture combined with opioids for cancer pain: a pilot pragmatic randomized controlled trial. Acupunct Med 2022;40(2):133–41. Available from: https://doi.org/10.1177/09645284211056016.

[30] Ge L, He Y, Wu D, et al. Acupuncture for cancer pain: an evidence-based clinical practice guideline. Chin Med 2022;17(1):8. Available from: https://doi.org/10.1186/s13020-021-00558-4.

[31] Weidenhammer W, Streng A, Linde K, et al. Acupuncture for chronic pain within the research program of 10 German health insurance funds—basic results from an observational study. Complement Ther Med 2007;15(4):238–46. Available from: https://doi.org/10.1016/j.ctim.2006.09.005.

[32] Crew KD, Capodice JL, Greenlee H, et al. Pilot study of acupuncture for the treatment of joint symptoms related to adjuvant aromatase inhibitor therapy in postmenopausal breast cancer patients. J Cancer Surviv 2007;1(4):283–91. Available from: https://doi.org/10.1007/s11764-007-0034-x.

[33] Ben-Arye E, Hausner D, Samuels N, et al. Impact of acupuncture and integrative therapies on chemotherapy-induced peripheral neuropathy: a multicentered, randomized controlled trial. Cancer 2022;128(20):3641–52. Available from: https://doi.org/10.1002/cncr.34422.

[34] Shalom-Sharabi I, Samuels N, Lev E, et al. Impact of a complementary/integrative medicine program on the need for supportive cancer care-related medications. Support Care Cancer 2017;25(10):3181–90. Available from: https://doi.org/10.1007/s00520-017-3726-4.

[35] Alimi D, Rubino C, Pichard-Leandri E, et al. Analgesic effect of auricular acupuncture for cancer pain: a randomized, blinded, controlled trial. J Clin Oncol 2003;21(22):4120–6. Available from: https://doi.org/10.1200/jco.2003.09.011.

[36] Mao JJ, Ismaila N, Bao T, et al. Integrative medicine for pain management in oncology: society for integrative oncology - ASCO guideline. J Clin Oncol 2022;40(34):3998–4024. Available from: https://doi.org/10.1200/jco.22.01357.

[37] Richardson MA. Complementary and alternative therapy use in gynecologic oncology: implications for clinical practice. Gynecol Oncol 2002;84(3):360–2. Available from: https://doi.org/10.1006/gyno.2002.6612.

[38] Blom M, Lundeberg T. Long-term follow-up of patients treated with acupuncture for xerostomia and the influence of additional treatment. Oral Dis 2000;6(1):15–24. Available from: https://doi.org/10.1111/j.1601-0825.2000.tb00316.x.

[39] Johnstone PA, Niemtzow RC, Riffenburgh RH. Acupuncture for xerostomia: clinical update. Cancer 2002;94(4):1151–6. Available from: https://doi.org/10.1002/cncr.10348.

[40] Deng G, Hou BL, Holodny AI, et al. Functional magnetic resonance imaging (fMRI) changes and saliva production associated with acupuncture at LI-2 acupuncture point: a randomized controlled study. BMC Complement Med Ther 2008;8:371. Available from: https://doi.org/10.1186/1472-6882-8-37.

Chapter 26

Treating spiritual, emotional, and psychological aspects of pain

Sergej Nadalin[1,2] and Dragana Marić[3,4]
[1]*Department of Psychiatry, General Hospital "Dr. Josip Benčević", Slavonski Brod, Croatia,* [2]*Faculty of Medicine, Catholic University of Croatia, Zagreb, Croatia,* [3]*Faculty of Medicine, University of Belgrade, Belgrade, Serbia,* [4]*Clinic of Pulmonology, University Clinical Center of Serbia, Belgrade, Serbia*

Key learning objectives

- Appreciate and value the highly individualized nature of spirituality in each patient.
- Determine the possible role of spirituality in the patient's pain experience by exploring and honoring his values and beliefs.
- Acknowledge that diverse spiritual interventions are beneficial for patients' quality of life and well-being, but at the same time have limited effectiveness in alleviating cancer pain.
- Appreciate the importance of the psychology profession in palliative care as an addition and not as an alternative to physical treatments.
- Determine the role of psychological factors in pain experience, response to pain, and treatment.
- Acknowledge the role of psychosocial interventions in pain management, specifically in alleviating cancer pain.

Introduction

Active holistic care of individuals across all ages with serious health-related suffering due to severe illness, especially of those near the end of life, is imperative in contemporary palliative care [1]. In addition to meticulous symptom management, the scope of palliative care includes addressing emotional, psychological, existential, and social issues of patients (and their caregivers) during the course of a disease whose trajectory is emotionally draining and may terminate with death [2].

Spirituality and spiritual care are also recognized as vital aspects of palliative care, relevant for efficient palliative care delivery with the power to ameliorate negative symptom experiences [3]. Therefore, over the past 15 years, initiatives have been launched to integrate this dimension of care into research and clinical settings. To foster future evidence-based research on spiritual care and to improve staff competence and confidence, the European Association for Palliative Care (EAPC) established the Spiritual Care Taskforce in 2010, while the Global Network for Spirituality and Health was formed in the United States in 2013 [4]. Despite spirituality being regarded as a core dimension of palliative care, it has received little attention within the field of pain medicine. Reasons for this may include insufficient knowledge and representation of the concept of the bio-psychospiritual model of pain, lack of understanding of spirituality itself and its importance in pain treatment, as well as uncertainty about how it may be addressed [5].

The psychology profession supports the World Health Organization's emphasis that palliative care incorporates not only physical treatments but also assessment and intervention to ensure the psychological well-being of the person, their family members, and carers, as well as members of the palliative care interdisciplinary team [6]. The EAPC Task Force identified five areas of expertise that psychology can bring to a palliative care setting: psychological assessment, psychological therapy for individuals, families, and groups, consultation, supervision and staff support, education, and research [7].

Pain Management in Palliative Care for Patients With Cancer. DOI: https://doi.org/10.1016/B978-0-443-30050-9.00046-1

Although advanced cancer patients experience both psychological and physical pain, pain treatment is mostly targeted at relieving the physical symptoms [8,9]. Nevertheless, psychological treatments need to be considered as additions and not as alternatives to physical treatments for cancer-related pain in any phase of care [10,11]. Indeed, among cancer patients, unrecognized psychological factors have been shown to influence both the experience of pain and their response to pain (including seeking healthcare) and treatment [9,10]. For instance, psychological factors such as depression, anxiety, and a fear of catastrophes are associated with more severe pain [9,12]. Besides, evidence suggests that across the cancer continuum, unrelieved pain increases the risk for psychological distress [10].

Psychologists seem to be less often integrated in palliative care than their psychosocial colleagues [11,13] due to lack of adequate education, inadequate size of the palliative medicine-trained workforce, and lack of sufficient reimbursement for palliative care [14]. One study found that psychologists provided services to just 24% of palliative care inpatients and 19% of outpatients in a major comprehensive cancer center [13].

Spirituality and pain

The evolution of the view of pain has come a long way from a purely biological view of pain, through a dualistic view of pain that is either physical and therefore "real" or psychological and therefore "in the mind," to Engel's bio-psychosocial model, which provided a conceptual framework for better recognition of the role of the mind, as well as the influence of behavioral and environmental factors on pain [5,15,16]. Although the bio-psychosocial model has become the dominant concept for pain assessment and treatment within the palliative care community, possibly in response to the needs of patients at the end of life, proponents of the bio-psychospiritual model have emerged, which attempts to integrate a spiritual component with the bio-psychosocial model [5,17].

The bio-psychospiritual model essentially relies on the conceptualization of "total pain" and "total suffering" conceived in the 1950s by Dame Cicely Saunders and practiced since then within the palliative care community, understanding the suffering specific to terminal illnesses as multidimensional—simultaneously physical, psychological, emotional, social, spiritual, and/or existential in nature [18].

Spirituality is an important aspect of every person's life. Although the concept of spirituality is broad and not amenable to a universal definition, the definitions present in the literature successfully capture the vague and transcendent nature of spirituality. It is defined as a dynamic process whereby humans experience a connection with self, others, nature, and/or a higher power and sense of meaning [19]. Although spirituality and religion are often seen as the same thing, they are different, albeit interrelated domains, and yet they can overlap for some patients. It can be said that adherence to a religion is a manifestation of spirituality, as the deity is transcendent. But the sources of the transcendent can be very different for different persons, often linked to nature, music, art, family, and community [20,21]. Therefore spirituality is commonly conceptualized through those aspects of life that lie at the core of a person's identity, such as values, beliefs, activities, and relationships that provide meaning and purpose for life [17,22,23]. The concept of spiritual well-being also came from this understanding of spirituality. In this state, a person has a feeling of strength, peace, and comfort that arises from a sense of meaning and purpose, and a certain level of spiritual well-being is common to all [5,24].

Spirituality is important in any part of life, but it takes on special meaning when dealing with life-threatening diseases such as cancer, which is a source of complex suffering in which pain holds a particularly significant place. Although spirituality and spiritual well-being are well positioned within the doctrine of palliative care, the focus is mainly on spiritual distress during the end-of-life period and issues related to death and dying. At the same time, they receive relatively little attention in the field of cancer-related pain treatment [5].

Yet there is ample evidence that spirituality exerts a significant influence on physical and spiritual health and that a higher level of self-reported spirituality is associated with positive physical and psychological outcomes in patients with multiple cancer-related symptoms [25]. In the study by Miller et al., higher spirituality was significantly associated with lower severity of fatigue, depression, and sleep disturbance in the group of 200 solid tumor cancer patients undergoing chemotherapy. Also, higher spirituality was significantly associated with lower overall symptom interference. At the same time, spirituality was not significantly associated with pain severity, although results trended toward significance in the expected direction [26]. Previous studies have also shown that the role of spirituality in the experience of pain is not straightforward—one study found no significant relationship between spirituality and pain, and another one showed an association between higher spirituality and lower pain severity, interference, and total symptom score [27,28]. These mixed results initiated further research related to the influence of spirituality on pain, especially on the so-called pain-related catastrophizing, which includes negative cognitions and beliefs related to pain (e.g., perceived lack of control, helplessness, expectation of negative outcomes) [29]. In a mixed-methods study by Miller et al., in the quantitative part, spiritual well-being was directly negatively associated with pain-related catastrophizing and indirectly negatively associated with pain outcomes (intensity, pain-related distress, and interference). In the qualitative phase of the study, the findings revealed the complex and pervasive role of pain-related catastrophizing (pain "spirals," "mental anguish" that are difficult to disentangle from the painful experience itself), but it was also found that spirituality helps patients to overcome pain-related catastrophizing by providing a framework for interpreting and making meaning of the experience of pain. On the other hand, it was found that spirituality has significant limitations in the context of certain pain experiences, such as chronic, neuropathic, or particularly severe pain. It was concluded that although research has highlighted the importance of spirituality for some patients with cancer pain, limitations to the usefulness of spirituality for some patients have also been established [30]. Moreover, while spirituality can provide hope and comfort and offer a sense of purpose and determination in cancer-related suffering [31], one recent study found that spirituality was not significantly related to pain, but that resilience (the ability to adjust to losses by finding new opportunities in those very losses) was directly negatively correlated with pain [32]. Considering the differences in the importance that spirituality can have for different patients, depending on a person's values and beliefs, as well as the cultural environment, it is especially important to establish which patients could benefit from the inclusion of spirituality in the plan of pain management. Also, deepening the understanding of the nature of the relationship between spirituality and pain and exploring potential mediators (optimism, mindfulness, interoception, and pain-related catastrophizing) may be important in tailored pain assessment and pain management plans.

Spiritual care and pain management

Spiritual care is conceptualized as care that "responds to the needs of the human spirit when faced with trauma, ill health, and sadness" and strives to address broader life aspects such as meaning and purpose, relating these dimensions to oneself, surroundings, and the divine [33,34]. Although spiritual care should complement the physical, psychological, and social care within the holistic palliative care concept, it remains a largely neglected aspect of palliative care. Today, we know that spiritual interventions can reduce pain intensity and pain-related stress (for some patients), and patients approaching the end of their lives often find spiritual care desirable and valuable.

Given that higher spiritual well-being is associated with positive coping and adaptation strategies such as "fighting spirit" and the ability to find meaning in life despite the presence of pain [35,36], ways to assess the patient's spirituality have been investigated. The spiritual assessment should allow healthcare professionals to help patients identify and mobilize their spiritual resources while coping with pain and other debilitating symptoms. Initially, the spiritual assessment was carried out in the context of a broader assessment of the quality of life (QoL), and the instruments for the assessment of spirituality are either part of multidimensional questionnaires for the QoL assessment or are derived from them.

Several spiritual assessment tools have been used in clinical practice, such as the 12-item Functional Assessment of Chronic Illness Therapy–Spiritual Well-Being (FACIT-Sp-12) Scale [24], as well as the Faith, Importance and Influence, Community, and Address (FICA) Spiritual History Tool [20]. The FACIT-Sp-12 is the most widely used quantitative tool for measuring spiritual well-being among those with cancer. It is a part of the larger FACIT measurement system, and the FACIT-Sp-12 may or may not be used in conjunction with the Functional Assessment of Cancer Therapy–General (FACT-G) measure. It is a self-administered questionnaire containing 12 items rated on a 5-point Likert scale and three subdomains of spiritual well-being: peace, meaning, and faith. On the other hand, FICA, along with several other available tools such as FAITH, SPIRITual, or HOPE, uses a qualitative approach [37]. A tool for assessing spiritual history within a clinical context using an acronym, FICA, that identifies four components of the spiritual history: (1) faith or beliefs—questions around personal meaning, faith, and beliefs; (2) importance and influence—explores how important faith and beliefs are in terms of living with the illness; (3) community—strive to understand the importance of relationships, particularly whether they are in the context of a spiritual or religious community; and (4) address—estimation of how the person would like to have these issues addressed as part of their care [5,21].

The National Coalition for Hospice and Palliative Care recommends that spiritual history should be taken as an integral part of assessment at the initial patient encounter, with subsequent regular spiritual assessments together with other clinical assessments [38].

Spiritual care should be provided by practitioners of medical and radiation oncology, palliative care specialists, nurses, and social workers in the aforementioned structured manner because they have multiple contacts with cancer patients suffering from pain [20]. However, defining what constitutes "spiritual intervention" is a major challenge. Since the nature of spirituality is highly individual, any single therapeutic intervention may have substantial spiritual meaning for one person and be devoid of spirituality for another.

A recent systematic review on spiritual interventions for cancer pain included 11 studies of different designs to determine whether spiritual interventions have a role in the treatment of cancer pain and which offer the greatest possible benefit [39]. Types of spiritual interventions included dignity therapy, prayer-based therapy, electronic support groups, mindfulness-based art therapy, peer helping, mindfulness-based stress reduction, and spiritually focused music therapy. The authors concluded that spiritual interventions, taken together, had only a small benefit in alleviating cancer pain—three studies reported a statistically significant improvement in pain scores, and more importantly, none of the interventions reviewed worsened the participants' pain. At the same time, the spiritual interventions were well received and welcomed by the participants [39].

On the other hand, a recent in-depth review presented evidence-based statements concerning spiritual care: (1) incorporate spiritual care into care for patients with serious illness; (2) incorporate spiritual care education into the training of interdisciplinary teams caring for persons with serious illness; and (3) include specialty practitioners of spiritual care in the care of patients with serious illness [40]. The presented recommendations emphasize the role of spiritual care as an essential component of holistic palliative care and cancer pain management, underlining the importance of effective communication in exploring the patient's spirituality and establishing a spiritual connection with the patient and caregivers. A summary of recommendations for spiritual care in pain management is provided in Table 26.1.

Even though existing evidence cannot firmly support the use of spiritual interventions in cancer pain management, the highly individualized and transcendent nature of spirituality necessitates a broader perspective. Although the question is whether and to what extent we can influence the measurable pain experience through these interventions, the spiritual connection of patients and practitioners is beneficial on different levels, promoting the QoL of patients and caregivers and helping to find meaning and comfort during inevitable suffering [20].

Psychological factors in pain experience

According to the theory of pain published by Melzack and Wall in 1965, the experience of pain is a multidimensional phenomenon influenced by affect, previous experience, and cultural beliefs, in addition to sensory input [15]. The International Association for the Study of Pain (IASP) identified psychology as an integral aspect of the pain experience and included it thusly in its definition of pain [41]. By IASP, pain is defined as "an unpleasant sensory and emotional experience associated with actual or potential tissue damage or described in terms of such damage" [42].

Psychological factors have been shown to influence both the experience of pain and the response to pain treatment among cancer patients [10]. Psychological factors involved in the experience of pain can be grouped starting with awareness of pain (i.e., attention factors), then with evaluation of pain (i.e., cognitive factors), resulting feelings (i.e., emotional factors), and behaviors related to pain (i.e., coping strategies) [43].

Attention factor influences focus toward pain [43]. Evidence suggests that pain sensitivity decreases with engagement in attention-requiring tasks and with the use of environmental distractors [44–46]. On the other hand, greater attention to pain increases pain intensity, emotional distress, and psychosocial disability, as well as pain-related healthcare utilization in patients with chronic pain [46–48]. There are numerous studies suggesting effects of attention and distraction on the experience of pain [46,49–52]. A recent scoping review indicated a higher prevalence of attention-deficit hyperactivity disorder (ADHD) among youth with chronic pain and a higher prevalence of chronic pain in samples of youth with ADHD [50].

TABLE 26.1 Recommendations for spiritual care in pain management.

- Establish which patients could benefit from the inclusion of spirituality in the plan of pain management.
- Spiritual history should be taken as an integral part of assessment at the initial patient encounter with subsequent regular spiritual assessments.
- Spiritual assessment tools that can be used in clinical practice: FACIT-Sp-12, FICA, FAITH, SPIRITual, and HOPE.
- Spiritual care should be provided by practitioners of medical and radiation oncology, palliative care specialists, nurses, and social workers in a structured manner.
- Types of spiritual interventions: dignity therapy, prayer-based therapy, electronic support groups, mindfulness-based art therapy, peerhelping, mindfulness-based stress reduction, spiritually focused music therapy, etc.
- Incorporate spiritual care education into the training of interdisciplinary teams caring for cancer patients.
- Spiritual interventions are welcomed by the patients improving their quality of life but usually have only a small benefit in alleviating cancer pain.

One study showed that patients with fibromyalgia syndrome had poorer performance on the divided attention task, as well as that chronic pain and induced acute pain exerted different effects on attention [49]. Some novel studies found greater attention toward pain (new or ongoing) among survivors of cancer, specifically prior to regularly scheduled testing for recurrence [51,52]. Such experiences can significantly affect the fear of cancer recurrence among cancer survivors, which is a distressing concern that influences both physiological and psychological recovery [52].

Cognitive factors play a role in the interpretation of pain [43]. Evidence suggests that patients with positive attitudes regarding pain manifest better psychological and physical functioning in comparison to patients with negative attitudes regarding pain [53,54]. Positive attitudes have been associated with better utilization of medical services for pain management and better engagement in pain-reducing behaviors [55]. On the other hand, pain catastrophizing is a negative cognition related to the belief that the pain will result in the worst possible outcome [56]. It has been correlated to many adverse pain-related outcomes, including poor treatment response, unsatisfactory QoL, and high disability related to both acute and chronic pain [56,57]. There are studies indicating the disruption of cognitive processing in chronic pain syndromes, including fibromyalgia, migraine, chronic back pain, rheumatoid arthritis, diabetic neuropathy, osteoarthritis, chronic regional pain syndrome, peripheral neuropathic pain syndromes, and multiple sclerosis [46]. In these studies, major cognitive parameters encompassed attention, learning, memory, sustained concentration, processing speed, psychomotor ability, and executive function [46]. Importantly, one novel study evaluated cognitive impairment among patients experiencing cancer pain using the Functional Assessment of Cancer Therapy–Cognitive Scale and estimated its prevalence at 35.9%, significantly higher compared with the general population [58]. Other important factors contributing to cognitive impairment among patients with cancer pain in that study were age, smoking history, sleep quality, intensity of pain, and anxiety and depression [58].

Emotional factors may increase or decrease pain perception by affecting an individual's emotions [43]. A recent study on patients with chronic pain secondary to spinal cord injury showed that positive psychology exercises consisting of "flow experiences," kindness, gratitude, taking care of the body, spirituality, optimism, and so on can reduce pain intensity beyond the effects of pharmacological treatment [59]. On the other hand, anxiety, depression, fear, anger, guilt, frustration, and other negative emotions have been associated with higher pain levels [60]. Among negative emotions, the most frequently researched are anxiety and depression, both of which have been implicated to play a role in the chronification of pain [60]. One of the most influential models of chronic pain is the fear-avoidance model, which indicates that fear of pain and fear of reinjury may hamper recovery from an acute pain episode because anticipatory anxiety motivates avoidance behavior and inactivity [61,62]. There is a strong association between chronic pain and depression, though a long-standing debate is whether chronic pain causes depression or whether depression is a risk factor for developing chronic pain [62,63]. Importantly, results of a recent study suggest a strong association between the presence and severity of pain and distress symptoms such as anxiety and depression in patients diagnosed with various types of cancer [64]. In this study, the proportion of patients with anxiety and depression was 9.2% and 17.7% in patients with no pain and about 32.8% and 36.7% in patients with severe pain, respectively [64].

Coping strategies are behavioral techniques activated by individuals to reduce pain [43]. Positive coping refers to direct, rational, problem-focused solving, while negative coping refers to avoiding, withdrawing, or ignoring the problem [65]. One study investigated which coping strategies were of relevance for patients with spine-associated pain syndromes, fibromyalgia, polyarthritis, migraine/headache, chronic inflammatory bowel diseases, cancer-related pain, amputations, and pain associated with psychophysical exhaustion [66]. Their results indicated that positive attitudes together with higher age were significant predictors of patients' life satisfaction but none of the other adaptive coping strategies [66]. Recently, numerous studies on coping strategies have been conducted among women with breast cancer [67–69]. One study on coping strategies found that the most used coping strategies for pain in women with breast cancer were positive self-statements, relying on religion, acceptance of diagnosis, self-distraction from thinking about pain, positive reframing, and denial [68]. Furthermore, in women with breast cancer, emotion-focused strategies had significant influence on pain, while problem-focused solving strategies did not significantly affect pain experience [67]. One study on women who had undergone surgical treatment for breast cancer found that women choosing constructive coping strategies (i.e., fighting spirit and positive redefinition) obtained higher QoL scores, while greater reliance on destructive coping strategies (i.e., anxious preoccupation and helplessness) was associated with significantly worse QoL [69]. In all functioning domains, higher levels of pain were associated with poorer QoL and more severe symptoms associated with the disease and its treatment [69].

Psychological care and pain management

The goal of psychosocial interventions is to help persons to establish the emotional, social, and intellectual skills needed to live, learn, and work in the community with the least amount of professional support [70]. Psychosocial interventions are usually delivered by medical (psychiatrists and psychiatric nurses) and allied health (social workers, psychologists, and occupational therapists) specialties [71].

Psychosocial interventions for cancer patients encompass a wide range of interventions that help patients make life changing decisions, manage debilitating symptoms, confront impending mortality, and improve other patient outcomes [42,72]. In line with the bio-psychosocial model of pain, psychosocial interventions play a central role in the effective management of cancer-related pain [11,43].

Psychosocial interventions in cancer-related pain can be divided into two subgroups: skill-based interventions and education-based interventions [43]. In skill-based interventions, the patient has an active role and learns how to manage pain, while in education-based interventions, the patient acquires information about the disease and treatment from medical and allied health professionals [43].

Recent systematic reviews and metaanalyses including patients experiencing various advanced cancer types indicated statistically significant, though rather small, effects of psychosocial interventions on self-reported pain severity [73,74]. The most common psychosocial interventions were relaxation techniques, cognitive-behavioral psychotherapy, music therapy, mindfulness- and acceptance-based interventions, and supportive-expressive group psychotherapy [73]. While one metaanalysis indicated no evidence for the superiority of any of the intervention types [73], another metaanalysis, conducted more recently, suggested that interventions with a cognitive basis, such as relaxation techniques with guided imagery and mindfulness-based interventions, were the most effective [74]. Importantly, various psychosocial interventions, even brief ones (i.e., consisting of four sessions or less and lasting less than 21 days), have been shown to exert beneficial effects on QoL, emotional distress, and existential distress in advanced cancer patients receiving palliative treatment [75]. Specifically, the results of metaanalysis indicated small effects of brief psychological interventions on QoL and existential distress and medium effects on emotional distress [75]. This finding is of clinical relevance because high levels of psychological distress, such as depression, anxiety, and a fear of catastrophes, are also reportedly associated with more severe pain [9,12]. Furthermore, chronic pain poses a considerable challenge to individuals' well-being, leading to decreased QoL, limitations in daily functioning, and a higher reliance on healthcare services [76]. In metaanalyses the most frequently used questionnaires for assessment of the emotional distress were the Hospital Anxiety and Depression Scale, FACT-G, and the Quality-of-Life Concerns in the End of Life (QOLC-E). Furthermore, the existential distress was mostly assessed via FACIT-SP, Herth Hope Index, and QOLC-E [73].

Relaxation techniques are based on teaching a patient to focus on letting go of muscle tension through awareness and suggestion [10]. When used for cancer pain, they usually include guided imagery, a cognitive-behavioral mind–body intervention that is employed to manage pain, which affects and/or modifies the psychophysiological state of patients [77]. During guided imagery, patients are asked to evoke positive and pleasant images and to imagine them in the greatest possible detail, with the involvement of multiple senses [77]. Some studies indicate that the brain stimulation by guided imagery may induce the release of excitatory neurotransmitters, such as serotonin, as well as various endogenous opioid peptides, implicated in pain response [78,79].

Cognitive-behavioral therapy is a structured, goal-oriented type of psychotherapy that helps people to identify and change destructive or disturbing thoughts that have a negative influence on their behavior and emotions [80]. When used for chronic pain, such as cancer-related pain, cognitive-behavioral therapy engages the patient in finding coping strategies to allow the patient to gain a sense of control and mastery over pain and the suffering associated with pain [81,82]. Cognitive-behavioral therapy addresses maladaptive beliefs about pain and pain catastrophizing through formal use of cognitive restructuring [81]. Cognitive-behavioral therapy–based interventions have also been shown to have beneficial effects on QoL, depression, and anxiety in advanced cancer patients [73].

Music therapy is a therapeutic approach that uses music and/or elements of music, including sound, rhythm, and harmony, to help people improve their mental health and overall well-being interventions [83]. In cancer care, music therapy can alleviate physical symptoms such as pain and nausea, possibly by cognitive engagement and distraction [84]. Listening to music can help the release of endogenous opioids and dopamine, which supports music-induced analgesia and may also contribute to reducing the severity of fatigue [85,86]. Furthermore, evidence suggests that music therapy may reduce the severity of anxiety and depression, which have been shown to play an important role in the exacerbation of pain perception [84]. Please refer to the special Chapter 28 about music therapy to learn more.

TABLE 26.2 Recommendations for psychological care in pain management.

- Unrecognized psychological factors may negatively influence pain experience, response to pain, and treatment among cancer patients.
- Psychological care needs to be considered as an addition and not as an alternative to physical treatments for cancer-related pain.
- Establish which patients could specifically benefit from the inclusion of psychological care in the plan of cancer-related pain management.
- Psychosocial interventions can have a beneficial effect on self-rated pain severity of advanced cancer patients, though their effects seem to be small.
- Psychosocial interventions should be delivered by medical (psychiatrists and psychiatric nurses) and allied health specialties (social workers, psychologists, and occupational therapists).
- Types of psychosocial interventions: relaxation techniques, cognitive-behavioral psychotherapy, music therapy, mindfulness- and acceptance-based interventions, supportive-expressive group psychotherapy, etc.
- Incorporate psychological care education into the training of interdisciplinary teams caring for cancer patients.

Mindfulness- and acceptance-based interventions constitute a family of methods that aim at increasing present-moment awareness, nonjudgment, and values-based living [87]. Mindfulness- and acceptance-based interventions may be effective for managing cancer-related pain because they can moderate the patient's emotional appraisal of pain and other stressful events by disengagement from pain-related threats [88,89]. When patients practice such interventions, they become less cognitively focused on their pain; therefore, pain tolerance is enhanced [89]. Mindfulness-based pain analgesia has also been associated with extensive bilateral deactivation of the thalamus, a critical node involved in the descending inhibition to modulate nociceptive inputs at the dorsal horn of the spinal cord [90,91].

Supportive-expressive group psychotherapy aims to provide emotional, social, cognitive, and symptomatic support for people coping with advanced and terminal cancer [92]. Supportive-expressive group psychotherapy has been found effective in relieving cancer-related pain, emotional distress, depression, and anxiety [92,93]. Several studies found that supportive-expressive group psychotherapy that included hypnosis reduced the increase of pain experience in women with metastatic breast cancer over a time period of 1 year [73,94,95]. The results of one study indicate that among women who received the intervention, those who were highly hypnotizable reported significantly greater benefit from the hypnosis compared to those who were less hypnotizable, suggesting that greater hypnotizability may facilitate other beneficial effects of the intervention [95]. Therefore the authors concluded that teaching self-hypnosis to cancer patients may provide them with a tool they can apply to purposes beyond simply the control of pain [67].

A summary of recommendations for psychological care in pain management is provided in Table 26.2.

Future implications

Available evidence suggests that spiritual interventions are beneficial for some patients, but not for all, and that spiritual interventions do not affect certain types of pain (e.g., neuropathic, especially severe). Further high-quality research is needed to clarify and better understand the role of spirituality in the pain experience and also to determine the characteristics of patients and possibly the characteristics of pain itself, which would benefit from incorporating spiritual interventions into cancer-related pain management. Establishing differences in the spiritual needs of individuals of different ages, nationalities, and social and cultural backgrounds can complement this research. Moreover, given the diversity of spiritual interventions, exploring which types of spiritual interventions would be effective and suitable for specific patients or types of pain, while also taking into account the cultural environment, will be particularly challenging in developing evidence-based spiritual care.

Available evidence suggests that psychosocial interventions can have a beneficial effect on self-rated pain severity of advanced cancer patients, though their effects seem to be rather small. Further studies to determine which interventions are effective for which cancer types are warranted. More research is also needed to determine which interventions are more beneficial than others.

Conclusions

- Addressing spirituality, despite its undefined and transcendent nature, may play an important role in the experience of cancer-related symptoms, and a focus on spirituality appears to improve the QoL of cancer patients.
- There is a limited evidence base to support the use of spiritual interventions in the treatment of cancer pain. A small number of spiritual interventions have been investigated, and only a few have shown a quantitative benefit in pain

alleviation. Qualitative trials suggest that spiritual interventions can help in coping with cancer even if they do not modify the pain itself.
- Various psychological factors play an important role in the pain experience.
- Unrecognized psychological factors may negatively influence pain experience, response to pain, and treatment among cancer patients.
- Due to their beneficial effect on the pain severity of advanced cancer patients, psychosocial interventions should be incorporated into standards of care for oncology programs.
- Psychosocial interventions with a cognitive basis might be the most effective for reducing pain in cancer patients.

References

[1] Radbruch L, De Lima L, Knaul F, et al. Redefining palliative care – a new consensus-based definition. J Pain Symptom Manage 2020;60(4):754–64. Available from: https://doi.org/10.1016/j.jpainsymman.2020.04.027.

[2] Temel JS, Greer JA, El-Jawahri A, et al. Effects of early integrated palliative care in patients with lung and GI cancer: a randomised controlled study. J Clin Oncol 2017;35:834–41. Available from: https://doi.org/10.1200/JCO.2016.70.5046.

[3] Gijsberts MHE, Liefbroer AI, Otten R, et al. Spiritual care in palliative care: a systematic review of the recent European literature. Med Sci (Basel) 2019;7(2):25. Available from: https://doi.org/10.3390/medsci7020025.

[4] Quinn B, Connolly M. Spirituality in palliative care. BMC Palliat Care 2023;22(1):1. Available from: https://doi.org/10.1186/s12904-022-01116-x.

[5] Siddall PJ, Lovell M, MacLeod R. Spirituality: what is its role in pain medicine? Pain Med 2015;16:51–60. Available from: https://doi.org/10.1111/pme.12511.

[6] Domenico Borasio G. Translating the World Health Organization definition of palliative care into scientific practice. Palliat Support Care 2011;9(1):1–2. Available from: https://doi.org/10.1017/S1478951510000489.

[7] Jünger S, Payne S, Constantini A, et al. EAPC task force on education for psychologists in palliative care. Eur J Palliat Care 2010;17(2):84–7.

[8] Chiu TY, Hu WY, Lue BH, et al. Sedation for refractory symptoms of terminal cancer patients in Taiwan. J Pain Symptom Manage 2001;21(6):467–72. Available from: https://doi.org/10.1016/s0885-3924(01)00286-x

[9] Lee YP, Wu CH, Chiu TY, et al. The relationship between pain management and psychospiritual distress in patients with advanced cancer following admission to a palliative care unit. BMC Palliat Care 2015;2(14):69. Available from: https://doi.org/10.1186/s12904-015-0067-2.

[10] Syrjala KL, Jensen MP, Mendoza ME, et al. Psychological and behavioral approaches to cancer pain management. J Clin Oncol 2014;32(16):1703–11. Available from: https://doi.org/10.1200/JCO.2013.54.4825.

[11] Feldstain A. Psychosocial intervention in palliative care: what do psychologists need to know. J Health Psychol 2024;29:707–20. Available from: https://doi.org/10.1177/13591053231222848.

[12] Andersen KG, Kehlet H. Persistent pain after breast cancer treatment: a critical review of risk factors and strategies for prevention. J Pain 2011;12(7):725–46. Available from: https://doi.org/10.1016/j.jpain.2010.12.005.

[13] Ann-Yi S, Bruera E, Wu J, et al. Characteristics and outcomes of psychology referrals in a palliative care department. J Pain Symptom Manage 2018;56(3):344–51. Available from: https://doi.org/10.1016/j.jpainsymman.2018.05.022.

[14] Myint AT, Tiraphat S, Jayasvasti I, et al. Factors influencing the willingness of palliative care utilization among the older population with active cancers: a case study in Mandalay, Myanmar. Int J Environ Res Public Health 2021;18(15):7887. Available from: https://doi.org/10.3390/ijerph18157887.

[15] Melzack R, Wall PD. Pain mechanisms: a new theory. Science 1965;150(3699):971–9. Available from: https://doi.org/10.1126/science.150.3699.971.

[16] Engel GL. The need for a new medical model: a challenge for biomedicine. Science 1977;196(4286):129–36. Available from: https://doi.org/10.1126/science.847460.

[17] Edwards A, Pang N, Shiu V, et al. The understanding spirituality and the potential role of spiritual care in end-of-life and palliative care: a meta-study of qualitative research. Palliat Med 2010;24(8):753–70. Available from: https://doi.org/10.1177/0269216310375860.

[18] Rattner M. "Total pain": reverence and reconsideration. Front Sociol 2023;8:1286208. Available from: https://doi.org/10.3389/fsoc.2023.1286208.

[19] Steinhauser KE, Fitchett G, Handzo GF, et al. State of the science of spirituality and palliative care research part I: definitions, measurement, and outcomes. J Pain Symptom Manage 2017;54(3):428–40. Available from: https://doi.org/10.1016/j.jpainsymman.2017.07.028.

[20] Balducci L. Geriatric oncology, spirituality, and palliative care. J Pain Symptom Manage 2019;57(1):171–5. Available from: https://doi.org/10.1016/j.jpainsymman.2018.05.009.

[21] Puchalski C, Romer AL. Taking a spiritual history allows clinicians to understand patients more fully. J Palliat Med 2000;3(1):129–37. Available from: https://doi.org/10.1089/jpm.2000.3.129.

[22] Egan R, MacLeod R, Jaye C, et al. What is spirituality? Evidence from a New Zealand hospice study. Mortality 2011;16(4):307–24. Available from: https://doi.org/10.1080/13576275.2011.613267.

[23] Chochinov HM, Cann BJ. Interventions to enhance the spiritual aspects of dying. J Palliat Med 2005;8:S103–15. Available from: https://doi.org/10.1089/jpm.2005.8.s-103.

[24] Peterman AH, Fitchett G, Brady MJ, et al. Measuring spiritual well-being in people with cancer: the functional assessment of chronic illness therapy – spiritual well-being scale (FACIT-Sp). Ann Behav Med 2002;24(1):49–58. Available from: https://doi.org/10.1207/S15324796ABM2401_06.

[25] Delgado-Guay MO, Palma A, Duarte E, et al. Association between spirituality, religiosity, spiritual pain, symptom distress, and quality of life among Latin American patients with advanced cancer: a multicenter study. J Palliat Med 2021;24(11):1606–15. Available from: https://doi.org/10.1089/jpm.2020.0776.

[26] Miller M, Kwekkeboom K, Cherwin C. The role of spirituality in symptom experiences among adults with cancer. Support Care Cancer 2021;30(1):49–57. Available from: https://doi.org/10.1007/s00520-021-06399-z.

[27] Bai J, Brubaker A, Meghani SH, et al. Spirituality and quality of life in Black patients with cancer pain. J Pain Symptom Manage 2018;56(3):390–8. Available from: https://doi.org/10.1016/j.jpainsymman.2018.05.020.

[28] Miller M, Xu D, Lehto R, et al. Pain and spirituality outcomes among women with advanced breast cancer participating in a randomized controlled trial of foot reflexology. Oncol Nurs Forum 2021;48(1):31–43. Available from: https://doi.org/10.1188/21.ONF.31-43.

[29] Poulin PA, Romanow HC, Rahbari N, et al. The relationship between mindfulness, pain intensity, pain catastrophizing, depression, and quality of life among cancer survivors living with chronic neuropathic pain. Support Care Cancer 2016;24(10):4167–75. Available from: https://doi.org/10.1007/s00520-016-3243-x.

[30] Miller M, Speicher S, Hardie K, et al. The role of spirituality in pain experiences among adults with cancer: an explanatory sequential mixed methods study. Support Care Cancer 2024;32(3):169. Available from: https://doi.org/10.21203/rs.3.rs-3425339/v1.

[31] Alimujiang A, Lilah K, Weinsch A, et al. "I am not a statistic" ovarian cancer survivors' views of factors that influenced their long-term survival. Gynecol Oncol 2019;155(3):461–7. Available from: https://doi.org/10.1016/j.ygyno.2019.10.007.

[32] Arefian M, Asgari-Mobarake K, Fazilatpour M, et al. Proposing and evaluating a model of depression, stress, resilience and spirituality in relation to pain in women with breast cancer: investigating the mediating role of mindfulness. Eur J Oncol Nurs 2023;62:102268. Available from: https://doi.org/10.1016/j.ejon.2023.102268.

[33] Hamilton IJ, Morrison J, Macdonald S. Should GPs provide spiritual care? Br J Gen Pract 2017;67(665):573–4. Available from: https://doi.org/10.3399/bjgp17X693845.

[34] Puchalski C, Ferrell B, Virani R, et al. Improving the quality of spiritual care as a dimension of palliative care: the report of the consensus conference. J Palliat Med 2009;12:885–904. Available from: https://doi.org/10.1089/jpm.2009.0142.

[35] Jimenez-Fonseca P, Lorenzo-Seva U, Ferrando PJ, et al. The mediating role of spirituality (meaning, peace, faith) between psychological distress and mental adjustment in cancer patients. Support Care Cancer 2018;26(5):1411–8. Available from: https://doi.org/10.1007/s00520-017-3969-0.

[36] Sorajjakool S, Thompson KM, Aveling L, et al. Chronic pain, meaning, and spirituality: a qualitative study of the healing process in relation to the role of meaning and spirituality. J Pastoral Care Counsel 2006;60:369–78. Available from: https://doi.org/10.1177/154230500606000406.

[37] Blaber M, Jone J, Willis D. Spiritual care: which is the best assessment tool for palliative settings? Int J Palliat Nurs 2015;21(9):430–8. Available from: https://doi.org/10.12968/ijpn.2015.21.9.430.

[38] Ferrell BR, Twaddle ML, Melnick A, et al. National consensus project clinical practice guidelines for quality palliative care guidelines, 4th edition. J Palliat Med 2018;21(12):1684–9. Available from: https://doi.org/10.1089/jpm.2018.0431.

[39] Hindmarch T, Dalrymple J, Smith M, et al. Spiritual interventions for cancer pain: a systematic review and narrative synthesis. BMJ Support Palliat Care 2022;12(1):1–9. Available from: https://doi.org/10.1136/bmjspcare-2021-003102.

[40] Balboni TA, VanderWeele TJ, Doan-Soares SD, et al. Spirituality in serious illness and health. JAMA 2022;328(2):184–97. Available from: https://doi.org/10.1001/jama.2022.11086.

[41] Darnall BD. Psychological treatment for chronic pain: improving access and integration. Psychol Sci Public Interest 2021;22(2):45–51. Available from: https://doi.org/10.1177/15291006211033612.

[42] Raja SN, Carr DB, Cohen M, et al. The revised International Association for the Study of Pain definition of pain: concepts, challenges, and compromises. Pain 2020;161(9):1976–82. Available from: https://doi.org/10.1097/j.pain.0000000000001939.

[43] Ciucă A, Băban A. Psychological factors and psychosocial interventions for cancer related pain. Rom J Intern Med 2017;55(2):63–8. Available from: https://doi.org/10.1515/rjim-2017-0010.

[44] Sloan P, Hollins M. Attention and pain: are auditory distractors special? Exp Brain Res 2017;235(5):1593–602. Available from: https://doi.org/10.1007/s00221-017-4903-x.

[45] Hoegh M, Seminowicz DA, Graven-Nielsen T, et al. Delayed effects of attention on pain sensitivity and conditioned pain modulation. Eur J Pain 2019;23(10):1850–62. Available from: https://doi.org/10.1002/ejp.1458.

[46] Khera T, Rangasamy V. Cognition and pain: a review. Front Psychol 2021;21(12):673962. Available from: https://doi.org/10.3389/fpsyg.2021.673962.

[47] Herbert MS, Goodin BR, Pero ST, et al. Pain hypervigilance is associated with greater clinical pain severity and enhanced experimental pain sensitivity among adults with symptomatic knee osteoarthritis. Ann Behav Med 2014;48(1):50–60. Available from: https://doi.org/10.1007/s12160-013-9563-x.

[48] Lee KE, Ryu H, Chang SJ. The effect of pain catastrophizing on depression among older Korean adults with chronic pain: the mediating role of chronic pain interference and sleep quality. Int J Environ Res Public Health 2020;17(23):8716. Available from: https://doi.org/10.3390/ijerph17238716.

[49] Moore DJ, Meints SM, Lazaridou A, et al. The effect of induced and chronic pain on attention. J Pain 2019;20(11):1353–61. Available from: https://doi.org/10.1016/j.jpain.2019.05.004.

[50] Battison EAJ, Brown PCM, Holley AL, et al. Associations between chronic pain and attention-deficit hyperactivity disorder (ADHD) in youth: a scoping review. Children (Basel) 2023;10(1):142. Available from: https://doi.org/10.3390/children10010142.

[51] Yi JC, Syrjala KL. Anxiety and depression in cancer survivors. Med Clin North Am 2017;101(6):1099–113. Available from: https://doi.org/10.1016/j.mcna.2017.06.005.

[52] Krok D, Telka E, Falewicz A, et al. Total pain and fear of recurrence in post-treatment cancer patients: serial mediation of psychological flexibility and mentalization and gender moderation. J Clin Med 2024;13(7):1974. Available from: https://doi.org/10.3390/jcm13071974.

[53] Finan PH, Garland EL. The role of positive affect in pain and its treatment. Clin J Pain 2015;31:177–87. Available from: https://doi.org/10.1097/AJP.0000000000000092.

[54] Ecija C, Catala P, Velasco L, et al. When it hurts, a positive attitude may help. The moderating effect of positive affect on the relationship between walking, depression, and symptoms in women with fibromyalgia. Pain Manag Nurs 2022;23(6):767–75. Available from: https://doi.org/10.1016/j.pmn.2022.05.007.

[55] Jensen MP, Karoly P. Pain-specific beliefs, perceived symptom severity, and adjustment to chronic pain. Clin J Pain 1992;8(2):123–30. Available from: https://doi.org/10.1097/00002508-199206000-00010.

[56] Petrini L, Arendt-Nielsen L. Understanding pain catastrophizing: putting pieces together. Front Psychol 2020;16(11):603420. Available from: https://doi.org/10.3389/fpsyg.2020.603420.

[57] Simic K, Savic B, Knezevic NN. Pain catastrophizing: how far have we come. Neurol Int 2024;16(3):483–501. Available from: https://doi.org/10.3390/neurolint16030036.

[58] Zhu H, Lin R, Chen J, et al. Cognitive function and its associated factors among patients with cancer pain: a multicentre cross-sectional study in China. BMJ Open 2024;14(1):e071417. Available from: https://doi.org/10.1136/bmjopen-2022-071417.

[59] Müller R, Segerer W, Ronca E, et al. Inducing positive emotions to reduce chronic pain: a randomized controlled trial of positive psychology exercises. Disabil Rehabil 2022;44(12):2691–704. Available from: https://doi.org/10.1080/09638288.2020.1850888.

[60] Linton SJ, Shaw WS. Impact of psychological factors in the experience of pain. Phys Ther 2011;91(5):700–11. Available from: https://doi.org/10.2522/ptj.20100330.

[61] Vlaeyen JWS, Linton SJ. Fear-avoidance model of chronic musculoskeletal pain: 12 years on. Pain 2012;153(6):1144–7. Available from: https://doi.org/10.1016/j.pain.2011.12.009.

[62] Peters L. Emotional and cognitive influences on pain experience. Mod Trends Pharmacopsychiatry 2015;30:138–52. Available from: https://doi.org/10.1159/000435938.

[63] Linton J. A Transdiagnostic approach to pain and emotion. J Appl Biobehav Res 2013;18(2):82–103. Available from: https://doi.org/10.1111/jabr.12007.

[64] Arora A, Saini SK, Nautiyal V, et al. Cancer pain, anxiety, and depression in admitted patients at a tertiary care hospital: a prospective observational study. Indian J Palliat Care 2019;25(4):562–6. Available from: https://doi.org/10.4103/IJPC.IJPC_4_19.

[65] Jia H, Uphold CR, Wu S, et al. Health-related quality of life among men with HIV infection: effects of social support, coping, and depression. AIDS Patient Care STDS 2004;18(10):594–603. Available from: https://doi.org/10.1089/apc.2004.18.594.

[66] Büssing A, Osterman T, Neugebauer E, et al. Adaptive coping strategies in patients with chronic pain conditions and their interpretation of disease. BMC Public Health 2010;20(10):507. Available from: https://doi.org/10.1186/1471-2458-10-507.

[67] Khalili N, Farajzadegan Z, Mokarian F, et al. Coping strategies, quality of life and pain in women with breast cancer. Iran J Nurs Midwifery Res 2013;18(2):105–11.

[68] Czerw A, Religioni U, Deptała A. Assessment of pain, acceptance of illness, adjustment to life with cancer and coping strategies in breast cancer patients. Breast Cancer 2016;23(4):654–61. Available from: https://doi.org/10.1007/s12282-015-0620-0.

[69] Ośmiałowska E, Misiag W, Ciałkowska-Rysz A, et al. Coping strategies, pain, and quality of life in patients with breast cancer. J Clin Med 2021;10(19):4469. Available from: https://doi.org/10.3390/jcm10194469.

[70] Anthony W, Cohen M, Farkas M, et al. Psychiatric rehabilitation. In: Nonaka T, Ohashi H, editors. Translators. 2nd ed. Boston: Boston University; 2002. p. 83-103.

[71] Branjerdporn G, Gillespie GM, Liaw SY, et al. Development of an interprofessional psychosocial interventions framework. Int J Environ Res Public Health 2023;20(8):5495. Available from: https://doi.org/10.3390/ijerph20085495.

[72] Teo I, Krishnan A, Lee GL. Psychosocial interventions for advanced cancer patients: a systematic review. Psychooncology 2019;28:1394–407. Available from: https://doi.org/10.1002/pon.5103.

[73] Warth M, Zöller J, Kessler J, et al. Psychosocial interventions for pain management in advanced cancer patients: a systematic review and meta-analysis. Curr Oncol Rep 2020;22(1):3. Available from: https://doi.org/10.1007/s11912-020-0870-7.

[74] Ruano A, García-Torres F, Roldán JJ, et al. Psychological and non-pharmacologic treatments for pain in cancer patients: a systematic review and meta-analysis. J Pain Symptom Manage 2022;63(5):e505–20. Available from: https://doi.org/10.1016/j.jpainsymman.2021.12.021.

[75] Warth M, Kessler J, Koehler F, et al. Brief psychosocial interventions improve quality of life of patients receiving palliative care: a systematic review and meta-analysis. Palliat Med 2019;33(3):332–45. Available from: https://doi.org/10.1177/0269216318818011.

[76] Pandelani FF, Nyalunga SLN, Mthembu TG, et al. Chronic pain: the prevalence of chronic pain in patients attending Soshanguve Community Health Centre. Can J Pain 2023;8(1):2284152. Available from: https://doi.org/10.1080/24740527.2023.2284152.

[77] De Paolis G, Naccarato A, Cibelli F, et al. The effectiveness of progressive muscle relaxation and interactive guided imagery as a pain-reducing intervention in advanced cancer patients: a multicentre randomised controlled non-pharmacological trial. Complement Ther Clin Pract 2019;34:280–7. Available from: https://doi.org/10.1016/j.ctcp.2018.12.014.

[78] Blalock JE, Smith EM. Conceptual development of the immune system as a sixth sense. Brain Behav Immun 2007;21(1):23–33. Available from: https://doi.org/10.1016/j.bbi.2006.09.004.
[79] Wiech K, Tracey I. The influence of negative emotions on pain: behavioral effects and neural mechanisms. Neuroimage 2009;47(3):987–94. Available from: https://doi.org/10.1016/j.neuroimage.2009.05.059.
[80] Nakao M, Shirotsuki K, Sugaya N, et al. Cognitive-behavioral therapy for management of mental health and stress-related disorders: recent advances in techniques and technologies. Biopsychosoc Med 2021;15(1):16. Available from: https://doi.org/10.1186/s13030-021-00219-w.
[81] Sturgeon J. Psychological therapies for the management of chronic pain. Psychol Res Behav Manag 2014;10(7):115–24. Available from: https://doi.org/10.2147/PRBM.S44762.
[82] Lim JA, Choi SH, Park JS, et al. Cognitive-behavioral therapy for patients with chronic pain: implications of gender differences in empathy. Medicine (Baltimore) 2018;97(23):e10867. Available from: https://doi.org/10.1097/MD.0000000000010867.
[83] van der Steen JT, van Soest-Poortvliet MC, van der Wouden JC, et al. Music-based therapeutic interventions for people with dementia. Cochrane Database Syst Rev 2017;5(5):CD003477. Available from: https://doi.org/10.1002/14651858.CD003477.pub3.
[84] Trigueros-Murillo A, Martinez-Calderon J, Casuso-Holgado MJ, et al. Effects of music-based interventions on cancer-related pain, fatigue, and distress: an overview of systematic reviews. Support Care Cancer 2023;31(8):488. Available from: https://doi.org/10.1007/s00520-023-07938-6.
[85] Lunde SJ, Vuust P, Garza-Villarreal EA, et al. Music-induced analgesia: how does music relieve pain? Pain 2019;160(5):989–93. Available from: https://doi.org/10.1097/j.pain.0000000000001452.
[86] Sezgin MG, Bektas H. The effect of music therapy interventions on fatigue in patients with hematological cancers: a systematic review and meta-analysis of randomized controlled trials. Support Care Cancer 2022;30(11):8733–44. Available from: https://doi.org/10.1007/s00520-022-07198-w.
[87] Tatta J, Willgens AM, Palombaro KM. Mindfulness and acceptance-based interventions in physical therapist practice: the time is now. Phys Ther 2022;102(3):pzab293. Available from: https://doi.org/10.1093/ptj/pzab293.
[88] Brown CA, Jones AKP. Meditation experience predicts less negative appraisal of pain: electrophysiological evidence for the involvement of anticipatory neural responses. Pain 2010;150(3):428–38. Available from: https://doi.org/10.1016/j.pain.2010.04.017.
[89] Ngamkham S, Holden JE, Smith EL. A systematic review: mindfulness intervention for cancer-related pain. Asia Pac J Oncol Nurs 2019;6(2):161–9. Available from: https://doi.org/10.4103/apjon.apjon_67_18.
[90] Ab Aziz CB, Ahmad AH. The role of the thalamus in modulating pain. Malays J Med Sci 2006;13(2):11–8.
[91] Halassa MM, Chen Z, Wimmer RD, et al. State-dependent architecture of thalamic reticular subnetworks. Cell 2014;158(4):808–21. Available from: https://doi.org/10.1016/j.cell.2014.06.025.
[92] Spiegel D, Kogon M. Supportive-expressive group therapy and the terminally ill. In: Breitbart HM, editor. Handbook of psychiatry in palliative medicine: psychosocial care of the terminally ill. 3rd ed. Oxford, UK: Oxford University Press; 2023. p. 596–606.
[93] Ho RTH, Fong TCT, Lo PHY, et al. Randomized controlled trial of supportive-expressive group therapy and body-mind-spirit intervention for Chinese non-metastatic breast cancer patients. Support Care Cancer 2016;24:4929–37. Available from: https://doi.org/10.1007/s00520-016-3350-8.
[94] Goodwin PJ, Leszcz M, Ennis M, et al. The effect of group psychosocial support on survival in metastatic breast cancer. N Engl J Med 2001;345:1719–26. Available from: https://doi.org/10.1056/NEJMoa011871.
[95] Butler LD, Koopman C, Neri E, et al. Effects of supportive-expressive group therapy on pain in women with metastatic breast cancer. Health Psychol 2009;28(5):579–87. Available from: https://doi.org/10.1037/a0016124.

Chapter 27

Treating conditions associated with painful stimuli

Jelena Spasić
Clinic for Medical Oncology, Institute for Oncology and Radiology of Serbia, Belgrade, Serbia

Key learning objectives

- Cancer pain is often accompanied by other symptoms of the cancer as part of "symptom clusters." These accompanying symptoms can aggravate the pain experience by either worsening the level of chronic pain or presenting as breakthrough pain.
- The most common symptoms that occur with and can aggravate cancer pain are constipation, dyspnea, cough, nausea, vomiting, anxiety, depression, fatigue, and others. How often they occur depends on the type of cancer or comorbidity.
- Early recognition and appropriate therapy of these conditions are necessary for maintaining a good quality of life for patients with cancer.
- Multidisciplinary and multiprofessional approach to a patient with symptoms of advanced cancer is essential.

Introduction

Cancer pain rarely presents as a solitary symptom but is more often accompanied by other symptoms of the cancer, which can aggravate the pain experience by either worsening the level of chronic pain or presenting as breakthrough pain. Adequate pain assessment is crucial to providing the necessary treatment and must include all conditions that can add to the pain experience, whether related to the cancer, its treatment, or comorbidities. These conditions must all be assessed and treated as part of a comprehensive approach to treatment of patients presenting with cancer pain, using a strategy known as "treat the treatable, correct the correctable."

Cancer pain as a frequent part of symptom clusters in patients with cancer

Patients with cancer very rarely present with just one isolated symptom but are most often encumbered by numerous symptoms and complaints. These symptoms are due to the disease itself, due to cancer treatment, or comorbidities. They may or may not be interrelated. Consequently, these symptoms have a major effect on the patients' quality of life and adherence to treatment. In a prospective study by Grond on over 2200 patients reporting cancer pain, it was shown that approximately 17% of reported pain was due to oncologic treatment and about 10% due to other conditions unrelated to the cancer [1]. Historically, symptoms of cancer have been studied separately, and while this has led to advances in treating the particular symptom, often the treatment of just one symptom is not sufficient for improving the quality of life of the patients. Over the past 20–30 years, a new term of *symptom cluster* has been coined and is currently defined as two or more symptoms that are related, occur together, form a stable group, and are relatively independent of other clusters [2]. As one of the most frequent symptoms of cancer in general, pain is often a part of symptom clusters. One of the first studies that provided evidence of this was performed in Germany in the 1980s on more than 1600 consecutive patients with different cancers treated at a pain clinic. It was found that patients suffering from severe pain also reported more concomitant symptoms than patients who had mild pain, most commonly insomnia, vomiting, and anxiety. Also, and somewhat logically, patients receiving strong opioids had more concurrent symptoms than those receiving mild or no opioids ($P < .001$). Limitations of this study were that the severity of symptoms other than pain was not assessed, nor were standardized questionnaires used; however, it does provide important insights regarding cancer pain as part of symptom

Pain Management in Palliative Care for Patients With Cancer. DOI: https://doi.org/10.1016/B978-0-443-30050-9.00009-6

clusters [3]. The most common symptoms that occur with and can aggravate cancer pain are constipation, dyspnea, cough, nausea, vomiting, anxiety, depression, fatigue, and others. How often they occur is in relation to the type of cancer or comorbid disease.

Respiratory symptoms

Dyspnea

Dyspnea (breathlessness, shortness of breath) is defined as the subjective experience of breathing discomfort or difficulty breathing. It is a very common symptom of advanced cancer, present in over 50% of patients suffering from lung and head and neck cancer or from lung metastases from other cancers (breast, urogenital, lymphoma). Dyspnea becomes more severe with advancing cancer and toward the end of life. It is a complex symptom, accompanied by pain and psychological distress, anxiety, depression, and fear, which in turn worsen the subjective feeling of breathlessness [4]. Furthermore, it has been shown to be an independent risk factor for mortality in advanced cancer and is associated with increased risk for sudden death in patients receiving palliative care [5]. Chronic dyspnea is a multifactorial experience with physical, psychological, emotional, and social components and as such requires both a pharmacological and nonpharmacological approach [6].

The causes of dyspnea can be related to the cancer itself or anticancer therapy, comorbidities, concurrent medication, and other factors [6,7]. As with any symptom of advanced cancer, all of them should be assessed, and if found, reversible causes should be properly treated (e.g., pleural drainage for effusion, medication for exacerbation of chronic obstructive pulmonary disease [COPD], blood transfusion for anemia, etc.). Since dyspnea is a multidimensional symptom, all its dimensions need to be properly assessed (physical—severity; psychological—anxiety, unpleasantness, existential questions; social—its effect on social interactions and its effect on caregivers, etc.) by using one of several available scales of assessment for each domain of dyspnea. Also, since the subjective component is dominant, assessment should be done by the patients using patient-reported outcome measures (PROMs) wherever possible [6,7]. A major part of the assessment of dyspnea is distinguishing between dyspnea at rest, dyspnea on exertion, and terminal dyspnea, since some therapeutic approaches may differ.

Dyspnea causes much physical and psychological distress, so it is imperative that symptomatic treatment begin immediately, in parallel with investigating its cause. Even after the dominant cause has been identified, all treatment measures should be conducted concurrently. This includes treating the cause, nonpharmacological and pharmacological treatment, and referral to specialist palliative care services [7].

As with all symptoms, any underlying cause that can be reversed should be addressed and treated appropriately (correct the correctable). Certain nonpharmacological measures are recommended by all guidelines, even though there is little evidence from clinical trials as to their effect on relieving the symptoms of dyspnea. One of the simplest measures is using a hand-held fan in front of the face, which has been shown in a review by Qian et al.'s 10 studies on 344 patients (190 with cancer) to significantly improve breathlessness in six of those studies [8]. Also, in a review of interview data from 111 patients, most of them (91) reported some or substantial benefit, especially in terms of shorter recovery time after activity [9]. Breathing exercises and retraining, mobility aids, pulmonary rehabilitation (in certain cases), and ensuring adequate room conditions can be of help in certain patients with dyspnea. There is some evidence from small randomized trials and a systematic review by von Trott and colleagues on 190 patients with cancer and 347 patients with COPD that acupressure or acupuncture as complementary therapies have some short-term benefit on the severity of breathlessness [10–13].

Opioids are recommended as the first choice of pharmacological therapy for dyspnea. Their efficacy in the treatment of dyspnea is backed by the largest body of evidence from clinical trials, although not all in patients with cancer. The latest Cochrane report from 2016 [14] of 12 randomized clinical trials (RCTs) with 198 participants (of which 21% had cancer) showed the efficacy of systemic opioids on chronic dyspnea compared to placebo. A recent systematic review and metaanalysis by Takagi et al. from 2023 reviewed 12 RCTs with over 300 patients with cancer and the efficacy of systemic opioids on dyspnea and showed that opioids were more effective than placebo for dyspnea (standardized mean difference, –0.43; 95% confidence interval, –0.75 to –0.12). However, a significant difference was found only between systemic morphine and placebo, while no significant difference could be detected for other opioids. The authors conclude that there was a small number of patients treated with other opioids and that, even though a difference was found for systemic morphine, further larger randomized trials are needed in this setting [15]. There is also evidence from small, randomized, placebo-controlled trials that the use of immediate-release opioids prior to activity (showering, eating, taking a walk, etc.) can have some effect on reducing dyspnea on exertion [16]. The biggest concern when using opioids in this setting is the risk of respiratory depression; however, it has been shown that if lower doses of opioids are carefully titrated up against the level of dyspnea, there is no real risk of causing respiratory depression [17]. Short-acting benzodiazepines are frequently used in clinical practice, since dyspnea is often associated with anxiety. However, there is very little high-quality evidence in favor of or against their use to treat dyspnea in patients with cancer or other diseases. Midazolam continuous infusion, alone or in combination with morphine, may be used in carefully selected situations, for palliative sedation in end-of-life care, where it has been shown not to shorten expected survival while maintaining dignity and quality of life [6,7,18]. Corticosteroids should only be used in cases where dyspnea is primarily caused by an inflammatory or a structural cause, such as obstruction of big airways, carcinomatous lymphangitis, or mediastinal lymphadenopathy [6,7]. Standard supplemental oxygen is indicated only for patients with hypoxemia, when oxygen saturation (SaO_2) is below 90%. There is some evidence that a time-limited course of noninvasive mechanical ventilation could benefit certain patients [6,7].

As always, attention should also be given to possible adverse effects of these drugs, and they should be addressed appropriately.

The recommended indications and dosing of recommended drugs for treating dyspnea are presented in Table 27.1.

Cough

Cough is a common symptom of cancer but is often overlooked. Like pain, cough initially has a protective function in clearing the upper part of the tracheobronchial tree of mucus and foreign bodies [2,19]. When frequent and/or poorly controlled, it can cause much distress to a patient since it is physically exhausting and can cause or worsen pain, vomiting, fatigue, muscle strain, and influence sleep, eating, or breathing, or cause urinary incontinence, all of which adversely affect the patients' quality of life and social interactions. The prevalence of cough in patients with cancer is 50%–80% and is highest in patients with lung or head and neck cancer, up to 90%. Malignancies likely to cause cough include those of the airways, lungs, pleura, and mediastinum or cancers that metastasize to the thorax. Cough is also often caused by

TABLE 27.1 Principles of management of dyspnea.

Treat underlying cause(s)	**Chronic obstructive pulmonary disease/asthma/ other pulmonary disease/exacerbation of heart disease**	Optimize therapy (pulmonology/cardiology consultation)
	Pneumonia, bronchitis	Antimicrobial therapy
	Pleural/pericardial effusion	Drainage if appropriate
	Airway obstruction	Anticancer treatment; consider stenting
	Anticancer therapy-induced pulmonary toxicities	Withhold treatment; consider corticosteroids and opioids
	Anemia	Transfusion
Nonpharmacologic interventions	**Adequate position**	
	Air flow	Hand-held fan in front of the face
	Breathing techniques	Through pursed lips, diaphragmatic breathing, timed breathing
	Environment	Adequate room temperature and moisture
	Complementary therapies	Acupressure, acupuncture
Pharmacologic interventions	**Opioids (immediate-release morphine preferred option)**	Start with 1–5 mg Q4 h for titration Opioid-naive: 10–30 mg morphine equivalent per 24 h Opioid-tolerant: increase current dose by 25%–50% For acute dyspnea: 2–2.5 mg i.v. morphine (Q15 min)
	Corticosteroids (especially if structural or inflammatory cause)	8–16 mg of dexamethasone (or equivalent); i.v. or oral
	Benzodiazepines	For anxiety associated with dyspnea: lorazepam 1–2.5 mg p.o. Acute dyspnea: midazolam 2–5 mg s.c. or i.v. (continuous infusion for palliative sedation; consult palliative medicine team)
	Antidepressants	For anxiety associated with dyspnea
	Oxygen	Only if SaO_2 <90%

i.v., Intravenous; *p.o.*, per os (oral route); *s.c.*, subcutaneous.

noncancer conditions such as chronic lung and heart disease, chronic inflammation (e.g., cystic fibrosis), or acute infection, as well as concomitant medications (e.g., angiotensin-converting enzyme [ACE] inhibitors). Therefore, when evaluating cough in patients with cancer, these should not be overlooked [19,20] (Table 27.2).

A good history and physical examination are crucial. Intensity, quality, onset, frequency, and aggravating and alleviating factors should all be evaluated, as well as whether the cough is productive or nonproductive. It is important to evaluate its effect on pain and other accompanying symptoms, not forgetting its impact on activities of daily living and emotional distress. If the cough is productive, patients should be asked about the type, color, and amount of sputum and whether it is accompanied by fever. A complete physical examination is necessary, and further tests such as complete blood count and chemistry, chest X-ray (or CT scan in certain cases), ultrasound of the pleural space or heart, arterial blood oxygen saturation, and others can be helpful in identifying the cause of the cough [4,20,21].

The first step in the treatment of cough is treating or alleviating any potentially reversible cause (antimicrobial agents for acute infection, specific therapy for pulmonary and heart disease, or replacing the drug that had caused the cough, i.e., ACE inhibitors). Treatment of cancer is the next step. However, the effect of anticancer therapy on alleviating cough is usually not fast, so symptomatic treatment is usually used in combination. Also, it must be kept in mind that some anticancer therapy can worsen or cause cough (radiation therapy by causing pneumonitis or fibrosis, certain chemotherapy agents, or immunotherapy again by causing pneumonitis). There is scarce data from randomized trials regarding treatment of cough as a symptom. Pharmacological options are in the form of expectorants for wet cough and antitussives (cough suppressants) for dry cough, although antitussives do have their place in certain instances for treating wet cough as well.

Main approaches to treating cough in patients with cancer are presented in Table 27.3.

It is important to exercise caution in patients who are too frail and weak and cannot cough and expel the mucus, since accumulation of unexpelled mucus can worsen dyspnea [4,20,22,23]. Constipation as a side effect of opioids must not be overlooked and must be prevented and/or managed, as it can add to the distress of an already distressed patient [4,19,20,22,24,25].

Other symptoms caused by tumors in the thoracic cavity

Lung cancer and metastases to the thoracic cavity from other cancers can cause infiltration of the ribs, pleura, or thoracic vertebrae or present as Pancoast tumors. The dominant symptom of these tumors is pain, sometimes neuropathic in its nature, and Pancoast tumors also present with ptosis (Horner syndrome), weakness of the arm, and atrophy of arm muscles. In addition to adequate pain treatment, including corticosteroids, treating the underlying cancer (with radiation or systemic therapy) is necessary for alleviating these symptoms.

Gastrointestinal symptoms

Dysphagia

A specific symptom of difficulty swallowing and being unable to swallow (dysphagia and aphagia, respectively) or painful swallowing (odynophagia) is characteristic of involvement of the esophagus. Patients can also complain of feeling "food

TABLE 27.2 Causes of cough in patients with cancer.

Cancer	**Tumor or metastases:** endobronchial infiltration/obstruction, lung parenchyma infiltration, pleural effusion, pericardial effusion, carcinomatous lymphangitis, *tracheoesophageal fistula*
	Cancer treatment: chemotherapy (bleomycin, methotrexate, cyclophosphamide); radiation pneumonitis/fibrosis
Other causes	**Infection**
	Comorbidities: chronic obstructive pulmonary disease, asthma, heart failure, gastroesophageal reflux, neuromuscular disorders
	Medication: angiotensin-converting enzyme inhibitors, β-blockers
	Other: smoking, vocal cord paralysis, allergies

Source: Adapted from Refs. [4,19,21].

TABLE 27.3 Main approaches for treating cough in patients with cancer.

Treat underlying cause(s)	**Chronic obstructive pulmonary disease/ asthma/other pulmonary disease/ exacerbation of heart disease**	Optimize therapy (pulmonology/cardiology consultation)
	Pneumonia, bronchitis	Antimicrobial therapy
	Drug-induced	Consider replacing the drug in question (i.e., angiotensin-converting enzyme inhibitors)
	Cancer	Anticancer treatment
	Anticancer therapy–induced cough	Withhold treatment; consider corticosteroids and opioids
	Smoking	Council for smoking cessation
Nonpharmacologic interventions	**Cough suppression approaches**	Identifying cough triggers and avoiding them if possible
		Breathing exercises
		Speech pathology training
		Maintaining oral and laryngeal hygiene
		Adequate hydration
Pharmacologic interventions	**Antitussives (cough suppressants)**	**Codeine** 10–20 mg q6–8 h (if opioid-naive)
		Morphine Opioid-naive: 2 mg p.o. Q4 h; 5–10 mg Q4 h in refractory cough Opioid-tolerant: increase daily dose by 20%–25%
	Expectorants (mucolytics)	Inhalations with saline and other solutions
		Acetylcysteine, bromhexine, ambroxol hydrochloride (not available in all countries)
		Good hydration to potentiate effect of mucolytics

p.o., Per os (oral route).
Source: Adapted from Refs. [4,19,22].

being stuck in the throat or chest behind the breast bone," food coming back after swallowing (regurgitation), and heartburn. Very often, dysphagia is the first symptom caused by cancer in the thoracic cavity and leads to diagnosis. It can be caused by extramural (from the outside) compression of the esophagus by a mass in the mediastinum or by a tumor growing inside the esophageal tube, most often primary esophageal cancer. Also, anticancer treatment, specifically radiation therapy involving this area, can cause esophagitis, which presents with these symptoms. Prevalence of dysphagia ranges from about 17% in general to up to 70% of patients with head and neck cancer and 90% in patients suffering from esophageal cancer [26,27]. Patients with dysphagia can also develop malnutrition, dehydration, and aspiration pneumonia; usually have problems taking oral medication; and in general have a significantly reduced quality of life and higher rates of anxiety and depression.

Treatment of dysphagia is aimed at relieving symptoms and ensuring adequate nutrition. Cancer-directed treatment with systemic therapy, radiation therapy, or a combination of the two provides the most long-term effect on dysphagia as it restores the lumen of the esophagus and should be administered wherever possible. In many patients, before starting cancer-directed treatment or in parallel with it, adequate symptom relief is necessary. Dysphagia is often accompanied by pain while swallowing, sometimes nausea, and often heartburn or regurgitation. Opioids are usually mandated, and due to the nature of the symptom itself, the transdermal or liquid route of administration of analgesics is recommended. Administration of antiemetics and proton pump inhibitors is also indicated in certain patients. Corticosteroids have a place, as they have an antiedematous effect, especially if the cause of dysphagia is extramural compression or treatment-related esophagitis [26–28].

Mechanical interventions to restore the lumen of the esophagus are an important part of the treatment of dysphagia. They are also usually carried out before starting cancer-directed therapy but can be used in patients with incurable cancer. The most commonly used interventions are esophageal dilation and placing a stent inside the esophagus, while brachytherapy and photodynamic therapy are more rarely used. Esophageal dilation with a balloon or wire-guided polyvinyl bougies provides temporary relief of dysphagia until the effect of more definitive treatment can be accomplished. Since it provides a short-term benefit (repeat dilation is usually required every 2–4 weeks), it is not very commonly used in everyday clinical practice and is not indicated for patients with incurable disease. Sometimes dilation is performed prior to placing a stent, which has become the most commonly used mechanical intervention for relieving dysphagia. A stent provides rapid opening of the lumen and has the advantage of covering an area of esophago-tracheal fistula if present. Nowadays, self-expanding metal stents (SEMSs) are used, which are usually covered with an outer layer of semipermeable membrane. This outer layer prevents tumor growth inside the stent, but these stents are more prone to migrating [26,29]. The European Society of Gastrointestinal Endoscopy recommends placement of partially or fully covered SEMSs for palliation of malignant dysphagia over other mechanical interventions for sealing malignant tracheoesophageal or bronchoesophageal fistulas but does not recommend "SEMS placement as a bridge to surgery or before preoperative chemoradiotherapy because it is associated with a high incidence of adverse events" [26,29]. The other part of treating dysphagia is maintaining adequate nutrition. In an expert consensus statement on the management of dysphagia in patients with head and neck cancer, it is stated that "A comprehensive nutritional assessment of head and neck cancer (HNC) patients with dysphagia includes anthropometric measurements, biochemical measurements, and individualized nutrition requirements." It is critical to maintain some level of oral intake, as it has been shown that it significantly contributes to maintaining a better quality for life of these patients and reduces the psychological impact of dysphagia [30]. The first step for patients who are still able to pass some content is advising oral intake of pureed or liquid food or liquid meal supplements and teaching swallowing techniques, which can be helpful in certain patients. If oral intake is inadequate, nutritional supplementation via feeding tube may be necessary. The recommended way of supplementation is through a nasogastric/naso-enteric tube or percutaneous endoscopic gastrostomy tube, while total parenteral nutrition is discouraged and is reserved for rare and specific cases [31]. American Society of Parenteral and Enteral Nutrition and National Comprehensive Cancer Network (NCCN) guidelines both recommend pretreatment prophylactic feeding tube placement for HNC patients undergoing chemoradiation therapy who have severe weight loss, dehydration, dysphagia, or other conditions impacting safe and adequate oral intake. According to the same guideline, enteral nutrition can be ceased when patients with HNC consume 75% of their nutrition requirements orally for at least 3 days [28,31,32]. Approaching the treatment of dysphagia in patients with cancer requires a multidisciplinary and multiprofessional approach in a specialized institution in order to relieve symptoms and provide adequate nutritional intake.

Constipation

In the Oxford Concise Medical Dictionary, constipation is "a condition in which bowel evacuations occur infrequently, or in which the feces are hard and small, or where passage of feces causes difficulty or pain." According to the Rome III criteria, chronic constipation requires the presence of any two of the following symptoms for at least 12 weeks in the previous 12 months (not necessarily consecutively): straining during bowel movements, lumpy or hard stool, sensation of incomplete evacuation, sensation of anorectal blockage or obstruction, manual evacuation procedures to remove stool, and <3 bowel movements per week [33]. It is a very common symptom in patients with advanced cancer and one that causes significant distress, but observational studies suggest that it is still not very well managed. The incidence of constipation in patients with advanced cancer varies between 32% and 87%, which reflects differing populations of patients and differences in assessment and reporting, while it is found in up to 14% of the general population [24]. The etiology of constipation in patients with cancer is complex. It can be caused by the cancer itself, by medication, and by comorbidities and is most often multifactorial. In general, causes can be broadly grouped into organic and functional. A specific and important cause of constipation in patients with cancer is opioid-induced, which is discussed separately in another chapter. The most common causes are presented in Table 27.4.

Besides the objective symptom of infrequent bowel movements, symptoms of constipation are varied and related to the gastrointestinal tract but can be systemic if constipation is prolonged or not treated appropriately. The most common ones include abdominal pain, swelling, bloating, nausea, emesis, lack of appetite, early satiety, heartburn, sometimes headache, and general malaise. Also, local complications can occur, such as fecal impaction, overflow diarrhea, rectal prolapse, hemorrhoids, anal tears, rectal bleeding, and, in severe cases, gastrointestinal perforation. Patient perception and effect on quality of life should also be assessed, so the use of PROMs is encouraged. Besides careful history, which is usually

TABLE 27.4 Causes of constipation.

Organic	**Structural**	Abdominal or pelvic mass, peritoneal carcinomatosis, postsurgery adhesions, radiation fibrosis
	Iatrogenic/ medication	Antiemetics, antitussives, antidepressants, antacids, anticholinergics, iron supplements, chemotherapy (vinca alkaloids) Opioid-induced
	Metabolic	Hypercalcemia, hypokalemia, dehydration, uremia, hypothyroidism
	Comorbidities	Neuromuscular disorders, spinal cord involvement, autonomic system disorders, Parkinson's disease
	Pain	Cancer pain, anorectal, bone
Functional	**Diet**	Low in fiber, low fluid intake
	Age	
	Lack of activity	Incidence increases with age
	Environment	Access to toilets, need for assistance, cultural issues

Source: Adapted from Refs. [24,25,34].

sufficient for diagnosing constipation, a physical examination should be performed (although it is often unremarkable), a rectal examination is necessary to assess for fecal impaction, and in certain cases a diagnostic abdominal X-ray may be useful in assessing fecal loading. Examinations to determine and evaluate possible causes should be performed, and if treatable causes are found (e.g., metabolic disorders), they should be addressed initially [23,24,34].

The cornerstones of management of constipation are screening (for the symptom), prevention, assessment, intervention (therapy), and follow-up (reassessment) (Fig. 27.1).

Malignant bowel obstruction

Malignant bowel obstruction (MBO) is a severe complication of advanced cancer that occurs in approximately 3%–15% of patients and in up to 51% of patients with ovarian cancer and usually signifies a short prognosis [35]. Obstruction can be caused mechanically by the presence of an intraluminal tumor within the bowel or by extramural compression from peritoneal carcinomatosis, mesenteric and omental masses, adhesions, and fibrosis. Functional obstruction can be caused by tumor infiltration of the mesentery, nerves, or *celiac (or coeliac)/enteric plexus* and paraneoplastic syndromes and metabolic disorders, and it can also be exacerbated by medication such as opioids, some antiemetics, and intraperitoneal chemotherapy. It can be partial or complete, at a single point or more than one, and is more common in the small bowel. Gastric and intestinal secretions build up in the intestinal lumen and trigger an inflammatory response with edema and eventually bacterial overgrowth. This causes symptoms of abdominal pain and cramps, nausea, vomiting, distension, constipation, sometimes overflow diarrhea, eventually malnutrition. Diagnosis can be made from clinical signs and symptoms only; abdominal X-rays can be useful, but CT scans (with only i.v. contrast) are necessary for diagnosing the site and type of obstruction and planning surgery, if appropriate [36,37].

Before deciding on the appropriate treatment, a scrupulous assessment of expected survival is necessary. In patients with longer expected survival, surgery to remove the cause of obstruction is the primary treatment. In other patients, alleviating symptoms and ensuring adequate quality of life is the main objective [23,37]. All treatment should be administered via the sublingual, subcutaneous, transdermal, intravenous, and intramuscular routes (depending on drug formulations), since the oral route is usually contraindicated. Most treatment is aimed toward alleviating nausea and vomiting and pain. Recommended drugs are antiemetics, anticholinergics, antisecretory drugs (most often octreotide), analgesics, and corticosteroids, usually a combination of some or all of them, while always keeping in mind whether gut function can be maintained or not. Metoclopramide and other antiemetics with a prokinetic effect on gut motility should be avoided in complete obstruction, as they can lead to symptom worsening, but could have a place in treating incomplete obstruction. Fluid intake should be restricted to 500–1000 mL per 24 hour, for majority of patients, although individual approach is necessary. Nasogastric tube or venting gastrostomy can be considered in patients where gut function cannot be maintained as a way of alleviating vomiting [23,36–38].

Screening

- Assess in all cancer patients; encourage using patient-reported outcomes (PROs)

Prevention

- Increase fiber and fluid intake
- Exercise if appropriate
- Profilactic medication (laxative with opioids)
- Ensure access to toilets

Assessment and diagnosis

- Usual bowel movements (recent changes); date of last defecation; consistency of stool; presence/absence of urge to defecate; sensation of evacuation; blood or mucus in stool; necessity for manual evacuation
- Physical assessment: abdominal examination, auscultation, perineal inspection, digital rectal examination (DRE)
- Investigate potential metabolic and other causes
- Abdominal X-ray or CT scan if appropriate

Intervention (therapy)

- Correct the correctable
- **Laxatives:**
  - Osmotic: macrogol, lactulose, magnesium and sulfate salts
  - Stimulant: senna, bisacodyl and sodium picosulfate
  - Not recommended: bulk laxatives, liquid paraffin, detergent/stool softeners
- **Enemas and suppositories**:
  - First-line therapy in fecal impaction, otherwise if first-line therapy not effective
  - Contraindicated for patients with neutropenia or thrombocytopenia, paralytic ileus or intestinal obstruction, recent colorectal or gynecological surgery, recent anal or rectal trauma, severe colitis, inflammation or infection of the abdomen, toxic megacolon, undiagnosed abdominal pain or recent radiotherapy to the pelvic area

Follow-up

- If adequate, continue maintenance therapy
- If response not achieved, revisit treatment strategy, consider specialized intervention

FIGURE 27.1 Management of constipation.
Adapted from Refs. [24, 25, 34].

TABLE 27.5 Recommended treatment of malignant bowel obstruction (MBO).

Drug	Mechanism of action	Dosing (per 24 h)
Octreotide	Antisecretory, antiperistaltic	300–900 mcg; 100 mcg q8 h s.c. or continuous i.v. infusion
Anticholinergics Hyoscine butylbromide	Antisecretory, antiperistaltic	60–120 mg; 20 mg q6 h s.c. or continuous i.v. infusion
Corticosteroids	Antiinflammatory, antiedematous	Dexamethasone 8–16 mg i.v.
Antiemetics Metoclopramide Haloperidol Olanzapine	Prokinetic (only for partial MBO) Antagonist of dopamine receptors (DR) Antagonist of DR and 5HT3 receptors	30–40 mg i.v. + PRN 2.5 mg s.c. q6 h or PRN 5 mg s.l.
Pain medication		Parenteral/subcutaneous opioid titrated to pain and prior opioid exposure. Fentanyl transdermal option in opioid-tolerant patients with stable chronic opioid needs.

i.v., Intravenous; *s.c.*, subcutaneous; *s.l.*, sublingual; *PRN*, pro re nata (as needed).
Source: Adapted from Refs. [23,36,37].

The recommended drugs for treatment of MBO are listed in Table 27.5. Although the availability of specific substances listed may not be the same in different countries, the principles of management still remain.

Mucosal pain (mucositis) due to treatment

Mucositis is an acute toxicity of anticancer therapy (chemotherapy, radiation therapy, or targeted therapy) manifesting as inflammatory, ulcerative, and erosive lesions of gastrointestinal tract mucosa. Its incidence depends on the site of lesions and type of treatment. Oral mucositis is found in approximately 20%–40% of patients receiving conventional chemotherapy for solid tumors, over 80% of patients treated with radio(chemo)therapy for head and neck cancers, and about 80% of patients receiving high-dose chemotherapy prior to hematopoietic stem cell transplantation. New targeted agents that have entered clinical practice over the past several years can also cause mucositis, mostly oral, and it somewhat differs from the one caused by conventional chemotherapy [38,39]. Oral mucositis presents with pain, problems with speech, difficulty consuming food and liquids and swallowing, and in severe cases even requiring enteral nutrition. Nausea, vomiting, bloating, cramping, diarrhea, and sometimes anal pain are the most common symptoms of gastrointestinal mucositis. In both cases, there is an increased risk of infection. Mucositis is graded on a 5-point scale using the National Cancer Institute Common Terminology Criteria for Adverse Events and according to the extent of mucosal changes, while oral mucositis is also graded according to subjective discomfort [40]. Management of mucositis is aimed at prevention and control of symptoms in order to improve quality of life and ensure cancer treatment is delivered in a timely and uninterrupted manner.

All major oncology and supportive care societies have published guidelines for the management of mucositis, thus emphasizing its importance. The main points of these guidelines are summarized in Fig. 27.2. It is important to note that even though the availability of specific substances is not the same in different countries, the principles of management of mucositis still remain.

Other symptoms related to cancer pain

Symptoms of increased intracranial pressure

The central nervous system is often the site of primary or secondary tumors (metastases). Lung, breast, and rectal cancer and melanoma most commonly metastasize to the brain. Brain tumors are frequently surrounded by edema, which, simply

Prevention

- **Oral care and hygiene:** using soft toothbrush, regular flossing, using non-medicated mouth rinses, avoid irritants (dentures, smoking, hot and spicy food, sparkling drinks), natural preparations (honey)
- **Oral cryotherapy**
- **Benzodyazamine mouthwash**
- **Photobyomodulation therapy**

Treatment

- **Pain:** 2% lidocain mouthwash, 0.2% morphine mouthrinse, mucosal cooling agents; systemic opioids
- **Diarrheal:** loperamide (up to 16mg per day), octreotide (3x100mcg s.c.), sulphasalazine, probiotics
- **Antimicrobial agents**

Nutritional support

- Advice on type and preparation of food
- Liquid diet supplements
- Gastrostomy (jejunotstomy) tube
- Total parenteral nutrition (rare)

FIGURE 27.2 Principles of management of mucositis.
Adapted from Refs. [38–40]

by applying pressure to the brain or by impairing cerebrospinal fluid (CSF) circulation, causes various symptoms and significantly affects the patient's quality of life. These symptoms can present as specific neurological deficits that are related to the location of the tumor or unspecific symptoms related to increased intracranial pressure, such as headache, with or without nausea and vomiting, cognitive impairment, personality changes, and gait disturbances [41]. Headache is present in about 40%–60% of patients with brain tumors and is more likely to be caused by infratentorial and intraventricular tumors than supratentorial due to the more frequent disturbance in CSF flow. Also, elderly patients are less likely to develop headaches and more likely to present with confusion or memory loss as a symptom of increased intracranial pressure (ICP). In general, any new, atypical, and progressive headache, especially if accompanied by other symptoms (nausea, vomiting, seizures), is indicative of a brain tumor. These headaches are usually diffuse and moderate to severe and, as mentioned previously, most often accompanied by other symptoms [42,43]. Nausea caused by increased intracranial pressure is usually not constant and is most often relieved by vomiting, making it different from other cancer-related nausea. Seizures are present in about 20% of patients with brain tumors or metastases [43], very often as the initial presentation of the condition. In patients exhibiting these symptoms, a quick diagnosis is imperative for proper treatment. It is usually made by magnetic resonance imaging, although a CT scan can be sufficient in some cases.

Treatment of symptoms of increased ICP is similar and includes specific anticancer therapy such as surgery, stereotactic radiosurgery, or palliative radiation therapy if appropriate. However, since anticancer treatment takes some time to be efficacious, treatment of symptoms of increased ICP should be immediate and aggressive with antiedematous therapy. Steroids are the mainstay of antiedematous therapy of brain edema, since their effect on symptom relief is quick, they can be administered orally or intravenously, and they are inexpensive [42,43]. The most commonly used drug is dexamethasone since it has very little mineralocorticoid effect and thus very little effect on blood electrolyte levels and has the added benefit of controlling tumor-associated pain, limiting nausea and vomiting, and improving appetite in patients with cancer. Recommended doses of dexamethasone are between 4 and 16 mg (oral or i.v.) in a single daily dose. In a randomized study by Vecht et al., there was no significant difference in patients' condition between 4 and 8 mg of dexamethasone, nor between 4 and 16 mg [44]. Therefore the recommended dose of dexamethasone is the lowest one that provides meaningful relief of symptoms. The duration of steroid treatment must be assessed individually, with the recommendation to start tapering as soon as clinically indicated, since long-term use can lead to significant side effects. If steroids are given for longer than 4 weeks, prophylaxis of *Pneumocystis jiroveci* pneumonia should be considered [42]. Prophylaxis of gastrointestinal side effects and monitoring of blood glucose levels are also necessary. In certain patients, steroids do not provide complete relief from headache, so analgesic therapy must be added, usually with opioids. Mannitol is a commonly used hyperosmolal agent even though its effect on treating brain edema from cancer has not been systematically investigated. Mannitol is most widely available as a 20% solution and administered on a weight-based bolus dosing schedule: 0.5–2 g/kg. Due to a very well-described rebound phenomenon, that is, edema recurrence after treatment cessation, it is not recommended for prolonged use, over a period of time longer than 3–5 days, and should be restricted to patients requiring immediate antiedematous effect [45,46]. Anticonvulsants are indicated for treating seizures and should be given as secondary prophylaxis. Primary prophylaxis of seizures is not recommended in patients with brain tumors or metastases, since it has not been shown to prevent first-ever seizure events. Anticonvulsants that have few drug–drug interactions (especially with anticancer drugs) are the drugs of choice, such as levetiracetam, lamotrigine, and lacosamide (over phenytoin, carbamazepine, or valproic acid) [42,47].

It is important to note that the presence of brain edema by itself does not necessarily require immediate treatment if symptoms are not present [42,45].

Lymphedema

Lymphedema is defined as the "abnormal accumulation of interstitial fluid and fibroadipose tissues resulting from reduced lymph transport because of injury, infection, cancer/cancer-related treatment, or congenital abnormalities of the lymphatic system" [48]. It can be due to the tumor itself (bulky tumors or microscopic disease blocking lymph drainage), sometimes even as a first presentation of disease, or due to cancer treatments. Lymphedema caused by cancer treatment can develop weeks, months, and even years after treatment has been completed. Incidence greatly varies according to type of cancer and treatment and ranges from 5% to 83% in breast cancer. It has been shown that patients with cancer and lymphedema are more likely to have greater psychological distress (anxiety, depression, sexual dysfunction, and social distancing) and, in general, a poorer quality of life than patients without edema [49]. The most commonly seen lymphedema is in patients after surgery and/or radiation therapy for breast cancer that involved the axillary region. Over the past several years the incidence has lowered due to advances in surgery and sentinel node procedures, which allow for less invasive dissections of axillary and inguinal lymph nodes. Irradiation of the lymph nodes also increases the risk of developing lymphedema, as shown in a study by Warren et al., where patients with breast cancer who had irradiation of the axillae had a greater risk of lymphedema than those who only had irradiation to the chest wall or breast [50]. Radiation-induced lymphedema can be acute, which presents as skin redness and mild swelling and is usually short-lived and requires only skin care and sometimes medication for a short time. Late-onset lymphedema after radiation is most often irreversible.

Symptoms of lymphedema depend on its localization, but common ones are pain or discomfort, joint swelling, limited functionality of the affected limbs, and, in severe cases, inability to walk. Head and neck cancer lymphedema can present as swelling of the oral cavity, neck, or face; difficulty swallowing; voice changes; or the feeling of having an object lodged in the throat if the edema is located internally. Infections of the skin in the affected area are also common [48]. Cancer-related lymphedema is incurable, but treatment of lymphedema should be aimed at decreasing the swelling if possible, relieving symptoms of discomfort and pain, improving functional status, addressing body image issues, and in general improving quality of life. Setting realistic treatment goals is essential when planning treatment of lymphedema, life expectancy must be taken into consideration. The first stage is to reduce the volume of the affected limb with decongestive lymphedema therapy, and the second is to maintain the reduced volume. To reduce the volume of affected limbs, short-stretch inelastic bandages are recommended to be worn for 24–72 hours at a time and repeated, along with a specific exercise program. Diuretics are not recommended, as their effect on reducing volume is short term by removing water but not proteins from soft tissue, which can in the long run lead to exacerbation of lymphedema and fibrosis. Cellulitis is a major concern, so patients must be educated on skin care [48,51].

Cancer-related fatigue

Cancer-related fatigue (CRF) is defined as "a distressing, persistent, subjective sense of physical, emotional, and/or cognitive tiredness or exhaustion related to cancer or cancer treatment that is not proportional to recent activity and interferes with usual functioning" [52,53]. It is important to distinguish it from fatigue or tiredness in healthy people, in that it is not relieved by sleep or rest. It is the most common symptom of cancer, experienced by most patients, estimated to be present in approximately 40% of patients at diagnosis, 80%–90% during cancer treatment, and even after completion of treatment during survivorship, in 20%–50% of patients [54]. Data suggest that CRF is more prevalent in patients with cancer experiencing pain, since pain affects the patients' ability to exercise and their social interactions. In a cross-sectional study of 22 patients experiencing cancer pain in Taiwan, the authors found a correlation between pain and fatigue ($P < .01$), regardless of disease stage and treatment, and that patients with moderate and severe pain had more severe fatigue and quality of life than patients with mild pain [55].

CRF is a complex symptom with physical, emotional, and cognitive components, which should all be addressed during evaluation. Also, many factors can contribute to the development or exacerbation of CRF, such as anemia, other symptoms related to cancer as well as anticancer therapies, comorbidities, or adverse events caused by other drugs, and they also need to be evaluated and treated if possible. It is important that all patients with cancer be screened for the presence of CRF at the first oncological visit, during therapy at regular intervals, and at the end of therapy, possibly even into survivorship. There are several validated tools for CRF screening, such as the Numeric Rating Scale (NRS) (from 0 to 10) and the Brief Fatigue Inventory, which also includes evaluation of the impact of CRF on functional domains. Both these tools are examples of self-assessment, or patient-reported outcomes, which has been proved to be the most valid and accurate way to assess CRF [56,57]. If moderate-to-severe CRF is diagnosed (NRS ≥ 4), further evaluation is indicated and should include a focused fatigue history (start, duration, exacerbating and alleviating factors, influence on daily activities), evaluation of the underlying cancer and its treatment, evaluation of concurrent symptoms related or not related to cancer, comorbidities, concurrent medication, laboratory tests, and a mental status examination [58,59].

Management of CRF is complex, and data from randomized studies is lacking. As with other symptoms, causes that can be treated and reversed should be addressed first (e.g., anemia, adverse drug effects). Most guidelines recommend a combination of patient and caregiver education, physical activity, psychosocial intervention, and pharmacologic treatment, taking into account the general condition of the patient and treatment goals [58,59]. A metaanalysis by Mustian and colleagues in 2017 compared pharmaceutical, psychosocial, and exercise treatments for CRF. The results showed that exercise and exercise plus psychological interventions improved CRF during and after primary treatment, whereas pharmaceutical interventions did not, and that the effectiveness of these treatments depended on cancer stage, baseline treatment status, experimental treatment format, and fatigue measures [60]. There is data from RCTs and systematic reviews based on which physical exercise of moderate-intensity aerobic and functional resistance exercise can be recommended in patients with CRF who do not have cachexia. Even so, there is no prescription for exercise that can be used for all patients. Recommendations on the type and duration of exercise should be based on an individual basis [58,61,62].

As for pharmacologic treatments, there is very little high-quality evidence that could recommend certain drugs, since studies carried out in this setting were in different patient populations, had a small number of patients, used different fatigue assessment tools, and had a very high response of CRF to placebo. The best evidence exists for the short-term use of dexamethasone or methylprednisolone in metastatic cancer patients [63,64]. Other drugs, such as psychostimulants and antidepressants, are not recommended for routine practice [58,59].

Nonpharmacological measures are an important part of treating CRF. Information and counseling of patients and caregivers about ways to prevent or manage CRF, psychoeducation, and cognitive-behavioral therapy are recommended by all guidelines based on several studies, while yoga, acupuncture, and mind–body interventions can be an option to improve CRF [58].

All in all, CRF is a complex symptom, often associated with cancer pain, and should be assessed and treated in all patients with cancer at regular intervals, requiring a multidisciplinary and multiprofessional approach.

Conclusion

Cancer pain very rarely occurs as an isolated symptom but is most often accompanied by other symptoms related to the disease itself, its treatment, or comorbidities. These symptoms have complicated and multidirectional relationships; therefore accurately assessing and treating each of them is imperative for ensuring a good quality of life for patients with cancer. Multidisciplinary and multiprofessional cooperation is necessary, while adhering to guidelines and not forgetting the goals of care and patient preferences.

References

[1] Grond S, Zech D, Diefenbach C, et al. Assessment of cancer pain: a prospective evaluation in 2266 cancer patients referred to a pain service. Pain 1996;64(1):107–14. Available from: https://doi.org/10.1016/0304-3959(95)00076-3.

[2] Kim HJ, McGuire DB, Tulman L, et al. Symptom clusters: concept analysis and clinical implications for cancer nursing. Cancer Nurs 2005;28(4):270–82; quiz 283–4. Available from: https://doi.org/10.1097/00002820-200507000-00005.

[3] Grond S, Zech D, Diefenbach C, et al. Prevalence and pattern of symptoms in patients with cancer pain: a prospective evaluation of 1635 cancer patients referred to a pain clinic. J Pain Symptom Manage 1994;9(6):372–82. Available from: https://doi.org/10.1016/0885-3924(94)90174-0.

[4] Currow D, Ekstrom M. Management of respiratory symptoms in people with cancer. In: Olver I, editor. The MASCC textbook of cancer supportive care and survivorship. Cham, Switzerland: Springer International Publishing AG; 2018. p. 217–29.

[5] Hardy JR, Turner R, Saunders M, et al. Prediction of survival in a hospital-based continuing care unit. Eur J Cancer 1994;30(3):284–8. Available from: https://doi.org/10.1016/0959-8049(94)90242-9.

[6] Hui D, Bohlke K, Bao T, et al. Management of dyspnea in advanced cancer: ASCO guideline. J Clin Oncol 2021;39(12):1389–411. Available from: https://doi.org/10.1200/JCO.20.03465.

[7] Hui D, Maddocks M, Johnson M. ESMO guidelines committee. Management of breathlessness in patients with cancer: ESMO clinical practice and guidelines. ESMO Open 2020;5(6):e001038. Available from: https://doi.org/10.1136/esmoopen-2020-001038.

[8] Qian Y, Wu Y, de Moraes AR, et al. Fan therapy for the treatment of dyspnea in adults: a systematic review. J Pain Symptom Manage 2019;58(3):481–6. Available from: https://doi.org/10.1016/j.jpainsymman.2019.04.011.

[9] Luckett T, Phillips J, Johnson MJ. Contributions of a hand held FAN to self-management of chronic breathlessness. Eur Respir J 2017;50(2):1700262. Available from: https://doi.org/10.1183/13993003.00262-2017.

[10] Doğan N, Taşcı S. The effects of acupressure on quality of life and dyspnea in lung cancer: a randomized, controlled trial. Altern Ther Health Med 2020;26(1):49–56.

[11] Bauml J, Haas A, Simone CB, et al. Acupuncture for dyspnea in lung cancer: results of a feasibility trial. Integr Cancer Ther 2016;15(3):326–32. Available from: https://doi.org/10.1177/1534735415624138.

[12] Minchom A, Punwani R, Filshie J, et al. A randomised study comparing the effectiveness of acupuncture or morphine versus the combination for the relief of dyspnoea in patients with advanced non-small cell lung cancer and mesothelioma. Eur J Cancer 2016;61:102–10. Available from: https://doi.org/10.1016/j.ejca.2016.03.078.

[13] von Trott P, Oei SL, Ramsenthaler C. Acupuncture for breathlessness in advanced diseases: a systematic review and meta-analysis. J Pain Symptom Manage 2020;59(2):327–38. Available from: https://doi.org/10.1016/j.jpainsymman.2019.09.007.

[14] Barnes H, McDonald J, Smallwood N, et al. Opioids for the palliation of refractory breathlessness in adults with advanced disease and terminal illness. Cochrane Database Syst Rev 2016;3(3):CD011008. Available from: https://doi.org/10.1002/14651858.CD011008.pub2.

[15] Takagi Y, Sato J, Yamamoto Y, et al. Opioids for the management of dyspnea in cancer patients: a systematic review and meta-analysis. Int J Clin Oncol 2023;28(8):999–1010. Available from: https://doi.org/10.1007/s10147-023-02362-6.

[16] Johnson MJ, Hui D, Currow DC. Opioids, exertion, and dyspnea: a review of the evidence. Am J Hosp Palliat Care 2016;33(2):194–200. Available from: https://doi.org/10.1177/1049909114552692.

[17] Verberkt CA, van den Beuken-van Everdingen MHJ, Schols JMGA, et al. Respiratory adverse effects of opioids for breathlessness: a systematic review and meta-analysis. Eur Respir J 2017;50(5):1701153. Available from: https://doi.org/10.1183/13993003.01153-2017.

[18] Cherny NI. ESMO clinical practice guidelines for the management of refractory symptoms at the end of life and the use of palliative sedation. Ann Oncol 2014;25:iii143–52. Available from: https://doi.org/10.1093/annonc/mdu238.
[19] Bausewein C, Simon ST. Shortness of breath and cough in patients in palliative care. Dtsch Arztebl Int 2013;110(33–34):563–72. Available from: https://doi.org/10.3238/arztebl.2013.0563.
[20] Rajer M, Ahmedzai SH. Respiratory problems. In: Bosnjak SM, editor. ESMO handbook of supportive & palliative care. Lugano, Switzerland: ESMO Press; 2022. p. 101–11.
[21] Ahmedzai SH. Cough in cancer patients. Pulm Pharmacol Ther 2004;17(6):415–23. Available from: https://doi.org/10.1016/j.pupt.2004.09.015.
[22] Molassiotis A, Smith JA, Mazzone P, et al. Symptomatic treatment of cough among adult patients with lung cancer. Chest 2017;151(4):861–74. Available from: https://doi.org/10.1016/j.chest.2016.12.028.
[23] National Comprehensive Cancer Network. NCCN Clinical Practice Guidelines in Oncology: Palliative Care; Version 1.2026 [Internet]. Fort Washington (PA): National Comprehensive Cancer Network. Available from: https://www.nccn.org/professionals/physician_gls/pdf/palliative.pdf.
[24] Davies A, Leach C, Caponero R, et al. MASCC recommendations on the management of constipation in patients with advanced cancer. Support Care Cancer 2020;28(1):23–33. Available from: https://doi.org/10.1007/s00520-019-05016-4.
[25] Scotte F, Legeay K. Constipation. In: Bosnjak S, editor. ESMO handbook of supportive & palliative care. Lugano, Switzerland: ESMO Press; 2022. p. 78–80.
[26] Kuhn MA, Gillespie MB, Ishman SL, et al. Expert consensus statement: management of dysphagia in head and neck cancer patients. Otolaryngol Head Neck Surg 2023;168(4):571–92. Available from: https://doi.org/10.1002/ohn.302.
[27] Munoz-Schuffenegger P, Chu RWK, Wong RKS. Dysphagia, reflux, and hiccups. In: Olver I, editor. The MASCC textbook of cancer supportive care and survivorship. Cham, Switzerland: Springer International Publishing AG; 2018. p. 379–89.
[28] National Comprehensive Cancer Network. NCCN Clinical Practice Guidelines in Oncology: Head and neck cancers; Version 1.2026 [Internet]. Fort Washington (PA): National Comprehensive Cancer Network. Available from: https://www.nccn.org/professionals/physician_gls/pdf/head-and-neck.pdf.
[29] Spaander MCW, van der Bogt RD, Baron TH, et al. Esophageal stenting for benign and malignant disease: European Society of Gastrointestinal Endoscopy (ESGE) guideline - update 2021. Endoscopy 2021;53(7):751–62. Available from: https://doi.org/10.1055/s-0042-114210.
[30] Hutcheson KA, Bhayani MK, Beadle BM, et al. Eat and exercise during radiotherapy or chemoradiotherapy for pharyngeal cancers: use it or lose it. JAMA Otolaryngol Head Neck Surg 2013;139(11):1127–34. Available from: https://doi.org/10.1001/jamaoto.2013.4715.
[31] Boullata JI, Carrera AL, Harvey L, et al. ASPEN safe practices for enteral nutrition therapy. JPEN J Parenter Enteral Nutr 2017;41(1):15–103. Available from https://doi.org/10.1177/0148607116673053.
[32] Sadaps M, Bhatt A, Chevalier C, et al. A practical guide to the management of dysphagia in patients with metastatic esophageal cancer. Ann Esophagus 2018;1:2. Available from: https://doi.org/10.21037/aoe.2018.03.01.
[33] Drossman DA. Rome III: the new criteria. Chin J Dig Dis 2006;7(4):181–5. Available from: https://doi.org/10.1111/j.1443-9573.2006.00265.x.
[34] Larkin PJ, Cherny NI, La Carpia D, et al. ESMO guidelines committee. Diagnosis, assessment and management of constipation in advanced cancer: ESMO clinical practice guidelines. Ann Oncol 2018;29:iv111–25. Available from: https://doi.org/10.1093/annonc/mdy148.
[35] Cousins SE, Tempest E, Feuer DJ. Surgery for the resolution of symptoms in malignant bowel obstruction in advanced gynaecological and gastrointestinal cancer. Cochrane Database Syst Rev 2016;2016(1):CD002764. Available from: https://doi.org/10.1002/14651858.CD002764.pub2.
[36] Ripamonti CI, Guglielmo M. Malignant bowel obstruction. In: Bosnjak SM, editor. ESMO handbook of supportive & palliative care. Lugano, Switzerland: ESMO Press; 2022. p. 73–7.
[37] Madariaga A, Lau J, Ghoshal A, et al. MASCC multidisciplinary evidence-based recommendations for the management of malignant bowel obstruction in advanced cancer. Support Care Cancer 2022;30(6):4711–28. Available from: https://doi.org/10.1007/s00520-022-06889-8.
[38] Kuten-Shorrer M, Treister NS, Keefe DMK. Mucositis. In: Bosnjak S, editor. ESMO handbook of supportive & palliative care. Lugano, Switzerland: ESMO Press; 2022. p. 88–90.
[39] Lalla RV, Bowen JM. Mucositis (oral and gastrointestinal). In: Olver I, editor. ESMO handbook of supportive & palliative care. Cham, Switzerland: Springer International Publishing AG; 2018. p. 409–20.
[40] Elad S, Cheng KKF, Lalla RV, et al. MASCC/ISOO clinical practice guidelines for the management of mucositis secondary to cancer therapy. Cancer 2020;126(19):4423–31. Available from: https://doi.org/10.1002/cncr.33100.
[41] Roth P, Pace A, Le Rhun E, et al. Neurological and vascular complications of primary and secondary brain tumours: EANO-ESMO clinical practice guidelines for prophylaxis, diagnosis, treatment and follow-up. Ann Oncol 2021;32(2):171–82. Available from: https://doi.org/10.1016/j.annonc.2020.11.003.
[42] Le Rhun E, Guckenberger M, Smits M, et al. EANO–ESMO clinical practice guidelines for diagnosis, treatment and follow-up of patients with brain metastasis from solid tumours. Ann Oncol 2021;32(11):1332–47. Available from: https://doi.org/10.1016/j.annonc.2021.07.016.
[43] Cathcart-Rake E, Dronca R, Loprinzi CL. Central nervous system symptoms: headache, seizures, encephalopathy, and memory impairment. In: Olver I, editor. The MASCC textbook of cancer supportive care and survivorship. Cham: Springer; 2018. p. 529–42.
[44] Vecht CJ, Hovestadt A, Verbiest HB, et al. Dose-effect relationship of dexamethasone on karnofsky performance in metastatic brain tumors: a randomized study of doses of 4, 8, and 16 mg per day. Neurol Neth 1994;44(4):675–80. Available from: https://doi.org/10.1212/wnl.44.4.675.
[45] Halstead MR, Geocadin RG. The medical management of cerebral edema: past, present, and future therapies. Neurotherapeutics 2019;16(4):1133–48. Available from: https://doi.org/10.1007/s13311-019-00779-4.
[46] Esquenazi Y, Lo VP, Lee K. Critical care management of cerebral edema in brain tumors. J Intensive Care Med 2017;32(1):15–24. Available from: https://doi.org/10.1177/0885066615619618.

[47] Vecht C, Royer-Perron L, Houillier C, et al. Seizures and anticonvulsants in brain tumours: frequency, mechanisms and anti-epileptic management. Curr Pharm Des 2017;23(42):6464–87. Available from: https://doi.org/10.2174/1381612823666171027130003.

[48] Chaput G, Ibrahim M. Cancer-related lymphedema. Can Fam Physician 2023;69(10):691–5. Available from: https://doi.org/10.46747/cfp.6910691.

[49] Pyszel A, Malyszczak K, Pyszel K, et al. Disability, psychological distress and quality of life in breast cancer survivors with arm lymphedema. Lymphology 2006;39(4):185–92.

[50] Warren LEG, Miller C, Horick N, et al. The impact of radiation therapy on the risk of lymphedema after treatment for breast cancer: a prospective cohort study. Int J Radiat Oncol Biol Phys 2014;88(3):565–71. Available from: https://doi.org/10.1016/j.ijrobp.2013.11.232.

[51] Armer JM, Ostby PL, Ginex PK, et al. ONS guidelines™ for cancer treatment-related lymphedema. Oncol Nurs Forum 2020;47(5):518–38. Available from: https://doi.org/10.1188/20.ONF.518-538.

[52] Horneber M, Fischer I, Dimeo F, et al. Cancer-related fatigue: epidemiology, pathogenesis, diagnosis, and treatment. Dtsch Arztebl 2012;109(9):161–71. Available from: http://doi.org/10.3238/arztebl.2012.0161.

[53] ICD-11. International statistical classification of diseases and related health problems, 11th Revision; 2019.

[54] Roila F, Fatigoni S. Fatigue. In: Bozovic-Spasojevic I, Mountzios G, Wood J, editors. ESMO handbook of supportive and palliative care. Bosnjak SM: ESMO Press; 2022. p. 175–81.

[55] Liu D, Weng J-S, Ke X, et al. The relationship between cancer-related fatigue, quality of life and pain among cancer patients. Int J Nurs Sci 2023;10(1):111–6. Available from: https://doi.org/10.1016/j.ijnss.2022.12.006.

[56] Fisher MI, Davies C, Lacy H, et al. Oncology Section EDGE task force on cancer: measures of cancer-related fatigue – a systematic review. Rehabil Oncol 2018;36(2):93–105. Available from: https://doi.org/10.1097/01.REO.0000000000000124.

[57] Mendoza TR, Wang XS, Cleeland CS, et al. The rapid assessment of fatigue severity in cancer patients: use of the brief fatigue inventory. Cancer 1999;85(5):1186–96. Available from: https://doi.org/10.1002/(SICI)1097-0142(19990301)85:5<1186::AID-CNCR24>3.0.CO;2-N.

[58] Fabi A, Bhargava R, Fatigoni S, et al. Cancer-related fatigue: ESMO clinical practice guidelines for diagnosis and treatment. Ann Oncol 2020;31(6):713–23. Available from: https://doi.org/10.1016/j.annonc.2020.02.016.

[59] National Comprehensive Cancer Network. NCCN Clinical Practice Guidelines in Oncology: Cancer-Related Fatigue; Version 1.2026 [Internet]. Fort Washington (PA): National Comprehensive Cancer Network. Available from: https://www.nccn.org/professionals/physician_gls/pdf/fatigue.pdf.

[60] Mustian KM, Alfano CM, Heckler C, et al. Comparison of pharmaceutical, psychological, and exercise treatments for cancer-related fatigue: a meta-analysis. JAMA Oncol 2017;3(7):961–8. Available from: https://doi.org/10.1001/jamaoncol.2016.6914.

[61] Puetz TW, Herring MP. Differential effects of exercise on cancer-related fatigue during and following treatment: a meta-analysis. Am J Prev Med 2012;43(2):e1–4. Available from: https://doi.org/10.1016/j.amepre.2012.04.027.

[62] Schmitz KH, Courneya KS, Matthews C, et al. American College of Sports Medicine roundtable on exercise guidelines for cancer survivors. Med Sci Sports Exerc 2010;42(7):1409–26. Available from: https://doi.org/10.1249/MSS.0b013e3181e0c112.

[63] Paulsen O, Klepstad P, Rosland JH, et al. Efficacy of methylprednisolone on pain, fatigue, and appetite loss in patients with advanced cancer using opioids: a randomized, placebo-controlled, double- blind trial. J Clin Oncol 2014;32(29):3221–8. Available from: https://doi.org/10.1200/JCO.2013.54.3926.

[64] Yennurajalingam S, Frisbee-Hume S, Lynn Palmer J, et al. Reduction of cancer-related fatigue with dexamethasone: a double-blind, randomized, placebo-controlled trial in patients with advanced cancer. J Clin Oncol 2013;31(25):3076–82. Available from: https://doi.org/10.1200/jco.2012.44.4661.

Chapter 28

Pet therapy and music therapy

Ivana Kukec[1] and Sanja Pleština[2,3]

[1]*Department of Medical Oncology, Clinical Hospital Center Zagreb, Zagreb, Croatia,* [2]*Department of Pulmonology, University Hospital Centre Zagreb, Zagreb, Croatia,* [3]*Faculty of Medicine, University of Rijeka, Rijeka, Croatia*

Key learning objectives

- To introduce pet and music therapy as important complementary therapies targeting the total pain concept for persons with cancer.
- To provide evidence about pet and music therapy, which can alleviate cancer pain and provide other benefits for patients with cancer.
- To analyze the evidence from clinical trials on the efficacy of pet and music therapy in reducing pain and improving emotional well-being in patients with cancer.
- To introduce practical applications of the information about pet and music therapy in pain management.

Pet therapy

The human–animal bond dates from ancient times. In 1860, Florence Nightingale commented on the positive impact of small animals on those with chronic illnesses [1]. Pet therapy, or animal-assisted intervention (AAI), can be divided into animal-assisted therapy (AAT) and animal-assisted activities (AAA) [2]. AAT is a healing modality that involves a carefully coordinated interaction between a patient, a trained animal, and a handler to achieve specific therapeutic outcomes [3–5]. These outcomes include physical, social, emotional, and cognitive improvement [5,6]. AAT is more structured than general interactions with pets, aiming to match the patient's needs with a suitably trained animal. AAA are recreational activities with common pets that can improve the quality of life (QoL). Animals like dogs and cats are typically used, although other species, including horses and dolphins, may also be involved [5]. It is a complementary method that provides therapeutic interactions that can alleviate pain, improve emotional well-being, and improve patients' QoL with chronic diseases [2,4]. Some studies have found the impact of AAT on alleviating "total" pain in palliative care settings. This may include physical aspects of pain, as well as suffering components such as emotional factors, loss issues, fear of death, and spiritual concerns [7–9].

Numerous studies have explored the impact of pet therapy. Most of these studies have been conducted with therapy dogs; common dog breeds employed were Labrador Retrievers and mixes. Studies on AAI varied widely in design, including randomized controlled trials, observational studies with or without controls, and nonintervention surveys. Most studies were observational, often lacking a control group, and showed a broad range in the duration and frequency of AAI sessions, from just 1 session to 32 sessions, and lasting anywhere from 10 minutes to several hours [10]. Orlandi et al. conducted an observational study examining the impact of pet therapy on patients undergoing chemotherapy in an oncological day hospital. Two trained and veterinarian-approved dogs, groomed and prepared before entering the ward, were used for the therapy. Pet therapy sessions took place once a week for 25 weeks. Patients could choose to receive chemotherapy in the room where the pet therapy was held (experimental group, 89 patients) or in an adjacent room. The control group, consisting of 89 patients, underwent chemotherapy the following day in the same room but without pet therapy. The therapy included three 20-minute phases: watching the dog perform exercises with the trainer, interacting and playing with the dog, and finally holding or feeding the dog. The results showed a reduction in patient anxiety, heart rate, and blood pressure, along with a notable decrease in depression and a significant improvement in arterial oxygen saturation. The study highlighted the value of AAT in comprehensive cancer care while also recommending further

Pain Management in Palliative Care for Patients With Cancer. DOI: https://doi.org/10.1016/B978-0-443-30050-9.00024-2

research to develop standardized protocols [2]. An anecdotal study on pet therapy in palliative care, conducted with 19 patients, did not employ empirical measures but rather recorded phenomenological experiences in medical charts and records. The presence of a therapy dog was reported to "lighten the atmosphere," bringing a sense of "home" and "normalcy" to the hospital room. Patients frequently experienced pain relief and improved emotional well-being. Notably, AAT also helped reduce feelings of loneliness and depression, contributing further to pain alleviation. Interestingly, it was observed that the stress levels of medical staff were also lowered with the introduction of AAT. Engelman has also discussed two possible mechanisms behind these effects: the hypnoanalgesic effect, due to the relaxation induced by AAT, and the distraction effect, which diverts patients' attention from their pain [4]. Research further supports the idea that AAT can trigger a relaxation response, as studies have shown reductions in heart rate, respiration rate, blood pressure, and pupillary constriction, alongside increased peripheral skin temperature [5,11]. These physiological changes suggest that AAT helps shift the body from sympathetic nervous system activation to parasympathetic activation, fostering relaxation. This relaxation response may reduce both the emotional and sensory components of certain types of pain, resembling a form of "hypnoanalgesia" [12]. The presence of animals has been shown to reduce anxiety and stress levels in patients, which can lower the perception of pain, as anxiety and stress can exacerbate pain sensations. Interactions with therapy animals could lead to decreased levels of cortisol and increased endorphin release, both of which are associated with pain relief [11]. The physical act of petting or interacting with an animal can trigger the release of oxytocin, a hormone associated with bonding and relaxation. Oxytocin can reduce stress responses, lower blood pressure, and induce a state of calm, all of which contribute to pain relief. The release of endorphins during animal interaction could directly counteract pain signals, providing a natural form of pain relief [8].

Engaging with animals provides a pleasant distraction from pain and discomfort. This distraction can be particularly beneficial during stressful procedures or treatments, making them more tolerable for patients [4,5].

Interacting with animals can evoke positive emotions and enhance mood, which can influence the perception of pain, making it feel less intense. The bond formed with therapy animals can provide a sense of purpose and motivation, which can be empowering for patients. White et al. [13] found that women with breast cancer who participated in AAT reported improved emotional well-being and a sense of empowerment.

At the 2023 American Society of Clinical Oncology (ASCO) Conference, a small study was presented highlighting the benefits of additional supportive services for patients at a community hematology/oncology center. The prospective survey revealed that having a therapy dog visit significantly reduced patient anxiety and enhanced relaxation during chemotherapy and immunotherapy infusion sessions [14].

Despite the variability, some studies reported significant reductions in pain among pediatric patients with cancer. For instance, Silva and Osório observed marked decreases in self-reported pain levels [15], while Coakley and Mahoney noted improvements in mood and pain perception following therapy dog interactions [16]. Chuback also demonstrated that following the therapy dog visit, patients had lower distress and significant decreases in worry, tiredness, fear, sadness, and pain [17].

In a randomized controlled trial, Marcus et al. studied the effects of single dog-assisted therapy at a tertiary university chronic pain outpatient clinic. The study analyzed 295 therapy dog visits (235 with patients, 34 with accompanying family/friends, and 26 with staff) and 96 waiting room surveys (83 patients, 6 family/friends, and 7 staff) over 2 months. Significant improvements were reported in pain, mood, and emotional distress for patients who interacted with therapy dogs (pain from 6.66 to 5.74, $P < .0001$; anxiety from 3.92 to 2.14, $P < .0001$; sadness from 3.12 to 1.80, $P < .0001$; and irritability from 3.46 to 1.63, $P < .0001$, as self-reported on a 0–10 numeric scale). Clinically meaningful pain relief (a decrease of ≥2 points) was observed in 23% of patients after the therapy dog visit, compared to only 4% in the waiting room control. Similarly, significant improvements were noted in emotional distress and well-being among both family/friends and staff after therapy dog visits [18].

Carvalho et al. conducted a similar study, although they used different tools to measure psychological symptoms. This nonrandomized pilot study investigated the impact of a single session of dog-assisted therapy (cynotherapy) on reducing pain and emotional distress in oncology outpatients, compared to the typical waiting room experience (control). Over 10 months, 81 patients participated, with 41 in the cynotherapy group and 40 in the control group. The cynotherapy group showed greater improvements in pain (median difference score: –1.0 vs 0.0; $P = .037$), distress (median: –1.0 vs 0.0; $P = .017$), and depression (median: –1.0 vs 0.0; $P = .030$). Although not statistically significant, the proportion of patients with clinically relevant pain improvement (a reduction of at least 2 points) was nearly double in the cynotherapy group compared to controls (39% vs 20%, odds ratio = 2.53, 95% confidence interval, 0.86–8.02; $P = .088$). The average satisfaction rate was 9.3/10, and no negative events were reported. In conclusion, a single session of dog-assisted therapy can provide immediate relief in pain and distress for cancer patients in an outpatient setting, with high satisfaction and no adverse effects. The study underscores the potential clinical relevance of incorporating cynotherapy into cancer pain clinics and suggests further large-scale, focused research [19].

Additionally, Phung et al. found that AAI not only alleviated pain but also reduced anxiety and improved emotional well-being. These findings highlight the potential of AAIs to provide both psychological comfort and physical relief during intensive treatments such as chemotherapy and immunotherapy [10,20].

Another study, also among cancer patients undergoing chemotherapy, showed reductions in pain levels, fatigue, and emotional distress and improved the overall QoL for cancer patients. However, it noted the need for larger sample sizes to strengthen the findings [21].

There are no standardized guidelines for the use of pet therapy in oncology patients receiving treatment. There are a lot of variations in practice; some hospitals limit visits to domestic companion animals, and others have extensive dog therapy programs.

Veterinarians should be included early for the prevention of zoonotic disease—any pet in contact with a person with cancer should receive a routine annual veterinary examination, yearly stool examinations for *Salmonella*, *Campylobacter*, *Giardia*, and *Cryptosporidium* spp., and vaccines as indicated by the local veterinary authority [14,22].

For example, at the University Hospital Center Zagreb, a therapy dog named Topsy is involved in individual psychotherapy sessions with hematological patients in remission, under the guidance of an integrative psychotherapist. Topsy's presence helps reduce anxiety, pain, and feelings of loneliness while promoting relaxation, socialization, and emotional healing. The careful selection of patients and strict veterinary protocols mitigate potential risks, such as infection or allergic reactions, ensuring that therapy dogs like Topsy can safely participate in the therapeutic process (Fig. 28.1).

Music therapy

What is music? Sound is a mental image created by the brain in response to vibrating molecules. No known human culture today or at any time in the recorded past has been without music. Making music is not just an art; it is also a science. Listening to music, performing, and composing activate almost all areas of the brain and involve nearly every nervous subsystem.

In the 1970s, neurobiologists showed that connections of the inner ear lead to the auditory cortex and directly to the cerebellum. The function of the cerebellum is crucial for music and is also involved in emotions. Furthermore, it contains numerous connections with the amygdala (involved in remembering emotional experiences) and the frontal lobe (involved in planning and impulse control), so it is activated when people listen to music they like [23]. Avram Goldstein from Stanford University showed more than 50 years ago that the enjoyment of listening to music and the thrills elicited by music may be in some people attenuated by naloxone, a specific antagonist at the opiate (endorphin) receptor [24]. Koelsch et al. showed that 30 seconds of pleasant versus unpleasant music are sufficient to elicit activity changes in the hippocampus, which has receptors for endorphins and can produce opioids and release them into different regions of the brain and can also trigger the release of endorphins in the hypothalamus [25,26].

FIGURE 28.1 Personal photo showing Topsy, a therapy dog, assisting hematological patients in remission at the University Hospital Center Zagreb, reducing anxiety, pain, and loneliness. Used by author's permission.

Music is one of humanity's oldest medicines. Many cultures have developed their rich traditions for using sound and rhythm to ease suffering, promote healing, and calm the mind. Palliative care music therapy has been formally practiced since the late 1970s and affords a variety of benefits. Music therapy is based on the clinical and evidence-based use of music interventions within a therapeutic relationship to accomplish individualized treatment goals [27,28]. For an experience to be considered music therapy, four elements are needed: a music therapist, a client with a need, music, and a therapeutic relationship [28]. The education requirements to practice as a music therapist vary throughout the globe [29]. Four theoretical perspectives that support why many patients report reduced pain sensation after music therapy include the psychological relationship between music and pain, the psychophysiological theory, spinal mechanisms involved in pain modulation, and the role of endorphins [30].

An analysis of the music therapy literature yields numerous reports to support the role of music in the alleviation of pain in palliative care. Pain-reducing effects of music listening are well established, but clinical relevance is sometimes questionable. There are systematic reviews that have demonstrated important pain-reducing effects of music therapy [28,31–35], but most of the studies included in these systematic reviews addressed acute and procedural pain. Music characteristics (like tempo, harmony, melody, instrumentation, volume, etc.) are very randomly described in clinical studies. A metaanalysis evaluating music listening for pain management in adults suggests that music without lyrics is effective, especially for pain management of nonprocedural pain. Across different metaanalyses, the pain-reducing effects of music listening were inconsistent among the analyzed studies [33,36,37].

For her master's thesis in nursing, Longfield conducted a quasiexperimental study measuring the effects on mood and pain in hospice patients receiving music therapy [38,39]. Eight adult subjects diagnosed with cancer comprised their control group and received music therapy via taped recordings and a headset for 45 minutes a day for 5 days. The pretest and posttest measurements used the Short-Form McGill Pain Questionnaire (SFMPQ) and the Linear Analog Self-Assessment Scale (LASA). All subjects were prescribed narcotic pain medications and continued taking medications as prescribed during the study. There was a significant decrease in pain as measured by the SFMPQ, but further research with larger sample sizes is necessary.

Suzan van Veen made a systematic review of 22 nonpharmacological interventions feasible for nursing practice and adaptable to palliative care patients' needs. One of them included music therapy with promising results in pain reduction [40].

In the quantitative study, a comparative study was done on 14 cancer patients admitted for pain relief under the Department of Pain and Palliative Medicine, who had moderate-to-severe pain. Patients were allocated to the test group or control group nonrandomly, and the test group patients were subjected to listening to instrumental music therapy for 20 minutes (a combination of traditional and modern contemporary tunes) while the control was kept occupied by talking to them for 20 minutes. The Numeric Rating Scale (NRS) was used to assess the preinterventional and postinterventional pain scores. A statistically significant reduction was seen in the pain scores in the test group after music therapy ($P = .003$) and a statistically significant reduction in the postintervention pain scores in the test group compared to the control group (difference between groups $P = .034$), but again a study with a larger sample size should be undertaken to confirm this finding [41].

Two hundred inpatients at University Hospitals Case Medical Center were enrolled in the study from 2009 to 2011. Patients were randomly assigned to two groups: standard care alone or standard care with music therapy. A clinical nurse specialist administered pretests and posttests to assess the level of pain using an NRS as the primary outcome and the Functional Pain Scale as the secondary outcome. The intervention incorporated music therapist–guided autogenic relaxation and live music. A significantly greater decrease in NRS pain scores was seen in the music therapy group, and the mean change in Functional Pain Scale scores was significantly greater in the music therapy group [42,43].

In a pilot study with an ex postfacto design, a computerized database for music therapy in palliative care was utilized. The goals of the study were to describe a research tool to evaluate the use of the computerized database and to investigate the efficacy of music therapy. The database was designed working with a computer consultant and familiar computer programs. Several tools were used for assessment and data collection, and they consisted primarily of Visual Analog Scales using separate scales for different problems (i.e., mood, pain, anxiety, and shortness of breath). Over 1 year, 150 patients were seen, and 90 of their initial music therapy sessions were analyzed. Statistical significance was found for patient-rated scores for pain following music therapy. Although the author acknowledged that this study was limited because subjects were not randomly assigned and it used an ex postfacto design, the data indicated that music therapy is beneficial for people suffering from pain within the palliative care treatment model [44].

In another study, a palliative care team analyzed 198 patients with advanced, life-limiting illnesses. The music therapy group received 20-minute sessions performed by a music therapist. Each music therapy session was individualized. The control group received the same comfort measures as the music therapy group, but a music therapist was not present. The pain was assessed at preintervention and postintervention. Although 20 minutes were allotted, a few accounts of up to 50 minutes were noted. As a result, the timing of the postintervention score also varied. Blinding was broken for a few patients because participants revealed their group assignment to the research assistant [45].

In an Australian study, 25 patients were recruited from inpatient hospice services. The music session length varied from 20 to 40 minutes, and any sessions outside of the study's session duration were not included in the data. Therefore data could be missing that would affect the overall results. Only preintervention and postintervention scores were recorded, so only short-term effects were assessed. Both studies showed music therapy improved symptom burden, and a decrease in anxiety and pain was observed in patients who received music therapy, although blinding to intervention and additional therapy sessions were concerns [46].

A prospective cohort study of a total of 18 hospice patients, selected based on scores ≥4 on the revised Edmonton Symptom Assessment System items on pain, depression, anxiety, or well-being, participated in music therapy sessions provided by a board-certified music therapist over 2–3 weeks. Three to four music therapy sessions tailored to individual needs (lasting between 20 and 45 minutes) were conducted for each patient, after which they reported a reduction in symptom severity and emotional distress and an increase in QOL. Patient QOL was assessed using the LASA. These results are promising, supporting the benefits of music therapy for patients in a hospice setting [47].

A Turkish study aimed to investigate the effect of music on pain, anxiety, comfort, and functional capacity of cancer patients who received care in a palliative care unit. The population of this randomized controlled trial consisted of cancer patients hospitalized in the palliative care service between July 2018 and July 2019. The study included 60 patients (30 interventions/30 controls). The patients in the intervention group were given a total of six music sessions, 10 minutes each, with the Turkish classical music in maqams of their choice. There was a significant difference between the mean total pain scores, anxiety, comfort, and functional capacity scores of the patients in the intervention and control groups before and after music therapy. Music therapy decreased the level of pain [48].

Music therapy for pain management goes beyond listening to prerecorded music to distract or relax and includes individualized, active music engagement to optimize pain management. Clements-Cortés found that both live and recorded music provided in music therapy treatment resulted in statistically significant reductions in pain perception and the enhancement of physical comfort [49]. In hospice and palliative care, music therapists use methods such as songwriting, improvisation, guided imagery and music, lyric analysis, singing, instrument playing, and music therapy relaxation techniques [38]. Cynthia Lusson conducted various individualized music therapy sessions with an average duration of 34 minutes, from October 2022 to May 2023, involving 141 cancer patients in traditional hospitalization or chemotherapy day hospital settings. The numeric visual scale for pain was included before and after each session to evaluate the benefits of music therapy. The quantitative evaluation of symptom improvement proved challenging for various reasons, while the analysis of verbatim provided a more comprehensive assessment (beyond solely relying on numeric visual scales of pain) and can serve as an evaluation modality. A reduction of symptoms was found, but additional rigorous studies are essential [43,50]. Werner et al. showed that active sensorimotor synchronization to music (such as tapping) significantly amplifies the pain-reducing effects of music listening and has stronger analgesic effects than simply listening to music. In a sample of 59 participants, pain was applied to their fingernails using pressure algometry while either listening to music or undergoing a silent control period and either performing an active foot-tapping task or a passive control task. Importantly, the pain-reducing effect of the active tapping task was descriptively larger while listening to music than during silence. Actively tapping (compared to the passive control task, accompanied by either music or silence) is consistent with the well-established analgesic effects of distracting attention from a painful stimulus, but the fact that tapping in silence did not significantly reduce pain (compared with silence without tapping) indicates that the attentional distraction by the tapping cannot alone explain the pain-reducing effects of tapping to music. Emotional mechanisms are likely involved in the analgesic effects of sensorimotor synchronization to music, and social bonding emerges with music due to the activation of the endogenous opioid system with sensorimotor synchronization [51].

According to Guidelines for Integrative Medicine for Pain Management in Oncology developed by a multidisciplinary group of experts from the Society for Integrative Oncology and ASCO in 2022, the quality of evidence for music therapy for pain is low, and music therapy may be offered to patients experiencing surgical pain from cancer surgery (type: evidence-based, benefits outweigh harms; evidence quality: low; strength of recommendation: weak) [52].

Considerations significant to the use of music in pain relief also include the question if music, used inappropriately, can aggravate pain sensation. Music therapy has been proven as a safe and well-established intervention, but still, music-guided therapy is not indicated for people who are sedated, disoriented, or confused. Careful assessment of the patient's readiness for and ability to tolerate certain music therapy interventions is needed, and instrumental improvisation may not be appropriate in some hospital settings (as the volume of the music may be harmful to others in the immediate environment).

Potential application

Pet and music therapy can be practically applied in various healthcare clinical settings to enhance the QoL and alleviate pain for cancer patients:

1. In hospice and palliative care units for terminally ill patients.
2. At chemotherapy centers—oncology and hematology day hospitals and inpatient wards.

TABLE 28.1 Scientific evidence on pet therapy in cancer pain management.

Study/research	Study population	Findings
Orlandi et al. [2]	178 cancer patients undergoing chemotherapy in an oncological day hospital. Experimental group of 89 patients receiving pet therapy	Pet therapy significantly reduced patient anxiety, heart rate, and blood pressure, while also decreasing depression and improving arterial oxygen saturation.
Marcus et al. [18]	295 chronic pain patients (including patients with cancer) in a tertiary outpatient clinic. 235 patients had dog-assisted therapy, control group was waiting room only	Significant reduction in pain (6.66–5.74), anxiety, sadness, and irritability scores. Clinically meaningful pain relief observed in 23% of patients after dog therapy compared to only 4% in the control group.
Silva and Osório [15]	Pediatric oncology patients	AAT sessions resulted in marked decreases in self-reported pain levels and significant improvements in emotional well-being among pediatric patients.
Johnson et al. [21]	Adults undergoing chemotherapy and immunotherapy	Animal-assisted intervention helped alleviate pain, reduce anxiety, and improve emotional well-being, offering both physical and psychological relief.
Carvalho et al. [19]	81 cancer outpatients in a pilot study comparing dog-assisted therapy group and a control group	Dog-assisted therapy led to greater improvements in pain, emotional distress, and depression compared to controls. Satisfaction rate was 9.3/10, and no adverse events were reported.
Engelman [4]	Cancer patients in palliative care	AAT resulted in relaxation, lowered cortisol, increased oxytocin and endorphin release, reducing anxiety and contributing to pain alleviation.

AAT, Animal-assisted therapy.

TABLE 28.2 Scientific evidence on music therapy in cancer pain management.

Study/research	Study population	Findings
Gutgsell et al. [43]	200 palliative care patients	Music therapy significantly reduced numeric pain ratings and improved functional pain scores compared to controls.
Krishnaswamy and Nair [41]	14 cancer patients undergoing chemotherapy	Significant reductions in pain and anxiety were observed after 20-min music therapy sessions compared to talking sessions.
Hilliard [38]	Hospice and palliative patients	Visual Analog Scales used indicated marked reduction in pain levels following music therapy sessions.
Werner et al. [51]	59 participants exposed to pain via pressure on fingernails	Active sensorimotor synchronization to music significantly reduced pain, with effects stronger compared to passive listening, suggesting an endogenous opioid effect.
Düzgün and Karadakovan [48]	60 cancer patients in palliative care service (30 intervention/30 control)	Music therapy sessions led to significant reductions in pain, anxiety, and improved comfort and functional capacity among the intervention group.

Future implication

Complementary, transdisciplinary interventions such as pet and music therapy could benefit from:

- Expanded research: Further studies with larger sample sizes and diverse patient populations must validate the findings and establish robust evidence.
- Standardized protocols: Developing standardized protocols for implementing pet and music therapy in clinical settings will ensure consistency and maximize the therapeutic benefits for patients. This includes guidelines for selecting suitable therapy and integrating it into existing care plans.
- Technology integration: Exploring the use of technology, such as virtual reality interactions with therapy animals, could provide alternative options for patients unable to interact with live animals due to allergies or other constraints.

Conclusion

Pet and music therapy are valuable complementary therapies for managing cancer pain. The most important findings are given in Tables 28.1 and 28.2. Both modalities may reduce pain intensity and, in some cases, analgesic use. They provide emotional support and reduce anxiety and stress, which can lower the perception of pain, as anxiety and stress can exacerbate pain sensations; they decrease levels of cortisol and increase endorphin release, and they enhance mood and offer a pleasant distraction. The psychological and physiological mechanisms underpinning these benefits are supported by a growing body of research, underscoring the importance of incorporating pet and music therapy into comprehensive cancer care plans. Further studies and guidelines are needed to continue exploring the potential of pet and music therapy and to establish standardized protocols for their implementation in oncology settings. They can be effectively integrated into hospice care, chemotherapy centers, inpatient oncology, and other wards to enhance patient well-being and pain management.

References

[1] Nightingale F. Notes on Nursing: What It Is and What It Is Not [Internet]. New York: D. Appleton and Co.; 1860. Available from: https://digital.library.upenn.edu/women/nightingale/nursing/nursing.html; [accessed 2024].

[2] Orlandi M, Trangeled K, Mambrini A, et al. Pet therapy effects on oncological day hospital patients undergoing chemotherapy treatment. Anticancer Res 2007;27(6C):4301–3.

[3] Delta Society. Animal assisted therapy | Delta therapy dogs | Australia, <https://www.deltasociety.com.au/animal-assisted-therapy;> 2024.

[4] Engelman SR. Palliative care and use of animal-assisted therapy. Omega (Westport) 2013;67(1–2):63–7. Available from: https://doi.org/10.2190/OM.67.1-2.g.

[5] Braun C, Stangler T, Narveson J, et al. Animal-assisted therapy as a pain relief intervention for children. Complement Ther Clin Pract 2009;15(2):105–9. Available from: https://doi.org/10.1016/j.ctcp.2009.02.008.

[6] Jorgenson J. Therapeutic use of companion animals in health care. Image J Nurs Sch 1997;29(3):249–54. Available from: https://doi.org/10.1111/j.1547-5069.1997.tb00993.x.

[7] Gagnon J, Bouchard F, Landry M, et al. Implementing a hospital-based animal therapy program for children with cancer: a descriptive study. Can Oncol Nurs J 2004;14(4):217–22. Available from: https://doi.org/10.5737/1181912x144217222.

[8] Geisler AM. Companion animals in palliative care: stories from the bedside. Am J Hosp Palliat Care 2004;21(4):285–8. Available from: https://doi.org/10.1177/104990910402100411.

[9] Horowitz S. Animal-assisted therapy for inpatients: tapping the unique healing power of the human–animal bond. Altern Complement Ther 2010;16(6):339–43. Available from: https://doi.org/10.1089/act.2010.16603.

[10] Holder TRN, Gruen ME, Roberts DL, et al. A systematic literature review of animal-assisted interventions in oncology (part I): methods and results. Integr Cancer Ther 2020;19:1534735420943278. Available from: https://doi.org/10.1177/1534735420943278.

[11] Cole KM, Gawlinski A. Animal-assisted therapy: the human-animal bond. AACN Clin Issues 2000;11(1):139–49. Available from: https://doi.org/10.1097/00044067-200002000-00015.

[12] Patterson DR, Jensen MP. Hypnosis and clinical pain. Psychol Bull 2003;129(4):495–521. Available from: https://doi.org/10.1037/0033-2909.129.4.495.

[13] White JH, Quinn M, Garland S, et al. Animal-assisted therapy and counseling support for women with breast cancer: an exploration of patient's perceptions. Integr Cancer Ther 2015;14(5):460–7. Available from: https://doi.org/10.1177/1534735415580678.

[14] Curley PB, Patel A, Famoso J. Evaluating the impact of therapy dogs on patients with cancer. JCO Oncol Pract 2023;19(11_suppl.):278. Available from: https://doi.org/10.1200/OP.2023.19.11_suppl.278.

[15] Silva NB, Osório FL. Impact of an animal-assisted therapy programme on physiological and psychosocial variables of paediatric oncology patients. PLoS One 2018;13(4):e0194731. Available from: https://doi.org/10.1371/journal.pone.0194731.

[16] Coakley AB, Mahoney EK. Creating a therapeutic and healing environment with a pet therapy program. Complement Ther Clin Pract 2009;15(3):141–6. Available from: https://doi.org/10.1016/j.ctcp.2009.05.004.

[17] Chubak J, Hawkes R. Animal-assisted activities: results from a survey of top-ranked pediatric oncology hospitals. J Pediatr Oncol Nurs 2016;33(4):289–96. Available from: https://doi.org/10.1177/1043454215614961.

[18] Marcus DA, Bernstein CD, Constantin JM, et al. Animal-assisted therapy at an outpatient pain management clinic. Pain Med 2012;13(1):45–57. Available from: https://doi.org/10.1111/j.1526-4637.2011.01294.x.
[19] Silva Carvalho F, Carvalho S, Conde R, et al. Cynotherapy in cancer pain management: a pilot study. Pain Med 2021;22(12):3051–61. Available from: https://doi.org/10.1093/pm/pnab118.
[20] Phung A, Joyce C, Ambutas S, et al. Animal-assisted therapy for inpatient adults. Nursing 2017;47(1):63. Available from: https://doi.org/10.1097/01.NURSE.0000504675.26722.d8.
[21] Johnson RA, Meadows RL, Haubner JS, et al. Animal-assisted activity among patients with cancer: effects on mood, fatigue, self-perceived health, and sense of coherence. Oncol Nurs Forum 2008;35(2):225–32. Available from: https://doi.org/10.1188/08.ONF.225-232.
[22] Chan MM, Tapia Rico G. The "pet effect" in cancer patients: risks and benefits of human-pet interaction. Crit Rev Oncol Hematol 2019;143:56–61. Available from: https://doi.org/10.1016/j.critrevonc.2019.08.004.
[23] Levitin DJ. This is your brain on music: understanding a human obsession. London: Atlantic Books; 2008. Available from: https://www.perlego.com/book/117532/this-is-your-brain-on-music-understanding-a-human-obsession-pdf [accessed 28.08.24].
[24] Goldstein A. Thrills in response to music and other stimuli. Psychobiology 1980;8(1):126–9. Available from: https://doi.org/10.3758/BF03326460.
[25] Koelsch S, Fritz T, Cramon V, et al. Investigating emotion with music: an fMRI study. Hum Brain Mapp 2006;27(3):239–50. Available from: https://doi.org/10.1002/hbm.20180.
[26] Lathe R. Hormones and the hippocampus. J Endocrinol 2001;169(2):205–31. Available from: https://doi.org/10.1677/joe.0.1690205.
[27] Davis MP, Cancer pain. In: Olver I, editor. The MASCC textbook of cancer supportive care and survivorship. Springer; 2018. p. 17–35. Available from: https://doi.org/10.1007/978-3-319-90990-5_2.
[28] Bradt J, Dileo C, Shim M. Music interventions for preoperative anxiety. Cochrane Database Syst Rev 2013;2013(6):CD006908. Available from: https://doi.org/10.1002/14651858.CD006908.pub2.
[29] Clements-Cortés A, Klinck S. Fostering pre-professionals and learning experiences in end-of-life care contexts: music therapy internship supervision. Healthcare 2024;12(4):459. Available from: https://doi.org/10.3390/healthcare12040459.
[30] O'Callaghan CC. Pain, music creativity and music therapy in palliative care. Am J Hosp Palliat Care 1996;13(2):43–9. Available from: https://doi.org/10.1177/104990919601300211.
[31] Klassen JA, Liang Y, Tjosvold L, et al. Music for pain and anxiety in children undergoing medical procedures: a systematic review of randomized controlled trials. Ambul Pediatr 2008;8(2):117–28. Available from: https://doi.org/10.1016/j.ambp.2007.12.005.
[32] Lee JH. The effects of music on pain: a meta-analysis. J Music Ther 2016;53(4):430–77. Available from: https://doi.org/10.1093/jmt/thw012.
[33] McConnell T, Scott D, Porter S. Music therapy for end-of-life care: an updated systematic review. Palliat Med 2016;30(9):877–83. Available from: https://doi.org/10.1177/0269216316635387.
[34] Yinger OS, Gooding LF. A systematic review of music-based interventions for procedural support. J Music Ther 2015;52(1):1–77. Available from: https://doi.org/10.1093/jmt/thv004.
[35] Huang E, Huang J. Music therapy: a noninvasive treatment to reduce anxiety and pain of colorectal cancer patients – a systemic literature review. Medicina (Kaunas) 2023;59(3):482. Available from: https://doi.org/10.3390/medicina59030482.
[36] Kühlmann AYR, de Rooij A, Kroese LF, et al. Meta-analysis evaluating music interventions for anxiety and pain in surgery. Br J Surg 2018;105(7):773–83. Available from: https://doi.org/10.1002/bjs.10853.
[37] Martin-Saavedra JS, Vergara-Mendez LD, Pradilla I, et al. Standardizing music characteristics for the management of pain: a systematic review and meta-analysis of clinical trials. Complement Ther Med 2018;41:81–9. Available from: https://doi.org/10.1016/j.ctim.2018.07.008.
[38] Hilliard RE. Music therapy in hospice and palliative care: a review of the empirical data. Evid Based Complement Alternat Med 2005;2(2):173–8. Available from: https://doi.org/10.1093/ecam/neh076.
[39] Longfield V. The effects of music therapy on pain and mood in hospice patients [Unpublished. Master's thesis]. St. Louis, MO: Saint Louis University; 1995.
[40] van Veen S, Drenth H, Hobbelen H, et al. Non-pharmacological interventions feasible in the nursing scope of practice for pain relief in palliative care patients: a systematic review. Palliat Care Soc Pract 2024;18:26323524231222496. Available from: https://doi.org/10.1177/26323524231222496.
[41] Krishnaswamy P, Nair S. Effect of music therapy on pain and anxiety levels of cancer patients: a pilot study. Indian J Palliat Care 2016;22(3):307–11. Available from: https://doi.org/10.4103/0973-1075.185042.
[42] Gallagher LM, Lagman R, Bates D, et al. Perceptions of family members of palliative medicine and hospice patients who experienced music therapy. Support Care Cancer 2017;25(6):1769–78. Available from: https://doi.org/10.1007/s00520-017-3578-y.
[43] Gutgsell KJ, Schluchter M, Margevicius S, et al. Music therapy reduces pain in palliative care patients: a randomized controlled trial. J Pain Symptom Manage 2013;45(5):822–31. Available from: https://doi.org/10.1016/j.jpainsymman.2012.05.008.
[44] Horne-Thompson A, Grocke D. The effect of music therapy on anxiety in patients who are terminally ill. J Palliat Med 2008;11(4):582–90. Available from: https://doi.org/10.1089/jpm.2007.0193.
[45] Gallagher LM, Steele AL. Developing and using a computerized database for music therapy in palliative medicine. J Palliat Care 2001;17(3):147–54. Available from: https://doi.org/10.1177/082585970101700308.
[46] Zeng YS, Wang C, Ward KE, et al. Complementary and alternative medicine in hospice and palliative care: a systematic review. J Pain Symptom Manage 2018;56(5):781–94.e4. Available from: https://doi.org/10.1016/j.jpainsymman.2018.07.016.
[47] Estell MH, Whitford KJ, Ulrich AM, et al. Music therapy intervention to reduce symptom burden in hospice patients: a descriptive study. Am J Hosp Palliat Care 2024;42(1):102–11. Available from: https://doi.org/10.1177/10499091241237991.

[48] Düzgün G, Karadakovan A. Effect of music on pain in cancer patients in palliative care service: a randomized controlled study. Omega (Westport) 2024;88(3):1085–100. Available from: https://doi.org/10.1177/00302228211059891.

[49] Clements-Cortés A. The effect of live music vs. taped music on pain and comfort in palliative care. Kor J Music Ther 2011;13(1):105–21.

[50] Lusson C, Bourgouin M. Musical therapy in oncology: intervention modalities and patient lived experience. Presented at MASCC 2024;2024. Available from: https://simul-europe.com/2024/mascc/Files/2363.pdf.

[51] Werner LM, Skouras S, Bechtold L, et al. Sensorimotor synchronization to music reduces pain. PLoS One 2023;18(7):e0289302. Available from: https://doi.org/10.1371/journal.pone.0289302.

[52] Mao JJ, Ismaila N, Bao T, et al. Integrative medicine for pain management in oncology: society for integrative oncology–ASCO guideline. J Clin Oncol 2022;40:34. Available from: https://doi.org/10.1200/JCO.22.01357.

Chapter 29

Telemedicine, telerehabilitation, and online support communities

Ahmad Samir Alfaar[1,2,3] and Amir Radfar[4]

[1]*Al-Moriat for eHealth Solutions, Berlin, Germany,* [2]*Vistarlin Center for Integrated Medicine, Berlin, Germany,* [3]*Roya Eye Center, Stendal, Sachsen-Anhalt, Germany,* [4]*Former Faculty Member, College of Medicine, University of Central Florida, Orlando, FL, United States*

Key learning objectives

- Understand how technological advancements lead to new therapies and management plans in pain management for palliative cancer care.
- Evaluate the impact of technologies like telemedicine, virtual reality, and wearables on pain treatment in cancer patients.
- Discuss the legal and ethical considerations in using technology for pain management, including data privacy and informed consent.
- Learn how e-learning platforms and educational websites empower patients and caregivers in pain management and self-care.

A historical overview of telemedicine

Telemedicine's origins are deeply rooted in the desire to connect patients with healthcare providers remotely, an idea that began taking shape during the 19th century when telegraphs were used to communicate medical information. The modern form of telemedicine gained significant traction in the mid-20th century, spurred by the rapid development of telecommunications technology. It allowed healthcare providers to extend their reach without the substantial costs associated with building new facilities or expanding staff, optimizing resource use, and creating new business opportunities in telecommunications and medical devices. One notable early instance of telemedicine occurred in the 1960s, when Massachusetts General Hospital partnered with Logan International Airport to establish a hospital-based multispecialty telemedicine practice, allowing remote clinical examinations for travelers and airport staff [1–4].

This early use of telemedicine emphasized its potential to bridge healthcare access gaps, particularly in areas with limited resources. During the 1970s, telemedicine experienced significant growth, particularly through initiatives such as the Alaska Satellite Project (ATS-6), which employed satellite communications to deliver medical services to rural communities. This initiative highlighted how technology could overcome the healthcare challenges posed by geographical isolation [4]. Similarly, the Space Technology Applied to Rural Papago Advanced Health Care project, a collaboration between NASA and the Indian Health Service, integrated telemedicine into healthcare delivery for remote populations [5]. Telemedicine facilitated continuous patient monitoring and early intervention, crucial components of effective healthcare. The technology also allowed specialists to consult on cases from afar, significantly improving care quality for patients in isolated regions.

By the 1990s, the rise of the internet and rapid advances in telecommunications further enhanced telemedicine's capabilities. Videoconferencing technologies became more accessible, supporting real-time and direct consultations between healthcare providers and patients. These advancements laid the foundation for telemental health, teleradiology, and home telecare, demonstrating clear benefits like reduced travel and increased patient satisfaction [6]. The COVID-19 pandemic catalyzed large-scale adoption, preserving continuity for chronic and cancer care during lockdowns [3,7–9].

The rise of telerehabilitation and online support communities

Telerehabilitation, an extension of traditional telemedicine, emerged to expand rehabilitative care beyond conventional clinical settings. As video and sensor technologies matured in the late 1990s, home-based telerehabilitation enabled

Pain Management in Palliative Care for Patients With Cancer. DOI: https://doi.org/10.1016/B978-0-443-30050-9.00034-5

remote progress monitoring and proved comparable to in-person care for musculoskeletal and poststroke recovery [10,11]. The late 1990s saw telerehabilitation begin incorporating video conferencing and sensor technologies to remotely monitor physical progress. Early studies demonstrated that remote rehabilitation was as effective as in-person visits, especially in managing musculoskeletal disorders and poststroke recovery [12].

Traditionally, support groups met in person, but the internet revolution of the 1990s moved these gatherings online, expanding their reach and accessibility. Usenet newsgroups and CompuServe forums served as a platform for discussion in the 1980s, followed by online communities, such as MedHelp and those created by the American Cancer Society, which facilitated the sharing of experiences and information among patients. The anonymity of these platforms encouraged candid communication, allowing participants to discuss stigmatized issues openly [13,14]. By the 2010s, social media and dedicated forums became vital resources for patients, fostering emotional support, reducing feelings of isolation, and promoting treatment adherence [6,15].

Telemedicine in cancer care and palliative care

Early telemedicine initiatives in oncology enabled remote consultations for diagnostics, such as transmitting radiological images to specialists, facilitating timely decision-making [4,16,17]. Virtual platforms facilitated access to mental health resources and support networks, effectively addressing patients' emotional well-being during the challenging periods [18–20].

In palliative oncology, telemedicine supports timely analgesic titration, psychological care access, and home-based monitoring, reducing emergency visits while maintaining care quality [3]. Virtual consultations enable timely medication adjustments and evaluation of treatment effectiveness, and during the COVID-19 pandemic, telemedicine minimized exposure risks for vulnerable patients while maintaining care quality. This model effectively managed pain, reduced emergency visits, and improved patient satisfaction [3].

Beyond single encounters, combining modalities yields greater effect. The integration of telemedicine, telerehabilitation, and online support communities has led to a holistic approach for managing pain in palliative care. In this chapter, telepractice is used as an umbrella term for integrated telemedicine, telerehabilitation, and online psychosocial support. This approach ensures that patients undergoing complex treatments such as chemotherapy receive comprehensive support. For instance, a patient might be monitored remotely for side effects, engage in telerehabilitation exercises to maintain mobility, and access an online community for emotional support [21].

Integrated telepractice addresses physical function, symptom control, and psychosocial support within a single, patient-centered pathway [22–24].

Randomized controlled trials have demonstrated that telehealth interventions lead to similar or improved outcomes in symptom management and overall patient satisfaction [25–27]. Patient-reported outcome measures have also indicated improved quality of life and reduced symptom burden among palliative care patients using telehealth. However, effect sizes vary across populations and platforms, highlighting the need for multicenter, standardized trials [28].

Practical aspects

Building a successful program that integrates telepractice for pain management in palliative cancer care settings requires a deep understanding of technological, clinical, and psychosocial elements of care delivery. Implementation hinges on three pillars: clinical workflow, remote monitoring, and rehabilitation adherence.

Telemedicine for cancer pain management

The implementation of telemedicine involves using both synchronous communication technologies, such as video consultations, and asynchronous tools, like secure messaging and patient portals, to maintain ongoing care. Remote consultations enable healthcare providers to modify pain management plans based on real-time assessments, including patient-reported outcomes and physiological data from wearable sensors [29].

Telemedicine-based pain management often begins with an in-depth assessment of the patient's pain, conducted either in person or via a secure virtual platform. This initial assessment forms the foundation for a tailored treatment plan for cancer pain management, incorporating pharmacological strategies as well as lifestyle modifications [21,30]. Follow-up appointments are conducted virtually to evaluate treatment efficacy, discuss new symptoms, and adjust medications as needed. This model is particularly beneficial for patients in the advanced stages of cancer, who may face significant mobility challenges and experience considerable physical discomfort [3,31].

Remote patient monitoring is a critical component of telemedicine for pain management, particularly for tracking vital signs like heart rate, respiratory function, and other indicators that inform clinicians of the efficacy of the pain management strategy. Wearable biosensors provide continuous physiological data, allowing healthcare providers to intervene promptly when abnormalities are detected [32,33]. Telemedicine also enables the delivery of nonpharmacological pain management techniques. Cognitive-behavioral therapy (CBT), which is widely used to address anxiety and depression—common comorbidities in cancer patients—can be effectively administered through telehealth platforms. Telemedicine can also support nonpharmacological and psychological interventions, including remotely delivered CBT-based approaches, which may improve mental health and treatment adherence in cancer patients [31,32]. For cancer pain specifically, systematic reviews and meta-analyses support the benefit of telemedicine-based pain management interventions [21,26].

Telerehabilitation for palliative cancer patients

Symptom control pairs naturally with maintenance of function. Telerehabilitation focuses on maintaining physical function and mitigating complications related to decreased mobility in cancer patients. A successful telerehabilitation program starts with a personalized care plan, tailored to the patient's needs. This plan includes exercises designed to improve muscle strength, mobility, and coordination, which are facilitated by physiotherapists through live video sessions. Additional support is offered via instructional videos and progress-tracking apps, enabling patients to continue their exercises independently [34]. Wearable technologies, such as activity trackers and motion sensors, are extensively used in telerehabilitation to monitor patients' activities and provide real-time data on progress. This data helps therapists adjust the rehabilitation plan to optimize patient outcomes. Patients recovering from cancer-related surgeries can particularly benefit from telerehabilitation, as it allows them to regain strength and mobility without the burden of traveling to a clinic [3,35].

Wearables and motion sensors quantify adherence and progress, enabling data-driven progression without travel burden. Both healthcare providers and patients require detailed guidance on the technology involved, such as using wearable sensors and accessing video conferencing platforms. Healthcare providers should receive training not only on the technical aspects but also on best practices for engaging patients remotely—ensuring they maintain motivation and adherence to the rehabilitation regimen [36].

Online support communities

These communities provide vital peer support, allowing patients and families to share experiences and coping strategies—especially important in palliative care, where psychological challenges are significant. An effective online support platform should include interactive features such as forums, video chats, and live Q&A sessions. Healthcare professionals should moderate these spaces to ensure information is accurate, offer expert guidance, and intervene when users show signs of distress, maintaining a reliable and supportive community [6,37].

The integration of these support communities with broader telehealth services enables healthcare professionals to monitor discussions, gain insights into patients' emotional states, and provide timely interventions. This integration helps bridge the gap between medical treatment and psychosocial support, promoting a more holistic care model for palliative patients [38]. Artificial intelligence (AI)-driven technologies may support personalization by suggesting relevant discussions, coping strategies, or exercises based on a patient's engagement history, although clinical evidence in oncology and palliative-care support communities remains emerging (see "Future aspects" section) [18,39,40].

Integrating technology and ensuring practical implementation

The success of a telepractice program depends on the seamless integration of technology, data security, and an interoperable digital infrastructure. It is essential that electronic health records (EHRs) are incorporated into telehealth platforms, allowing healthcare providers easy access to patient records during remote consultations. The use of secure, interoperable telehealth platforms can improve access to patient records during remote consultations while supporting privacy, security, and workflow requirements [38,41,42–44].

Reliable connectivity, compatible hardware such as tablets and sensors, and responsive technical support are prerequisites for successful implementation.

Training and education are the foundation of a successful telehealth implementation strategy. Healthcare professionals need continuous training on advancements in telehealth technologies and best practices in virtual care. Patients, in turn, benefit from user-friendly guides and educational resources that empower them to take an active role in managing their health, ultimately enhancing their autonomy and adherence to care protocols [45].

Once live, feedback loops sustain quality improvement. Continuous feedback mechanisms are essential for refining telehealth programs. Patient satisfaction surveys, provider feedback sessions, and data analytics help identify gaps and areas for improvement, ensuring that the program evolves to meet the needs of both patients and healthcare providers.

Future aspects

The future of telepractice programs is rapidly evolving with emerging technologies that aim to make healthcare delivery more responsive, personalized, and inclusive. The following subsections highlight the key innovations shaping the future of cancer pain management.

Wearable technology and remote monitoring systems

Devices like biosensors, wristbands, and smart patches continuously track physiological markers such as heart rate, activity levels, and blood oxygen levels, providing a real-time window into the patient's health status. AI-driven platforms can analyze this data and offer immediate insights, enabling providers to adjust care plans proactively [46]. For patients undergoing chemotherapy, wearable devices facilitate real-time monitoring of critical symptoms like nausea, temperature, or pain. Data from these devices can be directly transmitted to healthcare systems, allowing for timely interventions in response to any alarming changes. Early studies suggest that these systems may help identify impending symptom deterioration, including potential pain exacerbations, and support more proactive rather than purely reactive management [47,48].

Artificial intelligence and machine learning for personalized care

AI algorithms can analyze patient data to help predict pain flares and other complications, allowing preemptive intervention and thereby improving outcomes. These algorithms can also recommend personalized treatment plans based on a patient's medical history, genetic information, and prior responses to therapy [4,49]. Machine learning (ML), a subset of AI, holds potential in refining diagnostic accuracy and clinical decision-making in telemedicine. By analyzing imaging data and other health indicators, ML can provide predictive insights that guide oncologists in managing pain and other symptoms, particularly for patients in underserved or remote areas [50].

Conversational agents and other AI-driven tools are also beginning to support remote symptom monitoring, medication reminders, and basic psychoeducation in oncology and palliative care; early pilot projects suggest good feasibility and generally positive attitudes among patients and clinicians, although robust evidence on clinical outcomes is still emerging [4,49,50].

Integration of blockchain for secure and interoperable data sharing

Blockchain technology offers a decentralized, secure framework for managing patient records among multiple healthcare systems. This technology can mitigate data fragmentation, giving healthcare professionals a unified view of a patient's medical history, which improves decision-making during remote consultations [51,52].

Robotics in telerehabilitation

Robot-assisted rehabilitation systems offer structured exercise programs tailored to a patient's progress, minimizing the risk of injury. These robotic devices can also provide detailed feedback to therapists, allowing them to adjust treatment protocols effectively and potentially reduce the need for some in-person visits [4].

Virtual reality for pain management

Virtual reality (VR) is being explored as a tool for pain management, particularly by providing immersive, engaging experiences that distract from pain and medical procedures. Systematic reviews indicate that VR can meaningfully reduce pain and anxiety in several clinical contexts, especially in pediatric populations, and similar approaches are now being adapted for cancer and palliative care settings [54]. Combining VR with techniques such as guided relaxation or hypnosis is an emerging strategy, but long-term outcomes and integration into routine telehealth workflows are still being evaluated [18,54].

Enhanced online support communities with artificial intelligence-driven personalization

The next generation of online support communities will be shaped by AI-driven personalization. AI chatbots and natural language processing tools can offer real-time responses and context-aware support, helping patients manage symptoms, access local resources, or simply connect with others facing similar challenges. This innovation aims to create a more supportive, individualized community experience [39,40,47].

Telehealth and genomics: toward precision oncology

Telemedicine is poised to play a key role in precision oncology, where treatments are increasingly tailored based on a patient's genetic profile. Advances in genomics allow highly individualized cancer treatment plans. Telemedicine facilitates remote genetic counseling and helps disseminate personalized treatment plans, making advanced oncology care accessible to patients regardless of geographic barriers [55].

Limitations

The above-mentioned innovations have enhanced accessibility, personalized care, and patient engagement. However, despite these benefits, several limitations hinder their optimal effectiveness. These challenges span technological, ethical, personal, and regulatory aspects. Addressing them is crucial to enhancing patient care and ensuring that the full potential of these technologies is realized.

In many jurisdictions, pandemic-era telehealth flexibilities (such as temporary licensing waivers, reimbursement parity, and originating-site waivers) were introduced to sustain access during COVID-19 but have since been modified or allowed to lapse. These policy shifts should be interpreted alongside broader access barriers, including language, broadband, and telerehabilitation infrastructure limitations [56–58]. This policy and reimbursement volatility has led some programs to scale back virtual visit capacity, raising concerns about renewed access barriers for patients who had benefited most from remote care. These trends underscore the need for clear, stable legislative and regulatory frameworks to preserve the gains achieved during the pandemic.

Technological limitations

One of the primary challenges lies in technological constraints, mainly bandwidth and connectivity issues. Reliable internet access is fundamental for effective telehealth services. Yet, in many regions—especially rural and underserved urban areas—broadband internet remains limited or nonexistent. This digital divide prevents many patients from benefiting from telemedicine services. Even where internet access is available, insufficient bandwidth can lead to poor video quality, dropped calls, and delayed data transmission, all of which compromise the quality of remote care [56–58].

Moreover, many patients, mostly the elderly or those with limited education, may struggle with using telehealth platforms effectively. The complexity of user interfaces and the need to operate devices such as smartphones, tablets, or wearable sensors can be overwhelming, leading to frustration, decreased engagement, and ultimately, suboptimal care outcomes [59].

Access to telemedicine often correlates with socioeconomic status. Patients from lower-income backgrounds may not be able to afford the necessary devices or internet services required for telehealth. This disparity exacerbates existing health inequalities, as those who might benefit most from remote care are least likely to access it. Cultural and language barriers further impede effective communication. The absence of culturally sensitive resources and interpreters can result in misunderstandings, reduced trust, and lower patient satisfaction [35].

Privacy and data security concerns

The transmission and storage of sensitive health information over digital platforms introduce risks of unauthorized access and data breaches. Cybersecurity threats can compromise patient confidentiality, potentially leading to the misuse of personal health information. Different jurisdictions have varying laws regarding patient data protection, making it challenging for telehealth providers operating across borders to maintain compliance [42–44].

Ethical considerations

Unequal access to telehealth services raises concerns about justice and equity in healthcare (see "Technological limitations" section). Additionally, telemedicine can inadvertently depersonalize the patient–provider relationship. The lack of physical presence may reduce the empathy and emotional support that are crucial in palliative care, potentially affecting patient satisfaction and emotional well-being [60–62].

Regulatory and legal issues

Licensing and credentialing present significant obstacles, as healthcare providers may face challenges with cross-state or international licensing when offering telemedicine services. Regulatory inconsistencies can limit the availability of qualified professionals to patients in different regions. Liability concerns also arise, as the legal framework for malpractice in telemedicine is still evolving. Providers may be uncertain about their liability in cases of misdiagnosis or technical failures, potentially limiting their willingness to offer comprehensive telehealth services [63,64].

Limitations in physical examination and diagnostics

Telemedicine cannot fully replicate the tactile and observational nuances of in-person examinations. Certain assessments require physical touch, palpation, or auscultation—elements not possible remotely. This limitation can lead to missed or delayed diagnoses and inadequate pain management. Relying solely on patient-reported symptoms without physical confirmation can affect diagnostic accuracy. Nonverbal cues and subtle clinical signs that might be evident during in-person visits may be overlooked in virtual consultations.

Personal and psychological aspects

While online support communities offer a platform for connection, they cannot fully substitute face-to-face interactions. Physical isolation may persist, potentially exacerbating feelings of loneliness and depression among palliative care patients. Moreover, telehealth often requires assistance from caregivers to manage technology and provide support during virtual consultations. This added responsibility can increase caregiver stress and burnout, affecting the overall care environment [65].

Limitations of artificial intelligence and machine learning in telemedicine

Algorithmic bias is a significant concern, as AI models are only as good as the data on which they are trained. Biases in data can lead to unequal treatment recommendations, potentially disadvantaging certain patient groups. Overreliance on technology may also diminish the clinician's role in decision-making and should be considered alongside the broader ethical, legal, and social issues of telemedicine discussed above [60–64].

Challenges with online support communities

Online support communities may circulate unverified or incorrect medical information, leading patients to adopt harmful practices based on peer advice without professional guidance. Privacy concerns are also prominent, as sharing personal experiences online raises the risk of exposing sensitive information to unintended audiences, leading to potential stigma or discrimination.

Further challenges

Telehealth services may pose logistical challenges. Compatibility issues between different EHR systems can hinder seamless data exchange, affecting continuity of care. Sustainability and cost are additional concerns. The initial setup and maintenance of telehealth infrastructure require significant investment. Ensuring long-term sustainability and cost-effectiveness is particularly challenging in resource-limited settings.

Conclusion

The evolution of telepractice programs has transformed pain management for palliative cancer care patients, redefining healthcare by overcoming geographical barriers and integrating personalized, holistic care. Telemedicine enables timely interventions and continuous monitoring, while telerehabilitation and online support communities address patients' physical, emotional, and psychological needs. Emerging technologies like AI, ML, VR, and blockchain promise further advancements, offering predictive analytics, immersive experiences, and secure data handling. However, challenges such as technological disparities, ethical concerns, and regulatory issues must be addressed to maximize these benefits.

Achieving the full potential of telepractice requires bridging technological gaps, establishing ethical standards, and fostering collaboration among healthcare providers, technologists, policymakers, and advocates. Prioritizing accessibility and preserving empathy will ensure that telepractice not only extends life but also enhances the quality of life for palliative care patients [53]. Ultimately, the future of pain management lies at the intersection of technological innovation and compassionate, patient-centered care.

References

[1] Breen GM, Matusitz J. An evolutionary examination of telemedicine: a health and computer-mediated communication perspective. Soc Work Public Health 2009;25(1):59–71. Available from: https://doi.org/10.1080/19371910902911206.

[2] Hemyari BA, Coffey N, Inman ZW, et al. Healthcare providers' experience using telehealth during the onset of the COVID-19 pandemic in a predominantly rural patient base: an online survey. Cureus 2023;15(3):e36260. Available from: https://doi.org/10.7759/cureus.36260.

[3] Perez J, Niburski K, Stoopler M, et al. Telehealth and chronic pain management from rapid adaptation to long-term implementation in pain medicine: a narrative review. Pain Rep 2021;6(1):e912. Available from: https://doi.org/10.1097/PR9.0000000000000912.

[4] Jude HD, Balas VE. Telemedicine technologies: big data, deep learning, robotics, mobile and remote applications for global healthcare. 1st ed. Cambridge (MA): Academic Press, an imprint of Elsevier; 2019.

[5] Freiburger G, Holcomb M, Piper D. The STARPAHC collection: part of an archive of the history of telemedicine. J Telemed Telecare 2007;13(5):221–3. Available from: https://doi.org/10.1258/135763307781458949.

[6] Gott M. Telematics for health: the role of telehealth and telemedicine in home and communities. Oxford, UK: Radcliffe Medical Press; 1995.

[7] Mann DM, Chen J, Chunara R, et al. COVID-19 transforms health care through telemedicine: evidence from the field. J Am Med Inform Assoc 2020;27(7):1132–5. Available from: https://doi.org/10.1093/jamia/ocaa072.

[8] Kinoshita S, Cortright K, Crawford A, et al. Changes in telepsychiatry regulations during the COVID-19 pandemic: 17 countries and regions' approaches to an evolving healthcare landscape. Psychol Med 2020;52(13):2606–13. Available from: https://doi.org/10.1017/s0033291720004584.

[9] Xie J, Prahalad P, Lee T, et al. Pediatric subspecialty adoption of telemedicine amidst the COVID-19 pandemic: an early descriptive analysis. Front Pediatr 2021;9:648631. Available from: https://doi.org/10.3389/fped.2021.648631.

[10] Rosen K, Patel M, Lawrence C, et al. Delivering telerehabilitation to COVID-19 inpatients: a retrospective chart review suggests it is a viable option. HSS J 2020;16(1_suppl.):64–70. Available from: https://doi.org/10.1007/s11420-020-09774-4.

[11] Alonazi A. Effectiveness and acceptability of telerehabilitation in physical therapy during COVID-19 in children: findings of a systematic review. Children 2021;8(12):1101. Available from: https://doi.org/10.3390/children8121101.

[12] Zakhary M, Cooper G, Herrera J, editors. Telemedicine for the musculoskeletal physical exam: a pocket guide. Cham, Switzerland: Springer International Publishing; 2023. Available from: https://doi.org/10.1007/978-3-031-16873-4.

[13] Cranen K, Drossaert CHC, Brinkman ES, et al. An exploration of chronic pain patients' perceptions of home telerehabilitation services. Health Expect 2012;15(4):339–50. Available from: https://doi.org/10.1111/j.1369-7625.2011.00668.x.

[14] Cottrell MA, Hill AJ, O'Leary SP, et al. Clinicians' perspectives of a novel home-based multidisciplinary telehealth service for patients with chronic spinal pain. Int J Telerehabil 2018;10(2):81–8. Available from: https://doi.org/10.5195/ijt.2018.6249.

[15] Reitzel M, Letts L, Lennon C, et al. Co-designing solutions to enhance access and engagement in pediatric telerehabilitation. Front Rehabil Sci 2023;4:1293833. Available from: https://doi.org/10.3389/fresc.2023.1293833.

[16] Cui F. Clinical applications of telemedicine services using a regional telemedicine platform for cancer treatment: a cross-sectional study. BMC Cancer 2024;24(1):808. Available from: https://doi.org/10.1186/s12885-024-12563-5.

[17] Guadamuz JS, Wang X, Royce TJ, et al. Sociodemographic inequities in telemedicine use among US patients initiating treatment in community cancer centers during the ongoing COVID-19 pandemic, 2020–2022. JCO Oncol Pract 2023;19(12):1206–14. Available from: https://doi.org/10.1200/OP.23.00144.

[18] Kumar SN, Zafar S, Babulak E, Alam MA, Siddiqui F, editors. Artificial intelligence in telemedicine: processing of biosignals and medical images. 1st ed. Boca Raton (FL): CRC Press; 2024. Available from: https://doi.org/10.1201/9781003307778.

[19] Kostin AA, Samsonov YuV. Evaluation of telemedicine experience: artificial intelligence in cancer care. City Healthcare 2024;5(2):115–24. Available from: https://doi.org/10.47619/2713-2617.zm.2024.v.5i2;115-124.

[20] Patel K, Gonzalez BD, Turner K, et al. Estimated carbon emissions savings with shifts from in-person visits to telemedicine for patients with cancer. JAMA Netw Open 2023;6(1):e2253788. Available from: https://doi.org/10.1001/jamanetworkopen.2022.53788.

[21] Chen W, Huang J, Cui Z, et al. The efficacy of telemedicine for pain management in patients with cancer: a systematic review and meta-analysis. Ther Adv Chronic Dis 2023;14:20406223231153097. Available from: https://doi.org/10.1177/20406223231153097.

[22] Latifi R, Doarn CR, Merrell RC, editors. Telemedicine, telehealth and telepresence: principles, strategies, applications, and new directions. Cham, Switzerland: Springer International Publishing; 2021. Available from: https://doi.org/10.1007/978-3-030-56917-4.

[23] Valenza-Peña G. Effects of telerehabilitation on pain and disability in patients with chronic neck pain: a systematic review and meta-analysis. Healthcare 2024;12(7):796. Available from: https://doi.org/10.3390/healthcare12070796.

[24] Jess M, Timm HU, Dieperink KB. Video consultations in palliative care: a systematic integrative review. Palliat Med 2019;33(8):942–58. Available from: https://doi.org/10.1177/0269216319854938.

[25] Chan RJ, Crichton M, Crawford-Williams F, et al. The efficacy, challenges, and facilitators of telemedicine in post-treatment cancer survivorship care: an overview of systematic reviews. Ann Oncol 2021;32(12):1552–70. Available from: https://doi.org/10.1016/j.annonc.2021.09.001.

[26] Buonanno P, Marra A, Iacovazzo C, et al. Telemedicine in cancer pain management: a systematic review and meta-analysis of randomized controlled trials. Pain Med 2023;24(3):226–33. Available from: https://doi.org/10.1093/pm/pnac128.

[27] Uemoto Y, Yamanaka T, Kataoka Y, et al. Efficacy of telemedicine using videoconferencing systems in outpatient care for patients with cancer: a systematic review and meta-analysis. JCO Clin Cancer Inform 2022;6:e2200084. https://doi.org/10.1200/CCI.22.00084.

[28] AlDossary SA, Martin-Khan M, Bradford N, et al. A systematic review of the methodologies used to evaluate telemedicine service initiatives in hospital facilities. Int J Med Inform 2017;97:171–94. Available from: https://doi.org/10.1016/j.ijmedinf.2016.10.012.

[29] Cruz-Lim EM, Co HC, Mendoza MJ, et al. Physicians' perceptions on the role of telemedicine in cancer care during and post-COVID-19 pandemic. Acta Med Philipp 2021;55(2):264–70. Available from: https://doi.org/10.47895/amp.v55i2.2836.
[30] Mathew S, Chamberlain C, Alvarez K, et al. Impact of a pharmacy-led pain management team on adults in an academic medical center. Hosp Pharm 2016;51(8):639–45. Available from: https://doi.org/10.1310/hpj5108-639.
[31] Koç ZE, Kaplan E, Tanrıverdi D. The effectiveness of telehealth programs on the mental health of women with breast cancer: a systematic review. J Telemed Telecare 2022;30(3):405–19. Available from: https://doi.org/10.1177/1357633x211069663.
[32] Banisafar A, Ahi Q, Mansouri A, et al. Comparison of the effectiveness of online narrative therapy and cognitive behavioral therapy on mental health and adherence to treatment in cancer patients with low psychological hardiness. J Assess Res Appl Couns 2023;5(5):100–7. Available from: https://doi.org/10.61838/kman.jarac.5.5.13.
[33] Gadzinski AJ, Dwyer EM, Reynolds J, et al. Interstate telemedicine for urologic cancer care. J Urol 2024;211(1):55–62. Available from: https://doi.org/10.1097/ju.0000000000003749.
[34] Matamala-Gomez M, Maisto M, Montana JI, et al. The role of engagement in teleneurorehabilitation: a systematic review. Front Neurol 2020;11:354. Available from: https://doi.org/10.3389/fneur.2020.00354.
[35] Dennett AM, Harding KE, Reimert J, et al. Telerehabilitation's safety, feasibility, and exercise uptake in cancer survivors: process evaluation. JMIR Cancer 2021;7(4):e33130. Available from: https://doi.org/10.2196/33130.
[36] Carpallo-Porcar B, Romo-Calvo L, Pérez-Palomares S, et al. Efficacy of an asynchronous telerehabilitation program in post-COVID-19 patients: a protocol for a pilot randomized controlled trial. PLoS One 2022;17(7):e0270766. Available from: https://doi.org/10.1371/journal.pone.0270766.
[37] Shahid S, Taylor EV, Cheetham S, et al. Key features of palliative care service delivery to indigenous peoples in Australia, New Zealand, Canada and the United States: a comprehensive review. BMC Palliat Care 2018;17(1):72. Available from: https://doi.org/10.1186/s12904-018-0325-1.
[38] Khairat S, Chourasia P, Muellers K, et al. Patient and provider recommendations for improved telemedicine user experience in primary care: a multi-center qualitative study. Telemed Rep 2023;4(1):21–9. Available from: https://doi.org/10.1089/tmr.2023.0002.
[39] Palanica A, Flaschner P, Thommandram A, et al. Physicians' perceptions of chatbots in health care: cross-sectional web-based survey. J Med Internet Res 2019;21(4):e12887. Available from: https://doi.org/10.2196/12887.
[40] Khandelwal S. Social companion chatbot for human communication using Ml and NLP. Int J Eng Appl Sci Technol 2023;8(1):321–4. Available from: https://doi.org/10.33564/ijeast.2023.v08i01.048.
[41] Kozikowski A, Shotwell J, Wool E, et al. Care team perspectives and acceptance of telehealth in scaling a home-based primary care program: qualitative study. JMIR Aging 2019;2(1):e12415. Available from: https://doi.org/10.2196/12415.
[42] Bunnell BE, Barrera JF, Paige SR, et al. Acceptability of telemedicine features to promote its uptake in practice: a survey of community telemental health providers. Int J Environ Res Public Health 2020;17(22):8525. Available from: https://doi.org/10.3390/ijerph17228525.
[43] Obaid OI, Salman SA. Security and privacy in IoT-based healthcare systems: a review. Mesopot J Comp Sci 2022;2022:29–39. Available from: https://doi.org/10.58496/mjcsc/2022/007.
[44] Sardi A, Rizzi A, Sorano E, et al. Cyber risk in health facilities: a systematic literature review. Sustainability 2020;12(17):7002. Available from: https://doi.org/10.3390/su12177002.
[45] Ha D, Nunnery MA, Klocko R, et al. Lung cancer survivors' views on telerehabilitation following curative intent therapy: a formative qualitative study. BMJ Open 2023;13(6):e073251. Available from: https://doi.org/10.1136/bmjopen-2023-073251.
[46] Fraser M, Gorely T, O'Malley C, et al. Does connected health technology improve health-related outcomes in rural cardiac populations? Systematic review narrative synthesis. Int J Environ Res Public Health 2022;19(4):2302. Available from: https://doi.org/10.3390/ijerph19042302.
[47] Tauben DJ, Langford DJ, Sturgeon JA, et al. Optimizing telehealth pain care after COVID-19. Pain 2020;161(11):2437–45. Available from: https://doi.org/10.1097/j.pain.0000000000002048.
[48] Chauhan HP, Patil K. Use of the wearable device in telehealth during COVID-19. J Pharm Res Int 2021:29–38. Available from: https://doi.org/10.9734/jpri/2021/v33i35b31895.
[49] Noorbakhsh-Sabet N, Zand R, Zhang Y, et al. Artificial intelligence transforms the future of health care. Am J Med 2019;132(7):795–801. Available from: https://doi.org/10.1016/j.amjmed.2019.01.017.
[50] Bi WL, Hosny A, Schabath MB, et al. Artificial intelligence in cancer imaging: clinical challenges and applications. CA Cancer J Clin 2019;69(2):127–57. Available from: https://doi.org/10.3322/caac.21552.
[51] Alrebdi N, Alabdulatif A, Iwendi C, et al. SVBE: searchable and verifiable blockchain-based electronic medical records system. Sci Rep 2022;12(1):266. Available from: https://doi.org/10.1038/s41598-021-04124-8.
[52] Hau YS, Lee HJ. A quantitative and qualitative review on the main research streams regarding blockchain technology in healthcare. Healthcare (Basel) 2021;9(3):266. Available from: https://doi.org/10.3390/healthcare9030247.
[53] Bienfait F, Petit MS, Pardenaud R, et al. Applying M-health to palliative care: a systematic review on the use of M-health in monitoring patients with chronic diseases and its transposition in palliative care. Am J Hosp Palliat Care 2019;37(7):549–64. Available from: https://doi.org/10.1177/1049909119885655.
[54] Eijlers R, Utens EMWJ, Staals LM, et al. Systematic review and meta-analysis of virtual reality in pediatrics: effects on pain and anxiety. Anesth Analg 2019;129(5):1344–53. Available from: https://doi.org/10.1213/ane.0000000000004165.
[55] Ambroggi M, Biasini C, Giovane CD, et al. Distance as a barrier to cancer diagnosis and treatment: review of the literature. Oncologist 2015;20(12):1378–85. Available from: https://doi.org/10.1634/theoncologist.2015-0110.
[56] Rodriguez JA, Saadi A, Schwamm LH, et al. Disparities in telehealth use among california patients with limited english proficiency. Health Aff (Millwood) 2021;40(3):487–95. Available from: https://doi.org/10.1377/hlthaff.2020.00823.

[57] Brunton L, Bower P, Sanders C. The contradictions of telehealth user experience in chronic obstructive pulmonary disease (COPD): a qualitative meta-synthesis. PLoS One 2015;10(10):e0139561. Available from: https://doi.org/10.1371/journal.pone.0139561.

[58] Leochico CFD, Espiritu AI, Ignacio SD, et al. Challenges to the emergence of telerehabilitation in a developing country: a systematic review. Front Neurol 2020;11:1007. Available from: https://doi.org/10.3389/fneur.2020.01007.

[59] Truong M, Yeganeh L, Cook O, et al. Using telehealth consultations for healthcare provision to patients from non-Indigenous racial/ethnic minorities: a systematic review. J Am Med Inform Assoc 2022;29(5):970–82. Available from: https://doi.org/10.1093/jamia/ocac015.

[60] Permatasari G, Basrowi RW. Telemedicine: a literature review on ethical, legal, and social issue (ELSI). Indones J Commun Occup Med 2023;3(2):123–9. Available from: https://doi.org/10.53773/ijcom.v3i2.99.123-9.

[61] Tedeschi C. Ethical, legal, and social challenges in the development and implementation of disaster telemedicine. Disaster Med Public Health Prep 2021;15(5):649–56. Available from: https://doi.org/10.1017/dmp.2020.118.

[62] Moghbeli F, Langarizadeh M. Application of ethics for providing telemedicine services and information technology. Acta Inform Med 2017;25(3):208. Available from: https://doi.org/10.5455/aim.2017.25.208-212.

[63] Andrade AO, Soares AB, Palis AA, et al. On the use of telemedicine in the context of COVID-19: legal aspects and a systematic review of technology. Res Biomed Eng 2021;38(1):209–27. Available from: https://doi.org/10.1007/s42600-021-00133-8.

[64] Kaplan B. Revisiting health information technology ethical, legal, and social issues and evaluation: telehealth/telemedicine and COVID-19. Int J Med Inform 2020;143:104239. Available from: https://doi.org/10.1016/j.ijmedinf.2020.104239.

[65] Těšinová JK, Dobiášová K, Dušek Z, et al. Development of telemedicine in the Czech Republic from patients' and other key stakeholders' perspective. Front Public Health 2023;11:1202182. Available from: https://doi.org/10.3389/fpubh.2023.1202182.

Chapter 30

Virtual reality and wearable devices

Kazuyuki Niki[1] and Ryouhei Ishii[2]

[1]*Department of Clinical Pharmacy Research and Education, Graduate School of Pharmaceutical Sciences, The University of Osaka, Suita, Osaka, Japan,* [2]*Department of Occupational Therapy, Graduate School of Rehabilitation Science, Osaka Metropolitan University, Osaka, Osaka, Japan*

Key learning objectives

Upon completion of this chapter, the reader will be able to:

- Identify the efficacy of virtual reality (VR) and wearable devices in palliative care.
- Understand the challenges and solutions in advancing the use of VR and wearable devices in palliative care.
- Identify the potential use of VR and wearable devices in palliative care.

Introduction

Virtual reality (VR) is a generic term for technology that works on the human sensory organs to artificially create an environment that is not real but feels substantially like reality. In recent years, as digital transformation (DX) progresses, VR is being used in various fields, including entertainment, education, and medicine. In the medical field, some VR content that has been approved by the FDA in the United States and the regulatory agencies in Europe is playing an important role in medical DX, along with smartphone applications. We have been promoting the medical application of VR mainly in palliative care [1].

On the other hand, wearable devices have also significantly impacted humanity by transforming health monitoring, social engagement, and human–computer interaction. These devices enable continuous health monitoring, which is particularly beneficial for chronic disease management and early detection of illnesses, as evidenced during the COVID-19 pandemic, where wearables helped identify patterns distinguishing healthy from infected individuals [2,3]. The integration of sensors in wearables allows for the tracking of vital signs, physical activity, and sleep, promoting a quantified self-culture that emphasizes personal improvement and social engagement through mechanisms like motivating empowerment and friendly rivalry [4,5]. Wearables also play a crucial role in augmented reality, enhancing user experiences and potentially altering human behavior by creating new interaction paradigms [6,7]. The convergence of wearable technology with fields like biotechnology and intelligent textiles suggests a future where wearables are seamlessly integrated into healthcare, including palliative care [8].

This chapter explores the application of both VR and wearable devices in palliative care, technologies that offer distinct yet potentially complementary approaches to symptom management, particularly pain. VR technologies often focus on providing acute, situational relief by leveraging immersive experiences to achieve effects like distraction from pain or generation of positive affective states. In contrast, wearable devices excel at enabling continuous, longitudinal monitoring of physiological parameters (e.g., vital signs, activity levels, and surface electromyogram [sEMG] signals related to pain) over extended periods. This continuous data stream can inform ongoing pain assessment, track treatment efficacy, and potentially predict exacerbations. Understanding these differing operational timescales and mechanisms—VR for acute intervention and wearables for longitudinal tracking and assessment—is crucial for appreciating their individual contributions and synergistic potential within integrated palliative care strategies. In this section, we will introduce a

Pain Management in Palliative Care for Patients With Cancer. DOI: https://doi.org/10.1016/B978-0-443-30050-9.00005-9

brief summary about previous studies on VR and wearable devices in palliative care and the findings and challenges we have obtained so far, as well as our prospects.

Efficacy of virtual reality in patients receiving palliative care

Although it is not uncommon for patients with terminal cancer to wish to "go back to a memorable place" or "go home," it can be difficult to achieve this goal for a variety of reasons [9]. We believe that if VR could be used to fulfill patients' wishes, even in a simulated manner, it would lead to an improvement in their quality of life. Therefore, we started using VR and conducted a clinical trial to verify its efficacy [1].

In this clinical trial, patients with terminal cancer admitted to palliative care units at two hospitals in Japan who gave their consent were given simulated outings to places of their choice using VR travel experience software (Google Earth VR). In principle, the patients operated the VR software, but medical staff assisted them as needed. Fig. 30.1 shows the patient during VR travel. Before and after VR travel, the Edmonton Symptom Assessment System–revised Japanese version (ESAS-r-J) [10] and the Numeric Rating Scale (NRS) for dizziness, headache, fun, and happiness were evaluated. ESAS is a rating instrument developed to help assess the nine symptoms frequently experienced by patients eligible for palliative care (pain, tiredness, drowsiness, nausea, lack of appetite, shortness of breath, depression, anxiety, and well-being) on an 11-point scale from 0 (no symptom) to 10 (worst symptom) for each symptom. ESAS-r-J is a completed validation of ESAS in Japanese. A total of 20 patients participated in the clinical trial. The mean age of patients was 72.3 years, and 18 (90%) patients had an Eastern Cooperative Oncology Group Performance Status ≥3. Significant improvements were observed before and after the VR travel with respect to pain, tiredness, drowsiness, shortness of breath, depression, anxiety, well-being, fun, and happiness. One patient each had a slight worsening of symptoms after VR travel for tiredness, drowsiness, nausea, and dizziness, but their post-VR scores were all mild (NRS ≤ 3), and no treatment such as antiemetic medication was necessary.

These results suggest that VR may temporarily improve symptoms in patients with terminal cancer without causing serious adverse reactions. In this study, the item with the largest effect size was depression, followed by anxiety, which also showed high values; thus VR travel may have had a particular impact on the mental health aspect. The effects of VR-based treatments for mental illness have been reported in multiple studies. VR has been studied across several psychiatric contexts, including persecutory delusions [11]. A feasibility study of a VR-based stress-management program in people with mood disorders reported improvements in stress-related variables, depression, and anxiety [12]. There is also a report [13] that an intervention using VR for depressed patients with a tendency toward self-criticism improved depression and self-criticism. What is noteworthy about this study is that in addition to improving mental symptoms, pain was also reduced by VR travel. These results and reports that the medial temporal lobe, including the hippocampus, is activated by "mental time travel (MTT)," which recalls pleasant memories from the past [14], and that activation of the hippocampus by MTT may be involved in the suppression of depressive-like behavior [15], may suggest that the activation of the reward system in the brain by the external stimulus of VR travel may have improved psychiatric symptoms and pain.

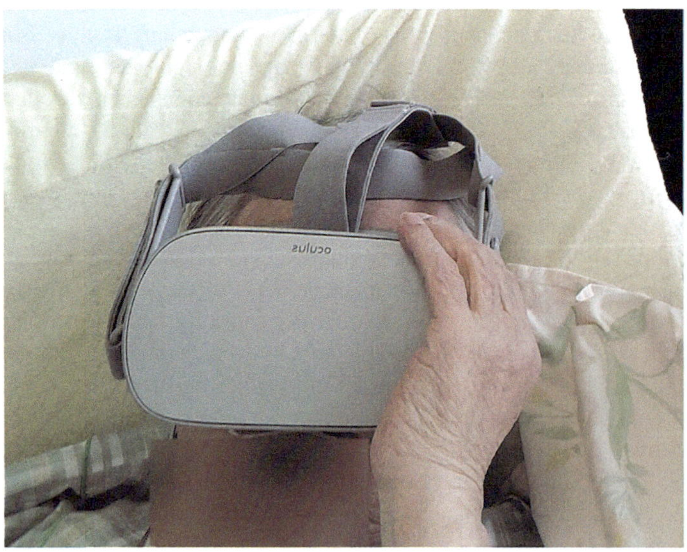

FIGURE 30.1 A patient using virtual reality in a palliative care setting.

With regard to VR-induced pain reduction, the gate-control theory is considered to be one of the possible mechanisms. That is, it has been reported that the immersive sensation created by VR distracts the patient from the stimulus (distraction effect), thereby increasing tolerance to painful stimuli [16–18]. A study compared the importance of the sense of immersion between 2D images and 360-degrees VR. Tashjian et al. [19] conducted a nonrandomized cohort study in which 100 patients with an NRS of 3 or higher were divided into two groups: 50 in the 2D group using a display and 50 in the VR group using an immersive VR head-mounted display. The intervention consisted of a single 15-minute session for both groups, with the 2D group viewing nature videos and the VR group experiencing the game named Pain RelieVR. Changes in pain before and after the intervention were assessed by NRS and it was found that pain was significantly reduced in the VR group compared to the 2D group (VR vs 2D group; change: –1.3 vs –0.6). Furthermore, 20 people (40%) in the 2D group showed a reduction in pain, while the number of people in the VR group was significantly higher, at 35 people (70%). Thus the sense of immersion created by VR is an important factor contributing to pain reduction. However, future studies from a mechanistic perspective are needed. Currently, we are working to resolve issues related to the use of VR for palliative care, which will be presented in the next section.

Potential applications: challenges and proposed solutions to the use of virtual reality in palliative care

The following are some questions we have received and challenges we have encountered while promoting the use of VR, some of which we are finding solutions to, and we present them together here.

Question 1: Which application should I choose?

Answer: There are applications for various genres and prices, but they all have reviews, so we try to choose from those with high ratings. Regardless of price, some free applications can be useful. While it is best to ask in advance what the patient would like and select applications in advance, it is also useful to prepare multiple genres (healing, travel, sports, etc.) so that we can respond to requests on the spot. In addition, since elderly patients are often bewildered by VR operations, we use applications that can be experienced simply by putting on VR goggles or applications that support multiplayer play (which we can operate on their behalf from outside) so that patients can enjoy VR even if they cannot operate it themselves.

Question 2: Are there any concerns about nausea or delirium?

Answer: Nausea, known as VR motion sickness, is the most worrisome adverse reaction; however, it can be addressed by selecting good image quality and adjusting the duration of the experience (less than 30 minutes). In addition, if patients experience nausea during the VR experience, they can recover by removing the goggles and resting for a while. Regarding delirium, we have not experienced any cases in which delirium appeared or worsened because of the VR experience, nor has it been reported in recent overseas studies [20,21].

Challenge 1: Dealing with the difficulty of VR operation and the weight of VR goggles.

It is desirable for elderly people and busy medical professionals to be able to use the VR system with intuitive and simple operations. In addition, although today's main VR goggles weigh around 500 g, many patients feel that they are heavy.

Solution: This challenge can be improved by using VR applications that are easy to operate (e.g., 360-degree video-viewing applications such as YouTube VR, world travel applications such as Wander, etc.) and head straps that adjust the goggles' center of gravity. However, modern VR goggles are developed primarily for gamers, and they have the disadvantage that they have multiple control buttons, which are difficult to operate until some degree of familiarity is achieved. In 2020, we introduced VR goggles that we thought were easy to use in palliative care [22]; however, the speed of VR goggles development since then has been extremely fast, and Oculus Go, the VR goggles we introduced as the easiest to use at the time, with few buttons and easy operation, was discontinued at the end of 2020. Therefore, we now use successor VR goggles such as Meta Quest 3.

Challenge 2: Realization of real-time communication in VR space.

When a patient requests a VR travel destination that cannot be viewed with the commercially available world travel application, such as within the patient's own home, we ask the patient's family to lend a 360-degree camera to take pictures, which are then viewed in the patient's hospital room as VR images. We frequently received requests to have conversations with people in the VR space; however, this was not possible with the conventional communication environment (4G) due to a time lag.

Solution: Real-time VR travel can now be realized by utilizing platforms that support high-speed communications such as 5G, which has become popular in recent years. However, the disadvantage is that there are very few such platforms, and they are very expensive to use. However, we have used it several times and feel that the ability to intuitively create the sensation of "being right there at the moment" is a major advantage over other platforms [22,23]. Now that the use of VR and the metaverse is becoming more widespread, we will discuss our thoughts on their future prospects in medicine in the next section.

Future implications: prospects for the future use of virtual reality and metaverse in medicine

With the current advancement of medical DX after the experience of the COVID-19 pandemic, it is expected that VR will be utilized more and more in the future. At present, the definition of the metaverse remains unclear, and in practice it is often treated as synonymous with VR. However, mutual interaction may be central to future metaverse-based approaches to distraction-related pain relief, especially if simple systems can be developed for older adults. Although this chapter focuses on VR in palliative care, VR and metaverse-based approaches may also be applicable in other settings, including noncancer pain and psychiatric disorders.

Enhancing palliative care through wearable devices

While the previous section focused on specific applications of VR based on our clinical experience, this section provides a broader overview of the diverse and rapidly developing field of wearable devices in palliative care. Wearable devices have emerged as a significant technological advancement in healthcare, offering capabilities for uninterrupted monitoring and data collection. In palliative care, these devices hold potential for improving patient outcomes by enabling remote monitoring, personalized care, and timely interventions. Wearable devices significantly enhance the quality of life for palliative care patients by providing continuous, noninvasive monitoring, which allows for timely interventions and personalized care.

Wearable devices have shown promise in palliative care, particularly for older patients, by enabling patient-centered care, treatment adjustments, and symptom management [24]. They have also been found to be feasible and accepted by patients, with the potential to predict health status deterioration and unplanned readmissions [25,26]. The use of wearable smartwatch technology has been explored to monitor symptoms in advanced illness [27], and a remote monitoring system has been found to be usable and acceptable in patients receiving palliative care at home [28]. Various types of wearable devices have already been implemented for various purposes and applications, and the following typology can be used to describe the most representative ones.

1. Remote monitoring and symptom management:
   Wearable devices enable ongoing observation of vital signs and physical activity, allowing healthcare providers to adjust treatments and manage symptoms more effectively [24,25,29,30]. These devices help in reducing the need for patients to travel to clinics, thus minimizing physical strain and improving quality of life [24,25].
2. Prognostic tools:
   Wearable devices can serve as prognostic tools by providing real-time data that helps in predicting patient outcomes and planning advanced care [26,29]. Data from wearables can correlate with clinical scales like the Karnofsky Performance Status and Palliative Performance Scale, aiding in better prognostication [26].
3. Patient-centered care:
   Wearables facilitate patient-centered care by enabling personalized treatment plans based on continuous data collection [24,31,32]. They support ecological momentary assessments (EMAs), which can capture real-time patient-reported outcomes and stress levels, enhancing personalized care [32].
4. Reduction in unplanned readmissions:

Persistent surveillance through wearables can help predict and prevent health status deterioration, thereby reducing unplanned hospital readmissions and emergency visits [29].

Studies indicate that wearable devices are feasible for use in palliative care settings, with high adherence rates and positive patient perceptions [29,32]. Patients and caregivers find value in the support provided by wearables, which can improve the overall care experience [32]. Despite these benefits, the potential role of digital health technology in palliative care, including wearable devices, is still being debated.

In the chapters that follow, we will delve into more specific examples of wearable devices in palliative care, their impact on clinical practice, and their future prospects.

Wearable devices, such as accelerometers, actigraphy devices, and smart patches, enable uninterrupted tracking of vital signs and physical activities, which is crucial for timely treatment adjustments and symptom control, particularly for conditions like dyspnea [24]. They facilitate remote monitoring, reducing the need for frequent clinic visits and allowing healthcare providers to make informed decisions based on real-time data, thus improving patient-centered care [24]. Wearable technology combined with deep learning models has shown higher accuracy in predicting survival outcomes for end-stage cancer patients compared to traditional prognostic tools, aiding in better decision-making and personalized care plans [34]. Additionally, wearable devices can trigger EMAs and electronic patient-reported outcomes, which are feasible and valued by patients and caregivers, enhancing the monitoring of stress and sleep disturbances [30]. They also help in predicting unplanned hospital readmissions by monitoring biomarkers like heart rate variability and step speed, which are significant indicators of health deterioration [25]. The use of wearable devices is associated with improved mental health, as they help reduce severe depression and anxiety levels among elderly patients [35]. Moreover, these devices support the transition from hospital to home care by providing ongoing assessment, which is essential for preventing rapid health deterioration and managing cancer-related fatigue [32]. Despite challenges like data security and device interoperability, the integration of wearable technology in palliative care holds promise for enhancing patient–professional relationships and ensuring holistic, dignified care [36]. Overall, wearable devices represent a significant advancement in palliative care, offering a blend of technological innovation and compassionate care.

Additionally, wearable devices equipped with sensors can measure pain-related physiometric signals and physical activity, providing objective data that can be correlated with pain scores to improve pain assessment and treatment [37]. Moreover, remote monitoring using wearables has shown feasibility and high patient acceptance, with devices like sensor-equipped bracelets and smartphones being used to track vital signs and activity levels, which can predict unplanned hospital readmissions and emergency visits [25,38]. These devices also facilitate the transition from hospital to home care by enabling continuous monitoring, thus reducing the need for frequent clinic visits and allowing for timely interventions when symptoms deteriorate [32].

Furthermore, recent advancements in wearable technology have opened up new possibilities for continuous pain monitoring using sEMG. The use of sEMG for pain assessment is particularly relevant in palliative care settings, where patients may have limited ability to self-report pain due to cognitive impairment, sedation, or other factors. Wearable devices integrated into Internet of Things (IoT) systems can monitor pain intensity through facial expressions using sEMG, providing an alternative to self-reporting, especially for patients unable to communicate their pain levels [39]. A feasibility study by Mieronkoski et al. demonstrated that sEMG combined with supervised machine learning methods can predict pain intensity from facial muscle activity, supporting the potential for objective, nonverbal pain assessment that could be valuable in palliative care settings where patients may be unable to reliably self-report pain [40]. While their initial results suggested potential in distinguishing between pain and nonpain states based on facial analysis, this evidence regarding automated systems incorporating sEMG for quantifiable pain assessment is still considered preliminary. Specific accuracy metrics were not detailed in the cited feasibility study, and larger studies are needed to rigorously quantify the sensitivity and specificity and validate the clinical utility of sEMG-based automated pain monitoring systems in end-of-life care. In other studies, researchers collected facial sEMG signals along with other physiological parameters from multiple facial muscles to explore continuous pain intensity monitoring using machine learning classification methods [41]. Further potential development of wearable systems measuring sEMG could have the potential to revolutionize pain management in palliative care by providing real-time, objective pain assessments that can inform treatment decisions and improve quality of life for patients.

Future directions for this technology include the development of more sophisticated algorithms that can distinguish between different types of pain (e.g., acute vs chronic) and the integration of sEMG data with other physiological and behavioral measures for a more comprehensive pain assessment [42]. As these technologies continue to evolve, they have the potential to significantly enhance pain management strategies in palliative care, ultimately improving patient comfort and quality of life during end-of-life care.

Overall, the integration of wearable technology in palliative care enhances patient-centered care, improves symptom management, and supports proactive health management.

Conclusions

Appropriate use of VR can be expected to temporarily improve physical and mental symptoms of patients with terminal cancer. Wearable devices offer significant benefits in palliative care by enabling continuous remote monitoring, improving symptom management, and supporting personalized care. They serve as valuable prognostic tools and help reduce the need for frequent clinic visits, thus enhancing the quality of life for patients. The feasibility and positive reception of these devices among patients and caregivers further underscore their potential in transforming palliative care practices. Improvement of specifications and systems, and further efficacy verification, are required in anticipation of the promotion of medical DX and the spread of the metaverse.

References

[1] Niki K, Okamoto Y, Maeda I, et al. A novel palliative care approach using virtual reality for improving various symptoms of terminal cancer patients: a preliminary prospective, multicenter study. J Palliat Med 2019;22(6):702–7. Available from: https://doi.org/10.1089/jpm.2018.0527.

[2] Yu S, Chen Z, Wu X. The impact of wearable devices on physical activity for chronic disease patients: findings from the 2019 Health Information National Trends Survey. Int J Environ Res Public Health 2023;20(1):887. Available from: https://doi.org/10.3390/ijerph20010887.

[3] Cosoli G, Scalise L, Poli A, et al. Wearable devices as a valid support for diagnostic excellence: lessons from a pandemic going forward. Health Technol (Berl) 2021;11(3):583–8. Available from: https://doi.org/10.1007/s12553-021-00540-y.

[4] Akdevelioglu D, Hansen S, Venkatesh A. Wearable technologies, brand community and the growth of a transhumanist vision. J Mark Manag 2021;38(5-6):569–604. Available from: https://doi.org/10.1080/0267257X.2021.2001559.

[5] Ajin Roch A, Karthik S, Arthi R. Dynamic programmable clock frequency using machine learning algorithms to reduce power consumption in wearables. Adv Intell Syst Comput 2021;1172:189–98. Available from: https://doi.org/10.1007/978-981-15-5495-7_19.

[6] Gaziani SA. The impacts of use of augmented reality wearable devices on human behaviour: a user perspective. EasyChair Prepr 2019;1(1):1–10.

[7] Rinaldi A, Becchimanzi C, Tosi F. Wearable devices and smart garments for stress management. Adv Intell Syst Comput 2018;795:1184–91. Available from: https://doi.org/10.1007/978-3-319-96071-5_92.

[8] Birringer J, Danjoux M. Wearable technology for the performing arts. Wearable Technol Med Health Care 2023:85–102. Available from: https://doi.org/10.1016/b978-0-12-819526-0.00005-9.

[9] Yamagishi A, Morita T, Miyashita M, et al. Pain intensity, quality of life, quality of palliative care, and satisfaction in outpatients with metastatic or recurrent cancer: a Japanese, nationwide, region-based, multicenter survey. J Pain Symptom Manage 2012;43(3):503–14. Available from: https://doi.org/10.1016/j.jpainsymman.2011.04.025.

[10] Yokomichi N, Morita T, Nitto A, et al. Validation of the Japanese version of the Edmonton Symptom Assessment System-Revised. J Pain Symptom Manage 2015;50(5):718–23. Available from: https://doi.org/10.1016/j.jpainsymman.2015.05.014.

[11] Freeman D, Bradley J, Antley A, et al. Virtual reality in the treatment of persecutory delusions: randomised controlled experimental study testing how to reduce delusional conviction. Br J Psychiatry 2016;209(1):62–7. Available from: https://doi.org/10.1192/bjp.bp.115.176438.

[12] Shah LBI, Torres S, Kannusamy P, et al. Efficacy of the virtual reality-based stress management program on stress-related variables in people with mood disorders: the feasibility study. Arch Psychiatr Nurs 2015;29(1):6–13. Available from: https://doi.org/10.1016/j.apnu.2014.09.003

[13] Falconer CJ, Slater M, Rovira A, et al. Embodying self-compassion within virtual reality and its effects on patients with depression. BJPsych Open 2016;2(1):74–80. Available from: https://doi.org/10.1192/bjpo.bp.115.002147.

[14] Milner B, Squire LR, Kandel ER. Cognitive neuroscience and the study of memory. Neuron 1998;20(3):445–68. Available from: https://doi.org/10.1016/s0896-6273(00)80987-3.

[15] Ramirez S, Liu X, Redondo RL, et al. Activating positive memory engrams suppresses depression-like behaviour. Nature 2015;522:335–9. Available from: https://doi.org/10.1038/nature14514.

[16] Wismeijer AAJ, Vingerhoets A. The use of virtual reality and audiovisual eyeglass systems as adjunct analgesic techniques: a review of the literature. Ann Behav Med 2005;30(3):268–78. Available from: https://doi.org/10.1207/s15324796abm3003_11.

[17] Hoffman HG, Patterson DR, Carrougher GJ, et al. The effectiveness of virtual reality pain control with multiple treatments of longer durations: a case study. Int J Hum-Comp Interact 2001;13(1):1–12. Available from: https://doi.org/10.1207/s15327590ijhc1301_1.

[18] Rutter CE, Dahlquist LM, Weiss KE. Sustained efficacy of virtual reality distraction. J Pain 2009;10(4):391–7. Available from: https://doi.org/10.1016/j.jpain.2008.09.016.

[19] Tashjian VC, Mosadeghi S, Howard AR, et al. Virtual reality for management of pain in hospitalized patients: results of a controlled trial. JMIR Ment Health 2017;4(1):e9. Available from: https://doi.org/10.2196/mental.7387.

[20] Ferguson C, Shade MY, Blaskewicz Boron J, et al. Virtual reality for therapeutic recreation in dementia hospice care: a feasibility study. Am J Hosp Palliat Care 2020;37(11):809–15. Available from: https://doi.org/10.1177/1049909120901525.

[21] Chirico A, Lucidi F, De Laurentiis M, et al. Virtual reality and music therapy as distraction interventions to alleviate anxiety and improve mood states in breast cancer patients during chemotherapy. J Cell Physiol 2020;235(6):5353–62. Available from: https://doi.org/10.1002/jcp.29422.

[22] Niki K, Okamoto Y, Ueda M. Response to Wang et al., virtual reality as a bridge in palliative care during COVID-19. J Palliat Med 2020;23(6):756–7. Available from: https://doi.org/10.1089/jpm.2020.0261.

[23] Niki K, Egashira S, Okamoto Y. A real-time virtual outing using virtual reality for a hospitalized terminal cancer patient who has difficulty going out: a case report. Front Virtual Real 2024;5(1):1269707. Available from: https://doi.org/10.3389/frvir.2024.1269707.

[24] Rada Sandic Spaho L, Uhrenfeldt L, Fotis T, et al. Wearable devices in palliative care for people 65 years and older: a scoping review. Digit Health 2023;9:20552076231181212. Available from: https://doi.org/10.1177/20552076231181212.

[25] Pavic M, Klaas VC, Theile G, et al. Mobile health technologies for continuous monitoring of cancer patients in palliative care aiming to predict health status deterioration: a feasibility study. J Palliat Med 2019;23(5):678–85. Available from: https://doi.org/10.1089/jpm.2019.0001.

[26] Huang Y, Kabir MA, Upadhyay U, et al. Exploring the potential use of wearable devices as a prognostic tool among patients in hospice care. Medicina (Kaunas) 2022;58(12):1824. Available from https://doi.org/10.3390/medicina58121824.

[27] Nwosu A, Quinn C, Samuels J, et al. Wearable smartwatch technology to monitor symptoms in advanced illness. BMJ Support Palliat Care 2018;8(2):237. Available from: https://doi.org/10.1136/bmjspcare-2017-001445.

[28] McCall K, Keen J, Farrer K, et al. Perceptions of the use of a remote monitoring system in patients receiving palliative care at home. Int J Palliat Nurs 2008;14(9):426–31. Available from: https://doi.org/10.12968/ijpn.2008.14.9.31121.

[29] Krey M, Mahadevan DA, Derungs K, et al. Wearable device technology in healthcare: exploring constraining and enabling factors. In: Proceedings of the 2020 international conference on healthcare informatics. Available from https://doi.org/10.1007/978-981-15-0637-6_1; 2020. p. 1–10.
[30] Schuler T, Back M, Clark K, et al. Wearable-triggered ecological momentary assessments are feasible in people with advanced cancer and their family caregivers: feasibility study from an outpatient palliative care clinic at a cancer center. J Palliat Med 2023;26(3):378–85. Available from: https://doi.org/10.1089/jpm.2022.0535.
[31] Hughes A, Shandhi M, Master H, et al. Wearable devices in cardiovascular medicine. Circ Res 2023;132(3):351–70. Available from: https://doi.org/10.1161/CIRCRESAHA.122.322389.
[32] Klaas VC. Monitoring outpatients in palliative care through wearable devices. 2018 [doctoral thesis]. Zurich: ETH Zurich. Available from: https://doi.org/10.3929/ethz-b-000299669.
[33] Mallick MK, Biser S, Haridas A, et al. Improving dyspnoea symptom control of patients in palliative care using a smart patch-a proof of concept study. Front Digit Health 2021;3:765867. Available from: https://doi.org/10.3389/fdgth.2021.765867.
[34] Bonacaro A, Rubbi I, Sookhoo D, et al. The use of wearable devices in preventing hospital readmission and improving the quality of life of chronic patients in the homecare setting: a narrative literature review. Prof Inferm 2019;72(2):143–51.
[35] Elkefi S, Asan O. Wearable devices' use in geriatric care between patient-centeredness and psychology of patients. Proc Int Symp Hum Factors Ergonom Health Care 2022;11(1):25–9. Available from: https://doi.org/10.1177/2327857922111025.
[36] Payne S, Tanner M, Hughes S. Digitisation and the patient-professional relationship in palliative care. Palliat Med 2020;34(4):441–3. Available from: https://doi.org/10.1177/0269216320911501.
[37] Leroux A, Rzasa-Lynn R, Crainiceanu CM, et al. Wearable devices: current status and opportunities in pain assessment and management. Digit Biomark 2021;5(1):89–102. Available from: https://doi.org/10.1159/000515576.
[38] Pavic M, Klaas VC, Theile G, et al. Feasibility and usability aspects of continuous remote monitoring of health status in palliative cancer patients using wearables. Oncology 2020;98(6):386–95. Available from: https://doi.org/10.1159/000507453.
[39] Yang G, Jiang M, Ouyang W, et al. IoT-based remote pain monitoring system: from device to cloud platform. IEEE J Biomed Health Inform 2018;22(6):1711–9. Available from: https://doi.org/10.1109/JBHI.2017.2776351.
[40] Mieronkoski R, Syrjälä E, Jiang M, et al. Developing a pain intensity prediction model using facial expression: A feasibility study with electromyography PLoS One 2020;15(7):e0235545. Available from: https://doi.org/10.1371/journal.pone.0235545.
[41] Jiang M, Mieronkoski R, Syrjälä E, et al. Acute pain intensity monitoring with the classification of multiple physiological parameters. J Clin Monit Comput 2019;33(3):493–507. Available from: https://doi.org/10.1007/s10877-019-00351-7.
[42] Lopez-Martinez D, Picard R. Multi-task neural networks for personalized pain recognition from physiological signals. In: 2017 Seventh international conference on affective computing and intelligent interaction workshops and demos (ACIIW), San Antonio, TX, USA; 2017, p. 181-4. Available from: https://doi.org/10.1109/ACIIW.2017.8272611.

Chapter 31

Nutrition in pain

Nataša Skočibušić[1,2,*] and Gordana Kenđel Jovanović[2,3,*]

[1]*Department of Clinical Pharmacology, University Hospital Center Rijeka, Rijeka, Croatia,* [2]*Faculty of Medicine, University of Rijeka, Rijeka, Croatia,* [3]*Department of Health Ecology, Teaching Institute of Public Health of Primorsko-goranska County, Rijeka, Croatia*

Key learning objectives

- Understanding nutritional support: Learning how nutritional interventions can impact pain management in cancer patients.
- Identifying nutrient deficiencies: Recognizing common nutrient deficiencies in cancer patients and their potential contribution to pain.
- Incorporating diet modifications: Understanding how dietary modifications can supplement conventional pain management strategies.
- Assessing individual needs: Learning to assess individual nutritional needs based on cancer type, treatment, and patient-specific factors.
- Collaborative care: Understanding the importance of a multidisciplinary approach involving nutritionists and cancer and pain management specialists (other specialists, psychologists, nurses, physiotherapists, and pharmacologists).

Introduction

Cancer treatments can alter patients' nutritional needs, leading to changes in eating patterns that may negatively affect digestion and food consumption. Chronic pain in patients on long-term opioid therapy may also lead to unhealthy eating habits, further exacerbating inflammation and neurosensitivity [1,2]. Nutrition interventions could potentially have a valuable role in managing pain by targeting pathways involved in inflammation and immune modulation. For example, a low-fat, low-sugar diet may reduce oxidative stress and prevent glial cell activation, contributing to pain management [2]. Certain nutrients, particularly plant-based phytonutrients, have antiinflammatory effects that may benefit cancer patients. These include anthocyanins, tannins, flavonoids, carotenoids, lycopene, and caffeine, which can modulate inflammation through various mechanisms and help regulate immune responses [3–5]. Foods rich in these nutrients include fruits, vegetables, whole grains, seeds, nuts, and beverages like tea and coffee. Additionally, these foods are high in dietary fiber, which supports gut microbiota health and the production of short-chain fatty acids with antiinflammatory properties [6,7]. One evidence-based dietary approach for improving cancer outcomes is the Mediterranean diet, which has been associated with lower cancer mortality and morbidity [8]. Rather than focusing on individual nutrients, this whole-diet approach could be more suitable for the majority of patients with cancer [9,10]. Incorporating behavior change strategies into nutrition interventions is essential for overcoming barriers to healthy eating in cancer patients. The behavior change model, which considers environment, policy, regulation, and clinical interventions, helps set specific dietary goals and facilitates successful dietary behavior change [11]. Nutrition professionals should use this model to support cancer patients in making sustainable dietary improvements, ultimately enhancing pain management and quality of life. A multidisciplinary approach, integrating personalized nutrition assessments and behavior change strategies, is essential for implementing effective nutrition interventions. Healthcare providers should collaborate to offer evidence-based guidance and continuous support for cancer patients in managing pain and improving their overall well-being (Table 31.1).

* Both authors equally contributed to this chapter.

Pain Management in Palliative Care for Patients With Cancer. DOI: https://doi.org/10.1016/B978-0-443-30050-9.00019-9

Mediterranean diet

- Fruits and vegetables: a wide variety, such as leafy greens, tomatoes, peppers, onions, berries, and seasonal fruits.
- Whole grains: whole wheat, brown rice, quinoa, oats, and barley.
- Healthy fats: primarily from olive oil, nuts, and seeds (e.g., almonds, walnuts, flax seeds, and chia seeds).
- Fish and seafood: fatty fish like salmon, mackerel, sardines, and shellfish.
- Legumes: lentils, chickpeas, beans, and green beans.
- Dairy: low-fat or fermented dairy products like Greek yogurt and cottage cheese.
- Nuts and seeds: walnuts, almonds, sunflower seeds, and flaxseeds.
- Herbs and spices: basil, oregano, thyme, rosemary, garlic, and cinnamon (used for reducing salt).
- Red meat: consumed occasionally in smaller portions.
- Water: the primary beverage for hydration.
- Wine: occasionally, in moderation, typically with meals (especially red wine, rich in polyphenols).

Impact of nutrition on pain management

The link between nutrition and pain management in cancer patients is complex. Tailored nutrition strategies targeting inflammation modulation, including education, could potentially help alleviate pain and improve the effectiveness of pain relief medications. Common challenges, such as mucositis, dysgeusia, nausea, vomiting, constipation, bloating, and dyspnea, require specific dietary interventions to alleviate discomfort, promote adequate nutrition, and significantly improve patient outcomes [12].

Mucositis

Mucositis, a painful inflammation of the mouth and gastrointestinal tract mucous membranes, often results from chemotherapy or radiation therapy, leading to ulcerations and infections [13] that severely hinder a patient's ability to eat and drink, resulting in malnutrition and dehydration. To reduce discomfort, patients should consume soft, bland foods like smoothies, soups, yogurt, and pureed vegetables, avoiding spicy, acidic, or rough foods [14]. Adequate hydration with nonacidic fluids is essential in managing mucositis. Recent studies suggest that natural honey may alleviate radiation-induced mucositis severity [15], although findings vary by study design, patient populations, and honey types. Potential risks, including hyperglycemia, should be considered, particularly in patients with diabetes or those receiving corticosteroids. Further research is necessary to standardize protocols and clarify honey's role in mucositis management [16].

TABLE 31.1 Antiinflammatory foods and key nutrients with antiinflammatory potential.

Food category	Examples	Key nutrients
Fruits	Berries (blueberries, strawberries), apples, oranges	Anthocyanins, flavonoids, vitamin C
Vegetables	Leafy greens (arugula, spinach, kale), broccoli, carrots	Carotenoids, phenolic acids, fiber
Whole grains	Brown rice, quinoa, oats, whole wheat	Fiber, antioxidants
Nuts and seeds	Walnuts, almonds, flaxseeds, chia seeds	Omega-3 fatty acids, fiber
Herbs and spices	Turmeric, ginger, garlic, cinnamon, black pepper	Curcumin, phenolic compounds, flavonoids, terpenes, antioxidants
Beverages	Green tea, black tea, coffee	Flavonoids, polyphenols
Legumes	Lentils, chickpeas, beans	Fiber, antioxidants, protein
Healthy fats	Olive oil, avocado	Omega-3 fatty acids, monounsaturated fats
Fish	Fatty fish (salmon, mackerel, sardines)	Omega-3 fatty acids (EPA, DHA)

DHA, Docosahexaenoic acid; *EPA*, eicosapentaenoic acid.

Glutamine supplementation has shown potential in reducing chemotherapy-induced mucositis and supporting mucosal healing. A suggested method involves rinsing with an aqueous solution containing 10 g of glutamine powder for 30–60 seconds, three times daily [17,18]. It is essential to recognize that gargling a glutamine solution is not the same as systemic supplementation, as it does not result in significant absorption. However, caution is advised for patients with renal impairment due to potentially elevated glutamine levels and for those with diabetes due to possible impacts on glucose metabolism. Further research is needed to confirm its safety and effectiveness in these groups [19,20].

Dysgeusia

Dysgeusia, often caused by chemotherapy, distorts taste and decreases appetite [21]. To manage this, patients can incorporate aromatic herbs, spices, cold foods, and flavor-enhancing sauces [22]. Trying various food combinations may also help find tolerable options during treatment.

Nausea and vomiting

Nausea and vomiting, common side effects of chemotherapy, can cause weight loss and dehydration [23]. Eating small, frequent meals, using ginger [23,24], and avoiding greasy or fried foods can help reduce symptoms. Staying hydrated with clear fluids is essential for preventing dehydration.

Constipation

Chemotherapy and opioid analgesics often cause constipation, worsening patients' quality of life [25]. Increasing dietary fiber from fruits, vegetables, and whole grains, along with proper hydration, is recommended, and tolerable physical activity can further promote gastrointestinal motility [26].

Bloating

Bloating, exacerbated by chemotherapy and medications, can negatively affect quality of life. FODMAPs—fermentable oligosaccharides, disaccharides, monosaccharides, and polyols—are short-chain carbohydrates that can be poorly absorbed in the small intestine, leading to fermentation by gut bacteria and subsequent gas production. Hence, a low-FODMAP diet may alleviate bloating [27]. Smaller, more frequent meals can reduce gastric distension and discomfort. Hot-water infusions of ginger, peppermint, or chamomile, and fermented foods containing probiotics are commonly used to improve digestion and reduce bloating, although robust evidence in cancer patients is limited [28,29]. Adequate hydration is essential, and carbonated beverages should be avoided [30,31].

Dyspnea

Dyspnea, or difficulty breathing, is common in cancer patients, especially in advanced stages. Adjusting the diet by increasing fat intake while reducing carbohydrates could potentially alleviate the respiratory burden [32,33]. A macronutrient ratio of 50%–55% fat, 15%–20% protein, and 30% carbohydrates may be used to help support energy needs and reduce carbon dioxide production.

Pharmacotherapy and nutritional support in palliative care

Pain can significantly impact nutritional intake, particularly in palliative care and at the end-of-life stage, when the focus shifts from aggressive nutritional interventions to comfort. Nutritional strategies aim to enhance quality of life rather than prolong it. Patients often experience reduced appetite, altered metabolism, and difficulties with digestion. Small, frequent meals or snacks of easy-to-digest foods that patients enjoy and nutrient-dense foods (e.g., shakes, mashed avocado) provide essential energy without overwhelming the body. Hydration through sips of water, ice chips, or flavored drinks helps maintain comfort and prevent dehydration [34]. In patients with advanced cancer, the use of percutaneous endoscopic gastrostomy (PEG) tubes may be necessary to ensure adequate nutritional intake when oral feeding is not feasible due to pain or other complications [34]. The management of nutrition in these patients requires careful consideration of their pain management regimen, as certain medications such as opioids may affect gastrointestinal function and nutrient absorption [35]. A multidisciplinary approach, integrating pharmacotherapy with tailored nutritional support, ensures the patient's comfort. Medications commonly used in cancer care, such as opioids, nonsteroidal antiinflammatory drugs (NSAIDs), corticosteroids, and paracetamol, often come with side effects that can complicate treatment and require careful management. Nutritional strategies are critical in mitigating these side effects and optimizing treatment efficacy. This holistic approach ensures that both pain management and nutritional needs are addressed in a coordinated, patient-centered manner.

Pain management and opioid-induced constipation

Opioids are central to pain management in palliative care but frequently cause constipation, a side effect that can worsen patients' discomfort. To manage opioid-induced constipation, increasing dietary fiber intake from fruits (e.g., apples, pears), vegetables (e.g., broccoli, carrots), and whole grains (e.g., oats, brown rice) is recommended. Adequate hydration and probiotic-rich foods like yogurt and kefir further support bowel health. Additionally, transdermal analgesics such as fentanyl may be influenced by a patient's nutritional status, particularly body fat composition (and hence depend on the patient's intake of fats), which can affect drug absorption, which affects drug absorption. Healthy fats from sources like avocado and nuts, along with omega-3-rich diets, may influence the pharmacokinetics of these drugs [35,36].

NSAIDs and gastrointestinal protection

NSAIDs, frequently used for pain and inflammation, can irritate the gastrointestinal tract and increase the risk of ulcers. To protect the stomach lining, patients should consume probiotic-rich foods and antioxidants (e.g., berries, leafy greens) that support gastric health. Additionally, avoiding alcohol and spicy foods and taking NSAIDs with food can help mitigate gastrointestinal irritation [24,37].

Paracetamol and liver health

Excessive paracetamol use can lead to liver toxicity, especially in patients with preexisting liver conditions. To support liver health, a diet rich in fruits, vegetables, and lean proteins (e.g., poultry, fish) is recommended. High-fiber whole grains, along with plant-based proteins like beans and lentils, support liver detoxification and function [38]. Hydration also plays a vital role in liver function, assisting with detoxification and overall metabolic health.

Antidepressants and slowed metabolism

Antidepressants, especially those affecting serotonin, can cause gastrointestinal side effects like slowed motility and altered appetite [39], leading to discomfort and nutritional issues. To manage these effects, patients can eat smaller, nutrient-rich meals, such as bananas, rice, applesauce, and toast. Omega-3-rich foods like fatty fish and flaxseeds may support mood and metabolic health [40]. Regular physical activity can also help improve gut motility and overall well-being.

Corticosteroids and metabolic management

Corticosteroids, often prescribed for inflammation and edema in cancer care, are associated with side effects like weight gain, hyperglycemia, and muscle wasting. Nutritional interventions should focus on high-protein diets to preserve muscle mass and reduce the catabolic effects of corticosteroids. Additionally, incorporating antiinflammatory foods, such as omega-3 fatty acids found in fish and flaxseeds, can help counteract the catabolic effects of corticosteroids [41], but more extensive studies are still needed to support these results. Hydration, particularly through water and herbal teas like dandelion, is essential for metabolic balance and overall well-being.

The most common side effects of corticosteroid therapy, along with their descriptions and recommended nutritional interventions, are outlined in Tables 31.2 and 31.3.

Potential clinical applications

Health professionals, working as a multidisciplinary team, can address nutrition-related risk factors for chronic pain, such as malnutrition [48], weight change [49], comorbidities [50], and abnormal biochemistry results pointing to nutrition deficiencies [51], reduced appetite, or gastrointestinal complaints [52]. Referring to dietitians for thorough assessments and personalized treatment plans provides dietary advice, evidence-based care, and tailored therapeutic diets [53]. If a dietitian is not available, consulting with an outsourced dietitian is advisable. Routine nutrition assessments for cancer patients help identify deficiencies and barriers to healthy eating. Questionnaire-based dietary screeners improve diet

TABLE 31.2 Nutritional strategies for managing corticosteroid therapy complications.

Side effects	Description	Nutritional strategies
Weight gain and metabolic changes	Corticosteroids can lead to significant weight gain due to increased appetite and altered metabolism	**Balanced diet** (whole foods, including fruits, vegetables, whole grains, and lean proteins, can help manage weight gain. Reducing the intake of high-calorie, low-nutrient foods is essential) **Portion control** (patients should be encouraged to eat smaller, more frequent meals to prevent excessive weight gain while still meeting their nutritional needs) **Physical activity** (regular physical activity, as tolerated)
Hyperglycemia	Corticosteroids can induce hyperglycemia, which poses risks for wound healing and increases the likelihood of infections	**Low glycemic index foods** (incorporating low glycemic index foods can help manage blood sugar levels. Foods such as whole grains, legumes, and nonstarchy vegetables should be prioritized) **Frequent monitoring** (regular monitoring of blood glucose levels and nutritional adjustments based on these readings) **Fiber-rich foods** (foods high in fiber, such as oats, beans, and fruits, can slow glucose absorption)
Osteoporosis and bone health	Long-term corticosteroid use is associated with an increased risk of osteoporosis and fractures due to decreased calcium absorption and altered bone metabolism	**Calcium and vitamin D** (dairy products, fortified plant-based milks, leafy greens, and fatty fish are excellent sources) **Limit caffeine and alcohol** **Weight-bearing exercises**
Gastrointestinal disturbances	Corticosteroids can lead to gastrointestinal side effects, including nausea, dyspepsia, and increased appetite, which may result in overeating and further weight gain	**Small, frequent meals** (eating smaller, more frequent meals can help manage gastrointestinal discomfort and prevent overeating) **Avoid irritating foods** (patients should avoid spicy, acidic, or high-fat foods that can exacerbate gastrointestinal symptoms. Instead, bland foods such as rice, bananas, and applesauce can be beneficial) **Hydration** (adequate hydration is needed due to possible modulation of electrolyte and water balance)
Psychological effects	Corticosteroids can lead to mood changes, anxiety, and insomnia, which can affect a patient's overall well-being and nutritional status	**Tryptophan-rich foods** (incorporating foods high in tryptophan, such as salmon, eggs, nuts, and spinach, can help enhance serotonin production, potentially improving mood and sleep quality) **Regular meal patterns** (establishing regular meal patterns can help stabilize mood and energy levels. Patients should be encouraged to eat at consistent times throughout the day) **Mindful eating** (practicing mindful eating techniques can help patients develop a healthier relationship with food and improve their overall mental health)

Source: Adapted from Refs. [41–46].

therapy effectiveness and pain management [12]. Also, the Behavior Change Model helps assist patients in changing dietary habits by assessing capabilities, opportunities, and motivation. With it, health professionals are guided in setting specific nutritional goals and successful dietary behavior changes [11,12]. Personalized dietary plans should meet nutritional needs, enhance quality of life, and consider cancer treatment interactions [12]. A multidisciplinary team, including doctors, dietitians, pharmacists, and other health professionals, is essential for integrating nutritional interventions into cancer-related pain management. Each team member should understand their role in nutrition therapy and follow standard practices [54]. Patients often recognize the importance of nutrition in cancer treatment but lack access to nutrition professionals. Effective communication and patient-centered consultations are crucial for raising awareness

TABLE 31.3 Dietary recommendations for management of cancer-related complications [47].

Symptom	Description	Cause	Discomfort	Nutritional management
Mucositis (mouth sores)	Inflammation and mucous membrane ulceration often occur in the mouth or throat.	Chemotherapy, radiation therapy to the head and neck, or certain medications.	Severe pain while eating, swallowing, and talking, making eating difficult.	**Soft, blended foods**: Pureed soups, smoothies, mashed potatoes/vegetables, scrambled eggs, and yogurt. **Cold or room temperature foods**: Cold smoothies, applesauce, and chilled pudding. **Avoid irritants**: Spicy, acidic, rough, or very hot foods (e.g., citrus, tomatoes, chili). **Hydration**: Plenty of fluids/oral rehydration solutions.
Bone pain	Deep, aching pain originating from bones, often due to cancer metastasis to bone tissue.	Cancer spread (commonly from breast, lung, or prostate cancer) to bones, leading to bone destruction.	Constant, localized pain that can worsen with movement or pressure.	**Calcium and vitamin D:** Dairy and fortified foods, leafy greens, and fatty fish (e.g., salmon) help maintain bone strength. **Antiinflammatory foods**: Omega-3-rich foods (e.g., fish, flaxseeds, nuts), turmeric, and ginger can help reduce bone pain associated with inflammation. **Protein:** Lean meats, fish, dairy, legumes, and tofu to support muscle mass and maintain bone integrity.
Neuropathic pain	Sharp, burning, or tingling pain due to nerve damage.	Chemotherapy-induced peripheral neuropathy, radiation-induced nerve damage, or direct tumor pressure on nerves.	Tingling, numbness, or pain, often in hands and feet, which can interfere with daily activities.	**B vitamins**: Particularly B6, B12, and folate, found in whole grains, eggs, and leafy greens, support nerve health. **Omega-3 fatty acids**: Antiinflammatory benefits from fish (e.g., salmon, sardines) and flaxseeds may help alleviate neuropathic pain. **Antioxidants**: Brightly colored fruits and vegetables (e.g., leafy greens, peppers, beets, tomatoes, carrots, oranges, and berries) can help reduce oxidative stress on nerves.
Constipation	Difficulty in passing stool, often leading to discomfort and abdominal pain.	Analgesics (especially opioids), dehydration, reduced physical activity, or bowel obstruction due to tumors.	Abdominal bloating, cramps, and straining during bowel movements.	**High-fiber foods**: Whole grains, legumes, fruits with skin (e.g., apples, pears), and vegetables like broccoli. **Hydration**: Adequate water intake (≥8 glasses/day) is essential for softening stools. **Probiotics**: Yogurt, kefir, and other fermented foods can promote gut health and regularity. **Avoid processed foods**: Refined sugars and processed foods can worsen constipation.
Fatigue	A persistent sense of exhaustion that is not relieved by rest.	Chemotherapy, radiation therapy, cancer-related anemia, malnutrition, or the cancer itself.	Profound physical and mental fatigue can contribute to depression, reduced appetite, and difficulty in managing daily activities.	**Small, frequent meals**: Eating nutrient-dense snacks rich in nuts, seeds, and fruit can provide sustained energy without overwhelming the digestive system. **Iron-rich foods**: Red meat, animal-sourced foods, spinach, and lentils help manage cancer-related anemia, a common cause of fatigue. Vitamin C enhances the absorption of iron. **Complex carbohydrates**: Whole grains (e.g., oats, quinoa) provide long-lasting energy and help stabilize blood sugar levels.

Nausea and vomiting	The sensation of needing to vomit or actual vomiting, often debilitating for patients.	Chemotherapy, radiation (especially to the abdomen), medications, or bowel obstruction.	Can lead to dehydration, weight loss, and reduced appetite, impacting nutrition and health.	**Small, frequent meals**: Eating smaller amounts of food more often can help prevent nausea. **Ginger**: Ginger tea or ginger candy can help settle the stomach. **Hydration**: Clear broths, electrolyte beverages, and herbal tea help maintain hydration without causing nausea. **Avoid strong smells**: Cold or room-temperature meals to minimize odors that might trigger nausea.
Anorexia and cachexia (loss of appetite and muscle wasting)	Reduced desire to eat, leading to severe weight loss and muscle wasting (cachexia).	Advanced cancer, inflammation, metabolic changes, and side effects from treatment.	Muscle weakness, severe fatigue, and impaired quality of life.	**High-calorie, high-protein foods**: Nutrient-dense foods like whole grains, seafood, lean meats, eggs, peas, beans, nuts, and full-fat dairy. **Small, frequent meals**: Eating small amounts throughout the day can help prevent feeling overwhelmed by large meals. **Supplements**: Oral nutritional supplements provide calories and protein in an easy-to-consume format. **Flavor enhancers**: Use herbs, spices, and sauces to make food more appealing.
Dyspnea (shortness of breath)	Difficulty in breathing or the sensation of not getting enough air.	Lung cancer, pleural effusion, anemia, pulmonary infections, or side effects of chemotherapy.	Anxiety, restlessness, and feeling of suffocation, leading to distress and fatigue.	**Small meals**: Eat smaller, nutrient-dense meals as large meals can pressure the diaphragm, making breathing harder. **Soft foods**: Soups and pureed or soft-textured foods (e.g., oatmeal, smoothies) may be easier to eat without causing breathlessness. **Avoid gas-producing foods**: Foods like beans, cabbage, and carbonated beverages can cause bloating, which may worsen dyspnea.
Ascites (fluid accumulation in the abdomen)	Buildup of fluid in the peritoneal cavity, causing swelling and discomfort.	Cancers of the liver, ovaries, or gastrointestinal tract.	Abdominal bloating, tightness, pain, and difficulty breathing or moving.	**Low-sodium diet**: Reducing sodium (processed foods, canned soups and foods, etc.) helps minimize fluid retention. **High-protein foods**: Maintain lean body mass with protein-rich foods as protein can help manage fluid balance. **Frequent small meals**: Helps avoid feeling overly full or bloated due to fluid buildup.
Headaches	Pain localized in the head, often due to pressure or inflammation.	Brain tumors, metastases to the brain, increased intracranial pressure, or chemotherapy.	Throbbing, pressure-like pain, which can be accompanied by nausea, vomiting, or visual disturbances.	**Hydration**: Adequate fluid intake (≥8 glasses/day) is important as dehydration can cause or exacerbate headaches. **Magnesium-rich foods**: Dark leafy greens, avocados, nuts, and seeds may help prevent or alleviate headaches, particularly in patients with magnesium deficiency. **Avoid triggers**: For some, foods like aged cheese, caffeine, or processed meats can trigger headaches, and should be avoided.

(Continued)

TABLE 31.3 (Continued)

Symptom	Description	Cause	Discomfort	Nutritional management
Lymphedema	Swelling, typically in the arms or legs, due to impaired lymphatic drainage.	Surgical removal of lymph nodes, radiation therapy, or cancer obstructing lymphatic flow.	Heaviness, tightness, and aching in the affected limb, along with reduced mobility.	**Antiinflammatory diet**: Increase intake of fruits, vegetables, omega-3 fatty acids, and whole grains to reduce inflammation and fluid retention. **Low-sodium diet**: Avoid foods high in salt (processed meats and other foods, canned soups and foods, etc.) to reduce water retention and swelling. **Hydration**: Staying well-hydrated helps maintain fluid balance and prevent swelling.
Skin irritation or radiation dermatitis	Redness, peeling, or painful sores on the skin.	Radiation therapy, particularly in areas like the chest, breast, or head and neck.	Burning, itching, or tenderness of the skin, making it difficult to wear clothing or tolerate movement.	**Hydrating foods**: Cucumber, watermelon, and broth-based soups can help keep the skin hydrated from the inside. **Vitamin E**: Foods like almonds, spinach, and sunflower seeds support skin repair and health. **Zinc-rich foods**: Whole grains, chickpeas, nuts, fish, and meat may help with wound healing and skin regeneration.
Abdominal pain	General discomfort or sharp pain in the abdomen.	Tumor pressure, obstruction of the intestines, liver metastases, or side effects from medications.	Can vary from cramping and bloating to severe, sharp pain that disrupts daily activities.	**Low-fiber diet**: If there is a risk of obstruction, reduce fiber intake (e.g. whole grains, plant foods) to minimize bowel irritation. **Easily digestible foods**: Soft, bland foods like broth, fruits/vegetables, and applesauce can be easier to digest. **Avoid gas-producing foods**: Reduce intake of beans, onions, carbonated drinks, and other foods that can lead to bloating.
Bladder and bowel incontinence	Loss of bladder or bowel control.	Tumors pressing on nerves controlling bladder/bowel functions, surgery, or radiation therapy.	Embarrassment, skin irritation, and discomfort from involuntary leakage.	**Fiber for bowel regulation**: Ensure a balanced fiber intake to regulate bowel movements, preventing constipation or diarrhea. **Adequate hydration**: Drinking enough fluids can reduce bladder irritation and help maintain normal bowel function. **Avoid bladder irritants**: Caffeine, alcohol, spicy foods, and acidic foods like citrus and tomatoes may worsen incontinence.
Depression and anxiety	Mental and emotional distress often manifests as sadness, hopelessness, and chronic worry.	The psychological impact of a cancer diagnosis, pain, fatigue, and the emotional toll of treatment.	Affects appetite, energy levels, and motivation to engage in daily activities, including eating.	**Mood-boosting nutrients**: Foods rich in omega-3 fatty acids (salmon, walnuts, seeds), folate (leafy greens), and magnesium (pumpkin seeds, dark chocolate) can help alleviate symptoms of depression. **Complex carbohydrates**: Whole grains, oats, and vegetables can help regulate blood sugar levels, which may influence mood stability. **Vitamin D**: Fortified dairy, eggs, fish.

Dysphagia (difficulty swallowing)	Trouble swallowing, leading to choking or discomfort when eating or drinking.	Head and neck cancers, esophageal cancer, or radiation therapy affecting the throat.	Can severely impact nutrition, leading to weight loss, malnutrition, and frustration with eating.	**Soft and pureed foods**: Smoothies, soups, mashed potatoes/vegetables, and yogurt are easier to swallow. **Thickened liquids**: Use commercial thickeners to make liquids easier to swallow and reduce the risk of aspiration. **High-calorie supplements**: Nutrient-dense oral supplements provide nutrition without needing to chew.
Muscle cramps	Sudden, involuntary muscle contractions causing pain.	Dehydration, electrolyte imbalances, or chemotherapy.	Sharp, intense pain that can temporarily immobilize the affected muscle.	**Magnesium and potassium**: Bananas, apples, avocado, starchy vegetables (potatoes, yams, carrots), legumes, spinach, meat and fish help prevent muscle cramps. **Hydration**: Adequate water intake is vital to prevent dehydration, which can contribute to muscle cramps. **Electrolyte balance**: Include soups, milk, leafy green vegetables, fruits, dried apricots and prunes, and dairy products.
Joint pain (arthralgia)	Stiffness, swelling, or pain in the joints.	Certain chemotherapy agents (like aromatase inhibitors) or cancer itself (e.g., in multiple myeloma or metastatic bone disease).	Aching joints that can limit mobility and affect daily functioning.	**Antiinflammatory foods**: Omega-3 fatty acids (e.g., fish, nuts, seeds), phytonutrients from green and leafy vegetables, fruits, berries, nuts, olive oil, spices and herbs (e.g. turmeric, ginger, Mediterranean herbs), green tea, and dark chocolate can help reduce inflammation in joints. **Hydration**: Adequate water intake (≥8 glasses/day) can help keep joints well-lubricated and manage joint pain.

Source: Adapted from Štimac D, Krznarić Ž, Vranešić Bender D, Obrovac Glišić M. editors. Dijetoterapija i klinička prehrana (Biblioteka Sveučilišni udžbenici). Zagreb, Croatia: Medicinska Naklada; 2014 and Vranešić Bender D. Dijetoterapija. Zagreb, Croatia: Medicinska Naklada; 2024.

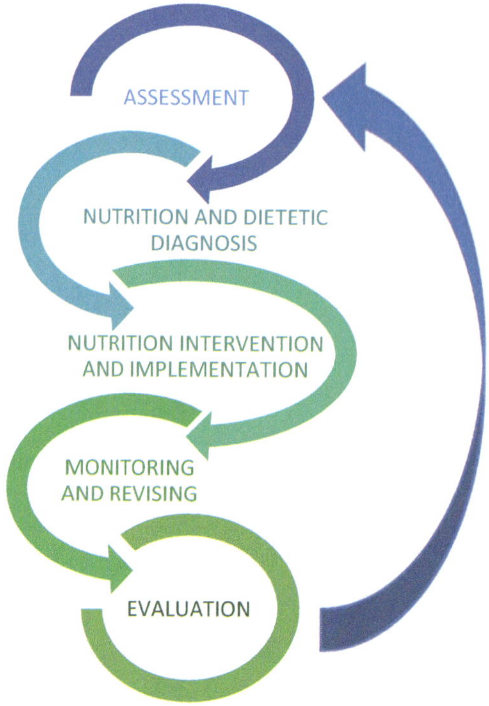

FIGURE 31.1 The process of nutrition and dietetic care of the cancer patient.

and encouraging informed decisions [54,55]. Continuous support and education on nutrition and lifestyle in pain management empower patients to participate actively in their care [56]. Education can be delivered in person and via leaflets, videos, or telehealth due to its effectiveness and acceptability [57]. Continuous support for cancer patients includes regular assessment of nutritional status and dietary behavior, adjusting interventions based on treatment response and current needs. This support is considered a fundamental right, ensuring patients' access to food and health [54,58].

Perspectives/opinions

Personalized dietary assessments and interventions can improve cancer patients' quality of life, support pain management, and may improve overall quality of life [34]. Currently, there is no consensus on standardized tools for nutrition screening or assessment in healthcare settings. The lack of standardized nutrition guidelines for managing pain in cancer patients and evidence-based data on dietary interactions with cancer treatments highlights the need for future guideline development [59]. Therefore creating evidence-based guidelines can standardize nutritional support in cancer care protocols specifically for cancer patients who experience pain.

Future implications

As the population of cancer survivors grows, healthcare providers must adapt to support those with persistent pain. Dietary interventions targeting pain's underlying causes can reduce costs and enhance patients' quality of life. This dynamic field requires continuous research, collaboration, and personalized care [59,60]. Developing evidence-based nutrition guidelines can improve pain management through a multidisciplinary approach. Personalized diets and education can reduce inflammation and improve well-being, addressing both nutritional and pain management needs [54]. More research is needed to integrate these approaches for chronic pain in cancer survivors, considering their varied nutritional needs based on cancer type, treatment, and lifestyle [2].

The nutrition and dietetic care process

The nutrition and dietetic care process for cancer patients is a systematic approach to providing personalized nutrition care (Fig. 31.1). By following this structured process, personalized dietetic care can significantly improve cancer patients' nutritional status, pain management, and overall treatment outcomes. This process helps evaluate outcomes and adjust diet plans. It includes assessment, diagnosis, intervention, implementation, monitoring, and evaluation [61]. A thorough initial assessment, including behavioral components, is essential for developing personalized interventions that promote better adherence to nutrition plans [62]. Early intervention through timely nutritional status assessment can help prevent malnutrition and delay cachexia, improving quality of life and treatment outcomes.

A comprehensive assessment is vital for effective dietetic interventions. It involves evaluating health, dietary habits, medical history, and sociodemographic factors like education and socioeconomic status. Psychological aspects such as depression or anxiety are also considered, emphasizing the need for tools like the motivation assessment of change (MAC2-A), food records, and dietary questionnaires (e.g., 24-hour recall). Screening tools like the Malnutrition Universal Screening Tool or Mini Nutritional Assessment-Short Form help identify nutritional risks.

The nutrition diagnosis phase identifies specific dietary problems, their causes, and symptoms by analyzing physical, psychological, and dietary data. This step bridges assessment and intervention by highlighting barriers to optimal nutrition.

Nutrition interventions are personalized strategies aimed at improving health and reducing pain. These include setting specific goals, patient education, counseling, and behavior change strategies to enhance adherence.

Monitoring and evaluation ensure that interventions remain effective. Regular assessments track progress, identify barriers, and guide necessary adjustments to meet individual needs. Finally, evaluating outcomes determines whether goals are achieved or if reassessment is needed to refine strategies for better results.

Conclusion

In cancer care, nutrition is an important part of managing pain and addressing treatment side effects. Nutritional strategies can significantly alleviate discomfort and improve patients' quality of life. Personalized dietary interventions support symptom management and enhance the efficacy of pain-relieving medications and other pharmacotherapies. As part of a comprehensive care plan, tailored nutrition can positively impact overall well-being and reduce treatment complications. Integrated, patient-centered nutrition represents a valuable tool in fostering holistic oncology care.

References

[1] Meleger AL, Froude CK, Walker J 3rd. Nutrition and eating behavior in patients with chronic pain receiving long-term opioid therapy. PM R 2014;6(1):7–12.e1. Available from: https://doi.org/10.1016/j.pmrj.2013.08.597.

[2] Nijs J, Tumkaya Yilmaz S, Elma Ö, et al. Nutritional intervention in chronic pain: an innovative way of targeting central nervous system sensitization? Expert Opin Ther Targets 2020;24(8):793–803. Available from: https://doi.org/10.1080/14728222.2020.1784142.

[3] Monjotin N, Amiot MJ, Fleurentin J, et al. Clinical evidence of the benefits of phytonutrients in human healthcare. Nutrients 2022;14(9):1712. Available from: https://doi.org/10.3390/nu14091712.

[4] Saleh HA, Yousef MH, Abdelnaser A. The anti-inflammatory properties of phytochemicals and their effects on epigenetic mechanisms involved in TLR4/NF-κB-mediated inflammation. Front Immunol 2021;12:606069. Available from: https://doi.org/10.3389/fimmu.2021.606069.

[5] Yahfoufi N, Alsadi N, Jambi M, et al. The immunomodulatory and anti-inflammatory role of polyphenols. Nutrients 2018;10(11):1618. Available from: https://doi.org/10.3390/nu10111618.

[6] Guo R, Chen LH, Xing C, et al. Pain regulation by gut microbiota: molecular mechanisms and therapeutic potential. Br J Anaesth 2019;123(5):637–54. Available from: https://doi.org/10.1016/j.bja.2019.07.026.

[7] Bjørklund G, Aaseth J, Doşa MD, et al. Does diet play a role in reducing nociception related to inflammation and chronic pain? Nutrition 2019;66:153–65. Available from: https://doi.org/10.1016/j.nut.2019.04.007.

[8] Monllor-Tormos A, García-Vigara A, Morgan O, et al. Mediterranean diet for cancer prevention and survivorship. Maturitas 2023;178:107841. Available from: https://doi.org/10.1016/j.maturitas.2023.107841.

[9] Guasch-Ferré M, Willett WC. The Mediterranean diet and health: a comprehensive overview. J Intern Med 2021;290(3):549–66. Available from: https://doi.org/10.1111/joim.13333.

[10] Rondanelli M, Faliva MA, Miccono A, et al. Food pyramid for subjects with chronic pain: foods and dietary constituents as anti-inflammatory and antioxidant agents. Nutr Res Rev 2018;31(1):131–51. Available from: https://doi.org/10.1017/S0954422417000270.

[11] Polak R, Dacey M, Phillips EM. Time for food – training physiatrists in nutritional prescription. J Rehabil Med 2017;49(2):106–12. Available from: https://doi.org/10.2340/16501977-2178.

[12] Brain K, Burrows TL, Bruggink L, et al. Diet and chronic non-cancer pain: the state of the art and future directions. J Clin Med 2021;10(21):5203. Available from: https://doi.org/10.3390/jcm10215203.

[13] Zheng Z, Zhao X, Zhao Q, et al. The effects of early nutritional intervention on oral mucositis and nutritional status of patients with head and neck cancer treated with radiotherapy. Front Oncol 2021;10:595632. Available from: https://doi.org/10.3389/fonc.2020.595632.

[14] Kusiak A, Jereczek-Fossa BA, Cichońska D, et al. Oncological-therapy related oral mucositis as an interdisciplinary problem-literature review. Int J Environ Res Public Health 2020;17(7):2464. Available from: https://doi.org/10.3390/ijerph17072464.

[15] Badders J, Coblens O, Ranasinghe V, et al. Medical honey in head and neck cancer. Cureus 2024;16(1):e52822. Available from: https://doi.org/10.7759/cureus.52822.

[16] UPMC Palliative and Supportive Institute. Honey for oral mucositis (The Tablet, Vol. 4, No. 6). Retrieved from https://dam.upmc.com/-/media/upmc/services/palliative-and-supportive-institute/documents/the-tablet-v4n6-honey-for-oral-mucositis.pdf; 2024.

[17] Gao Y, Yang H, Zhang X, et al. Effectiveness of glutamine oral care in reducing oral mucositis and improving oral health after neurosurgery: a randomized controlled trial with microbiome analysis. Med Sci Monit 2024;30:e942585. Available from: https://doi.org/10.12659/MSM.942585.

[18] Tsujimoto T, Yamamoto Y, Wasa M, et al. L-glutamine decreases the severity of mucositis induced by chemoradiotherapy in patients with locally advanced head and neck cancer: a double-blind, randomized, placebo-controlled trial. Oncol Rep 2015;33(1):33–9. Available from: https://doi.org/10.3892/or.2014.3564.

[19] Jafari-Vayghan H, Varshosaz P, Hajizadeh-Sharafabad F, et al. A comprehensive insight into the effect of glutamine supplementation on metabolic variables in diabetes mellitus: a systematic review. Nutr Metab 2020;17:80. Available from: https://doi.org/10.1186/s12986-020-00503-6.

[20] Wernerman J. Glutamine supplementation to critically ill patients? Crit Care 2014;18(2):214. Available from: https://doi.org/10.1186/cc13781.

[21] Silva PG, Barreto GA, Carlos AC, et al. Dysgeusia increases the risk for death and other side effects during antineoplastic systemic treatment for solid tumors: a cross-sectional study. Med Oral Patol Oral Cir Bucal 2024;29(3):e398–407. Available from: https://doi.org/10.4317/medoral.26389.

[22] Ghias K, Jiang Y, Gupta A. The impact of treatment-induced dysgeusia on the nutritional status of cancer patients. Clin Nutr Open Sci 2023;50:57–76. Available from: https://doi.org/10.1016/j.nutos.2023.06.004.

[23] Marx W, Kiss N, McCarthy AL, et al. Chemotherapy-induced nausea and vomiting: a narrative review to inform dietetics practice. J Acad Nutr Diet 2016;116(5):819–27. Available from: https://doi.org/10.1016/j.jand.2015.10.020.

[24] Raei M, Khanghah AG, Dadgari A, et al. Effect of herbal therapy to intensity chemotherapy-induced nausea and vomiting in cancer patients. Iran Red Crescent Med J 2013;15(2):101–6. Available from: https://doi.org/10.5812/ircmj.4392.

[25] Neves Duarte Lisboa I, Dantas De Sá Tinoco J, Dias Fernandes MIDC, et al. Constipation in chemotherapy patients: a diagnostic accuracy study. Asian Pac J Cancer Prev 2021;22(9):3017–21. Available from: https://doi.org/10.31557/APJCP.2021.22.9.3017.

[26] Wilson PB. Associations between physical activity and constipation in adult americans: results from the national health and nutrition examination survey. Neurogastroenterol Motil 2020;32(5):e13789. Available from: https://doi.org/10.1111/nmo.13789.

[27] Black CJ, Staudacher HM, Ford AC. Efficacy of a low FODMAP diet in irritable bowel syndrome: systematic review and network meta-analysis. Gut 2022;71(6):1117–26. Available from: https://doi.org/10.1136/gutjnl-2021-325214.

[28] Magro DO, Oliveira LMRD, Bernasconi I, et al. Effect of yogurt containing polydextrose, *Lactobacillus acidophilus* NCFM and *Bifidobacterium lactis* HN019: a randomized, double-blind, controlled study in chronic constipation. Nutr J 2014;13(1):75. Available from: https://doi.org/10.1186/1475-2891-13-75.

[29] Zwielehner J, Lassl C, Hippe B, et al. Changes in human fecal microbiota due to chemotherapy analyzed by TaqMan-PCR, 454 sequencing and PCR-DGGE fingerprinting. PLoS One 2011;6(12):e28654. Available from: https://doi.org/10.1371/journal.pone.0028654.

[30] Deleemans JM, Toivonen K, Reimer RA, et al. The chemo-gut study: a cross-sectional survey exploring physical, mental, and gastrointestinal health outcomes in cancer survivors. Glob Adv Health Med 2022;11:2164957X221145940 https://doiorg/10.1177/2164957X221145940.

[31] Simrén M, Törnblom H, Palsson OS, et al. Management of the multiple symptoms of irritable bowel syndrome. Lancet Gastroenterol Hepatol 2017;2(2):112–22. Available from: https://doi.org/10.1016/S2468-1253(16)30116-9.

[32] Crombeen AM, Lilly EJ. Management of dyspnea in palliative care. Curr Oncol 2020;27(3):142–5. Available from: https://doi.org/10.3747/co.27.6413.

[33] Nigro E, Perrotta F, Scialò F, et al. Food, nutrition, physical activity and microbiota: which impact on lung cancer? Int J Environ Res Public Health 2021;18(5):2399. Available from: https://doi.org/10.3390/ijerph18052399.

[34] Arends J, Bachmann P, Baracos V, et al. ESPEN guidelines on nutrition in cancer patients. Clin Nutr 2017;36(1):11–48. Available from: https://doi.org/10.1016/j.clnu.2016.07.015.

[35] Yılmaz ST, Elma Ö, Deliens T, et al. Nutrition/dietary supplements and chronic pain in patients with cancer and survivors of cancer: a systematic review and research agenda. Pain Physician 2021;24(4):335–44. Available from: https://doi.org/10.36076/ppj.2021.24.335.

[36] Alberts DS, Li Y, Wang L, et al. Fentanyl sublingual spray for breakthrough cancer pain in patients receiving transdermal fentanyl. Pain Manag 2016;6(3):175–86. Available from: https://doi.org/10.2217/pmt-2015-0009.

[37] Moore RA, Derry S, McQuay HJ, et al. Effects of food on pharmacokinetics of immediate release oral formulations of aspirin, dipyrone, paracetamol and NSAIDs – a systematic review. Br J Clin Pharmacol. 2015;80(3):381–8. Available from: https://10.1111/bcp.12628.

[38] Vranešić Bender D. Dijetoterapija. Zagreb, Croatia: Medicinska naklada; 2024.

[39] Oliva V, Lippi M, Paci R, et al. Gastrointestinal side effects associated with antidepressant treatments in patients with major depressive disorder: a systematic review and meta-analysis. Prog Neuropsychopharmacol Biol Psychiatry 2021;109:110266. Available from: https://doi.org/10.1016/j.pnpbp.2021.110266.

[40] Grosso G, Galvano F, Marventano S, et al. Omega-3 fatty acids and depression: scientific evidence and biological mechanisms. Oxid Med Cell Longev 2014;2014:313570. Available from: https://doi.org/10.1155/2014/313570.
[41] Lee M-K, Jeong HH, Kim M-J, et al. Nutrients against glucocorticoid-induced muscle atrophy. Foods 2022;11:687. Available from: https://doi.org/10.3390/foods11050687.
[42] Ravasco P. Nutrition in cancer patients. J Clin Med 2019;8(8):1211. Available from: https://doi.org/10.3390/jcm8081211.
[43] Hirano Y, Konishi T, Kaneko H, et al. Impact of prophylactic corticosteroid use on in-hospital mortality and respiratory failure after esophagectomy for esophageal cancer: nationwide inpatient data study in Japan. Ann Surg 2023;277(6):e1247–53. Available from: https://doi.org/10.1097/SLA.0000000000005502.
[44] Reuter M, Zamoscik V, Plieger T, et al. Tryptophan-rich diet is negatively associated with depression and positively linked to social cognition. Nutr Res 2021;85:14–20. Available from: https://doi.org/10.1016/j.nutres.2020.10.005.
[45] Muscular Dystrophy Association (MDA). Nutritional considerations while on corticosteroids. Retrieved September 20, 2024, from https://mdaquest.org/quest-article-nutritional-considerations-while-corticosteroids/; 2024.
[46] Esteves GP, Mazzolani BC, Smaira FI, et al. Nutritional recommendations for patients undergoing prolonged glucocorticoid therapy. Rheumatol Adv Pract 2022;6(2):rkac029. Available from: https://doi.org/10.1093/rap/rkac029.
[47] Štimac D, Krznarić Ž, Vranešić Bender D, Obrovac Glišić M, editors. Dijetoterapija i klinička prehrana (Biblioteka Sveučilišni udžbenici). Zagreb, Croatia: Medicinska Naklada; 2014.
[48] Skipper A, Coltman A, Tomesko J, et al. Position of the academy of nutrition and dietetics: malnutrition (undernutrition) screening tools for all adults. J Acad Nutr Diet 2020;120(4):709–13. Available from: https://doi.org/10.1016/j.jand.2019.09.011.
[49] Nicholson BD, Hamilton W, O'Sullivan J, et al. Weight loss as a predictor of cancer in primary care: a systematic review and meta-analysis. Br J Gen Pract 2018;68(670):e311–22. Available from: https://doi.org/10.3399/bjgp18X695801.
[50] Bruggink L, Hayes C, Lawrence G, et al. Chronic pain: overlap and specificity in multimorbidity management. Aust J Gen Pract 2019;48(10):689–92. Available from: https://doi.org/10.31128/AJGP-06-19-4966.
[51] Gunn J, Hill MM, Cotten BM, et al. An analysis of biomarkers in patients with chronic pain. Pain Physician 2020;23(1):E41–9.
[52] Drewes AM, Olesen AE, Farmer AD, et al. Gastrointestinal pain. Nat Rev Dis Primers 2020;6(1):1. Available from: https://doi.org/10.1038/s41572-019-0135-7.
[53] Dorner B, Friedrich EK. Position of the academy of nutrition and dietetics: individualized nutrition approaches for older adults: long-term care, post-acute care, and other settings. J Acad Nutr Diet 2018;118(4):724–35. Available from: https://doi.org/10.1016/j.jand.2018.01.022.
[54] Erickson N, Sullivan ES, Kalliostra M, et al. Nutrition care is an integral part of patient-centred medical care: a European consensus. Med Oncol 2023;40(4):112. Available from: https://doi.org/10.1007/s12032-023-01955-5.
[55] Lee SF, Brown T, Wyld D, et al. Investigating the dietary knowledge, attitudes, and beliefs of Australian patients with cancer. J Hum Nutr Diet 2023;36(3):612–21. Available from: https://doi.org/10.1111/jhn.13091.
[56] Prevost V, Delorme C, Grach MC, et al. Therapeutic education in improving cancer pain management: a synthesis of available studies. Am J Hosp Palliat Care 2016;33:599–612. Available from: https://doi.org/10.1177/1049909115586394.
[57] Adamse C, Dekker-Van Weering MG, van Etten-Jamaludin FS, et al. The effectiveness of exercise-based telemedicine on pain, physical activity and quality of life in the treatment of chronic pain: a systematic review. J Telemed Telecare 2018;24(8):511–26. Available from: https://doi.org/10.1177/1357633X17716576.
[58] Cardenas D. Ethical issues and dilemmas in artificial nutrition and hydration. Clin Nutr ESPEN 2021;41:23–9. Available from: https://doi.org/10.1016/j.clnesp.2020.12.010.
[59] Kristensen MB, Wessel I, Ustrup KS, et al. Nutrition screening and assessment tools for patients with cancer and survivors of cancer: a systematic review protocol. BMJ Open 2020;10(10):e037844. Available from: https://doi.org/10.1136/bmjopen-2020-037844.
[60] Tümkaya Yılmaz S, Malfliet A, Elma Ö, et al. Diet/nutrition: ready to transition from a cancer recurrence/prevention strategy to a chronic pain management modality for cancer survivors? J Clin Med 2022;11(3):653. Available from: https://doi.org/10.3390/jcm11030653.
[61] The British Dietetic Association Model and process for nutrition and dietetic practice. Retrieved September 25, 2024, from https://www.bda.uk.com/practice-and-education/nutrition-and-dietetic-practice/professional-guidance/model-and-process-for-dietetic-practice.html; 2020.
[62] Kohlenberg-Müller K, Ramminger S, Kolm A, et al. Nutrition assessment in process-driven, personalized dietetic intervention - the potential importance of assessing behavioural components to improve behavioural change: results of the EU-funded IMPECD project. Clin Nutr ESPEN 2019;32:125–34. Available from: https://doi.org/10.1016/j.clnesp.2019.03.017.

Part IV

Management of pain in specific cancers

Chapter 32

Head and neck cancers

Nikola Milašević
Clinic for Medical Oncology, Institute for Oncology, Clinical Centre of Montenegro, Podgorica, Montenegro

Key learning objectives

- Give an overview of epidemiology, incidence, pathogenesis, and prognosis for head and neck squamous cell carcinoma (HNSCC).
- Address the most common symptoms in patients being treated for HNSCC.
- Explore the impact of different treatment modalities on survival, pain reduction, and the quality of life in patients with HNSCC.
- Evaluate new strategies for pain relief for future studies.

Epidemiology, etiology, pathogenesis, and prognosis of patients with head and neck cancer

Head and neck squamous cell carcinoma (HNSCC) encompasses a variety of tumors of origin in the lip, oral cavity, hypopharynx, oropharynx, nasopharynx, or larynx, which represent more than 90% of patients with head and neck cancer (HNC), which is why HNSCC will be the focus of the chapter. They most often occur between the ages of 65 and 69. It is the sixth most common malignancy in the world, accounting for around 5% of all cancer cases and responsible for an estimated 1%–2% of all cancer deaths. The most common HNC sites are the oral cavity and larynx (incidence rates 3.9 and 2.3 per 100,000, respectively) [1]. Although tobacco- and alcohol-associated HNSCC incidence has declined in some settings, HPV-associated oropharyngeal squamous cell carcinoma has increased in the United States and other high-income countries, changing the epidemiology of HNSCC [1].

While genetic causes are uncommon, patients with Fanconi anemia have an over 500-fold increased risk of developing squamous cell carcinoma of the head and neck [2]. However, tobacco consumption and alcohol remain the main risk factors of HNSCC, which is why incidence is expected to rise in Eastern Europe and China, while infection with human papillomavirus (HPV) is one of the major factors associated with the growth of oropharyngeal tumors. The majority of patients have locally advanced or metastatic disease (stage III or IV) at the time of diagnosis. Locally advanced disease includes large tumors or tumors that have spread to regional lymph nodes but have not yet metastasized to distant organs [3,4]. About 10% of HNC cases have distant metastases at the time of diagnosis. However, this rate can increase to 20%–30% or more in cases of recurrence, especially after failure of initial treatments such as surgery, radiation, or chemotherapy [3]. The risk of advanced or metastatic disease is higher for certain primary tumor locations, such as the hypopharynx and nasopharynx, compared to other areas such as the oral cavity or larynx. For example, hypopharyngeal carcinoma is often detected at a more advanced stage, with distant metastases present in 25%–30% of cases [3]. HPV-positive oropharyngeal carcinomas tend to present at an advanced stage but have a better prognosis compared to HPV-negative cancers. Although HPV-positive cancers are often locally advanced, they have a lower rate of distant metastases (about 10%) compared to HPV-negative tumors [4].

Pain Management in Palliative Care for Patients With Cancer. DOI: https://doi.org/10.1016/B978-0-443-30050-9.00032-1

Cancer characteristics which influence the pain in head and neck squamous cell carcinoma

Patients with HNSCC share the common characteristics of upper aerodigestive malignancies, advanced age, a history of tobacco and alcohol abuse, and multiple cardiopulmonary comorbidities, which predispose them to severe adverse respiratory outcomes [5]. Moreover, potential toxicities from chemotherapy and radiotherapy (RT) have been shown to significantly increase rates of dysphagia, aspiration, pneumonia, and pain in this population [6]. The symptoms of HNC can be diverse and include sores in the mouth or throat, lumps in the neck, as well as swelling of the jaw. When it comes to changes in the mouth, they are most often nonhealing sores, which are painful and cause difficulty in swallowing, possibly also in breathing, and can appear as white or red spots on the gums, tongue, or mucous membrane of the mouth. Changes can also be asymptomatic or cause hoarseness, pain, and ringing in the ears or hearing problems as the first symptom of the disease. The complaints mentioned may be the consequences of less serious conditions, but the existence of a malignant process should always be ruled out [6].

The pain associated with HNSCC is predominantly associated with the location of the primary site and results from tissue destruction, nerve compression, and inflammation [7]. Oral and pharyngeal cancers often interfere with oral intake, as they are associated with dysphagia (difficulty swallowing) and odynophagia (painful swallowing). This pain is usually sharp or burning and can be made worse by eating, talking, or swallowing. This makes eating a challenge, especially after treatments such as radiation therapy, which further damages the mucosa, exacerbating mucositis and leading to severe pain [7]. Cancer of the larynx often leads to pain that is localized in the throat and may spread to the ears (otalgia). This pain is usually dull and persistent and may worsen during phonation or swallowing [7]. On the other hand, nasal and sinus cancer can lead to facial pain and headaches, often due to sinus obstruction or invasion of adjacent structures [7]. Tumor changes in the paranasal cavities and nasal cavity can also lead to chronic infections and nosebleeds, frequent headaches, facial swelling, eye problems, or jaw pain [6]. When it comes to salivary gland malignancy, there may be swelling under the chin or around the jaw, numbness or paralysis of the facial muscles, or pain in the face, chin, or neck that occurs periodically.

HNC primarily spreads to regional lymph nodes, especially the cervical region. Distant metastases of the HNC are in general less common compared to other malignancies but are more common in nasopharyngeal and hypopharyngeal cancer. In the presence of aggressive metastatic tumors such as adenoid cystic carcinoma, basaloid squamous cell carcinoma, and neuroendocrine carcinomas, an extensive metastatic work-up is recommended [3]. However, distant metastases can occur in lungs and less commonly in bones and liver. Pain associated with metastasis may include bone metastases, which can cause deep, stabbing, and often refractory localized pain that worsens with movement or palpation. It can also present with pathological fractures, especially in weight-bearing bones and spinal cord compression, and it can cause hypercalcemia. Bone pain caused by metastases may be due to multiple mechanisms, including release of chemical mediators, elevated intra osseous pressure, and periosteal elevation. Metastatic bone disease is a well-known cause of severe cancer-related pain, which impairs the patient's mobility, significantly reduces the quality of life (QoL), and requires a combination of pharmacological and nonpharmacological treatments [8]. Metastases in the lungs are usually not associated with pain, but they can cause dyspnea and chest discomfort, and in the case of involvement of the pleural tissue, they can cause localized pain, which increases with breathing and movement. Liver metastases are less common in patients with squamous cell carcinoma of the head and neck, but they can lead to impaired liver function and contribute to various symptoms such as nausea, anorexia, and general weakness, making pain management difficult. If liver metastases are present or if the patient has underlying liver disease, the metabolism of certain analgesics (such as acetaminophen or some opioids) may be impaired. This requires careful dose adjustment to avoid hepatotoxicity and drug accumulation [9].

Patient characteristics which influence the pain in head and neck squamous cell carcinoma

Comorbidities play a major role in HNSCC, as it is often occurring in smokers and alcohol drinkers. Many patients in this age group have preexisting comorbidities such as cardiovascular disease, chronic obstructive pulmonary disease, and diabetes. These comorbidities can significantly affect the treatment of pain [1,5]. Furthermore, kidney damage is common in elderly patients, especially those receiving nephrotoxic chemotherapy agents such as cisplatin. Decreased renal clearance may require dose adjustments of opioids and other analgesics to prevent toxicity. For example, morphine, which is excreted by the kidneys, can accumulate in patients with renal dysfunction, necessitating the use of alternatives such as fentanyl or hydromorphone, which are metabolized differently [10]. This is covered in more detail in another chapter of this book. An additional issue for HNSCC patients is the frequent presence of dysphagia, which limits oral analgesic use, necessitating alternatives like transdermal fentanyl. Beyond tumor-related pain, treatments introduce additional challenges for peroral use of medications.

Pain associated with treatment of head and neck squamous cell carcinoma

Different types of HNC treatments often make the pain worse. For example, RT is associated with the development of mucositis (painful inflammation of the mucous membrane) [6]. This condition often leads to severe pain in the mouth, which requires aggressive pain management with local anesthetics, systemic analgesics (such as opioids), and occasionally using a nasogastric tube to feed the patient [6]. Surgical resection of a tumor can also cause postoperative, neuropathic pain due to nerve damage and sometimes chronic pain due to scar tissue or nerve entrapment. Gabapentinoids like pregabalin are commonly used to treat neuropathic pain, often in combination with other analgesics to enhance efficacy. Pregabalin is frequently combined with other agents such as opioids, tricyclic antidepressants, serotonin–norepinephrine reuptake inhibitors, and nonsteroidal antiinflammatory drugs. Combining different mechanisms targets multiple pathways involved in neuropathic pain, potentially leading to better symptom control and the possibility of lower doses of individual drugs, reducing side effects [11]. The use of chemotherapy, especially platinum derivatives and taxanes, can lead to the onset or worsening of chemotherapy-induced peripheral neuropathy (CIPN), which is characterized by numbness, tingling, or pain in the hands and feet [11]. Cisplatin-based regimens are particularly associated with peripheral neuropathy (in 30%–50% of the patients who complete the full course of treatment, and in 10% of patients, it causes disabling symptoms) and hearing loss (in 40%–80% of patients). [12] Adequate treatment of these tumors can lead to the reduction or disappearance of the disease and reduction of symptoms directly caused by malignancy. Details about the use of adjuvant analgesics and treatment of neuropathy are covered in other chapters of this book.

Management of advanced head and neck squamous cell carcinoma with systemic treatment

HNSCC is often diagnosed at an advanced, inoperable stage of the disease. Systemic treatment is the standard of care for the treatment of metastatic and inoperable HNC, and the progress in systemic treatment that has been made over the past few years using anti-epidermal growth factor receptor (EGFR) therapy and immunotherapy, alone or in combination, has led to an improvement in overall survival (OS), progression-free survival, and improvements in QoL [13–17].

Historically, platinum-based chemotherapy (e.g., cisplatin or carboplatin) in combination with 5-fluorouracil (5-FU) has been the standard regimen for metastatic and unresectable HNC [13]. This approach, although effective in controlling disease progression, is associated with significant toxicity, including nausea, vomiting, neuropathy, and myelosuppression [12]. Cetuximab, an EGFR inhibitor, has been a major breakthrough for patients with unresectable or metastatic HNC. The EXTREME study, the results of which were published in 2008 for the first time, showed an improvement in median OS when cetuximab was combined with cisplatin and 5-FU, compared to the use of chemotherapy only, with a median OS of 10.1 months compared to 7.4 months. The objective response rate (ORR), which represents the significant reduction of tumor burden often associated with reduction of pain, was 36% in the EXTREME protocol group, compared to 20% in the control group, with a significant improvement in patients' QoL. The most common grade 3 or 4 adverse events in the chemotherapy alone and cetuximab groups were anemia (19% and 13%, respectively), neutropenia (23% and 22%), and thrombocytopenia (11% in both groups) [13].

The immune checkpoint inhibitors have revolutionized the treatment of recurrent, inoperable, or metastatic HNSCC over the past few years. Pembrolizumab (anti-programmed cell death protein 1 [PD-1]), in combination with chemotherapy, is an approved first-line therapy for the treatment of patients with inoperable or metastatic HNSCC, either regardless of programmed death-ligand 1 (PD-L1) biomarker values (according to National Comprehensive Cancer Network guidelines), or in the population of patients with positive CPS ≥1 (according to European Medicines Agency [EMA] indications) [14]. In the subgroup of patients diagnosed with Combined Positive Score (CPS) values >20 in tumor tissue, pembrolizumab is preferred as monotherapy because of the pronounced effectiveness of immunotherapy, and its use in combination with chemotherapy in that patient group does not add substantial benefit, but primarily toxicity [6]. Following progression on chemotherapy, and if immunotherapy was not previously used, two immunotherapeutic agents (pembrolizumab and nivolumab) can also be applied [13–16].

However, the use of PD-1 antibodies is often moved from the second to the first line of treatment following the results of KEYNOTE-048, a phase III study that compared the use of pembrolizumab, alone or in combination with chemotherapy, versus the EXTREME protocol (cetuximab with chemotherapy) in the first-line treatment for recurrent/metastatic HNSCC. Pembrolizumab alone improved OS over cetuximab with chemotherapy in tumors with CPS >20 (median 14.9 vs 10.7 months, hazard ratio (HR) 0.61), and in the population with CPS greater than or equal to 1 (12.3 vs 10.3 months, HR 0.78), and was noninferior in the overall population (11.6 vs 10.7 months, HR 0.85). Pembrolizumab with chemotherapy improved OS compared to cetuximab with chemotherapy in the overall population as well (13.0 vs 10.7 months, HR 0.77 [95% confidence interval, 0.63–0.93], $P = .0034$), in the population with CPS ≥20 (14.7 vs 11.0, HR 0.60) and in the population with a CPS ≥1 (13.6 vs 10.4, HR 0.65) [15].

If not given in the first line, pembrolizumab could also be applied in the second-line treatment following KEYNOTE-040, the results of which were published in 2017. This phase III trial compared pembrolizumab monotherapy versus standard-of-care control (methotrexate, docetaxel, or cetuximab) in second-line treatment in patients with HNSCC who had progressed to first-line platinum-based chemotherapy. Pembrolizumab improved OS compared with chemotherapy, with a median OS of 8.4 months versus 6.9 months, with an HR of 0.80. The ORR was 14.6% for pembrolizumab, and responses were generally profound and durable. Pembrolizumab also had a better safety profile compared to previous standard of care, with fewer grade 3–4 adverse events, which contributed to a better QoL for patients [16]. Both of these studies confirmed pembrolizumab as the standard of care in recurrent/metastatic HNSCC, especially in patients with high PD-L1 expression, due to its survival benefit and manageable safety profile [15,16]. The CheckMate 141 trial published in 2017 was a phase III study examining the efficacy of nivolumab monotherapy (PD-1 inhibitor) versus the standard of care (chemotherapy) in second-line HNSCC patients who had progressed to first-line platinum-based chemotherapy. Nivolumab improved OS to 7.7 months compared with 5.1 months with standard chemotherapy in platinum-resistant cases. One-year OS was 36% in the nivolumab group versus 16% in the chemotherapy-treated control group. The ORR was 13.3% in the nivolumab group versus 5.8% in the standard therapy group. Nivolumab significantly improved the patient's QoL and reduced physical symptoms and improved social functioning compared to the use of chemotherapy [17].

Impact of systemic treatment on quality of life

The mentioned EXTREME, KEYNOTE-048, and KEYNOTE-040 studies, in addition to evaluating the effectiveness and safety of these therapies, also evaluated the QoL and the treatment of pain, as one of the most common and severe symptoms of the disease. The appearance and severity of these symptoms were not always related to the primary tumor but also to numerous side effects of the cancer therapy they were receiving. Assessment of pain and QoL in the EXTREME study using the European Organisation for Research and Treatment of Cancer (EORTC) Quality of Life Questionnaire Core 30 (QLQ)-C30 and the HNC module, EORTC QLQ-H&N35, showed that the addition of cetuximab led to symptom relief in terms of difficulty swallowing and pain, which are critical aspects for QoL in these patients. However, no significant differences were found when assessing the overall QoL between the groups, probably due to the toxicity burden caused by the therapy [13,18]. Cetuximab has been associated with a high incidence of skin rash and stomatitis, which can cause discomfort and pain. Skin reactions usually manifest as an acne-like rash and require local or systemic interventions (e.g., corticosteroids, antibiotics). Stomatitis requires symptomatic treatment with mouthwashes, local anesthetics, and sometimes cetuximab dose modification [13]. The cytostatic agent cisplatin, which is a key component of the EXTREME regimen, is a well-known cause of peripheral neuropathy. Polyneuropathy occurs in 30%–50% of patients treated with cisplatin, especially in cumulative doses [12]. Treatment strategies for cisplatin-induced neuropathy include reducing the dose or switching to carboplatin, which has a lower incidence of neuropathy, or using neuropathic pain medications, such as the antiepileptic gabapentin or the antidepressant duloxetine. Neuropathic pain can also be treated with tricyclic antidepressants or opioids, depending on the severity [11]. QoL assessment in the KEYNOTE 048 study was performed using the EORTC QLQ-C30 and EORTC QLQ-H&N35 questionnaires. Patients on pembrolizumab, especially in the monotherapy group, showed a slower deterioration in QoL compared to those receiving the EXTREME protocol. Combining pembrolizumab with chemotherapy reduced pain and swallowing difficulties, and baseline EORTC QLQ-C30 Global Health Status QoL scores in the overall HRQoL population were well balanced across treatment groups with no significant difference [19]. When it comes to immunotherapy, pembrolizumab may be associated, in rare cases, with side effects that can cause pain, such as arthralgia (joint pain) in less than 2% of patients and myalgia (muscle pain) that occurs in 12% of patients. These side effects are well managed with symptomatic therapy, such as corticosteroid therapy, and in rare cases with other immunosuppressants [15,16].

Along with pembrolizumab, chemotherapy is often a critical part of treatment; however, despite an increase in efficiency, it contributes to the appearance of neuropathy and stomatitis. Peripheral neuropathy has been reported in approximately 30%–50% of patients, which is consistent with the frequency of cisplatin-based regimens, as is stomatitis in comparison with the application of the EXTREME protocol and it affected about 28% of patients [15]. As we have already stated, stomatitis is one of the frequent side effects of chemotherapy and anti-EGFR therapy, so correct and successful treatment of stomatitis can significantly improve the QoL of patients and thus their compliance and maintenance of the continuity of the use of oncological treatment. Stomatitis is treated with oral hygiene measures, mouth rinses (e.g., saline or lidocaine), and topical corticosteroids. Severe cases may require systemic analgesics or dose modification of the harmful agent [13]. In addition to the KEYNOTE 048 study, in the KEYNOTE 040 study pembrolizumab was associated with better QoL compared with standard chemotherapy, including less pain and difficulty swallowing. The improvement of symptoms, including the reduction of pain, was associated with the reduction in tumor size, caused by the therapeutic effects of pembrolizumab, usually within 6–8 weeks from the start of the therapy [16]. The advantage of using pembrolizumab monotherapy, compared to combined therapy, is a significant reduction of side effects, which occur as a consequence of the use of cytotoxic therapy, especially the use of cisplatin and cetuximab, such as neuropathy, stomatitis, and skin reactions [13–16].

Surgery and radiotherapy in pain management of head and neck squamous cell carcinoma

Surgery and RT, or a combination of chemotherapy and RT, represent the mainstay of treatment of the localized disease. Their use in patients with advanced or metastatic disease is limited, although for patients with recurrent operable disease, surgery with or without RT or chemoradiotherapy represents a plausible option even for patients with limited metastatic spread [6]. In palliative care, however, RT has a significant role in improving the QoL, although there is little high-level evidence and a lack of consensus to direct the selection of an optimal palliative radiation regimen. The treatment options include 30–32 Gy in 5–8 fractions, 24 Gy in 3 fractions, 20–25 Gy in 5 fractions, 30–40 Gy in 10 fractions, and 40–50 Gy in 16 fractions, while even regimens up to 72 Gy were evaluated [20]. Up to 85% response rate and 90% symptom relief were reported, although results dramatically differ between the trials, along with toxicity rates, which can result in up to 40% of grade 3 adverse effects [20]. A metaanalysis evaluating the efficacy of palliative RT schemes for locally advanced squamous cell carcinoma of the head and neck demonstrated that the stereotactic body radiotherapy (SBRT) or intensity-modulated radiation therapy approach yields significantly lower acute toxicity rates compared to conventional 2D or 3D RT, while the total biological effective doses (BED) of RT ($P = .001$), BED >60 Gy10 ($P = .001$), short course ($P = .01$), and SBRT ($P = .02$) were associated with a superior ORR [21].

Future implications in pain management and head and neck cancer therapy

The future of pain management is most likely to personalize the strategy of how to treat pain based on the genetic and molecular characteristics of the patient and the tumor. The identification of biomarkers that could show sensitivity to certain pain medications, as well as sensitivity to treatment-induced toxicity (e.g., neuropathy or stomatitis), may lead to the creation of protocols that would tell us how to treat pain better than we currently do. The metabolism of the drugs we administer is different for each patient, as well as the immune response they cause, so the personalization of therapy in the future will enable doctors to predict and alleviate, among other things, the painful side effects caused by this therapy. For example, paclitaxel is extensively metabolized in the liver, and cytochrome P450 enzymes play a key role in its biotransformation. Existing therapy, regardless of it being systemic therapy or analgesics, might target the same cytochrome enzyme, which might lead to higher toxicity or reduced effectiveness. Furthermore, many of the agents used in the treatment of HNSCC can also cause a painful syndrome that reduces food intake, leads to weight loss, and reduces the patient's QoL [13]. It is therefore necessary to work on the production of new molecules that would lead to similar efficiency as existing agents, with a more favorable profile of side effects of the drug, with reduction of pain and improvement of QoL.

When it comes to immunotherapeutics, such as checkpoint inhibitors, which can lead to side effects, such as arthralgia, myopathy, and neuropathy [15,16,22], research is underway to synthesize agents that would more precisely modulate the immune system in order to avoid such iatrogenic pain conditions. Also, it is necessary to improve symptomatic therapy, which is currently given in these conditions, such as corticosteroids and immunosuppressants, in order to balance the reduction of side effects and the impact of symptomatic therapy on the effectiveness of immunotherapy [23]. In this sense, the combination of immunotherapy with radioprotective agents or neuroprotective agents is being increasingly investigated in order to reduce inflammation and nerve damage in time, that is, to reduce the pain caused by the use of radiotherapy and chemotherapy, without jeopardizing the effectiveness of treatment [11].

A major problem in the treatment of HNC patients is the development of cisplatin-induced neuropathic pain (CIPN), most often with the use of certain cytostatics such as cisplatin or taxanes. This induced neuropathy is a significant cause of chronic pain, and new treatment modalities with neuroprotective agents are being developed to reduce or prevent CIPN in patients with HNC. Combining these neuroprotective strategies with chemotherapy could potentially provide pain relief while maintaining therapeutic efficacy [11,12].

Another way to reduce the pain caused by chemotherapy is the development of studies that investigate the use of cisplatin in smaller doses or its replacement with carboplatin. Also, the modification of the treatment protocol, which leads to the reduction in the dose or frequency of application of certain cytostatics, such as 5-FU or taxanes, leads to the reduction in the cumulative toxicity of these drugs. Great hopes are placed in nanotechnology, the technology of the future, for adequate and more precise delivery of pain medications to oncology patients. In this way, targeted systems would deliver painkillers directly to tumor cells or affected nerves, thus reducing systemic side effects while providing local pain control and improving efficacy [24].

What has been recognized as a necessity in the treatment of cancer patients during the past few years is a multimodal approach to pain treatment, which includes a combination of pharmacological, physical, psychological, social, and spiritual support. These approaches not only help with pain management but also address the psychological and emotional dimensions of chronic pain in patients with HNC. Early integration of palliative care as a parallel pathway in the treatment of HNC can improve patients' QoL, tolerability of specific cancer therapy, and better pain management. An increasing amount of data from clinical studies, which speaks about the advantage of applying early palliative interventions, emphasizes the need to include palliative care from the very beginning of a malignant disease. This modified approach results in more successful treatment of cancer-related pain syndrome, as well as reduced treatment-related toxicities, within a holistic approach to the cancer patient.

Conclusion

Patients with HNSCC can experience pain from tumor location, metastasis, and treatments like chemotherapy, radiation, and surgery. While the development of new systemic treatments, including anti-EGFR therapy (EXTREME study) and immunotherapy (KEYNOTE-048, CheckMate 141 trials), increased ORR, survival, and improved QoL compared to chemotherapy, a significant proportion of patients still experience pain, as well as cancer progression. Future directions include personalized medicine approaches, neuroprotective agents, and multimodal pain management strategies combining pharmacological and holistic care interventions.

References

[1] Licitra L, Karamouzis MV, editors. Head & neck cancers: essentials for clinicians. ESMO Press; 2017.

[2] Beddok A, Krieger S, Castera L, et al. Management of Fanconi anemia patients with head and neck carcinoma: diagnosis and treatment adaptation. Oral Oncol 2020;108:104816. Available from: https://doi.org/10.1016/j.oraloncology.2020.104816.

[3] Mehanna H, Beech T, Nicholson T, et al. Prevalence of human papillomavirus in oropharyngeal and non-oropharyngeal head and neck cancers—systematic review and meta-analysis of trends by time and region. Head Neck 2013;35(5):747–55. Available from: https://doi.org/10.1002/hed.22015.

[4] Argiris A, Karamouzis MV, Raben D, et al. Head and neck cancer. Lancet 2008;371(9625):1695–709. Available from: https://doi.org/10.1016/S0140-6736(08)60728-X.

[5] Bøje CR. Impact of comorbidity on treatment outcome in head and neck squamous cell carcinoma – a systematic review. Radiother Oncol 2014;110(1):81–90. Available from: https://doi.org/10.1016/j.radonc.2013.07.005.

[6] NCCN clinical practice guidelines in oncology (NCCN Guidelines®) head and neck cancers, version 5.2024, [October 24.10.24]. Available from: https://www.nccn.org/professionals/physician_gls/pdf/head-and-neck.pdf.

[7] Ye Y, Jensen DD, Viet CT, et al. Advances in head and neck cancer pain. J Dent Res 2022;101(9):1025–33. Available from: https://doi.org/10.1177/00220345221088527.

[8] Kapoor A, Kalwar A, Narayan S, et al. Analysis of bone metastasis in head and neck squamous cell carcinoma: experience of a regional cancer center. Clin Cancer Investig J 2015;4:206–10. Available from: https://doi.org/10.4103/2278-0513.148916.

[9] Soleimanpour H, Safari S, Shahsavari Nia K, et al. Opioid drugs in patients with liver disease: a systematic review. Hepat Mon 2016;16(4):e32636. Available from: https://doi.org/10.5812/hepatmon.32636.

[10] Dean M. Opioids in renal failure and dialysis patients. J Pain Symptom Manage 2004;28(5):497–504. Available from: https://doi.org/10.1016/j.jpainsymman.2004.02.021.

[11] Hu LY, Mi WL, Wu GC, et al. Prevention and treatment for chemotherapy-induced peripheral neuropathy: therapies based on CIPN mechanisms. Curr Neuropharmacol 2019;17(2):184–96. Available from: https://doi.org/10.2174/1570159X15666170915143217.

[12] Neife Aparecida Guimarães dos Santos S, Fontes RS, Pereira DR, et al. Overview of cisplatin-induced neurotoxicity and ototoxicity, and the protective agents. Food Chem Toxicol 2020;136:111079. Available from: https://doi.org/10.1016/j.fct.2019.111079.

[13] Vermorken JB, Mesia R, Rivera F, et al. Platinum-based chemotherapy plus cetuximab in head and neck cancer. N Engl J Med 2008;359(11):1116–27. Available from: https://doi.org/10.1056/NEJMoa0802656.

[14] European Medicines Agency. Keytruda [Internet]. Amsterdam: EMA; [cited 2026 May 20]. Available from: https://www.ema.europa.eu/en/medicines/human/EPAR/keytruda.

[15] Rischin D, Harrington KJ, Greil R, et al. Pembrolizumab alone or with chemotherapy versus cetuximab with chemotherapy for recurrent or metastatic head and neck squamous cell carcinoma (KEYNOTE-048): a randomised, open-label, phase 3 study. Lancet 2019;394(10212):1915–28. Available from: https://doi.org/10.1016/S0140-6736(19)32591-7.

[16] Cohen EEW, Soulières D, Le Tourneau C, et al. Pembrolizumab versus methotrexate, docetaxel, or cetuximab for recurrent or metastatic head and neck squamous cell carcinoma (KEYNOTE-040): a randomised, open-label, phase 3 study. Lancet 2019;393(10167):156–67. Available from: https://doi.org/10.1016/S0140-6736(18)31999-8.

[17] Ferris RL, Blumenschein G, Fayette J, et al. Nivolumab for recurrent squamous-cell carcinoma of the head and neck. N Engl J Med 2016;375(19):1856–67. Available from: https://doi.org/10.1056/NEJMoa1602252.

[18] Goerling U, Gauler T, Dietz A, et al. Quality of life of patients with head and neck cancer receiving cetuximab, fluorouracil, cisplatin comparing to cetuximab, fluorouracil, cisplatin, and docetaxel within the CEFCID trial. Oncol Res Treat 2022;45(6):319–25. Available from: https://doi.org/10.1159/000521415.

[19] Rischin D, Harrington KJ, Greil R, et al. Pembrolizumab alone or with chemotherapy for recurrent or metastatic head and neck squamous cell carcinoma: health-related quality-of-life results from KEYNOTE-048. Oral Oncol 2022;128:105815. Available from: https://doi.org/10.1016/j.oraloncology.2022.105815.

[20] Grewal AS, Jones J, Lin A. Palliative radiation therapy for head and neck cancers. Int J Radiat Oncol Biol Phys 2019;105(2):254–66. Available from: https://doi.org/10.1016/j.ijrobp.2019.05.024.

[21] Viani GA, Gouveia AG, Matsuura FK, et al. Assessing the efficacy of palliative radiation treatment schemes for locally advanced squamous cell carcinoma of the head and neck: a meta-analysis. Rep Pract Oncol Radiother 2023;28(2):137–46. Available from: https://doi.org/10.5603/RPOR.a2023.0021.

[22] Albarrán V, Chamorro J, Rosero DI, et al. Neurologic toxicity of immune checkpoint inhibitors: a review of literature. Front Pharmacol 2022;13:774170. Available from: https://doi.org/10.3389/fphar.2022.774170.

[23] Verheijden RJ, May AM, Blank CU, et al. Corticosteroids and other immunosuppressants for immune-related adverse events and checkpoint inhibitor effectiveness in melanoma. Eur J Cancer 2024;207:114172. Available from: https://doi.org/10.1016/j.ejca.2024.114172.

[24] Bhansali D, Teng SL, Lee CS, et al. Nanotechnology for pain management: current and future therapeutic interventions. Nano Today 2021;39:101223. Available from: https://doi.org/10.1016/j.nantod.2021.101223.

Chapter 33

Brain tumors

Ana Mišir Krpan[1,2] and Mihael Mišir[3,4]

[1]*Special Hospital Radiochirurgia Zagreb, Zagreb, Croatia,* [2]*Faculty of Dental Medicine and Health Osijek, University of Josip Juraj Strossmayer, Osijek, Croatia,* [3]*Department of Neurology and Neurosurgery, University Hospital Centre Osijek, Osijek, Croatia,* [4]*Faculty of Medicine Osijek, Josip Juraj Strossmayer University of Osijek, Osijek, Croatia*

Key learning objectives

- Recognize common mechanisms and clinical features of headache caused by primary brain tumors, brain metastases, and raised intracranial pressure.
- Identify red flags in patients with cancer-related headache that require urgent diagnostic evaluation and neuroimaging.
- Understand the role of MRI, CT, cerebrospinal fluid analysis, and specialist evaluation in diagnosing tumor-related headache.
- Describe treatment strategies for headache caused by brain tumors, including corticosteroids, analgesics, surgery, radiotherapy, and systemic therapy.

Introduction

According to the fifth edition of the World Health Organization (WHO) Classification of Tumors of the Central Nervous System (CNS), primary brain tumors are a heterogeneous group of neoplasms and range from very common benign tumors to highly invasive and virtually incurable tumors like glioblastoma (GBM) [1]. Brain metastases occur more often and can also be quite variable. These patients may have solitary or multiple brain metastases or leptomeningeal disease, and they may have a malignancy that is highly responsive or highly resistant to oncological treatment. In CNS tumors, there is sometimes no clear boundary between benign and malignant tumors. For example, a large skull base meningioma may cause pain, may require palliative treatment, and may ultimately lead to the same poor outcome as a malignant tumor.

The most common malignant primary brain tumor is GBM. The incidence rate increases with age, peaking at 75–79 years, and the median age at diagnosis is 64 years. The average annual age-adjusted incidence rate of GBM is 3.19 per 100,000 persons, and the overall prevalence is 9.23 per 100,000 persons [2]. The etiology of GBM remains unknown in most cases. Familial gliomas account for approximately 5% of malignant gliomas, and less than 1% of gliomas are associated with a known genetic syndrome.

The gold standard of treatment remains the treatment with surgery, after which chemoradiotherapy is applied, usually based on temozolomide. However, 5-year overall survival remains around 10% [3]. Despite the addition of both concomitant and adjuvant chemotherapy, the divergence in quality of life was only found in follow-up and only for social functioning, which was favored in the radiotherapy (RT) group. However, during subsequent evaluations, there was no significant difference in the quality–of–life evaluation, including pain assessment [4].

All brain tumors can be associated with headaches, and in some malignant brain tumors, like high-grade gliomas and metastases, pain control may be as important as oncological treatment. High-grade glioma patients, unlike the general cancer population, have specific symptoms in the end-of-life (EoL) phase. The majority of the patients experience headache, seizures, difficulty with swallowing and speech, progressive neurological deficits (especially motor deficits), incontinence, progressive cognitive deficits, and finally loss of consciousness. Symptoms differ in the early stage of the disease, during the course of the disease, or in the EoL stage. Pain is more pronounced at diagnosis and less commonly a challenge in EoL phase, compared to other tumors.

In brain tumor patients, pain can be caused by the tumor itself or induced by the treatment—surgery, RT, or systemic therapy.

Pain Management in Palliative Care for Patients With Cancer. DOI: https://doi.org/10.1016/B978-0-443-30050-9.00039-4

Headache caused by brain tumor

The headache may be attributed to a brain tumor if it has developed in close temporal relation with the development of the neoplasia, has significantly worsened in parallel with the worsening of the tumor, and/or has significantly improved following the successful treatment of the neoplasia.

A tumor headache is a consequence of pressure, traction, or infiltration of pain-sensitive intracranial structures as well as other structures of the head. A headache caused by a space-occupying tumor usually has at least one of the following characteristics: progressive, worse in the morning and/or when lying down, aggravated by Valsalva-like maneuvers and accompanied by nausea and/or vomiting. Headache can be an early sign and more frequent in fast-growing tumors like high-grade glioma because they tend to produce compression, and pain-sensitive structures do not have time to adapt to the direct or indirect pressure. The likelihood of headache is greater in young patients (including children), in patients with a history of primary headache, with rapid growth of the tumor, or with posterior fossa or midline localization of the tumor (International Classification of Headache Disorders) [5]. Hydrocephalus-causing tumors can cause headaches because they dislocate periventricular nociceptive structures. Headache is common in patients with leptomeningeal metastases, with incidence estimated between 32% and 75% throughout their disease course. The most common type of headache afflicting patients with leptomeningeal metastases, particularly early in the disease course, is generally attributed to meningeal irritation secondary to tumor cell and immunoinflammatory infiltrates within the subarachnoid space [6].

It is also important to define whether the headache is the "first or worst" or "new or different" [7]. Progression or deterioration of headache is a key feature. Symptoms of the first tumor headache vary in severity, but in most brain tumors, headache is initially absent or mild and tends to worsen over time. This progressive pattern of headache is often seen in high-grade glioma due to the mass effect or due to neuroendocrine secretion in pituitary adenomas.

Tumors can imitate and worsen primary headaches. In 23.5% of patients, tumor headaches also fulfill criteria for tension-type headache, and in 13.3%, for migraine without aura. On the other hand, patients with preexisting primary headaches (such as migraine and tension-type headache) have headaches more often than other patients with brain tumors (64% vs 38%) [8].

In patients with a brain tumor, headache is rarely isolated and is usually associated with some other neurological symptoms. In patients with raised intracranial pressure (ICP), headaches are more frequent in the bifrontal region and imitate moderate tension headaches [5,9–12]. In 42% of patients with infratentorial tumors, nausea and vomiting are present [8]. Headaches secondary to meningeal irritation tend to be intermittent, dull, or stabbing in character, and mild-to-moderate in intensity. They can be accompanied by mild photophobia and nausea. Classical meningeal sign, neck stiffness, is only present in 15% of patients [6]. If a headache worsens by exertion or coughing, sneezing, defecation, and bending over (or other Valsalva maneuvers), this symptom implies raised ICP, but according to clinical trials, it happens in only 5.1% of patients. Neurological deficits such as motor or sensory hemibody symptoms, dysphasia or dysarthria, visual field deficits, psychological changes, ataxia, paresthesia, epileptic seizures, anisocoria, and diplopia are the most important "red flags" mandating brain imaging.

In-depth evaluation of pain, which included headaches, is given in other chapters of this book.

Headache caused by brain tumor treatment

Another group of headaches in neuro-oncology patients is caused by oncological treatment with RT, systemic therapy, or other supportive therapy. RT can cause disruption of the blood–brain barrier (BBB), resulting in an increased permeability into the surrounding brain parenchyma. In one metaanalysis, 35% of the clinical studies included reported BBB disruption following RT [13]. Radiation-induced edema can develop following a single-fraction or multifraction stereotactic radiosurgery (SRS) for non–skull-base meningioma. Among analyzed studies that included only non–base of skull (non-BOS) tumors or separately grouped non-BOS tumors, symptomatic edema occurred in 5%–43% of patients and any edema occurred in 28%–50% [14].

Headaches are also a well-known side effect of many chemotherapeutic and immunotherapeutic treatments. These headaches lack localizing features on examination and have normal imaging. The patient must experience a few episodes of pain to meet the criteria for such a specific type of headache diagnosis. The temporal correlation to the start/change of the new therapy can be very instructive, especially in hormone-modulating therapies causing menstrual migraine-like headaches. The pain is mostly bilateral, with feelings of pressure and tightening and not throbbing.

Diagnostic considerations

A low threshold for investigation is suggested for any patient with a current or prior history of cancer, and all cancer patients with a new or changed headache pattern should be suspected to have brain metastasis unless otherwise evaluated. In addition to primary tumor or metastasis, hemorrhage into tumors, infarction, and infection could cause headache in patients with brain tumors. Magnetic resonance imaging (MRI) with and without contrast is a gold diagnostic standard [15]. In patients where MRI cannot be performed due to claustrophobia, metal implants and devices, or other technical issues, computed tomography with and without contrast should be done. If clinical symptoms and signs are compatible with leptomeningeal disease, but neuroimaging studies reveal negative findings, a cerebrospinal fluid (CSF) study including cytology should be considered. Due to the high rate (25%–50%) of false-negative CSF study analysis, a repeat of the study is required if clinical conditions are suggestive of tumors [16]. Additional examinations include an ophthalmological exam or evaluation by an ENT specialist and endocrinologist.

Treatment of headache caused by a brain tumor

General principles of treatment

The treatment of headache caused by primary or secondary brain tumors has two main components: an attempt to treat the cause of the pain by surgery, RT, or systemic therapy, or symptomatic treatment with antiedema therapy or analgesics.

In patients with headache and newly diagnosed intracranial tumors, antiedema therapy with corticosteroids should be started in addition to standard analgesic therapy. Further treatment modalities are directed toward surgery and tumor removal or reduction. Although postoperative headache remains in a significant percentage of patients, the most effective method is the removal of the pain cause [17]. If headache is the result of increased ICP, accompanied by nausea and vomiting, surgical intervention and tumor reduction should be strongly considered in order to reduce the mass effect, especially in patients with good life expectancy. For patients with a disturbance in the free flow of CSF through the ventricular system and an elevated ICP or hydrocephalus, procedures like ventriculostomy, ventricular shunting, or endoscopic third ventriculostomy could lead to symptom relief in the majority of patients.

Neurosurgical methods, evaluated in another chapter of this book, are the oncological treatment of choice for symptomatic relief, as RT takes several weeks or months for a full effect, so radiation is rarely used as a primary method of pain management. In fact, in patients with high-grade gliomas who initially have headaches, oncological treatment with radiation and chemotherapy can increase headaches due to BBB permeability, and therefore special caution and combined use of antiedema therapy are required before treatment can even begin [18]. If there are no symptoms of increased ICP, the prophylactic use of corticosteroids is not justified. In rare cases when there are no other treatment methods, such as in leptomeningeal carcinomatosis, whole-brain radiation can lead to pain reduction, especially if the tumor is radiosensitive. Other treatment modalities can influence quality of life and pain control. Tumor-treating fields therapy improves both progression-free and overall survival in patients with newly diagnosed GBM [19]. Although a positive effect on pain has been recorded, it is by no means a method of treating pain, especially not in a palliative setting [20].

For patients who have a headache and a past or present oncological disease, a low threshold for imaging must be applied. When brain tumors are detected, particularly with symptomatic edema, mass effect, raised ICP, or neurologic symptoms, corticosteroids should be the first line of treatment. Further treatment is considered depending on patient- and disease-related factors, available treatment modalities, and prognosis of the disease. A multidisciplinary approach is required and could include surgery, SRS, RT, or systemic therapy. Those with a less favorable prognosis need palliative care.

Oncology patients can also have a primary headache that is common in the healthy population and is not related to either the oncological disease or the therapy. It can also be present due to changes in lifestyle, for example, insomnia, anxiety, psychological stress, reduced movement, and physical activity. In addition to headaches, about 13%–25% of patients with brain tumors suffer from bodily pain. Its management is multimodal, and the approach taken is similar to pain management in patients with systemic cancers, including the WHO's sequential three-step analgesic ladder [18].

Nonpharmacologic and/or integrative interventions (e.g., massage, art or music therapy, acupuncture, relaxation, mindfulness, guided imagery) can be considered, but in brain tumor patients, reduced level of consciousness, decreased cognitive function, and lack of cooperation are present earlier in the course of the disease.

Systemic therapy agents in pain management

Corticosteroids

Corticosteroids are the mainstay of treatment for headaches in brain tumor patients. Dexamethasone has become the corticosteroid of choice in neuro-oncology because of its high potency, long half-life, and limited mineralocorticoid effects, but it can have significant adverse effects, evaluated in another chapter of this book [18]. Dexamethasone is a long-acting drug with a biologic half-life of 36–72 hours and therefore could be administered orally once a day [21]. Pharmacokinetics of oral versus intravenous (IV) dexamethasone show no significant difference in area under the curve [22]. Traditionally, for ease of prescribing, conversion of peroral (PO) to subcutaneous (SC) or intravenous (IV) dexamethasone was made on a 1:1 basis (e.g., 4 mg PO = 4 mg SC/IV). Vecht and colleagues conducted a randomized study of dexamethasone doses of 4, 8, and 16 mg/day. They concluded that administration of 4 mg dexamethasone per day for treatment of brain tumor edema results in the same degree of improvement as does administration of 16 mg/day. Toxic effects are dose dependent and, during a 4-week period, occurred more frequently in patients administered 16 mg dexamethasone per day [23].

Adverse events are increasing with higher doses and longer durations of treatment [24], and a downward titration of the dose (e.g., reducing dexamethasone by 1–2 mg every 3–5 days) is suggested whenever possible and when signs of neurologic stabilization are evident. Patients with mass effect that are prepared for RT need steroids for at least 24 hours before RT. For patients with a high risk of gastrointestinal (GI) side effects (i.e., perioperative patients, prior history of ulcers/GI bleed, receiving nonsteroidal antiinflammatory drugs [NSAIDs] or anticoagulation), H2 blockers or proton pump inhibitors are advised. Prophylactic treatment of *Pneumocystis jirovecii* pneumonia for patients undergoing long-term steroid therapy may be considered. When corticosteroids are associated with adverse events or unsatisfactory edema control, more effective therapy with bevacizumab could be introduced despite the lack of strong evidence from prospective randomized clinical trials [18].

Hyperosmolar agents

When the headache is severe and there are other signs of increased ICP, it is necessary to take additional measures to reduce the pressure. Guidelines for the acute treatment of cerebral edema suggest the addition of hyperosmolar agents (mannitol, hypertonic solution of 3% NaCl) with corticosteroids. Mannitol is widely used and very useful in decreasing ICP caused by edema, but optimal dosing and infusion rate are still debated. Mannitol-related complications are kidney injury, hypotension, electrolyte disturbances, a rebound effect after longer usage, and hyperglycemia [25]. The use of hyperosmolar agents in critical clinical situations merits close monitoring for adverse effects [26]. However, hyperosmotic therapy is not routinely recommended for all patients with brain tumor edema [27].

Analgesics and opioids

According to guidelines, NSAIDs, analgesics, and coanalgesics should be considered in the treatment of headache in patients with brain tumors, with the use of opioids for more significant pain [24]. Details regarding adverse effects of those different analgesics are given in specific chapters of the book. However, it is important to recognize a potential neurotoxicity induced by opioids, which can sometimes be misinterpreted as disease progression in patients with brain tumors. Special caution is required when using opioids in elderly patients with brain tumors due to an increased tendency to infections due to long-term use of corticosteroids and commonly found dehydration due to impaired consciousness or dysphagia [28].

Up to 85% of brain tumor patients have difficulty swallowing, particularly in the last weeks of life, both due to the inability to swallow (dysphagia) and in combination with decreased consciousness, which should influence the route by which analgesics are applied if necessary (SC injections or transdermal medications could be the preferred option) [24,29].

Anticancer medication

Temozolomide is the most commonly used drug in the GBM treatment; however, there is no effect on pain. Since temozolomide does not undergo hepatic metabolism and exhibits low protein binding, it is unlikely that it would affect the pharmacokinetics of other medicinal products. It is important to mention here that it can be safely administered with analgesics and majority of other drugs [30].

Conclusion

Atypical symptoms, changing of quality or intensity of primary headache, long-standing or worsening headache, and headache combined with neurological deficits are reasons for structural diagnostic procedures in any patient, especially in patients with a history of cancer. MRI with and without contrast is the gold standard. Headache is treated with analgesics, corticosteroids in the smallest possible dose for the shortest possible time, hyperosmolar agents in acute edema, and surgical intervention. As headaches in malignant primary or secondary brain tumors are usually caused by an expansive effect that leads to an increase in ICP, the most effective treatment is urgent decompression and surgical removal of the tumor. Oncological treatment with chemotherapy or RT, if successful, will ultimately lead to tumor shrinkage and pain reduction; however, the effect is slow. Antiedema therapy is used in all stages of the disease and is the main treatment in the palliative setting, although headache is not the dominant neurological symptom at the end of life. If analgesia or sedation is still required, morphine preparations may require other forms of drug administration, such as subcutaneously, due to other neurological symptoms such as dysphagia.

References

[1] Louis DN, Perry A, Wesseling P, et al. The 2021 WHO classification of tumors of the central nervous system: a summary. Neuro Oncol 2021;23(8):1231–51. Available from: https://doi.org/10.1093/neuonc/noab106.

[2] Ostrom QT, Patil N, Cioffi G, et al. CBTRUS statistical report: primary brain and other central nervous system tumors diagnosed in the United States in 2013–2017. Neuro Oncol 2020;22(12 Suppl. 2):iv1–96. Available from: https://doi.org/10.1093/neuonc/noaa200.

[3] Stupp R, Hegi ME, Mason WP, et al. European Organisation for Research and Treatment of Cancer Brain Tumour and Radiation Oncology Groups; National Cancer Institute of Canada Clinical Trials Group. Effects of radiotherapy with concomitant and adjuvant temozolomide versus radiotherapy alone on survival in glioblastoma in a randomised phase III study: 5-year analysis of the EORTC-NCIC trial. Lancet Oncol 2009;10(5):459–66. Available from: https://doi.org/10.1016/S1470-2045(09)70025-7.

[4] Taphoorn MJ, Stupp R, Coens C, et al. European Organisation for Research and Treatment of Cancer Brain Tumour Group; EORTC Radiotherapy Group; National Cancer Institute of Canada Clinical Trials Group. Health-related quality of life in patients with glioblastoma: a randomised controlled trial. Lancet Oncol 2005;6(12):937–44. Available from: https://doi.org/10.1016/S1470-2045(05)70432-0.

[5] Palmieri A, Valentinis L, Zanchin G. Update on headache and brain tumors. Cephalalgia 2021;41(4):431–7. Available from: https://doi.org/10.1177/0333102420974351.

[6] Wilcox JA, Estrera R, Boire A. The spectrum of headache in leptomeningeal metastases: a comprehensive review with clinical management guidelines. Curr Pain Headache Rep 2023;27:695–706. Available from: https://doi.org/10.1007/s11916-023-01180-9.

[7] Hainer BL, Matheson EM. Approach to acute headache in adults. Am Fam Physician 2013;87(10):682–7.

[8] Valentinis L, Tuniz F, Valent F, et al. Headache attributed to intracranial tumors: a prospective cohort study. Cephalalgia 2010;30:389–98. Available from: https://doi.org/10.1111/j.1468-2982.2009.01970.x.

[9] Suwanwela N, Phanthumchinda K, Kaoropthum S. Headache in brain tumor: a cross-sectional study. Headache 1994;34:435–8. Available from: https://doi.org/10.1111/j.1526-4610.1994.hed3407435.x.

[10] Do TP, Remmers A, Schytz HW, et al. Red and orange flags for secondary headaches in clinical practice: SNNOOP10 list. Neurology 2019;92:134–44. Available from: https://doi.org/10.1212/WNL.0000000000006697.

[11] Goffaux P, Fortin D. Brain tumor headaches: from bedside to bench. Neurosurgery 2010;67:459–66. Available from: https://doi.org/10.1227/01.NEU.0000372092.96124.E6.

[12] Schankin CJ, Ferrari U, Reinisch VM, et al. Characteristics of brain tumor-associated headache. Cephalalgia 2007;27:904–11. Available from: https://doi.org/10.1111/j.1468-2982.2007.01368.x.

[13] Hart E, Odé Z, Derieppe MPP, et al. Blood-brain barrier permeability following conventional photon radiotherapy – a systematic review and meta-analysis of clinical and preclinical studies. Clin Transl Radiat Oncol 2022;35:44–55. Available from: https://doi.org/10.1016/j.ctro.2022.04.013.

[14] Milano MT, Sharma M, Soltys SG, et al. Radiation-induced edema after single-fraction or multifraction stereotactic radiosurgery for meningioma: a critical review. Int J Radiat Oncol Biol Phys 2018;101(2):344–57. Available from: https://doi.org/10.1016/j.ijrobp.2018.03.026.

[15] Shekhar K, Chakra V. Assessment of diagnostic efficacy of CT scan and MRI in detecting brain tumors. Int J Med Res Prof 2018;4(1):511–4. Available from: https://doi.org/10.21276/ijmrp.2018.4.1.109.

[16] Cho S, Chu MK. Headache in brain tumors. Neurol Clin 2024;42(2):487–96. Available from: https://doi.org/10.1016/j.ncl.2023.12.004.

[17] Ravn Munkvold BK, Sagberg LM, Jakola AS, et al. Preoperative and postoperative headache in patients with intracranial tumors. World Neurosurg 2018;115:e322–30. Available from: https://doi.org/10.1016/j.wneu.2018.04.044.

[18] Koekkoek JAF, van der Meer PB, Pace A, et al. Palliative care and end-of-life care in adults with malignant brain tumors. Neuro Oncol 2023;25(3):447–56. Available from: https://doi.org/10.1093/neuonc/noac216.

[19] Stupp R, Taillibert S, Kanner A, et al. Effect of tumor-treating fields plus maintenance temozolomide vs maintenance temozolomide alone on survival in patients with glioblastoma: a randomized clinical trial. JAMA 2017;318(23):2306–16. Available from: https://doi.org/10.1001/jama.2017.18718. Erratum in: JAMA. 2018 May 1;319(17):1824. https://doi.org/10.1001/jama.2018.3431.

[20] Zhu JJ, Demireva P, Kanner AA, et al. Health-related quality of life, cognitive screening, and functional status in a randomized phase III trial (EF-14) of tumor treating fields with temozolomide compared to temozolomide alone in newly diagnosed glioblastoma. J Neurooncol 2017;135(3):545–52. Available from: https://doi.org/10.1007/s11060-017-2601-y.

[21] Liu D, Ahmet A, Ward L, et al. A practical guide to the monitoring and management of the complications of systemic corticosteroid therapy. Allergy Asthma Clin Immunol 2013;9(1):30. Available from: https://doi.org/10.1186/1710-1492-9-30.

[22] Spoorenberg SM, Deneer VH, Grutters JC, et al. Pharmacokinetics of oral vs. intravenous dexamethasone in patients hospitalized with community-acquired pneumonia. Br J Clin Pharmacol 2014;78(1):78–83. Available from: https://doi.org/10.1111/bcp.12295.

[23] Vecht CJ, Hovestadt A, Verbiest HBC, et al. Dose-effect relationship of dexamethasone on Karnofsky performance in metastatic brain tumors: a randomized study of doses of 4, 8, and 16 mg per day. Neurology 1994;44(4):675–80. Available from: https://doi.org/10.1212/WNL.44.4.675.

[24] Pace A, Dirven L, Koekkoek JAF, et al. European Association for Neuro-Oncology (EANO) guidelines for palliative care in adults with glioma. Lancet Oncol 2017;18(6):e330–40. Available from: https://doi.org/10.1016/S1470-2045(17)30345-5.

[25] Kim JH, Jeong H, Choo YH, et al. Optimizing mannitol use in managing increased intracranial pressure: a comprehensive review of recent research and clinical experiences. Korean J Neurotrauma 2023;19(2):162–76. Available from: https://doi.org/10.13004/kjnt.2023.19.e25.

[26] Cook AM, Morgan Jones G, Hawryluk GWJ, et al. Guidelines for the acute treatment of cerebral edema in neurocritical care patients. Neurocrit Care 2020;32(3):647–66. Available from: https://doi.org/10.1007/s12028-020-00959-7.

[27] Esquenazi Y, Lo VP, Lee K. Critical care management of cerebral edema in brain tumors. J Intensive Care Med 2017;32(1):15–24. Available from: https://doi.org/10.1177/0885066615619618.

[28] Godwin B, Frank C, Molnar F, et al. Identification and management of opioid-induced neurotoxicity in older adults. Can Fam Physician 2022;68(4):269–70. Available from: https://doi.org/10.46747/cfp.6804269.

[29] Pace A, Di Lorenzo C, Guariglia L, et al. End of life issues in brain tumor patients. J Neurooncol 2009;91:39–43. Available from: https://doi.org/10.1007/s11060-008-9670-x.

[30] Temodal, INN-temozolomide: summary of product characteristics. Amsterdam: European Medicines Agency; 2024. Available from: https://www.ema.europa.eu/en/documents/product-information/temodal-epar-product-information_en.pdf.

Chapter 34

Lower gastrointestinal cancers, gastrointestinal stromal tumors, and neuroendocrine neoplasms

Borislav Belev[1,2]
[1]*Department of Medical Oncology, Clinical Hospital Center Zagreb, Zagreb, Croatia,* [2]*Faculty of Medicine, University of Zagreb, Zagreb, Croatia*

Key learning objectives

- Describe the major pain-related clinical features of colorectal cancer, anal cancer, gastrointestinal stromal tumors, and neuroendocrine neoplasms.
- Understand how tumor location, obstruction, metastatic disease, and treatment-related complications contribute to pain in lower gastrointestinal cancers.
- Identify the role of surgery, radiotherapy, systemic therapy, targeted therapy, and supportive care in symptom and pain control.
- Recognize the importance of individualized, multidisciplinary management for patients with gastrointestinal malignancies and cancer-related pain.

Colorectal cancer

Epidemiology

According to global data, colorectal cancer (CRC) is the third most common cancer and the second most common cause of cancer mortality worldwide [1,2]. The highest age-standardized mortality rate was estimated for Europe (30.4 per 100,000 person-years) and was higher for men than women [1,2]. Early-onset CRC becomes an emerging trend, and a shift to younger age may include different reasons such as genetic predispositions, exposure to environmental, and various lifestyle-related factors (such as hyperlipidemia, obesity, alcohol consumption, etc.) [3,4].

Diagnosis

Diagnosis of CRC includes colonoscopy with the histopathological examination, computerized tomography (CT), magnetic resonance, and positron emission tomography/CT (PET/CT), with laboratory testing including tumor marker carcinoembryonic antigen, which is crucial before the initiation of the treatment. The majority of CRCs are adenocarcinoma, and this chapter will focus on these cancers.

Treatment of localized disease

In general, the majority of operable colon cancers are treated surgically, and adjuvant chemotherapy is applied for patients with higher-stage disease, depending on the patient's condition and comorbidities. Neoadjuvant approaches, however, including immunotherapy, are increasingly being recognized as potential treatment avenues. On the other hand, while some rectal cancers can be treated only endoscopically, locally advanced rectal cancers, even when they are operable, are usually treated with neoadjuvant chemo, radiotherapy (RT), or chemoradiotherapy before operation. Adjuvant treatment can also be applied after operation, usually for colon cancer, rarely for rectal ones. Several novel biomarkers, particularly the presence of microsatellite instability (MSI), hold crucial information and can guide treatment choices, even in early CRC, but are absolutely necessary in the metastatic setting [5–7].

Pain Management in Palliative Care for Patients With Cancer. DOI: https://doi.org/10.1016/B978-0-443-30050-9.00038-2

Treatment of metastatic disease

Recent advances in molecular profiling of metastatic CRC facilitate the ability to direct treatments to the biological features of the tumor. It is an important step toward personalized treatment, but cures remain uncommon. Although patients live longer, symptomatic treatment becomes increasingly important.

However, approximately one-third of patients with metastatic CRC have oligometastatic disease, either synchronous or metachronous, and the treatment intention in these cases can be curative, with 5-year overall survival (OS) rates of 35%–67% after metastasectomy [8]. Most prospective studies have concentrated on patients included in noncurative systemic therapy trials [9,10], some on patients undergoing liver resection [11,12], and a few on cytoreductive surgery [13] or resection of multiorgan metastases [14]. Health-related quality of life (HRQoL) is also an important outcome measure, along with progression-free survival (PFS) and OS, not only for curative but also for noncurative diseases, including palliative chemotherapy [15,16]. HRQoL is a multidimensional concept including physical, functional, social, and emotional components [17] and can be measured with different tools, cancer-specific or nonspecific questionnaires [18,19].

However, management of the majority of metastatic colorectal cancers (mCRCs) is palliative, although new treatments are available and include targeted therapy and checkpoint inhibitors, which may significantly prolong OS. Generally speaking, combining various modalities in patient-tailored treatment strategies enables an optimal outcome whenever possible, which means that evaluating patients on a multidisciplinary team (MDT) consisting of different experts in the field is crucial. Along with its effects on PFS and OS, the effect of systemic therapy on quality of life (QoL) and pain management is not negligible, especially in patients with bulky disease and in earlier lines of therapy [20].

The effect of first-line chemotherapy on quality of life

Pain in CRCs is present in up to 70% of the patients, with up to 40% of the patients experiencing neuropathic pain and almost up to a quarter of patients suffering from painful colonic obstructions. The most likely cause of pain is direct pressure via metastases, most commonly the liver, and could be both visceral and nociceptive. However, several chemotherapy options can also cause neuropathic pain (oxaliplatin), pain due to mucositis (5-fluorouracil), or cramps and diarrhea (irinotecan) [21]. Pain in CRC is very complex in its pathogenesis, demanding different treatment approaches, and systemic therapy can primarily help with the pain associated with tumor burden, which is a nociceptive component of the pain [22].

There are several treatment options for the treatment of first-line mCRC, which depend on tumor sidedness, molecular profile, and patient characteristics. FOLFOXIRI + bevacizumab can be given regardless of mutation status, producing response rates of up to 69%, with a median OS of 30.2 months. Importantly, even in unresectable settings, 39% of patients were able to convert to surgically amenable disease [23].

Another trial demonstrated that, despite a significant reduction in tumor burden following treatment with chemotherapy and bevacizumab, there was an actual increase in pain ($P = .003$), along with nausea and vomiting ($P = .003$), diarrhea ($P = .021$), and decreased appetite ($P = .003$). On the other hand, there was an improvement in emotional function ($P = .009$), cognitive function ($P = .033$), and perception of body image ($P = .026$), all of which were observed after 6 months of treatment. However, such results might also suggest the natural progression of the disease [24]. A similar randomized CHARTA trial tested the addition of irinotecan to FOLFOX/bevacizumab in the treatment of first-line mCRC and met its primary end point of an improved PFS rate at 9 months [16]. There was no significant difference in the median Global Health Status (GHS) score at weeks 0, 8, 16, and 24 between the two total population treatment groups. No significant difference was observed in either clinical group comparing both treatment arms; however, the GHS mean values showed a trend toward improvement at a clinically relevant difference during therapy for patients treated with FOLFOXIRI/bevacizumab among clinical group 3 patients (patients with no symptoms or risk of rapid deterioration) (>10 points, $P = .185$). Clinically relevant changes in GHS score (at least 10 points) were similar between the two groups, showing an improvement in 44.3% and 39.7% ($P = .629$) and deterioration in 21.5% and 26.9% ($P = .461$) for the control group compared to the experimental group, respectively. This study highlighted the importance of clinical stratification in guiding the implementation of specific protocols [16]. However, longer follow-up demonstrates a decrease in QoL regardless of the protocol used, primarily due to the natural progression of the disease [25].

In patients with rat sarcoma virus oncogene (RAS), rapidly accelerated fibrosarcoma oncogene (RAF) wild-type (WT) disease, FOLFOX-panitumumab (an anti-epidermal growth factor receptor [EGFR] antibody) is another option, which produces a similar overall response rate, disease control rate, PFS, and OS to FOLFOXIRI-bevacizumab. While the latter group has a higher incidence of chemotherapy-related adverse effects, grade 3 or 4 (59% vs 46%, $P = .04$), the treatments produce different adverse effects, with skin rash and hypomagnesemia more common in the panitumumab arm, and arterial hypertension and thromboembolic incidents more common in the bevacizumab arm [26]. However, despite a high rate of adverse effects, FOLFOXIRI-bevacizumab produces a similar global QoL score to FOLFOX-bevacizumab (59.8 vs 58.8; $P = .726$) [27].

What does adding anti-EGFR add to chemotherapy? In the trial CRYSTAL, adding cetuximab (another anti-EGFR antibody) to first-line FOLFIRI significantly improved outcomes in patients with Kirsten rat sarcoma virus oncogen (KRAS) WT mCRC [28]. However, there were no significant differences in GHS/QoL ($P = .12$) and social functioning scores ($P = .43$) between the two treatment arms [29]. What was of concern, skin toxicity in patients receiving cetuximab, did not significantly alter QoL scales. Another phase 3 trial, TAILOR (evaluating cetuximab to FOLFOX-4), demonstrated similar results in QoL analysis [30].

While there is an issue and dilemma on whether to put patients on chemo holiday or continue the treatment with maintenance, the data showed that treatment with panitumumab combinations resulted in an early deterioration of QoL during induction with partial recovery during maintenance, showing the importance of a quality follow-up of the patients and the effect of maintenance therapy on QoL [31].

The effect of immunotherapy and targeted therapy on symptoms

For a rare group of patients with MSI, immune checkpoint inhibitors remain the key. KEYNOTE-177 showed that treatment with pembrolizumab did not reach a median OS during follow-up, compared with 36.7 months with chemotherapy (hazard ratio [HR] 0.74; 95% confidence interval [CI] 0.53–1.03; $P = .036$). Serious adverse events attributed to study treatment occurred in 25 (16%) patients in the pembrolizumab group and in 41 (29%) patients in the chemotherapy group, suggesting a more favorable side effect profile compared to standard chemotherapy [32].

On the other hand, CheckMate 8HW demonstrated that 24-month PFS was 72% with nivolumab plus ipilimumab as compared with 14% with chemotherapy. QoL data, including the pain scales, were also positive, with the new combination surpassing the prespecified threshold for meaningful improvements starting at week 13 in the nivolumab-plus-ipilimumab group [33].

While additional data from CheckMate 8HW is awaited, a phase 2 trial with the same regimen and in a similar population demonstrated a response rate (RR) of 69%, while a total of 84% of patients had a reduction in tumor burden from baseline. The 24-month PFS was 73.6%, while the 12-month PFS was 76.4%, suggesting that patients who achieve at least 1 year without disease recurrence have a chance for long-term remission. QoL appeared largely stable over the treatment period, although pain levels were lower for the majority of the 115-week follow-up compared to the baseline [34].

For patients with B-Raf protooncogene (BRAF)-mutated tumors, a potential option is a combination of chemotherapy with encorafenib (BRAF inhibitor) and cetuximab. Compared to the standard of care, the new combination improved the response rate (60.9% vs 40.0%, odds ratio [OR] 2.4), although immature OS data were available. However, the HR was 0.47 in the primary analysis, with 1-year survival rates of 79.5% versus 66.1%. No QoL data or pain scales are available, but the incidence of adverse effects was similar between the two arms [35].

While current data do not support first-line palliative treatment of human epidermal growth factor receptor 2 (HER2)+ patients with targeted treatments, several new positive drugs, such as trastuzumab with tucatinib, showed remarkable promise. In the chemotherapy-refractory setting, the combination resulted in an almost 40% response rate, with three patients even achieving a complete response, although no QoL data were available. New studies are awaited, but data from other cancer settings suggest a potential for HER2-targeted therapies even in earlier stages, while the quality of the data and its impact on pain are also awaited [36].

The effect of later therapy lines on symptoms

While several effective options exist in the first-line setting, second-line chemotherapy is associated with significantly lower responses compared to first-line therapy, although this depends widely on tumor laterality, molecular profile, and patient characteristics [37].

A HRQoL from second-line chemotherapy was undertaken by researchers in the PRODIGE18 trial. Patients treated with chemotherapy and bevacizumab reported more frequent diarrhea but no other relevant impairment of patients' HRQoL between the two treatment arms. Pain scores showed reduction at the 8–10-week period but increased after the 20th week, demonstrating that while initial response to chemotherapy can lower the pain burden, natural disease progression and increase in pain can be expected after several months of second-line chemotherapy with a median time-until-definite deterioration just above 5–6 months regardless of the regimen used [38].

Following the progression of the first two lines of chemotherapy, treatment with trifluridine tipiracil and bevacizumab could yield the highest responses (6% partial response, median OS just under 11 months) [39] with QoL data from the real world showing relatively stable levels of pain during the trifluridine tipiracil treatment [40]. However, treatment with regorafenib, fruqintinib, or rechallenge/reintroduction are all valid but palliative options [5].

The effect of local therapy on symptoms

Along with systemic therapy, advanced or metastatic rectal cancer treated with palliative radiation can result in a 79% reduction of pain after 1 month with a median duration of 4.5 months [41]. Radiation can also be used in conjunction with chemotherapy, with one study reporting a 100% RR of pain relief, with a median duration of 7.8 months; both numbers were significantly higher than those with RT alone [42].

While palliative surgery could be an option for select patients, maximal tumor debulking in combination with palliative systemic therapy produced no difference in HRQoL compared with palliative systemic therapy alone [43]. On the other hand, a Finnish study evaluated HRQoL data in curatively treated mCRC, with or without systemic treatment, demonstrating that HRQoL remained stable during first-line to later-line treatments but declined with tumor progression. They showed that symptom burden was lower after treatment and remained slightly higher but stable throughout all phases of systemic therapy [44].

Although metastasectomy is the aim in every curative intention of liver-only mCRC, local ablative treatment (LAT) is an acceptable alternative therapy [45]. The CLOCC trial compared systemic therapy with or without radiofrequency ablation and demonstrated that adding a local treatment might prolong OS in patients with nonresectable CRC liver metastases [46]. It is unknown whether any LAT technique is superior to others. The ultimate aim of LAT is a longer, disease-free interval, as well as improved QoL [47–49].

Conclusion

In summary, most systemic therapies for mCRC are palliative.

However, the advancement of immunotherapy and development of LATs may result in long-term complete remission for a significant number of patients. For other patients, a substantial RR could be achieved in over 60% of cases. Although data on QoL remain limited, the associated reduction in tumor burden may be clinically meaningful. This is relevant in both first- and second-line treatment settings, where a considerable proportion of patients may experience pain relief, particularly during the initial months following treatment initiation. RT and surgery are viable options for reducing symptom burden, but an MDT must be involved in such decision-making.

However, some patients with CRC experience persistent discomfort reflecting oxaliplatin-related neuropathy, and so far, there is no unique established treatment to prevent and effectively treat chemotherapy-induced peripheral neuropathy (CIPN) [50]. Nevertheless, pharmacological treatment options include opioids, nonopioids, and adjuvant analgesics; doses and schedules for all of these medications are available in other chapters of this book [51].

About 24% of patients with CRC suffer from painful colonic obstruction, with 30-day postoperative mortality reaching 40% in end-stage disease. Image-guided percutaneous cecostomy/colostomy presents a high rate of technical success and can be used as an alternative to open and laparoscopic surgery; this is also covered in the chapter on surgical approaches [52–54].

Anal cancer

Anal cancer is a rare cancer usually associated with human papillomavirus and most commonly squamous cell in origin; the majority of patients are women. Diagnosis typically includes endoscopy with histopathology and a pelvic magnetic resonance imaging (MRI) with additional CT, MRI, or PET/CT scans to exclude metastatic disease, with a gynecological examination recommended in women [55].

Localized anal cancer can be cured by concomitant chemoradiation and is very effective in localized disease, which is specific for this disease [56,57]. Surgery is usually reserved for salvage therapy in recurrent disease, as well as for local complications [58].

Pain in anal cancer patients can occur usually due to local or metastatic tumor burden, with metastasis commonly involving the liver, lung, extrapelvic lymph nodes, peritoneum, and bones [59]. Capsular pain from liver metastasis is typically sharp and located in the right upper quadrant. Patients with liver capsular pain typically respond to nonsteroidal antiinflammatory drugs and corticosteroids, although regular doses of opioids are the mainstay of pain management. Side effects, toxicities, and tolerance might increase as the liver function worsens, and analgesic modifications are frequently necessary. Invasion of the porta hepatis, which results in referred pain from the liver to the shoulder via the phrenic nerve, is common but often minimized by clinicians. In some selected cases, even liver resection can be attempted to reduce the tumor burden [60,61].

Intensive treatment, particularly for locoregional disease, can also result in long-term toxicities, including pain. Patients with a higher T stage or a tumor relapse (compared to disease remission) showed significantly higher symptom burden, demonstrating the effect of tumor size on QoL. Older patients also suffered more often from pain compared to younger ones. On the other hand, treatment with intensity-modulated RT resulted in several improvements compared to 3D RT, including lower scores on the symptom scale for pain (−35.9 points) [62].

Locally advanced disease/relapses may cause similar symptomatic manifestations such as rectal cancer, and local RT may provide some relief; data on long-term QoL of cancer patients confirmed enormous deficits in these functions. In comparison to a healthy cohort, cancer patients suffered more often from depression and other issues regarding their emotional well-being [63,64].

The standard of care for metastatic anal cancer remains the chemotherapy combination of carboplatin and paclitaxel, which achieved 8.1 months of median PFS compared to 5.7 months reached for the cisplatin and 5-fluorouracil combination. The OS was 20 and 12.3 months, respectively (HR 2.00, $P = .014$). Importantly, for symptom management, the combination of paclitaxel and carboplatin achieved an overall RR of 59%, with 12.8% of complete responses. Unfortunately, the QoL assessment was limited due to poor compliance. However, the authors note that the Global Health Status score remained unchanged for carboplatin plus paclitaxel at 24 weeks, suggesting that this chemotherapy combination can at least halt the rapid deterioration expected from disease progression [65]. Second-line treatment includes checkpoint immunotherapy; studies demonstrated responses in up to around a quarter of patients, although there is a lack of data on QoL and pain reduction in later lines [66,67].

Interesting data also suggested that higher levels of pain were significantly associated with shorter survival, although multivariate analysis did not confirm the findings. Interestingly, the authors found no differences between patients with and without distant metastases in most areas of QoL studied [68].

Integration of palliative and analgesic therapy is necessary due to very intensive upfront treatment [60]. An interdisciplinary approach is necessary because oncologists alone sometimes cannot manage the complexity of this demanding clinical situation. Today, it is well established that earlier integration of palliative care, especially in the outpatient setting, is preferable and may minimize hospitalizations at the end of life. On the contrary, hospitalization at the end of life for patients with cancer negatively impacts a patient's QoL [60,69].

Gastrointestinal stromal tumors

Gastrointestinal stromal tumors (GIST), though rare, are the most common mesenchymal tumors of the gastrointestinal (GI) tract. The most common site of GISTs is the stomach, followed by the small intestine. Usually, GISTs are clinically represented by bleeding, or they are found incidentally during surgery or diagnostic procedures. A full radiological and endoscopic staging must be completed before treatment initiation. Rarely, GIST may present as a bulky disease, causing a compressive effect on surrounding tissues. In such cases, the disease is symptomatic, and the introduction of neoadjuvant treatment leads to the alleviation and improvement of symptoms and overall condition [70].

The nature of GISTs became better understood with the identification of KIT expression and c-KIT mutations. Thereafter, platelet-derived growth factor receptor alpha (PDGFRA) mutations were also recognized as oncogenic drivers in the majority of GISTs. Although most GISTs have either a mutation of KIT or PDGFRA kinase as driver mutations, approximately 10% of GISTs do not harbor these and are collectively grouped as KIT/PDGFRA-WT GIST, which usually respond more poorly to treatment.

Targeted therapy is the mainstay of systemic treatment for GIST [70,71]. Imatinib is still the "gold standard" of first-line treatment for the majority of patients due to its durable efficacy and favorable safety, except for patients with specific mutations [72]. Patients experiencing progressive disease on imatinib are suggested to receive second-line sunitinib, with a median PFS of approximately 6–9 months [70]. Regorafenib is currently established as a third-line therapy, with a longer median PFS of approximately 4–5 months compared to the placebo arm [73]. Ripretinib, a switch-control tyrosine kinase inhibitor (TKI), by dual binding to both the switch pocket and the activation loop, locks the kinase in an inactivated state. The phase III trial (INVICTUS) confirmed the role of ripretinib as a fourth-line treatment, based on a median PFS of over 6 months compared to only 1 month in the placebo arm. Mutation analysis is relevant for evaluating the efficacy of approved TKIs in patients with advanced GIST. Patients with an exon 11 mutation appear to be sensitive to standard-dose imatinib (400 mg daily), whereas those with an exon 9 mutation are less responsive to this dose [74]. Although most GISTs respond well to molecular therapy, they eventually develop resistance. The median time to progression is approximately 2 years [70]. Although a change of systemic therapy remains a mainstay of treatment, in very selective cases, surgery may show some benefit in alleviating symptoms [75].

The impact of TKIs on HRQoL in patients with GIST has been demonstrated in several studies. While one trial showed that the effect of imatinib on global health status did not vary significantly between baseline and 1 year of treatment, other researchers demonstrated an improvement in HRQoL in 25% of patients after 18 months of treatment, as well as a significant reduction in pain compared to the placebo. Over a third of patients on sunitinib in a second-line setting also showed an improvement in global QoL, while no such data were demonstrated for regorafenib. On the other hand, ripretinib treatment demonstrated maintenance of stable HRQoL compared to deterioration in the placebo group [76]. However, due to a low number of patients on further lines of TKIs, it is not reliable to evaluate the effect of these treatments on HRQoL.

The perspective on disease and the phase of treatment is undoubtedly crucial. In a cross-sectional analysis, it was found that the score for pain was the highest in the second or subsequent lines of treatment, similar to that in a palliative setting, which is significantly higher than in the first-line treatment or an adjuvant setting. Importantly, TKIs themselves can cause pain, particularly abdominal or muscle pain or stomatitis-associated pain, which must be closely monitored [70,76].

Neuroendocrine neoplasms

Neuroendocrine neoplasms (NENs), especially gastroenteropancreatic-NENs (GEP-NEN), are unusual and relatively rare neoplasms that present numerous clinical challenges. They characteristically synthesize, store, and secrete a variety of peptides and neurotransmitters, which can lead to the development of distinct clinical syndromes, including the carcinoid syndrome; however, many remain clinically silent until presentation with a mass effect. The majority of neuroendocrine tumors (NETs) have an indolent course [77]. Staging includes a CT scan of the thorax, abdomen, and pelvis (with MRI as a valid alternative), as well as specific imaging modalities such as somatostatin receptor PET/CT or MRI. Operable tumors are usually treated with surgical procedures, usually with no adjuvant treatment. For metastatic or inoperable disease, multiple new treatments have been approved in recent years, primarily based on evidence of tumor growth inhibition and disease progression delay [78].

Nevertheless, NETs are associated with a wide range of symptoms, from excessive hormone production to tumor burden. The most commonly reported symptoms are fatigue, diarrhea, flushing, and pain [79], which can have a significant impact on daily living and QoL [80]. Patient-reported outcomes should play a pivotal role in selecting treatment options in NET, as symptoms can be debilitating. In recent years, HRQoL measurements have gained increasing relevance in clinical trials [81]. A key characteristic of NENs is their functional activity and the production of endocrine syndromes, which can sometimes be more threatening to patients than the tumor burden itself. If there is pain, it may be caused by a fibrosis of the mesenteric folds, which leads to compression or fibrosis of the mesenteric duplicature. Tumors producing serotonin usually produce such an effect (predominantly mid-gut NETs) [82,83]. Pancreatic or lung NETs may have a mass effect, with a compressive effect on surrounding tissues, vessels, and other structures [84,85]. There are hormone-related symptoms and tumor burden–related symptoms (Table 34.1), and corrections of all these symptoms with systemic or locally ablative treatment can be associated with an increase in QoL, as well as the control of pain.

The management strategy for NEN is very complex and demands an interdisciplinary approach and centers of excellence to achieve better outcomes.

Treatment with lanreotide (120 mg subcutaneously per month) for advanced somatostatin receptor–positive NETs was associated with significantly prolonged PFS (HR, 0.47). However, despite the significant effect on reducing the progression, there were no significant between-group differences in QoL or OS, although there was a trend toward pain reduction in the lanreotide arm (–0.68 [standard error: 2.60]). However, the authors used the change from baseline to the last postbaseline value available, which limits the evaluation of the nuances and different time points [86].

Similar results were obtained for octreotide (long-acting, 30 mg intramuscularly every month) in a similar patient population, with a significantly longer time to tumor progression (HR = 0.34; 95% CI, 0.20–0.59; $P = .000072$) compared with placebo. Importantly, a more detailed effect on QoL was evaluated. Octreotide treatment was associated with a significantly longer time to deterioration for fatigue, insomnia, and pain (not reached vs 18.2; $P = .0435$), showing the analgesic effect of octreotide, along with its antitumor effect [87].

Everolimus also remains a potential option for advanced nonfunctional NETs, as it was associated with a 52% reduction in the estimated risk of progression or death compared to placebo [88]. However, no relevant differences were noted

TABLE 34.1 Symptoms in neuroendocrine tumors.

Hormone-related symptoms	Diarrhea, flushing, fatigue, loss of appetite, dyspnea, palpitations, weight loss
Tumor burden-related symptoms	Abdominal pain, abdominal distension, ascites, jaundice, compression of adjacent organs

between the everolimus and placebo groups for QoL parameters at any of the examined time points (weeks 8–48), although separate data were not provided for pain scores [89].

A significant change in treatment paradigm was achieved with the use of 177Lu-DOTA-TATE plus long-acting octreotide, which extended median PFS by 14 months in patients with grade 2 or 3 well-differentiated, advanced gastroenteropancreatic (GEP)-NEN compared to long-acting octreotide alone in treatment-naive patients [90,91]. The previous trial with 177Lu-DOTA-TATE in the second-line demonstrated a longer time to deterioration of QoL parameters for global health status (HR, 0.406), physical functioning (HR, 0.518), role functioning (HR, 0.580), fatigue (HR, 0.621), pain (HR, 0.566), diarrhea (HR, 0.473), disease-related worries (HR, 0.572), and body image (HR, 0.425). Such data demonstrate that along with longer survival, treatment with 177Lu-DOTA-TATE also results in a significant analgesic effect, even in a second-line setting [92].

Conclusion

Lower GI cancers represent a heterogeneous group of cancers. For metastatic settings, chemotherapy with biological therapy remains key for colon adenocarcinoma (except for immunotherapy in a small group of patients), while TKIs are a mainstay for GIST treatment and somatostatin analogs for NENs. Significant response and tumor reduction can be achieved but are predominantly reported for first-line settings. Emerging molecular and cell-based therapies are poised to transform the treatment landscape for lower GI cancers, with advances such as pan-RAS inhibitors, antibody-drug conjugates, and mKRAS-targeted vaccines showing promise for overcoming resistance and potentially resulting in more sustained and durable responses.

References

[1] Bray F, Laversanne M, Sung H, et al. Global cancer statistics 2022: GLOBOCAN estimates of incidence and mortality worldwide for 36 cancers in 185 countries. CA Cancer J Clin 2024;74(3):229–63. Available from: https://doi.org/10.3322/caac.21834.

[2] Siegel RL, Wagle NS, Cercek A, et al. Colorectal cancer statistics, 2023. CA Cancer J Clin 2023;73:233–54. Available from: https://doi.org/10.3322/caac.21772.

[3] Howren A, Sayre EC, Loree JM, et al. Trends in the incidence of young-onset colorectal cancer with a focus on years approaching screening age: a population-based longitudinal study. J Natl Cancer Inst 2021;113:863–8. Available from: https://doi.org/10.1093/jnci/djaa220.

[4] Murphy N, Campbell PT, Gunter MJ, et al. Unraveling the etiology of early-onset colorectal cancer. J Natl Cancer Inst 2021;113(5):505–6. Available from: https://doi.org/10.1093/jnci/djaa165.

[5] National Comprehensive Cancer Network. NCCN Clinical Practice Guidelines in Oncology (NCCN Guidelines®): colon cancer. Plymouth Meeting (PA): National Comprehensive Cancer Network, Inc.; 2024. Available from: https://www.nccn.org/professionals/physician_gls/pdf/colon.pdf.

[6] National Comprehensive Cancer Network. NCCN Clinical Practice Guidelines in Oncology (NCCN Guidelines®): rectal cancer, Version 2.2025. Plymouth Meeting (PA): National Comprehensive Cancer Network, Inc.; 2025. Available from: https://www.nccn.org/professionals/physician_gls/pdf/rectal.pdf.

[7] Wilkinson N. Management of rectal cancer. Surg Clin North Am 2020;100:615–28. Available from: https://doi.org/10.1016/j.suc.2020.02.014.

[8] Osterlund P, Salminen T, Soveri LM, et al. Repeated centralized multidisciplinary team assessment of resectability, clinical behavior, and outcomes in 1086 Finnish metastatic colorectal cancer patients (RAXO): a nationwide prospective intervention study. Lancet Reg Health Eur 2021;3:100049. Available from: https://doi.org/10.1016/j.lanepe.2021.100049.

[9] De Falco V, Napolitano S, Roselló S, et al. How we treat metastatic colorectal cancer. ESMO Open 2020;4:e000813. Available from: https://doi.org/10.1136/esmoopen-2020-000813.

[10] Dekker E, Tanis PJ, Vleugels JLA, et al. Colorectal cancer. Lancet 2019;394:1467–80. Available from: https://doi.org/10.1016/S0140-6736(19)32319-0.

[11] Hasegawa K, Takahashi M, Ohba M, et al. Perioperative chemotherapy and liver resection for hepatic metastases of colorectal cancer. J Hepatobiliary Pancreat Sci 2012;19(5):503–8. Available from: https://doi.org/10.1007/s00534-012-0509-7.

[12] Moris D, Tsilimigras DI, Machairas N, et al. Laparoscopic synchronous resection of colorectal cancer and liver metastases: a systematic review. J Surg Oncol 2019;119(1):30–9. Available from: https://doi.org/10.1002/jso.25313.

[13] Adam R, Kitano Y, Abdelrafee A, et al. Debulking surgery for colorectal liver metastases: foolish or chance? Surg Oncol 2020;33:266–9. Available from: https://doi.org/10.1016/j.suronc.2020.02.008.

[14] Stewart CL, Warner S, Ito K, et al. Cytoreduction for colorectal metastases: liver, lung, peritoneum, lymph nodes, bone, brain. When does it palliate, prolong survival, and potentially cure? Curr Probl Surg 2018;55(9):330–79. Available from: https://doi.org/10.1067/j.cpsurg.2018.08.004.

[15] Hamidou Z, Chibaudel B, Hebbar M, et al. Time to definitive health-related quality of life score deterioration in patients with resectable metastatic colorectal cancer treated with FOLFOX4 versus sequential dose-dense FOLFOX7 followed by FOLFIRI: the MIROX randomized phase III trial. PLoS One 2016;11:e0157067. Available from: https://doi.org/10.1371/journal.pone.0157067.

[16] Schmoll HJ, Mann J, Meinert F, et al. Efficacy and quality of life for FOLFOX/bevacizumab +/−irinotecan in first-line metastatic colorectal cancer—final results of the AIO CHARTA trial. Br J Cancer 2024;130:233–41. Available from: https://doi.org/10.1038/s41416-023-02496-4.

[17] Färkkilä N, Sintonen H, Saarto T, et al. Health-related quality of life in colorectal cancer. Colorectal Dis 2013;15:e215–22. Available from: https://doi.org/10.1111/codi.12143.

[18] Ramsey SD, Berry K, Moinpour C, et al. Quality of life in long term survivors of colorectal cancer. Am J Gastroenterol 2002;97:1228–34. Available from: https://doi.org/10.1111/j.1572-0241.2002.05694.x.

[19] Pate A, Lowery J, Kilbourn K, et al. Quality of life and the negative impact of comorbidities in long-term colorectal cancer survivors: a population-based comparison. J Cancer Surviv 2020;14:653–9. Available from: https://doi.org/10.1007/s11764-020-00876-w.

[20] Cervantes A, Adam R, Roselló S, et al. Metastatic colorectal cancer: ESMO clinical practice guideline for diagnosis, treatment and follow-up. Ann Oncol 2023;34(1):10–32. Available from: https://doi.org/10.1016/j.annonc.2022.10.003.

[21] Zielinska A, Włodarczyk M, Makaro A, et al. Management of pain in colorectal cancer patients. Crit Rev Oncol Hematol 2021;157:103–22. Available from: https://doi.org/10.1016/j.critrevonc.2020.103122.

[22] Cherny NI, Fallon MT, Kaasa S, Portenoy RK, Currow DC, editors. Chronic cancer pain syndromes. Oxford textbook of palliative medicine. 6th ed. Oxford: Oxford University Press; 2021.

[23] Tomasello G, Petrelli F, Ghidini M, et al. FOLFOXIRI plus bevacizumab as conversion therapy for patients with initially unresectable metastatic colorectal cancer: a systematic review and pooled analysis. JAMA Oncol 2017;3(7):e170278. Available from: https://doi.org/10.1001/jamaoncol.2017.0278.

[24] Moisuc DC, Marinca MV, Matei AM, et al. The impact of bevacizumab and chemotherapy on quality of life in metastatic colorectal cancer patients. Healthcare (Basel) 2023;11(4):591. Available from: https://doi.org/10.3390/healthcare11040591.

[25] Yamazaki K, Nagase M, Tamagawa H, et al. Randomized phase III study of bevacizumab plus FOLFIRI and bevacizumab plus mFOLFOX6 as first-line treatment for patients with metastatic colorectal cancer (WJOG4407G). Ann Oncol 2016;27(8):1539–46. Available from: https://doi.org/10.1093/annonc/mdw206.

[26] Pietrantonio F, Fucà G, Rossini D, et al. FOLFOXIRI-bevacizumab or FOLFOX-panitumumab in patients with left-sided RAS/BRAF wild-type metastatic colorectal cancer: a propensity score-based analysis. Oncologist 2021;26(4):302–9. Available from: https://doi.org/10.1002/onco.13642.

[27] Quidde J, Schmoll H-J, Garlipp B, et al. Impact of FOLFOXIRI and bevacizumab (bev) compared to FOLFOX and bev on health related quality of life (HRQOL) in patients with metastatic colorectal cancer (MCRC): analysis of the CHARTA-AIO 0209 trial. J Clin Oncol 2017;35:3544. Available from: https://doi.org/10.1200/JCO.2017.35.15_suppl.3544.

[28] Van Cutsem E, Koehne CH, Hitre E, et al. Cetuximab and chemotherapy as initial treatment for metastatic colorectal cancer. N Engl J Med 2009;360(14):1408–17. Available from: https://doi.org/10.1056/NEJMoa0805019.

[29] Láng I, Köhne CH, Folprecht G, et al. Quality of life analysis in patients with KRAS wild-type metastatic colorectal cancer treated first-line with cetuximab plus irinotecan, fluorouracil and leucovorin. Eur J Cancer 2013;49:439–48. Available from: https://doi.org/10.1016/j.ejca.2012.08.023.

[30] Liu T, Li J, Xu J, et al. Quality of life (QoL) analyses in patients with RAS wild-type (wt) metastatic colorectal cancer (mCRC) treated with first-line FOLFOX-4 ± cetuximab in the phase 3 TAILOR trial. Ann Oncol 2017;28:v201. Available from: https://doi.org/10.1093/annonc/mdx393.118.

[31] Raimondi A, Di Maio M, Morano F, et al. Health-related quality of life in patients with RAS wild-type metastatic colorectal cancer treated with panitumumab-based first-line treatment strategy: a prespecified secondary analysis of the Valentino study. Eur J Cancer 2020;135:230–9. Available from: https://doi.org/10.1016/j.ejca.2020.04.048.

[32] Diaz LA Jr, Shiu KK, Kim TW, et al. Pembrolizumab versus chemotherapy for microsatellite instability-high or mismatch repair-deficient metastatic colorectal cancer (KEYNOTE-177): final analysis of a randomised, open-label, phase 3 study. Lancet Oncol 2022;23(5):659–70. Available from: https://doi.org/10.1016/S1470-2045(22)00197-8.

[33] Andre T, Elez E, van Cutsem E, et al. Nivolumab plus ipilimumab in microsatellite-instability–high metastatic colorectal cancer. N Engl J Med 2024;391(21):2014–26. Available from: https://doi.org/10.1056/NEJMoa2402141.

[34] Lenz HJ, van Cutsem E, Limon ML, et al. First-line nivolumab plus low-dose ipilimumab for microsatellite instability-high/mismatch repair-deficient metastatic colorectal cancer: the phase II CheckMate 142 study. J Clin Oncol 2022;40:161–70. Available from: https://doi.org/10.1200/JCO.21.01015.

[35] Kopetz S, Yoshino T, Van Cutsem E, et al. Encorafenib, cetuximab and chemotherapy in BRAF-mutant colorectal cancer: a randomized phase 3 trial. Nat Med 2025;31:901–8. Available from: https://doi.org/10.1038/s41591-024-03443-3.

[36] Strickler JH, Cercek A, Siena S, et al. Tucatinib plus trastuzumab for chemotherapy-refractory, HER2-positive, RAS wild-type unresectable or metastatic colorectal cancer (MOUNTAINEER): a multicentre, open-label, phase 2 study. Lancet Oncol 2023;24(5):496–508. Available from: https://doi.org/10.1016/S1470-2045(23)00150-X.

[37] Fernández Montes A, Martínez Lago N, Covela Rúa M, et al. Efficacy and safety of FOLFIRI/aflibercept in second-line treatment of metastatic colorectal cancer in a real-world population: prognostic and predictive markers. Cancer Med 2019;8(3):882–9. Available from: https://doi.org/10.1002/cam4.1903.

[38] Bertaut A, Touchefeu Y, Blanc J, et al. Health-related quality of life analysis in metastatic colorectal cancer patients treated by second-line chemotherapy, associated with either cetuximab or bevacizumab: the PRODIGE 18 randomized phase II study. Clin Colorectal Cancer 2022;21(2):e49–61. Available from: https://doi.org/10.1016/j.clcc.2021.09.001.

[39] Prager GW, Taieb J, Fakih M, et al. Trifluridine–tipiracil and bevacizumab in refractory metastatic colorectal cancer. N Engl J Med 2023;388(18):1657–67. Available from: https://doi.org/10.1056/NEJMoa2214963.

[40] Kröning H, Göhler T, Decker T, et al. Effectiveness, safety and quality of life of trifluridine/tipiracil in pretreated patients with metastatic colorectal cancer: real-world data from the noninterventional TACTIC study in Germany. Int J Cancer 2023;153(6):1227–40. Available from: https://doi.org/10.1002/ijc.34603.

[41] Chia D, Lu J, Zheng H, et al. Efficacy of palliative radiation therapy for symptomatic rectal cancer. Radiother Oncol 2016;121(2):258–61. Available from: https://doi.org/10.1016/j.radonc.2016.06.023.

[42] Ito Y, Ohtsu A, Ishikura S, et al. Efficacy of chemoradiotherapy on pain relief in patients with intrapelvic recurrence of rectal cancer. Jpn J Clin Oncol 2003;33(4):180–5. Available from: https://doi.org/10.1093/jjco/hyg036.

[43] Bakkerus L, Buffart LM, Buffart TE, et al. Health-related quality of life in patients with metastatic colorectal cancer undergoing systemic therapy with or without maximal tumor debulking. J Natl Compr Canc Netw 2023;21(10):1059–1066.e5. Available from: https://doi.org/10.6004/jnccn.2023.7050.

[44] Lehtomäki K, Stedt HP, Osterlund E, et al. Health-related quality of life in metastatic colorectal cancer patients treated with curative resection and/or local ablative therapy or systemic therapy in the Finnish RAXO-study. Cancers 2022;14(7):1713. Available from: https://doi.org/10.3390/cancers14071713.

[45] Yoo GS, Rim CH, Cho WK, et al. Korean Cancer Association Oligometastasis Working Group. Metastasis-directed local therapy of hepatic oligometastasis from colorectal cancer and future perspective in radiation therapy. Cancer Res Treat 2023;55(3):707–19. Available from: https://doi.org/10.4143/crt.2022.1599.

[46] Macbeth F, Farewell V, Treasure T, et al. Local treatment of unresectable colorectal liver metastases: results of a randomized phase II trial. JNCI: J Natl Cancer Inst 2017;109(10):djx149. Available from: https://doi.org/10.1093/jnci/djx149.

[47] Miller SR, Chang DT. Local-regional therapy for oligometastatic colorectal cancer. Cancer J 2024;30(4):272–9. Available from: https://doi.org/10.1097/PPO.0000000000000729.

[48] Lee BM, Chang JS, Koom WS, et al. Importance of local ablative therapies for lung metastases in patients with colorectal cancer. Ann Surg 2023;278(1):e173–8. Available from: https://doi.org/10.1097/SLA.0000000000005466.

[49] Hitchcock KE, Romesser PB, Miller ED. Local therapies in advanced colorectal cancer. Hematol Oncol Clin North Am 2022;36(3):553–67. Available from: https://doi.org/10.1016/j.hoc.2022.02.007.

[50] Kang L, Tian Y, Xu S, et al. Oxaliplatin-induced peripheral neuropathy: clinical features, mechanisms, prevention and treatment. J Neurol 2021;268:3269–82. Available from: https://doi.org/10.1007/s00415-020-09942-w.

[51] Vieira C, Brás M, Fragoso M, et al. Opioids for cancer pain and its use under particular conditions. A narrative review. Acta Med Port 2019;32:388–99. Available from: https://doi.org/10.20344/amp.10500.

[52] Paice JA, et al. AAPT diagnostic criteria for chronic cancer pain conditions. J Pain 2017;18:233–46. Available from: https://doi.org/10.1016/j.jpain.2016.10.020.

[53] Tran M, Grant M, Clayton J, et al. Advance care decision making and planning. Aust J Gen Pract 2018;47:753–7. Available from: https://doi.org/10.31128/AJGP-06-18-4613.

[54] Pisano M, Zorcolo L, Merli C, et al. 2017 WSES guidelines on colon and rectal cancer emergencies: obstruction and perforation. World J Emerg Surg 2018;13:36. Available from https://doi.org/10.1186/s13017-018-0192-3.

[55] National Comprehensive Cancer Network. (2025). NCCN Clinical Practice Guidelines in Oncology (NCCN Guidelines®) – anal carcinoma, Version 4.2025. Retrieved from: https://www.nccn.org/professionals/physician_gls/pdf/anal.pdf

[56] Eng C, Ciombor KK, Cho M, et al. Anal cancer: emerging standards in a rare disease. J Clin Oncol 2022;40(24):2774–88. Available from: https://doi.org/10.1200/JCO.21.02566.

[57] Osborne MC, Maykel J, Johnson EK, et al. Anal squamous cell carcinoma: an evolution in disease and management. World J Gastroenterol 2014;20:13052–9. Available from: https://doi.org/10.3748/wjg.v20.i36.13052.

[58] Giunta EF, Bregni G, Hendlisz A, et al. Anal squamous cell carcinoma: standards of care, new data and ongoing clinical trials. Curr Opin Oncol 2021;33(4):372–7. Available from: https://doi.org/10.1097/CCO.0000000000000748.

[59] Yellu M, Deeb A, Olowokure O. Overview of recent trends in the management of metastatic anal cancer. World J Oncol 2015;6(1):311–5. Available from: https://doi.org/10.14740/wjon866w.

[60] Ferrell BR, Temel JS, Temin S, et al. Integration of palliative care into standard oncology care: American Society of Clinical Oncology clinical practice guideline update. J Clin Oncol 2017;35:96–112. Available from: https://doi.org/10.1200/JCO.2016.70.1474.

[61] Goldner M, Platoff R, Betances A, et al. Role of metastasectomy for liver metastasis in stage IV anal cancer. Am J Surg 2021;221(4):832–8. Available from: https://doi.org/10.1016/j.amjsurg.2020.08.016.

[62] Sauter C, Peeken JC, Borm K, et al. Quality of life in patients treated with radiochemotherapy for primary diagnosis of anal cancer. Sci Rep 2022;12:4416. Available from: https://doi.org/10.1038/s41598-022-08525-1.

[63] Welzel G, Hägele V, Wenz F, et al. Quality of life outcomes in patients with anal cancer. Strahlenther Onkol 2011;187:175–82. Available from: https://doi.org/10.1007/s00066-010-2175-5.

[64] Arndt V, Koch-Gallenkamp L, Jansen L, et al. Quality of life in long-term and very long-term cancer survivors versus population controls in Germany. Acta Oncol 2017;56:190–7. Available from: https://doi.org/10.1080/0284186X.2016.1266089.

[65] Rao S, Sclafani F, Eng C, et al. International rare cancers initiative multicenter randomized phase II trial of cisplatin and fluorouracil versus carboplatin and paclitaxel in advanced anal cancer: interAAct. J Clin Oncol 2020;38(22):2510–8. Available from: https://doi.org/10.1200/JCO.19.03266.

[66] Morris VK, Salem ME, Nimeiri H, et al. Nivolumab for previously treated unresectable metastatic anal cancer (NCI9673): a multicentre, single-arm, phase 2 study. Lancet Oncol 2017;18(4):446–53. Available from: https://doi.org/10.1016/S1470-2045(17)30104-3.

[67] Marabelle A, Cassier PA, Fakih M, et al. Pembrolizumab for previously treated advanced anal squamous cell carcinoma: results from the non-randomised, multicohort, multicentre, phase 2 KEYNOTE-158 study. Lancet Gastroenterol Hepatol 2022;7(5):446–54. Available from: https://doi.org/10.1016/S2468-1253(21)00382-4.

[68] Frank F, Hecht M, Loy F, et al. Differences in and prognostic value of quality of life data in rectal cancer patients with and without distant metastases. Healthcare (Basel) 2020;9(1):1. Available from: https://doi.org/10.3390/healthcare9010001.

[69] Valvo F, Ciurlia E, Avuzzi B, et al. Cancer of the anal region. Crit Rev Oncol Hematol 2019;135:115–27. Available from: https://doi.org/10.1016/j.critrevonc.2018.12.007.

[70] Casali PG, Blay JY, Abecassis N, et al. Gastrointestinal stromal tumours: ESMO-EURACAN-GENTURIS clinical practice guidelines for diagnosis, treatment and follow-up. Ann Oncol 2022;33(1):20–33. Available from: https://doi.org/10.1016/j.annonc.2021.09.005.

[71] Hirota S, Isozaki K, Moriyama Y, et al. Gain-of-function mutations of c-kit in human gastrointestinal stromal tumors. Science 1998;279:577–80. Available from: https://doi.org/10.1126/science.279.5350.577.

[72] Demetri GD, von Mehren M, Blanke CD, et al. Efficacy and safety of imatinib mesylate in advanced gastrointestinal stromal tumors. N Engl J Med 2002;347:472–80. Available from: https://doi.org/10.1056/NEJMoa020461.

[73] Demetri GD, Reichert P, Kang YK, et al. Efficacy and safety of regorafenib for advanced gastrointestinal stromal tumours after failure of imatinib and sunitinib (GRID): an international, multicentre, randomised, placebo-controlled, phase 3 trial. Lancet 2013;381:295–302. Available from: https://doi.org/10.1016/S0140-6736(12)61857-1.

[74] Blay JY, Serrano C, Heinrich MC, et al. Ripretinib in patients with advanced gastrointestinal stromal tumours (INVICTUS): phase 3 trial. Lancet Oncol 2020;21:923–34. Available from: https://doi.org/10.1016/S1470-2045(20)30168-6.

[75] Qiu HB, Zhou ZG, Feng XY, et al. Advanced gastrointestinal stromal tumor patients benefit from palliative surgery after tyrosine kinase inhibitors therapy. Medicine (Baltimore) 2018;97(2):e9097. Available from: https://doi.org/10.1097/MD.0000000000009097.

[76] van de Wal D, Elie M, Le Cesne A, et al. Health-related quality of life and side effects in gastrointestinal stromal tumor (GIST) patients treated with tyrosine kinase inhibitors: a systematic review of the literature. Cancers 2022;14:1832. Available from: https://doi.org/10.3390/cancers14071832.

[77] Meeker A, Heaphy C. Gastroenteropancreatic endocrine tumors. Mol Cell Endocrinol 2014;386(1-2):101–20. Available from: https://doi.org/10.1016/j.mce.2013.07.015.

[78] Detjen K, Hammerich L, Özdirik B, et al. Models of gastroenteropancreatic neuroendocrine neoplasms: current status and future directions. Neuroendocrinology 2021;111(3):217–36. Available from: https://doi.org/10.1159/000509864.

[79] Fernandes CJ, Leung G, Eads JR, et al. Gastroenteropancreatic neuroendocrine tumors. Gastroenterol Clin North Am 2022;51(3):625–47. Available from: https://doi.org/10.1016/j.gtc.2022.06.002.

[80] Pavel M, Öberg K, Falconi M, et al. Gastroenteropancreatic neuroendocrine neoplasms: ESMO Clinical Practice Guidelines for diagnosis, treatment and follow-up. Ann Oncol 2020;31(7):844–60. Available from: https://doi.org/10.1016/j.annonc.2020.03.304.

[81] Pearman TP, Beaumont JL, Cella D, et al. Health-related quality of life in patients with neuroendocrine tumors: an investigation of treatment type, disease status, and symptom burden. Support Care Cancer 2016;24:3695–703. Available from: https://doi.org/10.1007/s00520-016-3189-z.

[82] Grillo F, Albertelli M, Malandrino P, et al. Prognostic effect of lymph node metastases and mesenteric deposits in neuroendocrine tumors of the small bowel. J Clin Endocrinol Metab 2022;107(12):3209–21. Available from: https://doi.org/10.1210/clinem/dgac326.

[83] Madani A, Thomassen I, van Gestel YRBM, et al. Peritoneal metastases from gastroenteropancreatic neuroendocrine tumors: incidence, risk factors and prognosis. Ann Surg Oncol 2017;24(8):2199–205. Available from: https://doi.org/10.1245/s10434-016-5734-x.

[84] Ma ZY, Gong YF, Zhuang HK, et al. Pancreatic neuroendocrine tumors: a review of serum biomarkers, staging, and management. World J Gastroenterol 2020;26(19):2305–22. Available from: https://doi.org/10.3748/wjg.v26.i19.2305.

[85] Iyoda A, Azuma Y, Sano A. Neuroendocrine tumors of the lung: clinicopathological and molecular features. Surg Today 2020;50(12):1578–84. Available from: https://doi.org/10.1007/s00595-020-01988-7.

[86] Caplin ME, Pavel M, Ćwikła JB. Lanreotide in metastatic enteropancreatic neuroendocrine tumors. N Engl J Med 2014;371(3):224–33. Available from: https://doi.org/10.1056/NEJMoa1316158.

[87] Rinke A, Neary MP, Eriksson J, et al. Health-related quality of life for long-acting octreotide versus placebo in patients with metastatic midgut neuroendocrine tumors in the phase 3 PROMID trial. Neuroendocrinology. 2019;109(2):141-51. Available from https://doi.org/10.1159/000499469.

[88] Yao JC, Fazio N, Singh S, et al. Everolimus for the treatment of advanced, non-functional neuroendocrine tumours of the lung or gastrointestinal tract (RADIANT-4): a randomised, placebo-controlled, phase 3 study. Lancet 2016;387(10022):968–77. Available from: https://doi.org/10.1016/S0140-6736(15)00817-X.

[89] Pavel ME, Singh S, Strosberg JR, et al. Health-related quality of life for everolimus versus placebo in patients with advanced, non-functional, well-differentiated gastrointestinal or lung neuroendocrine tumours (RADIANT-4): a multicentre, randomised, double-blind, placebo-controlled, phase 3 trial. Lancet Oncol 2017;18(10):1411–22. Available from: https://doi.org/10.1016/S1470-2045(17)30471-0.

[90] Singh S, Halperin D, Myrehaug S, et al. [^{177}Lu]Lu-DOTA-TATE plus long-acting octreotide versus high-dose long-acting octreotide for the treatment of newly diagnosed, advanced grade 2-3, well-differentiated, gastroenteropancreatic neuroendocrine tumours (NETTER-2): an open-label, randomised, phase 3 study. Lancet 2024;403(10446):2807–17. Available from: https://doi.org/10.1016/S0140-6736(24)00701-3.

[91] Strosberg J, El-Haddad G, Wolin E, et al. NETTER-1 Trial Investigators. Phase 3 trial of ^{177}Lu-dotatate for midgut neuroendocrine tumors. N Engl J Med 2017;376(2):125–35. Available from: https://doi.org/10.1056/NEJMoa1607427.

[92] Strosberg J, Wolin E, Chasen B, et al. NETTER-1 Study Group. Health-related quality of life in patients with progressive midgut neuroendocrine tumors treated with ^{177}Lu-dotatate in the phase III NETTER-1 trial. J Clin Oncol 2018;36(25):2578–84. Available from: https://doi.org/10.1200/JCO.2018.78.5865.

Chapter 35

Hepatobiliary cancers

Juraj Prejac[1,2]
[1]*Department of Oncology, University Hospital Centre Zagreb, Kišpatićeva, Zagreb, Croatia,* [2]*School of Dental Medicine, University of Zagreb, Gundulićeva, Zagreb, Croatia*

Key learning objectives

- Understand the epidemiology, risk factors, and clinical presentation of hepatobiliary cancers.
- Recognize the unique challenges in diagnosing and managing pain in patients with advanced liver disease and hepatobiliary tumors.
- Evaluate the evolving landscape of systemic therapies, including the impact of immunotherapy and combination regimens on survival and quality of life in hepatobiliary cancers.
- Identify the limitations and risks of pharmacological pain management in liver impairment, and explore alternative or adjunctive interventions such as nerve blocks, radiotherapy, and bisphosphonates for bone metastasis.
- Appreciate the importance of multidisciplinary and early palliative care integration to optimize symptom control, patient comfort, and communication in advanced hepatobiliary malignancies.

Liver cancer

Liver cancer ranks as the sixth most prevalent cancer globally, with approximately 906,000 new cases and 830,000 deaths reported in 2020. It also stands as the third leading cause of cancer-related mortality worldwide [1]. The most common primary liver cancer is hepatocellular carcinoma (HCC), which constitutes approximately 80%–90% of all cases [1,2]. It arises from hepatocytes and is associated with well-known risk factors, including infection with hepatitis B and C viruses, alcohol intake, aflatoxin-contaminated foods, and metabolic syndrome [1,3]. Globally, HCC primarily arises in the context of chronic liver disease and cirrhosis, accounting for approximately 80% of cases. In Western countries, about 90% of HCC cases occur in individuals with an underlying cirrhotic condition [4]. However, the remaining 10%–20% occur in patients with noncirrhotic livers. This dual pattern underscores the complexity of HCC's pathogenesis with various etiological agents interacting, producing a diverse array of mechanisms for HCC development, including interactions between genetic, environmental, and immunological factors [5,6].

Anatomical considerations, cirrhosis, and end-stage liver disease

A visceral capsule, called the Glisson capsule, envelops the liver and consists of interstitial tissue [7]. It is continuous with the interstitial spaces and matrix surrounding the portal triads. Besides its role in fluid exchange, the capsule has a structural role. It encloses the parenchyma and shields the liver from physical impact while maintaining elasticity, allowing daily hepatocyte fluctuations in size [8]. Glisson capsule plays a role in lymph drainage, as the superficial lymph network resides within the capsule, covering both the anterior and posterior surfaces. The anterior surface drains primarily to phrenic lymph nodes connecting to the mediastinal and internal mammary lymphatic networks, while the posterior surface drains to hilar lymph nodes, including those around the cystic duct, common bile duct, hepatic artery, peripancreatic, pericardial, and celiac regions [9,10]. To accommodate this variability in size, the normal liver capsule is thin and well-innervated and contains abundant elastic fibers [7]. The primary mechanical difference between the liver and its capsule in disease states is due to their initial properties and composition. The liver is highly cellular and relatively soft, with a limited extracellular matrix. During fibrosis, the matrix within the liver becomes cross-linked and highly organized, significantly increasing the tissue's rigidity. This phenomenon is thought to be influenced by changes in the capsule matrix structure [11].

Pain Management in Palliative Care for Patients With Cancer. DOI: https://doi.org/10.1016/B978-0-443-30050-9.00041-2

In fibrosis, which is common in the majority of patients with liver cancer, changes occur that are secondary to the underlying parenchymal disease. This is more pronounced in patients with cirrhosis, who exhibit significant development of capsular fibrosis [11]. Cirrhosis is the final, irreversible stage of chronic liver disease, marked by the disruption of the liver's structure and the creation of regenerative nodules. Those who have cirrhosis but have not yet experienced significant complications are said to have compensated cirrhosis [12]. However, when complications arise—like variceal bleeding, fluid accumulation in the abdomen (ascites), or infection of the abdominal cavity (spontaneous bacterial peritonitis)—the condition progresses to decompensated cirrhosis. Advanced liver disease, liver failure, and decompensated cirrhosis, due to the typically permanent nature of these health issues, are often referred to as "end-stage liver disease" (ESLD) [13]. ESLD is characterized by the liver's inability to fulfill the body's physiological requirements and represents the final phase of chronic liver disease, which can arise from various causes and is marked by severe liver damage, impaired function, and failure. The clinical manifestations of ESLD can take several forms, including decompensated HCC. Unlike other terminal conditions, liver transplantation offers a unique possibility of a cure for such irreversible liver diseases. Nevertheless, a significant number of patients are ineligible for transplantation owing to various clinical and social considerations [14].

Patients with ESLD and liver cancer are known to experience significant discomfort and considerable distress. The various complications associated with chronic liver disease can significantly diminish the health-related quality of life (HRQoL), and this decline tends to escalate as the disease progresses [15]. Patients with chronic liver disease exhibit a reduced HRQoL compared to patients with no liver disease, which mirrors the HRQoL levels found in individuals with chronic obstructive pulmonary disease or congestive heart failure. Among those with cirrhosis, HRQoL is somewhat better preserved in patients suffering from cholestatic liver diseases than in those with hepatocellular conditions. Additionally, advancing age has a slight negative impact on the physical dimension of HRQoL [15].

Liver pain

The signs and symptoms of liver cancer often include pain, fatigue, weight loss, and the appearance of ascites, which significantly affect the patient's quality of life (QoL) [16,17]. The pain is usually localized in the upper right quadrant, in the area of the liver, and is difficult to locate accurately, with no specific point of maximum intensity [18]. The liver can be palpated with enlarged and irregular contours. Upper quadrant pain is a common symptom in patients with ESLD, with a prevalence ranging from 30% to 79%, according to a metaanalysis by Peng et al. [19], significantly affecting the QoL. Pain in that area can be etiologically related to the liver and surrounding structures, such as the vascular and biliary systems.

The neural control of liver function is complex and not fully understood, but the liver receives parasympathetic and sympathetic innervation. Nerve fibers originate from the celiac plexus, lower thoracic ganglia, right phrenic nerve, and the vagus nerves. The vagus nerves split into anterior (left) and posterior (right) branches as they travel from the thorax to the abdomen. The anterior vagus further divides into a cephalic and a hepatic branch, passing through the lesser omentum to provide parasympathetic innervation to the liver. Sympathetic innervation mainly comes from the celiac plexus and thoracic splanchnic nerves [9].

Liver cancer can lead to two types of pain: parietal or visceral [17,20]. Visceral pain is caused by the pressure exerted by tumors in the spaces between organs. Since slow, unmyelinated C-fibers typically mediate visceral pain, it is more diffuse and presents as dull, aching, and persistent discomfort or cramp-like sensations [21]. Parietal pain, transmitted by fast, myelinated Aδ-fibers, is caused by stretching of the liver capsule and is sharp and localized. Furthermore, patients with advanced liver disease may also experience ascites, which can result in abdominal and lower back pain.

Systemic therapy for hepatocellular carcinoma

The landscape of HCC treatment has evolved significantly over the past decade. Initially, sorafenib was the primary treatment, offering limited survival benefits. Lenvatinib later emerged as a more effective option, with a notable 24% objective response rate (ORR) [22]. However, the advent of immune checkpoint inhibitors marked a paradigm shift in HCC management. The FDA-approved combination of atezolizumab and bevacizumab, following the IMbrave150 trial results, has set a new standard for first-line treatment, demonstrating improved survival rates and a higher ORR [23]. The combination of atezolizumab and bevacizumab resulted in an improvement of median overall survival (OS) of 19.2 months over 13.4 months with sorafenib (hazard ratio [HR] 0.66, 95% confidence interval [CI] 0.52–0.85) [24]. It also showed a significant reduction in the risk of symptom deterioration in patients with cancer compared to sorafenib, according to the European Organisation for Research and Treatment of Cancer (EORTC) Quality of Life Questionnaire Core 30 (QLQ-C30) and Quality of Life Questionnaire for Hepatocellular Carcinoma (QLQ-HCC18) scales. The HRs indicate a lower probability of symptom worsening, including appetite loss, fatigue, diarrhea, and pain, which suggests improved QoL for patients undergoing treatment. A combination of atezolizumab plus bevacizumab reduced the risk of pain based on EORTC QLQ-C30 generic cancer symptom scales (HR 0.46, 95% CI 0.34–0.62) and two of three EORTC QLQ-HCC18 scales (HR 0.65, 95% CI 0.46–0.92) [25]. Subsequent trials like HIMALAYA have continued to build on these advancements, showcasing the potential of combination therapies in improving patient outcomes in advanced HCC. The combination of durvalumab and tremelimumab resulted in a significantly prolonged time to worsening of abdominal pain (HR 0.61, 95% CI 0.47–0.80). This was also observed with durvalumab monotherapy (HR 0.67, 95% CI 0.52–0.87) [26].

Analgesics and liver impairment

The liver metabolizes most drugs used for analgesia, and liver function significantly affects drug metabolism. The management of pain in patients with cirrhosis is a complex issue due to the altered pharmacokinetics and increased sensitivity to drugs. The impairment of drug metabolism positively correlates with the grade of liver dysfunction [27,28]. Acetaminophen, found in various over-the-counter and prescription drugs and widely used for its analgesic and antipyretic properties, has a well-documented risk profile concerning hepatotoxicity. The risk is heightened with acute overdoses, but even long-term use of therapeutic doses, particularly in advanced liver disease or alcohol use, can lead to significant liver damage [29,30]. For short-term or single-dose administration, up to 3–4 g per day may be permissible [31]. However, the dose must be corrected for patients with liver damage, where doses higher than 2 g are not recommended [32–34]. This is particularly important in individuals with chronic hepatitis C virus (HCV) infection, where the liver's compromised state may exacerbate the drug's hepatotoxic effects [35,36]. The absence of formal guidelines for acetaminophen use in HCV patients underscores the need for cautious application of expert recommendations and heightened vigilance in monitoring liver function tests, especially alanine aminotransferase levels, to prevent potential liver injury.

Nonsteroidal antiinflammatory drugs (NSAIDs) and opioids are commonly used analgesic drugs, which are evaluated in depth in other chapters. However, their use in patients with cirrhosis often has several specific limitations. Due to their mechanism of action, which involves prostaglandin inhibition, NSAIDs can lead to acute renal failure and gastrointestinal bleeding [33,34]. The essential role of prostaglandins and nitric oxide in maintaining gastrointestinal mucosal integrity means that NSAIDs can disrupt these protective mechanisms, increasing the risk of gastritis and peptic ulcer disease [37]. Additionally, NSAIDs may exacerbate the risk of variceal and nonvariceal upper gastrointestinal bleeding in patients with portal hypertension by reducing platelet aggregation [32,33].

Due to the risk of complications, maintaining renal function is vital in patients with cirrhosis and portal hypertension. Arterial vasodilation in splanchnic circulation can reduce effective blood volume and organ perfusion, activating the renin–angiotensin system, renal vasoconstriction, and increased cardiac output [38]. The vasodilatory effects of prostaglandins maintain renal homeostasis, and by inhibiting prostaglandin synthesis, NSAIDs can adversely affect renal function. The impact varies based on the severity of liver disease and the NSAID's potency in prostaglandin synthesis inhibition [33]. Therefore, systemic use of NSAIDs should generally be avoided in patients with cirrhosis. Opioids should be used with caution, if feasible, avoided in patients with a history of encephalopathy or substance abuse [31]. Potential nephrotoxicity of celecoxib in this population has not been proven, but data from more extensive trials with longer follow-up are lacking [32]. Alternative analgesics such as anticonvulsants (e.g., gabapentin and pregabalin) may be considered for neuropathic pain management in patients with advanced liver disease, as they undergo little to no transformation in the liver. Since these drugs are primarily excreted via the kidneys, doses require adjustment for renal failure [31,32].

Opioids are another class of drugs commonly prescribed for pain. In liver dysfunction, opioid clearance is decreased since the liver is the main organ for opioid metabolism [31,32]. This increases the bioavailability and accumulation of the drug and its metabolites, leading to more severe adverse reactions [32,39]. This may be further aggravated by hypoalbuminemia, increasing toxicity for albumin-bound drugs [28]. In patients with liver failure, opioids can indeed cause a range of complications, including sedation and encephalopathy [39]. In patients with cirrhosis, the half-life of both intravenous and oral morphine and its metabolites is increased [40]. Therefore it is recommended to use extended dosing intervals. Immediate-release formulations are preferred, and sustained-release forms should be avoided [34]. Unlike morphine, tramadol is a partial opioid agonist and also acts as a serotonin and norepinephrine reuptake inhibitor, which contributes to its profile of causing less respiratory depression and having a lower potential for abuse compared to other opioids [32]. However, in cirrhotic patients, there is a significant increase in plasma drug concentration, and the elimination half-life of tramadol is extended [41]. Extended-release formulations are contraindicated in severe hepatic impairment [32].

Hence, it is crucial for healthcare providers to balance the therapeutic benefits of analgesic drugs against the risks of hepatotoxicity and to tailor dosing regimens carefully, particularly in vulnerable populations such as those with existing liver conditions or concurrent alcohol use.

Bone metastases in liver cancers

Bone ranks as the third most common site for metastases from all types of tumors, following the lungs and liver. Bone metastases (BM) are the leading cause of bone pain in patients with cancer [42]. They can lead to serious health complications, including hypercalcemia, which is an abnormally high calcium level in the blood that can weaken bones, create kidney stones, and interfere with how the heart and brain function [43]. Decreased mobility and QoL and increased risk of fractures and neurovascular compression are also significant concerns.

Managing BM is a complex challenge in palliative care, where the goal is to alleviate symptoms and improve the QoL for patients with limited life expectancy. Treatment for BM is typically multifaceted, using medication and radiotherapy (RT) to relieve pain, stabilize the skeletal structure, reduce calcium levels, and improve the patient's functional abilities. In some cases, surgery may be necessary to repair fractures or decompress nerve compression [44].

HCC is known to metastasize to the bone in roughly 15% of cases. This significant rate of bone involvement highlights the importance of monitoring for skeletal metastases in patients with HCC [45,46]. The axial skeleton, especially the thoracic and lumbar vertebrae, is often impacted by BM from HCC. Similar to other cancers that spread to bones, widespread skeletal damage from multiple metastases usually indicates a dire prognosis [43,47]. Targeted local treatments often manage isolated lesions. Despite the severity of these conditions, complications such as pathological fractures and hypercalcemia are relatively rare occurrences [43,48]. From a palliative perspective, the recognition of hypercalcemia as a potential (though rare) cause of new or worsening pain is critical—especially in patients with advanced HCC and known or suspected bone involvement. Timely diagnosis and management of hypercalcemia (e.g., with bisphosphonates or denosumab, hydration, and corticosteroids) can lead to rapid symptomatic relief, reduce opioid requirements, and improve QoL. In a patient with HCC and new-onset diffuse pain, weakness, or altered mental status, calcium levels need to be monitored—especially if BM are known or suspected. It may be rare, but identifying it early can make a big difference in comfort and care planning. BM from HCC may be characterized as hypervascular soft tissue masses, which often lead to the destruction of the bone cortex, sometimes extending into surrounding muscles and fat [43,47,49]. These conditions are more resistant to RT because more cancer cells survive postradiation treatment than BM without soft tissue involvement [43,50].

Bisphosphonates are a class of drugs that have been widely used in the treatment of osteoporosis due to their ability to inhibit bone resorption and increase bone strength, thereby reducing the risk of fractures. They are extensively evaluated in another chapter of this book [51,52]. While bisphosphonates are established in treating various osteotropic cancers, their application in HCC BM is limited, with limited case reports and retrospective studies available [43]. Zoledronic acid has been found beneficial for pain management in patients with BM from HCC. Monthly administration of 4 mg of zoledronic acid significantly reduced pain measured using a Visual Analog Scale. Furthermore, usage of painkillers was reduced [53]. In clinical practice, pain reduction in this way can have a significant impact on drug prescription, considering that the use of opioid analgesics is limited in patients with significant liver damage [31].

External beam radiation therapy (EBRT) is a well-established treatment for pain relief in patients with BM, and it is also specifically evaluated in another chapter of this book. Research indicates that EBRT can provide significant pain relief, with a systematic review showing a pooled overall pain response among patients. However, while dose escalation may enhance pain relief, it does not translate to improved survival outcomes [49]. The equivalence of single and multifraction RT regimens regarding pain relief and control of symptoms suggests that shorter courses of RT could be beneficial [49,54–56]. By reducing the duration of palliative RT, patients can spend less time in treatment settings and have more comfort, potentially at home or in a hospice environment [56]. Although this approach aligns with the principles of palliative care, prioritizing patient comfort and QoL, the duration of relief may be shorter [49].

Biliary tract cancer (cholangiocarcinoma)

Cholangiocarcinoma is a rare malignancy that develops in the bile ducts connecting the liver to the gallbladder and small intestine. It can occur anywhere along the biliary tract, from the intrahepatic bile canaliculi to the point where the common bile duct enters the duodenum at the duodenal ampulla. Cholangiocarcinoma is categorized based on its anatomical origin: intrahepatic cholangiocarcinoma (iCCA) or extrahepatic cholangiocarcinoma (eCCA). The latter is divided into perihilar cholangiocarcinoma (pCCA) and distal cholangiocarcinoma (dCCA). Adenocarcinomas account for over 95% of cholangiocarcinomas [57,58]. Cholangiocarcinoma often arises without genetic predisposition and, unlike HCC, without a clear etiology, especially in Western countries [59]. However, certain risk factors are associated with developing cholangiocarcinoma, such as parasitic infections, primary sclerosing cholangitis, biliary tree calculi, cirrhosis, chronic hepatitis B, and cystic biliary lesions [57,60].

Diagnosis is often challenging in its early stages due to a lack of specific symptoms. The most common symptom for pCCA and dCCA types is obstructive jaundice, which results from blockages in the biliary tract. iCCA usually presents jaundice at more advanced stages. Other symptoms indicating advanced disease can include fatigue, abdominal pain, general discomfort, nausea, loss of appetite, and weight loss. Regular ultrasonography surveillance in cirrhosis patients can aid in the early detection of iCCA, which is otherwise often diagnosed incidentally or during imaging for other conditions [60]. Surgery is the only potentially curative option. However, only a minority of patients are diagnosed early due to a lack of specific symptoms. For those with metastatic or locally advanced disease, systemic therapies are the primary treatment modalities [58]. Recent advances in the treatment of cholangiocarcinoma include the addition of durvalumab, an anti-programmed death-ligand 1 monoclonal antibody, to gemcitabine- and cisplatin-based chemotherapy, with an ORR of 26.7% compared to 18.7%, respectively [61]. This resulted in significantly longer progression-free survival (HR 0.75, 95% CI 0.63–0.89) and OS (HR 0.80, 95% CI 0.66–0.97) with no difference in patient-reported symptoms, including pain or functioning, using EORTC QLQ-C30, EORTC QLQ-BIL21, or Global Health Status/QoL questionnaires [61,62]. That being said, a study by Wu et al. on a heterogeneous group of patients with bile duct or gallbladder cancer, pancreatic cancer (PC), and ampullary cancer reported that immunotherapy reduced opioid consumption and improved pain management compared to other treatment modalities [63].

One of the more common complications of cholangiocarcinoma is biliary obstruction, which clinically manifests as jaundice. It is more common in pCCA and dCCA, and studies indicate that as many as 70%–90% of cases of eCCA present with some degree of obstruction [64]. In contrast, iCCA is less likely to cause biliary obstruction, so it often remains asymptomatic longer. Obstructive jaundice is usually managed by placing a plastic or metal biliary stent. Biliary stenting is performed either via endoscopic retrograde cholangiopancreatography (ERCP) or percutaneously. This approach enables bile drainage and alleviates obstruction [65]. Alternatively, percutaneous transhepatic drainage (PTD) can be performed. Compared to ERCP, PTD is a more invasive method, as the drainage catheter is inserted externally through the liver parenchyma. The presence of the drain itself, its displacement, and the resulting pain can significantly impair the patient's QoL [66,67]. Complications associated with these two procedures include pancreatitis, bleeding, and infections in the case of ERCP [68], and pericatheter bile leakage, catheter occlusion, and drain dislodgement in the case of PTD [69].

Pancreatic cancer

The pancreas is a glandular organ with a complex relationship to neighboring structures. It plays a crucial role in digestion (exocrine pancreas) and blood sugar regulation (endocrine pancreas). Its position in the abdomen places it close to several vital organs and vessels, reflecting its integral role in the body's overall function. The pancreas is anatomically divided into head, body, and tail. Its head rests against the inferior vena cava and the renal vein, nestled within the curvature of the duodenum's C loop. Extending toward the spleen, the tail of the pancreas reaches the splenic hilum [70].

PC is one of the most lethal malignancies with rising incidence, especially in Western countries. Based on Global Cancer Observatory (GLOBOCAN) data, PC ranks 12th in incidence worldwide [71], with a high mortality-to-incidence ratio and a 5-year survival rate of approximately 10% [72]. The reason for the high mortality rate is that it is typically diagnosed at an advanced stage due to its subtle early symptoms. Most of these cancers are pancreatic ductal adenocarcinomas, which form in the lining of the glandular structures of the pancreas [73].

Pathophysiology of pancreatic cancer pain

PC-related pain presents as pain in the upper abdominal region, which can spread posteriorly and laterally, and radiate to the back area between the scapulae. It is a common symptom present in 72%–100% of patients with PC [74,75]. Less commonly, it may be diffuse within the abdomen or referred to the lower abdominal quadrants. The pain is often described as dull, aching, gnawing, or spasmodic. Its intensity can vary throughout the day, influenced by factors like food intake and position, for example, further exacerbated by lying down. Additionally, back pain associated with PC pain may worsen when the patient is lying down and improve when sitting forward [75]. The occurrence of pain could potentially indicate cancer spreading to other surrounding organs or distant metastases [76,77]. Another hallmark of PC is unexplained weight loss, often correlating with the onset of pain. Keep in mind that because PC is frequently diagnosed late, a significant percentage of patients (20%–30%) experience moderate-to-severe pain at the time of diagnosis [78].

The underlying mechanisms leading to this pain include regional invasion and pressure on surrounding nerves as the tumor grows [79]. Pain severity tends to increase with disease progression. The high prevalence of pain experienced by patients with PC is due to its neurotropic properties with a high propensity for perineural invasion [77]. The development of pain syndrome in PC is multifactorial. Cancer cells damage the neuronal sheath, exposing the nerves to other noxious stimuli. Furthermore, cancer growth induces the growth of nerves along with the neovascularization of tumors that contribute to the development of pain [80].

Psychosocial aspect

Given the strong association between PC and pain, distinguishing whether other symptoms are cancer-related or pain-related becomes challenging. Cancers in general, and especially aggressive diseases such as PC, can significantly impact an individual's physiology, often due to stress responses [81]. This stress frequently triggers adaptive behaviors that can manifest across various systems, including endocrine, immune, cognitive, emotional, and physiological domains [81,82]. In addition to an increase in treatment side effects, stress negatively impacts maintaining chemotherapy dose intensity, preserving body mass, and preserving the general condition of the patient [83].

Numerous studies have reported a notably elevated prevalence of depressed mood and anxiety among PC patients. Rates of depression in PC surpass the rates observed in other cancers with similar prognoses, even leading to an increase in the risk of suicide. Overall, symptom burden is substantial in this population, encompassing issues like disrupted sleep, fatigue, and nausea/vomiting associated with obstruction or delayed gastric emptying [83]. All of the above can significantly affect the perception and intensity of pain.

Systemic therapy of pancreatic cancer

Treatment options vary depending on the stage and location of the tumor and may include surgery, chemotherapy, radiation therapy, and targeted drug therapies. Based on trial results from PRODIGE 4/ACCORD 11 and MPACT clinical trials, FOLFIRINOX and nab-paclitaxel plus gemcitabine became therapies of choice in the first-line setting [84,85].

In the PRODIGE 4/ACCORD 11 trial, the FOLFIRINOX regimen resulted in an improved median OS of 11.1 months compared to 6.8 months in the gemcitabine group (HR 0.57, 95% CI 0.45–0.73; $P < .001$). The ORR was also significantly higher, 31.6% in the FOLFIRINOX group, compared to 9.4% in the gemcitabine group ($P < .001$) [84]. Similarly, in the MPACT trial, nab-paclitaxel plus gemcitabine resulted in an increased median OS of 8.5 months compared to 6.7 months for the gemcitabine monotherapy group (HR 0.72, 95% CI 0.62–0.83) with an ORR of 23% versus 7%, respectively ($P < .001$) [85].

Both regimens are considered the standard of care in the first-line setting and positively influence QoL. For example, in PRODIGE 4/ACCORD 11, using the EORTC QLQ-C30, chemotherapy reduced the pain sensation in both treatment arms. The effect of pain reduction was more pronounced in the FOLFIRINOX study arm [86].

Analgesic strategies for pancreatic cancer pain

Opioid medications are still the cornerstone of the management of pain associated with PC. Pain relief can often be effectively achieved with opioid analgesics alone. For patients experiencing persistent nausea and vomiting, oral medications are challenging to take. Hence, transdermal patches can be used for pain control in opioid-tolerant patients [87], considering the amount of fatty tissue necessary for absorption. Furthermore, due to the tumor's proximity to the celiac plexus, the pain may be neuropathic. Hence, the usual pain management strategies, using opiates or NSAIDs, are not always effective, necessitating the consideration of "adjuvant" medications such as gabapentin, pregabalin, or duloxetine [88,89].

Many studies have demonstrated the effectiveness of local procedures targeting the celiac plexus or splanchnic nerves [90–93]. The celiac plexus innervates the liver, gallbladder, stomach, pancreas, spleen, kidneys, small intestine, and the first two-thirds of the large intestine [94,95]. It is a complex of nerves located in front of the aorta at the level of the first lumbar vertebra. Although it innervates many upper abdominal organs, the plexus block is usually reserved for PC pain refractory to high-dose analgesic therapy [96,97]. Percutaneous bolus injection of local anesthetic, known as celiac plexus block, or neurolysis by chemically destroying the neural pathways or ganglia, can block the synaptic pathways for pain relief. It can lead to improved patient outcomes, including reduced pain levels and decreased reliance on opioids [94,97]. The celiac plexus block uses imaging techniques such as ultrasound or CT scan for guidance. Apart from percutaneous injection, it can be performed during endoscopic ultrasonography or surgically, with posterior and anterior paraaortic approaches being the most common [97]. The celiac plexus block is considered a highly effective therapeutic method for managing chronic abdominal pain, especially in the case of PC. However, in one single-center retrospective case–control study on patients with PC, the celiac plexus block was negatively associated with survival, although some other independent factors may have contributed to this result, including patient selection [98]. Similarly, neurolysis of splanchnic nerves originating from Th5 to L2, leading sympathetic fibers to the celiac plexus, can be performed using 95% alcohol [99]. Using this technique, the pain relief appears to be most significant in the first 3 months, with decreased opioid use present for 5 months following the procedure.

Alternatively, a short course of RT to alleviate pain and reduce the need for opioid analgesics is also feasible [100,101]. However, achieving the desired effect takes longer compared to the celiac plexus block. Furthermore, when deciding on the above procedures, it should be considered that none of them have a proven benefit in terms of OS.

Conclusion

In summary, palliative management of patients with hepatobiliary tumors is a multidisciplinary approach to optimize QoL. Continuous, clear communication among patients, family members, and healthcare providers is a cornerstone of this care. Importantly, palliative care can be integrated early in managing life-limiting conditions alongside other treatments that may extend life and are not solely reserved for the end-of-life stage [102].

References

[1] Sung H, Ferlay J, Siegel RL, et al. Global cancer statistics 2020: GLOBOCAN estimates of incidence and mortality worldwide for 36 cancers in 185 countries. CA Cancer J Clin 2021;71(3):209–49. Available from: https://doi.org/10.3322/caac.21660.

[2] Thomas London W, Petrick JL, McGlynn KA. Liver cancer. In: Thun M, Linet MS, Cerhan JR, Haiman CA, Schottenfeld D, editors. Cancer epidemiology and prevention. 4th ed. Oxford University Press; 2018. p. 635–60. ISBN: 9780190238667.

[3] Llovet JM, Zucman-Rossi J, Pikarsky E, et al. Hepatocellular carcinoma. Nat Rev Dis Primers 2016;2:16018. Available from: https://doi.org/10.1038/nrdp.2016.18.

[4] Forner A, Reig M, Bruix J. Hepatocellular carcinoma. Lancet 2018;391(10127):1301–14. Available from: https://doi.org/10.1016/S0140-6736(18)30010-2.

[5] Singal AG, Kanwal F, Llovet JM. Global trends in hepatocellular carcinoma epidemiology: implications for screening, prevention and therapy. Nat Rev Clin Oncol 2023;20:864–84. Available from: https://doi.org/10.1038/s41571-023-00825-3.

[6] Suresh D, Srinivas AN, Kumar DP. Etiology of hepatocellular carcinoma: special focus on fatty liver disease. Front Oncol 2020;10:601710. Available from: https://doi.org/10.3389/fonc.2020.601710.

[7] Stecco C, Sfriso MM, Porzionato A, et al. Microscopic anatomy of the visceral fasciae. J Anat 2017;231(1):121–8. Available from: https://doi.org/10.1111/joa.12617.

[8] Sinturel F, Gerber A, Mauvoisin D, et al. Diurnal oscillations in liver mass and cell size accompany ribosome assembly cycles. Cell 2017;169:651–63. Available from: https://doi.org/10.1016/j.cell.2017.04.015.

[9] Abdel-Misih SR, Bloomston M. Liver anatomy. Surg Clin North Am 2010;90(4):643–53. Available from: https://doi.org/10.1016/j.suc.2010.04.017.

[10] Skandalakis JE, Skandalakis LJ, Skandalakis PN, et al. Hepatic surgical anatomy. Surg Clin North Am 2004;84(2):413–35, viii. Available from: https://doi.org/10.1016/j.suc.2003.12.002.

[11] Llewellyn J, Fede C, Loneker AE, et al. Glisson's capsule matrix structure and function is altered in patients with cirrhosis irrespective of aetiology. JHEP Rep 2023;5(9):100760. Available from: https://doi.org/10.1016/j.jhepr.2023.100760.

[12] Ginès P, Krag A, Abraldes JG, et al. Liver cirrhosis. Lancet 2021;398(10308):1359–76. Available from: https://doi.org/10.1016/S0140-6736(21)01374-X.

[13] Chen T, Chen G, Wang G, et al. Expert consensus on the diagnosis and treatment of end-stage liver disease complicated by infections. Hepatol Int 2024;18(3):817–32. Available from: https://doi.org/10.1007/s12072-023-10637-3.

[14] Sanchez W, Talwalkar JA. Palliative care for patients with end-stage liver disease ineligible for liver transplantation. Gastroenterol Clin North Am 2006;35(1):201–19. Available from: https://doi.org/10.1016/j.gtc.2005.12.007.

[15] Younossi ZM, Boparai N, Price LL, et al. Health-related quality of life in chronic liver disease: the impact of type and severity of disease. Am J Gastroenterol 2001;96(7):2199–205. Available from: https://doi.org/10.1111/j.1572-0241.2001.03956.x.

[16] Zhu AX. Hepatocellular carcinoma: are we making progress? Cancer Invest 2003;21(3):418–28. Available from: https://doi.org/10.1081/cnv-120018233.

[17] Kumar M, Panda D. Role of supportive care for terminal stage hepatocellular carcinoma. J Clin Exp Hepatol 2014;4(Suppl. 3):S130–9. Available from: https://doi.org/10.1016/j.jceh.2014.03.049.

[18] Sun VC, Sarna L. Symptom management in hepatocellular carcinoma. Clin J Oncol Nurs 2008;12(5):759–66. Available from: https://doi.org/10.1188/08.CJON.759-766.

[19] Peng JK, Hepgul N, Higginson IJ, et al. Symptom prevalence and quality of life of patients with end-stage liver disease: a systematic review and meta-analysis. Palliat Med 2019;33(1):24–36. Available from: https://doi.org/10.1177/0269216318807051.

[20] Christian-Miller N, Frenette C. Hepatocellular cancer pain: impact and management challenges. J Hepatocell Carcinoma 2018;5:75–80. Available from: https://doi.org/10.2147/JHC.S145450.

[21] Boezaart AP, Smith CR, Chembrovich S, et al. Visceral versus somatic pain: an educational review of anatomy and clinical implications. Reg Anesth Pain Med 2021;46(7):629–36. Available from: https://doi.org/10.1136/rapm-2020-102084.

[22] Kudo M, Finn RS, Qin S, et al. Lenvatinib versus sorafenib in first-line treatment of patients with unresectable hepatocellular carcinoma: a randomised phase 3 non-inferiority trial. Lancet 2018;391(10126):1163–73. Available from: https://doi.org/10.1016/S0140-6736(18)30207-1.

[23] Finn RS, Qin S, Ikeda M, et al. Atezolizumab plus bevacizumab in unresectable hepatocellular carcinoma. N Engl J Med 2020;382(20):1894–905. Available from: https://doi.org/10.1056/NEJMoa1915745.

[24] Cheng AL, Qin S, Ikeda M, et al. Updated efficacy and safety data from IMbrave150: atezolizumab plus bevacizumab vs. sorafenib for unresectable hepatocellular carcinoma. J Hepatol 2022;76(4):862–73. Available from: https://doi.org/10.1016/j.jhep.2021.11.030.

[25] Galle PR, Finn RS, Qin S, et al. Patient-reported outcomes with atezolizumab plus bevacizumab versus sorafenib in patients with unresectable hepatocellular carcinoma (IMbrave150): an open-label, randomised, phase 3 trial. Lancet Oncol 2021;22(7):991–1001. Available from: https://doi.org/10.1016/S1470-2045(21)00151-0.

[26] Abou-Alfa GK, Chan SL, Kudo M, et al. Phase 3 randomized, open-label, multicenter study of tremelimumab and durvalumab as first-line therapy in patients with unresectable hepatocellular carcinoma: HIMALAYA. J Clin Oncol 2022;40:379. Available from: https://doi.org/10.1200/JCO.2022.40.4_suppl.379.

[27] Elbekai RH, Korashy HM, El-Kadi AO. The effect of liver cirrhosis on the regulation and expression of drug metabolizing enzymes. Curr Drug Metab 2004;5(2):157–67. Available from: https://doi.org/10.2174/1389200043489054.

[28] Verbeeck RK. Pharmacokinetics and dosage adjustment in patients with hepatic dysfunction. Eur J Clin Pharmacol 2008;64(12):1147–61. Available from: https://doi.org/10.1007/s00228-008-0553-z.

[29] Watkins PB, Kaplowitz N, Slattery JT, et al. Aminotransferase elevations in healthy adults receiving 4 grams of acetaminophen daily: a randomized controlled trial. JAMA 2006;296(1):87–93. Available from: https://doi.org/10.1001/jama.296.1.87.

[30] Larson AM, Polson J, Fontana RJ, et al. Acetaminophen-induced acute liver failure: results of a United States multicenter, prospective study. Hepatology 2005;42(6):1364–72. Available from: https://doi.org/10.1002/hep.20948.

[31] Chandok N, Watt KD. Pain management in the cirrhotic patient: the clinical challenge. Mayo Clin Proc 2010;85(5):451–8. Available from: https://doi.org/10.4065/mcp.2009.0534.

[32] Ma J, Björnsson ES, Chalasani N. The safe use of analgesics in patients with cirrhosis: a narrative review. Am J Med 2024;137(2):99–106. Available from: https://doi.org/10.1016/j.amjmed.2023.10.022.

[33] Malespin MH. Risk of nonsteroidal anti-inflammatory drugs and safety of acetaminophen in patients with advanced liver disease. Clin Liver Dis (Hoboken) 2018;12(3):85–8. Available from: https://doi.org/10.1002/cld.737.

[34] Rakoski M, Goyal P, Spencer-Safier M, et al. Pain management in patients with cirrhosis. Clin Liver Dis (Hoboken) 2018;11(6):135–40. Available from: https://doi.org/10.1002/cld.711.

[35] Myers RP, Shaheen AA, Li B, et al. Impact of liver disease, alcohol abuse, and unintentional ingestions on the outcomes of acetaminophen overdose. Clin Gastroenterol Hepatol 2008;6(8):918–25. Available from: https://doi.org/10.1016/j.cgh.2008.02.053.

[36] Nguyen GC, Sam J, Thuluvath PJ. Hepatitis C is a predictor of acute liver injury among hospitalizations for acetaminophen overdose in the United States: a nationwide analysis. Hepatology 2008;48(4):1336–41. Available from: https://doi.org/10.1002/hep.22536.

[37] Ferguson CB, Mitchell RM. Nonvariceal upper gastrointestinal bleeding: standard and new treatment. Gastroenterol Clin North Am 2005;34(4):607–21. Available from: https://doi.org/10.1016/j.gtc.2005.08.003.

[38] García-Pagán JC, Gracia-Sancho J, Bosch J. Functional aspects on the pathophysiology of portal hypertension in cirrhosis. J Hepatol 2012;57:458. Available from: https://doi.org/10.1016/j.jhep.2012.03.007.

[39] Soleimanpour H, Safari S, Shahsavari Nia K, et al. Opioid drugs in patients with liver disease: a systematic review. Hepat Mon 2016;16(4):e32636. Available from: https://doi.org/10.5812/hepatmon.32636.

[40] Hasselström J, Eriksson S, Persson A, et al. The metabolism and bioavailability of morphine in patients with severe liver cirrhosis. Br J Clin Pharmacol 1990;29(3):289–97. Available from: https://doi.org/10.1111/j.1365-2125.1990.tb03638.x.

[41] Lee CR, McTavish D, Sorkin EM. Tramadol. A preliminary review of its pharmacodynamic and pharmacokinetic properties, and therapeutic potential in acute and chronic pain states. Drugs 1993;46(2):313–40. Available from: https://doi.org/10.2165/00003495-199346020-00008.

[42] Zhu XC, Zhang JL, Ge CT, et al. Advances in cancer pain from bone metastasis. Drug Des Devel Ther 2015;9:4239–45. Available from: https://doi.org/10.2147/DDDT.S87568.

[43] Longo V, Brunetti O, D'Oronzo S, et al. Bone metastases in hepatocellular carcinoma: an emerging issue. Cancer Metastasis Rev 2014;33(1):333–42. Available from: https://doi.org/10.1007/s10555-013-9454-4.

[44] Reale C, Turkiewicz AM, Reale CA. Antalgic treatment of pain associated with bone metastases. Crit Rev Oncol Hematol 2001;37(1):1–11. Available from: https://doi.org/10.1016/s1040-8428(99)00066-9.

[45] Uchino K, Tateishi R, Shiina S, et al. Hepatocellular carcinoma with extrahepatic metastasis: clinical features and prognostic factors. Cancer 2011;117(19):4475–83. Available from: https://doi.org/10.1002/cncr.25960.

[46] Uka K, Aikata H, Takaki S, et al. Clinical features and prognosis of patients with extrahepatic metastases from hepatocellular carcinoma. World J Gastroenterol 2007;13(3):414–20. Available from: https://doi.org/10.3748/wjg.v13.i3.414.

[47] Kim SU, Kim DY, Park JY, et al. Hepatocellular carcinoma presenting with bone metastasis: clinical characteristics and prognostic factors. J Cancer Res Clin Oncol 2008;134(12):1377–84. Available from: https://doi.org/10.1007/s00432-008-0410-6.
[48] Kashima M, Yamakado K, Takaki H, et al. Radiofrequency ablation for the treatment of bone metastases from hepatocellular carcinoma. AJR Am J Roentgenol 2010;194(2):536–41. Available from: https://doi.org/10.2214/AJR.09.2975.
[49] Rim CH, Park S, Yoon WS, et al. Radiotherapy for bone metastases of hepatocellular carcinoma: a hybrid systematic review with meta-analyses. Int J Radiat Biol 2023;99(3):419–30. Available from: https://doi.org/10.1080/09553002.2022.2094020.
[50] Buijs JT, van der Pluijm G. Osteotropic cancers: from primary tumor to bone. Cancer Lett 2009;273(2):177–93. Available from: https://doi.org/10.1016/j.canlet.2008.05.044.
[51] Cremers S, Drake MT, Ebetino FH, et al. Pharmacology of bisphosphonates. Br J Clin Pharmacol 2019;85(6):1052–62. Available from: https://doi.org/10.1111/bcp.13867.
[52] Drake MT, Clarke BL, Khosla S. Bisphosphonates: mechanism of action and role in clinical practice. Mayo Clin Proc 2008;83(9):1032–45. Available from: https://doi.org/10.4065/83.9.1032.
[53] Montella L, Addeo R, Palmieri G, et al. Zoledronic acid in the treatment of bone metastases by hepatocellular carcinoma: a case series. Cancer Chemother Pharmacol 2010;65(6):1137–43. Available from: https://doi.org/10.1007/s00280-009-1122-6.
[54] Chow E, Harris K, Fan G, et al. Palliative radiotherapy trials for bone metastases: a systematic review. J Clin Oncol 2007;25(11):1423–36. Available from: https://doi.org/10.1200/JCO.2006.09.5281.
[55] Chow R, Hoskin P, Schild SE, et al. Single vs multiple fraction palliative radiation therapy for bone metastases: cumulative meta-analysis. Radiother Oncol 2019;141:56–61. Available from: https://doi.org/10.1016/j.radonc.2019.06.037.
[56] Schmid RK, Johnstone CA, Robbins JR. Palliative radiation for bone metastases from hepatocellular carcinoma: practice patterns and the amount of remaining life spent receiving treatment. Ann Palliat Med 2022;11(6):1900–10. Available from: https://doi.org/10.21037/apm-21-2657.
[57] Menon G, Garikipati SC, Roy P. Cholangiocarcinoma. StatPearls. Treasure Island (FL): StatPearls Publishing; 2024. Available from: https://www.ncbi.nlm.nih.gov/books/NBK560708/.
[58] Vogel A, Bridgewater J, Edeline J, et al. Biliary tract cancer: ESMO clinical practice guideline for diagnosis, treatment and follow-up. Ann Oncol 2023;34(2):127–40. Available from: https://doi.org/10.1016/j.annonc.2022.10.506.
[59] Khan SA, Tavolari S, Brandi G. Cholangiocarcinoma: epidemiology and risk factors. Liver Int 2019;39(Suppl. 1):19–31. Available from: https://doi.org/10.1111/liv.14095.
[60] Banales JM, Marin JJG, Lamarca A, et al. Cholangiocarcinoma 2020: the next horizon in mechanisms and management. Nat Rev Gastroenterol Hepatol 2020;17(9):557–88. Available from: https://doi.org/10.1038/s41575-020-0310-z.
[61] Oh DY, Ruth He A, Qin S, et al. Durvalumab plus gemcitabine and cisplatin in advanced biliary tract cancer. NEJM Evid 2022;1(8):EVIDoa2200015. https://doi.org/10.1056/EVIDoa2200015.
[62] Burris HA, 3rd, Okusaka T, Vogel A, et al. Durvalumab plus gemcitabine and cisplatin in advanced biliary tract cancer (TOPAZ-1): patient-reported outcomes from a randomised, double-blind, placebo-controlled, phase 3 trial. Lancet Oncol 2024;25(5):626–35. Available from: https://doi.org/10.1016/S1470-2045(24)00082-2.
[63] Wu X, Qin F, Zhang Q, et al. Immunotherapy improved cancer related pain management in patients with advanced hepato-pancreatic biliary cancers: a propensity score-matched (PSM) analysis. Front Oncol 2022;12:914591. Available from: https://doi.org/10.3389/fonc.2022.914591.
[64] Liu Y, Wang Y, Yu Y, et al. Comparison of clinical characteristics and mortality risk between patients with cholangiocarcinoma: a retrospective cohort study. Front Surg 2023;9:1037310. Available from: https://doi.org/10.3389/fsurg.2022.1037310.
[65] Wang H, Jiao Y, Ma Q, et al. Overview of endoscopic biliary stenting in malignant obstructive jaundice. World J Gastrointest Surg 2025;17(2):103378. Available from: https://doi.org/10.4240/wjgs.v17.i2.103378.
[66] Moole H, Dharmapuri S, Duvvuri A, et al. Endoscopic versus percutaneous biliary drainage in palliation of advanced malignant hilar obstruction: a meta-analysis and systematic review. Can J Gastroenterol Hepatol 2016;2016:4726078. Available from: https://doi.org/10.1155/2016/4726078.
[67] Robson PC, Heffernan N, Gonen M, et al. Prospective study of outcomes after percutaneous biliary drainage for malignant biliary obstruction. Ann Surg Oncol 2010;17(9):2303–11. Available from: https://doi.org/10.1245/s10434-010-1045-9.
[68] Szary NM, Al-Kawas FH. Complications of endoscopic retrograde cholangiopancreatography: how to avoid and manage them. Gastroenterol Hepatol (N Y) 2013;9(8):496–504.
[69] Antalek M, Patel ME, Knight GM, et al. Adverse events after percutaneous transhepatic biliary drainage: a 10-year retrospective analysis. J Vasc Interv Radiol 2025;36(4):564–572.e1. Available from: https://doi.org/10.1016/j.jvir.2024.12.022.
[70] Talathi SS, Zimmerman R, Young M. Anatomy, abdomen and pelvis, pancreas. StatPearls. Treasure Island (FL): StatPearls Publishing; 2024. Available from: https://www.ncbi.nlm.nih.gov/books/NBK532912/.
[71] Ferlay J, Ervik M, Lam F, et al. Global cancer observatory: cancer today. Lyon, France: International Agency for Research on Cancer; 2024. Available from https://gco.iarc.who.int/today [accessed 11.06.25].
[72] Ilic I, Ilic M. International patterns in incidence and mortality trends of pancreatic cancer in the last three decades: a joinpoint regression analysis. World J Gastroenterol 2022;28(32):4698–715. Available from: https://doi.org/10.3748/wjg.v28.i32.4698.
[73] Hu JX, Zhao CF, Chen WB, et al. Pancreatic cancer: a review of epidemiology, trend, and risk factors. World J Gastroenterol 2021;27(27):4298–321. Available from: https://doi.org/10.3748/wjg.v27.i27.4298.
[74] Bond-Smith G, Banga N, Hammond TM, et al. Pancreatic adenocarcinoma. BMJ 2012;344:e2476. Available from: https://doi.org/10.1136/bmj.e2476.

[75] Paice JA, Mulvey M, Bennett M, et al. AAPT diagnostic criteria for chronic cancer pain conditions. J Pain 2017;18(3):233–46. Available from: https://doi.org/10.1016/j.jpain.2016.10.020.

[76] D'Haese JG, Hartel M, Demir IE, et al. Pain sensation in pancreatic diseases is not uniform: the different facets of pancreatic pain. World J Gastroenterol 2014;20(27):9154–61. Available from: https://doi.org/10.3748/wjg.v20.i27.9154.

[77] Bapat AA, Hostetter G, Von Hoff DD, et al. Perineural invasion and associated pain in pancreatic cancer. Nat Rev Cancer 2011;11(10):695–707. Available from: https://doi.org/10.1038/nrc3131.

[78] Caraceni A, Portenoy RK. Pain management in patients with pancreatic carcinoma. Cancer. 1996;78(3 Suppl):639–53. Available from: https://doi.org/10.1002/(SICI)1097-0142(19960801)78:3<639::AID-CNCR45>3.0.CO;2-X.

[79] Carvajal G. Pancreatic cancer related pain: review of pathophysiology and intrathecal drug delivery systems for pain management. Pain Physician 2021;24(5):E583–94.

[80] di Mola FF, di Sebastiano P. Pain and pain generation in pancreatic cancer. Langenbecks Arch Surg 2008;393(6):919–22. Available from: https://doi.org/10.1007/s00423-007-0277-z.

[81] Pyter LM. The influence of cancer on endocrine, immune, and behavioral stress responses. Physiol Behav 2016;166:4–13. Available from: https://doi.org/10.1016/j.physbeh.2015.09.031.

[82] Batty GD, Russ TC, Stamatakis E, et al. Psychological distress in relation to site specific cancer mortality: pooling of unpublished data from 16 prospective cohort studies. BMJ 2017;356:j108. Available from: https://doi.org/10.1136/bmj.j108.

[83] Mazzella Ebstein AM, Joseph SJ, Hernandez M. Psychological stress and pancreatic cancer patients: a qualitative systematic review protocol. JBI Evid Synth 2020;18(3):576–82. Available from: https://doi.org/10.11124/JBISRIR-D-18-00006.

[84] Conroy T, Desseigne F, Ychou M, et al. FOLFIRINOX versus gemcitabine for metastatic pancreatic cancer. N Engl J Med 2011;364(19):1817–25. Available from: https://doi.org/10.1056/NEJMoa1011923.

[85] Von Hoff DD, Ervin T, Arena FP, et al. Increased survival in pancreatic cancer with nab-paclitaxel plus gemcitabine. N Engl J Med 2013;369:1691–703. Available from: https://doi.org/10.1056/NEJMoa1304369.

[86] Gourgou-Bourgade S, Bascoul-Mollevi C, Desseigne F, et al. Impact of FOLFIRINOX compared with gemcitabine on quality of life in patients with metastatic pancreatic cancer: results from the PRODIGE 4/ACCORD 11 randomized trial. J Clin Oncol 2013;31(1):23–9. Available from: https://doi.org/10.1200/JCO.2012.44.4869.

[87] Lohse I, Brothers SP. Pathogenesis and treatment of pancreatic cancer related pain. Anticancer Res 2020;40(4):1789–96. Available from: https://doi.org/10.21873/anticanres.14133.

[88] Lussier D, Huskey AG, Portenoy RK. Adjuvant analgesics in cancer pain management. Oncologist 2004;9(5):571–91. Available from: https://doi.org/10.1634/theoncologist.9-5-571.

[89] Paice JA, Portenoy R, Lacchetti C, et al. Management of chronic pain in survivors of adult cancers: American Society of Clinical Oncology clinical practice guideline. J Clin Oncol 2016;34(27):3325–45. Available from: https://doi.org/10.1200/JCO.2016.68.5206.

[90] Amr YM, Makharita MY. Neurolytic sympathectomy in the management of cancer pain-time effect: a prospective, randomized multicenter study. J Pain Symptom Manage 2014;48(5):944–56. Available from: https://doi.org/10.1016/j.jpainsymman.2014.01.015.

[91] Lou S. Endoscopic ultrasound-guided celiac plexus neurolysis to alleviate intractable pain caused by advanced pancreatic cancer. Surg Laparosc Endosc Percutan Tech 2019;29(6):472–5. Available from: https://doi.org/10.1097/SLE.0000000000000683.

[92] Arcidiacono PG, Calori G, Carrara S, et al. Celiac plexus block for pancreatic cancer pain in adults. Cochrane Database Syst Rev 2011;2011(3):CD007519. Available from: https://doi.org/10.1002/14651858.CD007519.pub2.

[93] Stefaniak T, Basinski A, Vingerhoets A, et al. A comparison of two invasive techniques in the management of intractable pain due to inoperable pancreatic cancer: neurolytic celiac plexus block and videothoracoscopic splanchnicectomy. Eur J Surg Oncol 2005;31(7):768–73. Available from: https://doi.org/10.1016/j.ejso.2005.03.012.

[94] Cornman-Homonoff J, Holzwanger DJ, Lee KS, et al. Celiac plexus block and neurolysis in the management of chronic upper abdominal pain. Semin Intervent Radiol 2017;34(4):376–86. Available from: https://doi.org/10.1055/s-0037-1608861.

[95] Vig S, Bhan S, Bhatnagar S. Celiac plexus block – an old technique with new developments. Pain Physician 2021;24(5):379–98.

[96] Ambai VT, Singh V, Boorman DW, et al. Celiac plexus neurolysis for abdominal cancers: going beyond pancreatic cancer pain. Pain Rep 2021;6(1):e930. Available from: https://doi.org/10.1097/PR9.0000000000000930.

[97] John RS, Dixon B, Hendrix JM, et al. Celiac plexus block. StatPearls. Treasure Island (FL): StatPearls Publishing; 2024. Available from: https://www.ncbi.nlm.nih.gov/books/NBK531469.

[98] Fujii-Lau LL, Bamlet WR, Eldrige JS, et al. Impact of celiac neurolysis on survival in patients with pancreatic cancer. Gastrointest Endosc 2015;82(1):46–56.e2. Available from: https://doi.org/10.1016/j.gie.2014.12.036.

[99] Dong D, Zhao M, Zhang J, et al. Neurolytic splanchnic nerve block and pain relief, survival, and quality of life in unresectable pancreatic cancer: a randomized controlled trial. Anesthesiology 2021;135(4):686–98. Available from: https://doi.org/10.1097/ALN.0000000000003936.

[100] Ceha HM, van Tienhoven G, Gouma DJ, et al. Feasibility and efficacy of high dose conformal radiotherapy for patients with locally advanced pancreatic carcinoma. Cancer. 2000;89(11):2222–9. Available from: https://doi.org/10.1002/1097-0142(20001201)89:11<2222::aid-cncr10>3.0.co;2-v.

[101] Morganti AG, Trodella L, Valentini V, et al. Pain relief with short-term irradiation in locally advanced carcinoma of the pancreas. J Palliat Care 2003;19(4):258–62. Available from: https://doi.org/10.1177/082585970301900407.

[102] World Health Organization. Palliative care, <https://www.who.int/news-room/fact-sheets/detail/palliative-care>; 2020 [accessed 19.08.24].

Chapter 36

Esophageal and gastric cancers

Majana Soče[1,2] and Marin Golčić[3,4]
[1]Department of Medical Oncology, Clinical Hospital Center Zagreb, Zagreb, Croatia, [2]University of Applied Health Sciences, University of Zagreb, Croatia, [3]Tumor Clinic, Clinical Hospital Center Rijeka, Rijeka, Croatia, [4]Faculty of Medicine, University of Rijeka, Rijeka, Croatia

Key learning objectives

- Understand the epidemiology, risk factors, and diagnostic procedures for esophageal and gastric cancers.
- Evaluate the role of systemic therapies, including chemotherapy, immunotherapy, and targeted therapy, in the treatment of advanced esophageal and gastric cancers.
- Recognize the impact of modern systemic treatments on survival outcomes, quality of life, and pain reduction in metastatic settings.
- Explore the value of radiotherapy and stenting to palliate pain in patients with advanced disease.

Epidemiology of esophageal and gastric cancers

Esophageal and gastric cancers are major contributors to cancer mortality worldwide, with gastric cancer responsible for 5.6% of all new cases and 7.7% of mortality and esophageal cancer (often divided into esophageal squamous cell carcinoma [ESCC] and esophageal adenocarcinoma [EAC]), contributing to 3.1% of new cases and 5.5% of mortality. Gastric cancer is more common in men and is particularly common in Central Asian countries, East Asia, and East Europe, while esophageal cancer depends on the histology—the ESCC is most commonly found in Asia and Africa, compared to EAC, most commonly found in higher-income Western countries [1].

Esophageal cancer risk factors include tobacco and alcohol consumption, obesity, chronic reflux (Barrett's esophagus) (particularly for EAC), and diet and nutrients. Hereditary cancer is rare, although some of known hereditary cancer predisposition syndromes are tylosis—focal non epidermolytic (palmoplantar keratoderma or Howel–Evans syndrome), a very rare autosomal dominant syndrome—and Bloom syndrome and Fanconi anemia—both rare autosomal recessive disorders [2]. The majority of gastric cancers (95%) are adenocarcinomas, which are classified anatomically (cardia/proximal or noncardia/distal) and histologically (diffuse or intestinal). The diffuse type occurs mostly in high-risk countries, while the intestinal type is related to environmental factors such as *Helicobacter pylori* infection and tobacco consumption. Smoking, obesity, high-salt diet, and processed meat are considered risk factors as well. About 3%–5% of gastric cancers are associated with hereditary predisposition syndromes, such as hereditary diffuse gastric cancer (an autosomal dominant syndrome characterized by young age onset of developing gastric cancer), Lynch syndrome, Peutz–Jeghers syndrome, and familial adenomatous polyposis [3].

Patients with esophageal and gastric cancers frequently present at advanced stages, where metastasis contributes heavily to disease burden and pain [1,4,5]. It is estimated that up to 60% of patients present with metastatic disease at diagnosis. Most common sites of metastasis of esophageal cancer include liver, lung, and lymph nodes (all above 50%), compared to gastric cancer which most commonly metastasize to lymph nodes, liver, and peritoneum (all above 50%). Pain from esophageal and gastric cancers can be nociceptive, neuropathic, or mixed, most commonly due to compression and expansion, although cancer treatment itself can cause significant pain (especially oxaliplatin). The detailed evaluation of how to treat pain caused by cancer treatment and which analgesics are preferred are evaluated in a different chapter of this book. Liver metastases are the most common site of metastasis for both cancers and are characterized by dull, persistent pain in the right upper quadrant, often worsening with movement or breathing. Bone metastases often occur as sharp, localized pain that may worsen at night and are more often found in EAC compared to ESCC (35% vs 16%); a similar

Pain Management in Palliative Care for Patients With Cancer. DOI: https://doi.org/10.1016/B978-0-443-30050-9.00001-1

difference is reported for diffuse gastric carcinoma compared to the intestinal type (30% vs 15%). Bone metastases are usually localized in the most vascularized skeletal parts, and although commonly painful, up to a third of the patients with metastasis found on imaging studies did not complain of bone pain [5–7]. While lung metastases are less common in gastric cancer, when evaluating patients with pleuritic pain, often exacerbated by deep breathing, it is also important to rule out both pulmonary embolism and lung metastasis [8]. Peritoneal metastases, especially common in gastric cancer, are frequently associated with severe abdominal pain and ascites and are a prognostic marker of poor survival [9].

Diagnosis of esophageal and gastric cancers

Diagnosis of esophageal and gastric cancers involves a combination of clinical evaluation, endoscopic procedures, laboratory analysis, and imaging studies. Upper gastrointestinal endoscopy is the gold standard for diagnosing these cancers, allowing direct visualization of the tumor and biopsy for histopathological examination. Endoscopic ultrasound can be used to assess tumor depth and nodal involvement, while computed tomography (CT) and positron emission tomography (PET) scans are employed to detect metastatic disease and to stage the cancer. PET/CT, when available, is commonly used to detect unsuspected distant metastases during staging and could rule out a futile radical operation in patients with widespread disease, or be used to evaluate response to treatment. In gastric cancer, diagnostic laparoscopy can be used to detect occult metastases, primarily for clinically resectable gastric cancer at high risk for peritoneal spread. Positive peritoneal cytology is associated with a poor prognosis and is considered as metastatic disease, even in the absence of other metastatic sites [10]. The accurate staging of the disease is essential for planning appropriate treatment, including the distinction between localized and metastatic disease, and the location of primary tumor.

One crucial distinction between gastric cancer and esophageal cancer is the Siewert classification, which has been used to plan treatment since its proposal in the 1980s. Siewert classification divides tumors into Siewert I if epicenter was in the distal esophagus, Siewert II if cancers were immediately arising at the gastroesophageal junction (GEJ), and Siewert III if the location was subcardial. Besides the choice of treatment, studies have shown that the Siewert classification is a prognostic factor in itself, with a rise in tumor diameter and the presence of perineural and vascular invasion as the tumor becomes more distal. Unsurprisingly, overall survival (OS) was worse in Siewert type III (2.62 years) compared to type I cancers (4.96 years) [11].

Treatment of localized disease in esophageal and gastric cancers

In patients with localized esophageal and gastric cancers, the main goal is to try to cure the patient. Surgery is the primary treatment modality for resectable disease, often after neoadjuvant chemotherapy [10,12]. Esophagectomy was for a long time the mainstay of treatment for localized esophageal cancer, although modern approach favors neoadjuvant chemoradiotherapy with trials such as CROSS showing improved OS compared to surgery alone (10-year survival benefit: 38% vs 25%, hazard ratio [HR] 0.60, $P < .05$) [13]. Total or subtotal gastrectomy is performed for localized gastric cancer. Neoadjuvant and adjuvant chemotherapy, as demonstrated in the FLOT trial, has shown to improve survival outcomes compared to patients undergoing only gastrectomy (OS 50 months vs 35 months, HR 0.77, $P < .05$) [14].

Systemic therapy in esophageal cancer

Systemic therapy in esophageal squamous cell carcinoma

For ESCC, the treatment landscape relied on platinum-based chemotherapy combined with fluoropyrimidines as the foundational regimen up until recently, with the traditional chemotherapy regimen demonstrating an OS of 10 months and an objective response rate (ORR) of 30% [10,15]. However, the advances in the understanding of immune system and cancer interactions led to the new group of drugs, called the programmed death-1 (PD-1) inhibitors, which work by blocking the interaction between tumor cells and immune cells, reducing the tumor suppression of T cells. PD-1 inhibitors such as pembrolizumab proved efficacy in various tumors leading to increased but also sustained responses, although they often require weeks and months to achieve full anticancer activity, unlike chemotherapy, which can result in rapid response. The efficacy of PD-1 inhibitors was proven in various cancers, and trials such as KEYNOTE-590 produced a new standard of care in esophageal cancer as well. The trial tested combination of pembrolizumab plus chemotherapy and demonstrated an improved ORR of 45% and a median OS of 12.6 months. Not only that, but the response duration of 2 years or longer was significantly higher in the pembrolizumab group (18% vs 6%) [15]. This difference is important, as there is almost a 15% higher ORR with immunotherapy, which represents the proportion of patients whose tumors show significant shrinkage in response to treatment. A higher ORR is often associated with pain reduction as the tumor burden decreases, which can potentially reduce pressure on surrounding organs and nerves and lead to pain reduction; however, it must be noted that the treatment itself can result in toxicity and pain. The quality-of-life study in the KEYNOTE-590 trial showed that patients receiving pembrolizumab had significantly better results in pain subscale (−4.78) compared to only chemotherapy (−1.85), compared to baseline, showing a correlation between ORR and better pain scores [15,16].

In another trial, CheckMate 648, the combination of another anti-PD-1 checkpoint inhibitor nivolumab with chemotherapy demonstrated an ORR of 47%, very similar to previous data with pembrolizumab [17]. The trial only tested ESCC, unlike Keynote-590 [15], which had a wider cohort of patients. The quality-of-life analysis did show lower pain levels in patients treated with immunotherapy combined with chemotherapy, although the difference did not reach significance [17,18]. The CheckMate 648 trial also had a study arm with a combination of nivolumab and ipilimumab, another checkpoint inhibitor. Although that arm had a slightly lower response rate (RR) (around 35%) than the combination arm, it still produced viable survival benefits, suggesting that a nonchemotherapy option is possible as well in selected patients [17]. Despite no statistically significant differences reported in quality of life between the two arms, it was maintained for the nivolumab-based regimens, with fewer patients receiving dual checkpoint inhibition being bothered by treatment adverse effects compared to chemotherapy-based treatments. Furthermore, significantly higher response rates produced by immunotherapy could lead to fewer symptoms [17].

A variety of other checkpoint inhibitors such as sintilimab, toripalimab, camrelizumab, or tiragolumab with atezolizumab recently demonstrated positive results as well in the first-line treatment of ESCC. This shows that immunotherapy combined with chemotherapy now appears to be the mainstay of metastatic ESCC treatment and leads not only to longer OS but also to a higher ORR, which could potentially reduce pain, although QoL data are limited [10].

Later-line treatment of ESCC relies on chemotherapy in majority of cases, usually taxane-, irinotecan-, or oxaliplatin-based. However, trials such as RATIONALE 302 demonstrated that for patients with metastatic ESCC, immunotherapy with tislelizumab, an anti-PD-1 antibody, can achieve a significant response in a subset of patients, even in later lines of treatment. Tislelizumab produced longer OS compared to chemotherapy (8.6 vs 6.3 months, HR 0.70, $P < .05$), and importantly, higher ORR (20.3% vs 9.8%), with fewer patients experiencing a grade 3 or higher treatment–related adverse effects. Unfortunately, no quality-of-life or pain scale data were reported, but higher response rate and lower adverse effects could potentially lead to lower pain levels, at least in some of the patients [19].

Systemic therapy in esophageal adenocarcinoma

EAC was historically treated with a combination of platinum compounds and fluorouracil. Physicians used standard chemotherapy with platinum compounds and fluorouracil, which demonstrated an OS of 9–11 months and an ORR of 25%–30% [15,20]. However, new advances in the field of immunotherapy have significantly altered the response and increased the survival.

The majority of trials usually evaluated EAC, particularly of GEJ together with patients with gastric cancer due to similarities in behaviors, as it appears that those two share more biological similarities than EAC and ESCC. The KEYNOTE-590 trial is one of the rare modern trials that evaluates both ESCC and EAC. While the whole group of patients derived benefit from the addition of pembrolizumab (HR 0.73), the dominant factor behind the benefit was ESCC (the majority of patients in the trial), or patients with a combined positive score (CPS) >10 [15]. While CheckMate 648 only evaluated ESCC patients, the trial CheckMate 649 evaluated gastric adenocarcinoma, GEJ adenocarcinoma, and EAC, where an addition of nivolumab to regular chemotherapy was analyzed. The trial was positive for the whole population in terms of OS (HR 0.80), although a particular benefit was found for patients with a CPS >5 (HR 0.71). Importantly, for the whole population the response rate was 60% in the nivolumab arm, compared to 45% in the chemotherapy-alone arm, with 12% versus 7% of complete responders, that is, the patients with no signs of disease following the treatment course. The trial also evaluated the quality-of-life parameter via the Functional Assessment of Cancer Therapy—General (FACT-G) questionnaire, which also included pain analysis. Patients in the nivolumab plus chemotherapy group had decreased risk of symptom deterioration than the chemotherapy-alone group while on treatment [18,20].

Recent advances have integrated targeted agents like trastuzumab for human epidermal growth factor receptor 2 (HER2)-positive cases (predominantly in GEJ cancer), significantly enhancing outcomes in those patients. The ToGA trial was the one which established trastuzumab as a standard treatment in HER2-positive esophagogastric cancer, achieving an ORR of 47% versus 35% with chemotherapy alone. Complete response was also more likely reported in the group with trastuzumab (5% vs 2%), with similar rates of stable disease (32% vs 35%). It needs to be emphasized that in the study, only 20% of the patients had GEJ cancers, with the rest having a gastric cancer. However, the addition of trastuzumab had a positive and similar effect on both cancer types (HR 0.67 and HR 0.76 for OS), although the statistical significance was achieved only for gastric cancers, most likely due to the larger cohort [21]. The rapid tumor response is important, especially for those patients experiencing obstructive symptoms due to primary or metastatic lesions.

Quality-of-life analysis from these studies showed that patients receiving trastuzumab, pembrolizumab, or nivolumab experienced not only improved survival but also significant reductions in pain and better functional scores compared to those receiving chemotherapy alone, showing that modern cancer therapy is also important in multimodal approach to pain management [20–22].

Systemic therapy in gastric cancer

The treatment of metastatic gastric cancer also relies heavily on systemic chemotherapy, with increasing integration of targeted and immunotherapy agents. The standard chemotherapy regimen for metastatic gastric cancer typically included platinum salts and fluoropyrimidines, with an OS of 8–10 months and an ORR of 25%–30% [20,21].

As already mentioned, the patients with HER2 positivity are eligible to receive treatment with trastuzumab combined with chemotherapy, based on the results of the ToGA trial. The majority of gastric cancer patients had an intestinal histologic type (77%), and only 9% had a pure diffuse histology. As mentioned above, gastric cancer patients receiving trastuzumab exhibited an HR of 0.74 for OS (median OS 13.8 vs 11.1 months vs chemotherapy alone), with better results for nondiffuse histological type and with no previous lines of chemotherapies. As mentioned, although a significant response rate was reported with trastuzumab treatment, which could be associated with pain reduction, although stomatitis (22% vs 15%) and abdominal pain (22% vs 19%) were more often reported in the trastuzumab group. On the other hand, constipation was more common in the chemotherapy-only group (32% vs 26%) [21]. Additional data further support the use of trastuzumab in pain management during the treatment of gastric cancer, with median time to 10% definite deterioration of 9.4 versus 6.8 months ($P < .0001$) for patients treated with trastuzumab compared to chemotherapy alone [22].

Recent data supports the treatment of patients with HER2-positive gastric or GEJ cancer with a combination of PD-1 and HER2 blockade. KEYNOTE-811 evaluated patients with metastatic GEJ adenocarcinoma and HER2 positivity; one cohort was treated with pembrolizumab, trastuzumab, and chemotherapy and another with trastuzumab and chemotherapy. The study demonstrated a median OS of 20 months for the combined treatment arm and 16.9 months for the trastuzumab-only arm. While in interim analysis OS difference did not reach significance, final data demonstrated an improvement in OS (HR 0.80, $P = .004$) [23]. An ORR of 72.6% was noted in the combined therapy arms (vs 59.8% for trastuzumab-only arm), suggesting that a significant impact on pain could be expected, although no data on quality of life were published at the time of writing of this book [24].

On the other hand, patients without HER2 positivity had no other option besides chemotherapy until recently. Previously mentioned trial CheckMate 649 evaluated gastric adenocarcinoma, GEJ adenocarcinoma, and EAC without HER2 positivity [20], demonstrating an improved quality of life and OS, particularly for patients with a CPS above 5, and demonstrating an ORR of 60%. Another cohort of the trial evaluated combined nivolumab–ipilimumab, which did not improve OS in the whole population but showed significant improvement in patients with microsatellite instability (ORR 70% and HR 0.28 for OS) [20].

On the other hand, trials with pembrolizumab such as KEYNOTE-062 tested chemotherapy, chemotherapy plus pembrolizumab, and monotherapy pembrolizumab in patients with locally advanced or metastatic HER2-negative gastric or GEJ adenocarcinoma. Initial results showed a noninferiority of pembrolizumab to chemotherapy for OS in patients with a CPS of 1 or greater but prolonged OS in patients with a CPS >10 (17.4 vs 10.8 months, HR 0.69, $P < .05$). Surprisingly, the combination arm was not superior to chemotherapy [25]. On the other hand, KEYNOTE-859 only tested combination chemotherapy, showing an improvement in median OS compared to the placebo (13.9 vs 11.5 months, HR 0.78, $P < .05$) and an improvement in ORR of 51% versus 42%. Importantly, the trial evaluated quality of life via the European Organisation for Research and Treatment of Cancer Quality of Life Questionnaire Core 30 (EORTC QLQ-C30). Time to true deterioration was similar between groups in QLQ-C30 Global Health Status/Quality of Life and favored pembrolizumab versus placebo in the EORTC QLQ—Stomach (EORTC QLQ-STO22) Pain Scale (HR 0.76). Such results suggest that ORR can translate to an improvement in pain scales [26].

However, not all patients are eligible for immunotherapy or have HER2 receptor positivity. For those patients, several new medications are options. For patients with Claudin 18.2 positivity and advanced gastric cancer, a monoclonal antibody targeting Claudin 18.2, zolbetuximab, was tested in combination with standard chemotherapy. Data demonstrate improvements in ORR (42.5% vs 40.3%) and OS (14.4 vs 12.1 months, HR 0.77) when zolbetuximab is added [27]. Importantly, although the difference in ORR is not as significant as in some of the other previously mentioned trials, the analysis of combined phase 3 clinical trials demonstrated maintenance in measured quality of life, a numerically longer time to deterioration of abdominal pain and discomfort was reported for zolbetuximab arm. Particularly in the SPOTLIGHT trial, a greater proportion of patients had improved scores for abdominal pain and discomfort in the zolbetuximab arm (50.0%) compared with the placebo arm (39.2%) at week 49 [28].

Bemarituzumab, a targeted therapy against fibroblast growth factor receptor 2b (FGFR2), is another promising therapeutic option. A phase 2 study in advanced non-HER2-positive gastric/GEJ cancer (FIGHT) demonstrated an improvement in OS when chemotherapy was combined with bemarituzumab (19.2 vs 13.5 months, HR 0.77, although 95% confidence interval [CI] was 0.52–1.14). However, for patients with FGFR2b-positive gastric cancer in ≥10% of tumor cells, HR for OS was 0.52 (95% CI 0.31–0.85) with an ORR of 56.5% versus 36.5%. Although no substantial quality-of-life data were available, there was a similar percentage of patients reporting abdominal pain and sensory neuropathy between the two groups [29].

Later lines of systemic cancer therapies have substantially worse results compared to first-line therapies, with chemotherapies with taxanes or irinotecan being the mainstay of the treatment. The RAINBOW trial evaluated ramucirumab combined with paclitaxel in advanced gastric or GEJ adenocarcinoma. The results demonstrated improved efficacy (OS 9.6 vs 7.4 months, HR 0.807, $P < .05$) over chemotherapy alone, with an ORR of 28% versus 16%. Importantly, quality of life was maintained during treatment with ramucirumab [30,31]. Ramucirumab, a vascular endothelial growth factor receptor 2 (VEGFR-2) inhibitor, works by inhibiting angiogenesis, which not only helps reduce tumor growth but could potentially also alleviate pain related to tumor-induced vascular changes.

For patients not treated with immunotherapy in the first line, there is evidence of efficacy in later lines, although the results are more modest. For example, ATTRACTION-2 studied nivolumab in pretreated gastric or GEJ cancer patients, demonstrating an ORR of 11.2% and a modest increase in OS (5.2 vs 4.1 months, HR 0.61, $P < .05$). Although modest, this response could lead to improved pain management for some patients, particularly those with symptomatic metastases. Although patient-reported outcomes on quality of life are not available, the incidence of severe treatment-related adverse events was low, and symptomatic adverse events were observed equally in both groups [32]. This suggests that even modest tumor shrinkage with immunotherapy can occur in later-line setting and could potentially be associated with pain reduction.

A great potential for later-line treatment could be reached by combining HER2 antibodies with chemotherapy as proven by the DESTINY-Gastric01 trial, which showed that treatment with trastuzumab deruxtecan, an antibody–drug conjugate, in pretreated HER2-positive advanced gastric cancer could result in a longer OS (12.5 vs 8.4 months, HR 0.59) and a staggering 51% ORR (compared to 14% in physician choice group). No quality-of-life data were available, but such tumor shrinkage would likely result even in a short-term pain reduction [33]. While other treatment regimens have also demonstrated an increase in OS in later lines, such as trifluridine-tipiracil and regorafenib, their effect on ORR is modest [34].

Regardless, quality-of-life data from these newer therapies suggest significant improvement in symptom relief, including pain reduction, further emphasizing the benefits of integrating targeted therapy in systemic management of gastric cancer.

Radiotherapy and stenting in pain control

Radiotherapy is a crucial tool in the palliative management of pain in metastatic esophageal and gastric cancers, particularly for patients with bone metastasis, bleeding, or obstructive lesions. A recent metaanalysis demonstrates that various different radiotherapy doses and variations exist for palliative treatment of gastric cancer, but response rates for bleeding, pain, and obstruction symptoms were 74%, 67%, and 68%, respectively. The majority of studies analyzed had a relatively small number of patients, ranging from 15 to 115, and most had a follow–up of only 3–4 months, which limits long-term evaluation of the response. However, one of the trials analyzed reported a median duration of pain relief of 7.8 months. A low biologically equivalent dose seems to be similarly effective [35], and especially for bone metastases, a single 8 Gy fraction completely resolved pain in 23% of the patients after 3 months from single fraction, with overall efficacy of treatment of 62% [36]. A single fraction 8 Gy could also be used for gastric cancer bleeding, particularly in patients with a high palliative prognostic index score. However, regimens of 20 Gy/5 fractions, 30 Gy/10 fractions, and 36 Gy/12 fractions were commonly used as well [35,37]. Importantly, palliative gastric radiotherapy could also lead to lower use of transfusion units [38]. Compared to pain results, median duration of response for obstruction ranged from 3.2 to 6.2 months. Toxicity was rare and occurred only in 15% of the patients treated with radiotherapy alone [35].

Chemoradiotherapy is usually used predominantly in esophageal cancer, as a neoadjuvant or definite treatment. However, the treatment is usually difficult for patients with numerous adverse effects. Data show that neoadjuvant chemoradiotherapy reduces quality of life, particularly toward the last cycle of chemoradiotherapy; however, most symptoms were restored 4–8 weeks following the cessation. Not only that, but odynophagia scores actually improved compared to baseline levels, showing that neoadjuvant chemoradiotherapy can reduce pain following a brief increase [39].

Most patients with dysphagia subsequently suffer from weight loss and malnutrition. Complete, or near-complete obstruction, is a frequent reason for hospitalization in this patient population. Surgical palliation has high morbidity and mortality. Instead, various other palliative approaches are available, all of which are relatively effective. Endoscopic placement of stents is now widely used for mechanical relief of dysphagia. Self-expanding metallic stents (SEMSs) are today the most frequently used stents, as they require no prior dilation, and they are successful in more than 95% of cases. SEMSs are very valuable for esophageal perforations or tracheoesophageal fistulae, helping to avoid rapid deterioration and death. A frequent complication is repeated dysphagia due to tissue ingrowth or overgrowth, food bolus, or migration and requires reintervention. Other complications are bleeding and perforation. Also, stenting has been associated with pain; according to the literature, pain may be experienced in up to 60% of patients [40]. Surgical treatment of malignant bowel obstruction and a more detailed overview of radiotherapy options are evaluated in different chapters of this book.

Conclusion

In conclusion, advancements in systemic therapies, including chemotherapy, immunotherapy, and targeted agents, have significantly improved the management of esophageal and gastric cancers. These treatments not only extend survival but also alleviate pain, enhancing patients' quality of life. The ongoing evolution in personalized medicine, along with integrated supportive interventions, holds the potential to further refine treatment outcomes, offering hope for more effective and compassionate cancer care. However, introduction to early palliative care and treatment of patients in a multidisciplinary setting remain the keystone in both pain and oncological management.

References

[1] Sung H, Ferlay J, Siegel RL, et al. Global cancer statistics 2020: GLOBOCAN estimates of incidence and mortality worldwide for 36 cancers in 185 countries. CA Cancer J Clin 2021;71(3):209–49. Available from: https://doi.org/10.3322/caac.21660.

[2] Domper Arnal MJ, Ferrández Arenas A, Lanas Arbeloa A. Esophageal cancer: risk factors, screening and endoscopic treatment in Western and Eastern countries. World J Gastroenterol 2015;21(26):7933–43. Available from: https://doi.org/10.3748/wjg.v21.i26.7933.

[3] Ko KP. Risk factors of gastric cancer and lifestyle modification for prevention. J Gastric Cancer 2024;24(1):99–107. Available from: https://doi.org/10.5230/jgc.2024.24.e10.

[4] Bray F, Laversanne M, Weiderpass E, et al. The ever-increasing importance of cancer as a leading cause of premature death worldwide. Cancer 2021;127(16):3029–30. Available from: https://doi.org/10.1002/cncr.33587.

[5] Verstegen MHP, Harker M, van de Water C, et al. Metastatic pattern in esophageal and gastric cancer: influenced by site and histology. World J Gastroenterol 2020;26(39):6037–46. Available from: https://doi.org/10.3748/wjg.v26.i39.6037.

[6] Shah MA, Kennedy EB, Catenacci DV, et al. Treatment of locally advanced esophageal carcinoma: ASCO guideline. J Clin Oncol 2020;38(23):2677–94. Available from: https://doi.org/10.1200/JCO.20.00866.

[7] Selvaggi G, Scagliotti GV. Management of bone metastases in cancer: a review. Crit Rev Oncol Hematol 2005;56(3):365–78. Available from: https://doi.org/10.1016/j.critrevonc.2005.03.011.

[8] Reamy BV, Williams PM, Odom MR. Pleuritic chest pain: sorting through the differential diagnosis. Am Fam Physician 2017;96(5):306–12.

[9] Messager M, Lefevre JH, Pichot-Delahaye V, et al. The impact of perioperative chemotherapy on survival in patients with gastric signet ring cell adenocarcinoma: a multicenter comparative study. Ann Surg 2011;254(5):684–93. Available from: https://doi.org/10.1097/SLA.0b013e3182352647.

[10] National Comprehensive Cancer Network (NCCN). NCCN clinical practice guidelines in oncology: esophageal and esophagogastric junction cancers, v4, <https://www.nccn.org/professionals/physician_gls/pdf/esophageal.pdf>; 2021.

[11] Curtis NJ, Noble F, Bailey IS, et al. The relevance of the Siewert classification in the era of multimodal therapy for adenocarcinoma of the gastro-oesophageal junction. J Surg Oncol 2014;109(3):202–7. Available from: https://doi.org/10.1002/jso.23484.

[12] van Hagen P, Hulshof MC, van Lanschot JJ, et al. Preoperative chemoradiotherapy for esophageal or junctional cancer. N Engl J Med 2012;366(22):2074–84. Available from: https://doi.org/10.1056/NEJMoa1112088.

[13] Eyck BM, van Lanschot JJB, Hulshof MCCM, et al. Ten-year outcome of neoadjuvant chemoradiotherapy plus surgery for esophageal cancer: the randomized controlled CROSS trial. J Clin Oncol 2021;39(18):1995–2004. Available from: https://doi.org/10.1200/JCO.20.03614.

[14] Al-Batran SE, Homann N, Pauligk C, et al. Perioperative chemotherapy with fluorouracil plus leucovorin, oxaliplatin, and docetaxel versus fluorouracil or capecitabine plus cisplatin and epirubicin for locally advanced, resectable gastric or gastro-oesophageal junction adenocarcinoma (FLOT4): a randomised, phase 2/3 trial. Lancet 2019;393(10184):1948–57. Available from: https://doi.org/10.1016/S0140-6736(18)32557-1.

[15] Sun JM, Shen L, Shah MA, et al. Pembrolizumab plus chemotherapy versus chemotherapy alone for first-line treatment of advanced oesophageal cancer (KEYNOTE-590): a randomised, placebo-controlled, phase 3 study. Lancet 2021;398(10302):759–71. Available from: https://doi.org/10.1016/S0140-6736(21)01234-4.

[16] Mansoor W, Joo S, Norquist JM, et al. Health-related quality-of-life analysis from KEYNOTE-590: pembrolizumab plus chemotherapy versus chemotherapy for advanced esophageal cancer. Oncologist. 2024;29(10):e1324–35. Available from: https://doi.org/10.1093/oncolo/oyae087. Erratum in: Oncologist. 2024 Aug 5;29(8):e1100. https://doi.org/10.1093/oncolo/oyae179

[17] Doki Y, Ajani JA, Kato K, et al. Nivolumab combination therapy in advanced esophageal squamous-cell carcinoma. N Engl J Med 2022;386(5):449–62. Available from: https://doi.org/10.1056/NEJMoa2111380.

[18] Moehler M, Xiao H, Blum SI, et al. Health-related quality of life with nivolumab plus chemotherapy versus chemotherapy in patients with advanced gastric/gastroesophageal junction cancer or esophageal adenocarcinoma from CheckMate 649. J Clin Oncol 2023;41(35):5388–99. Available from: https://doi.org/10.1200/JCO.23.00170.

[19] Shen L, Kato K, Kim SB, et al. RATIONALE-302 Investigators. Tislelizumab versus chemotherapy as second-line treatment for advanced or metastatic esophageal squamous cell carcinoma (RATIONALE-302): a randomized phase III study. J Clin Oncol 2022;40(26):3065–76. Available from: https://doi.org/10.1200/JCO.21.01926.

[20] Janjigian YY, Shitara K, Moehler M, et al. First-line nivolumab plus chemotherapy versus chemotherapy alone for advanced gastric, gastro-oesophageal junction, and oesophageal adenocarcinoma (CheckMate 649): a randomised, open-label, phase 3 trial. Lancet 2021;398(10294):27–40. Available from: https://doi.org/10.1016/S0140-6736(21)00797-2.

[21] Bang YJ, Van Cutsem E, Feyereislova A, et al. Trastuzumab in combination with chemotherapy versus chemotherapy alone for treatment of HER2-positive advanced gastric or gastro-oesophageal junction cancer (ToGA): a phase 3, open-label, randomized controlled trial. Lancet 2010;376(9742):687–97. Available from: https://doi.org/10.1016/S0140-6736(10)61121-X.

[22] Satoh T, Bang YJ, Gotovkin EA, et al. Quality of life in the trastuzumab for gastric cancer trial. Oncologist 2014;19(7):712–9. Available from: https://doi.org/10.1634/theoncologist.2014-0058.

[23] Rha SY, Kawazoe A, Bai Y, et al. Final overall survival for the phase III, KEYNOTE-811 study of pembrolizumab plus trastuzumab and chemotherapy for HER2+ advanced, unresectable or metastatic G/GEJ adenocarcinoma. Ann Oncol 2024;35(Suppl. 4):S1453–4. Available from: https://doi.org/10.1016/j.annonc.2024.10.154.

[24] Janjigian YY, Kawazoe A, Bai Y, et al. Pembrolizumab plus trastuzumab and chemotherapy for HER2-positive gastric or gastro-oesophageal junction adenocarcinoma: interim analyses from the phase 3 KEYNOTE-811 randomised placebo-controlled trial. Lancet 2023;402(10418):2197–208. Available from: https://doi.org/10.1016/S0140-6736(23)02033-0.

[25] Shitara K, Van Cutsem E, Bang YJ, et al. Efficacy and safety of pembrolizumab or pembrolizumab plus chemotherapy vs chemotherapy alone for patients with first-line, advanced gastric cancer: The KEYNOTE-062 phase 3 randomized clinical trial. JAMA Oncol 2020;6(10):1571–80. Available from: https://doi.org/10.1001/jamaoncol.2020.3370.

[26] Rha SY, Oh DY, Yañez P, et al. Pembrolizumab plus chemotherapy versus placebo plus chemotherapy for HER2-negative advanced gastric cancer (KEYNOTE-859): a multicentre, randomised, double-blind, phase 3 trial. Lancet Oncol 2023;24(11):1181–95. Available from: https://doi.org/10.1016/S1470-2045(23)00515-6.

[27] Shah MA, Shitara K, Ajani JA, et al. Zolbetuximab plus CAPOX in CLDN18.2-positive gastric or gastroesophageal junction adenocarcinoma: the randomized, phase 3 GLOW trial. Nat Med 2023;29(8):2133–41. Available from: https://doi.org/10.1038/s41591-023-02465-7.

[28] Lordick F, Van Cutsem E, Shitara K, et al. Health-related quality of life in patients with CLDN18.2-positive, locally advanced unresectable or metastatic gastric or gastroesophageal junction adenocarcinoma: results from the SPOTLIGHT and GLOW clinical trials. EClinicalMedicine 2024;9(8):103663. Available from: https://doi.org/10.1016/j.esmoop.2024.103663.

[29] Wainberg ZA, Kang YK, Lee KW, et al. Bemarituzumab as first-line treatment for locally advanced or metastatic gastric/gastroesophageal junction adenocarcinoma: final analysis of the randomized phase 2 FIGHT trial. Gastric Cancer 2024;27(3):558–70. Available from: https://doi.org/10.1007/s10120-024-01466-w.

[30] Wilke H, Muro K, Van Cutsem E, et al. Ramucirumab plus paclitaxel versus placebo plus paclitaxel in patients with previously treated advanced gastric or gastro-oesophageal junction adenocarcinoma (RAINBOW): a double-blind, randomised phase 3 trial. Lancet Oncol 2014;15(11):1224–35. Available from: https://doi.org/10.1016/S1470-2045(14)70420-6.

[31] Mehta R, Kommalapati A, Kim RD. The impact of ramucirumab treatment on survival and quality of life in patients with gastric cancer. Cancer Manag Res 2020;12:51–7. Available from: https://doi.org/10.2147/CMAR.S199827.

[32] Kang YK, Boku N, Satoh T, et al. Nivolumab in patients with advanced gastric or gastroesophageal junction cancer refractory to, or intolerant of, at least two previous chemotherapy regimens (ATTRACTION-2): a randomized, double-blind, placebo-controlled, phase 3 trial. Lancet 2017;390(10111):2461–71. Available from: https://doi.org/10.1016/S0140-6736(17)31827-5.

[33] Shitara K, Bang YJ, Iwasa S, et al. Trastuzumab deruxtecan in previously treated HER2-positive gastric cancer. N Engl J Med 2020;382(25):2419–30. Available from: https://doi.org/10.1056/NEJMoa2004413.

[34] National Comprehensive Cancer Network (NCCN). Gastric cancer, Version 4.2023 [Internet]. Plymouth Meeting (PA): National Comprehensive Cancer Network; 2023 [cited 2024 Oct 21]. Available from: https://www.nccn.org/professionals/physician_gls/pdf/gastric.pdf.

[35] Tey J, Soon YY, Koh WY, et al. Palliative radiotherapy for gastric cancer: a systematic review and meta-analysis. Oncotarget 2017;8(15):25797–805. Available from: https://doi.org/10.18632/oncotarget.15554.

[36] Shuja M, Elghazaly AA, Iqbal A, et al. Efficacy of 8 Gy single fraction palliative radiation therapy in painful bone metastases: a single institution experience. Cureus 2018;10(1):e2036. Available from: https://doi.org/10.7759/cureus.2036.

[37] Sekii S, Saito T, Kosugi T, et al. Who should receive single-fraction palliative radiotherapy for gastric cancer bleeding?: an exploratory analysis of a multicenter prospective observational study (JROSG 17-3). Clin Transl Radiat Oncol 2023;42:100657. Available from: https://doi.org/10.1016/j.ctro.2023.100657.

[38] Katano A, Yamashita H. The impact of palliative radiation therapy on patients with advanced gastric cancer: results of a retrospective cohort study. Cureus 2022;14(12):e32971. Available from: https://doi.org/10.7759/cureus.32971.

[39] Noordman BJ, Verdam MGE, Onstenk B, et al. Quality of life during and after completion of neoadjuvant chemoradiotherapy for esophageal and junctional cancer. Ann Surg Oncol 2019;26(13):4765–72. Available from: https://doi.org/10.1245/s10434-019-07779-w.

[40] Alzanbagi A, Qureshi LA, Ahmed I, et al. Self-expanding metallic stents for palliation of esophageal cancer: a single center experience from Saudi Arabia. Cureus 2022;14(12):e32096. Available from: https://doi.org/10.7759/cureus.32096.

Chapter 37

Lung and thoracic cancers

Bojan Zarić[1,2], Daliborka Bursać[1,2], and Vladimir Stojšić[1,2]
[1]*Faculty of Medicine, University of Novi Sad, Novi Sad, Serbia,* [2]*Institute for Pulmonary Diseases of Vojvodina, Sremska Kamenica, Serbia*

Key learning objectives

- Learn basic epidemiological data, clinical characteristics, and treatment principles of lung cancer.
- Gain insight into the specifics of non–small-cell lung cancer and the ways in which systemic therapy improves quality of life (QoL) and pain.
- Gain insight into the specifics of small-cell lung cancer and the ways in which systemic therapy improves QoL and pain.
- Gain insight into the specifics of malignant mesothelioma and thymic neoplasms and the ways in which systemic therapy improves QoL and pain.
- Understand future therapeutic options and their impact on QoL.

Epidemiological data, clinical characteristics, and treatment principles of lung cancer

Lung cancer remains the leading cause of cancer-related deaths worldwide. In 2023, over 230,000 new cases of lung cancer were reported in the United States, while over 120,000 patients died from the disease. Lung cancer remains most common in populations older than 60 years, but recent data show a shift toward younger populations. The data on gender distribution, originating from different parts of the world and reflecting economic status, vary. While in more developed countries the male-to-female ratio remains at 1.5:1, in low-income countries it rises to 6:1 in favor of the male gender. It is established that active smoking and exposure to tobacco smoke remain the most significant risk factors for lung cancer development. Smoking, especially active smoking, increases the risk for the development of lung cancer by up to 30 times compared with never-smokers. However, other factors cannot be neglected; exposure to air pollution, radon gas, industrial vapors in the automotive and metallurgy industries, and heavy metals are also important risk factors. Although genetic predisposition is not fully elucidated, accumulating evidence indicates that specific genetic factors may act as contributors to lung cancer susceptibility [1].

Lung cancer is a malignant disease originating from respiratory epithelial cells, which defines its histology, clinical presentation, and response to therapy. Traditionally, it is divided into two major groups: non–small-cell lung cancer (NSCLC) and small-cell lung cancer (SCLC). Despite both falling within the broader category of lung cancer, these two entities are examined separately because of their distinct clinical characteristics and treatment strategies [2,3].

Lung adenocarcinoma is the most common peripheral type of lung cancer, remaining clinically silent for longer periods, while thoracic pain due to involvement of the parietal pleura might be one of the first symptoms. Squamous cell lung cancer is believed to be centrally located in most cases; thus, the leading symptoms include cough, dyspnea, and hemoptysis. However, the symptoms and signs of lung cancer originate from the growth of the primary tumor (size, relationship with surrounding structures), regional growth (spread to the mediastinum, pleura, or regional mediastinal structures), systemic spread (distant metastases, e.g., brain, liver, bones, adrenal glands), and paraneoplastic syndromes. The main issue is that these symptoms are not characteristic, and the first symptom often occurs at an advanced stage, while 5%–10% of patients may remain asymptomatic; pain is a symptom in over 50% of lung cancer patients [4–6].

Diagnosis of lung cancer

Diagnostic algorithms require imaging guidance, commonly by computed tomography (CT) scans, followed by tissue sampling for definitive diagnosis. The diagnosis of lung cancer requires cytological or pathological confirmation of

Pain Management in Palliative Care for Patients With Cancer. DOI: https://doi.org/10.1016/B978-0-443-30050-9.00047-3

lung cancer. Main diagnostic procedures include bronchoscopy, transthoracic needle biopsy or puncture, thoracentesis (with sampling of pleural fluid for smears or cytoblock), or surgery. More advanced (minimally) invasive diagnostic techniques are complementary and include autofluorescence videobronchoscopy, narrow band imaging videobronchoscopy, endobronchial ultrasound (radial or linear), endoesophageal ultrasound, and video-assisted thoracic surgery (VATS) [7–9].

Therapeutic decisions are based on the stage of the tumor, and staging might require additional diagnostic workup, including positron emission tomography-CT scans, bone scintigraphy, or magnetic resonance imaging. Once the staging procedures are completed and the histology of the tumor is known, the staging is performed in accordance with the TNM classification, currently in its ninth edition. The stage of lung cancer further dictates the therapeutic approach. Diagnostic workup also requires molecular testing of the tumor tissue [10,11].

Treatment of lung cancer

Once the stage is established, a decision regarding resectability is usually debated by the multidisciplinary tumor board (MDT). Traditionally, stages from I to IIIA are regarded as surgically resectable; however, during the last several years, there has been growing evidence for a perioperative approach, which includes a combination of immunotherapy and chemotherapy prior to surgery, followed by consolidation immunotherapy. In some cases, potentially resectable tumors may benefit from neoadjuvant therapy—such as chemotherapy, immunotherapy, targeted therapy, or a combination of these approaches—to reduce tumor burden and enable a less extensive surgical resection. This neoadjuvant approach precedes surgery and might be effective, depending on the therapeutic approach, in 20%–70% of the cases. However, operability is another issue significantly influencing the ability for successful surgical treatment. Operability of the patient refers to the functional ability of the patient to be surgically treated, mainly regarding comorbidities and performance status. Performance status can be assessed by different measurement scales, but one of the most used is the Eastern Cooperative Oncology Group (ECOG) Scale. Patients fit for surgery are usually classified as ECOG 0 or ECOG 1 performance status [12–14].

Treatment of advanced or metastatic stage NSCLC depends on histological subtype, presence/absence of driver mutations, presence of other biomarkers (such as programmed death-ligand 1 [PD-L1]), performance status, and comorbidities. Additional diagnostic workup in advanced and metastatic stages requires molecular testing and usually the presence of both the molecular tumor board and MDTs. Driver mutations are most detected in lung adenocarcinoma and are the cornerstone of targeted (molecular) therapies. The European Society for Medical Oncology (ESMO) recommends testing for at least 11 mutations in metastatic lung adenocarcinoma, which are targetable with adequate drugs, although the recommendation for testing is constantly expanding. Tumors not harboring targetable mutations are, depending on ECOG performance status (PS), suitable for either immunotherapy alone, or a combination of immunotherapy and chemotherapy ± antivascular endothelial growth factor (anti-VEGF) inhibitors [15].

SCLC, as a separate entity of lung cancer, is difficult to treat, with a median overall survival (OS) of slightly over 1 year. Surgical treatment might be successful for the very early stage (stage I), but the prospective data on its true impact on OS are scarce. The addition of immunotherapy to chemotherapy for more advanced stages prolongs progression-free survival (PFS), but its effect on OS is clinically limited. Other thoracic malignancies, such as mesothelioma and thymic cancer, are rare diseases with limited options for treatment, which include surgery, chemotherapy, and radiotherapy [16,17].

Historically, in lung cancer, radiotherapy was most commonly used in a palliative setting; however, new approaches and techniques demonstrate a curative potential for a small subset of patients. Stereotactic body radiation therapy uses targeted radiation to the tumor with minimal effect on surrounding tissue, enabling good results of treatment in inoperable early-stage NSCLC, almost comparable to surgery. This approach also contributes to better survival outcomes in oligometastatic disease in patients undergoing targeted therapy or immunotherapy. External beam radiotherapy is, however, irreplaceable in the treatment of lung cancer emergencies such as superior vena cava syndrome or penetration into the spinal canal. It also provides a good therapeutic strategy for pain management when combined with systemic therapy and palliative care. However, the total impact of external beam radiotherapy on survival outcomes in advanced-stage (or even metastatic) lung cancer remains debatable. Recently, concurrent (or sequential) chemo-radiotherapy in stage III lung cancer, followed by durvalumab (if previous treatment accomplished at least stable disease), has been shown to benefit PFS and provides a good sign to improve OS in this stage of the disease [15,16,18].

Treatment of advanced-stage and metastatic nonsquamous cell lung cancer (lung adenocarcinoma)

Targeted therapy

For patients with lung adenocarcinoma in advanced or metastatic stages with tumors harboring driver mutations, targeted therapy is the standard of care. Each individual mutation corresponds to a subset of treatment options in first, second, or later lines of the treatment, keeping in mind that salvage therapy might be chemotherapy.

For cancers with epidermal growth factor receptor (EGFR) driver mutations (deletion in exon 19 or L858R mutation in exon 21), the current standard of care per ESMO guidelines is first-line osimertinib. These mutations occur in up to 10% of adenocarcinomas in the Caucasian population and over 45% of the East-Asian population. The OS of patients treated with osimertinib exceeded 38 months, with 54% of patients alive at 3 years and 28% of patients who are still receiving osimertinib at 3 years. The response rate is also remarkable and is around 80%, indicating that in many patients there can be a significant reduction in symptoms. This was shown to be true in the quality-of-life (QoL) analysis of the FLAURA trial, where there were key symptom improvements from baseline in both treatment arms, with a statistically significant difference favoring osimertinib for chest pain (−6.84 vs −3.88; $P = .021$) [17].

First- or second-generation EGFR tyrosine kinase inhibitors (TKIs) are also available but less efficient in terms of survival options. These drugs include erlotinib, gefitinib, afatinib, and dacomitinib, but also possible combinations of erlotinib with bevacizumab and erlotinib plus ramucirumab. Upon each progression, it is highly recommended to provide either tissue or liquid rebiopsy for the identification of secondary mutations. Possible second-line treatments include amivantamab (in case of exon 20 insertion or C-mesenchymal-epithelial transition factor [cMET] amplification); there is also an option of a combo therapy in the second line (atezolizumab + bevacizumab + platinum chemotherapy) and a platinum-based doublet if found appropriate. The main characteristics of targeted therapy in EGFR-mutated lung adenocarcinoma are a rapid response rate and relatively good duration of response (DoR). Because most treatments are administered orally, patient adherence remains high, with a marked improvement in the quality of life compared to the previous standard of care. Third-generation EGFR TKIs such as osimertinib and lazertinib maintain high activity in cerebrospinal fluid, providing significant central nervous system (CNS) penetration and good intracranial control of the disease, thus providing control for one of the main causes of serious QoL impairment in these patients. The significant response achieved from these drugs could provide substantial control of the cancer-related pain, especially for peripherally located adenocarcinomas affecting the parietal pleura. Rapid response rates also provide quick symptom relief in terms of dyspnea and cough [15,18]. So, concurrent targeted therapy, pain medication, and supportive care at the beginning of the systemic treatment are the best options for the patient.

The assessment of QoL has emerged as a pivotal end point in the treatment of advanced NSCLC, particularly in the context of targeted therapies for EGFR-mutant tumors. First- and second-generation EGFR TKIs—erlotinib, gefitinib, and afatinib—not only extend PFS but also have demonstrable effects on patient-reported outcomes (PROs), assessed through validated tools such as the European Organisation for Research and Treatment of Cancer Quality of Life Questionnaire Core 30 (EORTC QLQ-C30) and its lung cancer-specific module QLQ-LC13, as well as the Functional Assessment of Cancer Therapy–Lung (FACT-L). Erlotinib has shown consistent QoL benefits in multiple trials. In the BR.21 study, which evaluated erlotinib versus placebo in previously treated advanced NSCLC, statistically significant improvements were observed in global QoL, physical functioning, and symptom control. Specifically, EORTC QLQ-C30 scores demonstrated significant benefits in cough, dyspnea, and pain reduction. QoL improvement was achieved in approximately 44% of patients on erlotinib compared to 22% on placebo ($P < .001$) [17–21].

For anaplastic lymphoma kinase (ALK) translocation-positive lung adenocarcinomas, occurring in 1%–3% of all lung adenocarcinomas, the first line of treatment is equally effective with third-generation drugs (lorlatinib) and second-generation drugs (alectinib and brigatinib), while crizotinib, as a first-generation drug, is mostly abandoned in the first line along with ceritinib (due to higher toxicity). Lorlatinib might have a clinical advantage in the first line when neuroprotective activity and CNS control of the disease are paramount. ALK translocations are characteristically more often found in younger patient populations and nonsmokers, making this type of disease challenging to treat. Upon progression on first-line treatment, platinum, pemetrexed, or the atezolizumab–bevacizumab–paclitaxel–carboplatin combination might be successful. The same principles of efficacy found for EGFR TKI are generally true for ALK TKIs: rapid response rate, solid DoR, good neuroprotective effectiveness, and control of the disease in the brain. Common adverse events (AEs) for both groups of TKIs (EGFR and ALK) are diarrhea, rash, gastrointestinal, AEs, and some characteristic laboratory discrepancies (abnormal liver function tests). In most of the cases, AEs are Common Terminology Criteria for Adverse Events (CTCAE) grade I–II, not requiring special treatment; however, if AEs present in grade III or higher, they require dose reduction and/or specific therapeutic activities targeting specific AEs [15,18].

Alectinib and lorlatinib have redefined standards of care, not only by improving survival outcomes but also by positively impacting health-related QoL. The ALEX trial, a phase III study comparing alectinib to crizotinib in treatment-naive ALK-positive advanced NSCLC, demonstrated a significant improvement in PFS with alectinib (median PFS 34.8 months vs 10.9 months; HR 0.43, 95% confidence interval [CI] 0.32–0.58). Alectinib also reduced the cumulative incidence of CNS progression (12% vs 45% at 12 months). QoL assessments using the EORTC QLQ-C30 and QLQ-LC13 showed that alectinib delayed time to deterioration in global health status and key lung cancer symptoms such as cough and dyspnea, with patients reporting better overall well-being and fewer treatment-related side effects such as nausea and visual disturbances compared to crizotinib [22]. In contrast, lorlatinib, a highly potent third-generation ALK inhibitor with broad CNS activity, was evaluated in the CROWN trial, which randomized treatment-naive ALK-positive NSCLC patients to receive lorlatinib or crizotinib. The trial met its primary end point with a significant improvement in PFS (not reached vs 9.3 months; HR 0.28, 95% CI 0.19–0.41) and a markedly lower rate of CNS progression (3% vs 33% at 12 months). However, the QoL profile of lorlatinib is more nuanced. While patients benefited from improved intracranial disease control and prolonged systemic response, QoL data from the EORTC QLQ-C30 revealed a mixed picture: patients receiving lorlatinib reported stability or improvement in lung cancer symptoms but also experienced higher rates of cognitive and mood-related AEs, such as speech disturbances, memory impairment, and mood alterations. These neurocognitive effects were reflected in declines in the cognitive functioning subscale and increased reporting of emotional distress [23]. Taken together, both alectinib and lorlatinib offer substantial QoL benefits through superior disease control and intracranial efficacy. Alectinib is often associated with a more favorable tolerability and QoL profile, especially in long-term use, while lorlatinib provides unmatched CNS protection at the cost of a distinct neurocognitive toxicity profile that necessitates proactive monitoring and supportive care strategies. Thus QoL considerations, alongside efficacy, CNS disease burden, and patient-specific factors, play a pivotal role in selecting the optimal first-line ALK inhibitor [24,25,26].

ROS1 translocation is usually confirmed in approximately 1% of lung adenocarcinomas, typically in younger patients and nonsmokers or light smokers. The first line is again reserved for ROS1 TKIs (crizotinib, entrectinib, and repotrectinib). While cross-trial comparisons should be interpreted with caution, numerically better results in survival outcomes are observed with entrectinib and repotrectinib. The second line of treatment, if previously treated with ROS1 TKI, is either a platinum-based doublet or a clinical trial [15,18].

Other targetable changes include BRAF V600 mutation (treated with the dabrafenib–trametinib combination in the first line) and rearranged during transfection (RET) translocation (available first-line options include pralsetinib or selpercatinib). Neurotrophic tyrosine receptor kinase (NTRK) translocation should be treated in the first line with entrectinib or larotrectinib, human epidermal growth factor receptor 2 (HER2) mutation with trastuzumab deruxtecan, EGFR exon 20 insertion with amivantamab, and MET exon 14 skipping mutation with platinum doublet ± capmatinib or tepotinib monotherapy, while Kirsten rat sarcoma virus oncogene homologue (KRAS) G12C–mutated tumors could be treated with either KRAS inhibitors (sotorasib, adagrasib) or immunotherapy, either in mono- or in combo-regimen. All these mutations are most commonly mutually excluded [15,18].

Although many actionable driver alterations in NSCLC are treated with oral or targeted agents, the efficacy, CNS activity, toxicity profiles, resistance mechanisms, and optimal sequencing differ substantially by molecular subtype and by individual drug. Entrectinib is a potent, CNS-penetrant inhibitor of ROS1 and NTRK kinases, approved for the treatment of advanced NSCLC harboring ROS1 rearrangements, as well as NTRK gene fusions across solid tumors. Its efficacy and QoL impact have been evaluated in an integrated analysis of three phase I and II trials—ALKA-372-001, STARTRK-1, and STARTRK-2—which collectively enrolled patients with ROS1-positive NSCLC and NTRK fusion-positive tumors. In ROS1-positive NSCLC (n = 53, efficacy-evaluable), entrectinib achieved an objective response rate (ORR) of 77% (95% CI, 63.3–88.2), with a median PFS of 19.0 months (95% CI, 12.2–36.6). Among patients with baseline CNS metastases, the intracranial response rate was 55%, underscoring its CNS activity. In the broader NTRK fusion-positive solid tumor population (including NSCLC), entrectinib showed an ORR of 57% and durable responses with a median DoR of 10.4 months. These results established entrectinib as a first-line or early-line option in this molecularly defined patient population. QoL was assessed using validated patient-reported outcome tools, primarily the EORTC QLQ-C30 and QLQ-LC13. Patients reported rapid and sustained improvements in cough, dyspnea, and pain, aligning with the early tumor shrinkage and symptom control seen clinically. Notably, global health status and emotional functioning improved within the first few cycles of therapy. Time to definitive deterioration in QoL domains was generally prolonged, with a median exceeding 9 months across most subscales, indicating maintenance of daily functioning during therapy. Importantly, entrectinib's tolerability profile—characterized by predominantly low-grade side effects such as dysgeusia, fatigue, and weight gain—contributed to sustained QoL, with fewer dose-limiting toxicities or treatment discontinuations compared to older ROS1-targeted therapies. In summary, entrectinib demonstrates robust systemic and intracranial efficacy in ROS1- and NTRK-driven NSCLC, with a favorable impact on patient-reported QoL. Its ability to control both extracranial and CNS disease while preserving functional status makes it a valuable therapeutic option, particularly in patients with symptomatic or neurologically complex disease [18,27].

It is important to emphasize that these treatments are generally well tolerated, compliant, and manageable in terms of toxicity. Symptom relief is usually quick and sustainable; however, the progression is unavoidable due to secondary resistance, and patient surveillance and control should be maintained at a high level. In patients with poor performance status, the use of targeted therapies is still debatable; while the toxicity profiles of the drugs might not be a common problem, the efficacy in terms of survival outcomes and the cost could be an obstacle to initiation of the treatment.

Treatment of non–oncogene-addicted lung adenocarcinoma

The treatment of non–oncogene-addicted lung adenocarcinoma depends on the ECOG performance status and PD-L1 status. If the patients are categorized as ECOG 0–2 and have high expression of PD-L1 ≥50%, immunotherapy is a preferable therapeutic choice. In these cases, pembrolizumab, atezolizumab, or cemiplimab are indicated. Despite the significant clinical success in a subset of patients, immune checkpoint inhibitor (ICI) monotherapies are often associated with delayed clinical responses; therefore, in patients with a high disease burden, early initiation of palliative care and symptomatic management should be considered alongside anticancer treatment. Hence, pain therapy and symptomatic relief for dyspnea, hemoptysis, and cough should be initiated along with the systemic treatment, or in many cases earlier, to improve the condition of the patient. Immunotherapy leads to increased rates of immune-related AEs (irAEs), such as hypothyroidism, hyperthyroidism, immune-related pneumonitis, hepatitis, myocarditis, and other autoimmune diseases. As such, a wider multidisciplinary team, including an endocrinologist, gastroenterologist, dermatologist, and immunologist, is needed to follow, discuss, and treat the patient accordingly [28,29].

Immunotherapy with ICIs has revolutionized the treatment landscape of NSCLC, offering durable responses and meaningful improvements in QoL for subsets of patients. In pivotal phase III trials, such as KEYNOTE-024 and KEYNOTE-042, pembrolizumab monotherapy in PD-L1 high (≥50%) tumors demonstrated superior ORRs of 44.8% and 27.0%, respectively, compared to chemotherapy, with significantly delayed time to deterioration in symptoms such as dyspnea, cough, and chest pain (measured by EORTC QLQ-C30 and QLQ-LC13) [30,31]. Similarly, IMpower110 showed that atezolizumab improved ORR (38.3% vs 28.6%) and maintained global health status for longer durations than platinum-based chemotherapy [32]. In CheckMate 227, nivolumab plus ipilimumab improved OS compared to chemotherapy, with QoL analyses showing sustained improvements in fatigue, physical functioning, and emotional well-being [33]. Importantly, time to first and definitive deterioration was consistently delayed with immunotherapy across trials, reinforcing its role not only in prolonging survival but also in maintaining or improving daily functioning and symptom burden. While immune-related AEs can affect QoL and may, in some cases, be long-lasting or even permanent, they generally occur less frequently and are less cumulative than the chronic toxicities associated with chemotherapy. Collectively, these findings underscore that immunotherapy, particularly in biomarker-selected populations, confers both oncologic and patient-centered benefits, cementing its status as a cornerstone in lung cancer management.

However, once the disease progresses, even those patients would be considered for treatment with platinum-based chemotherapy. With limited effect on survival outcomes, chemotherapy brings its own AEs: hematological (neutropenia, thrombocytopenia, anemia, and pancytopenia), gastrointestinal (nausea and vomiting), hair loss, appetite loss, and cachexia. These AEs require broad and immediate action, especially when appearing in CTCAE grade ≥3. Colony-stimulating factors, transfusions of thrombocytes and erythrocytes, antiemetics, appetite boosters, and high-energy food supplements, along with hepatoprotective drugs and other supportive medications, must be readily available. If tumors have any expression of PD-L1 under 50%, a combination of immunotherapy plus platinum-based chemotherapy or even two immunotherapeutics with chemotherapy is the treatment of choice. That approach accomplishes faster control of symptoms related to the disease, pain, dyspnea, and cough relief, but also an increased rate of previously described AEs originating from both immunotherapy and chemotherapy [15,18].

Since this is a treatment of adenocarcinoma, some of the specific drugs are used in combination. Pemetrexed, as an antifolate agent, while providing sustainable efficacy, brings a completely different subset of AEs, mainly mucositis, diarrhea, and mucosa-related disorders requiring dose reduction, cycle interruptions, and rapid supportive care treatment. The addition of anti-VEGFs such as bevacizumab, ramucirumab, and nintedanib to immunotherapy and platinum-based chemotherapy is not without increased risk of worsening hypertension, kidney disease, and potentially fatal bleeding (hemoptysis). During the treatment of patients with combinations including anti-VEGF, a cardiologist, nephrologist, and pulmonologist (bronchoscopist) must be a part of the MDT. It is common to include IO–IO combinations in platinum-based chemotherapy; approved combinations include anti-PD-L1/anti-PD1 and anti-cytotoxic T-lymphocyte-associated antigen 4 (CTLA-4) combos (nivolumab + ipilimumab, durvalumab + tremelimumab), and in those scenarios, in addition to a more rapid effect on cancer symptoms, higher grades of irAEs are to be expected, so the coordination with patients and their families should be close and proactive. When the first line fails and the disease progresses, in many cases, the best supportive care will be the only option [15,18].

The recommended second line is usually chemotherapy without a platinum component or immunotherapy if not received in the first line, which has limited efficacy in terms of survival outcomes and often serves as salvage therapy. This also refers to the patients with ECOG 3–4, where best supportive care is the only viable treatment option for the majority of patients. Immunotherapy combination regimens have demonstrated substantial improvements in both clinical efficacy and health-related QoL in patients with advanced NSCLC. In the pivotal CheckMate 227 trial, the dual checkpoint blockade of nivolumab plus ipilimumab achieved an ORR of 35.9% in patients with PD-L1 ≥1% and 27.0% in those with PD-L1 <1% [33]. Importantly, QoL analyses using EORTC QLQ-C30 and QLQ-LC13 showed significantly longer time to confirmed deterioration in global health status, fatigue, and physical functioning compared to chemotherapy (median time to deterioration not reached vs 2.3 months for global health status; $P < .01$), highlighting the clinical and symptomatic benefits of the regimen. In CheckMate 9LA, which combined nivolumab plus ipilimumab with two cycles of chemotherapy, ORR was 38%, and QoL was preserved throughout the treatment, with delayed worsening in symptoms such as dyspnea, pain, and appetite loss [34]. The EQ-5D-3L index score remained stable over time, and fewer patients reported severe declines in mobility and self-care compared to those receiving standard chemotherapy. Similarly, the POSEIDON trial evaluated durvalumab plus tremelimumab with chemotherapy and reported an ORR of 38.8%, with improved OS and manageable toxicity [35]. QoL outcomes, assessed with EORTC QLQ-C30, demonstrated delayed deterioration in physical functioning (HR 0.76), fatigue (HR 0.73), and pain (HR 0.70) compared to chemotherapy alone. Notably, in all trials, combination immunotherapy delayed the time to first and definitive deterioration across multiple domains, reinforcing its ability to not only extend life but also preserve patient-reported well-being during treatment. While immune-related AEs were more frequent, they were generally manageable and did not offset the overall QoL gains observed with these regimens [18,30,36,37].

Treatment of advanced and metastatic stage squamous cell lung cancer

Treatment of advanced and metastatic stage squamous cell lung cancer is similar to the treatment of nononcogene-addicted adenocarcinoma, except drugs reserved for adenocarcinoma only (e.g., pemetrexed or bevacizumab). The first line in high expressers is reserved for immunotherapy monotherapy (atezolizumab, pembrolizumab, and cemiplimab). Recommended therapy in case of expression of PD-L1 1%–49% is a combination of immunotherapeutics with platinum-based chemotherapy [16,18].

Small-cell lung cancer: impact and role of systemic therapy in symptom and pain relief

SCLC is characterized by its rapid growth, leading to early dissemination and widespread metastasis [16]. Although SCLC often shows a strong initial response to chemotherapy and radiotherapy, treatment resistance and relapse are very common. The selection of therapy for SCLC depends on the disease stage, the patient's overall health status, ECOG performance status, and the presence of other health conditions or comorbidities.

Traditionally, SCLC has been classified into two main stages: limited disease (LD), where tumors are localized to one side of the chest and confined to a region that can be treated within a single radiation field, and extensive disease (ED), where tumors have spread beyond the primary site, including distant metastases [11].

For the treatment of SCLC, both the National Comprehensive Cancer Network and ESMO recommend a combination of platinum-based chemotherapy (cisplatin or carboplatin) with etoposide as the first-line treatment for both limited and ED stages. These guidelines suggest concurrent chemotherapy and radiotherapy for patients with limited-stage SCLC (LD SCLC), while surgical resection is rarely performed except in select cases without distant metastases. Prophylactic cranial irradiation is recommended for patients who respond well to initial therapy to reduce the risk of brain metastases, although the impact on OS is limited [16,38,39]. Recent studies and emerging treatments, such as immunotherapy (atezolizumab, durvalumab), are incorporated into ongoing clinical practice to improve outcomes for patients with SCLC, particularly in cases with extensive or recurrent disease [40–42].

Small-cell lung cancer and pain

Recent advances in the treatment of SCLC have modestly improved outcomes in a disease historically marked by rapid progression, high symptom burden, and limited therapeutic innovation. The incorporation of ICIs into first-line therapy represents a significant paradigm shift. In the IMpower133 trial, the addition of atezolizumab to carboplatin and etoposide in treatment-naive extensive-stage SCLC improved median OS to 12.3 months versus 10.3 months with chemotherapy alone (HR 0.70; 95% CI, 0.54–0.91) and increased PFS to 5.2 months versus 4.3 months [43]. The ORR was 60.2%. Similarly, in the CASPIAN trial, durvalumab plus platinum-etoposide resulted in a median OS of 13.0 months compared to 10.3 months (HR 0.73; 95% CI, 0.59–0.91), with an ORR of 67.9% [44]. Beyond survival benefits, both trials demonstrated meaningful QoL preservation, as assessed by patient-reported outcome measures including the EORTC QLQ-C30 and QLQ-LC13. Notably, pain control emerged as a significant area of benefit: in IMpower133, the time to definitive deterioration in pain symptoms was significantly delayed in the atezolizumab arm compared to chemotherapy alone (HR 0.76; 95% CI, 0.58–0.98), with patients reporting slower worsening in chest pain and pain-related interference with daily activities. In CASPIAN, durvalumab similarly prolonged the time to pain progression, with numerical improvements in analgesic use and pain interference domains, despite no formal statistical testing for these end points. These findings indicate that immune-based combinations not only prolong life but also provide clinically relevant symptom palliation, particularly in pain management, which is crucial for maintaining functional independence and dignity in a rapidly debilitating disease such as SCLC. Ongoing studies are investigating additional immunomodulatory agents, including anti-T-cell immunoglobulin and ITIM domain (TIGIT) antibodies, poly (ADP-ribose) polymerase (PARP) inhibitors, and T-cell engagers, with a continued emphasis on integrating survival benefit and symptom control—including pain—as coprimary therapeutic goals to enhance both clinical and patient-centered outcomes in SCLC [40–42].

Lung cancer–related and therapy-related pain

Local growth in the chest cavity might be the cause of the thoracic pain. Direct spread to surrounding structures might bring invasion of the esophagus, causing odynophagia; this pain is often followed by vomiting or regurgitation of esophageal contents into the lungs (manifesting as pneumonia). The pain due to compression of the esophagus increases during the swallowing of solid food. In case of invasion of the spinal nerves or invasion of the vertebrae, pain irradiation through the innervation path of the affected nerve is characteristic and might imitate the pain in some infectious diseases (such as herpes zoster). In case of peripheral growth of the tumor, the affection of well-innervated parietal pleura causes pain, often related to the breathing process and commonly worsening during cough. If there is an infiltration of local nerve plexuses, one of the first symptoms of lung cancer could be shoulder pain. Abdominal pain in lung cancer due to direct invasion of the diaphragm is commonly associated with hiccups and might imitate myocardial infarction [45–48].

But in the majority of cases of lung cancer-associated pain, the real cause is metastases. Bone metastases and fractures are the underlying cause of high-intensity pain and always require additional diagnostics (PET-CT or bone scintigraphy). Abdominal pain due to hepatic metastases is quite often followed by an increase in hepatic enzymes or jaundice due to biliary obstruction. The cause of abdominal pain may be metastatically involved adrenal glands compressing surrounding structures.

The pain in lung cancer is a consequence of either direct action of the tumor, local spread in the thoracic cavity, or distant metastases, and usually requires a tailored approach by a palliative care specialist and the team of supportive care. Due to different types, intensities, and underlying causes of the pain, oncologists are highly dependent on the other molecular tumor board (MTB) members for choosing the correct therapy and the correct time.

One of the most complicated forms of lung cancer–related pain is observed within the paraneoplastic syndrome. The most probable causes are interleukins, prostaglandins, or other proinflammatory cytokines, which could not be detected or diagnosed, and clinicians should be aware that this type of idiopathic pain can occur [45–48].

Malignant pleural mesothelioma

Malignant pleural mesothelioma (MPM) is not so common but is an aggressive type of cancer arising from mesothelial cells of the pleura. The vast majority of patients with MPM have at some point been directly or indirectly exposed to asbestos fibers. The onset of the disease is gradual, with a latency period estimated at approximately 20–40 years after initial exposure. The incidence of MPM is higher in males than in females, with rates per 100,000 persons of 1.7 and 0.4 in Europe (for males and females, respectively). Mesothelioma is very rare below the age of 50 years, with a median age at diagnosis of 76 years, and because of that is usually recognized as a disease of the elderly [49,50].

MPM is often diagnosed in a late stage of disease when curative treatment is not possible, and the remaining life expectancy is substantially reduced. Pleural mesothelioma has a poor prognosis. The 5-year OS is less than 5% in untreated cases, and the average survival is approximately 1 year after diagnosis, depending on numerous factors such as age, present comorbidities, histopathologic subtype, stage of disease, and previously received therapies [51,52].

Symptoms of MPM may be present for a long time before the diagnosis is made. The two most common symptoms are dyspnea and chest pain. Other symptoms include fatigue, loss of appetite, weight loss, and sweating. Hemoptysis can also occur, but is less common compared with lung cancer. Local tumor invasion can cause superior vena cava obstruction, dysphagia, or laryngeal nerve palsy. More than 70% of patients have pleural effusion at presentation. In early-stage disease, pleural effusion is the main source of dyspnea and shortness of breath. During the natural course of the disease, as MPM extends around the pleural surface and surrounds the lungs, dyspnea is a result of restricted respiratory movement. Although almost all patients with MPM are symptomatic, some patients might be asymptomatic at presentation and be detected by the imaging obtained for a different reason [53].

The diagnostic approach should begin with an accurate medical history and physical examination. Standard workup is composed of general laboratory blood tests, chest X-ray, CT scan of the chest and abdomen, thoracentesis with examination of the pleural effusion, and thoracoscopy with confirmatory biopsy. According to the recent guidelines, to secure a pathological diagnosis, pleural biopsies should be done by VATS, which carries a sensitivity of 95%, specificity of 100%, and negative predictive value of 94%. VATS is also superior for the assessment of the pleural cavity for staging purposes and performing a talc poudrage. When thoracoscopy is not feasible or contraindicated, CT-guided or ultrasound-guided biopsy is a good alternative [54].

Staging of MPM is challenging, specifically in the accurate assessment of tumor size and nodal involvement. The current eighth revision of the Union for International Cancer Control TNM staging system for mesothelioma is based on the updated International Association for the Study of Lung Cancer mesothelioma staging project [55].

Mesothelioma-induced pain and systemic therapy for advanced diseases

Treatment strategy in patients with MPM should be discussed within an MDT and determined according to the stage of the disease, histology, age, PS, comorbidities, and the patient's preferences. The role of surgery in MPM is still debatable and unclear, but is usually conducted as a radical removal of all visible disease or as a tissue-sparing, debulking procedure. Extrapleural pneumonectomy (EPP) is a radical procedure in which a surgeon removes the lung, pleura, pericardium, and diaphragm. The goal of this procedure is to eradicate all macroscopically visible tumor tissue. The Mesothelioma and Radical Surgery (MARS) trial is so far the only randomized trial to assess EPP in MPM. With some practical issues during the course of the study, analysis of clinical outcome data revealed potential harm associated with EPP. As a result of this trial and several smaller observational trials also suggesting potential harm of EPP, this procedure has been largely abandoned. Extended pleurectomy with decortication is a less radical, lung-sparing procedure in which the visceral and parietal pleura are removed. This procedure has no intent for macroscopic complete resection and therefore is usually engaged together with other treatment modalities [56].

Systemic treatment options for MPM are comprised of chemotherapy, targeted therapy, and immunotherapy. ICIs are also being evaluated in the first-line setting in the phase III CheckMate 743 trial. MPM patients were enrolled in a combination of nivolumab plus ipilimumab for up to 2 years versus a cisplatin/pemetrexed or carboplatin/pemetrexed combination for up to six cycles. Nivolumab plus ipilimumab significantly prolonged OS in comparison to chemotherapy, and these data put this immunotherapy combination as a new first-line option for unresectable MPM. The landmark CheckMate 743 trial established nivolumab plus ipilimumab as a new first-line treatment standard for unresectable MPM. In this randomized, open-label phase III study, 605 patients were assigned to receive either nivolumab (3 mg/kg every 2 weeks) plus ipilimumab (1 mg/kg every 6 weeks) or platinum-based chemotherapy (cisplatin or carboplatin plus pemetrexed). The combination immunotherapy arm demonstrated a significant improvement in OS, the primary end point, with a median OS of 18.1 months versus 14.1 months in the chemotherapy arm (HR 0.74; 95% CI, 0.61–0.89; $P = .002$). This benefit was particularly pronounced in the nonepithelioid histologic subtype, where median OS was 18.1 months with immunotherapy versus 8.8 months with chemotherapy (HR 0.46). While PFS—a secondary end point—was similar between arms (6.8 vs 7.2 months), the DoR was notably longer in the immunotherapy group (median DoR not reached vs 6.7 months), reflecting the durability of benefit. Importantly, QoL outcomes, assessed using the EORTC QLQ-C30 and LC13 modules, favored immunotherapy. Patients receiving nivolumab plus ipilimumab experienced a delayed time to deterioration in global health status, physical functioning, and key symptoms, including fatigue and dyspnea, compared to those receiving chemotherapy. Pain control was also superior, with fewer patients in the immunotherapy arm reporting moderate-to-severe chest pain during treatment [57].

Despite immune-related AEs, overall treatment tolerability was acceptable, with fewer discontinuations due to toxicity in the immunotherapy arm (15% vs 20%). These findings support dual checkpoint inhibition as a strategy that not only improves survival in MPM, particularly in nonepithelioid disease, but also preserves or enhances QoL—an especially critical consideration in a disease associated with profound symptom burden and limited curative options [57]. For the second-line approach upon progression of the disease, we still do not have clear evidence for the best agent. Phase II trials support the use of vinorelbine or gemcitabine/cisplatin doublet chemotherapy, but there are a number of ongoing trials of ICIs in second- or subsequent-line settings [58].

Radiotherapy has limited use in the treatment of MPM patients, and it could potentially be used as a part of a multimodality approach prophylactically to prevent subcutaneous metastases, or it can be given palliatively for symptom control to relieve symptoms, particularly in cases of chest wall invasion, nerve root involvement, or painful cutaneous metastases [59].

Palliative care and symptom management are essential and very important treatment aspects of MPM. Chest pain, as one of the most common symptoms of MPM, comes from either the pleural effusion or the tumor. The pain itself is generally heavy and dull, and it tends to worsen as the mesothelioma progresses, especially if invasion and infiltration of the chest wall occur. Besides that, neuropathic pain from intercostal nerve involvement, as well as bone pain due to rib invasion, may occur. If there is a presence of parietal pleura irritation, patients might feel pleuritic pain, which is less common. On the other hand, many of the treatment protocols used in MPM treatment can also cause pain. Surgery causes pain during the recovery period, and depending on the extent of the surgical procedure, the pain might last for days or weeks. Chemotherapy used in MPM treatment, as in other cancers, can cause uncomfortable side effects, including pain. Pain is caused by mouth sores, commonly seen with pemetrexed use, but also caused by digestive issues or chemotherapy-induced neuropathy [60].

Thymic epithelial tumors

Thymic epithelial tumors (TETs) belong to a group of rare neoplasms that originate from thymic epithelial cells, with the peak of the disease incidence in the sixth and seventh decades of life. In accordance with the World Health Organization [61] histologic classification, TETs are further classified into thymomas (TMs) and thymic carcinomas. In one-third of all cases, clinically speaking, TETs are asymptomatic. When patients are presented with symptomatic disease, usually they have mass effect symptoms or autoimmune disease symptoms such as myasthenia gravis, or patients might have some immunodeficiencies, mainly hypogammaglobulinemia. Symptoms from a mediastinal mass usually occur because of compression of the trachea, bronchus, or lung and may result in chest pain, dyspnea, and cough [62,63].

The cornerstone of radiological diagnostics is contrast-enhanced CT. Among all anterior mediastinal masses, TETs account for almost one-third of cases, and when we have a patient with some autoimmune disorder and an anterior mediastinal mass, the diagnosis of thymoma is highly likely [64,65]. Histopathological analysis is not required if the diagnosis is most likely and complete surgical resection is possible; however, if complete surgical resection is not achievable, then biopsy is recommended. For staging purposes, the TNM staging system and Masaoka-Koga staging are most extensively used [66,67].

The treatment strategy for patients with TETs is still surgical resection of the neoplasm, especially for early-stage disease. Postoperative radiotherapy in TMs is recommended for more aggressive TETs, while the use of adjuvant chemotherapy might be considered for stages II–IV, but only after evaluation by a multidisciplinary board [68–72]. In inoperable cases, neoadjuvant chemotherapy should be offered. The most frequently used chemotherapy regimens are the CAP protocol (cisplatin, doxorubicin, cyclophosphamide) and PE protocol (cisplatin, etoposide). If the patient, after neoadjuvant chemotherapy, is not eligible for surgical treatment, sequential radiotherapy is recommended (60–66 Gy in 30–33 fractions).

For metastatic TETs, palliative chemotherapy is indicated with the CAP protocol as a preferred first-line treatment, but other cisplatin-based chemotherapy combinations are also feasible options. Upon disease progression, the second-line chemotherapy options are paclitaxel plus carboplatin or gemcitabine plus capecitabine. New advances, including anti-PD-1 immunotherapy, are being developed, and the KEYNOTE-158 basket trial evaluated pembrolizumab for cancers with microsatellite instability, which also included patients with thymic carcinoma and reported an ORR of 34.3%, with a median OS of 23.5 months [73]. The JAVELIN Solid Tumor trial, investigating avelumab also in a cohort with thymoma and thymic carcinoma, showed modest efficacy but raised concerns due to high rates of autoimmune toxicity in thymoma [74]. While formal QoL data were limited in these trials, case series and PROs suggest that immunotherapy responders often experienced symptomatic improvement, particularly in chest pain and dyspnea, reflecting tumor control near mediastinal structures. In contrast, standard chemotherapy regimens—such as CAP for thymoma and carboplatin-paclitaxel for thymic carcinoma—achieve ORRs of 50%–70% and PFS of 8–11 months, with predictable but more cumulative toxicities. QoL assessments from retrospective and small prospective cohorts show that platinum-based regimens may lead to early tumor-related pain relief, particularly in bulky or invasive tumors, but often result in declines in physical functioning and fatigue due to cumulative toxicity. Novel approaches, including multikinase inhibitors (e.g., lenvatinib in the REMORA trial), have shown promising results with ORRs of ~38% in thymic carcinoma and a manageable toxicity profile [75]. While pain outcomes were not primary end points, reduction in tumor-related chest pain and dysphagia has been reported in responders, particularly with lenvatinib and pembrolizumab. Overall, the therapeutic landscape in TETs remains limited by rarity and immune sensitivity, but emerging targeted and immune-based therapies offer hope not only for prolonging survival but also for improving symptom burden and QoL, especially in relapsed or refractory disease [76–82].

Conclusion and further perspectives

The discovery of new biomarkers, along with new therapies, will undoubtedly prolong the survival outcomes but also make the symptoms more chronic. It is therefore highly advisable to have MTB readily available and engaged from the very moment of diagnosis. Artificial intelligence (AI), the use of clinical decision-making AI-based tools, machine learning tools for molecular diagnostics and prognostication, and comprehensive genome profiling are more or less a necessity for the near future of lung cancer diagnostics and treatment. Development of a new class of drugs, including antibody-drug conjugates, will open new possibilities in the treatment of even the most resilient tumors of the lung [83–86].

For SCLC, future treatment strategies are likely to focus on improving outcomes through a combination of novel therapies that target the underlying mechanisms of the disease. Future research is exploring biomarker-guided treatments, which could help identify patients most likely to respond to specific therapies, allowing for more personalized treatment approaches. For advanced and metastatic SCLC, treatments will continue to focus on palliative care, including interventions aimed at managing symptoms such as pain, fatigue, and respiratory difficulties. In summary, while challenges remain due to the rapid progression and resistance of SCLC, the ongoing research into novel therapies, including targeted treatments and immunotherapy, holds significant promise for improving outcomes in the future [87].

References

[1] Siegel RL, Miller KD, Wagle NS. Cancer statistics. CA Cancer J Clin 2023;73(1):17–48. Available from: https://doi.org/10.3322/caac.21763.

[2] Mott TF. Lung cancer: management. FP Essent 2018;464:27–30.

[3] Wadowska K, Bill-Lula I, Śliwińska-Mossoń M. Genetic markers in lung cancer diagnosis: a review. Int J Mol Sci 2020;21(13):4569. Available from: https://doi.org/10.3390/ijms21134569.

[4] Collins LG, Haines C, Perkel R, et al. Lung cancer: diagnosis and management. Am Fam Physician 2007;75(1):56–63.

[5] Nasim F, Sabath BF. Lung cancer. Med Clin North Am 2019;103(3):463–73. Available from: https://doi.org/10.1016/j.mcna.2018.12.006.

[6] Prado MG, Kessler LG, Au MA, et al. Symptoms and signs of lung cancer prior to diagnosis: case-control study using electronic health records from ambulatory care within a large US-based tertiary care centre. BMJ Open 2023;13(4):e068832. Available from: https://doi.org/10.1136/bmjopen-2022-068832.

[7] Zaric B, Stojsic V, Sarcev T, et al. Advanced bronchoscopic techniques in diagnosis and staging of lung cancer. J Thorac Dis 2013;5(4):S359–70. Available from: https://doi.org/10.3978/j.issn.2072-1439.2013.05.15.

[8] Zarogoulidis P, Matthaios D, Kosmidis C, et al. Effective early diagnosis for NSCLC: an algorithm. Expert Rev Respir Med 2021;15(11):1437–45. Available from: https://doi.org/10.1080/17476348.2021.1969916.

[9] Zaric B, Kovacevic T, Stojsic V, et al. New technologies in diagnostic bronchoscopy – an age of meta-analyses. Expert Rev Med Devices 2016;13(9):789–91. Available from: https://doi.org/10.1080/17434440.2016.1218276.

[10] Detterbeck FC, Woodard GA, Bader AS. The proposed ninth edition TNM classification of lung cancer. Chest 2024;166(4):882–95. Available from: https://doi.org/10.1016/j.chest.2024.05.026.

[11] Rami-Porta R, Nishimura KK, Giroux D, et al. The International Association for the Study of Lung Cancer lung cancer staging project: proposals for revision of the TNM stage groups in the forthcoming (ninth) edition of the TNM classification for lung cancer. J Thorac Oncol 2024;19(7):1007–27. Available from: https://doi.org/10.1016/j.jtho.2024.02.011.

[12] Alduais Y, Zhang H, Fan F, et al. Non-small cell lung cancer (NSCLC): a review of risk factors, diagnosis, and treatment. Medicine (Baltimore) 2023;102(8):e32899. Available from: https://doi.org/10.1097/MD.0000000000032899.

[13] Alexander M, Kim SY, Cheng H. Update 2020: management of non-small cell lung cancer. Lung 2020;198(6):897–907. Available from: https://doi.org/10.1007/s00408-020-00407-5.

[14] Oken MM, Creech RH, Tormey DC, et al. Toxicity and response criteria of the Eastern Cooperative Oncology Group. Am J Clin Oncol 1982;5(6):649–55.

[15] Remon J, Soria JC, Peters S. Early and locally advanced non-small-cell lung cancer: an update of the ESMO clinical practice guidelines focusing on diagnosis, staging, systemic and local therapy. Ann Oncol 2021;32(12):1637–42. Available from: https://doi.org/10.1016/j.annonc.2021.08.1994.

[16] Dingemans AC, Früh M, Ardizzoni A. ESMO Guidelines Committee. Small-cell lung cancer: ESMO clinical practice guidelines for diagnosis, treatment and follow-up. Ann Oncol 2021;32(7):839–53. Available from: https://doi.org/10.1016/j.annonc.2021.03.207.

[17] Ramalingam SS, Vansteenkiste J, Planchard D. FLAURA Investigators. Overall survival with osimertinib in untreated, EGFR-mutated advanced NSCLC. N Engl J Med 2020;382(1):41–50. Available from: https://doi.org/10.1056/NEJMoa1913662.

[18] Hendriks LE, Kerr K, Menis J, et al. ESMO Guidelines Committee. Non-oncogene-addicted metastatic non-small-cell lung cancer: ESMO clinical practice guideline for diagnosis, treatment and follow-up. Ann Oncol 2023;34(4):358–76. Available from: https://doi.org/10.1016/j.annonc.2022.12.013.

[19] Cho BC, Ahn MJ, Kang JH, et al. Lazertinib versus gefitinib as first-line treatment in patients with EGFR-mutated advanced non-small-cell lung cancer: results from LASER301. J Clin Oncol 2023;41(26):4208–17. Available from: https://doi.org/10.1200/JCO.23.00515.

[20] Passaro A, Wang J, Wang Y, et al. MARIPOSA-2 Investigators. Amivantamab plus chemotherapy with and without lazertinib in EGFR-mutant advanced NSCLC after disease progression on osimertinib: primary results from the phase III MARIPOSA-2 study. Ann Oncol 2024;35(1):77–90. Available from: https://doi.org/10.1016/j.annonc.2023.10.117.

[21] Bezjak A, Tu D, Seymour L, et al. Symptom improvement in lung cancer patients treated with erlotinib: quality-of-life analysis of the National Cancer Institute of Canada Clinical Trials Group Study BR.21. J Clin Oncol 2006;24(24):3831–7. Available from: https://doi.org/10.1200/JCO.2006.05.8073.

[22] Mok T, Camidge DR, Gadgeel SM, et al. Updated overall survival and final progression-free survival data for patients with treatment-naive advanced ALK-positive non-small-cell lung cancer in the ALEX study. Ann Oncol 2020;31(8):1056–64. Available from: https://doi.org/10.1016/j.annonc.2020.04.478.

[23] Shaw AT, Bauer TM, de Marinis F, et al. First-line lorlatinib or crizotinib in advanced ALK-positive lung cancer. N Engl J Med 2020;383(21):2018–29. Available from: https://doi.org/10.1056/NEJMoa2027187.

[24] Cooper AJ, Sequist LV, Lin JJ. Third-generation EGFR and ALK inhibitors: mechanisms of resistance and management. Nat Rev Clin Oncol 2022;19(8):499–514. Available from: https://doi.org/10.1038/s41571-022-00639-9. Epub 2022 May 9. Erratum in: Nat Rev Clin Oncol 2022 Nov;19(11):744. Available from: https://doi.org/10.1038/s41571-022-00680-8.

[25] Camidge DR, Dziadziuszko R, Peters S, et al. Updated efficacy and safety data and impact of the EML4-ALK fusion variant on the efficacy of alectinib in untreated ALK-positive advanced non-small cell lung cancer in the global phase III ALEX study. J Thorac Oncol 2019;14(7):1233–43. Available from: https://doi.org/10.1016/j.jtho.2019.03.007.

[26] Shaw AT, Bauer TM, de Marinis F, et al. First-line lorlatinib or crizotinib in advanced ALK-positive lung cancer. N Engl J Med 2020;383:2018–29. Available from: https://doi.org/10.1056/NEJMoa2027187.

[27] Doebele RC, Drilon A, Paz-Ares L, et al. Entrectinib in patients with advanced or metastatic NTRK fusion-positive solid tumours: integrated analysis of three phase 1–2 trials. Lancet Oncol 2020;21(2):271–82. Available from: https://doi.org/10.1016/S1470-2045(19)30691-6.

[28] Riely GJ, Wood DE, Aisner DL, et al. Non-small cell lung cancer, version 4.2026, NCCN clinical practice guidelines in oncology. J Natl Compr Canc Netw 2026;24(4):e260017. Available from: https://doi.org/10.6004/jnccn.2026.0017.

[29] Thompson JA, Schneider BJ, Brahmer J, et al. NCCN guidelines® insights: management of immunotherapy-related toxicities, version 2.2024. J Natl Compr Canc Netw 2024;22(9):582–92. Available from:https://doi.org/10.6004/jnccn.2024.0057.

[30] Reck M, Rodriguez-Abreu D, Robinson AG, et al. Updated analysis of KEYNOTE-024: pembrolizumab versus platinum-based chemotherapy for advanced non-small-cell lung cancer with PD-L1 tumor proportion score of 50% or greater. J Clin Oncol 2019;37(7):537–46. Available from: https://doi.org/10.1200/JCO.18.00149.

[31] Mok TSK, Wu Y-L, Kudaba I, et al. Pembrolizumab versus chemotherapy for previously untreated, PD-L1-expressing, locally advanced or metastatic non-small-cell lung cancer (KEYNOTE-042): a randomised, open-label, controlled, phase 3 trial. Lancet 2019;393(10183):1819–30. Available from: https://doi.org/10.1016/S0140-6736(18)32409-7.

[32] Herbst RS, Giaccone G, de Marinis F, et al. Atezolizumab for first-line treatment of PD-L1–selected patients with NSCLC. N Engl J Med 2020;383(14):1328–39. Available from: https://doi.org/10.1056/NEJMoa1917346.

[33] Hellmann MD, Paz-Ares L, Bernabe Caro R, et al. Nivolumab plus ipilimumab in advanced non–small-cell lung cancer. N Engl J Med 2019;381(21):2020–31. Available from: https://doi.org/10.1056/NEJMoa1910231.

[34] Paz-Ares L, Ciuleanu T-E, Lee JS, et al. Nivolumab plus ipilimumab with two cycles of chemotherapy versus chemotherapy as first-line treatment for metastatic non-small-cell lung cancer (CheckMate 9LA): an international, randomised, open-label, phase 3 trial. Lancet Oncol 2021;22(2):198–211. Available from: https://doi.org/10.1016/S1470-2045(20)30641-0.

[35] Johnson ML, Cho BC, Luft A, et al. Durvalumab tremelimumab chemotherapy versus chemotherapy alone in first-line metastatic NSCLC: the POSEIDON randomized phase III trial. J Clin Oncol 2023;41(14):2579–93. Available from: https://doi.org/10.1200/JCO.22.00975.

[36] Siciliano MA, Caridà G, Ciliberto DD, et al. Efficacy and safety of first-line checkpoint inhibitors-based treatments for non-oncogene-addicted non-small-cell lung cancer: a systematic review and meta-analysis. ESMO Open 2022;7(3):100465. Available from: https://doi.org/10.1016/j.esmoop.2022.100465.

[37] Siringo M, Baena J, Bote de Cabo H, et al. Future perspectives in the second line therapeutic setting for non-oncogene addicted non-small-cell lung cancer. Cancers (Basel) 2023;15(23):5505. Available from: https://doi.org/10.3390/cancers15235505.

[38] National Comprehensive Cancer Network. NCCN clinical practice guidelines in oncology: small cell lung cancer, Version 3.2025. 2024 Oct 29. <https://www.nccn.org/professionals/physician_gls/pdf/sclc.pdf>.

[39] Simone CB II, Bogart JA, Cabrera AR, et al. Radiation therapy for small cell lung cancer: an ASTRO clinical practice guideline. Pract Radiat Oncol 2020;10(3):158–73. Available from: https://doi.org/10.1016/j.prro.2020.02.009.

[40] Megyesfalvi Z, Gay CM, Popper H, et al. Clinical insights into small cell lung cancer: tumor heterogeneity, diagnosis, therapy, and future directions. CA Cancer J Clin 2023;73(6):620–52. Available from: https://doi.org/10.3322/caac.21785.

[41] Tariq S, Kim SY, Monteiro de Oliveira Novaes J, et al. Management of small cell lung cancer. Lung 2021;199(6):579–87. Available from: https://doi.org/10.1007/s00408-021-00486-y.

[42] Mestdagh F, Steyaert A, Lavand'homme P. Pain management: a narrative review of current concepts, strategies, and techniques. Curr Oncol 2023;30(7):6838–58. Available from: https://doi.org/10.3390/curroncol30070500.

[43] Horn L, Mansfield AS, Szczęsna A, et al. First-line atezolizumab plus chemotherapy in extensive-stage small-cell lung cancer. N Engl J Med 2018;379(23):2220–9. Available from: https://doi.org/10.1056/NEJMoa1809064.

[44] Paz-Ares L, Dvorkin M, Chen Y, et al. Durvalumab plus platinum–etoposide versus platinum–etoposide in first-line treatment of extensive-stage small-cell lung cancer (CASPIAN): a randomised, controlled, open-label, phase 3 trial. Lancet 2019;394(10212):1929–39. Available from: https://doi.org/10.1016/S0140-6736(19)32222-6.

[45] Mercadante S, Vitrano V. Pain in patients with lung cancer: pathophysiology and treatment. Lung Cancer 2010;68(1):10–5. Available from: https://doi.org/10.1016/j.lungcan.2009.11.004.

[46] Potter J, Higginson IJ. Pain experienced by lung cancer patients: a review of prevalence, causes and pathophysiology. Lung Cancer 2004;43(3):247–57. Available from: https://doi.org/10.1016/j.lungcan.2003.08.030.

[47] Zhang B, Li X, Ma Z, et al. Prevalence and management of pain in lung cancer patients in northern China: a multicenter cross-sectional study. Thorac Cancer 2022;13(11):1684–90. Available from: https://doi.org/10.1111/1759-7714.14444.

[48] Tan VS, Tjong MC, Chan WC, et al. Pain and interventions in stage IV non-small cell lung cancer: a province-wide analysis. Curr Oncol 2023;30(3):3461–72. Available from: https://doi.org/10.3390/curroncol30030262.

[49] Bibby AC, Tsim S, Kannelakis N, et al. Malignant pleural mesothelioma: an update on investigation, diagnosis and treatment. Eur Respir Rev 2016;25:472–86. Available from: https://doi.org/10.1183/16000617.0063-2016.

[50] Brims F. Epidemiology and clinical aspects of malignant pleural mesothelioma. Cancers (Basel) 2021;13(16):4194. Available from: https://doi.org/10.3390/cancers13164194.

[51] van Meerbeeck JP, Scherpereel A, Surmont VF, et al. Malignant pleural mesothelioma: the standard of care and challenges for future management. Crit Rev Oncol Hematol 2011;78(2):92–111. Available from: https://doi.org/10.1016/j.critrevonc.2010.04.004.

[52] Carbone M, Adusumilli PS, Alexander Jr HR, et al. Mesothelioma: scientific clues for prevention, diagnosis, and therapy. CA Cancer J Clin 2019;69(5):402–29. Erratum in: CA Cancer J Clin 2020 Jul;70(4):313-4.https://doi.org/10.3322/caac.21607. Available from: https://doi.org/10.3322/caac.21572.

[53] Roberts ME, Neville E, Berrisford RG, et al. BTS Pleural Disease Guideline Group. Management of a malignant pleural effusion: British Thoracic Society pleural disease guideline 2010. Thorax 2010;65(2):32–40. Available from: https://doi.org/10.1136/thx.2010.136994.

[54] Scherpereel A, Opitz I, Berghmans T, et al. ERS/ESTS/EACTS/ESTRO guidelines for the management of malignant pleural mesothelioma. Eur Respir J 2020;55:1900953. Available from: https://doi.org/10.1183/13993003.00953-2019.

[55] Pleural mesothelioma. In: Brierly JD, Gospodarowicz MK, Wittekind C, editors. TNM classification of malignant tumours. Chichester, West Sussex/Hoboken, New Jersey: John Wiley & Sons, Inc.; 2017. p. 113–4. <http://mom.gov.az/resources/content_files/ebook/TNM-Classification-of-Malignant-Tumours-8th-edition.pdf>

[56] Treasure T, Lang-Lazdunski L, Waller D, et al. Extra-pleural pneumonectomy versus no extra-pleural pneumonectomy for patients with malignant pleural mesothelioma: clinical outcomes of the mesothelioma and radical surgery (MARS) randomised feasibility study. Lancet Oncol 2011;12:763–72. Available from: https://doi.org/10.1016/S1470-2045(11)70149-8.

[57] Baas P, Scherpereel A, Nowak AK, et al. First-line nivolumab plus ipilimumab in unresectable malignant pleural mesothelioma (CheckMate743): a multicentre, randomised, open-label, phase 3 trial. Lancet 2021;397(10272):375–86. Available from: https://doi.org/10.1016/S0140-6736(20)32714-8.

[58] Fennell DA, Casbard AC, Porter C, et al. A randomized phase II trial of oral vinorelbine as second-line therapy for patients with malignant pleural mesothelioma. J Clin Oncol 2021;39(15):8507. Available from: https://doi.org/10.1200/JCO.2021.39.15_suppl.8507.

[59] Macleod N, Price A, O'Rourke N, et al. Radiotherapy for the treatment of pain in malignant pleural mesothelioma: a systematic review. Lung Cancer 2014;83(2):133–8. Available from: https://doi.org/10.1016/j.lungcan.2013.11.004.

[60] Saunders JA, Ashton M, Hall C, et al. Pain management in patients with malignant mesothelioma: challenges and solutions. Lung Cancer (Auckl) 2019;10:37–46. Available from: https://doi.org/10.2147/LCTT.S192558.

[61] WHO. Thoracic tumours: WHO classification of tumours. 5th ed. vol. 5. Lyon, France: International Agency for Research on Cancer; 2020.

[62] Hsu CH, Chan JK, Yin CH, et al. Trends in the incidence of thymoma, thymic carcinoma, and thymic neuroendocrine tumor in the United States. PLoS One 2019;14:e022719. Available from: https://doi.org/10.1371/journal.pone.0227197.

[63] IASLC. Thoracic tumours. 5th ed. Lyon, France: IARC Press; 2021.

[64] Carter BW, Marom EM, Detterbeck FC, et al. Approaching the patient with an anterior mediastinal mass: a guide for clinicians. J Thorac Oncol 2014;9(2):102–9. Available from: https://doi.org/10.1097/JTO.0000000000000294.

[65] Kaira K, Endo M, Abe M, et al. Biologic correlation of 2-[^{18}F]-fluoro-2-deoxy-D-glucose uptake on positron emission tomography in thymic epithelial tumors. J Clin Oncol 2010;28(23):3746–53. Available from: https://doi.org/10.1200/JCO.2009.27.4662.

[66] Detterbeck FC, Nicholson AG, Kondo K, et al. The Masaoka-Koga stage classification for thymic malignancies: clarification and definition of terms. J Thorac Oncol 2011;6(3):1710–6. Available from: https://doi.org/10.1097/JTO.0b013e31821e8cff.

[67] Detterbeck FC, Ströbel P, Marino M, et al. The IASLC/ITMIG thymic epithelial tumors staging project: proposal for an evidence-based stage classification system for the forthcoming (8th) edition of the TNM classification of malignant tumors. J Thorac Oncol 2014;9(2):65–72. Available from: https://doi.org/10.1097/JTO.0000000000000290.

[68] Bhora FY, Chakraborty D, Lee HJ. The ITMIG/IASLC thymic epithelial tumors staging project: a proposed lymph node map for thymic epithelial tumors in the forthcoming 8th edition of the TNM classification of malignant tumors. J Thorac Oncol 2014;9(2):88–96. Available from: https://doi.org/10.1097/JTO.0000000000000293.
[69] Tosi D, Damarco F, Franzi S, et al. Outcomes of extended surgical resections for locally advanced thymic malignancies: a narrative review. Gland Surg 2022;11:611–21. Available from: https://doi.org/10.21037/gs-21-642.
[70] Utsumi T, Shiono H, Kadota Y, et al. Postoperative radiation therapy after complete resection of thymoma has little impact on survival. Cancer 2009;115:5413–20. Available from: https://doi.org/10.1002/cncr.24618.
[71] Hamaji M, Shaw RR, Burt BM. A meta-analysis of postoperative radiotherapy for thymic carcinoma. Ann Thorac Surg 2017;103:1668–75. Available from: https://doi.org/10.1016/j.athoracsur.2016.12.042.
[72] Falkson CB, Boucher A, Brundage M. The management of thymoma: a systematic review and practice guideline. J Thorac Oncol 2009;4(7):911–9. Available from: https://doi.org/10.1097/jto.0b013e3181a4b8e0.
[73] Marabelle A, Le DT, Ascierto PA, et al. Efficacy of pembrolizumab in patients with noncolorectal high microsatellite instability/mismatch repair–deficient cancer: results from the phase II KEYNOTE-158 study. J Clin Oncol 2020;38(1):1–10. Available from: https://doi.org/10.1200/JCO.19.02105.
[74] Heery CR, O'Sullivan-Coyne G, Madan RA, et al. Avelumab for metastatic or locally advanced previously treated solid tumours (JAVELIN Solid Tumor): a phase 1a, multicohort, dose-escalation trial. Lancet Oncol 2017;18(5):587–98. Available from: https://doi.org/10.1016/S1470-2045(17)30239-5.
[75] Sato J, Satouchi M, Itoh S, et al. Lenvatinib in patients with advanced or metastatic thymic carcinoma (REMORA): a multicentre, phase 2 trial. Lancet Oncol 2020;21(6):843–50. Available from: https://doi.org/10.1016/S1470-2045(20)30162-5.
[76] Girard N, Lal R, Wakelee H, et al. Chemotherapy definitions and policies for thymic malignancies. J Thorac Oncol 2011;6(7 Suppl. 3):S1749–55. Available from: https://doi.org/10.1097/JTO.0b013e31821ea5f7.
[77] Girard N, Mornex F, Van Houtte P. Chemotherapy and targeted agents for thymic malignancies. Expert Rev Anticancer Ther 2012;12:685–95. Available from: https://doi.org/10.1586/era.12.29.
[78] Okuma Y, Sato M, Hosomi Y, et al. Key components of chemotherapy for thymic malignancies: a systematic review and pooled analysis for anthracycline-, carboplatin- or cisplatin-based chemotherapy. J Cancer Res Clin Oncol 2015;141:323–31. Available from: https://doi.org/10.1007/s00432-014-1800-6.
[79] Loehrer PJ Sr, Samuels BL, Weick JK. Octreotide alone or with prednisone in patients with advanced thymoma and thymic carcinoma: an Eastern Cooperative Oncology Group phase II trial. J Clin Oncol 2004;22(2):293–9. Available from: https://doi.org/10.1200/JCO.2004.02.047.
[80] Lattanzio R, La Sorda R, Facciolo F. Thymic epithelial tumors express vascularendothelial growth factors and their receptors as potential targets of antiangiogenic therapy: a tissue micro array-based multicenter study. Lung Cancer 2014;85(2):191–6. Available from: https://doi.org/10.1016/j.lungcan.2014.05.010.
[81] Giaccone G, Kim C, Thompson J. Pembrolizumab in patients with thymic carcinoma: a single-arm, single-centre, phase 2 study. Lancet Oncol 2018;19:347–55. Available from: https://doi.org/10.1016/S1470-2045(18)30062-7.
[82] Katsuya Y, Hishima T, Horinouchi H, et al. Single-arm, multicentre, phase II trial of nivolumab for unresectable or recurrent thymic carcinoma: PRIMER study. Eur J Cancer 2019;113:78–86. Available from: https://doi.org/10.1016/j.ejca.2019.03.012.
[83] Danielsen AV, Aasvang EK, Klepstad P, et al. Pain trajectories and neuropathic pain symptoms following lung cancer surgery: a prospective cohort study. Eur J Pain 2024;28(8):1343–55. Available from: https://doi.org/10.1002/ejp.2265.
[84] Pei Q, Li Y, Luo Y, et al. Artificial intelligence in clinical applications for lung cancer: diagnosis, treatment and prognosis. Clin Chem Lab Med 2022;60(12):1974–83. Available from: https://doi.org/10.1515/cclm-2022-0291.
[85] Avanzo M, Stancanello J, Pirrone G. Radiomics and deep learning in lung cancer. Strahlenther Onkol 2020;196(10):879–87. Available from: https://doi.org/10.1007/s00066-020-01625-9.
[86] Meng FT, Jhuang JR, Peng YT, et al. Predicting lung cancer survival to the future: population-based cancer survival modeling study. JMIR Public Health Surveill 2024;10:e46737. Available from: https://doi.org/10.2196/46737.
[87] Desai P, Takahashi N, Kumar R, et al. Microenvironment shapes small-cell lung cancer neuroendocrine states and presents therapeutic opportunities. Cell Rep Med 2024;5(6):101610. Available from: https://doi.org/10.1016/j.xcrm.2024.101610.

Chapter 38

Breast cancer

Lazar Popović[1,2], Maja Popović[1,2], and Gorana Matovina Brko[2]
[1]*Faculty of Medicine, University of Novi Sad, Novi Sad, Serbia,* [2]*Oncology Institute of Vojvodina, Sremska Kamenica, Serbia*

Key learning objectives

- Understand the heterogeneity of metastatic breast cancer (MBC) and its impact on prognosis and therapeutic strategies.
- Recognize the role of targeted therapies and immunotherapies in improving survival and quality of life (QoL) for hormone receptor-positive, human epidermal growth factor receptor 2+, and triple-negative breast cancer subtypes.
- Evaluate the significance of pain as a symptom in MBC and the need for individualized, multidisciplinary pain management.
- Evaluate the importance of incorporating patient-reported outcomes, particularly pain and QoL, in clinical decision-making for MBC.

Introduction

Breast cancer is the most common malignant disease of women. According to the World Health Organization data from the year 2022, breast cancer ranks first in terms of incidence (58.7 per 100,000) and mortality rate (17 per 100,000) of all malignant diseases of women [1]. The incidence of breast cancer is constantly increasing; however, over time the mortality is decreasing as a result of the implementation of screening programs, health education measures, improvement of diagnostic procedures, and also more effective therapeutic measures that are implemented in the treatment of this disease [2]. According to United Kingdom Cancer Research data, survival from breast cancer in Great Britain has almost doubled in the last 40 years (from 40% to 78%), and particularly great progress has been achieved in the treatment of early breast cancer, which today is often curable, as patient 10-year survival rate is up to 96% [3]. Unlike early breast cancer, metastatic breast cancer (MBC) is still generally an incurable disease, with a median survival of 2–3 years. However, outcomes vary significantly by subtype, and recent advances in treatment have led to substantial improvements in survival [4]. Most MBCs do not occur de novo (only 6%–40%, depending on the region) but follow the treatment of early breast cancer [2]. Recurrent MBC is more difficult to manage, driven in part by endocrine resistance and clonal evolution [5]. According to the available data, it is estimated that in 2015, half a million women died from MBC and that this number will be around 800,000 in 2030, worldwide [4].

Breast cancer is a heterogenous disease, and survival is significantly influenced by the breast cancer subtype. It is divided into several distinct subtypes according to immunohistochemical expression of estrogen receptor (ER), progesterone receptor (PR), and overexpression/amplification of human epidermal growth factor receptor 2 (HER2). Intrinsic subtypes identified from these expression profiles play a key role in guiding treatment decisions, and include ER +/PR+ (luminal A and luminal B), HER2 enriched (HER2+), and triple-negative breast cancer (TNBC, lacking all three receptors) [6,7]. Despite the adjuvant treatment of early breast cancer and the high survival rates of patients, a significant number of them still experience local relapse or metastatic disease. Distant metastases occur in up to 25% of HER2+ and triple-negative tumors, 20% of patients with luminal B tumors, and 9.5% of patients with the luminal A subtype [8,9]. According to the US Surveillance, Epidemiology, and End Results Program data, in patients with stage IV disease for whom the tumor subtype was known, 72.7% of patients were found to be HR+/HER2−, 12.2% were triple-negative (HR−/HER2−), 10.3% were HR+/HER2+, and 4.6% were HR−/HER2+ [10]. Hormone receptor-positive (HR+) MBC has a median OS of 37.0 months, while patients with HR-negative MBC have a median overall survival (OS) of 9.9 months [7]. Breast cancer most commonly metastasizes to the bone, liver, lungs, and brain, with these sites of spread closely linked to

Pain Management in Palliative Care for Patients With Cancer. DOI: https://doi.org/10.1016/B978-0-443-30050-9.00053-9

patient survival outcomes [11]. Patients with HR+ Luminal A (51.1%) and Luminal B (44.7%) breast cancers are most often diagnosed with bone metastasis. TNBC patients have a higher probability of brain metastasis [12]. Patients with brain metastases or metastases at multiple sites generally have relatively poor prognostic outcomes compared with lung and other visceral metastases [13]. With the improvement of survival in early breast cancer, there has been an increase in liver and central nervous system (CNS) metastases and a decline in bone metastases, suggesting that early breast cancer therapies seem to have led to an evolution in tumor biology in subsequent metastatic disease [5,14]. Since MBC remains an incurable disease for the majority of patients, the primary goal of treatment is palliative, aimed at prolonging survival while minimizing toxicity and preserving an acceptable quality of life (QoL) for patients [15]. Today, therapy for MBC involves the use of hormonal therapy, as monotherapy or in combination with targeted therapies, antibody-drug conjugates (ADC), immunotherapy, targeted therapies, and chemotherapy.

Pain, however, remains one of the most debilitating symptoms of MBC. It influences the functional status of the patient as well as QoL. In advanced breast cancer, pain can be a consequence of the disease spreading to bones, muscles, or neural structures or the release of inflammatory cytokines. On the other hand, pain can be a consequence of cancer treatment: chemotherapy-induced neuropathic pain or radiation-induced microvascular changes and nerve compressions. Pain can be chronic as well as acute and breakthrough [16]. More details are given in appropriate chapters of this book.

Pain severity is associated with the current metastatic site, with patients with visceral and bone metastases generally reporting more severe pain levels and greater pain interference than those with bone-only or visceral-only metastases [17]. Successful pain management in MBC patients requires a detailed evaluation of both patient and pain. The management should be tailored according to each patient and the actual cause of pain. Early recognition and comprehensive pain assessment are crucial for effective pain management [18].

The goal of specific breast cancer therapies is to prolong progression-free survival (PFS) and OS of breast cancer patients, without significant disturbance of QoL but also with improvement of pain symptoms and general health status of MBC patients.

Hormone receptor-positive HER2-negative metastatic breast cancer

All guidelines for the treatment of metastatic hormone receptor-positive human epidermal growth factor receptor 2–negative (HR+/HER2–) breast cancer today primarily recommend the use of hormonal therapy. Treatment depends on the type and duration of the previous hormonal adjuvant therapy and the time elapsed from the end of the hormonal therapy to the time of the recurrence of the disease [14,19]. Endocrine therapy (ET) with cyclin-dependent kinase (CDK)4/6 inhibitors is the first-line choice for the majority of metastatic HR+/HER2– breast cancer patients [20]. This combination has an advantage over chemotherapy and is recommended even in the presence of visceral disease, and it is not recommended only if there are signs of a visceral crisis, that is, if there is suspicion or proof that endocrine resistance has occurred [14,19]. A visceral crisis is not only the presence of visceral metastases but also indicates a significantly compromised organ function, which requires faster-acting and more time-efficient therapy, especially if after progression it will not be possible to apply the next line of therapy [21–23]. Three drugs from the category of CDK4/6 inhibitors—palbociclib, ribociclib, and abemaciclib—are approved in the first- and second-line therapies of metastatic HR+/HER2– breast cancer in combination with hormonal therapy [19,21,23]. By introducing these drugs in the first line in combination with an aromatase inhibitor, the patient's PFS was extended by about 10 months for all three drugs [24–26]. The objective response rate was higher for the combination of drugs in all three studies, reaching 40.7% up to 49.1% [24–27]. The improvement in OS was reported at 3.5 years for combination with ribociclib versus ET alone: 70% versus 46% (HR 0.71) [28].

In the second line of treatment, palbociclib, ribociclib, and abemaciclib are recommended in combination with fulvestrant, as well as in the first line in patients whose disease progression occurred during adjuvant hormonal therapy [19,21,23]. All three drugs, when combined with fulvestrant, improve PFS in comparison with fulvestrant alone [28–30]. Although clinical trials investigating combinations of ET and all three CDK4/6 inhibitors have met their primary end point of PFS, OS as a secondary end point was significantly improved only in combinations with ribociclib and abemaciclib [31,32].

The majority of modern trials with CDK4/6 inhibitors (CDK4/6i) evaluated QoL using questionnaires that also assessed pain, such as the European Organization for Research and Treatment of Cancer Quality of Life Questionnaire Core 30 (EORTC QLQ-C30), the Brief Pain Inventory (BPI), and its short form (BPI-sf). Pain as well as QoL were reported in pooled analyses or even as a secondary end point of the trials [33,34]. A recent review of the literature reports the available evidence about health-related QoL (HRQoL) of breast cancer patients receiving CDK4/6i both in clinical studies and in a real-world setting. The results show that the addition of a CDK4/6i to ET does not worsen patient HRQoL, with a positive trend toward pain improvement [35].

Subanalyses and pooled analyses mostly reveal a numerically longer time to deterioration (TTD) and a sustained deterioration of pain among patients in the experimental arms, as well as greater pain reduction in the experimental arms compared with the placebo arms, with clinically meaningful decreases in pain scores [34,36].

It was noticed that the combination of ET and CDK4/6 inhibitors was as effective as chemotherapy in visceral crisis for the majority of patients [20]. The RIGHT Choice trial confirmed the superiority of ribociclib and ET in comparison to combination chemotherapy. This trial also included patients with a definition of visceral crisis, pointing out that endocrine-based therapy is as effective as chemotherapy even in this subgroup of patients [37]. Similar response rate (RR) in chemotherapy versus ribociclib were reported: 61.8% versus 66.1%, with better tolerability of ribociclib over combination chemotherapy [37]. The RIGHT Choice changed the previous collective wisdom that breast cancer patients with visceral disease need chemotherapy.

The mainstay of ET treatment in MBC, especially in the first line of treatment, are aromatase inhibitors (AIs). AI-induced musculoskeletal symptoms (AIIMS) are a common side effect of ET in breast cancer and may cause adjuvant ET discontinuation. AIIMS are a consequence of the abrupt suppression of estrogen levels and consist of AI-induced bone loss and AI-induced arthralgias. AI-induced arthralgias are reported in up to 50% of postmenopausal patients [38].

AIIMS may present with arthralgia, myalgia, arthritis, or stiffness, and it may be the reason for discontinuation of the treatment in approximately 25% of the patients with early breast cancer. AIIMS symmetrically affect the fingers, hands, wrists, elbows, shoulders, knees, and ankles and can manifest in various ways, such as diffuse joint and muscle pain and stiffness, carpal tunnel syndrome, tenosynovitis, trigger finger, and reduced grip strength [39].

A recent metaanalysis of seven randomized clinical trials reported that arthralgia, myalgia, back pain, and bone pain were decreased and less reported in MBC patients treated with AI and CDK4/6i combination in comparison to AI monotherapy. The mechanism behind this pain reduction in the two-drug combination is not yet completely elucidated [40].

The optimal sequence of endocrine therapies, either alone or in combination with targeted agents, is not known [19]. In recent years several new groups of targeted therapies and combinations with hormonal therapy have emerged. Some of these drugs and combinations have become the standard of treatment. The optimal treatment sequence with all available options is yet to be defined. Here we will only briefly mention some of the treatments available.

Experts agree that the patients progressing on the first-line CDK4/6 with ET do not seem to benefit from prolongation of CDK4/6 treatment [20]. However, guidelines recommend the continuation of hormone therapy in the following lines of disease treatment, sequentially, in patients in whom there is no doubt or proof of the existence of endocrine resistance. Chemotherapy is recommended after several lines of ET therapy, that is, in case of endocrine resistance, most often as monotherapy [14,19]. The mechanisms of endocrine resistance are numerous. Resistance can exist at the very beginning of treatment, when the relapse of the disease occurs in the first months after the start of treatment, or it can develop over time [41].

Elacestrant is the first approved selective estrogen degrader that has been investigated in the second/third line of treatment of patients with activating estrogen receptor 1 (ESR1) mutation. In the EMERALD trial elacestrant prolonged PFS in previously treated patients in both the overall population (HR 0.70) and those with the ESR1 mutation (HR 0.55) [42]. The overall RR with elacestrant was 27.3% in patients evaluable for the response, while QoL was maintained between treatment arms in this trial [43].

Alpelisib is a phosphoinositide 3-kinase alpha-specific inhibitor approved for the treatment of phosphatidylinositol-4,5-bisphosphate 3-kinase catalytic subunit alpha (PIK3CA)-mutated HR+/HER2− advanced breast cancer. Alpelisib proved effective in combination with fulvestrant versus fulvestrant plus placebo in the SOLAR-1 trial, reaching median progression-free survival (mPFS) 11 versus 5.7 months (HR 0.65); however, there was no statistically significant OS benefit. In patients with PIK3CA mutations, overall response rates (ORRs) of 26.6% and 12.8% were reported [44,45]. Patient-reported outcomes (PROs) in the SOLAR-1 trial showed no statistical difference in the deterioration of overall Global Health Status (GHS)/QoL between the alpelisib plus fulvestrant arm and the placebo plus fulvestrant arm, despite the known side effects of alpelisib (diarrhea, appetite loss, hyperglycemia, nausea or vomiting, and fatigue) [46].

One of the potential mechanisms of resistance in HR+ breast cancer is activation of the mammalian target of rapamycin (mTOR). On the other hand, inhibition of mTOR can enhance the efficacy of ET. The BOLERO-2 trial managed to show significant mPFS prolongation and higher ORR with the mTOR inhibitor everolimus and exemestane versus exemestane alone; however, no OS benefit was recorded [47,48]. The median time to definitive deterioration (TDD) in HRQoL was 8.3 months with combination therapy versus 5.8 months with ET (hazard ratio [HR], 0.74) [49].

In patients with HR+/HER2− advanced breast cancer and PIK3CA/RAC(Rho family)-alpha serine/threonine-protein kinase 1 (AKT1) phosphatase and tensin homolog (PTEN) alterations, capivasertib is another new treatment option. In the CAPItello-291 trial, combination therapy with capivasertib and fulvestrant achieved prolonged mPFS in comparison to fulvestrant alone. Prolongation was recorded in both the overall population and the altered AKT signaling pathway. The objective response rate was 22.9% for the combination versus 12.2% for fulvestrant. OS data are not available yet [50]. PROs from CAPItello-291 show that the capivasertib–fulvestrant combination delayed TTD of GHS/QoL and maintained HRQoL (except symptoms of diarrhea). Pain outcomes were not reported [51].

The poly (ADP-ribose) polymerase (PARP) inhibitors olaparib and talazoparib are treatment options for germline (g) breast cancer gene (BRCA)1/2-mutated patients with metastatic HR+/HER2− cancer. Their approval was a result of the EMBRACA and OlympiAD trials that showed significant improvement in median PFS in gBRCA1/2-mutated patients in comparison to treatment of physician choice (PCT): 8.6 versus 5.6 months (HR for disease progression or death, 0.54) in the EMBRACA trial and 7.0 versus 4.2 months (HR for disease progression or death, 0.58) in the OlympiAD trial [52,53]. OS benefit was not reached in both trials [54,55]. Both the talazoparib arm in the EMBRACA trial and the olaparib arm in the OlympiAD trial achieved a confirmed higher ORR (62.6% and 59.9%, respectively) in comparison to chemotherapy (27.2% and 28.8%) [52,55].

The EMBRACA trial demonstrated a statistically significant greater delay in TTD in multiple cancer-related and breast cancer-specific symptoms and in GHS/QoL, favoring talazoparib over PCT (HR, 0.38; 24.3 vs 6.3 months, respectively; $P < .0001$) [56].

Trastuzumab-deruxtecan (T-Dxd) is approved for previously treated HER2-low HR+ MBC. In DESTINY-Breast-04, the HR+ cohort treated with T-Dxd versus physicians' choice of chemotherapy achieved an mPFS of 10.1 versus 5.4 months (HR 0.51) and a median overall survival (mOS) of 23.9 versus 17.5 months (HR 0.64) [57]. Data analyses of PRO in the DESTINY-Breast-04 trial showed that median TDD for GHS/QoL was delayed with T-DXd versus physician's choice (11.4 vs 7.5 months, respectively; HR, 0.69). Median TDD for pain symptoms was 16.4 versus 6.1 months (HR 0.40) for T-Dxd versus physicians' choice of chemotherapy, favoring T-Dxd for all prespecified PROs as well. DESTINY-Breast-06 investigated the efficacy of treatment in HER2-ultralow patients: in terms of pain, T-DXd significantly reduced the risk of deterioration per QLQ-C30 [58].

HER2+ breast cancer

HER2+ breast cancer accounts for about 20% of all breast cancers. Amplification of HER2 protein on the surface of malignant cells is a predictor of poor prognosis. In the past, precisely because of the previously mentioned fact, this subtype of cancer was characterized by a high rate of relapse and mortality [59,60]. In the last two decades, the prognosis of metastatic HER2+ breast cancer has completely changed due to the implementation of anti-HER2 therapy. The first drug that made a revolution was the monoclonal antibody trastuzumab, which led to a significant benefit in terms of OS and QoL for patients suffering from this subtype of the disease [61]. Then in 2012, after the results of the CLEOPATRA clinical study, the first line of treatment for metastatic HER2+ breast cancer was established: pertuzumab (monoclonal antibody), trastuzumab, and docetaxel (PHT), which is still the standard in the first line of treatment. This study compared trastuzumab, pertuzumab, and docetaxel versus trastuzumab, placebo, and docetaxel [62]. Dual anti-HER2 therapy significantly prolonged OS and time to disease progression, with a high overall RR of 86%. After an 8-year follow-up, the median survival of these patients is 57.1 months, and the reduction in the risk of death with dual anti-HER2 blockade and chemotherapy is 37% [63]. Additionally, a QoL analysis demonstrated that a double anti-HER2 blockade (trastuzumab and pertuzumab) in combination with docetaxel does not significantly affect the QoL and significantly prolongs the time until the recurrence of symptoms caused by breast cancer, primarily cancer pain [64].

Real-world data indicates that while the combination significantly improves PFS and OS, it also can cause a wide range of side effects, including muscle and back pain and joint pain. Taxane-induced acute pain syndrome (TAPS) is a common side effect of taxane-based chemotherapy. It occurs 2–3 days post taxane chemotherapy, and it is associated with reduced QoL and increased analgesic requirements [65].

Therefore up to 6–8 cycles of taxane combination are usually used, with continuation of anti-HER2 therapy only [14]. Details about TAPS are given in another chapter of this book.

Until 2013, the standard in the second line of treatment for patients with this subtype of MBC was the combination of lapatinib (a tyrosine kinase inhibitor) and capecitabine chemotherapy [66]. Another important breakthrough in the treatment of this disease occurred in 2013, after the results of the EMILIA study, which positioned the first antibody-drug conjugate, ado-trastuzumab emtasine (TDM-1), and since then this has been the standard second line of treatment for this disease for many years. In this study, TDM-1 was compared with lapatinib in combination with capecitabine. TDM-1 led to a significant prolongation of life, but the QoL was also significantly improved, and the time to worsening of disease symptoms was also significantly prolonged [67].

Antibody-drug conjugate trastuzumab deruxtecan led to a further improvement in prognosis and prolongation of life in these patients. Trastuzumab deruxtecan was first investigated in the DESTINY-Breast01 study in patients who were previously treated with multiple lines of therapy. After the results of this study, where 60% of patients achieved a complete or partial response, it was approved in this indication [68]. After that, the DESTINY-Breast03 study was created, which compared the effectiveness of TDM-1 and T-Dxd in the population of patients who were previously treated with one therapeutic line. It was the results of this study that led to a new standard second-line treatment for metastatic HER2+ breast cancer. Trastuzumab deruxtecan significantly prolonged time to disease progression and OS [69]. Subanalysis of the results of this study related to QoL and disease symptoms showed that T-Dxd is superior to TDM-1 in all examined indicators of disease symptom control and QoL (pain, physical, emotional, and social functioning) [70]. When it comes to an early relapse of the disease, that is, when the relapse occurs within the first 6 months after the completion of postoperative therapy, and the patients were treated with trastuzumab and pertuzumab in the preoperative or postoperative regimen, the therapy of choice in the first line of treatment for metastatic disease is T-Dxd.

The current recommended third-line therapy is tucatinib (a tyrosine kinase inhibitor) in combination with trastuzumab and capecitabine. This therapeutic option was approved after the published results of the HER2CLIMB study. This study compared tucatinib in combination with trastuzumab and capecitabine versus placebo, trastuzumab, and capecitabine, demonstrating a benefit, particularly for intracranial metastases [71]. There are now data from real-world studies that indicate the effectiveness of this therapeutic option even after progression to T-Dxd [71,72]. This is important, as symptoms of CNS metastases significantly disturb the QoL and shorten survival; most often, they are the cause of death.

After the proven benefit of T-Dxd in systemic disease control precisely because of the above, the DESTINY-Breast012 study was created, which included patients who were previously treated regardless of whether or not they have CNS metastases and whether they have active or stable CNS metastases. The primary objective of the study was the response rate to this therapy. The results showed that the intracranial response rate in patients with CNS metastases was 64%, which, in addition to the local treatment of CNS metastases, contributed to a significant control of disease symptoms and prolongation of OS. This is now the preferred second-line treatment option in all patients regardless of whether they have CNS disease [73].

It is known that about 50% of patients with HER2+ breast cancer have HR+ disease, the so-called triple-positive disease. On the other hand, we have a subgroup of patients who, due to their general condition and associated diseases, are not candidates for standard first-line treatment. For this reason, studies were created that examined the effectiveness of hormone therapy in combination with anti-HER2 therapy. Data from real clinical practice even indicated a better treatment outcome if anti-HER2 therapy is combined with hormonal therapy in triple-positive breast cancer compared to anti-HER2 therapy combined with chemotherapy [74]. The SYSUCC-002 study proved the noninferiority of the combination of anti-HER2 therapy with ET compared to the combination of anti-HER2 therapy with chemotherapy in terms of time to disease progression. On the other hand, ET is less toxic than chemotherapy [75]. Studies examining T-Dxd, tucatinib, and TDM-1 in various combinations in earlier lines of therapy are ongoing, the results of which will likely further change the treatment landscape for this disease and further improve treatment outcomes for this disease.

Triple-negative breast cancer

TNBC accounts for about 10%–15% of all subtypes of breast cancer. This subtype of breast cancer has a very aggressive course, characterized by a high rate of relapse after treatment of early breast cancer, as well as a high mortality rate [76]. In contrast to HR+ and HER2+ breast cancer, which can be treated with targeted therapy, in TNBC, which is HR– and HER2–, chemotherapy was the only therapeutic option for a long time [77]. However, this changed with immunotherapy trials following the knowledge about the presence of tumor-infiltrating lymphocytes and programmed death-ligand 1 (PD-L1) expression on immune cells in the tumor microenvironment, as well as tumor cells [78,79]. After numerous trials, the first-line treatment for metastatic TNBC has been established as pembrolizumab (PD-1 checkpoint inhibitor) in combination with chemotherapy (paclitaxel, nab-paclitaxel or gemcitabine in combination with carboplatin). The choice of chemotherapy depends on the time that has passed since the treatment of early breast cancer. However, this therapeutic option is used only in patients with a combined positive score (CPS) $\geq$10 [80]. A CPS represents PD-L1 expression on both tumor cells and immune cells in the tumor microenvironment [81]. According to literature data, PD-L1 expression is present in 20%–42% of patients with metastatic TNBC [82]. This therapeutic option significantly prolonged the OS of patients with this very difficult disease to treat [80].

The analysis of the QoL of the patients included in the study showed no difference between groups in TTD in QLQ-C30 GHS/QoL, emotional functioning, or physical functioning. Side effects of immunotherapy are a consequence of the immune system attacking healthy tissues, leading to immune-related adverse events (irAEs), which can present with various symptoms, affecting various organs. Pain can be a symptom of certain irAEs: musculoskeletal pain, myositis, polymyalgia rheumatica, arthritis, and tenosynovitis [83]. This is evaluated in detail in another chapter of this book.

Other than irAEs, immune checkpoint inhibitors also can cause transient pain enhancement or flare pain following administration, usually in joints or the primary tumor location. Data on this is, however, limited, and the detailed mechanism of this side effect is unknown [84].

In patients with TNBC, about 20% are carriers of gBRCA1 or gBRCA2 mutations, which was the basis for the investigation of targeted therapy, PARP inhibitors [85]. The before-mentioned OlympiAD study examined olaparib and included metastatic TNBC patients. The results are given in previous paragraphs and demonstrated an improvement in QoL as well as the control of disease symptoms, as discussed previously in the HR+/HER– section. PARP inhibitors are the therapy of choice in the first line of treatment of metastatic TNBC if the tumor is PD-L1 negative and the gBRCA1 or gBRCA2 mutation is proven [56].

ADCs contributed to further progress and improvement of treatment outcomes for patients with TNBC. Sacituzumab–govitecan is a combination of a monoclonal antibody that targets trophoblast surface antigen 2 and a topoisomerase I inhibitor linked via a linker. Following the results of the ASCENT study, sacituzumab–govitecan was approved for the treatment of metastatic TNBC in a population of patients previously treated with at least two lines of therapy [86]. The introduction of all previously mentioned therapeutic options significantly prolonged the OS of patients with this subtype of breast cancer, compared to the results of chemotherapy in the treatment of patients with metastatic TNBC, when survival was up to 12 months. The time until the symptoms of the disease worsen was also significantly extended. Median time to first clinically meaningful worsening was longer for sacituzumab–govitecan than for standard chemotherapy for physical functioning (22.1 vs 12.1 weeks, $P < .001$), role functioning (11.4 vs 7.1 weeks, $P < .001$), fatigue (7.7 vs 6.0 weeks, $P < .05$), and pain (21.6 vs 9.9 weeks, $P < .001$) [87].

The previously mentioned T-Dxd was also studied in the population of patients with low HER2 protein expression in comparison with chemotherapy chosen by the investigator (HER2 1+ or HER2 2+, but with a negative result of in situ hybridization). In this study, mostly HR+ patients were included, but 11.3% of HR– patients were also included, which is practically TNBC. The cohort of patients with HR– disease who were treated with T-Dxd had a benefit in terms of time to disease progression and OS [57]. There are no head-to-head studies comparing T-Dxd and sacituzumab–govitecan, but according to the results of the studies, if we have patients with low expression of the HER2 protein, the therapy of choice would be T-Dxd, and if the expression is not present at all, the therapy of choice would be sacituzumab–govitecan [88].

Today, studies are underway that examine combinations of immunotherapy, ADC, and targeted therapy, the results of which will further change the picture of the treatment of this aggressive disease and significantly improve the QoL of these patients.

Conclusion

The management of MBC has evolved substantially, with tailored therapies improving survival and QoL across subtypes. Advances in targeted and immunotherapies, especially for HR+, HER2+, and TNBC, have not only extended life but also delayed the onset and worsening of pain, a key determinant of patient well-being. Continued integration of PRO into clinical practice is essential for optimizing treatment strategies. Despite progress, MBC remains incurable, and ongoing research into novel agents and pain management strategies is critical to further enhance patient outcomes.

References

[1] Ervik M, Lam F, Laversanne M, et al. Trends in cancer incidence and mortality rates. Lyon, France: International Agency for Research on Cancer. Available from: https://gco.iarc.who.int/overtime; 2004 [accessed 15.04.25].

[2] Cancer Stat Facts: Female Breast Cancer. National Cancer Institute, Available from: <https://seer.cancer.gov/statfacts/html/breast.html>; 2025 [accessed 07.02.25].

[3] Breast cancer survival statistics. Cancer Research UK, Available from: <https://www.cancerresearchuk.org/health-professional/cancer-statistics/statistics-by-cancer-type/breast-cancer>; 2025 [accessed 07.02.25].

[4] Cardoso F, Spence D, Mertz S, et al. Global analysis of advanced/metastatic breast cancer: decade report (2005–2015). Breast 2018;39:131–8. Available from: https://doi.org/10.1016/j.breast.2018.03.002.
[5] Holzel D, Eckel R, Bauerfeind I, et al. Improved systemic treatment for early breast cancer improves cure rates, modifies metastatic pattern and shortens post-metastatic survival: 35-year results from the Munich cancer registry. J Cancer Res Clin Oncol 2017;143(9):1701–12. Available from: https://doi.org/10.1007/s00432-017-2428-0.
[6] Goldhirsch A, Wood WC, Coates AS, et al. Strategies for subtypes—dealing with the diversity of breast cancer: highlights of the St. Gallen international expert consensus on the primary therapy of early breast cancer 2011. Ann Oncol 2011;22(8):1736–47. Available from: https://doi.org/10.1093/annonc/mdr304.
[7] Valachis A, Carlqvist P, Ma Y, et al. Overall survival of patients with metastatic breast cancer in Sweden: a nationwide study. Br J Cancer 2022;127(4):720–5. Available from: https://doi.org/10.1038/s41416-022-01845-z.
[8] Hwang KT, Kim J, Jung J, et al. Impact of breast cancer subtypes on prognosis of women with operable invasive breast cancer: a population-based study using SEER database. Clin Cancer Res 2019;25(6):1970–9. Available from: https://doi.org/10.1158/1078-0432.CCR-18-2782.
[9] Van Maaren MC, De Munck L, Strobbe LJA, et al. Ten-year recurrence rates for breast cancer subtypes in the Netherlands: a large population-based study. Int J Cancer 2019;44(2):263–72. Available from: https://doi.org/10.1002/ijc.31914.
[10] Howlader N, Altekruse SF, Li CI, et al. US incidence of breast cancer subtypes defined by joint hormone receptor and HER2 status. J Natl Cancer Inst 2014;106(5):dju055. Available from: https://doi.org/10.1093/jnci/dju055.
[11] Chen MT, Sun HF, Zhao Y, et al. Comparison of patterns and prognosis among distant metastatic breast cancer patients by age groups: a SEER population-based analysis. Sci Rep 2017;7(1):9254. Available from: https://doi.org/10.1038/s41598-017-10166-8.
[12] Fan JH, Zhang S, Yang H, et al. Molecular subtypes predict the preferential site of distant metastasis in advanced breast cancer: a nationwide retrospective study. Front Oncol 2023;13:978985. Available from: https://doi.org/10.3389/fonc.2023.978985.
[13] De Ieso PB, Schick U, Rosenfelder N, et al. Breast cancer brain metastases – a 12 year review of treatment outcomes. Breast 2015;24(4):426–33. Available from: https://doi.org/10.1016/j.breast.2015.03.007.
[14] Gennari A, André F, Barrios CH, et al. ESMO clinical practice guideline for the diagnosis, staging and treatment of patients with metastatic breast cancer. Ann Oncol 2021;32(12):1475–95. Available from: https://doi.org/10.1016/j.annonc.2021.09.019.
[15] Higgins MJ, Wolff AC. Therapeutic options in the management of metastatic breast cancer. Oncology (Williston Park) 2008;22(6):614–23; discussion 623, 627–9.
[16] Alhazmi LSS, Bawadood MAA, Aljohani AMS, et al. Pain management in breast cancer patients: a multidisciplinary approach. Cureus 2021;13(6):e15994. Available from: https://doi.org/10.7759/cureus.15994.
[17] Wood R, Mitra D, de Courcy J, et al. Patient-reported pain severity, pain interference and health status in HR+/HER2- advanced/metastatic breast cancer. ESMO Open 2017;2(3):e000227. Available from: https://doi.org/10.1136/esmoopen-2017-000227.
[18] Swarm RA, Abernethy AP, Anghelescu DL, et al. Adult cancer pain. J Natl Compr Canc Netw 2013;11:992–1022. Available from: https://doi.org/10.6004/jnccn.2013.0119.
[19] National Comprehensive Cancer Network (NCCN). (NCCN guidelines) Breast cancer. NCCN Clinical Practice Guidelines in Oncology, Version 1. 2025; 2025. Available from: https://www.nccn.org/professionals/physician_gls/pdf/breast.pdf.
[20] Ditsch N, Untch M, Fasching PA, et al. ABC7 consensus: assessment by a german group of experts. Breast Care 2024;19(2). Available from: https://doi.org/10.1159/000535997.
[21] Rugo HS, Rumble RB, Macrae E, et al. Endocrine therapy for hormone receptor–positive metastatic breast cancer: American Society of Clinical Oncology guideline. J Clin Oncol 2016;34(25):3069–103. Available from: https://doi.org/10.1200/JCO.2016.67.1487.
[22] Cardoso F, Paluch-Shimon S, Schumacher-Wulf E, et al. 6th and 7th International consensus guidelines for the management of advanced breast cancer (ABC guidelines 6 and 7). Breast 2024;76:103756. Available from: https://doi.org/10.1016/j.breast.2024.103756.
[23] Cardoso F, Senkus E, Costa A, et al. 4th ESO-ESMO international consensus guidelines for advanced breast cancer (ABC 4). Ann Oncol 2018;29(8):1634–57. Available from: https://doi.org/10.1093/annonc/mdy192.
[24] Turner NC, Ro J, André F, et al. PALOMA3 study group. palbociclib in hormone-receptor-positive advanced breast cancer. N Engl J Med 2015;373(3):209–19. Available from: https://doi.org/10.1056/NEJMoa1505270.
[25] Hortobagyi GN, Stemmer SM, Burris HA. Updated results from MONALEESA-2, a phase III trial of first-line ribociclib plus letrozole versus placebo plus letrozole in hormone receptor-positive, HER2-negative advanced breast cancer. Ann Oncol 2018;29(7):1541–7. Available from: https://doi.org/10.1093/annonc/mdy155.
[26] Goetz MP, Toi M, Campone M, et al. MONARCH 3: abemaciclib as initial therapy for advanced breast cancer. J Clin Oncol 2017;35(32):3638–46. Available from: https://doi.org/10.1200/JCO.2017.75.6155.
[27] Tripathy D, Im SA, Colleoni M, et al. Ribociclib plus endocrine therapy for premenopausal women with hormone-receptor-positive, advanced breast cancer (MONALEESA-7): a randomised phase 3 trial. Lancet Oncol 2018;19(7):904–15. Available from: https://doi.org/10.1016/S1470-2045(18)30292-4.
[28] Im SA, Lu YS, Bardia A, et al. Overall survival with ribociclib plus endocrine therapy in breast cancer. N Engl J Med 2019;381(4):307–16. Available from: https://doi.org/10.1056/NEJMoa1903765.
[29] Cristofanilli M, Turner NC, Bondarenko I, et al. Fulvestrant plus palbociclib versus fulvestrant plus placebo for treatment of hormone-receptor-positive, HER2-negative metastatic breast cancer that progressed on previous endocrine therapy (PALOMA-3): final analysis. Lancet Oncol 2016;17(4):425–39. Available from: https://doi.org/10.1016/S1470-2045(15)00613-0.

[30] Sledge Jr GW, Toi M, Neven P, et al. MONARCH 2: abemaciclib in combination with fulvestrant in women with HR+/HER2- advanced breast cancer who had progressed while receiving endocrine therapy. J Clin Oncol 2017;35(25):2875–84. Available from: https://doi.org/10.1200/JCO.2017.73.7585.

[31] Hortobagyi GN, Stemmer SM, Burris HA, et al. Overall survival with ribociclib plus letrozole in advanced breast cancer. N Engl J Med 2022;386(10):942–50. Available from: https://doi.org/10.1056/NEJMoa2114663.

[32] Goetz MP, Toi M, Huober J, et al. Abemaciclib plus a nonsteroidal aromatase inhibitor as initial therapy for HR+, HER2- advanced breast cancer: final overall survival results of MONARCH 3. Ann Oncol 2024;35(8):718–27. Available from: https://doi.org/10.1016/j.annonc.2024.04.013.

[33] Janni W, Alba E, Bachelot T, et al. First-line ribociclib plus letrozole in postmenopausal women with HR+, HER2– advanced breast cancer: tumor response and pain reduction in the phase 3 MONALEESA-2 trial. Breast Cancer Res Treat 2018;169:469–79. Available from: https://doi.org/10.1007/s10549-017-4658-x.

[34] Kaufman P, Toi M, Neven P, et al. Health-related quality of life in MONARCH 2: abemaciclib plus fulvestrant in hormone receptor-positive, HER2-negative advanced breast cancer after endocrine therapy. Oncologist 2020;15(2):243–51. Available from: https://doi.org/10.1634/theoncologist.2019-0551.

[35] Di Lauro V, Barchiesi G, Martorana F, et al. Health-related quality of life in breast cancer patients treated with CDK4/6 inhibitors: a systematic review. ESMO Open 2022;7(6):100629. Available from: https://doi.org/10.1016/j.esmoop.2022.100629.

[36] Fasching PA, Bardia A, Nusch A, et al. Pooled analysis of patient (pt)-reported quality of life (QOL) in the MONALEESA (ML)-2, -3, and -7 trials of ribociclib (RIB) plus endocrine therapy (ET) to treat hormone receptor–positive, HER2-negative (HR+/HER2–) advanced breast cancer (ABC). Ann Oncol 2020;31(350):351. Available from: https://doi.org/10.1016/j.annonc.2020.08.378.

[37] Lu YS, Mahidin EIBM, Azim H, et al. Final results of RIGHT choice: ribociclib plus endocrine therapy versus combination chemotherapy in premenopausal women with clinically aggressive hormone receptor-positive/human epidermal growth factor receptor 2-negative advanced breast cancer. J Clin Oncol 2024;42(23):2812–21. Available from: https://doi.org/10.1200/JCO.24.00144.

[38] Beckwée D, Leysen L, Meuwis K, et al. Prevalence of aromatase inhibitor-induced arthralgia in breast cancer: a systematic review and meta-analysis. Support Care Cancer 2017;25:1673–86. Available from: https://doi.org/10.1007/s00520-017-3613-z.

[39] Gupta A, Henry NL, Loprinzi CL. Management of aromatase inhibitor-induced musculoskeletal symptoms. JCO Oncol Pract 2020;16(11):733–9. Available from: https://doi.org/10.1200/OP.20.00113.

[40] Andrikopoulou A, Fiste O, Liontos M, et al. Aromatase and CDK4/6 inhibitor-induced musculoskeletal symptoms: a systematic review. Cancers (Basel) 2021;13(3):465. Available from: https://doi.org/10.3390/cancers13030465.

[41] Clarke R, Tyson JJ, Dixon JM. Endocrine resistance in breast cancer – an overview and update. Mol Cell Endocrinol 2015;418(3):220–34. Available from: https://doi.org/10.1016/j.mce.2015.09.035.

[42] Bidard FC, Kaklamani VG, Neven P, et al. Elacestrant (oral selective estrogen receptor degrader) versus standard endocrine therapy for estrogen receptor-positive, human epidermal growth factor receptor 2-negative advanced breast cancer: results from the randomized phase III EMERALD trial. J Clin Oncol 2023;41(23). 3962. Available from: https://doi.org/10.1200/JCO.23.01239.

[43] Cortes J, Bidard FC, Bardia A, et al. EMERALD trial analysis of patient-reported outcomes (PROs) in patients with ER+/HER2– advanced or metastatic breast cancer (mBC) comparing oral elacestrant vs standard of care (SoC) endocrine therapy. ESMO Open 2023;8(1):101377. Available from: https://doi.org/10.1016/j.esmoop.2023.101377.

[44] André F, Ciruelos E, Rubovszky G, et al. Alpelisib for PIK3CA-mutated, hormone receptor-positive advanced breast cancer. N Engl Med 2019;380(20):1929–40. Available from: https://doi.org/10.1056/NEJMoa1813904.

[45] André F, Ciruelos EM, Juric D, et al. Alpelisib plus fulvestrant for PIK3CA-mutated, hormone receptor-positive, human epidermal growth factor receptor-2-negative advanced breast cancer: final overall survival results from SOLAR-1. Ann Oncol 2021;32(2):208–17. Available from: https://doi.org/10.1016/j.annonc.2020.11.011.

[46] Ciruelos EM, Rugo HS, Mayer IA, et al. Patient-reported outcomes in patients with PIK3CA-mutated hormone receptor-positive, human epidermal growth factor receptor 2-negative advanced breast cancer from SOLAR-1. J Clin Oncol 2021;39(18):2005–15. Available from: https://doi.org/10.1200/JCO.20.01139.

[47] Yardley DA, Noguchi S, Pritchard KI, et al. Everolimus plus exemestane in postmenopausal patients with HR(+) breast cancer: BOLERO-2 final progression-free survival analysis. Adv Ther 2013;30(10):870-84. Available from: https://doi.org/10.1007/s12325-013-0060-1. Epub 2013 Oct 25. Erratum in: Adv Ther. 2014;31(9):1008-9.

[48] Piccart M, Hortobagyi GN, Campone M, et al. Everolimus plus exemestane for hormone-receptor-positive, human epidermal growth factor receptor-2-negative advanced breast cancer: overall survival results from BOLERO-2. Ann Oncol 2014;25(12):2357–62. Available from: https://doi.org/10.1093/annonc/mdu456.

[49] Burris HA 3rd, Lebrun F, Rugo HS, et al. Health-related quality of life of patients with advanced breast cancer treated with everolimus plus exemestane versus placebo plus exemestane in the phase 3, randomized, controlled, BOLERO-2 trial. Cancer 2013;119(10):1908–15. Available from: https://doi.org/10.1002/cncr.31887.

[50] Turner NC, Oliveira M, Howell SJ, et al. Capivasertib in hormone receptor-positive advanced breast cancer. N Engl J Med 2023;388(22):2058–70. Available from: https://doi.org/10.1056/NEJMoa2214131.

[51] Oliveira M, Rugo HS, Howell SJ, et al. CAPItello-291 study group. Capivasertib and fulvestrant for patients with hormone receptor-positive, HER2-negative advanced breast cancer (CAPItello-291): patient-reported outcomes from a phase 3, randomised, double-blind, placebo-controlled trial. Lancet Oncol. 2024;25(9):1231–44. Available from: https://doi.org/10.1016/S1470-2045(24)00373-5.

[52] Litton JK, Rugo HS, Ettl J, et al. Talazoparib in patients with advanced breast cancer and a germline BRCA mutation. N Engl J Med 2018;379(8):753–63. Available from: https://doi.org/10.1056/NEJMoa1802905.

[53] Robson M, Im SA, Senkus E, et al. Olaparib for metastatic breast cancer in patients with a germline BRCA mutation. N Engl J Med 2017;377(6):523–33. Available from: https://doi.org/10.1056/NEJMoa1706450.

[54] Litton JK, Hurvitz SA, Mina LA, et al. Talazoparib versus chemotherapy in patients with germline BRCA1/2-mutated HER2-negative advanced breast cancer: final overall survival results from the EMBRACA trial. Ann Oncol 2020;31(11):1526–35. Available from: https://doi.org/10.1016/j.annonc.2020.08.2098.

[55] Robson ME, Tung N, Conte P, et al. OlympiAD final overall survival and tolerability results: olaparib versus chemotherapy treatment of physician's choice in patients with a germline BRCA mutation and HER2-negative metastatic breast cancer. Ann Oncol 2019;30(4):558–66. Available from: https://doi.org/10.1093/annonc/mdz012.

[56] Ettl J, Quek RGW, Lee KH, et al. Quality of life with talazoparib versus physician's choice of chemotherapy in patients with advanced breast cancer and germline BRCA 1/2 mutation: patient-reported outcomes from the EMBRACA phase III trial. Ann Oncol 2018;29(9):1939–47. Available from: https://doi.org/10.1093/annonc/mdy257.

[57] Modi S, Jacot W, Yamashita T, et al. DESTINY-Breast04 trial investigators. Trastuzumab deruxtecan in previously treated HER2-low advanced breast cancer. N Engl J Med 2022;387(1):9–20. Available from: https://doi.org/10.1056/NEJMoa2203690.

[58] Hu X, Curigliano G, Yonemori K, et al. Effects of trastuzumab deruxtecan (T-DXd) vs choice of chemotherapy (TPC) on patient-reported outcomes (PROs) in hormone receptor-positive, HER2-low or HER2-ultralow metastatic breast cancer (mBC): results from DESTINY-Breast06, Barcelona. Ann Oncol 2024:LBA22. Available from: https://doi.org/10.1016/j.annonc.2024.08.2260.

[59] Cronin KA, Harlan LC, Dodd KW, et al. Population-based estimate of the prevalence of HER-2 positive breast cancer tumors for early stage patients in the US. Cancer Invest 2010;28(9):963–8. Available from: https://doi.org/10.3109/07357907.2010.496759.

[60] Harper KL, Sosa MS, Entenberg D, et al. Mechanism of early dissemination and metastasis in Her2+ mammary cancer. Nature 2016;540(7634):588-92. Available from: https://doi.org/10.1038/nature20609. Epub 2016 Dec 14. Erratum in: Nature. 2018;553(7688):366. doi: 10.1038/nature24666.

[61] Slamon DJ, Leyland-Jones B, Shak S, et al. Use of chemotherapy plus a monoclonal antibody against HER2 for metastatic breast cancer that overexpresses HER2. N Engl J Med 2001;344(11):783-92. Available from: https://doi.org/10.1056/NEJM200103153441101.

[62] Baselga J, Cortés J, Kim SB, et al. CLEOPATRA Study Group. Pertuzumab plus trastuzumab plus docetaxel for metastatic breast cancer. N Engl J Med 2012;366(2):109–19. Available from: https://doi.org/10.1056/NEJMoa1113216.

[63] Swain SM, Miles D, Kim SB, et al. Pertuzumab, trastuzumab, and docetaxel for HER2-positive metastatic breast cancer (CLEOPATRA): end-of-study results from a double-blind, randomised, placebo-controlled, phase 3 study. Lancet Oncol 2020;21(4):519-30. Available from: https://doi.org/10.1016/S1470-2045(19)30863-0.

[64] Cortés J, Baselga J, Im YH, et al. Health-related quality-of-life assessment in CLEOPATRA, a phase III study combining pertuzumab with trastuzumab and docetaxel in metastatic breast cancer. Ann Oncol 2013;24(10):2630-5. Available from: https://doi.org/10.1093/annonc/mdt274.

[65] Asthana R, Zhang L, Wan BA, et al. Pain descriptors of taxane acute pain syndrome (TAPS) in breast cancer patients – a prospective clinical study. Support Care Cancer 2020;28(2):589–98. Available from: https://doi.org/10.1007/s00520-019-04845-7.

[66] Geyer CE, Forster J, Lindquist D, et al. Lapatinib plus capecitabine for HER2-positive advanced breast cancer. N Engl J Med 2006;355:2733–43. Available from: https://doi.org/10.1056/NEJMoa064320.

[67] Verma S, Miles D, Gianni L, et al. Trastuzumab emtansine for HER2-positive advanced breast cancer. N Engl J Med 2012;367(19):1783–91. Available from: https://doi.org/10.1056/NEJMoa1209124.

[68] Modi S, Saura C, Yamashita T, et al. DESTINY-Breast01 investigators. Trastuzumab deruxtecan in previously treated HER2-positive breast cancer. N Engl J Med 2020;382(7):610–21. Available from: https://doi.org/10.1056/NEJMoa1914510.

[69] Hurvitz SA, Hegg R, Chung WP, et al. Trastuzumab deruxtecan versus trastuzumab emtansine in patients with HER2-positive metastatic breast cancer: updated results from DESTINY-Breast03, a randomised, open-label, phase 3 trial. Lancet 2023;401(10371):105–17. Available from: https://doi.org/10.1016/S0140-6736(22)02420-5.

[70] Curigliano G, Dunton K, Rosenlund M, et al. Patient-reported outcomes and hospitalization data in patients with HER2-positive metastatic breast cancer receiving trastuzumab deruxtecan or trastuzumab emtansine in the phase III destiny-breast03 study. Ann Oncol 2023;34(7):569–77. Available from: https://doi.org/10.1016/j.annonc.2023.04.516.

[71] Curigliano G, Mueller V, Borges V, et al. Tucatinib versus placebo added to trastuzumab and capecitabine for patients with pretreated HER2+ metastatic breast cancer with and without brain metastases (HER2CLIMB): final overall survival analysis. Ann Oncol 2022;33(3):321–9. Available from: https://doi.org/10.1016/j.annonc.2021.12.005.

[72] Frenel JS, Zeghondy J, Guérin-Charbonnel C, et al. Tucatinib Combination Treatment After Trastuzumab-Deruxtecan in Patients With ERBB2-Positive Metastatic Breast Cancer. JAMA Netw Open 2024;7(4):e244435. Available from: https://doi.org/10.1001/jamanetworkopen.2024.4435.

[73] Kabraji S, Ni J, Sammons S, et al. Preclinical and clinical efficacy of trastuzumab deruxtecan in breast cancer brain metastases. Clin Cancer Res 2023;29(1):174–82. Available from: https://doi.org/10.1158/1078-0432.CCR-22-1138.

[74] Statler AB, Hobbs BP, Wei W. Real-world treatment patterns and outcomes in HR+/HER2+ metastatic breast cancer patients: a national cancer database analysis. Sci Rep 2019;9(18126):9. Available from: https://doi.org/10.1038/s41598-019-54402-9.

[75] Hua X, Bi XW, Zhao JL, et al. Trastuzumab plus endocrine therapy or chemotherapy as first-line treatment for patients with hormone receptor-positive and HER2-positive metastatic breast cancer (SYSUCC-002). Clin Cancer Res 2022;28(4):637–45. Available from: https://doi.org/10.1158/1078-0432.CCR-21-3435.

[76] Foulkes WD, Smith IE, Reis-Filho JS. Triple-negative breast cancer. N Engl J Med 2010;363(20):1938–48. Available from: https://doi.org/10.1056/NEJMra1001389.
[77] Lehmann BD, Jovanović B, Chen X, et al. Refinement of triple-negative breast cancer molecular subtypes: implications for neoadjuvant chemotherapy selection. PLoS One 2016;11(6):e0157368. Available from: https://doi.org/10.1371/journal.pone.0157368.
[78] Salgado R, Denkert C, Demaria S, et al. The evaluation of tumor-infiltrating lymphocytes (TILs) in breast cancer: recommendations by an International TILs Working Group 2014. Ann Oncol 2015;26(2):259–71. Available from: https://doi.org/10.1093/annonc/mdu450.
[79] Mittendorf EA, Philips AV, Meric-Bernstam F, et al. PD-L1 expression in triple-negative breast cancer. Cancer Immunol Res 2014;2(4):361–70. Available from: https://doi.org/10.1158/2326-6066.CIR-13-01277.
[80] Cortes J, Rugo HS, Cescon DW, et al. Pembrolizumab plus chemotherapy in advanced triple-negative breast cancer. N Engl J Med 2022;387(3):217–26. Available from: https://doi.org/10.1056/NEJMoa2202809.
[81] De Marchi P, Leal LF, Duval da Silva V, et al. PD-L1 expression by tumor proportion score (TPS) and combined positive score (CPS) are similar in non-small cell lung cancer (NSCLC). J Clin Pathol 2021;74(11):735–40. Available from: https://doi.org/10.1136/jclinpath-2020-206832.
[82] Rozenblit M, Huang R, Danziger N, et al. Comparison of PD-L1 protein expression between primary tumors and metastatic lesions in triple negative breast cancers. J Immunother Cancer 2020;8(2):e001558. Available from: https://doi.org/10.1136/jitc-2020-001558.
[83] Brahmer JR, Long GV, Hamid O, et al. Safety profile of pembrolizumab monotherapy based on an aggregate safety evaluation of 8937 patients. Eur J Cancer 2024;199:113530. Available from: https://doi.org/10.1016/j.ejca.2024.113530.
[84] Saito Y, Sato J, Takeshima T, et al. Transient worsening of pain after administration of immune checkpoint inhibitors – a case series. In Vivo 2024;38(2):944–8. Available from: https://doi.org/10.21873/invivo.13524.
[85] Hahnen E, Hauke J, Engel C, et al. Germline mutations in triple-negative breast cancer. Breast Care (Basel) 2017;12(1):15–9. Available from: https://doi.org/10.1159/000455999.
[86] Bardia A, Hurvitz SA, Tolaney SM, et al. ASCENT clinical trial investigators. Sacituzumab govitecan in metastatic triple-negative breast cancer. N Engl J Med 2021;384(16):1529–41. Available from: https://doi.org/10.1056/NEJMoa2028485.
[87] Loibl S, Loirat D, Tolaney SM, et al. Health-related quality of life in the phase III ASCENT trial of sacituzumab govitecan versus standard chemotherapy in metastatic triple-negative breast cancer. Eur J Cancer 2022;178:23–33. Available from: https://doi.org/10.1016/j.ejca.2022.10.003.
[88] Schlam I, Tolaney SM, Tarantino P. How I treat HER2-low advanced breast cancer. Breast 2023;67:116–23. Available from: https://doi.org/10.1016/j.breast.2023.01.005.

Chapter 39

Gynecological tumors

Nevenka Lukovac Janjić
Clinic for Medical Oncology, Institute for Oncology, Clinical Centre of Montenegro, Podgorica, Montenegro

Key learning objectives

- Knowledge of the epidemiology, risk factors, and diagnostics of gynecological cancers.
- Knowledge of the role of systemic therapy, including chemotherapy, immunotherapy, targeted therapy, and radiotherapy in the treatment of gynecological cancers
- Impact of modern systemic treatments on survival, quality of life, and pain in metastatic stages of the disease.

Introduction

Gynecological cancers encompass cancers affecting the cervix uteri, corpus uteri, ovaries, fallopian tubes, vulva, and vagina. According to the GLOBOCAN report, the incidence and mortality vary significantly between countries but contribute to a significant percentage of incidence and mortality among cancer-related deaths among women worldwide [1]. This chapter will evaluate three of the most common gynecological cancers—cervical, endometrial, and ovarian—in more detail, examining potential treatment options and the most common pain syndromes associated with them, as well as strategies for managing these conditions.

Cervical cancer

Cervical cancer is the fourth most common cancer in women worldwide, with about 660,000 new cases in 2022. In the same year, about 94% of the 350,000 deaths caused by cervical cancer occurred in low- and middle-income countries. The highest incidence and mortality rates from cervical cancer are in sub-Saharan Africa, Central America, and Southeast Asia. Regional differences in the burden of cervical cancer are associated with inequalities in access to vaccination, screening, and treatment services, and risk factors including HIV prevalence [2].

The World Health Organization (WHO) recognizes three categories of epithelial tumors of the cervix: squamous, glandular (adenocarcinoma), and other epithelial tumors, including adenosquamous carcinoma, neuroendocrine tumors, and undifferentiated carcinoma. About 70%–80% of cervical cancer is squamous cell carcinoma, while adenocarcinoma accounts for 20%–25%. Other forms of cervical cancer are rare [2].

Human papillomavirus (HPV) infection is the most important etiological factor of cervical cancer. HPV 16/18 serotypes cause at least two-thirds of cervical cancer. HPV vaccines have a major impact on cervical and HPV-related cancers in men and women and could potentially lead to an annual reduction of 90% in cervical cancer, 85% in vaginal cancer, 87% in vulvar cancer, 92% in anal cancer, and 85% in penile cancers [3].

Cervical cancer is clinically staged based on tumor size, vaginal or parametrial involvement, bladder/rectal infiltration, and the presence of distant metastases. The diagnostic algorithm involves the use of computed tomography (CT) to detect pathological lymph nodes, while magnetic resonance imaging (MRI) can determine tumor size, degree of stroma penetration, parametrial involvement, vaginal extension, and corpus extension [4]. Positron emission tomography(PET)/CT has the potential to accurately determine the extent of disease, particularly in lymph nodes that are not macroscopically enlarged and in distant places, with high sensitivity and specificity [5].

Tumor risk assessment includes tumor size, stage, depth of tumor invasion, lymph node status, lymphovascular space invasion (LVSI), and histological subtype. The status and number of lymph nodes are the most important prognostic factors [6].

Pain Management in Palliative Care for Patients With Cancer. DOI: https://doi.org/10.1016/B978-0-443-30050-9.00037-0

The primary treatment for cervical cancer is surgical treatment, depending on the stage of the disease, according to the International Federation of Gynecology and Obstetrics (FIGO) and the tumor, node, metastasis (TNM) classification. The earliest stage of microinvasive cervical cancer (stage IA1) without LVSI can be successfully treated with conization or simple preservation trachelectomy for fertility [7]. A simple hysterectomy may be offered if the patient does not want to preserve fertility. In stage IA1 with LVSI, surgical evaluation of pelvic lymph nodes should be discussed with the patient, including the sentinel lymph node (SLN).

Five-year relative survival rates are based on statistical information from the database administered by the National Cancer Institute. The database does not group cancers according to FIGO stages but instead categorizes them into localized, regional, and distant stages, with 5-year survival rates of 91% for localized, 62% for regional, and around 19% for distant stages [8].

In patients with FIGO stages IA2, IB, and IIA, radical hysterectomy with bilateral lymph node dissection (with or without SLN) is the standard treatment if the patient does not want to preserve fertility. Open abdominal radical hysterectomy is generally preferred/standard for most patients requiring radical hysterectomy, while minimally invasive approaches require careful selection, counseling, and adherence to current gynecologic-oncology guidelines [9]. As radiotherapy (RT) and surgery are equally effective in early stages, surgery should only be considered in patients with earlier stages (up to FIGO IIA) [9,10]. Chemoradiotherapy is the treatment of choice for locally advanced cervical cancer and has been the standard for patients with FIGO IB2-IVA disease for nearly two decades [11].

The gold standard for the treatment of metastatic or recurrent programmed death-ligand 1 (PD-L1)-negative cervical cancer is the doublet of cisplatin–paclitaxel with bevacizumab, a monoclonal antibody that inhibits vascular endothelial growth factor, a key parameter of tumor angiogenesis in the endothelium. This milestone was due to GOG protocol 240 meeting its primary end point with the chemotherapy plus bevacizumab arm resulting in significantly improved overall survival (OS) compared with chemotherapy alone (16.8 vs 13.3 months; hazard ratio [HR], 0.765; P = .0068) [12]. Improvements in OS and progression-free survival (PFS) attributed to the incorporation of bevacizumab into the treatment of advanced cervical cancer were not accompanied by any significant deterioration in health-related quality of life. Patients responding to antiangiogenesis therapy who maintain an acceptable quality of life could be suitable for progression to treatment with other novel therapies that might confer additional benefit [13].

However, immunotherapy has led to changes in the standard of the treatment of this disease, so today the combination of pembrolizumab + paclitaxel and cisplatin/carboplatin with bevacizumab is the preferred first-line regimen in metastatic or recurrent cervical cancer, particularly for PD-L1-positive tumors [14]. A phase III KEYNOTE-826 study evaluated the efficacy and safety of adding pembrolizumab versus placebo to platinum-based chemotherapy, with or without bevacizumab, as first-line treatment for persistent, recurrent, or metastatic cervical cancer. In the PD-L1 combined positive score (CPS) ≥1 (n = 548), all-comer (n = 617), and CPS ≥10 (n = 317) populations, median OS with pembrolizumab-chemotherapy versus placebo-chemotherapy was 28.6 versus 16.5 months (HR for death, 0.60 [95% confidence interval (CI), 0.49–0.74]), 26.4 versus 16.8 months (HR, 0.63 [95% CI, 0.52–0.77]), and 29.6 versus 17.4 months (HR, 0.58 [95% CI, 0.44–0.78]), respectively. The incidence of grade 3 or higher adverse events was 82.4% with pembrolizumab-chemotherapy and 75.4% with placebo-chemotherapy. These results show that pembrolizumab plus chemotherapy, with or without bevacizumab, continued to provide clinically meaningful improvements in OS for patients with persistent, recurrent, or metastatic cervical cancer. The objective response rate (ORR) remained higher in the pembrolizumab group (ORR, 68.5% vs 50.9%), with a greater number of complete responses to therapy (CR, 25.6% vs 14.5%). The median depth of response in the pembrolizumab group was 19.2 months (1.31–40.91), longer compared to 10.4 months (1.51–40.71) in the chemotherapy group. The addition of pembrolizumab to chemotherapy, with or without bevacizumab, did not negatively affect health-related quality of life. The significant response rate holds promise that a reduction in tumor burden could lead to improved pain control [15].

Endometrial cancer

In developed countries, endometrial cancer is more common than cervical cancer, but viewed globally, it is in second place. Endometrial cancer is the second most common and the fourth leading cause of death worldwide among gynecological malignancies, as recorded in 2020 [16]. The global incidence reported in 2020 was 417,367, accounting for about 2.2% of newly diagnosed cancer cases in that year, making it the sixth most commonly diagnosed cancer in the world. In 2020, approximately 97,370 deaths were attributed to endometrial cancer, accounting for 1% of all cancer deaths [16].

The incidence of endometrial cancer has been increasing in recent years and is estimated to increase by 50% worldwide by 2040. Incidence and mortality vary between countries. It can be seen that the incidence is higher in high-income or developed countries than in low- and middle-income countries [16,17].

The precise etiology of endometrial cancer is unknown, but prolonged exposure to estrogen is thought to be a significant causative factor. Therefore, women who have early menarche and delayed menopause have a high risk of developing endometrial cancer. Other risk factors include nulliparity, obesity, comorbidities such as hypertension and diabetes, estrogen-only hormone replacement therapy, and genetic factors such as Lynch syndrome and Cowden syndrome. The highest incidence occurs in the 55–65-year age group. The most common histological form is the endometrioid type, which accounts for 80% of all cases of endometrial cancer. Less common subtypes are mucinous carcinoma, serous carcinoma, clear cell carcinoma, and squamous cell carcinoma [18–20].

The diagnosis of endometrial cancer is made on the basis of clinical and gynecological examination, including transvaginal ultrasound. Depth of myometrial invasion and cervical stromal invasion are important aspects of endometrial cancer staging. Dynamic contrast-enhanced MRI is an excellent method for assessing these features, with an accuracy of up to 98% and 90% for the assessment of myometrial and cervical stromal invasion, respectively [21]. Abdominal and thoracic CT scans should be considered to investigate the presence of extrapelvic disease. Fluorodeoxyglucose (FDG)-PET-CT shows high specificity and positive predictive value for detecting distant metastases [22].

Surgery is the primary method of treatment for most localized tumors. Total hysterectomy with bilateral salpingo-oophorectomy with or without lymph node dissection is the most commonly performed operation. Postoperative external RT or vaginal brachytherapy depends on the stage, degree, and LVSI of the tumor [23].

The systemic therapy recommendation includes immunotherapy and chemotherapy-based combination regimens as preferred, first-line options for recurrent disease. The NRG-GY018, a randomized, phase III trial, evaluated the benefits of pembrolizumab when combined with a carboplatin/paclitaxel regimen in 816 patients with stage III or IVA endometrial carcinoma with measurable disease, or stage IVB or recurrent disease of any histologic subtype, except for carcinosarcoma [24]. The patients were stratified based on the mismatch repair (MMR)-deficient (dMMR) or MMR-proficient (pMMR) status of their tumors. The 1-year PFS was 74% versus 38% in the dMMR cohort for the triplet regimen versus the chemotherapy arm, respectively (HR, 0.30; 95% CI, 0.19–0.48; $P<.001$). In pMMR tumors, the median PFS was 13.1 months in the pembrolizumab arm versus 8.7 months in the chemotherapy arm (HR, 0.54; 95% CI, 0.41–0.71; $P<.001$) [24]. Another phase III, randomized trial (RUBY) showed benefits of adding dostarlimab to the carboplatin/paclitaxel regimen in 494 patients with stage III or IV or recurrent disease, including all histologies. At 24 months, OS was 71.3% versus 56% (HR, 0.69; 95% CI, 0.54–0.89) in the dostarlimab arm versus the chemotherapy arm, respectively. Significantly more benefits were observed in patients with dMMR/microsatellite instability-high (MSI-H) tumors with PFS of 61.4% versus 15.7% (HR, 0.32; 95% CI, 0.17–0.63; $P<.001$) in the triplet versus the doublet therapy arms, respectively. Based on the results from the NRG-GY018 and RUBY trials, the National Comprehensive Cancer Network Panel has added pembrolizumab/carboplatin/paclitaxel and dostarlimab/carboplatin/paclitaxel as category 1, preferred, first-line treatment options for recurrent endometrial carcinoma. The addition of immunotherapy provides an option for long-term remission or even cure in some patients, while the response rate hovers around 70% and offers significant potential to reduce pain in those patients whose pain is caused by a tumor mass [25,26].

Ovarian cancer

Ovarian cancer was the third most common gynecological cancer in the world in 2020. Ovarian cancer is often diagnosed at a late stage, making this malignancy the deadliest gynecological cancer. The prognosis of ovarian cancer is usually poor, with a 5-year survival rate of only 17% for patients at an advanced stage [27]. The risk of ovarian cancer increases in women with age, along with family history, familial cancer syndrome, and breast cancer susceptibility gene (BRCA) mutations [28]. Reproductive risk factors for ovarian cancer include having children later or never having a full-term pregnancy, reaching menopause at an older age, and taking hormone therapy after menopause [29].

At the time of diagnosis, most women have advanced disease [30]. Despite surgery and chemotherapy, more than 70% of patients experience progression. Epithelial ovarian cancer (EOC) accounts for up to 95% of ovarian cancers [31,32]. EOC is further classified into five distinct histopathologic subtypes: low-grade serous, high-grade serous, mucinous, clear cell, and endometrioid ovarian carcinomas. High-grade serous ovarian carcinoma (HGSOC) is the most common subtype of EOC, accounting for over 70% of EOC. Different histological types of EOC have a different cellular origin, a different mutational spectrum, and therefore a different prognosis. Even within one histological type, different molecular subtypes with different prognoses can be found [33-35].

There is still a lack of reliable diagnostic methods that can detect disease early and are suitable for screening [36]. For the past two decades, the conventional treatment for EOC has been optimal cytoreductive surgery and platinum-based chemotherapy. Although approximately 80% of patients respond to first-line chemotherapy, relapse rates and chemo-resistance are still high [37,38]. Staging classifications are used to describe the extent of disease spread and to determine treatment and prognosis. Ovarian cancer is typically staged according to the FIGO system, American Joint Committee on Cancer, or TNM classification. The FIGO staging system is the most significant prognostic indicator [39,40].

Adjuvant platinum-based chemotherapy significantly prolongs OS and PFS in patients with early-stage EOC and is generally recommended for FIGO stages I–IIB, either paclitaxel–carboplatin or carboplatin alone (six cycles) [41]. For patients receiving paclitaxel–carboplatin, a minimum of three cycles is recommended, except for high-grade serous carcinoma (HGSC)/EOC or any stage IC-II, regardless of histotype, for which six cycles are suggested. Systemic chemotherapy with paclitaxel + carboplatin or cisplatin is used for advanced stages after maximum cytoreduction or without it, with the inclusion of antiangiogenic therapy (bevacizumab) with chemotherapy and in maintenance therapy [41].

Two phase 3 randomized trials, GOG-0218 and ICON7, tested the effects of adding bevacizumab during first-line platinum-based combination chemotherapy and as single-agent maintenance therapy after first-line chemotherapy [42,43]. Results across both studies showed that patients with features associated with poor prognosis tend to derive a greater benefit from the addition of bevacizumab [42]. Interim analyses of data from GOG 0218 showed that bevacizumab improved OS in patients with stage IV disease and in patients with ascites, another high-risk group; however, the final analyses did not show a correlation between OS and ascites in the bevacizumab-containing arms. For ICON7, although after long-term follow-up there were no significant effects of bevacizumab on PFS or OS for the total population, subgroup analyses identified a high-risk group for which bevacizumab improved both PFS (median PFS for arm 1 vs arm 2: 10.5 vs 16.0 months; HR, 0.73 [95% CI, 0.61–0.88]; $P = .001$) and OS (median OS for arm 1 vs arm 2: 30.2 vs 39.7 months; HR, 0.78 [95% CI, 0.63–0.97]; $P = .03$) [43]. This high-risk group included those with either stage IV, inoperable stage III, or suboptimally debulked (residual disease >1 cm) stage III. Exploratory analyses suggest that stage may be more important than the extent of residual disease for identifying patients who may benefit from bevacizumab. Response rates in the bevacizumab group were almost 70%, indicating that significant tumor control and potential for pain reduction could be achieved with this combination [43,44].

Up to 50% of HGSC/EOC are detected as homologous recombination deficiency (HRD)-positive by current tests. This includes 15%–20% of patients with a germline BRCA1/2 mutation. BRCA1/2 mutations and HRD are pathways involved in homologous recombination repair of DNA double-strand breaks. These mutations cause the cell to lose the ability to repair DNA damage caused by external insults, specifically those to the double helix, resulting in a condition defined as HRD. This occurrence favors tumor initiation, growth, and evolution. HRD positivity, with or without BRCA1/2 mutations, is a well-established predictive factor of response to poly (ADP-ribose) polymerase (PARP) inhibitors. The incorporation of PARP as maintenance therapy after first-line chemotherapy, alone or in combination with bevacizumab, has ushered in a new era in the first-line treatment of advanced HGSOC/EOC [44,45].

PARP inhibitors were approved after trials such as SOLO-1 demonstrated a remarkable improvement in PFS with single-agent olaparib versus placebo as maintenance therapy for patients with a germline or somatic BRCA1/2 mutation who had a CR/PR after first-line platinum-based chemotherapy. The risk of progression or death was 70% lower, with a median PFS of 13.8 months for the placebo group, and the median PFS for olaparib had not been reached after a median follow-up of 41 months; OS data are also immature [46].

The phase III PAOLA-1/ENGOT-ov25 trial evaluated the addition of maintenance olaparib to bevacizumab in patients with newly diagnosed, advanced ovarian cancer who were candidates to receive bevacizumab in combination with first-line platinum-based chemotherapy and who were unselected by biomarker or surgical status [47]. Maintenance olaparib plus bevacizumab provided a significant PFS benefit compared with placebo plus bevacizumab in the overall PAOLA-1 population (median PFS, 22.1 vs 16.6 months; HR, 0.59; 95% CI, 0.49–0.72; $P < .001$). In prespecified subgroup analyses, the greatest PFS benefit occurred with olaparib plus bevacizumab versus placebo plus bevacizumab in patients who tested positive for HRD (median PFS, 37.2 vs 17.7 months; HR, 0.33; 95% CI, 0.25–0.45) and in patients with a tumor BRCAm (median PFS, 37.2 vs 21.7 months; HR, 0.31; 95% CI, 0.20–0.47). While maintenance therapy does not directly reduce pain, in some patients it will help prevent the tumor recurrence, which in turn will prevent the need for analgesic therapy [47].

Pain syndrome in gynecological cancers

Patients diagnosed with advanced gynecological malignancies, including cervical cancer, endometrial cancer, and ovarian cancer, experience varying degrees of physical pain. This pain significantly impacts the psychological and physiological well-being of patients, ultimately compromising the efficacy of treatment. Consequently, the appropriate diagnosis and management of pain are of utmost importance to improve patient outcomes and reduce suffering [48].

The pain syndromes that accompany gynecological cancer most often arise as a consequence of the growth of the primary tumor or metastases but can also occur as a result of treatment (surgery, chemotherapy, immunotherapy, hormonal therapy, or RT). Finally, patients with cancer may develop pain that is entirely unrelated to the cancer or its treatment [49,50].

Malignant ascites is defined as the accumulation of fluid containing cancer cells in the abdomen. The pathophysiology of ascitic fluid accumulation is multifactorial and may result from the obstruction of veins and lymphatic vessels, oncotic shifts, or increased vascular permeability due to the diffuse release of cytokines [51]. Malignant ascites may also arise as a result of direct invasion of the peritoneum (peritoneal carcinomatosis). Ovarian cancer is the most common gynecological malignancy that causes ascites. Large amounts of ascites in patients with EOC usually indicate the presence of peritoneal metastases; therefore, ascites is found in most patients (89%) with advanced disease (FIGO stages III and IV). However, the absence of ascites may not exclude malignancy, since ascites is rarely (17%) seen in early disease (FIGO stages I and II) and is absent in nearly half of borderline tumors. Unlike primary OC, recurrent disease is not strongly associated with ascites and is found in 38% of patients with recurrent OC [52]. Patients with malignant ascites have abdominal distension, pain, dyspnea, nausea, anorexia, and vomiting. In more than 50% of patients diagnosed with malignant ascites, this fluid accumulation has caused the first signs and symptoms of their underlying malignancy. Due to the numerous potential etiologies that can cause ascites, paracentesis is necessary to determine if a malignant source is present. In these cases, cytology is diagnostic in 50%–60% of cases [52].

Treatment strategies for malignant ascites aim to alleviate symptoms, similar to those for patients with nonmalignant ascites. Initial interventions include the use of diuretics and therapeutic paracentesis [52]. When these measures are unsuccessful, a peritoneal drain can be placed, with relief in 83%–100% of cases [53]. Peritoneovenous shunt (PVS) involves one-way valves that allow flow from the peritoneum to the vena cava and is a more invasive alternative to extraperitoneal drainage. Although few data are comparing the two interventions, PVS has been shown to have a higher patency rate. There is a theoretical risk of disseminated intravascular coagulation in patients with PVS due to shunting of ascitic proteins/tissue factor, in the vena cava [54].

Malignant bowel obstruction (MBO) is broadly defined as mechanical, functional, and/or radiological obstruction of the intestinal tract in the presence of known primary or metastatic intraabdominal, retroperitoneal, or pelvic malignancy. Globally, the prevalence of MBO is estimated to range from 3% to 15% in cancer patients, although the exact extent is difficult to quantify due to the heterogeneity of definitions and available data. Primary malignancies of the colon, ovary, stomach, pancreas, and uterus are most often associated with MBO, with metastases of extra abdominal origin playing a more minor role [55]. Retrospective series report that the prevalence of MBO in gynecological malignancies is high and occurs in about 28%–51%. MBO in women with advanced gynecologic cancer is common and a major clinical challenge, as EOC is the dominant cause of MBO [56].

MBO can generally be characterized by a clinical syndrome with nausea, vomiting, and abdominal pain in the presence of known primary and/or metastatic malignancy in the abdomen, with or without correlating radiographic evidence of obstruction. From the report of the Clinical Protocol Subcommittee of the International Conference on MBO on the establishment of clinical trials for MBO, the following definition was proposed to ensure an accurate diagnosis for comparison of outcomes: (1) clinical evidence of bowel obstruction by history, physical examination, or radiographic examination; (2) bowel obstruction beyond the ligament of Treitz; (3) intraabdominal primary cancer with incurable disease; or (4) nonintraabdominal primary cancer with clear intraperitoneal disease [56]. However, the exact definition varies in the literature, contributing to the diagnostic and therapeutic challenges associated with this condition [55,57].

The pathophysiology of MBO may be directly related to extraluminal compression and/or intraluminal obstruction by a tumor mass, benign etiologies (such as adhesions or scars), motility disorders (e.g., tumor infiltration of the mesentery or pharmacological side effects), or treatment consequences of the underlying malignancy (radiation-induced enteritis). Often, the etiology is multifactorial, making it challenging to predict the true causes of MBO accurately. Due to the heterogeneity in presentation, causes, and patient factors among those with MBO, therapeutic options must be tailored to each patient and their specific treatment goals [58]. It is important to note that 36%–49% of MBO resolve with aggressive medical treatment alone; however, a significant proportion of them will recur (72%) [55].

Pharmacologic therapies are often used to relieve symptoms of MBO, primarily nausea, vomiting, abdominal pain, and dehydration. They are broadly categorized into analgesics (e.g., opioids), antiemetics, prokinetic agents, anticholinergics, and antisecretory agents. Granisetron (a 5-HT3 antagonist) has been studied in a clinical trial, and the findings suggest a significant reduction in the severity of nausea and frequency of vomiting when used in combination with dexamethasone and haloperidol. Systemic corticosteroids in the treatment of acute MBO, as reported in the Cochrane database, revealed a nonsignificant trend suggesting that intravenous dexamethasone may contribute to the resolution of MBO without affecting OS. The prokinetic agent metoclopramide has been shown to be effective in relieving nausea and vomiting, as well as restoring intestinal transit, in partial MBO. However, it is currently not recommended for use in complete obstruction due to the risk of ischemia and perforation [58,59].

MBO is the most common cause of surgical consultation in the palliative setting [60]. A variety of surgical procedures may be offered depending on the preoperative and intraoperative assessment, including colostomy/ileostomy creation, bypass with primary anastomosis, placement of a gastrostomy tube for ventilation, and/or resection of the disease.

Bone metastasis occurs rarely and is reported in the literature with an incidence ranging from 2% to 6% [61]. Generally, bone metastasis occurs in the axial skeleton and is a significant contributor to the cancer pain in patients with metastatic cancer [62]. Metastatic cancer–induced bone pain is caused by inflammation and bone remodeling or by damage to or dysfunction of the nervous system due to tumor invasion [63,64]. Bone pain caused by metastases is most often a combination of neuropathic and nociceptive pain [64,65]. The details about bone metastasis and the treatment algorithm are given in other chapters of this book.

Fungating tumors are associated with advanced cancers, including primary, secondary, and recurrent malignant diseases. They are characterized by rapid proliferation, ulceration, necrosis, and skin erosion, causing pain, hemorrhage, excessive drainage, and malodor [66]. Primarily, cervical carcinoma spreads through a direct local extension to paracervical tissue and pelvic organs (vagina, pelvic wall, urinary bladder, and rectum) and lymphatic embolization [66,67]. The probable mechanism of cutaneous metastasis is an alteration of the lymphatic flow. The deep lymphatic pathways are obstructed by the tumor, resulting in the shunting of lymphatic flow to the cutaneous lymphatics [66,68]. Skin metastasis is usually associated with generalized dissemination, locoregional recurrence, and a rapidly progressive disease process that signals poor prognosis, with an average reported survival of 3–8.5 months. Only palliative treatment can be given in the form of the extirpation of skin lesions, followed by RT or chemotherapy [69,70].

Herpes zoster occurs when the varicella-zoster virus reactivates from its dormant state during periods of reduced cell-mediated immunity, causing a characteristic rash and substantial pain that can significantly impact quality of life. Cancer patients may experience cell-mediated immunosuppression, resulting from chemotherapy, psychological stress, or physical trauma of surgery or RT, putting them at a greater risk of herpes zoster. Postherpetic neuralgia (PHN) is the most common complication of herpes zoster. Treatments target symptom control but are often inadequate at relieving pain and are ineffective in a significant number of patients with PHN [71].

Patients with gynecological cancer may have a high risk of polymicrobial infections due to the presence of physiological microflora resulting from the anatomical proximity of the lower intestine and urinary tract. In certain infections, such as deep tissue abscess, cellulitis in patients with a chronic fistular tract, the presence of a large necrotic tumor, and a history of multiple recent infections, the likelihood of polymicrobial infection remains high [72].

Neoplastic plexopathy usually develops as a late-stage complication of cancer, usually when the tumor has locally or regionally advanced. Cervical, brachial, and lumbosacral plexus (LSP) are the most frequently encountered neoplastic plexopathies in clinical practice [73]. Neoplasms are the most common cause of lumbar plexopathy. Plexopathy is part of the initial presentation of cancer in approximately 15% of patients. Involvement of the plexus by cancer is seen in about 1% of cancers [73]. LSP is located near many abdominal and pelvic organs, including the bladder, colon, ovaries, and cervix, and may be involved as an infiltrative or compressive process [74].

The effects of radiation can be seen up to 6 months after the start of radiation [75]. Mechanisms of injury include direct toxic effects on axons and on the vasa nervorum, with secondary microinfarction of the nerve, with dose-dependent escalation [76]. Concurrent chemotherapy may act synergistically to increase the risk of radiation plexopathy. Clinical syndromes may manifest with the involvement of the upper plexus (L1–L4), lumbosacral trunk (L4–L5), and lower plexus (S1–2 and S4). Plexopathy is typically unilateral but may be bilateral in 25% of patients [76]. Pain is the dominant symptom (98%) in lumbosacral plexopathy. LSP usually begins with leg pain, which is soon followed by numbness and weakness. The pain tends to be dull with a superimposed sharp pain in a radicular distribution. The pain is worse when the patient lies on their back but may be aggravated by prolonged movement, sitting, or the Valsalva maneuver. If the iliopsoas muscle is involved, the patient may tend to rest with the hips flexed (as seen in meningitis). Pain may involve the costovertebral region, radiating to the upper thigh, in lesions affecting the upper lumbar plexus. The pain involves the lower back, buttocks, thighs, or hips when the lower lumbar plexus is involved [77].

Lymphedema of the lower extremities is a chronic, often irreversible condition that affects many patients treated for gynecologic malignancies, with reported rates of up to 70% in selected populations. Lymph from the gynecological organs primarily drains into three lymph nodes: the pelvic, para-aortic, and inguinofemoral nodes. These basins are often sampled or completely excised as part of the surgical treatment of gynecological malignancies. Overall, the incidence of treatment-related lymphedema is about 25%, but it can be as high as 70% in some patient populations [78]. There are two consistent contributors to lower extremity lymphedema in these patients: lymphadenectomy and radiation therapy. Lymphadenectomy—defined as the complete excision of a lymph node basin—directly disrupts the normal return of lymphatic fluid from the lower extremities. In general, the risk of lymphedema is proportional to the number of lymph nodes sampled, with excision of specific lymph nodes and lymph node basins considered to pose a higher risk [79]. SLN mapping alone has been shown to reduce the risk of lower extremity lymphedema to less than 10% in gynecologic malignancies. Radiation-induced lymphedema is thought to be secondary to lymph node and lymphatic vessel sclerosis, scarring, and subsequent impedance of upstream lymphatic flow [78].

Chemotherapy-induced peripheral neuropathy (CIPN) can often be registered in gynecological cancer as agents that are used to treat all three mentioned types of gynecological cancers are taxanes (paclitaxel and docetaxel) and platinum salts (cisplatin and carboplatin). All of these can damage the structures of the nervous system and cause a range of neuropathies, including large and small fiber neuropathies, sensory and/or motor neuropathies, and cranial and autonomic neuropathies, collectively referred to as CIPN. The effects of chemotherapy on the nervous system vary among different drug classes, depending on the specific physical and chemical properties of the drug used, as well as its individual or cumulative doses. The prevalence of CIPN depends on the agent, with reported rates highest with platinum-based drugs (70%–100%) and taxanes (11%–87%) [80]. Toxicity can occur either with a high single dose or after cumulative exposure. Recent studies reported that the prevalence of CIPN was highest in the first month after chemotherapy and reduces in prevalence after 3 and 6 months [81].

Clinically, CIPN manifests as a deficit in sensory, motor, and/or autonomic functions of varying intensity. Sensory symptoms typically develop first, affecting the feet and hands, and usually present as a classic "glove and stocking" neuropathy. Symptoms include numbness, tingling, altered sense of touch, impaired vibration, and paresthesias and dysesthesias triggered by touch and exposure to hot or cold temperatures. Painful sensations often occur, including spontaneous burning, shooting, or electric shock-like pain, as well as mechanical or thermal allodynia or hyperalgesia [82]. Sensory disturbances occur more frequently than autonomic symptoms, which usually include orthostatic hypotension, constipation, and altered sexual or urinary function [83].

RT is widely used to treat malignant diseases located in the pelvis [84]. Although RT delivery techniques are continuously improving, external beam RT and brachytherapy can lead to adverse effects depending on the radiation field and the dose applied to the organs. Local complications of the urinary bladder or rectum are commonly reported after pelvic RT [85]. The molecular effects of RT can be divided into immediate cell death and late-onset injury [86]. Clinical symptoms caused by the immediate effects of RT are often self-limiting. However, the delayed molecular impact of RT can lead to organ-specific adverse effects such as bleeding, reduced capacity, ulceration, or ischemia, including RT-induced chronic painful neuropathy, all of which can result in pain. The syndrome encompassing these organ-specific adverse effects after pelvic RT is known as pelvic radiation disease and includes radiation proctitis, cystitis, urethritis, osteitis, lumbar plexopathy, lymphedema, and sexual dysfunction [86]. More detailed information about CIPN and RT effects is given in Chapter 44.

Management of pain syndrome in gynecological cancer

About 70% of patients with advanced gynecological malignancies report some degree of cancer-related pain. This pain may result from primary tumor growth, metastatic disease, compression or invasion of surrounding tissues, and treatment-related complications. Cancer pain is common, often severe, and frequently difficult to manage, requiring ongoing assessment and individualized pharmacologic intervention [87-89].

Pain should be evaluated at the initial examination and reassessed throughout treatment and follow-up. There are various assessment tools, the use of which should depend on the patient's physical condition, cultural background, and level of consciousness [89]. More details are available in Chapter 5.

The nature of pain experienced by patients with gynecological cancer is generally dull and of a moderate degree. The pain predominantly affects the pelvic cavity and adjacent tissues, including the hypogastrium, perineum, and waist. Additionally, several patients experience pain in more than two sites. Cancer-related pain is a burdensome issue for patients and may lead to various adverse outcomes, such as interference with daily activities, hindrance of social function, and impairment of cognitive function [90]. The consequences of cancer-related pain can negatively impact patients' quality of life, leading to pessimism, anxiety, fear, and depression, resulting in patients' societal and economic dependence. Furthermore, pain may affect patients' sleep quality, leading to adverse emotions and reduced disease resistance, which is potentially detrimental to tumor treatment and symptom relief.

The chronification process of pain in gynecological cancer patients is complex and involves the interaction of various mechanisms. Firstly, ongoing tissue damage and inflammation can lead to the sensitization of peripheral nociceptors, resulting in increased pain sensitivity and decreased pain thresholds. This can lead to a self-perpetuating cycle of pain, as continuous stimulation of nociceptors can alter their excitability and trigger the release of pro inflammatory mediators, thereby intensifying the pain response [91,92].

Secondly, the central nervous system (CNS) can undergo neuroplastic changes in response to chronic pain. These changes can lead to altered pain perception, making pain feel more intense and difficult to control. Neuroplastic changes can occur at various levels of the CNS, including the spinal cord and brain, and can involve altered activity in pain-processing regions and disrupted inhibitory pathways [93,94].

Thirdly, psychosocial factors such as anxiety, depression, and stress can also contribute to the chronicification of pain in gynecological cancer patients. These factors can affect pain perception and increase pain sensitivity, as well as impair coping strategies and quality of life. The relationship between pain and psychosocial factors is bidirectional, with pain exacerbating psychosocial symptoms and psychosocial symptoms exacerbating pain [95,96]. Effective pain management strategies in these patients may involve a multifaceted approach that targets peripheral and central sensitization, as well as psychosocial factors, and are paramount for ensuring the well-being of patients [97,98].

Pain management aims to optimize analgesic efficacy while minimizing adverse drug reactions and avoiding inappropriate drug administration, with a shift from an anesthesiologist-based treatment model to a nurse-based care model [99,100]. Pharmacological interventions remain the foundation of clinical management for gynecological cancers, although complementary therapies and other interventions may be acceptable in certain conditions [101,102]. More details about adjuvants or coanalgesics and topical analgesics are given in detail in another chapter of this book [103–105]. Additionally, RT, which is commonly used in gynecological tumors and can achieve pain control in as many as 60%–80% of patients, is evaluated in detail in Chapter 19 of the book [106–110].

Furthermore, while the three-step WHO analgesic method forms the basis of pain management, recommending that treatment begin with nonopioids and then proceeding to opioid-based therapy, nonpharmacological and integrative management therapies, including massage, music, or acupuncture, are also options [111–116]. Such an approach also includes creating a comfortable treatment environment, regular psychoeducation and psychosocial interventions if necessary, health education, and cognitive-behavioral intervention [117–120].

Future development

Multiple innovative pathways are evolving gynecological cancer care into an individualized approach based on a distinct molecular profile. Artificial intelligence and precision medicine are transforming diagnosis and treatment selection, with AI-driven models showing promise in differential diagnosis and molecular profiling to guide personalized therapies. Advanced immunotherapies are expanding beyond checkpoint inhibitors to include chimeric antigen receptor (CAR) T cells (CAR-T) therapies targeting specific tumor antigens, such as mesothelin, in ovarian cancer, although challenges remain in solid tumor applications. Liquid biopsy technologies utilizing multi omics approaches are showing potential for early detection, and nanotechnology-based drug delivery systems are being developed to enhance therapeutic targeting while reducing systemic toxicity. Combination strategies integrating PARP inhibitors with immunotherapy are also showing an interesting potential for ovarian cancers.

Conclusion

Gynecological cancers encompass a significant burden of disease globally, with cervical, endometrial, and ovarian cancers representing major challenges in women's health. The integration of modern therapeutic approaches has transformed treatment outcomes. Immunotherapy has emerged as a game-changer, particularly with pembrolizumab-based combinations showing remarkable improvements in survival for cervical and endometrial cancers. PARP inhibitors have revolutionized ovarian cancer management, especially in patients with homologous recombination deficiencies. Pain management remains a complex process, requiring multimodal approaches that combine pharmacological, interventional, and supportive therapies. This chapter emphasizes that approximately 70% of patients with advanced gynecological malignancies experience varying degrees of cancer pain, significantly impacting quality of life. An enhanced understanding of pain mechanisms and the development of targeted interventions have improved patient outcomes and comfort during treatment.

References

[1] Sung H, Ferlay J, Siegel RL, et al. Global cancer statistics 2020: GLOBOCAN estimates of incidence and mortality worldwide for 36 cancers in 185 countries. CA Cancer J Clin 2021;71:209–49. Available from: https://doi.org/10.3322/caac.21660.

[2] Stelzle D, Tanaka LF, Lee KK, et al. Estimates of the global burden of cervical cancer associated with HIV. Lancet Glob Health 2021;9(2):e161–9. Available from: https://doi.org/10.1016/S2214-109X(20)30459-9.

[3] Serrano B, de Sanjosé S, Tous S, et al. Human papillomavirus genotype attribution for HPVs 6, 11, 16, 18, 31, 33, 45, 52 and 58 in female anogenital lesions. Eur J Cancer 2015;51:1732–41. Available from: https://doi.org/10.1016/j.ejca.2015.06.001.

[4] Wagenaar HC, Trimbos JB, Postema S, et al. Tumor diameter and volume assessed by magnetic resonance imaging in the prediction of outcome for invasive cervical cancer. Gynecol Oncol 2001;82:474–82. Available from: https://doi.org/10.1006/gyno.2001.6267.

[5] Patel CN, Nazir SA, Khan Z. 18F-FDG PET/CT of cervical carcinoma. AJR Am J Roentgenol 2011;196:1225–33. Available from: https://doi.org/10.2214/AJR.10.5084.

[6] Kim SM, Choi HS, Byun JS. Overall 5-year survival rate and prognostic factors in patients with stage IB and IIA cervical cancer treated by radical hysterectomy and pelvic lymph node dissection. Int J Gynecol Cancer 2000;10:305–12. Available from: https://doi.org/10.1046/j.1525-1438.2000.010004305.x.

[7] Dittrich R, Lotz L, Hackl J. Fertilitätserhalt bei Krebserkrankungen. Frauenarzt 2014;55:240–6.

[8] American Cancer Society. Survival rates for cervical cancer, <https://www.cancer.org/cancer/types/cervical-cancer/detection-diagnosis-staging/survival.html>; 2025 [accessed 16.03.2026].

[9] Geetha P, Nair MK. Laparoscopic, robotic and open method of radical hysterectomy for cervical cancer: a systematic review. J Minim Access Surg 2012;8:67–73. Available from: https://doi.org/10.4103/0972-9941.97584.

[10] Whitney CW, Sause W, Bundy BN. Randomized comparison of fluorouracil plus cisplatin versus hydroxyurea as an adjunct to radiation therapy in stage IIB–IVA carcinoma of the cervix with negative para-aortic lymph nodes: a Gynecologic Oncology Group and Southwest Oncology Group study. J Clin Oncol 1999;17:1339–48. Available from: https://doi.org/10.1200/JCO.1999.17.5.1339.

[11] Peters WA 3rd, Liu PY, Barrett RJ 2nd. Concurrent chemotherapy and pelvic radiation therapy compared with pelvic radiation therapy alone as adjuvant therapy after radical surgery in high-risk early-stage cancer of the cervix. J Clin Oncol 2000;18:1606–13. Available from: https://doi.org/10.1200/JCO.2000.18.8.1606.

[12] Tewari KS, Sill MW, Penson RT, et al. Final overall survival of the phase III randomised trial of chemotherapy with and without bevacizumab for advanced cervical cancer: an NRG oncology/gynecologic oncology group study. Lancet. 2017;390(10103):1654–63. Available from https://doi.org/10.1016/S0140-6736(17)31607-0

[13] Tewari KS, Monk BJ. Invasive cervical cancer. In: DiSaia PJ, Creasman WT, editors. Clinical gynecologic oncology. 8th ed. Philadelphia: Mosby; 2012.

[14] Abu-Rustum NR, Campos SM, Amarnath S, et al. Cervical cancer, version 2.2026, NCCN clinical practice guidelines in oncology. J Natl Compr Canc Netw 2025;23(12):549–73. Available from: https://doi.org/10.6004/jnccn.2025.0057.

[15] Monk BJ, Colombo N, Tewari KS, et al. First-line pembrolizumab + chemotherapy versus placebo + chemotherapy for persistent, recurrent, or metastatic cervical cancer: final overall survival results of KEYNOTE-826. J Clin Oncol 2023;41(36):5505–11. Available from: https://doi.org/10.1200/JCO.23.00914.

[16] Torre LA, Islami F, Siegel RL, et al. Global cancer in women: burden and trends. Cancer Epidemiol Biomarkers Prev 2017;26:444–57. Available from: https://doi.org/10.1158/1055-9965.EPI-16-0858.

[17] Tanvir I, Riaz S, Hussain A, et al. Hospital-based study of epithelial malignancies of endometrial cancer frequency in Lahore, Pakistan, and common diagnostic pitfalls. Pathol Res Int 2014;2014:179384. Available from: https://doi.org/10.1155/2014/179384.

[18] Creasman WT, Odicino F, Maisonneuve P, et al. Carcinoma of the Corpus uteri. Int J Gynaecol Obstet 2006;95(Suppl. 1):S105–43. Available from: https://doi.org/10.1016/S0020-7292(06)60031-3.

[19] Brooks RA, Fleming GF, Lastra RR, et al. Current recommendations and recent progress in endometrial cancer. CA Cancer J Clin 2019;69(4):258–79. Available from: https://doi.org/10.3322/caac.21561.

[20] Feinberg J, Albright B, Black J, et al. Ten-year comparison study of type 1 and 2 endometrial cancers: risk factors and outcomes. Gynecol Obstet Invest 2019;84(3):290–7. Available from: https://doi.org/10.1159/000493132.

[21] Peungjesada S, Bhosale PR, Balachandran A. Magnetic resonance imaging of endometrial carcinoma. J Comput Assist Tomogr 2009;33:601–8. Available from: https://doi.org/10.1097/RCT.0b013e31818d4279.

[22] Gee MS, Atri M, Bandos AI. Identification of distant metastatic disease in uterine cervical and endometrial cancers with FDG PET/CT: analysis from the ACRIN 6671/GOG 0233 multicenter trial. Radiology 2018;287:176–84. Available from: https://doi.org/10.1148/radiol.2017170963.

[23] Oaknin A, Bosse TJ, Creutzberg CL, et al. Endometrial cancer: ESMO clinical practice guideline for diagnosis, treatment and follow-up. Ann Oncol 2022;33(9):860–77. Available from: https://doi.org/10.1016/j.annonc.2022.05.009.

[24] Eskander RN, Sill MW, Beffa L, et al. Pembrolizumab plus chemotherapy in advanced endometrial cancer. N Engl J Med 2023;388(23):2159–70. Available from: https://doi.org/10.1056/NEJMoa2302312.

[25] Powell MA, Bjørge L, Willmott L, et al. Overall survival in patients with endometrial cancer treated with dostarlimab plus carboplatin–paclitaxel in the randomized ENGOT-EN6/GOG-3031/RUBY trial. Ann Oncol 2024;35(8):728–38. Available from: https://doi.org/10.1016/j.annonc.2024.05.546.

[26] Abu-Rustum NR, Campos SM, Amarnath S, et al. NCCN Guidelines® Insights: Uterine Neoplasms, Version 3.2025. J Natl Compr Canc Netw. 2025;23(8):284–91. Available from: https://doi.org/10.6004/jnccn.2025.0038.

[27] Baldwin LA, Huang B, Miller RW, et al. Ten-year relative survival for epithelial ovarian cancer. Obstet Gynecol 2012;120:612–8. Available from: https://doi.org/10.1097/AOG.0b013e318264f794.

[28] Momenimovahed Z, Tiznobaik A, Taheri S, et al. Ovarian cancer in the world: epidemiology and risk factors. Int J Womens Health 2019;11:287–99. Available from: https://doi.org/10.2147/IJWH.S197604.

[29] Salehi F, Dunfield L, Phillips KP, et al. Risk factors for ovarian cancer: an overview with emphasis on hormonal factors. J Toxicol Environ Health B Crit Rev 2008;11:301–21. Available from: https://doi.org/10.1080/10937400701876095.

[30] Torre LA, Trabert B, DeSantis CE, et al. Ovarian cancer statistics, 2018. CA Cancer J Clin 2018;68:284–96. Available from: https://doi.org/10.3322/caac.21456.

[31] Prat J. Ovarian carcinomas: five distinct diseases with different origins, genetic alterations, and clinicopathological features. Virchows Arch 2012;460:237–49. Available from: https://doi.org/10.1007/s00428-012-1203-5.

[32] Lheureux S, Braunstein M, Oza AM. Epithelial ovarian cancer: evolution of management in the era of precision medicine. CA Cancer J Clin 2019;69:280–304. Available from: https://doi.org/10.3322/caac.21559.

[33] Pentheroudakis G, Pavlidis N. Serous papillary peritoneal carcinoma: unknown primary tumour, ovarian cancer counterpart or a distinct entity? A systematic review. Crit Rev Oncol Hematol 2010;75:27–42. Available from: https://doi.org/10.1016/j.critrevonc.2009.10.003.

[34] Bignotti E, Tassi RA, Calza S, et al. Gene expression profile of ovarian serous papillary carcinomas: identification of metastasis-associated genes. Am J Obstet Gynecol 2007;196:245.e1–11. Available from: https://doi.org/10.1016/j.ajog.2006.10.874.

[35] Kujawa KA, Lisowska KM. Ovarian cancer–from biology to clinic. Postepy Hig Med Dosw (Online) 2015;69:1275–90. Available from: https://doi.org/10.5604/17322693.1184451.

[36] Coleman RL, Monk BJ, Sood AK, et al. Latest research and treatment of advanced-stage epithelial ovarian cancer. Nat Rev Clin Oncol 2013;10:211. Available from: https://doi.org/10.1038/nrclinonc.2013.5.

[37] Cortez AJ, Tudrej P, Kujawa KA, et al. Advances in ovarian cancer therapy. Cancer Chemother Pharmacol 2018;81:17–38. Available from: https://doi.org/10.1007/s00280-017-3501-8.

[38] Lalwani N, Prasad SR, Vikram R, et al. Histologic, molecular, and cytogenetic features of ovarian cancers: implications for diagnosis and treatment. Radiographics 2011;31:625–46. Available from: https://doi.org/10.1148/rg.313105066.

[39] Colombo N, Sessa C, du Bois A, et al. ESMO-ESGO consensus conference recommendations on ovarian cancer: pathology and molecular biology, early and advanced stages, borderline tumours and recurrent disease. Gynecol Oncol 2019;30(5):672–705. Available from: https://doi.org/10.1093/annonc/mdz062.

[40] Chan JK, Tian C, Fleming GF, et al. The potential benefit of 6 vs. 3 cycles of chemotherapy in subsets of women with early-stage high-risk epithelial ovarian cancer: an exploratory analysis of a Gynecologic Oncology Group study. Gynecol Oncol 2010;116(3):301–6. Available from: https://doi.org/10.1016/j.ygyno.2009.10.073.

[41] NCCN. NCCN Clinical practice guidelines in oncology: ovarian cancer (epithelial ovarian cancer/fallopian tube cancer/primary peritoneal cancer), Version 3.2025. Fort Washington (PA): NCCN. Available from: https://www.nccn.org/professionals/physician_gls/pdf/ovarian.pdf [accessed 05.12.25].

[42] Burger RA, Brady MF, Bookman MA, et al. Incorporation of bevacizumab in the primary treatment of ovarian cancer. N Engl J Med 2011;365:2473–83. Available from: https://doi.org/10.1056/NEJMoa1104390.

[43] Oza AM, Cook AD, Pfisterer J, et al. Standard chemotherapy with or without bevacizumab for women with newly diagnosed ovarian cancer (ICON7): overall survival results of a phase 3 randomised trial. Lancet Oncol 2015;16:928–36. Available from: https://doi.org/10.1016/S1470-2045(15)00086-8.

[44] González Martín A, Oza AM, Embleton AC, et al. Exploratory outcome analyses according to stage and/or residual disease in the ICON7 trial of carboplatin and paclitaxel with or without bevacizumab for newly diagnosed ovarian cancer. Gynecol Oncol 2019;152:53–60. Available from: https://doi.org/10.1016/j.ygyno.2018.08.036.

[45] Pujade-Lauraine E, Ledermann JA, Selle F, et al. Olaparib tablets as maintenance therapy in patients with platinum-sensitive, relapsed ovarian cancer and a BRCA1/2 mutation (SOLO2/ENGOT-Ov21): a double-blind, randomised, placebo-controlled, phase 3 trial. Lancet Oncol 2017;18:1274–84. Available from: https://doi.org/10.1016/S1470-2045(17)30469-2.

[46] DiSilvestro P, Colombo N, Scambia G, et al. Efficacy of maintenance olaparib for patients with newly diagnosed advanced ovarian cancer with a BRCA mutation: subgroup analysis findings from the SOLO1 Trial. J Clin Oncol. 2020;38(30):3528–37. Available from http://doi.org/10.1200/JCO.20.00799.

[47] Ray-Coquard I, Pautier P, Pignata S, et al. Olaparib plus bevacizumab as first-line maintenance in ovarian cancer. N Engl J Med 2019;381:2416–28. Available from: https://doi.org/10.1056/NEJMoa1911361.

[48] Rivera Dominguez A, Mora Jurado A, Garcia de la Oliva A, et al. Gynecological pelvic pain as emergency pathology. Radiologia 2017;59(2):115–27. Available from: https://doi.org/10.1016/j.rx.2016.09.010.

[49] Jacox A, Carr DB, Payne R, et al. Management of cancer pain. In: Rockville MD, editor. Clinical practice guideline No. 9. AHCPR publication No. 94-0592. Agency for Health Care Policy and Research, Public Health Service, U.S. Department of Health and Human Services. Rockville, Maryland; 1994.

[50] Cherny NI, Portenoy RK. Cancer pain: principles of assessment and syndromes. In: Wall PD, Melzack R, editors. Textbook of Pain. Edinburgh: Churchill Livingstone; 1994. p. 787–823.

[51] Ahmed N, Stenvers KL. Getting to know ovarian cancer ascites: opportunities for targeted therapy-based translational research. Front Oncol 2013;3:256. Available from: https://doi.org/10.3389/fonc.2013.00256.

[52] Cerne K, Kobal B. Ascites in advanced ovarian cancer. In: Devaja O, Papadopoulos A, editors. Ovarian cancer - from pathogenesis to treatment. London, UK. InTech; 2018. Available from: http://dx.doi.org/10.5772/intechopen.72698.

[53] Ilgen JS, Marr A. Cancer emergencies: the acute abdomen. Emerg Med Clin North Am 2009;27(3):381–99. Available from: https://doi.org/10.1016/j.emc.2009.04.006.

[54] Faught W, Kirkpatrick JR, Krepart GV, et al. Peritoneovenous shunt for palliation of gynecologic malignant ascites. J Am Coll Surg 1995;180(4):472–4.

[55] Tuca A, Guell E, Martinez-Losada E, et al. Malignant bowel obstruction in advanced cancer patients: epidemiology, management, and factors influencing spontaneous resolution. Cancer Manag Res 2012;4:159–69. Available from: https://doi.org/10.2147/CMAR.S29297.

[56] Anthony T, Baron T, Mercadante S. Report of the clinical protocol committee: development of randomized trials for malignant bowel obstruction. J Pain Symptom Manage 2007;34:S49–59. Available from: https://doi.org/10.1016/j.jpainsymman.2007.04.011.

[57] Krouse RS. Malignant bowel obstruction. J Surg Oncol 2019;120:74–7. Available from: https://doi.org/10.1002/jso.25451.

[58] Cousins SE, Tempest E, Feuer DJ. Surgery for the resolution of symptoms in malignant bowel obstruction in advanced gynaecological and gastrointestinal cancer. Cochrane Database Syst Rev 2016;2016:CD002764. Available from: https://doi.org/10.1002/14651858.CD002764.pub2.
[59] Tuca A, Roca R, Sala C. Efficacy of granisetron in the antiemetic control of nonsurgical intestinal obstruction in advanced cancer: a phase II clinical trial. J Pain Symptom Manage 2009;37:259–70. Available from: https://doi.org/10.1016/j.jpainsymman.2008.01.014.
[60] Badgwell BD, Smith K, Liu P. Indicators of surgery and survival in oncology inpatients requiring surgical evaluation for palliation. Support Care Cancer 2009;17:727–34. Available from: https://doi.org/10.1007/s00520-008-0554-6.
[61] Sahinler I, Erkal H, Akyazici E, et al. Endometrial carcinoma and an unusual presentation of bone metastasis: a case report. Gynecol Oncol 2001;82:216–8. Available from: https://doi.org/10.1006/gyno.2001.6206.
[62] Mantyh PW. Bone cancer pain: from mechanism to therapy. Curr Opin Support Palliat Care 2014;8(2):83–90. Available from: https://doi.org/10.1097/SPC.0000000000000048.
[63] Fallon M, Giusti R, Aielli F, et al. Management of cancer pain in adult patients: ESMO clinical practice guidelines. Ann Oncol 2018;29 (Suppl. 4):191. Available from: https://doi.org/10.1093/annonc/mdy152.
[64] Sindhi V, Erdek M. Interventional treatments for metastatic bone cancer pain. Pain Manag 2019;9(3):307–15. Available from: https://doi.org/10.2217/pmt-2018-0073.
[65] Mercadante S, Villari P, Ferrera P, et al. Optimization of opioid therapy for preventing incident pain associated with bone metastases. J Pain Symptom Manage 2004;28(5):505–10. Available from: https://doi.org/10.1016/j.jpainsymman.2004.02.024.
[66] Adderley UJ. Topical agents and dressings for fungating wounds. Cochrane Database Syst Rev 2014:CD003948. Available from: https://doi.org/10.1002/14651858.CD003948.pub3.
[67] Agrawal A, Yau A, Magliocco A, et al. Cutaneous metastatic disease in cervical cancer: a case report. J Obstet Gynaecol Can 2010;32:467. Available from: https://doi.org/10.1016/S1701-2163(16)34501-7.
[68] Imachi M, Tsukamoto N, Kinoshita S, et al. Skin metastasis from carcinoma of the uterine cervix. Gynecol Oncol 1993;48:349–54. Available from: https://doi.org/10.1006/gyno.1993.1061.
[69] Khurana R, Singh S. Isolated cutaneous metastasis to thigh from cancer cervix – fourteen years after curative radiotherapy. Internet J Gynecol Obstet 2009;11:1.
[70] Grabiec M, Małgorzata W, Andrzej M. Multiple skin metastases to vulva from carcinoma of the cervical stump. Ginekol Pol 2010;81:140–3.
[71] Dworkin RH, Schmader KE, Goldstein EJC. Treatment and prevention of postherpetic neuralgia. Clin Infect Dis 2003;36:877–82. Available from: https://doi.org/10.1086/368196.
[72] Safdar A, Rolston K, Bennettrobert GL, et al. Management of infections in patients with gynecologic malignancy. 2016. Available from: https://obgynkey.com/management-of-infections-in-patients-with-gynecologic-malignancy/#ch29ref2.
[73] Lee EQ. Nervous system metastases from systemic cancer. Contin (Minneap Minn) 2015;21(2):415–28. Available from: https://doi.org/10.1212/01.CON.0000464178.81957.18.
[74] Bourque PR, Sampaio ML, Warman-Chardon J, et al. Neurolymphomatosis of the lumbosacral plexus and its branches: case series and literature review. BMC Cancer 2019;19(1):1149. Available from: https://doi.org/10.1186/s12885-019-6365-y.
[75] Dyck PJ, Thaisetthawatkul P. Lumbosacral plexopathy. Contin (Minneap Minn) 2014;20(5):1343–58. Available from: https://doi.org/10.1212/01.CON.0000455877.60932.d3.
[76] Jaeckle KA. Neurologic manifestations of neoplastic and radiation-induced plexopathies. Semin Neurol 2010;30(3):254–62. Available from: https://doi.org/10.1055/s-0030-1255219.
[77] Jaeckle KA, Young DF, Foley KM. The natural history of lumbosacral plexopathy in cancer. Neurology 1985;35(1):8–15. Available from: https://doi.org/10.1212/WNL.35.1.8.
[78] Cormier JN, Askew RL, Mungovan KS, et al. Lymphedema beyond breast cancer: a systematic review and meta-analysis of cancer-related secondary lymphedema. Cancer 2010;116:5138–49. Available from: https://doi.org/10.1002/cncr.25458.
[79] Abu-Rustum NR, Barakat RR. Observations on the role of circumflex iliac node resection and the etiology of lower extremity lymphedema following pelvic lymphadenectomy for gynecologic malignancy. Gynecol Oncol 2007;106:4–5. Available from: https://doi.org/10.1016/j.ygyno.2007.03.026.
[80] Park SB, Goldstein D, Krishnan AV, et al. Chemotherapy-induced peripheral neurotoxicity: a critical analysis. CA Cancer J Clin 2013;63:419–37. Available from: https://doi.org/10.3322/caac.21204.
[81] Seretny M, Currie GL, Sena ES, et al. Incidence, prevalence, and predictors of chemotherapy-induced peripheral neuropathy: A systematic review and meta-analysis. Pain. 2014;155:2461–70. Available from https://doi.org/10.1016/j.pain.2014.09.020.
[82] Kolb NA, Smith AG, Singleton JR, et al. The association of chemotherapy-induced peripheral neuropathy symptoms and the risk of falling. JAMA Neurol 2016;73:860–6. Available from: https://doi.org/10.1001/jamaneurol.2016.0383.
[83] Mols F, van de Poll-Franse LV, Vreugdenhil G, et al. Reference data of the European Organisation for Research and Treatment of Cancer (EORTC) QLQ-CIPN20 questionnaire in the general Dutch population. Eur J Cancer 2016;69:28–38. Available from: https://doi.org/10.1016/j.ejca.2016.09.020.
[84] American Cancer Society. Cancer facts & figures, <https://www.cancer.org/research/cancer-facts-statistics/all-cancer-facts-figures/cancer-facts-figures-2020.html> [accessed 15.10.2025].
[85] Wu W, He X, Li S, et al. Pain nursing for gynecologic cancer patients. Front Oncol 2023;13:1205553. Available from: https://doi.org/10.3389/fonc.2023.1205553.
[86] Browne C, Davis NF, Mac Craith E, et al. A narrative review on the pathophysiology and management for radiation cystitis. Adv Urol 2015;2015:346812. Available from: https://doi.org/10.1155/2015/346812.

[87] Plow EB, Pascual-Leone A, Machado A. Brain stimulation in the treatment of chronic neuropathic and non-cancerous pain. J Pain 2012;13(5):411–24. Available from: https://doi.org/10.1016/j.jpain.2012.02.001.

[88] Juganavar A, Joshi KS. Chronic pelvic pain: a comprehensive review. Cureus 2022;14(10):e30691. Available from: https://doi.org/10.7759/cureus.30691.

[89] Boz I, Teskereci G. Experiences of gynecologic oncology nurses regarding caring behaviors: a hermeneutic phenomenological study. J Psychosom Obstet Gynaecol 2021;42(1):29–39. Available from: https://doi.org/10.1080/0167482X.2020.1726316.

[90] Rodriguez C, Ji M, Wang HL, et al. Cancer pain and quality of life. J Hosp Palliat Nurs 2019;21(2):116–23. Available from: https://doi.org/10.1097/NJH.0000000000000507.

[91] Latremoliere A, Woolf CJ. Central sensitization: a generator of pain hypersensitivity by central neural plasticity. J Pain 2009;10(9):895–926. Available from: https://doi.org/10.1016/j.jpain.2009.06.012.

[92] Hooten WM. Chronic pain and mental health disorders: shared neural mechanisms, epidemiology, and treatment. Mayo Clin Proc 2016;91(7):955–70. Available from: https://doi.org/10.1016/j.mayocp.2016.04.029.

[93] Yang S, Chang MC. Chronic pain: structural and functional changes in brain structures and associated negative affective states. Int J Mol Sci 2019;20(13):3130. Available from: https://doi.org/10.3390/ijms20133130.

[94] Ji RR, Nackley A, Huh Y, et al. Neuroinflammation and central sensitization in chronic and widespread pain. Anesthesiology 2018;129(2):343–66. Available from: https://doi.org/10.1097/ALN.0000000000002130.

[95] Cillessen L, Johannsen M, Speckens AEM, et al. Mindfulness-based interventions for psychological and physical health outcomes in cancer patients and survivors: a systematic review and meta-analysis of randomized controlled trials. Psychooncology 2019;28(12):2257–69. Available from: https://doi.org/10.1002/pon.5214.

[96] Campos ACP, Antunes GF, Matsumoto M, et al. Neuroinflammation, pain and depression: an overview of the main findings. Front Psychol 2020;11:1825. Available from: https://doi.org/10.3389/fpsyg.2020.01825.

[97] Ji RR, Chamessian A, Zhang YQ. Pain regulation by non-neuronal cells and inflammation. Science 2016;354(6312):572–7. Available from: https://doi.org/10.1126/science.aaf8924.

[98] Littell RD, Kumar A, Einstein MH, et al. Advanced communication: a critical component of high quality gynecologic cancer care: a Society of Gynecologic Oncology evidence based review and guide. Gynecol Oncol 2019;155(1):161–9. Available from: https://doi.org/10.1016/j.ygyno.2019.07.026.

[99] Li D, Gao L, Ren LY, et al. Knowledge and attitudes regarding cancer pain management among oncology nurses in China. J Int Med Res 2021;49(1): 300060520979448. https://doi.org/10.1177/0300060520979448.

[100] Hagedorn JM, Pittelkow TP, Hunt CL, et al. Current perspectives on spinal cord stimulation for the treatment of cancer pain. J Pain Res 2020;13:3295–305. Available from: https://doi.org/10.2147/JPR.S263857.

[101] Kwon JH. Overcoming barriers in cancer pain management. J Clin Oncol 2014;32(16):1727–33. Available from: https://doi.org/10.1200/JCO.2013.52.4827.

[102] Paice JA, Portenoy R, Lacchetti C, et al. Management of chronic pain in survivors of adult cancers: American Society of Clinical Oncology clinical practice guideline. J Clin Oncol 2016;34(27):3325–45. Available from: https://doi.org/10.1200/JCO.2016.68.5206.

[103] Dworkin RH, O'Connor AB, Backonja M, et al. Pharmacologic management of neuropathic pain: evidence-based recommendations. Pain 2007;132(3):237–51. Available from: https://doi.org/10.1016/j.pain.2007.08.033.

[104] Moulin DE, Clark AJ, Gilron I, et al. Pharmacological management of chronic neuropathic pain – consensus statement and guidelines from the Canadian Pain Society. Pain Res Manag 2007;12(1):13–21. Available from: https://doi.org/10.1155/2007/730785.

[105] Tramèr MR, Carroll D, Campbell FA, et al. Cannabinoids for control of chemotherapy induced nausea and vomiting: quantitative systematic review. BMJ 2001;323(7303):16–21. Available from: https://doi.org/10.1136/bmj.323.7303.16.

[106] Araujo AM, Gómez M, Pascual J, et al. Treatment of pain in the oncology patient. An Sist Sanit Navar 2004;27(Suppl. 3):63–75.

[107] Chow E, Zeng L, Salvo N, et al. Update on the systematic review of palliative radiotherapy trials for bone metastases. Clin Oncol 2012;24:112–24. Available from: https://doi.org/10.1016/j.clon.2011.11.004.

[108] Goblirsch MJ, Zwolak PP, Clohisy DR. Biology of bone cancer pain. Clin Cancer Res 2006;12:6231s–5s. Available from: https://doi.org/10.1158/1078-0432.CCR-06-0682.

[109] Goblirsch M, Mathews W, Lynch C, et al. Radiation treatment decreases bone cancer pain, osteolysis and tumor size. Radiat Res 2004;161:228–34. Available from: https://doi.org/10.1667/RR3108.

[110] Wu JSY, Wong RKS, Lloyd NS, et al. Radiotherapy fractionation for the palliation of uncomplicated painful bone metastases—an evidence-based practice guideline. BMC Cancer 2004;4:71. Available from: https://doi.org/10.1186/1471-2407-4-71.

[111] Stjernsward J. WHO cancer pain relief programme. Cancer Surv 1988;7(1):195–208.

[112] Deng G. Integrative medicine therapies for pain management in cancer patients. Cancer J 2019;25(5):343–8. Available from: https://doi.org/10.1097/PPO.0000000000000399.

[113] Vitzthum LK, Riviere P, Murphy JD. Managing cancer pain during the opioid epidemic-balancing caution and compassion. JAMA Oncol 2020;6(7):1103–4. Available from: https://doi.org/10.1001/jamaoncol.2020.0779.

[114] Staley SA, Tucker KR, Clark LH. The role of obesity in the development and management of gynecologic cancer. Obstet Gynecol Surv 2020;75(5):308–16. Available from: https://doi.org/10.1097/OGX.0000000000000779.

[115] Kim CH, Lefkowits C, Holschneider C, et al. Managing opioid use in the acute surgical setting: a society of gynecologic oncology clinical practice statement. Gynecol Oncol 2020;157(3):563–9. Available from: https://doi.org/10.1016/j.ygyno.2020.01.024.

[116] Mitchell MT. Dancing with deterrents: understanding the role of abuse-deterrent opioid formulations and naloxone in managing cancer pain. Oncologist 2019;24(12):1505–9. Available from: https://doi.org/10.1634/theoncologist.2019-0340.

[117] Beesley VL, Price MA, Butow PN, et al. Physical activity in women with ovarian cancer and its association with decreased distress and improved quality of life. Psychooncology 2011;20(11):1161–9. Available from: https://doi.org/10.1002/pon.1834.
[118] Tucker K, Staley SA, Clark LH, et al. Physical activity: impact on survival in gynecologic cancer. Obstet Gynecol Surv 2019;74(11):679–92. Available from: https://doi.org/10.1097/OGX.0000000000000731.
[119] Sheinfeld Gorin S, Krebs P, Badr H, et al. Meta-analysis of psychosocial interventions to reduce pain in patients with cancer. J Clin Oncol 2012;30(5):539–47. Available from: https://doi.org/10.1200/JCO.2011.37.0437.
[120] Syrjala KL, Donaldson GW, Davis MW, et al. Relaxation and imagery and cognitive-behavioral training reduce pain during cancer treatment: a controlled clinical trial. Pain 1995;63(2):189–98. Available from: https://doi.org/10.1016/0304-3959(95)00039-U.

Chapter 40

Urologic cancers

Andrej Žist
Division of Medical Oncology, Institute of Oncology Ljubljana, Ljubljana, Slovenia

Key learning objectives

- Learn basic epidemiological data on urologic cancers.
- Gain insight into the specifics of prostate cancer and the ways systemic therapy improve quality of life (QoL) and pain.
- Gain insight into the specifics of bladder cancer and the ways systemic therapy improve QoL and pain.
- Gain insight into the specifics of renal cancer and the ways systemic therapy improve QoL and pain.
- Have an understanding of how future systemic agents will affect symptom management in urologic cancers.

Epidemiology of urologic cancers

According to the most recent global data, urologic cancers as a group represent 14% of all new cancer diagnoses in 2022 [1]. Prostate cancer represents half of all new urologic cancers diagnosed yearly, being the fourth most common malignancy, after lung, breast, and colorectal cancers. Global incidence and cancer patterns differ according to world region and Human Development Index (HDI). Out of countries for which data are available, prostate cancer ranks as the most frequent malignancy in men in almost two-thirds of countries (118 of 185), solidifying its role as a major public health problem. The widespread availability of prostate-specific antigen (PSA) testing in the late 1980s and early 1990s led to a rapid increase in incidence in the majority of countries. In contrast, trends in mortality show declines largely in high-resource countries and increased prostate cancer mortality in lower-resource settings [2,3].

After prostate cancer, bladder cancer is the second most common urologic cancer and ninth in overall incidence, with three out of four cases occurring in male patients [1]. Over the last 15 years incidence of bladder cancer has been declining in most parts of the world, although regional differences persist (e.g., rising incidence in Eastern Europe). Overall, the same trends are observed in men and women, with the exception of Western Europe, where incidence is increasing for women, likely reflecting the smoking epidemic in women in current years [4,5]. Regarding mortality, bladder cancer ranked 13th in 2022 and 9th among men. Three out of four deaths due to bladder cancer occur in men, with age-standardized mortality rates being highest in Northern Africa and Europe [4].

Kidney cancer ranks 9th in incidence and 16th in mortality worldwide, with a significant variance depending on geography, gender, and age. Incidence is generally higher in Europe and North America as compared to Asia, Africa, or Latin America and the Caribbean. It is hypothesized that this difference is largely due to wider access to abdominal imaging, although this does not completely explain the wide margin between different locations [6]. While worldwide incidence is slowly increasing, mortality has been slowly declining, especially in Western countries. The advent of new systemic therapies with a dramatic effect on survival and overdiagnosis of small renal masses of indolent oncologic potential are hypothesized as the main reasons for this trend [7,8].

Testicular and penile cancers are rare malignancies with a generally favorable prognosis, representing 0.4% and 0.2% of all cancer diagnoses per year worldwide, respectively [1]. Testicular cancer, although a rare cancer, still represents the most common cancer in young males between the ages of 14 and 44 years in countries with a high or very high HDI. Incidence of testicular cancer varies geographically from less than one affected individual per 100,000 males in Africa and Asia to 9.9 affected individuals per 100,000 males in Norway and other parts of Northern Europe [9,10]. Overall survival (OS) is excellent, with 5-year survival rates of 95% and cancer-specific survival at 15 years for patients with initially stage I disease above 99% [10,11]. Penile cancer is a rare disease, accounting for <1% of newly diagnosed cancers in men.

Pain Management in Palliative Care for Patients With Cancer. DOI: https://doi.org/10.1016/B978-0-443-30050-9.00033-3

Incidence is higher in low- and middle-income countries, lower in high-income countries, and negligible in Islamic countries and Israel due to religious neonatal circumcision. Long-term survival is favorable with 5-year OS of 90%, but it dramatically worsens with local or distant metastasis [12].

Prostate, bladder, and kidney cancers represent more than 95% of all urologic cancers and will be discussed in-depth in this chapter.

Prostate cancer

Specifics of prostate cancer

Prostate cancer is diagnosed at the median age of 70 years and rarely before 55 years of age [13,14]. Screening with blood serum biomarker PSA can be useful in detecting asymptomatic localized cancer, which is potentially curable, but comes at a risk of overdiagnosis and overtreatment of clinically indolent disease [15,16]. Standard diagnostic tools for detecting prostate cancer include, besides PSA testing, digitorectal examination and multiparametric magnetic resonance imaging (mpMRI) of the prostate. A prostate biopsy is used to confirm the presence of cancer in the gland. Additional approaches include mpMRI-guided transrectal or transperineal biopsies [17,18]. The standard for disease staging represents conventional imaging using contrast-enhanced abdominopelvic CT and technetium-99m bone scan. Novel molecular imaging methods with gallium-68 prostate-specific membrane antigen (PSMA) positron emission tomography (PET)-CT have higher sensitivity and specificity for locoregional and distant spread and can aid in identifying metastatic disease [19]. After staging, around 75% of patients have localized disease, 15% have locally advanced, and 10% have metastatic cancer, with 5-year survival rates up to 99% for localized and 30%–40% for metastatic disease [17,20]. For localized disease, patients are further divided into risk groups using different assessment tools. Standard treatment options include radical prostatectomy with/without lymphadenectomy, radiotherapy (external beam radiotherapy [EBRT] or brachytherapy) with/without transient androgen deprivation therapy (ADT), or active surveillance [17]. Currently almost 50% of patients in the low-risk group can safely be managed with active surveillance, delaying local therapy and related complications without compromising long-term survival [21–23]. Besides disease characteristics, patient age and comorbidities play an important role in selecting the most appropriate treatment, as up to 30% of patients can be older than 80 years. Active treatment of localized prostate cancer is offered to patients with an expected survival of 10–15 years depending on their age at diagnosis and comorbidities [24–26].

Prostate cancer–induced pain and systemic therapy for advanced disease

A total of 5%–10% of patients with prostate cancer have distant metastasis at diagnosis, with an additional 5% relapsing after local treatment. Common sites of metastasis include bone (84%), distant lymph nodes (10.6%), visceral metastases (around 10%), and brain and dura (<2%) [17,27]. Bone metastases commonly affect the axial skeleton and can be located in the vertebrae (69%), pelvic bones (41%), long bones (25%), and the skull (25%). Metastases can cause skeletal-related events (SRE) (pathological fracture, radiotherapy or surgery to the bone, spinal cord compression, hypercalcemia; SREs), which are the major causes of morbidity and mortality in prostate cancer patients. SREs can cause pain and negatively affect physical activity and quality of life (QoL) [28,29]. Concurrently, ADT for prostate cancer induces reduction of bone mineral density, which can lead to osteoporosis and its complications [30].

Prostate cancer–induced bone pain initially presents as dull aches, but with disease progression, it can become constant and severe, being intensified during movement. Breakthrough pain has a prevalence of 75% in patients with bone metastases and typically presents with acute, piercing, and very severe pain without an obvious trigger. Management of pain follows the World Health Organization's (WHO) three-step analgesic ladder with new agents being introduced or rotated according to the severity of pain. Adjuvant agents can be added to alleviate the neuropathic component [29,31], while bone-modifying agents (BMAs) such as bisphosphonates (e.g., zoledronic acid) or receptor activator of nuclear factor kappa-B ligand (RANKL) inhibitors (denosumab) can also be used to prevent SREs and pain progression [32]. Recent trials of BMAs in metastatic castration-resistant prostate cancer (mCRPC) focused on de-escalation of treatment from 4- to 12-weekly regimens. Initial results suggest that de-escalation is a feasible and reasonable option for mCRPC patients [33]. While widespread use of BMAs in everyday practice led to significantly lower incidence of SREs, an unequivocal effect on improvement of bone pain is not established [34,35]. A Cochrane metaanalysis showed little to no difference in pain responses to bisphosphonates versus placebo or no treatment, and none of the trials included reported pain responses to denosumab [36].

A recent systematic review and metaanalysis of patients with solid tumors comparing conventional external beam radiotherapy (cEBRT) with stereotactic body radiotherapy (SBRT) showed similar overall pain response with either technique in patients with painful bone metastases. Overall pain after 1 month improved for 52% (95% confidence interval [CI], 43%–61%) of patients treated with cEBRT and 62% (95% CI, 49%–75%) with SBRT. However, complete pain response after 6 months was achieved in only 18% (95% CI, 12%–24%) of patients after cEBRT, compared to 48% (95% CI, 39%–58%) after SBRT, suggesting that SBRT may be more effective in some patient subgroups [37]. A comparison between single-fraction radiotherapy and multifraction radiotherapy shows similar complete pain response rates (RRs) (23% vs 24%), but a higher rate of retreatment with single-fraction treatment (20% vs 8%, $P<.01$) [38]. According to recent guidelines, single-fraction radiotherapy (8 Gyx1) is recommended for uncomplicated bone metastases in patients with limited life expectancy or poor functional status to offer effective, rapid pain relief in a logistically convenient way, while multifraction radiotherapy is used for long-lasting pain control and in cases of complicated bone metastases (e.g., spinal cord compression, large symptomatic lesions, weight-bearing bone metastases) [39].

Metastatic prostate cancer encompasses the states of metastatic castration-sensitive prostate cancer (mCSPC) and mCRPC. For the past 80 years, ADT has been the cornerstone of treatment of mCSPC, which translated to a median survival expectancy of 10–15 years for men with recurrent disease after local therapy, but only 2–3 years when presenting with primary metastatic disease. After progression on ADT, patients enter the mCRPC state, which is characterized by a higher symptom burden and median survival of only 2–3 years [17,40]. In the last 30 years, major advances in systemic therapy have led to improvement not only in survival but also in QoL. Initial treatments for mCRPC with mitoxantrone did not show an OS benefit but were effective in improving QoL and were implemented into clinical practice more than two decades ago [41]. Work continued with the milestone study of docetaxel, which was the first systemic therapy to also improve OS in mCRPC patients in the early 2000s, quickly becoming a standard of care [42]. The 2010s marked the coming of novel androgen receptor pathway inhibitors (ARPIs), first to the mCRPC state and later to the mCSPC state, improving OS with a more favorable safety profile as compared to taxane chemotherapy (Tables 40.1 and 40.2) [45,46,48–50]. Novel approaches include exploitation of distinct genetic alterations that govern the pathogenesis of metastatic prostate cancer and a resurgence of radionuclides in the treatment of metastatic prostate cancer. Monotherapy or drug combinations with poly (ADP-ribose) polymerase (PARP) inhibitors in patients with somatic or germline mutations in genes of homologous recombination repair, especially in BRCA genes, have proven to be efficacious in these high-risk patients, but with only limited improvements in symptom control [51,52]. Radium-223 was the first targeted alpha therapy approved by the US Food and Drug Administration (FDA) and European Medicines Agency (EMA). Compared to placebo, it improved OS, time to first SRE, and QoL irrespective of docetaxel use [53]. Radium-223 accumulates only in bone and penetrates less than 0.1 mm into soft tissue, which minimizes toxicity and makes it tolerable for the elderly patient population. In contrast, beta-emitting radionuclides, including strontium-89, samarium-153, phosphorus-32, and others, can penetrate 5–10 mm into soft tissue, but their use in metastatic prostate cancer does not improve OS [29]. Recently, beta-emitting lutetium-177 coupled to a small molecule ligand PSMA-617 targeting PSMA was FDA and EMA approved on the basis of the phase 3 VISION trial. Treatment with Lu-177-PSMA-617 provided superior pain responses and improved OS [54].

TABLE 40.1 Selected trials in mCRPC stage.

Trial/author	Drugs	No. of pts.	OS, months	*P*-value	Pain, %	*P*-value	QoL, %	*P*-value
TAX 327, Tannock et al. [42]	Docetaxel 3w + P vs docetaxel 1w + P vs mitoxantrone + P	1006 (335 vs 334 vs 337)	18.9 vs 17.4 vs 16.5	.009; .36	35 vs 31 vs 22	.01; .08	22 vs 23 vs 13	.009; .005
TROPIC, De Bono et al. [43]	Cabazitaxel + P vs mitoxantrone + P	755 (378 vs 377)	15.1 vs 12.7	<.0001	9.2 vs 7.7	.63	No data	–
COU-AA-301, Logothetis et al. [44]	Abiraterone + P vs placebo + P	1195	15.8 vs 11.2	<.0001	45.0 vs 28.8	.0005	No data	–
AFFIRM, Fizazi et al. [45]	Enzalutamide vs placebo	1199 (800 vs 399)	18.4 vs 13.6	<.001	45 vs 7	.0079	42 vs 15	<.0001
PREVAIL, Loriot et al. [46]	Enzalutamide vs placebo	1717 (872 vs 845)	32.4 vs 30.2	<.001	58 vs 16	<.0001	40 vs 23	<.0001

OS, Overall survival; *P*, prednisone; *Pain*, pain improvement from baseline, %; *QoL*, quality-of-life improvement from baseline, %.

TABLE 40.2 Selected trials in metastatic castration-sensitive prostate cancer stage.

Trial/author	Drug	No. of pts.	OS, months	*P*-value	Pain, months	*P*-value	QoL, months	*P*-value
CHAARTED, Sweeney et al. [47]	Docetaxel + androgen deprivation therapy (ADT) vs ADT	790 (397 vs 393)	57.6 vs 44.0	<.001	No data	–	No data	–
LATITUDE, Chi et al. [48]	Abiraterone + P + ADT vs placebo + ADT	1199 (597 vs 602)	53.3 vs 36.5 mo	<.0001	10.2 vs 6.5	0.001	12.9 vs 8.3	0.032
ARCHES, Armstrong et al. [49]	Enzalutamide + ADT vs placebo + ADT	1150 (574 vs 576)	NR (HR 0.66)	<.001	14.09 vs 11.1	0.032	11.14 vs 8.38	0.007
TITAN, Chi et al. [50]	Apalutamide + ADT vs placebo + ADT	1052 (525 vs 527)	NR vs 52.2	<.0001	19.3 vs 12.0	0.17	9.0 vs 9.2	0.32

HR, Hazard ratio; *NR*, not reached; *OS*, overall survival; *P*, prednisone; *Pain*, time to pain deterioration; *QoL*, time to quality-of-life deterioration.

TABLE 40.3 Selected trials with novel agents and radionuclides in mCRPC stage.

Trial/author	Drug	No. of pts.	OS, months	*P*-value	Pain, months	*P*-value	QoL, months	*P*-value
PROFOUND, Hussain et al. [51]	Olaparib vs physician choice	Cohort A: 245 (162 vs 83)	19.1 vs 14.7	.02	Not reached (NR) vs 9.92	.019	5.55 vs 3.65	.49
PROPEL, Clarke et al. [52]	Olaparib + abiraterone vs placebo + abiraterone	796 (399 vs 397)	42.1 vs 34.7	.0544	NR (hazard ratio 1.06)	.75	No data	–
VISION, Sartor et al. [54]	177Lu-PSMA-617 + standard care vs standard care	831 (551 vs 280)	15.3 vs 11.3	<.001	5.9 vs 2.2	<.001	9.7 vs 2.4	<.001
ALSYMPCA, Parker et al. [53]	Radium-223 dichloride vs placebo	921 (614 vs 307)	14.9 vs 11.3	<.001	[a]30.2 vs 20.1	.01	[b]24.6 vs 16.1	.02

OS, Overall survival; *P*, prednisone; *Pain*, time to pain deterioration; *QoL*, time to quality-of-life deterioration.
[a] *Pain improvement from baseline, %.*
[b] *QoL improvement from baseline, %.*

With the multitude of treatment possibilities for advanced prostate cancer, optimal sequencing of treatments is an ongoing issue. For mCSPC, doublet therapy with ADT plus ARPI represents the current standard. An upfront triplet regimen incorporating the addition of docetaxel is reserved only for selected patients with aggressive disease and the ability to tolerate additional toxicity [55]. Upon progression to mCRPC, genomic and molecular markers help tailor optimal treatment sequences, giving priority to targeted treatments with a more favorable efficacy/toxicity profile over traditional taxane chemotherapy [56]. For older and frail patients, their individual health status represents the cornerstone for decision-making when choosing treatment strategies. Patients at risk may benefit from a comprehensive geriatric assessment, with geriatric interventions helping to establish an appropriate and effective treatment plan [57] (Table 40.3).

Future implications

Patient selection on the basis of molecular biomarkers and combinations of systemic therapy will guide future systemic therapy options. While combinations provide better efficacy, this usually comes at the price of higher toxicity and polypharmacy for the patients. New and emerging treatment options will have to be carefully evaluated, and their benefit should be against the added toxicity. Novel biomarkers will define patient groups with higher risk, where intensified therapy is justified, and groups where standard therapy or even de-escalation will be possible.

Bladder cancer

Specifics of bladder cancer

The mean age at diagnosis of bladder cancer is 73 years [58]. Painless, visible hematuria is the most common presentation of bladder cancer in about 75% of cases. About 5% present with microscopic hematuria, which can correspond to a lower stage of disease [59]. Cystoscopy is the gold standard for diagnosis of bladder cancer, followed by transurethral resection of bladder tumor (TURBT) [60]. Depending on the depth of invasion of the tumor into the bladder wall, tumors are divided into non–muscle-invasive bladder cancer (NMIBC) and muscle-invasive bladder cancer (MIBC), with the latter requiring more aggressive management. Standard staging includes abdominopelvic CT imaging with CT urography, including the upper urinary tract. Multiparametric MRI has better resolution for soft tissues compared to CT and can be considered for additional local staging. Chest CT is recommended for MIBC but is not necessary for NMIBC. Fluorodeoxyglucose (FDG) PET-CT is not standard but can have a role when metastatic disease is suspected [59,61].

NMIBC represents 80% of all diagnosed bladder cancer and can effectively be managed with regular TURBT and intravesical instillations of systemic agents (e.g., Bacillus Calmette–Guerin [BCG] vaccine) for patients at a higher risk of disease recurrence. Five-year OS is favorable, from 95% for pTis stage to 70% for pT1 stage [59,62]. The gold standard of treatment for MIBC is radical cystectomy with pelvic lymphadenectomy and urinary diversion. The choice of type of urinary diversion depends on patient and disease factors and can involve formation of an ileal conduit and orthotopic neobladder [59]. Selected patients can be candidates for bladder-preserving trimodality treatment (TMT) comprising maximal TURBT followed by concurrent chemoradiotherapy. TMT is an acceptable alternative for patients who wish to retain their bladder or are unfit for radical cystectomy and may be more effective in disease localized only to the bladder [63]. Five-year recurrence-free survival following surgery ranges from 73% to 85% for disease localized to the bladder, dropping to 56%–58% with tumor invasion into surrounding tissue (pT3–4) and to 33%–35% with regional lymph node metastases (N+) [64,65]. Perioperative systemic therapy with either neoadjuvant or adjuvant therapy can improve survival [66–68]. For patients who are not candidates for cisplatin-based chemotherapy prior to or after surgery or have residual after surgery, adjuvant immunotherapy with nivolumab can be an option [69]. Comorbidities represent a major obstacle in the delivery of systemic therapy. Up to 25% of bladder cancer patients present with chronic kidney disease, which is a limiting factor for cisplatin-based chemotherapy. Cisplatin itself is nephrotoxic, and adequate renal function is required for safe and effective treatment [70–72].

Bladder cancer–induced pain and systemic therapy for advanced disease

The most common sites of metastases in bladder cancer are lymph nodes (26%–90%), bones (35%–47%), lungs (20%–45%), liver (13%–47%), and peritoneum (16%–19%). Bone metastases are a common cause of bladder cancer–associated pain and primarily appear in the spine. Other common sites include the pelvis, ribs, skull, and proximal long bones of the upper and lower extremities [73]. Therapeutic management includes the use of opioids according to the WHO ladder and the addition of supportive medication. Use of zoledronic acid and denosumab is recommended by current guidelines with the aim of reducing and delaying SREs, but evidence for efficacy in urothelial cancer is limited [74]. A phase 2 trial with 40 bladder cancer patients showed efficacy of zoledronic acid in decreasing risk of SREs and improving pain [75]. Surgery can be used to relieve pain caused by vertebral body fractures, and EBRT is effective in controlling pain due to bone metastases. In addition to relieving bone pain, EBRT can be used to relieve pain caused by infiltration of adjacent pelvic structures or from ureteral obstruction, the other two major causes of bladder cancer-related pain [76]. A recent trial of short-course radiotherapy in patients with pelvic tumors, which included bladder cancer patients, showed a 92.9% overall pain response and improvement in QoL indices for well-being, fatigue, and ability to perform daily activities in 64%, 36%, and 48% of patients, respectively [77].

Survival for metastatic bladder cancer is poor with 5-year survival ranging from 5% to 8% and OS in everyday practice around 15 months. Systemic therapy in bladder cancer is underutilized, with almost 50% of patients not receiving any treatment for metastatic disease. Barriers for systemic chemotherapy include performance status, comorbidities, and toxicity of cisplatin-based chemotherapy, which, until recently, was the treatment of choice in first-line [78]. Cisplatin-based chemotherapy became the standard of care in the early 1990s with the methotrexate + vinblastine + doxorubicin + cisplatin regimen. Subsequent trials established the gemcitabine + cisplatin regimen as a less toxic alternative with similar efficacy. Due to the toxicity profile of cisplatin (nephrotoxicity, nausea/vomiting, neurotoxicity, ototoxicity, and myelosuppression), only 30% of patients in everyday practice are suitable candidates [59,78,79]. From 2015, immune-checkpoint inhibitors (ICIs) came into regular practice for the treatment of advanced solid tumors. ICIs became available as second-line treatment following platinum-based chemotherapy, as first-line treatment for cisplatin-ineligible patients, or as maintenance therapy following first-line platinum-based chemotherapy. Durable responses could be seen in 20%–25% of patients, but at the expense of immune-related adverse events, which can occur in up to 40% of patients [59,80–82]. The latest advancements in the field are antibody-drug conjugates between cytotoxic chemotherapy and an antibody against target receptors on bladder cancer cells, such as enfortumab vedotin (EV), and targeted therapy with pan-fibroblast growth factor receptor inhibitor erdafitinib. Recently, a phase 3 clinical trial evaluating the combination of EV and pembrolizumab against the standard combination of gemcitabine and platinum as first-line treatment for advanced urothelial cancer showed prolonged OS by 15.4 months with the combination, almost doubling the survival in the experimental arm (31.5 vs 16.1 months, hazard ratio [HR] 0.47, 95% CI 0.38–0.58, $P < .00001$). Patients treated with the combination also had better objective responses to treatment (67.7% vs 44.4%, $P < .00001$). The number of adverse events was equally distributed between the two arms (97% vs 95.6%), with patients on the combination experiencing more serious adverse events (27.7% vs 19.6%). Cooccurrences of peripheral sensory neuropathy, skin rash, alopecia, eye disorders, and hyperglycemia in the combination arm and hematotoxicity, nausea, and fatigue in the platinum-based chemotherapy arm stood out [83]. The combination of EV with pembrolizumab was the first treatment to surpass cisplatin-based chemotherapy and represents a new standard for the treatment of advanced urothelial cancer. As this combination is also significantly less nephrotoxic, it marks a major milestone in the treatment of advanced bladder cancer (Table 40.4).

Future implications

The best way to improve a symptom is to treat the underlying cause, and in bladder cancer, new drugs will be doing just that. Improved responses with novel drug combinations will have an immediate effect on the management of pain and other symptoms of advanced bladder cancer. As preliminary data show similar efficacy of novel drug combinations in early-stage bladder cancer, this could lead to improved cure rates and fewer patients with metastatic bladder cancer.

TABLE 40.4 Selected trials in metastatic bladder cancer.

Trial/author	Drug	No. of pts.	OS, months	*P*-value	ORR, %	*P*-value	QoL, months	*P*-value
Von der Maase et al. [79]	GC vs MVAC	405 (203 vs 202)	14.0 vs 15.2	.66	49.4 vs 45.7	.51	No data	–
KEYNOTE-045, Bellmunt et al. [81]	Pembrolizumab vs physician choice chemotherapy	542 (270 vs 272)	10.3 vs 7.4	.002	21.1 vs 11.4	.001	3.5 vs 2.3	.008
JAVELIN 100, Powles et al. [82]	Maintenance avelumab + BSC vs BSC	700 (350 vs 350)	21.4 vs 14.3	.001	–	–	Not reached vs 13.8	.2
THOR, Loriot et al. [84]	Erdafitinib vs physician choice chemotherapy	266 (133 vs 130)	12.1 vs 7.8	.005	45.6 vs 11.5	<.001	No data	–
EV-302, Powles et al. [83]	EV + pembrolizumab vs gemcitabin+ cisplatin/ carboplatin	886 (442 vs 444)	31.5 vs 16.1	<.001	67.7 vs 44.4	<.001	5.9 vs 3.2	NS

BSC, Best supportive care; *EV*, enfortumab vedotin; *GC*, gemcitabine + cisplatin; *MVAC*, methotrexate + vinblastine + doxorubicin + cisplatin; NS, nonsignificant; *ORR*, objective response rate, %; *OS*, overall survival; *QoL*, time to quality-of-life deterioration, months.

Kidney cancer

Specifics of kidney cancer

Kidney cancer is usually diagnosed at an age of 64 years [85]. The classical triad seen in kidney cancer (hematuria, flank pain, and flank mass) is seen in less than 15% of cases, with the majority of tumors being found incidentally on abdominal imaging due to other reasons before any symptoms [85,86]. Optimal disease staging includes contrast-enhanced CT of the chest, abdomen, and pelvic region to determine primary tumor assessment and regional and distant spread [85,87]. Initial assessment of patients with metastatic disease should include laboratory testing needed to determine the patients' prognostic group criteria. Patients can be divided into good (median survival 43.2 months), intermediate (median survival 22.5 months), or poor prognosis groups (median survival 7.8 months) [88]. Pathohistological verification of malignancy is obtained by core biopsy or with nephrectomy and, in 80% of cases, reveals clear-cell renal cell carcinoma. The remaining 20% are collectively referred to as non–clear-cell renal cell carcinoma and comprise a histologically diverse group, with the majority being papillary renal cell carcinoma [89,90]. At diagnosis, two-thirds of patients present with localized disease and a 5-year survival rate of 93%. Locally advanced disease can be present in around 20% of patients and primary metastatic disease in 15%, with 5-year survival of 73% and 12%, respectively [91].

Standard treatment for localized disease is surgery, with partial nephrectomy indicated for patients with T1 disease. If partial nephrectomy is not feasible, open or robot-assisted laparoscopic nephrectomy is indicated [85,89]. Adjuvant therapy with pembrolizumab is the first systemic therapy for kidney cancer to improve disease-free survival and OS and can be offered to patients at a higher risk of disease recurrence after surgery [92]. If patients are deemed medically inoperable, ablative therapies (cryotherapy, radiofrequency ablation) can be performed percutaneously using ultrasound or CT guidance and with up to 90% efficacy [93]. Stereotactic ablative body radiotherapy (SABR) is a noninvasive treatment option in elderly, medically inoperable renal cancer patients with excellent disease control (5-year cancer-specific survival 92%) and a low rate of local failure (5-year local failure 5%) [94].

Systemic therapy for advanced kidney cancer

When taking into account recurrence after local therapy, around 40% of all kidney cancer patients develop distant metastases [91,95]. Most common sites of metastasis include lung (45%–55%), bone (30%–38%), lymph node (19%–22%), liver (19%–21%), adrenal glands (9%–11%), and brain (3%–10%). Also, kidney cancer is known for its atypical pattern of metastases, which can occur in various soft tissues in the body (small and large intestine, pancreas, kidney, pleura, skin, thyroid, testis, nasopharynx, vagina) [96–100]. For patients with primary metastatic kidney cancer, upfront cytoreductive nephrectomy (CN) is suggested to reduce paraneoplastic syndromes and improve QoL [101]. Initial randomized trials also showed an improvement in OS [102,103], while recent trials did not confirm an OS benefit with the addition of upfront CN for patients treated with novel systemic therapy for metastatic disease [89,104,105]. Currently CN may be considered in patients with low-volume single-organ metastatic disease with a large primary tumor or for patients with a complete response to upfront systemic therapy [89,104,105]. Local therapies for the treatment of kidney cancer metastases and their effect on symptom control also remain debatable. In retrospective studies, OS in patients after surgical metastasectomy was improved in the range from 36 to 142 months compared to 8 to 27 months with no metastasectomy [106]. SABR can represent an alternative to surgery in oligometastatic kidney cancer [107,108]. While retrospective data suggest metastases-directed therapy can improve OS, its effect on QoL is not well documented, and the risk of morbidity, especially with surgery, has to be balanced with possible gains. Therefore candidates for metastases-directed therapy should have a multidisciplinary evaluation by an experienced team to provide the best shared decision-making.

In the mid-2000s treatment with tyrosine kinase inhibitors (TKIs) became the first effective therapy for metastatic kidney cancer. Treatment with TKIs sunitinib or sorafenib offered RR (11%–31%) and improved median OS (15–26 months) but at the cost of a challenging toxicity profile. While standard chemotherapy can produce higher-grade acute toxicity following application of drugs, TKIs lead to lower-grade toxicity but for longer periods of time due to the continuous regimens of drug administration, which could negatively impact patients' QoL [109–111]. Supportive care to manage fatigue, hand–foot syndrome, skin rashes, stomatitis and other mucosal inflammation, diarrhea, and hypertension became crucial for the effective delivery of these potent drugs [112,113]. First-generation TKIs are no longer the first choice of treatment in newly diagnosed metastatic kidney cancer but are still regularly used in later treatment lines [89].

TABLE 40.5 Selected trials in metastatic kidney cancer.

Trial/author	Drug	No. of pts	OS, months	*P*-value	ORR, %	*P*-value	QoL, months	*P*-value
CheckMate 214, Motzer et al. [115]	Nivolumab + ipilimumab vs sunitinib	1096 (550 vs 546)	55.7 vs 38.4	<.001	39.3 vs 32.4	.0055	4.2 vs 2.1	Sig.
CheckMate 9ER, Choueiri et al. [118]	Nivolumab + cabozantinib vs sunitinib	651 (323 vs 328)	46.5 vs 36.0	.001	55.7 vs 27.7	<.001	19.4 vs 7.0	.0001
KEYNOTE-426, Rini et al. [119]	Pembrolizumab + axitinib vs sunitinib	861 (432 vs 429)	45.7 vs 40.1	<.0001	60.4 vs 39.6	<.001	25 vs NR, HR 1.4	Sig.
[a]CLEAR, Motzer et al. [120]	Pembrolizumab + lenvatinib vs sunitinib	1069 (355 vs 357)	53.7 vs 54.3	.0424	71.3 vs 36.7	<.0001	30.9 vs 27.0	.0081
RENOTORCH, Yan et al. [121]	Toripalimab + axitinib vs sunitinib	421 (210 vs 211)	NR vs 26.8	.0186	56.7 vs 30.8	<.0001	No data	–

HR, Hazard ratio; *NR*, not reached; *ORR*, objective response rate; *OS*, overall survival; *QoL*, time to quality-of-life deterioration; *Sig.*, statistically significant.
[a] *Data on third arm of trial with lenvatinib + everolimus not reported.*

The next milestone in the treatment of metastatic kidney cancer followed in the mid-2010s with the introduction of ICIs; survival data show that between 15% and 20% of patients treated with ICIs as monotherapy or in combination can have durable responses and long-term survival [114–117]. As in bladder cancer, immune-related adverse events can occur with the use of ICIs in kidney cancer patients, with a higher incidence with the use of ICI combinations [80].

Currently the standard of treatment of newly diagnosed metastatic cancer is combination therapy with either ICI–ICI or TKI–ICI combination with newer TKIs [89]. Several combinations are available with similar median OS (range 46–54 months) and differences in RR and toxicity profile. In general, TKI–ICI combinations offer higher RR (range 56%–71%) but at the cost of higher toxicity (grade ≥3 in 67%–82%) and higher discontinuation rates due to adverse events (11%–37%). In comparison, the ICI–ICI combination nivolumab–ipilimumab has an RR of 39% and an incidence of grade ≥3 toxicity of 48%, but with the possibility for durable responses as already discussed (Table 40.5) [115,116,118–126].

Future implications

Combination systemic therapy has transformed the treatment landscape for metastatic kidney cancer in the last 5 years. As projected in the past [85], 5-year survival rates are nearing the 50% mark, which is a fourfold improvement from the time of initial treatment for metastatic renal cancer 25 years ago. Future improvements in the field of disease stratification on the basis of molecular markers will help match patients/disease subtypes with the most efficacious therapy as multiple biomarker-driven trials are running.

Conclusion

Systemic treatment of urologic cancer has made tremendous improvements in the past 25 years. In prostate cancer, the treatment with ARPIs led to an almost uniformly good response to treatment and long-lasting remissions, also improving the time to deterioration of pain or deterioration of QoL. For disease refractory to hormonal agents, taxane chemotherapy remains a viable and effective option in providing symptom relief, and newer treatments with beta-emitting radionuclides like 177-lutetium are slowly finding their way into everyday clinical practice.

Treatment of kidney cancer is based on TKIs in monotherapy or in combination with ICIs, but patients have to be able to tolerate the added toxicity. Local treatments (SABR, surgical resection) of primary tumors in the metastatic setting or solitary metastases must also not be neglected as an effective tool to deal with symptomatic metastases (e.g., painful bone metastases) or symptomatic, large primary tumors with paraneoplastic symptoms.

Lastly, treatment options for metastatic bladder cancer have also improved, with the novel combination of EV and pembrolizumab introduced in 2023. With improved RR and a better toxicity profile (primarily lower nephrotoxicity), this is not just another treatment option but a major stepping stone on which further improvements in the field will be made.

References

[1] Bray F, Laversanne M, Sung H, et al. Global cancer statistics 2022: GLOBOCAN estimates of incidence and mortality worldwide for 36 cancers in 185 countries. CA Cancer J Clin 2024;74(3):229–63. Available from: https://doi.org/10.3322/caac.21834.

[2] Center MM, Jemal A, Lortet-Tieulent J, et al. International variation in prostate cancer incidence and mortality rates. Eur Urol 2012;61(6):1079–92. Available from: https://doi.org/10.1016/j.eururo.2012.02.054.

[3] Wong MCS, Goggins WB, Wang HHX, et al. Global incidence and mortality for prostate cancer: analysis of temporal patterns and trends in 36 countries. Eur Urol 2016;70(5):862–74. Available from: https://doi.org/10.1016/j.eururo.2016.05.043.

[4] van Hoogstraten LMC, Vrieling A, van der Heijden AG, et al. Global trends in the epidemiology of bladder cancer: challenges for public health and clinical practice. Nat Rev Clin Oncol 2023;20:287–304. Available from: https://doi.org/10.1038/s41571-023-00744-3.

[5] Teoh JYC, Huang J, Ko WYK, et al. Global trends of bladder cancer incidence and mortality, and their associations with tobacco use and gross domestic product per capita. Eur Urol 2020;78(6):893–906. Available from: https://doi.org/10.1016/j.eururo.2020.09.006.

[6] Bukavina L, Bensalah K, Bray F, et al. Epidemiology of renal cell carcinoma: 2022 update. Eur Urol 2022;82:529–42. Available from: https://doi.org/10.1016/j.eururo.2022.08.019.

[7] Sung WW, Ko PY, Chen WJ, et al. Trends in the kidney cancer mortality-to-incidence ratios according to health care expenditures of 56 countries. Sci Rep 2021;11(1):1479. Available from: https://doi.org/10.1038/s41598-020-79367-y.

[8] Bhindi B, Thompson RH, Lohse CM, et al. The probability of aggressive versus indolent histology based on renal tumor size: implications for surveillance and treatment. Eur Urol 2018;74(4):489–97. Available from: https://doi.org/10.1016/j.eururo.2018.06.003.

[9] Znaor A, Lortet-Tieulent J, Jemal A, et al. International variations and trends in testicular cancer incidence and mortality. Eur Urol 2014;65:1095–106. Available from: https://doi.org/10.1016/j.eururo.2013.11.004.

[10] Cheng L, Albers P, Berney DM, et al. Testicular cancer. Nat Rev Dis Prim 2018;4:29. Available from: https://doi.org/10.1038/s41572-018-0029-0.

[11] Kier MG, Lauritsen J, Mortensen MS, et al. Prognostic factors and treatment results after bleomycin, etoposide, and cisplatin in germ cell cancer: a population-based study. Eur Urol 2017;71(2):290–8. Available from: https://doi.org/10.1016/j.eururo.2016.09.015.

[12] Thomas A, Necchi A, Muneer A, et al. Penile cancer. Nat Rev Dis Prim 2021;7(1):11. Available from: https://doi.org/10.1038/s41572-021-00246-5.

[13] Zahed H, Feng X, Sheikh M, et al. Age at diagnosis for lung, colon, breast and prostate cancers: an international comparative study. Int J Cancer 2024;154(1):28–40. Available from: https://doi.org/10.1002/ijc.34671.

[14] Clark R, Vesprini D, Narod SA. The effect of age on prostate cancer survival. Cancers (Basel) 2022;14(17):4149. Available from: https://doi.org/10.3390/cancers14174149.

[15] Ilic D, Djulbegovic M, Jung JH, et al. Prostate cancer screening with prostate-specific antigen (PSA) test: a systematic review and meta-analysis. BMJ 2018;362:k3519. Available from: https://doi.org/10.1136/bmj.k3519.

[16] Hayes JH, Barry MJ. Screening for prostate cancer with the prostate-specific antigen test: a review of current evidence. JAMA 2014;311:1143–9. Available from: https://doi.org/10.1001/jama.2014.2085.

[17] Rebello RJ, Oing C, Knudsen KE, et al. Prostate cancer. Nat Rev Dis Prim 2021;7(1):9. Available from: https://doi.org/10.1038/s41572-020-00243-0.

[18] Ahdoot M, Wilbur AR, Reese SE, et al. MRI-targeted, systematic, and combined biopsy for prostate cancer diagnosis. N Engl J Med 2020;382(10):917–28. Available from: https://doi.org/10.1056/NEJMoa1910038.

[19] Hofman MS, Lawrentschuk N, Francis RJ, et al. Prostate-specific membrane antigen PET-CT in patients with high-risk prostate cancer before curative-intent surgery or radiotherapy (proPSMA): a prospective, randomised, multicentre study. Lancet 2020;395(10231):1208–16. Available from: https://doi.org/10.1016/S0140-6736(20)30314-7.

[20] Siegel RL, Giaquinto AN, Jemal A. Cancer statistics, 2024. CA Cancer J Clin 2024;74(1):12–49. Available from: https://doi.org/10.3322/caac.21820.

[21] Timilshina N, Komisarenko M, Martin LJ, et al. Factors associated with discontinuation of active surveillance among men with low-risk prostate cancer: a population-based study. J Urol 2021;206(4):903–13. Available from: https://doi.org/10.1097/JU.0000000000001903.

[22] Hamdy FC, Donovan JL, Lane JA, et al. Fifteen-year outcomes after monitoring, surgery, or radiotherapy for prostate cancer. N Engl J Med 2023;388(17):1547–58. Available from: https://doi.org/10.1056/NEJMoa2214122.

[23] Donovan JL, Hamdy FC, Lane JA, et al. Patient-reported outcomes 12 years after localized prostate cancer treatment. NEJM Evid 2023;2(4):EVIDoa2300018. Available from: https://doi.org/10.1056/EVIDoa2300018.

[24] Piccirillo JF, Vlahiotis A, Barrett LB, et al. The changing prevalence of comorbidity across the age spectrum. Crit Rev Oncol Hematol 2008;67(2):124–32. Available from: https://doi.org/10.1016/j.critrevonc.2008.01.013.

[25] Williams GR, Mackenzie A, Magnuson A, et al. Comorbidity in older adults with cancer. J Geriatr Oncol 2016;7:249–57. Available from: https://doi.org/10.1016/j.jgo.2015.12.002.

[26] Benzo RM, Moreno PI, Fox RS, et al. Comorbidity burden and health-related quality of life in men with advanced prostate cancer. Support Care Cancer 2023;31(8):496. Available from: https://doi.org/10.21203/rs.3.rs-2572781/v1.

[27] Gravestock P, Somani BK, Tokas T, et al. A review of modern imaging landscape for prostate cancer: a comprehensive clinical guide. J Clin Med 2023;12(3):1186. Available from: https://doi.org/10.3390/jcm12031186.

[28] Baldessari C, Pipitone S, Molinaro E, et al. Bone metastases and health in prostate cancer: from pathophysiology to clinical implications. Cancers (Basel) 2023;15(5):1518. Available from: https://doi.org/10.3390/cancers15051518.

[29] Smith AE, Muralidharan A, Smith MT. Prostate cancer induced bone pain: pathobiology, current treatments and pain responses from recent clinical trials. Discov Oncol 2022;13(1):108. Available from: https://doi.org/10.1007/s12672-022-00569-z.

[30] Rachner TD, Coleman R, Hadji P, et al. Bone health during endocrine therapy for cancer. Lancet Diabetes Endocrinol 2018;6:901–10. Available from: https://doi.org/10.1016/S2213-8587(18)30047-0.

[31] Fallon M, Giusti R, Aielli F, et al. Management of cancer pain in adult patients: ESMO clinical practice guidelines. Ann Oncol 2018;29:iv166–91. Available from: https://doi.org/10.1093/annonc/mdy152.

[32] Coleman R, Hadji P, Body JJ, et al. Bone health in cancer: ESMO clinical practice guidelines. Ann Oncol 2020;31(12):1650–63. Available from: https://doi.org/10.1016/j.annonc.2020.07.019.

[33] Clemons M, Ong M, Stober C, et al. A randomised trial of 4- versus 12-weekly administration of bone-targeted agents in patients with bone metastases from breast or castration-resistant prostate cancer. Eur J Cancer 2021;142:132–40. Available from: https://doi.org/10.1016/j.ejca.2020.08.019.

[34] Mitchell AP, Meza AM, Panageas KS, et al. Real-world use of bone modifying agents in metastatic, castration-resistant prostate cancer. Prostate Cancer Prostatic Dis 2023;26(1):126–32. Available from: https://doi.org/10.1038/s41391-022-00573-y.

[35] Jakob A, Zahn MO, Nusch A, et al. Real-world patient-reported outcomes of breast cancer or prostate cancer patients receiving antiresorptive therapy for bone metastases: final results of the PROBone registry study. J Bone Oncol 2022;33:100420. Available from: https://doi.org/10.1016/j.jbo.2022.100420.

[36] Macherey S, Monsef I, Jahn F, et al. Bisphosphonates for advanced prostate cancer. Cochrane Database Syst Rev 2017;12:CD006250. Available from: https://doi.org/10.1002/14651858.CD006250.pub2 [accessed 22.10.25].

[37] Bindels BJJ, Mercier C, Gal R, et al. Stereotactic body and conventional radiotherapy for painful bone metastases: a systematic review and meta-analysis. JAMA Netw Open 2024;7(2):E2355409. Available from: https://doi.org/10.1001/jamanetworkopen.2023.55409.

[38] Rich SE, Chow R, Raman S, et al. Update of the systematic review of palliative radiation therapy fractionation for bone metastases. Radiother Oncol 2018;126(3):547–57. Available from: https://doi.org/10.1016/j.radonc.2018.01.003.

[39] Alcorn S, Cortés ÁA, Bradfield L, et al. External beam radiation therapy for palliation of symptomatic bone metastases: an ASTRO clinical practice guideline. Pract Radiat Oncol 2024;14(5):377–97. Available from: https://doi.org/10.1016/j.prro.2024.04.018.

[40] Desai K, McManus JM, Sharifi N. Hormonal therapy for prostate cancer. Endocr Rev 2021;42:354–73. Available from: https://doi.org/10.1210/endrev/bnab002.

[41] Tannock IF, Osoba D, Stockler MR, et al. Chemotherapy with mitoxantrone plus prednisone or prednisone alone for symptomatic hormone-resistant prostate cancer: a Canadian randomized trial with palliative end points. J Clin Oncol 1996;14(6):1756–64. Available from: https://doi.org/10.1200/JCO.1996.14.6.1756.

[42] Tannock IF, de Wit R, Berry WR, et al. Docetaxel plus prednisone or mitoxantrone plus prednisone for advanced prostate cancer. N Engl J Med 2004;351(15):1502–12. Available from: https://doi.org/10.1056/NEJMoa040720.

[43] De Bono JS, Oudard S, Ozguroglu M, et al. Prednisone plus cabazitaxel or mitoxantrone for metastatic castration-resistant prostate cancer progressing after docetaxel treatment: a randomised open-label trial. Lancet 2010;376(9747):1147–54. Available from: https://doi.org/10.1016/S0140-6736(10)61389-X.

[44] Logothetis CJ, Basch E, Molina A, et al. Effect of abiraterone acetate and prednisone compared with placebo and prednisone on pain control and skeletal-related events in patients with metastatic castration-resistant prostate cancer: exploratory analysis of data from the COU-AA-301 randomised trial. Lancet Oncol 2012;13(12):1210–7. Available from: https://doi.org/10.1016/S1470-2045(12)70473-4.

[45] Fizazi K, Scher HI, Miller K, et al. Effect of enzalutamide on time to first skeletal-related event, pain, and quality of life in men with castration-resistant prostate cancer: results from the randomised, phase 3 AFFIRM trial. Lancet Oncol 2014;15(10):1147–56. Available from: https://doi.org/10.1016/S1470-2045(14)70303-1.

[46] Loriot Y, Miller K, Sternberg CN, et al. Effect of enzalutamide on health-related quality of life, pain, and skeletal-related events in asymptomatic and minimally symptomatic, chemotherapy-naive patients with metastatic castration-resistant prostate cancer (PREVAIL): results from a randomised, phase 3 trial. Lancet Oncol 2015;16(5):509–21. Available from: https://doi.org/10.1016/S1470-2045(15)70113-0.

[47] Sweeney CJ, Chen YH, Carducci M, et al. Chemohormonal therapy in metastatic hormone-sensitive prostate cancer. N Engl J Med 2015;373(8):737–46. Available from: https://doi.org/10.1056/NEJMoa1503747.

[48] Chi KN, Protheroe A, Rodríguez-Antolín A, et al. Patient-reported outcomes following abiraterone acetate plus prednisone added to androgen deprivation therapy in patients with newly diagnosed metastatic castration-naive prostate cancer (LATITUDE): an international, randomised phase 3 trial. Lancet Oncol 2018;19(2):194–206. Available from: https://doi.org/10.1016/S1470-2045(17)30911-7.

[49] Armstrong AJ, Azad AA, Iguchi T, et al. Improved survival with enzalutamide in patients with metastatic hormone-sensitive prostate cancer. J Clin Oncol 2022;40(15):1616–22. Available from: https://doi.org/10.1200/JCO.22.00193.

[50] Chi KN, Chowdhury S, Bjartell A, et al. Apalutamide in patients with metastatic castration-sensitive prostate cancer: final survival analysis of the randomized, double-blind, phase III TITAN study. J Clin Oncol 2021;39(20):2294–303. Available from: https://doi.org/10.1200/JCO.20.03488.

[51] Hussain M, Mateo J, Fizazi K, et al. Survival with olaparib in metastatic castration-resistant prostate cancer. N Engl J Med 2020;383(24):2345–57. Available from: https://doi.org/10.1056/NEJMoa2022485.

[52] Clarke NW, Armstrong AJ, Thiery-Vuillemin A, et al. Abiraterone and olaparib for metastatic castration-resistant prostate cancer. NEJM Evid 2022;1(9):EVIDoa2200043. Available from: https://doi.org/10.1056/EVIDoa2200043.

[53] Parker C, Nilsson S, Heinrich D, et al. Alpha emitter radium-223 and survival in metastatic prostate cancer. N Engl J Med 2013;369(3):213–23. Available from: https://doi.org/10.1056/NEJMoa1213755.

[54] Sartor O, de Bono J, Chi KN, et al. Lutetium-177–PSMA-617 for metastatic castration-resistant prostate cancer. N Engl J Med 2021;385(12):1091–103. Available from: https://doi.org/10.1056/NEJMoa2107322.

[55] Azad AA, Kostos L, Agarwal N, et al. Combination therapies in locally advanced and metastatic hormone-sensitive prostate cancer. Eur Urol 2025;87(4):455–67. Available from: https://doi.org/10.1016/j.eururo.2025.01.010.

[56] Francini E, Agarwal N, Castro E, et al. Intensification approaches and treatment sequencing in metastatic castration-resistant prostate cancer: a systematic review. Eur Urol 2025;87(1):29–46. Available from: https://doi.org/10.1016/j.eururo.2024.09.008.

[57] Boyle HJ, Alibhai S, Decoster L, et al. Updated recommendations of the International Society of Geriatric Oncology on prostate cancer management in older patients. Eur J Cancer 2019;116:116–36. Available from: https://doi.org/10.1016/j.ejca.2019.04.031.

[58] Lin W, Pan X, Zhang C, et al. Impact of age at diagnosis of bladder cancer on survival: a surveillance, epidemiology, and end results-based study 2004-2015. Cancer Control 2023;30:10732748231152322. Available from: https://doi.org/10.1177/10732748231152322.

[59] Dyrskjøt L, Hansel DE, Efstathiou JA, et al. Bladder cancer. Nat Rev Dis Prim 2023;9(1):58. Available from: https://doi.org/10.1038/s41572-023-00468-9.

[60] Babjuk M, Burger M, Capoun O, et al. European Association of Urology Guidelines on non–muscle-invasive bladder cancer (Ta, T1, and carcinoma in situ). Eur Urol 2022;81:75–94. Available from: https://doi.org/10.1016/j.eururo.2021.08.010.

[61] Hensley PJ, Panebianco V, Pietzak E, et al. Contemporary staging for muscle-invasive bladder cancer: accuracy and limitations. Eur Urol Oncol 2022;5:403–11. Available from: https://doi.org/10.1016/j.euo.2022.04.008.

[62] Ślusarczyk A, Zapała P, Zapała Ł, et al. Cancer-specific survival of patients with non-muscle-invasive bladder cancer: a population-based analysis. Ann Surg Oncol 2023;30(12):7892–902. Available from: https://doi.org/10.1245/s10434-023-14051-9.

[63] Pieretti A, Krasnow R, Drumm M, et al. Complications and outcomes of salvage cystectomy after trimodality therapy. J Urol 2021;206(1):29–36. Available from: https://doi.org/10.1097/JU.0000000000001696.

[64] Madersbacher S, Hochreiter W, Burkhard F, et al. Radical cystectomy for bladder cancer today – a homogeneous series without neoadjuvant therapy. J Clin Oncol 2003;21(4):690–6. Available from: https://doi.org/10.1200/JCO.2003.05.101.

[65] Stein JP, Lieskovsky G, Cote R, et al. Radical cystectomy in the treatment of invasive bladder cancer: long-term results in 1,054 patients. J Clin Oncol 2001;19(3):666–75. Available from: https://doi.org/10.1200/JCO.2001.19.3.666.

[66] Pfister C, Gravis G, Flechon A, et al. Perioperative dose-dense methotrexate, vinblastine, doxorubicin, and cisplatin in muscle-invasive bladder cancer (VESPER): survival endpoints at 5 years in an open-label, randomised, phase 3 study. Lancet Oncol 2024;25(2):255–64. Available from: https://doi.org/10.1016/S1470-2045(23)00587-9.

[67] Vale CL. Adjuvant chemotherapy in invasive bladder cancer: a systematic review and meta-analysis of individual patient data Advanced Bladder Cancer (ABC) Meta-analysis Collaboration. Eur Urol 2005;48:189–201. Available from: https://doi.org/10.1016/j.eururo.2005.04.005.

[68] Burdett S, Fisher DJ, Vale CL, et al. Adjuvant chemotherapy for muscle-invasive bladder cancer: a systematic review and meta-analysis of individual participant data from randomised controlled trials. Eur Urol 2022;81:50–61. Available from: https://doi.org/10.1016/j.eururo.2021.09.028.

[69] Bajorin DF, Witjes JA, Gschwend JE, et al. Adjuvant nivolumab versus placebo in muscle-invasive urothelial carcinoma. N Engl J Med 2021;384(22):2102–14. Available from: https://doi.org/10.1056/NEJMoa2034442.

[70] Jiang DM, Gupta S, Kitchlu A, et al. Defining cisplatin eligibility in patients with muscle-invasive bladder cancer. Nat Rev Urol 2021;18:104–14. Available from: https://doi.org/10.1038/s41585-020-00404-6.

[71] Nguyen C, Ghodoussipour S, Winter M, et al. Chronic kidney disease and radical cystectomy for bladder cancer: perioperative and oncologic outcomes in 1,214 patients. Urol Oncol 2022;40(8):381.e9–381.e16. Available from: https://doi.org/10.1016/j.urolonc.2022.04.010.

[72] Barone B, Finati M, Cinelli F, et al. Bladder cancer and risk factors: data from a multi-institutional long-term analysis on cardiovascular disease and cancer incidence. J Pers Med 2023;13(3):512. Available from: https://doi.org/10.3390/jpm13030512.

[73] Shinagare AB, Ramaiya NH, Jagannathan JP, et al. Metastatic pattern of bladder cancer: correlation with the characteristics of the primary tumor. AJR Am J Roentgenol 2011;196(1):117–22. Available from: https://doi.org/10.2214/AJR.10.5036.

[74] Witjes JA, Bruins HM, Cathomas R, et al. European Association of Urology Guidelines on muscle-invasive and metastatic bladder cancer: summary of the 2020 guidelines. Eur Urol 2021;79:82–104. Available from: https://doi.org/10.1016/j.eururo.2020.03.055.

[75] Zaghloul MS, Boutrus R, El-Hossieny H, et al. A prospective, randomized, placebo-controlled trial of zoledronic acid in bony metastatic bladder cancer. Int J Clin Oncol 2010;15(4):382–9. Available from: https://doi.org/10.1007/s10147-010-0074-5.

[76] Yi L, Ai K, Li X, et al. Bone metastasis in bladder cancer. J Personal Med 2023;13(1):54. Available from: https://doi.org/10.3390/jpm13010054.

[77] Farina E, Macchia G, Siepe G, et al. Palliative short-course radiotherapy in advanced pelvic cancer: a phase II study (SHARON project). Anticancer Res 2019;39(8):4237–42. Available from: https://doi.org/10.21873/anticanres.13585.

[78] Swami U, Grivas P, Pal SK, et al. Utilization of systemic therapy for treatment of advanced urothelial carcinoma: lessons from real world experience. Cancer Treat Res Commun 2021;27:100325. Available from: https://doi.org/10.1016/j.ctarc.2021.100325.

[79] von der Maase H, Hansen SW, Roberts JT, et al. Gemcitabine and cisplatin versus methotrexate, vinblastine, doxorubicin, and cisplatin in advanced or metastatic bladder cancer: results of a large, randomized, multinational, multicenter, phase III study. J Clin Oncol 2000;18(17):3068–77. Available from: https://doi.org/10.1200/JCO.2000.18.17.3068.

[80] Johnson DB, Nebhan CA, Moslehi JJ, et al. Immune-checkpoint inhibitors: long-term implications of toxicity. Nat Rev Clin Oncol 2022;19:254–67. Available from: https://doi.org/10.1038/s41571-022-00600-w.

[81] Bellmunt J, de Wit R, Vaughn DJ, et al. Pembrolizumab as second-line therapy for advanced urothelial carcinoma. N Engl J Med 2017;376(11):1015–26. Available from: https://doi.org/10.1056/NEJMoa1613683.

[82] Powles T, Park SH, Voog E, et al. Avelumab maintenance therapy for advanced or metastatic urothelial carcinoma. N Engl J Med 2020;383(13):1218–30. Available from: https://doi.org/10.1056/NEJMoa2002788.

[83] Powles T, Valderrama BP, Gupta S, et al. Enfortumab vedotin and pembrolizumab in untreated advanced urothelial cancer. N Engl J Med 2024;390(10):875–88. Available from: https://doi.org/10.1056/NEJMoa2312117.

[84] Loriot Y, Matsubara N, Park SH, et al. Erdafitinib or chemotherapy in advanced or metastatic urothelial carcinoma. N Engl J Med 2023;389(21):1961–71. Available from: https://doi.org/10.1056/NEJMoa2308849.

[85] Hsieh JJ, Purdue MP, Signoretti S, et al. Renal cell carcinoma. Nat Rev Dis Primers 2017;3:17009. Available from: https://doi.org/10.1038/nrdp.2017.9.

[86] Moore SC, Chow WH, Schatzkin A, et al. Physical activity during adulthood and adolescence in relation to renal cell cancer. Am J Epidemiol 2008;168(2):149–57. Available from: https://doi.org/10.1093/aje/kwn102.

[87] Nicaise EH, Yildirim A, Sheth S, et al. Cytoreductive surgery, systemic treatment, genetic evaluation, and patient perspective in a young adult with metastatic renal cell carcinoma. CA Cancer J Clin 2024;74(4):323–38. Available from: https://doi.org/10.3322/caac.21835.

[88] Heng DYC, Xie W, Regan MM, et al. External validation and comparison with other models of the international metastatic renal-cell carcinoma database consortium prognostic model: a population-based study. Lancet Oncol 2013;14(2):141–8. Available from: https://doi.org/10.1016/S1470-2045(12)70559-4.

[89] Powles T, Albiges L, Bex A, et al. Renal cell carcinoma: ESMO clinical practice guideline for diagnosis, treatment and follow-up. Ann Oncol 2024;35(8):P692–706. Available from: https://doi.org/10.1016/j.annonc.2024.05.537.

[90] Moch H, Amin MB, Berney DM, et al. The 2022 World Health Organization classification of tumours of the urinary system and male genital organs—part a: renal, penile, and testicular tumours. Eur Urol 2022;82:458–68. Available from: https://doi.org/10.1016/j.eururo.2022.06.016.

[91] Padala SA, Barsouk A, Thandra KC, et al. Epidemiology of renal cell carcinoma. World J Oncol 2020;11(3):79–87. Available from: https://doi.org/10.14740/wjon1279.

[92] Choueiri TK, Tomczak P, Park SH, et al. Overall survival with adjuvant pembrolizumab in renal-cell carcinoma. N Engl J Med 2024;390(15):1359–71. Available from: https://doi.org/10.1056/NEJMoa2312695.

[93] El Dib R, Touma NJ, Kapoor A. Cryoablation vs radiofrequency ablation for the treatment of renal cell carcinoma: a meta-analysis of case series studies. BJU Int 2012;110(4):510–6. Available from: https://doi.org/10.1111/j.1464-410X.2011.10885.x.

[94] Siva S, Ali M, Correa RJM, et al. 5-year outcomes after stereotactic ablative body radiotherapy for primary renal cell carcinoma: an individual patient data meta-analysis from IROCK (the International Radiosurgery Consortium of the Kidney). Lancet Oncol 2022;23(12):1508–16. Available from: https://doi.org/10.1016/S1470-2045(22)00656-8.

[95] Alzubaidi AN, Sekoulopoulos S, Pham J, et al. Incidence and distribution of new renal cell carcinoma cases: 27-year trends from a statewide cancer registry. J Kidney Cancer VHL 2022;9(2):7–12. Available from: https://doi.org/10.15586/jkcvhl.v9i2.219.

[96] Bianchi M, Sun M, Jeldres C, et al. Distribution of metastatic sites in renal cell carcinoma: a population-based analysis. Ann Oncol 2012;23(4):973–80. Available from: https://doi.org/10.1093/annonc/mdr362.

[97] Semenescu LE, Kamel A, Ciubotaru V, et al. An overview of systemic targeted therapy in renal cell carcinoma, with a focus on metastatic renal cell carcinoma and brain metastases. Curr Issues Mol Biol 2023;45:7680–704. Available from: https://doi.org/10.3390/cimb45090485.

[98] Wei H, Miao J, Cui J, et al. The prognosis and clinicopathological features of different distant metastases patterns in renal cell carcinoma: analysis based on the SEER database. Sci Rep 2021;11(1):17822. Available from: https://doi.org/10.1038/s41598-021-97365-6.

[99] Meyer CP, Sun M, Karam JA, et al. Complications after metastasectomy for renal cell carcinoma—a population-based assessment. Eur Urol 2017;72(2):171–4. Available from: https://doi.org/10.1016/j.eururo.2017.03.005.

[100] Adashek JJ, Aydin AM, Kim P, et al. The role of metastasectomy in the treatment of metastatic renal cell carcinoma. AME Med J 2019:4. Available from: https://doi.org/10.21037/amj.2019.06.02.

[101] Kumada N, Iinuma K, Kubota Y, et al. Impact of cytoreductive nephrectomy in the management of metastatic renal cell carcinoma: a multicenter retrospective study. Diseases 2024;12(6):122. Available from: https://doi.org/10.3390/diseases12060122.

[102] Flanigan RC, Salmon SE, Blumenstein BA, et al. Nephrectomy followed by interferon alfa-2b compared with interferon alfa-2b alone for metastatic renal-cell cancer. N Engl J Med 2001;345(23):1655–9. Available from: https://doi.org/10.1056/NEJMoa003013.

[103] Mickisch GHJ, Garin A, Van Poppel H, et al. Radical nephrectomy plus interferon-alfa-based immunotherapy compared with interferon alfa alone in metastatic renal-cell carcinoma: a randomised trial. Lancet 2001;358(9286):966–70. Available from: https://doi.org/10.1016/s0140-6736(01)06103-7.

[104] Méjean A, Ravaud A, Thezenas S, et al. Sunitinib alone or after nephrectomy in metastatic renal-cell carcinoma. N Engl J Med 2018;379(5):417–27. Available from: https://doi.org/10.1056/NEJMoa1803675.

[105] Li KP, He M, Wan S, et al. Comparison of upfront versus deferred cytoreductive nephrectomy in patients with metastatic renal cell carcinoma receiving systemic therapy: a systematic review and meta-analysis. Int J Surg 2023;109(10):3178–88. Available from: https://doi.org/10.1097/JS9.0000000000000591.

[106] Ouzaid I, Capitanio U, Staehler M, et al. Surgical metastasectomy in renal cell carcinoma: a systematic review. Eur Urol Oncol 2019;2:141–9. Available from: https://doi.org/10.1016/j.euo.2018.08.028.

[107] Ali M, Mooi J, Lawrentschuk N, et al. The role of stereotactic ablative body radiotherapy in renal cell carcinoma. Eur Urol 2022;82:613–22. Available from: https://doi.org/10.1016/j.eururo.2022.06.017.

[108] Zaorsky NG, Lehrer EJ, Kothari G, et al. Stereotactic ablative radiation therapy for oligometastatic renal cell carcinoma (SABR ORCA): a meta-analysis of 28 studies. Eur Urol Oncol 2019;2:515–23. Available from: https://doi.org/10.1016/j.euo.2019.05.007.

[109] Negrier S, Escudier B, Lasset C, et al. Recombinant human interleukin-2, recombinant human interferon alfa-2a, or both in metastatic renal-cell carcinoma. N Eng J Med 1998;338:1272–8. Available from: https://doi.org/10.1056/NEJM199804303381805.

[110] Motzer RJ, Hutson TE, Tomczak P, et al. Sunitinib versus interferon alfa in metastatic renal-cell carcinoma. N Engl J Med 2007;356:115–24. Available from: https://doi.org/10.1056/NEJMoa065044.

[111] Escudier B, Eisen T, Stadler WM, et al. Sorafenib in advanced clear-cell renal-cell carcinoma. N Engl J Med 2007;356(2):125–34. Available from: https://doi.org/10.1056/NEJMoa060655.

[112] Deng H, Li M, Wu Q, et al. A 2/1 sunitinib dosing schedule provides superior antitumor effectiveness and less toxicity than a 4/2 schedule for metastatic renal cell carcinoma: a systematic review and meta-analysis. Front Oncol 2020;10:313. Available from: https://doi.org/10.3389/fonc.2020.00313.

[113] Brown JE, Royle KL, Gregory W, et al. Temporary treatment cessation versus continuation of first-line tyrosine kinase inhibitor in patients with advanced clear cell renal cell carcinoma (STAR): an open-label, non-inferiority, randomised, controlled, phase 2/3 trial. Lancet Oncol 2023;24(3):213–27. Available from: https://doi.org/10.1016/S1470-2045(22)00793-8.

[114] Motzer RJ, Escudier B, McDermott DF, et al. Nivolumab versus everolimus in advanced renal-cell carcinoma. N Engl J Med 2015;373(19):1803–13. Available from: https://doi.org/10.1056/NEJMoa1510665.

[115] Motzer RJ, Tannir NM, McDermott DF, et al. Nivolumab plus ipilimumab versus sunitinib in advanced renal-cell carcinoma. N Engl J Med 2018;378(14):1277–90. Available from: https://doi.org/10.1056/NEJMoa1712126.

[116] Motzer RJ, McDermott DF, Escudier B, et al. Conditional survival and long-term efficacy with nivolumab plus ipilimumab versus sunitinib in patients with advanced renal cell carcinoma. Cancer 2022;128(11):2085–97. Available from: https://doi.org/10.1002/cncr.34180.

[117] George S, Larkin J, Chepynoga K, et al. 729 Poster session long-term survivorship rates among patients with previously treated advanced renal cell carcinoma receiving nivolumab. Immuno-Oncology Technol 2023;20:100575. Available from: https://doi.org/10.1016/j.iotech.2023.100575.

[118] Choueiri TK, Powles T, Burotto M, et al. Nivolumab plus cabozantinib versus sunitinib for advanced renal-cell carcinoma. N Engl J Med 2021;384(9):829–41. Available from: https://doi.org/10.1056/NEJMoa2026982.

[119] Rini BI, Plimack ER, Stus V, et al. Pembrolizumab plus axitinib versus sunitinib for advanced renal-cell carcinoma. N Engl J Med 2019;380(12):1116–27. Available from: https://doi.org/10.1056/NEJMoa1816714.

[120] Motzer RJ, Porta C, Eto M, et al. Lenvatinib plus pembrolizumab versus sunitinib in first-line treatment of advanced renal cell carcinoma: final prespecified overall survival analysis of CLEAR, a phase III study. J Clin Oncol 2024;42(11):1222–8. Available from: https://doi.org/10.1200/JCO.23.01569.

[121] Yan XQ, Ye MJ, Zou Q, et al. Toripalimab plus axitinib versus sunitinib as first-line treatment for advanced renal cell carcinoma: RENOTORCH, a randomized, open-label, phase III study. Ann Oncol 2024;35(2):190–9. Available from: https://doi.org/10.1016/j.annonc.2023.09.3108.

[122] Cella D, Motzer RJ, Suarez C, et al. Patient-reported outcomes with first-line nivolumab plus cabozantinib versus sunitinib in patients with advanced renal cell carcinoma treated in CheckMate 9ER: an open-label, randomised, phase 3 trial. Lancet Oncol 2022;23(2):292–303. Available from: https://doi.org/10.1016/S1470-2045(21)00693-8.

[123] Plimack ER, Powles T, Stus V, et al. Pembrolizumab plus axitinib versus sunitinib as first-line treatment of advanced renal cell carcinoma: 43-month follow-up of the phase 3 KEYNOTE-426 study. Eur Urol 2023;84(5):449–54. Available from: https://doi.org/10.1016/j.eururo.2023.06.006.

[124] Bedke J, Rini BI, Plimack ER, et al. Health-related quality of life analysis from KEYNOTE-426: pembrolizumab plus axitinib versus sunitinib for advanced renal cell carcinoma. Eur Urol 2022;82(4):427–39. Available from: https://doi.org/10.1016/j.eururo.2022.06.009.

[125] Cella D, Grünwald V, Escudier B, et al. Patient-reported outcomes of patients with advanced renal cell carcinoma treated with nivolumab plus ipilimumab versus sunitinib (CheckMate 214): a randomised, phase 3 trial. Lancet Oncol 2019;20(2):297–310. Available from: https://doi.org/10.1016/S1470-2045(18)30778-2.

[126] Motzer R, Porta C, Alekseev B, et al. Health-related quality-of-life outcomes in patients with advanced renal cell carcinoma treated with lenvatinib plus pembrolizumab or everolimus versus sunitinib (CLEAR): a randomised, phase 3 study. Lancet Oncol 2022;23(6):768–80. Available from: https://doi.org/10.1016/S1470-2045(22)00212-1.

Chapter 41

Sarcomas

Pampina Pilavaki[1], Poorni Priya Jaganathan[2], Robin Jones[3], and Anastasia Constantinidou[4]

[1]*Barking, Havering and Redbridge University Hospitals NHS Trust, London, United Kingdom,* [2]*Chelsea and Westminster Hospital NHS Foundation Trust, London, United Kingdom,* [3]*Royal Marsden NHS Foundation Trust, London, United Kingdom,* [4]*Medical School, University of Cyprus, Nicosia, Cyprus; Bank of Cyprus Oncology Centre, Nicosia, Cyprus*

Key learning objectives

- Understand the burden and complexity of pain in patients with soft tissue and bone sarcomas.
- Recognize the need for active pain screening, validated assessment tools, and regular review of analgesic requirements in sarcoma care.
- Describe the role of multimodal pain management, including anticancer therapy, analgesics, interventional approaches, and nonpharmacological strategies.
- Explain how chemotherapy and radiotherapy may contribute to symptom relief and pain control in selected patients with advanced sarcoma.

Introduction

Sarcomas represent a rare and heterogeneous group of malignancies arising from the mesenchymal tissue, accounting for approximately 1% of all adult malignancies [1–3]. Soft tissue sarcomas (STS) and bone sarcomas (BS) comprise the two main categories of sarcomas with an estimated incidence in Europe of 4–5 cases/100,000 per year and 0.8–0.9 cases/100,000 per year, respectively [4,5]. More than 100 distinct entities of sarcomas are identified according to the latest World Health Organization (WHO) classification [6]. This diversity is further complicated by several clinical manifestations, encompassing differences in age of onset, anatomical locations, and genetic predisposition [7]. Due to the complexity of sarcomas, a multidisciplinary approach at a specialist sarcoma center is essential for the optimal management and best patient outcome [4,8].

For patients diagnosed with local/locoregional sarcomas, the standard of practice is surgical excision of the primary tumor with clear margins [4]. Additional radiotherapy (RT) and/or chemotherapy may be considered in selected cases with nonmetastatic disease in the adjuvant or neo-adjuvant setting [4,5]. Despite initial surgical management, approximately half of patients with localized intermediate- or high-grade STS will develop progression of disease and metastases associated with dismal prognosis [9,10].

Patients with metastatic sarcomas are managed with systemic treatment; however, RT and surgery may also have a role in advanced disease [4,5]. Since the early 1970s, first-line treatment has comprised anthracycline-based chemotherapy [4,11,12]. There are some exceptions to this approach, which depend on the subtype of sarcoma and the characteristics of the patient [4,5]. Additional lines of therapy include gemcitabine plus docetaxel [13], or dacarbazine combinations [14], high-dose ifosfamide [15], trabectedin [16], pazopanib [17], and eribulin [18]. Targeted therapies and new therapeutic strategies such as immunotherapy and multireceptor tyrosine kinase inhibitors have been investigated in patients with different types of sarcoma, but their application in daily clinical practice remains limited [19,20]. Despite advances in surgical and medical treatment, for the majority of sarcoma subtypes, the median overall survival in the metastatic setting ranges from 16 to 20 months [21–23].

In the advanced setting, the therapeutic goals include prolonging survival, stabilizing or reducing tumor growth, alleviating disease-related symptoms such as pain, and preserving symptom control for an extended duration of time [24–26]. Patients diagnosed with advanced sarcomas exhibit a clinically significant symptom burden, particularly pain [27–29]. Despite the substantial burden of symptoms, the extent to which systemic therapy diminishes disease-related symptoms and affects health-related quality of life (HRQoL) has rarely been assessed or integrated as the primary end point of clinical trials [25]. In general, to achieve symptom control, the general principles of cancer pain management are applied to patients with sarcomas [27,30]. The key strategies for pain management in advanced sarcomas are described in Table 41.1.

Pain Management in Palliative Care for Patients With Cancer. DOI: https://doi.org/10.1016/B978-0-443-30050-9.00045-X

TABLE 41.1 Key strategies for pain management in palliative care: insights for sarcoma patients.

- Pain is the most common disease-related physical symptom in patients with advanced sarcomas.
- Pain in sarcoma patients is a complex problem requiring a holistic and multidisciplinary approach.
- Various treatment modalities, including primary anticancer therapy, analgesic drugs, interventional analgesic therapies, and nonpharmacological approaches, can be used for optimal pain management in sarcomas.
- Opioid analgesics remain the mainstay of analgesic therapy.
- Response to chemotherapy may assist with pain control, but further investigation is warranted.
- Radiotherapy may improve pain in advanced sarcomas, but further investigation is required to identify the optimal fractionation regimen.

The burden of pain in sarcomas in clinical practice

Pain in patients with sarcomas is a complex problem, frequently correlating with mass effect on the surrounding anatomy, including tumor invasion, compression of adjacent structures, neuropathic mechanisms, and treatment-related complications [31]. Importantly, pain constitutes the most common physical symptom in patients with advanced sarcomas, necessitating prompt, continued management [27–29]. The nature and intensity of pain may vary significantly amongst patients, requiring a personalized plan of care [32].

Gough et al. conducted a retrospective study analyzing 81 patients to investigate the symptom burden prior to the start of a new treatment decision and the role of palliative care in patients with advanced STS [27]. First-line chemotherapy was given to 50 patients, second-line to 28 patients, third-line to 15 patients, and fourth-line to 7 patients. Eight patients were referred for a phase I clinical trial, and 48 patients received the best supportive care, which indicates no further active treatment recommended by the treating sarcoma team. Palliative RT was given to 18 patients (22%), aiming to reduce primary tumor size, provide analgesia, treat brain metastasis, and manage spinal cord compression. At all treatment decisions/stages of disease, the most common symptom was pain, which was reported by 50% of patients prior to the start of first-line chemotherapy (20% of them reported uncontrolled pain), 82% prior to second-line therapy, and 79% prior to best supportive care (48% were documented as having uncontrolled pain). The WHO three-step analgesic ladder was followed for pain management. Specifically, 86% of the patients required a regular "step 1" analgesic such as paracetamol, 64% required a "step 2" analgesic such as codeine, 64% used a regular "step 3" analgesic such as morphine sulfate, and 28% of the patients were given an adjunctive agent specified for neuropathic pain such as gabapentin. Overall, patients with advanced STS demonstrate a significant, progressive symptom burden, which commonly includes pain.

In another study, a cross-sectional survey performed in a sarcoma center explored symptom prevalence and severity in patients with locally advanced inoperable or metastatic STS undergoing different therapeutic approaches, including first-line palliative chemotherapy, active surveillance pre–first-line and post–first-line palliative chemotherapy, and palliative care alone [28]. One hundred and thirteen patients were included in the study; 42 started first-line palliative chemotherapy, 27 were on active surveillance pre–first-line chemotherapy, 24 were on active surveillance post–first-line chemotherapy, and 20 patients received palliative care alone. The patient-reported Memorial Symptom Assessment Scale-Short Form (MSAS-SF) was used to assess symptom burden. Symptom prevalence, severity, and MSAS-SF subscales were documented prior to the start of a new treatment. The median overall number of reported symptoms was 11, and the most common physical symptom was pain (prevalence of 77% among all patients; 95% CI 68–84), which caused "quite a bit" or "very much" distress in 24% of those cases. The lowest symptom burden was reported in the active surveillance pre–first-line chemotherapy group and increased through active surveillance post–first-line chemotherapy, first-line chemotherapy, and the palliative care group (most symptomatic), which likely reflects stages of STS "disease progression." The prevalence of pain in each treatment group was 74% (95% CI 55–87) in the active surveillance pre–first-line chemotherapy group, 79% (95% CI 64–88) in the first-line chemotherapy group, 71% (95% CI 50–85) in the active surveillance post–first-line chemotherapy group, and 85% (95% CI 63–97) in the palliative care group.

Another retrospective study investigated the patterns of physical symptoms in patients with locally advanced/metastatic sarcomas and described the palliative treatment used in the last 2 weeks of life [29]. Forty-six patients (37 STS and 9 BS) were included in the study. Pain was the most frequently reported physical symptom, affecting 43 patients (93%), with 41 patients (98%) requiring the administration of opioids. Specifically, 36 patients (88%) were treated with injectable opioids, 4 patients (10%) with oral opioids, and 1 patient (2%) with transdermal opioids. Morphine was the most frequently used opioid, which was administered to 35 patients (86%). Thirty-three patients (79%) received nonsteroidal antiinflammatory drugs and acetaminophen. Antidepressants and anticonvulsants were administered to 21% of the patients (nine patients), in conjunction with opioids, for the management of neuropathic pain. Additionally, palliative chemotherapy was administered in 21 patients (46%) during the last 2 weeks of life, and oral therapy was continued until a median of

5.6 days before death. Twelve patients proceeded with local RT, mostly to manage pain and symptoms secondary to brain metastasis. Sedative treatment was given to five patients (29%) due to uncontrolled pain within the last 2 weeks of life. Specifically, intravenous (IV) midazolam was administered to most patients. No difference in opioid mean dose and pain control could be identified between the types of sarcoma (STS vs BS), primary lesion location (trunk vs extremity), gender, and age. The conclusion was that the high burden of pain in patients with musculoskeletal sarcomas at the end of life is profound, and effective pain management is paramount.

Furthermore, a cross-sectional study explored the prevalence of pain in sarcoma patients attending outpatient clinics in a tertiary cancer referral center [31]. A total of 149 patients (including STS and BS) were recruited in the study, of whom 22% had metastatic disease. Patients who experienced pain in the previous 7 days had further assessment of the characteristics and management of the pain using pain assessment tools (Brief Pain Inventory [BPI], the Self-report Leeds Assessment of Neuropathic Symptoms and Signs [S-LANSS], and the Pain Management Index [PMI]). Within the preceding week, 53% of the patients reported pain, and 25% had significant pain. Inadequately controlled pain and neuropathic pain were reported in 63% and 36% of those with pain, respectively. Potential risk factors associated with the presence and severity of pain were further explored; however, age, gender, tumor type, and the type of cancer therapy were not implicated as predictors. The investigators concluded that patients diagnosed with sarcomas should be actively screened for pain, and their analgesic requirements should be regularly reviewed.

Overall, patients with advanced sarcomas commonly experience pain, which is the most common disease-related symptom in these patients. Importantly, a substantial number of patients suffer from inadequate pain control. The character and severity of pain may vary significantly from one patient to another. However, from the limited studies performed, no risk factors related to the presence and intensity of pain have been identified. Given the high burden of pain in patients diagnosed with sarcoma and the detrimental impact this can have on quality of life (QoL), prompt and active screening and assessment of pain, with regular review of analgesic requirements, are essential to optimize management.

Basic principles in cancer pain management

Continuous assessment and management of pain is necessary to improve patients' QoL and reduce distress and should be an integral part of the care provided to cancer patients [28,30]. Effective treatment of cancer-related pain starts with efficient evaluation of pain [33]. A fundamental component of pain assessment is the prompt and regular self-reporting of pain intensity using validated tools [30,34].

To facilitate a holistically minded model of care, an integrated approach to pain management should be followed, incorporating different treatment modalities, which may include primary anticancer therapy, analgesic drugs, interventional analgesic therapies, and a variety of other nonpharmacological approaches [30,32,35]. A balance between the benefits and burdens of the treatment plan should be sought in relation to the goals of care [32]. Crucially, patients should be encouraged to demonstrate an active role in their management and communicate with their medical team about their symptoms, the efficacy of treatment, and the side effects [30]. Pharmacological treatments are the cornerstone of managing cancer-related pain [35], with opioid analgesics remaining the mainstay of analgesic therapy [30]. Details about opioid use and other pharmacological treatments are given in other chapters of this book. It is important to maintain a balance between pain alleviation, whilst minimizing adverse effects, and the type and dose of analgesic medications should be adjusted accordingly to achieve this goal [30]. These basic principles can be applied in patients with sarcomas when indicated [27,36,37].

Role of chemotherapy and radiotherapy in pain control in sarcomas

Chemotherapy is the mainstay of treatment for the majority of non-gastrointestinal stromal tumor (non-GIST) advanced sarcomas; it is administered with palliative intent, aiming to improve quantity and QoL [38]. The presence of symptoms, especially pain, and those resulting from the mass effect of space-occupying sarcomas influence the decision-making regarding treatment [28,39]. Other factors that may affect the therapeutic approach in sarcoma patients are the histological subtype, the grade, the performance status, the goals of care, patients' comorbidities, and potential toxicities. Anticancer therapy may be the cause of pain, but it may also assist with controlling cancer-related pain [30]. However, there are limited studies supporting the analgesic effect of chemotherapy in patients with sarcomas. Several studies have been performed investigating the QoL in sarcoma patients [8,26,40–48], but pain control has not been the primary aim of those studies, though it has been investigated as a component of QoL assessment [49].

Granda-Cameron et al. conducted a pilot study including 11 patients to investigate the symptom distress and QoL in newly diagnosed patients with sarcoma (8 STS, 3 BS) receiving chemotherapy [49]. Patients completed the Edmonton Symptom Assessment Scale and the Functional Assessment of Cancer Therapy–General at baseline and on days 1, 15, and

21 of their chemotherapy treatment. Overall, during chemotherapy, increasing symptom distress and reduced QoL over time were reported. The most commonly reported symptoms during chemotherapy included anxiety, poor well-being, lack of appetite, drowsiness, and depression, while fatigue was the most prevalent and pervasive symptom reported in most cycles. Patients who proceeded with second-line therapy experienced higher levels of fatigue. Additionally, fatigue increased over time in Caucasians and patients with osteosarcoma but was not reported in African Americans and patients with STS. Symptom distress had an increasing trend over time in older patients (>48-year-old), except pain. However, pain increased over time among younger patients (<48-year-old) and women. Furthermore, in younger patients, symptom distress did not show an increase (tired, depressed, and anxious subscales) or decrease (appetite subscale) over time. Given that the anxiety and depression scores of younger patients did not change over time, the increase in pain scores could be partially explained by patients' perception of pain and its impact on their daily activities. The findings of this study should be interpreted cautiously given the small sample size, but clearly further investigation is warranted to validate such results.

Gough et al. performed a mixed-methods longitudinal study to examine what constitutes HRQoL, whether the most commonly used HRQoL assessment tool measures this, and to what extent HRQoL and its components change during and after chemotherapy [38]. Sixty-six patients with locally advanced, inoperable, or metastatic STS were enrolled in the study, of whom 42 were commencing first-line palliative chemotherapy and 24 were under surveillance postchemotherapy completion, having achieved at least stable disease. Serial European Organisation for Research and Treatment of Cancer Quality of Life Questionnaire Core 30 (EORTC QLQ-C30) questionnaires were used, and nested-qualitative semistructured interviews with a subsample of participants were performed. According to the study, HRQoL components were classified into three domains: the physical, the psychological, and the social. The participants recognized five main factors, including pain, as important components of HRQoL. These factors are examined within the EORTC QLQ-C30, whilst other factors such as loss of independence are not. For the patients receiving chemotherapy, the mean overall EORTC QLQ-C30 HRQoL score deteriorated from baseline (60.4) to the primary evaluation point (50.2) (change of –10.2, *t*-test –2.70, $P = .01$), which is consistent with patients' qualitative accounts. Regarding patients under surveillance, baseline overall HRQoL scores were higher compared to those of the chemotherapy group at baseline, which is suggestive of a correlation with chemotherapy response and longer-term improvement in HRQoL. The presence and control of physical symptoms were of great importance in maintaining HRQoL, as indicated by both the HRQoL scores and qualitative accounts. Interestingly, in the chemotherapy group, fatigue worsened after cycle 2 (baseline 41.7–52.8; change of +11.1, *t*-test +2.51, $P < .05$), but pain (baseline 41.5–32.1; change –9.4, *t*-test –2.06, $P < .05$) and sleep disturbance (baseline 43.1–28.5; change –14.6, *t*-test –3.05, $P < .05$) were improved. Overall, the EORTC QLQ-C30 measures some but not all components of HRQoL in sarcoma patients, which indicates the necessity to develop an STS-specific module within the EORTC QLQ-C30 or the creation of a new STS-specific HRQoL tool. Also, first-line chemotherapy with palliative intent improves specific symptoms that are prevalent and affect HRQoL and therefore may improve HRQoL.

Another longitudinal study explored the change or stability in sarcoma patients' QoL and attempted to identify the distinct trajectories of change from the diagnostic to treatment phase [50]. Additionally, demographic and clinical predictors of QoL during treatment were investigated. QoL of 36 patients with nonadvanced primary BS and STS was evaluated using the EORTC QLQ-C30 (Portuguese version) at the time of diagnosis (baseline) and again at 3–6 months after the beginning of treatment (neoadjuvant or adjuvant chemotherapy, RT, or both). Generally, patients experienced diminished global health/QoL, poorer functioning, more symptoms, and more financial difficulties compared to the general population, both in the diagnostic and treatment phases. During treatment, patients reported worse physical functioning and more fatigue, nausea, vomiting, and appetite loss but a better overall QoL and less pain compared to the time of diagnosis. The better global health/QoL during treatment may be explained by the improvement of pain, whilst reduction of pain could be a result of the combination of pain management through analgesic medication and reduction of tumor size in response to treatment [50].

An important treatment modality for the management of sarcoma patients is RT. Clinical practice guidelines address the indications for RT, sequencing of local therapies, appropriate dose, and planning techniques for the management of primary, operable, localized STS in adults [51]. RT is recommended for patients with extremity and superficial truncal STS who are at high risk of local recurrence of resected tumor. Additionally, RT demonstrates a significant role in the management of sarcoma patients with nonresectable tumor, oligometastatic disease, or for palliation [52]. The application of RT is considered an effective approach to control pain or other local complications in sarcomas. Focal pain associated with mass effect or local destruction by a tumor, especially bone pain, can be amenable to palliative RT [32]. However, available data regarding the optimal dose, fractionation schedule, efficacy, and side effects specifically for sarcoma patients are limited [52,53]. Therefore dose and fractionation follow the general rules for palliative RT established for different body sites regardless of histology, which leads to a variety of fractionation concepts [52]. The Royal College of Radiologists' *Radiotherapy dose fractionation guidance* (fourth edition) recommends that the selection of dose fractionation schedules should be individualized, and different dose fractionation regimens for palliative purposes for sarcoma patients are proposed [54]. Moreover, selected sarcoma patients with local disease may obtain more durable local control with higher total doses [54]. The European Society for Medical Oncology (ESMO) recommends external beam radiation therapy, 8 Gy single dose, for cancer patients suffering from painful bone metastases, whilst patients with recurrent bone pain after previous irradiation should be offered reirradiation with a further dose of 8 Gy [30]. Additionally, ESMO recommends that stereotactic body radiation therapy should be considered, preferably within clinical trials, for patients with oligometastases, good performance status, and well-controlled primary sites [30].

Soyfer et al. presented their findings of the use of hypofractionated RT for palliation of sarcoma metastases [55]. Seventeen patients with symptomatic metastatic sarcomas, including STS and BS, who experienced either pain or discomfort were included in the study. A short and intensive course of RT was followed; specifically, 39 Gy was given in 13 fractions of 3 Gy/day, five times a week. Durable pain control was reported in 12 out of 15 cases treated for gross metastases, and the treatment was well tolerated. Acute side effects included grade 1 skin toxicity, and late side effects included skin pigmentation and induration of irradiated soft tissues.

Tween et al. performed a retrospective study aiming to assess the effectiveness of RT in providing symptomatic improvement for advanced sarcomas [53]. One hundred and five patients were included in the study with 137 treated lesions (114 STS and 23 BS), located in 15 different anatomical sites. In total, 25 different dose fractionation schedules were used, including total doses of 8–60 Gy in 1–28 fractions, whilst 20 Gy in 5 fractions was the most common fractionation, accounting for 26%. The most common indication for RT was pain, experienced by 74 patients. Symptomatic improvement was reported in 70% of STS and 55% of BS at 3 months post-RT. The response rate was higher in patients with STS compared to BS; specifically, all patients with angiosarcoma and leiomyosarcoma reported an improvement in symptoms, while 25% of patients with osteosarcomas and chondrosarcomas experienced an improvement. Efficacy of different fractionation schedules was investigated; for comparison purposes, each schedule was converted to biological effective dose (BED), and an alpha–beta ratio (α/β) of four for tumor was used. An increase in symptomatic response rate was observed with increasing BED up to 50 Gy, but the response rate did not improve further beyond this point. Large-volume, multicenter, prospective studies are essential to better understand the most appropriate dose and fractionation regimes for different sarcoma subtypes.

Conclusion

Sarcomas constitute a rare group of malignancies characterized by significant heterogeneity. Patients with advanced disease demonstrate poor prognosis and suffer from a high symptom burden, with pain reported as the most common physical symptom in the advanced setting. The characteristics of pain vary significantly among patients, necessitating a personalized approach to each patient. Prompt, effective, and continued assessment and management of pain should be an integral part of the care provided to patients with sarcomas. Improvement of cancer-related pain in sarcomas is likely multifactorial. Different treatment modalities may be required, such as analgesic drugs, interventional analgesic therapies, and nonpharmacological approaches (i.e., psychological therapy and rehabilitative interventions). Response to chemotherapy and/or RT may also contribute to pain control in sarcomas, but further research is required to investigate the extent to which chemotherapy contributes to pain management and to identify the optimal dose fractionation regimen for palliation for the different sarcoma subtypes.

References

[1] Hui JYC. Epidemiology and etiology of sarcomas. Surg Clin North Am 2016;96(5):901–14. Available from: https://doi.org/10.1016/j.suc.2016.05.005.

[2] Sbaraglia M, Bellan E, Dei Tos AP. The 2020 WHO classification of soft tissue tumours: news and perspectives. Pathologica 2021;113(2):70–84. Available from: https://doi.org/10.32074/1591-951X-213.

[3] Stiller CA, Trama A, Serraino D, et al. Descriptive epidemiology of sarcomas in Europe: report from the RARECARE project. Eur J Cancer 2013;49(3):684–95. Available from: https://doi.org/10.1016/j.ejca.2012.09.011.

[4] Gronchi A, Miah AB, Dei Tos AP, et al. Soft tissue and visceral sarcomas: ESMO–EURACAN–GENTURIS Clinical Practice Guidelines for diagnosis, treatment and follow-up☆. Ann Oncol 2021;32(11):1348–65. Available from: https://doi.org/10.1016/j.annonc.2021.07.006.

[5] Strauss SJ, Frezza AM, Abecassis N, et al. Bone sarcomas: ESMO–EURACAN–GENTURIS–ERN PaedCan Clinical Practice Guideline for diagnosis, treatment and follow-up. Ann Oncol 2021;32(12):1520–36. Available from: https://doi.org/10.1016/j.annonc.2021.08.1995.

[6] WHO Classification of Tumours Editorial Board. Soft Tissue and Bone Tumors. WHO Clarification of Tumors. 5th ed. vol. 3. Lyon: International Agency for Research on Cancer; 2020.

[7] Arifi S, Constantinidou A. Editorial: precision medicine in sarcomas: the road to an effective biomarker-driven-care strategy. Front Oncol 2023;13:1194593. Available from: https://doi.org/10.3389/fonc.2023.1194593.

[8] Andritsch E, Beishon M, Bielack S, et al. ECCO essential requirements for quality cancer care: soft tissue sarcoma in adults and bone sarcoma. A critical review. Crit Rev Oncol Hematol 2017;110:94–105. Available from: https://doi.org/10.1016/j.critrevonc.2016.12.002.

[9] Bajpai J, Susan D. Adjuvant chemotherapy in soft tissue sarcomas…Conflicts, consensus, and controversies. South Asian J Cancer 2016;5(1):15–9. Available from: https://doi.org/10.4103/2278-330X.179687.

[10] Dangoor A, Seddon B, Gerrand C, et al. UK guidelines for the management of soft tissue sarcomas. Clin Sarcoma Res 2016;6(1):20. Available from: https://doi.org/10.1186/s13569-016-0060-4.

[11] Meyer M, Seetharam M. First-line therapy for metastatic soft tissue sarcoma. Curr Treat Options Oncol 2019;20(1):6. Available from: https://doi.org/10.1007/s11864-019-0606-9.

[12] O'Bryan RM, Luce JK, Talley RW, et al. Phase II evaluation of adriamycin in human neoplasia. Cancer 1973;32(1):1–8. Available from: https://doi.org/10.1002/1097-0142(197307)32:1<1::aid-cncr2820320101>3.0.co;2-x.

[13] Maki RG, Wathen JK, Patel SR, et al. Randomized phase II study of gemcitabine and docetaxel compared with gemcitabine alone in patients with metastatic soft tissue sarcomas: results of sarcoma alliance for research through collaboration study 002 [corrected]. J Clin Oncol 2007;25(19):2755–63. Available from: https://doi.org/10.1200/jco.2006.10.4117.

[14] García-Del-Muro X, López-Pousa A, Maurel J, et al. Randomized phase II study comparing gemcitabine plus dacarbazine versus dacarbazine alone in patients with previously treated soft tissue sarcoma: a Spanish Group for Research on Sarcomas study. J Clin Oncol 2011;29(18):2528–33. Available from: https://doi.org/10.1200/jco.2010.33.6107.

[15] Martin-Liberal J, Alam S, Constantinidou A, et al. Clinical activity and tolerability of a 14-day infusional ifosfamide schedule in soft-tissue sarcoma. Sarcoma 2013;2013:1–6. Available from: https://doi.org/10.1155/2013/868973.

[16] Demetri GD, von Mehren M, Jones RL, et al. Efficacy and safety of trabectedin or dacarbazine for metastatic liposarcoma or leiomyosarcoma after failure of conventional chemotherapy: results of a phase III randomized multicenter clinical trial. J Clin Oncol 2016;34(8):786–93. Available from: https://doi.org/10.1200/JCO.2015.62.4734.

[17] van der Graaf WT, Blay JY, Chawla S, et al. Pazopanib for metastatic soft-tissue sarcoma (PALETTE): a randomised, double-blind, placebo-controlled phase 3 trial. Lancet 2012;379(9829):1879–86. Available from: https://doi.org/10.1016/S0140-6736(12)60651-5.

[18] Demetri GD, Schöffski P, Grignani G, et al. Activity of eribulin in patients with advanced liposarcoma demonstrated in a subgroup analysis from a randomized phase III study of eribulin versus dacarbazine. J Clin Oncol 2017;35(30):3433–9. Available from: https://doi.org/10.1200/JCO.2016.71.6605.

[19] Fleuren EDG, Vlenterie M, van der Graaf WTA. Recent advances on anti-angiogenic multi-receptor tyrosine kinase inhibitors in osteosarcoma and Ewing sarcoma. Front Oncol 2023;13. Available from: https://doi.org/10.3389/fonc.2023.1013359.

[20] Pilavaki P, Panagi M, Arifi S, et al. Exploring the landscape of immunotherapy approaches in sarcomas. Front Oncol 2023;12. Available from: https://doi.org/10.3389/fonc.2022.1069963.

[21] Billingsley KG, Burt ME, Jara E, et al. Pulmonary metastases from soft tissue sarcoma: analysis of patterns of disease and postmetastasis survival. Ann Surg 1999;229(5):602–10.; discussion 610–2 Available from: https://doi.org/10.1097/00000658-199905000-00002.

[22] Ryan CW, Merimsky O, Agulnik M, et al. PICASSO III: A phase III, placebo-controlled study of doxorubicin with or without palifosfamide in patients with metastatic soft tissue sarcoma. J Clin Oncol 2016;34(32):3898–905. Available from: https://doi.org/10.1200/JCO.2016.67.6684.

[23] Tap WD, Wagner AJ, Schöffski P, et al. Effect of doxorubicin plus olaratumab vs doxorubicin plus placebo on survival in patients with advanced soft tissue sarcomas: the ANNOUNCE randomized clinical trial. JAMA 2020;323(13):1266–76. Available from: https://doi.org/10.1001/jama.2020.1707.

[24] Schöffski P, Cornillie J, Wozniak A, et al. Soft tissue sarcoma: an update on systemic treatment options for patients with advanced disease. Oncol Res Treat 2014;37(6):355–62. Available from: https://doi.org/10.1159/000362631.

[25] Younger E, Jones RL, Desar IME, et al. Health-related quality Of Life In patients with advanced Soft TIssue sarcomas treated with Chemotherapy (The HOLISTIC study): protocol for an international observational cohort study. BMJ Open 2020;10(6):e035171. Available from: https://doi.org/10.1136/bmjopen-2019-035171.

[26] Younger E, Jones RL, den Hollander D, et al. Priorities and preferences of advanced soft tissue sarcoma patients starting palliative chemotherapy: baseline results from the HOLISTIC study. ESMO Open 2021;6(5):100258. Available from: https://doi.org/10.1016/j.esmoop.2021.100258.

[27] Gough NJ, Smith C, Ross JR, et al. Symptom burden, survival and palliative care in advanced soft tissue sarcoma. Sarcoma 2011;2011:325189. Available from: https://doi.org/10.1155/2011/325189.

[28] Gough N, Koffman J, Ross JR, et al. Symptom burden in advanced soft-tissue sarcoma. J Pain Symptom Manage 2017;53(3):588–97. Available from: https://doi.org/10.1016/j.jpainsymman.2016.10.357.

[29] Kawashima H, Ariizumi T, Yamagishi T, et al. Symptom burden and end-of-life palliative treatments during the last two weeks of life in patients with advanced musculoskeletal sarcoma. J Palliat Med 2019;22(8):908–14. Available from: https://doi.org/10.1089/jpm.2018.0415.

[30] Fallon M, Giusti R, Aielli F, et al. Management of cancer pain in adult patients: ESMO Clinical Practice Guidelines. Ann Oncol 2018;29:iv166–91. Available from: https://doi.org/10.1093/annonc/mdy152.

[31] Kuo PY, Yen JT, Parker GM, et al. The prevalence of pain in patients attending sarcoma outpatient clinics. Sarcoma 2011;2011:1–6. Available from: https://doi.org/10.1155/2011/813483.

[32] Portenoy RK. Treatment of cancer pain. Lancet 2011;377(9784):2236–47. Available from: https://doi.org/10.1016/S0140-6736(11)60236-5.

[33] Portenoy RK, Ahmed E. Cancer pain syndromes. Hematol Oncol Clin North Am 2018;32(3):371–86. Available from: https://doi.org/10.1016/j.hoc.2018.01.002.
[34] Fink RM, Brant JM. Complex cancer pain assessment. Hematol Oncol Clin North Am 2018;32(3):353–69. Available from: https://doi.org/10.1016/j.hoc.2018.01.001.
[35] Paice JA, Ferrell B. The management of cancer pain. CA Cancer J Clin 2011;61(3):157–82. Available from: https://doi.org/10.3322/caac.20112.
[36] Rodríguez SM, Yamak Altinpulluk E, Oliver-Fornies P, et al. Postoperative analgesia and palliative care of PENG block for terminal osteosarcoma. J Palliat Care 2022;37(4):469–70. Available from: https://doi.org/10.1177/08258597221121319.
[37] Chen YC, Lin FS, Sun WZ. Percutaneous cervical cordotomy for managing pain in terminal malignant sarcoma and the accompanied shared decision-making: a case report. Asian J Anesthesiol 2019;57(4):125–7. Available from: https://doi.org/10.6859/aja.201912_57(4).0003.
[38] Gough N, Koffman J, Ross JR, et al. Does palliative chemotherapy really palliate and are we measuring it correctly? A mixed methods longitudinal study of health related quality of life in advanced soft tissue sarcoma. PLoS One 2019;14(9):e0210731. Available from: https://doi.org/10.1371/journal.pone.0210731.
[39] Sleijfer S, Judson I, Demetri GD. How to decide whether to offer and use 'nonstandard' therapies in patients with advanced sarcomas and gastrointestinal stromal tumors: global variations in clinical practice, assessment, and access to therapies in diseases with limited incidence and data. Am Soc Clin Oncol Educ Book 2012(32):645–50. Available from: https://doi.org/10.14694/edbook_am.2012.32.304.
[40] Eichler M, Hentschel L, Richter S, et al. The Health-Related Quality of Life of Sarcoma Patients and Survivors in Germany—Cross-Sectional Results of a Nationwide Observational Study (PROSa). Cancers (Basel) 2020;12(12):3590. Available from: https://doi.org/10.3390/cancers12123590.
[41] Hinds PS, Gattuso JS, Billups CA, et al. Aggressive treatment of non-metastatic osteosarcoma improves health-related quality of life in children and adolescents. Eur J Cancer 2009;45(11):2007–14. Available from: https://doi.org/10.1016/j.ejca.2009.04.020.
[42] den Hollander D, Van der Graaf WTA, Fiore M, et al. Unravelling the heterogeneity of soft tissue and bone sarcoma patients' health-related quality of life: a systematic literature review with focus on tumour location. ESMO Open 2020;5(5):e000914. Available from: https://doi.org/10.1136/esmoopen-2020-000914.
[43] Le Cesne A, Blay JY, Cupissol D, et al. A randomized phase III trial comparing trabectedin to best supportive care in patients with pre- treated soft tissue sarcoma: T-SAR, a French Sarcoma Group trial. Ann Oncol 2021;32(8):1034–44. Available from: https://doi.org/10.1016/j.annonc.2021.04.014.
[44] Poveda A, López-Pousa A, Martín J, et al. Phase II Clinical Trial With Pegylated Liposomal Doxorubicin (CAELYX®/Doxil®) and Quality of Life Evaluation (EORTC QLQ-C30) in Adult Patients With Advanced Soft Tissue Sarcomas: a study of the Spanish Group for Research in Sarcomas (GEIS). Sarcoma 2005;9(3–4):127–32. Available from: https://doi.org/10.1080/13577140500287024.
[45] Reichardt P, Leahy M, Garcia Del Muro X, et al. Quality of life and utility in patients with metastatic soft tissue and bone sarcoma: the sarcoma treatment and burden of illness in North America and Europe (SABINE) study. Sarcoma 2012;2012:1–11. Available from: https://doi.org/10.1155/2012/740279.
[46] Calaminus G, Jenney M, Hjorth L, et al. Quality of Life of Patients With Osteosarcoma in the European American Osteosarcoma Study-1 (EURAMOS-1): development and implementation of a questionnaire substudy. JMIR Res Protoc 2019;8(8):e14406. Available from: https://doi.org/10.2196/14406.
[47] Calaminus G, Jenney M, Hjorth L, et al. Quality of life (QoL) in osteosarcoma: first results of the presurgery treatment period of EURAMOS-1 (NCT00134030). J Clin Oncol 2010;28(15_suppl.):10062. Available from: https://doi.org/10.1200/jco.2010.28.15_suppl.10062.
[48] Day JR, Miller B, Loeffler BT, et al. Patient reported quality of life in young adults with sarcoma receiving care at a sarcoma center. Front Psychol 2022;13:871254. Available from: https://doi.org/10.3389/fpsyg.2022.871254.
[49] Granda-Cameron C, Hanlon AL, Lynch MP, et al. Experience of newly diagnosed patients with sarcoma receiving chemotherapy. Oncol Nurs Forum 2011;38(2):160–9. Available from: https://doi.org/10.1188/11.onf.160-169.
[50] Paredes T, Pereira M, Moreira H, et al. Quality of life of sarcoma patients from diagnosis to treatments: predictors and longitudinal trajectories. Eur J Oncol Nurs 2011;15(5):492–9. Available from: https://doi.org/10.1016/j.ejon.2011.01.001.
[51] Salerno KE, Alektiar KM, Baldini EH, et al. Radiation therapy for treatment of soft tissue sarcoma in adults: executive summary of an ASTRO clinical practice guideline. Pract Radiat Oncol 2021;11(5):339–51. Available from: https://doi.org/10.1016/j.prro.2021.04.005.
[52] Roeder F. Radiation therapy in adult soft tissue sarcoma—current knowledge and future directions: a review and expert opinion. Cancers (Basel) 2020;12(11):3242. Available from: https://doi.org/10.3390/cancers12113242.
[53] Tween H, Peake D, Spooner D, et al. Radiotherapy for the palliation of advanced sarcomas—the effectiveness of radiotherapy in providing symptomatic improvement for advanced sarcomas in a single centre cohort. Healthcare (Basel) 2019;7(4):120. Available from: https://doi.org/10.3390/healthcare7040120.
[54] Radiotherapy dose fractionation. 4th ed. London: The Royal College of Radiologists; 2024. Available from: https://www.rcr.ac.uk/media/0d4mrqi5/radiotherapy-dose-fractionation-fourth-edition.pdf.
[55] Soyfer V, Corn BW, Kollender Y, et al. Radiation therapy for palliation of sarcoma metastases: a unique and uniform hypofractionation experience. Sarcoma 2010;2010:927972. Available from: https://doi.org/10.1155/2010/927972.

Chapter 42

Melanomas

Davorin Herceg[1,2], Luka Simetić[1], and Viktor Šabarić[1]
[1]Department of Medical Oncology, Clinical Hospital Center Zagreb, Zagreb, Croatia, [2]Faculty of Dental Medicine, University of Zagreb, Zagreb, Croatia

Key learning objectives

- Epidemiological trends of cutaneous-type melanoma and differentiation from other types of melanoma (mucosal melanoma and uveal melanoma).
- Determine the success of systemic treatment for cutaneous melanoma.
- High impact of pain in patients with metastatic melanoma (data from cohort studies).
- Impact of nonsteroidal antiinflammatory drugs on checkpoint immunotherapy.
- Define the association between the activation of the programmed cell death-1 /programmed death-ligand 1 pathway and pain in melanoma patients.

Introduction

Melanoma is a malignant tumor originating from melanocytes, which are more abundant in the skin (90%) and in the middle layer of the eye (the uvea) but can also be found in other organs, such as the inner ear, meninges, heart, and mucosal tissues such as the intestines, vagina, and paranasal sinuses. Uveal and mucosal melanomas occur more rarely (1%–5%) and exhibit different biological behavior based on a different genetic basis. The incidence of skin melanoma has continuously risen, particularly in Caucasian populations: 25/100,000 in Europe, 30/100,000 in the United States, and 60/100,000 in Australia and New Zealand [1]. Melanoma mortality also steadily increased until 2013. Mortality rates in highly developed Western countries decreased by 18% within 3 years [2], at least partially due to the success of novel systemic therapies, such as checkpoint immunotherapy and targeted therapy, used in adjuvant and metastatic settings [3].

Melanoma arises from a complex interaction of environmental and hereditary factors. UV ray exposure is the most important environmental risk factor for melanoma development. It is connected with several oncogenic aberrations (mutations like v-raf murine sarcoma viral oncogene homolog B1 [BRAF], neuroblastoma RAS viral oncogene homolog [NRAS], and KIT proto-oncogene receptor tyrosine kinase [KIT]). Phenotypic risk factors are a lighter skin type and blue eyes (Nordic phenotype), a high nevus count, and the presence of atypical nevi. Familial melanoma occurs in at least two first-degree relatives as a result of germline mutations (cyclin-dependent kinase inhibitor 2A [CDKN2A], melanocortin 1 receptor [MC1R], and BRCA1-associated protein 1 [BAP1]) or shared environmental risk [4]. In the prevention of melanoma, the key is to avoid exposure to UV rays. Sun-protective behavior includes avoiding the sun during peak insolation times, as well as regular use of sunscreen and protective clothing, hats, and sunglasses. Screening programs reveal a higher number of thin melanomas, but as the incidence of thick melanomas does not decrease at the same time, there is limited evidence that screening programs reduce melanoma-specific mortality [5].

Mucosal melanoma risk factors are poorly understood, and the pathogenesis remains unclear. The molecular landscape includes a lower incidence of BRAF oncogene mutations but a high frequency of KIT mutations, suggesting a distinct genetic origin compared to cutaneous melanoma [6].

Melanoma signs are typically referred to by the "ABCDE" acronym (asymmetry, irregular or blurred borders, color changes, diameter more than 6 mm, and "evolving" changes in size, color, etc.). Pain and pruritus are not significant signs of primary cutaneous melanoma, but an itching mole can sometimes be a sign of evolving melanoma. Typical signs of uveal melanoma are visual disturbances. Pain is a rare sign of uveal melanoma (<1%) and can be attributed to secondary

Pain Management in Palliative Care for Patients With Cancer. DOI: https://doi.org/10.1016/B978-0-443-30050-9.00031-X

glaucoma or tumor-related necrosis, uveal effusion, choroiditis, or localized scleritis [7]. The symptoms of mucosal melanoma can vary depending on the location of the cancer. In contrast to cutaneous melanoma, pain, along with bleeding and the formation of a mass lesion, is a common sign of primary mucosal melanoma.

Primary melanoma is typically managed with wide excision, and in cutaneous melanoma, sentinel lymph node biopsy (SLNB) or complete lymph node dissection (CLND) in case of nodal disease is performed [8]. However, even these procedures can result in significant long-term pain. Chronic postoperative pain is suggested to be neuropathic in 31%–66% of patients with SLNB and 82%–89% of patients with CLND. In patients with SLNB, 2% reported at least moderate pain, and 2% reported that the pain had an impact on their quality of life (QoL). After CLND, 12% reported at least moderate pain, and 14% reported that pain had an impact on their QoL 3–48 months after surgery [9].

Cutaneous melanoma with a high risk of recurrence of local disease or metastasis (stages IIB–C, IIIB–D, and resected stage IV) is treated with adjuvant therapy, either with BRAF/mitogen-activated protein kinase (MEK) targeted therapy or immunotherapy with programmed cell death-1 (PD-1) checkpoint immunotherapy [10–14]. Recently, neoadjuvant therapy has also been developed in the treatment of clinically evident stage III melanoma [15,16].

Treatment of metastatic melanoma

Fifteen years ago, stage IV melanoma had an extremely poor prognosis; even with chemotherapy, the 2-year survival was less than 10%. However, treatment outcomes have improved dramatically after the introduction of targeted therapy and immunotherapy for the treatment of stage IV cutaneous melanoma. The combination of BRAF and MEK inhibitors acts by inhibiting kinases further downstream in the mitogen-activated protein kinase (MAPK) cascade, also known as the Ras–Raf–MEK–ERK pathway. This prevents some of the activated MAPK signaling and provides more potent and durable inhibition of extracellular signal-regulated kinase (ERK) signaling. In targeted therapy, three combinations of BRAF/MEK inhibitors (vemurafenib + cobimetinib, dabrafenib + trametinib, and encorafenib + binimetinib) have been registered, with small differences in efficacy but with less toxicity in the last combination (encorafenib + binimetinib). All three combinations achieve a significant response in the first year of treatment, but due to the emergence of acquired resistance, the median overall survival (mOS) is between 25 and 33 months, and about 35% of patients have a 5-year survival. Importantly, all three combinations of medications have similar clinical effectiveness, with over 90% of patients having control of the disease and over 60% showing a response rate, usually within the first 6 months of therapy. Unsurprisingly, this results in significant improvement in QoL and pain scales, showing that rapid and significant control of pain can be achieved using BRAF/MEK inhibitors alone [17–19]. Data from the trials also show that visceral metastases (79%) are more common than nonvisceral (21%) [18], which can result in pain due to nociceptive, neuropathic, or mixed mechanisms.

Checkpoint immunotherapy (anti-PD-1, anti-cytotoxic T-lymphocyte–associated protein 4 [CTLA-4] and anti-lymphocyte-activation gene 3 [LAG-3]), which works by inhibiting the interaction between tumors and immune cells, can produce durable survival in a subset of responders. The anti-PD-1 inhibitors pembrolizumab and nivolumab have superior efficacy in comparison to ipilimumab (an anti-CTLA-4 inhibitor), with a higher response rate of 33%–44%. After 10 years, 34.0% of patients who received pembrolizumab remained alive [20,21]. On the other hand, long-term follow-up data suggest that the most efficacious treatment option is the combination immunotherapy (ipilimumab + nivolumab), which yielded the highest response rate and the longest survival (10-year survival rate of 43% and mOS of 71.9 months) but at the cost of toxicity (grade 3 and 4 adverse events in 55% of patients) [22]. However, despite higher toxicity, nivolumab + ipilimumab and nivolumab alone maintained health-related quality of life (HRQoL), and no deterioration was observed over time compared with ipilimumab. No changes in HRQoL were observed for patients who discontinued treatment for any reason in the later follow-up. Pain was evaluated using the EuroQol 5-Dimension (EQ-5D) questionnaire, but no clinically meaningful changes were observed in any group while on treatment at any time point [23].

The next immunotherapy combination, nivolumab + relatlimab (an anti-LAG-3 inhibitor), has a significantly lower rate of grade 3 or 4 treatment-related adverse events (TRAE) (18.9%) and improved efficacy regarding progression-free survival at 12 months compared to nivolumab (47.7% vs 36%). Surprisingly, despite the significant difference in grade 3 or 4 TRAE between the two arms, there was no significant difference in QoL scores. While this could be due to the limitations of questionnaires, it could also signal that patients adjust to side effects over time, and benefits may offset side effects. However, only the combination of nivolumab and ipilimumab managed a response rate close to 60%, which is expected to be achieved with BRAF/MEK inhibitors. [24]. The selection of an appropriate therapy in the first line of stage IV melanoma is challenging. According to results from the SECOMBIT and DREAMseq trials, ipilimumab + nivolumab is the preferable first-line therapy option for patients with advanced BRAF-mutant melanoma, as initiation of the dual checkpoint blockade yields significantly longer survival despite a similar early response rate of 60%. BRAF/MEK inhibitors do tend to work more quickly than immunotherapy and remain an option for the treatment of BRAF-mutated patients with metastases in critical locations, where there are contraindications to immunotherapy (e.g., transplanted patients) or where a quick response is required. Also, while no robust data exist, due to the mechanism of action, it is expected that for patients in pain, BRAF/MEK inhibition could yield a quicker response. No markers are sufficiently robust for decision-making in a metastatic melanoma setting beyond the BRAF status. Unfortunately, there are no data comparing QoL and pain scores in the DREAMseq or SECOMBIT trials, which could directly compare the effect of the drugs on pain scores and potential pain management [25,26].

Pain and melanoma

According to some accounts, around 20% of the global adult population suffers from pain, and the pain burden in people diagnosed with cancer is even more pronounced. Up to 59% of cancer patients receiving treatment are affected by pain, which impoverishes their physical performance and emotional status. Due to increasing cancer survivorship and demographic shifts, cancer prevalence worldwide is on the rise, highlighting the need for well-being in cancer survivors [27]. So far, there have been mostly small-sample, single-cancer studies regarding QoL in community-dwelling cancer survivors.

Joshy et al. gathered data from surveys completed by 16,053 cancer survivors and 106,345 patients without cancer in Australia. Through a questionnaire, bodily pain and high-impact pain (defined as pain sufficient to interfere with daily activities) were evaluated for 13 cancer types, including melanoma, and stratified according to cancer stage, time since diagnosis, and recent treatment. According to the study, a slightly higher percentage of cancer survivors, 34.9%, compared to 31.3% of participants without cancer, reported bodily pain (hazard ratio [HR] 1.07), and 15.9% versus 13.1% (HR 1.13), respectively, reported high-impact pain. Furthermore, high-impact pain and bodily pain were reported more often in patients with metastatic disease, patients diagnosed within 2 years, and patients who had cancer treatment in the previous month. In melanoma patients, the occurrence of high-impact pain and bodily pain was similar to participants without cancer. However, less invasive procedures could be used to cure earlier-stage melanoma (e.g., skin excision) compared to breast or prostate cancer surgery, as it is important to emphasize the time interval of the study (2012–15), which suggests that novel therapy options such as immunotherapy were seldom used [27].

In the era of immunotherapy in cancer treatment, especially for melanoma, there is surging interest in the effects of different classes of concomitant drugs, such as analgesics, considering their possible immunomodulatory effects on immunotherapy's potency of action. Some laboratory and clinical studies have already demonstrated paracetamol's effect on the expansion of regulatory T cells, which resulted in shorter survival in cancer patients receiving immunotherapy [28]. With nonsteroidal antiinflammatory drugs (NSAIDs) being the first step in the WHO pain management ladder, Kennedy et al. conducted an analysis of the prognostic and predictive value of NSAIDs in the European Organisation for Research and Treatment of Cancer 1325/KEYNOTE-054 phase III trial of pembrolizumab versus placebo in 1019 adult patients with stage IIIA (sentinel node cancer deposit size >1 mm), IIIB, or IIIC melanoma postregional lymphadenectomy. Although some prior laboratory evidence demonstrated that NSAIDs could reduce activation of T cells, the results of this study clearly showed no negative effects of NSAIDs on recurrence-free survival and distant metastasis-free survival. Moreover, in patients receiving adjuvant pembrolizumab, concomitant use of NSAIDs was not related to an increased number of immune-related adverse events (irAEs) [29].

Apart from successfully being used to treat multiple types of cancer, including melanoma, checkpoint inhibitors affecting the PD-1/programmed death-ligand -1 (PD-L1) pathway could also have a role in pain regulation. In one study, Chen et al. [30] determined that in naive mice, exogenous application of PD-L1 induced analgesia. On the other hand, blockade of endogenous PD-L1 and PD-1 signaling resulted in hyperalgesia. In addition, PD-L1 alleviated chronic pain, including nerve injury–induced neuropathic pain and bone cancer pain in rodents, through peripheral and central actions and downgraded formalin-induced acute inflammatory pain. Furthermore, activation of PD-1 by PD-L1 inhibited action potential induction in mouse and human dorsal root ganglion neurons, which resulted in analgesic effects through both peripheral and central mechanisms. It is also significant that inhibition of the PD-1/PD-L1 pathway in melanoma-inoculated mice resulted in the aggravation of pain symptoms. Finally, the study showed that, along with immunosuppression, the PD-1/PD-L1 axis could also suppress the nociceptive system, wherein, through PD-L1 expression, tumor cells generate a susceptible environment for tumor growth. In another study involving 516 patients affected with cancer pain, Zhang et al. [31] showed that soluble PD-1 (s-PD-1) content in peripheral blood increased and PD-1 expression on T lymphocytes decreased with increasing pain severity. The proposed mechanism is that the increase in s-PD-1 results in augmented binding with PD-L1, and lower PD-1 levels decrease the possibility of binding with PD-L1, both resulting in diminishing the analgesic function of the PD-1/PD-L1 pathway and aggravation of pain. In addition, the intensity of interaction between PD-1 on the surface of T- and B-cell membranes and PD-L1 on the surface of vascular endothelial cells, tissue cells, and monocytes decreased significantly, so that the activated T cells could not promptly obtain negative regulatory signals, resulting in excessive activation and proliferation of T cells, autologous tissue damage, production of inflammatory factors, and aggravation of pain. The study's data analysis showed a particularly high PD-1-positive expression rate in melanoma patients, which could explain the lack of cancer pain in patients with early-stage melanoma, thereby evading prompt diagnosis [31].

Immune checkpoint inhibitors (ICIs) induce potentially severe and even lethal side effects called irAEs, which are considered off-target tissue damage and involve essentially every organ system. A wide range of irAEs have been described, including hypophysitis, thyroiditis, myocarditis, pneumonia, pancreatitis, hepatitis, nephritis, adrenal insufficiency, enteritis, and skin rash. Although the underlying mechanisms of irAEs are not fully understood, they include aberrant cytotoxic T-cell activation, increased autoantibody production, direct molecular mimicry, inflammatory cytokine production, complement-mediated inflammation, impaired regulatory T-cell function, and other mechanisms. Rheumatic irAEs are important in the context of bone and joint pain. In patients with metastatic melanoma, up to 5% developed arthritis, and the same frequency was noted for myalgia [32].

An increasing body of evidence suggests that ICI therapy could potentially leave a permanent effect on HRQoL in treated melanoma patients. Through various questionnaires, several studies have examined the connection between ICI therapy and HRQoL in both metastatic and adjuvant settings. Patients consistently reported deterioration in physical, cognitive, social, and emotional functioning. The most common long-term symptoms were fatigue, worry, trouble sleeping, and joint and muscle aches. Symptoms were more pronounced in patients who received combination ICI therapy, apart from muscle ache, which was more prevalent in the group that received only PD-1/PD-L1 therapy. Age and gender differences did not significantly affect HRQoL or overall health outcome, although musculoskeletal symptoms were more common in women, and deterioration in social functioning was more associated with younger patients [33–35].

The rationale behind the use of the combination of BRAF and MEK inhibitors is to prevent the aberrant activation of mutated BRAF in the context of the RAF–MEK–ERK pathway. The toxicity profiles of BRAF inhibitors in patients with metastatic disease are similar: rash, joint pain, and fatigue are recorded with overlapping incidence rates, though there are a few exceptions, such as the excessive sensitivity to sunlight caused by vemurafenib, or hyperkeratosis or dysesthesia, which seem to occur more frequently, particularly with encorafenib monotherapy. MEK inhibitors, on the other hand, mostly display rash, diarrhea, peripheral edema, and fatigue. Although a direct comparison between the three drug combinations is not possible, indirect comparisons from individual studies are feasible. One of the most common side effects associated with pain is arthralgia. Milder arthralgia occurs with vemurafenib and cobimetinib therapy in over 30% of cases, while the incidence is lower with dabrafenib + trametinib (15%) and encorafenib + binimetinib (25%). When considering grade 3–4 arthralgia in therapies with dabrafenib + trametinib and encorafenib + binimetinib, arthralgia is registered in 1% of cases, rising to 4% with vemurafenib + cobimetinib. According to these data, it is important for elderly patients with rheumatic or degenerative joint disorders to choose a BRAF/MEK combination with the lowest incidence of arthralgia. In a metaanalysis aimed at indirectly comparing the different safety profiles of BRAF and MEK inhibitors in metastatic melanoma settings, no correlation was found between the incidence of side effects and median progression-free survival ($r = 0.15$), response rate ($r = -0.10$), or complete response rate ($r = -0.09$). There was a slight positive correlation between the incidence of side effects and mOS ($r = 0.22$) [36].

Future research

Future research in the field of melanoma therapy should focus on improving diagnostic accuracy using better prognostic and predictive tools, which include classic clinicopathological characteristics of melanoma combined with multiomic analyses. Immunotherapy with PD-1 inhibitors is the foundation of modern systemic treatment for melanoma. However, there is also evidence by Chen et al. [30] in animal studies with a previously unrecognized role of PD-L1 as an endogenous pain inhibitor and neuromodulator. The information that PD-L1 can have high potency in suppressing activities of human nociceptive neurons suggests that local targeting of PD-L1/PD-1 signaling in sensory neurons may lead to the development of novel analgesics. Future studies are necessary to investigate the signaling mechanisms by which PD-L1/PD-1 regulates synaptic transmission and synaptic plasticity in the spinal cord and brain. Additionally, research on large clinical databases (real-world data) is needed to further characterize pain as an irAE.

Conclusion

Pain prevalence as a common, debilitating, and feared symptom varies by cancer type and stage and is reported in a slightly higher percentage even in cancer survivors, including melanoma patients. Nowadays, the leading role in the systemic treatment of advanced melanoma is played by immunotherapy. There is an important relationship between pain therapy with analgesics and immunotherapy, and on the other hand, checkpoint inhibitors can also exhibit pain modulation. Regarding the effects of NSAIDs, the data from the KEYNOTE-054 study clearly showed no negative effects of NSAIDs on recurrence-free survival and distant metastasis-free survival [29]. In animal models, PD-L1 has been shown to act as a neuromodulator of pain sensitivity via both peripheral and central mechanisms. PD-L1 could also control chronic pain by suppressing T-cell activation and proinflammatory responses. Exogenous application of PD-L1 induced analgesia, whereas blockade of endogenous PD-L1 and PD-1 signaling via soluble programmed death-ligand 1 (sPD-L1), PD-1 antibodies, or PD-1 deletion resulted in hyperalgesia [30]. The results of a clinical study involving 516 patients showed that the expression of PD-1 on T-cell surfaces decreased, and peripheral soluble programmed death (sPD) content increased with increasing degrees of cancer pain [31], suggesting a potential relationship between PD-1/PD-L1 signaling and cancer pain.

References

[1] Ferlay J, Ervik M, Lam FC. Cancer statistics for year 2020. International Agency for Research on cancer. https://gco.iarc.fr/today; 2020 [accessed 01.12.23].

[2] Berk-Krauss J, Stein JA, Weber J, et al. New systemic therapies and trends in cutaneous melanoma deaths among US Whites, 1986–2016. Am J Public Health 2020;110:731–3. Available from: https://doi.org/10.2105/AJPH.2020.305567.

[3] Swetter SM, Johnson D, Albertini MR, et al. NCCN guidelines® insights: melanoma: cutaneous, version 2.2024. J Natl Compr Canc Netw 2024;22(5):290–8. Available from: https://doi.org/10.6004/jnccn.2024.0036.

[4] Long GV, Swetter SM, Menzies AM, et al. Cutaneous melanoma. Lancet 2023;402:485–502. Available from: https://doi.org/10.1016/S0140-6736(23)00821-8.

[5] Ferris LK, Saul MI, Lin Y, et al. A large skin cancer screening quality initiative: description and first-year outcomes. JAMA Oncol 2017;3(8):1112–5. Available from: https://doi.org/10.1001/jamaoncol.2016.6779.

[6] Ma Y, Xia R, Ma X, et al. Mucosal melanoma: pathological evolution, pathway dependency and targeted therapy. Front Oncol 2021;11:702287. Available from: https://doi.org/10.3389/fonc.2021.702287.

[7] Rishi P, Shields CL, Khan MA, et al. Headache or eye pain as the presenting feature of uveal melanoma. Ophthalmology 2013;120:1946–7. Available from: https://doi.org/10.1016/j.ophtha.2013.06.015.

[8] Moncrieff MD, Lo SN, Scolyer RA, et al. Evaluation of the indications for sentinel node biopsy in early-stage melanoma with the advent of adjuvant systemic therapy: an international, multicenter study. Ann Surg Oncol 2022;29(9):5937–45. Available from: https://doi.org/10.1245/s10434-022-11761-4.

[9] Slagelse C, Petersen KL, Dahl JB, et al. Persistent postoperative pain and sensory changes following lymph node excision in melanoma patients: a topical review. Melanoma Res 2014;24(2):93–8. Available from: https://doi.org/10.1097/CMR.0000000000000041.

[10] Larkin J, Del Vecchio M, Mandalá M, et al. Adjuvant nivolumab versus ipilimumab in resected stage III/IV melanoma: 5-year efficacy and biomarker results from CheckMate 238. Clin Cancer Res 2023;29:3352–61. Available from: https://doi.org/10.1158/1078-0432.CCR-22-3145.

[11] Eggermont AMM, Blank CU, Mandalà M, et al. Adjuvant pembrolizumab versus placebo in resected stage III melanoma (EORTC 1325-MG/KEYNOTE-054): distant metastasis-free survival results from a double-blind, randomised, controlled, phase 3 trial. Lancet Oncol 2021;22(5):643–54. Available from: https://doi.org/10.1016/S1470-2045(21)00065-6.

[12] Luke JJ, Ascierto PA, Khattak MA, et al. Pembrolizumab versus placebo as adjuvant therapy in resected stage IIB or IIC melanoma: final analysis of distant metastasis-free survival in the phase III KEYNOTE-716 study. J Clin Oncol 2024;42(14):1619–24. Available from: https://doi.org/10.1200/JCO.23.02355.

[13] Kirkwood JM, Del Vecchio M, Weber J. Adjuvant nivolumab in resected stage IIB/C melanoma: primary results from the randomized, phase 3 CheckMate-76K trial. Nat Med 2023;29:2835–42. Available from: https://doi.org/10.1038/s41591-023-02583-2.

[14] Dummer R, Hauschild A, Santinami M, et al. Five-year analysis of adjuvant dabrafenib plus trametinib in stage III melanoma. N Engl J Med 2020;383:1139–48. Available from: https://doi.org/10.1056/NEJMoa2005493.

[15] Versluis JM, Menzies AM, Sikorska K, et al. Survival update of neoadjuvant ipilimumab plus nivolumab in macroscopic stage III melanoma in the OpACIN and OpACIN-neo trials. Ann Oncol 2023;37:423–30. Available from: https://doi.org/10.1016/j.annonc.2023.01.004.

[16] Long GV, Blank CU, Amaria RN. Long-term survival with neoadjuvant therapy in melanoma: updated pooled analysis from the International Neoadjuvant Melanoma Consortium (INMC). Ann Oncol 2024;35(suppl_2):S1232. Available from https://doi.org/10.1016/j.annonc.2024.08.2282.

[17] Dummer R, Ascierto PA, Gogas HJ. Encorafenib plus binimetinib versus vemurafenib or encorafenib in patients with BRAF-mutant melanoma (COLUMBUS): a multicentre, open-label, randomised phase 3 trial. Lancet Oncol 2018;19(5):603–15. Available from: https://doi.org/10.1016/S1470-2045(18)30142-6.

[18] Robert C, Grob JJ, Stroyakovskiy D. Five-year outcomes with dabrafenib plus trametinib in metastatic melanoma. N Engl J Med 2019;381:626–36. Available from: https://doi.org/10.1056/NEJMoa1904059.

[19] Ascierto PA, Dréno B, Larkin J, et al. 5-Year outcomes with cobimetinib plus vemurafenib in BRAFV600 mutation-positive advanced melanoma: extended follow-up of the coBRIM study. Clin Cancer Res 2021;27:5225–35. Available from: https://doi.org/10.1158/1078-0432.CCR-21-0809.

[20] Robert C, Long GV, Brady B. Five-year outcomes with nivolumab in patients with wild-type BRAF advanced melanoma. J Clin Oncol 2020;38(33):3937–46. Available from: https://doi.org/10.1200/JCO.20.00995.

[21] Long GV, Carlino MS, McNeil C, et al. Pembrolizumab versus ipilimumab for advanced melanoma: 10-year follow-up of the phase III KEYNOTE-006 study. Ann Oncol 2024;35(12):1191–9. Available from: https://doi.org/10.1016/j.annonc.2024.08.2330.

[22] Wolchok JD, Chiarion-Sileni V, Rutkowski P. Final, 10-year outcomes with nivolumab plus ipilimumab in advanced melanoma. N Engl J Med 2025;392(1):11–22. Available from: https://doi.org/10.1056/NEJMoa2407417.

[23] Schadendorf D, Larkin J, Wolchok J, et al. Health-related quality of life results from the phase III CheckMate 067 study. Eur J Cancer 2017;82:80–91. Available from: https://doi.org/10.1016/j.ejca.2017.05.031.

[24] Tawbi HA, Schadendorf D, Lipson EJ. Relatlimab and nivolumab versus nivolumab in untreated advanced melanoma. N Engl J Med 2022;386:24–34. Available from: https://doi.org/10.1056/NEJMoa2109970.

[25] Atkins MB, Lee SJ, Chmielowski B. Combination dabrafenib and trametinib versus combination nivolumab and ipilimumab for patients with advanced BRAF-mutant melanoma: the DREAMseq trial-ECOG-ACRIN EA6134. J Clin Oncol 2023;41(2):186–97. Available from: https://doi.org/10.1200/JCO.22.01763.

[26] Ascierto PA, Mandalá M, Ferrucci PF. Sequencing of ipilimumab plus nivolumab and encorafenib plus binimetinib for untreated BRAF-mutated metastatic melanoma (SECOMBIT): a randomized, three-arm, open-label phase II trial. J Clin Oncol 2023;41(2):212–21. Available from: https://doi.org/10.1200/JCO.21.02961.

[27] Joshy G, Khalatbari-Soltani S, Soga K, et al. Pain and its interference with daily living in relation to cancer: a comparative population-based study of 16,053 cancer survivors and 106,345 people without cancer. BMC Cancer 2023;23(1):774. Available from: https://doi.org/10.1186/s12885-023-11214-5.

[28] Bessede A, Marabelle A, Guégan JP, et al. Impact of acetaminophen on the efficacy of immunotherapy in cancer patients. Ann Oncol 2022;33(9):909–15. Available from: https://doi.org/10.1016/j.annonc.2022.05.010.

[29] Kennedy OJ, Glassee N, Kicinski M, et al. Prognostic and predictive value of non-steroidal anti-inflammatory drugs in the EORTC 1325/KEYNOTE-054 phase III trial of pembrolizumab versus placebo in resected high-risk stage III melanoma. Eur J Cancer 2024;201:113585. Available from: https://doi.org/10.1016/j.ejca.2024.113585.

[30] Chen G, Kim YH, Li H, et al. PD-L1 inhibits acute and chronic pain by suppressing nociceptive neuron activity via PD-1. Nat Neurosci 2017;20(7):917–26. Available from: https://doi.org/10.1038/nn.4571.

[31] Zhang J, Zhang H, Luo Y. Association between activation of the programmed cell death-1 (PD-1)/programmed death-ligand 1 (PD-L1) pathway and pain in patients with cancer. Med Sci Monit 2019;25:1275–82. Available from: https://doi.org/10.12659/MSM.912632.

[32] Dang QM, Watanabe R, Shiomi M. Rheumatic immune-related adverse events due to immune checkpoint inhibitors—a 2023 update. Int J Mol Sci 2023;24:5643. Available from: https://doi.org/10.3390/ijms24065643.

[33] Looman EL, Cheng PF, Lai-Kwon J, et al. Health-related quality of life in survivors of advanced melanoma treated with anti-PD1-based immune checkpoint inhibitors. Cancer Med 2023;12(11):12861–73. Available from: https://doi.org/10.1002/cam4.5967.

[34] Egeler M, Lai-Kwon J, Tissier R, et al. Real-world health-related quality of life outcomes for patients with resected stage III/IV melanoma treated with adjuvant anti-PD1 therapy. Eur J Cancer 2024;200:113601. Available from: https://doi.org/10.1016/j.ejca.2024.113601.

[35] Boekhout AH, Rogiers A, Jozwiak K, et al. Health-related quality of life of long-term advanced melanoma survivors treated with anti-CTLA-4 immune checkpoint inhibition compared to matched controls. Acta Oncol 2021;60(1):69–77. Available from: https://doi.org/10.1080/0284186X.2020.1818823.

[36] Garutti M, Bergnach M, Polesel J, et al. BRAF and MEK inhibitors and their toxicities: a meta-analysis. Cancers (Basel) 2023;15(1):141. Available from: https://doi.org/10.3390/cancers15010141.

Chapter 43

Hematological malignancies

Velizar Stefanov Shivarov
Department of Experimental Research, Medical University Pleven, Pleven, Bulgaria

Key learning objectives

- To acknowledge the fact that hematological malignancies (HMs) encompass a large group of entities with diverse prognoses and to understand the general principles of their current classification.
- To acknowledge the fact that HM patients are underrepresented in the palliative care facilities, but pain is a frequent symptom in HMs during end-of-life care.
- To understand that various types of pain syndromes can be encountered in HM patients in palliative settings.
- To understand that pharmacological treatment of pain in HM patients follows the general principles for management of cancer-associated patients but needs to be adapted on an individual basis depending on the type of underlying blood cancer and the fitness of the patient.
- To gain insight into the recent advances in two frequent HMs, such as diffuse large B-cell lymphoma and multiple myeloma, with special emphasis on chimeric antigen receptor T-cell therapies.

Hematological malignancies

Contemporary classification of hematological malignancies

The latest fifth edition of the World Health Organization (WHO-HAEM5) classification of hematolymphoid tumors follows the principles set with the 2008 edition to define entities based on clinicopathological principles [1,2]. The main principle to classify those entities is based on the postulated normal counterpart cell and to assign entities to main lineages (myeloid, lymphoid, or histiocytic/dendritic) has been preserved since the 2008 edition [3]. The classification recognizes the instances of lineage infidelity or plasticity and the fact that recurrent genetic abnormalities can lead to either myeloid or lymphoid malignancies, which are recognized as separate categories [2]. Another level of classification principle is to recognize precursor neoplasms (acute leukemias) versus more mature neoplasms (e.g., chronic leukemias, myeloproliferative neoplasms, and lymphomas). Classification of mature myeloid malignancies further utilizes genetic features and preservation of effective erythropoiesis to define myeloproliferative versus myelodysplastic syndromes [2]. On the other hand, classification of mature lymphoid malignancies is primarily based on clinical presentation and the postulated normal counterpart. Classification categorizes separately B-cell lymphoid proliferations and lymphomas versus T-cell and NK-cell lymphoid proliferations and lymphomas [1]. Of note, plasma cell neoplasms and other diseases with paraproteins are classified not as a separate category but rather as a subcategory under B-cell lymphoid proliferations and lymphomas [1]. For the first time the WHO-HAEM5 introduces a separate category of stroma-derived neoplasms of lymphoid tissue, which encompasses entities of mesenchymal origin specific for lymph nodes or spleen, such as hamartomas and sarcomas [1]. Finally, the classification defines a group of inherited genetic syndromes associated with hematolymphoid tumors, such as Fanconi anemia and Bloom syndrome [2].

As some of the WHO-HAEM5–recognized entities are quite rare, often for epidemiological purposes, various entities are reported collectively under general categories such as Hodgkin lymphoma, non-Hodgkin lymphomas (NHLs), multiple myeloma (MM), acute myeloid leukemia, acute lymphoid leukemia, chronic lymphocytic leukemia, chronic myeloid leukemia (CML), and other leukemias [4]. These broad categories and individual entities may have quite diverse prognoses, with overall survival (OS) ranging from decades to several weeks or months in untreated cases and therefore have different impacts on global health measures. Globally, in 2022, NHLs accounted for approximately 250,000 death

Pain Management in Palliative Care for Patients With Cancer. DOI: https://doi.org/10.1016/B978-0-443-30050-9.00021-7

cases, leukemias for approximately 305,000, and MM for approximately 121,000 [5]. According to other studies, deaths from NHLs were slightly declining between 2015 and 2019, but those from leukemias and MM were gradually but consistently increasing over the same period [6].

Furthermore, in spite of the improvement in OS for patients with hematological malignancies (HMs), another recent analysis shows that index cancer diagnosis or diagnosis of a second malignancy remains the major cause of death in almost all entities, most prominently for patients with acute leukemias. Therefore, management of HMs in the palliative setting is a common clinical scenario with some specific challenges [7–9]. Some reports, however, suggest that patients with HMs were underrepresented in the inpatient palliative care units and hospices [10,11]. The possible explanation for this is that a large number of such patients remain on active oncological treatment even in very advanced stages of diseases and that end-of-life (EoL) care for hematological malignancy patients is often delivered at home [12]. The rapidly evolving field of clinical drug development often modifies significantly and in an unpredictable fashion the individual prognosis of the HM patients [9]. This prognostic uncertainty is recognized as one of the major challenges in the management of EoL HM patients, and therefore hematologists in general underrecognize the versatility of palliative care options [9]. Frequent encounters with the healthcare system for continuous maintenance therapy, such as transfusion support, pose significant extra operational and emotional burdens to HM patients and their caregivers [11]. Indeed, such burdensome care can be one of the leading causes for enrollment of patients in specialized palliative home care programs [12]. On the other hand, a recent study suggested that pain was the second most common reason for enrolling patients in such programs, further emphasizing the importance of pain management for patients with hematologic malignancies in palliative care settings [12].

Pain syndromes in hematological malignancies

Types of pain syndromes

Pain syndromes observed in patients with HM follow the general classification of pain syndromes in cancer patients [13–15]. In terms of onset and duration, pain syndromes can be classified as acute and chronic. Acute pain syndromes are relatively rare in HM and especially in patients in palliative settings. They might be associated with specific diagnostic and therapeutic procedures and generally respond well to short-term nonopioid analgesia. Acute pain syndromes in HM palliative settings that are directly related to tumors are usually due to bone fractures (e.g., vertebral), acute obstructions (bowel obstruction or vena cava syndrome), and deep vein thromboses (DVTs). Bone fractures and acute obstruction are caused by bulky tumor masses (more frequently lymphoid), and the treatment of choice is local debulking through external beam radiotherapy, usually in single doses for rapid pain relief [14]. However, in EoL settings, depending on the general fitness of the patient, fractionated radiotherapy can be the preferred approach [14]. DVTs and pulmonary embolism in palliative settings are usually managed through anticoagulant therapy and rarely with vascular surgery [14]. Acute pain syndromes related to antineoplastic or supportive care with granulocyte colony-stimulating factor (G-CSF) therapy are not typical for palliative settings and will not be discussed further, as they are reviewed in-depth elsewhere [13,16].

Typically HM patients in palliative settings can experience chronic pain syndromes. They can be classified as nociceptive (visceral or somatic), neuropathic, or mixed-type pain syndromes [13,14]. HM-related somatic pain is caused by soft tissue and bone involvement by the tumor mass. Bone invasion can be oligofocal or multifocal, manifesting with pathological fractures, compression fractures, lytic lesions, and hyperproliferative foci. Pathological and vertebral compression fractures can be particularly common in patients with heavily pretreated advanced diseases such as MM. They can also be caused by paravertebral proliferation of extraosseous masses in patients with advanced lymphomas or acute lymphoblastic leukemia. Skin and soft tissue involvement with necrotic painful lesions are more common for advanced T-cell NHLs and some myeloid neoplasms, such as blastic plasmacytoid dendritic cell neoplasms.

Tumor-related visceral pain syndromes are commonly described as diffuse sensations of increased pressure that cannot be specifically localized [14]. They might be due to chronic bowel obstruction or bulky abdominal tumor masses. They are probably less common in HM patients than in solid cancer patients but might be of particular concern for patients with advanced myeloproliferative neoplasms such as myelofibrosis or splenic lymphomas, which are characterized by excessive splenomegaly. In patients with advanced HM, one can also expect chronic visceral pain due to chronic intestinal obstruction due to bulky tumor masses. Similarly, bulky masses can cause ureteral obstructions with unilateral or bilateral hydronephrosis and associated typical pain syndromes.

Neuropathic pain syndromes in HM patients are caused by direct invasion of neural structures by the tumor mass. This is more commonly observed in lymphoid malignancies. Leptomeningeal invasion is more commonly expected in acute lymphoblastic leukemias and can present with nonspecific headaches and other symptoms of intracranial hypertension, such as nausea, vomiting, and seizures. Radiculopathies, plexopathies, and peripheral mononeuropathies can also be observed in HM, and the presentation would depend on the localization and size of the tumor mass and its relationship to adjacent neural structures. Paraneoplastic pain syndromes are more rarely observed in HMs than in solid tumors and can present as subacute sensory gangliopathy or axonopathies. A well-known example of painful sensoropathies are immune-mediated polyneuropathies in MM and other plasma cell disorders such as Polyneuropathy, Organomegaly, Endocrinopathy, Monoclonal gammopathy, and Skin changes (POEMS) syndrome [17].

A fairly large subgroup of HM patients can develop painful peripheral neuropathies due to chemotherapy or targeted therapy. The classes of drugs used in HMs that are typically associated with such neuropathic pain are vinca alkaloids (e.g., vincristine and vinblastine), proteasomal inhibitors (e.g., bortezomib and carfilzomib), and immunomodulatory drugs (e.g., thalidomide and lenalidomide) [16]. Novel antibody-drug conjugates that are increasingly used in HM (e.g., brentuximab vedotin) frequently use microtubule formation inhibitors, and neuropathic pain is one of their common side effects [18]. It is expected that more patients with such neuropathic pain syndromes will eventually be seen in the palliative settings in the future. Furthermore, because of the severe immunosuppression during treatment of HM, some patients may experience herpes zoster reactivation both during active treatment and during EoL care. Along with the acute herpetic pain, some patients may suffer long term from postherpetic neuralgia, which is sometimes severe [16].

Finally, unique for HM patients are painful symptoms due to hematopoietic stem cell transplantation (HSCT). Acute pain syndromes that might be encountered can be visceral pains due to mucositis and veno-occlusive disease or acute graft-versus-host disease (GVHD) or bone pain due to G-CSF administration [16]. However, the HSCT-related pain syndrome that is more relevant to palliative settings is the pain in severe refractory chronic GVHD associated with gastrointestinal symptoms [16].

General principles of management of pain syndromes in hematological malignancies

Severity of pain in HM patients can be assessed directly by using standardized and widely used tools such as the Visual Analog Scale, Numerical Analog Scale, and Facial Expression Scale [14]. One can also apply the common terminology criteria for adverse events to grade pain syndromes associated with anticancer therapy [16]. Furthermore, pain sensation subjective rating can also be part of specific quality-of-life (QoL) questionnaires developed and validated for some HMs such as the Myelofibrosis Symptom Assessment Form [19] and disease-specific Functional Assessment of Cancer Therapy (FACT) questionnaires (e.g., FACT-Leu, FACT-MM, and FACT-Lym) [20]. Assessment of QoL using such questionnaires is a frequent efficacy end point in clinical trials in malignant hematology, and in EoL settings, improvement in QoL is often a primary efficacy measure [21,22].

Pharmacological management of pain in HM, including in palliative settings, follows the general principles for pain management of cancer patients according to the WHO ladder for analgesic therapy [13,14,23]. Specific dosing schedules of common analgesics used for HM patients are beyond the scope of this chapter. They are summarized in a number of excellent reviews, which can be used as practical guidelines [13,14,23]. Over the coming years it is expected that more HM patients in palliative settings will receive active therapy with novel small molecule inhibitors or immunotherapies. Some of those patients will experience pain syndromes requiring pharmacological therapy. In all such cases the treating physicians would be advised to refer to the current prescribing information in order to account for any possible drug–drug interactions with analgesic drugs when planning the overall management of the HM patient toward the EoL. Nonpharmacological approaches might be of particular value in heavily pretreated HM patients receiving palliative care, although they have not been formally investigated [14]. Psychological support and some cognitive therapy approaches have proven beneficial during the EoL period of HM patients [24]. Finally, there exist reports that complementary care measures such as acupuncture might be of value in pain syndromes associated with proteasome inhibitors, suggesting that alternative approaches deserve further exploration in the clinical settings [25].

Recent therapeutic advances in selected relapsed/refractory hematological malignancies

As discussed above, the hallmark of HMs is their biological and clinical heterogeneity with currently recognized more than 200 entities. It is important to remember that virtually all HMs are considered primary systemic diseases, and localized radiotherapy with curative intent is possible in only a minority of stage I indolent lymphomas, and therefore the majority of cases of HM require systemic chemotherapy/immunotherapy or targeted therapy. As discussed above, pain management may become a significant issue in HMs in heavily pretreated relapsed/refractory cases. For the purposes of this textbook, it is assumed that any therapeutic advances in those settings may have some potential therapeutic effect and thereby mitigate the risk of developing pain syndromes in palliative settings. It is beyond the scope of this chapter to cover all recent therapeutic advances in two major entities, which are common and may lead to pain syndromes in advanced disease settings as discussed above. We will briefly discuss below the main recent phase 3 trial results in relapsed/refractory diffuse large B-cell lymphoma (DLBCL) and MM.

Diffuse large B-cell lymphoma

Basic facts on diffuse large B-cell lymphoma with focus on contemporary therapeutic options for advanced disease

DLBCL encompasses a number of mature B-cell lymphomas with medium- to large-sized morphology [1]. Several variants are currently recognized based on morphology and molecular and genetic features [1]. Most cases are categorized as DLBCL, not otherwise specified (NOS). DLBCL, NOS accounts for 25%–30% of adult cases of NHLs worldwide and is relatively more frequent in developing countries [26]. The median age at presentation is between 60 and 70 years. The usual presentation is with palpable enlarged lymph node(s) with or without constitutional symptoms. Diagnosis is made based on excisional/core lymph node biopsy confirming typical histology and immunohistochemical features.

Initial staging is based on whole-body imaging techniques such as CT or positron emission tomography (PET)/CT [8]. In certain cases, particularly if PET/CT is used, bone marrow biopsy, often used for initial staging, can be omitted [8]. Prognostic stratification is based on the original international prognostic index for aggressive lymphomas or any of its subsequent versions [27]. The current standard of first-line therapy for fit patients is chemo-immunotherapy with the R-CHOP-21 regimen (cyclophosphamide, doxorubicin, vincristine, and prednisone every 21 days). Addition of rituximab to CHOP has been shown to lead to 90% 6-year OS and 80% progression-free survival (PFS) in younger patients below 60 years [28]. Similar outcomes were reported for patients ≥60 years; 3-year OS was 67%, and 3-year failure-free survival was 53% [29,30]. In some more aggressive DLBCL subtypes in fit patients, it is preferable to use as first-line therapy a more aggressive regimen such as dose-adjusted EPOCH-R (DA-EPOCH-R) (etoposide, prednisone, vincristine, cyclophosphamide, doxorubicin, and rituximab) [31]. Polatuzumab vedotin (an anti-CD79b antibody-drug conjugate) in combination with rituximab, cyclophosphamide, doxorubicin, and prednisone (pola-R-CHP regimen) [32] is also approved in Europe and the United States [32].

In spite of the effective first-line treatment with R-CHOP, which is curative in the majority of cases, it is estimated that 10%–15% of patients do not respond adequately to initial treatment, and up to 35% of patients will eventually relapse at some point after achieving a complete response. Prognosis and choice of subsequent treatment vary depending on whether the disease was refractory to initial systemic therapy and on the interval from initial treatment until relapse, with the initial 12-month period being a landmark. In some developed countries, medically, patients with a first early relapse may preferentially be offered to go for chimeric antigen receptor T-cell (CAR-T) therapy with either axicabtagene ciloleucel (axi-cel) [33] or lisocabtagene maraleucel (liso-cel) [34], as both products have recently been shown to be superior to autologous stem cell transplantation (ASCT). The use of CAR-T cell therapy in patients with DLBCL at first relapse has been enabled by advances in the management of treatment-related toxicities, particularly cytokine release syndrome (CRS) and immune effector cell-associated neurotoxicity syndrome (ICANS), which were major concerns in early studies of B-cell malignancies (Table 43.1). However, in most countries the treatment of choice for medically fit DLBCL patients after the first relapse would be 2–3 cycles of intensive chemotherapy followed by ASCT. This approach has been shown to increase overall response rate to 64% with 5-year event-free survival of 46% [39].

Approximately 50% of patients with diffuse large B-cell lymphoma (DLBCL) experience relapse after achieving remission, whether following treatment for primary refractory disease or after a first relapse [37]. Under these circumstances the proportion of patients who are medically fit to undergo intensive salvage chemotherapy is substantially low. Such patients who have not received CAR-T therapy will be offered such if locally available. Indeed, the first regulatory-approved use of autologous CAR-T cell products in DLBCL was in these settings. We have summarized the data from the pivotal phase 2 studies, which led to the approval of three products: axi-cel, tisagenlecleucel, and liso-cel (Table 43.1). A viable option is also antibody-based therapy with novel products such as tafasitamab and polatuzumab vedotin [40]. Bispecific antibodies are also actively explored as an option for such patients [41]. However, allogeneic stem cell therapy is to be offered only to patients who achieve a complete response to salvage therapy. For unfit DLBCL patients in second relapse, the options are even more limited, aiming to relieve symptoms and prolong life with minimal toxicity. Those would include bispecific antibody therapies such as glofitamab and epcoritamab. Both products are CD20 × CD3 bispecific antibodies leading to complete responses in up to 40% of patients [42,43]. Being T-cell–engaging agents, those bispecific antibodies confer significant risk for life-threatening CRS and/or ICANS [42,43]. Therefore, their

TABLE 43.1 Pivotal clinical studies with chimeric antigen receptor T-cell therapy for diffuse large B-cell lymphoma (DLBCL) patients in second relapse.

Trial	Indication	Allocation to therapy	End points	Outcomes
ZUMA-1 [35]	Patients with large B-cell lymphomas (LBCLs) with refractory or relapsed disease within 12 months of autologous stem cell transplantation (ASCT)	For phase 2 part of the study 111 patients enrolled. 101 received axicabtagene ciloleucel	Primary: objective response rate (ORR); secondary: RFS; OS; duration of response (DOR); safety	ORR 82% Median DOR 11.1 months CRS in 93% of patients ICANS in 64% of patients 5-year OS 42.6%
JULIET [36,37]	Patients with relapsed/refractory DLBCL after or ineligible for ASCT	In phase 2 settings, 111 patients received infusion with tisagenlecleucel	Primary: best overall response rate; secondary: DOR, OS, safety	ORR 52% CRS in 58% of patients ICANS in 11% of patients Median OS 11.1 months 60.4% of responders maintained response for 40.3 months
TRANSCEND NHL 001 [38]	Patients with relapsed/refractory LBCL after previous stem cell transplantation	269 patients received infusion of lisocabtagene maraleucel under seamless study design	Primary: rate of adverse events, the probability of dose-limiting toxicities, and the objective response rate; secondary: rate of CR, DOR, PFS, OS	ORR 73% CR rate 53% CRS in 42% of patients ICANS in 30% of patients Median DOR 23.1 months Median OS 27.3 months

CRS, Cytokine release syndrome; *DOR*, duration of response; *ICANS*, immune effector cell-associated neurotoxicity syndrome; *ORR*, objective response rate; *OS*, overall survival; *PFS*, progression-free survival; *RFS*, relapse-free survival.

administration in frail, unfit patients should be considered after a careful risk–benefit ratio assessment. Finally, low-intensity chemotherapy or even pulses with dexamethasone may lead to some short-lived responses in some patients in palliative settings.

Future directions in diffuse large B-cell lymphoma management

A significant proportion of DLBCL patients can be cured with the standard of care R-CHOP regimen. It is expected that CAR-T cell-based therapy will become the standard therapeutic approach for patients in first relapse. This is expected to be facilitated by the development of off-the-shelf allogeneic CAR-T cell products, which will drastically reduce cost and improve patients' access. Further avenues of development are precise risk stratification of patients and eventually devising therapeutic strategies based on individual molecular characteristics of the lymphoma, as demonstrated recently by the development of the venetoclax, ibrutinib, prednisone, obinutuzumab, and lenalidomide regimen in relapsed/refractory DLBCL [44].

Multiple myeloma

Basic facts on multiple myeloma with focus on contemporary therapeutic options for advanced disease

MM is the major entity in the continuum of plasma cell neoplasms [1,45]. It accounts for approximately 10% of all HM and is more frequent in males as well as in people of African descent [46]. The median age at diagnosis is 65 years. Currently, the diagnosis is made based on the revised International Myeloma Working Group criteria [45]. In general, they are based on the confirmation of a clonal plasma cell population of at least 10% in the bone marrow or histologically proven plasmacytoma. Further to that, patients with MM must have a myeloma-defining event (i.e., any of the following: hypercalcemia, renal failure, anemia, or lytic bone lesions) or any of three specific biomarkers (i.e., clonal bone marrow plasma cells ≥60%, serum-free light-chain [FLC] ratio ≥100 [provided the involved FLC level is ≥100 mg/L], and >1 focal lesion on magnetic resonance imaging [MRI] studies) [45].

The natural history of the disease is inevitably lethal because of the development of end-stage organ failure. However, in the last two decades the prognosis of MM patients has dramatically improved, and the OS for transplant-eligible MM patients is currently estimated to be more than 10 years outside of clinical trials [47]. Traditionally, risk stratification was based mainly on tumor burden, but recent risk stratification approaches aim to account for both tumor burden and cytogenetic features of the disease, as exemplified by the revised international staging system (RISS) [48]. Notably, due to recent advances with current therapeutic approaches, patients with high-risk cytogenetics (classified as stage III disease according to RISS) who are eligible for ASCT can reach a median survival rate of approximately 8 years [47].

Indeed, several classes of new medications have been incorporated into MM management in the last two decades, including proteasomal inhibitors, immunomodulatory agents, targeted immunotherapies, bispecific antibodies, and CAR-T cell therapies [47,48]. However, front-line therapeutic strategy is still devised based on the patient's fitness to undergo consolidation therapy with ASCT after initial induction therapy. In general the goal is to maximize response with the first 3–4 induction cycles so that the likelihood for an optimal stem cell harvest is preserved. A plethora of options exist for induction regimens in such patients, but the quadruplet regimens with daratumumab, bortezomib, lenalidomide, and dexamethasone (Dara-VRd) or with isatuximab, bortezomib, lenalidomide, and dexamethasone (Isa-VRd) are considered superior to any triplet regimen in terms of rates and depth of responses [47,48]. The availability of such regimens is still not universal, and therefore triplet regimens are still much more commonly used as induction therapy. VRd is probably the most widely used, followed by daratumumab, lenalidomide, and dexamethasone (DRd). Additional options in consideration depending on comorbidities are bortezomib, cyclophosphamide, and dexamethasone (VCd), or VCd + daratumumab (Dara-VCd), or even bortezomib, thalidomide, and dexamethasone (VTd). All these regimens can be used also in transplant-ineligible patients along with classical melphalan-based regimens such as bortezomib, melphalan, and prednisone (VMP) [47]. Timing of ASCT postinduction can be either immediately after completing four cycles of induction or be postponed to a later date depending on logistics and the patient's preferences. In any case, ASCT-based consolidation is still considered a standard of care for transplant-eligible patients, as it adds approximately 12 months to the OS of those patients [47]. Furthermore, it is considered standard of care for all patients to receive post-ASCT or postinduction (for transplant-ineligible) maintenance therapy. The drug of choice for this therapy is lenalidomide, but its optimal duration remains uncertain [49].

Eventually all MM patients will relapse. The optimal therapy in first or subsequent relapses is uncertain and will depend on several main variables, including time since last induction therapy, response to previous therapy, overall performance status, tumor burden and aggressiveness at relapse, and the locally available therapeutic options. The general principle is to select a triplet regimen not containing drugs that were used to achieve or maintain a response to previous induction therapies. Recent reviews of main clinical trial results in these settings are readily available [47]. We would like to focus briefly on recent phase III study results in relapsed MM with CAR-T cell therapies, which showed striking efficacy in heavily pretreated MM patients.

The currently approved autologous CAR-T cell products, including idecabtagene vicleucel (ide-cel) and ciltacabtagene autoleucel (cilta-cel), are targeting the surface B-cell maturation antigen (BCMA) (Table 43.2). In the phase 3 KarMMa-3 study, ide-cel showed highly significant improvement in PFS and overall response rate in triple-refractory MM patients versus standard regimens (see Table 43.2) [50]. Notably, the safety profile of ide-cel was typical for CAR-T cell therapies [50]. CRS was reported in 84% of ide-cel–treated patients, with grade 3 and higher events in only 5%. ICANS of any grade was reported in 18% of these patients, with 3% having grade 3 [50].

Similar impressive efficacy was reported for cilta-cel in the phase 3 CARTITUDE-4 study in lenalidomide-refractory MM patients [51]. Cilta-cel was superior in terms of PFS and response rates versus standard regimens pomalidomide, bortezomib, and dexamethasone (PVd) or daratumumab, pomalidomide, and dexamethasone (DPd) [51]. The safety profile was similar to ide-cel. CRS was observed in 95% of cilta-cel–treated patients, with just 4% developing grade 3 or 4 CRS [51]. ICANS was reported in 21% of patients, but 9% had a grade 3 or higher event [51].

Based on these results, CAR-T cell therapy in MM is emerging as a highly effective therapy for patients not only in late relapses but also in first relapse. Their widespread use, however, may be hindered by cost and specific comorbidities in elderly patients [47].

Success in late relapses of MM is further fostered by positive results with belantamab mafodotin, which is an anti-BCMA anti-body–drug conjugate with monomethyl auristatin-F [52]. It showed efficacy either as mono- or combination therapy in heavily pretreated relapsed/refractory MM patients [53,54]. Finally, three bispecific antibodies have been recently approved for relapsed/refractory MM patients. Teclistamab [55] and elranatamab target BCMA [56], whereas talquetamab targets G-protein–coupled receptor family C group 5 member D (GPRC5D) surface protein [57]. Expectedly, as T-cell–engaging therapy, they are associated with CRS, but its rate and severity are lower than with CAR-T cell products discussed above [55–57], but still require careful management.

Future directions in multiple myeloma management

After decades of being considered an incurable disease, MM is now on the verge of unprecedented improvement in management because of the recent advances in immunotherapy. There is a reasonable chance to expect that in the near

TABLE 43.2 Registration of phase 3 clinical studies with chimeric antigen receptor T-cell therapy for relapsed/refractory multiple myeloma (MM) patients.

Indication	Allocation to therapy	End points	Outcomes
Patients with relapsed/refractory MM after two to four previous regimens (including immunomodulatory agents, proteasome inhibitors, and daratumumab) and who had disease refractory to the last regimen	In phase 3 settings 132 patients were randomized to standard therapy and 254 patients were randomized to ide-cel	Primary: PFS as assessed by independent response committee Key secondary: response rate (defined as a partial response or better) assessed by the independent response committee, and overall survival	Median PFS 13.3 months for ide-cel versus 4.4 months for standard therapy ($P<.001$) Overall response rate was 71% for ide-cel versus 42% for standard therapy ($P<.001$)
Patients with relapsed/refractory MM with lenalidomide resistance who had one to three lines of previous therapy, including a proteasome inhibitor and an immunomodulatory drug.	In phase 3 settings 211 patients were randomized to standard therapy and 208 patients were randomized to cilta-cel	Primary: PFS key secondary CR or better response rate, ORR, minimal residual disease (MRD) negativity, OS, and worsening of patient-reported symptoms	Median PFS was not reached for cilta-cel group versus 11.8 months for standard therapy ($P<.001$) Overall response rate was 84.6% for cilta-cel versus 67.3% for standard therapy ($P<.001$)

MM, Multiple myeloma; *MRD*, minimal residual disease; *ORR*, overall response rate; *OS*, overall survival; *PFS*; progression-free survival.

future CAR-T cell therapies or bispecific antibodies will be integrated in the management of MM patients at first relapse or even as a consolidation therapy. Of course, many questions will need to be addressed in light of such opportunities, for example, what would be the optimal frontline therapy if immunotherapy is readily available? Will ASCT still be the standard of care or not? Will maintenance therapy be needed at all? What would be the optimal sequencing of immunotherapies?, and so on. The hope is that with optimal management of such a vast range of therapeutics, some MM patients will actually be definitively cured at the expense of minimal toxicity.

Conclusion

HMs are an extremely heterogeneous group of malignancies with variable underlying pathogenesis, clinical course, therapeutic options, and clinical outcomes. Most of them have been considered incurable in the past, but in the last decades we have witnessed dramatic improvement in the OS in several HMs, such as CML, most lymphomas, and MM. This success was due to the development of targeted therapies with small molecule inhibitors and monoclonal antibodies. We expect further improvements in the clinical outcomes due to wider use of CAR-T cell therapies and bispecific antibodies in a number of indications such as DLBCL and MM. However, in many cases the underlying HM will remain incurable and will be the cause of death. This raises specific issues with the EoL management in HM patients. Although not as frequent as in solid cancers, pain syndromes in HM patients may significantly worsen their QoL toward the EoL. Therefore, pain management requires wider recognition in the hematological community as an integral part of the EoL care from HM patients. Optimal pain management plans for HM patients should take into account some specifics of the underlying hematological disease and of the general condition of the patient. Formation of multidisciplinary teams between hematologists, oncologists, palliative care physicians, physical therapy specialists, and psychologists will be useful in devising the optimal pain management strategy for such patients. Rapid developments in the specific therapies for HM patients require up-to-date information regarding potential drug–drug interactions to be taken into account when planning pain management in palliative settings.

References

[1] Alaggio R, Amador C, Anagnostopoulos I, et al. The 5th edition of the World Health Organization classification of haematolymphoid tumours: lymphoid neoplasms. Leukemia 2022;36(7):1720–48. Available from: https://doi.org/10.1038/s41375-022-01620-2.

[2] Khoury JD, Solary E, Abla O, et al. The 5th edition of the World Health Organization classification of haematolymphoid tumours: myeloid and histiocytic/dendritic neoplasms. Leukemia 2022;36(7):1703–19. Available from: https://doi.org/10.1038/s41375-022-01613-1.

[3] Vardiman JW, Thiele J, Arber DA, et al. The 2008 revision of the World Health Organization (WHO) classification of myeloid neoplasms and acute leukemia: rationale and important changes. Blood 2009;114(5):937–51. Available from: https://doi.org/10.1182/blood-2009-03-209262.

[4] Kocarnik JM, Compton K, Dean FE, et al. Cancer incidence, mortality, years of life lost, years lived with disability, and disability-adjusted life years for 29 cancer groups from 2010 to 2019 a systematic analysis for the global burden of disease study 2019. JAMA Oncol 2022;8(3):420–44. Available from: https://doi.org/10.1001/jamaoncol.2021.6987.

[5] Bray F, Laversanne M, Sung H, et al. Global cancer statistics 2022: GLOBOCAN estimates of incidence and mortality worldwide for 36 cancers in 185 countries. CA Cancer J Clin 2024;74(3):229. Available from: https://doi.org/10.3322/caac.21834.

[6] Zhang N, Wu J, Wang Q, et al. Global burden of hematologic malignancies and evolution patterns over the past 30 years. Blood Cancer J 2023;13(1). Available from: https://doi.org/10.1038/s41408-023-00853-3.

[7] Sant M, Minicozzi P, Mounier M, et al. Survival for haematological malignancies in Europe between 1997 and 2008 by region and age: results of EUROCARE-5, a population-based study. Lancet Oncol 2014;15(9):931–42. Available from: https://doi.org/10.1016/S1470-2045(14)70282-7.

[8] Cheson BD, Fisher RI, Barrington SF, et al. Recommendations for initial evaluation, staging, and response assessment of Hodgkin and non-Hodgkin lymphoma: the Lugano classification. J Clin Oncol 2014;32(27):3059–68. Available from: https://doi.org/10.1016/10.1200/JCO.2013.54.8800.

[9] Gerlach C, Alt-Epping B, Oechsle K. Specific challenges in end-of-life care for patients with hematological malignancies. Curr Opin Support Palliat Care 2019;13(4):369–79. Available from: https://doi.org/10.1097/SPC.0000000000000470.

[10] Alt-Epping B, Wulf G, Nauck F. Palliative care for patients with hematological malignancies – a case series. Ann Hematol 2011;90(5). Available from: https://doi.org/10.1007/s00277-010-1057-3.

[11] Beaussant Y, Daguindau N, Chauchet A, et al. Hospital end-of-life care in haematological malignancies. BMJ Support Palliat Care 2018;8(3):314–24. Available from: https://doi.org/10.1136/bmjspcare-2017-001446.

[12] Kaiser F, Rudloff LV, Vehling-Kaiser U, et al. Palliative home care for patients with advanced haematological malignancies—a multicenter survey. Annals of Hematology 2017;96(9):1557–62. Available from: https://doi.org/10.1007/s00277-017-3045-3.

[13] Niscola P, Arcuri E, Giovannini M, et al. Pain syndromes in haematological malignancies: an overview. Hematol J 2004;5(4):293–303. Available from: https://doi.org/10.1038/sj.thj.6200530.

[14] Niscola P, Tendas A, Scaramucci L, et al. Pain in malignant hematology. Expert Rev Hematol 2011;4(1):81–93. Available from: https://doi.org/10.1586/ehm.10.79.

[15] Caraceni A, Shkodra M. Cancer pain assessment and classification. Cancers 2019;11(4). Available from: https://doi.org/10.3390/cancers11040510.

[16] Sharan K. Essentials of interventional cancer pain management; 2019. pp. 123–131.

[17] Dispenzieri A. POEMS syndrome: 2011 update on diagnosis, risk-stratification, and management. Am J Hematol 2011;86(7):591. Available from: https://doi.org/10.1002/ajh.22050.

[18] Mariotto S, Tecchio C, Sorio M, et al. Clinical and neurophysiological serial assessments of brentuximab vedotin-associated peripheral neuropathy. Leuk Lymphoma 2019;60(11):2806–9. Available from: https://doi.org/10.1080/10428194.2019.1605068.

[19] Mesa RA, Schwager S, Radia D, et al. The Myelofibrosis Symptom Assessment Form (MFSAF): an evidence-based brief inventory to measure quality of life and symptomatic response to treatment in myelofibrosis. Leuk Res 2009;33(9):1199–203. Available from: https://doi.org/10.1016/j.leukres.2009.01.035.

[20] Cella DF, Tulsky DS, Gray G, et al. The functional assessment of cancer therapy scale: development and validation of the general measure. J Clin Oncol 1993;11(3):570–9. Available from: https://doi.org/10.1200/jco.1993.11.3.570.

[21] Allart-Vorelli P, Porro B, Baguet F, et al. Haematological cancer and quality of life: a systematic literature review. Blood Cancer J 2015;5(4):e305. Available from: https://doi.org/10.1038/bcj.2015.29.

[22] Mercieca-Bebber R, King MT, Calvert MJ, et al. The importance of patient-reported outcomes in clinical trials and strategies for future optimization. Patient Relat Outcome Meas 2018;9:353–67. Available from: https://doi.org/10.2147/prom.s156279.

[23] Geyer HL, Gazelka H, Mesa R. How I treat pain in hematologic malignancies safely with opioid therapy. Blood 2020;135(26):2354. Available from: https://doi.org/10.1182/blood.2019003116.

[24] El-Jawahri A, Nelson AM, Gray TF, et al. Palliative and end-of-life care for patients with hematologic malignancies. J Clin Oncol 2020;38(9):944–53. Available from: https://doi.org/10.1200/JCO.18.02386.

[25] Bao T, Lao L, Medeiros M, et al. Improvement of painful bortezomib-induced peripheral neuropathy following acupuncture treatment in a case series of multiple myeloma patients. Med Acupunct 2012;24(3):181–7. Available from: https://doi.org/10.1089/acu.2011.0868.

[26] Shivarov V, Dimitrov J, Guenova M, et al. Lymphoid neoplasms in Bulgaria according to the WHO classification (2001). The experience of the National Center of Haematology. Haematologica 2005;90(12 Suppl.):ELT07.

[27] Ruppert AS, Dixon JG, Salles G, et al. International prognostic indices in diffuse large B-cell lymphoma: a comparison of IPI, R-IPI, and NCCN-IPI. Blood 2020;135(23):2041–8. Available from: https://doi.org/10.1182/BLOOD.2019002729.

[28] Pfreundschuh M, Trümper L, Osterborg A, et al. CHOP-like chemotherapy with or without rituximab in young patients with good-prognosis diffuse large-B-cell lymphoma: 6-year results of an open-label randomised study of the MabThera International Trial (MInT) Group. Lancet Oncol 2011;12(11):1013–22. Available from: https://doi.org/10.1016/S1470-2045(11)70235-2.

[29] Coiffier B, Lepage E, Briere J, et al. Chop chemotherapy plus rituximab compared with chop alone in elderly patients with diffuse large-B-cell lymphoma. N Engl J Med 2002;346(4):235–42. Available from: https://doi.org/10.1056/NEJMoa011795.

[30] Habermann TM, Weller EA, Morrison VA, et al. Rituximab-CHOP versus CHOP alone or with maintenance rituximab in older patients with diffuse large B-cell lymphoma. J Clin Oncol 2006;24(19):3121–7. Available from: https://doi.org/10.1200/JCO.2005.05.1003.

[31] Wilson WH, Dunleavy K, Pittaluga S, et al. Phase II study of dose-adjusted EPOCH and rituximab in untreated diffuse large B-cell lymphoma with analysis of germinal center and post-germinal center biomarkers. J Clin Oncol 2008;26(16):2717–24. Available from: https://doi.org/10.1200/JCO.2007.13.1391.

[32] Tilly H, Morschhauser F, Sehn LH, et al. Polatuzumab vedotin in previously untreated diffuse large B-cell lymphoma. N Engl J Med 2022;386(4):351–63. Available from: https://doi.org/10.1056/NEJMoa2115304.

[33] Locke FL, Miklos DB, Jacobson CA, et al. Axicabtagene ciloleucel as second-line therapy for large B-cell lymphoma. N Engl J Med 2022;386(7):640–54. Available from: https://doi.org/10.1056/NEJMoa2116133.

[34] Kamdar M, Solomon SR, Arnason J, et al. Lisocabtagene maraleucel versus standard of care with salvage chemotherapy followed by autologous stem cell transplantation as second-line treatment in patients with relapsed or refractory large B-cell lymphoma (TRANSFORM): results from an interim analysis of an open-label, randomised, phase 3 trial. Lancet 2022;399(10343):2294–308. Available from: https://doi.org/10.1016/S0140-6736(22)00662-6. Erratum in: Lancet 2022 Jul 16;400(10347):160. https://doi.org/10.1016/S0140-6736(22)01207-7.

[35] Neelapu SS, Locke FL, Bartlett NL, et al. Axicabtagene ciloleucel CAR T-cell therapy in refractory large B-cell lymphoma. N Engl J Med 2017;377(26):2531–44. Available from: https://doi.org/10.1056/NEJMoa1707447.

[36] Schuster SJ, Bishop MR, Tam CS, et al. Tisagenlecleucel in adult relapsed or refractory diffuse large B-cell lymphoma. N Engl J Med 2019;380(1):45. Available from: https://doi.org/10.1056/NEJMoa1804980.

[37] Schuster SJ, Tam CS, Borchmann P, et al. Long-term clinical outcomes of tisagenlecleucel in patients with relapsed or refractory aggressive B-cell lymphomas (JULIET): a multicentre, open-label, single-arm, phase 2 study. Lancet Oncol 2021;22(10):1403–15. Available from: https://doi.org/10.1016/S1470-2045(21)00375-2.

[38] Abramson JS, Palomba ML, Gordon LI, et al. Two-year follow-up of lisocabtagene maraleucel in relapsed or refractory large B-cell lymphoma in TRANSCEND NHL 001. Blood 2024;143(5):404–16. Available from: https://doi.org/10.1182/blood.2023020854.

[39] Philip T, Guglielmi C, Hagenbeek A, et al. Autologous bone marrow transplantation as compared with salvage chemotherapy in relapses of chemotherapy-sensitive non-hodgkin's lymphoma. N Engl J Med 1995;333(23):1540–5. Available from: https://doi.org/10.1056/NEJM199512073332305.

[40] Duell J, Abrisqueta P, Andre M, et al. Tafasitamab for patients with relapsed or refractory diffuse large B-cell lymphoma: final 5-year efficacy and safety findings in the phase II L-MIND study. Haematologica 2024;109(2):553. Available from: https://doi.org/10.3324/haematol.2023.283480.

[41] Cassanello G, Luna de Abia A, Falchi L. Trial watch: bispecific antibodies for the treatment of relapsed or refractory large B-cell lymphoma. Oncoimmunology 2024;13(1):2321648. Available from: https://doi.org/10.1080/2162402X.2024.2321648.

[42] Hutchings M, Mous R, Clausen MR, et al. Dose escalation of subcutaneous epcoritamab in patients with relapsed or refractory B-cell non-Hodgkin lymphoma: an open-label, phase 1/2 study. Lancet 2021;398(10306):1157–69. Available from: https://doi.org/10.1016/S0140-6736(21)00889-8.

[43] Dickinson MJ, Carlo-Stella C, Morschhauser F, et al. Glofitamab for relapsed or refractory diffuse large B-cell lymphoma. N Engl J Med 2022;387(24):2220–31. Available from: https://doi.org/10.1056/NEJMoa2206913.

[44] Melani C, Lakhotia R, Pittaluga S, et al. Combination targeted therapy in relapsed diffuse large B-cell lymphoma. N Engl J Med 2024;390(23):2143–55. Available from: https://doi.org/10.1056/NEJMoa2401532.

[45] Rajkumar SV, Dimopoulos MA, Palumbo A, et al. International Myeloma Working Group updated criteria for the diagnosis of multiple myeloma. Lancet Oncol 2014;15(12):e538–48. Available from: https://doi.org/10.1016/S1470-2045(14)70442-5.

[46] Siegel RL, Giaquinto AN, Jemal A. Cancer statistics, 2024. CA Cancer J Clin 2024;74(1):12–49. Available from: https://doi.org/10.3322/caac.21820.

[47] Rajkumar SV. Multiple myeloma: 2024 update on diagnosis, risk-stratification, and management. Am J Hematol 2024;99(9):1802–24. Available from: https://doi.org/10.1002/ajh.27422.

[48] Palumbo A, Avet-Loiseau H, Oliva S, et al. Revised international staging system for multiple myeloma: a report from international myeloma working group. J Clin Oncol 2015;33(26):2863–9. Available from: https://doi.org/10.1200/JCO.2015.61.2267.

[49] McCarthy PL, Holstein SA, Petrucci MT, et al. Lenalidomide maintenance after autologous stem-cell transplantation in newly diagnosed multiple myeloma: a meta-analysis. J Clin Oncol 2017;35(29):3279–89. Available from: https://doi.org/10.1200/JCO.2017.72.6679.

[50] Rodriguez-Otero P, Ailawadhi S, Arnulf B, et al. Ide-cel or standard regimens in relapsed and refractory multiple myeloma. N Engl J Med 2023;388(11):1002–14. Available from: https://doi.org/10.1056/NEJMoa2213614.

[51] San-Miguel J, Dhakal B, Yong K, et al. Cilta-cel or standard care in lenalidomide-refractory multiple myeloma. N Engl J Med 2023;389(4):335–47. Available from: https://doi.org/10.1056/NEJMoa2303379.

[52] Offidani M, Corvatta L, Morè S, et al. Belantamab mafodotin for the treatment of multiple myeloma: an overview of the clinical efficacy and safety. Drug Des Devel Ther 2021;15:2401–15. Available from: https://doi.org/10.2147/DDDT.S267404.

[53] Dimopoulos MA, Hungria VTM, Radinoff A, et al. Efficacy and safety of single agent belantamab mafodotin versus pomalidomide plus low-dose dexamethasone in patients with relapsed or refractory multiple myeloma (DREAMM-3): a phase 3, open-label, randomised study. Lancet Haematol 2023;10(10):e801–12. Available from: https://doi.org/10.1016/S2352-3026(23)00243-0.

[54] Dimopoulos MA, Beksac M, Pour L, et al. Belantamab mafodotin, pomalidomide, and dexamethasone in multiple myeloma. N Engl J Med 2024;391(5):408–21. Available from: https://doi.org/10.1056/NEJMoa2403407.

[55] Moreau P, Garfall AL, van de Donk NWCJ, et al. Teclistamab in relapsed or refractory multiple myeloma. N Engl J Med 2022;387(6):495–505. Available from: https://doi.org/10.1056/NEJMoa2203478.

[56] Lesokhin AM, Tomasson MH, Arnulf B, et al. Elranatamab in relapsed or refractory multiple myeloma: phase 2 MagnetisMM-3 trial results. Nat Med 2023;29(9):2259. Available from: https://doi.org/10.1038/s41591-023-02528-9.

[57] Chari A, Minnema MC, Berdeja JG, et al. Talquetamab, a T-cell-redirecting GPRC5D bispecific antibody for multiple myeloma. N Engl J Med. 2022;387(24):2232–44. Available from: https://doi.org/10.1056/NEJMoa2204591.

Part V

Special situations in pain treatment

Chapter 44

Treating pain resulting from cancer treatment

Krešimir Blažičević[1] and Mario Nalbani[2]
[1]*University Hospital Centre Zagreb, Zagreb, Croatia,* [2]*General Hospital Dubrovnik, Dubrovnik, Croatia*

Key learning objectives

- Get acquainted with the basic principles of chronic pain syndromes that arise from oncologic treatments.
- Understand the selection of key medications and nondrug options for different symptoms.
- Recognize the role of prevention and structured follow-up.

Introduction

With the widespread development of oncology in general and new treatment options being introduced in clinical practice faster than ever, significant progress has been made in terms of improving long-term outcomes in patients with malignant diseases. As oncology patients live longer, some patients face consequences from cancer treatment for many years, even after the end of the treatment itself. All of those challenges are summed up in the term “cancer survivorship,” and it is increasingly discussed and researched [1].

Although a growing number of patients now survive beyond 5 years after a cancer diagnosis, there is still no universally accepted definition of “cancer survivor” or “cancer survivorship.” Moreover, many individuals who experience relapse or develop a second primary cancer have already undergone treatments such as surgery, radiotherapy, or systemic therapy, and may continue to face effects from previous interventions even before the beginning of new treatments [2–4]. The cancer diagnosis by itself is not the only thing that matters in cancer survivorship; the emphasis is on the physical, psychological, social, and economic challenges experienced by people and caregivers after a cancer diagnosis [5].

One of the most common side effects and residual effects of cancer treatment is pain. In this chapter, we will try to emphasize some of the most common pain syndromes after cancer treatment. Antineoplastic treatments have a high propensity of causing adverse effects, and a large proportion of them can be long-lasting, extending significantly longer after the end of treatment. Chemotherapy, radiotherapy, and surgical treatment are commonly related to pain as a long-term consequence of treatment. Although the prevalence of pain in cancer survivors is less than in those receiving active treatment or those in a palliative setting, it is still rather high, with an estimated 35.8% of patients with chronic pain [6]. Some authors suggested classification by type of treatment as well as by the system that is affected [7].

Postsurgery chronic pain syndromes

Postmastectomy pain syndrome

Postsurgical syndromes are probably the most underestimated pain syndromes in everyday clinical practice. Postsurgical chronic pain is mainly associated with thoracic surgeries, breast cancer surgery, and pelvic surgeries, but any form of surgery can cause chronic pain.

Postmastectomy pain syndrome (PMPS) presents as neuropathic pain with typical characteristics: burning, tingling, stinging, and hyperesthesia. It is mainly defined as pain in the anterior/lateral chest wall, axilla, and medial upper arm that persists more than 3–6 months after surgery. The overall incidence of PMPS pain following a breast cancer operation ranges from 20% to 72%. This variability results from different study populations and PMPS definitions between data sets [8].

Pain Management in Palliative Care for Patients With Cancer. DOI: https://doi.org/10.1016/B978-0-443-30050-9.00006-0

Established risk factors for PMPS include known preoperative pain, severe acute postoperative pain, younger age, preoperative anxiety, and higher body mass index. The type of surgery is also linked to the prevalence of PMPS, and it has been discovered that the more invasive breast operations, the more likely is damage to sensory nerves. Systemic treatments, including taxanes, platinum agents, vinca alkaloids, and aromatase inhibitors (AIs), can also contribute to the development of chronic pain. Axillary radiation is another risk factor due to radiation fibrosis and chronic inflammation that cause nerve entrapment [9].

The best treatment for PMPS is prevention primarily through limiting nerve injury and working on preexisting psychological problems. Local or regional blocks in the perioperative period and intraoperative use of regional anesthesia, such as paravertebral blocks or pectoral nerve blocks, showed promising results in reducing PMPS [10,11].

Physical therapy is crucial in treating PMPS, helping improve mobility, reduce stiffness, and prevent muscle atrophy in the affected arm. Therapeutic exercises, including stretching, strengthening, and range-of-motion, can help patients regain functionality and decrease pain perception. A metaanalysis of 18 trials confirmed that exercise is beneficial in reducing pain severity in PMPS [12]. Additionally, psychological therapies like cognitive-behavioral therapy (CBT) are beneficial as they address pain-related distress and anxiety, which can exacerbate the experience of pain [13].

However, for patients with advanced disease and persistent PMPS, the use of neuropathic adjuvants may be necessary, although opioids such as oral morphine or oxycodone may be necessary. In palliative care, breakthrough doses of immediate-release opioids are often used in combination with scheduled extended-release formulations to maintain consistent analgesia.

Cancer-related pelvic pain

Cancer-related pelvic pain is a common but often underrecognized consequence of cancer and its treatment in survivors of colorectal, prostate, and gynecological malignancies. It is a complex condition that arises from multiple overlapping mechanisms, including direct tumor invasion, treatment-induced tissue damage, postsurgical complications (such as disturbance of pelvic nerve plexuses), postradiation therapy fibrosis, and neuropathic sequelae. The resulting pain can be nociceptive, neuropathic, or a combination of both, with a profound impact on functional capacity. Management of cancer-related pelvic pain in this population requires a tailored, multimodal approach. Pharmacologic therapies such as duloxetine, gabapentinoids, or tricyclic antidepressants (TCAs) can provide relief for neuropathic components. Antiinflammatory agents and topical lidocaine may benefit localized nociceptive symptoms. Each of those drugs is evaluated in separate chapters of this book. Pelvic floor physical therapy is crucial, particularly for patients with dyspareunia or myofascial dysfunction. Ultimately, early identification and individualized treatment of cancer-related pelvic pain can restore function, alleviate suffering, and improve long-term quality of life in this growing group of patients [14].

Postradical neck dissection pain syndrome

Radical neck dissection is the principal treatment modality in head and neck cancers when curative intention is the goal, but it is also done in some skin cancers. When we talk about chronic pain after radical neck dissection, there is rarely an occurrence of chronic pain at the exact surgical site in the neck; mostly, we refer to subsequent shoulder dysfunction and pain [15]. Radical neck dissection includes neck lymphadenectomy, as well as removal of adjacent structures. Accessory nerve sacrifice is considered a critical factor in determining postoperative shoulder function and is associated with subsequent denervation of the trapezius muscle, leading to inability of shoulder girdle elevation alongside shoulder pain, with a significant detrimental effect on quality of life. However, sometimes pain can occur even if the accessory nerve is spared. The actual incidence of chronic pain after radical neck dissection is not well known, but it is estimated that a significant proportion of patients can experience chronic pain after radical dissection [16,17]. Although there is not a lot of data, early physical therapy could potentially be helpful in reducing periscapular symptoms [17]. Acupuncture can be beneficial for alleviating chronic pain compared to other treatments (physical therapy, analgesia, and/or antiinflammatory drugs) as per results of a randomized trial that found significant reduction in pain, dysfunction, and xerostomia [18]. Local subcutaneous botulinum toxin injection showed a significant chronic pain reduction in a small nonrandomized trial. In patients with metastatic head and neck cancers undergoing palliative care, opioid therapy remains a cornerstone of treatment, especially when nonsteroidal antiinflammatory drugs (NSAIDs) or muscle relaxants are insufficient [19].

Phantom pain after limb amputation

The majority of limb amputee patients are of vascular or trauma cause, while only less than 2% are caused by cancer [20]. Most cancer amputee patients are attributed to sarcoma surgery, predominantly young patients, and amputation can sometimes be curative. Phantom limb pain is clinically defined as the perception of pain or discomfort in a limb that no longer exists and can present with a variety of symptoms. In some patients, phantom pain can gradually fade over the course of months, but some patients suffer for years after amputation. Treatment for phantom limb pain is difficult and thus warrants a multidisciplinary approach. While classical analgesics are sometimes prescribed, there is limited evidence to support their effectiveness. Antidepressants have shown some benefits as well, although the evidence is conflicting [21]. Nonpharmacologic measures have emerged as a potentially more effective approach with transcutaneous electrical nerve stimulation (TENS), mirror therapy, and spinal cord stimulation (SCS), all valid options in some cases. Novel technological advancements like virtual and augmented reality are a form of mirror therapy that demonstrated some success as well [22,23]. For cancer patients, opioids are sometimes used in palliative settings to manage severe residual pain unresponsive to adjuvant or nonpharmacological therapies. Controlled-release morphine or methadone may be considered due to their dual analgesic and N-methyl-D-aspartate receptor-modulating properties [21].

Radiation-induced chronic pain syndromes

Chronic pain, as a consequence of ionizing radiation, is pain caused by delayed local damage to the nervous system, bones, or other soft tissues in the field of radiotherapy given to treat the primary tumor or metastases [24]. The onset of pain after radiotherapy ranges from several months to several years after the end of radiotherapy. It has been shown that chronic pain after radiotherapy is associated with higher doses of radiation, high doses per fraction, and after combination protocols. Technological advancements in radiotherapy and optimized dosing schedules have reduced the frequency of chronic effects of radiation [25]. Prevalence of pain when radiotherapy is used as a curative treatment is estimated to be 33%, and is less prevalent than in patients who are on treatment or those with advanced or metastatic disease [26].

The clinical features, in addition to pain, include weakness and sensory changes. In most cases, pain is mild to moderate; rarely is the pain severe. With late occurrences of pain after irradiation, the first step should always be the exclusion of disease recurrence, which is often challenging, and early recognition is paramount. Although there are several chronic pain entities as a consequence of radiotherapy, here we will mainly focus on the most common chronic neuropathic pain syndromes. Nervous tissue is especially vulnerable to ionizing damage, but it is not the sole cause of pain. Radiotherapy-induced fibrosis also contributes to neuropathic pain [27].

Radiation-induced neuropathy and radiation-induced fibrosis

Fibrosis and neuropathy, following radiotherapy, occur more often after several years but are often progressive and irreversible. The underlying pathophysiologic mechanism is not fully understood, and it is assumed that the key factors include nerve compression by indirect radiation-induced fibrosis, direct axonal damage, demyelination, and microvascular injury causing ischemia [25,28].

Radiation-induced brachial plexopathy, following breast cancer irradiation, is the best described syndrome, occurring in less than 1% of patients (with doses of 50 Gy in 2 Gy fractions) but with possible devastating sequels [28]. It manifests as pain in the upper extremities with paresthesia and swelling. In more severe cases it can lead to paralysis.

Radiation-induced lumbosacral plexopathy commonly affects the lumbosacral plexus after irradiation for pelvic, colon, or testicular cancers. Typically, it occurs at least a year after treatment and is manifested as asymmetric bilateral lower-extremity weakness, while pain is an uncommon symptom [28]. If there is pain, treatment with nonopioid analgesics is suggested, although benzodiazepines, TCAs, and antiepileptics can potentially be helpful. Pain control can also be achieved with pregabalin [29]. In addition to pharmacological treatment, physiotherapy can provide functional improvement and prevent joint complications, especially if the underlying mechanism appears to be radiation fibrosis.

Radiation-induced myelopathy is seen in patients receiving radiotherapy for cervical and upper thoracic regions. Clinical manifestations depend on location and subsequent spinal cord injury and often include sensory or motor deficits. Corticosteroids could be used in the treatment; however, their use must be based on clinical judgment [25].

Radiation fibrosis syndrome is a common late-occurring complication of radiotherapy. It is characterized as a progressive fibrotic tissue sclerosis in the irradiation field [30]. Pain is usually managed with NSAIDs. If the neuropathic component is emphasized, pregabalin or gabapentin is recommended, as well as duloxetine or TCAs. An important part of treatment is rehabilitation and physical therapy, which can improve functional status and increase muscle strength, therefore maintaining adequate functionality and quality of life. When radiation fibrosis or neuropathy progresses in advanced disease, opioids can provide essential relief, particularly for deep tissue or nerve entrapment syndromes not responding to conventional adjuvants [25,30,31].

Other radiation-induced chronic pain syndromes

Other than neuropathic syndromes and radiation fibrosis, other late consequences of radiotherapy that can contribute to pain are dermatitis, mucositis, enteritis, and lymphedema. The majority of patients who are treated with radiotherapy will eventually develop radiation-induced dermatitis, typically in the first few months of treatment [32]. Rarely, acute dermatitis progresses to the chronic phase with chronic wounds and skin necrosis. On the contrary, chronic radiation-induced dermatitis develops months after radiation exposure and can lead to permanent skin damage. It can lead to fibrosis, atrophy, pigmentation changes, as well as secondary cutaneous malignancies. Fibrosis leads to limitation of movement, contractures, and pain. Due to skin sensitivity in the face, neck, and chest region, it is especially susceptible to radiation damage [33].

Prevention is one of the most important aspects of minimizing skin damage and includes good skin hygiene, moisturizers, adequate loose-fitting clothes, and sun exposure avoidance. As for pain management, NSAIDs are often used, with adjuvant analgesics per clinical judgment.

Mucositis is a common side effect in patients with head and neck cancers treated with radiotherapy, especially those with curative intent (higher irradiation doses) or concomitant with chemotherapy. As with dermatitis, prevention is even more emphasized. Good oral hygiene in the first place, mouthwash with normal saline or alkali, and benzydamine oral rinse are recommended. Opioids can also be recommended if pain progresses [34].

Pain syndromes after systemic oncological treatment and during long-term adjuvant therapy

Chemotherapy-induced peripheral neuropathy

Introduction

Chemotherapy-induced peripheral neuropathy (CIPN) is one of the most common adverse effects following cancer treatment. It ranges from mild to severe but is commonly a long-lasting side effect whose burden impacts the quality of life and work productivity, profoundly affecting quality of life. CIPN can be caused by diverse chemotherapeutic drugs that can damage sensory, autonomic, or motor nerves. Approximately 30%–60% of patients who receive neurotoxic chemotherapy develop CIPN [35]. With more time elapsed, the percentage of patients affected decreases.

To evaluate the prevalence of CIPN, researchers have used different clinician-assessed or patient-reported questionnaires because the gold-standard electromyoneurography is painful and invasive. The most extensive metaanalysis included 31 studies in which 4179 patients with different tumor types were treated with commonly used chemotherapeutic drugs. The highest prevalence (68%) of CIPN was observed within 1 month of chemotherapy. The prevalence dropped to 60% at 3 months after chemotherapy and was reduced to 30% at 6 months after chemotherapy [36].

The most important etiological factor for CIPN is the type and schedule of chemotherapy agents used. The most common chemotherapy agents that cause CIPN are platinum agents, taxanes, and vinca alkaloids, proteasome inhibitors, immunomodulatory drugs, and epothilones [36]. New drugs, most notably antibody-drug conjugates, which essentially bring higher doses of chemotherapy, more specifically to tumors, are also not immune to causing CIPN.

Although these agents are neurotoxic due to various pathological mechanisms, they mainly cause symmetrical sensory "glove and stocking" neuropathies, predominantly peripheral neuropathy that may be associated with significant pain. Chemotherapy is not perfectly targeted to cancer cells, so it causes cytotoxic changes in normal tissues, including those of the nervous system. Peripheral nerves originating from neurons in the dorsal root ganglion are particularly susceptible to damage as the blood–brain barrier does not protect them. The mechanism of CIPN differs between chemotherapies and is mostly multifactorial. For example, platinum-induced neurotoxicity is caused by crosslinking DNA and interfering with its repair and replication, leading to apoptosis of neurons in the dorsal root ganglia. Taxanes, conversely, work through microtubule disruption, leading to mitotic arrest, again mainly in the dorsal root ganglia. Both agents cause axonal degeneration, mitochondrial damage, increased reactive oxygen species, altered calcium homeostasis, and changes in ion channel expression on the cell level. Understanding each chemotherapy's specific molecular mechanism of neurotoxicity could help develop agent-specific preventive measures [37].

Patient characteristics are also important. Clinical risk factors include neuropathy at baseline, smoking, abnormal creatinine clearance, and specific sensory changes during chemotherapy. Also, prior neuropathy, for example, diabetic neuropathy, is important. Many single-nucleotide polymorphisms are associated with a range of proteins, including voltage-gated sodium channels, Schwann cell function–related proteins, receptors for cell surface collagen, receptors involved in neuronal apoptosis, neuronal crest cell development, and an enzyme involved in pyruvate metabolism [38].

Treatment of chemotherapy-induced peripheral neuropathy

Most treatments for CIPN are symptomatic, not causal, and no single approach is universally applicable. Several pharmacological and nonpharmacological treatments have shown promise in clinical trials and are recommended by international guidelines. Among these options, duloxetine, gabapentin, pregabalin, and tricyclic antidepressants (TCAs) are all commonly used and regarded as potential therapeutic approaches, with some emerging evidence for the use of acupuncture. Each of these is evaluated in depth in different chapters of the book, but we will give an overview in this chapter.

Duloxetine, a serotonin–norepinephrine reuptake inhibitor, is considered the most effective pharmacological treatment for CIPN based on current evidence. By increasing the levels of serotonin and norepinephrine, duloxetine modulates the pain perception pathways in the central nervous system. In a key randomized controlled trial, duloxetine reduced pain scores by an average of 0.73 points on a 10-point scale compared to placebo, representing a clinically meaningful improvement for many patients. About 60% of patients experience pain relief, and approximately 30% significantly reduce symptoms per one researcher [39], although metaanalysis did not demonstrate a significant effect compared to placebo [38]. Despite limited evidence, duloxetine is recommended as a first-line treatment for CIPN by major international guidelines. In clinical practice, duloxetine should be initiated at 30 mg once daily for the first 7 days, then increased to 60 mg once daily if tolerated. In severe CIPN (pain score ≥7), combining duloxetine with a second-line agent like pregabalin is a potential option. However, early referral to a pain specialist and multidisciplinary approach should be warranted for complex cases. This stepwise titration helps reduce side effects: nausea, dry mouth, somnolence, and dizziness. Some patients also report fatigue, insomnia, and constipation, although these tend to decrease over time. Rare but more severe side effects, such as hepatotoxicity, may occur, particularly in patients with preexisting liver conditions [40].

Gabapentin and pregabalin are anticonvulsants that have shown some benefit in managing CIPN. These drugs act by binding to voltage-gated calcium channels, reducing the release of excitatory neurotransmitters and alleviating neuropathic pain. Gabapentin is typically started at a dose of 300 mg at bedtime, titrated up to a maximum of 1800–3600 mg daily in divided doses. In palliative patients with neuropathic cancer pain, gabapentin is started at 100–300 mg at bedtime and titrated every 3 days as tolerated. For frail patients or those with renal impairment, use a slower titration schedule (e.g., 100 mg increments weekly) and monitor for sedation and dizziness. Conversely, pregabalin is usually started at 75 mg twice daily, with the possibility of increasing to 150 mg twice daily and a maximum of 300 mg twice daily. Common side effects of both medications include dizziness, sedation, peripheral edema, and weight gain. Pregabalin tends to have a faster onset of action but may also cause more pronounced sedation. Both drugs can increase the risk of falls, particularly in older patients [29,35]. Studies suggest that 25%–40% of patients may experience some reduction in pain, but the results are inconsistent. Major international guidelines do not recommend these drugs as first-line treatments, though they may be used off-label as second-line options, particularly when duloxetine is ineffective or contraindicated [41].

TCAs, such as amitriptyline, nortriptyline, and desipramine, are another class of drugs used for managing neuropathic pain. These medications work by inhibiting serotonin and norepinephrine reuptake, enhancing pain modulation pathways. Amitriptyline is typically started at 10–25 mg at bedtime, titrating up to 75–150 mg daily if needed. Nortriptyline and desipramine can also be used with similar dosing strategies. However, TCAs have significant side effects, including dry mouth, constipation, urinary retention, blurred vision, orthostatic hypotension, and sedation. These anticholinergic effects are more pronounced with amitriptyline, making it less suitable for older adults or those with cardiovascular issues. The effectiveness of TCAs in CIPN is generally modest. Around 20%–30% of patients may experience some pain relief, but the high incidence of side effects often limits their use. When compared to duloxetine or gabapentin, TCAs are typically less well-tolerated and less effective for CIPN treatment. Major oncological guidelines do not recommend TCAs, considering their limited efficacy and higher side effect profile than other options [42].

Topical creams could also be a beneficial option for localized relief without the systemic side effects of pills. However, these may offer localized relief in select cases but are not part of guideline-endorsed first-line management. There are many creams on the market, but the most widely used is capsaicin cream, which is derived from chili peppers. Capsaicin reduces the substance P, a neurochemical involved in transmitting pain signals. When applied topically, it can help reduce the sensation of pain. However, it may cause an initial burning or stinging sensation, which tends to lessen with continued use. Lidocaine cream is a local anesthetic that temporarily numbs the area where it is applied. It is often used in patients with CIPN to reduce pain and discomfort, providing short-term relief by blocking nerve signals. Menthol-based creams offer a cooling sensation that can alleviate the discomfort of burning or tingling sensations caused by neuropathy.

Compounded cream combines several active ingredients, such as gabapentin, ketamine, and amitriptyline. These multiingredient creams aim to target various pain pathways to offer broader relief, with various results [42].

Acupuncture is increasingly used in supportive care, and small trials suggest potential benefit in CIPN. Acupuncture regimens for CIPN typically involve sessions two to three times per week, lasting 30–45 minutes each, with treatment duration depending on individual patient response. Treatment courses usually last from 4 to 8 weeks. Patients treated with acupuncture report moderate to significant pain and sensory function improvements, although more trials are needed to establish high-quality data. While the evidence supporting its use is weak, it is widely considered safe and potentially beneficial as part of a multimodal approach, especially when duloxetine is not available or effective. Both American Society of Clinical Oncology and National Comprehensive Cancer Network recognize the benefit of acupuncture but primarily for AI-related joint pain, general cancer pain, musculoskeletal pain, and fatigue [40,43–45].

Several other nonpharmacological treatments are explored for the management of CIPN, and while they may not be as well-researched as pharmacological treatments, they provide valuable options, especially for patients looking for alternatives to medications. Exercise has emerged as a promising option for managing CIPN. Regular physical activity has been shown to reduce pain and enhance overall quality of life in patients with neuropathy. Aerobic exercise combined with resistance training promotes nerve regeneration and improves muscle strength. However, the exact mechanisms through which exercise improves CIPN symptoms are still understudied. While the evidence is not as robust as for some pharmacological treatments, the low risk of side effects and the overall health benefits make exercise a recommended option in many guidelines. Another approach is TENS, a therapy that uses low-voltage electrical currents to provide pain relief by stimulating peripheral nerves. The electrical impulses are thought to block pain signals and encourage the release of natural pain-relieving chemicals, such as endorphins. However, studies on TENS in CIPN have produced mixed results, with some trials showing benefit and others finding little to no effect. As a result, TENS is not recommended as a first-line treatment for CIPN but may be considered as an adjunct therapy, especially in patients looking for noninvasive, drug-free options [46]. Physical therapy is another nonpharmacological treatment that can be beneficial for patients with CIPN. Physical therapy programs focus on improving strength, coordination, and mobility through tailored exercises. By helping to improve nerve function and muscle strength, physical therapy can enhance patients' quality of life and reduce the risk of falls, which is a common concern in those with severe neuropathy [46]. In addition to the therapies mentioned, there is ongoing research into novel nonpharmacological treatments such as neuromodulation, including SCS, which aims to relieve pain by delivering electrical impulses directly to the spinal cord. While this is an emerging area, initial studies have shown promise in relieving neuropathic pain, though further research is needed [46].

Chronic pain caused by aromatase inhibitors

AIs are a class of drugs commonly used in the treatment of hormone receptor-positive breast cancer. These medications block the enzyme aromatase, which converts androgens into estrogen. A significant side effect associated with the use of AIs is chronic pain. AI, for example, anastrozole, letrozole, and exemestane, reduce estrogen production, leading to various musculoskeletal symptoms, commonly referred to as AI-associated musculoskeletal syndrome (AIMSS). This syndrome manifests as joint stiffness, pain, and muscle discomfort [47].

Although the exact mechanism through which these drugs cause pain is not fully understood, estrogen is believed to play a protective role in maintaining joint and bone health. When estrogen levels drastically fall, the joints and bones become more susceptible to inflammation and damage, leading to pain. The most common complaints involve pain in the "small joints" like the hand and wrists, but it can also affect the knee, back, and shoulder. Patients often describe the pain as severe enough to limit their daily activities, such as walking, dressing, and even sleeping, with some even mentioning objects falling out of their hands. Since endocrine therapy is taken for at least 5 years, this chronic pain affects patients' adherence to therapy, as many patients choose to discontinue treatment due to discomfort, leading to worse treatment results. Not all patients who take AI will experience chronic pain. Some predisposing factors have been identified: postmenopausal women with a preexisting history of osteoarthritis, obesity, vitamin D deficiency, and low levels of physical activity before starting treatment. Some studies suggest that letrozole is associated with a higher incidence of musculoskeletal symptoms than anastrozole or exemestane, though these differences are not always significant across different studies. Genetic predispositions may also play a role, as individual variations in drug metabolism and response to estrogen depletion may influence how the body reacts to these medications [48].

The chronic pain associated with AIs significantly impacts both the physical and emotional well-being of patients. In addition to the physical limitations caused by joint stiffness and pain, patients also experience psychological distress. Chronic pain not only affects patients' ability to perform daily tasks but can also hinder their motivation to engage in physical activity, which is crucial for overall health and cancer recovery. Physical inactivity may, in turn, exacerbate the symptoms, creating a cycle of pain and limited mobility [49]. Managing chronic pain in patients taking AIs is challenging and includes a multimodal approach. NSAIDs are often the first line of treatment for managing joint and muscle pain. These medications help reduce inflammation, a contributor to pain and stiffness. For many patients, NSAIDs provide only moderate relief, but relief enough to function normally. Unfortunately, long-term use of NSAIDs carries additional health risks, making this approach impractical for longer periods. For patients who do not experience sufficient relief with NSAIDs, more potent analgesics such as opioids can be prescribed, often for short-term management of acute pain flare-ups. In some cases, local corticosteroid injections into the affected joints can provide substantial pain relief, particularly when inflammation is concentrated in one or two areas. These injections reduce inflammation directly at the source, alleviating joint stiffness and discomfort. However, the use of corticosteroids is usually limited to infrequent applications due to potential side effects such as cartilage damage and systemic effects with prolonged use. The role of antidepressants and anticonvulsants is yet unclear, as they are mostly prescribed for neuropathic pain, which, from a pathophysiological point of view, is not the primary mechanism of pain in AIMSS [47].

Physical therapy is an important part of managing musculoskeletal pain associated with AI. Physical therapy can increase joint mobility, improve muscle strength, and reduce stiffness. These programs often involve stretching exercises, range-of-motion activities, and strength training [50]. Exercise, particularly low-impact aerobic activities like walking, swimming, or cycling, has been shown to reduce joint pain and stiffness in AI users. Additionally, strength training exercises, particularly those that target muscles around affected joints, can enhance stability and reduce the load on the joints, thereby minimizing pain. Additionally, acupuncture has emerged as a promising complementary therapy for most chronic pain, including AIMSS. Due to the small sample size of trials, which were mostly nonrandomized, it is not clear whether this is a placebo effect or an actual acupuncture effect [51]. Lifestyle modifications, such as maintaining a healthy weight and ensuring adequate vitamin D and calcium intake, are also important for managing AIMSS since obesity is a known risk factor for joint pain, and weight loss has been shown to reduce the severity of musculoskeletal symptoms in AI-treated patients. Furthermore, vitamin D supplementation may benefit patients with low bone density or those at risk for osteoporosis, as it helps support bone health and may alleviate some of the bone-related pain associated with AI use [52]. If none of the above approaches help, giving a drug holiday or switching to another AI or other endocrine therapy is always an option.

Medication-related osteonecrosis of the jaws

Adjuvant treatment with preventive bisphosphonates or denosumab to diminish the negative effects on bone loss and protect bones from metastasis is often applied in breast cancer patients, along with cancers that metastasize to bone, such as prostate cancers. The most severe possible complication of bisphosphonate use is osteonecrosis of the jaw (ONJ). Bisphosphonates work by inhibiting osteoclast activity, reducing bone resorption, and increasing bone density, while denosumab inhibits a protein called RANK ligand (RANKL). By blocking RANKL, it prevents the maturation, function, and survival of osteoclasts [53]. However, prolonged use of bisphosphonates and denosumab, especially in higher doses commonly administered to cancer patients, is associated with the development of ONJ. This condition is characterized by the death of bone tissue in the jaw, leading to areas of exposed bone, infection, and significant pain [54]. ONJ primarily occurs after invasive dental procedures. The compromised ability of the bone to heal and regenerate due to the long-term suppression of osteoclasts results in necrotic bone tissue that fails to repair itself, causing persistent pain. This pain is often chronic and can severely affect daily activities such as eating, speaking, and sleeping, resulting in a diminished quality of life [55].

Chronic pain is the most debilitating symptom of ONJ. Pain is usually deep and often exacerbated by jaw movement or infection. Many cancer patients experience anxiety and depression as a result of their ongoing pain, making pain management a crucial component of postcancer care.

Risk factors for developing ONJ are duration and dosage of bisphosphonate or denosumab therapy, concurrent use of corticosteroids, and preexisting dental conditions [56]. Prevention is the best treatment. Preventive measures include pretreatment dental screening with comprehensive dental assessments. Any necessary dental procedures, such as extractions or implants, should be completed before starting treatment to minimize the risk of ONJ. Patient education on maintaining oral hygiene, avoiding invasive dental procedures during the treatment, and recognizing early symptoms of ONJ such as pain, swelling, or exposed bone is also important. We should not forget to advise patients to undergo routine dental examination not only at the beginning but also throughout antiresorptive therapy to monitor oral health and address any issues early [57].

Once ONJ develops, treatment focuses on managing symptoms, promoting healing, and preventing further complications. Early referral to maxillofacial or dental specialists is critical to guide surgical or advanced dental intervention where needed. The management of ONJ is typically categorized into conservative, medical, and surgical approaches, depending on the severity of the condition. Conservative management includes chlorhexidine mouth rinses to reduce oral bacteria and infection risk. In cases where infection is present, systemic antibiotics can be used to control the spread of infection. Pain management with NSAIDs and analgesics is essential for managing discomfort in ONJ patients. Medical management involves temporary or permanent discontinuation of bisphosphonates or denosumab. Surgical management includes sequestrectomy, which involves the removal of dead bone in cases where necrosis is localized. In more severe cases, resection of the affected bone followed by reconstruction using vascularized bone grafts or alloplastic materials may be necessary. Soft tissue flaps, such as mucosal or muscle flaps, can cover exposed bone and promote healing. In refractory cases of ONJ in metastatic cancer patients, opioid treatment is essential for sustained pain control [58,59].

Immunotherapy-related chronic pain syndromes

Checkpoint inhibitor immunotherapy (ICI) has revolutionized perspectives and outcomes in many cancers, as indications for its use are becoming more frequent and in a greater number of different tumor sites. In simple terms, checkpoint inhibitors are monoclonal antibodies that act by blocking immunosuppressive signaling by inhibiting binding of cancer cells or antigen-presenting cells to T-cell surface molecules (called immune checkpoints), which leads to resetting antitumor immune response [60].

Characteristic side effects are immunologically mediated (immune-related adverse events [IrAE]), resulting from damage caused by activation of the immune system and consequent inflammation. IrAE can affect practically every organ system [61].

Although the majority of IrAE occur early in the course of treatment, some adverse events occur later, after the treatment is discontinued. Fortunately, late occurrences are very rare, with an estimated incidence of about 5% [62]. Chronic IrAE, defined as those persisting for more than 12 weeks after discontinuation of ICI, turned out to be more frequent than previously thought, as reported in retrospective analysis. The most common chronic IrAE are endocrinopathies and rheumatologic/rheumatological events [63]. Metaanalyses estimate the incidence of arthralgia reported in clinical trials is 1%–43% and arthritis 1%–7% [64,65]. In a single-center prospective study, it is estimated that 6.6% of patients treated with ICI are referred to rheumatologists, and the incidence of inflammatory arthritis was 3.8% [65]. Rheumatic immune-related syndromes resemble well-known mixed connective tissue disorders like rheumatoid arthritis, polymyalgia rheumatica, Sjögren syndrome, and others. Pathogenesis of rheumatic IrAE is still not fully understood, but it is suggested that there is a possible overlapping mechanism with classical rheumatological disease [61].

Inflammatory arthritis is the most common rheumatic IrAE, and other less common syndromes include polymyalgia rheumatica/giant cell arteritis, inflammatory myopathies (dermatomyositis and polymyositis), and others. Inflammatory arthritis is defined as an inflammatory joint pain resulting from ICI treatment [65]. Unlike classical rheumatoid arthritis, a majority of patients who develop inflammatory arthritis are seronegative for rheumatoid factor or anticitrullinated protein antibodies, although some seropositive cases were reported [66]. It is manifested as joint pain with tenderness, swelling, and warmth. Inflammatory arthritis can vary in clinical picture with both small and large joint involvement, as well as oligoarthritis or polyarthritis (with the latter being more common), and can range from mild to severe [66,67]. ICI-induced arthritis progresses to erosive joint destruction early, earlier than classical rheumatoid arthritis. Diagnosis is based on clinical history and physical examination. In cancer survivors, new onset of pain should always prompt appropriate evaluation with the aim of excluding relapse of the disease.

When there is a mild form of arthritis with a low symptom burden that does not influence everyday functional status, NSAID/paracetamol is the preferred option. Oral NSAIDs are mainstay, but some patients might benefit with topical NSAID (e.g., diclofenac gel, ibuprofen gel). Treatment should be carried out for as short as possible with the lowest possible dose until control or cessation of symptoms is achieved. In patients where NSAIDs are not recommended or in cases where symptom control cannot be achieved or a prompter response is needed, low-dose corticosteroids are recommended. Prednisone is a preferred option with starting doses of 10–20 mg daily [67,68]. In cases of moderate-to-severe arthritis with multiple joints affected and impaired functionality, moderate-to-high doses of oral corticosteroid are recommended, with doses from 40 mg daily up to 1 mg/kg daily [68,69]. As soon as the symptoms improve, tapering should continue for 4–6 weeks. In patients with only a few joints affected or in large joints, intraarticular corticosteroid injections can be recommended, in a similar manner as in rheumatoid arthritis. If unable to taper corticosteroid dose to <10 mg daily, conventional disease-modifying antirheumatic drugs (DMARDs—methotrexate, sulfasalazine, hydroxychloroquine) should be included in therapy in consultation with a rheumatologist. Also, in patients who develop side effects or those at risk of developing side effects, DMARDs can be considered even in earlier grades of arthritis. In severe cases, with persistent arthritis, refractory to corticosteroids (or who are unable to wean to low-dose) or DMARDs, treatment should be escalated to biological therapy to prevent long-term negative impact on joints and to preserve or restore functionality [68].

Data regarding the influence of DMARDs on survival or cancer relapse are scarce, although some reports have not found a negative impact. The effect of long-term high-dose immunosuppression or biological therapy with non-tumor necrosis factor inhibitors is not known. Careful use of therapy is crucial, and consultation with rheumatology specialists is often necessary.

Symptom and pain control in rheumatological IrAE is very important since its impact on quality of life can be aggravating. Pain management should always be individualized and carried out in a multidisciplinary approach with rheumatological consultation, even in early phases of arthritis. In patients with advanced disease experiencing chronic inflammatory or musculoskeletal pain due to immune-related mechanisms, opioids may be considered when NSAIDs and corticosteroids fail to provide sufficient relief. Palliative protocols emphasize the judicious use of opioids in these scenarios to ensure symptom relief and preserve patient function and dignity.

Conclusion

This chapter underscores the importance of addressing chronic pain in cancer survivors, given its significant impact on quality of life long after active treatment concludes. Chronic pain following surgery, radiotherapy, chemotherapy, and systemic treatments like AIs and bisphosphonates remains a critical issue in survivorship care. Each treatment modality presents unique pain challenges, requiring tailored interventions that address both physical and psychological aspects. Effective pain management involves a multidisciplinary approach, incorporating pharmacological treatments such as NSAIDs, opioids, antidepressants, and anticonvulsants alongside nonpharmacological therapies like physical therapy, acupuncture, and CBT. Preventative strategies, such as preoperative analgesia and early physical therapy, play a crucial role in reducing the incidence of chronic pain syndromes, particularly postsurgical pain. Patient education on pain management and proactive symptom monitoring can empower survivors to seek timely interventions. In the palliative setting, the emphasis shifts toward timely symptom relief, often involving systemic opioids, nerve blocks, or palliative radiotherapy rather than preventative modalities. Collaborating with specialists from various fields ensures a holistic approach that considers each patient's unique experience and treatment history. Ultimately, focusing on chronic pain management improves survivorship outcomes and enhances their overall well-being.

References

[1] Zebrack B. Cancer survivorship-a framework for quality cancer care. J Natl Cancer Inst 2024;116(3):352–5. Available from: https://doi.org/10.1093/jnci/djad266.

[2] Davies A, Dégi CL, Aapro M, et al. Free from cancer: achieving quality of life for all cancer patients and survivors. European Cancer Organisation; 2020. Available from: https://www.europeancancer.org/resources/publications/reports/free-from-cancer.html.

[3] Rowland JH, Kent EE, Forsythe LP, et al. Cancer survivorship research in Europe and the United States: where have we been, where are we going, and what can we learn from each other? Cancer 2013;119(Suppl. 11):2094–108. Available from: https://doi.org/10.1002/cncr.28060.

[4] National Cancer Institute [cited 2025 Jun 4]. Available from: https://www.cancer.gov/publications/dictionaries/cancer-terms/def/survivor.

[5] National Cancer Institute [cited 2025 Jun 4]. Available from: https://www.cancer.gov/publications/dictionaries/cancer-terms/def/cancer-survivor.

[6] Snijders RAH, Brom L, Theunissen M, et al. Update on prevalence of pain in patients with cancer 2022: a systematic literature review and meta-analysis. Cancers 2023;15(3):591. Available from: https://doi.org/10.3390/cancers15030591.

[7] Glare PA, Davies PS, Finlay E, et al. Pain in cancer survivors. J Clin Oncol 2014;32(16):1739–47. Available from: https://doi.org/10.1200/JCO.2013.52.4629.

[8] Gärtner R, Jensen MB, Nielsen J, et al. Prevalence of and factors associated with persistent pain following breast cancer surgery. JAMA 2009;302(18):1985–92. Available from: https://doi.org/10.1001/jama.2009.1568.

[9] Steegers MA, Wolters B, Evers AW, et al. Effect of axillary lymph node dissection on prevalence and intensity of chronic and phantom pain after breast cancer surgery. J Pain 2008;9(9):813–22. Available from: https://doi.org/10.1016/j.jpain.2008.04.001.

[10] Zhang Y, Liu T, Zhou Y, et al. Analgesic efficacy and safety of erector spinae plane block in breast cancer surgery: a systematic review and meta-analysis. BMC Anesthesiology 2021;21(1):59. Available from: https://doi.org/10.1186/s12871-021-01277-x.

[11] Stillman M. Clinical approach to patients with neuropathic pain. Cleve Clin J Med 2006;73(8):726–36. Available from: https://doi.org/10.3949/ccjm.73.8.726.

[12] Kannan P, Lam HY, Ma TK, et al. Efficacy of physical therapy interventions on quality of life and upper quadrant pain severity in women with post-mastectomy pain syndrome: a systematic review and meta-analysis. Qual Life Res 2022;31(4):951–73. Available from: https://doi.org/10.1007/s11136-021-02926-x.

[13] Daniels S. Cognitive behavior therapy for patients with cancer. J Adv Pract Oncol 2015;6(1):54–6.

[14] Hwang S, Clark M. Cancer-related pelvic pain. In: Gulati A, Hung JC, editors. Essentials of interventional cancer pain management. Cham, Switzerland: Springer; 2019. Available from: https://doi.org/10.1007/978-3-319-99684-4_42.

[15] Talmi YP, Horowitz Z, Pfeffer MR, et al. Pain in the neck after neck dissection. Otolaryngol Head Neck Surg 2000;123(3):302–6. Available from: https://doi.org/10.1067/mhn.2000.104946.

[16] Gane EM, Michaleff ZA, Cottrell MA, et al. Prevalence, incidence, and risk factors for shoulder and neck dysfunction after neck dissection: a systematic review. Eur J Surg Oncol 2017;43(7):1199–218. Available from: https://doi.org/10.1016/j.ejso.2016.10.026.

[17] Espitalier F, Testelin S, Blanchard D, et al. Management of somatic pain induced by treatment of head and neck cancer: postoperative pain. Guidelines of the French Oto-Rhino-Laryngology–Head and Neck Surgery Society (SFORL). Eur Ann Otorhinolaryngol Head Neck Dis 2014;131(4):249–52. Available from: https://doi.org/10.1016/j.anorl.2014.05.002.

[18] Pfister DG, Cassileth BR, Deng GE, et al. Acupuncture for pain and dysfunction after neck dissection: results of a randomized controlled trial. J Clin Oncol 2010;28(15):2565–70. Available from: https://doi.org/10.1200/JCO.2009.26.9860.

[19] Wittekindt C, Liu WC, Preuss SF, et al. Botulinum toxin A for neuropathic pain after neck dissection: a dose-finding study. Laryngoscope 2006;116(7):1168–71. Available from: https://doi.org/10.1097/01.mlg.0000217797.05523.75.

[20] Caruso M, Harrington S. Prevalence of limb loss and limb difference in the United States: implications for public policy [White paper]. Amputee Coalition; 2024. Available from: https://www.amputee-coalition.org/wp-content/uploads/2024/05/Prevalence-of-Limb-Loss-and-Limb-Difference-in-the-United-States_Implications-for-Public-Policy.pdf.

[21] Culp CJ, Abdi S. Current understanding of phantom pain and its treatment. Pain Physician 2022;25(7):E941–57.

[22] Hanyu-Deutmeyer AA, Cascella M, Varacallo M. Phantom limb pain. StatPearls [Internet]. Treasure Island (FL): StatPearls Publishing; 2023. 2024 Jan.

[23] Kaur A, Guan Y. Phantom limb pain: a literature review. Chin J Traumatol 2018;21(6):366–8. Available from: https://doi.org/10.1016/j.cjtee.2018.04.006.

[24] Bennett MI, Kaasa S, Barke A, et al. IASP taskforce for the classification of chronic pain. The IASP classification of chronic pain for ICD-11: chronic cancer-related pain. Pain 2019;160(1):38–44. Available from: https://doi.org/10.1097/j.pain.0000000000001363.

[25] Karri J, Lachman L, Hanania A, et al. Radiotherapy-specific chronic pain syndromes in the cancer population: an evidence-based narrative review. Adv Ther 2021;38(3):1425–46. Available from: https://doi.org/10.1007/s12325-021-01640-x.

[26] van den Beuken-van Everdingen MH, de Rijke JM, Kessels AG, et al. Prevalence of pain in patients with cancer: a systematic review of the past 40 years. Ann Oncol 2007;18(9):1437–49. Available from: https://doi.org/10.1093/annonc/mdm056.

[27] Brown M, Farquhar-Smith P. Pain in cancer survivors; filling in the gaps. Br J Anaesth 2017;119(4):723–36. Available from: https://doi.org/10.1093/bja/aex202.

[28] Delanian S, Lefaix JL, Pradat PF. Radiation-induced neuropathy in cancer survivors. Radiother Oncol 2012;105(3):273–82. Available from: https://doi.org/10.1016/j.radonc.2012.10.012.

[29] Jiang J, Li Y, Shen Q, et al. Effect of pregabalin on radiotherapy-related neuropathic pain in patients with head and neck cancer: a randomized controlled trial. J Clin Oncol 2019;37(2):135–43. Available from: https://doi.org/10.1200/JCO.18.00896.

[30] Purkayastha A, Sharma N, Sarin A, et al. Radiation fibrosis syndrome: the evergreen menace of radiation therapy. Asia Pac J Oncol Nurs 2019;6(3):238–45. Available from: https://doi.org/10.4103/apjon.apjon_71_18.

[31] DiFrancesco T, Khanna A, Stubblefield MD. Clinical evaluation and management of cancer survivors with radiation fibrosis syndrome. Semin Oncol Nurs 2020;36(1):150982. Available from: https://doi.org/10.1016/j.soncn.2019.150982.

[32] Salvo N, Barnes E, van Draanen J, et al. Prophylaxis and management of acute radiation-induced skin reactions: a systematic review of the literature. Curr Oncol 2010;17(4):94–112. Available from: https://doi.org/10.3747/co.v17i4.493.

[33] Bray FN, Simmons BJ, Wolfson AH, et al. Acute and chronic cutaneous reactions to ionizing radiation therapy. Dermatol Ther 2016;6(2):185–206. Available from: https://doi.org/10.1007/s13555-016-0120-y.

[34] Mallick S, Benson R, Rath GK. Radiation induced oral mucositis: a review of current literature on prevention and management. Eur Arch Otorhinolaryngol 2016;273(9):2285–93. Available from: https://doi.org/10.1007/s00405-015-3694-6.

[35] Burgess J, Ferdousi M, Gosal D, et al. Chemotherapy-induced peripheral neuropathy: epidemiology, pathomechanisms and treatment. Oncol Ther 2021;9(2):385–450. Available from: https://doi.org/10.1007/s40487-021-00168-y.

[36] Seretny M, Currie G, Sena E, et al. Incidence, prevalence, and predictors of chemotherapy-induced peripheral neuropathy: a systematic review and meta-analysis. Pain 2014;155(12):2461–70. Available from: https://doi.org/10.1016/j.pain.2014.09.020.

[37] Starobova H, Vetter I. Pathophysiology of chemotherapy-induced peripheral neuropathy. Front Mol Neurosci 2017;10:174. Available from: https://doi.org/10.3389/fnmol.2017.00174.

[38] Chow R, Novosel M, So O, et al. Duloxetine for prevention and treatment of chemotherapy-induced peripheral neuropathy (CIPN): systematic review and meta-analysis. BMJ Support Palliat Care 2023;13(1):27–34. Available from: https://doi.org/10.1136/spcare-2022-003815.

[39] Smith EM, Pang H, Cirrincione C, et al. Effect of duloxetine on pain, function, and quality of life among patients with chemotherapy-induced painful peripheral neuropathy: a randomized clinical trial. JAMA. 2013;309(13):1359–67. Available from: https://doi.org/10.1001/jama.2013.2813.

[40] Loprinzi CL, Lacchetti C, Bleeker J, et al. Prevention and management of chemotherapy-induced peripheral neuropathy in survivors of adult cancers: ASCO guideline update. J Clin Oncol 2020;38:3325–48. Available from: https://doi.org/10.1200/JCO.20.01399.

[41] Wang C, Chen S, Jiang W, et al. Treatment for chemotherapy-induced peripheral neuropathy: a systematic review of randomized controlled trials. Front Pharmacol 2022;13:1080888. Available from: https://doi.org/10.3389/fphar.2022.1080888.

[42] Desforges AD, Hebert CM, Spence AL, et al. Treatment and diagnosis of chemotherapy-induced peripheral neuropathy: an update. Biomed Pharmacother 2022;147:112671. Available from: https://doi.org/10.1016/j.biopha.2022.112671.

[43] Li K, Giustini D, Seely D. A systematic review of acupuncture for chemotherapy-induced peripheral neuropathy. Curr Oncol 2019;26(2):e147–54. Available from: https://doi.org/10.3747/co.26.4261.

[44] Mao JJ, Ismaila N, Bao T, et al. Integrative medicine for pain management in oncology: society for integrative oncology–ASCO guideline. J Clin Oncol 2022;40(34):3998–4024. Available from: https://doi.org/10.1200/JCO.22.01357.

[45] National Comprehensive Cancer Network. NCCN Clinical Practice Guidelines in Oncology: Cancer-Related Fatigue, Version (Fatigue guideline PDF), <https://www.nccn.org/patients/guidelines/content/PDF/fatigue-patient.pdf>; 2024 [accessed 04.12.25].

[46] Zhang X, Wang A, Wang M, et al. Non-pharmacological therapy for chemotherapy-induced peripheral neurotoxicity: a network meta-analysis of randomized controlled trials. BMC Neurol 2023;23(1):433. Available from: https://doi.org/10.1186/s12883-023-03485-z.

[47] Henry NL, Giles JT, Stearns V. Aromatase inhibitor-associated musculoskeletal symptoms: etiology and strategies for management. Oncology (Williston Park) 2008;22(12):1401–8.

[48] Sestak I, Cuzick J, Sapunar F, et al. Risk factors for joint symptoms in patients enrolled in the ATAC trial: a retrospective, exploratory analysis. J Clin Oncol 2008;9(9):866–72. Available from: https://doi.org/10.1016/S1470-2045(08)70182-7.

[49] Morales L, Pans S, Paridaens R, et al. Debilitating musculoskeletal pain and stiffness with letrozole and exemestane: associated tenosynovial changes on magnetic resonance imaging. Breast Cancer Res Treat 2007;104(1):87–91. Available from: https://doi.org/10.1007/s10549-006-9394-6.

[50] Yu K, Portes P, Morris GS, et al. The role of exercise in aromatase inhibitor-induced arthralgia [Internet]. PM R 2024;16(12):1406–16. Available from: https://doi.org/10.1002/pmrj.13193.

[51] Mao JJ, Xie SX, Farrar JT, et al. A randomized trial of electro-acupuncture for arthralgia related to aromatase inhibitor use. Eur J Cancer 2014;50(2):267–76. Available from: https://doi.org/10.1016/j.ejca.2013.09.022.

[52] Eastell R, Hannon RA, Cuzick J, et al. Effect of an aromatase inhibitor on bmd and bone turnover markers: 2-year results of the Anastrozole, Tamoxifen, Alone or in Combination (ATAC) trial (18233230). J Bone Miner Res 2006;21(8):1215–23. Available from: https://doi.org/10.1359/jbmr.060508.

[53] Hanley DA, Adachi JD, Bell A, et al. Denosumab: mechanism of action and clinical outcomes. Int J Clin Pract 2012;66(12):1139–46. Available from: https://doi.org/10.1111/ijcp.12022.

[54] Marx RE. Pamidronate (Aredia) and zoledronate (Zometa) induced avascular necrosis of the jaws: a growing epidemic. J Oral Maxillofac Surg 2003;61(9):1115–7. Available from: https://doi.org/10.1016/S0278-2391(03)00720-1.

[55] Dimopoulos MA, Kastritis E, Anagnostopoulos A, et al. Osteonecrosis of the jaw in patients with multiple myeloma treated with bisphosphonates: evidence of increased risk after treatment with zoledronic acid. Haematologica 2006;91(7):968–71.

[56] Shannon J, Shannon J, Modelevsky S, et al. Bisphosphonates and osteonecrosis of the jaw. J Am Geriatr Soc 2011;59(12):2350–5. Available from: https://doi.org/10.1111/j.1532-5415.2011.03713.x.

[57] Singh M, Gonegandla GS. Bisphosphonate-induced osteonecrosis of the jaws (BIONJ). J Maxillofac Oral Surg 2020;19(2):162–7. Available from: https://doi.org/10.1007/s12663-019-01211-2.

[58] Vescovi P, Merigo E, Meleti M, et al. Surgical approach and laser applications in BRONJ osteoporotic and cancer patients. J Osteoporos 2012;2012:585434. Available from: https://doi.org/10.1155/2012/585434.

[59] Domschke C, Schuetz F. Side effects of bone-targeted therapies in advanced breast cancer. Breast Care (Basel) 2014;9(5):332–6. Available from: https://doi.org/10.1159/000368844.

[60] Darvin P, Toor SM, Sasidharan Nair V, et al. Immune checkpoint inhibitors: recent progress and potential biomarkers. Exp Mol Med 2018;50(12):1–11. Available from: https://doi.org/10.1038/s12276-018-0191-1.

[61] Johnson DB, Nebhan CA, Moslehi JJ, et al. Immune-checkpoint inhibitors: long-term implications of toxicity. Nat Rev Clin Oncol 2022;19(4):254–67. Available from: https://doi.org/10.1038/s41571-022-00600-w.

[62] Owen CN, Bai X, Quah T, et al. Delayed immune-related adverse events with anti-PD-1-based immunotherapy in melanoma. Ann Oncol 2021;32(7):917–25. Available from: https://doi.org/10.1016/j.annonc.2021.03.204.

[63] Patrinely JR Jr, Johnson R, Lawless AR, et al. Chronic immune-related adverse events following adjuvant anti-PD-1 therapy for high-risk resected melanoma. JAMA Oncol 2021;7(5):744–8. Available from: https://doi.org/10.1001/jamaoncol.2021.0051. Erratum in: JAMA Oncol. 2021 May 1;7(5):785. Available from: https://doi.org/10.1001/jamaoncol.2021.1616.

[64] Cappelli LC, Gutierrez AK, Bingham CO 3rd, et al. Rheumatic and musculoskeletal immune-related adverse events due to immune checkpoint inhibitors: a systematic review of the literature. Arthritis Care Res 2017;69(11):1751–63. Available from: https://doi.org/10.1002/acr.23177.

[65] Ghosh N, Tiongson MD, Stewart C, et al. Checkpoint inhibitor-associated arthritis: a systematic review of case reports and case series. J Clin Rheumatol 2021;27(8):e317–22. Available from: https://doi.org/10.1097/RHU.0000000000001370.

[66] Calabrese LH, Calabrese C, Cappelli LC. Rheumatic immune-related adverse events from cancer immunotherapy. Nat Rev Rheumatol 2018;14(10):569–79. Available from: https://doi.org/10.1038/s41584-018-0074-9.

[67] Dang QM, Watanabe R, Shiomi M, et al. Rheumatic Immune-related Adverse Events Due To Immune Checkpoint Inhibitors-a 2023 Update. Int J Mol Sci 2023;24(6):5643. Available from: https://doi.org/10.3390/ijms24065643.

[68] Jeurling S, Cappelli LC. Treatment of immune checkpoint inhibitor-induced inflammatory arthritis. Curr Opin Rheumatol 2020;32(3):315–20. Available from: https://doi.org/10.1097/BOR.0000000000000701.

[69] Schneider BJ, Naidoo J, Santomasso BD, et al. Management of immune-related adverse events in patients treated with immune checkpoint inhibitor therapy: ASCO guideline update. J Clin Oncol 2021;39(36):4073–126. Available from: https://doi.org/10.1200/JCO.21.01440. Erratum in: J Clin Oncol. 2022 Jan 20;40(3):315. Available from: https://doi.org/10.1200/JCO.21.02786.

Chapter 45

Treating pain in liver and renal insufficiency

Andrej Belančić[1] and Marijana Vučković[2]
[1]Department of Basic and Clinical Pharmacology and Toxicology, Faculty of Medicine, University of Rijeka, Rijeka, Croatia, [2]Department of Internal Medicine, Division of Nephrology and Dialysis, University Hospital of Split, Split, Croatia

Key learning objectives

- Understand the impact of liver and renal insufficiency on the pharmacokinetics and pharmacodynamics of analgesics in palliative care.
- Identify the mechanisms by which liver and renal insufficiency alter drug metabolism, distribution, and excretion.
- Evaluate appropriate opioid and nonopioid analgesics for use in patients with varying degrees of organ dysfunction.
- Implement evidence-based dose adjustments for analgesics based on the estimated glomerular filtration rate and Child–Pugh scores.
- Recognize the risks associated with inappropriate dosing and develop strategies for safe pain management in this vulnerable population.

Introduction

In palliative care, pain management must be effective to alleviate suffering and enhance quality of life [1]. Pain management in patients with hepatic and/or renal insufficiency is complex because of altered pharmacokinetics (PK) and pharmacodynamics (PD) of analgesics. Without effective dosage adjustment, these patients face high risks of drug accumulation or toxicity, thus requiring proper understanding of its pathophysiology and how it affects drug metabolism [1–13].

The purpose of this chapter is to discuss the pharmacological basis for managing pain in patients with liver and renal insufficiency in a palliative care context. As such, it explains how the metabolism and clearance of the analgesic drugs are affected by organ malfunction as well as provides guidance on selection thereof based on the estimated glomerular filtration rate (eGFR) and Child–Pugh scores.

Liver and renal insufficiency in palliative care

Defining liver and renal insufficiency

Liver insufficiency

Liver insufficiency refers to a condition where the liver fails to carry out essential hepatic functions such as drug metabolism, protein synthesis (e.g., albumin and blood clotting factors), and detoxification of ammonia and bilirubin. The condition is often measured based on the Child–Pugh class, comprising five clinical parameters: bilirubin levels, albumin

Pain Management in Palliative Care for Patients With Cancer. DOI: https://doi.org/10.1016/B978-0-443-30050-9.00017-5

concentrations, international normalized ratio, presence of ascites, and hepatic encephalopathy. Based on this scoring system, liver insufficiency is categorized into three classes: A (mild), B (moderate), and C (severe) [1,4,5,9–12,14].

Renal insufficiency

The condition is characterized by a diminished eGFR, leading to excess accumulation of waste products as well as drugs excreted by the kidneys. It is classified on the basis of stages, which range from mild renal impairment with a glomerular filtration rate (GFR) of 60–89 mL/min/1.73 m^2 (stage 2) to end-stage kidney disease marked by a GFR of less than 15 mL/min/1.73 m^2 or the need for dialysis [2,3,6–8,13].

Pathogenesis and metabolism in liver and renal insufficiency

Liver insufficiency pathogenesis

Common causes in palliative care include hepatic metastases, primary liver cancers, chemotherapy-induced hepatotoxicity, viral hepatitis, and cirrhosis. The liver's reduced functional capacity leads to decreased synthesis of albumin (affecting drug binding) and impaired cytochrome P450 (CYP450) enzyme activity, which is crucial for the metabolism of many opioids and nonopioid analgesics. CYP3A4 and CYP2D6, the primary enzymes involved in the metabolism of opioids such as oxycodone and methadone, are particularly affected [1,4,5,9–12,15,16].

Renal insufficiency pathogenesis

Common causes include chronic kidney disease due to diabetes or hypertension and acute kidney injury from sepsis, nephrotoxic drugs, or dehydration. Impaired renal function reduces the clearance of drugs and their metabolites, especially those with active metabolites such as morphine-6-glucuronide (M6G) from morphine. Accumulation of such metabolites can cause neurotoxicity (e.g., agitation, myoclonus, seizures) if not appropriately managed [2,3,6–8,13,17].

Impact on pharmacological pain management

Liver and renal insufficiency significantly alter the PK and PD of analgesics:

Liver insufficiency

Reduces drug clearance, increases bioavailability due to reduced first-pass metabolism, and prolongs the half-life of many drugs. For example, the bioavailability of morphine can increase by 50% in severe liver disease, necessitating dose reductions [1,4,5,9–12,14].

Renal insufficiency

Impairs drug excretion, leading to the accumulation of both parent drugs and active metabolites. In patients with eGFR <30 mL/min/1.73 m^2, the half-life of drugs such as gabapentin may increase up to fivefold, requiring significant dose reductions or alternative therapies [2,3,6–8,13,18].

Choice of analgesics based on organ function

Opioids

Morphine: This opiate should not be used in severe renal impairment (eGFR < 30 mL/min/1.73 m^2) due to high accumulation of M6G and morphine-3-glucuronide (M3G). In Child–Pugh B, doses should be reduced by 50% due to decreased clearance [1–14].

Hydromorphone: For moderate renal impairment (eGFR 30–60 mL/min/1.73 m^2), hydromorphone is the preferred choice over morphine but still requires dose adjustment. In severe renal disease, dose reduction at least by 50% is recommended [2,3,6–8,13,19].

Fentanyl: An excellent option in severe renal failure due to its minimal renal excretion and no active metabolites. In liver disease, the dose can usually be maintained unless severe liver failure (Child–Pugh C) is present, where dose titration may be needed [1–13].

Methadone: This opioid should be used with caution due to its intricate PK that involve CYP3A4, CYP2B6, and CYP2D6, among other enzymes. Dose adjustments should be made cautiously, with a 25%–50% reduction in Child–Pugh B and C and frequent monitoring for QT interval prolongation [1,4,5,9–12,15,16].

Nonopioids

Acetaminophen: Reduced doses (≤2 g/day in Child–Pugh A and B, avoid in Child–Pugh C) are usually safe in liver insufficiency. Although generally considered safe in patients with chronic kidney disease, acetaminophen should be used cautiously in individuals with significantly reduced renal function (eGFR < 30 mL/min/1.73 m²). Prolonged use in this population may increase the risk of further deterioration in kidney function [1–5,7–13,20–25].

Nonsteroidal antiinflammatory drugs (NSAIDs): Avoid in both liver and renal insufficiency due to risks of gastrointestinal bleeding, nephrotoxicity, and impaired platelet function [1–5,7–13].

Gabapentin and pregabalin: A dose reduction in renal impairment may be necessary; reduce by 50% if eGFR is 30–60 mL/min/1.73 m², and reduce by 75%–90% if eGFR is <30 mL/min/1.73 m². Although these drugs are not metabolized by the liver, caution should still be exercised in patients with severe liver disease [1–5,7–13,18,26].

Pharmacological considerations in pain management

Pharmacokinetic and pharmacodynamic changes in liver and renal insufficiency

The PK and PD of analgesics are significantly altered in patients with liver and renal insufficiency, making pain management in this population particularly challenging.

Liver insufficiency

Among other functions, the liver takes a central role in drug metabolism, mainly via the CYP450 enzyme system. In liver insufficiency, there is a decrease in hepatic blood flow, impaired hepatocyte function, and diminished enzyme activity, all affecting drug metabolism. For instance, clearance of drugs metabolized by CYP3A4 (such as fentanyl) or CYP2D6 (such as codeine and tramadol) is notably decreased. The impaired hepatic synthesis of albumin in liver insufficiency also decreases the binding of highly protein-bound drugs (such as methadone and fentanyl), leading to higher free drug concentrations, potentially increasing both efficacy and toxicity. Due to decreased first-pass metabolism, there is also an increase in bioavailability for orally administered drugs, which therefore necessitates dose reduction. Liver failure impairs both phase I metabolic pathways and phase II conjugation processes, including glucuronidation, sulfation, and acetylation. In combination with reduced hepatic blood flow, these alterations can prolong the half-lives of drugs such as morphine (metabolized via glucuronidation to M6G) and paracetamol (undergoes glucuronidation and sulfation). Thus, in patients with moderate-to-severe liver disease (Child–Pugh B or C), dose reductions of on average 25%–50% for most drugs are often necessary. For example, while issues with morphine clearance are pronounced in patients with kidney failure, morphine clearance may also decrease by up to 50% in patients with Child–Pugh C cirrhosis, while its active metabolite M6G accumulates, increasing the risk of sedation, respiratory depression, and neurotoxicity [1,4,5,9–12,14].

Renal insufficiency

The kidneys play a major role in excreting hydrophilic drugs as well as their metabolites. In renal insufficiency, particularly when eGFR is <30 mL/min/1.73 m², there is a decreased elimination of renally cleared drugs and their active metabolites, leading to prolonged half-lives and accumulation. For example, gabapentin, which is almost exclusively excreted unchanged by the kidneys, has a half-life that can increase from 5–7 hours in healthy individuals to 52 hours in patients with severe renal impairment (eGFR < 15 mL/min/1.73 m²). Similarly, in extreme situations the half-life of hydromorphone can increase from 2–3 to 10–20 hours, with significant accumulation of its neurotoxic metabolite, hydromorphone-3-glucuronide. In renal insufficiency, the PD response can also be altered due to uremia-related changes in receptor sensitivity and increased blood–brain barrier permeability, which may enhance drug effects, particularly for centrally acting analgesics. The doses of these medications are usually adjusted based on the level of renal function deterioration, denoted by eGFR. For instance, morphine should be avoided in patients who have severe renal impairment (eGFR < 30 mL/min/1.73 m²), while those medicines with minimal urine excretion such as fentanyl and methadone remain the best choice [2,3,6–8,13].

A practical summary of the main pharmacological considerations and drug adjustments in liver and renal dysfunction is shown in Fig. 45.1.

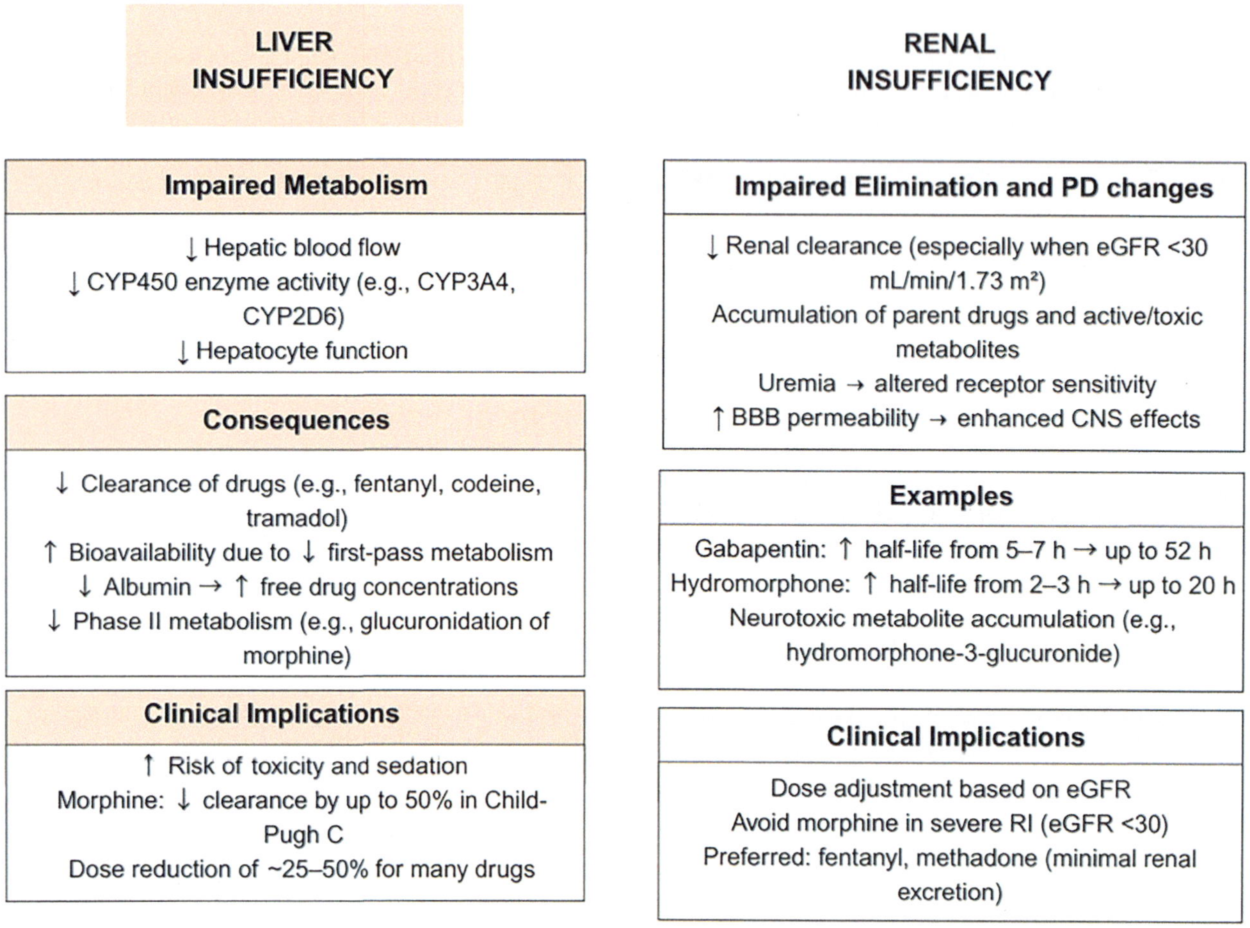

FIGURE 45.1 Pharmacokinetic and pharmacodynamic changes in liver and renal insufficiency.

Importance of dose modifications for quality of life

Dose modification is crucial not only for preventing adverse events but also for optimizing quality of life in palliative patients. Adjusting analgesic regimens to reduce the risk of drug toxicity and adverse effects (such as sedation, nausea, and constipation) directly correlates with improved pain management and overall patient well-being. Inadequate pain control or excessive side effects can diminish functional capacity, cognitive function, and the ability to participate in daily activities, leading to a significant reduction in quality of life.

Moreover, deprescribing or reducing unnecessary medications while optimizing the doses of essential analgesics can minimize polypharmacy risks, including drug–drug interactions and cumulative toxicity. This personalized approach enhances safety, comfort, and overall satisfaction in palliative care, aligning with the goals of maintaining dignity and maximizing the patient's remaining life quality [2,3,9,11].

Opioid analgesics: detailed pharmacological considerations and dose modifications

Morphine

Morphine is a strong opioid often used in palliative care, but it is extensively metabolized in the liver to M6G and M3G. M6G is primarily responsible for analgesic effects but can accumulate in renal impairment, leading to increased toxicity risks. In Child–Pugh B, it is reasonable to start with 50% of the usual dose, with further dose reductions based on clinical response and side effects. In renal insufficiency, morphine should be avoided if eGFR is <30 mL/min/1.73 m² due to the risk of M6G accumulation and neurotoxicity. For mild-to-moderate renal impairment (eGFR 30–60 mL/min/1.73 m²), morphine should be used with caution at a reduced dose and with extended dosing intervals (e.g., every 6–8 hours instead of every 4 hours) [1–14].

Hydromorphone

Hydromorphone is metabolized in the liver to hydromorphone-3-glucuronide, which can accumulate in renal failure and cause neuroexcitatory effects. In patients with moderate renal impairment (eGFR 30–60 mL/min/1.73 m^2), consider reducing the dose by 25%–50%. In severe renal impairment (eGFR < 30 mL/min/1.73 m^2), reduce the dose by 50%–75% or consider alternative opioids such as fentanyl. For liver insufficiency, reduce the initial daily dose by 25%–50% in Child–Pugh B and by 50% (or avoid) in Child–Pugh C [1–13,19].

Fentanyl

Fentanyl is a synthetic opioid with high lipophilicity and is metabolized primarily by CYP3A4. It has minimal active metabolites and undergoes minimal renal excretion, making it a preferred opioid in patients with severe renal impairment (eGFR < 30 mL/min/1.73 m^2). In case of liver insufficiency, fentanyl might accumulate as its metabolism reduces. Mild or moderate liver impairment (Child–Pugh A or B) might not require dosage adjustments. Severe liver failure (Child–Pugh C), on the other hand, necessitates close monitoring and a careful titration because this can lead to overdosing effects [1–13].

Methadone

Methadone undergoes metabolism by CYP450 enzymes, particularly CYP3A4, CYP2B6, and CYP2D6, producing no active metabolites. Thus it is appropriate for individuals with kidney disease, although consider starting with lower dose and with more regular follow-ups in patients with severe kidney failure. Methadone's half-life varies substantially, ranging from 15 to 60 hours, and may elicit QT prolongation; hence, it needs careful administration. For patients with moderate liver insufficiency (Child–Pugh B), start at 50% of the usual dose and titrate slowly. Avoid methadone or use it with extreme caution in low doses in severe liver insufficiency (Child–Pugh C), monitoring QTc closely [1,4,5,9–12,15,16].

Buprenorphine

Buprenorphine is a partial mu-agonist with a ceiling effect for respiratory depression, making it relatively safe in both renal and hepatic impairment. Minimal renal excretion suggests no dose adjustment is needed for renal failure. In patients with moderate liver dysfunction, dose adjustment is usually not needed, but in patients with severe liver failure (Child–Pugh C), a 25%–50% decrease in the dose should be considered and signs of accumulation carefully monitored [1–13,27–30].

Nonopioid analgesics and adjuvants

Acetaminophen

Acetaminophen is primarily metabolized in the liver by glucuronidation and sulfation, with a small proportion undergoing CYP450-mediated oxidation to a toxic metabolite, N-acetyl-p-benzoquinone imine (NAPQI), which is detoxified by glutathione. In liver insufficiency, the production of NAPQI may be increased, and glutathione stores depleted, raising the risk of hepatotoxicity. Therefore limit the dose to 2 g/day in patients with Child–Pugh A or B and generally avoid in Child–Pugh C. In mild and moderate renal impairment, acetaminophen is generally safe, but prolonged use should be avoided in patients with severe renal failure due to potential accumulation [1–5,7–13,20–25].

NSAIDs (e.g., ibuprofen and naproxen)

In both liver and renal insufficiency, NSAIDs should generally be avoided because they can be nephrotoxic to an extent that is nonnegligible, they carry a significant risk of causing gastrointestinal bleeding, and they decrease the synthesis of protective prostaglandins that could make the condition of renal function even worse. They are also known to indirectly trigger or worsen hepatic encephalopathy among individuals who are at risk due to liver disease [1–5,7–13].

Gabapentin and pregabalin

These are commonly used adjuvants for neuropathic pain but require significant dose adjustments in renal impairment due to renal excretion. For mild-to-moderate renal impairment (eGFR 30–60 mL/min/1.73 m^2), reduce the dose by 50%. In severe renal impairment (eGFR < 30 mL/min/1.73 m^2), reduce the dose by 75% or more. Despite not being metabolized by the liver, they should be used cautiously in patients with severe liver impairment [1–5,7–13,18,26].

Multimodal analgesia strategies

Multimodal analgesia is defined as the use of multiple drugs with varying mechanisms to manage pain while reducing side effects associated with a high dose of a single agent. For instance, in order to achieve pain management in palliative patients who have kidney and liver insufficiency, it is possible to combine an adjuvant such as gabapentin or a corticosteroid with a reduced-dose opioid. The key is to balance efficacy with safety by considering the PK and PD properties of each medication in the context of organ dysfunction.

In patients with severe liver or renal insufficiency, carefully titrated combinations such as low-dose methadone with gabapentin (adjusted for renal function) or low-dose fentanyl with acetaminophen (adjusted for liver function) can provide adequate pain relief while minimizing adverse effects. However, in these circumstances, a multidisciplinary and individualized approach with regular assessments and adjustments based on patient response, changes in organ function, and side effects, is crucial [1,2,8,31] (Tables 45.1 and 45.2).

TABLE 45.1 Examples of dose adjustments for opioid analgesics based on diminished estimated glomerular filtration rate (eGFR) and Child–Pugh scores.

Medication	Usual dose	eGFR 30–60 mL/min	eGFR < 30 mL/min	Child–Pugh A	Child–Pugh B	Child–Pugh C
Morphine	10 mg every 4 h	5–7.5 mg every 6 h	Avoid if possible	7.5–10 mg every 4 h	5 mg every 8 h	Avoid
Hydromorphone	4 mg every 4–6 h	2–3 mg every 6–8 h	1–2 mg every 8–12 h	3 mg every 6–8 h	2 mg every 8–12 h	Avoid or reduce by 50%
Fentanyl	For example, 25 mcg/h patch	No adjustment needed	No adjustment needed	No adjustment needed	No adjustment needed	Titrate with caution
Methadone	2.5–10 mg every 8–12 h	No adjustment needed	Consider starting with 2.5 mg every 24 h	5 mg every 12 h	2.5 mg every 12 h	Start with 1 mg every 24 h
Buprenorphine	5 mcg/h patch	No adjustment needed	No adjustment needed	No adjustment needed	No dose adjustment	Reduce dose by 50%

eGFR, Estimated glomerular filtration rate.
Patients with an eGFR below 30 mL/min or classified as Child–Pugh C should be managed by a multidisciplinary team, with treatment decisions tailored to the individual patient.

TABLE 45.2 Examples of dose adjustments for nonopioid analgesics and adjuvants based on diminished estimated glomerular filtration rate (eGFR) and Child–Pugh scores.

Medication	Normal dose	eGFR 30–60 mL/min	eGFR < 30 mL/min	Child–Pugh A	Child–Pugh B	Child–Pugh C
Acetaminophen	500–1000 mg every 4–6 h	500 mg every 4–6 h	500 mg every 8–12 h	500 mg every 4–6 h, up to 2 g	500 mg every 6 h, up to 2 g	Avoid if possible
Gabapentin	300–1200 mg 3 times daily	100–300 mg 3 times daily	300 mg once daily	No adjustment needed	Use with caution	Use with caution
Pregabalin	75 mg twice daily	75 mg once daily	25–50 mg once daily	No adjustment needed	Use with caution	Use with caution
NSAIDs (Ibuprofen)	200–400 mg every 6–8 h	Avoid	Avoid	Avoid	Avoid	Avoid

eGFR, Estimated glomerular filtration rate; *NSAID*, nonsteroidal antiinflammatory drug.
Patients with an eGFR below 30 mL/min or classified as Child–Pugh C should be managed by a multidisciplinary team, with treatment decisions tailored to the individual patient.

Practical applications in clinical settings

Developing patient-centered pain management plans

Creating an individualized pain management plan for patients with liver or renal insufficiency involves several key steps [8,10,32]:

- **Assessment of organ function:** Regularly assess liver function (Child–Pugh score) and renal function (eGFR).
- **Drug selection and dosing:** Choose analgesics with PK properties suitable for the patient's degree of organ dysfunction. Start at lower doses and titrate slowly.
- **Monitoring and reassessment:** Use validated pain scales, assess for side effects, and adjust doses accordingly. Monitor liver enzymes and renal parameters periodically.
- **Patient and family education:** Ensure the patient and their family understand the goals of pain management, potential side effects, and the importance of adherence to the prescribed regimen.

Monitoring and adjusting treatment

Routine monitoring includes clinical assessments (e.g., pain scores, sedation scales), laboratory tests (e.g., liver enzymes, serum creatinine), and specific drug levels if needed (e.g., methadone). Treatment should be adjusted promptly in response to any signs of toxicity, deterioration in organ function, or changes in pain severity [8,10,32].

Practical barriers in clinical practice

Despite best practice, practical constraints often hinder the optimal management of pain in patients with liver and renal insufficiency. Routine ECG monitoring for methadone-induced QT prolongation may not be feasible due to limited access to electrocardiography, and this is particularly true in community settings or where resources are limited. Similarly, the availability of palliative care or specialist consultation outside of tertiary centers may be limited. Time pressures in overcrowded outpatient clinics and patient polypharmacy are not conducive to individualized care. In practice, teleconsultations could be initiated to seek timely advice from specialists, and simplified monitoring regimens could be prioritized for lower-risk patients. Whenever the opportunity arises, high-risk patients should be consulted by cardiologists, nephrologists, or clinical pharmacologists. Analgesic regimens can be reviewed and simplified in advance, and validated dosing tables based on eGFR and Child–Pugh scores can be used for added safety and speed in these rather complex situations [8,10,32].

Special considerations in pain management

Prevention of adverse events through dose modifications

Failure to appropriately adjust analgesic doses can result in a wide range of adverse events, including opioid-induced neurotoxicity, respiratory depression, and severe gastrointestinal side effects such as constipation and nausea. In patients with liver or renal insufficiency, the accumulation of drugs or their active metabolites can lead to serious complications such as encephalopathy, sedation, myoclonus, and even death. For example, in liver impairment, failure to adjust doses of acetaminophen can lead to acute liver failure, while in renal impairment, unmodified doses of morphine can cause severe opioid toxicity due to metabolite accumulation [8,10,32].

Dose modifications and deprescribing to enhance quality of life

Effective dose modification strategies that minimize side effects while achieving adequate pain control directly impact quality of life. For palliative patients, reducing the burden of symptoms such as drowsiness, dizziness, and gastrointestinal discomfort allows for greater participation in meaningful activities, improved mood, and enhanced social interactions. Deprescribing or adjusting medications, especially in the context of polypharmacy, helps to mitigate the risk of drug–drug interactions and ensures a more tailored and patient-centered approach to care [1,2,8,31].

Other challenges in pain assessment

Patients with liver or renal insufficiency may have altered mental status due to hepatic encephalopathy or uremic confusion, complicating pain assessment. For these patients, adapt traditional pain assessment tools (such as the Visual Analog Scale) by involving caregivers or using observational scales such as the Pain Assessment in Advanced Dementia (PAINAD) scale. The PAINAD scale evaluates pain based on breathing, facial expression, body language, and

TABLE 45.3 Common nonverbal indicators of pain in patients with encephalopathy or uremia.

Observation	Possible interpretation
Facial grimacing	Localized or systemic pain
Restlessness or agitation	Discomfort, pain, or emerging delirium
Moaning or groaning	Painful stimulus or unrelieved discomfort
Guarding or withdrawal	Procedural or movement-related pain
Sudden changes in vital signs (e.g., heart rate or blood pressure)	Autonomic response to unrecognized pain
Resistance to care or touch	Possible localized pain or hypersensitivity
Decreased interaction or alertness	Sedation masking pain or overwhelming discomfort

Patients with encephalopathy or uremia should be managed by a multidisciplinary team, with the treatment approach and decisions tailored to the individual patient.

consolability and is particularly suited for patients with limited communication ability. Another useful tool is the Behavioral Pain Scale, often used in intensive care settings, which assesses facial expression, upper limb movement, and compliance with ventilation as proxies for pain intensity. Clinicians must be vigilant for atypical presentations of pain, such as agitation, restlessness, moaning, facial grimacing, or guarding in patients unable to verbalize discomfort (Table 45.3). Involving caregivers who are familiar with the patient's baseline behavior is essential in detecting subtle changes that may indicate pain [8,10,32].

Multidisciplinary and integrative approaches

A multidisciplinary approach facilitates and enhances pain management in patients with liver and renal insufficiency. Palliative care specialists, hepatologists, nephrologists, clinical pharmacologists, pharmacists, and nurses play an essential role in this process. Consequently, the integration of nonpharmacological interventions such as physiotherapy can be particularly useful in the care of patients with ascites-related pain, as it improves posture, mobility, and diaphragmatic breathing to reduce abdominal pressure.

Cognitive-behavioral therapy may be particularly useful for patients with pain-related anxiety and can be explored in primary care or through online platforms that offer structured cognitive-behavioral therapy. A synergistic approach with pharmacotherapy may improve outcomes and reduce drug-related complications [8,10,32].

Future implications and directions

There is a need for more studies on the safety and efficacy of newer analgesics in liver and renal insufficiency, as well as on the role of pharmacogenetics in personalized medicine. Novel analgesics with safer profiles are being developed. PK modeling and AI tools may provide more accurate dosing recommendations. Greater use of individualized dosing regimens and multimodal approaches should be pursued to improve outcomes in palliative care patients.

Conclusions

- Prioritize patient safety through evidence-based dose adjustments.
- Regularly reassess pain and adjust treatment based on changing organ function.
- Consider the patient's overall goals of care in all pain management decisions.
- Stay current with guidelines and emerging research. Use validated assessment tools and consider specialist consultation for complex cases.

References

[1] Rogal SS, Hansen L, Patel A, et al. AASLD practice guidance: palliative care and symptom-based management in decompensated cirrhosis. Hepatology 2022;76(3):819–53. Available from: https://doi.org/10.1002/hep.32378.

[2] Aronoff GR, Bennett WM, Berns JS, et al. Drug prescribing in renal failure: dosing guidelines for adults and children. 5th ed. American College of Physicians; 2007.

[3] Golightly LK, Teitelbaum I, Kiser TH, editors. Renal pharmacotherapy: dosage adjustment of medications eliminated by the kidneys. New York, USA: Springer Science; 2013.

[4] Hamilton JP. Pain management in liver disease. Gastroenterol Hepatol 2023;19(6):355–8.

[5] Chandok N, Watt KD. Pain management in the cirrhotic patient: the clinical challenge. Mayo Clin Proc 2010;85(5):451–8. Available from: https://doi.org/10.4065/mcp.2009.0534.

[6] Coluzzi F, Caputi FF, Billeci D, et al. Safe use of opioids in chronic kidney disease and hemodialysis patients: tips and tricks for non-pain specialists. Ther Clin Risk Manag 2020;16:821–37. Available from: https://doi.org/10.2147/TCRM.S262843.

[7] Niscola P, Scaramucci L, Vischini G, et al. The use of major analgesics in patients with renal dysfunction. Curr Drug Targets 2010;11(6):752–8. Available from: https://doi.org/10.2174/138945010791170879.

[8] Koncicki HM, Unruh M, Schell JO. Pain management in CKD: a guide for nephrology providers. Am J Kidney Dis 2017;69(3):451–60. Available from: https://doi.org/10.1053/j.ajkd.2016.08.039.

[9] Bosilkovska M, Walder B, Besson M, et al. Analgesics in patients with hepatic impairment: pharmacology and clinical implications. Drugs 2012;72(12):1645–69. Available from: https://doi.org/10.2165/11635500-000000000-00000.

[10] Rakoski M, Goyal P, Spencer-Safier M, et al. Pain management in patients with cirrhosis. Clin Liver Dis (Hoboken) 2018;11(6):135–40. Available from: https://doi.org/10.1002/cld.711.

[11] Dwyer JP, Jayasekera C, Nicoll A. Analgesia for the cirrhotic patient: a literature review and recommendations. J Gastroenterol Hepatol 2014;29(7):1356–60. Available from: https://doi.org/10.1111/jgh.12560.

[12] Gupta NK, Lewis JH. Review article: the use of potentially hepatotoxic drugs in patients with liver disease. Aliment Pharmacol Ther 2008;28(9):1021–41. Available from: https://doi.org/10.1111/j.1365-2036.2008.03822.x.

[13] Davison SN. Clinical pharmacology considerations in pain management in patients with advanced kidney failure. Clin J Am Soc Nephrol 2019;14(6):917–31. Available from: https://doi.org/10.2215/CJN.05180418.

[14] Hasselström J, Eriksson S, Persson A, et al. The metabolism and bioavailability of morphine in patients with severe liver cirrhosis. Br J Clin Pharmacol 1990;29(3):289–97. Available from: https://doi.org/10.1111/j.1365-2125.1990.tb03638.x.

[15] Novick DM, Kreek MJ, Arns PA, et al. Effect of severe alcoholic liver disease on the disposition of methadone in maintenance patients. Alcohol Clin Exp Res 1985;9(4):349–54. Available from: https://doi.org/10.1111/j.1530-0277.1985.tb05558.x.

[16] Chalabianloo F, Høiseth G, Vold JH, et al. Impact of liver fibrosis and clinical characteristics on dose-adjusted serum methadone concentrations. J Addict Dis 2023;41(1):53–63. Available from: https://doi.org/10.1080/10550887.2022.2057140.

[17] Kreek MJ, Schecter AJ, Gutjahr CL, et al. Methadone use in patients with chronic renal disease. Drug Alcohol Depend 1980;5(3):197–205. Available from: https://doi.org/10.1016/0376-8716(80)90180-5.

[18] Blum RA, Comstock TJ, Sica DA, et al. Pharmacokinetics of gabapentin in subjects with various degrees of renal function. Clin Pharmacol Ther 1994;56(2):154–9. Available from: https://doi.org/10.1038/clpt.1994.118.

[19] Durnin C, Hind ID, Ghani SP, et al. Pharmacokinetics of oral immediate-release hydromorphone (Dilaudid IR) in subjects with moderate hepatic impairment. Proc West Pharmacol Soc 2001;44:83–4.

[20] Berg KJ, Djøseland O, Gjellan A, et al. Acute effects of paracetamol on prostaglandin synthesis and renal function in normal man and in patients with renal failure. Clin Nephrol 1990;34(6):255–62.

[21] Forrest JA, Clements JA, Prescott LF. Clinical pharmacokinetics of paracetamol. Clin Pharmacokinet 1982;7(2):93–107. Available from: https://doi.org/10.2165/00003088-198207020-00001.

[22] Martin U, Temple RM, Winney RJ, et al. The disposition of paracetamol and the accumulation of its glucuronide and sulphate conjugates during multiple dosing in patients with chronic renal failure. Eur J Clin Pharmacol 1991;41(1):43–6. Available from: https://doi.org/10.1007/BF00280104.

[23] Prescott LF, Speirs GC, Critchley JA, et al. Paracetamol disposition and metabolite kinetics in patients with chronic renal failure. Eur J Clin Pharmacol 1989;36(3):291–7. Available from: https://doi.org/10.1007/BF00558162.

[24] Benson GD. Acetaminophen in chronic liver disease. Clin Pharmacol Ther 1983;33(1):95–101. Available from: https://doi.org/10.1038/clpt.1983.14.

[25] Hayward KL, Powell EE, Irvine KM, et al. Can paracetamol (acetaminophen) be administered to patients with liver impairment? Br J Clin Pharmacol 2016;81(2):210–22. Available from: https://doi.org/10.1111/bcp.12802.

[26] Randinitis EJ, Posvar EL, Alvey CW, et al. Pharmacokinetics of pregabalin in subjects with various degrees of renal function. J Clin Pharmacol 2003;43(3):277–83. Available from: https://doi.org/10.1177/0091270003251119.

[27] Elkader A, Sproule B. Buprenorphine: clinical pharmacokinetics in the treatment of opioid dependence. Clin Pharmacokinet 2005;44(7):661–80. Available from: https://doi.org/10.2165/00003088-200544070-00001.

[28] Filitz J, Griessinger N, Sittl R, et al. Effects of intermittent hemodialysis on buprenorphine and norbuprenorphine plasma concentrations in chronic pain patients treated with transdermal buprenorphine. Eur J Pain 2006;10(8):743–8. Available from: https://doi.org/10.1016/j.ejpain.2005.12.001.

[29] Hand CW, Sear JW, Uppington J, et al. Buprenorphine disposition in patients with renal impairment: single and continuous dosing, with special reference to metabolites. Br J Anaesth 1990;64(3):276–82. Available from: https://doi.org/10.1093/bja/64.3.276.

[30] Summerfield RJ, Allen MC, Moore RA, et al. Buprenorphine in end stage renal failure. Anaesthesia 1985;40(9):914. Available from: https://doi.org/10.1111/j.1365-2044.1985.tb11063.x.

[31] Christian-Miller N, Frenette C. Hepatocellular cancer pain: impact and management challenges. J Hepatocell Carcinoma 2018;5:75–80. Available from: https://doi.org/10.2147/JHC.S145450.

[32] Klinge M, Coppler T, Liebschutz JM, et al. The assessment and management of pain in cirrhosis. Curr Hepatol Rep 2018;17(1):42–51. Available from: https://doi.org/10.1007/s11901-018-0389-7.

Chapter 46

Pain management in the intensive care unit and emergency department

Zeeshan M. Rizwan[1] and Srdjan Gavrilović[2,3]
[1]*Department of Pharmacy, Mayo Clinic, Rochester, MN, United States,* [2]*Faculty of Medicine, University of Novi Sad, Novi Sad, Serbia,* [3]*Institute for Pulmonary Diseases of Vojvodina, Sremska Kamenica, Serbia*

Key learning objectives

- To understand different factors in evaluating pain in the intensive care unit (ICU) and emergency department (ED) for palliative care cancer patients.
- To identify effective pharmacological strategies for managing pain for palliative care cancer patients.
- To evaluate nonpharmacological interventions suitable for ICU/ED settings.
- To discuss the future directions and ongoing research in palliative care pain management within these settings.

Introduction

Pain is one of the most common and debilitating symptoms experienced by cancer patients, affecting their functional status and quality of life. Effective pain management in the intensive care unit (ICU) and emergency department (ED) is critical due to the high prevalence of moderate-to-severe pain among this population, especially those in the late stages. Prompt and effective pain management in the ICU and ED can significantly reduce suffering, lower the risk of complications, and enhance treatment outcomes. In addition, proper pain control helps alleviate emotional distress, increase patient comfort, and may shorten hospital stays and reduce readmissions [1,2]. Pain management in critically ill palliative care cancer patients has several unique aspects. The International Association for the Study of Pain defines pain as "an unpleasant sensory and emotional experience associated with or resembling actual or potential tissue damage" [3]. This definition emphasizes the need for a multifaceted approach to pain management in these patients. A comprehensive evaluation is required to assess the impact of pain on the patient's psychological, social, spiritual, and existential domains in order to ensure a holistic treatment strategy [4]. Pain in cancer patients is frequently multifactorial, necessitating the inclusion of psychological distress, substance abuse, and comorbidities in treatment planning [5]. Effective pain management combines pharmacological and nonpharmacological treatments, and complex pain issues frequently necessitate the collaboration of a multidisciplinary team.

Assessment of pain

Intensive care unit

Effective pain management in the ICU begins with a thorough assessment. For patients who can self-report, tools such as the Numeric Rating Scale (NRS) and Visual Analog Scale (VAS) are utilized [6]. However, many ICU patients are nonverbal due to sedation, mechanical ventilation, or altered consciousness. In these cases, observational tools such as the Behavioral Pain Scale and the Critical-Care Pain Observation Tool (CPOT) assess physical indicators such as grimacing, muscle tension, and restlessness. A more detailed evaluation of pain diagnosis is provided in Chapter 5 [6–8].

Pain Management in Palliative Care for Patients With Cancer. DOI: https://doi.org/10.1016/B978-0-443-30050-9.00028-X

Emergency department

In the ED, pain assessment is critical, especially for palliative care cancer patients. It should encompass the location, quality, intensity, onset, duration, and frequency of pain, as well as factors that relieve or exacerbate it. Acute pain can be measured with the NRS, VAS, and the Faces Pain Scale–Revised. For chronic pain common in palliative care, multidimensional assessments such as the Brief Pain Inventory and the Short-Form McGill Pain Questionnaire measure pain intensity, impact on daily activities, and various pain types [6,9]. Neuropathic pain assessments use scales such as the Leeds Assessment of Neuropathic Symptoms and Signs and the Neuropathic Pain Scale to ensure precise management [10]. For nonverbal patients, observational scales such as Face, Legs, Activity, Cry, Consolability; Pain Assessment in Advanced Dementia; and CPOT help identify and manage pain through indicators such as facial expressions and body movements [6–8]. A comprehensive assessment facilitates a seamless transition to a multifaceted treatment approach integrating nonpharmacological and pharmacological strategies.

Nonpharmacologic treatment

Nonpharmacologic interventions are crucial for managing pain in palliative care cancer patients, both in the ICU and ED. These techniques include music therapy, massage, and relaxation methods, which are generally low-cost, safe, and simple to implement. However, their use in emergencies is limited, and they are primarily feasible after patient stabilization [11]. Clinicians proactively identify and eliminate specific pain triggers by repositioning uncomfortable medical equipment, modifying painful procedures, or adjusting patient positioning. Addressing anxiety through guided imagery or involving social workers to support family communication, including evaluating thoughts, emotions, perceptions, beliefs, and attitudes, significantly impacts patient suffering [12,13]. Clearly communicating details about potentially stressful medical procedures, such as invasive tests or surgeries, and establishing targeted postoperative pain management strategies can greatly improve patient comfort. In addition, minimizing environmental stressors, such as using noise-canceling headphones, dimming lights, or ensuring personal space along with the presence of family and spiritual care, further supports pain management in the ICU and ED [12]. This holistic approach ensures that both the physical and emotional needs of end-of-life patients are addressed comprehensively in both ICU and ED settings.

Pharmacologic treatment

Intensive care unit

In the ICU, opioid analgesics are central to the pharmacological management of pain for palliative care patients. Common choices include morphine, oxycodone, hydromorphone, fentanyl, and methadone, selected based on factors such as patient-specific characteristics including renal and hepatic function, potential adverse effects, and available routes of administration [1,2,12,14]. For example, morphine is commonly used but may be less suitable for patients with renal failure due to the risk of metabolite accumulation leading to toxicity. Alternatives such as fentanyl, which have inactive metabolites and are less reliant on renal excretion, may be more appropriate in these cases. The practice of rotating opioids is also implemented to manage tolerance that might develop with long-term opioid use, aiming to maintain efficacy and minimize adverse effects such as sedation, respiratory depression, and gastrointestinal upset. Nonopioid analgesics such as nonsteroidal antiinflammatory drugs (NSAIDs) and acetaminophen or paracetamol may be used adjunctively to enhance pain relief and reduce opioid requirements, though their use must be judicious to avoid adverse effects, particularly in compromised patients [15]. In addition, ketamine, an N-methyl-D-aspartate receptor antagonist, is considered for refractory pain due to its analgesic properties at subanesthetic doses, particularly when traditional opioids are insufficient or inappropriate. The details on the dosing, formulations, adverse effects, and monitoring are given in the appropriate chapters of this book [16].

Emergency department

In the ED, pain is usually managed with targeted interventions following the WHO three-step pain ladder, starting with nonopioid analgesics for mild pain, progressing to weak opioids for moderate pain, and escalating to strong opioids for severe pain. However, analgesic selection should also be guided primarily by pain severity, prior opioid exposure, organ function, and need for rapid onset [15]. NSAIDs and acetaminophen are typically the first line of treatment for mild-to-moderate pain due to their fast-acting properties. As pain intensity increases to moderate levels, weaker opioids such as codeine and tapentadol can be introduced, potentially in combination with nonopioid analgesics to better manage the pain and underlying conditions. However, as codeine and tapentadol are only available in oral formulations, their onset of action is relatively slower and their suitability for rapid titration in the ED/ICU is limited. For severe pain, strong opioids such as morphine, hydromorphone, fentanyl, and oxycodone become necessary [17]. These are chosen based on their rapid onset and effectiveness, crucial in the ED setting where pain management needs to be quick and effective. The administration of these opioids may be adjusted based on the patient's response and the presence of any side effects such as respiratory depression or systemic reactions.

In addition, management strategies must consider emergent issues such as opioid-induced hyperalgesia, where increased opioid use paradoxically exacerbates the pain. In such cases, strategies include rotating opioids or adjusting dosages, alongside the potential use of ketamine, which has been shown to be effective for refractory pain without the risk of opioid-induced complications [16]. In managing breakthrough pain, rapid-onset opioids such as intravenous fentanyl are essential for quick relief. Continuous reassessment of pain and symptoms is crucial, adapting to changes in the patient's condition and aligning with the overarching goals of palliative care, which focus on comfort and quality of life. A detailed overview of the specific medications and their dosages can be found in Table 46.1.

Review of current research and clinical trials

Pain is a prevalent and significant symptom among patients presenting to the ED and ICU. Effective pain management is crucial, particularly for palliative care patients, necessitating an approach that aligns with their care goals and respects their end-of-life preferences. In oncology, increasing evidence underscores the link between quality of life and survival with timely and comprehensive palliative care, particularly when pain management is ineffective [17].

It is important to avoid potential pain triggers and interventions that might increase discomfort, one of the most common complaints, with global studies reporting prevalence rates ranging from 9% to 73% depending on the population and setting [19–25]. Azizoddin et al., in a prospective observational cohort study, highlighted that up to 40% of cancer patients reporting pain may not receive optimal analgesia, reflecting inadequate treatment [26]. Ineffective pain control has been associated with increased ED visits, as demonstrated by Elsayem et al. [27], and Adhikari et al.'s extensive study on outpatient pain management patterns in cancer patients, utilizing data from the national ambulatory medical care survey (2006–18), revealed significant undertreatment of pain: 61% did not receive pain medication, 24% were prescribed nonopioid analgesics, and 14% received opioids [28]. Beck et al., in a secondary analysis of a prospective cohort study, reported that cancer patients with preexisting chronic pain experienced higher pain severity despite receiving more opioid analgesics compared to those without chronic pain [29]. These findings underscore the critical role of outpatient pain management in oncology, potentially preventing one-fifth to almost half of ED visits related to cancer pain [30–32].

A multicenter cohort study evaluating 1075 adult cancer patients across 18 EDs in the United States found that opioid administration in the ED correlated with higher hospital admission and 30-day readmission rates. Severe pain at presentation was associated with increased 30-day mortality, though this risk was mitigated by adjusting for clinical factors, emphasizing the heightened risk for patients requiring opioids [33]. In a recent prospective study on cancer patients presenting with moderate-to-severe pain in the ED, biopsychosocial factors such as pain catastrophizing and anxiety were significantly associated with worse pain outcomes. In addition, outpatient opioid use independently correlated with increased pain severity and interference, suggesting potential benefits from enhanced pharmacologic and behavioral interventions [26].

Defining and aligning care goals in the ED with patient wishes, medical standards, and ethical norms is essential. Apostol et al.'s pilot cohort study demonstrated that integrating goals-of-care meetings for advanced cancer patients at risk of critical-care interventions may reduce ICU admissions [34]. The integration of palliative care into emergency medicine can ease decision-making burdens and facilitate expert-led discussions on care goals, thereby enhancing patient management in these critical settings [35,36].

Most published studies are limited by factors such as reliance on database queries, lack of detailed pain-related information, small sample sizes, focus on specific cancer subtypes, single-center settings, homogeneous participant groups, and missing or incomplete data.

Guidelines

There are several guidelines with a special focus on pain treatment in palliative or cancer patients in ED and ICU:

The United States Best Practice Guidelines for Primary Palliative Care in the ED, established by an expert panel from six medical societies, provided evidence-based recommendations and consensus-driven guidance for frontline ED clinicians in screening, assessing, and managing palliative care needs. The guidelines address comprehensive care strategies, including opioid-tolerant pain management and symptoms induced by opioids, aiming to optimize patient outcomes and care transitions [35].

TABLE 46.1 Summary of suggested drugs for pain management and dosing [2,12,16,18].

Pain severity	Medication	Description	Route of administration	Dosage
Mild to moderate	Acetaminophen	Nonopioid analgesic	IV/PO	Standard dosing: Up to 1 g every 6 h
	Ibuprofen	Nonsteroidal antiinflammatory drug (NSAID), reduces inflammation and mild pain	IV/PO	Up to 800 mg every 6–8 h
	Ketorolac	NSAID, for moderate pain	IV/IM	15–30 mg every 6 h, max 120 mg/day, max 5 days. Dose needs adjustment if patient <50 kg or ≥65 years
	Codeine[a]	Opioid analgesic for moderate pain	PO	30–60 mg every 4–6 h as needed
	Tapentadol[a]	Opioid analgesic, inhibits norepinephrine reuptake	PO	50–100 mg every 4–6 h as needed
Severe	Morphine	Mainstay for severe pain management and breakthrough pain in both ICU and ED	IV	2–10 mg every 3–4 h as needed
	Hydromorphone	Mainstay for severe pain management and breakthrough pain in both intensive care unit and ED	IV	0.3–1.5 mg every 3–4 h as needed
	Fentanyl	Mainstay for severe pain management and breakthrough pain in both intensive care unit and ED	IV	50–100 μg every 0.5–2 h as needed
	Oxycodone[a]	Immediate-release formulation	PO	5–10 mg every 4–6 h as needed
Adjuvants	Ketamine	Effective in hyperalgesia, chronic, neuropathic, and cancer pain	IV	0.1–0.3 mg/kg/h infusion
	Gabapentin[a]	For neuropathic pain	PO	Start at 100–300 mg up to 3 times daily
	Pregabalin[a]	For neuropathic pain	PO	Start at 75 mg once or twice daily
Opioid-related complications	Management of opioid-induced hyperalgesia	Adjusting opioid types or doses, adding ketamine	Varies	Adjust based on symptoms
	Breakthrough pain	Managed by rapid-onset opioids such as fentanyl	IV	Adjusted to patient's needs

ED, Emergency department; *ICU*, intensive care unit; *IM*, intramuscular; *IV*, intravenous; *PO*, peroral.
Therapeutic approach should be individualized and multidisciplinary, taking into account liver and kidney function frailty, prior medication use, and potential drug interactions. For detailed discussion, refer to relevant chapters in this book.
[a] *Due to their availability exclusively as oral formulations, these drugs are primarily used in patients after initial pain stabilization.*

The European Society for Emergency Medicine guidelines, published in 2020, offer structured recommendations for acute pain management in emergency settings across Europe. They cover assessment, nonpharmacological and pharmacological therapies, and specific considerations for prehospital and hospital care. Recent advancements include a diagnostic algorithm for identifying breakthrough cancer pain, which should complement detailed clinical evaluations and insights from patients and caregivers to optimize management [37].

The recommendations from 10 German medical societies emphasize the integration of palliative care into intensive care medicine for both incurable cancer and nononcological patients. The focus extends beyond pain management to improving the quality of life and symptom control, recognizing the urgent need for supportive care for patients and their highly stressed relatives [38].

The Clinical Practice Guidelines and Consensus Statements on Pain Management in Critically Ill End-of-Life Patients: A Systematic Review, authored by an international expert group, offers recommendations covering pain assessment, pharmacologic and nonpharmacologic pain management strategies, and opioid and analgesic dosing for terminal ICU patients, among other topics. These findings aim to foster international consensus on the dignified management of end-of life pain within ICU settings [12].

Future implications and conclusion

In both the ICU and ED, effective palliative care integrates a philosophy aimed at improving the physical, mental, spiritual, and psychosocial well-being of patients and their families. As palliative care has evolved from focusing primarily on end-of-life scenarios to encompassing comprehensive supportive care throughout a patient's disease trajectory, there is a growing need to apply these practices early in the disease course. This early integration has demonstrated benefits such as symptom alleviation, enhanced patient and caregiver quality of life, improved understanding of illness and prognosis, reduced healthcare costs, and potentially improved survival rates [2,25]. Future implications for ICU and ED settings include developing and implementing innovative models that facilitate early integration of palliative care. These models should address existing disparities in access to palliative care, particularly in underserved populations, and focus on overcoming barriers at the patient, provider, and system levels [39,40]. In addition, there is a need for ongoing research to improve the quality of care and understand multilevel factors that contribute to disparities in palliative care utilization among cancer patients [41].

Advancing palliative care in the ICU and ED involves more than just managing acute symptoms; it requires a holistic approach that considers all aspects of a patient's well-being throughout their medical journey. The future of palliative care in these critical settings depends on the development of new care models that ensure early and equitable access to palliative services, aiming to improve both the quality of care and clinical outcomes. Such efforts should be multidisciplinary and inclusive of various stakeholders, including nursing, medicine, pharmacy, psychology, and social work, to ensure comprehensive and effective implementation across the healthcare spectrum.

References

[1] Nordness MF, Hayhurst CJ, Pandharipande P. Current perspectives on the assessment and management of pain in the intensive care unit. J Pain Res 2021;14:1733–44. Available from: https://doi.org/10.2147/JPR.S256406.

[2] Serra S, Spampinato MD, Riccardi A, et al. Pain management at the end of life in the emergency department: a narrative review of the literature and a practical clinical approach. J Clin Med 2023;12(13):4357. Available from: https://doi.org/10.3390/jcm12134357.

[3] Raja SN, Carr DB, Cohen M, et al. The revised international association for the study of pain definition of pain: concepts, challenges, and compromises. Pain 2020;161(9):1976–82. Available from: https://doi.org/10.1097/j.pain.0000000000001939.

[4] Bennett MI, Kaasa S, Barke A, et al. IASP Taskforce for the Classification of Chronic Pain. The IASP classification of chronic pain for ICD-11: chronic cancer-related pain. Pain 2019;160(1):38–44. Available from: https://doi.org/10.1097/j.pain.0000000000001363.

[5] Syrjala KL, Jensen MP, Mendoza ME, et al. Psychological and behavioral approaches to cancer pain management. J Clin Oncol 2014;32(16):1703–11. Available from: https://doi.org/10.1200/JCO.2013.54.4825.

[6] Puntillo K, Nelson JE, Weissman D, et al. Palliative care in the ICU: relief of pain, dyspnea, and thirst–a report from the IPAL-ICU Advisory Board. Intensive Care Med 2014;40(2):235–48. Available from: https://doi.org/10.1007/s00134-013-3153-z.

[7] Crellin DJ, Harrison D, Santamaria N, et al. Systematic review of the Face, Legs, Activity, Cry and Consolability scale for assessing pain in infants and children: is it reliable, valid, and feasible for use? Pain 2015;156(11):2132–51. Available from: https://doi.org/10.1097/j.pain.0000000000000305.

[8] Warden V, Hurley AC, Volicer L. Development and psychometric evaluation of the Pain Assessment in Advanced Dementia (PAINAD) scale. J Am Med Dir Assoc 2003;4(1):9–15. Available from: https://doi.org/10.1097/01.JAM.0000043422.31640.F7.

[9] Tan G, Jensen MP, Thornby JI, et al. Validation of the brief pain inventory for chronic nonmalignant pain. J Pain 2004;5(2):133–7. Available from: https://doi.org/10.1016/j.jpain.2003.12.005.

[10] Bennett M. The LANSS Pain Scale: the Leeds Assessment of Neuropathic Symptoms and Signs. Pain 2001;92(1):147–57. Available from: https://doi.org/10.1016/S0304-3959(00)00482-6.

[11] Coelho A, Parola V, Cardoso D, et al. Use of non-pharmacological interventions for comforting patients in palliative care: a scoping review. JBI Database System Rev Implement Rep 2017;15(7):1867–904. Available from: https://doi.org/10.11124/JBISRIR-2016-003204.

[12] Durán-Crane A, Laserna A, López-Olivo MA, et al. Clinical practice guidelines and consensus statements about pain management in critically Ill end-of-life patients: a systematic review. Crit Care Med 2019;47(11):1619–26. Available from: https://doi.org/10.1097/CCM.0000000000003975.

[13] Sallnow L, Smith R, Ahmedzai SH, et al. Report of the Lancet Commission on the value of death: bringing death back into life. Lancet 2022;399(10327):837–84. Available from: https://doi.org/10.1016/S0140-6736(21)02314-X.

[14] Mularski RA, Puntillo K, Varkey B, et al. Pain management within the palliative and end-of-life care experience in the ICU. Chest 2009;135(5):1360–9. Available from: https://doi.org/10.1378/chest.08-2328.

[15] Nabal M, Librada S, Redondo MJ, et al. The role of paracetamol and nonsteroidal anti-inflammatory drugs in addition to WHO Step III opioids in the control of pain in advanced cancer. A systematic review of the literature. Palliat Med 2012;26(4):305–12. Available from: https://doi.org/10.1177/0269216311428528.

[16] Riccardi A, Guarino M, Serra S, et al. Narrative review: low-dose ketamine for pain management. J Clin Med 2023;12(9):3256. Available from: https://doi.org/10.3390/jcm12093256.

[17] Swarm RA, Paice JA, Anghelescu DL, et al. Adult cancer pain, version 3.2019, NCCN clinical practice guidelines in oncology. J Natl Compr Canc Netw 2019;17(8):977–1007. Available from: https://doi.org/10.6004/jnccn.2019.0038.

[18] Fallon M, Giusti R, Aielli F, et al. Management of cancer pain in adult patients: ESMO clinical practice guidelines. Ann Oncol 2018;29(Suppl. 4):iv166–91. Available from: https://doi.org/10.1093/annonc/mdy152.

[19] Haltia O, Hirvonen OM, Saarto T, et al. Impact of the regional palliative care pathway on emergency room visits and hospitalizations. Anticancer Res 2021;41(3):1701–6. Available from: https://doi.org/10.21873/anticanres.14934.

[20] Crock B, Islam MR, Subramaniam S. Emergency department utilisation by palliative patients in a regional Australian setting. Am J Hosp Palliat Care 2022;39(8):956–61. Available from: https://doi.org/10.1177/10499091211055903.

[21] Green E, Ward S, Brierley W, et al. "They shouldn't be coming to the ED, should they?": a descriptive service evaluation of why patients with palliative care needs present to the emergency department. Am J Hosp Palliat Care 2017;34(10):984–90. Available from: https://doi.org/10.1177/1049909116676774.

[22] Hsu H-S, Wu T-H, Lin C-Y, et al. Enhanced home palliative care could reduce emergency department visits due to non-organic dyspnea among cancer patients: a retrospective cohort study. BMC Palliat Care 2021;20(1):42. Available from: https://doi.org/10.1186/s12904-021-00713-6.

[23] Nordt SP, Ryan JM, Kelly D, et al. Palliative care patient emergency department visits at tertiary university-based emergency department in Ireland. Am J Emerg Med 2023;66:76–80. Available from: https://doi.org/10.1016/j.ajem.2023.01.035.

[24] Verhoef M-J, de Nijs E, Horeweg N, et al. Palliative care needs of advanced cancer patients in the emergency department at the end of life: an observational cohort study. Support Care Cancer 2020;28(3):1097–107. Available from: https://doi.org/10.1007/s00520-019-04906-x.

[25] Díez-Concha JF, Gómez-García DM, Agudelo JA, et al. Assessing the impact of palliative care admission of end-of-life cancer adults. Palliat Care Soc Pract 2023;17:26323524231198545. Available from: https://doi.org/10.1177/26323524231198545.

[26] Azizoddin DR, Beck M, Flowers KM, et al. Psychological evaluation of patients with cancer presenting to the emergency department with pain: independent predictors of worse pain severity, interference, and higher hourly opioid administration. JCO Oncol Pract 2022;18(10):e1648–60. Available from: https://doi.org/10.1200/OP.22.00142.

[27] Elsayem AF, Merriman KW, Gonzalez CE, et al. Presenting symptoms in the emergency department as predictors of intensive care unit admissions and hospital mortality in a comprehensive cancer center. J Oncol Pract 2016;12(5):e554–63. Available from: https://doi.org/10.1200/JOP.2015.009019.

[28] Adhikari B, Shrestha N, Bhatt VR. A real-world study of the quality of outpatient pain management in patients with cancer. J Cancer Surviv 2025;19(6):2101–9. Available from: https://doi.org/10.1007/s11764-024-01621-3.

[29] Beck M, Schreiber KL, Wilson JM, et al. A secondary analysis: the impact of pre-existing chronic pain among patients with cancer presenting to the emergency department with acute pain. Support Care Cancer 2024;32(2):129. Available from: https://doi.org/10.1007/s00520-024-08314-8.

[30] Alsirafy SA, Raheem AA, Al-Zahrani AS, et al. Emergency department visits at the end of life of patients with terminal cancer: pattern, causes, and avoidability. Am J Hosp Palliat Care 2016;33(7):658–62. Available from: https://doi.org/10.1177/1049909115581819.

[31] Roy M, Halbert B, Devlin S, et al. From metrics to practice: identifying preventable emergency department visits for patients with cancer. Support Care Cancer 2021;29(7):3571–5. Available from: https://doi.org/10.1007/s00520-020-05874-3.

[32] Salama H, Omer MH, Shafqat A, Binahmed A, et al. Avoidable emergency department visits among palliative care cancer patients: novel insights from Saudi Arabia and the Middle East. BMC Palliat Care 2024;23(1):60. Available from: https://doi.org/10.1186/s12904-024-01389-4.

[33] Coyne CJ, Reyes-Gibby CC, Durham DD, et al. Cancer pain management in the emergency department: a multicenter prospective observational trial of the Comprehensive Oncologic Emergencies Research Network (CONCERN). Support Care Cancer 2021;29(8):4543–53. Available from: https://doi.org/10.1007/s00520-021-05987-3.

[34] Apostol CC, Waldfogel JM, Pfoh ER, et al. Association of goals of care meetings for hospitalized cancer patients at risk for critical care with patient outcomes. Palliat Med 2015;29(4):386–90. Available from: https://doi.org/10.1177/0269216314560800.

[35] Loffredo AJ, Chan GK, Wang DH, et al. United States best practice guidelines for primary palliative care in the emergency department. Ann Emerg Med 2021;78(5):658–69. Available from: https://doi.org/10.1016/j.annemergmed.2021.05.021.
[36] Ouchi K, Aaronson EL, Engel KG, et al. Emergency palliative care: nongoal concordant care and time-limited trials in the emergency department. J Palliat Med 2024;27(6):823–6. Available from: https://doi.org/10.1089/jpm.2023.0371.
[37] Idrissi SH, Coffey F, Doblas V, et al. Guidelines for the management of acute pain in emergency situations (Report). Aartselaar (Belgium): European Society for Emergency Medicine; 2020. Available from: https://www.eusem.org/images/EUSEM_EPI_GUIDELINES_MARCH_2020.pdf.
[38] Michels G, Schallenburger M, Neukirchen M. Recommendations on palliative care aspects in intensive care medicine. Crit Care 2023;27(1):355. Available from: https://doi.org/10.1186/s13054-023-04622-3.
[39] Downar J, Hua M, Wunsch H. Palliative care in the intensive care unit: past, present, and future. Crit Care Clin 2023;39(3):529–39. Available from: https://doi.org/10.1016/j.ccc.2023.01.007.
[40] Rosa WE, McDarby M, Buller H, et al. Palliative care clinician perspectives on person-centered end-of-life communication for racially and culturally minoritized persons with cancer. Cancers 2023;15(16):4076. Available from: https://doi.org/10.3390/cancers15164076.
[41] Sanders JJ, Temin S, Ghoshal A, et al. Palliative care for patients with cancer: ASCO guideline update. J Clin Oncol 2024;42(19):2336–57. Available from: https://doi.org/10.1200/JCO.24.00542.

Chapter 47

Treating pain in children and adolescents

Jelena Roganović[1,2] and Nuša Matijašić Stjepović[1]

[1]*Department of Pediatric Oncology and Hematology, Children's Hospital Zagreb, Zagreb, Croatia,* [2]*Faculty of Biotechnology and Drug Development, University of Rijeka, Rijeka, Croatia*

Key learning objectives

- To comprehend the basic principles of pediatric palliative care, including the World Health Organization guidelines for the treatment of chronic pain in children and adolescents with life-threatening, life-limiting, and terminal conditions.
- To name the most commonly used pain assessment tools in children and adolescents and choose appropriate ones depending on the type of pain, child's age, and developmental stage.
- To understand the mechanism of action of opioids used in children and adolescents with advanced cancer, routes of administration, and routinely applied starting doses.
- To identify available adjuvant drugs for pain control in children and adolescents with end-stage cancer, along with their applicability and side effects.
- To get insight into the most frequently used interventional and nonpharmacological procedures for pain relief in pediatric oncological patients at terminal life phases.

Introduction

Treatment novelties and advanced supportive care have led to immense progress in pediatric oncology, with 5-year survival rates nowadays exceeding 80% [1]. Moreover, new therapeutic options may shift cancer from an acute to a chronic illness. Despite these encouraging facts, approximately 4 million children with malignancies require pediatric palliative care (PPC) worldwide [2,3]. As defined by the World Health Organization (WHO), PPC integrates multidisciplinary approaches (drugs, anesthetic interventions, and integrative therapies) to prevent and relieve suffering in children with life-threatening, life-limiting, and terminal conditions [4]. In 2022, new international standards for PPC were published, once again highlighting the importance of alleviating distressing symptoms as a fundamental component of care [5]. Regardless of a deep understanding of the basic principles of PPC, physicians still feel insecure when delivering PPC to cancer patients, not only while addressing their physical needs but also, expectedly, feel even less confident in meeting children's emotional and spiritual needs [6].

Pain remains one of the leading symptoms in children and adolescents with advanced-stage malignancies [7–11], following fatigue and reduced mobility [7]. It is also the most distressing symptom, as perceived by parents [8]. This unpleasant sensation, along with psychological, spiritual, and emotional anguish, leads to end-of-life (EoL) suffering [12]. Pain severely diminishes the quality of life (QoL) of terminally ill children and their families. One of the core principles of PPC states that severe, uncontrolled pain is regarded as a medical emergency requiring immediate intervention [13]. Adequate pain control sets the ground for proper terminal at-home care, as the goal in PPC.

However, some studies suggest that even in resource-rich countries, pain is undertreated in as many as 70% of pediatric cancer cases [7,9,14], and almost one-fifth of patients endure severe chronic pain [15]. This may be the result of the complex pain pathophysiology in oncological patients. Additionally, pain in children differs tremendously from that in adults due to diverse physiology and variances in pain perception depending on the cognitive stage of the child [16,17]. Numerous psychological and emotional factors affect the child's comprehension of pain and can stimulate their response to the sensation. The pain pathway changes during growth. Neurobiological studies demonstrated that nociceptors start functioning at the 20th week of gestation. Their density on the body's surface is higher in children than in adults. The

Pain Management in Palliative Care for Patients With Cancer. DOI: https://doi.org/10.1016/B978-0-443-30050-9.00018-7

amount of neuromediators is higher as well, which could potentially lead to an increased sensitivity to pain in childhood. Hence, painful signals could be more intense and last longer in earlier life, although the mechanism is not fully elucidated. Due to the plasticity of a child's neurological system, prolonged or repeated pain enhances the risk of neuron malfunction and death [17]. Besides, pain assessment in young, nonverbal, and/or intellectually disabled children is challenging.

Opioids are the backbone of palliative pain treatment in children and adults with cancer. The high-quality research on opioid use in pediatric palliative oncology is scarce. Regarding the dosing regimen, the adopted strategy is dose escalation until pain is properly controlled or serious side effects occur. The misconceptions on opioid use, such as provoking addiction or hastening death through respiratory arrest, unfortunately act as barriers for satisfactory pain relief in young patients [13–16,18–21]. The multimodal approach to pain in children at terminal phases is complex and includes interventional and nonpharmacological approaches. Given the multifactorial etiology and individual response across patients, pain treatment must be tailored to suit each child [16].

Pain assessment in children and adolescents

For successful relief, pain must be first detected, localized, and quantified, and its cause determined. Upon treatment initiation, continuous reassessment is required. The major obstacles regarding pain evaluation in pediatric patients are the inability of young children to verbalize sensations and the insufficient clinicians' familiarity with currently available pain scales [22–27]. At present, there is no pain assessment tool validated exclusively for the PPC setting [27].

Preschool children are mostly incapable of self-reported pain measurement; they are typically unable to clearly communicate their pain, often do not completely understand what is being asked of them, and may have a tendency to simulate, exaggerate, or deny pain. In addition, parents/caregivers influence the way young children express their feelings. Therefore, in those younger than 6 years, Behavioral Pain Scales, based on observing the child, are mandatory. Some tools require monitoring of physiological indices, such as heart rate, blood pressure, oxygen saturation, and/or sweating [22–25]. As chronicity leads to adaptation, biological signs become more applicable for acute pain, while behavioral changes remain the key instrument in evaluating chronic discomfort [25].

A recent systematic review identified 65 different observational pediatric pain assessment scales [26]. The most commonly applied ones are the following: Premature Infant Pain Profile (PIPP); Échelle de Douleur et d'Inconfort du Nouveau-né (EDIN)/Neonatal Pain and Discomfort Scale (NPDS); Douleur Aiguë du Nouveau-né (DAN); Neonatal Infant Pain Scale (NIPS); Neonatal Facial Coding System (NFCS); Evaluation Enfant Douleur (EVENDOL); Face, Legs, Activity, Cry, Consolability (FLACC); revised Face, Legs, Activity, Cry, Consolability (r-FLACC); Children's Hospital of Eastern Ontario Pain Scale (CHEOPS); Douleur Enfant Gustave Roussy (DEGR); Hétéro-évaluation de la douleur chez l'enfant (HEDEN); Non-Communicating Children's Pain Checklist–Revised (NCCPC-R); Pediatric Pain Profile (PPP); and the COMFORT Scale [22,23,28–39]. The age distribution for scales used in acute and prolonged pain is displayed in Table 47.1.

While most of the above mentioned instruments are used to quantify acute and procedural pain (PIPP, DAN, NIPS, NFCS, EVENDOL, FLACC, CHEOPS) [22,28,30–34], three of them are tailored to objectify prolonged pain and discomfort: EDIN/NPDS, DEGR, and HEDEN [22,23,29,35,36]. EDIN/NPDS was developed on preterm infants as the first tool to evaluate protracted pain in neonates. It comprises five behavioral indicators: facial activity, body movements, quality of sleep, contact with nurses, and consolability [29]. DEGR is an observational scoring system specifically developed to assess pain in children aged 2–6 years with cancer. It consists of 15 items distributed in three main categories: behaviors specific to pain, psychomotor inertia, and anxiety. Each item is rated depending on its severity (0–4) so that theoretical values of DEGR range from 0 to 60. The scale's items are summarized in Table 47.2 [23]. HEDEN is DEGR's shorter version, as the original takes about 23 minutes to complete [22,36].

The gold standard in children older than 6 years is self-reported rating, such as through Visual Analog Scale, Numeric Rating Scale, Verbal Rating Scale, Faces Pain Scale (FPS) and Faces Pain Scale–Revised (FPS-R), Poker Chip Tool, Memorial Symptom Assessment Scale for children aged 7–12 years, Rainbow Pain Scale, and Symptom Screening in Pediatrics Tool [22,25,27,40–45]. Some self-assessment tools, such as the Wong–Baker FACES Scale, may be used in younger children, above the age of 3 years [46]. FPS-R is recommended in PPC [27]. It consists of six faces aligned horizontally, with the one on the far left being neutral and the one on the far right representing maximal pain [41].

When choosing the appropriate tool, clinicians must also consider the child's developmental abilities, regardless of the chronological age [25]. Few scales seem reliable in children with cognitive and neurological impairments: R-FLACC, NCCPC, and PPP, while COMFORT is advised in sedated and unconscious patients [22,34,37–39].

TABLE 47.1 Observational pain scales for acute and prolonged pain according to child's age.

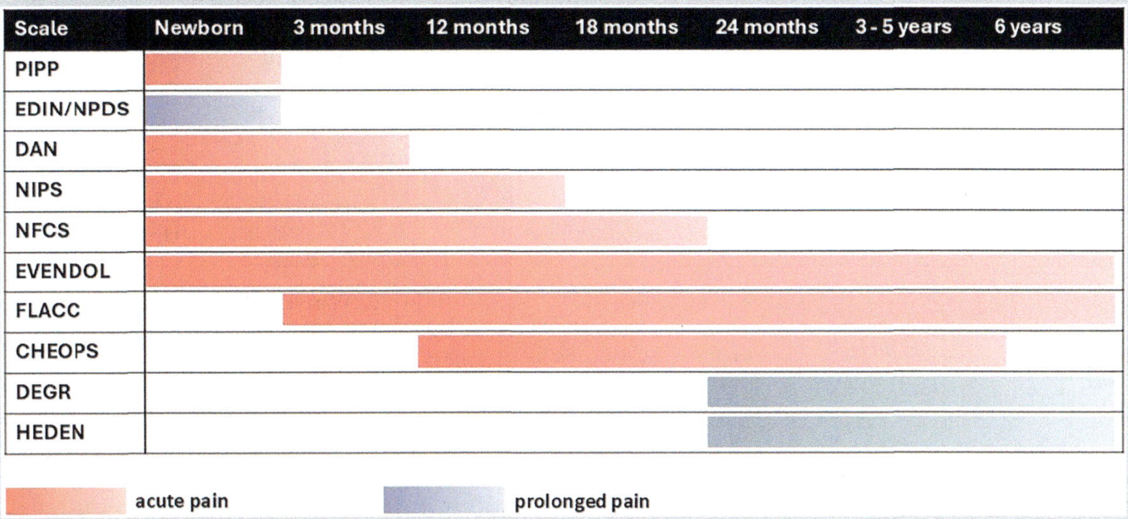

CHEOPS, Children's Hospital of Eastern Ontario Pain Scale; *DAN*, Douleur Aiguë du Nouveau-né; *DEGR*, Douleur Enfant Gustave Roussy; *EDIN/NPDS*, Échelle de Douleur et d'Inconfort du Nouveau-né or Neonatal Pain and Discomfort Scale; *EVENDOL*, Evaluation Enfant Douleur; *FLACC*, Face, Legs, Activity, Cry, Consolability; *HEDEN*, Hétero Evaluation Douleurenfant; *NFCS*, Neonatal Facial Coding System; *NIPS*, Neonatal Infant Pain Scale; *PIPP*, Premature Infant Pain Profile.

Pharmacological treatment of pain

Multimodal analgesia is the benchmark of pharmacological pain relief in terminally ill children with cancer. The analgesic ladder, proposed by the WHO in 1986, remains an essential tool for everyday clinical practice, despite its possible limitations. The original ladder consisted of three steps: the first one addressed mild pain (nonsteroidal antiinflammatory drugs and acetaminophen), while the second and third steps addressed moderate and severe pain by using weak or potent opioids, with or without nonopioid analgesics and adjuvants [47,48].

Opioids

General aspects, classification, dosing, and side effects

Opioids are the mainstay of EoL pain treatment in pediatric cancer patients. Trained professionals must deliver them according to the opioid stewardship [4]. Bakir et al. reported that 90% of pediatric oncological patients treated in the algology clinic eventually used opioids, while dual opioid therapy was prescribed to almost one-third of them [15]. Children with solid tumors (ST) outside the central nervous system (CNS) are more likely to require multiple opioids, higher opioid doses, and opioid rotation [20,49]. The most common weak opioids include codeine and tramadol, while the standard potent ones are morphine, hydrocodone, methadone, fentanyl, oxycodone, buprenorphine, tapentadol, hydromorphone, and oxymorphone [48]. In 2012, the WHO guidelines introduced a new, two-step ladder strategy for managing pediatric cancer pain, bypassing weak opioids, namely codeine. At the time, there were no formal comparative studies between the two approaches, but it seemed that the benefits of attaining pain control by potent opioids (primarily morphine) outweighed benefits of using codeine/tramadol, drugs with unpredictable risks in children [50]. The two-step pain-relieving ladder was later examined in the study of Abdel-hadi and coworkers—low-dose morphine was associated with lower pain scores, fewer drug changes, and shorter duration of pain, although side effects were more common compared to weak opioids [51].

The complexity of pharmacodynamics and pharmacokinetics of opioids, and ethical issues concerning their usage in children has led to the fact that applicability and dosages for many opioid drugs were extrapolated from adult studies [18]. The fundamental principles of opioid therapy include titrating doses to efficacy and opioid rotation when tolerance or dose-limiting toxicities occur [14,52]. Dosing should be weight-based up to 50 kg, above which adult dosing is applied. In case the body mass index is >30, ideal weight is to be considered [53]. In PPC, opioid doses frequently reach 10–100 times

TABLE 47.2 *Douleur Enfant Gustave Roussy*/Gustave Roussy Child Pain Scale (all items scored from 0 to 4).

Behaviors specific to pain	Psychomotor inertia	Anxiety
1. Unnatural postures • Avoidance of certain painful positions or adoption of a particular position; when the child is sitting or lying down with no physical activity	**1. Resignation** • Resignation to everything that happens, loss of protest, or resistance; during an unpleasant situation (e.g., venipuncture)	**1. Tenseness** • The degree of nervous tension in the child's body
2. Protection of painful areas • Continuous avoidance of all contact with painful areas	**2. Withdrawal** • "Withdrawing into the shell"	**2. Hostility** • The child's hostility toward those around
3. Expressing pain • The way in which the child verbalizes the pain, either spontaneously, or when asked	**3. Lack of expression** • The ability of the child to register and express feelings by the tone of voice, eyes, and facial expression; when the child is active (e.g., during games)	**3. Wariness at being moved** • When the child is being moved around he/she is worried, demonstrates how he/she wants to be moved, resists, or holds onto the adult's hand
4. Indicating painful areas • The way the child locates pain, either spontaneously, or when asked	**4. Lack of interest in surroundings** • The child's energy for interaction with the environment	**4. Crying easily** • The extent to which the child cries
5. Pain avoidance when moving • Spontaneous avoidance of all movements or parts of the body; during sequences of movements (e.g., walking)	**5. Slowness and paucity of movement** • Slowness, restriction, and stiffness, the trunk and large joints are particularly motionless; in relation to a normal child's movements	
6. Reactions to examination of the painful area • Resistance, pulling away, or emotional reactions		

Source: Created after Gauvain-Piquard A, Rodary C, Rezvani A, Serbouti S. The development of the DEGR(R): a scale to assess pain in young children with cancer. Eur J Pain 1999;3(2):165–176.

the standard starting doses [14,19,54]. Compared to patients with hematological malignancies, children with ST tend to have a more difficult-to-manage terminal-phase pain [49], resulting dominantly from tumor spread to the spine and major nerves [54]. The aggressive opioid plan signifies high oral doses, defined as >20 mg/kg/day morphine equivalent dose (MED) [20], and massive opioid infusions (>3 mg/kg/h of MED) [54]. Reported infusion rates may exceed 500 mg/kg/h, as described in an infant with an isolated metastasis in the periaqueductal gray matter, part of the brain stem linked to pain sensation [54]. However, such dosing is rare, requires expert oversight, and intensive monitoring. Interestingly, Vallero et al. discovered that children with CNS tumors require less analgesia compared to other tumors. In their research on 39 neuro-oncological patients, slightly above half of them received morphine at EoL, and none suffered from uncontrolled pain [55]. Table 47.3 shows usual starting doses of main opioid analgesics in children, according to the WHO guidelines [50].

Pain medications should be administered to children in the simplest, most effective, and least painful way. Routes of administration are oral, intravenous (IV), subcutaneous (SC), transdermal, rectal, intrathecal, or epidural, with the percentage of IV applications rising in terminally ill children approaching death. The oral route is the most convenient and effective, and the least expensive one. However, patients in terminal phases commonly manifest dysphagia, and clinicians prefer the IV route when pain is uncontrolled and precise titration is rapidly needed [13,18,20]. New routes of delivery (intranasal, buccal, and sublingual) are being investigated in PPC for drugs such as fentanyl [18,56,57]. For chronic pain, slow-release opioids should be taken regularly, at predetermined intervals—"by the clock" [48].

Opioid patient-controlled analgesia (PCA) is becoming increasingly present in the field of palliative oncology. It is proven to be safe and efficacious in children with various conditions, including terminal cancer [49,58]. At the start of PCA therapy, pulse oximetry is mandatory, and advised monitoring comprises respiratory and heart rate, blood pressure,

TABLE 47.3 Starting doses of morphine, fentanyl, and methadone for persisting pain in children (WHO 2012 guidelines).

Morphine	Fentanyl	Methadone
Oral (prolonged-release formulation) • **Child**: **200–800 mcg/kg** every **12 h**	*Continuous IV infusion* • **Neonate** or **infant**: initial IV bolus of **1–2 mcg/kg** (over 3–5 min) followed by **0.5–1 mcg/kg/h** • **Child**: initial IV bolus of **1–2 mcg/kg** (over 3–5 min), followed by **1 mcg/kg/h**	*Oral, SC, or IV* • **Child**: initially **100–200 mcg/kg every 4 h** for the first 2–3 doses, then **100–200 mcg/kg** every **6–12 h**; maximum of 5 mg per starting dose; IV methadone must be administered slowly over 3–5 min
Continuous intravenous (IV) infusion • **Neonate**: initial IV bolus **25–50 mcg/kg** over at least 5 min, followed by **5–10 mcg/kg/h** • **Infant 1–6 months**: initial IV bolus **100 mcg/kg** over at least 5 min, followed by **10–30 mcg/kg/h** • **Infant** or **child 6 months–12 years**: initial IV bolus **100–200 mcg/kg** over at least 5 min, followed by **20–30 mcg/kg/h**	*Transdermal patch* • **Child ≥2 years** (opioid tolerant on at least 45–60 mg of oral morphine equivalent per day; stable pain management): use a ratio of 45 mg of oral morphine being equivalent to 12.5 mcg/h in patch dosage, 90 mg morphine to 25 mcg/h patch.	
Continuous subcutaneous (SC) infusion • **Infant 1–3 months**: **10 mcg/kg/h** • **Infant** or **child 3 months–12 years**: **20 mcg/kg/h**		

and pain, sedation, and nausea scores every 2 hours. The obligatory condition for at-home PCA is an on-call specialist available 24 hours a day, 7 days a week. SC PCA is believed to be as efficacious as IV [13]. The largest study on opioid PCA in pediatric palliative oncology showed a significant increase in opioid requirements in the last 2 weeks of life. It also found that younger patients (<13 years) have a higher opioid consumption on the last day of life, compared to adolescents and young adults (AYA), presumably as a reflection of increased pain perception by caregivers in the time of imminent death [58].

Children display a variety of combinations of different μ-opioid subtypes due to alternative splicing. This contributes to the individual variance of therapeutic response and incomplete cross-tolerance among opioids. Therefore opioid rotation is a widespread practice in instances of intolerable side effects and inadequate pain control [59]. Not surprisingly, children with malignancies, who often undergo prolonged opioid therapy, sometimes require multiple opioid switches to achieve optimal analgesia with acceptable tolerability [20,52]. The favored rotation is morphine to fentanyl [52], and understanding of dose equivalency to morphine is essential [18,60]. A considerable proportion of children on major opioids continue to receive simple analgesia (step 1 WHO), which provides a synergistic effect as mechanisms of action differ [20].

Studies on children are continuously reporting that opioids' adverse effects are neither as common nor as severe as traditionally believed, which will hopefully eliminate opiophobia among physicians and contribute to adequate control of severe EoL pain. Even potent opioids are generally safe when appropriately monitored, with up to 90% of patients reporting no serious side effects [15]. Among the negative effects encountered, the most frequent are nausea, vomiting, pruritus, constipation, urinary retention, sweating, and sedation [13,15,19,61–67]. Laxatives and antiemetics should be given prophylactically. Urinary retention is generally resolved by calming measures, wet towel application, and the sound of running water, but in some instances cholinomimetics or intermittent catheterization is helpful. Pruritus is one of the main reasons for opioid rotation in children, which is also helpful for hallucinations and confusion [13].

Weak opioids

Historically, **codeine** was the number one oral opioid used in children. It is a prodrug (3-methyl-morphine) without intrinsic analgesic activity, metabolized by cytochrome P450 2D6 (CYP2D6) into morphine. In ultrarapid metabolizers, high morphine levels accumulate, potentially leading to apnea [18,68–71]. Fatalities in children, reported almost exclusively after tonsillectomy and/or adenoidectomy [69–71], led to prohibition of codeine in children younger than 12 and a cautious and restricted use in those above this age limit [18,68,72–74].

Tramadol, a synthetic analog of codeine, acts through at least two complementary mechanisms: [1] as a centrally acting μ-opioid and [2] as an inhibitor of serotonin and noradrenalin reuptake [18,68,75]. Its potency is only one-tenth to one-fifth of that of morphine, and it is more nauseating than other weak opioids [76]. The pharmacodynamics of tramadol is also affected by the CYP2D6 activity. Few tramadol-related deaths have been described in children, urging the US Food and Drug Administration (FDA) to further restrict its use in children since 2017 [18,68,74,75]. Nevertheless, studies suggest that tramadol is still an extensively prescribed weak opioid in pediatric cancer practice, particularly where morphine is not available [15,18,61]. Some large pediatric centers questioned the FDA's decision, as it was based on three deaths in half a century following administration of high-concentration drops [14]. New evidence supports this opinion. Ornstein et al. reported not only excellent relief of mucositis-associated pain with tramadol monotherapy but also a safe profile of the drug [77].

Strong opioids

Morphine is the medicine of choice for moderate-to-severe pain in PPC, according to the WHO [4,50]. It is the most prescribed major opioid in children dying of cancer [15,20]. The drug is cheap and widely available in developed countries, but concerns on its lack in some parts of the world, specifically sub-Saharan Africa, are worrisome [78]. The use of morphine as a continuous infusion to treat EoL pain in pediatric oncology was described since the early 1980s [79]. Continuous morphine infusion is a feasible analgesic option, although, it is important to consider dose-limiting side effects (e.g., apnea, hallucinations) and administer prophylactic laxatives and antiemetics [21]. Oral morphine is available in both immediate- and extended-release formulations, the latter to be taken at regular intervals and the first one to treat breakthrough pain on demand. Long-acting morphine is ideally suited for PPC settings; it is proven to be efficacious, safe, and especially convenient for at-home terminal care in children and adolescents [80]. Morphine is metabolized by hepatic enzymes and cleared mostly renally and partially by biliary excretion. Clearance rates mature with age and reach that of adults by 6–12 months of age [81]. In children aged 1 year and older, drug doses calculated per body weight exceed those in adults due to higher elimination and a greater fraction of total body water [62,82]. Dosing in children has various recommendations. The study conducted on children with cancer set the starting dose of 1.5–2.0 mg/kg/day to provide adequate plasma concentrations (>12 ng/mL) [82]. Pharmacokinetic parameters of morphine show, however, a great variability among individuals, so a range of plasma concentrations are found during continuous infusions. Table 47.3 demonstrates initial doses of morphine for opioid-naive patients, according to the 2012 WHO guidelines on persisting pain in children. The dosage must be adjusted to an effective level, with a maximum dosage increase of 50% in 24 hours. Experienced physicians may prescribe a 100% daily increase but with close monitoring of the child. Breakthrough pain is to be treated with boluses of immediate-release oral, IV, or SC morphine. They are given as 5%–10% of the daily amount, as frequently as needed. In cases of moderate or severe renal impairment, doses must be reduced by 25% or 50%, respectively [50]. **Hydromorphone** is a semisynthetic analog of morphine, with a 5–7.5 times higher potency. Compared to morphine, it is a μ-receptor agonist with low affinity to κ- or δ-receptors [13]. Starting dose of hydromorphone in children is 30–80 mcg/kg (maximum 2 mg per dose), given orally every 3–4 hours, or 15 mcg/kg per dose for SC/IV use, given slowly over several minutes, every 3–6 hours [50].

Other strong opioids to consider in PPC include methadone (for neuropathic and nonresponsive nociceptive pain) [13,14,18,21,63–65,83,84], fentanyl (in stable pain situations as a transdermal patch, especially when oral opioids cause severe constipation) [13,18,21,57,66,85,86], buprenorphine (first choice in patients with renal failure) [13,18,21,67,87–89], oxycodone [13,15,21], and tapentadol [90,91].

Methadone is a multimechanistic drug, proven excellent for advanced pain management, but still underutilized in children. The main causes are a wide variety of interindividual responses, differences in half-life (being shorter in young people), and complex equianalgesic conversion in children, making it challenging to titrate and requiring close follow-up [14,64,84]. In 2018, Mott et al. suggested guidelines for methadone administration in children. The daily conversion ratio of morphine to methadone is complex and depends on multiple factors; however, it can generally be approximated as 10:1 for pain control in patients receiving <1000 mg of oral morphine and 20:1 for those receiving >1000 mg. Maximum starting daily doses are 40 or 60 mg for children weighing <15 or >15 kg, respectively [64]. The adverse effect profile in the pediatric population is favorable [63,64]. Prolongation of QTc was described [21,92], but Mott et al. advised omitting ECG monitoring in situations of terminal disease and/or intractable pain [64]. Table 47.3 shows starting methadone doses according to the WHO 2012 guidelines. After initial dosing, such as for morphine, doses may be escalated but maximally by 50% daily. Sometimes, 2–3 days after the effective drug level is achieved, doses must be reduced temporarily by 50% to prevent adverse effects of methadone accumulation [50].

Fentanyl and **buprenorphine** are both available for transdermal use. Fentanyl is a pure μ-opioid agonist, appropriate for children in whom pain control is already obtained (relief for at least 24 hours). Due to its slow absorption and long half-life, it is not suitable as a starting drug [13,66,86]. In those <10 years of age, doses per kilogram are higher than in older patients, an observation similar to that of morphine consumption, explained by increased clearance rate in younger children [85]. Rapid-onset fentanyl, in the form of sublingual tablets, lozenges, or intranasal spray, can be used for breakthrough pain in children as little as 13 kg [57]. Initial dosing of intravenous fentanyl as well as conversion to patch doses are shown in Table 47.3 [50]. Buprenorphine is a powerful lipophilic synthetic opioid. The drug is of great clinical interest owing to different application routes, long duration of action, metabolism largely independent of renal function, and antihyperalgesic effect obtained through κ-receptor antagonism. The latter may contribute to its favorable suppression of neuropathic pain [13]. It has long been available for parenteral and sublingual use but is now mainly administered as a matrix patch formulation [67]. The transdermal route is preferred in PPC, as it is simple to use and feasible in patients having difficulty swallowing [88]. Ruggiero et al. demonstrated that transdermal buprenorphine used in the setting of pediatric cancer pain significantly improved the QoL of patients, including sleeping and activity [67]. Associated side effects are prevailingly nausea, vomiting, and local skin reactions [88]. In most European countries, transdermal buprenorphine is available in patches having drug release rates of 35, 52.5, and 70 μg/h, corresponding to a daily dose of 0.8, 1.2, and 1.6 mg, respectively. The patches have a duration of 3 days and can be conveniently cut into halves or quarters. Formulations with a duration of 7 days are available in the United States and United Kingdom. Experience suggests that transdermal buprenorphine is even stronger than previously believed, leading to a change of conversion ratio (transdermal buprenorphine:oral morphine) from 1:75 to 1:110–115 [89].

Oxycodone, a semisynthetic opioid of high bioavailability, has approximately double the analgesic potency of morphine. It is available in the form of intravenous solution and immediate- or extended-release oral preparations [13]. Studies in children rely on data from its use for postsurgical and fracture-related pain [93,94]. Negative effects are common, mostly drowsiness [93]. Initial recommended doses of oxycodone for pediatric age group are 0.05–0.15 mg/kg every 4–6 hours [50].

Tapentadol is one of the latest approved opioid analgesics and the first one to undergo a formal pediatric development program designed to fulfill requirements of the FDA and the Pediatric Committee of the European Medicines Agency. It acts through two synergistic mechanisms: as a μ-opioid receptor agonist and a norepinephrine reuptake inhibitor. The use of tapentadol in children was tempting from the start, when it displayed effective pain relief for acute and chronic pain in adults, as well as better gastrointestinal tolerability compared to other opioids. Furthermore, it showed predictable pharmacokinetics. In the European Union it was authorized in 2018 as a new treatment option for moderate-to-severe acute pain in children from 2 to 18 years [90,91]. In 2021, favorable pharmacokinetics and safety were confirmed even in children younger than 2 years [95]. While investigating efficacy and safety of prolonged release tapentadol in children, Howard et al. used starting doses of approximately 1.25–1.5 mg/kg given every 12 hours, and the most common side effects were nausea and constipation [90].

Adjuvants

When opioids do not accomplish the desired effect on pain relief or cause serious adverse effects, alternative drugs are needed to alleviate EoL pain. Coanalgesics are prescribed in roughly one-fifth of pediatric cancer patients, again, more commonly in those with ST [15]. The careful selection of an analgesic regimen should be based on all contributory factors, including the underlying etiology of pain. Neuropathic pain, for instance, is particularly challenging to manage, which can require the use of adjuvants such as gabapentinoids, tricyclic antidepressants, serotonin–norepinephrine reuptake inhibitors, low-dose ketamine, alpha-agonists, and lidocaine [18,96]. Corticosteroids and bisphosphonates may be used to decrease bone pain, while muscle spasms are relieved by benzodiazepines [97–99].

Gabapentin and **pregabalin** have revolutionized the management of difficult-to-treat pain and are nowadays used by most pediatric palliative specialists. Gabapentinoids, along with **tricyclic antidepressants**, are considered first-line therapy for intractable neuropathic pain that emerges from chemotherapy-related neuronal injury, ST spread to nerves, phantom limb sensations, and complex combination of factors at the EoL (secondary hyperalgesia, peripheral and central sensitization, wind-up phenomena, etc.). The theory behind their use is the inhibition of excitation within the CNS. Dosing is incremental to avoid side effects such as dizziness and somnolence [100–102]. The starting dose of gabapentin is 10 mg/kg/day and escalates to 30 mg/kg/day over a few days to a week, divided into three daily consumptions. Pregabalin is commenced at a dose of 1–2 mg/kg/day, which can be gradually titrated upwards until effective analgesia is achieved (final dose 150–300 mg/day) [100,102,103].

Ketamine's main advantage lies in reversing opioid-induced hyperalgesia through N-methyl-D-aspartate (NMDA) receptor antagonization. These receptors are implicated in increasing pain stimulus, as their activation leads to a state of hyperexcitability. Ketamine can be given by almost every route, with intravenous being the most accepted. Recently, intranasal and sublingual formulations have become available, providing rapid absorption [104]. Low-dose or subanesthetic ketamine infusions are defined as those up to 1 mg/kg/h. Initial infusion rates range from 0.01 to 0.2 mg/kg/h, with the standard escalation rate of 0.02 mg/kg/h. Side effects are mainly psychotropic (nightmares, hallucinations, drowsiness, and mood disorders) [105]. Simultaneous administration of benzodiazepines (e.g., lorazepam) may eliminate these negative drug-induced episodes [106]. Moreover, ketamine acts as an antidepressant, so its function on mood and global perception of EoL pain in PPC is to be investigated [107,108].

Lidocaine downregulates pain transmission through blockage of voltage-gated ion channels of neurons in the spinal cord and inhibition of NMDA receptors. Pain relief is felt beyond elimination of the drug, contributing to the modifying effect on pain wind-up [109]. In 2020, a relevant prospective, randomized study, conducted on children after a major spinal surgery, proved the positive effect of perioperative lidocaine administration on pain relief and sooner convalescence. After a bolus of 1.5 mg/kg over 30 minutes, continuous infusion was delivered at a rate of 1 mg/kg/h [110]. Knowledge on the analgesic use of lidocaine infusions in children with cancer is limited and relies on case reports or retrospective analyses [109,111–113]. Anghelescu et al. confirmed the opioid-sparing effect of systemic lidocaine in pediatric patients with malignancies, as well as the good tolerability of the drug in the observed group. Only one patient experienced hypotension that required discontinuation of infusion, but contributing factors might have been hypothyroidism and an elevated body mass index (possible overdosing as compared to ideal body mass) [112].

Bisphosphonates were not recommended in the WHO 2012 guidelines as coanalgesics, as there were no systematic reviews or randomized controlled trials on their use in the age group, while serious adverse events such as jaw osteonecrosis were described in adults [50]. However, in a small retrospective study conducted on children with metastatic

TABLE 47.4 Most commonly used adjuvants for treatment of end-of-life pain in pediatric cancer patients.

Drug class	Representatives	Indication for use	Side effects
Gabapentinoids	Gabapentin Pregabalin	Neuropathic pain	Drowsiness, dizziness, mood changes, fatigue, blurred vision, hoarseness, lower back pain, flu-like symptoms
Tricyclic antidepressants	Amitriptyline Nortriptyline	Neuropathic pain	Drowsiness, blurred vision, xerostomia, constipation, diarrhea, nausea/vomiting, unusual taste, urine retention
Serotonin–norepinephrine reuptake inhibitors	Duloxetine	Neuropathic pain	Drowsiness, fatigue, headache, blurred vision, xerostomia, constipation, nausea/vomiting, loss of appetite, increased sweating
Alpha-agonists	Clonidine	Neuropathic pain	Dizziness, fatigue, constipation, xerostomia, headache, insomnia
Anesthetics	Ketamine	Neuropathic pain	Dizziness, hallucinations, nightmares, mood disorders, disorientation, confusion, nausea/vomiting
	Lidocaine	Neuropathic pain	Drowsiness, dizziness, headaches, tingling or numbness around mouth, metallic taste, nausea, palpitations, arrhythmia
Corticosteroids	Dexamethasone	Neuropathic pain, bone pain, inflammation	Mood changes, insomnia, headache, gastrointestinal bleeding, hyperglycemia, palpitations, hypertension, weight gain, edema
Bisphosphonates	Zoledronic acid Pamidronate	Bone pain	Flu-like symptoms, bone pain/myalgia/arthralgia, hypocalcemia, hypophosphatemia, nausea, constipation, diarrhea, hypotension, osteonecrosis of the jaw
Benzodiazepines	Diazepam	Muscle spasms	Dizziness, ataxia, muscle weakness, confusion, xerostomia, constipation

bone disease, zoledronic acid, a third-generation bisphosphonate, showed satisfactory pain relief and a safe drug profile [114]. Another retrospective study confirmed a significant decrease in opioid consumption in 35 children with primary bone tumors or metastases after administration of intravenous bisphosphonates (zoledronate and pamidronate) [115].

Customarily used adjuvants, indications for their use, and most frequently reported side effects are presented in Table 47.4.

Interventional pain management

While numerous interventional approaches are successfully implemented in treating cancer-related pain in adults, they are still underutilized in the pediatric population. Cuviello et al. reported a low application rate (only 6.5%) of these procedures in the largest to date known study concerning regional pain management in EoL pediatric cancer care [116]. The reasons behind this are the traditional fear of procedural complications and the distress that invasive techniques pose to children, particularly in terminal phases. Therefore invasive strategies in children are reserved for limited situations when all other treatment options, previously mentioned, fail. New data brings evidence on benefits of invasive pain-relieving techniques at EoL in children, including efficacy, safety, and the possibility of good outpatient pain control. Children with ST are more often exposed to interventional pain management due to greater tumor burden and the release of neuroimmune mediators causing nociceptive pain [116,117].

Interventional pain management includes regional anesthesia and analgesia and rarely used techniques such as radiofrequency ablation (RFA), cordotomy, and vertebral augmentation [117]. Regional pain control is obtained by drug delivery in proximity to a peripheral nerve (femoral, sciatic, axillary, supraclavicular, infraclavicular, or interscalene) plexus, or through central neuraxial blockades (epidural or intrathecal). The technique is employed either as a single shot or via catheter, as repetitive injections, or as continuous infusion. Catheter placement is preferred as it provides a long-term solution. Regional approaches are also used for neurodestruction with alcohol, referred to as neurolytic interventions. [116–121].

Ropivacaine and bupivacaine are routinely used anesthetics for continuous nerve blocks, sometimes in conjunction with opioids (fentanyl, morphine, and hydromorphone) [116]. The use of intrathecal clonidine is also described. Clonidine exerts its effect through agonism of alpha-2 receptors on the dorsal horn of the spinal cord and activation of descending inhibitory noradrenergic pathways [119]. Single-shot blocks are performed prevailingly on the celiac plexus, and drugs administered include anesthetics (lidocaine, bupivacaine), ethanol, and corticosteroids [116]. Celiac plexus neurolysis is delivered to patients with uncontrollable visceral pain caused by abdominal tumors [122,123].

Adverse outcomes associated with regional techniques comprise postpuncture headache, respiratory depression, tachyphylaxis, tachycardia, bleeding, infections, and interference with urination and defecation [116,121]. According to new findings, undesired effects of regional pain interventions are much less common than previously believed [116,124]. Infections and bleeding can be minimized by tunneled catheter placement in accordance with the patient's bloodwork (platelets >50,000/μL and absolute neutrophil count >1500/μL) [116]. In addition, the widespread use of ultrasound mitigates the risks of bleeding, nerve injury, and catheter dislodgement [117,124,125].

Spinal cordotomy (open anterolateral or percutaneous) involves lesioning of the spinothalamic tract to interrupt pain afferents and exert contralateral analgesia [126]. RFA uses electric current to generate heat and causes thermal necrosis of various structures, including nerves. In pediatric oncology, it is dominantly used in the management of painful osteoid osteoma. Vertebral augmentation through balloon kyphoplasty has long been used as a treatment for compression fracture pain in adults, while the experience in children is scarce [117,127]. Details on these infrequently used procedures for palliative pain relief in children extend the purpose of this chapter.

Integrative medicine and nonpharmacological methods

Pain of any etiology is considered a biopsychosocial phenomenon, and palliative pain, or "total pain," especially, comprises a multidimensional experience that unifies physical, psychological, social, spiritual, and practical dimensions [128,129]. Therefore treating symptoms at EoL in children should not be restricted to pharmacological measures alone. Complementary and alternative medicine (CAM) is defined as all medical practices, products, and even medical systems that are normally not a part of conventional medicine. Integrative medicine (IM) is a coordinated approach that combines evidence-based complementary and conventional strategies [130,131]. The recognition of the value of IM has led many healthcare systems to incorporate these modalities [132,133]. According to Schütze et al., the prevalence of CAM use in children with malignancies at the palliative period is 43%, with the most common one being homeopathy [130].

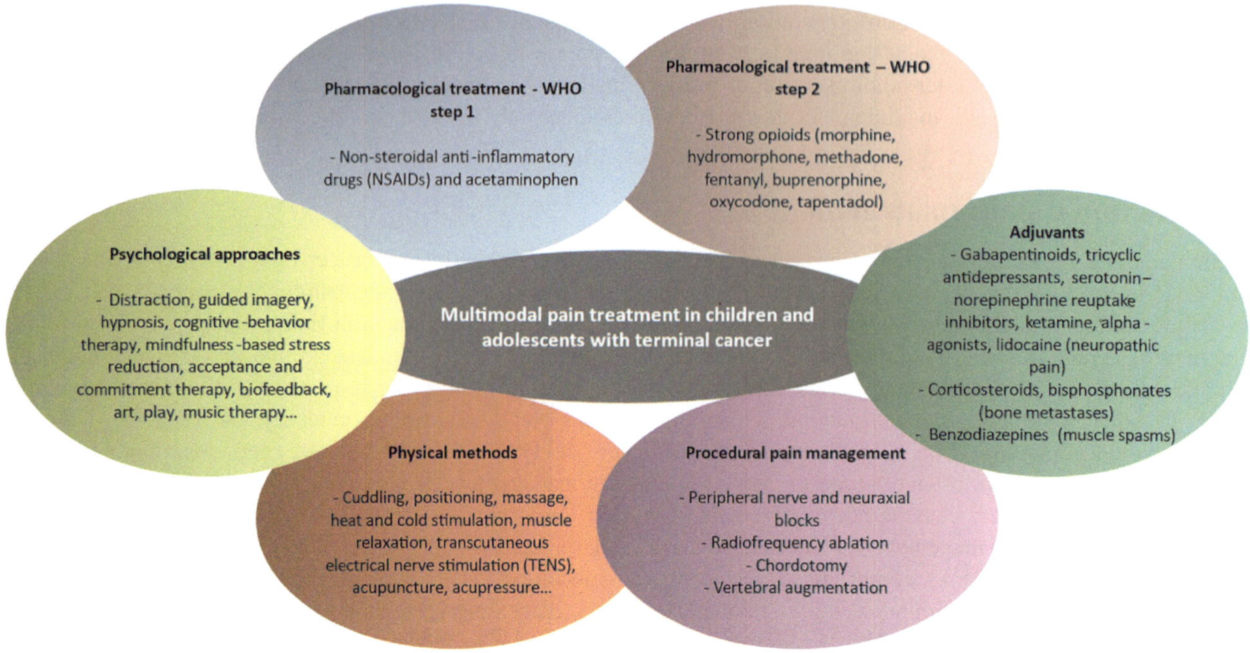

FIGURE 47.1 Multimodal pain treatment.

Nondrug interventions are standardly divided into physical (cuddling, positioning, massage, heat and cold stimulation, muscle relaxation, transcutaneous electrical nerve stimulation, acupuncture, acupressure, etc.) and psychological therapies (distraction, guided imagery, hypnosis, cognitive-behavior therapy, mindfulness-based stress reduction, acceptance and commitment therapy, biofeedback, art, play, music therapy, etc.) [134]. A recent literature review of nonpharmacological treatments of pain in children with cancer disclosed that adjunctive methods alleviated not only pain but also other undesired symptoms such as nausea, vomiting, and fatigue, consequently leading to improved levels of energy, sleep, and mood [135]. Out of 11 studies, only two reported adverse events, none of which were severe–mild bruising and local bleeding after acupuncture [136], and aggressiveness in some children due to the fear of an upcoming procedure for therapeutic massage. The latter could be mitigated with the presence of parents/caregivers during complementary methods, particularly for younger children [137]. Patients' biological and psychological development influences their engagement with integrative approaches and potential outcomes. Younger children are more likely to choose creative therapies such as dance and music, while AYA patients prefer mind–body therapies, massage, and acupuncture [138].

New nonpharmacological methods are being implemented in clinical practice. One of them is virtual reality as a form of distraction, a strategy in which pain is suppressed by consciously shifting attention to a more pleasant activity or thought. Traditionally, guided imagery was a typical distraction tool that involved envisioning peaceful settings to calm the sympathetic nervous system and reduce pain and anxiety. Virtual reality is a 3D, interactive technology that creates an artificial environment and engages multiple senses. It has been used in pediatrics for the last two decades, mostly for procedural pain [139].

The scheme of the multimodal approach to EoL pain treatment in children and adolescents with cancer is shown in Fig. 47.1.

Further implications

The expansion of education and clinical training programs on pain management in PPC is needed, most importantly to root out deeply established fear of opioids. Robust clinical validation for pain assessment tools in PPC is essential to timely recognize and objectify pain in children with terminal cancer. Advancing knowledge on opioid pharmacogenetics is expected to establish a personalized pain treatment plan, with the aim to avoid therapy failure and minimize side effects. Addition of appropriate adjuvants and interventional approaches merits consideration in situations where opioid escalation does not produce adequate analgesia. Nondrug modalities should be implemented more often in terminally ill children with cancer, as studies regarding their use in other types of pain in childhood show great efficacy.

Conclusion

- Pain is still one of the most common symptoms felt by a large proportion of children and adolescents with terminal cancer, regardless of advances in PPC, including available guidelines on chronic pain management. Inadequate analgesia severely diminishes QoL of patients and their families.
- Pain physiology and perception differ tremendously in children compared to adults, and assessment is challenging. Despite numerous available pediatric pain scales, no instrument has been validated for the setting of oncological PPC exclusively.
- Opioids are the mainstay of the treatment, even though opiophobia and inadequate knowledge on dose titration and rotation frequently lead to poorly controlled pain. Morphine is the number one drug recommended by the WHO. Drugs should be given by the most feasible route, at predetermined intervals, with boluses for breakthrough pain.
- Neuropathic and bone-related pain are especially challenging to treat and often require adjuvants, most frequently gabapentinoids and tricyclic antidepressants for neuropathic pain and corticosteroids and bisphosphonates for bone metastases. Old drugs, such as ketamine and lidocaine, are establishing their way as pharmacological options for complex, refractory pain.
- Invasive procedures are well tolerated but unfortunately underutilized in children. Interventions such as peripheral or neuraxial blocks may significantly alleviate pain in circumstances of inadequate pain relief despite opioid escalation.
- Nonpharmacological measures, comprising physical and psychological approaches, should be included in the multidimensional treatment plan, tailored to suit each child.
- No child should die in pain!

References

[1] Botta L, Gatta G, Capocaccia R, et al. Long-term survival and cure fraction estimates for childhood cancer in Europe (EUROCARE-6): results from a population-based study. Lancet Oncol 2022;23(12):1525–36. Available from: https://doi.org/10.1016/S1470-2045(22)00637-4.

[2] Connor SR, Downing J, Marston J. Estimating the global need for palliative care for children: a cross-sectional analysis. J Pain Symptom Manage 2017;53(2):171–7. Available from: https://doi.org/10.1016/j.jpainsymman.2016.08.020.

[3] Benini F, Avagnina I, Giacomelli L, et al. Pediatric palliative care in oncology: basic principles. Cancers (Basel) 2022;14(8):1972. Available from: https://doi.org/10.3390/cancers14081972.

[4] World Health Organization. Integrating palliative care and symptom relief into paediatrics: a WHO guide for health-care planners, implementers and managers. Geneva: World Health Organization; 2018. Available from: https://iris.who.int/handle/10665/274561.

[5] Benini F, Papadatou D, Bernadá M, et al. International standards for pediatric palliative care: from IMPaCCT to GO-PPaCS. J Pain Symptom Manage 2022;63(5):e529–43. Available from: https://doi.org/10.1016/j.jpainsymman.2021.12.031.

[6] McNeil MJ, Godfrey A, Loggetto P, et al. Physician perceptions of and barriers to pediatric palliative care for children with cancer in Brazil. JCO Glob Oncol 2023;9:e2300057. Available from: https://doi.org/10.1200/GO.23.00057.

[7] Jalmsell L, Kreicbergs U, Onelöv E, et al. Symptoms affecting children with malignancies during the last month of life: a nationwide follow-up. Pediatrics 2006;117(4):1314–20. Available from: https://doi.org/10.1542/peds.2005-1479.

[8] Hedén L, Pöder U, von Essen L, et al. Parents' perceptions of their child's symptom burden during and after cancer treatment. J Pain Symptom Manage 2013;46(3):366–75. Available from: https://doi.org/10.1016/j.jpainsymman.2012.09.012.

[9] Wolfe J, Grier HE, Klar N, et al. Symptoms and suffering at the end of life in children with cancer. N Engl J Med 2000;342(5):326–33. Available from: https://doi.org/10.1056/NEJM200002033420506.

[10] Hongo T, Watanabe C, Okada S, et al. Analysis of the circumstances at the end of life in children with cancer: symptoms, suffering and acceptance. Pediatr Int 2003;45(1):60–4. Available from: https://doi.org/10.1046/j.1442-200x.2003.01668.x.

[11] Madden K, Magno Charone M, Mills S, et al. Systematic symptom reporting by pediatric palliative care patients with cancer: a preliminary report. J Palliat Med 2019;22(8):894–901. Available from: https://doi.org/10.1089/jpm.2018.0545.

[12] Cherny NI, Coyle N, Foley KM. Suffering in the advanced cancer patient: a definition and taxonomy. J Palliat Care 1994;10(2):57–70.

[13] Zernikow B, Michel E, Craig F. Pediatric palliative care: use of opioids for the management of pain. Paediatr Drugs 2009;11(2):129–51. Available from: https://doi.org/10.2165/00148581-200911020-00004.

[14] Friedrichsdorf SJ. From tramadol to methadone: opioids in the treatment of pain and dyspnea in pediatric palliative care. Clin J Pain 2019;35(6):501–8. Available from: https://doi.org/10.1097/AJP.0000000000000704.

[15] Bakır M, Rumeli Ş, Pire A. Multimodal analgesia in pediatric cancer pain management: a retrospective single-center study. Cureus 2023;15(9):e45223. Available from: https://doi.org/10.7759/cureus.45223.

[16] World Health Organization. Guidelines on the management of chronic pain in children. Geneva: World Health Organization; 2020. Available from: https://iris.who.int/handle/10665/337999.

[17] Pancekauskaitė G, Jankauskaitė L. Paediatric pain medicine: pain differences, recognition and coping acute procedural pain in paediatric emergency room. Medicina (Kaunas) 2018;54(6):94. Available from: https://doi.org/10.3390/medicina54060094.

[18] Downing J, Jassal SS, Mathews L. Pediatric pain management in palliative care. Pain Manag 2015;5(1):23–35. Available from: https://doi.org/10.2217/pmt.14.45.

[19] Siden H, Nalewajek. High dose opioids in pediatric palliative care. J Pain Symptom Manage 2003;25(5):397–9. Available from: https://doi.org/10.1016/s0885-3924(03)00071-x.

[20] Hewitt M, Goldman A, Collins GS. Opioid use in palliative care of children and young people with cancer. J Pediatr 2008;152(1):39–44. Available from: https://doi.org/10.1016/j.jpeds.2007.07.005.

[21] Hall EA, Hagemann TM, Shelton CM. A narrative review of pain in pediatric oncology: the opioid option. Paediatr Drugs 2024;26:565–96. Available from: https://doi.org/10.1007/s40272-024-00640-y.

[22] Beltramini A, Milojevic K, Pateron D. Pain assessment in newborns, infants, and children. Pediatr Ann 2017;46(10):e387–95. Available from: https://doi.org/10.3928/19382359-20170921-03.

[23] Gauvain-Piquard A, Rodary C, Rezvani A, et al. The development of the DEGR(R): a scale to assess pain in young children with cancer. Eur J Pain 1999;3(2):165–76. Available from: https://doi.org/10.1053/eujp.1999.0118.

[24] O'Rourke D. The measurement of pain in infants, children, and adolescents: from policy to practice. Phys Ther 2004;84(6):560–70. Available from: https://doi.org/10.1093/ptj/84.6.560.

[25] Duffy EA, Dias N, Hendricks-Ferguson V, et al. Perspectives on cancer pain assessment and management in children. Semin Oncol Nurs 2019;35(3):261–73. Available from: https://doi.org/10.1016/j.soncn.2019.04.007.

[26] Andersen RD, Langius-Eklöf A, Nakstad B, et al. The measurement properties of pediatric observational pain scales: a systematic review of reviews. Int J Nurs Stud 2017;73:93–101. Available from: https://doi.org/10.1016/j.ijnurstu.2017.05.010.

[27] Chan AY, Ge M, Harrop E, et al. Pain assessment tools in paediatric palliative care: a systematic review of psychometric properties and recommendations for clinical practice. Palliat Med 2022;36(1):30–43. Available from: https://doi.org/10.1177/02692163211049309.

[28] Ballantyne M, Stevens B, McAllister M, et al. Validation of the premature infant pain profile in the clinical setting. Clin J Pain 1999;15(4):297–303. Available from: https://doi.org/10.1097/00002508-199912000-00006.

[29] Debillon T, Zupan V, Ravault N, et al. Development and initial validation of the EDIN scale, a new tool for assessing prolonged pain in preterm infants. Arch Dis Child Fetal Neonatal Ed 2001;85(1):F36–41. Available from: https://doi.org/10.1136/fn.85.1.f36.

[30] Carbajal R, Paupe A, Hoenn E, et al. DAN: une échelle comportementale d'évaluation de la douleur aiguë du nouveau-né [APN: evaluation behavioral scale of acute pain in newborn infants]. Arch Pediatr 1997;4(7):623–8. Available from: https://doi.org/10.1016/s0929-693x(97)83360-x.

[31] Hudson-Barr D, Capper-Michel B, Lambert S, et al. Validation of the pain assessment in neonates (PAIN) scale with the neonatal infant pain scale (NIPS). Neonatal Netw 2002;21(6):15–21. Available from: https://doi.org/10.1891/0730-0832.21.6.15.

[32] Grunau RE, Oberlander T, Holsti L, et al. Bedside application of the neonatal facial coding system in pain assessment of premature neonates. Pain 1998;76(3):277–86. Available from: https://doi.org/10.1016/S0304-3959(98)00046-3.

[33] Fournier-Charrière E, Tourniaire B, Carbajal R, et al. EVENDOL, a new behavioral pain scale for children ages 0 to 7 years in the emergency department: design and validation. Pain 2012;153(8):1573–82. Available from: https://doi.org/10.1016/j.pain.2012.02.024.

[34] Malviya S, Voepel-Lewis T, Burke C, et al. The revised FLACC observational pain tool: improved reliability and validity for pain assessment in children with cognitive impairment. Paediatr Anaesth 2006;16(3):258–65. Available from: https://doi.org/10.1111/j.1460-9592.2005.01773.x.

[35] Gauvain-Piquard A, Rodary C, Rezvani A, et al. Pain in children aged 2-6 years: a new observational rating scale elaborated in a pediatric oncology unit–preliminary report. Pain 1987;31(2):177–88. Available from: https://doi.org/10.1016/0304-3959(87)90034-0.

[36] Marec-Berard P, Gomez F, Combet S, et al. HEDEN pain scale: a shortened behavioral scale for assessment of prolonged cancer or postsurgical pain in children aged 2 to 6 years. Pediatr Hematol Oncol 2015;32(5):291–303. Available from: https://doi.org/10.3109/08880018.2015.1005324.

[37] Breau LM, McGrath PJ, Camfield CS, et al. Psychometric properties of the non-communicating children's pain checklist-revised. Pain 2002;99(1-2):349–57. Available from: https://doi.org/10.1016/s0304-3959(02)00179-3.

[38] Hunt A, Goldman A, Seers K, et al. Clinical validation of the paediatric pain profile. Dev Med Child Neurol 2004;46(1):9–18. Available from: https://doi.org/10.1017/s0012162204000039.

[39] van Dijk M, Peters JM, van Deventer P, et al. The COMFORT behavior scale: a tool for assessing pain and sedation in infants. Am J Nurs 2005;105(1):33–6. Available from: https://doi.org/10.1097/00000446-200501000-00019.

[40] Bieri D, Reeve RA, Champion DG, et al. The faces pain scale for the self-assessment of the severity of pain experienced by children: development, initial validation, and preliminary investigation for ratio scale properties. Pain 1990;41(2):139–50. Available from: https://doi.org/10.1016/0304-3959(90)90018-9.

[41] Hicks CL, von Baeyer CL, Spafford PA, et al. The faces pain scale-revised: toward a common metric in pediatric pain measurement. Pain 2001;93(2):173–83. Available from: https://doi.org/10.1016/S0304-3959(01)00314-1.

[42] Hester NO, Foster R, Kristensen K. Measurement of pain in children: Generalizability and validity of the pain ladder and the poker chip tool, Advances in Pain Research and Therapy. In: Fields HL, Dubner R, Cervero F, editors. New York: Raven Press; 1990. p. 79–84.

[43] Collins JJ, Devine TD, Dick GS, et al. The measurement of symptoms in young children with cancer: the validation of the memorial symptom assessment scale in children aged 7-12. J Pain Symptom Manage 2002;23(1):10–6. Available from: https://doi.org/10.1016/s0885-3924(01)00375-x.

[44] Mahon P, Holsti L, Siden H, et al. Using colors to assess pain in toddlers: validation of "the rainbow pain scale"-a proof-of-principle study. J Pediatr Oncol Nurs 2015;32(1):40–6. Available from: https://doi.org/10.1177/1043454214555197 [published correction appears in J Pediatr Oncol Nurs 2016;33(6):460. Available from: https://doi.org/10.1177/1043454216671089].

[45] Dupuis LL, Johnston DL, Baggott C, et al. Validation of the symptom screening in pediatrics tool in children receiving cancer treatments. J Natl Cancer Inst 2018;110(6):661–8. Available from: https://doi.org/10.1093/jnci/djx250.

[46] Hockenberry MJ, Wilson D. Wong's essentials of pediatric nursing. St. Louis: Mosby; 2009.

[47] Ventafridda V, Saita L, Ripamonti C, et al. WHO guidelines for the use of analgesics in cancer pain. Int J Tissue React 1985;7(1):93–6.

[48] Anekar AA, Hendrix JM, Cascella M. WHO analgesic ladder. StatPearls. Treasure Island (FL): StatPearls Publishing; 2024. Available from: https://www.ncbi.nlm.nih.gov/books/NBK554435/.

[49] Hoell JI, Warfsmann J, Balzer S, et al. End-of-life care in children with hematologic malignancies. Oncotarget 2017;8(52):89939–48. Available from: https://doi.org/10.18632/oncotarget.21188.

[50] World Health Organization. Persisting pain in children package: WHO guidelines on the pharmacological treatment of persisting pain in children with medical illnesses. World Health Organization; 2012. Available from: https://iris.who.int/handle/10665/44540.

[51] Abdel-Hadi S, Ghazaly MM, Mohamed MA, et al. Impact of pain management using the WHO analgesic ladder in children with cancer in South Egypt Cancer Institute, Assiut University. SECI Oncol 2014;2014(1):2–8. Available from: https://doi.org/10.18056/seci2014.1.

[52] Drake R, Longworth J, Collins JJ. Opioid rotation in children with cancer. J Palliat Med 2004;7(3):419–22. Available from: https://doi.org/10.1089/1096621041349590.

[53] Norris S, Minkowitz S, Scharbach K. Pediatric palliative care. Prim Care 2019;46(3):461–73. Available from: https://doi.org/10.1016/j.pop.2019.05.010.

[54] Collins JJ, Grier HE, Kinney HC, et al. Control of severe pain in children with terminal malignancy. J Pediatr 1995;126(4):653–7. Available from: https://doi.org/10.1016/s0022-3476(95)70370-5.

[55] Vallero SG, Lijoi S, Bertin D, et al. End-of-life care in pediatric neuro-oncology. Pediatr Blood Cancer 2014;61(11):2004–11. Available from: https://doi.org/10.1002/pbc.25160.

[56] Banala SR, Khattab OK, Page VD, et al. Intranasal fentanyl spray versus intravenous opioids for the treatment of severe pain in patients with cancer in the emergency department setting: a randomized controlled trial. PLoS One 2020;15(7):e0235461. Available from: https://doi.org/10.1371/journal.pone.0235461.

[57] Coombes L, Burke K, Anderson AK. The use of rapid onset fentanyl in children and young people for breakthrough cancer pain. Scand J Pain 2017;17:256–9. Available from: https://doi.org/10.1016/j.sjpain.2017.07.010.

[58] Anghelescu DL, Snaman JM, Trujillo L, et al. Patient-controlled analgesia at the end of life at a pediatric oncology institution. Pediatr Blood Cancer 2015;62(7):1237–44. Available from: https://doi.org/10.1002/pbc.25493.

[59] Brennan MJ. The clinical implications of cytochrome p450 interactions with opioids and strategies for pain management. J Pain Symptom Manage 2012;44(6 Suppl.):S15–22. Available from: https://doi.org/10.1016/j.jpainsymman.2012.08.012.

[60] Mazur A, Radziewicz Winnicki I, Szczepański T. Pain management in children. Ann Agric Environ Med 2013;1:28–34.

[61] Zernikow B, Smale H, Michel E, et al. Paediatric cancer pain management using the WHO analgesic ladder–results of a prospective analysis from 2265 treatment days during a quality improvement study. Eur J Pain 2006;10(7):587–95. Available from: https://doi.org/10.1016/j.ejpain.2005.09.002.

[62] Zernikow B, Lindena G. Long-acting morphine for pain control in paediatric oncology. Med Pediatr Oncol 2001;36(4):451–8. Available from: https://doi.org/10.1002/mpo.1109.

[63] Davies D, DeVlaming D, Haines C. Methadone analgesia for children with advanced cancer. Pediatr Blood Cancer 2008;51(3):393–7. Available from: https://doi.org/10.1002/pbc.21584.

[64] Mott C, Sarpal A, Moss K, et al. Methadone for analgesia in children with life-limiting illness: experience from a tertiary children's health service. Children (Basel) 2018;5(7):86. Available from: https://doi.org/10.3390/children5070086.

[65] Hall EA, Sauer HE, Habashy C, et al. Methadone for cancer pain in pediatric end-of-life care. Am J Hosp Palliat Care 2021;38(8):914–9. Available from: https://doi.org/10.1177/1049909120963641.

[66] Hiyama E, Yamaguchi S, Okawa K, et al. An open-label study of the pharmacokinetics and tolerability of once-a-day fentanyl citrate patch in Japanese pediatric and adolescent patients with cancer pain. Clin Drug Investig 2021;41(12):1087–98. Available from: https://doi.org/10.1007/s40261-021-01097-4.

[67] Ruggiero A, Coccia P, Arena R, et al. Efficacy and safety of transdermal buprenorphine in the management of children with cancer-related pain. Pediatr Blood Cancer 2013;60(3):433–7. Available from: https://doi.org/10.1002/pbc.24332.

[68] Rieder MJ, Jong G. The use of oral opioids to control children's pain in the post-codeine era. Paediatr Child Health 2021;26(2):120–7. Available from: https://doi.org/10.1093/pch/pxaa133.

[69] Ciszkowski C, Madadi P, Phillips MS, et al. Codeine, ultrarapid-metabolism genotype, and postoperative death. N Engl J Med 2009;361(8):827–8. Available from: https://doi.org/10.1056/NEJMc0904266.

[70] Kuehn BM. FDA: No codeine after tonsillectomy for children. JAMA 2013;309(11):1100. Available from: https://doi.org/10.1001/jama.2013.2403.

[71] Kelly LE, Rieder M, van den Anker J, et al. More codeine fatalities after tonsillectomy in North American children. Pediatrics 2012;129(5):e1343–7. Available from: https://doi.org/10.1542/peds.2011-2538.

[72] Ono S, Ono Y, Koide D, et al. Relationship between severe respiratory depression and codeine-containing antitussives in children: a nested case-control study. J Epidemiol 2020;30(3):116–20. Available from: https://doi.org/10.2188/jea.JE20180224.

[73] European Medicines Agency (EMA). Codeine-containing medicines – referral, Available from: <https://www.ema.europa.eu/en/medicines/human/referrals/codeine-containing-medicines>; 2013 [accessed 01.07.24].

[74] US Food and Drug Administration. FDA drug safety communication: FDA restricts use of prescription codeine pain and cough medicines and tramadol pain medicines in children; recommends against use in breastfeeding women, Available from: <https://www.fda.gov/downloads/Drugs/DrugSafety/UCM553814>; 2017 [accessed 01.07.24].

[75] Rodieux F, Vutskits L, Posfay-Barbe KM, et al. When the safe alternative is not that safe: tramadol prescribing in children. Front Pharmacol 2018;9:148. Available from: https://doi.org/10.3389/fphar.2018.00148.

[76] Hain RD, Miser A, Devins M, et al. Strong opioids in pediatric palliative medicine. Paediatr Drugs 2005;7(1):1–9. Available from: https://doi.org/10.2165/00148581-200507010-00001.

[77] Yaffe Ornstein M, Stocki D, Levin D, et al. Tramadol treatment for chemotherapy-induced mucositis pain in children. J Pediatr Hematol Oncol 2022;44(2):e487–92. Available from: https://doi.org/10.1097/MPH.0000000000002003.

[78] Harding R, Simms V, Penfold S, et al. Availability of essential drugs for managing HIV-related pain and symptoms within 120 PEPFAR-funded health facilities in East Africa: a cross-sectional survey with onsite verification. Palliat Med 2014;28(4):293–301. Available from: https://doi.org/10.1177/0269216313498637.

[79] Miser AW, Miser JS, Clark BS. Continuous intravenous infusion of morphine sulfate for control of severe pain in children with terminal malignancy. J Pediatr 1980;96(5):930–2. Available from: https://doi.org/10.1016/s0022-3476(80)80585-3.

[80] Goldman A, Bowman A. The role of oral controlled-release morphine for pain relief in children with cancer. Palliat Med 1990;4(4):279–85. Available from: https://doi.org/10.1177/026921639000400405.

[81] van Lingen RA, Simons SH, Anderson BJ, et al. The effects of analgesia in the vulnerable infant during the perinatal period. Clin Perinatol 2002;29(3):511–34. Available from: https://doi.org/10.1016/s0095-5108(02)00018-0.

[82] Hunt A, Joel S, Dick G. Population pharmacokinetics of oral morphine and its glucuronides in children receiving morphine as immediate-release liquid or sustained-release tablets for cancer pain. J Pediatr 1999;135(1):47–55. Available from: https://doi.org/10.1016/s0022-3476(99)70326-4.

[83] Anghelescu DL, Faughnan LG, Hankins GM, et al. Methadone use in children and young adults at a cancer center: a retrospective study. J Opioid Manag 2011;7(5):353–61. Available from: https://doi.org/10.5055/jom.2011.0076.

[84] Fife A, Postier A, Flood A, et al. Methadone conversion in infants and children: retrospective cohort study of 199 pediatric inpatients. J Opioid Manag 2016;12(2):123–30. Available from: https://doi.org/10.5055/jom.2016.0324.

[85] Hunt A, Goldman A, Devine T, et al. FEN-GBR-14 Study Group. Transdermal fentanyl for pain relief in a paediatric palliative care population. Palliat Med 2001;15(5):405–12. Available from: https://doi.org/10.1191/026921601680419456.

[86] Finkel JC, Finley A, Greco C, et al. Transdermal fentanyl in the management of children with chronic severe pain: results from an international study. Cancer 2005;104(12):2847–57. Available from: https://doi.org/10.1002/cncr.21497.

[87] Attinà G, Ruggiero A, Maurizi P, et al. Transdermal buprenorphine in children with cancer-related pain. Pediatr Blood Cancer 2009;52(1):125–7. Available from: https://doi.org/10.1002/pbc.21736.

[88] Haupt TS, Smyth M, Gregoire MC. A scoping review of transdermal buprenorphine use for non-surgical pain in the pediatric population. Cureus 2019;11(10):e5954. Available from: https://doi.org/10.7759/cureus.5954.

[89] Attinà G, Romano A, Triarico S. Transdermal buprenorphine for pain management in children. Drugs Context 2021;10:2021-6-1. Available from: https://doi.org/10.7573/dic.2021-6-1.

[90] Howard RF, Radic T, Sohns M, et al. Tapentadol prolonged release for long-term treatment of pain in children. J Pain Res 2020;13:3157–70. Available from: https://doi.org/10.2147/JPR.S272751.

[91] Eerdekens M, Radic T, Sohns M, et al. Outcomes of the pediatric development plan of tapentadol. J Pain Res 2021;14:249–61. Available from: https://doi.org/10.2147/JPR.S290487.

[92] Madden K, Park M, Liu D, et al. The frequency of QTc prolongation among pediatric and young adult patients receiving methadone for cancer pain. Pediatr Blood Cancer 2017;64(11):e26614. Available from: https://doi.org/10.1002/pbc.26614.

[93] Ali S, Manaloor R, Johnson DW, et al. An observational cohort study comparing ibuprofen and oxycodone in children with fractures. PLoS One 2021;16(9):e0257021. Available from: https://doi.org/10.1371/journal.pone.0257021.

[94] Czarnecki ML, Jandrisevits MD, Theiler SC, et al. Controlled-release oxycodone for the management of pediatric postoperative pain. J Pain Symptom Manage 2004;27(4):379–86. Available from: https://doi.org/10.1016/j.jpainsymman.2003.08.007.

[95] Eissa A, Tarau E, Beuter C, et al. Tapentadol for the treatment of moderate-to-severe acute pain in children under the age of two years. J Pain Res 2021;14:229–48. Available from: https://doi.org/10.2147/JPR.S269530.

[96] Splinter W. Novel approaches for treating pain in children. Curr Oncol Rep 2019;21(2):11. Available from: https://doi.org/10.1007/s11912-019-0766-6.

[97] Bunn M. Top 10 drugs in paediatric palliative care. Malawi Med J 2008;20(4):133–4.

[98] Moody K, Baig M, Carullo V. Alleviating terminal pediatric cancer pain. Children (Basel) 2021;8(3):239. Available from: https://doi.org/10.3390/children8030239.

[99] Eghbali-Fatourechi G. Bisphosphonate therapy in pediatric patients. J Diabetes Metab Disord 2014;13(1):109. Available from: https://doi.org/10.1186/s40200-014-0109-y.

[100] Anghelescu DL, Tesney JM. Neuropathic pain in pediatric oncology: a clinical decision algorithm. Paediatr Drugs 2019;21(2):59–70. Available from: https://doi.org/10.1007/s40272-018-00324-4.

[101] Drake R, Prael G, Phyo Y, et al. Gabapentin for pain in pediatric palliative care. J Pain Symptom Manage 2024;67(3):212–22.e1 https://doi.org/10.1016/j.jpainsymman.2023.11.011.

[102] Butkovic D, Toljan S, Mihovilovic-Novak B. Experience with gabapentin for neuropathic pain in adolescents: report of five cases. Paediatr Anaesth 2006;16(3):325–9. Available from: https://doi.org/10.1111/j.1460-9592.2005.01687.x.

[103] Vondracek P, Oslejskova H, Kepak T, et al. Efficacy of pregabalin in neuropathic pain in paediatric oncological patients. Eur J Paediatr Neurol 2009;13(4):332–6. Available from: https://doi.org/10.1016/j.ejpn.2008.06.011.

[104] Bali A, Dang AK, Gonzalez DA. Clinical uses of ketamine in children: a narrative review. Cureus 2022;14(7):e27065. Available from: https://doi.org/10.7759/cureus.27065.

[105] Courade M, Bertrand A, Guerrini-Rousseau L, et al. Low-dose ketamine adjuvant treatment for refractory pain in children, adolescents and young adults with cancer: a pilot study. BMJ Support Palliat Care 2022;12(e5):e656–63. Available from: https://doi.org/10.1136/bmjspcare-2018-001739.

[106] Finkel JC, Pestieau SR, Quezado ZM. Ketamine as an adjuvant for treatment of cancer pain in children and adolescents. J Pain 2007;8(6):515–21. Available from: https://doi.org/10.1016/j.jpain.2007.02.429.

[107] Mandal S, Sinha VK, Goyal N. Efficacy of ketamine therapy in the treatment of depression. Indian J Psychiatry 2019;61(5):480–5. Available from: https://doi.org/10.4103/psychiatry.IndianJPsychiatry_484_18.

[108] Benini F, Congedi S, Giacomelli L. Refractory symptoms in paediatric palliative care: can ketamine help? Drugs Context 2021;10:2021-2-5. Available from: https://doi.org/10.7573/dic.2021-2-5.

[109] Gibbons K, DeMonbrun A, Beckman EJ, et al. Continuous lidocaine infusions to manage opioid-refractory pain in a series of cancer patients in a pediatric hospital. Pediatr Blood Cancer 2016;63(7):1168–74. Available from: https://doi.org/10.1002/pbc.25870.

[110] Batko I, Kościelniak-Merak B, Tomasik PJ. Lidocaine as an element of multimodal analgesic therapy in major spine surgical procedures in children: a prospective, randomized, double-blind study. Pharmacol Rep 2020;72(3):744–55. Available from: https://doi.org/10.1007/s43440-020-00100-7.

[111] Rav E, Sheth R, Ahmad AH. Systemic lidocaine infusions for pediatric patients with cancer-related pain. Children (Basel) 2022;9(12):1934. Available from: https://doi.org/10.3390/children9121934.

[112] Anghelescu DL, Morgan KJ, Frett MJ, et al. Lidocaine infusions and reduced opioid consumption – retrospective experience in pediatric hematology and oncology patients with refractory pain. Pediatr Blood Cancer 2021;68(11):e29215. Available from: https://doi.org/10.1002/pbc.29215.

[113] Massey GV, Pedigo S, Dunn NL, et al. Continuous lidocaine infusion for the relief of refractory malignant pain in a terminally ill pediatric cancer patient. J Pediatr Hematol Oncol 2002;24(7):566–8. Available from: https://doi.org/10.1097/00043426-200210000-00015.

[114] August KJ, Dalton A, Katzenstein HM, et al. The use of zoledronic acid in pediatric cancer patients. Pediatr Blood Cancer 2011;56(4):610–4. Available from: https://doi.org/10.1002/pbc.22681.

[115] Anghelescu DL, Pankayatselvan V, Nguyen R, et al. Bisphosphonate use in pediatric oncology for pain management. Am J Hosp Palliat Care 2019;36(2):138–42. Available from: https://doi.org/10.1177/1049909118793114.

[116] Cuviello A, Cianchini de la Sota A, Baker J, et al. Regional blocks for pain control at the end of life in pediatric oncology. Front Pain Res (Lausanne) 2023;4:1127800. Available from: https://doi.org/10.3389/fpain.2023.1127800.

[117] Le-Short C, Katragadda K, Nagda N, et al. Interventional pain management for the pediatric cancer patient: a literature review. Children (Basel) 2022;9(3):389. Available from: https://doi.org/10.3390/children9030389.

[118] Argun G, Çayırlı G, Toğral G, et al. Are peripheral nerve blocks effective in pain control of pediatric orthopedic tumor surgery? Eklem Hastalik Cerrahisi 2019;30(1):46–52. Available from: https://doi.org/10.5606/ehc.2019.62395.

[119] Whyte E, Lauder G. Intrathecal infusion of bupivacaine and clonidine provides effective analgesia in a terminally ill child. Paediatr Anaesth 2012;22(2):173–5. Available from: https://doi.org/10.1111/j.1460-9592.2011.03672.x.

[120] Mele C. Intrathecal pain pump for pediatric perverse oncological pain. J Pediatr Nurs 2021;56:99–100. Available from: https://doi.org/10.1016/j.pedn.2020.10.012.

[121] Rork JF, Berde CB, Goldstein RD. Regional anesthesia approaches to pain management in pediatric palliative care: a review of current knowledge. J Pain Symptom Manage 2013;46(6):859–73. Available from: https://doi.org/10.1016/j.jpainsymman.2013.01.004.

[122] Baek SW, Erdek MA. Time-dependent change in pain threshold following neurolytic celiac plexus block. Pain Manag 2019;9(6):543–50. Available from: https://doi.org/10.2217/pmt-2019-0021.

[123] Anghelescu DL, Guo A, Morgan KJ, et al. Pain outcomes after celiac plexus block in children and young adults with cancer. J Adolesc Young Adult Oncol 2018;7(6):666–72. Available from: https://doi.org/10.1089/jayao.2018.0035.

[124] Baizhanova A, Zhailauova A, Sazonov V. Regional anesthesia for pain control in children with solid tumors – a review of case reports. Front Pediatr 2024;11:1275531. Available from: https://doi.org/10.3389/fped.2023.1275531.

[125] Merella F, Canchi-Murali N, Mossetti V. Erratum to 'General principles of regional anaesthesia in children' [BJA Education 19 (2019) 342-348]. BJA Educ 2020;20(1):32. Available from: https://doi.org/10.1016/j.bjae.2019.10.001.

[126] Steel D, Kirkman MA, Thompson DNP, et al. Open thoracic anterolateral cordotomy for pain relief in children: report of 2 cases. J Neurosurg Pediatr 2017;20(3):278–83. Available from: https://doi.org/10.3171/2017.5.PEDS17119.

[127] Hoashi JS, Thomas SM, Goodwin RC, et al. Balloon kyphoplasty for managing intractable pain in pediatric pathologic vertebral fractures. J Pediatr Orthop 2017;37(4):e286–91. Available from: https://doi.org/10.1097/BPO.0000000000000886.

[128] Liossi C, Howard RF. Pediatric chronic pain: biopsychosocial assessment and formulation. Pediatrics 2016;138:e20160331. Available from: https://doi.org/10.1542/peds.2016-0331.

[129] Ong CK, Forbes D. Embracing Cicely Saunders's concept of total pain. BMJ 2005;331:576–7. Available from: https://doi.org/10.1136/bmj.331.7516.576-d.

[130] Schütze T, Längler A, Zuzak TJ, et al. Use of complementary and alternative medicine by pediatric oncology patients during palliative care. Support Care Cancer 2016;24(7):2869–75. Available from: https://doi.org/10.1007/s00520-016-3097-2.

[131] NCCAM. Complementary, alternative, or integrative health: what's in a name? Available from: http://nccam.nih.gov/health/whatiscam; 2013.

[132] Shafto K, Gouda S, Catrine K, et al. Integrative approaches in pediatric palliative care. Children (Basel) 2018;5(6):75. Available from: https://doi.org/10.3390/children5060075.

[133] Deng G. Integrative medicine therapies for pain management in cancer patients. Cancer J 2019;25(5):343–8. Available from: https://doi.org/10.1097/PPO.0000000000000399.

[134] El Geziry A, Toble Y, Al Kadhi F, et al. Non-pharmacological pain management. In: Shallik N editor. Pain management in special circumstances. IntechOpen; London (UK). 2018. Available from: https://doi.org/10.5772/intechopen.79689.

[135] Paes TV, Silva-Rodrigues FM, Ávila LKD. Non-pharmacological methods for pain management in pediatric oncology: evidences in literature. Braz J Cancerol 2021;67(2):e-031027. Available from: https://doi.org/10.32635/2176-9745.RBC.2021v67n2.1027.

[136] Chokshi SK, Ladas EJ, Taromina K, et al. Predictors of acupuncture use among children and adolescents with cancer. Pediatr Blood Cancer 2017;64(7):e26424. Available from: https://doi.org/10.1002/pbc.26424.

[137] Çelebioğlu A, Gürol A, Yildirim ZK, et al. Effects of massage therapy on pain and anxiety arising from intrathecal therapy or bone marrow aspiration in children with cancer. Int J Nurs Pract 2015;21(6):797–804. Available from: https://doi.org/10.1111/ijn.12298.

[138] Yun H, Romero SAD, Record B, et al. Utilization of integrative medicine differs by age among pediatric oncology patients. Pediatr Blood Cancer 2019;66(6):e27639. Available from: https://doi.org/10.1002/pbc.27639.

[139] Hoag JA, Karst J, Bingen K, et al. Distracting through procedural pain and distress using virtual reality and guided imagery in pediatric, adolescent, and young adult patients: randomized controlled trial. J Med Internet Res 2022;24(4):e30260. Available from: https://doi.org/10.2196/30260.

Chapter 48

Treating pain in geriatric patients with cancer

Muhammad Atif Waqar[1] and Tomi Kovačević[2,3]

[1]Section of Palliative Medicine, Department of Oncology, The Aga Khan University Hospital, Karachi, Pakistan, [2]Faculty of Medicine, University of Novi Sad, Novi Sad, Serbia, [3]Institute for Pulmonary Diseases of Vojvodina, Sremska Kamenica, Serbia

Key learning objectives

- **Recognize the unique challenges presented in geriatric cancer pain management:** Identify the multifactorial nature of cancer-related pain in older adults, including the influence of comorbidities, cognitive decline, frailty, and age-associated physiological changes.
- **Implement comprehensive pain assessments and develop effective pain management plan tailored for older adults:** Utilize validated, age-appropriate tools, especially in patients with cognitive impairment and/or communication barriers and make appropriate individualized strategy for pain management.
- **Understand the changes in pharmacokinetics and pharmacodynamics as a consequence of normal aging:** Integrate geriatric principles into the pharmacological management of pain.
- **Incorporate nonpharmacologic and interventional therapies:** Integrate physical therapy, psychological support, radiotherapy, and nerve blocks as essential components of a multimodal approach to pain control in the elderly.
- **Promote interdisciplinary approaches:** Recognize the importance of multidisciplinary collaboration, routine reassessment, and the need for age-inclusive decision-making in pain management.

Introduction

Over half of cancer diagnoses occur in individuals aged 70 or older, a figure projected to rise as life expectancy increases [1]. As the population ages, the incidence of cancer in older adults increases [2]. Data shows that as of 2013, individuals over 60 years of age comprised 12% of the global population—a proportion expected to reach 22% (over 2 billion people) by 2050 [3–6]. According to the World Health Organization (WHO), the global population of older adults is expected to triple in the next four decades [3,6].

Chronic cancer pain is a prevalent and often undertreated symptom among older adults with cancer due to cancer itself, its treatment, and age-related comorbidities. Pain prevalence increases significantly in >65-year-old patients with cancer [7]. Cancer pain in older patients is frequently multifactorial, often with preexisting chronic pain exacerbated by cancer therapies [4]. Apart from cancer-related and complex cancer pain, pain in this population may stem from other existing diseases and chronic noncancer conditions, such as osteoarthritis or osteoporosis [8,9]. Noncancer chronic pain conditions are common in older adults, affecting 40%–70% of the ambulatory elderly population and further complicating pain management in those with cancer [4].

Unfortunately, pain is underdiagnosed in geriatric patients, especially if they have concurrent neurocognitive disorders. Pain is often underreported by older adults and clinicians often overlook pain assessments because they are overwhelmed by the multiple comorbid conditions that warrant their medical attention. Barriers to effective pain management in this population include fears of medication side effects (e.g., sedative effect that could threaten their functional capacity and independence), atypical pain presentations, and concerns about polypharmacy and opioid use (e.g., fear of addiction) [10,11]. The ability to tolerate severe pain also decreases with age, thus placing older adults in a very vulnerable position of suffering [10]. Despite high symptom burden, older patients may underreport pain due to cognitive impairment, depression, or a belief that pain is a

Pain Management in Palliative Care for Patients With Cancer. DOI: https://doi.org/10.1016/B978-0-443-30050-9.00023-0

normal aspect of aging [8,12]. Additionally, physiological, psychological, and social changes in aging significantly affect pain perception and response to treatment [12]. Even when pain may be identified, clinicians are very hesitant and reluctant to address the symptom effectively in geriatric patients due to fear of polypharmacy, lack of adequate knowledge and expertise in analgesic strategies, and fear of utilizing opioids in this population [11,13].

Unrecognized and/or poorly managed pain increases suffering for both patients and their caregivers [5,12]. Poor pain control is associated with functional decline, emotional distress, and diminished quality of life (QoL), necessitating holistic, multidisciplinary care [4,5]. Geriatric patients with cancer often face unique challenges, especially concerning pain management. A comprehensive assessment of older cancer patients experiencing pain should evaluate contributing comorbidities, emotional and spiritual distress, disability, and frailty. It is also essential to utilize validated, age-appropriate pain and symptom assessment tools [12].

Effective pain control in older adults requires an interdisciplinary strategy combining pharmacologic and nonpharmacologic interventions, with a central aim of reducing suffering and enhancing QoL [12]. This population presents unique challenges due to age-related physiological changes, multiple comorbidities, polypharmacy, and decreased functional status. Older adults with cancer are particularly vulnerable and require careful evaluation and individualized analgesic titration [14]. The development of the geriatric oncology as subspecialty reflects the need to address the unique challenges posed by aging-related physiological changes that affect cancer pathophysiology and response to treatment [4].

Because of aging, there are numerous physiological changes that occur in the body. These, among other, include a loss of integrated physiological homeostasis, loss of ability to adapt to internal and external stressors, decrease in drug metabolism and drug clearance, decrease in glomerular filtration rate and renal elimination, and unpredictable pharmacokinetics and pharmacodynamics of drugs [11]. For example, skin changes can alter the distribution of transdermal drugs; liver and renal changes, increased body fat along with decrease of body water, and reduction of lean body mass can lead to adverse effects due to delayed elimination and/or increased concentration of drugs. [13].

Pain management in geriatric patients with cancer requires a tailored approach considering various factors. Individualized analgesic regimens must consider altered pharmacokinetics, cognitive status, comorbidities, and polypharmacy. A team-based approach is crucial to achieving optimal pain relief while minimizing adverse effects. Substantial progress has been made in cancer pain management for older adults, yet further improvement is needed [4].

Pain assessment for elderly

Effective cancer pain management in older adults begins with thorough and routine pain screening. However, this population presents unique challenges, including sensory deficits, cognitive impairment, communication difficulties, and age-related changes in pain perception [14,15]. These factors complicate both the recognition and the accurate measurement of pain severity [12,14,16]. Older adults frequently underreport pain due to misconceptions, including beliefs that pain is a normal part of aging, a sign of disease progression, or that treatment will not be beneficial [4]. A comprehensive and thorough pain assessment is essential to formulate an effective analgesic plan for geriatric patients. The domains that are crucial include assessment of pain intensity, location, quality, and impact on physical and psychological functioning. There are various pain assessment tools that have been specifically validated for older adults, considering their cognitive and communication limitations. Pain intensity can be assessed using one of the validated pain assessment tools consistently through disease trajectory (Table 48.1).

Comprehensive geriatric assessment (CGA) is crucial for tailoring pain management strategies. It encompasses physical health, functional status, cognition, emotional well-being, social support, and spiritual needs. Tools such as the Geriatric Pain Measure, Mini-Mental State Examination, and cross-cultural WHODAS 2.0 can provide further insights [12,14,17–22].

Multidimensional, individualized approach that integrates patient preferences, functional status, and goals of care is essential for optimal pain assessment in the geriatric oncology population.

Pharmacological treatment

Pharmacological management of cancer pain in older adults requires a careful, individualized approach. Physiological changes in aging affect the pharmacokinetics and pharmacodynamics of analgesic drugs, necessitating adjusted dosing strategies and vigilant monitoring [9]. Age-related changes in drug metabolism, increased prevalence of comorbidities, and polypharmacy all contribute to heightened vulnerability to adverse drug reactions [9,11,15].

Geriatric patients often have multiple medical comorbidities, syndromes (e.g., delirium, dementia, failure to thrive, frailty) falls, malnutrition, and functional dependence. Medications are prescribed for most of their comorbid conditions and/or syndromes leading to polypharmacy.

TABLE 48.1 Pain assessment tools for older adults with cancer.

Tool name	Target population	Assessment type
Numeric Rating Scale	Cognitively intact	Self-report (0–10 scale)
Verbal Descriptor Scale	Mild-to-moderate cognitive impairment	Self-report (descriptive terms)
Faces Pain Scale–Revised	Mild-to-moderate cognitive impairment	Self-report (facial expression scale)
Pain Assessment in Advanced Dementia	Advanced dementia	Observational (behavioral cues)
Comprehensive Geriatric Assessment	Older frail adults	Multidisciplinary (multidimensional)
Checklist of Nonverbal Pain Indicators	Cognitive impairment	Observational (nonverbal indicators)
Doloplus-2 Scale	Severe cognitive impairment	Observational (multidomain)
Non-Communicative Patient's Pain Assessment Instrument (NOPPAIN)	Severe cognitive impairment	Observational (caregiver-administered)
Geriatric Pain Measure	Older adults with chronic pain	Self-report (multidimensional)
Mini-Mental State Examination	All older adults (cognitive screening)	Cognitive screening tool
WHODAS 2.0	Geriatric population (disability assessment)	Self-report or proxy (functional domains)

Table developed by the authors based on the literature [9,11,16–22].

Polypharmacy increases the risk of drug toxicity, sedation, delirium, injurious falls, and noncompliance in geriatric patients. Drug–drug interactions and the potential for adverse effects in older adults must be carefully considered when prescribing analgesics [11]. The American Geriatrics Society Beers Criteria is a list of potentially inappropriate medications that may result in harm or have a high risk of adverse events if prescribed in geriatric patients [23]. This list helps assist clinicians in critically reviewing certain groups of medications that should be avoided or used with caution in specific circumstances to avoid iatrogenic complications. Analgesic medications (including opioids and NSAIDs), skeletal muscle relaxants, antiseizure medications, and antidepressants are included in the Beers Criteria list [24].

The general principles of analgesic therapy remain similar to younger adults but with heightened caution. Drug choices should consider physiological changes in absorption, distribution, metabolism, and excretion. For example, aging leads to reduced renal and hepatic clearance, decreased lean body mass, and altered protein binding, all of which can modify drug pharmacokinetics [12,13].

Initiation of pharmacologic therapy should involve shared decision-making that incorporates patient goals, functional status, and the overall treatment plan. An emphasis should be placed on minimizing adverse effects while maintaining QoL [16]. The WHO analgesic ladder remains the foundation of pain management in elderly, progressing from nonopioids (e.g., paracetamol, NSAIDs) to weak opioids and then strong opioids for severe pain [25,26]. For older adults, the approach is typically "start low and go slow," titrating doses carefully to balance efficacy and tolerability [9,27]. Although opioids remain central to cancer pain control, older adults may have increased sensitivity to these agents. Side effects such as constipation, sedation, and delirium are more common and must be closely monitored. Morphine, for instance, may accumulate in renal dysfunction due to active metabolites, whereas fentanyl or oxycodone may be preferred in certain clinical contexts [11]. Opioid rotation, dose adjustments, and the use of adjuvant analgesics (e.g., duloxetine and gabapentinoids for neuropathic pain) may improve tolerability. Nonopioid agents such as acetaminophen are commonly used but must be dosed cautiously in patients with hepatic insufficiency. NSAIDs, although potentially effective, are associated with higher risks of gastrointestinal and renal toxicity in this population and should be used sparingly and for short durations [4,9,15].

Ultimately, pharmacologic treatment of cancer pain in older adults must be guided by a CGA, close monitoring, and regular reassessment to ensure optimal balance between pain relief and medication safety. Practical advice for most commonly used pain medication in older patients with cancer is shown in Table 48.2.

TABLE 48.2 Practical advice for commonly used pain medications in older patients with cancer.

	Favorable drug profile	Prescription recommendation	Caution
Acetaminophen (paracetamol)	• Could be sufficient for cancer pain in older patients as monotherapy	• In patients without liver dysfunction: ceiling dose is 3 g/day in frail patients, >80 years of age and regular alcohol users: ceiling dose is 2 g/day	• Monitor for liver impairment
NSAID	• Recommended for shortest durations possible at lowest effective dose with addition of a gastroprotective agent	• Considered for short course to opioids for uncontrolled metastatic bone pain in a patient undergoing radiation therapy	• Frequently result in undesirable gastrointestinal side effects in older patients • Should be used with extreme caution and on a case-by-case basis
Tramadol		• Start dose: 12.5–25 mg • Dose increase until expected effect • Ceiling dose 300 mg/day in 2–3 divided doses • Ceiling dose in frail or renal impairment: 200 mg/day	• Dose adjustment for renal impairment • Avoid in patients at risk for seizures • Risk of respiratory depression and sedation if associated with other CNS depressants • Increased risk of delirium compared with other opioids • Risk of serotonin syndrome
Morphine	• Good option for titration of opioid treatment	• Initiate with the lowest possible dose and with 25% reduction for patients beginning at age >60 and with 50% beginning at age >80 • Slowly titrate and monitor closely during dose-titration phase • Increase by 30%–50% per day • Determine daily dose requirement after 3–7 days • Use of laxatives is recommended	• The risk–benefit ratio of each opioid should be carefully weighed, keeping in mind the patient's physiological status such as renal function and absorption issues • Additional laxatives may need to be employed • Consider methylnaltrexone for refractory constipation • Dementia, dehydration, and use of other CNS medication are additional risk factors for oversedation and confusion • Risk for falls is increased
Hydromorphone	• Preferred over morphine in patients with mild-to-moderate renal and hepatic dysfunction • Decreased risk of delirium compared with other opioids • Potentially better efficacy in patients with cancer pain and with sleep disturbance	• Starting dose of short-acting formulation 0.2–0.5mg orally every 3 to 4 h • Dose requirement after 3–7 days	• Dose reduction in renal impairment • Accumulation of a neurotoxic (excitatory) metabolite may become a concern in older adults with severe renal insufficiency
Oxycodone	• Low dose associated with less alteration of cognitive function • Can be useful for neuropathic pain treatment • Is considered as a less constipating opioid	• Starting Dose: (>60 to 75 years): 1.25–2.5mg (short-acting formulation) • After 3–7 days, determine daily dose requirement • Frail/Geriatric Starting Dose (>80 years): 1.25mg (short-acting formulation) or switching to an ultra-low liquid titration	• Observe for headache, drowsiness, dizziness, nausea, and vomiting • Pharmacokinetics is largely affected by age and renal function
Fentanyl	• In short-acting transmucosal form offers rapid oral mucosal absorption and rapid pain relief • Transdermal (TD) administration of long-acting available	• Avoided as a frontline option in opioid naive patients • Initial dose is determined by titration with immediate-release opioid	• Delayed absorption rate is in patients with increased adipose tissue • Avoid in cachectic older patients with cancer due to insufficient adipose tissue • Careful use and close monitoring in severe renal and/or hepatic impairment • *Erythema at application-site is possible* • In case of fever absorption can be increased risking of overdose
Methadone	• Lack of data in older cancer population • Can be effective in cases of neuropathic pain	• Not recommended as a first-line opioid • Starting dose: at the low end of the dosing range, in elderly or frail patients, start as low as 1mg orally once per day	• Risk of apnea at high doses • Be aware of possible QT interval prolongation • Preferable to be initiated under close, specialized supervision

Buprenorphine	• Can be used in case of renal failure • Low addiction potential • Lower risk of nausea, vomiting, and constipation • Demonstrates antidepressive and antianxiety effects • Not affecting cognitive functions when TD dose applied	• TD patch: start with 5 mcg/h • Increase if needed every 7 days • Monitor closely at the starting dose and titration	• Due to risk for falls. monitor for possible CNS side effects and hypotension • Monitor for skin changes at the site of application
• Anticonvulsants • Gabapentin and pregabalin	• Effective for treatment of neuropathic pain and as adjuvant treatment	• Gabapentin • Start dose: 100 mg/day (preferable at bedtime) • Increase slowly until desired effect • Doses >300 mg/day are associated with adverse effects • Monitor closely due to possible CNS side effects • Pregabalin • Start dose: 25–50 mg (preferable at bedtime) • Increase slowly until desired effect or 150 mg/day in 2–3 divided doses	• Increased risk of serious, life-threatening respiratory depression and accidental injuries (e.g., falls) • Sedation and dizziness might limit the use in older adults
Tricyclic Antidepressants (TCA)	• Potentially inappropriate medication, use with caution • Not frontline option for older patients • Causes or exacerbates Syndrome of Inappropriate Antidiuretic Hormone Secretion (SIADH), hyponatremia, sedation and orthostatic hypotension	• Nortriptyline: 10–25 mg/day • Amitriptyline: 10–25 mg once daily at bedtime	• Can cause atrioventricular heart block and orthostatic hypotension; therefore cardiac monitoring is recommended especially for patients with history of cardiac disease • Clinically significant hyponatremia and falls are increased in older adults
• SNRI: duloxetine and venlafaxine	• Usually well-tolerated and therefore a good option for seniors • Effective as adjuvant treatment of neuropathic pain • Cardiovascular and anticholinergic effects are less common than in TCA	• Duloxetine • Start dose: 20–30 mg/day • Gradual titration over 2–4 weeks until desired effect • A daily dose of 30 mg/day could be effective • Venlafaxine • Start dose: 37.5 mg/day • Gradual titration over 2–4 weeks until desired effect	• Duloxetine should be avoided in patients with liver dysfunction, severe renal impairment, and alcohol abusers • Be aware of potential exacerbation syndrome of inappropriate antidiuretic hormone secretion (SIADH) or hyponatremia • Risk of falls • Observe possible suicidal ideation and suicidal tendencies
Corticosteroids	• Option for adjuvant pain treatment	• To be considered for short periods before and/or after radiotherapy to avoid pain flares • Tailored dosage according to patient's clinical status	• Be aware of possible worsening of frailty, increased risk of falls, insomnia, confusion. and delirium
Cannabinoids	• Weak quality of research data • Less promising in cancer pain management		• Be aware of possible serious adverse cardiovascular effects even in patients with low cardiac risk, in older adults
Bisphosphonates	• As adjuvant treatment of painful bone metastases	• When baseline creatinine clearance is 30–60 mL/min dose reduction is recommended • In case of severe renal impairment zoledronic acid is not recommended	• Before administering a dental review to prevent osteonecrosis of the jaw is preferable

Table developed by the authors based on the literature [9,11,27–34].

Nonpharmacologic and interventional therapies

Nonpharmacologic strategies play a vital role in a comprehensive pain management plan for older adults with cancer and offer meaningful benefits in terms of pain relief, improved mood, and overall QoL, and should not be overlooked, in the care of older adults with cancer pain. These approaches are often used as adjuncts to medications, with the goal of improving function, reducing distress, and enhancing QoL. Nonpharmacologic modalities include physical therapy, occupational therapy, cognitive-behavioral therapy, acupuncture, massage, and relaxation techniques. These interventions can reduce reliance on medications, especially in frail individuals at high risk for drug toxicity [11,16].

Psychosocial interventions are crucial in managing the emotional and existential dimensions of cancer pain. Supportive counseling, psychotherapy, hypnosis, imaginative relaxation, and spiritual care should be considered based on patient needs and preferences, and with the goal to ease mental symptoms affecting the sole pain perception [9,12]. Psycho oncological interventions can be a very important nonpharmacologic intervention to address complex cancer pain in older adults. Through psychoeducation, geriatric patients can understand what pain is together with its psychological meaning and patients potentiated to take self-control of the pain [9].

Rehabilitative therapies are particularly important in mitigating the impact of pain on mobility and independence. Physical therapy can address musculoskeletal contributors to pain and support safe ambulation, while occupational therapy can assist with activities of daily living and promote functional independence [11]. Incorporating regular physical therapy and physical activity may be beneficial in lowering of pain scores and improving performance status, gait, and balance stability, but the data are not clear and conclusive for older patients with cancer [35]. For bedbound and functionally compromised frail geriatric patients, passive range of motion exercises, repositioning, gentle massage, and intentional touch may help to decrease the pain experience [36].

Interventional pain management techniques, including nerve blocks, spinal analgesia, and neuromodulation, may be considered for refractory pain not adequately controlled with systemic pharmacological therapy. Procedures should be selected cautiously, considering comorbidities, frailty, and patient goals [27]. Additionally, transcutaneous electrical nerve stimulation units, occupational therapy, heat and cold therapy, acupuncture, guided imagery, aromatherapy, hypnosis, biofeedback, and pet therapy are all evidence-based modalities that can also reduce pain and can be beneficial for geriatric patients depending on comorbidities.

Palliative radiation therapy, specifically single-fraction radiation for uncomplicated painful osseous metastasis, remains a highly effective modality for palliation of localized bone metastases or tumor-related nerve compression. It should be considered in patients with focal pain and good performance status [4,9,37].

Education and support

Providing detailed patient education and allowing shared decision-making is essential in developing a comprehensive pain management strategy for older adults with cancer. Geriatric patients must be taught how to use pain medications effectively. Counseling should be performed to explain the rationale for dosing schedules and the importance of breakthrough analgesic medication. Both geriatric patients and their caregivers should be thoroughly educated about pain management strategies, potential side effects of the prescribed medications, and the importance of adherence to the analgesic treatment regimen. Providing partner-guided pain management training to caregivers can improve social and psychological functioning and decrease discomfort in older adults. A pain diary can also serve as a useful tool for monitoring the intake of prescribed pain medication.

Special considerations in cognitive impairment and frailty

Cognitive impairment, including mild cognitive decline and dementia, is common in older adults with cancer and poses a significant barrier to adequate pain assessment and management. These patients may have diminished capacity to self-report pain, leading to reliance on observational tools and caregiver input for accurate assessment [27,34].

Geriatric patients with cognitive impairment pose unique challenges since they are not able to participate effectively in complex pain assessment. Comorbidities such as dementia can further obscure pain assessment and necessitate careful observation for nonverbal signs of discomfort [4].

Pain perception in dementia is complex and varies with disease type and stage. While individuals with Alzheimer's disease may retain the capacity to experience pain, their ability to communicate and interpret pain is often compromised [12]. When self-report is not feasible, clinicians should interpret changes in behavior, function, appetite, sleep, and social interaction as potential indicators of pain [13]. In cases where cognitive impairment complicates pain reporting, clinical signs such as grimacing, agitation, withdrawal, altered sleep or appetite, and autonomic changes should be closely monitored [12,34]. Healthcare professionals must also recognize the role of caregivers, who can provide critical insights into behavioral changes suggestive of pain. For patients with cognitive impairment, clinicians must monitor for behavioral and physiological indicators of distress and utilize validated observational tools such as the Hurley Discomfort Scale formally known as the Discomfort Scale for Dementia of the Alzheimer's Type (DS-DAT), the Pain Assessment in Advanced Dementia (PAINAD), Checklist of Nonverbal Pain Indicators (CNPI), Doloplus-2 Scale, and the Non-Communicative Patient's Pain Assessment Instrument (NOP-PAIN) to support diagnosis and management of pain [4,11,18–21,27,34]. These scales utilize nonverbal cues and behaviors such as facial expressions, body movements, changes in activity and interpersonal interactions, and vocalization to help guide the clinician regarding pain assessment in cognitively impaired older adults. Facial expressions, vocalizations, agitation, and physiological signs such as tachycardia or diaphoresis can aid in diagnosing pain, particularly in patients with advanced cognitive decline [27]. Furthermore, cognitive impairment can increase the risk of delirium, especially in the context of poorly controlled pain or opioid initiation. Delirium, in turn, can mask or mimic pain, complicating both diagnosis and management. Clinicians must balance pain relief with the risk of neurotoxicity, particularly when using centrally acting medications [16].

Frailty, characterized by diminished physiologic reserves and vulnerability to stressors, is associated with increased prevalence and intensity of pain. Pain in frail older adults can lead to further deterioration in function, increased fall risk, depression, and reduced QoL [1,27]. Management strategies in this population must be highly individualized, prioritizing safety and functional preservation. The presence of frailty should prompt dose reductions of analgesics, consideration of nonpharmacologic options, and frequent reassessment. In addition, a careful review of polypharmacy and comorbidities is essential to minimize adverse drug interactions [11].

In summary, pain management in older adults with cognitive impairment and frailty demands a nuanced, multidisciplinary approach. Regular use of validated tools, caregiver involvement, attention to behavioral cues, and a cautious pharmacologic strategy are essential to optimizing outcomes in this high-risk population.

Future implications

Addressing the significant gaps in knowledge related to pain prevalence, pathophysiology, and optimal management strategies is one of the main cornerstones in pain management of elderly patients. This includes trials that incorporate frailty indices, cognitive assessments, and patient-reported outcomes relevant to the elderly, taking into account all ethical dilemmas.

As the field of geriatric oncology continues to evolve, prioritizing compassionate, tailored pain care for older adults with cancer is not only a clinical imperative but a fundamental component of equitable, patient-centered medicine. Therefore better education of healthcare professionals on geriatric pain, development of age-adapted clinical guidelines, and the integration of geriatrics and palliative medicine expertise into oncology practice is vital.

Future research in geriatric pain management should also be focused on the integration of technology, particularly utilizing technology-based interventions for remote monitoring and management of pain in geriatric patients with cancer.

Conclusion

Despite advances in cancer therapies and palliative care, pain remains underrecognized and undertreated in older adults. The coexistence of comorbidities, frailty, and functional decline in elderly complicates both assessment and treatment, underscoring the importance of tools tailored to cognitive status and functional ability. Effective pain management in older adults with cancer requires a comprehensive, individualized, and multidisciplinary approach that considers the unique physiological, cognitive, and psychosocial complexities of aging. Practical applications of pain management in geriatric patients with cancer include regular comprehensive geriatric and pain assessment using validated tools tailored to cognitive status. Pharmacologic treatment must be initiated with caution, emphasizing the principles of "start low and go slow," and should be accompanied by close monitoring for efficacy and adverse effects. Nonpharmacologic and interventional strategies offer critical adjuncts that support functional independence and holistic well-being, especially in frail and cognitively impaired individuals. Pain often coexists with psychological and spiritual distress, further emphasizing the need for holistic, patient-centered care.

AI disclosure

During the preparation of this work the author(s) used ChatGPT in order to assist with proofreading, formatting, and grammar checking. After using this tool/service, the author(s) reviewed and edited the content as needed and take(s) full responsibility for the content of the publication.

References

[1] Lopez Alarcon D, Villegas Estevez F. Expert consensus on the management of breakthrough cancer pain in older patients. A Delphi study. J Geriatr Oncol 2019;10:643–53. Available from: https://doi.org/10.1016/j.jgo.2019.03.012.
[2] Russo MM, Sundaramurthi T. An overview of cancer pain: epidemiology and pathophysiology. Semin Oncol Nurs 2019;35:223–8. Available from: https://doi.org/10.1016/j.soncn.2019.04.002.
[3] WHO. Ageing and health, Available from: <https://www.who.int/news-room/fact-sheets/detail/ageing-and-health>.
[4] Finnerty D, O'Gara A, Buggy DJ. Managing pain in the older cancer patient. Curr Oncol Rep 2019;21:100. Available from: https://doi.org/10.1007/s11912-019-0854-7.
[5] Torres A, Garcia-Valdes P, Henriquez-Santos G. Geriatric emergencies in older adults with cancer: why older patients need a special approach. Gac Mex Oncol 2023;22(1):41–51. Available from: https://doi.org/10.24875/j.gamo.22000126.
[6] Santos J, Martins S, Azvedo L, et al. Pain as a risk factor for suicidal behavior in older adults: a systematic review. Arch Gerontol Geriatr. 2020;87:104000. Available from: https://doi.org/10.1016/j.archger.2019.104000.
[7] Domenichiello AF, Ramsden CE. The silent epidemic of chronic pain in older adults. Prog Neuropsychopharmacol Biol Psychiatry 2019;93:284–90. Available from: https://doi.org/10.1016/j.pnpbp.2019.04.006.
[8] Brunello A, Achene-Djaballah S. Prevalence of pain in patients with cancer aged 70 years or older: a prospective observational study. J Geriatr Oncol 2019;10:637–42. Available from: https://doi.org/10.1016/j.jgo.2019.01.005.
[9] Pysz-Waberski D, Sadurska J, Kędzierska-Jamróz J, et al. Multidisciplinary management of chronic pain in elderly oncology patients. Contemp Oncol (Pozn) 2022;26(3):157–64. Available from: https://doi.org/10.5114/wo.2022.119466.
[10] Yoon S, Scarton L, Duckworth L, et al. Pain, symptom distress, and pain barriers by age among patients with cancer receiving hospice care: comparison of baseline data. J Geriatr Oncol 2021;12:1068–75. Available from: https://doi.org/10.1016/j.jgo.2021.04.008.
[11] Brant JM. Assessment and management of cancer pain in older adults: strategies for success. Asia Pac J Oncol Nurs 2018;5:248–53. Available from: https://doi.org/10.4103/apjon.apjon_11_18.
[12] Delgado-Guay M, Bruera E. Management of pain in the older person with cancer. Oncology (Williston Park) 2008;22(1):55–61.
[13] Cope D. Cancer pain management considerations in older adults. Semin Oncol Nurs 2019;25:274–8. Available from: https://doi.org/10.1016/j.soncn.2019.04.008.
[14] Thota R, Seshadri R. Assessment of pain and analgesic prescriptions in older adults. Cancer Res Stat Treat 2024;7(1):144–5. Available from: https://doi.org/10.4103/crst.crst_412_23.
[15] Urban D, Cherny N, Catane R. The management of cancer pain in the elderly. Crit Rev Oncol Hematol 2010;73(2):176–83. Available from: https://doi.org/10.1016/j.critrevonc.2009.03.008.
[16] Malec M, Levine S. How do I best treat pain in my older patient with cancer? J Geriatr Oncol 2019;10:841–4. Available from: https://doi.org/10.1016/j.jgo.2019.05.009.
[17] Welsh TJ, Gordon AL, Gladman JR. Comprehensive geriatric assessment–a guide for the non-specialist. Int J Clin Pract 2014;68(3):290–3. Available from: https://doi.org/10.1111/ijcp.12313.
[18] Blozik E, Stuck AE, Niemann S, et al. Geriatric Pain Measure short form: development and initial evaluation. J Am Geriatr Soc 2007;55(12):2045–50. Available from: https://doi.org/10.1111/j.1532-5415.2007.01474.x.
[19] Snow AL, Weber JB, O'Malley KJ, et al. NOPPAIN: a nursing assistant-administered pain assessment instrument for use in dementia. Dement Geriatr Cogn Disord 2004;17(3):240–6. Available from: https://doi.org/10.1159/000076446.
[20] Pautex S, Herrmann FR, Michon A, et al. Psychometric properties of the Doloplus-2 observational pain assessment scale and comparison to self-assessment in hospitalized elderly. Clin J Pain 2007;23(9):774–9. Available from: https://doi.org/10.1097/AJP.0b013e318154b6e3.
[21] Paulson CM, Monroe T, Mion LC. Pain assessment in hospitalized older adults with dementia and delirium. J Gerontol Nurs 2014;40(6):10–5. Available from: https://doi.org/10.3928/00989134-20140428-02.
[22] Folstein MF, Folstein SE, McHugh PR. Mini-mental state". A practical method for grading the cognitive state of patients for the clinician. J Psychiatr Res 1975;12:189–98. Available from: https://doi.org/10.1016/0022-3956(75)90026-6.
[23] By the 2023 American Geriatrics Society Beers Criteria® Update Expert Panel. American Geriatrics Society 2023 updated AGS Beers Criteria® for potentially inappropriate medication use in older adults. J Am Geriatr Soc 2023;71(7):2052–81. Available from: .https://doi.org/10.1111/jgs.18372.
[24] American Geriatrics Society Panel on Pharmacological Management of Persistent Pain in Older Persons. Pharmacological management of persistent pain in older persons. J Am Geriatr Soc 2009;57(8):1331–4. Available from https://doi.org/10.1111/j.1532-5415.2009.02376.x.
[25] Anekar AA, Hendrix JM, Cascella M. WHO analgesic ladder. StatPearls [Internet]. Treasure Island (FL): StatPearls Publishing; 2023.
[26] WHO. WHO guidelines for the pharmacological and radiotherapeutic management of cancer pain in adults and adolescents. Geneva, Switzerland: World Health Organization; 2018.

[27] El Hachem G, Oliviera Rocha F, Pepersack T, et al. Advances in pain management for older patients with cancer. Ecancermedicalscience 2019;13:980. Available from: https://doi.org/10.3332/ecancer.2019.980.

[28] Paice JA, Bohlke K, Barton D, et al. Use of opioids for adults with pain from cancer or cancer treatment: ASCO Guideline. J Clin Oncol 2023;41(4):914–30. Available from: https://doi.org/10.1200/JCO.22.02198.

[29] Ang S, Sidharthan S, Lai W, et al. Cannabinoids as a potential alternative to opioids in the management of various pain subtypes: benefits, limitations, and risks. Pain Ther 2023;12:355–75. Available from: https://doi.org/10.1007/s40122-022-00465-y.

[30] Beedham W, Sbai M, Allison I, et al. Cannabinoids in the older person: a literature review. Geriatrics 2020;5(1):2. Available from: https://doi.org/10.3390/geriatrics5010002.

[31] Freo U, Ruocco C, Valerio A, et al. Paracetamol: A review of guideline recommendations. J Clin Med 2021;10:3420. Available from: https://doi.org/10.3390/jcm10153420.

[32] Nathan R, Mupamombe C, Elibol J, et al. Assessing efficacy and use patterns of medical cannabis for symptom management in elderly cancer patients. Am J Hosp Palliat Care 2023;40(4):368–73. Available from: https://doi.org/10.1177/10499091221110217.

[33] Mercadante S, Arcuri E. Pharmacological management of cancer pain in the elderly. Drugs Aging 2007;24(9):761–76. Available from: https://doi.org/10.2165/00002512-200724090-00004.

[34] Giovannini S, Coraci D, Brau F, et al. Neuropathic pain in the elderly. Diagnostics 2021;11:613. Available from: https://doi.org/10.3390/diagnostics11040613.

[35] Plinsinga ML, Singh B, Rose GL, et al. The effect of exercise on pain in people with cancer: a systematic review with meta-analysis. Sports Med 2023;53(9):1737–52. Available from: https://doi.org/10.1007/s40279-023-01862-9.

[36] Stöckigt B, Suhr R, Sulmann D, et al. Implementation of intentional touch for geriatric patients with chronic pain: a qualitative pilot study. Complement Med Res 2019;26:195–204. Available from: https://doi.org/10.1159/000496063.

[37] O'Donovan A, Morris L. Palliative radiation therapy in older adults with cancer: age-related considerations. Clin Oncol (R Coll Radiol) 2020;32(11):766–74. Available from: https://doi.org/10.1016/j.clon.2020.06.011.

Chapter 49

Treating pain in end-of-life care

Mamak Tahmasebi[1] and Hedieh Matinrad[2]

[1]*Palliative Medicine Unit, Radiation Oncology Department, Tehran University of Medical Sciences, Tehran, Iran,* [2]*UCSF Division of Palliative Medicine, San Francisco, CA, United States*

Key learning objectives

In the provision of compassionate palliative and end-of-life care, the management of pain plays a crucial role. The skillful alleviation of pain has the power to transform the final days of those afflicted with terminal illnesses, offering solace and tranquility as they approach their inevitable departure from the world. When it comes to navigating the realm of pain management at life's threshold, there are key considerations that demand our attention.

- Describe the common etiologies and types of pain in patients with cancer at the end of life.
- List frequently used treatment modalities for pain at the end of life, including pharmacologic and nonpharmacologic modalities.
- Appreciate the multidimensional elements of a person's experience and perception of pain.

Epidemiology

Understanding the epidemiology of cancer pain in end-of-life care is essential for creating effective strategies to manage pain and ensure that patients can maintain their quality of life and dignity during their final chapter. While "end of life" can encompass a spectrum of time, a period on the order of the final days to weeks or short months of life is how it is used in the literature and clinical setting.

Research indicates that the prevalence of cancer-related pain is strikingly high among patients nearing the end of life, with a prevalence of 66.4% in people with advanced, metastatic, or terminal disease [1].

Despite the existence of effective treatment options for managing this distressing symptom, there persists a troubling trend in undertreatment. A comprehensive analysis conducted by Deandrea et al. [2] highlighted that nearly one in two patients facing cancer pain did not receive adequate pain relief.

Cancer pain at the end of life can be caused by various factors, including the primary tumor, metastatic lesions, treatment-related side effects, or a combination of these factors. Neuropathic pain, bone pain, and visceral pain are common types of cancer pain reported by patients [3].

Several risk factors have been associated with a higher likelihood of experiencing cancer pain at the end of life, including advanced disease stage, presence of metastases, specific cancer types (e.g., lung, head and neck, and gastrointestinal cancers), and certain treatments (e.g., chemotherapy, radiation therapy) [4,5].

Uncontrolled cancer pain can not only have a profound negative impact on a patient's quality of life, contributing to physical, psychological, and spiritual distress, but also worsen caregiver burden physically and psychologically [6].

Of utmost importance is a comprehensive assessment of the patient's unique experience with pain, conducted diligently and consistently over time. This evaluation should delve beyond physical discomfort to explore its nuanced qualities, such as location, intensity, and underlying factors that may exacerbate or alleviate suffering. It is essential to recognize that the experience of pain is not only physical but also social, psychological, and spiritual—a concept referred to as total pain [7].

Pain Management in Palliative Care for Patients With Cancer. DOI: https://doi.org/10.1016/B978-0-443-30050-9.00035-7

Pathophysiology of cancer pain at the end of life

The pathophysiology of cancer-related pain at the end of life is multifactorial, making its management challenging. A thorough understanding of these processes is essential for creating multimodal strategies to address a patient's unique pain experience [3].

As tumors grow and metastasize, they can invade and compress surrounding tissues, bones, nerves, and organs, leading to nociceptive pain. Tumors also release inflammatory agents such as prostaglandins and cytokines, heightening nociceptor sensitivity and perpetuating an inflammatory cascade that worsens pain. Metastatic bone lesions cause severe pain due to increased pressure, inflammatory mediator release, and pathological fractures [8].

Neuropathic pain mechanisms include nerve compression or infiltration by tumors affecting peripheral nerves or the spinal cord. Chemotherapeutic agents, such as platinum-based drugs and taxanes, can cause chemotherapy-induced peripheral neuropathy and subsequent neuropathic pain. Radiation therapy may also damage nerves, leading to similar symptoms [9].

Central sensitization plays a critical role in chronic cancer pain by amplifying pain signals in the central nervous system. Prolonged nociceptive input can result in a "wind-up" phenomenon, where repeated stimulation of nociceptors increases sensitivity in dorsal horn neurons, intensifying the pain response [10].

Beyond physical factors, psychological, emotional, social, and spiritual aspects also influence pain perception. Anxiety, depression, and social isolation can magnify pain, while fear and negative thoughts further contribute to overall distress. Addressing these dimensions is crucial for effective pain management at the end of life [11].

Management

Knowing a detailed account of past medication and treatment history is an important first step in pain management at the end of life. This includes a comprehensive record of both current and past analgesic medications, meticulously noting doses, routes, and schedules, the effectiveness of these medications, any resulting side effects, and identifying potential barriers to adherence. Eliciting the history of nonpharmaceutical interventions undertaken in pursuit of pain relief is equally important [7,12].

Beyond medical history taking, a patient's and family's preferences, pain management goals, cultural beliefs that shape their outlook on healing and suffering, and openness to various treatment modalities are essential in our care of the whole person [13].

In end-of-life care, special attention should be given to anticipatory prescribing—providing medications ahead of time for anticipated symptoms such as pain ensures quick relief when it is most needed and helps prevent unnecessary suffering. It is essential to remain vigilant by regularly reassessing and adjusting treatment plans. Monitoring changes in pain levels and evaluating the effectiveness of medications are vital for maintaining proper control over pain management.

In addition, during the end-of-life phase, it is paramount that medications not aimed at improving quality of life in this phase be evaluated and deprescribed, as polypharmacy is commonly observed. Polypharmacy not only increases the risk of delirium but also leads to a misdirection of focus on ingestion of medications instead of preservation of time and focus on comfort and quality of life; as an example, statins, whose benefit is over many years through the prevention of cardiovascular events, generally no longer have a role at the end-of-life phase and could be safely deprescribed [14,15].

Important to remember that while the Numeric Rating Scale for pain assessment (0–10 scale) is often used with patients who can communicate, at the end of life, there is often a time when a patient is no longer able to verbally communicate their pain. At this point, relying on other signs such as grimacing and moaning and including the use of pain scales that do not rely on verbal communication, such as the Face, Legs, Activity, Cry, Consolability (FLACC) scale, is helpful to assess discomfort and treat it adequately. The FLACC scale was originally used to assess discomfort in children and has been adopted for pain assessment in nonverbal adults. A score of 0–2 is given for each category, 0 indicating relaxed and comfortable appearing, and 2 indicating most uncomfortable appearing; scores are added up for a total score. A total score of 0–3 is considered mild discomfort, 4–6 moderate pain, and 7–10 severe pain or discomfort [16].

The management of pain in people with terminal cancer requires a comprehensive approach that encompasses both pharmaceutical and nonpharmaceutical interventions. The effective alleviation of pain is vital in ensuring a sense of dignity and quality of life during the final stages of existence [17].

Pharmacological treatment

Opioid analgesics represent the cornerstone in the pharmacological toolbox aimed at managing moderate-to-severe cancer-associated pain. Availability of specific opioids varies by setting, and dosing is dependent on whether they have used opioids previously for pain relief or if they are opioid naive. Common opioids used in end-of-life care include morphine, hydromorphone, fentanyl, and oxycodone [12].

Proper dosing, along with careful consideration regarding the route by which these medications are administered (oral, parenteral/subcutaneous injection, or transdermal), is essential for optimal pain control. In particular, the route of administration often depends on the setting of care at the end of life and medication availability. As patients lose the ability to safely swallow, transdermal, sublingual, subcutaneous, and intravenous routes become modes of administration. For families caring for a patient at home, oral, sublingual, and transdermal administration are more accessible, while more training is required for home subcutaneous opioid administration. If high-concentration oral morphine is available (20 mg/mL), a very small volume can be given orally, some of which will absorb in the oral mucosa while the rest will trickle down the esophagus, without significant aspiration risk given the exceedingly small volumes used [18]. Short-acting oral opioids take effect within ~1 hour and can be safely administered in the last hours to days of life as frequently as every 1–2 hours as needed for pain when focus is exclusively on ensuring comfort. Intravenous/subcutaneous administration has an onset of effect within 10–15 minutes. In the last hours to days at the end of life, intravenous and subcutaneous opioids can be given every 15 minutes as needed to ensure comfort. Opioid dosing is highly dependent on a patient's prior opioid needs and tolerance. However, starting doses for opioids of oral morphine equivalent of 10–15 mg are typically safe and well tolerated for moderate-to-severe cancer-related pain. Incremental dose increases to find an effective dose with tolerable side effects will be needed. While morphine should be avoided in the setting of significant renal failure, cautious use of lower doses can be an acceptable approach if other alternative medications are not available [18,19].

In instances where satisfactory levels of pain relief are not achieved despite appropriate dose and frequency escalation, or where intolerable side effects emerge, rotation between different types of opioid agents may be considered [19].

Use of opioids has a stigma attached to it, given its historic and current use as a substance to achieve euphoria and cause addiction. It is critical that we educate patients and families that, in the setting of end of life, the use of opioids is for the relief of pain and discomfort. Given the short time that opioids will be used, concern for addiction development is low, although dependence can still occur. This is difficult to stringently study, but is a widely shared expert opinion, including the opinion held by the authors [3,20].

Supplementary agents, such as anticonvulsants (such as gabapentin and pregabalin), antidepressants (such as serotonin–norepinephrine reuptake inhibitors such as duloxetine and tricyclic antidepressants such as amitriptyline), and corticosteroids, can be employed in conjunction with opioids to amplify pain alleviation and target neuropathic or bone-related discomfort. Using all these adjuvant medications depends on the estimated prognosis, as it may take days or weeks for some agents to become effective [21].

For mild-to-moderate pain management, nonopioid analgesics such as nonsteroidal antiinflammatory drugs and acetaminophen may be utilized, particularly when paired with adjuvant analgesics or opioids [21].

In certain instances, interventional methods such as nerve blocks, epidural analgesia, or neurolytic procedures could be considered for persistent pain that does not respond well to conventional treatment approaches or for specific types of pain syndromes [19].

Nonpharmaceutical management

Within nonpharmaceutical management, various interventions can be utilized to alleviate pain and improve overall well-being. The utility of some methods may depend on the patient's cognitive status, engagement, and prognosis. Psychological and behavioral strategies such as cognitive-behavioral therapy, mindfulness-based techniques, and psychoeducation serve to guide individuals in managing their pain perception, coping mechanisms, and emotional distress [7,13].

Physical modalities, including heat/cold therapy, massage, acupuncture, and transcutaneous electrical nerve stimulation, offer nonpharmaceutical avenues for pain relief and increased comfort [7,13].

In addition, radiation therapy is a potential solution for managing pain associated with bone metastases or specific tumor sites. Palliative radiation treatment has shown efficacy in controlling pain levels in these cases [11].

Beyond the physical aspect of pain management lies the importance of spiritual and emotional support. Addressing existential concerns, providing emotional guidance, and involving chaplains or counselors can facilitate greater alleviation of suffering while enhancing overall well-being. This holistic approach recognizes the interconnected nature of physical discomfort with psychological distress and offers a comprehensive framework for improving quality of life amidst difficult circumstances [7,17].

In the field of complementary and alternative therapies, healing modalities such as music therapy, aromatherapy, and other nontraditional methods may be integrated into a patient's care regimen in alignment with their personal openness and cultural values [13].

The successful alleviation of pain in patients at the end of life with terminal cancer demands an approach that is centered on the individual, drawing upon a diverse array of medical and holistic strategies customized to suit their unique needs, social and cultural values, and preferences at the end of life. It is imperative that continuous evaluation and refinement of the treatment protocol take place as the patient's circumstances shift and their experience of pain transforms over time.

At last, there are times when, despite available holistic management for pain, a person's suffering is not alleviated. This is when shared decision-making with the patient and/or family and the medical team can be done to provide relief through sedation instead, via the use of benzodiazepines such as lorazepam or midazolam. However, this should be considered once attempts at adequate treatment of pain have been exhausted [22].

Conclusion

Treatment of a patient's pain at the end of life is important not only for the alleviation of that person's suffering but also to support their family members and caregivers. Understanding the type of pain experienced and its etiology can guide clinicians' choice of pharmacologic treatment, though the agents of choice for moderate-to-severe cancer-related pain at the end-of-life care are opioids. A person's experience of pain, however, is influenced by many factors, including psychological, social, and spiritual/existential factors. Adequate treatment of pain requires consideration and addressing of these factors, when possible, within an interdisciplinary team. The route of administration of medications for pain is dependent on the setting, though parenteral and subcutaneous opioids have a more rapid onset and, as such, can be given more frequently. In the home setting, low-volume oral liquid short-acting opioids or subcutaneous opioids can be used to treat pain in the last hours to days of life when patients are often unable to awaken or swallow pills. Eliciting a patient's and family's wishes as it relates to the end of life is important, particularly as some hold cultural or religious values connected to the experience of pain or suffering; some patients may opt for more awareness in the balance of pain control and wakefulness, and this may change with time.

References

[1] van den Beuken-van Everdingen MH, Hochstenbach LM, Joosten EA, et al. Update on prevalence of pain in patients with cancer: systematic review and meta-analysis. J Pain Symptom Manag 2016;51(6):1070–90. Available from: https://doi.org/10.1016/j.jpainsymman.2015.12.340.

[2] Deandrea S, Montanari M, Moja L, et al. Prevalence of undertreatment in cancer pain. A review of published literature. Ann Oncol 2008;19(12):1985–91.

[3] Dans M, Kutner JS, Agarwal R, et al. NCCN guidelines® insights: palliative care, version 2.2021. J Natl Compr Canc Netw 2021;19(7):780–8. Available from: https://doi.org/10.6004/jnccn.2021.0033.

[4] van den Beuken-van Everdingen MH, de Rijke JM, Kessels AG, et al. Prevalence of pain in patients with cancer: a systematic review of the past 40 years. Ann Oncol 2007;18(9):1437–49. Available from: https://doi.org/10.1093/annonc/mdm056.

[5] Isaac T, Stuver SO, Davis RB, et al. Incidence of severe pain in newly diagnosed ambulatory patients with stage IV cancer. Pain Res Manag 2012;17(5):347–52. Available from: https://doi.org/10.1155/2012/542354.

[6] Chi N-C, Demiris G, Pike KC, et al. Pain management concerns from the hospice family caregivers' perspective. Am J Hosp Palliat Care 2018;35(4):601–11.

[7] Hamza Habib M, Zheng J, Kozlov E, et al. Top ten tips palliative care clinicians should know about total pain. J Palliat Med 2026;29(2):237–44. Available from: https://doi.org/10.1089/jpm.2025.0159.

[8] Bohren Y, Cachemaille M, Timbolschi ID, et al. Understanding the physiopathology of pain pathways for a practical approach of cancer pain management. Cardiovasc Intervent Radiol 2025;48(7):899–906. Available from: https://doi.org/10.1007/s00270-024-03920-9.

[9] Sarezky J, Sachs G, Elinzano H, et al. Cancer and peripheral nerve disease. Clin Geriatr Med 2021;37(2):289–300. Available from: https://doi.org/10.1016/j.cger.2021.01.003.

[10] Herrero JF, Laird JM, López-García JA. Wind-up of spinal cord neurones and pain sensation: much ado about something? Prog Neurobiol 2000;61(2):169–203. Available from: https://doi.org/10.1016/s0301-0082(99)00051-9.

[11] Paice JA, Ferrell B. The management of cancer pain. CA Cancer J Clin 2011;61(3):157–82. Available from: https://doi.org/10.3322/caac.20112.

[12] Bruera E, Paice J. Cancer pain management: safe and effective use of opioids. Am Soc Clin Oncol Educ Book 2015;1(35):e593–9. Available from: https://doi.org/10.14694/EdBook_AM.2015.35.e593.

[13] Mao JJ, Ismaila N, Bao T, et al. Integrative medicine for pain management in oncology: Society for Integrative Oncology-ASCO guideline. J Clin Oncol 2022;40(34):3998–4024. Available from: https://doi.org/10.1200/JCO.22.01357.

[14] Elsten EECM, Pot IE, Geijteman ECT, et al. Recommendations for deprescribing of medication in the last phase of life: an international Delphi study. J Pain Symptom Manage 2024;68(5):443–55.e2. Available from: https://doi.org/10.1016/j.jpainsymman.2024.07.029.
[15] Scott IA, Hilmer SN, Reeve E, et al. Reducing inappropriate polypharmacy: the process of deprescribing. JAMA Intern Med 2015;175(5):827–34. Available from: https://doi.org/10.1001/jamainternmed.2015.0324.
[16] Voepel-Lewis T, Zanotti J, Dammeyer JA, et al. Reliability and validity of the face, legs, activity, cry, consolability behavioral tool in assessing acute pain in critically ill patients. Am J Crit Care 2010;19(1):55–61. Available from: https://doi.org/10.4037/ajcc2010624.
[17] Janjan N. Improving cancer pain control with NCCN guideline-based analgesic administration: a patient-centered outcome. J Natl Compr Canc Netw 2014;12(9):1243–9. Available from: https://doi.org/10.6004/jnccn.2014.0122.
[18] Caraceni A, Hanks G, Kaasa S, et al. Use of opioid analgesics in the treatment of cancer pain: evidence-based recommendations from the EAPC. Lancet Oncol 2012;13(2):e58–68. Available from: https://doi.org/10.1016/S1470-2045(12)70040-2.
[19] Swarm RA, Paice JA, Anghelescu DL, et al. Adult cancer pain, version 3.2019, NCCN clinical practice guidelines in oncology. J Natl Compr Canc Netw 2019;17(8):977–1007. Available from: https://doi.org/10.6004/jnccn.2019.0038.
[20] Pinkerton R, Mitchell G, Hardy J. Stringent control of opioids: sound public health measures, but a step too far in palliative care? Curr Oncol Rep 2020;22(4):34. Available from: https://doi.org/10.1007/s11912-020-0900-5.
[21] Portenoy RK. Treatment of cancer pain. Lancet 2011;377(9784):2236–47. Available from: https://doi.org/10.1016/S0140-6736(11)60236-5.
[22] Blinderman CD, Billings JA. Comfort care for patients dying in the hospital. N Engl J Med 2015;373(26):2549–61. Available from: https://doi.org/10.1056/NEJMra1411746.

Chapter 50

Treating pain in war-stricken areas

Farzana Khan[1,2]

[1]*Fasiuddin Khan Research Foundation, Dhaka, Bangladesh,* [2]*Knowledge Mobilisation Hub, NIHR Global Health Research Unit on Respiratory Health (RESPIRE), Usher Institute, University of Edinburgh, United Kingdom*

Key learning objectives

- Describe the scope of pain-related suffering among cancer patients in conflict- and disaster-stricken areas, especially among refugees and displaced populations.
- Identify the major barriers to effective pain relief and palliative care in low-resource humanitarian settings, including legal, logistical, and cultural challenges.
- Evaluate the psychological and ethical burdens experienced by frontline healthcare workers delivering care in crisis zones.
- Apply principles of culturally responsive care to better address pain in diverse humanitarian populations.
- Recommend strategies to integrate pain management and palliative care into emergency health protocols and build health system resilience.
- Implement a modular training framework to improve self-preparedness and pain management competency among humanitarian health workers.

Introduction

In conflict zones, war-stricken regions, and humanitarian crises, the burden of pain, especially among patients with cancer or terminal illnesses, is immense and often neglected. This neglect underscores the urgent relevance of palliative care in humanitarian contexts. Palliative care is a critical component of humanitarian healthcare that addresses physical, emotional, psychosocial, and spiritual suffering. Yet, access to essential pain management remains limited due to logistical, ethical, cultural, and policy-related barriers. The suffering is especially dire for refugees, internally displaced persons, and populations in fragile or failing health systems. Pain management for patients with cancer in humanitarian settings is especially emphasized due to the challenges of delivering palliative care in low-resource environments. Recent COVID-19 pandemic additionally exposed systemic inadequacies and proposed recommendations for integrating pain relief into global humanitarian health agendas.

Pain is not just a symptom but a condition that dehumanizes and exacerbates vulnerability. Studies have documented high levels of pain-related suffering in refugee populations [1], especially in protracted crisis settings such as Cox's Bazar, Bangladesh, which houses Rohingya refugees [2,3]. Despite the known burden, pain management is rarely prioritized in humanitarian responses [4,5].

Cancer-related pain is particularly acute in these settings. Refugees and displaced populations with cancer often receive fragmented or no care at all due to their uncertain legal status, resource scarcity, and disrupted health infrastructure [4,6,7]. There is a growing body of research that underlines the unique challenges faced by refugee cancer patients in countries such as Lebanon, Turkey, and Jordan, which have absorbed large numbers of displaced persons from Syria and Palestine [4,5,7,8]. These patients experience compounded trauma from both the disease and displacement, often lacking access to opioids or even basic pain relief [9].

Healthcare providers (HCPs) in crisis settings face emotional and moral distress when unable to alleviate suffering [10–12]. Frontline providers report psychological burdens associated with their inability to offer dignified end-of-life care. Expatriate and national staff alike experience burnout and moral distress, which the COVID-19 pandemic further intensified [10,13,14].

Pain Management in Palliative Care for Patients With Cancer. DOI: https://doi.org/10.1016/B978-0-443-30050-9.00044-8

Barriers to effective pain management for patients with cancer in humanitarian contexts

Barriers to pain relief in humanitarian contexts are multifaceted. Refugees with cancer are doubly marginalized. Studies from Turkey and the Middle East show that refugee cancer patients often present late, are excluded from national cancer care programs, and have little continuity in their treatment [4,5,7,8]. This results in unmanaged symptoms, disease progression, and preventable suffering. Evidence from Lebanon and Jordan indicates that national health systems struggle to absorb the cancer care needs of refugee populations, particularly in radiation therapy, chemotherapy, and pain control [4,6]. Global oncology institutions have urged systemic reforms to ensure equitable cancer care access for displaced populations [6].

Most barriers to effective pain management in humanitarian crises can be grouped as follows:

1. **Limited opioid availability and regulation**: Strict regulatory environments and limited HCP training hinder access to opioids. In Bangladesh, for instance, providers cite legal concerns and lack of knowledge as major deterrents to prescribing opioids [15].
2. **Systemic and logistical challenges**: Disrupted supply chains, limited cold storage, and fragmented care models reduce the availability of pain relief medications [6,12,16].
3. **Neglect of noncommunicable diseases**, including cancer: Humanitarian response priorities often overlook noncommunicable diseases [4–6].
4. **Insufficient training for frontline workers**: Frontline workers often lack training in recognizing and managing complex pain and distress [10,13,17].
5. **Cultural and ethical considerations**: Perceptions around death, suffering, and opioid use vary significantly across cultures and can influence both patient receptiveness and provider willingness to offer palliative care [18].
6. **Policy gaps**: Humanitarian health guidelines historically focused on curative and life-saving interventions, often excluding palliative care [10,16]. However, frameworks such as the World Health Organization's integration of palliative care into primary care and the Sphere Handbook now advocate for its inclusion [16,19].

The COVID-19 pandemic compounded humanitarian suffering and highlighted weaknesses in emergency preparedness [2,14,20]. It underscored the need for better training, guidelines, and the institutionalization of palliative care protocols in humanitarian settings. It also amplified the need for culturally competent care, communication, and empathy [18,21].

Proposed modular training: training on self-preparedness, pain management, and palliative care in humanitarian settings

As already mentioned, a modular training framework needs to be implemented to enhance self-preparedness and clinical competencies for HCPs in crisis settings. Training in pain management during crises constitutes a critical intervention that should be prioritized at the onset of such events, as it can partially mitigate the sudden and substantial disruptions in healthcare provision. One of the main objectives of these programs for HCPs is building capacity of care for themselves while caring for others in prolonged or high-risk humanitarian settings.

Modular training must include modules with learning content, objectives, activities, and expected outcomes. Evaluation metrics across all modules implies pretraining and posttraining assessments (knowledge, attitude, confidence); objective structured clinical examination-style evaluations during simulations and follow-up surveys on field impact after deployment.

Proposed modular training for pain management and palliative care in humanitarian settings suggests following modules:

Module 1. **Foundations of pain and palliative care in crisis contexts**: To provide participants with an understanding of the scope and ethical necessity of pain and palliative care in humanitarian settings, particularly for cancer and trauma patients.

Module 2. **Clinical pain assessment and basic interventions**: To train HCPs in assessing pain and delivering pharmacological and nonpharmacological interventions using minimal resources.

Module 3. **Self-preparedness for crisis work**: To build the capacity of HCPs to care for themselves while caring for others in prolonged or high-risk humanitarian settings.

Module 4. **Culturally responsive care and communication**: To enable participants to deliver care that respects local cultures, religions, and social norms, especially in multicultural and refugee contexts.

Module 5. **Team-based palliative practice and leadership**: To strengthen collaborative team function, delegation, and interdisciplinary coordination in fast-moving emergency environments.

Module 6. **Systems and advocacy for sustainable pain relief**: To equip participants with tools to advocate for system-level changes in palliative care delivery and supply chain resilience.

Module 7. **Simulation and real-world integration**: To provide participants with hands-on crisis simulations to practice integrated pain care, team communication, and self-care under pressure (Table 50.1).

TABLE 50.1 Proposed modular training for pain management and palliative care in humanitarian settings: module content, activities, and expected outcomes.

	Content	Activities	Outcomes
Module 1	• Definitions of palliative care and humanitarian pain management. • Epidemiology of pain in crises: trauma, cancer, and chronic illnesses. • Overview of global standards (e.g., WHO, Sphere Handbook) and recent developments in cancer care for refugees [4,6,16,17]. • Ethical principles in humanitarian palliative care: dignity, equity, and nonabandonment [12,14].	• Interactive lecture with real-world case examples (Rohingya, Syrian refugees, Turkey). • Small group discussions on ethical dilemmas and triage decisions.	• Clear understanding of humanitarian palliative principles. • Familiarity with global frameworks and cancer-specific burdens in displaced populations.
Module 2	• Pain types and assessment tools for adults and children (including language/cultural adaptation). • Opioid pharmacology and pain relief ladder in low-resource settings [15]. • Barriers to opioid access and strategies to overcome them (regulatory and logistical reforms). • Nonpharmacological interventions for pain and comfort [17,22].	• Pain assessment simulations using vignettes. • Role-play scenarios to practice low-resource interventions.	• Increased clinical confidence in recognizing and managing pain. • Ability to deliver effective pain care within resource-constrained environments.
Module 3	• Concepts of self-preparedness and emotional resilience [13]. • Recognizing and managing burnout, moral distress, and secondary trauma. • Self-assessment tools and wellness planning. • Lessons from past crises (e.g., COVID-19, Gaza, Lebanon) on frontline distress [10].	• Guided reflection and journaling. • Development of a personalized self-care plan. • Peer mentoring and support group formation.	• Improved ability to manage stress and sustain clinical effectiveness. • Structured approach to psychological preparedness.
Module 4	• Understanding cultural humility and patient values [18]. • Communication strategies across language and belief systems. • Managing grief and end-of-life practices in diverse communities. • Addressing stigma in cancer and pain perception among displaced populations [7].	• Case-based analysis of cultural miscommunications. • Training on the use of interpreters and cultural mediators.	• Enhanced trust-building and cultural alignment in care delivery. • Fewer conflicts due to misunderstandings in palliative contexts.
Module 5	• Roles and responsibilities in palliative care teams. • Team communication tools (e.g., Situation, Background, Assessment, Recommendation [SBAR], check-ins). • Leadership under pressure and supporting junior staff.	• Simulation of multidisciplinary team meetings in a refugee setting. • Conflict-resolution role-plays.	• Stronger team dynamics and reduced role ambiguity. • Clear communication channels in the field.
Module 6	• Navigating regulatory environments for opioid availability. • Emergency preparedness planning for medication stockpiles. • Engaging donors, governments, and International Non-Governmental Organizations (INGOs) for policy reform [6,20]. • Leveraging community health networks and digital solutions.	• Advocacy pitch practice: building a case for opioid inclusion in relief packs. • Policy scenario planning workshops.	• Practical advocacy skills and system-mapping awareness. • Action plans for policy influence and resource mobilization.
Module 7	• Full-spectrum case simulations: war-injured patient with cancer pain in refugee camp. • Simulated breakdowns: sudden loss of drug supply, family conflicts, or interpreter failure. • Debriefing for emotional and operational feedback.	• Multistation rotating crisis simulation. • Peer and trainer feedback loops.	• Confident application of knowledge in real-life scenarios. • Reinforced muscle memory for crisis decision-making.

WHO, World Health Organization.

Conclusion

Integrating advanced pain control and palliative care into humanitarian responses is crucial but fraught with challenges. These include limited resources, access to medication, cultural specificity, systemic neglect, and the emotional toll on healthcare workers. The COVID-19 pandemic highlighted the urgency of these efforts, reinforcing the need for enhanced training, better resource allocation, and embedded palliative protocols. Especially for refugees and displaced patients with cancer, humanitarian healthcare systems must be restructured to alleviate suffering and uphold dignity. Future research must deepen our understanding of these populations' needs and develop actionable strategies for integrating palliative care into all phases of humanitarian intervention.

Recommendations toward integrated solutions

Addressing pain in war-affected settings requires systemic change, which includes:

- **Inclusion of pain management and palliative care in emergency health protocols**: Guidelines must prioritize symptom management and pain relief as part of emergency care standards [16,19].
- **Capacity building**: Training HCPs in opioid use, communication skills, and palliative care principles is essential [15,20].
- **Cultural sensitivity**: Understanding the cultural dimensions of suffering and care can facilitate better communication and patient-centered care [18].
- **Resource allocation**: Ensuring sustainable access to essential medications, diagnostics, and staffing in conflict-affected regions must become a priority [4,6,12].

References

[1] Doherty M, Power L, Petrova M, et al. Illness-related suffering and need for palliative care in Rohingya refugees and caregivers in Bangladesh: a cross-sectional study. PLoS Med 2020;17(3):e1003011. Available from: https://doi.org/10.1371/journal.pmed.1003011.

[2] Nouvet E, Sivaram M, Bezanson K, et al. Palliative care in humanitarian crises: a review of the literature. J Int Humanit Action 2018;3(1):5. Available from: https://doi.org/10.1186/s41018-018-0033-8.

[3] Doherty M, Khan F. Neglected suffering: the unmet need for palliative care in Cox's Bazar. London: World Child Cancer; 2018.

[4] El Saghir NS, Charara RN, Kreidieh FY, et al. Cancer care for refugees and displaced populations: Middle East conflicts and global natural disasters. Am Soc Clin Oncol Educ Book 2018(38):433–40. Available from: https://doi.org/10.1200/EDBK_201365.

[5] Hassan MA, Maniam A. Refugees have cancer too. Ecancermedicalscience 2024;18:ed134. Available from: https://doi.org/10.3332/ecancer.2024.ed134.

[6] Ilbawi A, Elzinga G, Allemani C, et al. Cancer care for refugees: time to invest in people and systems. Lancet Oncol 2020;21(5):604–5. Available from: https://doi.org/10.1016/S1470-2045(20)30163-7.

[7] Güney S, Doğan ÖÇ, Bağçivan G. Cancer care for refugees in Türkiye: challenges and achievements. Semin Oncol Nurs 2025;41(3):151879. Available from: https://doi.org/10.1016/j.soncn.2025.151879.

[8] Jozef S, Dibusova L, Mihalik J, et al. Neglected population of cancer patients – migrants and refugees. Eur J Cancer 2015;51(3):160–1. Available from: https://doi.org/10.1016/S0959-8049(16)30472-5.

[9] Guo P, Alajarmeh S, Alarja G, et al. Compounded trauma: a qualitative study of the challenges for refugees living with advanced cancer. Palliat Med 2021;35(5):916–26. Available from: https://doi.org/10.1177/02692163211000236.

[10] Schneider M, Chappuis F, Pautex S. How do expatriate health workers cope with needs to provide palliative care in humanitarian emergency assistance? A qualitative study. Palliat Med 2018;32(10):1567–74. Available from: https://doi.org/10.1177/0269216318794091.

[11] Hunt M, Chénier A, Bezanson K, et al. Moral experiences of humanitarian health professionals caring for patients who are dying or likely to die in a humanitarian crises. J Int Humanit Action 2018;3(1):12. Available from: https://doi.org/10.1186/s41018-018-0040-9.

[12] Hunt M, Nouvet E, Chénier A, et al. Addressing obstacles to the inclusion of palliative care in humanitarian health projects: a qualitative study of humanitarian health professionals' and policy makers' perceptions. Confl Health 2020;14(1):70. Available from: https://doi.org/10.1186/s13031-020-00314-9.

[13] Khan F, Azad TB, Bhuyian S, et al. Exploring the self-preparedness of frontline healthcare workers in a low- and middle-income country from a humanitarian context during the COVID-19 pandemic: a constructivist grounded theory study. Front Public Health 2023;11:1043050. Available from: https://doi.org/10.3389/fpubh.2023.1043050.

[14] Radbruch L, Knaul FM, de Lima L, et al. The key role of palliative care in response to the COVID-19 tsunami of suffering. Lancet 2020;395(10235):1467–9. Available from: https://doi.org/10.1016/S0140-6736(20)30964-8.

[15] Khan F, Ahmad N, Iqbal M, et al. Physicians knowledge and attitude of opioid availability, accessibility and use in pain management in Bangladesh. Bangladesh Med Res Counc Bull 2014;40(1):18–24. Available from: https://doi.org/10.3329/bmrcb.v40i1.20324.

[16] The Sphere Handbook: Humanitarian Charter and Minimum Standards in Humanitarian Response. 4th ed. Geneva: Sphere Association; 2018.

[17] World Health Organization. Quality of care in fragile, conflict-affected and vulnerable settings: taking action. Geneva: World Health Organization; 2020.

[18] Schuster-Wallace CJ, Nouvet E, Rigby I, et al. Culturally sensitive palliative care in humanitarian action: lessons from a critical interpretive synthesis of culture in palliative care literature. Palliat Support Care 2022;20(4):582–92. Available from: https://doi.org/10.1017/S1478951521000894.

[19] World Health Organization. Integrating palliative care and symptom relief into primary health care. Geneva: World Health Organization; 2018.

[20] Daubman BR, Khan F, Slater SE, et al. Save lives and relieve suffering: the twin imperatives of humanitarian response and the role of palliative care. Front Oncol 2023;13:1120380. Available from: https://doi.org/10.3389/fonc.2023.1120380.

[21] Van Der Cingel M. Compassion and professional care: exploring the domain. Nurs Philos 2009;10(2):124–36. Available from: https://doi.org/10.1111/j.1466-769X.2009.00397.x.

[22] Wang Y, Aaron R, Attal N, et al. An update on non-pharmacological interventions for pain relief. Cell Rep Med 2025;6(2):101940. Available from: https://doi.org/10.1016/j.xcrm.2025.101940.

Chapter 51

Ethical dilemmas in pain management

Mamak Tahmasebi[1] and Ehsan Shamsi Gooshki[2,3]
[1]*Palliative Medicine Unit, Radiation Oncology Department, Tehran University of Medical Sciences, Tehran, Iran,* [2]*Monash Bioethics Centre, Monash University, Melbourne, VIC, Australia,* [3]*Medical Ethics and History of Medicine Research Centre, Tehran University of Medical Sciences, Tehran, Iran*

Key learning objectives

- The importance of the distinction between pain and suffering.
- The moral significance of the alleviation of pain and suffering.
- The important ethical issues in the context of cancer pain management.
- The role of the market influence in cancer pain management.

Introduction

The issue of moral challenges in controlling pain in cancer patients is both important and complex. As emphasized by the World Health Organization Expert Committee on Cancer Pain Relief and Active Supportive Care (1990), "Freedom from pain should be seen as a right of every cancer patient, and access to pain therapy as a measure of respect for this right" (p. 10). Yet, despite this recognition, the realization of this right often involves navigating numerous ethical dilemmas [1,2].

The importance of the distinction between pain and suffering

The sensation of pain is an embodiment of physical discomfort, a signal alerting the body to potential harm or injury. It is a primal response triggered by stimuli perceived as threatening to the intricate web of tissues and nerves that compose our corporeal form. Pain, though unwelcome in its presence, plays an essential role in the human experience; it stands as a sentinel warning us to steer clear of danger or prompting us to address any looming peril [3,4].

On the contrary, suffering is a complex and profound ordeal that delves into the realms of psyche and emotions. It encapsulates the array of detrimental mental conditions that can manifest in reaction to pain, as well as other harrowing circumstances such as bereavement, deprivation, unease, desolation, or crises regarding existence itself. Suffering entails the cognitive and emotional digestion of anguish or other adversities, frequently culminating in sensations of torment, hopelessness, or a perception of futility in one's predicament [5,6].

Although pain and suffering often intertwine, they are not interchangeable concepts. It is conceivable to endure physical anguish without enduring substantial emotional torment, as evidenced by practices such as stoicism, mindfulness, or successful pain relief techniques. Conversely, one may undergo profound suffering in the absence of actual physical pain, exemplified by experiences of grief, depression, or fear of death [7].

The differentiation between pain and suffering holds significant importance as it underscores the notion that easing human affliction cannot be achieved solely through eradicating physical discomfort. Rather, attending to the emotional, mental, and philosophical aspects of our existence is frequently deemed equally, if not more vital in diminishing the overall burden of human anguish [5,7].

The moral significance of alleviation of pain and suffering management

Among the oldest moral norms in medical practice is the ethical obligation to relieve pain—an enduring expression of the principle of beneficence. While this duty remains a cornerstone of professional care, its application has been significantly

Pain Management in Palliative Care for Patients With Cancer. DOI: https://doi.org/10.1016/B978-0-443-30050-9.00042-4

shaped by advances in medical science and technology, which have expanded both the possibilities and the ethical complexities of effective pain management. Scholars such as Peter Singer contend that prioritizing the relief of pain and suffering should be central to ethical deliberation [8].

The act of easing pain and alleviating suffering can also be perceived as a way to honor the inherent dignity and autonomy of individuals. Unabated agony has the potential to diminish one's capacity to lead a purposeful and independent existence. It can be argued that without effective pain control, obtaining valid informed consent becomes difficult—if not impossible. Informed consent requires the patient's focused attention to receive, process, and understand the disclosed information. This process depends on establishing a foundation for clear, mutual communication, which is often unattainable when a patient is in significant pain. Severe pain can impair decision-making capacity by affecting cognitive functioning or by exerting a form of internal coercion, thereby undermining the voluntariness of the patient's consent. Moreover, failure to manage pain runs counter to the principles of patient-centered care, which is increasingly emphasized in contemporary clinical practice.

Numerous ethical paradigms, such as virtue ethics and care ethics, place great emphasis on the moral significance of compassion and the obligation to aid those experiencing adversity. The act of alleviating pain and suffering is viewed as a tangible expression of this moral excellence [9].

Uncontrolled pain can wield profound and adverse effects, manifesting in impaired cognitive capacities, diminished efficiency, societal estrangement, and the gravest of all outcomes—thoughts of self-destruction. Mitigating this suffering is imperative to avert such catastrophic consequences [10-12].

Within the realm of healthcare and governmental regulations, justice requires fairness in the distribution of resources, access for individuals and communities to pain management tools, and a nondiscriminatory approach in allocating related resources. This principle emphasizes a profound moral obligation to alleviate human pain and suffering that stands at the core of all deliberations regarding a justified approach to decision-making processes and responsible advancement of novel treatments and interventions [13,14].

The deep significance of alleviating suffering and anguish is firmly grounded in core ethical values such as promoting general well-being, honoring human dignity, demonstrating empathy, and averting harm. A careful exploration of these ethical dimensions holds the power to inform not only individual actions but also wider societal approaches toward addressing pain and suffering [5,15].

Jamie Mayerfeld stated pain and suffering is not simply a burden for those enduring it, but rather universally undesirable [16]. The alleviation of suffering benefits the sufferer individually and enhances the collective state of existence in our world. It is also important to recognize the role of social expectations regarding pain control. Pain is often the chief complaint that drives individuals to seek care at hospitals and healthcare facilities, as it is one of the most immediate and tangible signs of illness or infirmity. As such, the effective alleviation of pain is not only a clinical responsibility but also a societal expectation. When healthcare professionals respond promptly and compassionately to pain, it reinforces public trust in the medical profession. In this sense, the ability to manage pain effectively plays a crucial role in maintaining the moral credibility of healthcare systems.

Cancer is a critical area where pain control and management play a significant role, as they profoundly impact patients' quality of life, emotional well-being, and treatment outcomes. Cancer-related pain—stemming from tumor progression, treatment side effects, or comorbid conditions—is often chronic and debilitating. Effective pain management not only alleviates physical suffering but also reduces psychological distress, enhances mobility and daily functioning, and enables patients to better tolerate and adhere to cancer therapies. Furthermore, it improves patients' ability to actively engage in treatment decisions, which is particularly important given that cancer can be either treatable or terminal. This is especially crucial in end-of-life decision-making, where comfort, dignity, and patient preferences take on heightened importance. Overall, effective pain management strengthens patient-centered care and facilitates shared decision-making across all stages of the disease.

Several important ethical issues arise in the context of cancer pain management

Autonomy and informed consent

The regard for patient autonomy and the necessity of obtaining their informed consent hold utmost importance in the realm of cancer pain management, where intricate treatment choices are frequently at play. Patients must be afforded a detailed disclosure of relevant information and to a level that their understanding of the potential pitfalls, advantages, and alternative courses available in the arena of pain alleviation strategies could be ensured. Special consideration should be directed toward individuals with compromised decision-making capabilities, notably those grappling with cognitive deficiencies or nearing life's end [13,17].

While the significance of informed consent is well-established in medical ethics literature, it can be challenging to uphold in certain circumstances—particularly in cultural contexts where families play a central role in medical decision-making. In some cases, families may ask the healthcare team to withhold specific information from a competent patient, even when the patient has been diagnosed with terminal cancer. Another complicating factor is the assessment of a patient's decision-making capacity. Factors such as fluctuating consciousness, severe pain or distress, and death anxiety can interfere with a clear evaluation of a patient's ability to make informed decisions. This assessment is a critical milestone, as it informs the next steps in care, including when and how to involve surrogate decision-makers in the process.

Balancing benefit and harm

In the delicate balance between benefit and harm, healthcare providers are tasked with a weighty responsibility. They must meticulously assess the advantages of pain management interventions in comparison to the lurking dangers of adverse reactions, unforeseen complications, and iatrogenic harm. This undertaking becomes particularly daunting when confronting potent analgesics or invasive procedures fraught with perilous outcomes.

At the heart of this critical decision-making process lies the principle of nonmaleficence, which requires avoiding preventable harms as much as possible. This guiding principle is paramount in navigating the complex terrain of pain management where every action taken must be executed with utmost precision and consideration for potential consequences. In such moments of heightened stakes and grave uncertainties, a provider's unwavering commitment to preventing harm elevates their practice to an elevated sense of duty and moral obligation toward those entrusted in their care [13,18].

Beneficence, as another core ethical principle, requires maximizing potential benefits while minimizing possible harms. A key challenge lies in aligning the medical understanding of benefit—typically grounded in an objective account of health and clinical well-being—with the patient's or their surrogate decision-maker's subjective perception of well-being. This tension highlights the importance of considering what constitutes a benefit not only from a clinical standpoint but also from the patient's personal, cultural, and existential perspective. In end-of-life care especially, what is considered beneficial may extend beyond prolonging life to include comfort, dignity, and alignment with the patient's values and preferences.

In the context of cancer treatment, the risk of addiction resulting from opioid use is an important factor in risk–benefit assessment. In the era of the opioid epidemic, there arises a pressing need to reexamine the prudent administration and consumption of powerful analgesics in the treatment of cancer-related pain. The key to successfully treading this fine line lies in conducting thorough evaluations of patients, ensuring their full comprehension of potential risks through education initiatives, and embracing innovative approaches to pain management that veer away from traditional pharmaceutical interventions. By adhering to these foundational principles underscored by awareness and vigilance against opioid-related complications, we can aspire toward a more refined strategy that addresses both physical anguish and safeguarding against misuse tendencies within this realm of complex ailment management [19,20].

Equitable access to pain management

The issue of equitable access to pain management is one that plagues the healthcare system, especially when it comes to effective cancer pain relief. This discrepancy in access, particularly among marginalized communities, raises profound questions of morality and equity.

Various factors come into play here: socioeconomic status, race, ethnicity, geographical location, and insurance coverage all intermingle to determine a patient's ability to receive comprehensive pain care. Addressing these disparities is essential in guaranteeing fair and just pain management for all individuals battling cancer. The quest for equality in this realm remains paramount in providing optimal care for every patient in need [21–23].

End-of-life considerations

In the twilight of life, a shifting perspective on pain management emerges. The focus transitions from pursuit of cure to the compassionate provision of palliative care and alleviation of pain and suffering. Here lies a realm where ethical dilemmas loom large—grappling with decisions regarding potentially life-shortening analgesics, navigating the difficult path of withholding or withdrawing life-sustaining treatments, and considering interventions aimed solely at providing comfort and enhancing the quality of life. These profound considerations underscore the delicate balance between respecting individual autonomy and ensuring that every aspect of end-of-life care is imbued with dignity and compassion [24,25]. In addition to its inherent sensitivity, end-of-life decision-making is particularly important because it often involves deeply held beliefs and values shared by all parties involved—including healthcare providers and patients or their families—who may conscientiously object to certain options. These decisions frequently touch on ethical, cultural, religious, and personal dimensions that can create moral conflict. Moreover, legal constraints tend to be more pronounced in end-of-life contexts, especially when decisions involve hastening death or prolonging life, making the process even more complex and requiring careful ethical and legal consideration.

Voluntary assisted dying (VAD) (which may include euthanasia and physician-assisted suicide, depending on the jurisdiction) is a request in some societies in end of life. Although the question around permissibility of VAD has a more ethical nature, the social, legal, and political aspects of it cannot be ignored. Healthcare professionals play a critical role in shaping the discourse and policies related to VAD, highlighting the need for further research and dialogue in this complex and sensitive area [26]. The main ethical issues in VAD practice are:

- **Autonomy and informed consent**: Respecting patients' rights to make decisions about their own lives, ensuring they are fully informed about their options and the implications of choosing euthanasia or VAD.
- **Moral and religious perspectives:** Balancing the sanctity of life with the quality of life, considering different moral and religious views on the permissibility of ending life to alleviate suffering.
- **Legal and regulatory issues:** Navigating the varying legal statuses of euthanasia and VAD across different regions and ensuring strict regulations to prevent abuse.
- **Potential for abuse:** Protecting vulnerable populations from coercion or abuse and implementing safeguards to ensure decisions are voluntary and informed.
- **Impact on healthcare providers:** Addressing the influence of social trust to healthcare professionals and healthcare system, and the moral distress healthcare providers may experience, and providing support systems to help them cope with the ethical challenges.
- **End-of-life care:** Emphasizing the importance of high-quality palliative care to alleviate suffering and potentially reduce the demand for euthanasia and VAD, and adopting a holistic approach to address physical, emotional, psychological, and spiritual suffering.
- **Terminal sedation (palliative sedation):** Involving the use of sedatives to relieve intractable symptoms in terminally ill patients until death; this is also known as palliative sedation. This practice is typically reserved for patients with refractory symptoms that cannot be managed by other means. The primary goal is to alleviate suffering, not to hasten death. Ethical distinctions between VAD which involves a deliberate act to end life at the patient's request, and terminal sedation which aims to relieve suffering without the explicit intention of causing death should be considered [27].

Clinician burnout and moral distress

In the realm of healthcare, the burden of managing intricate and severe cancer pain can take a heavy toll on clinicians, leading to burnout and moral distress. This predicament is heightened when medical caregivers find themselves impotent in assuaging the distress of their patients, particularly when confronted with obstacles to prescribing opioids. Accessibility and availability of appropriate medication and enough knowledge about handling opiophobia and opioid side effects would be the key to success in cancer pain management. Consequently, attending to the well-being of clinicians and offering robust support systems becomes imperative in sustaining ethical, empathetic, and efficacious cancer pain management [28].

To navigate these profound ethical quandaries, it is crucial to adopt a comprehensive approach that revolves around patients' needs at its core. Embracing open communication, fostering shared decision-making processes, and carefully assessing the distinct circumstances and values of each patient and their loved ones must take center stage in this multidisciplinary endeavor [29,30].

Role of the market influence in cancer pain management

The market's influence in the realm of healthcare, specifically within the domains of pain alleviation and costly medical treatments, necessitates a nuanced examination of issues such as fairness, availability, and the delicate interplay between efficacy and societal obligation [19].

Equity and access in the realm of pain management are a matter of great concern. The landscape is marred by prohibitively high prices, making certain procedure-based treatments, such as advanced interventional techniques and cutting-edge pharmaceuticals, an elusive luxury for a significant portion of patients [31]. Consequently, a stark inequality emerges where only individuals with substantial financial means or comprehensive insurance plans have the privilege to avail themselves of these sophisticated avenues for pain relief [32].

The government's involvement and regulatory measures in society hold a crucial importance. Governments have the power to intervene in cases of market failures, ensuring fairness and societal well-being. This can be achieved through initiatives such as universal healthcare coverage, implementing price controls, providing subsidies for vital pain management services, and regulatory oversight [33,34].

As new frontiers in pain management beckon with promises of relief and healing, the shadow of rising healthcare expenses looms ominously over those seeking solace (e.g., price of fentanyl patch compared to methadone pills). A bitter dichotomy emerges as innovation clashes with accessibility—a dilemma faced by policymakers and health institutions alike. Striking a harmonious chord between fostering groundbreaking advancements and preserving fair access becomes an intricate challenge that demands keen discernment and unwavering resolve [35].

The significance of embracing a multidisciplinary strategy for the management of pain cannot be overstated. By enlisting a holistic approach that integrates both pharmacological and nonpharmacological interventions, it becomes feasible to tackle the intricate biopsychosocial aspects of pain effectively. This method holds promise in diminishing dependency on expensive, intervention-based therapies, thereby offering a more sustainable solution to alleviate suffering [36].

In essence, the market's impact on healthcare, especially within the realm of costly, procedure-centric pain management, is one defined by a nuanced equilibrium between fostering innovation, guaranteeing universal reach, and upholding a viable economic structure. Those in positions of policymaking authority, as well as healthcare practitioners and invested parties alike, face the intricate task of weighing these opposing forces to confront the complexities associated with pain management with both impartiality and efficacy [19].

Conclusion

The affliction of pain proves to be a malevolent force upon those subjected to its cruel grasp. In the realm of cancer sufferers, this torment is coupled with heightened emotional turmoil and significant impediments on day-to-day functionality. Left unattended, this agonizing sensation may ultimately prompt pleas for VAD while also precipitating needless hospitalizations and emergency room visits. Indeed, pain bears ill consequences beyond the individualistic scope. Thus it is imperative that steps are taken toward the prevention and alleviation of cancer-related suffering not only as a means of ameliorating the plight of those directly affected but as a moral imperative to foster righteousness within our broader societal framework.

Acknowledgment

We acknowledge the use of OpenAI's ChatGPT (version GPT-4) for assistance with language editing and clarification of the text. This tool was employed to enhance the clarity, grammar, and readability of the manuscript, while all intellectual content, analysis, and conclusions remain the responsibility of the authors.

References

[1] WHO Expert Committee on Cancer Pain Relief and Active Supportive Care & World Health Organization. Cancer pain relief and palliative care: report of a WHO expert committee [meeting held in Geneva from 3 to 10 July 1989]. Geneva: World Health Organization; 1990. Available from: https://iris.who.int/items/803bf528-2d21-4435-8ba1-d2c483df4d62.

[2] World Health Organization. Narcotic and psychotropic drugs: achieving balance in national opioids control policy: guidelines for assessment. World Health Organization; 2000. Available from: https://iris.who.int/items/a89149c2-962d-4f41-87bd-7d643c554236.

[3] Loeser JD, Melzack R. Pain: an overview. Lancet 1999;353(9164):1607–9. Available from: https://doi.org/10.1016/S0140-6736(99)01311-2.

[4] Breivik H, Borchgrevink PC, Allen SM, et al. Assessment of pain. Br J Anaesth 2008;101(1):17–24. Available from: https://doi.org/10.1093/bja/aen103.

[5] Kahn DM, Steeves RH. The experience of suffering: conceptual clarification and theoretical definition. J Adv Nurs 1986;11(6):623–31. Available from: https://doi.org/10.1111/j.1365-2648.1986.tb03379.x.

[6] Morse JM, Beres MA, Spiers JA, et al. Identifying signals of suffering by linking verbal and facial cues. Qual Health Res 2003;13(8):1063–77. Available from: https://doi.org/10.1177/1049732303256401.
[7] Cassell EJ. The nature of suffering and the goals of medicine. N Engl J Med 1982;306(11):639–45. Available from: https://doi.org/10.1056/NEJM198203183061104.
[8] Singer P. Practical ethics. New York: Cambridge University Press; 1979.
[9] Noddings N. Caring: a feminine approach to ethics and moral education. Berkeley, California: University of California Press; 1984.
[10] Eccleston C, Crombez G. Pain demands attention: a cognitive-affective model of the interruptive function of pain. Psychol Bull 1999;125(3):356–66. Available from: https://doi.org/10.1037/0033-2909.125.3.356.
[11] Turk DC, Monarch ES. Biopsychosocial perspective on chronic pain. In: Turk DC, Gatchel RJ, editors. Psychological approaches to pain management: a practitioner's handbook. 2nd ed. Washington, D.C: Guilford Press; 2002.
[12] Chiulli DL. Suicide risk identification and mitigation in patients with chronic pain prescribed opioid medication. Ment Health Clin 2024;14(6):328–33. Available from: https://doi.org/10.9740/mhc.2024.12.328.
[13] Beauchamp TL, Childress JF. Principles of biomedical ethics. 8th ed. New York: Oxford University Press; 2019.
[14] World Health Organization. WHO guidelines for the pharmacological and radiotherapeutic management of cancer pain in adults and adolescents. Geneva; 2018.
[15] Burns JH, Hart HLA, Burns JH, Rosen F, editors. An introduction to the principles of morals and legislation. Oxford: Clarendon Press; 1996.
[16] Mayerfeld J. Suffering and moral responsibility. Oxford, UK: Oxford University Press; 1999.
[17] Faden RR, Beauchamp TL, King NMP. A history and theory of informed consent. New York: Oxford University Press; 1986.
[18] Varkey B. Principles of clinical ethics and their application to practice. Med Princ Pract 2021;30(1):17–28. Available from: https://doi.org/10.1159/000509119.
[19] Volkow ND, McLellan AT. Opioid abuse in chronic pain—misconceptions and mitigation strategies. N Engl J Med 2016;374(13):1253–63. Available from: https://doi.org/10.1056/NEJMra1507771.
[20] Brennan F, Carr DB, Cousins M. Pain management: a fundamental human right. Anesth Analg 2007;105(1):205–21. Available from: https://doi.org/10.1213/01.ane.0000268145.52345.55.
[21] Institute of Medicine (US) Committee on Advancing Pain Research, Care, and Education. Relieving pain in America: a blueprint for transforming prevention, care, education, and research. National Academies Press; 2011. Available from: https://doi.org/10.17226/13172.
[22] Shavers VL, Bakos A, Sheppard VB. Race, ethnicity, and pain among the US adult population. J Health Care Poor Underserved 2010;21(1):177–220. Available from: https://doi.org/10.1353/hpu.0.0255.
[23] Burgess DJ, van Ryn M, Crowley-Matoka M, et al. Understanding the provider contribution to race/ethnicity disparities in pain treatment: insights from dual process models of stereotyping. Pain Med 2006;7(2):119–34. Available from: https://doi.org/10.1111/j.1526-4637.2006.00105.x.
[24] Steinhauser KE, Clipp EC, McNeilly M, et al. In search of a good death: observations of patients, families, and providers. Ann Intern Med 2000;132(10):825–32. Available from: https://doi.org/10.7326/0003-4819-132-10-200005160-00011.
[25] Quill TE, Abernethy AP. Generalist plus specialist palliative care—creating a more sustainable model. N Engl J Med 2013;368(13):1173–5. Available from: https://doi.org/10.1056/NEJMp1215620.
[26] Blaschke S-M, Schofield P, Taylor K, et al. Common dedication to facilitating good dying experiences: qualitative study of end-of-life care professionals' attitudes towards voluntary assisted dying. Palliat Med 2019;33(6):562–9. Available from: https://doi.org/10.1177/0269216318824276.
[27] Gilbertson L, Savulescu J, Oakley J, et al. Expanded terminal sedation in end-of-life care. J Med Ethics 2023;49(4):252–60. Available from: https://doi.org/10.1136/jme-2022-108511.
[28] Rushton CH, Batcheller J, Schroeder K, et al. Burnout and resilience among nurses practicing in high-intensity settings. Am J Crit Care 2015;24(5):412–20. Available from: https://doi.org/10.4037/ajcc2015291.
[29] Papadimos TJ, Maldonado Y, Tripathi RS, et al. An overview of end-of-life issues in the intensive care unit. Int J Crit Illn Inj Sci 2011;1(2):138–46. Available from: https://doi.org/10.4103/2229-5151.84801.
[30] Chen L, Chu H, Wang H, et al. Effects of communication, trust, and respect on shared decision-making: insights from healthcare providers in Chinese public hospitals. Front Med (Lausanne) 2025;12:1577276. Available from: https://doi.org/10.3389/fmed.2025.1577276.
[31] Callahan D. Setting limits: medical goals in an aging society. Washington D.C: Georgetown University Press; 1995.
[32] Garrison LP, Mansley EC, Abbott TA, et al. Good research practices for measuring drug costs in cost-effectiveness analyses: a societal perspective: the ISPOR drug cost task force report—part II. Value Health 2010;13(1):8–13. Available from: https://doi.org/10.1111/j.1524-4733.2009.00660.x.
[33] Berwick DM, Nolan TW, Whittington J. The triple aim: care, health, and cost. Health Aff 2008;27(3):759–69. Available from: https://doi.org/10.1377/hlthaff.27.3.759.
[34] Porter ME. What is value in health care? N Engl J Med 2010;363(26):2477–81. Available from: https://doi.org/10.1056/NEJMp1011024.
[35] Lakdawalla DN, Doshi JA, Garrison LP, et al. Defining elements of value in health care-a health economics approach: an ISPOR special task force report [3]. Value Health 2018;21(2):131–9. Available from: https://doi.org/10.1016/j.jval.2017.12.007.
[36] Gatchel RJ, Peng YB, Peters ML, et al. The biopsychosocial approach to chronic pain: scientific advances and future directions. Psychol Bull 2007;133(4):581–624. Available from: https://doi.org/10.1037/0033-2909.133.4.581.

Chapter 52

The role of the nurse in pain management

Imatullah Akyar
Faculty of Nursing, Hacettepe University, Ankara, Türkiye

Key learning objectives

- Understand the complexity and multidimensional nature of pain experienced by patients in palliative care and its impact on patients and their families.
- Recognize the importance of a comprehensive patient assessment and the value of a tailored, individualized approach in optimizing pain relief and functional outcomes.
- Identify the goals of pain and symptom management within the context of palliative care.
- Describe the pivotal role of nurses in the interdisciplinary palliative care team.

Introduction

In palliative care, pain management is crucial for improving the quality of life of patients with cancer, and nurses play an essential role in managing their patients' pain [1,2]. As pain and suffering are among the most bothersome symptoms experienced by patients with cancer, nurses should be involved in all steps of cancer pain management [3]. The American Nurses Association's position statement on the ethical responsibility to manage pain and suffering emphasizes that nurses should provide clinically excellent care, which includes awareness of clinical indications, mutual identification of goals for pain management, interprofessional collaboration, and understanding of professional standards for the assessment and management of different types of pain [4]. The goal of pain management is to decrease pain intensity and increase functionality, performance status, and comfort for patients and caregivers [5].

In the management of pain, guidelines recommend screening for pain at every engagement with a healthcare professional, assessing the components of the pain, agreeing on a tailored pain management plan, and regularly reviewing and evaluating cancer pain treatment outcomes [6]. Nurses, who are the first point of contact and who spend the maximum time with patients and families, are responsible for pain screening and assessment, patient education and monitoring, and pain treatment [1,4] (Table 52.1).

Nurses' role in pain assessment

Pain assessment is the central pillar of pain management. Nurses are responsible for conducting assessments to understand and evaluate the pain experienced by patients. They should initiate communication on pain at every engagement with the patient [6]. One example of screening can include posing a key screening question of "What has been your worst pain in the last 24 hours on a scale of 0–10? Where 0 is no pain, and 10 is the worst imaginable" or using different patient-reported tools to understand patients' pain experience [7]. When evaluating pain, it is important to gain insight into aspects of pain such as intensity, location, and physical, psychological, socioeconomic, and spiritual factors underlying the total pain experience and its broader impact on patients' well-being during pain history. In addition to pain history, extending the assessment to include medical and psychosocial history, physical examination, and laboratory tests [8] helps nurses to identify the classification, pathophysiology, etiology, and mechanisms of pain that guide and direct the management plan [8,9].

Nurses should be cognizant of several crucial principles required to assess pain effectively [1,7–14]. First, the patients must always be believed, as pain is a personal experience that occurs when and where the person says it does. Since most

Pain Management in Palliative Care for Patients With Cancer. DOI: https://doi.org/10.1016/B978-0-443-30050-9.00015-1

TABLE 52.1 Role of nurses in palliative care pain management.

Focus area	Nurses' role	
Assessment	• Use validated tools based on the patient's status. • Perform routine and ongoing reassessments. • Involve caregivers for nonverbal patients.	• Include physical, emotional, spiritual, and psychosocial dimensions.
Communication	• Practice open, honest, and compassionate communication.	• Use active listening.
Care planning	• Collaborate with the team and the patient/family to create an individualized pain management plan. • Document specific pain goals and evaluate outcomes regularly.	• Align goals with patient values, prognosis, and preferences.
Intervention (pharmacological & nonpharmacological)	• Administer medications per clinical guidelines. • Monitor for effectiveness, breakthrough pain, and side effects. • Educate patients on medication use, risks, and reporting side effects. • Advocate for timely dose adjustments.	• Integrate nonpharmacological interventions. • Tailor interventions based on cognitive/physical ability and preferences. • Encourage participation and assess effectiveness. • Document outcomes of interventions.
Education	• Educate patients/families about pain mechanisms, treatment options, and side effect management. • Use simple language and teach-back method to confirm understanding.	• Provide materials in various formats (written, digital). • Involve caregivers.
Monitor	• Reassess pain 30–60 min postintervention, or even earlier with parenteral analgesia. • Track changes in functionality, sleep, and quality of life.	• Report any new or worsening symptoms to the team promptly.
Advocacy	• Speak up for patient preferences, especially when patients cannot speak for themselves. • Advocate for effective and timely pain control.	• Escalate care to specialists when needed. • Ensure ethical and culturally sensitive pain management decisions.
Documentation	• Document pain scores, interventions, responses, education, and side effects.	• Use consistent terminology to support continuity of care.

patients report multiple pain experiences, each pain should be assessed separately, and communication about pain must always be encouraged, even if pain is not always manifested. It is also recommended to expand pain assessment with additional approaches such as development of an understanding of the diagnosis and effects of cancer treatment, understanding the meaning of pain for the patient and family, discussing the patient's typical coping responses to stress or pain, and describing the changes in mood that have occurred as a result of the pain. In addition, discussing the patient's knowledge of, curiosity about, preferences for, and expectations about pain management methods and explaining the patient's concerns about using controlled substances such as opioids or anxiolytics could all add value to pain treatment.

In patients' pain assessment, particular attention should be paid to any discrepancies between patients' verbal descriptions of pain and their behavior and appearance. Assessments need to be performed at regular intervals, in response to changes in pain intensity, after each pain intervention, during the peak effect of medication/nonpharmacological intervention, and after modification of the pain management plan. In addition to pain assessment, adequate assessment also covers reassessment of the patient and the ability to support, inform, communicate, and understand the patient and family's needs.

Pain intensity and treatment outcomes should be assessed regularly and consistently with some of the methods in detail evaluated in other chapters in this book. While using the Visual Analog Scale or Numeric Rating Scale related to the worst pain in the last 24 hours is the most common means of pain evaluation, there is no consensus on a standard tool. However, validated tools should be used in pain assessment. The Expert Working Group of the European Association for Palliative Care [15] recommends using multidimensional tools like the Brief Pain Inventory and the McGill Pain Questionnaire for assessment. In patients with cognitive impairment, limited communicative skills, or in older adults who limit self-report of pain, observation of pain-related behaviors and discomfort is an important part in pain assessment [16]. Particularly in those patients, nurses need to be aware of potential causes of pain, observe patient behaviors and utilize behavioral pain assessment tools, solicit proxy reporting of pain, and evaluate behavior/activity changes. Possible nonverbal pain behaviors can be assessed through evaluation of *facial expressions* (frown, wince, sad or frightened look, wrinkled forehead, closed or tightened eyelids, rapid blinking, clenched teeth or jaw), *body movements* (restless, agitated, jittery, fidgeting, pacing, rocking, constant or intermittent shifting of position, withdrawing), *protective mechanisms* (bracing, guarding, rubbing or massaging a body part, splinting, clutching or holding onto side rails, bed, tray table, or affected area during movement), *verbalizations* (saying common phrases such as "help me," "leave me alone," "don't touch me," "ouch," cursing, verbally abusive, praying out loud), *vocalizations* (sighing, moaning, crying, whining, screaming, chanting, breathing heavily), *mental status changes* (confusion, disorientation, irritability, distress, depression), and *changes in activity patterns, routines, or interpersonal interactions* (decreased appetite, sleep alterations, decreased social activity participation, change in ambulation, immobilization, aggression, combativeness, resisting care) [1,9,17].

Nurses' role in pain management plan

Nurses should manage and monitor pain through individualized pharmacological and nonpharmacological strategies aligned with patient goals, educate and support patients and families, advocate for timely interventions, and ensure continuity of care through clear documentation and team collaboration. Nurses' role begins with the development of a holistic and individualized management plan based on a thorough assessment. A tailored approach to each patient is critical for managing pain effectively [4]. The pain management plan should include pharmacological, interventional, and nonpharmacological interventions, adapted to the patient's overall condition, prognosis, preferences, and goals of care. The details on potential management plans are given in other chapters of this book.

Regardless of the intervention used, it is important to provide clear, jargon-free explanations of medications, their purpose, dosing schedules, side effects, and when to report problems. One of the options nurses could apply is to use teach-back methods to verify understanding, as it is essential for patients to participate actively in pain management and to take an active role in decision-making and follow-up planning [10].

In addition, reassessing pain and side effects regularly is crucial. Nurses should document findings and communicate with prescribers if adjustments are needed. This is particularly true for patients with changes in kidney/liver function, cognition, or new symptoms. Nurses must monitor whether treatments align with the patient's goals, assess response to treatment, identify the need for supplemental analgesics, evaluate medication-related adverse effects, and understand drug–drug and drug–disease interactions [1].

In order to evaluate pain following an intervention, a fixed reassessment time could be used (e.g., 30–60 minutes postintervention). If no relief is observed, nurses should initiate escalation protocols or multidisciplinary consults (e.g., palliative care, pain specialist). They must also follow up on indicators of functionality, sleep quality, emotional state, and patients'/caregivers' reports of the impact of pain on the quality of life [8]. If the pain is not adequately relieved within a reasonable amount of time after applying standard analgesic guidelines and interventions, consulting team members is also the nurses' responsibility [4]. Nurses should help patients and families understand prescriptions and educate them on appropriate medication usage [4].

In order to improve compliance, written instructions, visual aids, and practical demonstrations could be provided (e.g., using a pillbox, applying topical patches). The importance of adherence and tracking proper medication use can be supported through pain diaries or digital apps, while written material can be instrumental in helping patients comprehend nonpharmacological interventions, particularly after long consultations with significant amount of new information. Such education can be provided to patients through nurse-led consultations in which motivational interviewing and patient coaching can be used as a strategy [4,8]. The education supports adherence and encourages self-management [18]. It should cover:

- Different treatment approaches.
- Purpose and use of each medication.
- Expected outcomes and potential risks associated with a specific medication.
- Side effects and their management.
- Importance and methods of follow-up reporting (such as patient diaries or apps).
- Basic pain physiology and its consequences [4].

In every treatment plan, it is important to assess patient readiness and preferences and to tailor strategies to the individual's capacity and comfort level. Self-management education provides knowledge and information on lifestyle, psychology, social support, navigation and coordination, and medical decision-making [19]. Nurses should use multiple, individualized strategies, including cognitive and behavioral strategies [20].

Addressing patient fears or misconceptions (e.g., opioid addiction, tolerance) using evidence-based information and empathetic communication is also paramount. Self-management strategies also encourage patients to make independent use of nonpharmacological interventions [21]. Even though pharmacological interventions remain central, the multifactorial etiology of pain, the potential for adverse effects, and misconceptions about analgesics can impede pain management [22].

One of the options to improve compliance and overall pain management is to integrate nonpharmacological techniques into the daily routine, which can be used as an adjuvant to pharmacological treatment or as a stand-alone approach. Such techniques could reduce the dose of the analgesics required to control pain, thus minimizing the adverse effects of pharmacological therapies and helping patients gain a sense of control, manage fatigue, stress, and anxiety, and enhance physical activity, functional capacity, and quality of life [8,19,22,23]. These interventions also help with individualized pain management by building and sustaining therapeutic relationships, recognizing the diversity of patients' needs, developing person-centered care, and incorporating significant others [24].

Nonpharmacological interventions are mainly classified into two categories: cognitive-behavioral and physical interventions [23]. Cognitive-behavioral interventions include relaxation, imagery, distraction, music, prayer, hypnosis, mindfulness meditation, psychotherapy, and education, and physical interventions include acupuncture, heat/cold application, exercise, massage, reflexology, positioning, and transcutaneous electrical nerve stimulation [1,25]. Cognitive-behavioral interventions help patients by providing information and support to develop problem-solving and coping skills, enabling them to manage pain and solve pain-related problems, and by distracting attention, providing a greater sense of control and improved mood. Physical interventions, on the other hand, effectively decrease pain, which is at least partially explained by the gate control theory; a nonpainful sensation, such as touch, pressure, or vibration, activates nerves that do not transmit pain signals (called nonnociceptive fibers) and can interfere with signals from pain fibers [8,21].

For every intervention, it is also important to match interventions to the phase of illness and patient engagement level. For advanced stages, nurses should prioritize comfort-enhancing strategies, while for earlier phases, structured education with active coaching could be offered. In a metaanalysis conducted by Park and Lee [22], psychoeducational, physical, and music interventions were effective in pain management; all interventions reduced barriers to pain management and coping and improved other physical and emotional symptoms. Due to short-term effectiveness and ease of participation, music and physical interventions provided an option for end-of-life care patients. On the other hand, psychoeducational interventions, which require greater active involvement, should be provided on time to educate patients in the early stage of cancer diagnosis [22]. Whenever possible, nurses should use tailored nonpharmacological educational interventions, which are more effective than standardized ones [21].

While evidence-supported interventions are the basis of pain management, nurses should remain flexible to include patient-preferred practices when they are safe and meaningful. Some behavioral interventions could include dance movement therapy, exercise, expressive writing, hypnosis, relaxation/progressive muscle relaxation, guided imagery, yoga, body-based interventions, acupressure, foot massage, massage, energy-based interventions, Reiki, and therapeutic touch [21,26]. Despite the lack of proven efficacy in large patient trials, some of these options could be meaningful for a particular patient; massage and mindful breathing were identified as best-evidence nonpharmacological interventions to support pain management in palliative care patients by Veen et al. [26].

Engaging with family and caregivers is also an important step in providing a proper pain management plan. Nurses should consider patient and caregiver abilities (including fatigue level, interest, cognition, and other factors) when selecting nonpharmacological interventions and engage family members and significant others in the planning and delivery of the pain management plan [24,26].

Future implications

In palliative care settings, nurses are pivotal in managing multifaceted and dynamic pain experiences, yet they frequently face significant hurdles, including time constraints and resource shortages. Studies consistently highlight that high patient-to-nurse ratios and competing clinical demands often impede comprehensive pain assessments and frequent reevaluations that are essential in palliative care [27,28]. Insufficient access to specialized training and limited interdisciplinary collaboration restrict nurses' ability to apply pharmacologic and nonpharmacologic interventions effectively, whereas institutional barriers undermine optimal palliative pain management [29]. Targeted strategies, such as enhanced staffing models, institutional investment in educational initiatives, and policy reforms empowering nursing autonomy, can address these challenges and help nurses manage pain and deliver patient-centered care effectively.

For effective pain management, nurses should be supported with knowledge and institutional policies to screen and assess pain effectively and in a timely manner and to use self-management techniques. More rigorous research is needed to characterize the prescription of nonpharmacologic interventions in terms of timing, feasibility, duration, and patient involvement. Research should also focus on patient follow-up and family engagement.

Conclusion

The prevalence of pain with advanced cancer is very high, and most patients experience moderate-to-severe pain daily, leading to serious compromises in health, loss of independence, loss of mobility, and a decrease in quality of life. Pain management aims to relieve pain, improve functionality, and improve quality of life. Nurses need to build rapport with patients and families, continue to develop knowledge and skills, and stay updated about new techniques for pain management to control pain effectively and alleviate patient suffering.

References

[1] Ferrell BR. Physical aspects of care: pain and gastrointestinal symptoms. USA: Oxford University Press; 2015.
[2] Sekse RJT, Hunskår I, Ellingsen S. The nurse's role in palliative care: a qualitative meta-synthesis. J Clin Nurs 2018;27(1-2):e21–38. Available from: https://doi.org/10.1111/jocn.13912.
[3] De Munter J, Dodlek N, Khmaladze A, et al. The role of cancer nurses in cancer-related pain management in Europe. Palliat Care Soc Pract 2023;17(1):1–12. Available from: https://doi.org/10.1177/26323524231216996.
[4] Coyne P, Mulvenon C, Paice JA. American Society for Pain Management Nursing and Hospice and Palliative Nurses Association position statement: pain management at the end of life. Pain Manag Nurs 2018;19(1):3–7. Available from: https://doi.org/10.1016/j.pmn.2017.10.019.
[5] Culgin JL, Duffy CM, Levitan LL. Acute, chronic, oncologic, and end-of-life pain management in primary care. In: Buttaro TM, Polgar-Bailey P, Sandberg-Cook J, Trybulski J, editors. Primary care: interprofessional collaborative practice. USA: Elsevier Health Sciences; 2021.
[6] Chapman EJ, Edwards Z, Boland JW, et al. Practice review: evidence-based and effective management of pain in patients with advanced cancer. Palliat Med 2020;34(4):444–53. Available from: https://doi.org/10.1177/0269216319896955.
[7] Fallon M, Giusti R, Aielli F, et al. ESMO Guidelines Committee. Management of cancer pain in adult patients: ESMO clinical practice guidelines. Ann Oncol 2018;29(Suppl. 4):iv166–91. Available from: https://doi.org/10.1093/annonc/mdy152.
[8] Can G. Oncology nursing. Türkiye: Nobel Medicine Publishing; 2019.
[9] Kitchen R, Faull C, Russell S, Wilson J, editors. Handbook of palliative care. USA: John Wiley & Sons; 2024.
[10] Alghamdi FA, Alahmed ZA, Sowadi SY, et al. Nursing role and strategy in pain management in palliative care. Int J Community Med Public Health 2024;11(1):398–402. Available from: https://doi.org/10.18203/2394-6040.ijcmph20233830.
[11] Carbonara L, Casale G, De Marinis MG, et al. Adherence to ESMO guidelines on cancer pain management and their applicability to specialist palliative care centers: an observational, prospective, and multicenter study. Pain Pract 2024;25(1):e13418. Available from: https://doi.org/10.1111/papr.13418.
[12] Charak S, Thattil RG, Srivastava CM. Assessment and management of pain in palliative care. London: IntechOpen; 2021.
[13] Ferrell BR, Paice JA. Oxford textbook of palliative nursing. USA: Oxford University Press; 2019.
[14] Wessex Palliative Physicians. Palliative care handbook a good practice guide, <https://www.ruh.nhs.uk/For_Clinicians/departments_ruh/Palliative_Care/documents/palliative_care_handbook.pdf>; 2019 [accessed 02.10.24].
[15] Caraceni A, Cherny N, Fainsinger R, et al. Pain measurement tools and methods in clinical research in palliative care: recommendations of an Expert Working Group of the European Association of Palliative Care. J Pain Symptom Manage 2002;23(3):239–55. Available from: https://doi.org/10.1016/S0885-3924(01)00409-2.
[16] Herr K, Coyne PJ, Ely E, et al. Pain assessment in the patient unable to self-report: clinical practice recommendations in support of the ASPMN 2019 position statement. Pain Manag Nurs 2019;20(5):404–17. Available from: https://doi.org/10.1016/j.pmn.2019.07.005.
[17] Larson S, Laures E, Seo M, et al. Evidence-based pain assessment in nonverbal palliative care patients. Pain Manag Nurs 2024;25(2):152–9. Available from: https://doi.org/10.1016/j.pmn.2023.12.005.
[18] Zhang J, Chan DN, Liu X, et al. Effects of self-management interventions for cancer patients with pain: a systematic review of randomised controlled trials. J Clin Nurs 2023;32(17–18):5652–67. Available from: https://doi.org/10.1111/jocn.16669.
[19] Russo MM, Sundaramurthi T. An overview of cancer pain: epidemiology and pathophysiology. Semin Oncol Nurs 2019;35(3):223–8. Available from: https://doi.org/10.1016/j.soncn.2019.04.002.
[20] van Dongen SI, De Nooijer K, Cramm JM, et al. Self-management of patients with advanced cancer: a systematic review of experiences and attitudes. Palliat Med 2020;34(2):160–78. Available from: https://doi.org/10.1177/0269216319883976.

[21] Eaton LH, Brant JM, McLeod K, et al. Nonpharmacologic pain interventions: a review of evidence-based practices for reducing chronic cancer pain. Clin J Oncol Nurs 2017;21(3):54–70. Available from: https://doi.org/10.1188/17.CJON.S3.54-70.

[22] Park YJ, Lee MK. Effects of nurse-led nonpharmacological pain interventions for patients with cancer: a systematic review and meta-analysis. J Nurs Scholarsh 2022;54(4):422–33. Available from: https://doi.org/10.1111/jnu.12750.

[23] Syrjala KL, Jensen MP, Mendoza ME, et al. Psychological and behavioral approaches to cancer pain management. J Clin Oncol 2014;32(16):1703–11. Available from: https://doi.org/10.1200/JCO.2013.54.4825.

[24] Munkombwe WM, Petersson K, Elgán C. Nurses' experiences of providing nonpharmacological pain management in palliative care: a qualitative study. J Clin Nurs 2020;29(9–10):1643–52. Available from: https://doi.org/10.1111/jocn.15232.

[25] Swarm RA, Paice JA, Anghelescu DL, et al. Adult cancer pain, version 3.2019, NCCN clinical practice guidelines in oncology. J Natl Compr Cancer Netw 2019;17(8):977–1007. Available from: https://doi.org/10.6004/jnccn.2019.0038.

[26] van Veen S, Drenth H, Hobbelen H, et al. Non-pharmacological interventions feasible in the nursing scope of practice for pain relief in palliative care patients: a systematic review. Palliat Care Soc Pract 2024;18(1):1–23. Available from: https://doi.org/10.1177/26323524231222496.

[27] Dakka FJ. Nurses barriers to evidence-based practice in palliative care: a systematic review. SAGE Open Nurs 2022;8(1):1–8. Available from: https://doi.org/10.1177/23779608221142957.

[28] Gerber K, Willmott L, White B, et al. Barriers to adequate pain and symptom relief at the end of life: a qualitative study capturing nurses' perspectives. Collegian 2022;29(1):1–8. Available from: https://doi.org/10.1016/j.colegn.2021.02.008.

[29] Barroso-Castaño P, Cabrera-Jaime S, Huertas-Zurriaga A, et al. Exploring the roles and regulatory challenges of advanced practice nurses in acute pain management: A scoping review. J Adv Nurs 2025;0:1–52. Available from: https://doi.org/10.1111/jan.17027.

Chapter 53

Implementation of guidelines in pain management

Auxiliadora Fernández López[1,2], Luis Felipe Calle Cruz[3], and Marin Golčić[4,5]

[1]*Home Support Team for Palliative Care, Virgen del Rocío University Hospital, Seville, Spain,* [2]*Faculty of Medicine, University of Seville, Seville, Spain,* [3]*Primary Care Centre of Castilleja de la Cuesta, Seville, Spain,* [4]*Tumor Clinic, Clinical Hospital Center Rijeka, Rijeka, Croatia,* [5]*Faculty of Medicine, University of Rijeka, Rijeka, Croatia*

Key learning objectives

- Identify the main barriers to effective implementation of pain management guidelines in palliative cancer care.
- Understand the importance of education and training initiatives for healthcare providers and patients to overcome misconceptions and improve pain management practices.
- Recognize the role of systematic, multidisciplinary approaches and organizational strategies in facilitating guideline adherence and improving clinical outcomes.
- Explore future directions in pain management implementation, including the integration of technology and early palliative care interventions to enhance patient-centered care.

Introduction

Despite advances in cancer treatment and palliative care, uncontrolled pain is a significant issue. The epidemiology of such pain was covered by other chapters in this book. Many organizations have tried to systematically approach pain treatment and standardize pain management and diagnosis, resulting in an abundance of different guidelines, many of which are specific for patients with cancer. Some of the most important pain treatment guidelines are published by the World Health Organization, European Society for Medical Oncology, National Comprehensive Cancer Network, and American Society of Clinical Oncology, among others. However, in this chapter, we primarily aim to understand and evaluate the barriers and propose improvements for the implementation of pain treatment guidelines in everyday practice, and not to evaluate specific guidelines [1–4].

Challenges in implementation of guidelines in pain treatment for palliative care patients with cancer arise due to physician and organizational or patient-related factors. While some barriers might be subjective due to attitudes and misconceptions, others may clearly be objective such as workforce-related constraints and regulatory or financial issues. Regardless, any approach must be multifaceted and multidimensional, as correcting a single issue alone would not significantly reduce the barriers to proper implementation of pain management guidelines or adequately alleviate pain in patients with cancer.

Importance of pain management guidelines

Clinicians need information continuously and dynamically and use both their clinical experience (personal clinical expertise) and external clinical scientific evidence. However, the discretion in clinical decision-making and the degree of uncertainty lead to a variability in the response of the care process in practice when faced with a health problem [5]. Hence, evidence-based medicine and regularly updated clinical practice guidelines are required to systematically provide equal and scientifically based pain management. This is especially emphasized by the fact that pain is one of the most feared symptoms in cancer patients [6].

Pain Management in Palliative Care for Patients With Cancer. DOI: https://doi.org/10.1016/B978-0-443-30050-9.00040-0

However, the complexity in therapeutic, diagnostic, and prognostic decision-making has increased, requiring paradigm shifts and new approaches, especially in the management of the production, implementation, and management of scientific knowledge. In a way, we can say that clinical practice guidelines represent an effort to adopt scientific advances into clinical experience through regular clinical practice and are a fundamental tool for therapeutic dialogue with the patient that allows shared decision-making to be achieved. However, development and updating of clinical practice guidelines for pain management require substantial resources and necessarily require a multidisciplinary group of experts in the field of oncology, pain medicine, and palliative care. The guidelines should strive to improve quality of life, standardize treatment, provide education and training, identify pain early, allow for a multidisciplinary approach, personalization of treatment, monitoring and follow-up, and improve the cost-effectiveness of actions. The quality of the levels of evidence and the degree of recommendation form the basis of evidence-based clinical practice guidelines, which allow the clinician to determine the validity and applicability in clinical practice of published research. However, despite the availability of such guidelines, there are several main barriers in implementation in everyday clinical practice.

Barriers in implementation of pain guidelines

Physician-related challenges

One of the most important healthcare provider–related challenges in implementing pain management guidelines in everyday clinical practice is insufficient education and training. Despite decades of research documenting unmet global needs, pain management continues to be a low priority in health professional curricula even in high-resource countries [7]. Many healthcare professionals learn pain management only through limited and informal curriculum in clinical settings that may perpetuate a culture of stigma and inadequate pain care practices [8]. Hence, without formal education, providers may rely on anecdotal experiences or outdated practices rather than evidence-based guidelines. This knowledge gap could be partially responsible for lack of understanding of the benefits of palliative care and appropriate timing for referrals [9].

Another significant issue is the stigma and misconceptions associated with palliative care. Even where there is adequate knowledge about palliative care, unfavorable attitudes among healthcare providers, discomfort when addressing palliative care requirements, and associating palliative care exclusively with end-of-life care can create obstacles to timely referrals. Physicians' discomfort discussing palliative care with patients and families, desire to maintain hope, uncertainty about prognosis, emotional labor in calling for palliative care consultation, and time constraints are the leading causes of delay in initiating appropriate pain management. Many physicians are also resistant to sharing their traditional roles with palliative care teams, creating barriers to interdisciplinary care that is essential for effective pain management [10–13].

Furthermore, when multiple physicians are involved in a patient's care, differences of opinion about the direction of care can create inconsistency in pain management approaches. Conflicts may arise among physicians about the compatibility of palliative care with curative therapy, and differences of opinion regarding the timing or eligibility of patients for specialized pain management services further complicate care delivery. This is especially problematic for countries with no palliative care specialties or lack of mobile palliative care teams and hospices. In such settings, lack of specialty primarily focused on symptom control might be a primary barrier to proper implementation of pain guidelines in everyday practice. The lack of standardization in pain assessment and management practices also emphasizes this issue as without consistent approaches to pain assessment, implementing guideline-based treatment algorithms becomes complicated and it becomes increasingly difficult to objectively evaluate the success of any pain intervention or implementation of pain guidelines [14–17].

Healthcare system and regulatory constraints

In order for a healthcare system to achieve a successful implementation of pain treatment guidelines in everyday practice, many key steps are involved: identifying specific guideline recommendations to implement, measuring current practice, identifying gaps between current practice and recommendations, developing implementation strategies, addressing barriers, implementing changes, and evaluating outcomes. However, in order to accomplish these steps, a strong regulatory framework and resource allocation are required. Without clear implementation strategies and dedicated resources, guidelines may remain as documents without meaningful impact on clinical practice or patient outcomes [18,19].

The regulatory environment surrounding pain management, particularly regarding opioid prescriptions, presents significant barriers as well. The opioid crisis has led to increased scrutiny and regulation of opioid prescribing, creating an additional layer of complexity in pain management for palliative care patients. While these regulations aim to prevent misuse and addiction, they sometimes create unintended barriers to appropriate pain management for cancer patients who legitimately need opioid medications [20,21]. Clinicians may hesitate to increase opioid dosages when appropriate because of misconceptions about opioids causing respiratory depression, addiction, or decreased life expectancy. This regulatory environment can create a negative effect on prescribing practices that adversely affect cancer patients with severe pain, particularly those who require increasing doses due to disease progression or tolerance [22,23].

Resource limitations are also a major problem. There could be a resource limitation in the supply of crucial drugs such as opioids, particularly in large or remote areas of coverage. Limited resources are also often the cause of inadequate workforce and lack of proper hospital infrastructure. High workloads with complexity in care, low rates of referrals, and lack of partnering community service providers further complicate guideline implementation [24,25]. This is particularly true for a significant portion of cancer patients for whom the standard analgesic therapy fails to provide effective pain relief or causes intolerable side effects. These complex cases often require interventional techniques such as peripheral nerve blocks, neuraxial analgesia, sympathetic blocks, or neurolytic blocks which require specialized training, equipment, and infrastructure that may not be available in all settings, creating disparities in access to comprehensive pain management [9,18,26,27].

Additional significant systemic barrier, particularly in some countries, is the fragmentation of healthcare systems. Cancer patients frequently receive care across multiple settings and from various providers, creating discontinuities that hinder consistent pain management. In palliative care this is even more true as in high-resource settings, there is a lack of a bridging care system from hospital to home. A skewed distribution of palliative services could result in limited access, particularly for patients in underserved or rural areas [9,28].

Patient and family-related challenges

Lastly, one of the main challenges of implementing proper pain management and adhering to pain management guidelines is the misconceptions and fears held by the patients and their families. There are still many people who believe that palliative care is only for end-of-life patients or that it is the same as hospice care. Family members may not be willing to accept palliative care referrals or may feel offended by the suggestion because of these negative connotations. Wrong but common interchangeable use of terms such as "palliative care" and "end-of-life care" contributes to these misconceptions and creates barriers to acceptance of appropriate pain management strategies. Yet, without proper access to early palliative care, there is also limited focus on management of pain and other symptoms [29–31].

Associating palliative care with end-of-life care could influence many patients and families to consider palliative care as a last resort, which can lead to unrealistic expectations about what palliative care can achieve. If these expectations are not met, family or patients may express dissatisfaction with care, which can also lead to worse adherence to treatment recommendations [32].

Cultural beliefs, personal preferences, and communication barriers significantly impact the implementation of pain management guidelines as well, with some patients seeing pain treatment as giving up or a necessary component of fighting against cancer, particularly with opioid use. Cultural attitudes toward pain expression, stoicism, and the use of certain medications may influence patients' willingness to report pain or adhere to recommended treatments. In addition, language barriers between healthcare providers and patients can lead to inadequate pain assessment and management. Previous disappointing experiences with pain management, fears about addiction or side effects, and concerns about being labeled as "drug-seeking" may also affect engagement with pain management strategies [16,33–35].

Strategies to overcome implementation challenges

Education and training initiatives

Improving education and training for healthcare providers is essential for effective guideline implementation, while educating the general public is a prerequisite for removing misconceptions toward palliative care and symptom management. Pain education should be integrated into the training or preemployment education of all health professionals. Educational strategies for physicians to develop comfort with palliative care patients and training in communication skills are particularly important. Training should focus on developing strategies and techniques to deal with difficult situations, breaking bad news, interacting with family members and coworkers in challenging situations, and collaborating with other organizations. Organizing workshops, seminars, and online courses, and providing decision support systems with implementation of clinical software, alerts, and reminders could all be used for training. Furthermore, financial incentives and providing additional resources and feedback could all incentivize education of physicians.

Targeted educational interventions for patients, families, and communities regarding the potential benefits of palliative care services can help address misconceptions and improve acceptance of pain management approaches. Clearly distinguishing between palliative care and end-of-life care in communications with patients and families may help reduce negative associations with palliative care services. Involving receptive family members in explaining aspects of care and engaging religious or spiritual leaders may help gain the confidence of patients and families. Healthcare providers should reconsider and update their approach to discussing palliative care and pain management, emphasizing that palliative care can be beneficial for patients at any stage of serious illness, not just at the end of life [9,36–40].

Systematic implementation approaches

While initial structured implementation processes that include identifying specific guideline recommendations and developing a palliative and symptom-management team might require substantial investments, in the long run, such policy is cost-saving. This process should include measuring current practice to identify gaps between recommendations and practice, determining resources required for implementation, and evaluating potential benefits for patients. By following a structured approach, organizations can more effectively translate guideline recommendations into clinical practice improvements and monitor their impact on patient outcomes. This process requires strong institutional support, the availability of adequate resources, continuous training, and the active participation of professionals in the development of the guidelines. It also requires a clear plan, an analysis of the gaps between clinical practice and the ideal, tailored communication strategies for each target group, decisions about which implementation methods work best in each system context, evaluation of cost-effectiveness and opportunity costs, assessment of impact, and, finally, adaptation to the local environment and circumstances [19,41].

Improving interdisciplinary communication and coordination can enhance the implementation of pain management guidelines. Providing palliative care skill training to all healthcare team members helps improve coordination in pain management, which can often require an interdisciplinary approach. This is particularly important for developing clear protocols for referrals and consultations, which can enhance timely access to appropriate pain management interventions [42,43]. Shared decision-making approaches involving patients, families, and healthcare providers from various disciplines help ensure that pain management strategies align with patient preferences and goals of care. This collaborative approach acknowledges the complex nature of pain and the need for multidimensional assessment and management strategies [44].

Future directions for implementation research

Future strategies for the improvement of pain management should continue to focus on the implementation of a multidisciplinary approach and improving the regulatory approach, developing both early and terminal (late) palliative care, particularly improving access to palliative care and symptom management in underdeveloped areas. Particularly important is the integration of technology in pain management, including virtual reality, wearables, and artificial intelligence. Technology offers promising solutions for enhancing guideline implementation and monitoring pain management outcomes; it can both help in early diagnosis of the pain and also follow the response to pain alleviation interventions by following objective parameters such as pulse, saturation, sweating, and other parameters. Electronic health records can incorporate decision support tools that guide clinicians through evidence-based pain assessment and management protocols. Mobile applications can additionally facilitate remote monitoring of pain and symptoms, allowing for timely adjustments to pain management strategies. The development and validation of more objective pain assessment tools could also enhance consistency in pain assessment and management. As these technologies evolve, guidelines should incorporate recommendations for their appropriate use and integration into comprehensive pain management approaches [45,46].

Conclusion

Implementing pain management guidelines in palliative cancer care requires navigating complex challenges across healthcare systems, organizations, providers, patients, and families. Despite the availability of evidence-based guidelines, significant gaps remain between recommendations and practice, resulting in inadequate pain management for many cancer patients. Main causes of such barriers include systemic/regulatory barriers, patient misconceptions, and physician-related issues. As we continue to develop and refine pain management guidelines for palliative cancer care, equal attention must be given to implementation strategies that address these multifaceted barriers. Only through comprehensive approaches that target barriers at all levels can we hope to bridge the gap between evidence-based recommendations and clinical practice, ultimately improving pain management and quality of life for cancer patients in palliative care.

References

[1] World Health Organization. WHO guidelines for the pharmacological and radiotherapeutic management of cancer pain in adults and adolescents. Geneva: World Health Organization; 2019.

[2] Fallon M, Giusti R, Aielli F, et al. Management of cancer pain in adult patients: ESMO clinical practice guidelines. Ann Oncol 2018;29(Suppl. 4):iv166–91. Available from: https://doi.org/10.1093/annonc/mdy152.

[3] National Comprehensive Cancer Network. NCCN clinical practice guidelines in oncology: adult cancer pain, version 3.2024. Plymouth Meeting, PA: National Comprehensive Cancer Network; 2024. Available from: https://www.nccn.org/guidelines/guidelines-detail?category=3&id=1413.

[4] Mao JJ, Ismaila N, Bao T, et al. Integrative medicine for pain management in oncology: society for integrative oncology-ASCO guideline. J Clin Oncol 2022;40(34):3998–4024. Available from: https://doi.org/10.1200/JCO.22.01357.

[5] Sackett DL. Evidence-based medicine. Semin Perinatol 1997;21(1):3–5. Available from: https://doi.org/10.1016/s0146-0005(97)80013-4.

[6] LeMay K, Wilson KG, Buenger U, et al. Fear of pain in patients with advanced cancer or in patients with chronic noncancer pain. Clin J Pain 2011;27(2):116–24. Available from: https://doi.org/10.1097/AJP.0b013e3181f3f667.

[7] Mezei L, Murinson BB. Pain education in North American medical schools. J Pain 2011;12(12):1199–208. Available from: https://doi.org/10.1016/j.jpain.2011.06.006.

[8] Bond M. Pain education issues in developing countries and responses to them by the International Association for the Study of Pain. Pain Res Manag 2011;16(6):404–6. Available from: https://doi.org/10.1155/2011/654746.

[9] Agrawal US, Sarin J, Bakhshi S, et al. Challenges and opportunities in providing palliative care services to children with a life-limiting illness: a systematic review. Natl Med J India 2022;35(5):284–9. Available from: https://doi.org/10.25259/NMJI_349_2022.

[10] Formagini T, Poague C, O'Neal A, et al. "When I Heard the Word Palliative": obscuring and clarifying factors affecting the stigma around palliative care referral in oncology. JCO Oncol Pract 2022;18(1):e72–9. Available from: https://doi.org/10.1200/OP.21.00088.

[11] Zhang S, Li Z, Yue P. Healthcare providers' attitudes and associated factors on palliative care referral: a qualitative systematic review and meta-aggregation. J Clin Nurs 2024;33(9):3355–80. Available from: https://doi.org/10.1111/jocn.17160.

[12] Salins N, Ghoshal A, Hughes S, et al. How views of oncologists and haematologists impacts palliative care referral: a systematic review. BMC Palliat Care 2020;19(1):175. Available from: https://doi.org/10.1186/s12904-020-00671-5.

[13] Zuniga-Villanueva G, Ramirez-GarciaLuna JL, Weingarten K. Factors associated with knowledge and comfort providing palliative care: a survey of pediatricians in Mexico. J Palliat Care 2019;34(2):132–8. Available from: https://doi.org/10.1177/0825859719827021.

[14] Kaye EC, Gattas M, Kiefer A, et al. Provision of palliative and hospice care to children in the community: a population study of hospice nurses. J Pain Symptom Manage 2019;57(2):241–50. Available from: https://doi.org/10.1016/j.jpainsymman.2018.10.509.

[15] Tong W, Murali KP, Fonseca LD, et al. interpersonal conflict between clinicians in the delivery of palliative and end-of-life care for critically ill patients: a secondary qualitative analysis. J Palliat Med 2022;25(10):1501–9. Available from: https://doi.org/10.1089/jpm.2021.0631.

[16] Gilron I, Carr DB, Desjardins PJ, et al. Current methods and challenges for acute pain clinical trials. Pain Rep 2019;4(3):e647. Available from: https://doi.org/10.1097/PR9.0000000000000647.

[17] Xu X, Chen H, Xu J, et al. The status and challenges of pain assessment in hospitalized patients: a cross-sectional study. Pain Manag Nurs 2024;26(2):e124–30. Available from: https://doi.org/10.1016/j.pmn.2024.11.003.

[18] De Hert S, de Paula-Garcia N. Implementation of guidelines in clinical practice; barriers and strategies. Curr Opin Anesthesiol 2024;37(2):155–62. Available from: https://doi.org/10.1097/ACO.0000000000001344.

[19] Guideline Central. Guideline implementation overview: 7 steps to improve clinical guideline adherence at your organization - guideline central [Internet]. Guideline central. 2023 [cited 2025 Jun 20]. Available from: https://www.guidelinecentral.com/insights/guideline-implementation-overview-7-steps-to-improving-clinical-guideline-adherence-at-your-organization/.

[20] Manjiani D, Paul DB, Kunnumpurath S, et al. Availability and utilization of opioids for pain management: global issues. Ochsner J 2014;14(2):208–15.

[21] Webster LR, Grabois M. Current regulations related to opioid prescribing. PM R 2015;7(Suppl. 11):S236–47. Available from: https://doi.org/10.1016/j.pmrj.2015.08.011.

[22] Ebbert JO, Philpot LM, Clements CM, et al. Attitudes, beliefs, practices, and concerns among clinicians prescribing opioids in a large academic institution. Pain Med 2018;19(9):1790–8. Available from: https://doi.org/10.1093/pm/pnx140.

[23] Im DD, Chary A, Condella AL, et al. Emergency department clinicians' attitudes toward opioid use disorder and emergency department-initiated buprenorphine treatment: a mixed-methods study. West J Emerg Med 2020;21(2):261–71. Available from: https://doi.org/10.5811/westjem.2019.11.44382.

[24] Gourd E. Inadequate government funding for UK hospices affects palliative care. Lancet Oncol 2024;25(6):697–8. Available from: https://doi.org/10.1016/S1470-2045(24)00253-5.

[25] Shi Z, Du M, Zhu S, et al. Factors influencing accessibility of palliative care: a systematic review and meta-analysis. BMC Palliat Care 2025;24(1):80. Available from: https://doi.org/10.1186/s12904-025-01704-7.

[26] Salikhanov I, Connor SR, Kunirova G, et al. Challenges for developing palliative care services in resource- limited settings of Kazakhstan. Public Health Rev 2023;44:1605672. Available from: https://doi.org/10.3389/phrs.2023.1605672.

[27] Fumić Dunkić L, Hostić V, Kustura A. Palliative treatment of intractable cancer pain. Acta Clin Croat 2022;61(Suppl. 2):109–14. Available from: https://doi.org/10.20471/acc.2022.61.s2.14.

[28] Alizadeh Z, Rohani C, Rassouli M, et al. Bridging the gap and developing a home-based palliative care model for cancer patients. Asia Pac J Oncol Nurs 2024;11(11):100583. Available from: https://doi.org/10.1016/j.apjon.2024.100583.

[29] Shen MJ, Wellman JD. Evidence of palliative care stigma: the role of negative stereotypes in preventing willingness to use palliative care. Palliat Support Care 2019;17(04):374–80. Available from: https://doi.org/10.1017/S1478951518000834.

[30] Alcalde J, Zimmermann C. Stigma about palliative care: origins and solutions. Ecancermedicalscience 2022;16:1377. Available from: https://doi.org/10.3332/ecancer.2022.1377.

[31] Murray SA, Murray IR. End of life care still not living up to public and doctors' expectations. BMJ 2016;353:i2188. Available from: https://doi.org/10.1136/bmj.i2188.

[32] von Blanckenburg P, Riera Knorrenschild J, Hofmann M, et al. Expectations, end-of-life fears and end-of-life communication among palliative patients with cancer and caregivers: a cross-sectional study. BMJ Open 2022;12(5):e058531. Available from: https://doi.org/10.1136/bmjopen-2021-058531.

[33] Givler A, Bhatt H, Maani-Fogelman PA The importance of cultural competence in pain and palliative care. National Library of Medicine. StatPearls Publishing; 2023. Available from: https://www.ncbi.nlm.nih.gov/books/NBK493154/.

[34] Peacock S, Patel S. Cultural influences on pain. Rev Pain 2008;1(2):6–9. Available from: https://doi.org/10.1177/204946370800100203.

[35] Dressler G, Cicolello K, Anandarajah G. "Are They Saying It How I'm Saying It?" A qualitative study of language barriers and disparities in hospice enrollment. J Pain Symptom Manage 2021;61(3):504–12. Available from: https://doi.org/10.1016/j.jpainsymman.2020.08.019.

[36] Watt-Watson J, Hogans BB. Current status of pain education and implementation challenges - International Association for the Study of Pain (IASP) current status of pain education and implementation challenges | IASP [Internet]. Available from: https://www.iasp-pain.org/resources/fact-sheets/current-status-of-pain-education-and-implementation-challenges/; 2021.

[37] Artioli G, Bedini G, Bertocchi E, et al. Palliative care training addressed to hospital healthcare professionals by palliative care specialists: a mixed-method evaluation. BMC Palliat Care 2019;18:88. Available from: https://doi.org/10.1186/s12904-019-0476-8.

[38] van Zuilekom I, Thiesen – van Staveren J, Dericks-Issing M, et al. Optimizing palliative care education nationwide: a practice example from The Netherlands. Palliat Care Soc Pract 2024;18 26323524241298288. Available from: https://doi.org/10.1177/26323524241298288.

[39] Chi NC, Barani E, Fu YK, et al. Interventions to support family caregivers in pain management: a systematic review. J Pain Symptom Manage 2020;60(3):630–56. Available from: https://doi.org/10.1016/j.jpainsymman.2020.04.014.

[40] Campanella P, Lovato E, Marone C, et al. The impact of electronic health records on healthcare quality: a systematic review and meta analysis. Eur J Public Health 2016;26(1):60–4. Available from: https://doi.org/10.1093/eurpub/ckv122.

[41] Sheridan PE, LeBrett WG, Triplett DP, et al. Cost savings associated with palliative care among older adults with advanced cancer. Am J Hosp Palliat Care 2021;38(10):1250–7. Available from: https://doi.org/10.1177/1049909120986800.

[42] Dudley N, Ritchie CS, Rehm RS, et al. Facilitators and barriers to interdisciplinary communication between providers in primary care and palliative care. J Palliat Med 2019;22(3):243–9. Available from: https://doi.org/10.1089/jpm.2018.0231.

[43] Thiel M, Harden K, Brazier LJ, et al. Improving the interdisciplinary clinical education of a palliative care program through quality improvement initiatives. Palliat Med Rep 2020;1(1):270–9. Available from: https://doi.org/10.1089/pmr.2020.0092.

[44] Rabben J, Vivat B, Fossum M, et al. Shared decision-making in palliative cancer care: a systematic review and metasynthesis. Palliat Med 2024;38(4):406–22. Available from: https://doi.org/10.1177/02692163241238384.

[45] Leroux A, Rzasa-Lynn R, Crainiceanu C, et al. Wearable devices: current status and opportunities in pain assessment and management. Digit Biomark 2021;5(1):89–102. Available from: https://doi.org/10.1159/000515576.

[46] Abd-Elsayed A, Robinson CL, Marshall Z, et al. Applications of artificial intelligence in pain medicine. Curr Pain Headache Rep 2024;28(4):229–38. Available from: https://doi.org/10.1007/s11916-024-01224-8.

Chapter 54

Coping with cancer pain: families, caregivers, and medical staff

Irina Filip
Diplomate of the American Board of Psychiatry and Neurology, Avicenna Medical and Clinical Research Institute, Mission Viejo, CA, United States

Key learning objectives

- Understand pain's psychological impact on patients, families, and medical staff.
- Identify effective coping strategies and stress reduction techniques for managing emotional pain.
- Explain the relationship between emotional well-being and pain management.
- Evaluate research trials related to coping mechanisms and burnout prevention.
- Apply practical strategies in clinical settings to support families and staff.

Introduction

Pain management is a crucial component of care for patients with advanced cancer, especially those receiving palliative treatment. Cancer pain is not only a physical symptom but also a multidimensional experience that carries significant emotional and psychological burdens for patients and those around them [1]. Family caregivers, who often provide the primary support outside the healthcare team, are deeply affected by the patient's suffering. While caregiving can strengthen bonds and create meaning, it also exposes families to anticipatory grief, anxiety, helplessness, and emotional fatigue. Over time, these stressors can lead to compassion fatigue and burnout [2,3]. Medical staff, continually confronted with pain and suffering, are similarly vulnerable. Prolonged exposure contributes to emotional exhaustion, moral distress, and reduced professional efficacy. Resilience has been identified as a key protective factor, yet burnout remains highly prevalent among oncology and palliative care clinicians, with consequences for both their well-being and the quality of patient care [4–6]. Taken together, these dynamics highlight that cancer pain is a multidimensional challenge—impacting patients, families, and medical professionals alike. Addressing it requires not only effective symptom control but also integrated coping strategies and systemic support to protect the emotional health of all those involved.

Understanding the psychological effect of pain

Exploration of the emotional toll on families witnessing a loved one in pain

Witnessing a family member battle cancer can be one of the most painful experiences. Emotional responses range from anxiety, helplessness, uncertainty about outcomes, fear of treatment unpredictability, anticipatory grief, and guilt for not being able to provide more help [2]. The prolonged stress of caregiving may lead to emotional fatigue, physical exhaustion, mood disorders, and somatic symptoms. As responsibilities intensify, family members may divert attention from other relationships, leading to interpersonal conflict, isolation, and loneliness [3].

Stress and burnout among medical staff

Medical staff exposed to the ongoing suffering of patients face high emotional and physical demands, often resulting in compassion fatigue and burnout. The emotional exhaustion of staff is multifaceted, involving feelings of being overwhelmed, disengagement from responsibilities, and loss of empathy over time [6]. Prolonged empathic engagement

Pain Management in Palliative Care for Patients With Cancer. DOI: https://doi.org/10.1016/B978-0-443-30050-9.00014-X

across entire shifts may contribute to empathy erosion [7]. In addition, medical personnel face stress from caring for multiple patients while meeting documentation standards. Somatic symptoms such as headaches, gastrointestinal issues, muscle tension, and physical strain are frequently reported [3]. Moral distress, discussed in the review by Zanatta et al., highlights how systemic limitations such as institutional policies can exacerbate ethical tensions and stress [4]. Over time, such distress may lead to psychiatric disorders, including anxiety, depression, and insomnia [5].

Effective coping strategies

Techniques for families/caregivers to manage emotional distress

Structurally organized support groups and psychoeducation are known to assist families in coping with emotional trauma. There are many workshops and seminars in the community designed specifically for family members and caregivers of patients with terminal illnesses, and they provide practical advice on stress management and communication skills.

Group therapy provides a safe space to share and process one's struggles, fears, and frustrations, which can be comforting and make the caregivers feel less isolated. Family members feel validated and supported by relating to others' experiences in similar circumstances. They are encouraged to articulate their needs and concerns, improving their communication skills. Hearing about the success stories of others, caregivers may feel empowered and develop better resilience. The caregivers can learn new strategies and acquire new skills in caring for the loved ones during group interactions reducing the social isolation.

Encouraging caregivers to take breaks often, accept help from others, and maintain their well-being by engaging in self-care activities and relaxation practices promote their emotional well-being. The caregivers should feel supported, heard, understood, and valued by the family members to feel fulfilled in what they do. Encouraging families to celebrate small periods of remission and cherish meaningful moments with their loved ones also reinforces positive connections.

Psychotherapists can help families identify the emotional aspects of their problems, family conflicts, and communication issues, which are essential for stress management and emotional health maintenance. Individual or family therapy also creates an environment for venting emotions, acquiring new coping strategies, and improving emotional health in general. In addition, many palliative care programs offer specialized counseling or therapy not only for the patient but also for the family members, even after the loss of a loved one [2,8].

Effective communication: maintaining a sense of normalcy

A family member's diagnosis of cancer is an emotionally taxing experience and tends to leave the family members stunned and in dismay. Family members may feel reluctant to initiate difficult conversations or know what to say, worrying they would offend the patient. Effective communication is vital for patients and their families navigating the cancer journey [9].

Family members should feel comfortable sharing their emotions with others, voicing their needs and expectations, and discussing with the other family members how to better assist the patient in furthering their treatment. Family members can maintain a sense of normality by keeping up with daily routines and activities, continuing to socialize with family and friends, and accepting help from others when offered [2].

Open and honest communication helps family members and caregivers to understand the diagnosis, treatments, and evolution of the disease, preparing them emotionally and practically to navigate through the patient's cancer journey [9]. Celebrating small treatment successes and periods of remission reinforces the concept that one can still find joy in life despite the challenges of the anticipated prognosis.

Education and understanding of the disease process

The families of patients with cancer and the caregivers are more likely to provide effective and practical support to their loved ones if they have a good understanding of the patient's diagnosis, prognosis, and nature of the pain. To better understand the patient's condition, they can accompany patients to the doctor's appointments and prepare a list of questions in advance to ensure they are being answered. They can also meet with the oncologist or other healthcare team members to discuss the patient's diagnosis and prognosis and request written summaries of medical consultations [8,9].

By gaining knowledge of what to expect, family members can recognize potential medication side effects, newly emerging symptoms or pain triggers, pain levels, and the effectiveness of pain relief treatments, and they can engage more effectively with healthcare providers to address them appropriately.

Family members can also access websites or renowned organizations, such as the American Cancer Society or National Cancer Institute, to learn more about different types of cancer, their treatment, and prognosis. Data shows that being informed makes the family less afraid of uncertainty and empowers them for what is coming ahead, making them more equipped to assist the patient through cancer treatment [2].

Resilience techniques

The medical staff working with cancer patients is under tremendous stress triggered by job demands and compassion fatigue. Resilience has been shown as a protective factor against adversity and emotional trauma. It consists of the ability to adjust and adapt to traumatic and challenging circumstances posed by exposure to pain, suffering, and death. Resilience develops from the interplay of genetics, personality traits, and environmental factors. The more resilient a person is, the more likely he is to overcome the adverse effects of stress and adversity or to "bounce back" faster if thrown into deep emotional states [4,10].

Resilience training programs for palliative care providers are usually conducted through retreats and encompass self-care and relaxation practices. Psychological education for death and grief awareness, cognitive-behavior therapy, meditation, mindfulness-based stress reduction, and skill building are among research-based methods to prevent burnout.

By practicing meditation and mindful techniques, healthcare providers exposed to pain and suffering can better regulate their emotions. Research findings show that mindfulness techniques bring attention to the moment, reinstating a sense of control in the present moment and distracting the focus from the future, which is unknown and anxiety-producing.

The practice of meditation can also assist in developing a calm and focused mind, which can be beneficial in dealing with job roles. Learning how to set up professional emotional boundaries by limiting emotional involvement can protect healthcare providers caring for patients in pain from compassion fatigue and mental exhaustion. Providers should regularly check in with their emotions, trying to detach if they feel drawn too deeply into patients' suffering. This emotional awareness helps providers remain empathic without becoming overwhelmed [10].

Through "guided imagery," a widely used complementary relaxation technique, clients imagine themselves in a peaceful place that distracts them from stressful situation. By visiting this mental space often, the body's physiological responses are lowered, the sympathetic system is inhibited, and the body releases endorphins and feels more relaxed.

Biofeedback is another effective technique in which clients learn how to relax mentally and take control of their bodily functions, such as heart rate, breathing, temperature, and body tension. They can see tangible outcomes, which enhances their sense of control over emotions and feelings. In addition to guided imagery—which has been shown in randomized trials to reduce anxiety and stress responses in cancer patients [11]—biofeedback interventions, including heart-rate-variability training [12] and device-assisted feedback in chemotherapy patients [13], have also demonstrated benefits in improving coping skills.

Current studies have found that healthcare professionals in palliative care settings can mitigate burnout by improving communication and receiving emotional support from peer support groups. Most palliative care and oncology programs have daily or weekly debriefing sessions where providers discuss cases and allow them to express their emotions and concerns in a safe and supportive environment, learning coping strategies from each other. This alleviates the stress and pressure of making independent treatment decisions. After debriefing with colleagues or using mindfulness techniques, healthcare providers can use reflective practices to gain a new perspective on their emotional responses [14].

Healthcare institutions should provide free individual counseling sessions to employees through employee assistance programs [10]. In these sessions, providers can process their experiences with a trained therapist and learn and apply personalized coping strategies. Focusing on self-care, such as new or old hobbies, physical exercise, or socializing, can give healthcare providers purpose and validation outside the professional world, making them cope better with work challenges.

Another source of stress among healthcare providers is working additional hours and having to cover for the providers who are out, which increases their patient load and responsibilities. This type of stress can be alleviated by providing adequate medical staff on duty and standby support in case of emergencies. Healthcare professionals should have enough time to rest between shifts [4].

Practical applications in clinical settings

Many healthcare facilities have already adopted the multidisciplinary pain management team model, which includes social workers, therapists, and pain specialists who can make tailored recommendations based on each patient's needs.

The facilities could offer mindfulness-based interventions for doctors, nurses, or other palliative care team members. This creates a safe space to practice mindfulness and stress reduction techniques to mitigate burnout and improve healthcare workers' job satisfaction [10].

Palliative care facilities can organize weekly family meetings as part of a structured family support program. In these meetings, trained therapists can introduce and practice stress management techniques to provide emotional support and increase the emotional resilience of the family members [14].

Future implications

There are several promising directions for palliative care in the near future. Personalized medicine (also known as precision medicine) may optimize treatment timing, minimize medication-related adverse effects, improve diagnostic accuracy, and reduce aggressive end-of-life interventions [15]. This approach may also enhance treatment outcomes and predictability, support decision-making, and lessen decisional burden for caregivers and families.

In addition, with the emergence of the COVID-19 pandemic, a trend was noted in delivering medical care using online technology. Given the loss of function in patients with advanced cancer and mobility issues, the consideration of using online counseling, psychiatric and pain management visits, or using mental health apps to reach the patients and their families in their environment and provide them with coping tools and strategies is highly considered.

Another future direction in palliative care aims to apply "trauma-informed practices" in this field. Patients or their families may experience preexisting posttraumatic stress disorder symptoms caused by adverse childhood events, war trauma, trauma caused by displacement in conflict, trauma from discrimination and racism, or healthcare-induced trauma resulting from painful surgeries, procedures, adverse medication effects, intensive care unit admissions, and emergency room visits.

These additional stressors add to the diagnosis of cancer and may reactivate memories of prior traumas, which can interfere with the patient's acceptance of palliative care treatments. "Trauma-informed care" does not directly address each source of the trauma. Instead, it focuses on building trust in the therapeutic relationship, which is meant to increase treatment compliance, decrease symptom burden, and improve the quality of life in this patient population [16].

Summary

Palliative care for patients with cancer pain is inherently complex and must extend beyond the patient to include families and caregivers. Coping strategies are essential, as they help families manage the emotional and practical burdens of caregiving while supporting the patient. For medical staff, fostering resilience and preventing burnout are critical to maintaining a healthy workforce and ensuring consistent, high-quality care. Ongoing research and adaptation in clinical practice are needed to strengthen support systems for both families and healthcare professionals.

References

[1] Kelley AS, Morrison RS. Palliative care for the seriously ill. N Engl J Med 2015;373(8):747–55. Available from: https://doi.org/10.1056/NEJMra1404684.

[2] Northouse LL, Katapodi MC, Song L, et al. Interventions with family caregivers of cancer patients: meta-analysis. CA Cancer J Clin 2010;60(5):317–39. Available from: https://doi.org/10.3322/caac.20081.

[3] Stajduhar KI. Burdens of family caregiving at the end of life. Clin Invest Med 2013;36(3):E121–6. Available from: https://doi.org/10.25011/cim.v36i3.19722.

[4] Zanatta F, Maffoni M, Giardini A. Resilience in palliative healthcare professionals: a systematic review. Support Care Cancer 2020;28(3):971–88. Available from: https://doi.org/10.1007/s00520-019-05194-1.

[5] Rotenstein LS, Ramos MA, Rosales RC, et al. Prevalence of burnout among physicians: a systematic review. JAMA 2018;320(11):1131–50. Available from: https://doi.org/10.1001/jama.2018.12777.

[6] Pereira SM, Fonseca AM, Carvalho AS. Burnout in palliative care: a systematic review. Nurs Ethics 2011;18(3):317–26. Available from: https://doi.org/10.1177/0969733011398092.

[7] Hojat M, Vergare MJ, Maxwell K, et al. The devil is in the third year: a longitudinal study of erosion of empathy in medical school. Acad Med. 2009;84(9):1182–91. Available from: https://doi.org/10.1097/ACM.0b013e3181b17e55.

[8] Hannon B, Swami N, Krzyzanowska MK, et al. Satisfaction with oncology care among patients with advanced cancer and their caregivers. Qual Life Res 2013;22(9):2341–9. Available from: https://doi.org/10.1007/s11136-013-0371-3.

[9] Mazor KM, Beard RL, Alexander GL, et al. Patients' and family members' views on patient-centered communication during cancer care. Psychooncology 2013;22(11):2487–95. Available from: https://doi.org/10.1002/pon.3317.

[10] Back AL, Steinhauser KE, Kamal AH, et al. Building resilience for palliative care clinicians: an approach to burnout prevention based on individual skills and workplace factors. J Pain Symptom Manage 2016;52(2):284–91. Available from: https://doi.org/10.1016/j.jpainsymman.2016.02.002.

[11] Charalambous A, Giannakopoulou M, Bozas E, et al. A randomized controlled trial for the effectiveness of progressive muscle relaxation and guided imagery as anxiety reducing interventions in breast and prostate cancer patients undergoing chemotherapy. Evid Based Complement Alternat Med 2015;2015:270876. Available from: https://doi.org/10.1155/2015/270876.

[12] Spada GE, Masiero M, Pizzoli SFM, et al. Heart rate variability biofeedback in cancer patients: a scoping review. Behav Sci (Basel) 2022;12(10):389. Available from: https://doi.org/10.3390/bs12100389.

[13] Mi N, Zhang L, Sun H, et al. A randomized controlled trial of mindfulness meditation combined with an intelligent biofeedback device in pancreatic cancer patients receiving chemotherapy. Brain Behav 2024;14(12):e70197. Available from: https://doi.org/10.1002/brb3.70197.

[14] Yongpraderm S, Inpithuk P, Wongprom I. Resilience-building in palliative care professionals: scoping review. BMJ Support Palliat Care 2026;16:1–10. Available from: https://doi.org/10.1136/spcare-2024-005144.

[15] Sampetrean A, Aldea M, Mateus C. Precision medicine's new frontier: Integrating palliative care at the right time. ESMO Open 2023;8(5):101632. Available from: https://doi.org/10.1016/j.esmoop.2023.101632.

[16] de Groot JM, Fehon DC, Calman L, et al. Trauma-informed palliative care is needed: a call for implementation and research. Palliat Med 2023;37(10):1470–3. Available from: https://doi.org/10.1177/02692163231206998.

Glossary of Terms

Acetaminophen A centrally acting analgesic used for mild-to-moderate nociceptive cancer pain; functions at least partially through the inhibition of CNS cyclooxygenase-2 (COX-2).

Acupuncture Integrative modality activating endorphin release and gate control via needle stimulation at meridians, reducing chemotherapy-induced nausea and neuropathic pain.

Adenocarcinoma Malignancy of glandular epithelial origin (e.g., lung, breast, colorectal), often with propensity for visceral and bone metastases.

Adjuvant analgesics Nonopioid pharmacologic agents used adjunctively for cancer pain syndromes (also known as coanalgesics).

Advance care planning (ACP) A structured process involving patients and caregivers to align further disease management with documented goals of care (often via advance directives).

ALK mutation Oncogenic ALK gene rearrangements that promote tumor growth.

Allodynia A neuropathic phenomenon where nonnoxious stimuli (e.g., light touch) cause pain, often due to central sensitization.

American Society of Clinical Oncology (ASCO) Professional organization representing physicians across all oncology subspecialties who care for people with cancer and that publishes oncology guidelines and informs evidence-based cancer treatment and palliative frameworks.

Androgen deprivation therapy (ADT) Endocrine therapy reducing testosterone to control prostate cancer progression and pain.

Animal-assisted therapy (AAT) Adjunctive intervention employing trained animals to attenuate cancer pain perception via limbic system modulation.

Anorexia Cytokine-mediated appetite loss in cancer cachexia that exacerbates fatigue and pain sensitivity.

Antibody-drug conjugate (ADC) A targeted cancer therapy that links a monoclonal antibody to a potent cytotoxic drug, delivering the drug directly to cancer cells while minimizing damage to healthy tissue.

Anticonvulsants A class of drugs primarily used to treat seizures, which also play an adjuvant role in neuropathic pain.

Antidepressants A class of medication used to treat depression and anxiety disorders, which also play an adjuvant role in pain management.

Ascites Pathologic peritoneal fluid accumulation that can cause visceral pain and dyspnea.

Background pain Ongoing pain that is relatively constant and persists over time in patients with chronic or palliative conditions.

Barbiturates GABAergic sedatives used for refractory seizures but also for deep palliative sedation in end-of-life care.

Benzodiazepines A class of sedatives used to manage anxiety, insomnia, and muscle spasms, with possible adjuvant benefits for pain.

Best supportive care (BSC) A multidimensional evidence-based approach optimizing relief of pain and other cancer burdens, providing the best possible quality of life in incurable cancer.

Biologically effective dose (BED) A calculated radiation dose that accounts for both the total dose delivered and the effect of dose fractionation, allowing comparison of different radiotherapy schedules.

Bisphosphonates A class of drugs inhibiting the activity of osteoclasts used to slow down or prevent bone loss and reduce skeletal-related events.

Bone-seeking agents Radiopharmaceuticals delivering targeted beta or alpha emissions to osteoblastic metastases, providing bone pain palliation.

BRCA Tumor suppressor genes (BRCA1/2) whose mutations elevate breast, ovarian, and prostate cancer risk.

BRAF/MEK inhibitors Targeted agents used to treat patients with BRAF mutation in cancer, most commonly melanoma.

Brachytherapy Localized radiation modality delivering a high dose of radiation directly to the cancer cells by placing the radioactive source inside or near the tumor, with minimal effects to surrounding tissue.

Breakthrough dose Supplemental dosing of pain medication (most commonly opioids) used to manage transient cancer pain flares.

Breakthrough pain Acute, severe pain flares that overcome analgesia of background pain.

C fibers Unmyelinated nerve fibers that conduct pain signals slowly, typically associated with dull, aching, or burning pain.

Calciferol A fat-soluble vitamin essential for bodily functions such as bone health.

Cannabinoids A class of compounds found in the *Cannabis* plant, which can provide symptomatic relief for a variety of painful conditions.

CAR T-cell therapy A type of cancer immunotherapy in which a patient's T cells are modified to identify and attack cancer cells.

Central sensitization An increased responsiveness of the central nervous system to stimulation, leading to increased pain sensitivity.

CDK4/6 inhibitors A class of cancer drugs that inhibit the function of cyclin-dependent kinases 4 and 6 (CDK4/6), slowing the growth of cancer cells.

Checkpoint inhibitor (CPI) A type of immunotherapy drug that helps the immune system recognize and attack cancer cells by blocking proteins that suppress immune responses.

Chemoradiotherapy Cancer treatment that combines chemotherapy and radiotherapy.

Chemotherapy A type of cancer treatment that uses drugs to stop the growth of rapidly dividing cells, including cancer cells.

Chemotherapy-induced peripheral neuropathy (CIPN) Nerve damage caused by certain cancer drugs, leading to symptoms such as numbness, tingling, burning pain, and sensory disturbance.

Child–Pugh scores A system that assesses the severity of chronic liver disease and helps estimate prognosis.

Cholecalciferol A crucial nutrient that allows for absorption of calcium (known as vitamin D3).

Cognitive-behavioral therapy (CBT) Structured psychotherapy targeting maladaptive pain–related cognitions (via corticolimbic modulation).

Computed tomography (CT) An imaging technique that uses X-rays and computer processing to create detailed cross-sectional pictures of the body.

Comprehensive geriatric assessment (CGA) A multidomain assessment of geriatric overall health and well-being.

COX enzymes Enzymes involved in the production of prostaglandins—chemicals that mediate inflammation, pain, and fever in the body.

CYP isoenzymes A family of enzymes mainly responsible for the oxidative metabolism of different drugs.

Demoralization syndrome Existential distress syndrome marked by hopelessness, loss of purpose, and amplified pain perception in advanced cancer.

Deprescribing Structured discontinuation of medications with unfavorable risk-benefit in terminal illness, reducing polypharmacy complications such as drug–drug interactions and sedation.

Early palliative care Integration of palliative services at the beginning of terminal illness diagnosis, improving quality of life and even survival.

Eastern Cooperative Oncology Group (ECOG) Scale Performance status functional assessment scaled 0–5 used to guide treatment management.

Edmonton Symptom Assessment System (ESAS) Validated symptom score scaled 0–10 assessing the most common cancer burdens (e.g., pain, fatigue, anxiety, and others).

EGFR mutation A genetic change in the *EGFR* gene that can lead to cancer growth (especially in lung cancer).

End-of-life (EoL) care Healthcare that focuses on providing comfort and support to patients who are in the last stages of their lives.

Equianalgesic dosing A method to determine equivalent pain-relieving doses when switching between different opioids.

Estimated glomerular filtration rate (eGFR) A calculation that estimates how well the kidneys are filtering waste from the blood.

European Society for Medical Oncology (ESMO) A professional organization dedicated to advancing cancer care through education, clinical guidelines, research, and policy across Europe and globally.

European Society for Radiotherapy and Oncology (ESTRO) A professional organization that promotes innovation, research, and education in the field of radiation oncology in Europe.

External beam radiation therapy (EBRT) A type of cancer treatment using high-energy beams to target and destroy cancer cells.

FLACC scale Behavioral pain assessment for nonverbal patients scaled 0–10, developed for use in pediatric and noncommunicative cancer patients.

Functional Assessment of Cancer Therapy (FACT) Validated quality-of-life instrument assessing symptom burden in patients with cancer.

Gamma-aminobutyric acid (GABA) receptor A receptor in the nervous system targeted by sedatives to produce calming and muscle-relaxing effects.

Glutamate A major excitatory neurotransmitter involved in transmitting pain signals in the nervous system.

Guided imagery Cognitive-behavioral technique using visualization to modulate pain perception and anxiety through corticolimbic pathways.

Hazard ratio (HR) A statistical measure used to compare the risk of an event occurring at any time point between two groups.

Holistic care An approach integrating biological, psychological, social, and spiritual support for comprehensive care of patients.

Hormonal therapy A form of treatment involving the use of hormones or hormone-altering drugs to treat various conditions, primarily to inhibit tumor growth and reduce metastatic pain.

Hospice care Comprehensive palliative care prioritizing comfort and quality of life provided at home or in specialized facilities for patients with <6 months of prognosis.

Hyperalgesia Exaggerated pain sensitivity from central or peripheral mechanisms (including opioid-induced hyperalgesia via mu-receptor neuroplasticity).

Hypnotics A class of medications primarily used to induce and maintain sleep.

Immune-related adverse events (irAEs) Autoimmune toxicities from CPI often managed with corticosteroids.

Immunotherapy A type of cancer treatment that boosts the body's natural defenses—primarily through T-cell responses to help reduce and stop cancer growth.

Integrated Palliative Care Outcome Scale (IPOS) A brief, validated multidimensional tool for simple and comprehensive assessment of patients' needs and the quality of care, which is suitable for use in diverse healthcare settings and can be completed by patients or healthcare professionals.

International Federation of Gynecology and Obstetrics (FIGO) Global authority providing staging criteria for gynecologic cancers, facilitating prognostication and palliative analgesic planning.

Lymphedema A condition that involves the buildup of lymph fluid most commonly in the arms and legs, leading to the swelling of the tissue.

Magnetic resonance imaging (MRI) An imaging technique that uses strong magnetic fields and radio waves to produce detailed images of internal body structures.

Malignant bowel obstruction (MBO) Tumor-induced intestinal blockage usually causing crampy visceral pain and nausea.

Metastatic castration-resistant prostate cancer (mCRPC) An advanced stage of prostate cancer that has spread to other parts of the body and has progressed on initial hormone therapies.

Metastatic castration-sensitive prostate cancer (mCSPC) Prostate cancer that is spreading to other parts of the body but still responds to ADT.

Metastatic spinal cord compression (MSCC) Compression of the spinal cord usually due to the pressure from bone metastases.

Microsatellite instability (MSI) A condition of genetic hypermutability resulting from defects in mismatch repair, commonly found in gastrointestinal cancers.

Minimally invasive procedures (MIPs) Medical interventions performed through small incisions or natural body openings using specialized instruments (e.g., to reduce pain), allowing faster recovery time compared to traditional surgical interventions.

Monoclonal antibody A laboratory-made protein designed to bind to a specific target on cells, commonly used in cancer treatment to mark cancer cells for immune attack, block growth signals, or deliver toxins directly to tumors.

mTOR inhibitors A class of drugs that target the mammalian target of rapamycin (mTOR) protein, inhibiting tumor growth.

Multidimensional pain scales Assessment instruments measuring multiple dimensions of cancer pain, including intensity, quality, and interference with function (e.g., Brief Pain Inventory).

Multidisciplinary team A group of healthcare professionals from different fields collaborating to provide comprehensive care tailored to the individual patient's needs.

Multimodal pain management Combining multiple types of therapies (e.g., pharmacological and nonpharmacological) to achieve better pain relief and reduce side effects.

N-methyl-D-aspartate (NMDA) antagonist A class of drugs that block NMDA receptors, resulting in a reduction of excitatory neurotransmission (used for analgesia, anesthesia, and management of certain neuropathic or refractory pain states).

National Comprehensive Cancer Network (NCCN) US-based alliance publishing high-level, evidence-based cancer care guidelines integrating multimodal pain management, palliative care, and targeted therapy algorithms.

Nerve blocks Regional anesthetic injections aiming to interrupt pain transmission.

Neuropathic pain Pain caused by nerve damage or dysfunction, often described as burning, shooting, tingling, or numbing.

Nociceptive pain Pain from tissue damage or inflammation, typically described as aching or throbbing.

Nonpharmacologic pain management Pain management techniques that do not involve medication (e.g., massage, cognitive-behavioral therapy, or acupuncture).

Nonverbal Pain Scale (NVPS) Behavioral tool assessing pain in noncommunicative patients via facial expressions, movement, and vital signs.

Nonsteroidal antiinflammatory drugs (NSAIDs) A class of medications used to manage mild-to-moderate pain by blocking the production of chemicals in the body, primarily prostaglandins.

Numeric Rating Scale (NRS) Patient-reported 0–10 pain scale for pain assessment.

OLDCARTS Mnemonic pain assessment tool for comprehensive cancer pain assessment including Onset, Location, Duration, Characteristics, Aggravating/Relieving factors, Timing, Severity.

Oncolytic virus therapy A form of cancer therapy in which engineered viruses are used to lyze tumor cells while stimulating immune responses.

On-demand palliative care A model in which specialist palliative care is provided based on emerging patient needs rather than through routine scheduled visits.

Opioids Opioids are natural, semisynthetic, or synthetic compounds that act on opioid receptors and are primarily used for analgesia.

Opioid-induced constipation (OIC) Constipation resulting from mu-receptor-mediated GI dysmotility from chronic opioid use.

Opioid-induced hyperalgesia (OIH) Paradoxical pain amplification from prolonged use of opioids due to central neuroplasticity.

Opioid rotation Switching between opioids to improve analgesic efficacy in cancer pain and/or avoid adverse effects.

Opioid-sparing effect The use of nonopioid drugs such as NSAIDs or acetaminophen to reduce the required dose of opioids in pain management.

Opioid use disorder (OUD) Maladaptive opioid behavior causing distress or dysfunction.

OPQRSTUV Mnemonic comprehensive cancer pain assessment including Onset, Provocation, Quality, Region, Severity, Timing, Understanding, Values.

Oral mucositis A painful inflammation and ulceration of the mucous membranes in the mouth, often as a result of chemotherapy or radiotherapy.

Overall survival (OS) Time from treatment initiation to death, often used as a primary end point in cancer trials.

Pain scales Tools used to measure and monitor a patient's pain intensity.

Palliative Performance Scale (PPS) A tool used in palliative care to measure a patient's functional status and predict disease progression, based on ambulation, activity, self-care, intake, and consciousness.

Palliative sedation Ethically guided use of sedatives at end of life to relieve refractory suffering (must follow institutional and legal protocols).

PARP inhibitors A class of drugs that block DNA repair in BRCA-mutated cancers.

PD-1/PD-L1 Proteins involved in immune checkpoints that help cancer cells to remain hidden from immune defenses. Blocking these proteins can help the immune system to target and destroy cancer cells.

Performance status Functional capacity of cancer patients usually assessed with ECOG PS (scaled 0–5) or Karnofsky PS (scaled 0%–100%), which help in determining the eligibility criteria for cancer therapies.

Perineural invasion The invasion of cancer cells into the perineural space.

P-glycoprotein A cellular transporter protein that affects drug absorption and elimination, which plays a role in drug interactions with opioids.

Physical therapy Rehabilitative intervention using exercise, manual techniques, and movement education to address pain from deconditioning, neuropathy, or musculoskeletal dysfunction.

PIK3CA inhibitors A class of drugs that are used to treat certain cancers by inhibiting the PI3K/AKT/mTOR pathway, a crucial pathway for cell proliferation.

Polypharmacy The concurrent use of multiple medications (five or more), increasing the risk of adverse drug events, interactions, and other negative health outcomes.

Positron emission tomography (PET) An imaging technique that uses radioactive tracers to show metabolic activity in tissues, often used to detect cancer, evaluate cancer spread, and monitor treatment response.

Primary palliative care (PPC) Basic palliative care services provided by nonspecialist clinicians such as general practitioners and nurses.

Progression-free survival (PFS) Time from the start of the treatment to disease progression or death.

Quality of life (QoL) in cancer A measure assessing physical, emotional, social, and functional well-being of individuals living with cancer.

Radiodermatitis Skin inflammation and damage caused by exposure to ionizing radiation, often seen as a side effect of cancer radiotherapy.

Radium-223 (Ra-223) dichloride Alpha-emitting radiopharmaceutical used for bone pain in mCRPC via osteoblastic targeting.

Reflexology Manual pressure therapy that can reduce cancer pain and anxiety.

Response rate (RR) Proportion of patients with partial or complete tumor shrinkage following treatment.

Samarium-153 (Sm-153) lexidronam Beta-emitting radiopharmaceutical primarily used for targeting osteoblastic metastases.

Scrambler therapy Noninvasive cutaneous neuromodulation using electrical impulses to modulate afferent nerve activity in conditions such as chemotherapy-induced peripheral neuropathy (CIPN).

Secondary palliative care Specialist-level palliative care delivered by trained experts in palliative care for patients with complex or advanced needs.

Self-expandable metal stent (SEMS) A tubular metal device that expands after placement to keep a hollow structure, such as the esophagus, bile duct, or airway, open when it is narrowed or blocked.

Serotonin–norepinephrine reuptake inhibitors (SNRIs) Dual-acting antidepressants often used to treat neuropathic cancer pain.

Serotonin syndrome Potentially fatal adverse reaction from serotonergic excess, presenting with clonus, hyperreflexia, and altered mental status caused by medications that build up high levels of serotonin in the body.

Shared decision-making A collaborative framework in which clinicians, patients, and caregivers align treatment options with patient goals and values in cancer pain care.

Single-fraction radiotherapy The administration of a single high dose of radiation instead of multiple smaller doses over a longer period of time.

Skeletal-related events (SREs) Pathologic complications from bone metastases (e.g., fractures, spinal compression), causing pain and morbidity.

SOCRATES Mnemonic pain assessment tool: Site, Onset, Character, Radiation, Associations, Time course, Exacerbating/Relieving factors, Severity.

Somatic pain Type of nociceptive pain that arises from the musculoskeletal system (bones, muscles, ligaments, tendons, and joints).

Spiritual pain A form of suffering rooted in existential distress, such as loss of meaning or questions about the purpose of illness and life.

Stereotactic body radiation therapy (SBRT) Precision radiotherapy delivering high-dose radiation to localized lesions.

Substance P A neuropeptide that enhances the transmission of pain signals from peripheral nerves to the spinal cord.

Supportive care An integrative approach in oncology, managing pain and other symptoms in cancer alongside disease-directed treatments and its side effects, often used interchangeably with palliative care, emphasizing multimodal strategies and quality of life.

Systemic radiotherapy A type of radiation therapy that circulates radioactive drugs throughout the body to target and kill cancer cells.

Targeted therapy A type of cancer treatment that uses molecular guided agents designed to interfere with specific molecules involved in tumor growth and progression.

Transcutaneous electrical nerve stimulation (TENS) A form of noninvasive electrical stimulation that utilizes electric currents to modulate peripheral nerve activity and provide pain relief.

Tertiary palliative care Highly specialized care offered in academic centers, focusing on complex symptom management, clinical research, and education.

TNM A cancer staging system describing tumor size and extent (T), involvement of regional lymph nodes (N), and presence of distant metastasis (M).

Topical pain drugs Medications applied directly to mucosal surfaces or skin, which offer local analgesia with fewer systemic side effects.

Total pain A multidimensional concept encompassing physical, psychological, social, and spiritual aspects of patients with cancer, requiring integrative, multimodal palliative strategies.

Transcranial magnetic stimulation (TMS) A neuromodulation technique using magnetic fields to stimulate brain regions, explored as a treatment for chronic pain.

Transdermal patch (TDP) A medicated adhesive patch that, when placed on the skin, releases a specific dose of medication into the bloodstream.

Tyrosine kinase inhibitors (TKIs) A class of molecules that selectively block the activity of tyrosine kinases, which can slow or stop the growth of certain cancers.

Virtual reality (VR) Immersive, computer-generated environments, which can reduce cancer pain and anxiety via distraction and corticolimbic modulation.

Visceral pain A type of pain that originates from the internal organs and is often described as deep, dull, cramping, or poorly localized.

Visual Analog Scale (VAS) A patient-reported pain intensity assessment tool that measures pain on the continuous 0–10 cm line.

Voluntary assisted dying (VAD) The process by which a terminally ill and mentally competent adult can be prescribed medication by a medical professional to end their life (not universally available and legal status varies across countries).

World Health Organization (WHO) analgesic ladder A three-step framework developed by the WHO for escalating pain management from nonopioids to strong opioids, now expanded to include adjuvants and interventional approaches.

Xerostomia Dry mouth resulting from salivary gland damage, commonly after head and neck radiotherapy, causing oral pain and dysphagia.

25-Hydroxyvitamin D The main circulating form of vitamin D; deficiency contributes to various conditions, including osteomalacia and bone fragility.

Note This glossary provides concise clinical definitions aligned with current best practices and guidelines pertinent to cancer-related pain management with the assistance of an AI tool. It does not aim to serve as a universal oncology or palliative medicine glossary, and terminology outside the core focus of this book has been intentionally excluded.

Index

Note: Page numbers followed by "*f*" and "*t*" refer to figures and tables, respectively.

N

T

U

V

W

Y

Z

9780443300509